Encyclopedia of Astronomy and Astrophysics

ENCYCLOPEDIA

OF

ASTRONOMY

AND

ASTROPHYSICS

Robert A. Meyers, Editor
TRW, Inc.

Steven N. Shore, Scientific Consultant
New Mexico Institute of Mining and Technology
and DEMIRM, Observatoire de Meudon

Academic Press, Inc.
Harcourt Brace Jovanovich, Publishers
San Diego New York Berkeley Boston
London Sydney Tokyo Toronto

This work is a derivative from the *Encyclopedia of Physical Science and Technology*
edited by Robert A. Meyers, copyright © 1987 by Academic Press, Inc.

ACADEMIC PRESS, INC.
San Diego, California 92101

United Kingdom Edition published by
ACADEMIC PRESS LIMITED
24-28 Oval Road, London NW1 7DX

Library of Congress Cataloging-in-Publication Data

Encyclopedia of astronomy and astrophysics / Robert A. Meyers, editor,
   Steven N. Shore, scientific editor.
       p.   cm.
     ISBN   0-12-226690-0   (alk. paper)
       1. Astronomy—Dictionaries. 2. Astrophysics—Dictionaries.
   I. Meyers, Robert A. (Robert Allen), Date. II. Shore, Steven N.
   QB14.E53   1988
   520′.3′21—dc19   88-24062
                                                         CIP

PRINTED IN THE UNITED STATES OF AMERICA
88 89 90 91   9 8 7 6 5 4 3 2 1

CONTENTS

CONTRIBUTORS

ALBEE, ARDEN L. *California Institute of Technology*. Lunar Rocks.

ANDERSON, JAMES L. *Stevens Institute of Technology*. Relativity, General.

BARR, L. D. *National Optical Astronomy Observatories*. Optical Telescopes.

BODENHEIMER, PETER. *Lick Observatory, University of California, Santa Cruz*. Stellar Structure and Evolution.

BRANCH, DAVID. *University of Oklahoma*. Supernova 1987A; Supernovae.

BURATTI, BONNIE J. *JPL, California Institute of Technology*. Moon (Astronomy); Planetary Satellites, Natural.

CARRUTHERS, GEORGE R. *Naval Research Laboratories*. Astronomy, Ultraviolet Space.

CRANNELL, CAROL JO. *NASA Goddard Space Flight Center*. Solar Physics.

DAVIS, JR., RAYMOND. *University of Pennsylvania*. Neutrino Astronomy; Neutrino Observational Astronomy (in part).

DERMOTT, STANLEY F. *Cornell University*. Solar System, General.

DYKLA, JOHN J. *Loyola University of Chicago; Theoretical Asrophysics Group, Fermilab*. Cosmology.

GEHRZ, ROBERT D. *University of Minnesota*. Astronomy, Infrared (in part).

GRASDALEN, GARY L. *University of Wyoming*. Astronomy, Infrared (in part).

GRINDLAY, JONATHAN E. *Harvard University*. X-Ray Astronomy.

GULKIS, SAMUEL. *California Institute of Technology*. Radio Astronomy, Planetary.

HACKWELL, JOHN A. *University of Wyoming*. Astronomy, Infrared (in part).

HOOTS, FELIX R. *Air Force Space Command*. Celestial Mechanics.

HUCHRA, JOHN P. *Harvard-Smithsonian Center for Astrophysics*. Galactic Structure and Evolution.

LANZEROTTI, L. J. *AT&T Bell Laboratories*. Solar–Terrestrial Physics.

LESTER, JOHN B. *University of Toronto*. Stellar Spectroscopy.

LEVINE, JOEL S. *NASA Langley Research Center*. Planetary Atmospheres.

MCFADDEN, LUCY-ANN. *University of Maryland*. Primitive Solar System Objects: Asteroids and Comets.

MCGERVEY, JOHN D. *Case Western Reserve University*. Relativity, Special.

MANN, ALFRED K. *University of Pennsylvania*. Neutrino Observational Astronomy (in part).

MEYER, PETER. *University of Chicago*. Cosmic Radiation.

MICHEL, F. CURTIS. *Rice University*. Pulsars.

OSTRO, STEVEN J. *California Institute of Technology*. Planetary Radar Astronomy.

REEDY, ROBERT C. *Los Alamos National Laboratory*. Meteorites, Cosmic Ray Record.

RUSSELL, C. T. *University of California, Los Angeles*. Solar System, Magnetic and Electric Fields.

SAENZ, RICHARD A. *California Polytechnic State University*. Black Holes (Astronomy) (in part).

SHAPIRO, STUART L. *Cornell University*. Black Holes (Astronomy) (in part).

SHIPMAN, HARRY L. *University of Delaware*. Astrophysics.

SHORE, STEVEN N. *New Mexico Institute of Mining and Technology; DEMIRM, Observatoire de Meudon; Radioastronomie, Ecole Normale Superieure*. Binary Stars; Dark Matter in the Universe; Neutron Stars; Quasars; Radio Astronomy; Star Clusters; Variable Stars.

TRYON, EDWARD P. *The City University of New York*. Cosmic Inflation.

WHITE, R. STEPHEN. *University of California, Riverside*. Gamma-Ray Astronomy.

YORK, DONALD G. *University of Chicago*. Space, Interstellar Matter.

FOREWORD

Astrophysics has changed more fundamentally in the past decade than perhaps any other discipline in the physical sciences and certainly more than at any time in its history.

For the first time, the sky throughout the electromagnetic spectrum stands as a surveyed territory. The successes in space observation of the past two decades form an impressive list. *UHURU, SAS-3, HEAO-A, EINSTEIN, COS-B, EXOSAT, TENMA,* and *GINGA* have revealed the gamma-ray and X-ray sky and have permitted detailed high-resolution spectroscopy and imaging, as well as high-time-resolution monitoring of selected sources. *OAO-2, COPERNICUS, ANS, TD1,* and *IUE* have performed photometric all-sky surveys and acted as high-resolution spectroscopic observatories. *IRAS* has surveyed and imaged the far infrared sky for the first time, discovering new components to the interstellar medium and spurring the study of global star formation processes.

In ground-based technology, the past decade has seen the explosive growth in the use of optical and infrared linear solid-state array imaging detectors, such as charge coupled devices (CCDs), whose sensitivities and stability exceed any detectors previously available. Numerous large optical and infrared observatories are now scattered around the world, allowing for almost 24-hr monitoring of celestial phenomena. New millimeter observatories, such as Nobayama and IRAM, have discovered a wealth of new molecules and structures in the interstellar medium. Radio interferometry, exemplified by the Very Large Array, has made high-resolution imaging possible and practical at centimeter wavelengths.

The expansion of computer technology has made it possible to simulate physical processes and test theoretical models of radiative, hydrodynamic, gravitational, and plasma phenomena in extremely complex astrophysical contexts with increasing ease. Entirely new areas have been opened as a result of the merger of high-energy physics, general relativity, and cosmology.

These changes in technology have opened unprecedented research opportunities in virtually all areas of astrophysics. For the first time, it is possible to obtain a complete picture, at all wavelengths, of the wide range of energetic phenomena displayed by cosmical environments and objects. With these new observations, however, come a large number of new theoretical questions about the physical processes that structure these objects. In a rapidly evolving field, it is essential to have a basic reference to which the researcher can turn when new questions present themselves, and to which laypeople can turn to learn more about the background of new discoveries.

This next decade should see an even more dramatic increase in our observational capabilities. The Hubble Space Telescope, Infrared Space Observatory, Cosmic Background Explorer, the *Galileo* mission to Jupiter, the *Phobos* Mars orbiter, and the *HIPPARCHOS* astrometric all-sky survey satellite are among the space projects that should be operating routinely during the 1990s. Ground-based optical and infrared interferometry, though just now beginning, should see rapid expansion and application. Several millimeter arrays are either operating, under construction, or planned for the next ten years. The Very Long Baseline Array is now coming on line for milliarcsecond imaging at centimeter wavelengths. Ground-based optical telescope technology is in a state of rapid development and several 8- to 10-m active mirror telescopes are now under construction. In short, this next decade represents a new era in the history of astronomy. It seems fitting at the transition to present this volume to a public to whom many of these topics will be active areas of research and contemplation in the years soon to come.

Steven N. Shore

PREFACE

Astronomy is that critical science that deals with celestial bodies and the observation and interpretation of the radiation received either directly by Earth-based observatories and instrument arrays or by transmission from satellites and deep space probes from the component parts of the universe.

Humankind strives to answer such cosmic questions as: Is the universe finite or infinite? Is it bounded or unbounded? Is it similar or dissimilar in composition to Earth and the Solar System? Is the universe immutable in its total structure, or does it evolve with the passage of time?

The *Encyclopedia of Astronomy and Astrophysics* has been prepared in response to a well-defined need for a comprehensive, in-depth and yet concise treatment of the latest in fundamental astronomy and astrophysics theory, instrumentation, and observations of the Solar System, stars, and stellar systems. This unique publication is intended for the intelligent layperson, advanced placement high school student, bachelor's degree candidate or graduate student in the sciences or engineering, or practicing professional. Users of this *Encyclopedia* will include individuals, journalists, and educators who wish a better understanding of the advances in space theory that appear in the news on a nearly daily basis; scientists and engineers who work in allied interactive fields such as optics, communications, rocketry, navigation, controls, aeronautics, atmospheric sciences, earth sciences and oceanography, sensors, and relativistic physics; and researchers in astronomy and astrophysics in need of a ready reference at an appropriate scientific level. The encyclopedia format was chosen to allow two modes of retrieving information: a major subject entry and the usual detailed subject index at the end of the volume.

This *Encyclopedia* treats each subject area in a separate article, prepared by a recognized expert, that includes basic theory, derivations of mathematical relationships, recent history, status, and a forecast of future progress. Often this information is presented in photographic, line drawing, and graph format. The approximately 800-page volume contains more than 300 photographs and illustrations, 70 tables, 300 bibliographic entries, and 300 glossary entries. The authors are affiliated with such leading astronomical centers as the Jet Propulsion Laboratory, NASA, Kitt Peak National Observatory, and the Harvard Smithsonian Center for Astrophysics, as well as several universities, including the California Institute of Technology, the University of California, the University of Chicago, Harvard, and Cornell.

The articles average 20 pages in length and contain a glossary of terms specific to the subject covered. The glossaries will facilitate understanding by readers not familiar with the subject matter covered. Further, each article begins with an outline and an introductory definition of the subject, which together with the glossary allow the reader to understand the specific portions of the article containing the desired information. This approach is not found in any other book on astronomy.

The articles cover, in an interlocking manner, the five components of astronomy and astrophysics: (1) fundamentals, instrumentation, techniques, and observations; (2) cosmic rays; (3) the Solar System; (4) stars; and (5) stellar systems, galactic and extragalactic objects, and the universe. Fundamentals are presented in articles titled "Astronomy, Infrared"; "Astronomy, Ultraviolet Space"; "Astrophysics"; "Celestial Mechanics"; "Gamma-Ray Astronomy"; "Neutrino Astronomy"; "Neutrino Observational Astronomy"; "Optical Telescopes"; "Planetary Radar Astronomy"; "Radio Astronomy"; "Radio Astronomy, Planetary"; "Relativity, General"; "Relativity, Special"; and "X-Ray Astronomy." Cosmic rays are covered in articles titled "Cosmic Radiation" and "Meteorites,

Cosmic Ray Record." The Solar System is described in the articles "Lunar Rocks"; "Moon (Astronomy)"; "Planetary Atmospheres"; "Planetary Satellites, Natural"; "Primitive Solar System Objects: Asteroids and Comets"; "Solar Physics"; "Solar System, General"; "Solar System, Magnetic and Electric Fields"; and "Solar–Terrestrial Physics." Stars are presented in articles titled "Binary Stars"; "Black Holes (Astronomy)"; "Neutron Stars"; "Pulsars"; "Quasars"; "Stellar Structure and Evolution"; "Supernova 1987A"; "Supernovae"; and "Variable Stars." The cosmos is presented in the articles "Cosmic Inflation"; "Cosmology"; "Dark Matter in the Universe"; "Galactic Structure and Evolution"; "Space, Interstellar Matter"; and "Star Clusters."

Robert A. Meyers

ASTRONOMY, INFRARED

Robert D. Gehrz   *University of Minnesota*
Gary L. Grasdalen
John A. Hackwell   *University of Wyoming*

GLOSSARY

Background-limited incident power (BLIP): Background radiation level at which shot noise due to the random arrival rate of background photons equals detector noise.

Background subtraction: Cancellation of background signal from the telescope and sky by spatial chopping to produce source and reference beams.

Beam switching: Exchanging the source and reference beams by driving the telescope. The procedure cancels sky gradients along the spatial chopping direction and offset signals produced by the telescope.

Bipolar nebula: Object in which a toroidal disk obscures all radiation from an embedded star except that escaping from the polar regions of the disk.

Circumstellar shell, cocoon: Clouds of dust and gas that closely surround an embedded star.

Detector noise: Intrinsic noise of the detector.

Extinction and reddening: Extinction, the absorption of starlight as it passes through interstellar dust; reddening, the phenomenon that occurs when extinction decreases with increasing wavelength, making extinguished objects appear redder than their intrinsic color.

Far infrared (FIR): Spectral region from 40 to 1000 μm. The wavelength range of 200 μm–1000 μm is often called the *submillimeter* region.

Interstellar dust: Solid grains that lie in the general galactic medium or within higher density clouds.

Magnitude scale: Logarithmic scale of intensities upon which 5 magnitudes (mags) is a factor of 100; thus one mag is a factor of 2.512. Sources fainter than the zero point (0 mag) have positive magnitudes, and sources brighter than 0 mag have negative magnitudes. The scale can be defined at any wavelength.

Mass loss: Mass expelled from a star by outflow of gas and dust.

Noise-equivalent power (NEP): The power detected at a signal-to-noise level of 1 in 1 second.

Near infrared (NIR): Spectral region from 0.7 to 3 μm.

Sky noise: Noise produced by fluctuations in the sky background that cannot be removed by spatial chopping.

Thermal infrared (TIR): Spectral region where photon noise from the thermal radiation of the telescope and the earth's atmosphere limits the sensitivity of ground-based infrared detection systems; ordinarily the region from 3 to 40 μm.

Infrared (IR) astronomy is the study of radiation from the astrophysical environment at wavelengths between $\approx$0.7 and $\approx$1000 μm (1 μm = 10^{-6} m). Principal IR radiation mechanisms include emission from cool blackbody objects, emission from dust, 3K cosmic background radiation, continuum and line emission from gas, and nonthermal emission from energetic particles. Continuum and line absorption by gas and dust have a profound effect on the IR energy distributions of many astrophysical objects.

Infrared astronomy is central to many topics in stellar and extragalactic astronomy, because IR radiation is characteristically emitted by cool objects and because the relatively long IR rays can penetrate effectively through large amounts of obscuring material. It is also a primary tool for remote-sensing investigations of cold solar system objects and of components of planetary systems surrounding nearby stars. Topics uniquely addressed by IR astronomy are the nature of solar system objects, stages in the formation of stars and stellar systems, the physical conditions attending the death of stars, the contents of the material in the interstellar medium, the chemical evolution of galaxies, the physics of primordial galaxies, and the evolution of the universe. [See GALACTIC STRUCTURE AND EVOLUTION; STELLAR STRUCTURE AND EVOLUTION.]

I. History

Sir John Herschel demonstrated the existence of IR radiation in 1800 by dispersing solar light through a glass prism onto several thermometers. American Infrared Astronomy appears to have been born on 28 July 1878 in a hen-house in Rawlins, Wyoming, where Thomas Edison observed a total solar eclipse with an infrared detector of his own invention. During the early 1900s, measurements of the IR intensities of the moon, principal planets, and brightest stars were made by S. Pettit and E. Nicholson, who used primitive thermopile detectors. At this point, all the astronomical observations that were possible with existing infrared detector technology were complete.

It remained, in the early 1960s, for Frank J. Low to breathe new life into the infant field of IR astronomy with his invention of the low-noise gallium-doped germanium (Ga : Ge) bolometer. These detectors enabled astronomers to detect signals as weak as $\approx 10^{-15}$ watts in a few hours using standard astronomical telescopes—a sensitivity gain of a thousand over the primitive detectors used by earlier investigators. The 20 years since Low's invention have seen the evolution of solid-state IR detectors which,

mounted on space-based cryogenically cooled telescopes, may provide yet another thousand-fold increase in sensitivity. Today, IR astronomers address some of the most urgent questions in astrophysics.

II. Observational Techniques in Infrared Astronomy

Because IR radiation is invisible to the naked eye, the techniques used to study IR sources are quite different from those used in optical astronomy. Optical measurements can be made with photographic plates, photoelectric detectors, and charge injection device (CID) and charge coupled device (CCD) area detectors. Thermal, photoconductive, and photovoltaic detectors are used in the IR; there are no photographic emulsions that respond to wavelengths longer than ≈ 1 μm. Furthermore, the IR spectrum covers three decades (a factor of 1000 in frequency), whereas the entire visible spectrum covers only one octave (a factor of two in frequency)! Sensitive detection over the entire IR spectrum requires the use of several different techniques. [See STELLAR SPECTROSCOPY.]

A. GENERAL REQUIREMENTS FOR INFRARED OBSERVATIONS

Figure 1 shows the transmission characteristics of the earth's atmosphere from 1 to 1000 μm. Only a few windows exist through which the transmission is sufficient for consistent measurements from the ground. The best ground-based windows are centered at 1.25, 1.6, 2.3, 3.6, 5, 10.5, and 20 μm. It is apparent from Fig. 1 that terrestrial water vapor is the most important source of atmospheric absorption. Since the scale height (e-folding length) for water vapor is about 3000–5000 ft (0.9–1.5 km), dry, high-altitude mountain sites are required for the best IR transmission. At such sites, the 20-μm window is open most of the time, and the windows at 33, 350, and 1000 μm open under the driest conditions. The remaining regions of the IR spectrum are accessible only from airborne and spaceborne platforms. Airborne platforms are unable to rise above the CO_2 and O_3 absorptions, which occur high in the atmosphere (see Fig. 1).

Because the atmosphere and telescope are relatively warm ($T \approx 250$–300 K), they dominate the background radiation for ground-based ob-

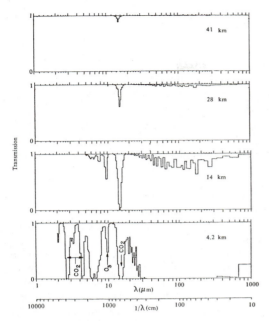

FIG. 1. Atmospheric transmission curve from 1 to 1000 μm at several altitudes, showing windows free of absorption by water vapor and other molecular constituents. (Figure courtesy of W. Traub.)

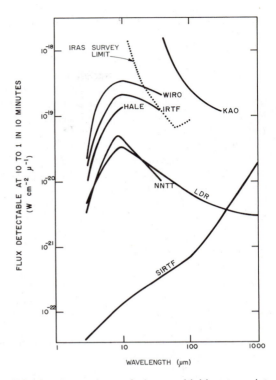

FIG. 2. Comparison of the sensitivities to point sources of several existing and proposed infrared observatories: Wyoming Infrared Observatory (WIRO, 2.34-m aperture), NASA Infrared Telescope Facility (IRTF, 3.05-m aperture), Hale 200-inch telescope (HALE, 5.08-m aperture), National New Technology Telescope (NNTT, 15-m aperture), NASA Kuiper Airborne Observatory (KAO, 0.92-m aperture), NASA Large Deployable Reflector (LDR, 10-m aperture), and NASA Space Infrared Telescope Facility (SIRTF, 0.8-m aperture). Sensitivity is given as a function of wavelength in flux detectable at a signal-to-noise ratio of 10 in 10 minutes of integration time. Flux limits for the NASA IRAS All-sky Infrared Survey are indicated by the dotted line. The Hale 5-m telescope is competitive only for observations at wavelengths ≤ 10 μm because of its low-altitude site.

servations. The radiation from warm objects is discussed in Section III,A. A state-of-the-art IR telescope is designed to minimize such emission, and the sky above an ideal observatory site is cold and transparent. The photometer beam must be as small as possible in angular extent to reduce the number of background photons that are accepted along with the radiation from an astronomical point source, so the atmosphere above the site must have excellent seeing (stability of the image of a point source as viewed through the atmosphere). The visual ($\lambda = 0.55$ μm) atmospheric seeing disks observed at the best ground-based sites are as small as 0.25 to 0.5 arcseconds for brief periods of time. Seeing-disk diameter is thought to scale as $\lambda^{-0.2}$, and atmosphere-limited image sizes should be smaller in the IR than in the visible. Ground-based IR telescopes with apertures as large as 10 to 15 m are therefore required to take full advantage of the best seeing beyond 10 and 20 μm.

For a wide range of observations, the most capable IR telescopes will be those placed in earth orbit. Their optics can be cooled to a few kelvins to eliminate instrumental background radiation. The entire IR spectrum is accessible.

Because there is no "seeing," diffraction-limited observations with extremely large apertures can be contemplated, and the only background radiation is that from the natural sources in the astrophysical environment. On the other hand, these facilities are exceedingly costly, and their requirements for cryogenic cooling may render their missions rather short-lived. The relative sensitivities of a number of existing and proposed IR astronomical observatories are compared in Fig. 2.

B. DETECTORS, FILTERS, AND DEWAR DESIGN

1. Detectors

State-of-the-art infrared astronomical measurements are made with several types of detector. The most popular broadband detector longward of ≈ 5 μm is still the Ga : Ge bolometer introduced by Low. The temperature of the device increases as it absorbs incident infrared photons. When operated at 1 K, the electrical resistance of the doped germanium crystal is roughly proportional to the fourth power of its temperature. Small changes in the detector resistance, which are proportional to the incident power, are measured through a pair of electrical leads. Since the thermal noise of the detector is proportional to the square root of its temperature, these cooled detectors are many times more sensitive than detectors of similar characteristics operated at room temperature (≈ 300 K). The bolometer is typically painted black so that it absorbs IR radiation efficiently. Since the bolometer must be operated in the ac mode for background removal and noise bandwidth reduction, the thermal time constant of the detector (proportional to the mass of the detector and inversely proportional to the conductivity of its leads to the thermal bath) must be kept as small as possible. In general, this dictates that the sensitive bolometer element be kept physically small. In practice, such detectors are cubes measuring about 0.008″ to 0.015″ on a side with leads to the thermal bath measuring 0.0005″ or 0.001″ in diameter by 0.10″ long.

At the longer IR wavelengths ($\lambda \geq 350$ μm), it is necessary to bond the sensitive detection element to a large-area, low-thermal-mass substrate that will efficiently absorb the radiation and heat the bolometer element. Such devices are called *composite bolometers*.

A bolometer's optimum operating point is achieved at the BLIP condition. It has proven difficult to eliminate all sources of current noise in bolometers, so they do not meet the BLIP condition in the NIR. The most sensitive 1-K bolometers currently available have an NEP of $\approx 5 \times 10^{-15}$ W Hz$^{-1/2}$. Recently developed bolometers for operation at liquid-He^3 temperatures (≤ 0.3 K) using closed-cycle refrigeration systems are capable of achieving NEPs as low as 10^{-16} W Hz$^{-1/2}$. This is the detector of choice for the FIR and for moderate-resolution spectroscopy in the TIR.

Photoconductive and photovoltaic solid-state detectors are also widely used in infrared astronomy. Both types rely on the production of carriers within the detector by the photoelectric effect. In photoconductors, the absorption of a photon results in the excitation of an electron or hole into the conduction band. Electrically, the effect of light is to increase the conductivity of the detector. Most photoconductors are Ge or Si crystals doped with small amounts of impurities. Since a minimum photon energy is required to move the carrier from the impurity level to the conduction band, photoconductors have a long-wavelength cutoff, beyond which they are completely insensitive. A bias voltage sweeps out those carriers that have been excited into the conduction band.

Photoconductors have an advantage over bolometers in that they can be operated at the BLIP condition for extremely low background levels, and the noise is limited only by the random arrival rate of photons from the source and background. Unfortunately, photoconductors have an additional noise factor of $2^{1/2}$ because of the random recombination of electron–hole pairs as the carriers are de-excited.

Photovoltaic detectors are diodes in which a carrier excited by a photon is "pushed" across the diode barrier. Thus the detector produces a current that is directly proportional to the photon flux. These detectors need no bias voltage, so there is no excess noise due to carrier recombination. The cutoff wavelength depends on the diode energy gap. InSb photovoltaic detectors, which are widely used in the 1–5-μm region, can achieve NEPs as low as 10^{-18} W Hz$^{-1/2}$. Some photoconductors, such as As : Si and In : Si, can reach comparable performance levels in the TIR if operated under very low background conditions.

Advances in solid-state technology have allowed industry to begin development of arrays of photoconductive and photovoltaic detectors for use under the low background conditions experienced with high-resolution IR spectroscopy and for observations throughout the IR spectrum from space-borne observatories. These arrays will soon be available in formats as large as 32×32, 64×64, and 128×128 pixels. They will usher in an era in which IR astronomers image moderately large fields of view in much the same way as optical astronomers image fields using CCD area detectors. IR area detectors can be employed in IR spectrometers as photographic film is deployed in optical spectrometers.

TABLE I.   The Wyoming Photometric System

λ (μm)	Δλ (μm)	Name	Zero magnitude calibration[a]			Detector
			W/cm² · μ	Janskys[b]	W/m² · Hz	
0.9	0.2	I	8.9×10^{-13}	2400	2.4×10^{-23}	InSb
1.2	0.2	J	3.3×10^{-13}	1600	1.6×10^{-23}	InSb
1.6	0.3	H	1.2×10^{-13}	1000	1.0×10^{-23}	InSb
2.2	0.4	K	4.2×10^{-14}	680	6.8×10^{-24}	InSb/bolo.
3.6	1.2	L	6.5×10^{-15}	280	2.8×10^{-24}	InSb/bolo.
4.9	0.7	M	2.0×10^{-15}	160	1.6×10^{-24}	InSb/bolo.
8.7	1.0	—	2.1×10^{-16}	54	5.4×10^{-25}	bolo.
10.0	5.8	N	1.2×10^{-16}	41	4.1×10^{-25}	bolo.
11.4	2.0	—	7.4×10^{-17}	32	3.2×10^{-25}	bolo.
12.6	0.8	—	4.9×10^{-17}	26	2.6×10^{-25}	bolo.
19.5	5.8	Q	8.7×10^{-18}	11	1.1×10^{-25}	bolo.
23.0	6	—	4.5×10^{-18}	8	8.0×10^{-26}	bolo.
33.0	22[c]	Z	1.1×10^{-18}	4	4.0×10^{-26}	bolo.

[a] There are 10% disagreements between the absolute calibrations of different photometric systems.
[b] 1 Jansky = 10^{-26} W/m² · Hz.
[c] Bandpass is limited by the actual atmospheric transmission at the time of observation.

2. Filters and Dispersive Devices

IR measurements may be classified as either photometric or spectroscopic. Photometry involves the use of broad bandpasses ($30 \geq \lambda/\Delta\lambda \geq 1$), whereas spectroscopy is performed at higher resolutions ($10^6 \geq \lambda/\Delta\lambda \geq 30$). For photometric applications, individual thin-film interference filters mounted in a slide or wheel are placed in the incident radiation beam. Several standard sets of photometric filters are currently in use by IR astronomers to cover the atmospheric windows in the NIR and TIR. The photometric system in use at the Wyoming IR Observatory is given in Table I. Table II gives the magnitudes of bright primary standard stars on the Wyoming photometric system. Broad-band filters for the longer IR wavelengths ($\lambda \geq 30$ μm) are difficult to fabricate and usually have several elements. A diffusing element is used to scatter short-wavelength radiation out of the beam, while additional interference and short/long-pass elements may be required to define the desired bandpass. Moderate-resolution spectroscopy ($200 \geq \lambda/\Delta\lambda \geq 30$) is ordinarily accomplished using circular variable thin-film

TABLE II.   Magnitudes of Selected Infrared Standard Stars on the Wyoming Photometric System[a]

λ μm	Δλ μm	Name	Standard Stars						
			α Lyr	α Boo	α Tau	β Peg	β And	β Gem	μ UMa
0.9	0.2	I	0.00	−1.67	−1.31	−0.40	−0.19	−0.11	+0.81
1.2	0.2	J	−0.01	−2.18	−1.84	−1.10	−0.82	−0.49	−0.09
1.6	0.3	H	−0.01	−2.92	−2.71	−2.08	−1.76	−1.06	−0.69
2.2	0.4	K	−0.02	−3.02	−2.86	−2.26	−1.92	−1.14	−0.88
3.6	1.2	L	−0.03	−3.14	−2.98	−2.46	−2.09	−1.21	−1.03
4.9	0.7	M	−0.03	−3.00	−2.81	−2.30	−1.89	−1.12	−0.78
8.7	1.0	—	−0.03	−3.16	−2.98	−2.46	−2.04	−1.22	−0.93
10.0	5.8	N	−0.03	−3.15	−2.97	−2.51	−2.06	−1.19	−0.95
11.4	2.0	—	−0.03	−3.21	−3.05	−2.57	−2.14	−1.22	−1.04
12.6	0.8	—	−0.03	−3.23	−3.07	−2.59	−2.05	−1.19	−1.04
19.5	5.8	Q	−0.03	−3.20	−3.16	−2.80	−2.11	−1.24	−1.01
23.0	6	—	−0.03	−3.20	−3.16	−2.80	−2.11	−1.24	−1.01
33.0	22	Z	0.0	−3.2	−3.2	−2.8	−2.1	−1.2	−1.0

[a] There are small variations between the standard magnitudes used by different observers. The magnitudes for the 23-μm and 33-μm bands are approximate.

interference filters (CVFs), for which the wave-length transmitted is a function of angle. These CVFs are rotated in front of the detector to scan the wavelength interval of interest. Typical examples of the IR-transmitting substrates used for IR interference filters are sapphire for the NIR and germanium or silicon for TIR applications. CVFWs with good transmission cannot be produced for wavelengths beyond 25 μm.

Still higher spectral resolutions require the use of dispersive optical elements such as reflection gratings or prisms constructed of IR-transmitting materials such as NaCl and BaF_2. The highest spectral resolutions ($10^6 \geq \lambda/\Delta\lambda \geq 10^4$) are achieved using Michelson and Fabry–Perot interferometers and heterodyne detection techniques.

3. IR Instrument Design

IR detectors must be operated at low temperatures and shielded against background radiation from the telescope, the surroundings, and the detection system components. As much of the IR detection system as possible must be enclosed within a cryogenic dewar vessel (Fig. 3). Incident radiation from the telescope focal plane is introduced into the dewar through an IR transmitting window. Background from the telescope and its surroundings is minimized by imaging the telescope pupil within the dewar and placing a cold pupil stop at this point (see Fig. 4). Thus, as seen by the detector, the telescope entrance pupil is surrounded by a cold, black wall that emits no radiation. Emission from filters, dispersive elements, and other optical components can be eliminated by placing them within the cold enclosure and thermally clamping them to the cold bath. It is generally adequate to cool the optical elements to liquid nitrogen temperatures. A typical 1-K detection system contains a helium vessel surrounded by a liquid nitrogen temperature radiation shield. The temperature of the liquid helium bath is lowered using a high-throughput mechanical pumping system.

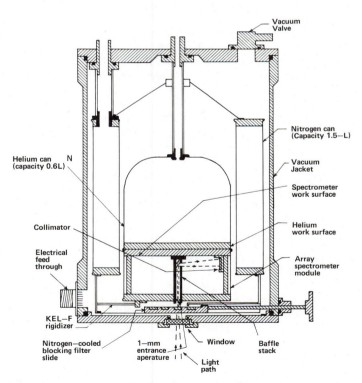

FIG. 3. Cross section of a Wyoming Infrared Observatory liquid nitrogen jacketed liquid helium dewar. [Courtesy of R. D. Gehrz and the *Publications of the Astronomical Society of the Pacific.* Gehrz, R. D., Hackwell, J. A., and Smith, J. R. (1976). *P.A.S.P.* **88,** 971.]

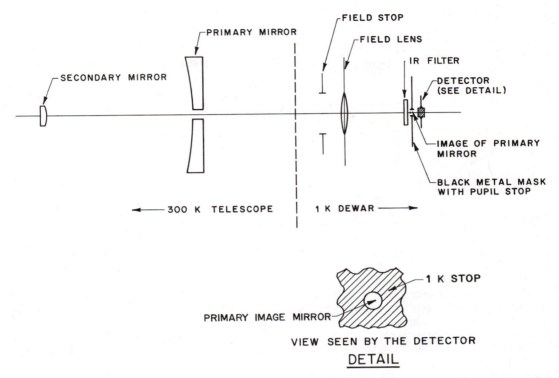

FIG. 4. Schematic diagram illustrating the cold-baffling of the entrance pupil in a typical infrared detection system.

C. BACKGROUND REMOVAL

Background radiation from the telescope and sky can be 10^6 to 10^7 times larger than the faintest signals detected in the TIR. To remove the effects of background, it is necessary to spatially modulate the background and use synchronous ac detection techniques. Typically, this is accomplished by rocking the secondary mirror of the telescope back and forth at a frequency between 5 and 20 cycles/sec. This rocking motion alternately places two patches of sky upon the detector. When a source of interest is placed in one of the two sky patches, the signal upon the detector from that patch (the source beam) is equal to the background signal plus the source signal. The signal in the other patch (reference beam) is due to background alone. If the background signals from the two sky patches are equal, then the difference in the signals from the source and reference beams will equal the signal from the source of interest (see Fig. 5).

The typical observing procedure is to demodulate the source signal electronically (subtract the signals produced in the two beams to remove the background signal) with the source placed in one of the beams and to integrate it for 5 or 10 sec. The relative positions of the source and reference beams are then reversed by beam switching, and the source signal is again integrated for the same length of time. The true source strength is given by the average source signal for these two operations. Any gradients due to spatial variations in the sky background along the chopping direction and any fixed ac signals produced by background signal from the telescope are removed by beam switching.

Although background cancellation is most effectively accomplished when the secondary mirror of the telescope is used as the spatial modulation element, schemes to cancel background have been attempted using mechanical devices near the focal plane of the telescope. However, focal plane choppers introduce excess noise, because they cannot be constructed to maintain the mechanical tolerances required to provide the detector with a stable view of the background radiation from telescope components.

In the TIR, fluctuations in the sky background can be spatially large or rapid enough so that

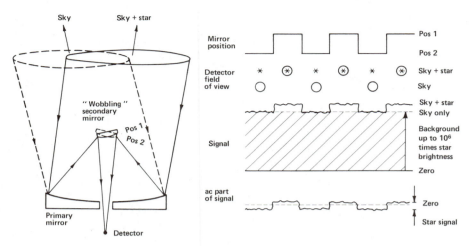

FIG. 5.   Schematic representation of background cancellation using a chopping secondary mirror.

they cannot be entirely compensated by the background cancellation technique described above. The resulting sky noise may dominate the sensitivity of the telescope and detection system combination. Sky noise is generally associated with poor sky conditions such as high and rapidly fluctuating water vapor content, nearly invisible high-altitude cirrus clouds, local blowing dust and snow, ground fog, and insects flying across the telescope aperture. Avoidance of sky noise is a major factor in the selection of prime ground-based IR observatory sites.

When the background level from the sky and telescope is highly stable, especially in the NIR, it is possible to integrate in the dc mode. Background in these cases is removed by occasionally observing a reference patch of blank sky.

D. GROUND-BASED INFRARED OBSERVATORIES

Ground-based IR observatories have telescopes that are specially designed to have low thermal emission and are situated at high-altitude sites characterized by low water vapor, clear sky, and excellent seeing conditions. A state-of-the-art IR telescope (Fig. 6) has low emissivity optics, a chopping secondary mirror for background cancellation, and extensive computer control capabilities. The latter are required to point the telescope and maintain its tracking on objects that are invisible to optical guiding techniques. Also, because the atmosphere does not scatter solar TIR radiation, it is possible to observe in the IR 24 hours a day. IR telescopes are therefore required to point and

track the telescope during daylight hours when optical tracking reference sources are obscured by scattered light.

The most advanced IR telescopes currently in use are in the 2- to 4-m aperture class, but seeing

FIG. 6.   The 2.34-m Cassegrain Wyoming Infrared Telescope. (Photograph courtesy of L. R. Shaw.)

conditions at the best sites in the world suggest that 10–15-m class telescopes can produce significant sensitivity gains in the IR. The Keck Telescope (Caltech/University of California) and the proposed National New Technology Telescope (NNTT) will have 10 and 15-m equivalent apertures, respectively, which would revolutionize high-spatial-resolution observations in the TIR region of the spectrum.

In addition to the single-aperture telescopes just described, several projects are planed that would enable the realization of high-spatial-resolution measurements in the TIR and FIR using the interferometrically combined optical outputs of several widely separated small-aperture telescopes.

E. Airborne and Space IR Facilities

All ground-based IR telescopes have very small fields of view; all-sky surveys at optical, radio, and infrared wavelengths are required to select target objects. In the IR, the pioneering all-sky surveys were made with sounding rockets launched by the IR group at the Air Force Geophysical Laboratories in the 1970s. These were followed by the highly successful NASA IR Astronomical Satellite (IRAS), which surveyed 95% of the sky in 11 months from 12 to 100 μm and produced a very detailed picture of the global characteristics of the IR heavens. The IRAS survey catalog contains about 250,000 infrared sources including stars, regions of star formation, and galaxies. In addition to the catalog, the IRAS data are available as all-sky images in four wavelength bands centered at 12, 25, 60, and 100 μm. Survey redundancy gives information about source variability on timescales of hours, weeks, and months. The Cosmic Background Explorer (COBE) mission, to be launched in the late 1980s, will study the spectrum and isotropy of the diffuse radiation of the universe at wavelengths between 1 μm and 9.6 mm. This experiment will increase our knowledge about both the 3-K cosmic background radiation and the diffuse IR backgrounds associated with the solar neighborhood and the Galaxy.

A number of observational experiments and facilities have already been developed to exploit the advantages offered by airborne and space-based IR astronomy. The Kuiper Airborne Observatory, a 0.92-m telescope mounted in a C141 aircraft, has allowed IR astronomers to obtain regular observations in the FIR and at TIR

FIG. 7. The proposed Stratospheric Observatory for Infrared Astronomy (SOFIA) is to be a 3.5-meter-class telescope mounted in a modified Boeing 747 aircraft. (Photograph courtesy of NASA.)

wavelengths obscured from the ground by water vapor. Similar observations have been made with a 0.3-m telescope flown in a Lear jet. Several balloon experiments have yielded successful observations in the FIR.

Several proposed new facilities promise to advance the state of infrared astronomy significantly over the next several decades. The Space IR Telescope Facility (SIRTF), which will be launched in the 1990s, will enable IR astronomers to obtain the first sensitive high-resolution images of the IR sky. Imaging and spectroscopic capabilities in TIR and submillimeter astronomy will be greatly enhanced by the Stratospheric Observatory for Infrared Astronomy, SOFIA (Fig. 7), a 3.5-m telescope mounted in a 747 aircraft, and the Large Deployable Reflector (LDR), an orbiting 10-m dish. SOFIA and LDR will enable sensitive observations of a host of molecular and atomic transitions, which are expected to elucidate the physical conditions in and dynamics of cool interstellar clouds in our own and other galaxies.

III. Physical Mechanisms Detected in IR Astronomy

A. Blackbody Radiation

The detection, measurement, and elucidation of the nature of sources that emit like blackbodies is a principal objective of IR astronomy. Wein's law states that the wavelength of maximum emission for a blackbody of temperature T is $\lambda T \approx 2670$ μm K. Thus spectrophotometric measurements in the 1- to 1000-μm wavelength range are sensitive to emissions from blackbodies whose temperatures range from about 3 to 3000 K. At the low-temperature end, one observes emission from the 3-K cosmic background and cold (10–30 K) interstellar dust. Cool dust embedded in molecular clouds and planetary objects has temperatures between 30 and 500 K. Circumstellar dust, planetary objects, and cool stellar companion stars emit at temperatures between 100 and 2000 K. Finally, low-luminosity and very cool, high-luminosity stars emit in the 2000–5000 K range.

B. Dust Emission

Dust can be identified by characteristic infrared spectral features when the emission is from optically thin dust grains. For example, silicate-rich materials (those with SiO_2 bonds) such as olivine and enstatite exhibit characteristic 10- and 20-μ emission features caused by the stretching and bending molecular vibrational modes. Diatomic molecules, such as SiC, exhibit only the stretching mode vibrations at 10 μm. Small iron and graphite grains emit a featureless continuous spectrum whose shape is determined largely by the grain size distribution.

Detailed spectral examination of the shapes, strengths, and doppler shifts of dust features can reveal the chemical composition and abundance of the dust as well as information about the dynamics and physical conditions of the environment in which the dust resides. When grains of any type become large, or when the shell in which the grains are contained has sufficient optical depth, the dust spectrum is black, and specific information about the composition and physical state of the grain material cannot generally be obtained. However, the physical properties of condensed ices and silicate materials in large solid bodies such as asteroids can be studied using NIR reflectance spectroscopy.

C. Gas Emission

There are two conditions under which gas will emit a continuum in the infrared. The first of these, free–free emission or bremsstrahlung, arises in ordinary thermal plasmas. Electrons passing near a positively charged nucleus or ion will be accelerated and emit radiation. Typically this process is important for gas at temperatures near 10,000 K. The process is rarely seen in its pure form, since most material also contains dust, whose emission will also be present. In the early development of most novae, before any dust has condensed, the expanding gas shell exhibits a nearly pure free–free spectrum (see Fig. 17).

The second process of continuum emission, synchrotron or magnetobremsstrahlung, is due to the spiral motion of relativistic electrons in a magnetic field. This process is important in su-

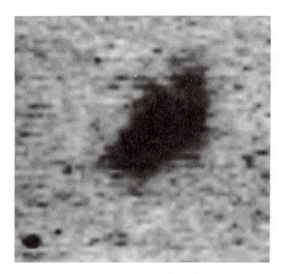

FIG. 8.   Crab Nebula at 2 μm, showing distribution of synchrotron (magneto-bremsstrahlung) emission. [Courtesy of G. L. Grasdalen and the *Publications of the Astronomical Society of the Pacific*. Grasdalen, G. L. (1979). *P.A.S.P.* **91**, 436.]

pernova remnants, such as the Crab Nebula (Fig. 8), and appears to be operating in the nuclei of active galaxies and quasistellar objects.

Unlike continuum emission, which requires rather special conditions, line emission in the infrared is characteristic of most phases of the interstellar medium. Collisional excitation is a common mode of line emission excitation. If an atom or molecule is raised to an excited state by a collision with an electron, ion, atom, or molecule, the density in the interstellar medium is usually so low that it will radiate the energy away as a photon rather than have it taken up in another collision. Thus the radiative rates for the line emission are not relevant, and lines that are "forbidden" in the laboratory are easily produced in the interstellar medium. The lines of molecular hydrogen are a good example of this process. Under laboratory conditions these lines are very difficult to produce, yet the pure rotation and vibrational–rotational spectra of molecular hydrogen have been observed in the low-density gas around both young and old stars (Figs. 9 and 10).

Line emission is diagnostic of the physical conditions in the gas, including temperature, density, chemical composition, and excitation conditions. For the lines to appear in the infra-

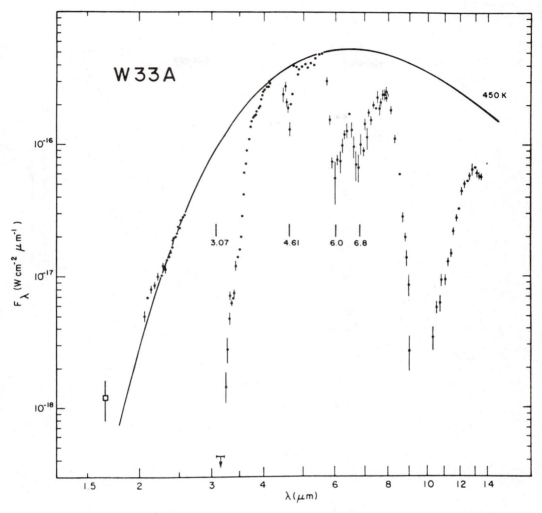

FIG. 9. IR energy distribution of W33A showing absorption features characteristic of cocoon enshrouded objects in regions of star formation. [Reprinted courtesy of B. T. Soifer and *The Astrophysical Journal,* published by the University of Chicago Press; copyright 1979 by The American Astronomical Society. Soifer, B. T., Puetter, R. C., Russell, R. W., Willner, S. P., Harvey, P. M., and Gillett, F. C. (1979). *Ap. J.* **232**, 153.]

red, the energy level differences must be quite small. In general, this means that for atoms and ions the transitions involved are between fine structure levels. The wavelengths shift with the degree of ionization from the far infrared for neutral species (CI at 610 μm), to the thermal infrared for moderately ionized species (SIV at 10.5 μm), to the near infrared for highly ionized species (SiIX at 3.9 μm). Energy levels are also close together near the ionization limit. Under most conditions these levels are not highly populated. However, in dense plasmas, recombina-

tion from the next ionization stage may provide enough transitions for observation. The hydrogen series $n \to 3$ (Paschen), $n \to 4$ (Brackett), $n \to 5$ (Pfund), $n \to 6$ (Humphreys), and $n \to 7$ have all been observed (Fig. 10).

Emission from rotational transitions of molecular species such as H_2O, OH, NH_3, HCl, and CO are believed to be significant sources of cooling in the long TIR, FIR, and submillimeter spectral regions. These lines can be used to trace the dynamical and physical conditions in molecular clouds and interstellar gas.

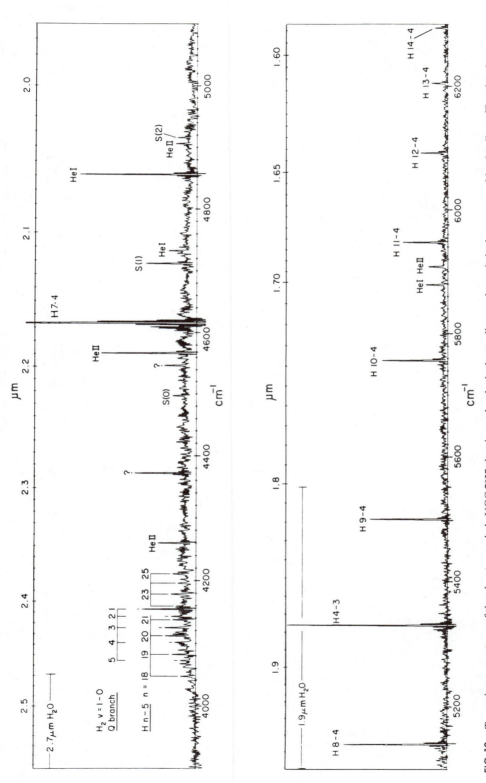

FIG. 10. Two-micron spectrum of the planetary nebula NGC 7027 showing molecular hydrogen lines and atomic hydrogen recombination lines. [Reprinted courtesy of U. Fink and *The Astrophysical Journal*, published by the University of Chicago Press; copyright 1981 by The American Astronomical Society, Smith, H. A., Larson, H. P., and Fink, U. (1981). *Ap. J.* **244**, 835.]

D. Absorption Processes

Continuous absorption of infrared radiation by atomic or molecular processes requires very large column densities found only in stellar atmospheres. For stellar temperatures below about 7000 K, the dominant source of opacity in the infrared is the free–free spectrum of H^-. At higher temperatures, the dominant source of opacity is the free–free spectrum of hydrogen (Kramers opacity). Electron scattering (Thomson opacity) dominates at the highest temperatures.

In the interstellar medium, the continuum opacity is produced by dust grains. The scattering and absorption of light by small particles produces a continuous opacity which decreases with wavelength. In the NIR, this decrease goes roughly as the first power of the wavelength, whereas ultimately, beyond the last resonances and where the wavelength is long compared to the grain size, the opacity decreases as the fourth power of the wavelength.

There may be resonances in the material composing the dust grains, which appear as broad absorption features. The most prominent of these is the feature near 9.7 μm, which has been identified as due to silicate grains (Fig. 9). This feature is much broader than those normally seen in single-mineral silicates. This additional breadth has been interpreted as indicating that the interstellar silicate grains are not stoichiometric crystalline structures but are rather amorphous. Within interstellar clouds, additional absorption features appear. A feature near 3.07 μm has been interpreted as due to water and ammonia ices (Fig. 9). Additional absorption features have been seen between 4 and 7 μm, again against very heavily obscured objects (Fig. 9). The chemical identification of these features is still uncertain.

The production of absorption lines, unlike the production of emission lines, does depend on the radiative rate coefficients of the atoms or molecules. Since the energy differences are small, only the high-lying states of ions or molecules have sufficiently large rate coefficients. Transitions involving low-lying terms are usually "forbidden" and therefore not seen in absorption. Thus atomic and ionic absorption lines appear only in stellar atmospheres, where the densities and temperatures are high enough to produce substantial populations in the high states. Molecules, on the other hand, have their pure rotation and rotation–vibration transitions in the infrared. Thus one of the most prominent absorption systems in the infrared is the vibrational–rotational spectrum of CO. This is seen in cool stellar atmospheres and as absorption features produced by the material in the interstellar medium. Interstellar temperatures are so low that the thermal broadening of the lines is very small. Therefore, observations of interstellar molecular absorption require very high spectral resolution. $\Delta\lambda/\lambda \geq 10^4$ (Fig. 11).

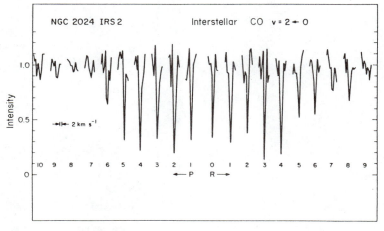

FIG. 11. Interstellar CO fundamental absorption lines seen against NGC 2024 no. 2. [Reprinted courtesy of J. Black and *The Astrophysical Journal,* published by the University of Chicago Press; copyright 1984 by The American Astronomical Society. Black, J. H., and Willner, S. P. (1984). *Ap. J.* **279,** 673.

IV. Solar System Objects

Although the known planets, moons, and other solar systems objects were first discovered and studied by the visible sunlight they reflect, the key to unlocking may secrets of the solar system lies in studies of the infrared spectral region. This is apparent by considering that the temperature T of a solid body lying a distance r from the sun is approximately.

$$T = \{L(1 - A)/16\pi\sigma r^2\}^{1/4}$$

where $L = 4 \times 10^{33}$ erg sec^{-1} is the solar luminosity, $\sigma = 5.7 \times 10^{-5}$ is the Stefan–Boltzmann radiation constant, and A is the albedo (reflectance) of the body averaged over the solar spectrum. T can be as high as 1000 K, for dust grains on the verge of evaporation close to the sun, to 10–40 K, for objects outside the orbit of Pluto. Thus the thermal emission from solar system objects lies within the range from $\approx$3 to 300 μm. [*See* SOLAR SYSTEM, GENERAL.]

A. PLANETS

Thermal infrared spectrophotometric imaging of planets and moons gives insight into their surface composition and atmospheric properties. An example is provided by recent imaging studies of the lunar surface. The 20-μm image (Fig. 12) shows the spatial distribution of thermal radiation from the moon near the crater Aristarchus approximately 10 days past new moon.

Such images, examined as a function of lunar phase, give information about the lunar geological surface composition by revealing the rate at which different surface features cool. Large boulders, which are embedded in the deeper layers of the lunar soil, cool more slowly than does loose surface dust. Similar imaging studies are useful during eclipses. Thermal lunar infrared imaging is also an attractive method for searching for evidence of regions of recent lunar volcanism, which should tend to remain hotter than the surrounding terrain during the night phase of the lunar cycle. Although the existence of such regions has been suggested, none has yet been found. High-resolution spectroscopy of the lunar surface in the 7–14-μm spectral window has shown that the reststrahlen bands can be used to identify the surface distribution of various silicate rocks and sand upon the lunar surface. Yet another type of imaging study of the lunar surface 2-μm spectropolarimetric imaging of the reflected solar light, is capable of revealing information about the surface geology of the moon.

Thermal spectrophotometric images of the atmospheres of the giant planets reveal the distribution of methane, ethane, ammonia, and clouds. The strong band structure shown in the 5-μm image of Jupiter (Fig. 13) shows the cloud structure in the atmosphere, with the warmer and deeper layers being visible in the strong equatorial bands. Imaging of the thermal infra-

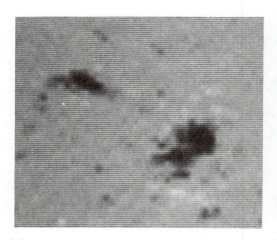

FIG. 12. Twenty-micrometer image of the moon, showing a warm region near the crater Aristarchus 10 days after new moon. The blackest regions have the highest intensity. At this time, the region is in darkness and is seen only in thermal emission. [Picture courtesy of Wyoming Infrared Observatory.]

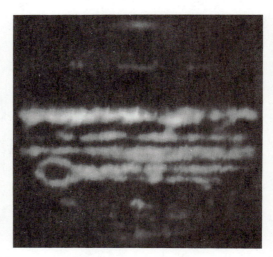

FIG. 13. Five-micrometer image of Jupiter showing cloud structure. [Courtesy of R. J. Terrile and *Science;* copyright 1977 by the AAAS. Terrile, R. J., Capps, R. W., Beckman, D. E., Becklin, E. E. Cruikshank, D. P., Beichman, C. A., Brown, R. H., and Westphal, J. A. (1977). *Science* **204**, 1007.]

red radiation from Saturn as a function of phase has revealed seasonal variations in the atmospheric pressure, temperature, cloud structure, and chemical composition. This information can be used to develop accurate theoretical models of the Saturnian atmosphere.

Full-disk infrared photometric measurements of Jupiter and Saturn have not yet satisfactorily answered the question of whether or not they have significant internal heat sources. Jupiter is an especially interesting case, because its mass is only slightly smaller than is believed necessary to sustain the nuclear fusion reactions that characterize a star.

Infrared studies can reveal the nature of the dust and rocks that comprise planetary ring structures, which appear to be a common feature of the giant planets. Ringlike structures are also seen around young, evolving stars. External galaxies are observed to have ringlike structures that are believed to have been produced during the gravitational collapse phase of galactic evolution. In this respect, solar system ring structures may be a model for many astrophysical processes. [See PLANETARY SATELLITES, NATURAL.]

B. COMETS, SATELLITES, AND ASTEROIDS

The chemical history of the solar system may be preserved in primitive solar system bodies such as comets, satellites, and asteroids. These objects contain unaltered materials that were trapped in them at the time of their formation; they are the Rosetta Stone of the solar system.

Infrared spectroscopy and polarimetry of such objects have led to an understanding of the chemical composition of comets and the composition and crystal structure of the minerals comprising asteroids and satellites.

Comets are particularly interesting because material from within their nuclei, which has been protected from the radiation environment of space since its entrapment, is released as the surface layers of the cometary nucleus are evaporated during perihelion passage. Infrared studies have already revealed the existence of small silicate dust particles in the cometary tail that is blown away from the comet by solar radiation pressure, and of large particles in the anti-tail which is drawn toward the sun by gravity (Fig. 14). Infrared studies of comet Halley suggest that the dust material spews from jets on the sunward side of the comet. Such jets were seen in the visible by the Giotto satellite. Infrared

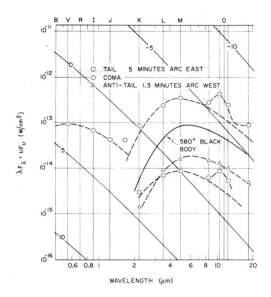

FIG. 14.   Energy distributions of the coma and tails of comet Kohoutek. The anti-tail, composed of large particles attracted toward the sun by gravity, has a blackbody energy distribution. The tail driven by solar radiation shows the emission features characteristic of small silicate grains. [Reprinted courtesy of E. P. Ney and *The Astrophysical Journal*, published by the University of Chicago Press; copyright 1974 by The American Astronomical Society. Ney, E. P. (1974). *Ap. J. (Letters)* **189**, L141.]

spectroscopy to detect ices in comets is a prime objective of the new infrared observatories.

Infrared studies have shown that the chemical composition of asteroids is a function of their distance from the sun. This discovery is fascinating because it is known that the elements in the collapsing and condensing solar system were spatially fractionated in such a way that the chemical compositions of the sun and individual planets are quite different. It is possible that asteroids retain a record of the chemical fractionation and spatial fragmentation history of the solar system.

C. THE ZODIACAL LIGHT

The existence of the zodiacal dust cloud has been known for many years through observations of the visible sunlight scattered by the dust. Recent infrared observations by AFGL sounding rockets and IRAS have mapped the thermal radiation from the dust itself and have given us insight into both the grain size distribution and the chemical composition. It has been shown that the Poynting–Robertson effect (the loss of orbital angular momentum because of the radiation pressure of the solar radiation field) should cause the solar gravitational field to sweep the zodiacal cloud into the sun at a very rapid rate. The material must be constantly replenished by asteroids and comets. IR spectroscopic studies of the chemical composition of the zodiacal cloud may reveal the sources of mass input to the cloud.

V. Infrared Observations of Stellar Sources

A. STELLAR SPECTRA

Generally, the continuum energy distribution of a stellar atmosphere approximates that of a blackbody at the effective temperature of the star. Thus the infrared continuum for most stars follows a Rayleigh–Jeans distribution, with the peak of the energy distribution being far to the blue of the infrared. Only for the coolest stars ($T \leq 3000$ K) does the peak of the energy distribution move into the NIR. For these cool stars, broad-band photometric studies have been very useful.

The primary opacity sources in these stars are H^- absorption and Rayleigh scattering by molecular hydrogen. The ratio of these two opacity sources is very sensitive to the surface gravity of the star. In low surface gravity stars (supergiants), the degree of ionization is much higher, so H^- opacity dominates, whereas in high surface gravity stars (dwarfs), scattering dominates. The deep minimum at 1.6 μm in the H^- opacity provides a powerful tool for diagnosing the surface gravity of late type stars. One problem in studies of metal-deficient stars has been the accurate diagnosis of their effective temperatures because the low metal content alters their visible colors. Since neutral and ionized metals have strong lines in the visible portion of the spectrum, these lines become much weaker as the abundances of the heavy species are reduced. By making observations in the infrared, where there are only a few weak lines, we can accurately measure the temperatures of these stars.

The infrared absorption lines from high-lying atomic and ionic states are weak because these states are sparsely populated. An additional condition that weakens IR lines is that the temperature never falls to zero in a stellar atmosphere. Typically, the temperature in the outermost layers of a stellar atmosphere is about 70%

of the temperature of the layers in which the continuum is formed. Thus for lines in the Rayleigh–Jeans portion of the energy distribution, the central depth of an absorption line cannot exceed $\approx 30\%$ of the continuum level. Because the continuum opacity rises toward longer wavelengths, the continuum formation point moves out in the stellar atmosphere, and the central depth of even the strongest lines decreases.

Even though the infrared atomic and ionic lines are not very numerous or strong in stellar spectra, high-resolution spectroscopy provides astronomers with a powerful tool for examining stellar magnetic fields. The energy difference due to Zeeman splitting of atomic and ionic energy levels depends on the Lande g of the level and the magnetic field. Since the breadth of stellar absorption lines is due to the doppler shift produced by thermal motions, the ratio of splitting to line width goes as the first power of the wavelength. Thus high-resolution observations at 2 μm are four to five times more sensitive to magnetic fields than observations at optical wavelengths.

The primary spectral features in cool stars are vibration–rotation spectra of molecules: CO, CN, and H_2O. Molecular spectra, particularly of CO, have proven useful for revealing the isotopic composition in cool stars. For ordinary atomic or ionic species, hyperfine splitting distinguishes different isotopes ($\lambda/\Delta\lambda \leq 10^4$). In molecular spectra, different atomic weights produce changes in the moment of inertia of the molecule, so that the rotational parameters are altered. Thus the bands heads of ^{13}CO are easily resolved from those due to ^{12}CO at resolutions $\lambda/\Delta\lambda \geq 10^3$ (Fig. 15).

The CO band is useful for studying very distant Galactic Stars. Because of interstellar dust, optical studies of Galactic stars are limited to a very small volume immediately around the sun. Since the interstellar medium is relatively transparent in the NIR, distant stars may be observed. The strong CO band at 2.4 μm can provide information of the surface temperature, surface gravity, and radial velocity of these stars.

B. STARS WITH CIRCUMSTELLAR SHELLS

Infrared spectrophotometry has revealed the existence of circumstellar material around many types of stars. The first TIR spectra of late-type stars, obtained by Gillett, Low, and Stein in

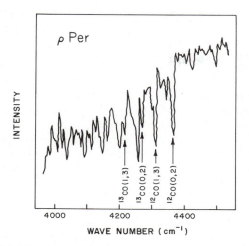

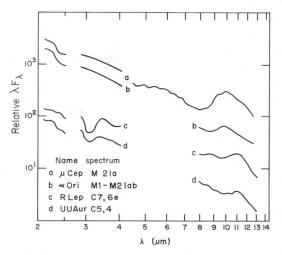

FIG. 15. CO bandheads in the star $\rho\alpha\alpha$Persei showing that isotopic species are resolved. [Courtesy of the *Astronomical Journal*. Johnson, H. L., and Mendez, M. E. (1970). *Astron. J.*, **75**, 785.]

FIG. 16. Infrared spectra of typical late-type stars which have infrared emission features. The M-stars have a 20-μm emission feature which is not shown here. [Courtesy of K. M. Merrill and the *Publications of the Astronomical Society of the Pacific*. Adapted from Merrill, K. M., and Stein, W. A. (1976). *P.A.S.P.* **88**, 285.]

1968 using a CVFW spectrometer and a bolometer, showed clear evidence of features in the 8–13-μm region. However, the calibrations were sufficiently uncertain that it was unclear whether there was a 9–12-μm feature in emission of 8.5- and 12.5-μm absorption bands. Woolf and Ney, in 1969, used infrared photometry to show that the feature was in emission and suggested that it was due to circumstellar silicate dust heated by optical and NIR radiation from the central star. Since the radiation is in excess of an extrapolation of the star's blackbody photospheric spectrum, the cooler emitting shell must have an area greater than that of the star. This argument is commonly used by infrared astronomers to deduce the presence of circumstellar material.

Figure 16 shows infrared spectra of typical late-type stars which have infrared emission features. The M-type stars, which have more oxygen than carbon (O > C), have circumstellar silicate dust that is similar in composition to many terrestrial rocks. In the carbon stars (C > O), the excess is due to SiC dust. These identifications were made by comparing the detailed shapes of the infrared emission features with the spectra of terrestrial materials. There is substantial evidence that this circumstellar dust and the gas in which it is embedded are driven away from the stars into interstellar space. It may be one of the ways in which the interstellar medium is enriched with heavy elements produced by nuclear reactions in stellar interiors.

Broad infrared emission features or "excesses" are not confined solely to late-type stars. Many hot, early-type stars have substantial infrared excesses, usually due to free–free emission from hot circumstellar gas. This phenomenon has been used to deduce the extent and density of hot gas around emission-line stars (Be, Of, and Wolf–Rayet stars). Surprisingly, some early-type stars, particularly certain WC stars (Wolf–Rayet stars with strong emission lines due to carbon), show emission features characteristic of circumstellar dust. Dust can apparently form even in the relatively hostile radiation environments around the hottest stars.

C. Dust Formation in Circumstellar Shells

Much effort has gone into studying the physical processes governing dust formation in circumstellar shells. Physical chemistry can predict condensation temperatures for different dust species if we assume that the gas is in chemical equilibrium. Relative elemental abundances can be determined from high-resolution spectra. Chemical equilibrium calculations imply that dust composed almost entirely of silicates will condense around M-stars for gas temperatures ≤ 1200 K. In carbon stars, carbon particles condense for $T \leq 1800$ K, and SiC particles form for $T \leq 1300$ K. The dependence upon the C to O ratio in the star's atmosphere occurs because the formation of CO in the cooling gas consumes either all of the oxygen or all of the carbon. This affects the chemistry of the shell and of subsequent dust condensation. Chemical equilibrium calculations are greatly simplified. In particular, they ignore the stellar radiation and the possibility that the gas can fall below the predicted condensation temperature (supersaturate) if no condensation nuclei are present. Nonetheless, their predictions are remarkably close to the observations. In all cases studied, once the gas temperature falls below the condensation temperature predicted by the chemical equilibrium calculations, dust formation quickly follows.

It is not clear how dust is able to form in the highly ionized environments surrounding very hot stars such as novae and emission-line stars. Perhaps the material is very clumpy and shields some of the matter from the highly energetic stellar radiation. The presence of dust around hot stars does show that, if heavy elements are present, dust will form as soon as the gas temperature and pressure allow it to, independent of the presence of ionizing radiation. This may explain why it is so common to observe dust around stars that are losing mass.

Most stars with IR excesses have been losing mass for a long time, so the circumstellar dust density has reached a "steady-state." Thus the observed infrared excesses do not vary with time, unless the luminosity of the star that is heating the dust changes. One important exception is dust formation around Galactic novae. These objects are white-dwarf stars that slowly accumulate matter on their surfaces, usually from a nearby companion star. When enough matter has accumulated, a brief runaway nuclear reaction occurs on the surface of the white dwarf, the luminosity of the star suddenly increases by a factor of up to 10^6, and about 10^{-6}–10^{-5} solar masses of gas are ejected from the star. Several of these novae (*nova* means *new star*) are discovered each year, many by amateur observers. Infrared studies show that dust forms around most novae 30 to 100 days after the explosion. The delay depends upon the velocity with which the gas is expelled from the star and the luminosity of the explosion, which determine the time after outburst when the gas temperature falls below the dust condensation temperature. All of the available heavy elements are converted into dust within 20–40 days of the onset of dust condensation. Dust forms even

though the nova is emitting large quantities of ultraviolet and soft X-ray radiation that could interfere with the chemical processes. In some cases, the dust is so thick that it blocks almost all ($\geq 99\%$) of the visible light emitted by the nova. Other dust shells intercept $\leq 1\%$ of the visible light, and the dust is seen only because it reradiates in the TIR the short-wavelength energy that it intercepts.

Figure 17 shows how the infrared energy distribution of a typical nova changes with time. The energy distribution for day 26 is characteristic of bremsstrahlung from a hot plasma. The onset of dust formation (day 34) is signaled by the appearance of a blackbody infrared excess that rises in intensity as the grains grow and intercept more radiation from the hot core. Maximum shell opacity occurs on day 75. Spectrum for day 3 gives the outburst luminosity, which is maintained by the hot core for several hundred days.

Carbon particles are the most common condensate in the circumstellar shells of Galactic novae. Two notable exceptions to this were Nova Aquilae 1982 and Nova Vulpeculae 1984 #2. Nova Aquilae 1982 showed a peculiar feature in its 8–13-μm spectrum, which fitted none of the spectra from either M-type or C-type stars and so is unlikely to be due to normal silicate or SiC dust. Nova Vulpeculae 1983 #2 showed

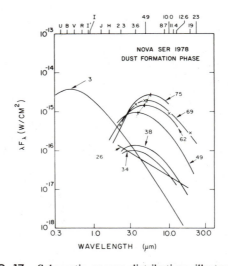

FIG. 17. Schematic energy distributions illustrating the temporal development of the dust shell of Nova Serpentis 1978. [Reprinted courtesy of R. D. Gehrz and *The Astrophysical Journal*, published by the University of Chicago Press; copyright 1980 by The American Astronomical Society. Gehrz, R. D., Grasdalen, G. L., and Hackwell, J. A. (1980). *Ap. J.* **237,** 855.]

very strong forbidden 12.8-μm [NeII] emission after 6 months and eventually formed silicate grains in its ejecta. More than two years after its eruption, this nova developed intense infrared coronal emission lines that dominated the luminosity of its shell. Because the gas emitted from novae is the result of nuclear reactions on the surface of a white dwarf, the relative abundances of the elements in the ejecta can be expected to be affected by material from the surface layers of the white dwarf. Peculiar chemical abundances resulting from an expulsion on the surface of a Mg–Ne–O-rich white dwarf may be the reason for the strange features seen in Nova Aquilae and Nova Vul.

D. Mass-Loss Processes

Infrared measurements of a star with an optically thin dust shell can be used to estimate the dust density. If these data are combined with high-resolution spectroscopy of the circumstellar gas, which gives an outflow velocity for the shell material, we can estimate the mass-loss rate of matter from the star. Table III summarizes the results of this technique as applied to late-type stars, early-type stars, and novae. Because most of the stars in this table have masses between 1 and 10 times that of the sun, it is clear that they cannot long continue to lose mass in this way. One to 10 million years is an upper limit to the length of these mass-loss phases, or about 1% of a stellar lifetime. These mass-loss phases probably represent the end of the evolution of most of the stars listed in Table III.

TABLE III. Typical Mass Loss Rates from Stars in the Galaxy

Type of Star	Approximate Mass-loss per Year (solar masses)	Estimated Number in Galaxy	Total Annual Mass Contribution to Galaxy (solar masses)
M-type variable	10^{-6}	2×10^6	2
C-type variable	10^{-6}	2×10^4	0.02
Late-type supergiant	10^{-5}	10^4	0.1
Emission-line star	10^{-7}	10^6	0.1
Planetary nebula	10^{-6}	10^4	0.01
Galactic nova	10^{-5}/outburst	40/year	0.0004
Supernova	25/outburst	1/100 years	0.25
Total	—	—	2.5

About 2.5 solar masses of gas and dust are ejected into the interstellar medium (ISM) each year by stars in our galaxy. It is presumably from this gas and dust that the giant molecular clouds, and ultimately new stars, will form. Because the stars that are ejecting material are at the endpoints of their lifetimes, their envelopes are enriched in the heavy elements that were produced in their interiors by stellar nuclear reactions during their main sequence evolution. This enriched material is added by the mass-loss process to the pre-existing ISM and can therefore be incorporated into new generations of stars at their formation time. It is believed that star formation converts a similar amount of interstellar material into new stars each year. Thus infrared astronomy helps us to follow the important phase of Galactic evolution that connects dying stars to forming ones.

VI. Observations of Regions of Star Formation

Infrared astronomy is a unique tool for studying star formation. Collapsing stars are obscured by dense circumstellar dust clouds. Furthermore, the sites of star formation are deeply embedded within the dustiest regions of the interstellar medium. Thus young stars must be detected by the thermal radiation emitted by their dust cocoons, through techniques that can probe deeply into the dusty areas in which they reside. Techniques that have proved to be especially useful are 2–20-μm spectrophotometric imaging of regions of star formation, FIR photometry and spectroscopy of the large-scale structure of molecular cloud cores, and NIR/TIR spectroscopic studies of compact embedded sources.

A. Star Formation in the Cores of Molecular Clouds

Star formation in our galaxy proceeds within the dense, warm cores of giant molecular clouds (GMCs). Although the large-scale morphology of these cold clouds is best mapped in the radio emissions of CO, the cloud cores themselves are strong emitters in the TIR and FIR. When the density of the matter in a GMC core is compressed enough so that the gravitational pressure in a localized region can exceed the cloud gas pressure, a stellar-mass sized condensation can collapse to form a single star (or small cluster of stars) surrounded by a disklike structure in which planetary condensations might form. Infrared spectroscopic studies of GMC cores have revealed the presence of silicate dust, graphite, and a number of organic molecules containing the elements that are the basis of life.

The collapse process is believed to occur in several distinctive phases, all of which emit primarily in the infrared. Localized condensations on the verge of collapse appear as unresolved or barely resolved knots in the molecular cloud core. Since the dust temperature is quite low at this time ($\leq$30–50 K), FIR (50–300 μm) mapping is most effective for revealing the initial stage of stellar formation. Molecular spectroscopy in the FIR can be used to study the chemical composition and physical conditions in the extended cool molecular cloud core and in localized condensations. Rotational transitions of the metal hydrides, for example, can be used to trace the metal abundance throughout the cloud core and give insight into the molecular chemistry in the cloud. Polarimetric measurements in the FIR can be used to determine the effects of the local interstellar magnetic field on the cloud core. Some of these studies have already been carried out using the KAO and balloon platforms. A deficiency of current FIR maps of GMC cores is their low spatial resolution. The smallest angle that can be resolved by a telescope of aperture D observing at a wavelength λ is given by the Rayleigh criterion, $\theta \approx 1.22\lambda/D$. For existing 1-m FIR telescopes, this angle is $\approx$90 arcseconds at 350 μm. Images in the NIR and TIR, on the other hand, can be obtained with spatial resolutions as high as 1 arcsecond using existing telescopes. Thus resolution of comparable details at 350 μm requires very-large-aperture telescopes or long-baseline interferometers. LDR and SOFIA will be especially useful for FIR imaging and spectroscopic studies of regions of star formation.

B. PROTOSTARS

Once gravitational pressure has exceeded the gas pressure in a localized condensation, the cloud will undergo a gravitational free-fall collapse, during which the evolution of the stellar system is governed by gravity alone. This phase, called the *protostellar phase,* is extremely short-lived. The free-fall time is directly proportional to cloud mass and ranges from about 10^3 to 10^6 years for clouds with masses in the range from 0.1 to 100 solar masses. Again, the cool temperature (T $\leq$ 100–200 K) of the collapsing dusty

cloud renders it observable only in the TIR and FIR. Astronomers have not yet positively identified a star in the true protostellar collapse phase. Condensations detected in nearby molecular clouds by IRAS have been suggested as protostellar objects, but this hypothesis remains to be proven by supporting observations with other facilities. TIR imaging of GMC cores using area detectors on the largest existing and proposed telescopes should identify a number of protostars. Spectroscopic and temporal studies of spatially resolved protostellar objects are crucial to our understanding of early stellar evolution. It is believed that the spatial morphology of planetary systems, the spatial fractionation of elements, and the complex chemistry of such systems originate during this phase.

C. THE DISTRIBUTION OF GAS AND DUST

Once the central density of a collapsing protostar becomes high enough that collisions between the gas molecules can thermalize the energy of the free fall, the internal energy of the system begins to rise. When the contraction has generated sufficient pressure and heat in the central core, nuclear burning will begin. At this point, the central condensation is a true star, even though it may still be deeply embedded within and visually obscured by the remnant dust and gas cocoon of the collapsing protocloud. Infrared astronomy provides the only way to observe the spatial distribution, chemical makeup, and dynamical motions of such systems.

As the nuclear burning phase develops, the central star begins to dissipate the remnants of the collapsed protocloud. Heat from the star vaporizes the dust nearest the star, and radiation pressure from the star, acting in conjunction with a stellar wind of high energy particles, blows the remaining dust away. These processes can rapidly accelerate dust grains to supersonic velocities within the circumstellar region, so that the gas phase of the protocloud gets dragged along by the dust grains. In this manner, a young star can dispose entirely of its placental material within a few tens of millions of years. The FIR provides many spectroscopic indications of the warm molecular gas produced by this outflow. Particularly important are the high-J rotation transitions of CO, which are excited in molecular gas over the temperature range from 100 to 2000 K.

Molecular cloud cores containing luminous

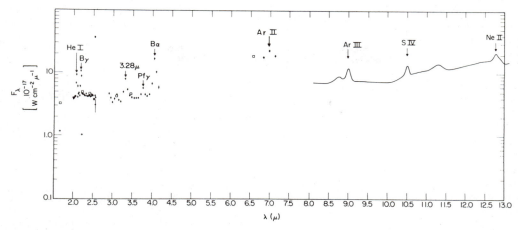

FIG. 18.   Infrared spectrum of W3 showing recombination and fine-structure emission lines present in young stellar sources embedded in regions of star formation. [Reprinted courtesy of T. Herter and *The Astrophysical Journal*, published by the University of Chicago Press; copyright 1981 by The American Astronomical Society. Herter, T., Pipher, J. L., Helfer, H. L., Willner, S. P., Puetter, R. C., Rudy, R. J., and Soifer, B. T. (1981). *Ap. J.* **244,** 511.]

young stars that have dissipated portions of their circumstellar cocoons can emit a complex spectrum of gas and dust emission over an extended spatial area (see Figs. 9 and 18). If the young stars are massive, they are hot enough emit a substantial ionizing photon flux, which gives rise to an extended ionized hydrogen (HII) region in the surrounding cloud core. The infrared spectrum of the core is characterized by free–free emission from the hot gas, by emission lines from the various atomic and molecular species contained within the gas phase, and emission from the dust grains that are heated by the stellar radiation and hot gas (Fig. 18).

Evolved massive, young stars near the edge of a GMC appear in a relatively clear field, and the adjacent molecular material shows HII emission from a "blister" caused by the expansion of the ionizing radiation from the young star into the GMC.

Infrared imaging and spectroscopy have shown that many luminous young stars embedded in molecular cloud cores are bipolar nebulae. There is evidence for high energy outflow of the gas in the bi-conical lobes. These observations suggest that star formation is accompanied by the collapse of high density disk-like structures that may be sites for the formation of planetary systems. Infrared images of the core of the S106 molecular cloud similarly reveal the presence of a luminous young star embedded in a cold rotating disk and show that a high-energy outflow is occurring in the bipolar lobes lying

above and below the disk structure. The molecular cloud core of Mon R-2 shows a blister HII region and several condensations that are sites where massive stars have recently formed (Fig. 19).

Low-mass stars do not become hot enough to ionize any gas, but the dust surrounding these

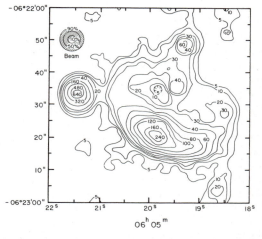

FIG. 19.   Mon R-2 at 20 μm. The isophotes reveal a "blister" HII region with embedded condensations. Contour levels are in units of 10^9 Jy sr^{-1}. [Reprinted courtesy of J. Hackwell and *The Astrophysical Journal*, published by the University of Chicago Press; copyright 1982 by The American Astronomical Society. Hackwell, J. A., Grasdalen, G. L., and Gehrz, R. D. (1982). *Ap. J.* **252,** 250.]

objects reflects stellar radiation and can be distinguished because the reflection polarizes the infrared light. Such sources are often obscured by cold dust in dark clouds and must be sought by penetrating the material with imaging techniques in the NIR.

D. PHYSICS OF REGIONS OF STAR FORMATION

Infrared astronomy has already had a profound impact upon our understanding of star formation. In particular, the energetic outflows emanating from luminous young stars discovered by radio and IR techniques were not predicted by theory. IR imaging has also revealed the morphology of disklike structures in many young stars. Today, IR astronomy is in a unique position to address several other crucial questions.

The onset of star formation process is poorly understood. It is widely believed that shock compression of GMC material is required to precipitate stellar collapse. Mechanisms proposed include shocks produced by supernova events, cloud–cloud collisions, the high-energy outflows from young stars themselves, and sonic waves propagating through galaxies as they evolve. Infrared imaging and spectroscopics studies of regions of star formation in our own and other galaxies will reveal the motions, geometries, and excitation temperatures associated with the star formation process.

The thermal properties of GMCs suggest that the clouds should be in free gravitational collapse. The implied star formation rate from freely collapsing clouds is several orders of magnitude higher than the rate actually inferred from the amount of interstellar material remaining in the Galaxy. Star formation appears to be self-regulating; large-scale dynamical motions within the clouds show that sufficient energy is being imparted to them to impede the star formation rate. It seems possible that the high-energy bipolar outflows emanating from young stars recently formed within a GMC core return sufficient energy to the clouds to power the large-scale motions that inhibit excessive star formation. Infrared imaging and dynamical studies through infrared spectroscopy should reveal the detailed processes whereby the outflow energy is coupled back into the GMC material.

Finally, high-resolution NIR and TIR imaging of regions of star formation with powerful new facilities such as NNTT, SIRTF, and LDR might be expected to reveal the elusive protostellar phase of the collapse. The extraordinarily high spatial and spectral resolution available with NNTT and LDR might allow us to actually investigate the dynamical and chemical processes that occur during the formation of planetary systems.

VII. Dust and Gas in the Interstellar Medium

Infrared astronomy is a unique tool for investigating the dusty and gaseous material pervading the space between the stars.

A. THE INTERSTELLAR EXTINCTION LAW

A close look at the Milky Way in Cygnus with the naked eye reveals a dark band down the center: the Great Rift. This rift does not represent a real absence of stars in the midplane of the Galaxy but is due to the extinction of starlight by interstellar dust grains. This extinction makes it difficult to observe stars in the Galactic plane at distances of more than 1 or 2 kiloparsecs, our local "swimming hole," as Walter Baade called it. These distances are not large compared with the 30-kiloparsec diameter of the Galaxy or the distance of 10 kiloparsecs to the Galactic center.

In the infrared, the extinction by the interstellar dust is much reduced. In magnitudes, the opacity at 2 μm is 10 times smaller than the opacity at 0.5 μm. Thus a star extinguished by 20 magnitudes at 0.5 μm (Fig. 20), which is not an exceptionally large extinction for a distant star in the Galactic plane, will appear 10^{-8} times fainter because of interstellar extinction. At 2

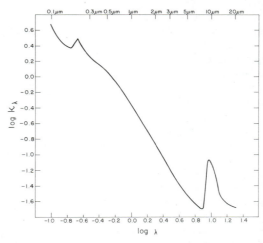

FIG. 20.   The interstellar reddening law is illustrated. The logarithm of the interstellar opacity (normalized at 0.55 μm) is plotted against the logarithm of the wavelength. [Adapted from Savage, B. D., and Mathis, J. S. (1979). *Ann. Rev. Ast. and Ap.* **17,** 73; and from Merrill, K. M. (private communication).]

μm, the same dust will extinguish the light from the star by a factor of only 0.16, a ratio of nearly 10^7 in the relative transmission at the two wavelengths! At longer wavelengths, the relative transmission is even larger. This illustrates the tremendous power of infrared astronomy to study objects that are heavily obscured by interstellar dust. The obscuration may be due to extinction by dust within molecular clouds, or simply because of the great distances light from distant stars must traverse within the Galactic plane. Because radiation from late-type stars peaks in the 1–2-μm region, surveys of the Galactic plane at these wavelengths are the only way to determine the distribution of stars in our own Galaxy.

B. IR OBSERVATIONS OF THE ISM

Studies of the wavelength dependence of interstellar extinction, often called the *interstellar reddening law,* provide important information about the composition and size distribution of interstellar dust. Details of the interstellar extinction curve depend upon the Galactic longitude, but the approximately λ^{-1} law seems to be universal in the NIR. The 10-μm silicate absorption is seen where the total extinction is high: toward the Galactic center, in the spectrum of the heavily extinguished supergiant star VI Cyg #12, and for stars embedded in molecular clouds (Fig. 11). This silicate dust may have its origin in the mass-loss from late-type M-stars discussed in Section V,D above. Many stars embedded in molecular clouds also show a 3.1-μm absorption feature due to ice, but this is not usually seen in stars extinguished only by the normal interstellar medium. Probably the ice accretes onto dust grains that are protected from ultraviolet and visible radiation inside dense molecular clouds but is sputtered from grains in less dense interstellar space.

Measurements of the infrared interstellar extinction law require observations of reddened stars whose intrinsic energy distributions are well known. The interstellar contribution is deduced by comparing the observed (apparent) and intrinsic energy distributions of such stars. Reddened stars culled from surveys at visible wavelengths are generally not extinguished heavily enough in the infrared to permit accurate determination of the infrared portion of the interstellar reddening law. The IRAS survey contains about 250,000 sources, most of them stars. Astronomers will be able to select samples of highly reddened stars from this survey for ground-based follow-up studies of the infrared extinction law.

Some progress in understanding infrared reddening by the ISM has been made by studying distant luminous stars detected by their radio OH maser emission, which easily penetrates the ISM. The distances to these stars can be estimated by means of their radial velocities determined from radio observations, and their infrared energy distributions can be measured. The NIR extinction and the depth of the 10-μm silicate band appear to correlate with increasing distance for these OH/IR stars. The interstellar opacity implied by measurements of these objects is 2 to 4 times higher than that implied by optical observations of distant supergiants. Either optical observations are biased by selection of stars that lie in directions of relatively low extinction or some of the extinction in OH/IR stars is circumstellar.

It is tempting to think of interstellar dust as uniformly permeating the Galactic plane, with the density falling at higher Galactic latitudes, and with a few embedded clumps representing molecular clouds. But the IRAS results have completely changed the way in which astronomers think of the distribution of interstellar matter. Interstellar dust is heated by the stellar radiation field to temperatures between 5 and 50 K; thus we expect the dust extinction to turn to emission at FIR wavelengths. Figure 21, which

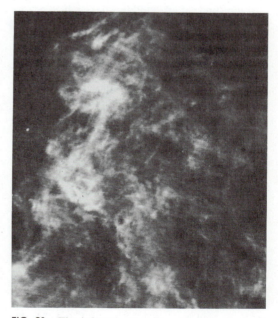

FIG. 21. The infrared cirrus near the Galactic north pole as mapped at 100 μm by IRAS. (IRAS survey picture courtesy of NASA.)

is a view of the sky near the north Galactic pole
at 100 μm, shows the irregular distribution of the
infrared emission by interstellar dust. Because
of their resemblance to earthly cirrus clouds,
these structures are called *infrared cirrus*.

FIR/submillimeter spectroscopy, only now
coming into its own, is proving to be an enormously useful tool for tracing the cool gaseous
components of the ISM condensed into molecular clouds. Observations of high-J molecular
transitions, especially those of the metal hydrides, may trace the metal abundance throughout the Galaxy and yield new information about
interstellar chemistry. Recent measurements
with KAO suggest that FIR fine-structure transitions can be useful for tracing the atomic and
ionic constituents of both dense and diffuse
clouds in the ISM.

VIII. Infrared Observations of Galaxies

A. THE CENTER OF OUR GALAXY

Because the infrared is transmitted through
the interstellar medium, we can observe the center of our Galaxy in the IR directly. In the NIR,
the dominant structure is the concentration of
stars toward the center of the Galaxy (Fig. 22).
These stars are cool giants in the last stages of
their life. Detailed studies of their chemical compositions and radial velocities have yet to be undertaken. The chemical history of the material
near the Galactic center may be very different
from that of the material in the solar neighbor-

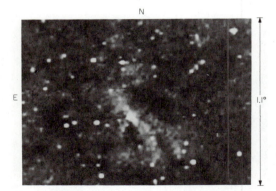

FIG. 22. Wide-angle view of the Galactic center at
2.2 μm. [Courtesy of E. E. Becklin and the *Publications of the Astronomical Society of the Pacific*. Becklin, E. E., and Neugebauer, G. (1978). *P.A.S.P.* **90**,
657.]

hood. Stellar velocities reveal the gravitational
structure of the Galactic center region. Since
stellar velocities are not easily influenced by
anything except gravity, they therefore provide
the most direct measure of the mass distribution
in the Galactic center.

The observed properties of our Galactic center are similar to, though less energetic than,
those of active galactic nuclei and quasars. The
energetic activity occurring near the center of
our Galaxy is manifested by radio, TIR, FIR, X-
ray, and even gamma ray emission. Although
the total energy involved is only $1-3 \times 10^7$ solar
luminosities, we can study this activity at much
higher spatial resolution in our Galaxy than in
any other. This provides a unique opportunity to
unravel the processes that may be occurring in
other more violent galactic nuclei. In the TIR,
the Galactic center region is dominated by a set
of filaments and small sources. These are hot,
gaseous filaments containing warm dust, which
correspond quite closely to a similar set of filaments seen in the radio regime. TIR line emission is also seen from the hot gas in the filaments. High-resolution spectral studies using
Fabry–Perot interferometers show that the gas
has velocities of 100–200 km sec^{-1}. If one accepts these velocities as due to gravitation from
a central object, then there must be 5 million
solar masses with the inner 2 parsecs of the galaxy! The energy source for the ionization and
heating of these filaments has yet to be discovered, but likely candidates are massive black
holes and major bursts of star formation. Attention currently centers upon the nonthermal radio
point source IRS 16.

In the FIR, we observe a large cloud of cool
material 2 by 4 parsecs in size surrounding the
Galactic center (Fig. 23). The total luminosity of
this cloud is 2×10^6 solar luminosities. Apparently this is a disk of material heated by the
luminous nuclear object. The dust density actually decreases close to the nucleus. The role of
this cool material in the evolution of the Galactic
center is not yet well understood.

B. ACTIVE GALACTIC NUCLEI

The term *active galactic nucleus* means any
galactic nucleus that emits nonstellar light. The
objects involved span an enormous range of
types. They include objects like our own Galactic center, Radio galaxies, Seyfert galaxies, and
exotic objects like BL Lac and quasars.

There are two major types of infrared bright

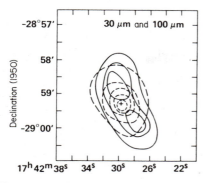

FIG. 23. FIR map of the Galactic center showing the energy distributions at 30 and 100 μm. [Reprinted courtesy of E. E. Becklin and *The Astrophysical Journal*, published by the University of Chicago Press; copyright 1982 by The American Astronomical Society. Becklin, E. E., Gatley, I., and Werner, M. W. (1982). *Ap. J.* **258**, 135.]

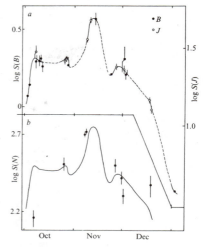

FIG. 24. Outburst of the BL Lac object AO 0235 + 164 showing rapid large-amplitude variations at 0.44 μm (B), 1.25+ μm (J), and 10 μm (N). [Courtesy of G. H. Rieke and *Nature*. Rieke, G. H., Grasdalen, G. L., Kinman, T. D., Hintzen, P., Wills, B. J., and Wills, D. (1976). *Nature* **260**, 754.]

galactic nuclei. One class is dominated in the NIR and TIR by nonthermal emission, which is diagnosed by spectral shape and rapid variations in brightness. The most violent examples of such variability are the BL Lac objects, whose luminosity may change by a factor of two in a few days. The size of these objects must be incredibly small, less than 10 light days across. The total energy involved in an outburst of a BL Lac object is enormous; the outburst of the object AO 0224 + 164 in 1975 (Fig. 24) released an energy equivalent to a rest mass of ≈1 solar mass. Similar processes appear to be operating in QSOs and Type I Seyfert nuclei.

The second class of active galactic nuclei has recently undergone a major burst of star formation very near the nucleus. In some nearby cases, this region of star formation can even be resolved: for example, NGC 253 (Fig. 25). This is diagnosed by the presence of dust emission in the TIR, thermal radio flux from gas ionized by young hot stars and in some cases by young supernova remnants. At even moderate distances from us, these central regions can only be marginally resolved. For example, the Seyfert galaxy NGC 1068 at a distance of 16 megaparsecs is only just resolved at 10 μm by the Palomar 5-m telescope. Beyond this distance it becomes very difficult, on the basis of spatial information alone, to distinguish the two types of nuclear activity.

Observations of the continuum energy distribution through the TIR and FIR into the radio regime are useful in distinguishing between thermal and nonthermal processes. These observa-

tions are extremely difficult with present-day equipment but have been very valuable in demonstrating the nonthermal character of a number of distant sources. Why some galaxies have undergone such massive bursts of star formation in their inner regions remains an enigma. Interaction with another galaxy is a strong possibility in some cases.

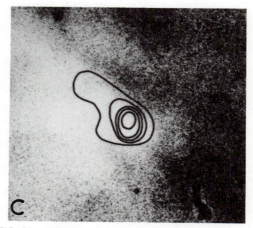

FIG. 25. TIR flux contours superimposed on an optical image-tube picture of NGC 253. [Reprinted courtesy of G. H. Rieke and *The Astrophysical Journal*, published by the University of Chicago Press; copyright 1975 by The American Astronomical Society. Rieke, G. H., and Low, F. J. (1975). *Ap. J.* **197**, 17.]

C. Star Formation in Other Galaxies and Galactic Evolution

Many important questions about star formation can be addressed only by IR studies of entire galaxies. Such studies are often unfeasible for our own Galaxy. Sensitive ground-based NIR and TIR measurements require small (1–30 arcseconds) photometer beams; the time required to map the star formation process throughout our Galaxy with such systems is prohibitive. Furthermore, heavy obscuration by the interstellar dust in our Galactic plane (see Section VII) prevents infrared observations of star formation regions much more distant than the Galactic center. About 2×10^4 new IR bright galaxies, including starburst galaxies and active galactic nuclei, have been found by the IRAS survey. These will provide fertile ground for studying the star formation process.

Infrared studies of the morphology of star formation on a galactic scale in other galaxies can provide insight into the shock-front patterns associated with regions of star formation, the manner in which high-energy outflows from active star formation regions interact with the adjacent interstellar medium, the efficiency and rate of the process as a function of position within a galaxy, and the way in which star formation has proceeded in galaxies of different types. These studies are particularly important, because the rates of star formation and stellar evolution are thought to control the chemical evolution of galaxies. This evolution proceeds as nuclear fusion within stars transforms the primordial hydrogen- and helium-rich material of a galaxy into heavy elements.

The NIR is especially suitable for the study of galaxies. During galactic formation, a large fraction of the mass is converted rapidly into low-mass, cool stars which burn their nuclear fuel very slowly. For this reason, most of the stars in galaxies are nearly as old as the galaxy itself ($\approx 10^{10}$ years); all of the ancient, massive, fast-burning stars have already died. For low-mass stars, the final phase of evolution is the red giant phase. It is apparent that the oldest stellar populations, including the cool luminous red giants

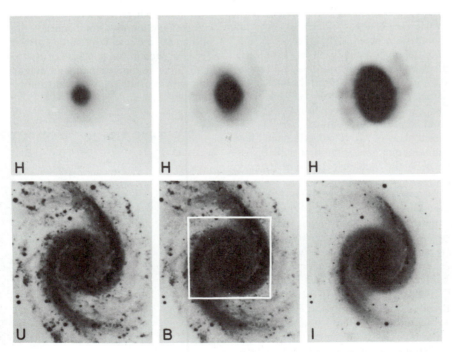

FIG. 26.   Infrared and optical images of NGC 1566 showing that a barred structure becomes prominent at infrared wavelengths. The top row shows three different contrast levels of the same H (1.6 μm) data. Several distinct stellar populations representing different epochs of star formation in the evolution of this galaxy appear to be represented. [Reprinted courtesy of J. Hackwell and *The Astrophysical Journal*, published by the University of Chicago Press; copyright 1983 by The American Astronomical Society. Hackwell, J. A., and Schweizer, F. (1983). *Ap. J.* **265**, 643.]

that represent the death phase of the oldest among them, emit most of their energy in the NIR. Thus optical and ultraviolet studies of galaxies are dominated by the hot young stars with very small mass-to-light ratios, whereas infrared studies reveal the old, low-luminosity populations that have high mass-to-light ratios. Observations in the NIR should trace the distribution of mass in a galaxy more faithfully than observations at any other wavelength.

The beautiful spiral patterns seen in optical photographs of disk galaxies are thought to arise from density waves in the disks, which compress the interstellar medium. The compression initiates star formation along the crests of the waves. NIR observations, because they trace the mass distribution, are a powerful tool for the study of the density waves. Such studies have already shown much higher wave amplitudes in the disk than were expected on theoretical grounds. Tracing the true mass distribution in NIR studies offers the best hope for revealing the sources of excitation of density waves. Several recent infrared imaging observations of other galaxies demonstrate the enormous power of such NIR studies. Figure 26 shows infrared imaging observations of the southern galaxy NGC 1566 that reveal different spiral patterns as a function of wavelength. Presumably, different stellar populations, and by inference different epochs in the star formation history of the galaxy, are being observed. Similar NIR images of other galaxies have shown that the rate of star formation in spiral galaxies like our own is directly proportional to the density of the interstellar gas.

The evolution of the elements is controlled by the star formation history in a galaxy. Primeval material, processed through stars, is contaminated by the ashes of nuclear burning. The mix of heavy elements produced depends on the mass of the star. Element abundances depend in a complex way on the number of previous generations of stars and on their mass distribution. The composition of the interstellar medium may influence the mass distribution of young stars. To comprehend fully the evolution of galaxies requires another level of sophistication in our understanding of the star formation process. By observing the mass distribution in the NIR and the star formation processes in the TIR and FIR, astronomers hope to approach this problem directly.

The star formation process may contribute to the anomalous and high energetic output of peculiar galactic systems such as Markarian 171

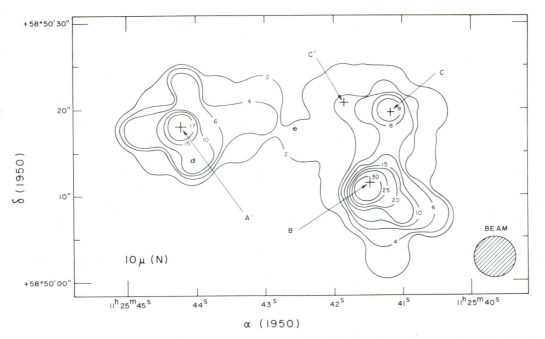

FIG. 27.   A 10-μm map of the peculiar galaxy Mkn 171 (NGC 3690) showing starburst regions. [Reprinted courtesy of R. D. Gehrz and *The Astrophysical Journal,* published by the University of Chicago Press; copyright 1983 by The American Astronomical Society. Gehrz, R. D., Sramek, R. A., and Weedman, D. W. (1983). *Ap. J.* **267**, 551.

(Fig. 27). Such galaxies appear to support enormous bursts of star formation in which the formation rate greatly exceeds that ordinarily inferred for star formation regions within our own Galaxy. These star bursts may indicate a transient phase in galactic evolution. Alternatively, they may be indicative of collisional interactions among several galaxies or between a galaxy and its local intergalactic environment. Presumably, shock fronts associated with the collisions, generated in the interstellar gas of the galaxies, produce exceedingly luminous bursts of star formation. The peculiar galaxy Arp 220 has a luminosity approaching that of a quasar and may be too energetic to be explained by star bursts alone. A massive energetic source in the galactic nucleus may be involved. Infrared imaging and spectroscopic studies of such systems in the NIR and TIR are capable of elucidating these physical processes.

D. Protogalaxies

As we look to larger and larger distances, we are observing objects as they existed in the past. In principle, it is possible to observe the evolution of the universe directly. The expansion of the universe conveniently sorts different epochs by their redshift Z, where

$$Z - 1 = \lambda(obs)/\lambda(emitted) = R(obs)/R(emitted)$$

and the values of R are measures of the size of the universe when the light was observed and first emitted.

Current thinking about the origin of galaxies places their formation in the radius R of the universe somewhere between $R(now)/R(then) = Z + 1 = 5$ and 1000. The lower limit arises from the observation of QSOs at redshifts of $Z = 4$. If QSOs, as is widely thought, are extreme examples of active galactic nuclei, this means that galaxies were present at that epoch. Detailed spatial observations of the cosmic background radiation show that galaxies had not yet formed at epochs corresponding to $Z \approx 1000$. This has led to the hypothesis that galaxies formed at $Z \approx 10$.

It is expected that, in the early phases of galactic evolution, the systems will be very luminous, since the first generation of stars will contain a very large number of very massive, luminous stars. These luminous stars will produce most of their energy in the UV portion of the spectrum ($\lambda \leq 0.1$ μm). If the stars are surrounded by clouds of hydrogen gas, the transfer of radiation through the gas will degrade the ultraviolet light into Lyman-alpha photons. For redshifts beyond 10, this luminosity will be shifted into the NIR. Thus the NIR becomes the logical place to search for galaxies in the earliest star forming phase.

Acknowledgments

The authors acknowledge C. Beichman, R. Bessey, J. A. Eddy, F. J. Low, K. Matthews, T. L. Murdock, E. P. Ney, R. Terrile, and H. A. Thronson, Jr. for contributing useful information. The many individuals who allowed us to reproduce their work are cited in the figure captions.

Bibliography

Gehrz, R. D., Black, D. C., and Solomon, P. M. (1984). *Science* **224**, 823.
Habing, H. J., and Israel, F. P. (1979). *Annu. Rev. Astron. Astrophys.* **17**, 345.
Kessler, M. F., and Phillips, J. P., eds. (1984). "Galactic and Extragalactic Infrared Spectroscopy." Reidel Publ., Dordrecht, Netherlands.
Low, F. J., and Rieke, G. H. (1974). *Methods Exp. Phys.* **12A**, 415–462.
Rieke, G. H., and Lebofsky, M. J. (1979). *Annu. Rev. Astron. Astrophys.* **17**, 477.
Savage, B. D., and Mathis, J. S. (1979). *Annu. Rev. Astron. Astrophys.* **17**, 73.
Setti, G., and Fazio, G. G., eds. (1978). "Infrared Astronomy," Proc. NATO Adv. Study Inst. Reidel Publ., Dordrecht, Netherlands.
Stein, W. A., and Soifer, B. T. (1983). *Annu. Rev. Astron. Astrophys.* **22**, 177.
Wynn-Williams, C. G. (1982). *Annu. Rev. Astron. Astrophys.* **20**, 597.
Wynn-Williams, C. G., and Cruikshank, D. P. (1981). "Infrared Astronomy," Proc. IAU Symp. No. 96. Reidel Publ., Dordrecht, Netherlands.

ASTRONOMY, PLANETARY RADAR—*SEE* PLANETARY RADAR ASTRONOMY

ASTRONOMY, ULTRAVIOLET SPACE

George R. Carruthers   *Naval Research Laboratory*

GLOSSARY

Blackbody radiator: Object that absorbs or emits radiation with 100% efficiency at all wavelengths and whose emission spectrum follows the laws derived by Max Planck and others.

Extinction: Attenuation of radiation by the processes of scattering, pure absorption, or both.

Extreme ultraviolet: Wavelength range below 1000 Å, extending to the X-ray wavelength range below 100 Å.

Far ultraviolet: Wavelength range below 2000 Å, usually the range 1000–2000 Å.

Flux distribution: Radiated energy, in energy units or number of photons per unit area or for the total object, per second versus wavelength in a region of the electromagnetic spectrum.

Ionosphere: Region of a planetary atmosphere in which a significant portion of the atoms or molecules are ionized (i.e., stripped of one or more electrons).

Middle ultraviolet: Wavelength range 2000–3000 Å.

Resonance transition: Spectral transition of an atom, molecule, or ion that is between the ground state (state of lowest energy) and an excited state (of higher energy), with corresponding emission or absorption of a discrete wavelength of electromagnetic radiation.

Subordinate transition: Spectral transition between two excited states.

Ultraviolet (UV) space astronomy is the study of extraterrestrial objects in the UV wavelength range of the electromagnetic spectrum. This portion of the spectrum provides information unavailable in other wavelength ranges, particularly on planetary and stellar atmospheres and the interstellar medium. However, because the UV spectrum is inaccessible to ground-based telescopes, UV astronomy must be conducted with instruments on rockets, earth satellites, and other space vehicles outside the earth's atmosphere.

I. Significance of the Research

One of the primary benefits of doing astronomy in space rather than at ground-based observations is the much wider range of the electromagnetic spectrum that is accessible to observation (Fig. 1). In particular, the entire UV and X-ray wavelength range below 3000 Å is absorbed by oxygen and ozone in the earth's atmosphere and hence is totally inaccessible to even the largest ground-based telescopes. To observe the sun, stars, and other celestial objects in this wavelength range, the instruments must be carried above the absorbing atmosphere by means of sounding rockets or space vehicles. [*See* ASTROPHYSICS.]

The UV spectral range is of importance to astronomy for a number of reasons. Two of the most significant are as follows. (1) The primary, or *resonance,* spectral transitions (those involving the ground state) of the most common atoms, ions, and molecules occur in the UV spectral range below 3000 Å; (2) very hot stars,

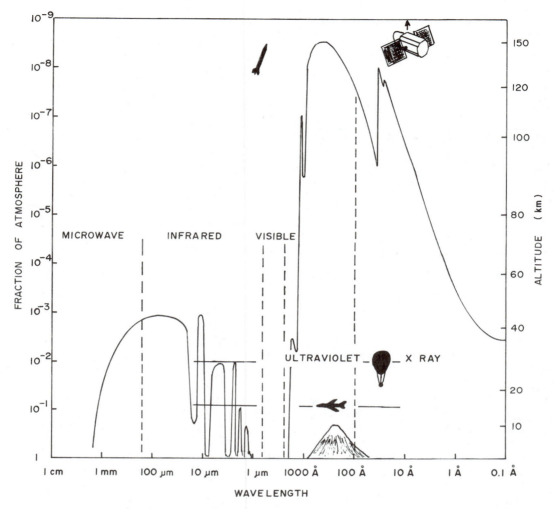

FIG. 1. Altitude and fraction of atmosphere remaining overhead that must be reached in order to observe one-half of the radiation coming from an extraterrestrial source versus wavelength in the electromagnetic spectrum.

having surface temperatures in excess of 10,000 K (vs. ~5800 K for our sun) emit much or most of their radiation in the ground-inaccessible UV. [*See* STELLAR SPECTROSCOPY.]

Hydrogen and helium are the most abundant elements in the universe, accounting for more than 99% of all atoms in stars, the interstellar medium, and the giant planets Jupiter and Saturn. Of the heavier elements, the most abundant in numerical order are oxygen, carbon, neon, and nitrogen. For all of these, in atomic or ionic form or in simple molecules such as H_2, N_2, and CO, the resonance spectral transitions occur in the far UV (below 2000 Å). The resonance transitions, to or from the ground state, are usually much stronger and more quantitatively useful for remote sensing of gas composition and physical state than are the subordinate transitions (between two excited states). This is particularly true for cool material, such as in the interstellar medium and in planetary atmospheres, where most of the atoms or molecules reside in the ground state. Therefore, observations in the far UV provide information on the composition and properties of astronomical objects that cannot be obtained at ground-accessible wavelengths.

Stars that are of much higher temperature than our sun (ranging from ~10,000 to more than 100,000 K) emit much or most of their radiation at wavelengths below the 3000-Å limit of ground-based observations. Therefore, it is difficult to determine, from ground-based observations alone, the total energy outputs and energy flux distributions of these stars. Although the flux distributions of hot stars have crude resemblances to those of classical blackbody radiators (in that the peak of the distribution shifts to shorter wavelengths as the temperature increases), more detailed theoretical models predict significant differences from blackbody distributions, particularly at the shortest wavelengths. Comparisons of observations with theory and accurate determinations of the "effective" temperatures of hot stars therefore requires accurate measurements in the far UV.

Even for cooler stars, such as our sun, UV measurements are important for detecting and characterizing energetic phenomena that occur in their outer atmospheres. For the sun, these include the high-temperature solar corona, solar flares, and other manifestations of solar activity. Also, solar UV radiation is largely responsible for the maintenance of the terrestrial and other planetary ionospheres and for exciting UV emissions observed in planetary atmospheres and in comets.

II. Instrumentation for Ultraviolet Space Astronomy

The instrumentation used for space UV astronomy is, in most respects, similar to that used in ground-based astronomy. The differences have to do, primarily, with the fact that UV radiation is not transmitted as efficiently by, or by as wide a variety of, transparent materials (used for windows, lenses, and other refractive optical elements) as are visible and near-infrared radiation. Also, particularly at wavelengths below 1200 Å, reflective coatings for mirrors and reflection gratings are not as efficient as at longer wavelengths. [See OPTICAL TELESCOPES.]

Categories of instruments can be defined by the types of measurements to be made, such as imagery, photometry (measurement of light intensity), spectroscopy, and spectrophotometry (measurement of intensity vs. wavelength in a spectrum). There is no sharp distinction between these categories, and they are not mutually exclusive; it is mainly a matter of emphasis. For example, imagery emphasizes high spatial resolution and simultaneous coverage of a wide field of view (in comparison with the size of the resolution element), but photometric information can also be obtained from images. Traditional spectroscopy emphasizes the detection, identification, and wavelength measurement of spectral features, whereas spectrophotometry emphasizes measurement of the intensity of spectral features or intensity versus wavelength in a continuum. Both can be accomplished with the same instrument if properly designed. Imagery in very narrow wavelength range (e.g., by the use of interference filters) can be considered a form of spectroscopy or spectrophotometry, and "imaging spectrographs" are those that retain spatial intensity resolution in one dimension while providing spectral intensity resolution in the transverse dimension.

As is true in ground-based astronomy, the type of instrumentation used depends to some extent on the object of study; for example, solar studies require different instrumentation than do observations of stars or nebulae. [See SOLAR PHYSICS.]

Since UV astronomy can be done only at very high altitudes or in space, its progress has been paced by the development of rocket vehicles for carrying the instruments above the atmosphere and of pointing controls for directing the instruments at the objects of interest with adequate accuracy and stability. The rate of progress in the various subfields of UV astronomy has therefore depended on both instrumentation and vehicular developments. Solar UV astronomy was the first subfield to be developed; because the sun is so much brighter than any other astronomical object, the requirements regarding both scientific instrumentation and rocket attitude control systems were less stringent than for observations of other celestial targets.

Solar UV astronomy began in the late 1940s with experiments flown by Naval Research Laboratory scientists on captured German V-2 rockets. The initial experiments were unpointed but nevertheless returned new and useful information on the intensity and spectral distribution of solar UV radiation. In the early 1950s, the development of instrument pointing controls for Aerobee sounding rockets imparted greater sensitivity and higher resolution to spectroscopy and photometry. The first measurements of stellar UV radiation were made from free-spinning Aerobee rockets in the late 1950s; the development of inertial attitude control systems in the early 1960s greatly improved the quality of data return.

The first satellite vehicles dedicated to space

astronomy were the *Orbiting Solar Observatories (OSOs)*, the first of which was launched in 1962, and the *Orbiting Astronomical Observatories (OAOs)*, the first successful launch of which was in 1968. The advent of the satellite observatories gave a tremendous improvement in available observing time, in comparison to the typical 5 min available in a sounding rocket flight. This allowed much more comprehensive studies of the sun and of stars and other celestial objects than possible with sounding rockets. However, the latter still remained useful for carrying out special-purpose investigations not suitable for the long-duration (but technologically more difficult) satellite observatories. Ultraviolet astronomical observations have also been carried out in several manned missions, including *Gemini*, *Apollo*, and *Skylab*. Current missions include the *Solar Maximum Mission (SMM)* and the *International Ultraviolet Explorer (IUE)* satellites. Most recently, solar observations were carried out during the Spacelab-2 space shuttle flight in July, 1985. Near future missions include celestial UV astronomy with the *Astro* shuttle payload and launch by the shuttle of the *Hubble Space Telescope*.

There are a large variety of instruments used for UV space astronomy, and in general, they are similar to instrumentation used in ground-based astronomy. In the UV, the number of optical components (mirrors, gratings, etc.) must be kept to a minimum because of the generally lower efficiencies of optical elements in the UV as compared with the visible. However, instruments now in use or planned for near future missions in UV astronomy are fully comparable in sensitivity, resolution, and other performance capabilities to similar visible-light instruments.

In UV astronomy, as in visible-light astronomy, photography has been a major recording technique for imagery and spectroscopy but is rapidly being replaced by electronic imaging detectors. These include image intensifiers and electrographic detectors (having final images recorded on film) and devices whose final output is an electronic signal (television-type detectors), such as the charge coupled device or devices based on microchannel-plate electron multiplier arrays.

We show here two examples of current UV astronomy instruments: the telescope and spectrographs used in the *IUE* satellite, launched in 1978 and still in operation (Fig. 2), and the Naval Research Laboratory's high-resolution telescope and spectrometer (HRTS), which has flown on several sounding rocket missions and on *Spacelab 2* (Fig. 3). The *Astro* UV astronomy payload (Fig. 4) consists of three separate instruments: the Johns Hopkins University's UV telescope for spectroscopy in the wavelength range 850–1850 Å, the Goddard Space Flight Center's ultraviolet imaging telescope for imagery in the range 1250–1800 Å, and the University of Wisconsin's UV photopolarimeter instrument for studies in the range 1400–3200 Å.

A number of UV imaging and spectrographic instruments are carried on the *Hubble Space Telescope* (Fig. 5), planned for a space shuttle launch. The *Hubble Space Telescope* has a collecting aperture of 2.4 m, more than twice that of the largest previous UV space astronomy telescope. It operates in the visible and UV wavelength ranges. Of particular importance is that it has diffraction-limited imaging performance. This allows it to achieve an angular resolution of better than 0.1 arc sec, which is 10 times better than that achieved in the visible by even the largest ground-based telescopes. This in turn provides an additional increment in detection sensitivity for point sources (such as stars), as well as much improved image quality. The *Hubble Space Telescope* scientific instruments incorporate detector concepts that are substantial improvements over those used in past and current satellite experiments. Even more advanced instrumentation and detector concepts are currently under development for use in future missions.

III. The Sun

The sun is the nearest star, the one of most practical importance to us, and the only one whose surface features can be observed in detail. The visible "surface" of the sun, known as the photosphere, emits a continuous spectrum of radiation whose intensity and spectral distribution crudely resemble those of a classical blackbody radiator having a temperature of ~5800 K. The intensity of solar radiation is a maximum in the visible (near 5000 Å wavelength) and decreases rapidly toward shorter wavelengths and less rapidly toward longer wavelengths. Thus, in the UV (particularly in the far UV, below 2000 Å) the solar photosphere is far less bright than in the visible.

However, the outer atmospheric layers of the sun, above the photosphere, have much higher

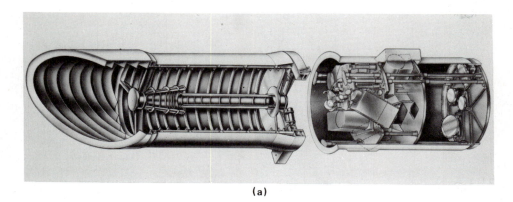

(a)

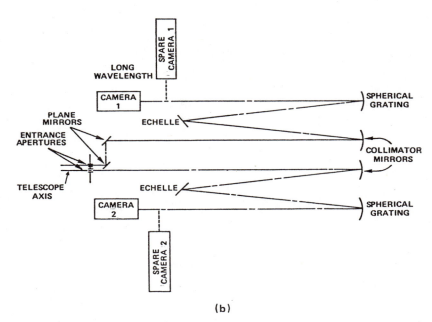

(b)

FIG. 2.   Instrumentation of the *International Ultraviolet Explorer* (*IUE*) satellite, launched in January 1978. (a) Cutaway diagrams of the 45-cm-aperture telescope (left) and the UV of spectrographs (right). (b) Simplified schematic diagram of the *IUE* spectrographs, as operated in high-resolution mode. Cameras 1 operate in the 1900–3200-Å wavelength range, and cameras 2 in the 1150–1950-Å range. Flat mirrors are placed in front of the echelle gratings for operation in low-resolution mode. (NASA illustrations.)

gas temperatures than does the photosphere. As seen in Fig. 6, the temperature rises very sharply in a "transition region" between the lower atmosphere (chromosphere) and the upper atmosphere (corona), from ~10,000 K in the upper chromosphere to more than 1,000,000 K in the corona. Highly active regions in the solar atmosphere, such as solar flares, can have much higher temperatures still. However, because of the very low gas densities in these regions, the energy content per unit volume is still very much less than in the photosphere. One of the most important current problems of solar physics is to explain how the outer atmosphere of the sun can be maintained at million-degree temperatures in contact with a 6000 K photosphere. Related problems include elucidation of the mechanism for the production and acceleration of the solar wind and of the mechanisms of, and physical processes occurring in, solar flares.

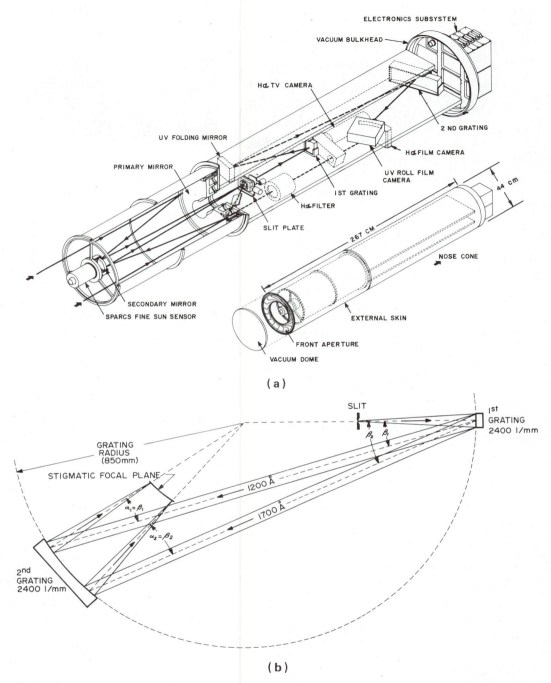

FIG. 3. Diagram of the Naval Research Laboratory's high-resolution telescope, and spectrograph, used in solar studies from rockets and *Spacelab 2*. (a). Diagram of the complete telescope and spectrograph. The optical parameters are as follows. Spatial: field of view, 0.5″ × 16′; resolution, <1″. Spectral: range, 1175–1715 Å and Hα; resolution, 0.05 Å. (b). Simplified schematic of the spectrograph. (Naval Research Laboratory illustrations.)

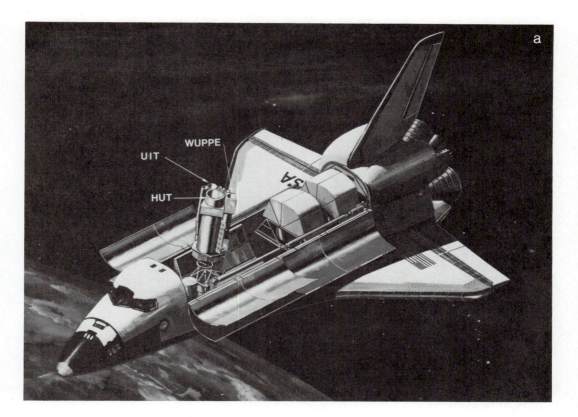

a

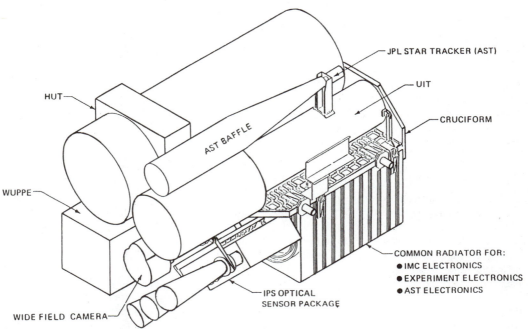

Astro Instruments Mounted to the IPS

(b)

FIG. 4. *Astro* shuttle payload for UV astronomy. (a) Artist's concept of the *Astro* cluster of three UV telescopes, mounted on the European Space Agency, instrument pointing system (IPS) in the shuttle payload bay. (b) Diagram of the *Astro* instrument cluster. (NASA illustrations.) HUT, Johns Hopkins University's ultraviolet telescope; UIT, Goddard Space Flight Center's ultraviolet imaging telescope; WUPPE, University of Wisconsin's UV photopolarimeter instrument.

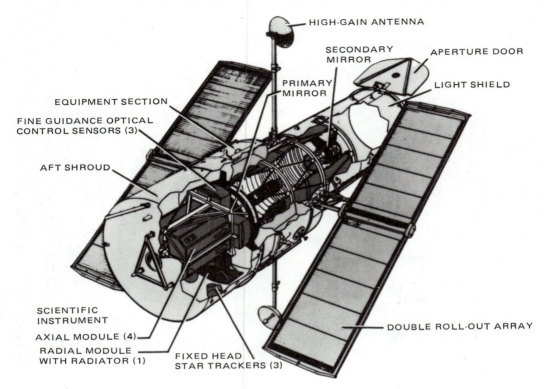

FIG. 5. Diagram of the 2.4-m-aperture *Hubble Space Telescope,* planned for launch by the space shuttle. The scientific instruments include imaging cameras, a high-speed photometer–polarimeter, a high-resolution spectrograph, and a faint-object spectrograph. (NASA photograph.)

In the visible and through the near and middle UV, the solar spectrum is a continuum with superimposed absorption lines (Fraunhofer lines), due to cool gas in the upper photosphere and lower chromosphere (the "temperature minimum" region in Fig. 6). In the far UV, however, the continuum fades away and is replaced by an emission line spectrum, produced by the hotter gas in the transition region and corona.

It is particularly noteworthy that, in the far and extreme UV, the outer atmosphere of the sun is the predominant source of radiation, due to its very high temperature. This constitutes one of the major advantages of short-wavelength observations of the sun: the ability to observe the high-temperature, active regions of the outer atmosphere without interference from the cooler (but, in the visible, far brighter) photosphere. Although (with the exception of sunspots) the solar disk appears almost featureless in white light, observations in far-UV emission lines (and also in narrow wavelength ranges centered on visible absorption lines, such as hydrogen Balmer α (Hα) at 6563 Å) reveal a great deal of detail and time variability in the structure of the solar atmosphere. Different emission or absorption features are produced by gas at different temperatures and hence at different spatial locations in the solar atmosphere. Therefore, observations from the visible to the X-ray wavelength range are needed to characterize the temperature, density, and time variations in the solar atmosphere. For example, imagery of the sun in the light of neutral hydrogen (Lyman α at 1216 Å) reveals gas having a characteristic temperature of ~10,000 K, whereas the light of ionized helium (He II) at 304 Å is emitted by gas with a temperature of ~80,000 K. Five-times-ionized oxygen (O VI) emission at 1032 and 1036 Å reveals material at a temperature of ~300,000 K. Nine-times-ionized magnesium (Mg X) emission near 625 Å is characteristic of material at a temperature near 1,600,000 K.

Studies of the sun in the UV and X-ray wavelength ranges are also of practical importance, because these radiations significantly influence the earth's upper atmosphere and are primarily responsible for the production and maintenance

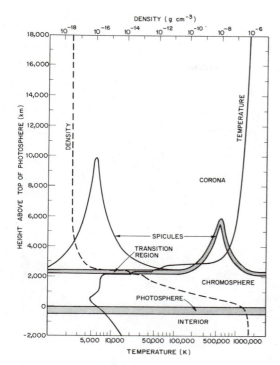

FIG. 6. Temperature and density in the solar atmosphere versus altitude. Note that a minimum temperature is reached just above the visible-light surface, or photosphere, and temperature rises sharply in the transition region between the chromosphere and corona. [Adapted from Eddy, J. A. (1979). "A New Sun," NASA SP-402. W. S. Government Printing Office, Washington, D.C.]

of the ionosphere (which is essential to long-distance radio communications). Ultraviolet radiation below 2000 Å dissociates molecular oxygen in the upper atmosphere, indirectly resulting in the formation of ozone (O_3). Lyman α radiation ionizes nitric oxide (NO), producing the lower ionosphere, whereas radiations of wavelength less than 1026, 911, and 796 Å ionize molecular oxygen, atomic oxygen, and molecular nitrogen, respectively.

As mentioned previously, the sun was the first extraterrestrial object to be observed in the ground-inaccessible UV, in part because of its brightness (even in the far UV, much greater than that of any other celestial object) and its relative ease of acquisition by primitive rocket pointing control systems. The first rocket observations were exploratory in nature, attempting to define the general nature and intensity distribution of the solar UV spectrum. Successive experiments gradually improved the spectral reso-

lution and photometric accuracy of the measurements and extended them toward shorter wavelengths. Also, once strong emission lines (such as hydrogen Lyman α at 1216 Å and ionized helium at 304 Å) were identified, imaging instruments (such as spectroheliographs) were used to obtain monochromatic images of the entire solar disk in these emissions.

The advent of long-duration satellite and *Skylab* observations allowed not only more detailed measurements, but studies of the time variations of the solar UV output and its correlations with other indications of solar activity, such as sunspots and structure of the white-light corona. Recent and near future observations, such as with the Naval Research Laboratory's HRTS instrument and with the planned solar optical telescope, will emphasize improved spatial and temporal resolution in the measurements. Also, it is increasingly clear that understanding the physical phenomena taking place in the solar atmosphere requires simultaneous or complementary measurements over a wide range of wavelengths, from the radio through the X-ray and γ-ray ranges.

Figure 7 is a spectrum of the sun, obtained in a sounding rocket flight, showing the transition of the solar spectrum from a continuum with superimposed absorption lines (Fraunhofer spectrum) to an emission line spectrum below 1800 Å. Measurements of this type are useful for measuring the total solar energy output in specific lines and wavelength ranges in the far and ex-

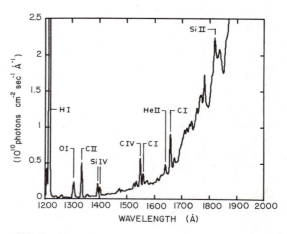

FIG. 7. Far-UV spectrum of the sun, obtained in a sounding rocket flight, showing the transition to an emission line spectrum below ~1800 Å. The effective resolution is 5 Å. (Courtesy of G. H. Mount, Naval Research Laboratory.)

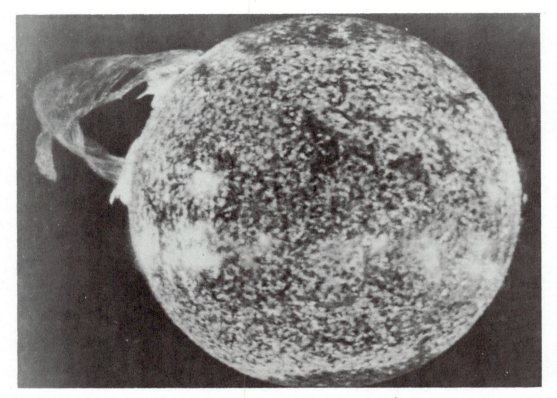

FIG. 8. Image of the sun in the light of ionized helium (He II) emission at 304-Å wavelength, showing mottled structure of the transition region, active regions associated with sunspots, and a giant eruptive prominence (at left). Obtained with the Naval Research Laboratory extreme-UV spectroheliograph from *Skylab*. (Naval Research Laboratory photograph.)

treme UV. Figure 8 is a monochromatic image of the entire sun in the light of ionized helium (304 Å), taken with the Naval Research Laboratory's extreme-UV spectroheliograph, which was flown on the U.S. *Skylab* space station in 1973–1974. Images of this type reveal the structure of the solar atmosphere in various temperature ranges.

Figure 9 is a spectrum of the sun taken in a sounding rocket flight of the Naval Research Laboratory's HRTS instrument. Unlike previous instruments, this combined high spectral resolution with high spatial resolution in the along-slit direction (an imaging spectrograph). Thus, new information on the spatial distributions and velocity structures of solar active regions was obtained, as were variations in the spectral intensity and temperature distributions.

IV. Planetary Atmospheres

Ultraviolet measurements are important for studies of planetary atmospheres, including the earth's upper atmosphere. This is due to the fact that the resonance absorption and emission spectral features fall primarily in the UV, making this spectral range much more sensitive for detection and measurement of the most common atmospheric gases than are other wavelength ranges. For example, O_2 absorbs strongly at wavelengths below 2000 Å; hence, UV spectroscopy in this wavelength range allows sensitive detection of O_2 in other planetary atmospheres (provided that one observes from above the earth's O_2-rich atmosphere). The hydrogen Lyman α emission line at 1216 Å, produced by scattering of solar Lyman α radiation by hydrogen atoms, is a very sensitive test for this gas in the outer atmospheres of planets and comets. [*See* PLANETARY ATMOSPHERES.]

In the earth's upper atmosphere (and, in planetary probe missions, the atmospheres of other planets) atmospheric composition can be measured *in situ* with mass spectrometers and related instrumentation. However, UV measurements provide a capability for *remote sensing* of

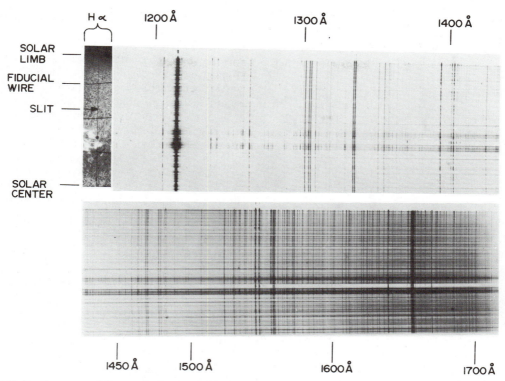

FIG. 9.   Spectrum of the sun obtained with the Naval Research Laboratory high-resolution telescope and spectrograph in a sounding rocket flight. This stigmatic (imaging) spectrograph reveals spatial variations across the disk of the sun (top to bottom; region covered by the spectrograph slit is shown at the left) simultaneously with the spectral variations (left to right). (Naval Research Laboratory photograph.)

atmospheric composition and its variation with altitude, geographic location, and time, which supplements the extends *in situ* measurements where available and can be applied to many objects not yet visited by spacecraft (e.g., in observations of other planets by spacecraft in near-earth orbit). In addition to atmospheric composition, UV measurements can remotely sense the atmospheric temperature structure and its spatial and temporal variations. Also, information on the fluxes, energies, and spatial distributions of incoming energetic particles (such as those that cause the earth's polar auroras) can be obtained.

Three basic types of UV observations applicable to planetary atmosphere studies are (1) observations of far-UV emission features, excited by solar UV radiation or by charged-particle impact (as in the earth's polar auroras); (2) observations of middle-UV absorption features, superimposed on the reflected solar spectrum, for example, observations of ozone in the earth's atmosphere or of sulfur dioxide in that of Venus;

and (3) observations of middle-UV or far-UV absorption features superimposed on the solar spectrum or a stellar UV spectrum when the sun or a star is viewed directly with a line of sight passing through the planetary atmosphere (solar or stellar occultation). All of these have been applied to studies of the earth's upper atmosphere; various combinations of these have been applied (in various wavelength ranges) to studies of the atmospheres of other planets. The pioneering observations of both the earth's upper atmosphere and those of other planets were made using sounding rocket vehicles. More recent observations of the other planets have been made both with earth-orbiting astronomical observatories (such as the *IUE* satellite) and from planetary flyby or orbiter spacecraft (such as the *Mariners* and *Voyagers*).

Figure 10 shows UV spectra of three planets: the earth, Venus, and Jupiter. It is seen that Lyman α (1216 Å) emission is present in all of these. The atmospheres of both Venus and Mars consists mainly of carbon dioxide, and hence

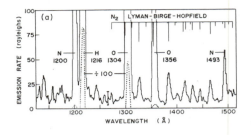

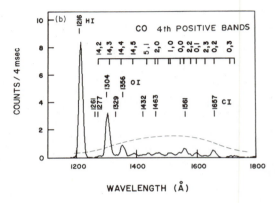

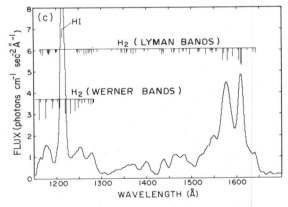

FIG. 10. Far-UV emission spectra of three planetary atmosphere. (a) Earth's upper atmospheric day airglow, observed in a sounding rocket flight. [Reprinted with permission from Takacs, P. Z. and Feldman, P. D. (1977). *J. Geophys. Res.* **82**, 5013.] (b) Day airglow of Venus, observed with the spectrometer on the *Pioneer Venus Orbiter*. [Reprinted with permission from Durrance, S. T. (1981). *J. Geophys. Res.* **86**, 9116.] (c) An aurora on Jupiter, observed with the *International Ultraviolet Explorer* satellite. [Reprinted with permission from Durrance, S. T. *et al.* (1982). *Geophys. Res. Lett.* **9**, 653.]

their UV spectra are similar, but there are subtle and significant differences. Atomic oxygen emissions (1304 and 1356 Å) are present in all but the spectrum of Jupiter, which shows only atomic and molecular hydrogen features. The earth spectrum is nearly unique in showing strong features of atomic and molecular nitrogen, although these have been detected in the UV spectrum of Titan, the largest satellite of Saturn. For our moon, the planet Mercury, and Ganymede (Jupiter's largest satellite), UV observations have set upper limits on atmospheric densities that are far lower than those established by ground-based measurements. As in the case of the sun, imagery of planetary atmospheres in individual spectral lines provides information on the altitude distributions of various gases, indirectly yielding information on temperature distributions. In addition, it provides information on the planetographic distributions of localized phenomena such as auroras. [*See* SOLAR SYSTEM, GENERAL.]

V. Comets

Comets are unique, in comparison with the planets and satellites, in many ways. Of particular significance here is that they are objects of very low mass (comparable to small asteroids), but are composed largely of volatile materials such as water ice. Thus, when they approach the sun while traveling along their highly eccentric orbits, a significant portion of their mass is "boiled off" to produce a gaseous halo, or coma, which is quite prominent in the UV as well as at other wavelengths. As in the case of planetary atmospheres, UV observations can provide important information on the volatile composition of comets. They also can be used to determine the vaporization rates of various cometary materials and to provide information on the physical interactions between the cometary atmosphere and solar UV radiation and the solar wind.

Hydrogen Lyman α and the OH molecular band emission near 3100 Å are the two most prominent spectral features in comets. These indicate that water is indeed the dominant volatile constituent of comets; other materials (which are responsible for the ground-accessible spectral features, such as CH, CN, C_2, and NH band emissions) are only minor constituents. Because of the small mass of the hydrogen atom, the cometary hydrogen coma is much larger than the coma revealed in heavier molecular emis-

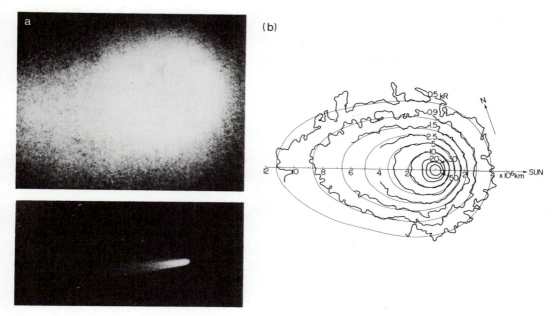

FIG. 11.   Imagery of Comet Kohoutek in hydrogen Lyman α (1216-Å) emission. (a). Comparison of far-UV (Lyman α) image of the comet, obtained by the Naval Research Laboratory in a January 1974 sounding rocket flight, with a visible-light image of the comet to the same scale. (Courtesy of P. Feldman, Johns Hopkins University). (b) Isointensity contour plots of the UV image are compared with a best-fit model (thin lines). For the conditions existing at the time, the hydrogen production rate Q is derived to be 6.2×10^{24} atoms per second.

sions (Fig. 11); it can be many times larger than the sun. Measurements of the Lyman α brightness distribution in this halo can be used to determine the hydrogen production rate and hence the vaporization rate of water and other hydrogen compounds.

Ultraviolet observations of Comet West in 1975 (Fig. 12) revealed that CO is also a prominent cometary constituent. Ultraviolet observations have also resulted in the detection of minor species not previously known in comets, including atomic and diatomic sulfur, ionized carbon, and the SH radical. Future, more sensitive observations and ones extending to shorter wavelengths are expected to reveal other constituents such as H_2, N_2, N, N^+, and O^+.

VI. The Stars

A. HOT STARS

Hot stars, meaning those having surface temperature in excess of 10,000 K (vs. ~5800 K for our sun), emit much or most of their radiation in the UV wavelength range below 3000 Å, the limit for ground-based observations. This was predicted theoretically, many years before the first space observations, from models of stellar atmospheres. It also was inferred for the very hottest stars (above 20,000 K) from observations of emission nebulae or ionized hydrogen (H II) regions; in these regions, ionization of hydrogen is produced by stellar radiation in the extreme UV, below 912-Å wavelength. Figure 13 compares ground-based observations and theoretical models for two stars: one a solar-type star of temperature 6000 K and one a very hot star of 30,000 K. The flux distributions have been adjusted to be equal in the visual range near 5500 Å. However, if the model fit for the hot star is extended into the UV, as shown in Fig. 13, it is seen that the vast majority of the total radiation output of the hot star is in the ground-inaccessible UV. Furthermore, although the flux distribution for the hot star is rising toward shorter wavelengths in the ground-accessible range, the shape of the distribution (or of the theoretical model curves) is insensitive to temperature for very hot stars. Hence, it is very difficult to determine temperature or total radiation output for very hot stars from ground-based observations.

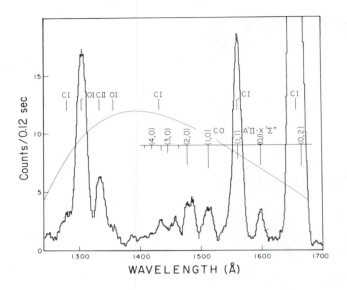

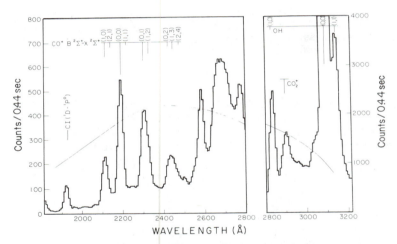

FIG. 12.   Far-UV spectrum of Comet West, obtained by Johns Hopkins University in a March 1975 sounding rocket flight. The smooth curve gives the relative sensitivity versus wavelength in each of the two spectrometer wavelength ranges. (Courtesy of P. Feldman, John Hopkins University).

The earliest sounding rocket observations of hot stars in the UV, in the early 1960s, indicated much lower UV fluxes than were predicted by the theoretical stellar atmosphere models. However, over the following years, improvements in both the observations and the models resulted in convergence of the two, so that now there is reasonable agreement between theory and observation. In particular, the improved models include the effects of the far-UV absorption lines (line blanketing), omitted in early models such as that in Fig. 13, and predict lower far-UV fluxes for a given temperature. Figure 14 shows a series of model flux distributions computed by E. Avrett of the Harvard–Smithsonian Center for Astrophysics.

The first large, self-consistent set of stellar UV spectrophotometric data was that obtained by the University of Wisconsin experiment on the *OAO-2* satellite. Figure 15 shows measured

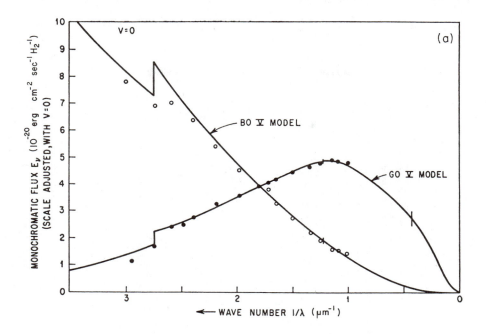

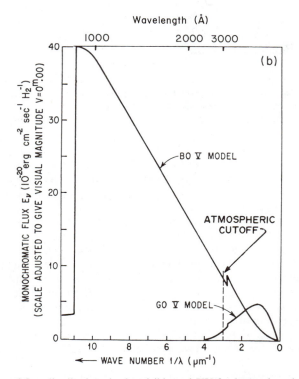

FIG. 13.   Comparison of flux distributions in the visible and UV for hot and cool stars. (a) Flux distributions in the ground-accessible wavelength range for a hot star (spectral type B0 V, temperature ~30,000 K) and for a cooler, solar-type star (spectral type G0 V, temperature ~6000 K), and best-fit theoretical models. The distributions have been adjusted to be equal in the visual band near 5500 Å. [●, β Com (corrected for line blanketing); ○, ν Ori (corrected for interstellar reddening]. (b) Extrapolation of the theoretical models to shorter wavelengths, showing the most of the energy output of the hot star is at wavelengths below 3000 Å, inaccessible from the ground. [Reprinted with permission from Code, A. D. (1960). *Astron. J.* **65**, 239.]

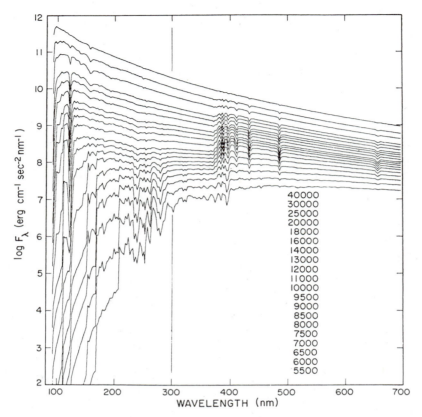

FIG. 14. Model atmosphere flux distributions, more recent than those in Fig. 13, taking into account absorption line blanketing. Effective temperatures for the models are as indicated, from top to bottom. [Reprinted with permission from Kurucz, R. L. (1979). *Astrophys. J. Suppl.* **40,** 1.]

flux distributions for four hot stars of different temperatures, a combination of UV observations by *OAO-2* with ground-based spectrophotometry. Similar data sets have been obtained with the European *TD-1* and *ANS* satellites and, more recently, these measurements have been extended to much fainter stars by the *IUE* satellite spectrometers. Current efforts include more accurate measurements in the range from 912 to 1200 Å, where there are still disagreements among various measurements and theory. Most stars are not observable at wavelengths below 912 Å, even from space, because this radiation is absorbed by atomic hydrogen in interstellar space.

Spectrometric observations of individual hot stars have been supplemented by imagery of starfields in moderate wavelength ranges in the UV. The advantage of these is that much fainter objects can be measured and that a very large number of objects can be recorded in a single exposure. Ultraviolet imagery is particularly

useful as a means of *surveying* large areas of the sky in order to detect and measure hot stars not previously known to exist (or at least not previously known to be hot). Figure 16 compares a far-UV image of the constellation Orion, obtained in a sounding rocket flight, with a visible-light image. The UV image shows, at a glance, the distribution of hot stars without confusion by the far more numerous (and, in some cases, brighter in visible light) cool stars.

Ultraviolet spectroscopy is useful not only for measuring the energy flux distributions of hot stars, but also for gaining insight into the properties and physical processes occurring in stellar atmospheres. This information is provided by the line features in the spectra. One of the first major discoveries in stellar UV spectroscopy was that the strong line features such as C IV (1550 Å) and Si IV (1400 Å) had what is known as a P Cygni profile: a very strong, broad absorption feature shifted to shorter wavelengths (relative to its laboratory, or rest, wavelength)

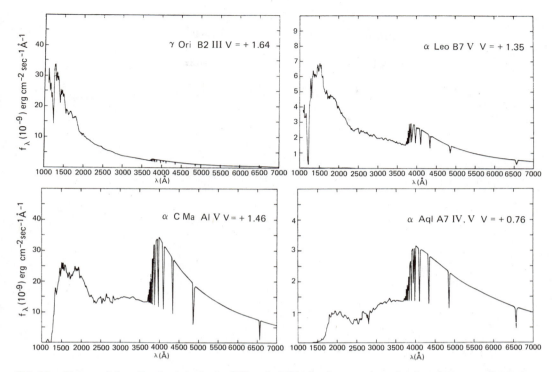

FIG. 15. Measured flux distributions in the UV and visible for four stars ranging in temperature from about 8000 K (type A7) to 20,000 K (type B2). The UV measurements are from the University of Wisconsin experiment package on *OAO-2*. (Courtesy of A. Code, University of Wisconsin.)

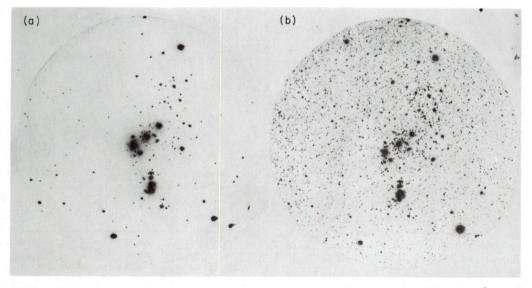

FIG. 16. Comparison of visible and UV imagery of the constellation Orion. (a). Far-UV (1230–2000 Å) image of Orion, obtained by the Naval Research Laboratory in a 1975 sounding rocket flight. (b). Visible-light image, to the same scale. (Hale Observatories photograph.)

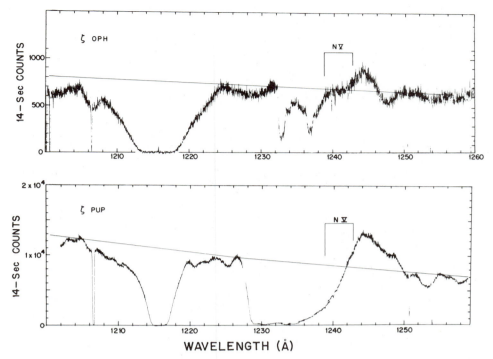

FIG. 17.   Far-UV high-resolution (0.05-Å) spectra of the stars ζ Ophiuchi (spectral type O9.5) and ζ Puppis (type O4) obtained with the Princeton University spectrometer on the *Copernicus* satellite. Of interest here are the absorption features due to interstellar atomic hydrogen at 1216 Å (considerably stronger, indicating a larger hydrogen column density, toward ζ Ophiuchi) and due to four-times-ionized nitrogen (N V) in the stellar atmospheres. The latter feature exhibits a P Cygni profile (much stronger in ζ Puppis), indicating high-velocity mass ejection. [Reprinted with permission from Morton, D. C. (1976) *Astrophys. J.* **203**, 386.]

combined with an emission feature at, or slightly longer than, the rest wavelength (Fig. 17). As shown in Fig. 18, the production of such a spectral feature can be explained as being due to a very strong *stellar wind*. The observations indicate that some very hot, luminous stars are losing mass at very high rates—of the order of 10^{-6} solar mass per year. Such high mass loss rates, many orders of magnitude greater than that associated with the solar wind, can significantly influence the evolution of a massive hot star. [*See* STELLAR STRUCTURE AND EVOLUTION.]

In addition, UV line spectra are useful because they contain the resonance absorption features of most of the common ions expected in stellar atmospheres and provide sensitive means of measuring temperature and relative abundances. Since, as discussed later, starlight can be absorbed or scattered by dust particles in interstellar space, measurements of absorption lines produced in the stellar atmosphere can, in

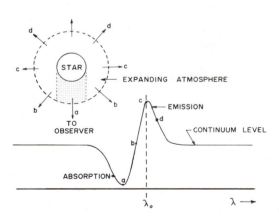

FIG. 18.   Illustration of the formation of a P Cygni profile (emission line near the "rest" wavelength, with blue-shifted absorption) in a spectral feature, resulting from high-velocity outflow of gas from a stellar atmosphere.

some cases, provide more useful temperature information than measurements of the continuum flux distribution.

B. COOL STARS

Cool stars, in the temperature range 3000–9000 K, emit most of their radiation in the wavelength range accessible from the ground (as does our sun). However, as for the sun, the UV wavelength range reveals energetic processes in the outer atmospheres of these stars. Objects of similar surface temperatures can exhibit quite different far-UV spectra, indicative of quite different conditions in their outer envelopes.

Although some pioneering observations of cool stars in the UV were obtained with sounding rockets, *OAO-2*, and *Copernicus*, a truly systematic study covering a wide range of cool star types had to await the higher sensitivity of the *IUE* satellite. Observations with *IUE* have revealed that some giant stars, such as Capella, are much more active than our sun (as are also some stars in close binary systems). Cool supergiant stars, on the other hand, do not show evidence of hot coronas; it is presumed that this is due to the presence of strong stellar winds, which deplete the energy that might otherwise go into heating. Cool class M dwarf stars, having temperatures of only 3000 K and being much less luminous than our sun, surprisingly show very active chromospheres and coronas—in some cases, with higher far-UV fluxes per unit area than our sun. These cool stars also sometimes exhibit flare activity far surpassing that of the sun. In eclipsing binary systems consisting of, for example, a hot star and a cool giant star, UV measurements can provide information on the outer atmosphere of the cool star by observing the hot star as it is eclipsed by the cool star.

VII. Interstellar Gas

The space between the stars is not empty, but contains highly rarefied gas and solid particles (dust). The density and temperature, and to some extent the composition, of this interstellar material are highly variable from place to place in our galaxy. On the average, the interstellar gas contains about one hydrogen atom per cubic centimeter, although it can range from 0.01 or less to more than 10^6 (in dense molecular clouds). It is this interstellar material from which new stars are formed. [*See* SPACE, INTERSTELLAR MATTER.]

The composition of the interstellar medium is believed to be similar to that of the sun and stars: Hydrogen accounts for ~90% of all atoms, and helium most of the remaining 10%. All heavier elements make up less than 1%. Much of the heavier element component is believed to be in the form of dust grains rather than in gaseous form, although the relative proportion is variable. Except in the close vicinity of hot and highly luminous stars, this interstellar medium is made evident only by its attenuation of the light of stars seen through it. The dust particles produce continuous attenuation, whereas the gas absorbs only in discrete spectral lines. Since the resonance transitions of most of the common elements occur in the far UV, the composition of the interstellar gas was only poorly known before the advent of space UV spectroscopy. Only relatively minor constituents of the interstellar gas, such as sodium and calcium, can be detected and measured by ground-based optical absorption spectroscopy. The far UV allows measurements of the major constituents, such as atomic and molecular hydrogen, and also atomic oxygen, carbon, and nitrogen.

A. INTERSTELLAR ATOMIC AND MOLECULAR HYDROGEN

The first observations of interstellar atomic hydrogen, by means of its Lyman α absorption line at 1216 Å, were made in sounding rockets as early as 1966. Molecular hydrogen was first observed in a 1970 rocket flight. Surveys of interstellar atomic hydrogen were made with the *OAO-2* satellite spectrometers. However, the instrument that produced the most comprehensive studies of the interstellar gas was the far-UV spectrometer provided by Princeton University, carried on the *Copernicus* satellite launched in 1972. This instrument observed in the wavelength range 912–3000 Å with spectral resolution as good as 0.05 Å. Figure 17 shows spectra of the stars ζ Ophiuchi and ζ Puppis, obtained with *Copernicus,* in the region of the Lyman α absorption line. It is apparent that there is much more atomic hydrogen in the line of sight to the former star than toward the latter.

The observations confirmed theoretical predictions that molecular hydrogen is formed from atomic hydrogen by "three-body recombination" on dust particles; here, the dust particle acts as a catalyst and takes up the energy of recombination of the two hydrogen atoms. The hydrogen molecules are broken up, or photodissociated, by UV radiation from hot stars in the interstellar medium outside of dense dust clouds. Thus, molecular hydrogen is the major form of interstellar hydrogen in dense dust clouds, whereas in the general interstellar medium hydrogen is primarily in atomic form.

B. OTHER CONSTITUENTS OF THE INTERSTELLAR GAS

The *Copernicus* satellite also observed many other species in the interstellar gas, including neutral and ionized carbon, neutral oxygen, nitrogen, and argon. The molecules CO and OH were also observed toward some stars. It was found that most of the heavier elements were less abundant in the interstellar gas, relative to hydrogen, than expected on the basis of relative elemental abundances in the sun and stars. It was conjectured that this "depletion" was the result of dust grain formation. Regions of the interstellar medium containing less than the usual proportion of dust showed less depletion than average, whereas the converse was true in lines of sight through dense dust clouds. *Copernicus* also discovered a very hot phase of the interstellar gas, revealing absorption lines of five-times-ionized oxygen (O VI) characteristic of temperatures near 300,000 K. This gas is probably heated by shock waves in the interstellar medium resulting from supernova explosions.

The *IUE* satellite, launched in 1978, does not have as high a spectral resolution as, or reach wavelengths as short as, the *Copernicus* spectrometer. However, in the range of wavelengths longer than 1200 Å, *IUE* is much more sensitive than *Copernicus*. Therefore, it observes much fainter and more distant stars (including some in nearby external galaxies, such as the Magellanic Clouds) and stars whose light is more severely attenuated by interstellar dust. It extended studies of high-temperature interstellar gas through observations of C IV, Si IV, and N V absorption lines. In observations of hot stars in the Magellanic Clouds, it detected interstellar absorption lines attributed (because of their Doppler shifts relative to locally produced interstellar absorption lines) to a hot "halo" of gas surrounding our galaxy and also perhaps to similar halos surrounding the Magellanic Clouds.

C. EMISSION NEBULAE

The *IUE* satellite also provided new information on the composition of interstellar gas through observations of emission nebulae (H II regions, planetary nebulae, and supernova remnants), in this case by observations of UV emission lines. Figure 19 shows *IUE* spectra of the Orion nebula (a typical H II region) and of the Cygnus Loop supernova remnant. In particular, the abundance of carbon (difficult to determine from ground-based observations) has been measured by observations of the strong features of C II (2325 Å), C III (1909 Å), and (in some cases) C IV (1550 Å). In these objects, the electron temperatures (and, in supernova remnants, shock wave velocities) have also been determined from observations in the far UV.

The H II regions are representative of gas from which stars have recently formed or are in the process of forming. Planetary nebulae, on the other hand, are formed from the outer envelopes of stars nearing the end points of their evolutionary life cycles. The collapsed, very hot core of a dying star provides the UV radiation that ionizes the nebula. Since the gas in the planetary nebula has been "processed" by the star, its composition differs from that of an H II region; however, it is the gas ejected by old stars that is recycled to form the next generation of stars. Supernova remnants are the result of much more violent disruptions of (more massive) stars; the shock waves produced by the stellar explosion excite emissions both from gas in the castoff envelope and in the surrounding interstellar gas, which is swept up in the expanding shell. Hence, the elemental compositions of supernova remnants are intermediate between those of H II regions and planetary nebulae.

VIII. Interstellar Dust

Dust particles in interstellar space make themselves known by both their attenuation and their reflection of starlight. In the visible range, the combined absorption and scattering (extinction) of starlight increases toward shorter wavelengths, so that stars seen through dust clouds appear both dimmer and redder in color than they would otherwise. In the close vicinity of bright stars, dust clouds are illuminated and appear as reflection nebulae.

Ultraviolet measurements of interstellar extinction and of reflection nebulae have the potential to provide additional information on the properties and composition of the dust particles. As mentioned, in the visible range, the extinction increases smoothly toward shorter wavelengths. However, early rocket measurements

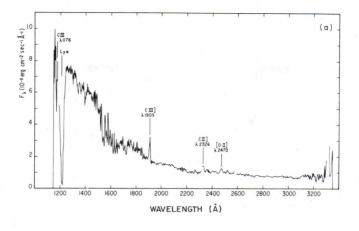

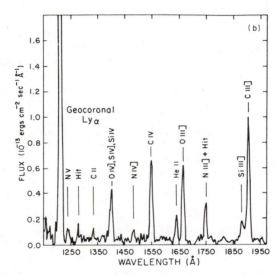

FIG. 19.   Far-UV spectra, obtained with *IUE*, of diffuse nebulae. (a) Orion nebula, a typical H II region, showing continuum due to dust scattering of UV starlight, with nebular emission lines. [Reprinted with permission from Torres-Peimbert, S. *et al.* (1980). *Astrophys. J.* **293,** 133.] (b) Cygnus Loop, a supernova remnant, showing emission lines due to shock-heated gas. [Reprinted with permission from Raymond, J. C. *et al.* (1980). *Astrophys. J.* **238,** 881.]

(confirmed by more comprehensive satellite observations), in which the spectra of reddened stars were compared with those of unreddened stars of the same type, showed that this trend does not continue indefinitely into the UV. Instead, the interstellar extinction versus wavelength relation reaches a peak near 2200 Å, following which the extinction *decreases* toward shorter wavelengths to ~1600 Å. Below this, the extinction again increases toward shorter wavelengths (see Fig. 20). The peak in the extinction curve near 2200 Å matches that expected theo-

retically to be produced by graphite particles. Hence, UV observations produced the first definite information on the composition of at least some of the interstellar dust particles. However, it has been found that the UV extinction curve is highly variable in both magnitude and shape in different directions in our galaxy and in the Magellanic Clouds. Hence, it appears that there must be at least three independent components of the interstellar dust, whose relative abundances vary in different ways with local conditions in the interstellar medium.

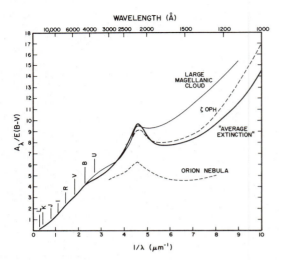

FIG. 20. Interstellar extinction versus wavelength in the UV and ground-accessible wavelength ranges. The letters indicate central wavelengths of ground-based photometric pass bands. Note large variations in the extinction curve in the UV for different regions of space.

Observations of the dust reflection spectrum of interstellar dust have also yielded some insights concerning the properties of interstellar dust particles; in particular it has been found that, in at least some reflection nebulae, the particles are very efficient scatterers of far-UV radiation. Except in the vicinity of the 2200-Å extinction peak, it appears that most of the interstellar extinction is due to scattering rather than to pure absorption.

Most galactic H II regions appear as reflection nebulae in the far UV, as illustrated by the *IUE* spectrum of the Orion nebula in Fig. 19. This is due to the relatively high dust densities (roughly proportional to the gas density) in these regions, combined with the presence of UV-bright stars (which both illuminate the dust and excite the UV and visible emission lines). However, the dust in H II regions is probably not typical of dust in other reflection nebulae (illuminated by somewhat cooler stars) and in the general interstellar medium, since the intense radiation fields in regions such as the Orion nebula can significantly modify the size distributions and compositions of the dust particles. This is evidenced by both their reflection spectra and their extinction curves, which differ from those observed in less excited regions of the interstellar medium.

IX. Extragalactic Objects

The external galaxies, like our own, are gigantic clusterings of stars (as many as 10^{11} to 10^{12}) and associated interstellar material. There is a wide variety of shapes and stellar content among external galaxies; many are spiral galaxies resembling our own, whereas others are spheroidal or elliptical and are devoid of obvious spiral patterns. [*See* GALACTIC STRUCTURE AND EVOLUTION.]

As in the case of our galaxy, UV observations are expected mainly to reveal the hot stellar content. However, an advantage in observing other galaxies is that the spatial distribution of the hot stellar population is seen at a glance, whereas this is much more difficult to determine for our galaxy, which we view from within. The variations in this distribution from one galaxy to the next, and in comparison with the distribution of cooler stars and of interstellar material in these galaxies, provide information on star formation rates and history and on the overall evolution of galaxies. The ability to detect and measure hot stars in the presence of a much larger number of cool stars by means of UV observation is even more important in studies of external galaxies than our own, since in the former individual stars often are not individually resolved and hence problems due to image overlap and confusion are more severe. Ultraviolet observations also provide more sensitive measurements of the interstellar material in observations of starlight extinction or reflection.

Some galaxies, notably the Seyfert galaxies and quasi-stellar objects (quasars), exhibit highly energetic activity in their central regions, which far transcends that which can be associated with even the most massive individual stars. The total luminosity of an active galactic nucleus can greatly exceed the total luminosity of the remainder of the galaxy; quasars are, in fact, by far the most luminous objects in the universe. Observations of these objects in the UV can add to the store of knowledge that is needed to acquire an understanding of these objects.

Photometric and spectrophotometric measurements of a number of external galaxies were obtained with *OAO-2* and subsequently were supplemental by more sensitive measurements with *IUE*. Ultraviolet images of a number of galaxies have been obtained in several rocket experiments and other investigations performed by

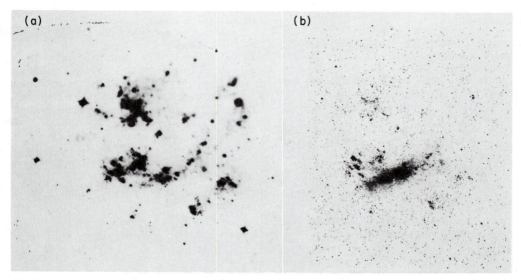

FIG. 21. Comparison of far-UV and visible imagery of the Large Magellanic Cloud, the nearest external galaxy. (a) Image in the wavelength range 1250–1600 Å (10 min) obtained with the Naval Research Laboratory S201 camera on *Apollo 16*. (b). Visible-light image, to the same scale. (Lick Observatory photograph.)

the Naval Research Laboratory and Goddard Space Flight Center (Figs. 21 and 22). The *Astro* ultraviolet imaging telescope and the imaging cameras on *Hubble Space Telescope* are expected to improve greatly the sensitivity and resolution of such imagery. The spectrographic instruments on *Hubble Space Telescope* and the Johns Hopkins ultraviolet telescope on *Astro* will provide more detailed spectrometric information on stars and nebulae in external galaxies, active galactic nuclei, and quasars. One of the

most important objectives of *Hubble Space Telescope* is to determine more accurately the distance sale of the universe; this it can do by observing galaxies at much greater distances with the same degree of detail as nearer galaxies are presently observed with ground-based telescopes. Ultraviolet observations will be an important aspect of these observations, because the most luminious hot stars are useful distance indicators, and these are much brighter in the UV than in the visible.

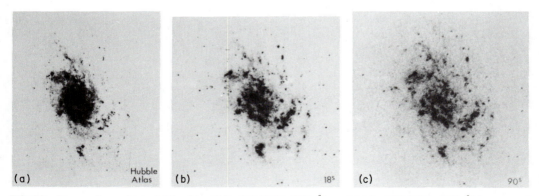

FIG. 22. Comparison of (a) visible, (b) middle-UV (1750–2750 Å), and (c) far-UV (1250–1750 Å) imagery of the nearby spiral galaxy M33. (Courtesy of T. Stecher, NASA-GSFC).

BIBLIOGRAPHY

Chapman, R. D., ed. (1981). "The Universe at Ultraviolet Wavelengths," NASA CP-2171. U.S. Natl. Aeronaut. Space Admin., Washington, D.C.

Code, A. D., ed. (1972). "The Scientific Results from the Orbiting Astronomical Observatory (OAO-2)." U.S. Natl. Aeronaut. Space Admin., Washington, D.C.

Cornell, J., and Gorenstein, P., ed. (1983). "Astronomy from Space." MIT Press, Cambridge, Massachusetts.

Eddy, J. A. (1979). "A New Sun: The Solar Results from Skylab," U.S. Natl. Aeronaut. Space Admin., Washington, D.C.

Hanle, P. A., and Chamberlain, V. D., eds. (1981). "Space Science Comes of Age." Smithsonian Institution Press, Washington, D.C.

Henbest, N., and Marten, M. (1983). "The New Astronomy." Cambridge Univ. Press, London and New York.

Kondo, L., Mead, J. M., and Chapman, R. D., eds. (1982). "Advances in Ultraviolet Astronomy," NASA CP-2238. U.S. Natl., Aeronaut. Space Admin., Washington, D.C.

Lundquist, C. A., ed. (1979). "Skylab's Astronomy and Space Sciences," NASA SP-404. U.S. Natl. Aeronaut. Space Admin., Washington, D.C.

Noyes, R. W. (1982). "The Sun, Our Star." Harvard Univ. Press, Cambridge, Massachusetts.

Spitzer, L., Jr. (1982). "Searching Between the Stars." Yale Univ. Press, New Haven, Connecticut.

ASTROPHYSICS

Harry L. Shipman *University of Delaware*

GLOSSARY

Active galactic nucleus: Nucleus of a galaxy producing energy that is not just starlight, usually in the form of high-speed charged particles, which interact with magnetic fields and gas to produce radio, optical, or X radiation.

Archeoastronomy: Interdisciplinary field combining astronomy, archeology, and sometimes anthropology, using the remains of ancient cultures to study what they knew about the sky.

Black hole: Object that has become so small that its gravity will not permit anything, even light, to escape.

Galaxy: Object consisting of a large number $(10^6–10^{12})$ of stars, held together by gravity.

Interstellar medium: Gas and dust located between the stars in a galaxy.

Main sequence: Phase of stellar evolution, which the sun is presently in, in which hydrogen fuses to form helium at the stellar center.

Planet: Object in which nuclear fusion plays a negligible role in its evolution; generally refers to an object the size of Jupiter or smaller.

Primeval fireball: Microwave radiation left over from the "big bang," the hot, early stage of the evolution of the universe.

Pulsar: Rotating neutron star, a 1-solar-mass $(1\text{-}M_\odot)$ object that is ~ 10 km across, in which radio and occasionally other electromagnetic radiation is emitted in regularly spaced bursts or pulses.

Quasar: Very active galactic nucleus in which radiation produced by the interaction of high-speed charged particles with magnetic fields and gas overwhelms the light from the galaxy.

Red giant star: Star in an advanced evolutionary state that is cooler and considerably larger than the sun.

White dwarf star: Final evolutionary state of a low-mass star, with a diameter of $\sim 10^4$ km, and a density of $\sim 10^7$ g/cm^3.

The scope of modern astrophysics is cosmic, ranging from the study of planets and smaller bodies within the solar system to the most distant galaxies and quasars at the edge of the observable universe. Will the universe expand forever? How do stars form? Why have planets developed in such fascinatingly different ways? Astrophysicists seek to answer these and other deep questions using sophisticated high-technology instruments to make measurements of all known types of electromagnetic radiation. The ability to make measurements from above the atmosphere has opened a number of additional windows in the electromagnetic spectrum and, in the 1960s, 1970s, and 1980s, has led to the discovery of a considerable number of unusual objects.

I. Scope of Modern Astronomy and Astrophysics

A. THE OLDEST SCIENCE—AND ONE OF THE YOUNGEST SCIENCES

The roots of astronomy and astrophysics probably extend back in time to the development of human consciousness. The first intelligent humans were, no doubt, impressed with the beauty of the star-speckled sky on dark nights and wondered how it was all put together, how it came to be. Early people satisfied their curiosity with myths, populating the sky with innumerable creatures, human, superhuman, and otherwise, who were believed to carry the sun across the sky on chariots, producing the regular celestial motions because of their own actions. This prescientific world view was eventually replaced by other cosmic models in which the universe inexorably grew larger and larger, and the place of humanity within the universe became, apparently, more and more insignificant.

Despite the ancient heritage of astronomy, however, the conception of the universe that most people believed in throughout most of human history was fundamentally wrong. It was only in the seventeenth and eighteenth centuries that we developed the necessary mathematical and physical tools that enabled us to understand the basic architecture of the solar system. And it was only in the late nineteenth century that we developed a fundamentally correct understanding of what the most abundant constituents of the universe, stars, were made of. We did not know how stars worked until the 1930s, when we learned enough about nuclear physics to model successfully the way nuclear reactions work in the center of the sun and other stars. After the dawn of the space age, when we could hurl telescopes above the atmosphere to measure radiation outside the narrow optical band of the spectrum, we discovered that the quiet, slow pace of stellar evolution was only a part of the cosmos and that the apparently serene universe is punctuated by astrophysical explosions of tremendous violence. Exploration of the planets and satellites in the solar system by robot probes has produced an explosive growth in our knowledge of other planetary bodies, leading some planetary scientists to call the 1960s and 1970s a "golden age" of planetary exploration.

B. Observations: Covering the Entire Electromagnetic Spectrum

Astrophysics differs from almost any other field of science and technology in that in this field a scientist cannot handle, modify, or control the objects being studied. Consequently, our entire understanding of the universe is dependent on our ability to measure the location, intensity, wavelength, time variation, and polar-

ization of various forms of electromagnetic radiation. Many advances in astronomy are directly attributable to advances in our ability to measure radiation with greater sensitivity (e.g., Galileo's use of the telescope) or at different wavelengths. [See OPTICAL TELESCOPES.]

It is true that there are a few cases in which physical objects from extraterrestrial bodies can be brought into the laboratory. Some examples are the moon rocks that were returned to terrestrial laboratories, rocks on Mars that were pushed around and chemically analyzed by remotely operated landers, and meteorites—rocks from other parts of the solar system that fall to the ground. Cosmic rays, high-energy particles from sources outside the solar system, are another source of extraterrestrial material. However, the opportunities to handle such material are quite rare. In the case of cosmic rays or meteorites, it is not at all straightforward to determine where the material is coming from. For these reasons, most of our understanding of the universe is based on measurements of electromagnetic radiation. [See COSMIC RADIATION.]

The electromagnetic spectrum covers a very wide range of energy, from long-wavelength radio waves with energies of 10^{-6} electron volts (eV) to ultrahigh-energy γ rays with energies of 10^6 eV and even beyond. As Fig. 1 illustrates, visible light is only a tiny piece in the middle of that broad range. The only other type of radiation that readily penetrates the earth's atmosphere is radio radiation, which was not widely exploited for astronomical purposes until after World War II.

The wavelength of any form of electromag-

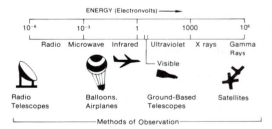

FIG. 1. The electromagnetic spectrum includes radiation of different wavelengths. Only radio waves and visible light can penetrate the earth's atmosphere; at other wavelengths, balloons, airplanes, space shuttles, space stations, or rockets are needed to carry telescopes high into or above the earth's atmosphere. [From Shipman, H. L. (1980). "Black Holes, Quasars, and the Universe," 2nd ed., copyright Houghton Mifflin, Boston.]

netic radiation is closely related to the temperature of the body that emits it; warm objects emit infrared radiation, with less energy than visible light, and very hot objects emit high-energy radiation such as X or γ rays. Because of the very limited extent of the optical window in the spectrum, it was stars and planets (which shine by the reflected light of one star, our sun) that were the focus of astrophysics before the 1960s. Our ability to open other regions of the electromagnetic spectrum has resulted in the discovery of a large number of new cosmic phenomena and has also dramatically improved our understanding of stellar astronomy. [*See* PLANETARY RADAR ASTRONOMY; RADIO ASTRONOMY, PLANETARY.]

The same object looks very different when seen in various parts of the electromagnetic spectrum. Figure 2 illustrates a schematic view of an extreme case, a cluster of galaxies, the largest structure in the universe. In visible light, what one sees are individual galaxies, collections of hundreds of billions of stars, each at a temperature of $\sim 10,000$ K and shining because nuclear reactions go on in the stellar core. For many years astrophysicists, limited to the optical window, thought that this picture showed all the important phenomena. We had a misleadingly incomplete picture of the phenomena in this cluster of galaxies, suspecting that in order to understand them, we would merely have to understand what the stars in the individual galaxies did. An X-ray picture of the same cluster of galaxies, however, shows a completely differ-

ent picture. One-tenth of the mass of a large cluster of galaxies is in the form of high-temperature (10^7–10^8 K) plasma, which permeates the intergalactic space between the galaxies in the cluster. We do not yet have γ-ray pictures of clusters of galaxies, but on the basis of what we know about other γ-ray sources we would expect to see a few active galactic nuclei, in which high-energy processes in the cores of a small fraction of galaxies are responsible for the emission of ultrahigh-energy radiation. [*See* GALACTIC STRUCTURE AND EVOLUTION.]

Working from the visible part of the spectrum to lower energies, an infrared picture of the same cluster of galaxies would show heated dust within the galaxies themselves, possibly including a very few unusual galaxies in which almost all of the emission comes out in the infrared part of the spectrum. (The existence of such infrared-emitting galaxies was unknown until the mid-1980s.) The radio radiation from a cluster of galaxies comes not from low-energy particles but from synchrotron radiation produced when energetic electrons, ejected from a galactic core at high speeds, spiral around magnetic field lines and send radio waves off into the universe. [*See* ASTRONOMY, INFRARED.]

It is this need to make measurements at many different wavelengths that generates the astrophysicist's seemingly insatiable appetite for more and more, bigger and bigger telescopes. Since astronomical observatories typically work in only one part of the electromagnetic spectrum, one observatory provides only a partial picture of a particular astronomical phenomenon. [*See* ASTRONOMY, ULTRAVIOLET SPACE; X-RAY ASTRONOMY.]

C. MODELING AND ASTROPHYSICAL THEORY

An understanding of an astronomical phenomenon is not limited to a description of what one sees in various parts of the electromagnetic spectrum, no matter how extensive the observations are. A true understanding of what is going on in a distant cosmic object usually requires an astrophysicist to develop some kind of conceptual and mathematical model for the phenomenon being studied. This modeling process involves using existing observations of an object to infer its fundamental nature—what it is made of, how it works, and how it evolves or changes through time. A good astrophysical model will then explain some previously puzzling measurements in some part of the spectrum or, better

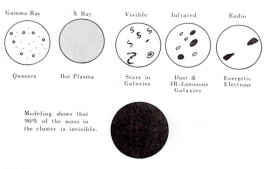

FIG. 2. At different wavelengths, a cluster of galaxies looks quite different. Astronomers must use information from all parts of the electromagnetic spectrum to form a complete picture, and even then modeling of the motion of the galaxies indicates that most of the mass in the cluster is invisible. [Adapted from Harwit, M., and Neal, V. (1985). "The Great Observatories for Space Astrophysics," NASA Astrophysics Division, Washington, D.C.]

yet, will suggest new measurements that could confirm or falsify the model.

Consider the cluster of galaxies in Fig. 2 as an example. Measurements of the velocities of a number of galaxies in clusters like this one indicated that the galaxies were traveling at surprisingly high speeds. A theoretical astrophysicist would ask how a cluster of galaxies like this hangs together through gravity, and can calculate what mass is needed in order to hold the cluster of galaxies together, given the speed of the galaxies in it. The answer turns out to be surprisingly large: Something like 90% of the mass in most clusters of galaxies is not emitting any radiation at all, in any parts of the electromagnetic spectrum that we see. Most of the material in this cluster of galaxies is "dark matter," and we do not know just what this dark matter is. It could, but probably does not, consist of prosaic things like planets wandering between the galaxies. Particle physicists speculate that exotic, undiscovered particles like axions or superstrings could account for the mass that is missing in clusters of galaxies.

D. HOW ASTROPHYSICAL OBJECTS WORK: THE ASTROPHYSICAL ZOO

The pervasive force of gravity, and the response of objects to it, is a unifying concept that is found in virtually all subfields of astrophysics, ranging from planetary science to cosmology. Figure 3 illustrates virtually the entire range of objects studied by astrophysicists, excluding very small objects that can be detected only indirectly, such as interstellar dust grains, 10^{-6} m across, which extinguish and redden the light from distant stars in the same way that dust in the earth's atmosphere makes the sunset red. Broadly speaking, the basic physical phenomena at work in various objects change as one looks at objects of different sizes, and one purpose of this illustration (and the paragraphs that follow) is to point out the different processes at work in objects of different sizes: inactive planets, active planets, stars and clusters of stars, and galaxies. The remainder of this article, which treats individual objects and classes of objects, can thus be put in context. [See SPACE, INTERSTELLAR MATTER.]

The universe is a fascinatingly complex place. The photographs and descriptions of various cosmic objects in this article will illustrate how different these cosmic objects can be. However, there is one basic similarity among all astrophysical objects, ranging from the tiniest asteroid to the largest cluster of galaxies. Everything in the universe is subject to the gravitational force, and the life cycle of every cosmic object is subject to one universal tendency: the orders from gravity

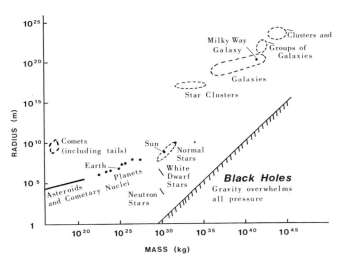

FIG. 3. Cosmic objects cover a wide range in mass and radius. Asteroids, planets, and normal stars have similar densities and roughly similar structures, where the force of gravity is opposed by pressure from normal matter. Star clusters, galaxies, and groups of galaxies consist of collections of widely separated stars and are fundamentally different.

that want to make that object smaller. If any cosmic object reaches a stage in its life cycle where those gravitational commands can no longer be resisted, it will collapse indefinitely, becoming a black hole. A black hole forms when a mass becomes so small that the gravitational force at its surface is so big that nothing, not even light, can get away from it. [See BLACK HOLES (ASTRONOMY).]

Many, if not most, cosmic objects have found a way to counteract the universal tendency towards collapse to oblivion. The strength of the gravitational force increases, generally, with the increasing mass of the object, up to the scale of the largest stars. Different sorts of pressures in the interiors of stars and planets provide the resistance to gravity in these objects, and consequently they go through different life cycles, ranging from the placid life histories of inactive planets through the violent and comparatively rapid evolution of massive stars. The next largest objects, clusters of stars, are fundamentally different in that their constituents interact only through the long-range gravitational force, and the principal reason they do not collapse is that the individual objects are orbiting in empty space and can trace out these orbits almost indefinitely. [See STAR CLUSTERS.]

It is our understanding of the forces at work in various cosmic objects and the power of the scientific method—which enables us to figure out how things that look like tiny dots, even in the largest telescopes, work—that have produced our contemporary concept of the cosmos. While much remains to be done, we can understand the basic makeup and evolution of many of the structures illustrated in Fig. 3. The basis for our understanding is the development of observational techniques that allow us to make precise measurements throughout the electromagnetic spectrum and the parallel development of analytical procedures that allow us to interpret what we see in quantitative terms.

II. Instruments and Techniques

A. OPTICAL TELESCOPES

The pursuit of astronomy requires that astronomers measure the position, intensity, wavelength, polarization state, and time variability of the electromagnetic radiation coming from astronomical sources. The difficulty we face is that the amount of energy coming from stars is mi-

croscopically small. In lifting this book a few centimeters off the desk in front of you, you expend about 1 joule (J) of energy. You would have to stare at the brightest star for a thousand years in order to accumulate 1 J of energy in the starlight that strikes your retina. By lifting this book from the desk, you use more energy than astronomers have measured since the invention of the telescope. Spectroscopists working in chemistry or physics laboratories can often buy a more powerful laser in order to generate more photons; astronomers do not have this option.

The science of astronomy—indeed, our understanding of the entire universe—made a tremendous leap forward when Galileo Galilei heard of a Dutch invention that made images appear to come closer, and turned the telescope toward the heavens. Many subsequent revolutions in astronomy have occurred when physicists, radio engineers, or rocket experts have devised new gadgets that open new windows on the universe. Galileo's discoveries were largely responsible for the acceptance of the Copernican model of the solar system, which banished humanity from the center of the universe.

In Galileo's time, the primary virtue of the telescope was that it enabled him to see more detail on objects within the solar system. The sun could not be regarded as a perfect object because it had spots on it. Venus showed phases, and Jupiter had a satellite system of its own, contradicting the geocentric model of the universe. Ever since then, telescope salesmen from post-Renaissance fairs to department stores have regarded "power" or magnification as one of the chief selling points of a telescope. These days, however, when a telescope is used for research purposes, the primary goal is to collect as much light as possible, not to magnify the images. Amateur astronomers who wish to photograph the wispy outer filaments of beautiful clouds of interstellar gas have the same goal. The tiny amounts of energy that reach us from everything in the cosmos (except the sun) cannot be increased, and so one strives to have a telescope with the largest aperture possible in order to gather the maximum number of photons.

Unfortunately, many of the most interesting astrophysical objects—the faint smudges that represent galaxies at the edge of the observable universe or the tiny glowing balls that are stars in their late, most exciting evolutionary stages—are feeble photon emitters indeed, and so astronomers are constantly striving to develop ways to collect more photons. In the late 1960s and 1970s, the power of large telescopes like the Mt. Palomar 200-in. was vastly increased by the development of sophisticated electronic detectors, which would record the impact of almost every photon on them, in contrast to photographic film or emulsions, which require ~100 photons to produce a detectable event. One cannot increase the efficiency of detectors beyond 100%, however, and so the quest for increased aperture became increasingly urgent by the 1980s. Engineering limitations have prevented the development of a traditional telescope (in which a single piece of glass focuses the light striking it to a point) much beyond the size of the 200-in.-diameter mirror of the Mt. Palomar telescope, built in the 1940s. A number of larger instruments using new technologies are planned for the 1990s. In all cases a key element in the design is that the supporting structure of the telescope is stressed slightly as the telescope points in different directions, and so the optical surface maintains its shape (with a tolerance of a fraction of the wavelength of light, or ~0.1 μm). [See OPTICAL TELESCOPES.]

In addition, the quality of the site at which a telescope is located can strongly affect its observing efficiency, partly because cloudy nights are a total loss and partly because atmospheric turbulence smears out the stellar images. Astronomers working in the infrared part of the spectrum require particularly high-altitude, dry sites like the summit of Mauna Kea, Hawaii (which, at an altitude of 14,000 ft, is much colder than the beaches at Waikiki). [See ASTRONOMY, INFRARED.]

B. SPECTROSCOPY: ANALYZING THE DATA

Often, what comes out of an optical telescope is a spectrum, or a graph showing the intensity of light as a function of wavelength. Figure 4 shows an example, a spectrum of a white dwarf star, a dying cinder, the remnant of a middle-sized star like the sun. What does the spectrum mean? The dips in the spectrum are called spectral lines, for historical reasons. Until 15 years ago, most spectra were recorded on film, with different wavelengths in different positions perpendicular to the length of the spectrograph slit. A place in the spectrum where the star emitted less light than at adjacent wavelengths shows up as a dark line on a photograph of a spectrum.

What produces these spectral lines, and what can one learn from analyzing them? Atoms go-

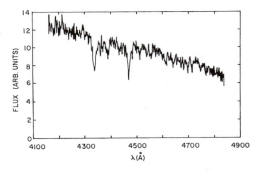

FIG. 4. Spectrum of a star, called PG2234 + 034, showing spectrum lines of hydrogen (at 4340 Å, the leftmost dip) and helium (with a weak feature at 4387 Å and a stronger dip at 4471 Å). Hydrogen and helium are the two most common elements in the universe.

ing from states of higher to lower energy emit photons at particular wavelengths, and atoms going from states of lower to higher energy absorb radiation, removing photons of particular wavelengths from the beam of starlight passing from the star's interior to the observer. The wavelength pattern depends on the element in question. The most easily recognized pattern at wavelengths of visible light is the Balmer series of hydrogen. In hydrogen, the wavelengths λ of successive lines in a series follow the relation $1/\lambda = R(1/n^2 - 1/m^2)$. For the Balmer series, n is 2, and R is the Rydberg constant (here equal to $\frac{1}{911}$ if the wavelength λ is in angstroms). Other elements have their own distinct pattern, and by recognizing this pattern in the spectrum of a star one can determine its chemical content. The star whose spectrum is shown in Fig. 4 contains both hydrogen and helium; the strong dip at 4340 Å comes from hydrogen, and the dip at 4471 Å and a weaker, barely detectable dip at 4387 Å come from helium.

Doing a quantitative analysis to derive the maximum information from the spectrum of a star (or, for that matter, any other astrophysical object) requires a bit more work. As light darts around through a stellar atmosphere, slowly making its way from the hot stellar interior to the cold depths of outer space, it is absorbed much more strongly at the wavelengths characteristic of the absorption pattern of the atoms in the stellar envelope. How much light is absorbed, however, depends on a number of complicating factors. If the element in question is mainly ionized in the star (where the heat in the envelope causes one or more electrons to be stripped from the atom), that element will not absorb much light at wavelengths characteristic of the neutral atom. Many spectrum lines in the visual part of the electromagnetic spectrum come from transitions where an atom goes from one excited electronic state to another, and the amount of light that is absorbed depends on how many atoms are in that excited state.

However, all of these processes that go on in a stellar envelope can in principle be modeled. Computer techniques developed in the early 1960s and still being refined follow the transfer of radiation through the outer layers of the star. The modeler can predict what sort of spectrum a star of given temperature, surface gravity, and chemical composition should have. One then matches this prediction to the real spectrum to determine the properties of stars. The spectra can directly or indirectly provide measurements of a star's temperature, surface gravity (GM/R^2), and chemical composition. For example, PG2234 + 034, the white dwarf whose spectrum is shown in Fig. 4, turns out to be composed mainly of helium, having one hydrogen atom for every 2500 helium atoms.

Similar modeling techniques can be used to analyze the spectra, in any wavelength region, of other astrophysical objects like the gas clouds, accretion disks, and quasars described in the following sections. Since these objects can be more complex than stars, the task of the modeler is more challenging, and the amount of quantitative information we can gain from the spectrum may be more limited. In addition, we can often gain additional information from analysis of other data such as an object's color, the polarization properties of the radiation it emits, and the time variations in its position in the sky or in the intensity of its radiation. Frequently, modelers use data from all over the electromagnetic spectrum in order to determine what a particular astrophysical object is.

C. Radio Telescopes

The other major wavelength region where electromagnetic radiation from the cosmos penetrates to the ground is the radio part of the spectrum. Radio radiation from the cosmos had not been expected, because cosmic objects are hot and radiate few photons in the low-energy, radio part of the electromagnetic spectrum. Karl Jansky of Bell Laboratories accidentally discovered radio emission from the center of the Milky Way galaxy in 1933, but for various reasons this discovery was not followed up until after World War II. The radio emission from the galactic center comes from electrons traveling at speeds close to the speed of light, spiraling around magnetic fields and emitting synchrotron radiation because the motion of the electrons is accelerated. Thus much of radio astronomy involves the study of active galactic nuclei. These are probably giant black holes surrounded by violent processes, where particles are accelerated to high energies. Similar types of radio emission from plasma processes are produced in the solar system. [*See* RADIO ASTRONOMY, PLANETARY.]

Other astrophysical objects that emit radio-frequency radiation include very cool gas clouds in interstellar space. Simple molecules can emit radio waves at specific frequencies, often called spectral lines in terminology borrowed from optical spectroscopy. Well over 50 different molecules have been discovered in interstellar space; in some cases, the species are rather exotic ones such as HC_9N, and in other cases they are familiar ones like water, ammonia, and carbon dioxide. These cool gas clouds often contain dust and are consequently impenetrable to the optical astronomer; the dust absorbs any light that might pass through the cloud.

Radio telescopes are built for the same purpose as optical telescopes: to gather light and concentrate it at a focus, where the intensity, polarization state, frequency, and time variability of radiation from a single source can be measured. Because the wavelength of radio radiation is relatively long, the precise polishing needed for the mirror (or lens) of an optical telescope is unnecessary, and the telescope can be made much larger. But diffraction limits the ability of a radio telescope to isolate the radio radiation from a particular source, and so a single telescope can neither locate nor map a source precisely. Radio astronomers use pairs of radio telescopes as interferometers in order to overcome this limitation. If you have an array of radio telescopes working together, each can be used in tandem with each of the others to provide a large number of interferometers, and the results can be combined to produce a fine-scale map with a resolution equal to or better than those of optical telescopes, as is indicated in some of the illustrations found later. The Very Large Array (VLA) of radio telescopes (Fig. 5) is an assembly of some 27 telescopes that can be moved on railroad tracks in order to map radio sources.

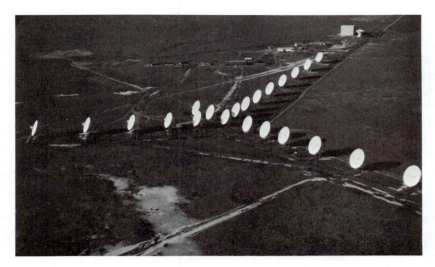

FIG. 5.   Very Large Array of radio telescopes, in which the telescopes, working together, build up a detailed picture of the radio sky. The antennas are on railroad tracks, and the finest detail is produced when they are moved out to the full length of the 21- and 19-km tracks. (Courtesy of the National Radio Astronomy Observatory, operated by Associated Universities, Inc., under contract with the National Science Foundation.)

D. TELESCOPES IN SPACE

But most of the electromagnetic spectrum cannot be observed from the ground. Our ability to send telescopes above the earth's atmosphere has opened up a number of other windows on the universe, and has produced a dramatic increase in the number of known types of cosmic phenomena. It is the hottest gas in the cosmos which emits X- and γ-ray photons, and space astronomers have discovered swirling disks and columns of hot gas in binary star systems where one of the stars is very compact, compressing and heating the gas that falls towards it. Space astronomy played a pivotal role in the discovery of the first black holes. The possibilities that have been suggested by the discoveries made so far have prompted some very ambitious plans, including the launch of the Space Telescope which is scheduled for the late 1980s or possibly the early 1990s. (Fig. 6).

The technological challenges of space astronomy can be formidable. Because it costs money (several thousand dollars a pound) to send something into space, the first telescopes to explore any part of the electromagnetic spectrum have typically been small ones. The data is collected in space and telemetered to the ground, to be analyzed later by the astronomer who asked for it. Designers of X- and γ-ray telescopes face the difficulty that normal optical systems, where

the radiation strikes optical surfaces head-on, will not work because these high-energy photons go right through common substances. (One gets around this difficulty by using collimators or grazing incidence telescopes to isolate photons from particular parts of the sky.) Our ability to meet these technological challenges, as well as the excitement generated by the discoveries of first-generation space instrumentation suggest a tremendous future for the field. The hope is that eventually the costs of working in space will decrease to the point that larger telescopes, kept operating for longer periods of time, will become economically as well as technologically feasible. [See ASTRONOMY, ULTRAVIOLET SPACE; GAMMA-RAY ASTRONOMY; X-RAY ASTRONOMY.]

E. PROBES TO THE PLANETS

For solar system studies, an opportunity presented by the space program is the ability to send experimental apparatus (and occasionally people) to other bodies in the solar system. Here, the instruments used are quite unlike those used in the rest of astronomy, because one can occasionally pick up, chemically analyze, and otherwise experiment with a piece of something elsewhere in the universe. The first planetary probes often had relatively simple instrumentation—cameras, possibly a simple mass

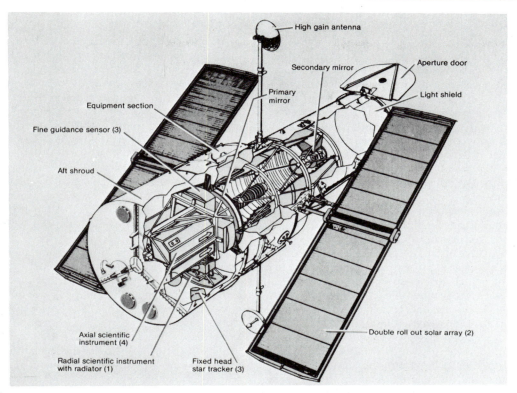

FIG. 6.   Artist's illustration shows NASA's Hubble space telescope with its exterior shroud cut away. Light strikes the primary mirror at lower left, bounces off the secondary mirror, and is then directed to the scientific instruments located behind the primary, at the lower left of the illustration. The other parts of the telescope are needed in order to point it, provide electrical power, and relay the images and spectra back to earth. (Courtesy of NASA.)

spectrograph, and particle detectors were a typical configuration. Later, more specialized instruments were devoted to a particularly intriguing property of a planet (such as the so-called life-detection experiments sent to Mars).

F. PARTICLES FROM SPACE: METEORITES, COSMIC RAYS, AND NEUTRINOS

The first material objects or particles that were known to come from space were meteorites, rocks orbiting in the inner part of the solar system that occasionally fall to earth. A few of these objects are known to have come from the region of the asteroid belt, between the orbits of mars and Jupiter. Most of the rest probably came from the same source, though there have been a few meteorites where mineralogical analysis possibly indicates a lunar or Martian origin. Analysis of these objects has enabled us to determine the age of the solar system (4.7 billion years), reconstruct the element-building events that took place just before the formation of the solar system, and determine the chemical state of the gas that condensed to form the solar system.

A second type of particle that comes from space consists of cosmic rays, high-energy charged particles that can be detected at the earth's surface with particle detectors. A major, well-publicized controversy in American physics in the early part of the twentieth century was the question of the composition of this radiation. It turns out to consist of atomic nuclei, mostly protons with mixtures of heavier nuclei ranging from helium on up. Cosmic rays with the lowest energies, those that only rarely penetrate the earth's atmosphere and are detectable at ground level, come from the sun. Higher energy cosmic rays come from sources somewhere in the Milky Way. Because these particles are charged, they follow curved paths in the magnetic fields that permeate interplanetary and interstellar space, and it is not easy to determine where they come from. [*See* METEORITES, COSMIC RAY RECORD.]

A third type of particle used in astrophysics is a rather improbable one: the neutrino. Neutrinos are neutral particles that interact with other matter only via the gravitational and weak nuclear interactions, and so they can travel through things like planets and stars very easily. Neutrinos emitted in nuclear reactions in the center of the sun travel right through the body of the sun without being stopped, and therefore they give us direct information on what happens at the solar center. A tank of perchloroethylene located far beneath the earth's surface in a gold mine in Lead, South Dakota, has detected neutrinos from the center of the sun, but the flux is one-quarter of that expected from solar models. It is not clear at present whether this is due to a fundamental deficiency in the solar model or whether it is a matter of detail, since this experiment is sensitive only to the very highest energy solar neutrinos emitted in a rare branch of the nuclear energy cycle that goes on in the solar core. [*See* NEUTRINO ASTRONOMY.]

These particles from the cosmos provide unique and valuable information of astrophysical interest. They are as much of an additional window on the universe as any other part of the spectrum is. However, there are limitations on what we can learn from them. These limitations may also apply to future detections of extraterrestrial particles or other types of radiation such as gravitational radiation or the exotic particles that underground proton decay detectors may have detected from the double star system Cygnus X-3. In some cases (e.g., cosmic rays and most meteorites) we have no direct information on the source of these particles. We can infer the origin of meteorites, from their composition, but then the amount of independent information that one learns is correspondingly reduced. Another limitation, which applies to solar neutrinos and meteorites as well, is that they come from only one place in the cosmos, and so information of broader astrophysical significance is missing.

III. Solar System

A. Golden Age of Planetary Exploration

Before 1960, solar system science was a very small subfield of astronomy practiced by a few courageous souls who all too often became involved in such unproductive debates as whether there were canals on Mars. Although ground-based telescopes could then, and still can, provide large amounts of very significant data on planets, comets, satellites, and asteroids, the interests of most astronomers lay elsewhere, in the realm of the stars and galaxies. Here, perhaps more than anywhere else in astronomy, the ability to escape from the earth's surface has generated a revolution in our approach to the field. The period from 1960 to 1980, during which probes first explored planets beyond earth, is sometimes referred to as the golden age of planetary exploration. The golden age either ended or went into limbo in 1980, when the budgetary constraints of the mid-1970s led to a significant slowing of the pace of solar system exploration. It remains to be seen whether the excitement of this period will continue.

The name "golden age," recalling the sixteenth and seventeenth centuries, when European explorers first mapped the surface of the globe, reflects the explosive growth in our knowledge of planetary surfaces. Before 1957, the only planet whose surface had been mapped to any extent was the moon (Fig. 7). Images returned from planetary probes revealed surprising features such as giant craters and volcanoes on Mars, eight active volcanoes on Jupiter's moon Io, continental-size plateaus on Venus, complex structure, as small as a few hundred meters, in the rings of Saturn, and rings around Jupiter, Uranus, and perhaps Neptune. Both the number and the surprising nature of these new discoveries produced an excitement among astronomers and the general public that rarely occurs in science. The properties of the planets are summarized in Table I. [*See* SOLAR SYSTEM, GENERAL.]

B. Inactive Bodies: Asteroids, Satellites, and Small Planets

To approach solar system objects in a way that goes beyond mere description, recall that the evolution of a planet, like the evolution of any other astronomical object, is governed by gravity. Since all planets have roughly the same density, the strength of the gravitational force increases with increasing planetary (or stellar) mass. Planets are small enough that the gravitational force that tends to compress them is quite weak. The rigidity of common materials is sufficient to resist gravitational compression. Temperatures at their centers never get sufficiently high to initiate the nuclear fusion of hydrogen

FIG. 7. Ground-based telescopic photograph of the surface of the moon, showing the large-impact crater Copernicus and secondary-impact craters around it. (Palomar Observatory Photograph.)

into helium, and so nuclear energy can be ignored in understanding planetary evolution.

The smallest planets have the dullest life cycles, since they are geologically inactive. Because their mass is small, thermal energy left over from gravitational contraction, as well as heat from radioactive decay, is also comparatively small. The envelope is so thin that whatever heat is generated leaks out fast, and so heat flow from their interior is presently negligible. In general their surfaces, once formed, remain virtually the same forever. An occasional minor

TABLE I. Planetary Properties

Planet	Distance from sun (AU)[a]	Orbital period	Equatorial radius (km)	Density (g/cm³)	Atmospheric composition	Interior composition
Mercury	0.387	87.97 d	2,439	5.43	Slight trace of H, He	Rock
Venus	0.723	224.7 d	6,050	5.27	CO_2	Rock
Earth	1.000	365.24 d	6,378	5.52	N_2, O_2	Rock
Mars	1.524	686.98 d	3,396	3.93	CO_2	Rock
Jupiter	5.202	11.86 yr	71,398	1.33	H_2, He, CH_4	H + He
Saturn	9.539	29.46 yr	60,330	0.69	H_2, He, CH_4	H + He
Uranus	19.18	84.01 yr	25,900	1.2	H_2, He, CH_4	H + He[b]
Neptune	30.06	164.8 yr	24,750	1.67	H_2, He, CH_4	H + He[b]
Pluto	39.44	247.7 yr	1,500	1	CH_4	CH_4 + (?), CO_2, H_2O

[a] One astronomical unit (AU) equals the semimajor axis of the earth's orbit, 1.49×10^8 km.
[b] Uranus and Neptune probably have cores of ice or rock that contain a substantial fraction of the planetary volume. While Jupiter and Saturn also have icy or rocky cores, they are much smaller, and the bulk composition of Jupiter and Saturn is similar to the composition of the sun.

change occurs whenever another, still smaller body, a passing meteorite or comet, hits the surface, producing a circular impact crater, such as those seen on the lunar surface (Fig. 7). With a few important exceptions, little else happens. Because these bodies are so inactive, the physical conditions that prevailed when they were formed, at the time the solar system was formed 4.5 billion years ago, play a greater role in shaping their present appearance, and so the tiniest bodies in the solar system are the most useful cosmic fossils.

The number of objects in the solar system that fit into this category of inactive bodies is quite large. The asteroid belt between Mars and Jupiter contains a couple of thousand objects whose orbit is known, and these are a small fraction of the total number. Except for an early era during which flows of molten rock (or magma) filled the lunar basins, the earth's moon has been inactive. Mercury, Pluto, and many planetary satellites also fall into the category of inactive bodies. These objects typically have radii of 2400 km (the radius of Mercury) or smaller. [*See* PLANETARY SATELLITES, NATURAL; MOON (ASTRONOMY); LUNAR ROCKS.]

C. SMALL BODIES WITH BIG NEIGHBORS: COMETS, RINGS, AND WEIRD PLANETARY SATELLITES

There are some important exceptions to the general rule that small planets are objects in which impact cratering is the only significant process that occurs on the planetary surface. If a small planet is quite close to some other body, other forces can intervene, often spectacularly. For example, comets, small chunks of ice ~10 km across, heat up, sublimate, and eject a great deal of gas and dust when they come close to the sun. Sunlight interacts with this gas and dust, producing the cometary display that is occasionally visible to the naked eye. Most comets follow orbits with extremely long periods, which take them far from the inner solar system, and so they are discovered very shortly before their closest approach to the sun. Of the comets with orbits contained within the solar system, following orbits with periods less than 100 yr, only Halley's comet is sufficiently bright to be visible to the unaided eye on most of its appearances. (In 1986, it did not pass very close to the earth, so that many people could not see it without binoculars.)

A spectacular example of small bodies being influenced by larger neighbors are the ring systems of the largest planets in the solar system. The tidal gravitational forces from giant planets like Jupiter and Saturn are sufficiently large to perturb or even destroy medium-sized satellites that orbit in their vicinity. Once an object comes closer to a giant planet than a certain limit (called the Roche limit), tidal forces will rip it apart (or may prevent it from forming in the first place), and what results is a swarm of tiny orbiting bodies circling the planet in a common orbit, which looks from a distance like a ring. The rings of Saturn were discovered, in a sense, by Galileo, though he could not make out their actual shape. Subsequently, rings were discovered by accident around Uranus, because they occulted the light from a distant star passing behind the planet. A search from the *Voyager* space probe discovered rings around Jupiter. Neptune apparently has some debris orbiting around it, but it is not clear whether this is part of a well-developed ring system. The structure of the rings around Saturn is quite complex (Fig. 8).

A less familiar example of the effects of tides involves the alteration of the surfaces of satellites of large planets. The innermost large satellite of Jupiter, Io, is so close to the giant planet that its interior is kept perpetually molten by tidal stresses from Jupiter and from other Jovian satellites. As a result this object is geologically more active than the earth, which is (apparently) the most active of the rocky planets in the solar system. The surface of one Saturn's satellites Enceladus is relatively crater free, also (apparently) a result of tidal forces modifying its appearance. Some other satellites in these systems (Europa and Ganymede) do not show the ancient, crater-strewn landscapes that one would expect on such small bodies. Many of the satellites of Uranus, photographed by *Voyager* in January 1986, show unusual and varied surface terrain, quite different from what one would expect of such small bodies were they completely isolated in space.

In any field that has developed as fast as planetary science, there will be objects that do not fit into the picture presented here, which attributes anything other than an impact crater on a small body to tidal interaction. Titan, the largest satellite of Saturn, has a thick atmosphere of nitrogen and methane, and may have methane lakes. Many of the small objects in the outer solar system are composed of ice rather than the rock that makes up the earth's moon, and the different thermal and mechanical properties of ice

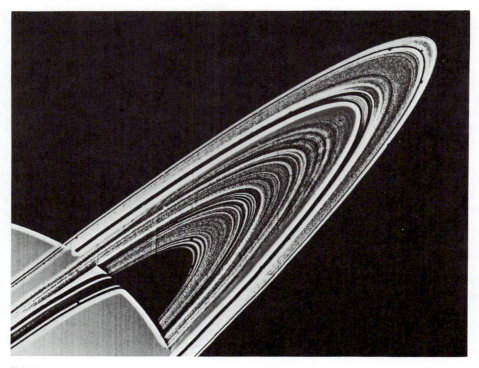

FIG. 8. The complex structure of the ring system of Saturn is evident in this photograph from *Voyager 1*. (NASA; photo provided courtesy of the National Space Science Data Center.)

may affect their evolution. Can a slushy subsurface layer account for the absence of craters on Europa? How is Pluto, which is composed largely of frozen methane rather than frozen water, different? Similar questions proliferate.

D. LARGE TERRESTRIAL PLANETS AND COMPARATIVE PLANETOLOGY

Once one starts to consider planets roughly the size of Mars, another planetary process enters the picture. Virtually all cosmic objects smaller than star clusters or galaxies (and possibly even those) are formed from the collapse of some kind of cloud of gas and dust. When something is formed from gravitational collapse, some heat is generated in the object's interior. Additional heat in the interior comes from radioactive decay. Once a planet becomes big enough, the rocky or icy mantle surrounding the hot core is sufficiently thick that the heat leakage can continue for a substantial fraction of the age of the solar system, in some cases until the present day. This heat flow from the interior produces such things as volcanism, plate tectonics, and the ejection of gases that form the atmospheres of the lightweight, rocky planets like the earth, Venus, and Mars. Of these three planets, Mars is the smallest, with a mass of one-tenth of an earth mass, and the other two are approximately equal in mass. They present a less than ideally equipped, but still serviceable cosmic laboratory in which to test various ideas about why planets are different from one another, including the idea that planetary mass is a fundamental determinant of planetary evolution. Such testing is called comparative planetology.

Since the mid-1960s, geologists have embraced the idea that plate tectonics, the motion of continental-sized plates past and into each other, is a major sculptor of the surface of the earth. The Himalayas exist because the Indian subcontinent is crashing into Asia. The ring of fire around the Pacific exists because the ocean bottom is crunching under the continental crust, or sliding past it in a jerky motion (as in the San Andreas fault in California). Do such processes occur on other planets?

On Mars and Venus, where one would expect more activity than on small objects like the earth's moon, there are signs of large-scale tectonics. There is a huge canyon stretching about one-quarter of the way around Mars, and this appears to be the result of stretching of the Martian crust by an enormous bulge in the northern hemisphere of the planet. Mars contains the largest volcanic mountain in the solar system, Olympus Mons, which is similar in shape to the Hawaiian islands but much larger. There are other, similar Martian shield volcanos. The surface of Venus is not as well mapped, since planetary radar astronomy is required to penetrate the cloud layer, but features characteristic of geological activity such as raised plateaus and volcanic mountains have been discovered in the 1980s.

Comparative planetology has been applied in other ways to explain, for example, the weather patterns on other planets. Mars has a very thin atmosphere, and its weather resembles in some ways that of terrestrial deserts. Venus's atmosphere is much thicker, and in addition the planet rotates only very slowly. The differing temperatures on Venus, the earth, and Mars are explained by differing degrees of greenhouse effect on the three planets. Mars is particularly puzzling in that channels on its surface, cut by running water (Fig. 9), could not have been formed when the atmosphere was as thin as it is now. Our suspicion is that at some time in the past, the climate was different and the atmosphere was thicker, perhaps because the polar caps were thinner. In this sense Mars is currently experiencing an ice age, reminiscent of the time on earth when ice sheets covered much of North America. [See PLANETARY ATMOSPHERES.]

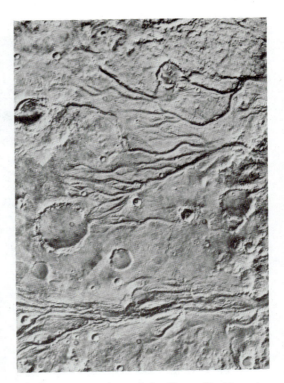

FIG. 9. *Viking 1* photo of the Chryse Planitia region of Mars, obtained from August 4 through 9, 1976. The channels were cut by flowing water a long time ago. (Courtesy of NASA.)

E. GAS GIANTS

Four of the planets in the solar system—Jupiter, Saturn, Uranus, and Neptune—are significantly larger than the rest. The radius of Jupiter is ~10 earth radii, and Neptune, the smallest of these gas giants, is still ~3.9 times the size of the earth. Instead of being made of solid ice or rock, as the smaller bodies are, they are made mainly of gas, hydrogen, and helium, with a composition more like the sun's than the earth's. All of them have a rocky core, and in the two smallest, Uranus and Neptune, the rocky core is a substantial part of the planet's volume.

At this writing, considerably less is known about these gas giants than about the smaller members of the Solar System. They are farther away from the earth, and it is more difficult for space probes to reach them; the first probe to have orbited (not just whizzed by) any of them is the *Galileo* probe to Jupiter, which may reach the giant planet in the 1990s. When we look at these planets, we do not see any solid surface layer; rather we see a fascinating, changing filigree of cloud patterns. In contrast to the cyclonic storms seen on the comparatively slowly rotating earth, these planets show band structure, where the high- and low-pressure areas have been stretched out so that planet-girdling winds dominate the planet's optical appearance. (From the viewpoint of comparative planetology, it is interesting that computer programs developed to simulate the earth's climate can also reproduce the banded structure on Jupiter.)

F. THE SUN AND SOLAR–TERRESTRIAL RELATIONS

The sun is a star, but because it is so close to us, we know much more about it than we would if it were just an ordinary star. We can photograph detail in its surface layers and in the layers above the surface, and so we know how strong a role magnetic fields play in shaping the structures of gas that pervade space near the sun. When these magnetic fields suddenly reconnect, or when some other plasma instability results in a sudden readjustment near the solar surface, a solar flare results, with the accompanying increase in brightness in the outer solar atmosphere, a solar radio burst, and a sudden emission of rapidly traveling charged particles. Even outside of periods of solar flares, less intense emissions of particles and radio waves are seen. [See SOLAR PHYSICS.]

These violent plasma events near the surface of the sun have effects on the earth's environment. The most obvious to the average person are the auroras, electrical discharges in the upper layer of the earth's atmosphere that occur because a burst of high-energy charged particles produced in a solar flare reaches the earth. Unexplored until the advent of the space program were the complicated structures of charged particles and magnetic fields surrounding the earth and other planets with magnetic fields. These fields trap high-speed charged particles into so-called radiation belts near the earth's surface, and solar flares can be followed by disturbances in the ionosphere, the earth's upper atmosphere, which can affect radio communications.

IV. Stars

A. BASIC PHYSICS OF STARS

The interiors of astronomical objects with masses significantly higher than planetary masses, past a limiting figure of 8% of the sun's mass (or ~80 Jupiter masses), can become sufficiently hot that nuclear fusion reactions within them can last for a long time. (The solar mass ($M_\odot$) is 1.989×10^{30} kg.) As a result, these objects are self-luminous stars rather than planets, which only reflect the light of their parent sun and, in the case of the largest ones, slowly leak away the gravitational energy that was released when they were formed. Normal stars like the sun work because heat pressure in their interiors counteracts the force of gravity and prevents them from collapsing. Because hydrogen is fusing to helium in the centers of these stars, releasing energy, the interior can remain hot in spite of the fact that the star is shining and pouring energy out into interstellar space.

The density of matter in ordinary stars like the sun is reasonably normal, similar to the density of water. When the nuclear energy sources run out, however, the stars reach the end of their life cycle. Eventually they must collapse, becoming much smaller. If their final mass is small enough, they can become very peculiar, compact stars, either white dwarf stars or neutron stars; but if their final mass is too large, they collapse indefinitely, becoming black holes, where an event horizon, a boundary from which no light or other objects can escape, surrounds them. [See STELLAR STRUCTURE AND EVOLUTION.]

B. OBSERVATIONAL STELLAR PROPERTIES

Stars look like pinpoints of light when seen through the telescope, and only in a few exceptional cases can astronomers use clever techniques to overcome the blurring of stellar images by the earth's atmosphere and resolve the visible disks of stars. Nevertheless, it is possible to measure a number of stellar properties, and for historical reasons the nomenclature of stellar astronomy is polluted by numerous anachronisms. The obvious way to measure stellar brightness, which is being used increasingly by astronomers working outside the optical region of the spectrum, is to measure the flux f of energy in a specified frequency range that crosses each square centimeter of telescope surface area, in watts per square meter per hertz (W/cm² Hz) for example. The first star catalog, compiled by the Greek astronomer Hipparchus, divided the stars into groups according to stellar brightness, with the brightest being first magnitude, the next brightest group being second magnitude, and so on down to sixth magnitude. Largely for historical reasons and partly because the magnitude scale is a convenient way of reckoning how prominent a star appears to be in the sky, it has persisted to the present day, but become precisely quantified: the stellar magnitude $m = -2.5 \log(f) +$ an additive constant, where the value of the constant varies depending on the waveband one is in.

Other interesting stellar properties include distance from earth, temperature, luminosity, and radius. The distance from earth is, for the closest stars, measured by triangulation, using the earth's orbital diameter as a baseline and detecting the shift in a star's position as it is seen from opposite ends of the orbit. The distance that corresponds to an angular displacement of 1 sec of arc with a 1-astronomical-unit (1-A.U.)

baseline (1 astronomical unit is the radius of the earth's orbit, or half of the maximum possible baseline) is called a parsec (1 pc = 3.08×10^{16} m); stars, on the average, are a little more than 1 pc apart. A star's temperature can be determined either by obtaining a spectrum of it and classifying the spectrum or by measuring its color and doing a more quantitative job, fitting an energy distribution of the star to models. The long arm of history is again apparent in the spectral classification. The original classification system used the letters of the alphabet, the stars with the strongest hydrogen lines coming first (the A stars). However, more quantitative determinations of stellar temperature and the elimination of redundant spectral classes has resulted in the spectral sequence, running from hottest to coolest, of O, B, A, F, G, K, M, other types being added for stars in particular evolutionary stages (such as the white dwarf stars, which are DO, DA, DB, DC, DQ, and DZ, with various subtypes) and the numerical indices being added to subdivide the classes, where the smallest numbers correspond to the hottest temperatures.

The quantities often of most interest when it comes to unscrambling the life cycle of a star are the star's mass, luminosity, and radius. The mass can be directly measured only in a few relatively favorable cases (~100) in which the star is in orbit around one or more other stars and Kepler's third law, relating the orbital period and the orbital size to the total mass of the system, can be applied. The luminosity is the star's total power, which can be measured in watts but is often expressed in units of the solar luminosity of 3.84×10^{26} W. It can be determined from the flux and the distance because of the inverse square relation between flux f, luminosity L, and distance d: $f = L/(4\pi)d^2$. (To see where this relation comes from, draw a sphere centered on the star with its radius equal to d, assume the star radiates isotropically, and divide the star's total power L by the area of the sphere to get the power crossing each unit surface area of the sphere.) This relation holds either in a specific part of the electromagnetic spectrum or across the whole spectrum. To determine the star's radius, measure or infer the total luminosity L and use the relation from thermodynamics between the absolute temperature T and the emissivity e of a black body, $e = \sigma T^4$, where σ is the Stefan–Boltzmann constant, 5.69×10^{-8} W/m^2 K^4. The luminosity of a star is then its surface area times e, so that $L = (4\pi)R^2\sigma T^4$.

These relations can be used to infer a number of stellar quantities; usually, one measures flux f, distance d, and temperature T and infers the luminosity and radius from the above relations. However, many important applications of these relations change the order of these procedures. Triangulation can be used only to determine distances to the closest stars, but if there is some indirect way of determining the luminosity of a particular star, one can determine its distance d by measuring its flux f. A particular type of variable star called a Cepheid variable follows a period–luminosity relation by which its luminosity can be determined from its pulsation period. These variables have been used as cosmic yardsticks; find a Cepheid variable in a distant galaxy, measure its flux and pulsation period, and you know how far away the galaxy is.

C. Stellar and Astronomical Nomenclature

Historical anachronisms are often responsible for the peculiar vocabulary of a scientific field, and astronomy is no exception. A star often has a number of names or designations. The brightest stars are identified by what constellation they are in. Constellations are patterns of stars in the skies; the names of many derive from ancient times, and the International Astronomical Union, the only organization that can sanction "official" astronomical names, divided the sky into 88 patches in 1930. Fainter stars are generally identified by catalog numbers, some of which are identified with a particular individual. Usage in the literature determines which designation is used most often. Star clusters and galaxies are also generally identified by catalog number, though a few of the brighter ones have acquired names over the years.

The names of solar system objects and their surface features have been established by tradition as officially sanctioned by the International Astronomical Union. The discoverer of a new planet, comet, or asteroid generally has the privilege of naming it. Custom dictates that mythological names be given to important permanent objects. The names of living astronomers are used only to identify ephemeral objects like comets or unimportant objects like minor lunar craters.

Table II lists the names and properties of a few representative stars. These stars were selected to represent the extremes of various stellar properties; most stars have masses and radii within an order of magnitude of the sun's. Some-

TABLE II.   Selected Stars[a]

Star name	Magnitude (visual)	Distance (pc)	Spectral class	Temperature (K)	Luminosity (sun = 1)	Radius (sun = 1)	Mass (sun = 1)
L726-8 B (UV Ceti)	13.10	2.60	M5.5	2,600	0.001	0.15	0.11
Betelgeuse	0.4	200e	M2	3,250	120,000e	1100e	10e
70 Ophiuchi A	4.24	4.93	K0	5,400	0.44	0.80	0.84
Procyon A	0.37	3.48	F5	6,510	7	2.1	1.77
Sirius A	−1.46	2.67	A0	9,970	25	1.67	2.14
Regulus	1.35	24.3	B7	12,210	270	3.6	3e
Epsilon Orionis	1.50	500e	B0	24,800	470,000e	37e	25e
Sirius B	8.4	2.67	DA2	27,000	0.027	0.008	1.05
H1504 + 65	16.2	800e	DZ0	160,000	6,000e	0.1e	1e

[a] An "e" indicates that a quantity is estimated; uncertainties vary from object to object but are probably roughly a factor of 2. The magnitude, radius, and luminosity of Betelgeuse are variable with time. Stars designated "A" or "B" are in binary systems, the A component being the brighter. [From Shipman, H. L. (1978). "The Restless Universe: an Introduction to Astronomy," Houghton Mifflin, Boston; updated as necessary from Popper, D. (1980). *Annu. Rev. Astron. Astrophys.* **18**, 115 (masses); Thejll, P., and Shipman, H. *Publ. Astron. Soc. Pacific* (in press) (Sirius B); Nousek, J., Shipman, H., Holberg, J., Liebert, J., Pravdo, S., Giommi, P., and White, N., (in press). *Astrophys. J.* (H1504 + 65).]

times stellar properties are displayed on a graph called a Hertzsprung–Russell diagram, where luminosity is plotted against temperature. (For historical reasons, luminosity is on the y axis, increasing upward, and temperature is on the x axis, plotted backward with temperature *decreasing* to the right.) However, tables like this one, or comparable graphs, can tell us only what stars are like, and the astrophysically interesting question of how they evolve has yet to be answered. Since stars can be analyzed in some detail, we understand stellar evolution better than we understand the evolution of many other astronomical objects, though many significant open questions still remain.

V. Life Cycles of Stars

A. STARBIRTH

Stars form when an interstellar gas cloud becomes unstable and collapses gravitationally. The space between the stars contains a great deal of gas in the form of clouds. The tiny black splotches in Fig. 10, silhouetted against the bright gas that is lit up by a nearby star, are tiny globules containing gas and dust. It is globules like these that can collapse to form stars. Several possible mechanisms can trigger starbirth by causing the gravitational collapse of small clouds of gas. One popular suggestion is that the explosion of a nearby supernova sends a shock wave through the interstellar medium, causing

the collapse. Isotopic analyses of meteorites indicate that a supernova might well have triggered the formation of the solar system.

The detailed processes leading to starbirth are largely unknown, partly because all of them take place inside a dark dust cloud from which light cannot escape. If one cannot see into a cloud, it is difficult to find out what is happening there, although infrared astronomers and radio astronomers studying line radiation from the molecules in the interstellar medium are beginning to make progress. There is no reason to believe, however, that when stars form objects of smaller mass—planets—are not formed as well. Recently, infrared astronomers have discovered one object with a (just barely) substellar mass, a companion of the nearby star Van Biesbroeck 8 (so it is called VB 8B). Several other systems contain disks of matter that may be similar to the disk that eventually formed the solar system.

Our current scenario for the formation of the solar system postulates that the cloud of gas that collapsed to form the sun was rotating somewhat and that the part of it that did not become incorporated into the sun formed a rotating disk called the solar nebula. Material in this disk condensed into solids, and then collision of the solids with one another, perhaps aided by the gravitational attraction between them, caused these solids to accrete and become planets. Planets in the inner solar system are composed primarily of rocky material, because the inner part of the solar nebula was hot, and rocks

FIG. 10.   Small globules of dust, seen silhouetted against a bright cloud of glowing gas called the Rosette nebula, are probably tiny interstellar cloudlets that are about to become stars. (Palomar Observatory photograph.)

rather than ice condense out of high-temperature gas. In the outer solar system, the colder gas condensed into ice, and the small objects out there, the satellites of Jupiter and Saturn, are predominantly icy.

B. STELLAR MIDDLE AGE: THE MAIN-SEQUENCE PHASE

Once a star has completed its gravitational contraction, it becomes hotter than 4×10^6 K in its center and nuclear fusion reactions can start. In these fusion reactions, four protons fuse to form a helium nucleus, getting rid of the two extra positive charges by emitting two positrons This nuclear rearrangement is one of the most efficient of all nuclear reactions, and so it can provide enough heat to keep a star like the sun in its present state for 10 billion years. More massive stars use their nuclear fuel much more rapidly, and their lifetimes are consequently shorter.

The reason that we believe this model of stellar evolution is correct is that the tools of stellar astronomy are reasonably well developed. Detailed computer models of the interior of a star follow the nuclear reactions and energy transfer through the star, predicting just how hot, large, and luminous a main-sequence star of a given mass should be. Observations of various stellar properties can confirm these predictions in detail. Hundreds of nearby stars fit the relation between temperature and luminosity that is predicted from computer models. Around 50 stars in binary systems, where one can use Kepler's laws to measure stellar masses, fit the predicted relation between stellar mass and luminosity. The rate at which stars age and leave the main sequence, as a function of mass, also agrees with the models. While there are several places where the data still do not fit into the picture, these are currently viewed as anomalies rather than indications of a major problem with the model.

C. Old Age: Red Giant Stars

Most of the stars in space have roughly the same radii. One can measure a star's radius by determining its luminosity and its temperature. Since each unit area of the star's surface radiates an amount of power equal to σT^4, where σ is the Stefan–Boltzmann constant, the total power radiated by the star is just its surface area times the power per unit area, or $4\pi R^2 \sigma T^4$. If L and T are known, R can be measured. In general, the most luminous stars are the hottest.

There are some conspicuous exceptions, however, which can be noticed even with the unaided eye in the nighttime sky. Five of the brightest stars in the sky are noticeably red, and these stars are bright because they are quite luminous. Quantitative measurements of their radii indicate that they are hundreds of times larger than the sun and that if their centers were placed where the sun is, the earth would be orbiting inside their surfaces. These types of stars are called red giant stars.

The physics of stellar evolution, confirmed by computer models, can explain the red giant phenomenon. A star stays on the main sequence until it runs out of hydrogen fuel. When nuclear fusion stops at the center, the central regions contract, reaching a considerably higher density. The star is no longer homogeneous, and the molecular weight of the material in the core, helium, is higher than the molecular weight of the hydrogen envelope. The high pressure gradient needed to support the high-molecular-weight core material ends up pushing the hydrogen envelope out into space, making it cooler. Even though the star has a higher luminosity than it did as a main-sequence star, the enormous increase in surface area means that it can cool down and still radiate away the energy that is generated in the central core.

D. White Dwarf Stars

However, red giant stars cannot last forever. They exist for a short period of time because there are still nuclear energy sources in the star's interior. In most red giant stars, hydrogen fuses to form helium in a shell surrounding the center, and in the center nuclear reactions involving heavier elements occur. Three helium nuclei can fuse to form a carbon atom, and carbon can capture a number of helium nuclei to form heavier elements like oxygen, neon, magnesium, and iron. Eventually, in a few hundred million years or so (depending on the star's mass), these energy sources are exhausted too, and the star must come to terms with gravity as it encounters its final fate.

The stars of lowest mass, stars that have an initial mass of less than ~ 8 $M_\odot$, end their red giant stage by shedding their outer envelopes, leaving a tiny, hot core. These outer envelopes are often ionized by the hot core and are visible as round, fluorescing gas clouds (Fig. 11). An excellent example of the illogic of astronomical nomenclature, these clouds are called planetary nebulae because in the telescope some of them look like planets. The hot core is a tiny ball that rapidly (in a few thousand years) shrinks to become no larger than the earth. A variety of physical processes that are not understood alter the chemical composition of the visible surface layers of these stars, and as they cool, radiating away the energy left from their earlier nuclear burning stages, their surface compositions probably change as radiative forces push elements to the surface and gravity causes the heavy ones to sink.

These stellar cores, called white dwarf stars, can last forever and simply cool. Their densities are $\sim 10^7$ g/cm^3, or ~ 100 tons/in.^{3}, and the nature of matter under such conditions is fundamentally different from that at ordinary densities. The atoms are forced so close together that the electron clouds from each atom would interpenetrate if they were allowed to do so. What happens is that the electrons become degenerate, and the uncertainty principle from quantum mechanics forces some of them to acquire momentum that exceeds the thermal momentum

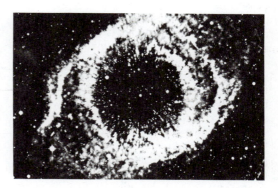

FIG. 11. Planetary nebula NGC 7293, also called the Helix nebula. The star in the center is about to become a white dwarf star, and the shell of gas was the outer envelope of this star when it was a red giant star. (Courtesy of National Optical Astronomy Observatories.)

they have just because they are hot. Quantitatively, the uncertainty principle is $\Delta p \, \Delta x > h$, and if the density of degenerate stuff is d, each electron is forced into a volume of space with a size of $\Delta x \sim d^{-1/3}$. Since the uncertainty in the electron's position is limited by the high density, the uncertainty principle requires that the electron's momentum also be limited, and the limiting momentum (called the Fermi momentum p_F) is $h/\Delta x$, and so $p_F \sim d^{1/3}$. To compress degenerate matter, increasing d, energy must be supplied to give the electrons a higher Fermi momentum, and so degenerate matter exerts a pressure called degeneracy pressure.

From the viewpoint of stellar evolution, the important thing about degeneracy pressure is that it is independent of temperature. Therefore, a star that is supported by degeneracy pressure, like a white dwarf star, can find a permanent refuge from the gravitational straitjacket. White dwarf stars cool, radiating away their internal heat but not shrinking. This configuration is the final resting place of low-mass stars, stars that begin their lives with less than 8 $M_\odot$.

E. NEUTRON STARS

Some stars are too massive to become white dwarf stars. If a white dwarf star is formed with a mass greater than a limiting value, called the Chandresekhar limit after its discoverer, it is unstable and collapses to become a smaller object. The numerical value of the Chandresekhar limit is 1.2–1.4 $M_\odot$ depending on chemical composition. Stellar remnants more massive than this can resist gravity when their centers become so dense that the electrons find a way around the uncertainty principle by combining with protons in atomic nuclei to form neutrons. When stellar matter reaches the almost unbelievably high density of 10^{14} g/cm^3, these neutrons exert a degeneracy pressure quite analogous to the electron degeneracy pressure in white dwarf stars. A ball of degenerate neutrons is called a neutron star, and typical neutron stars, with masses of $\sim$1.4 solar masses, have a radius of $\sim$10 km, being about the size of Manhattan Island. The properties of neutron stars depend on the equation of state, or pressure–density relation, of matter at these high densities, and so analysis of neutron stars may test our knowledge of the physics of dense matter.

Although neutron stars were discussed as purely theoretical objects in the 1930s, they remained an uninteresting theoretical toy until

they were discovered by chance in the late 1960s as pulsars. Jocelyn Bell, in Anthony Hewish's group at Cambridge University, was investigating the variability of radio sources on time scales of several minutes, a phenomenon produced by the passage of radio waves through the interplanetary medium. She noticed some peculiar radio sources, which, on further investigation, turned out to be regularly pulsating radio sources that emitted radio emission in precisely spaced bursts or pulses. The radio emission from these objects sounds like the ticking of a clock; the intensity varies irregularly, but the pulses come at precisely timed intervals.

Once a number of very fast pulsars were discovered, it became reasonably clear that pulsars were rotating neutron stars. The pulsing is produced because the magnetic poles of the neutron star, which (like the earth's magnetic poles) are not aligned with the rotational poles, are the source of a beam of radiation. When the star rotates, the beam sweeps around the sky. Someone at any point in space sees a pulse when the beam sweeps by. [See PULSARS.]

At least some, and probably all, pulsars are formed from the death of massive stars in supernova explosions. The core of the star is compressed to the point where a massive core, exceeding the Chandresekhar limit, rapidly collapses to neutron star densities. The energy released by this collapse blows the remainder of the star off into interstellar space, leaving an expanding cloud of gas called a supernova remnant. Optical (Fig. 12) and radio (Fig. 13) images of supernova remnants illustrate the variety of shapes that these remnants can assume. A su-

FIG. 12. A red-light photograph of the Crab nebula, Messier 1, shows the debris left over from a supernova that Chinese astronomers observed in 1054 A.D. (Palomar Observatory photograph.)

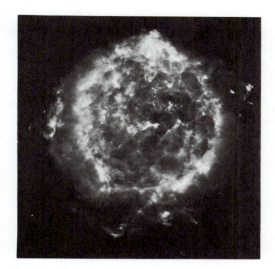

FIG. 13. Computer-produced radio image of the supernova remnant Cassiopeia A. Twenty minutes of computer time on a Cray supercomputer were required to produce this map, which was based on data obtained with the Very Large Array radio telescope (Fig. 5). The resolution in this image is superior to that in the best optical photographs. (Courtesy of the National Radio Astronomy Observatory, operated by Associated Universities, Inc., under contract with the National Science Foundation.)

pernova explosion goes off in a galaxy like ours approximately every 30 yr, but in many cases the dust between us and the supernova does not permit us to see it.

The analysis of pulsars has been difficult, largely because we do not understand precisely how pulse beams are produced. We have no equivalent of stellar spectroscopy to provide us with temperature, chemical compositions, and so forth, although X-ray telescopes scheduled to be orbited in the 1990s may be sensitive enough to see the thermal radiation from the hottest, most recently formed neutron stars. A good deal of information has come from measuring the arrival times of pulses. In one case, a pulsar and a nonpulsing neutron star are orbiting one another so tightly that the orbit is affected by the differences between Newtonian gravity and general relativity, and so pulsar timing measurements can confirm the prediction of general relativity that a system like this should lose energy by gravitational radiation.

F. BLACK HOLES

Just as there is a limiting mass for white dwarf stars, so there is a limiting mass for neutron stars. The exact numerical value of this mass is not well known since it depends on the equation of state of neutron star matter, but a reasonable value would be 2–3 $M_\odot$. If a star finishes its evolutionary cycle with a mass exceeding this value, its remnant core will collapse indefinitely, becoming a black hole.

Black holes, first mathematically described in 1916, were not identified until the 1970s, when a number of X-ray sources were tentatively identified as containing black holes. In a few binary systems, a visible star is being orbited by an invisible object that is accreting matter. In still fewer cases (three as of early 1986, called Cyg X-1, LMC X-3, and A0600-00), the mass of the invisible object is sufficiently large that it is probably not a neutron star but a black hole. Larger black holes may be found in active galactic nuclei (see Section VII,E).

G. STARS IN BINARY SYSTEMS

The story of stellar evolution that has been told to this point has presumed that the star evolves in isolation, expanding, contracting, and going through various nuclear burning stages in isolation from other stars. A large number of stars are found in double star systems, where two stars orbit one another in the same way that planets orbit the single sun. In cases where the stars orbit in paths that take them quite close to one another, the star that evolves faster (the one that was originally more massive) will not be able to become a red giant in the normal way, since there is not room in the binary system for a star with a huge envelope. Rather than swelling up and engulfing the other star, it will begin dumping mass onto its companion. The exchange of mass between the two stars continues in one way or another for the rest of the stars' combined life cycles and produces some unusual behavior in the evolution of this binary.

The most energetic and explosive phenomena that occur in the life cycles of interacting binaries occur in the very late stages, when one of the stars has become a compact object (white dwarf star, neutron star, or black hole). When mass is dumped onto a compact object, it gains a considerable amount of energy. A kilogram of material dropped onto a white dwarf star acquires an energy of $\sim 10^{14}$ J, comparable to the energy released in the Hiroshima nuclear bomb. The gravitational potential well in a neutron star is considerably deeper, and the gravitational energy of the infalling material is two orders of magnitude greater than in the white dwarf case. The exchange of mass in such interacting binaries leads to tremendous outbursts of radiation, where stars flare up, increasing their brightness by many orders of magnitude, in very short times ranging from minutes to hours. Such sudden increases of optical brightness have been known for a long time and are called nova explosions. Only recently have we monitored the X- and γ-ray sky and discovered that explosive surges in the high-energy brightness of these stars are relatively common.[*See* BINARY STARS.]

VI. Milky Way Galaxy

A. PHYSICS OF STAR CLUSTERS AND GALAXIES

There is a fundamental difference in density between the kind of objects discussed so far, with masses of 100 $M_\odot$ or less, and the larger objects discussed in the balance of this article. Except for stellar remnants (white dwarfs and neutron stars), stars and planets have mean densities within several orders of magnitude of ordinary substances that are familiar on earth. However, the mean density of even the densest star cluster is $\sim 10^{-14}$ g/cm^3, eight orders of magnitude less dense than the least dense star known, a huge, bloated red giant star.

The dividing line between stars and star clusters marks an important astrophysical transition between single objects and collections of single objects whose main interaction is gravitational. In objects the size of star clusters and larger, there are no longer easy ways for a gravitationally collapsing system to dissipate energy. Thus these objects do not have to deal with the threat of gravitational collapse in the same way that smaller systems do, because when gravity pulls a star toward the center of a typical star cluster, it will go right through the star cluster and come out the other side instead of crashing into something and staying there, forming a larger and larger central condensation.

B. STAR CLUSTERS

An examination of the wide diversity of astrophysical objects that are larger than stars indicates that the absence of nuclear energy, mechanical rigidity, or heat flow in them does not result in a uniformly simple collection of stellar systems. The smallest star clusters are those like the tiny Pleiades star cluster, the so-called seven sisters, which can be seen on autumn nights, rising in the east after dark. The star cluster actually contains more than a hundred members, most of which are too dim to be visible to the naked eye. In star clusters like the Pleiades, a few hundred stars are born together and follow the same orbit around the Milky Way galaxy for a few hundred million years until the tidal gravitational forces in the galaxy rip them apart. The star clusters of highest mass are the globular clusters, collections of 10^5 stars that orbit sufficiently far from the plane of the Milky Way that the tidal forces are weak, and they have lasted essentially for the entire age of the universe, some 10–20 billion years.

C. INTERSTELLAR GAS AND DUST

Another detectable, but not readily visible constituent of the Milky Way is gas and dust in the interstellar medium, about 5–10% of the total mass in the galaxy. The easiest way to see this gas is to look for visible gas clouds around very hot stars, sometimes called "nebulae" for historical reasons. Figure 14 shows one such nebula. The light one sees coming from the wispy gas clouds is produced by fluorescence. The gas absorbs ultraviolet ionizing radiation from the hot star in the middle of the nebula and reemits it in the optical part of the spectrum when electrons and ions recombine.

Indirect techniques have uncovered other forms of gas in the interstellar medium. Cold gas in the form of atomic hydrogen or other molecules can be detected because hydrogen atoms and complex molecules emit photons in the radio part of the spectrum. Most of the mass in the interstellar medium is in the form of such gas, and much of it is concentrated to giant molecular clouds, given their name because the easiest way to find them is to detect radio emission from the CO molecule. Most of the volume of space between the stars contains a very hot, low-density gas that was not detected until the 1970s. Another visible constituent of the interstellar medium is the dust, micrometer-sized particles of silicate minerals and graphite, seen when it absorbs the light from distant stars. When interstellar dust (or, for that matter, tiny dust of any sort) is in between the observer and a distant light source, the dust changes the color of the light because it absorbs more blue light than red light. Stars seen through an interstellar dust cloud look redder than they really are, in the same way that the sun looks red at sunset be-

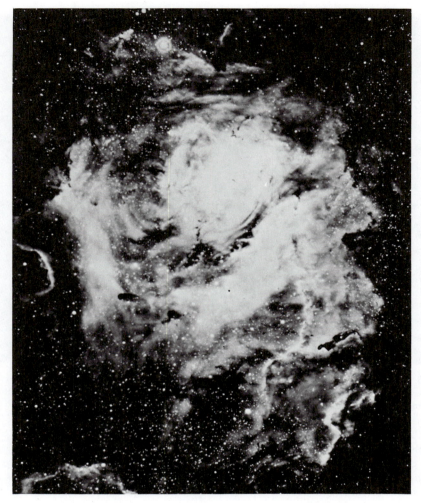

FIG. 14.   Messier 8, nicknamed the Lagoon nebula, is a huge cloud of glowing gas, lit
up by the newly formed stars in its center. The gas in the interstellar medium is most
easily seen in objects like these. (Courtesy of National Optical Astronomy Observatories.)

cause particles in the earth's atmosphere preferentially remove blue light from the sunbeam.

D. SPIRAL STRUCTURE

The major way in which stars are organized in
the universe consists of objects far larger than
star clusters: galaxies. Although the absorption
of starlight by interstellar dust has made it difficult to determine the structure of our own galaxy, a number of observational techniques,
mostly in the radio part of the spectrum, have
demonstrated that the Milky Way is a spiral galaxy, somewhat similar to the galaxy Messier 101

(Fig. 15), with spiral arms that are coiled more
tightly. The sun is located about two-thirds of
the way toward the edge of the visible spiral,
~10 kiloparsecs (kpc) (3×10^{20} m) from the center. The sun travels in a circular orbit around the
center, taking 250 million years to make one cir-
cuit.

Perpendicular to the plane of the spiral arms,
the Milky Way, like other spiral galaxies, is very
thin, only a few hundred parsecs thick (a small
percentage of the horizontal extent of the disk).
A small part of the very oldest material in the
Milky Way is part of a structurally different sub-
system of the galaxy, the spherical galactic halo.

FIG. 15.   The galaxy Messier 101 in optical radiation. The Milky Way galaxy is a spiral galaxy like this one, with the arms more tightly wound. The sun is located about two-thirds of the way out from the center. (Palomar Observatory photograph.)

Material in this halo travels along long, skinny, elliptical orbits that take it in toward the galactic center. The old material is easily distinguished because it does not share the circular orbital velocity of the solar system and thus is moving at a high velocity relative to the earth.

Why do spiral galaxies like the Milky Way have the shape that they do? The flatness of the galactic disk is readily attributed to the rotation of the protogalactic cloud that collapsed to form the disk of the Milky Way. The spiral arms are a bit more puzzling; two explanations for their shapes have been proposed. In one, these concentrations of young material are the result of spiral density waves that propagate through the gravitationally interacting material in the disk. Density waves are the result of the gravitational attraction between stars in the disk, causing concentration of material to form a stable pattern in the same way that waves on the surface of the ocean are a stable pattern. Density waves are regularly spaced; another example of density waves may be the regular spacings in the rings of Saturn (Fig. 8).

A second possible explanation for spiral structure is suggested by the fact that the structure is most prominently seen when young material is examined. A photograph like Fig. 15, taken in blue light, shows primarily the young, hot, short-lived, blue main-sequence stars. The idea here is that massive stars form, explode as supernovas, trigger the formation of stars in nearby regions of space, and end up setting off a chain reaction. Computer models suggest that when such chain reactions of star formation occur in rotating systems, spiral structure, not as regular as that produced by density waves, is the result. This idea is given the name "self-propagating star formation" model. It is likely that both the density wave process and self-propagating star formation play a role in shaping any galaxy. Another possible effect is the interaction of one galaxy with another.

E. GALACTIC NUCLEUS

The center of the Milky Way galaxy contains a fundamentally different beast from that found elsewhere. The first hint of the existence of high-energy phenomena in the galactic nucleus came from the pioneering radio observations of Karl Jansky, which indicated that the nucleus was a strong source of radio emission. It subsequently became clear that this radio emission was produced by high-speed electrons circling around magnetic fields, and the idea that the galactic nucleus contained a central, compact object that was the source of these electrons became popular.

Because visible light from the galactic center is absorbed by interstellar dust before it can reach us, we have to rely on infrared and radio observations. The structure of the central 2 kpc of the galaxy is very complex and still poorly understood. Material in the innermost spiral arm around the galactic center is not circling around the center but expanding away from it. At the very center itself there is a point source of radio radiation no longer than the solar system. Infrared measurements of Doppler shifts from material near the center suggest (but do not prove) that the central object is a giant black hole with a mass of ~1 million $M_\odot$.

F. DARK MATTER

Astrophysicists can analyze, model, and understand readily only matter that emits some sort of electromagnetic radiation. However, if we understand a particular stellar system quite well, we can map its gravitational fields by measuring the velocity of tracer objects moving around in it. Stars moving in low-mass galaxies will move more slowly than stars moving near high-mass objects.

Careful analysis of the motions of stars perpendicular to the galactic plane indicates that the mass in the disk of the galaxy, as indicated by its gravitational effects, is approximately twice as much as that which can be accounted for in the form of stars, stellar remnants, and interstellar matter. Said differently, about half of the material in the galactic disk is "dark matter." We have no clear idea of what this dark matter is. A popular suggestion is that it consists of a large number of very low mass stars or substellar objects. The substellar objects, in which nuclear reactions do not play a major role in heating the interior, would be about 10–50 times the mass of Jupiter and are called brown dwarfs (see Section V,A).

An even greater quantity of dark matter was found by astronomers measuring the velocities of molecular clouds located at great distances from the galactic center. If most of the mass of the Milky Way were contained in the visible disk, the orbital velocities of outlying objects would decrease in rough agreement with Kepler's third law. (The agreement is only approximate because the Milky Way is not a point mass.) However, measurements show that the velocities do not decrease at all with distance and therefore that the Milky Way Galaxy is embedded in a huge, spherical halo containing nearly 10 times as much mass as the visible disk with its spiral arms. The nature of the material making up the dark halo is unknown. It is remarkable that astrophysicists have been able to discover these forms of invisible dark matter, but it is sobering and a bit distressing to realize that virtually all of the mass in our home galaxy is in a totally unknown form.

VII. Galaxies and Quasars

A. SPIRAL, ELLIPTICAL, AND IRREGULAR GALAXIES

The spiral shape of the Milky Way galaxy is not the only shape that galaxies can have. Figure 16 illustrates some other types of ordinary spiral galaxies, designated by an "S" followed by a suffix 0 (zero), a, b, and c. The subclasses refer to, depending on the classifier, the looseness with which the spiral arms are wound or the predominance of the galactic nucleus in the optical image. Messier 101 (Fig. 15) is an Sc galaxy. In addition, there are "barred" spiral galaxies (designated SB) with a central bar crossing the nucleus of the galaxy; elliptical (E) galaxies, which are apparently featureless elliptical or circular collections of stars; and the irregulars (Irr), which have no apparent regular shape. The S0 galaxies, one of which is shown in Fig. 16, are almost like the ellipticals but show signs of a disk component as well. Figure 17 illustrates an irregular galaxy (at the right) and a spiral galaxy (at the center) in a small group of galaxies.

The variety of galactic shapes is correlated with other galactic properties (Table III). The loosely wound spirals (the Sc's) and the irregulars contain the greatest number of young blue stars and the greatest fraction of gas that forms young stars. In extreme contrast, the elliptical galaxies contain no gas at all, and their stellar population consists of a collection of very old stars that apparently were all formed at the same time. The colors of galaxies vary correspondingly: The spirals tend to be bluer than the ellipticals, since young stellar populations contain a

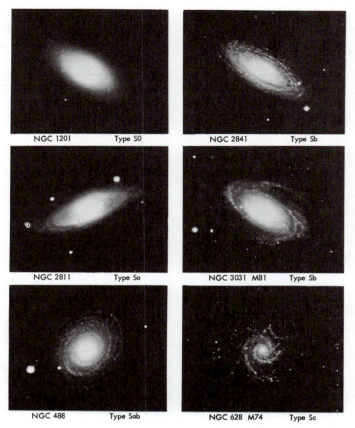

FIG. 16.   Six spiral galaxies showing the way in which they are classified according to the prominence of the nucleus and the shape of the spiral arms. (Courtesy of Mount Wilson and Las Campanas Observatories of the Carnegie Institution of Washington. Copyright Carnegie Institution of Washington.)

larger fraction of short-lived blue stars, which dominate the light output.

B. Galactic Evolution

Why do galaxies look the way they do? The correlation of galactic forms with galactic shape can be explained reasonably well in terms of a global picture of the evolution of the stellar populations within the Milky Way galaxy, which is then adapted for other galaxies. The fundamental origin of galactic shapes is a bit less clear.

The Milky Way contains two basic stellar populations: the disk population and the halo population. The evolution of the halo population is particularly simple: All the stars were formed at once, and the stars are slowly aging, the most massive stars becoming red giants first. In the disk population, however, the gas ejected by

stars as they evolve from red giants to stellar remnants (white dwarfs and neutron stars) is retained in a disk, in the form of the interstellar medium, and is available to form new stars. Much of the gas ejected from massive red giant stars is rich in heavy elements, and so the younger stars in spiral galaxies contain far more of such biologically important elements as carbon, nitrogen, oxygen, and iron than the old stars do. Halo population objects contain a much smaller proportion of heavy elements (although, surprisingly, the heavy-element content is not zero).

In the simplest conceptual scheme of galactic evolution, different types of galaxies contain these two populations in various mixtures, the ellipticals generally containing no disk population and the most loosely wound spirals containing no halo population. The evolution of the ellipticals is then quite simple, each star's

FIG. 17. Optical photograph of a small group of galaxies, the Messier 81 group. The largest galaxy (at the center) is a spiral, the blob on the right (Messier 82) is an irregular galaxy, and the two very small blobs at the right and lower left are small elliptical galaxies. North is at the top. (Palomar Observatory photograph.)

TABLE III. Some Selected Galaxies[a]

Name	Type	Luminosity (sun = 1)	Gas mass (sun = 1)	Total mass[b] (sun = 1)	Gas mass/total mass
NGC 4486	E0	8×10^{10}	Near 0	4×10^{12}	Near 0
NGC 205	E5	3×10^{8}	Near 0	1×10^{10}	Near 0
NGC 224	Sb	7×10^{10}	8×10^{9}	2×10^{11}	0.04
NGC 5457[c]	Sc	2×10^{10}	9×10^{9}	2×10^{11}	0.06
LMC	SBm	3×10^{9}	5×10^{8}	1×10^{10}	0.10
SMC	Im	7×10^{8}	5×10^{8}	2×10^{9}	0.32

[a] NGC numbers refer to the "New General Catalogue" of galaxies [see, e.g., de Vaucouleurs, G., de Vaucouleurs, A., and Corwin, H. (1976). "Second Reference Catalogue of Bright Galaxies," University of Texas Press, Austin]. LMC, large Magellanic Cloud; SMC, small Magellanic Cloud. One solar luminosity is 3.84×10^{26} W; 1 solar mass is 1.989×10^{30} kg.

[b] Burbidge, E. M., and Burbidge, G. R., The masses of galaxies; and Roberts, M. S., Radio observations of neutral hydrogen in galaxies, in "Galaxies and the Universe" (Sandage, A., Sandage, M., and Kristian, J., eds.). University of Chicago Press, Chicago, 1976.

[c] M101; see Fig. 15.

evolutionary time clock being set when the galaxy is formed, and the stars peeling off the main sequence at different times, according to their initial mass. In spirals the gas ejected by old stars is recycled to form new stars, and in some spirals the proportion of new stars is quite high. This simple picture can explain most of what we know about most galaxies, though there are still some outstanding puzzles, such as the ''starburst'' galaxies, where all of the stars seem to have been formed very recently. Another weird class of galaxies produces almost all of its luminosity in the infrared; the irregular galaxy on the right-hand side of Fig. 17 is one of those.

What is it that makes a galaxy become a spiral or an elliptical? Two factors appear to play a role. The intrinsic properties of a galaxy—its initial mass and angular momentum—are undoubtedly important, the fastest spinning galaxies becoming spirals and the slower ones becoming ellipticals and failing to form a disk. It is becoming increasingly clear, however, that the external environment of a galaxy also has an effect. Galaxies found in a small group, such as that illustrated in Fig. 17, tend to be spirals, only the smallest galaxies being ellipticals. Large elliptical galaxies tend to be found in the cores of much more concentrated clusters of galaxies, where interactions between galaxies are much more frequent. These correlations of galaxy morphology with environment are weak, and exceptions to the general trends are numerous. No explanation for this phenomenon has yet appeared.

C. Clusters of Galaxies

Like stars, galaxies are clustered together in space. Unlike stars, the distance between galaxies is only a modest multiple (say, 10–30) of their linear dimensions. A small group of galaxies is illustrated in Fig. 17. Many other small groups exist, such as the Local Group of galaxies, consisting of three large spirals (one of which is the Milky Way) along with more than a dozen smaller members.

The clustering of galaxies continues on all scales, ranging from small groups upward. Among the largest single objects in the universe are rich clusters of galaxies, containing of the order of 10^3 large galaxies and countless smaller ones. Many of the most important properties of clusters of galaxies have been alluded to already, in Section VII,A and in the discussion of Fig. 2 earlier. The dominant members of rich clusters of galaxies, found in the center of the clusters, are giant elliptical galaxies, with total masses one or two orders of magnitude greater than the mass of the Milky Way. It seems clear that this type of galaxy is found only in the rich clusters. There is an intracluster medium in the rich clusters containing ~10% of the total mass of the cluster, which was discovered when we photographed these clusters in the X-ray part of the electromagnetic spectrum for the first time.

D. Large-Scale Structure of the Universe and Dark Matter

Various heroic efforts to map the structure of the universe on the largest scales have indicated that the large clusters of galaxies are ordered in the universe as a whole and that the largest scale structures of any sort are long, filamentary superclusters of galaxies nearly 100 megaparsecs (Mpc) long (1 Mpc = 10^6 pc, or 3×10^{22} m). Our Local Group of galaxies may well be part of one such supercluster, which stretches out to the nearest large cluster of galaxies (the Virgo cluster), ~20 Mpc away, and beyond. These filamentary or flat superclusters are separated by giant spherical voids, where few galaxies appear to exist. The structure of the universe, according to this model, then resembles soap suds, where flat or filamentary soap films separate spherical air bubbles. The existence of this type of structure has been fairly well established in some directions in space, and it is a reasonable hypothesis that such a structure is universal.

Measurements of the dynamics of these large clusters and superclusters of galaxies have indicated rather definitively that the matter we can see in the galaxies themselves, even making allowance for their giant halos, is insufficient to bind the clusters gravitationally. Astronomers patiently measure the Doppler shifts of each galaxy in a cluster or supercluster and then allow for projection effects, since the Doppler shift measures only the galaxy's velocity toward or away from the earth. For clusters and superclusters of galaxies, this dark matter constitutes ~80% of the mass in the cluster. Dark matter, first encountered in the disk of the Milky Way galaxy and again in its halo, seems to be a pervasive constituent of the universe. The dynamics of the three major types of dark matter indicate reasonably clearly that the same type of stuff cannot be in all three places.

The origin of all of this structure is one of the major unsolved cosmological problems of the late 1980s. The grand unified theories of particle physics suggest that there may be a plethora of particles that interact only gravitationally or via the weak nuclear interaction with the rest of the universe. Such particles would not emit light. These particles would seem to be ideal candidates for the dark matter, and some early computer scenarios that tried to account for the large-scale structure of the universe suggested that if most of the mass in the universe were in the form of low-mass particles (with an energy equivalent of 10 to 20 eV, or a mass about $\frac{1}{50000}$ of the electron mass) the observed filamentary structure would be produced by the gravitational collapse of such objects. However, there is at this time no compelling scenario that produces a truly adequate explanation, and there is an additional concern that the number of possible particle properties is sufficiently large that any unambiguous connection between the large-scale structure of the universe and the world of particle physics may be difficult to establish firmly.

E. ACTIVE GALACTIC NUCLEI AND QUASARS

A surprising cosmological discovery in the early 1960s was that some apparently stellar, pointlike sources of light had enormously high Doppler shifts. The most logical explanation for these high redshifts was that these objects, eventually called quasars, were located at great distances from the earth and that the redshifts were produced by the expansion of the universe (which is itself discussed below). A problem with this explanation was that the brightest of these objects were as bright as ordinary, not too distant galactic stars. If these objects were truly at cosmological distances, the inferred luminosities were quite large, of the order of 10^{12} times the luminosity of the sun, or 10^{38} W, three orders of magnitude larger than the luminosity of a galaxy like the Milky Way. And yet the time scale with which these objects varied constrained them to be quite small, about the same size as the solar system (according to recent observations). The idea that something the size of the solar system could outshine an entire galaxy seemed rather outlandish, and the cosmological explanation of the quasar redshifts was questioned persistently. [See QUASARS.]

In the years since quasars were discovered, astronomers have gathered evidence that makes a compelling case for the cosmological origin of the redshifts of quasars. In this view, quasars are hyperactive nuclei of galaxies in which some of the same processes that are going on in the center of the Milky Way galaxy are occurring on a much more energetic scale. Quasars look like stars because the powerful emission from the tiny galactic nucleus overwhelms the light emitted by the stars in the surrounding galaxy. One of the most powerful pieces of evidence supporting the cosmological explanation of quasar redshifts was the discovery of fuzzy stellar emission surrounding some of the nearest quasars. Another piece of evidence, which also sheds some light on the nature of activity in galaxies, is the recognition that galaxies have active nuclei that put out nonthermal, nonstellar radiation on all energy scales ranging over seven orders of magnitude, from the very modest emission from the Milky Way to the most powerful quasars.

Radio images of active galaxies and quasars have, thanks to the power of the Very Large Array Telescope, demonstrated a fascinating variety of shapes, with one common theme: linear jets. Figure 18 shows a typical case. High-speed particles emitted in two directions by the galactic nucleus produce linear jets of radio emission when these particles spiral around magnetic fields. The blobs of radio emission at the end of the jets are produced, in this model, when the jets interact with the intergalactic medium sur-

FIG. 18.   Computer-produced radio image of the radio galaxy 3C 449. The galaxy itself is at the center of the picture, between the two long, skinny jets of radio emission. The blobs at the end of the jets represent the expansion of the jets outside the halo of this galaxy. (Copyright © 1978 by Associated Universities, Inc. Image produced by the Very Large Array Radio Telescope of the National Radio Astronomy Observatory, operated by Associated Universities, Inc., under contract with the National Science Foundation.)

rounding the galaxy. In some extreme cases, the jets are millions of parsecs ($\sim 10^{23}$ m) long, indicating that the alignment of the jets was preserved for 10^6 yr at least.

This line of evidence, along with many other indications, suggests that the fundamental energy source found in quasars is a larger version of the one postulated for the center of the Milky Way: a giant black hole, with a mass of up to 10^9 $M_\odot$ in the case of the most energetic objects. When this giant black hole swallows matter, energetic processes in the swirling disk of gas around the hole produce the ejection of narrow, collimated beams of high-energy particles. All the other phenomena associated with quasars ultimately derive energy from this central engine. The black hole model is not the only possibility; a giant supermassive star could in principle also do the job, but such massive stars are expected to last a very short time and soon collapse to form giant black holes or explode completely.

Because quasars are so luminous, they can be seen at great distances. The most distant quasar is located 15 billion light years from the earth. When we look at objects that are so far away, we are looking backward in time; the light from such a quasar took 15 billion years to get here. Examination of these objects can tell us what the universe was like 10^{10} years ago, and we find that it was somewhat different in that quasars were much more abundant than they are now. The difficulty of finding quasars much more distant than 15 billion light years suggests that before that time, quasars (and possibly galaxies) had not yet formed. These pieces of evidence are some of the many indications that we live in an evolutionary universe, which is not always the same and which does not last forever. The study of evolution on the largest scales is the domain of the field of cosmology.

VIII. Cosmology

A. The Expanding Universe

Astronomers, from the first cave dweller who gazed at the sky to the modern scientist using a complex telescope, probably became astronomers in order to figure out where the universe came from, how (or even whether) it evolves, and where it is all going. Much of astronomy, however, deals with the constituents of the universe rather than the universe itself. A superficial glance at the night sky suggests that the universe is unchanging, and it is relatively constant on time scales comparable to a human lifetime. The evidence for cosmic evolution is sufficiently subtle that it was not until very recently, in the 1920s, that we realized that the universe, as a whole, evolves. [See COSMOLOGY.]

The construction of large telescopes and spectrographs that, while abysmally inefficient by modern standards, were much larger and faster than any previous ones had enabled astronomers like Edwin Hubble to obtain spectra of very distant galaxies. Surprisingly, all these spectra showed a redshift, and the velocity of the galaxy increased with increasing distance. The relationship between galaxy velocity v and galaxy distance D, $v = HD$, is now known as Hubble's law in honor of its discoverer, and the proportionality constant H is known as Hubble's constant. About 50 years of observation have demonstrated the validity of Hubble's law, which describes quite well the average motion of galaxies.

The discovery of Hubble's law demonstrated the expansion of the universe. Its functional form does not mean that the earth is at the center of the universe. The expanding universe is like a huge jungle gym made of telescoping pipe or an expanding raisin cake, with children in the jungle gym or raisins in the cake representing galaxies. From the vantage point of any galaxy, all the other galaxies are moving away from it. [See COSMIC INFLATION.]

The size of Hubble's constant H is related to the age of the universe. If it is assumed that the expansion rate of the universe has not changed (observations show that, within a factor of two or so, this assumption is quite reasonable), then the time t it took the universe to expand to the point where a particular galaxy is at a distance D from the earth is given by the familiar relation between distance, velocity, and time $D = vt$. Combining this relation with Hubble's law gives $D = (HD)t$, or, rearranging, $t = 1/H$. The exact value of Hubble's constant H is not well known because a galaxy has to be quite far from the earth for the random motions of galaxies to be negligible compared with the expansion of the universe, and it is difficult to measure accurately the distance to such distant galaxies. However, a reasonable range for the value of Hubble's constant gives an age for the universe of between 10 and 20 billion years.

B. Proof of the Evolutionary Universe: Discovery of the Primeval Fireball

The most obvious explanation of Hubble's law is that everything in the universe was concentrated in a dense configuration at some time in the distant past. The expansion of this dense configuration, driven by heat, caused space to expand and everything in the universe to separate from everything else. This cosmological model, called the big bang model, predicts that there was a beginning to cosmic evolution at some definite time in the past, 10–20 billion years ago.

The idea of a universe that began at a definite time raises some philosophical questions, such as what preceded the big bang? For this and other reasons, a number of astronomers were reluctant to accept the big bang cosmology, and dreamed up other ways of producing Hubble's law in a nonevolving universe. Until the mid-1960s, when the Hubble relation was essentially the only cosmological fact available, such models were still viable. But in the mid-1960s a number of big bang fossils were found that have, in the minds of all but a very few astronomers, confirmed the standard picture of big bang cosmology.

The least ambiguous of these big bang fossils was a discovery that, like many astronomical discoveries, was the serendipitous result of what started out as a relatively uninteresting research project. In 1965, Arno Penzias and Robert Wilson were embarking on some projects that would make use of a horn-shaped radio antenna at Bell Laboratories. The unique feature of a horn antenna is that it can be calibrated very precisely. Their first chore was to identify all the sources of background noise in this antenna, so that when they pointed it at a radio source (or a communication satellite) they could know how much of the radio flux they received was coming from the source. There was a certain amount of background noise that they could not identify at first. It subsequently became clear that this excess radio noise was the echo of the big bang.

In the early stages of a hot big bang universe, there are a great many photons around, simply because the matter in the universe is hot. Once the universe is a million years old, it becomes un-ionized; these photons no longer interact with the matter in the universe, and they are progressively redshifted to the radio part of the spectrum by the time one reaches the present.

George Gamow, one of the early architects of the big bang cosmology, in fact predicted that a universal background noise of radio photons should exist (although his prediction was not the motivation for the Penzias and Wilson work). Particularly when measurements of the microwave background were made at many other wavelengths, confirming that the spectral shape of this background was indeed the blackbody curve one expected, the identification of the excess radio noise first identified by Penzias and Wilson with the primeval fireball radiation seemed quite secure.

C. Very Early Universe

Once the basic validity of the big bang cosmological model was accepted, astrophysicists began exploring its consequences and ramifications. In the very earliest stages of cosmic evolution, matter and photons interacted at very high energies. The simplest classical big bang model predicts that the temperature of the universe should increase without limit as one approaches the beginning at time $t = 0$. This classical model becomes invalid at a time of 10^{-43} sec, when the density of the universe is such that one would need a quantum theory of gravity in order to describe it; since we have no such quantum theory of gravity, this earliest of the early epochs has to be regarded as unexplored territory. Even at somewhat later epochs, however, something like 10^{-35} sec after the big bang, the energetic interactions of matter with photons and with itself present particle physicists with a way of exploring how matter interacts at energies far beyond the capacity of any particle accelerators to probe it. In the 1980s, exploration of the very early universe is one of the hottest frontiers of the field of particle physics.

One of the most exciting developments of this exploration was the "inflationary scenario," which offered promise of explaining some of the deepest cosmological mysteries. Because this period of cosmic evolution is only loosely coupled to any measurements we can make at present, it is often difficult to confirm any theoretical picture with observational data (hence the use of the word *scenario* rather than *model* or *theory*). In the inflationary scenario, the decay of extremely energetic particles is postulated to occur asymmetrically, particular parts of the universe becoming dominated by ordinary particles and other parts of the universe becoming dominated by antiparticles. A rapid expansion of the universe then means that the entire observable universe, all that we can see, is contained in one of these domains that was dominated by ordinary particles. This scenario accounts both for the homogeneity of the observable universe, where the temperature of the microwave background is almost exactly the same in any direction one looks, and for the absence of matter in the observable universe.

D. PRIMORDIAL NUCLEOSYNTHESIS

A second big bang fossil is provided by the abundance of the lightest elements in the universe, the various isotopes of hydrogen, helium, and lithium. During the first 20 min of cosmic evolution, the temperature and density of the universe were such that some simple nuclear reactions between protons and neutrons could occur. Classical big bang cosmology makes some quite definite predictions about the abundance of these light elements. For example, the abundance of the most common isotope of helium, ^{4}He, should be almost exactly 25% by mass for a wide variety of initial cosmological conditions. (This ratio was in fact established in the first 3 min of cosmic evolution, when the ratio of neutrons to protons was fixed.) Observations of the helium abundance of a wide variety of objects, appropriately interpreted to take helium production by stars into account, are in very good agreement with these predictions. Similar predictions of the abundances of deuterium and lithium are also in agreement with the big bang model.

After the synthesis of helium and other light elements in the first 20 min of cosmic evolution, not much of cosmological interest happened for a long time. When the universe was a million years old, ions and electrons recombined and the photons that were then around interacted with matter for the last time, beginning their journey through space and time that, for some of them, ended when they hit the radio telescope operated by Penzias and Wilson and provided us with one of the key clues to the puzzle of cosmic evolution. In the first few billion years of cosmic evolution, galaxies were formed. Somewhere between 5 and 15 billion years after the bang, a supernova explosion triggered the collapse of a gas cloud, giving birth to an apparently insignificant yellow star in the outskirts of a spiral galaxy. Before another few hundred million years had passed, chemical reactions in the primordial ocean of the third planet from that sun began, producing the phenomenon known as life. In another 5 billion years, the complex processes of evolutionary biology led to the development of intelligent creatures on that planet. They called the star the Sun, the galaxy the Milky Way, and in a comparatively brief amount of time these humans were able to figure out how the cosmos developed.

E. FUTURE OF COSMIC EVOLUTION

The universe is expanding, but each of the galaxies in it exerts an inexorable gravitational tug on each of the other galaxies, trying to pull the universe back together again. Will the expansion of the universe simply keep on going, or will these gravitational interactions dominate, bringing the expansion to a stop? This is one of the outstanding cosmological questions of the latter part of the twentieth century, and the answer is still unclear.

One can try to answer this question by simply taking account of all the matter in the universe and seeing whether it is sufficient to bring the expansion to a halt. A difficulty with this procedure is that there is far more dark matter in the universe than there is visible matter. Still, the dark matter can be detected because of its gravitational interactions, and the indications are that the universe does not contain enough matter to stop the expansion. Similar, and similarly tentative, conclusions come from a consideration of the density of the universe in its early evolutionary stages, which are inferred from measurements of the abundances of light nuclei produced in primeval nucleosynthesis.

Another way to ask this question is to use a variety of data to determine whether the expansion rate of the universe has changed. There is no evidence of any change, a change that would be necessary if the expansion were to come to a stop someday. However, the case for an open, ever-expanding universe remains tentative for these explorations involve pushing current telescope technology to the limits.

IX. Summary

The scope of astrophysics ranges from the study of comparatively nearby objects like planets to the consideration of the evolution of the entire universe and the measurement of photons from the most distant galaxies, billions of light years away. A few common themes run through the work of all types of astronomers, although the field is rapidly dividing into subspecialties. First is that with a few exceptions astronomers are limited to measuring the feeble trickle of photons and other forms of radiation that reach us from these distant objects, and that there is a continual push for more sophisticated ways to measure them. A second common theme is the role of gravity in cosmic objects; gravity is the dominant force governing the evolution of astronomical objects ranging from asteroids to clusters of galaxies. Gravity wants to squash these objects and make them black holes, and the evolution of all astronomical objects in some ways involves those tendencies that oppose that universal tendency.

The astrophysicist's ability to send telescopes above the earth's atmosphere in order to explore new wavelength regions in the electromagnetic spectrum has resulted in the discovery of a wide variety of new, exotic astrophysical objects in the latter part of the twentieth century. Volcanoes on other planets, pulsars, black holes, dark matter, quasars, and such were completely unknown as recently as 1955. Our basic picture of the evolution of the universe, the big bang cosmology, was not firmly established until the 1970s. The satellites of planets in the outer solar system were only tiny, mysterious dots until 1979, and the exploration of these objects continued through the 1980s. Because much of these additional insights are no older than the space program, astrophysics is currently in an explosive growth phase, with the consequence that much of our knowledge is continually being revised. There is no sign that the pace of astrophysical discoveries will slacken in the future.

BIBLIOGRAPHY

Allen, C. (1973). "Astrophysical Quantities," 3rd ed. Athlone Press, London.
"Annual Review of Astronomy and Astrophysics." Annual Reviews, Palo Alto.
Greenstein, George (1983). "Frozen Star," Freundlich/Scribners, New York, 1983.
Hartmann, William K. (1983). "Astronomy: the Cosmic Journey." Wadsworth, Palo Alto.
Pasachoff, Jay (1982). "Astronomy: From the Earth to the Universe." Saunders, New York.
Shipman, H. L. (1980). "Black Holes, Quasars, and the Universe," 2nd ed. Houghton Mifflin, Boston.
Trefil, J. S. (1985). "Space, Time, Infinity," Random House (Smithsonian Inst. Press), Washington.
Zombeck, M. (1982). "Handbook of Space Astronomy and Astrophysics," Cambridge Univ. Press, London.

BINARY STARS

Steven N. Shore   *New Mexico Institute of Mining and Technology and DEMIRM, Observatoire de Meudon*

GLOSSARY

Accretion disk: Structure formed when the material accreting onto a star has excess angular momentum, forming a circulating disk of material supported by internal pressure and heated by turbulent stresses.

Lagrangian points: Stable points in the orbit of a third body in a binary system; the inner Langrangian point, L_1, lies along the line of centers and marks the Roche limit for a tidally distorted star.

Main sequence: Phase of hydrogen core burning; first stable stage of nuclear processing and longest epoch in the evolution of a star.

Mass function: Method by which the mass of an unseen companion in a spectroscopic binary can be estimated using the radial velocity of the visible star and the orbital period of the binary.

Orbital parameters: Inclination, i, of the orbital plane to the line of sight; P, the period of revolution; e, the eccentricity of the orbit; q, the mass ratio.

Red giant: Stage of helium core burning; subsequent to the subgiant stage.

Subgiant: Stage of hydrogen shell burning, when the deep envelope initiates nuclear processing around an inert helium core produced by main sequence hydrogen core fusion. This is the transition stage in the expansion of the envelope from the main sequence to the red giant phase.

Units: Solar mass ($M_\odot$), 2×10^{23} g; solar radius ($R_\odot$), 7×10^{10} cm.

Binary stars are gravitationally bound stars, formed simultaneously, which orbit a common center of mass and evolve at the same time. These stars are formed with similar initial conditions, although often quite different masses. Visual binaries are both sufficiently separated in space and sufficiently near that their angular motion can be directly observed. Spectroscopic binaries are unresolved systems for which motion is detected using radial velocity variations of spectral lines from the component stars. If the orbital plane is inclined at nearly 90° to the line of sight, the components will display mutual eclipses. Depending on the orbital period and mass, the stars may either share a common envelope (contact), be in a state of unidirectional mass transfer between the components (semidetached), or evolve without mass transfer but with mutual gravitational perturbation (detached). In semidetached systems, depending on the rate of mass transfer and the nature of the accreting object, a hot accretion disk will be formed. If the companion star is gravitationally collapsed, being a neutron star or black hole, X-ray emission will result. Accretion onto white dwarf or neutron stars results in flash nuclear reactions that can trigger the nova event.

I. Historical Introduction

At the close of the eighteenth century, William Herschel argued that the frequency of close visual pairs was larger in any area of the sky than would be expected by chance. On this basis, it was suggested that binary stars—that is, physically gravitationally bound stellar systems—must exist. Prior to the discovery of Neptune, this was the most dramatic available demonstration of the universality of Newton's theory of gravitation. Herschel, Boole, and others extended this study to the demonstration that clustering is a general phenomenon of gravitational systems. The discovery of the wobble in the proper motion of Sirius led Bessel, in the 1840s, to argue for the presence of a low-mass, then unseen companion; it was discovered about 20 yr later by Clarke. Routine observations of visual binary star orbits were being made by the end of the century. For very low mass stars, the method is still employed in the search for planets through proper motion perturbations, much like the method by which Neptune was discovered.

About the same time as Herschel's original work, Goodricke observed the photometric variations in the star β Persei, also known as Algol. The variations, he argued, were due to the system being an unresolved, short-period binary system, with one star considerably brighter than the other but otherwise about the same physical size. He argued that the light variations were consistent with eclipses, and that were we able to resolve the system, we would see two stars orbiting a common center of mass with about a three-day period. The dramatic confirmation of this picture came with the discovery, by Hartmann and Vogel at the end of the last century, of radial velocity variations in this system consistent with the eclipse phenomenology. By midcentury, in part as a result of the work at Bamberg under Argelander, large-scale searches for variable stars began to produce very large samples of stars with β Persei-like behavior.

Most theoretical work on binary stars is the product of the past seventy years. Methods for the analysis of eclipses, based on light curve fitting by spheroids, were developed by Russell and Merrill in the first two decades of this century. Atmospheric eclipses were first discussed by Kopal in the 1930s. The study of mass transfer in binary systems was initiated largely by Struve in the mid-1930s, and the applications of orbital dynamics to the study of mass transfer began in the 1940s with Kuiper's study of β Lyrae. Using large-scale computer models, stellar evolution in binary star systems was first studied in detail in the 1960s by Paczynski and collaborators in Warsaw and Plavec and colleagues in Prague. Hydrodynamic modeling has only recently been possible using realistic physics and remains a most interesting area of study. Much recent work on binary star evolution and hydrodynamics has been spurred by the study of binary X-ray sources. Following the discovery of binarity for several classical novas, by Walker in the 1950s, the connection between low-mass X-ray binaries and cataclysms has been central to the study of evolution of stars undergoing mass exchange.

II. Some Preliminaries

The broadest separation between types of binary stars is on the basis of observing method. For widely separated systems, which are also close to us, the stars appear physically distinct on the sky. If the orbital periods are short enough, that is less than millennia, it is possible to determine the plane of the projected orbit by observing directly the motion of the stars. For those systems in which the luminosity ratios (and possibly mass ratios) are large, it is possible to obtain orbital characteristics for the two members by observing periodic wobbling in the proper motion of the visible member. Such methods are frequently employed in the search for planetary-like companions to high proper motion stars, that is, stars with large transverse velocities to the line of sight, such as Barnard's star. For systems of low proper motion and possibly long period, speckle interferometry, intensity, and Michelson interferometry, as well as lunar occultations, can be useful in separating components and at least determining luminosity ratios. Such methods, extended to the near infrared, have been recently employed in the search for brown dwarf stars, objects of Jupiter-sized mass.

When the system is unresolved, even at separations of several milliarcseconds, it is necessary to employ spectroscopic methods to determine the composition and motion of the constituent stars. These are the spectroscopic binaries, by far the largest group so far studied. Two methods of analysis, which are happily sometimes complementary, can be used—observation of radial velocity variations of the components and eclipse phenomena. [*See* STELLAR SPECTROSCOPY.]

A. SPECTROSCOPIC BINARY VELOCITY CURVES

Consider two stars in a circular orbit about the center of mass. Regardless of the perturbations, we can say that the separation of the two stars of mass M_1 and M_2 is $a = r_1 + r_2$ in terms of the separation of the stars from the center of mass. The individual radii are the distances of the components from the center of mass:

$$r_1/r_2 = M_2/M_1 \qquad (1)$$

The velocity ratio is given by $V_2/V_1 = M_1/M_2$ for a circular orbit, where V is the orbital velocity. By Kepler's law,

$$GM = \omega^2 a^3 \qquad (2)$$

where ω is the orbital frequency, given by $2\pi/P$, where P is the period and M is the total mass of the system. Now, assume that the inclination of the plane of the orbit to the observer is i, that the maximum *observed* radial velocity for one star is given by K, and that we observe only one of the stars. Then,

$$K^3 P/2\pi G = M_2^3 \sin^3 i/M^2 = f(M) \qquad (3)$$

The function $f(M)$ is called the *mass function* and depends only on observable parameters, the maximum radial velocity of one of the stars, and the period of the orbit. If M_1 is the mass of the visible star, $f(M)$ serves to delimit the mass of the unseen companion. If both stars are visible, then,

$$K_1/K_2 = M_2/M_1 \qquad (4)$$

independent of the inclination. Thus, if both stars can be observed, both the mass ratio and the individual masses can be specified to within an uncertainty in the orbital inclination using $f(M)$. The mass function permits a direct determination of stellar masses, independent of the distance to the stars. This means that we can obtain one of the most important physical parameters for understanding stellar evolution merely by a kinematic study of the stars.

If the orbit is eccentric, departures from simple sinusoidal radial velocity variations with time are seen. The eccentricity of the orbit also introduces another symmetry breaking factor into the velocity variations, because the angle between the observer and the line of apsides, the line that marks the major axis of each ellipse, determines the shape of the velocity variation curve with orbital phase.

B. Eclipsing Binary Light Curves

The orbital plane of eclipsing stars lies perpendicular to the plane of the sky. Depending on the relative sizes of the stars, the orbital inclination over which eclipses can occur is considerable, but in general only a small fraction of the known binary systems will be seen to undergo eclipses. The variation in light serves two purposes. It permits a determination of the relative radii of the stars, since the duration of ingress and the duration of eclipse depend on the difference in the sizes of the stars. That is, if Δt_1 is the total time between first and last contact, and Δt_2 is the duration of totality, assuming that the eclipse is annular or total, then,

$$\frac{\Delta t_1}{\Delta t_2} = \frac{r_g + r_s}{r_g - r_s} \qquad (5)$$

where r_s is the smaller and r_g is the greater radius, respectively. The diminution in light from the system depends on the relative surface brightness of the stars, which in turn depend on the surface (effective) temperature, T_{eff}. Eclipses will not be total if the two stars are not precisely in the line of sight, unless they differ considerably in radius, so that the mark of totality is that the light does not vary during the minimum in brightness.

C. Distortions in Photometric and Velocity Curves

Several effects have been noted that distort the light curve and can be used to determine more physical information about the constituent stars.

1. Reflection Effect

Light from one component of a close binary can be scattered from the photosphere and outer atmosphere of the other, producing a single sinusoidal variation in the system brightness outside of eclipse. This *reflection effect* is useful in checking properties of the atmospheres of the stars. If the illuminating star is of significantly higher temperature, it can also produce a local heating, which alters the atmospheric conditions of the illuminated star. Such an effect is especially well seen in X-ray sources, particularly HZ Her = Her X-1, which varies from an F-star to an A-star spectrum as one moves from the unilluminated side to the substellar point facing the X-ray source.

2. Photospheric Nonuniformities

A similar phenomenon has been noted in the RS CVn stars, where it is caused by the presence of large-scale, active magnetic regions, called *starspots*, on the stellar surfaces. Unlike reflection, these dark regions migrate with time through the light curve as the active regions move with the differential rotation of the stellar envelope, analogously to the motion of sunspots. Chemically peculiar magnetic stars also show departures from simple eclipse profiles, because of locally cooler photospheric patches, but these appear to be stable in placement on the stellar surface.

3. Circumstellar Material

The presence of disks or other circumstellar matter also distorts the light curves, and can alter the radial velocity variations as well. In several Algol systems, notably SX Cas and RX Cas, this is especially important. The timing of eclipses indicates a circular orbit, while the radial velocity variations are more like that of a highly eccentric one. The explanation lies in the fact that here is considerable optical depth in the matter in the orbit, which results in the atomic lines producing a distortion in the radial velocity variations. Many of the W Serpentis stars show this effect. It is most noticeable in eclipsing systems because these present the largest path length through material in the orbital plane. In some cases, atmospheric eclipses can also distort the lines because of stellar winds and convection cells intercepted by the line of sight. These motions, however, are generally small compared with the radial velocity and so alter the photometry (light curve instabilities during eclipse are well marked in the ζ Aur stars) but do not seriously affect the radial velocity determinations.

4. Ellipsoidal Distortions

If the stars are close enough together, their mutual gravitational influences raise tides in the envelopes, distorting the photospheres and producing a double sinusoidal continuous light variation outside of eclipse. Many of these systems also suffer from reflection effect distortion, so that there are many equivalent periods in these systems, depending on whether or not they eclipse.

Departures from symmetric minima should accompany expansion of the stars within their tidal surfaces. As the photosphere comes closer to the tidal-limiting radius, the *Roche limit,* the star becomes progressively more distorted from a symmetric ellipsoid and the photometric variations become more like sine curves. An additional feature is that as the stars become larger relative to the Roche limit they subtend a greater solid angle and display increasing reflection effect from the companion.

5. Limb Darkening

Stellar surfaces are not solid, and they have a continuous variation in surface brightness as one nears the limb. This effect, called *limb darkening,* is produced by the temperature gradient of the outer stellar atmosphere compared with the photospheres. The effect of limb darkening on light curves is to produce a departure from the behavior of simple, uniform spheres most notable in the softening of the points of contact during eclipse. It is one of the best ways available for measuring the temperature gradients of stellar atmospheres.

6. Apsidal Motion

The additional effect of the tidal distortion is that the stars are no longer simple point sources, but produce a perturbation on the mutual gravitational attraction. The departure of the gravitational potential from that of two point masses produces *apsidal motion,* the slow precession of the line connecting the two stars. This rate depends on the degree of distortion of the two stars, which in turn provides a measure of the degree of central concentration of the stars. Such information is an important input for stellar evolutionary models. One system that has been especially well studied is α Virginis (Spica). An ad-

ditional source of apsidal motion is the emission of angular momentum from the system, and the presence of a third body.

7. Third Light

Either because of circumstellar material in the orbital plane, which is not eclipsed but which scatters light from the binary components, or because of the presence of a faint third body in the system that is unresolved, some additional light may be present at a constant level in the eclipsing binary light curve. Frequently, high-resolution spectroscopy is able to reveal the lines of the companion star, as in Algol, but often it remains a problem to figure out the source for the nonphotospheric contributions to the light curve. This is simply added as an offset in the determinations of eclipse properties in most methods of light curve analysis.

III. Classification of Close Binary Systems

There are several distinctive classes of binary stars, distinguished nominally by their prototypes, usually the first observed or best known example of the phenomenology. In several cases, however, overlaps in the properties of the various systems make the prototypical separation confusing and less useful. Two main features distinguish classes of stars, the masses of the components and the separations. Alternatively, the period of the binary and the evolutionary status of the components are useful, and we shall use these alternately as needed.

The broad distinction among various binaries is whether the stars are physically separated by sufficient distance that the photospheres are distinct, in which case they are called *detached,* or have been significantly tidally distorted and may be in the process of mass transfer in some form. This latter class divides into those that have only one star transferring mass, the *semidetached,* and those with both stars immersed in a common envelope of gas that is mutually exchanged, the *contact* systems. This classification, first developed by Kuiper, has proven to be a most general taxonomic tool for distinguishing the various physical processes that occur at different stages in the evolution of close binaries. It is most important to note that a binary, depending on its initial period, mass, and mass ratio may pass through any or all of these stages at some time in its life. This is due to the expansion of stellar envelopes as stars evolve.

A. Contact Systems

If the two stars are sufficiently close at the beginning of their lifetimes, in the hydrogen core burning phase, they will form a common convective or radiative envelope. Such a star rapidly becomes surrounded by an optically thick cloak of gas, which radiates like a single photosphere, although more peanut shaped than spherical. Such stars have been observed among the W Ursa Majoris or W UMa stars. These are main sequence stars *only,* with mass ratios between 1 and 10, but in which the components have about $1M_\odot$. They have orbital periods of less than 12 hr. They also have analogs on the upper main sequence among the O stars, which are contact systems but for which the envelopes generally show surface temperature gradients and smaller optical depths. The W UMa stars are characterized by continuously variable light curves, indicative of extremely distorted photospheres. For such stars, as the system revolves, its projected surface area in the line of sight is continually changing, so the brightness of the system is never constant. The degree of contact, that is the fractional overlap of the stellar envelopes, varies from pointlike to over 20%.

There are several problems that have to be resolved in contact systems, the most notable being the Kuiper paradox. This paradox states that the main sequence radius for a star depends on its mass in a fashion different from that observed for the contact surface. In particular, the Roche limit is given by $R \sim M^{1/3}$, while for main sequence stars $R \sim M^{1/2}$. Thus, in order to satisfy both of these criteria simultaneously and maintain stable nuclear burning, the mass ratios for contact systems should be unity. This contradicts the observations, which show that the mass ratios can often reach extreme values, of order 0.1, and therefore these systems cannot be stable objects. The presence of common envelopes in contact systems argues for effective thermal coupling between the components, in spite of the very different input of energy from the two stars. In most W UMa stars, the variation of effective temperature from the two components is undetectable. Two models have been proposed for these systems in order to answer this apparent paradoxical behavior. One, the contact discontinuity model, assumes that there is a stable temperature inversion that lies above the lower luminosity star and that mass flows in the envelope with little thermal interaction with the underlying stars. The alternative model argues that the systems actually are not in strong contact for most of the time, and that the envelope undergoes a thermal relaxation on long time scales, which brings it into and out of thermal

contact at the L_1 point. The jury is still out in this matter, and further high-resolution observations of the line profiles and photometric behavior of these stars are needed to clarify the physics.

B. Semidetached Systems

If one star is sufficiently distended that its mass is no longer uniquely bound to that object and starts to flow toward the companion, because of tides, then the system is called *semidetached.* The prototype of such systems, Algol (β Persei), consists of a main sequence star of about $5M_\odot$ and a secondary star, which is about the same radius, is more evolved (in fact, a subgiant), and has a mass of about $0.8M_\odot$. These systems present a seemingly paradoxical circumstance, since here the more evolved star is the less massive star in the system. This *Algol paradox* is resolved, as we shall discuss in Section III, C, by the realization that mass loss and transfer have occurred in the system, and that the observed masses only weakly reflect the initial conditions of the system. In fact, the less massive star was at one time the more massive one, but it has been losing mass either by expulsion from the system as a whole or transfer to the companion, which has increased in mass with time.

C. Detached Systems

The majority of binary stars never arrive at the semidetached or contact phase, but evolve with their masses intact for their entire lives. These stars are called *detached* because they never come into contact with the Roche surface. Even in these systems, the presence of the companion can have interesting and important effects.

First of all, tidal locking can occur. Since stars are not solid objects, they can respond to the tide raised by the companion by deforming. The gravitational potential, varying over the surface, sets up circulation in a rotating star, which can be dissipative of the angular momentum of the binary and can also enforce corotation of the star with the orbital time scale. The effect has been well observed in the Earth–Moon system and is expected to be quite efficient in binary stars closer than periods of several weeks. The synchronism of the rotation and orbital periods generally slows the rotation to values well below that of field stars, producing main sequence stars that have rotation velocities typically two to three times slower than normal. For evolved stars, however, the process works in reverse. Normally, as a stellar envelope expands, the rotation velocity of the stellar surface is reduced. By the action of tides, the evolved star winds up rotating faster than would normally be expected. The characteristic generation of magnetic fields by dynamo processes has been implicated as the source of activity—enhanced X-ray and ultraviolet emission—in many giants among the RS CVn stars. Here, it is assumed that the strong tidal coupling between the components has spun up the members of the binary and enhanced the formation of strong surface magnetic fields. While the details are far from clear, the broad view and observational result is that the rapid rotation of evolved stars invariably leads to enhanced chromospheric and coronal emission lines and strong radio flaring behavior.

The wider the system, the smaller the coupling between the components. Eventually (and again the dividing line is fuzzy), the stars evolve more or less independently. However, some mass can be transfered from one star to another by stellar winds.

More evolved versions of the detached systems, the ζ Aurigae stars, are similar to the Algols in constitution. They possess main sequence hot stars, but the evolved star is the more massive one. Here, the primary is an evolved giant of late spectral type, but the system is so wide and the period so long that there is little direct interaction between the components. However, one system, ε Aur, remains an intriguing link between the Algols and the ζ Aur systems in having strong evidence for mass flows in the system and an extremely evolved hypergiant for the primary star. This system has the longest period of the class, about 27 yr; most of these binaries have periods of about 1 yr.

Among the massive stars, the Wolf–Rayet (WR) stars form an important parallel sequence of binary and nonbinary stars. These are generally very massive, with primary main sequence stars of more than $5M_\odot$ and secondaries that may have started out at $20–30M_\odot$ but have lost considerable portions of their initial masses. It is important to note here that the WR stars are discovered by their gross spectroscopic characteristics, extremely strong and broad emission lines, and that they are frequently not members of binary systems. Those of the WN sequence, which are nitrogen rich, appear to be binaries more frequently than those of the parallel WC sequence of carbon-rich stars. A main sequence group of OB stars, the OBN and OBC, may be related to the WR stars; this group also shows the same statistical separation between N-rich stars being binaries and C-rich being single stars.

Probably the most extreme examples of detached binaries are the *binary pulsars*, which consist of

magnetic neutron stars in systems with periods ranging from days to years, accompanied by evolved, often subdwarf, stars. These systems, because of the incredible accuracy achieved in pulse timing from the neutron star radio emission, offer the highest accuracy yet achieved in orbital determination. One system, 1913 + 16, has been cited as the prime example for orbital dynamics being altered by the emission of gravitational radiation by the revolving binary components.

D. SOME PROTOTYPE SUBCLASSES OF BINARY SYSTEMS

In this section, we summarize some general properties of important subclasses of close binary stars. Several of these have been discussed previously as well, in order to place them in a more physical context.

1. W Ursa Majoris Systems

These are main sequence contact binaries. They are typically low mass, of order $1–2M_\odot$, with orbital periods from about 2 hr to 1 day. The envelopes are distinguished as being in either radiative or convective equilibrium. The chief observational characteristics are that they show continuously variable light curves and line profile variations indicative of uniform temperature and surface brightness over both stars, although the mass ratio ranges from 0.1 to 1. Surface temperatures range from about 5500 to 8000 K. The lower mass systems are called W type, the higher mass systems are A type; the W-type envelopes are convective. Typical of this class are W UMa, TX Cnc, and DK Cyg. There may be massive analogs of this class, although the light curves are more difficult to interpret.

2. RS Canes Venaticorum Stars and Active Binaries

Close binaries with periods less than 2 weeks induce synchronous rotation via tidal coupling on time scales of order $10^8—10^9$ yrs. For main sequence stars, this generally results in slow rotation compared with that observed in single stars; for evolved stars, the opposite holds. The RS CVn and related stars show rapid rotation of a cool evolved star that displays enhanced dynamo activity. These stars are marked by exceptionally strong chromospheres and coronas, sometimes having UV and X-ray fluxes greater than 10^3 times that observed in normal G–K giants (cool giants). The photometric behavior of these systems is marked by the appearance of a dark wave, large active regions, which migrates through the light curve toward a decreasing phase, suggestive of differential rotation. The active stars have deep convective envelopes. Several subgiant systems, notably V471 Tau = BD + 16°516, have white dwarf companions, although most systems consist of detached subgiant or giant primaries and main sequence secondaries. With the exception of HR 5110, most of these systems are detached. Other representative members of this class are AR Lac, Z Her, and WW Dra. Typically, the mass ratios are very near unity, although this may be a selection effect.

An analog class, the FK Com stars, shows many of the RS CVn characteristics, especially enhanced chromospheric and coronal activity and rapid rotation, but it appears to be a class of single stars. The FK Com stars are argued to have resulted from the common envelope phase of an evolved system leading to accretion of the companion.

Both of these subclasses are especially notable as radio sources, often displaying long time duration (days to weeks) flares with energy releases of some 10^7 times that of the largest solar flares. The dMe stars are the low-mass analogs of these systems, although not all of these are binaries. It appears that the binarity is most critical in producing more rapid than normal rotation, which is responsible for the enhanced dynamo activity.

3. ζ Aurigae Stars and Atmospheric Eclipses

These systems consist of hot, main sequence stars, typically spectral type B, and highly evolved giants or supergiants with low surface temperatures (G or K giants). For several eclipsing systems, notably ζ Aur, 31 Cyg, and 32 Cyg, eclipses of the hot star can be observed through the giant atmosphere. The B star thus acts like a probe through the atmosphere of the giant during eclipse, almost like a CAT scan. Observations of photometric and spectroscopic fluctuations during eclipse provide a unique opportunity for studying turbulence in the envelopes of evolved stars. The systems are long period, although for several, notably 22 Vul, there is evidence for some interaction between the stars due to accretion of the giant wind by the main sequence star in the form of an accretion wake. The most extreme example of this subclass is ε Aur, a 27-yr-period binary with an unseen supergiant or hypergiant evolved cool star accompanied by an early-type companion.

4. Algol Binaries

These are the classic semidetached systems. They are marked by evidence for gas streams, distortions of the eclipse profiles due to instabilities in the circumstellar material on the time scale of several orbits, and sometimes enhanced radio and X-ray flaring activity of the more evolved star. For several systems, notably U Cep, the stream ejected from the giant hits the outer envelope of the accreting star and spins it up to very high velocity. For others, the stream circulates to form an accretion disk about the companion, which is heated both by turbulent viscosity and the impact of the stream in its periphery. These systems typically show inverted mass ratios, in that the more evolved star has the lower mass. They have masses ranging from about $1M_\odot$ each to greater than $5M_\odot$ for the constituent stars. Among the best examples of these systems are SX Cas, W Ser, and β Lyr.

5. Symbiotic Stars

These systems are so named because of the observation of strong emission lines from highly ionized species and cool absorption lines of neutral atoms and molecules in the same spectrum. They consist of a highly evolved cool giant or supergiant and either a main sequence, subdwarf, or collapsed companion. Several of the systems, notably R Aqr, show pulsating primary stars with periods of hundreds of days. The orbital periods are typically quite long, of order one year; the mass ratios have not generally inverted except in those systems where a white dwarf is established as a member. The ionizing source appears to be an accretion disk about the companion star, fed by a stellar wind and perhaps gas streams in the system. Radio and optical jets have been observed emanating from several systems, especially CH Cyg and R Aqr. These systems also display unstable light curves, presumably attributable to instabilities in the accretion disks. The most extreme examples of this class, having the longest periods and the most evolved red components, are the VV Cep stars.

6. Cataclysmics

These systems typically consist of low-mass main sequence stars of less than $1M_\odot$, and either white dwarf or neutron star companions. They display outbursts of the nova type when flash nuclear reactions release sufficient energy to expel the outer accreted layers off the surface of the collapsed star, or show unstable disks that appear to account for the dwarf novas. These systems will be discussed later at greater length. Among the best examples of this class are U Gem, SS Cyg, and OY Car for the dwarf novas; GK Per and DQ Her for the classical novas; and AM Her and CW 1103 + 254 for the magnetic white dwarf accreters. The low-mass X-ray binaries share many of the same characteristics without the extreme photometric variability, for example, Cyg X-2 and Sco X-1.

IV. Evolutionary Processes for Stars in Binary Systems

Normally, stars evolve from the main sequence, during which time their energy generation is via core hydrogen fusion, to red giants, when the star is burning helium in its core, at roughly constant mass. While stellar winds carry off some material during the main sequence stages of massive stars, most stars do not undergo serious alteration of their masses until the postgiant stages of their lives. This is only achieved when the escape velocity has been reduced to such an extent that the star can impart sufficient momentum to the outer atmosphere for a flow to be initiated. Envelope expansion reduces the surface gravity of the star, so that the radiative acceleration due to the high luminosity of stars in the postgiant stage, or the extreme heating affected by envelope convection in the outer stellar atmosphere, provides the critical momentum input. In a binary system, however, the star is no longer necessarily free to expand to any arbitrary radius. The presence of a companion fixes the maximum radius at which matter can remain bound to a star.[See STELLAR STRUCTURE AND EVOLUTION]

The companion exerts a tidal force on the star as it becomes more distended. The differential acceleration produced by the companion distorts the envelope of the evolving star until material can no longer be kept solely in the gravitational influence of the star. The maximum radius is set by the mass of the companion, the mass ratio, and the separation of the stars, and is called the *Roche limit*. It is given by

$$R_{RL} = G^{1/3}M^{1/3}P^{2/3}f(q) \qquad (6)$$

where $f(q)$ is a function of the mass ratio, given by a detailed analytic fit to the equipotential surfaces in the three-body problem as a function of the mass ratio q. An approximate form for this function is

$$f(q) \approx 0.4 - 0.2 \log q \qquad (7)$$

but other forms have been obtained. These functions are formal fits to the dependence of the radius of the L_1 point in the three-body problem as a function of mass ratio. Here, $f(q)$ is a slowly varying function of the mass ratio. Should the primary, that is more massive, star in a binary have a radius that exceeds this value, it will develop a cusp along the line of centers at the *inner Lagrangian point*, L_1. Nuclear processes continue in the stellar interior, driving the

increase in the envelope radius, so that even though the mass of the star is decreasing, the center of mass of the system shifts toward the companion star and the continued expansion of the primary causes the mass transfer to be maintained. Inexorably, the mass ratio will continue to increase until the star is so sufficiently stripped of matter that it becomes smaller than the instantaneous Roche lobe. At this point, the mass transfer stops.

The evolution of the system is determined by the fraction of the lost mass that is accreted by the second star and the fraction of both mass and angular momentum of the system that is lost through the outer contact surface. The loss of mass from one of the stars alters its surface composition as successive layers are peeled off. It is generally assumed that the star will appear as nitrogen enhanced, because the outer layers of the CNO-burning shell will be exposed to view if enough of the envelope is removed. The OBN and WN stars are assumed to be the result of such processing. In addition, the alteration of the mass of the star will produce a change in the behavior of turbulent convection in the envelope, although the details are presently very uncertain. The enhancement of turbulent mixing should be responsible for exposing the effects of nuclear processing of even deeper layers to view, but this has yet to be fully explored.

The behavior of the mass loser with time is significantly sensitive to whether the envelope is convective or radiative, that is, to whether the primary mode of energy transport is by mass motions or photons. In turn, these are sensitive to the temperature gradient. If the envelope is convective, the reduction in mass causes the envelope to expand. If radiative, the envelope will contract on mass being removed. Consequently, the mass transfer is unstable if the envelope is convective, and the star will continue to dump mass onto the companion until it becomes so reduced in mass that its envelope turns radiative. The instability, first described by Bath, may be responsible for the extreme mass transfer events seen in symbiotic stars and may also be implicated in some nova phenomena.

V. Mass Transfer and Mass Loss in Binaries

Mass transfer between components in a binary system takes place in two ways, by the formation of a stream or a wind. Either can give rise to an accretion disk, depending on the angular momentum of the accreting material. In the ζ Aur systems and in most WR binaries, the accretion is windlike. This also occurs in some highmass X-ray binaries (HMXRB), notably Cir X-1. In these, the accretion radius is given by the gravitational capture radius for the wind, which varies as

$$R \sim M/v_{wind}^2 \qquad (8)$$

where M is the mass of the accreting star and v_{wind} is the wind velocity at the gainer. The formation of an accretion wake has been observed in several systems, notably 22 Vul. The wake is accompanied by a shock. Should the star have a wind of its own, however, the material from the primary loser will be accelerated out of the system along an accretion cone, with little actually falling onto the lower mass component. Such wind–wind interaction is observed in Wolf–Rayet systems, notably V444 Cyg.

The formation of a stream is assured if the mass loss rate is low, and the star losing mass is in contact with the Roche surface. In this case, the L_1 point in the binary acts to funnel the mass into a narrow cone, which then transports both angular momentum and mass from the loser to the gainer.

The atmosphere of the mass losing star has a finite pressure, even though at the L_1 point the gravitational acceleration vanishes; thus, the mass loss becomes supersonic interior to the throat formed by the equipotentials, and a stream of matter is created between the stars. The fact that the center of mass is not the same as the center of force, that is L_1, means that the stream has an excess angular momentum when it is in the vicinity of the secondary or mass gaining star. It thus forms an accretion disk around the companion. However, if the ejection velocity is great enough, the size of the companion large enough compared with the separation of the stars, and the mass ratio small enough, the stream will impact the photosphere of the gainer. Instead of the formation of a stable disk, the stream is deflected by the stellar surface, after producing an impacting shock, with the consequence that the outer layers of the gainer are sped up to nearly the local orbital speed, also called the Keplerian velocity:

$$v_K = (GM_2/R_2)^{1/2} \qquad (9)$$

Some mass, the fraction is not well known, will also be lost through the outer Lagrangian point, L_3, on the rear side of the loser from the gainer along the line of centers. The mass of the system as a whole is therefore reduced. This means that the matter can also carry away angular momentum from the system. If the reduced mass of the system is given by

$$\mu = M_1 M_2/M \qquad (10)$$

then the angular momentum of the binary is

$$J = \mu a^2 \omega = G^{2/3} M_1 M_2 M^{-1/3} \omega^{-4/3} \qquad (11)$$

The change in the angular momentum of the system then produces a period change:

$$F_J = \frac{\delta M_1}{M_1}\left(1 - \frac{1}{3}\frac{M_1}{M}\right)$$
$$+ \frac{\delta M_2}{M_2}\left(1 - \frac{1}{3}\frac{M_2}{M}\right) + \frac{4}{3}\frac{\delta P}{P} \qquad (12)$$

for an amount F_J of angular momentum lost from the system and an amount δM of mass lost. Notice that the period evolution is very sensitive to both the amount of angular momentum lost and to the fraction of the mass lost from M_1, which is lost from the binary system.

The loss of angular momentum from the system is one of the currently unknown physical properties of various models. It is the most critical problem currently facing those studying the long-term behavior of the mass transfer in binaries, since it is the controlling factor in the orbital evolution. Two classes of models have been proposed, those in which much of the angular momentum of the stream is stored in the accretion disk or in the spun-up stellar envelope of the gainer and those in which the J is carried out of the system entirely. Magnetic fields can also act to transport angular momentum, and the degree of spin–orbit coupling between the components also affects the overall system evolution. As a result, much of the detailed behavior of mass exchanging or semi-detached systems is not yet fully understood.

VI. X-Ray Sources and Cataclysmics

The presence of a collapsed component in the system alters much of the observable behavior of binaries. In particular, the signature of mass accretion onto a white dwarf, neutron star, or black hole is X-ray emission. With the discovery of binary X-ray sources in the late 1960s, following the launch of UHURU, the first all-sky X-ray survey satellite, the field has rapidly grown. Early observations were interpreted as accretion onto white dwarf stars, but the physical details of the accretion processes onto specific compact objects have been refined so that it is now possible to distinguish many of the marks of a specific gainer by the observable behavior.

X-ray emission results from accretion onto a collapsed star because of the depth of its gravitational potential well. As the infalling matter traverses the accretion disk, it heats up because of collisions with rapidly revolving matter and radiates most of its energy away. If the disk is optically thick, this radiation will appear at the surface of the accretion disk as a local blackbody emitter at a temperature characteristic of the local heating rate for the matter in the disk. Since the material is slowly drifting inward, because of loss of angular momentum through viscosity-like interactions within the disk, the heating can be likened to that resulting from a turbulent medium that is capable of radiating away kinetic energy gained from infall. The mass distribution is not radially uniform, and the vertical extent of the disk is determined by pressure equilibrium in the z direction, so that the surface area and temperature vary as functions of distance from the central star. As a result, the flux merging from the surface and seen by a distant observer is an integrated one, summing up different regions of the disk which have different temperatures. The emergent spectrum of the material is not that of a blackbody, nor even very similar to a star. In general, it will appear to be a power law distribution with frequency, looking nonthermal but in fact reflecting the run of temperature and pressure in the disk.

A. ACCRETION DISK PROCESSES

When material is lost from the one star in a binary and flows into the vicinity of the companion, it may have sufficient excess angular momentum that it cannot accrete directly. Instead, it forms an accretion disk, a low-density circulating disk of gas that encircles the companion. Viscous processes then dominate the slow inward drift of matter in the disk onto the companion star. Two important effects come into play in these disks: the heating due to energy dissipation by the disk that produces observable radiation from the surface of the accretion disk and the dissipation of angular momentum.

In the current picture of such disks, the angular momentum is due to Keplerian motion, so that the rate of shear of the disk is $\sigma_{r\phi} = \partial\omega_K/\partial r$, where $\omega_K = (GM)^{1/2}r^{-3/2}$ is the Keplerian rotation frequency. Thus, the rate of energy dissipation, in the presence of viscosity, is the shear times the stress, $\varepsilon = T_{r\phi}\sigma_{r\phi}$. The greatest uncertainty currently rests in the specification of the stresses in the circulating matter. One common prescription is the α disk model, in which $T_{r\phi} = \alpha p$, where p is the pressure in the disk and α is a constant of order $0.1-1$. The local heating by dissipative processes thus creates a vertical temperature gradient. The radiation is transported to the surface of the disk and can be observed as a power law distribution of flux with distance from the central star, which translates to a power law in frequency. In other words, the disk looks to an outside observer as if it is a nonthermal spectrum, while in fact it is thermalized at every radial point, but the

distribution of the heating rate is governed by the shear and is thus a power law with radius.

The scale height of the disk, equivalent to its vertical thickness, is presumed to be due to the gas being in vertical hydrostatic equilibrium with respect to the tidal component of the gravitational acceleration directed toward the midplane:

$$\frac{dp}{dz} = -\rho\,\frac{GMz}{r^3} = -\rho\omega^2 z \qquad (13)$$

Taking the disk to be vertically isothermal gives

$$\rho(z) = \rho(0)e^{-z/2z_0^2} \qquad (14)$$

where $z_0 = a_s^2 v_\phi^{-2} r$ is the scale height for the disk. Here, a_s is the local sound speed, v_K is the Keplerian velocity, and $\rho(0)$ is the midplane density. The rate of energy dissipation balances the rate of energy generation by shearing stresses and governs the local temperature, which is presumed to be regulated by radiative losses from the surface of the disk. The accretion disk is thus self-luminous, tapping the turbulence in its interior, which is optically thick, and radiating the excess energy away as observable light. The greater the circulation velocity, the hotter the disk will be and the thinner it will appear. Thus, there is a radial temperature gradient in any disk, although the details of the viscous calculation remain to be clarified. It should be added that if the disk becomes sufficiently hot, nuclear processing may take place, resulting in additional energy release in the disk midplane in the innermost parts of the accretion disk. While not important for most accretion phenomena, such processes have recently been discussed for disks around black holes as well as in active galactic nuclei.

The total luminosity is determined by the rate of infall, since it is assumed that accretion is the primary source for emission of radiation, and the mean temperature is given by the shock at the surface of the gainer. As an approximate estimator, the temperature can be obtained from the virial theorem:

$$T_s = \frac{GM}{m_p k R} \approx 10^7\,\frac{M}{R}\,K \qquad (15)$$

where in the latter equation M and R are in solar units. The luminosity is given by the rate of accretion, so that

$$L = \frac{GM}{R}\dot{M} \qquad (16)$$

where $\dot{M}$ is the mass accretion rate. The maximum luminosity that a source of radiation can emit without driving some of the material off by radiation pressure, the so-called *Eddington limit*, is given by

$$L_{Edd} = 10^4 \frac{M}{M_\odot} L_\odot \qquad (17)$$

so that the accretion can only be stable as long as $L \leq L_{Edd}$. Most mass estimates for the gainer in X-ray binaries are limited by this luminosity.

The temperature of the emergent spectrum can serve as a guide to the nature of the accreting star. The lower the temperature, that is, the softer the X-ray spectrum, the lower the M/R ratio. White dwarfs have $M/R \approx 10^2$, neutron stars and black holes have $M/R \geq 10^5$ so that, in general, the more collapsed the gainer, the higher the temperature and the luminosity for the same rate of mass accretion.

Disk formation is also different in different types of systems. The gravitational potential for a white dwarf is lower, so the circulation velocity of the material is also lower. The efficiency of heating being reduced, the disks are cooler and the opacity is also higher. Consequently, the disks are sensitive to small fluctuations in the local heating rate and may be thermally unstable. This results in both flickering and alteration of the vertical structure of the disk with time. Also, the heating at the inner boundary of both neutron stars and white dwarfs differs from that seen for black holes, since there is actually no surface in the latter on which matter can accumulate. Consequently, neutron stars and white dwarfs are more sensitive to the accumulation of matter on their surfaces and can undergo flash nuclear processing, the initiation of nuclear fusion under low-pressure, upper boundary conditions, which serves to blow matter off their surfaces on time scales corresponding to the rate of expansion of the outer layers because of the release of nuclear binding energy. In effect, these stars behave like the cores of stars without the presence of the overlying matter, while black holes, which may go through some slow nuclear processing of matter as it falls into the central region, cannot initiate flash processes. Thus, novas and dwarf novas appear to be due to non–black hole systems.

The details of mass accretion can only be understood by detailed modeling of the accretion disk processes. These depend critically on the nature of the environment of the compact object, because the presence of strong magnetic fields appears to significantly alter the details of the accretion. A magnetic field exerts a pressure against the matter infalling to the surface and can serve to channel the matter to the poles of the accreter. In addition, it alters the pressure and density conditions at the interaction region in the plane of the accretion disk, forming an extended layer at large distance from the central star and can thus lower the effective temperature emerging from the shock region.

B. White Dwarf Accretion

The accreting star in most cataclysmics is a white dwarf, a compact star of about $1M_\odot$. With radii of order $0.01R_\odot$, these stars are not sufficiently collapsed to produce gamma emission and usually are imbedded in accretion disks dominated by boundary layer effects. These stars generally have weak magnetic fields compared with neutron stars, and while they produce pulsed emission because of rotation, they do not show many of the other effects of magnetic accretion, in particular rapid oscillations and periodic rapid outbursts in optical and X-ray.

As matter falls onto the surface of a white dwarf, it cools and compresses. The base of the accreting layer increases in density and temperature. If enough matter has accumulated, a pyconuclear or pressure-induced reaction takes place. Because the reaction zone is at the surface of the star, there is no matter overlying the nuclear burning zone to suppress the ejection of the rapidly heated material from the stellar surface; there is not sufficient weight above the layer to maintain steady burning. In addition, the underlying stellar envelope is degenerate, so the increased energy input from the nuclear reactions does not alter the pressure of the substratum. Such an event can also disperse the accretion disk for a time, and blow a shell out of the binary system. Such events will release most of the binding energy of the burning layer, typically of order 10^{45} erg, corresponding to about $10^{-8}M_\odot$ of material. The ejected matter will reach velocities of order 10^3 km sec^{-1}. This is the basis of current models of the *nova* event. The maximum brightening of these objects, many orders of magnitude less than that of supernovas, appears to be consistent with the small amount of matter lost from the system.

The accretion may also take place at the magnetic poles. In this case, because of the low mass accumulated, there will be no flash nuclear burning, instead it will be replaced by the slower release of gravitational energy. These systems, called the AM Her stars, are also called *polars* because they show rotationally modulated circular polarization in the emission lines formed above the magnetic poles of the star. The emission line strengths are also variable on the rotation time scale for the white dwarf, which is nearly synchronous with the orbital period.

C. Neutron Star Accretion

The presence of strong magnetic fields alters the inner boundary condition for the accretion disks of neutron stars. Here, the rotation of the magnetosphere of the star at the stellar rotation period is far lower than the Keplerian velocity of the infalling matter. Recent observations, primarily with EXO-SAT, have shown that the X-ray emission is modulated on short time scales, of order 0.1 sec, by the beating between the local Keplerian frequency and that of the stellar surface. The beating is not stable in time, but appears in the *quasi-periodic oscillations* or QPOs. These have been observed in a wide variety of low-mass X-ray binaries (LMXRB) with neutron star components as the gainers.

Variations in the rate of mass accretion, due to instabilities in the loser, may alter the placement of the boundary in radial distance from the neutron star and also affect the local heating. When the disk expands, it shields the inner part of the accretion disk from the line of sight and the magnetospheric boundary disappears from view. Recent arguments suggest that the brightness of the source is correlated with its spectrum, being *softer* when brighter, and it is during these bright phases that the QPOs disappear. When the source fades, the spectrum becomes harder and at this time the QPO of the X-ray emission is most apparent.

Several X-ray sources possess neutron stars as the compact accreter. Perhaps the best studied is Her X-1, which is an X-ray pulsar with a period of about 1 sec and which possesses an accretion disk that also shows an accretion disk instability on a 35-day period. Other examples are Cen X-3, Sco X-1, and LMC X-4.

D. BLACK HOLE ACCRETION

There is little difference between a neutron star and black hole from the point of view of X-ray generation. The main difference is that the inner boundary of the accretion disk is not solid. No flashes will occur from these systems because nuclear reactions do not occur on a stellar photosphere. However, some processing does take place in the inner region of the disk because of the extreme temperatures reached in the relativistic regime of the motion. The p–p interactions produce π mesons, which decay into gamma rays. These are in turn reprocessed by absorption and scattering into hard X-rays by the disk.

Other than periodic variations due to disk instabilities and precession of jets generated in the inner parts of the disk, black hole sources are not expected to show periodic, accretion-related behavior. The disks are unstable and flicker and also undergo occassional transient changes in brightness in X-ray and visible light, but otherwise the characteristics of these systems are easily distinguished from the QPO behavior of accreting neutron stars.

To date, LMC X-3, a binary in the Large Magel-lanic Cloud, and Cyg X-1 = HDE 226868 are the only X-ray binaries that require black hole components on the basis of orbital dynamics. One other object, SS 433, also appears to be a black hole accreter imbedded in a massive stellar wind and disk.

VII. Formation of Binary Systems

This is one of the major areas of study in stellar evolution theory, because at present, there are few good examples of pre-main-sequence binary stars so that the field remains dominated by theoretical questions. Protostellar formation begins with the collapse of a portion of an interstellar cloud, which proceeds to form a massive disk. Through viscosity and interaction with the ambient magnetic field, the disk slowly dissipates its angular momentum. If the disk forms additional self-gravitating fragments, they will collide as they circulate and accrete to form more massive structures. Models show that such disks are unstable to the formation of a few massive members, which then accrete unincorporated material and grow.

Classical results for stability analysis of rapidly rotating homogeneous objects point to several possible alternatives for the development of the core object. One is that the central star in the disk, if it is still rapidly rotating, may deform to a barlike shape, which can pinch off a low-mass component. The core may undergo spontaneous fission into fragments, which then evolve separately. Simulations show, however, that the fission scenario does not yield nearly equal mass fragments. Such systems more likely result either from early protostar fragmentation and disk accretion in the first stages of star formation or of coalescence of fragments during some intermediate stage of disk fragmentation before the cores begin to grow.

Binary star formation appears to be one of the avenues by which collapsing clouds relieve themselves of excess angular momentum, replacing spin angular momentum with orbital motion of the components. However, while the distribution of q may be a clue to the mechanism of formation, even this observational quantity is very poorly determined. The recent discovery of debris disks around several intermediate-mass, main sequence stars, especially β Pic and α Lyr = Vega, has fueled the speculation that planetary systems may be an alternative to the formation of binary stars for some systems. Statistical studies show that radial velocity variations are observed in many low-mass, solar-type stars, but the period and mass ratio distribution for these systems is presently unknown.

VIII. Concluding Remarks

Binary stars form the cornerstone of the understanding of stellar evolution. They are the one tool with which we can determine masses and luminosities for stars independent of their distances. They present theaters in which many fascinating hydrodynamic processes can be directly observed, and they serve as laboratories for the study of the effects of mass transfer on the evolution of single stars. By the study of their orbital dynamics, we can probe the deep internal density structures of the constituent stars. Finally, binary stars provide examples of some of the most striking available departures from quiescent stellar evolution. The development of space observation, using X-ray (TENMA, EXOSAT, EINSTEIN, and GINGA) and ultraviolet (International Ultraviolet Explorer–IUE) satellites, has in the past decade provided much information about the energetics and dynamics of these systems. The coming decade, especially with the planned launch of the Hubble Space Telescope and the Infrared Space Observatory (ISO), promises to be a fruitful one for binary star research.

BIBLIOGRAPHY

Batten, A. (1973). "Binary and Multiple Star Systems." Pergamon, Oxford.

Eggelton, P. P., and Pringle, J. E. eds. (1985). "Interacting Binary Stars." Reidel, Dordrecht, Netherlands.

Frank, J., King, A. R., and Raine, D. (1985). "Accretion Power in Astrophysics." Cambridge Univ. Press, London and New York.

Gehrels, T., and Matthews, M., eds. (1985). "Protostars and Protoplanets II." Univ. of Arizona Press, Tuscon.

Kenyon, S. J. (1986). "The Symbiotic Stars." Cambridge Univ. Press, London and New York.

Popper, D. M., Ulrich, R. K., and Plavec, M., eds. (1980). "Close Binary Systems: IAU Symp. 88." Reidel, Dordrecht, Netherlands.

Pringle, J. (1981). Accretion disks in astrophysics, *Annu. Rev. Astron. Astrophys.* **19**, 137.

Shu, F., and Lubow, S. (1981). Mass, angular momentum, and energy transfer in close binary systems, *Annu. Rev. Astron. Astrophys.* **19**, 277.

van der Kamp, P. (1986). "Dark Companions of Stars." Reidel, Dordrecht, Netherlands.

BLACK HOLES (ASTRONOMY)

Richard A. Saenz   *California Polytechnic State University*
Stuart L. Shapiro   *Cornell University*

GLOSSARY

Accretion disk: Disk of gaseous material, often captured from a normal binary companion star, which spirals onto a compact object. While falling through the gravitational field of the compact object, the gas may emit observable radiation.

Chandrasekhar mass: Maximum mass of a cold star that can support itself by internal pressure against gravitational collapse: ~1.4 solar masses.

Compact object: White dwarf, neutron star or black hole; one of the endpoint states of stellar evolution once thermonuclear burning ceases.

Cygnus X-1:   X-ray source in a binary system. The unseen compact companion star in this binary is a good black hole candidate.

Event horizon: Surface of a black hole; the boundary enclosing the region of spacetime that cannot communicate with the external universe. A one-way membrane across which infalling matter or light is captured but can never escape.

General relativity: Einstein's theory of gravitation. It identifies gravity with the curvature of spacetime and the source of curvature with mass-energy.

Gravitational redshift: Increase in the wavelength of light corresponding to a decrease in its frequency and energy as it climbs out of a gravitational field.

Hawking radiation: Thermal radiation emitted by an evaporating black hole.

Schwarzschild radius: Radius of a nonrotating, uncharged spherical black hole = $2GM/c^2$, where M is the mass of the hole, G the gravitational constant and c the speed of light. For a one-solar-mass black hole, this radius equals ~3 km.

Singularity: Region of infinite tidal force inside a black hole.

Spacetime: Four dimensional manifold composed of ordinary, three-dimensional space plus time. The presence of mass–energy causes the geometry of spacetime to be curved and non-Euclidean.

Tidal force: Variation in the gravitational force with spatial separation, causing an object free-falling in a gravitational field to be deformed.

A black hole is a region of spacetime enclosed by an event horizon; a region with gravity so strong that no matter, light, or signals of any kind can escape from its surface.

I. Describing a Black Hole

Long before the theory of gravity necessary to fully describe and understand black holes was developed (Einstein's Theory of General Relativity in 1915), Laplace had contemplated the possibility of a black hole. He noted in 1795 that as a consequence of Newton's theory of gravity, even light would be unable to escape from a sufficiently compact star.

Even after the publication of Einstein's papers and Karl Schwarzschild's (1916) exact solution for the gravitational field of a spherical mass, many leading physicists reeled from some of its inevitable implications, especially the possibility of a black hole. In 1935 Eddington called for "... a law of Nature to prevent the star from behaving in this absurd way." At about the same time Landau noted that "there exists in the whole quantum theory no cause preventing the system from collapsing to a point" and therefore concluded that there are "regions in which the laws of quantum mechanics (and therefore quantum statistics) are violated." [See COSMOLOGY; STELLAR STRUCTURE AND EVOLUTION.]

By 1939 Oppenheimer and Snyder had demonstrated that, as a consequence of Einstein's theory, a collapsing star could indeed eventually be cut off from all communication with the outside universe. Further work along these lines was left essentially until the 1960s. Our understanding of black holes has progressed rapidly since then. [See RELATIVITY, GENERAL.]

The black hole is the region of spacetime surrounded by an *event horizon*. As its name implies, we cannot see beyond the horizon since we can receive no messages emitted from its interior. The horizon is a one-way membrane: matter and light can cross the horizon and fall into a black hole, but once inside, the horizon cannot be crossed again—there is no escape from a black hole. Although the matter that creates the black hole is hidden within the horizon, objects in the outside universe can detect its presence. That presence is signaled by the gravitational field of the black hole. Far from the black hole the gravity is described by Newtonian theory: the more massive the gravitating object the greater its gravity, while the strength of the gravitational field diminishes as the square of the distance from the object.

Consider an astronaut who orbits close to the horizon of a black hole. He has a choice: turn around and be saved or explore closer and be lost to the outside universe forever. Once inside the horizon our curious astronaut would find that it is now impossible, even with the most powerful rocket engine, to prevent the increasingly rapid plunge toward the center of the black hole. Soon thereafter the astronaut is crushed to infinite density at the center of the black hole by the infinite gravitational tidal forces there. He thus joins the other matter making up the central *singularity* of the black hole. Because it lies beyond the horizon, this singularity is not visible to the outside universe and cannot send messages to the outside. Clothed by an event horizon, it is said to be causally disconnected from the exterior world. (It is conjectured that there are no "naked singularities" in general relativ-

ity.) Later we will report in more detail on the effects of the black hole on our intrepid astronaut.

How does one characterize a black hole? How do black holes differ? A black hole's dominant characteristic is its mass. The matter from which the black hole has formed plus any material that has been captured by the black hole determine its size and magnitude of its gravitational field. Its size is given by the *Schwarzschild radius* of the black hole

$$R_S = 2GM/c^2$$

where G is the universal gravitational constant, M the mass of the black hole and c the speed of light. For a black hole with the mass of the sun R_S is ~3 km. For a nonrotating, uncharged black hole the event horizon is located at R_S. Physically, the Schwarzschild radius measures, at any instant, the spherical surface area ($4\pi R_S^2$) at the black hole's event horizon.

In addition to its mass a black hole can have two other independent characteristics (but amazingly no others!). If the matter that collapses to form the black hole is rotating, the resulting black hole will be rotating. Such a black hole is said to have angular momentum. Similarly, if the matter making up the black hole is not electrically neutral, but is charged, the black hole will have an electrical charge, and will exert an electrical force on matter outside the horizon. (Since space is relatively abundant with charged particles of both signs and since the electrical force is relatively strong and far reaching, we expect that an initially charged black hole will quickly collect oppositely charged particles to electrically neutralize itself.) A black hole is completely characterized by these three attributes: its mass, angular momentum, and charge. All other distinguishing characteristics are lost. The black hole retains no other knowledge of the material from which it formed—matter or antimatter, leptons or baryons, strange or charmed particles, etc. J. A. Wheeler has summarized this situation with the aphorism "A black hole has no hair."

Consider now the effects of a black hole on the outside world. Focus on a nonrotating, uncharged black hole. It is the black hole's gravitational field that signals the black hole's presence to an outside observer. Far from the black hole Einstein's and Newton's theories make virtually identical predictions. In fact, Kepler's third law, which relates the distance from a central gravitating body (like the sun) to the period of an orbiting object (like the earth), can be used to find the mass of the central object (sun or black hole). The formula is

$$M = \frac{4\pi^2}{G} \frac{a^3}{T^2}$$

where T is the period of revolution, a the distance from the central to the orbiting body, and M the mass of a central body. The black hole's most important characteristic—its mass—can be measured without risking life or limb! [*See* CELESTIAL MECHANICS; SOLAR SYSTEM, GENERAL.]

Einstein's theory of general relativity tells us that gravity acts on all forms of energy. Objects with mass m have an associated rest mass energy E given by Einstein's famous special relativistic relationship

$$E = mc^2$$

where c is the speed of light. But even objects without mass, such as particles of light (called photons) have energy and are affected by gravity. Photons follow "curved" paths near a gravitating object similar to those of a comet that passes the sun. This effect, called light bending, was one of the first predictions of general relativity to be confirmed by solar system measurements. Near a black hole the gravitational field is so strong that photons can be put into circular orbits around the black hole (but only at a radius of $1.5R_S$). However, the orbit is unstable—the smallest disturbance will cause the photon to fall into or away from the black hole. Large-angle deflections of photons are also possible (see Fig. 1). A photon that approaches too close to the hole is, of course, captured. In fact, at a radius of $1.5R_S$, all photons emitted inward by a stationary source are captured; only those photons emitted outward can escape to large distances.

Just as a ball thrown from the surface of the earth slows as it rises, or gains speed as it falls, the local energy (rest mass energy plus kinetic energy) of anything moving near a black hole is also affected by gravity. The local energy of light is proportional to the photon's frequency (that is, the frequency of the light wave); high frequency means high energy. Since a photon has no rest mass, it cannot slow down (light always travels at the speed of light, $c = 300,000$ km/sec). However, the light's frequency drops as it moves away from the black hole, thereby decreasing the local energy. This effect is called the *gravitational redshift* (low frequency visible

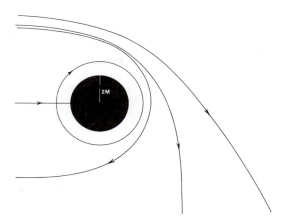

FIG. 1.   The curved paths followed by photons as they pass near a black hole.

light is red in color). Light falling toward a black hole is shifted to higher frequency or "blueshifted." [*See* RELATIVITY, SPECIAL.]

If our curious astronaut were to carry a laser pointed toward the earth as he coasts toward the horizon of a black hole, we on earth would see the color of the laser light change. Each succeeding photon we receive would have been emitted from a position closer to the horizon. As the photons travel away from the black hole they expend some of their energy to get away from the gravitational pull of the black hole. The closer to the horizon they originate, the more energy (frequency) they must lose before reaching us. Those emitted very near the horizon are heavily redshifted and reach us with almost no energy. Moreover, those photons emitted after the laser has crossed the horizon never get out! By contrast, a laser on earth, pointed at the astronaut, sends out photons that can always reach the astronaut. Each succeeding photon received by such an individual is received with a higher frequency than the last.

Another gravitational effect of prime importance to our erstwhile astronaut are the tidal forces exerted by the black hole. Because the effects of gravity are greater on objects closer to the black hole, any object of finite size will experience stronger gravity on the side closer to the black hole than on the more distant end. This difference is called the tidal force. Imagine our astronaut, falling feet first toward the black hole. As the horizon is approached the pull on the feet would begin to exceed that on the head to such a degree that the astronaut is elongated in the radial direction. Meanwhile, every particle of the astronaut's body is plunging toward the singularity at the black hole's center. Simple geometry tells us that because all radii meet at the center, the astronaut is also being compressed in the directions perpendicular to the astronaut's radial motion. The astronaut begins to look like very long spaghetti! Eventually the radial tidal force becomes so great, and the stretching forces so large, that the astronaut (and constituent molecules and atoms) are ripped apart. By the time the horizon is reached, only elementary particles remain. Upon reaching the singularity even these are crushed to infinite density!

Consider next the effects of a black hole's gravitational field on the passage of time. Let us return to light and the frequency of the wave associated with it, and use them as probes of time. The frequency is the measure of the number of wave crests (or whole wavelengths) that are emitted (or received, or pass by) each second. Comparing the rate at which waves pass with the rate of any other process of interest will give us their relative rates. The light thereby serves as a clock. Imagine now that a light wave is sent from an observer (perhaps our astronaut) sitting motionless (with the help of a rocket) near a black hole. Suppose it is arranged that when emitted, the light has a frequency equal to the astronaut's heartbeat rate. We receive the light signal far from the black hole, but the frequency at which we receive the signal is lower due to the gravitational redshift. We therefore conclude that the observer's heart is beating at a slower rate than the observer would measure. We can also discuss the observations made of us by the astronaut. The astronaut will receive any signal we send with a frequency higher (blueshifted) than the frequency at which we send the signal. The astronaut therefore concludes that our clock (and heart) is running faster than it would have had it been in the astronaut's vicinity. We can summarize all of this by saying that the rate at which a clock ticks, as viewed by a distant observer, is dependent on the relative strengths of the gravitational field at the point where the clock resides and the point where the observer resides. Clocks in a region of stronger gravity will be observed to tick at a slower rate than the clock the distant observer carries.

The angular momentum of a black hole is believed to play a large and important role in the astrophysics of black holes (see Sections II and V). The radius at which the horizon forms is determined by both the mass and the angular momentum of the black hole. More interestingly, there exists a region surrounding (and outside) the horizon, called the ergosphere. Any particle or observer inside the ergosphere is dragged along with the rotation of the black hole; there are no static observers allowed in the ergosphere. It is also possible to extract some of the black hole's rotational energy by sending a particle into the ergosphere. If the particle divides into two in the ergosphere with one piece falling into the black hole, the other may emerge with an energy greater than that of the original particle (this is called the Penrose mechanism).

II. Origin of Black Holes

The life of a star is a contest between the inward pull of gravity and the outward push exerted by gas pressure. Because a normal luminous star continuously loses energy from its surface it must generate heat at its center in order to sustain the necessary thermal pressure. Once the thermonuclear fuel in the star's core has been exhausted, the thermal pressure drops and the star has no choice but to contract. The final outcome of this gravitational collapse, the endpoint of stellar evolution, will depend on the mass of the star.

Relatively low-mass stars, like our sun, reach a state where a high density, metal-like state for the core can prevent further gravitational contraction. These objects are called *white dwarfs*. White dwarfs have a density a million times larger than the sun, but cannot have a mass greater than 1.4 times that of the sun. This limiting mass is called the Chandrasekhar mass. Heavier stars have a more complex, but faster evolution than lower-mass stars. When they reach a state of fuel exhaustion the weight of the star causes a rapid catastrophic collapse of the inner core, and possibly, an explosive shedding of the outer layers of the star. Such events are called supernovae. The fate of the collapsing core is determined principally by the mass of the parent star. For a star with a mass less than about ten to twenty times that of the sun, the collapse will be halted when the star reaches a density close to that of an atomic nucleus (about a billion times denser than a white dwarf). At these densities most of the matter in the core is in the form of neutrons—hence the name *neutron star*. At these densities, it is difficult to squeeze neutrons closer together. For a neutron star with a mass less than ~3 solar masses, this neutron pressure is sufficient to prevent further gravitational contraction. Neutron stars have radii of ~10 km.

If the mass of the parent star is too large (bigger than about ten to twenty solar masses), there is no escape from *total* gravitational collapse. There is no configuration of the matter, no outward pressure, that is able to halt the collapse of the core! Successive layers of the imploding core pass through the event horizon, as the matter collapses to the center forming a singularity. A black hole is born. Black holes resulting from the supernova process can easily have masses up to several tens of solar masses.

III. How to Detect a Black Hole

A. SEARCHING FOR BLACK HOLES

An isolated black hole would be virtually impossible to detect. It emits no light and is quite small (recall $R_S = 3$ km for a black hole with the mass of the sun). To identify a black hole we must therefore understand and observe the effects of the black hole's gravity on its external environment.

The most promising situation for detecting a black hole occurs when the black hole is in a binary star system in orbit about a normal companion star. A large fraction of the stars in our galaxy are in binary systems. The stars in such a system interact via their gravitational fields. Because of the small size of a black hole, the distance between the black hole and its orbiting companion can be quite small, and the gravitational effects on the normal companion star quite large. In what is a form of stellar cannibalism, the tidal forces can draw gaseous material from the normal star to the black hole. Because the stars are revolving about each other, the stellar material will form a swirling *accretion disk* of material that gradually spirals toward the black hole. (See Fig. 2.)

As it does, the gas is heated by compression and internal friction to temperatures as high as 10^6 K or more. Matter at such temperatures will radiate strongly at x-ray wavelengths. Observational evidence of a small but intense source of x-rays from a binary system could well be evidence for the existence of a black hole. Unfortunately, the above model is not much different if the companion to the normal star is a white dwarf or a neutron star instead of a black hole. In order to differentiate one compact source from another it is necessary to look closely at

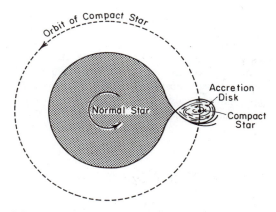

FIG. 2. Material captured from a normal companion forms an accretion disk and spirals toward the black hole giving rise to x-rays.

the x-ray emission and at the visible normal companion star. Rapid variations in the x-ray emission tell us that the emitting region is small. Variations on time scales of milliseconds would indicate an object less than 300 km in size. This would rule out white dwarfs but not neutron stars. It is therefore necessary to show that the unseen compact companion has a mass exceeding the mass limit for a neutron star (about 3 solar masses). In order to do this, the separation of the stars and their orbital period must be determined. Kepler's third law can then be used to give the masses (or at least a limit on the masses) of the two objects. If we can determine that the unseen compact object around which x-rays are being generated has a mass greater than 3 solar masses we would have strong evidence for the existence of a black hole.

B. BLACK HOLE CANDIDATES

Cygnus X-1 was the first x-ray source discovered in the constellation Cygnus by the x-ray satellite Uhuru. It is located in our galaxy about 8000 light years from earth and emits x-rays equal in power to ten thousand suns. The optical star companion to the x-ray source is a supergiant star, known as HDE 226868. HDE 226868 and the unseen compact object orbit each other every 5.6 days. Even with all the uncertainties associated with the observations of, and assumptions about HDE 226868, a lower mass limit of 3.4 solar masses for the compact object (with a more realistic lower limit of 5.3 solar masses) has been determined. Since both of these exceed the upper limits for neutron stars, we are led to the conclusion that Cygnus X-1 is very likely a black hole.

A more recently discovered candidate is an x-ray source discovered in the Large Magellenic Cloud, one of our galaxy's companion dwarf galaxies. The source is called LMC X-3. Estimates put the mass of the compact object at 10 solar masses. It is an excellent candidate for a black hole.

IV. Black Hole Evaporation

In the 1970s it was noted that the classical laws governing the interaction of a black hole with its surroundings could be summarized by a few statements quite reminiscent of the laws of thermodynamics. These were called the "laws of black hole dynamics." Steven Hawking proved a theorem, "the second law of black hole dynamics": in any interaction, the surface area of a black hole can never decrease, the area being proportional to the square of the mass of a nonrotating hole. For instance, this theorem puts strict upper limits on the amount of gravitational energy that can be radiated away when two black holes collide to form a larger black hole.

However, the connection with thermodynamics was not complete. A classical black hole cannot be in thermal equilibrium with another system. The classical black hole can absorb but not lose energy. Hawking was able to show that, when quantum mechanics is taken into account, a black hole *can* radiate energy—it can evaporate away its mass! In fact, he found that the emitted radiation has the same spectrum (the black body spectrum) as a hot object with a temperature given by

$$T = \frac{hc^3}{16\pi^2 kGM}$$

where h is Planck's constant, k Boltzmann's constant, and M the mass of the black hole; this expression holds for a nonrotating Schwarzschild black hole). For a one-solar-mass black hole T is only 10^{-7} K. Such a black hole has a negligible radiation loss, even over times as long as the age of the universe. However, if there exist black holes with masses that are much smaller, say 10^{15} g, (the mass of a large mountain), the temperature is about 10^{12} K.

Such a "mini" black hole may well have been formed by density fluctuations in the very early universe, and would only now be explosively radiating away its last bit of mass-energy. An appreciable fraction of this radiation would be in the form of gamma rays. The detection of such bursts would be a stunning confirmation of the ideas presented by Hawking.

V. Black Holes and Quasars

At the higher end of the mass scale lies another interesting prospect for black holes: explaining the energy mechanism in quasars and active galactic nuclei. Quasars were discovered in the early 1960s. Until that time their images on photographic plates had been overlooked, having been mistaken for not very interesting dim stars. Closer examination of their spectra indicated that the quasars are at cosmological distances from our own galaxy. If the quasars are at the distances indicated by their spectra, it is necessary to explain their enormous power output (vastly exceeding that of normal galaxies) emanating from regions often not any bigger than the solar system. It may well be that supermassive black holes supply the energy for all the various kinds of active galactic nuclei including quasars, radio galaxies and Seyfert galaxies. Possible origins for the supermassive black hole include the collapse of a supermassive star or the collapse of a dense cluster of neutron stars or stellar-mass black holes. [*See* QUASARS.]

Just as in the case of binary x-ray sources discussed previously, matter near the black hole would accrete onto a supermassive black hole. The complex interactions of the black hole's strong gravity, the electric and magnetic fields associated with the infalling matter, the angular momentum of the hole and of the accreting matter, can give rise to twin jets along the rotation axis, as well as to the synchrotron radiation and emission lines associated with active galactic nuclei. The mass of such a massive black hole need be on the order of 10^8 solar masses for it to give rise to observed power outputs of 10^{46} ergs/sec. [*See* GALACTIC STRUCTURE AND EVOLUTION.]

Even in our own galaxy there is observational evidence for a "small" (a few million solar masses) black hole at the center of the galactic nucleus. The evidence comes from infrared and radio observations of the motions of stars near the center of the galaxy. The high velocities of these stars, confined to a small region, indicate an unseen central mass of this size.

BIBLIOGRAPHY

Chandrasekhar, S. (1983). ''The Mathematical Theory of Black Holes.'' Oxford Univ. Press (Clarendon), London and New York.

Hawking, S. W. (1977). *Sci. Am.* **236**(1), 34–40.

Hawking, S. W., and Ellis, G. F. R. (1973). ''The Large Scale Structure of Space-Time.'' Cambridge Univ. Press, London and New York.

Misner, C. W., Thorne, K. S., and Wheeler, J. A. (1973). ''Gravitation.'' Freeman, San Francisco, California.

Penrose, R. (1972). *Sci. Am.* **226**(5), 38–46.

Shapiro, S. L., and Teukolsky, S. A. (1983). ''Black Holes, White Dwarfs and Neutron Stars: The Physics of Compact Objects.'' Wiley, New York.

Shipman, H. L. (1980). ''Black Holes, Quasars and the Universe,'' 2nd ed. Houghton Mifflin, Boston, Massachusetts.

Thorne, K. S. (1967). *Sci. Am.* **217**(5), 88–98.

Thorne, K. S. (1974). *Sci. Am.* **231**(6), 32–43.

CELESTIAL MECHANICS

Felix R. Hoots   *Air Force Space Command*

GLOSSARY

Angle of ascending node: Angle, measured eastward in the fundamental plane, between the ray from the attracting center to the vernal equinox and the ray from the attracting center to the ascending node; the latter ray will hereinafter be referred to as the line of ascending node.

Apocenter: Point in the orbit that is farthest from the central attracting body (called apogee for orbits about the earth and aphelion for orbits about the sun).

Argument of pericenter: Angle in the orbital plane, measured in the direction of motion, between the line of ascending node and the ray from the attracting center to pericenter; this ray will hereinafter be referred to as the pericentric line.

Celestial sphere: Imaginary sphere assumed to surround the central attracting body at an infinite distance from it.

Eccentricity: Dimensionless parameter that measures the degree of roundness of an orbit.

Equinox: Line in space through the centers of the earth and sun defined by the instant the sun appears to cross the earth's equator from south to north. This line intersects the celestial sphere in two points referred to as the vernal equinox and the autumnal equinox, respectively. If the sun's path projected on the celestial sphere is south to north as it passes through the equinox, that point is called the vernal equinox. The antipoint is called the autumnal equinox.

Fundamental plane: For planetary orbits, the plane of the earth's orbit about the sun called the ecliptic plane and, for orbits about a planet, the equatorial plane of that planet.

Inclination: Angle of tilt of the orbital plane with respect to the fundamental plane.

Line of nodes: Line of intersection of the orbital plane with the fundamental plane. The points of intersection of the orbital path with the fundamental plane are called the nodes. The point at which the orbiting body moves from south to north through the fundamental plane is called the ascending node. The antipoint is called the descending node.

Mean motion: Average angular rate with which an orbiting body sweeps through the central angle.

Pericenter: Point in the orbit that is closest to the central attracting body (called perigee for orbits about the earth and perihelion for orbits about the sun).

Perturbation: Any force that causes a small disturbance or deviation from the ideal two-body motion. Deviations proportional to the time are called secular perturbations, whereas oscillatory deviations are classified as periodic perturbations.

Semimajor axis: Half of the maximum distance across the ellipse.

True anomaly: Angle measured in the direction of motion from the pericentric line to a ray from the attracting center to the instantaneous position of the body in the orbit.

Two-body orbit: Ideal path followed by two bodies considered as point masses and moving only under their mutual gravitational attraction.

Celestial mechanics is the branch of astronomy concerned with the investigation of the motion of natural and artificial bodies in space. The specific application of this science to the motion of artificial satellites, space probes, and rockets is often called astrodynamics. Application to the forces and motions of natural celestial bodies and their dynamics is sometimes called dynamical astronomy. Such specific terminology is not universally or consistently employed, the term *celestial mechanics* often being restricted to the study of the dynamics of the solar system. However, since the dynamic and mathematical principles are universally applicable to the motion of bodies in space regardless of their origin, the term *celestial mechanics* will be taken to encompass all the branches just mentioned.

I. Historical Overview

A. EARLY ASTRONOMERS

Early celestial mechanics was concerned with the behavior of the major members of the solar system (the five visible planets, the sun, and the moon). About 3000 B.C., Babylonian priest-astrologers made crude observations of the motion of the moon and planets primarily in order to advise kings, regulate the calendar, and provide aids to travelers. They were also aware of a slow, apparent motion of the stars with respect to the passing of the first day of spring each year and invented the signs of the zodiac for tracking this precession. Neither they nor the Egyptians were able to fabricate a consistent explanation of the motion of the planets. This was due in part to their belief that the sun, moon, and planets moved on paths centered at the earth. [*See* SOLAR SYSTEM, GENERAL.]

This belief of earth-centered motion continued through the Greek period (beginning about 400 B.C.) and was further burdened by the philosophical assumptions that all motion was uniform and circular. For many years, the science of celestial mechanics was saddled with these philosophical restrictions and little progress was made. From the time of Aristotle (384–322 B.C.), it had been assumed that the planets moved along circular paths or combinations of smaller circular paths moving along larger ones (epicycles). Such explanations of planetary motion culminated in the work of Claudius Ptolemaus (A.D. 200), who represented the apparent planetary motion by epicycles centered at the earth.

Although arguments for a heliocentric (sun-centered) system had been raised as early as 270 B.C. by the Greek astronomer Aristarchus, the first consistent heliocentric theory was not put forth until the sixteenth century by Nicolaus Copernicus (1473–1543).

B. TYCHO BRAHE AND JOHANN KEPLER

In order to refine the theories of planetary motion, some new information was necessary. For many years, the Danish astronomer Tycho Brahe (1546–1601) had been meticulously collecting and recording accurate positional data for the planets. However, he was not particularly gifted in the areas of data analysis and theoretical mathematics. It is indeed fortunate that Johann Kepler (1571–1630) and Brahe happened to meet and begin working together only 18 months before Brahe's death. In contrast to Brahe, Kepler had great patience and mathematical insight and immediately recognized the wealth of analytical information waiting to be extracted from the data.

Kepler began his analysis with Brahe's observations of Mars, trying to explain them with some simple geometric model of the motion. Kepler's gift for theoretical speculation led him to three revolutionary assumptions that opened the door for the later developments of the fundamental laws of dynamics. The assumptions were that (1) Mars' orbit is a circle with the sun slightly off center, (2) the orbital motion takes place in a plane that is fixed in space, and (3) Mars does not necessarily move with uniform velocity along the circle. These fresh ideas immediately freed Kepler's analysis from the false constraints that had obstructed progress for more than 14 centuries. Since the observations were made from the earth, it was necessary to determine precisely the motion of the earth around the sun. By using the same three assumptions for the motion of the earth, Kepler was led to his first fundamental discovery. His computations showed that the earth did not move with uniform speed, but rather faster or slower inversely with the distance from the sun. This proportionality brought him to the general conclusion that a line joining the sun and a planet will sweep out equal areas in equal amounts of time. This discovery would later become known as Kepler's second law.

With all these assumptions, Kepler's analysis of the Mars data began to fall into place. All data points but two were consistent with the model.

This is where Kepler's patience, determination, and insight served him well. He decided that, in order to reconcile the data, the last of the Greek concepts of circular motion must be abandoned, and for more than a year he tried fitting various geometric curves to the data of Mars. At last, he tried the ellipse, all the data fit nearly perfectly, and Kepler's first law was born. The first two laws were published in 1609 in the *Astronomia Nova,* and a third law followed in 1619.

II. Fundamental Laws

A. KEPLER'S LAWS

A precise statement of Kepler's three laws is as follows. (1) The orbit of each planet is an ellipse with the sun at one focus of the ellipse; (2) the line joining the sun to the planet sweeps out equal areas within the ellipse in equal times; and (3) the squares of the periods of revolution of any two planets are in the same proportions as the cubes of their semimajor axes. An additional assumption made, although not stated as a separate law, was that the motion of each planet is confined to a plane that also contains the center of mass of the sun.

Today, we know that these laws are not exactly true. The first and second laws are true only for a two-body system in which the effects of planets other than the one to be modeled are ignored. The third law is true only with the additional assumption that the mass of the planet is negligible in comparison with the mass of the sun.

The ellipse of the first law is shown in Fig. 1.

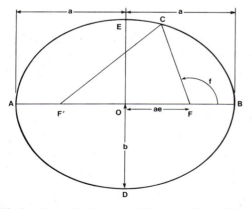

FIG. 1. Under idealized conditions, orbiting objects follow an ellipse with the central attracting body located at the focus *F*. See text for discussion.

It is a plane curve formed by the set of all points such that the sum of the distances from a point, say *C*, to two fixed points, *F* and *F'*, is a constant. Each of the two fixed points is called a focus of the ellipse, and it is at one focus, say *F*, that the central attracting body is situated. The point *B* on the orbit closest to the attracting body at *F* is called the pericenter, and the point *A*, which is farthest away, is called the apocenter. When specialized to a particular attracting body, the word *center* in pericenter and apocenter is usually altered to reflect the name of the body. For example, it is perihelion and aphelion for the sun, and it is perigee and apogee for the earth. The maximum distance across the ellipse is denoted by the line *AB* and is called the major axis. Half of this length is the semimajor axis, usually denoted by *a*. The line *DE* is called the minor axis; half of this length is called the semiminor axis and is denoted by *b*. The ratio of the focal length *OF* divided by the semimajor axis is a dimensionless parameter known as the eccentricity *e* of the ellipse. For nearly circular orbits, the offset of the focus from the center of the ellipse becomes small and the eccentricity approaches zero. For more elongated ellipses, the ratio of the distances and consequently the eccentricity approach a value of 1.

There are two other cases that were not considered by Kepler. In the limit as eccentricity becomes 1 or greater than 1, the orbits degenerate into a parabolic or hyperbolic shape, respectively. These geometric figures are open at one end and hence do not allow closed repeating motion. Some comets follow such trajectories.

For elliptic orbits, the size and shape of the orbit are completely specified by the two parameters *a* and *e*. The pericentric distance *q* can be computed from the quantities *a* and *e* by the formula $q = a(1 - e)$. It can be seen that the fractional size of the eccentricity determines the amount by which the pericentric distance is smaller than the semimajor axis. Likewise, the apocentric distance $Q = a(1 + e)$ is larger than the semimajor axis by an amount proportional to the eccentricity. For example, the planet Pluto with $e \simeq 0.25$ has perihelion distance 25% smaller and aphelion distance 25% larger than the semimajor axis of its orbit. Pluto has the most eccentric orbit of our planets, and Venus with $e \simeq 0.007$ has the most nearly circular orbit. The earth's orbit has $e \simeq 0.017$, resulting in less than a 2% variation in distance from the sun over the course of a year.

The second law states that the areas in Fig. 2

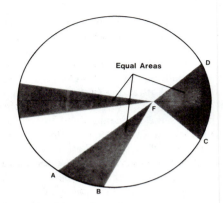

FIG. 2. The orbiting body moves in such a way that equal areas are swept out in equal amounts of time.

are all equal provided that equal times have elapsed between corresponding locations of the planets. That is, suppose a planet is at point A and 1 month later it is at point B. Now, at a later time, if the planet is at point C and 1 month later is at point D, then Kepler's second law states that areas ABF and CDF are equal. Since the points C and D are closer to F than A and B, and since the angular measure of CFD is larger than the angular measure of AFB, it can be concluded that the velocity at perihelion must be greater than elsewhere on the orbit. Likewise, the velocity at aphelion must be the least. The rate at which the planet sweeps through the central angle continually changes around the orbit. The average value of this rate is called the mean motion. It can be shown that the law of areas is true if and only if the force of attraction between the sun and a planet is directed along a line joining the two. Such a force is called a central force.

For the earth's orbit, each day a line from the sun to the earth sweeps out ~74 trillion square miles. For the month centered about perihelion passage, the earth would traverse through a central angle of ~30.6°, whereas for the month centered about aphelion passage, the angle traversed would be only ~28.6°. Because of the larger aphelion distance and the smaller perihelion distance, the differing angles traversed are counteracted and equal areas of ~2 quadrillion square miles are swept out in each of the 2 months.

The third law relates the semimajor axes of the planets to their periods of revolution around the sun. Since one-half of the sum of the pericentric and apocentric distances equals the length of the semimajor axis, the quantity a is usually considered the mean distance of the ob-

ject from the central attracting body. The orbital period P of planetary motion is related to the mean motion n by $P = 2\pi/n$. The third law states that the quantity P^2/a^3 is the same for all planets. Thus, for two planets labeled with subscripts 1 and 2, Kepler's third law gives $P_1^2/a_1^3 = P_2^2/a_2^3$. With this relationship, once the period of revolution of a planet is known, its mean distance from the sun can be computed relative to the known parameters of another planet. The earth's mean distance from the sun is called an astronomical unit (≈ 92.9 million miles) and is used as a standard unit of measure for the solar system. From observation, it is known that Jupiter's orbital period is 11.86 yr. With this value for P_2 and the values $a_1 = 1$ astronomical unit and $P_1 = 1$ yr for the earth, it is easy to compute $a_2 = 5.20$ astronomical units as the mean distance of Jupiter from the sun.

It is remarkable that Kepler was able to extract this empirical law from Brahe's observations. Later developments using sophisticated techniques in dynamics and calculus have produced the generalized form $P_1^2(m_0 + m_1)/a_1^3 = P_2^2(m_0 + m_2)/a_2^3$ for Kepler's third law, where m_0, m_1, and m_2 are the masses of the sun and two planets, respectively. When the masses of the planets are neglected in comparison with the mass of the sun, the generalized form reduces to the law originally stated by Kepler.

Kepler's laws may be considered the last significant purely kinematic contribution to celestial mechanics. His laws sought to describe the motion of the planets rather than to explain why they moved. On the other hand, Galileo (1564–1642) made a major contribution by introducing concepts of modern dynamics. His experiments revealed the fact that bodies maintain uniform motion (or are at rest) when no force acts on them. This was a radical departure from the Greek belief that rest was a natural state and any motion required force. Galileo's discovery was a major step forward and actually anticipated the essence of Newton's first two laws.

B. NEWTON'S LAWS

Without a doubt, the English scientist Isaac Newton (1642–1727) is considered the founder of the science of celestial mechanics, for celestial mechanics requires the simultaneous application of the laws of dynamics and the mathematics of calculus. It is Newton who gave us both.

After considerable analysis of Kepler's laws,

together with observational data on the motion of the moon, Newton formulated his three laws of motion, first published in his *Principia Mathematica* in 1687. The laws state that (1) any particle of matter will continue in a state of rest or a state of uniform motion in a straight line unless it is compelled by some external force to change that state, (2) the rate of change of the linear momentum (mass times velocity) of a particle is proportional to the force applied to the particle and takes place in the same direction as that force, and (3) the mutual actions of any two bodies are always equal and oppositely directed. The first law is really a corollary of the second, since if no force is applied, the momentum will not change and the velocity will remain constant.

As a consequence of these laws and with the insight provided by Kepler's laws, Newton proved that the force required to keep a planet in its orbit must obey his law of universal gravitation. This law states that two bodies attract one another with a force that is proportional to the product of the masses of the bodies, is inversely proportional to the square of the distance between the bodies, and is along the line connecting the bodies. This law is indeed universal in the sense that it is not limited to the effect of the sun on a planet. It includes the attraction of two stars, a planet on its satellite, or the earth on an apple. In order to apply the law in the last case, Newton showed that, at all points outside of a homogeneous, spherically symmetric rigid body, the gravitational force is the same as it would be were the total mass of the body concentrated at its center. Thus, to a good approximation, he could treat the sun, the planets, or the apple as a point mass. It was in deriving this result that Newton found it necessary to invent the branch of mathematics known as calculus.

Not only is the law of gravitation a universal one, but it is also unique. It has been shown that a force inversely proportional to the square of the distance is the only possible law of gravitation that both implies elliptical orbits and depends only on the distance.

III. Consequences of Newton's Laws

With Newton's laws in hand, it is not difficult to write the basic equations that must be solved to describe the motion of a body in space. The difficulties arise only in the solution of these equations and are of a mathematical nature. Progress in celestial mechanics since Newton's time has consisted almost wholly of the creation of mathematical techniques with which to contend with these equations.

While Kepler's laws give an empirical explanation of observed motion, Newton's laws provide the analytical tools to understand and derive such motion. According to the law of universal gravitation, the force of gravity is always directed toward the force center and is dependent only on the masses and separation distance. For such a force, it can be shown using Newton's second law that the motion must take place in a plane that is fixed in space and contains the force center. Although this is not given as a separate law, it was a necessary assumption made by Kepler in stating his laws. The constancy of the orbital plane immediately implies a constant rate of sweeping-out area, which is Kepler's second law. With a little more manipulation, Newton's second law yields a solution for the motion of the body in space. This is called two-body motion because it describes the motion of two isolated bodies moving under their mutual gravitational attraction. The solution not only predicts motion along an ellipse (Kepler's first law) but also allows motion along a parabola or hyperbola. The latter two are not closed curves but rather begin from infinitely far away, approach the central body passing a point of closest approach, and again recede to an infinite distance. Exactly which type of trajectory a body will follow is determined by the total amount of energy the body has when it begins the orbit. Finally, Kepler's third law follows from the constancy of sweeping-out area and the definition of mean motion.

In order to understand how an object can orbit a central body, it is useful to consider the following example. Imagine standing on a very high mountain on the earth (Fig. 3) and throwing a

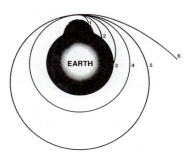

FIG. 3. By successively increasing the initial velocity, an earth orbit can be achieved.

ball horizontally out from the top of the mountain. Given a small amount of velocity, the ball might follow trajectory 1 and land partway down the mountainside. If the ball is given a somewhat greater velocity, it might follow trajectory 2 and not stop until the foot of the mountain. This is a somewhat unique and interesting situation. The ball will always fall downward and move outward in such a way that its trajectory closely follows the shape of the mountainside. That is, the perpendicular distance from the mountainside to the ball remains constant. By increasing the velocity of the ball, it is possible to achieve trajectory 3 and even trajectory 4, a circular orbit! Increasing the velocity further produces the elliptic trajectory 5, and an even further increase in velocity may result in a parabolic or hyperbolic trajectory 6, in which the ball will escape from the earth. Such trajectories are used to send space probes to the moon and planets.

In order to describe an orbit completely or to compare two orbits, some common reference frame is needed. The usual choice is a rectangular XYZ system. For planetary motion, the fundamental XY plane, called the ecliptic plane, is chosen as the orbital plane of the earth. The origin is the center of the sun and the Z axis is perpendicular to the earth's orbital plane. For orbits about a planet, the fundamental XY plane is chosen as the equatorial plane of the planet, the origin is the center of the planet, and the positive Z axis is perpendicular to the planet's equator. Both reference systems are chosen to be nonrotating by aligning the positive X axis with a relatively fixed point in space called the vernal equinox.

Usually the plane of the orbit will not lie in the XY plane. The orientation of the orbital plane in Fig. 4 can be described by two angles. The inclination i is the angle of tilt of the orbital plane

with respect to the fundamental XY plane. The angle of ascending node Ω gives the orientation with respect to the X axis of the line of intersection of the orbital plane and the fundamental plane. When the fundamental plane is the equatorial plane, this angle is called the right ascension of ascending node and is called the longitude of ascending node when using the ecliptic plane. This line of intersection is called the line of nodes of the orbit, and the point where the orbiting body ascends through the fundamental plane from south to north is called the ascending node.

To describe the orientation of the ellipse within the orbital plane, the angle measured in the direction of motion from the line of ascending node around to the pericentric line is used. This angle is the argument of pericenter ω. Finally, the true anomaly angle f locates the radius vector to the orbiting body with respect to the pericentric line.

In all there are six parameters associated with orbital motion. The semimajor axis and eccentricity describe the size and shape of the orbit. The angle of ascending node and inclination give the orientation of the orbital plane. And the argument of pericenter and true anomaly orient the ellipse within the plane and locate the orbiting body on the ellipse. Collectively, the six parameters, or some equivalent set thereof, are called the orbital elements.

For two-body motion, the true anomaly is the only parameter that varies. The remaining five orbital elements are constants. Thus, the ellipse remains fixed in space. It is much like a railroad track, and the orbiting body is like the train. The pathway is fixed, and only the position of the train can vary. Extending the analogy by putting hills in the railroad track (whose tops correspond to apocenter) and putting valleys in the railroad track (whose bottoms correspond to pericenter), it is easy to imagine that, if the engineer proceeds along with a constant throttle setting, the train will reach its maximum speed at the bottom of the valley and its minimum speed at the top of the hill. This is exactly how an orbiting body behaves as it traverses the ellipse.

The information contained in the six orbital elements is equivalent to knowing the three components of both position and velocity. There are equations for computing orbital elements from position and velocity and vice versa. The position and velocity give precise information about the instantaneous location and direction of motion of the body. On the other hand, the

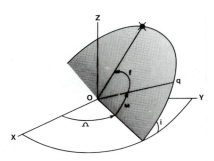

FIG. 4.   Orientation of the orbital plane and the orbiting body within the plane.

orbital elements provide a geometric picture of the pathway the body is following as well as its current location.

IV. Perturbations

A. GENERAL FORMULATION AND METHODS

The two-body solution is a concise and elegant description of idealized motion. However, in practice, it has only limited application. In general, the central body will be neither homogeneous nor perfectly spherical and thus its gravitational force will differ from Newton's universal law. In addition, other forces such as a third body may influence the motion. Such effects will generally cause small deviations from perfect two-body motion. These deviations are called perturbations, and it is with these that mathematicians have struggled for almost three centuries.

The force of a perturbation is usually small compared with the two-body force, but over a long period of time, the effect of a small perturbation can accumulate and cause significant deviations from two-body motion. Perturbative forces cause the orbital elements to change slowly from their two-body state. The rates at which they change can be expressed with a concise set of six variational equations known as the Lagrange planetary equations. These equations, obtained by Joseph Lagrange (1736–1813) for the purpose of studying the perturbations of planetary motion, are equally applicable to the study of perturbed motion of any orbiting body.

Given a set of perturbations to be considered, the perturbing forces are easily incorporated into the Lagrange variational equations. However, these equations must be solved to determine the dependence of the elements on time and thus ascertain the future evolution of the orbit. Since the time of Newton, many powerful analytical methods have been developed for attacking this problem. The ultimate goal is to obtain explicit expressions for the motion of the body or the state of its orbital elements as functions of time. Such solutions are called general perturbations solutions. In general, exact solutions can be obtained for only the simplest problems of celestial mechanics. In practice, one must resort to approximate solutions obtained by transformations of variables and series solution methods. Such solutions provide reasonable approximations of the true motion over

short time intervals and are of unquestionable value since they also provide insight into the general nature of the motion.

With the advent of high-speed electronic computers, another solution method became increasingly feasible. Special perturbations is a numerical method for solving equations. The computer can be used to carry out long, complicated, repetitive calculations very quickly and reliably. In contrast to general perturbations whose end product is a general prediction formula, special perturbations obtains a numerical result tailored to a specific problem input. However, because of the powerful computational capabilities of the computer, special perturbations can utilize more extensive force models and thus generally produce more accurate results. More recently, computers have been used to aid algebraic formula manipulation. Such tools have spawned several quite extensive general perturbations developments.

B. TYPES AND EFFECTS

Important secular perturbations of orbits of artificial satellites of the earth are due to the oblate shape of the earth. Because the earth was a hot molten mass in its early life, centrifugal forces arising from its rotation caused a deformation from a perfectly spherical shape to that of an approximate oblate spheroid having an equatorial radius ~21 km greater than its polar radius. This uneven mass distribution, or equatorial bulge, causes satellites to stray in both secular and, less importantly, periodic fashion from the idealized two-body motion. The size and character of such secular deviation are determined largely by the relative orientation (inclination) of the orbital plane to the equatorial bulge. The oblateness causes the right ascension of ascending node to decrease with time for orbits with inclination less than 90°. Thus, the orbital plane slowly rotates westward around the Z axis (see Fig. 4) while maintaining a constant inclination, and the line of nodes is said to regress. A typical time for a complete rotation is a few months. For inclinations greater than 90°, the right ascension of ascending node slowly increases with time, and the line of nodes is said to precess. For an inclination of exactly 90°, the equatorial bulge causes no motion of the orbital plane.

This perturbation property has been capitalized on in order to enhance the usefulness of certain scientific satellites. These satellites survey earth resources by photographic means. The photographic equipment has certain lighting conditions under which its operation is optimal. These conditions are determined by the orientation of the orbital plane with respect to a line joining the earth and sun. In Fig. 5, orientation A shows an orbital plane such that the satellite sees ideal noon lighting as it passes over the daylight side of the earth. However, 3 months later, at orientation B, the orbit has to have moved (its node must have precessed) by 90° in order to maintain the idealized lighting conditions. Over a complete year, the orbital plane has to precess slowly around the full 360°. By using general perturbations solutions, it is possible to compute that typically such motion will take place for an inclination of ~98°. Satellites that use this perturbation property are called sun-synchronous satellites.

Oblateness also causes a secular change in the argument of pericenter. The only stable orientation occurs for an inclination of ~63.4°, the so-called critical inclination. For inclinations less than this, the argument of pericenter slowly increases so that the location of pericenter will slowly precess around the orbit. For orbital inclinations greater than the critical inclination, the argument of pericenter slowly decreases with time. Again, a typical time for a complete transit around the orbit is a few months. Orbital planes with nearly critical inclination are very popular choices for those missions where it is desirable to maintain the low-altitude pericenter point over a particular latitude on the ground. If the satellite orbit is also such that it makes a whole number of revolutions per day, then the low point will be over a fixed ground location once per day. This is particularly useful for com-

manding and communicating from a single ground station.

The final secular effect of oblateness is in the mean motion of the satellite. For inclinations greater than ~54.7°, the mean motion will be slightly less than the two-body value and the satellite will take about a tenth of a minute longer to complete an orbit. For inclinations less than 54.7°, the mean motion will be slightly greater and the period will be shorter than for two-body motion. At the transition inclination between these regions, the period will be exactly the two-body period. In a similar manner, the sun is slightly oblate due to its rotation on its own axis. The oblateness causes analogous perturbations on the planetary orbits, but on a much larger time scale.

There are many other anomalies in the earth's gravitational field that cause satellite orbits to deviate from two-body motion. These effects are generally much smaller than the secular effects of oblateness and appear as small precessions or periodic variations about the two-body orbit.

However, there are cases known as resonance in which these effects can become important. As satellite altitude is increased, the orbital path becomes longer and, consequently, the orbital period increases. At an altitude of ~35,800 km above the earth, the orbital period is ~1 day, the same as the rotational period of the earth. For an inclination of zero degrees, the satellite seems to hover over a fixed point on the earth. Such satellites are said to be in synchronous orbits and are widely used for relaying communications. However, because such a satellite remains over the same location for a long period of time, local gravitational anomalies whose effects would normally be miniscule have time to produce a cumulative perturbation effect on the satellite. The satellite will start a slow drift away from its synchronous location and must be periodically maneuvered back to its desired longitude. If left uncorrected, the satellite will move into a resonant motion one mode of which is a slow period pendulumlike variation in longitude. Longitudes of 75° east and 105° west are stable points for synchronous satellites. All other longitudes require station-keeping maneuvers.

A second important perturbation of artificial satellites of the earth is caused by atmospheric drag. Assume that initially the satellite is in an elliptical orbit about the earth. As the satellite passes through the atmosphere, it has many collisions with molecules of air. In these collisions,

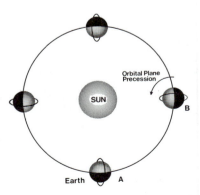

FIG. 5. Geometry of sun-synchronous satellites that use gravitational perturbations to maintain a fixed orbit–sun geometry.

a tiny bit of energy is taken from the satellite (mostly near perigee) and transferred to the air molecules. With less energy, the satellite cannot quite return to its original apogee and so, little by little, the apogee height decreases and the orbit becomes more and more circular (Fig. 6). Ultimately, the satellite will lose so much energy that it can no longer complete an orbit and will spiral into the earth. This happens to an average of more than one man-made object per day, but most of these completely disintegrate due to friction before ever reaching the earth's surface.

Many factors determine the rate at which a satellite orbit contracts. The atmospheric density and therefore the probability of satellite–molecule encounters decrease more or less exponentially as altitude increases. For satellites above ~1000 km, the density is sufficiently small that most remain aloft for hundreds of years. At the other end of the spectrum, a satellite orbiting at 100 km will generally not complete another revolution. The orbital contraction rate is directly proportional to the cross-sectional area and inversely proportional to the mass of the satellite. Other, more subtle factors such as sunspot activity also affect atmospheric density and, consequently, satellite orbital contraction rate. Because of the dynamic nature of the atmosphere, the drag perturbation is the most difficult to model and predict. A typical satellite at an altitude of 300 km will orbit for about 80–100 days before final impact, but the same satellite at 400 km will remain aloft for almost 2 yr. Such estimates obtained from general perturbations equations are important in es-

timating useful lifetimes of satellites or planning onboard fuel requirements for periodic reboosts.

The perturbative effect of a resisting medium was of interest long before the advent of artificial earth satellites. Encke (1791–1865) suggested that unexplained perturbations in the motion of Encke's comet might be accounted for by the passage of the comet through a diffuse medium surrounding the sun. Such an interplanetary medium continues to attract interest and serves as a possible explanation for the near circularity of planetary orbits in many theories of the formation and evolution of the solar system.

Gravitational forces due to bodies other than the primary are called third-body forces and can cause important perturbations in an orbit. To predict the motion of the moon about the earth with reasonable accuracy requires inclusion of the third-body force due to the sun. This is known as the main problem of lunar theory. One of the most famous problems in perturbations analysis occurred when observed variations in the motion of the planet Uranus could not be explained in terms of the then known six other planets. The English astronomer John Couch Adams and the French astronomer Jean Leverrier independently arrived at the conclusion that an eighth planet must cause the deviations. In 1845 and 1846, respectively, they computed the position of the unknown planet, observations were trained on the predicted spot, and eventually Neptune was discovered. For earth-orbiting satellites, the effect of an external body depends in a very complicated way on the relative distances to the earth and the third body as well as the geometric orientation of the two with respect to the satellite.

FIG. 6. Contraction and circularization of an orbit due to atmospheric drag.

V. Application to Artificial Satellites

A. ORBITAL COMPUTATIONS

Because of the close proximity to the attracting body, artificial earth satellites pose a particularly difficult problem in celestial mechanics. Unlike the planets and moons whose motions are much slower and more accurately long-term predictable, artificial satellite orbits evolve rapidly. Major perturbations such as oblateness and atmospheric drag must be modeled in order to predict adequately the motion for a few days. Locational data for a satellite must be frequently refined and updated. If not, one may not be able to later locate the satellite, for its orbit may have evolved away from the expected values. Unlike planets, which can be located visually and identified by physical characteristics, artificial satellites are often not visible and cannot be distinguished from one another by sight. Thus, positional information or orbital elements must be maintained on each artificial satellite from the time of its launch until its decay into the atmosphere. In 1986, for example, there were more than 6000 man-made objects orbiting the earth. This number includes operational payloads, rocket bodies, shrouds, debris, and so on, but each object is properly called a ''satellite'' of the earth. [*See* MOON (ASTRONOMY); PLANETARY SATELLITES, NATURAL.]

When a new satellite is launched, orbital elements must be established for the payload as well as the rocket body and other launch debris. An initial estimate of the orbit based on launch parameters is propagated downrange from the launch site. Observing sensors search the sky in and around the suspected trajectory. These observations are collected and used to compute an initial orbit. This is the same process used in astronomy when a new planet, moon, or comet is first observed. An orbit is computed that will fit through the observations. One of the earliest methods of initial orbit computation was given by Gauss (1777–1855). Since astronomical observations consisted of only angular directional measurements, these early methods are referred to as angles only initial orbit determinations. With the availability of radars for tracking artificial satellites, range information also became available. Several modifications and extensions of the Gauss method have been made to utilize this range information. Notable among these is the Herrick–Gibbs method. All of these methods are based on two-body mechanics and Taylor series expansions, and the end product is the two-body orbit that best matches the observations.

With a two-body orbit established, the future motion of the satellite can be approximately predicted for the next few revolutions, making possible the collection of further observations. All of these observations together represent many more data than are needed to determine the six orbital elements. This is called an overdetermined system. In order to utilize all the data, a fitting process such as least squares (invented by Gauss) is used to select mathematically the one orbit that best fits through all the available data. This process is called a differential correction,

and the resulting element set can now be used for predicting the future motion of the satellite.

Because of shortcomings in the physical models or of the mathematical solution or both, the true motion of the satellite will slowly stray from the predicted motion. This may take days, weeks, or months, but the orbital elements are time perishable and eventually must be redetermined to reflect the new state of the orbit. Again, observations are collected and a differential correction is made to produce new orbital elements to be used for future predictions. The entire process is an exercise in bootstrap celestial mechanics in which observations are used to determine orbital elements, which in turn are used to predict where to observe the satellite. Similar procedures are used to continually refine and update predicted positions for the members of the solar system.

B. ORBITAL GEOMETRY PROBLEMS

Orbital elements are used for much more than simply predicting future positions. There is a class of celestial mechanics problems that are highly geometric but whose components are basic orbits. For example, consider the problem of determining future observing opportunities for a radar site to view a given satellite. The radar site, being fixed on the earth's surface, follows a circular path in space whose center is on the rotational axis of the earth and whose plane is offset according to the latitude of the site. The satellite follows an approximate two-body el-

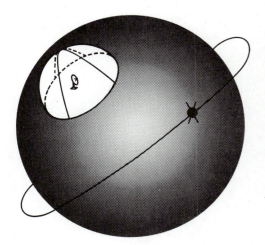

FIG. 7.   The satellite is visible to the radar only when the satellite is within the hemisphere of radar coverage.

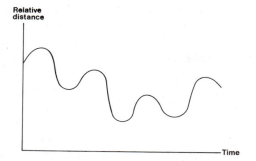

FIG. 8.   Variation with time of the relative distance between two satellites in different elliptic orbits.

lipse with focus at the center of the earth. The field of view of the radar is roughly defined by a hemispheric dome with base tangent to the surface of the earth with center of the base at the radar site and having radius equal to the range of the radar. Figure 7 illustrates this geometry. The problem to be solved involves finding the sequence of time intervals in which the satellite is within the hemisphere of radar coverage, a challenging problem in both geometry and dynamics.

Another example arises with the relative motion of two satellites. Here the problem is to find the sequence of times in which the relative distance between the two satellites achieves a local minimum. Typical variation of the relative distance with time is shown in Fig. 8. This is really a basic calculus problem where the distance function that is to be minimized is a complicated function of the position of the satellites in their respective orbits. Again, the combination of geometry and dynamics produces a challenging problem that currently has not yielded totally to analytical solution methods. Such problems are of extreme importance in planning a rendezvous or predicting possible future collisions.

Both these problems are further complicated by the fact that the orbital paths are perturbed ellipses. These examples are just a few of the many fascinating geometry/dynamics problems in celestial mechanics that have arisen and continue to surface daily because of our use of artificial satellites.

VI. Experimental Celestial Mechanics

A. Physical Experimentation

The field of experimental celestial mechanics has been made possible because of the tremen-

dous advances in computational capability of electronic computers and because of the successful launchings of earth satellites and space probes. In the past the celestial mechanics practitioner had to be content with observing the motion of the natural bodies of the solar system. Physical experimentation in astronomy was not possible. But now satellites can be launched into orbits specifically designed to obtain information about certain physical characteristics of a planet or moon. For example, by observing the rate at which the right ascension of ascending node moves, one can make a more accurate determination of the oblateness of the earth. Many of the irregularities in the earth's gravitational field have been determined from observation of the motion of satellites. To determine some of the smaller gravitational anomalies, clever orbits have been designed on the basis of analytical analysis of the equations of motion. By using orbits that very nearly make a whole number of revolutions per day, a near resonance situation is established whereby very small effects are amplified to many times their normal proportions. Such studies comprise a branch of celestial mechanics known as satellite geodesy.

Early models of the atmosphere of the earth were based on functions that decrease exponentially as altitude increases. By comparing observed contraction rates of satellite orbits with those predicted using the atmospheric models, improved values of atmospheric density can be extracted. Later atmospheric models include strong considerations of theoretical structure, but still rely on observed orbital contraction rates for proper calibration.

A high point for experimental celestial mechanics thus far has been the use of accelerometers on board satellites. These sensitive devices can measure the presence of any force on the satellite that is nongravitational. By recording the accelerations experienced, density values are obtained almost directly rather than being surmised from the long-term evolution of the orbit. Accelerometers are also used in geodesy. By including thrusting capability on the satellite, a compensating acceleration can be applied each time the accelerometer senses that a nongravitational force is influencing the motion of the satellite. In this way, the satellite can be made to follow an orbit that is a man-made simulation of drag-free motion. Then any perturbations in the orbit can be assumed to arise totally from gravitational effects and thus provide an excellent data base for satellite geodesy.

B. Numerical Experimentation

Another aspect of experimental celestial mechanics centers on the planning of scientific missions. In such instances there is some set of mission considerations to be optimized. These may include maximizing payload, minimizing fuel consumption, optimizing data collection opportunities for unmanned missions, and minimizing launch acceleration forces for manned flights. In each case, there are many trajectories that satisfy certain of these basic requirements, but the need is to pick the one that best satisfies all. There has been much success in using analytical methods to determine an optimum orbit for a scientific satellite of the earth. However, for more complicated problems such as sending a manned vehicle to the moon or a probe to a distant planet, the problem quickly exceeds the capabilities of present analytical techniques. This is where the computational power of the modern computer becomes indispensable. Once the governing forces have been included in the perturbation equations, the computer can be used to solve numerically for the future motion based on a given initial position and velocity. By systematically varying the initial values, the computer can be used to survey an enormous number of possible motions. The optimal solution can then be selected. An outstanding example of such numerical experimentation is the ongoing *Voyager* probes whose orbits not only have been selected for planetary flybys, but have also included considerations for near encounters with planetary moons.

Computer simulations have also been used to gain insight into some of the great unsolved problems in celestial mechanics. By computing many orbits over a wide spectrum of initial conditions, it has been possible to recognize regularities and patterns that give clues as to the general character of the analytical solution. Results of such studies include the discovery of the "horseshoe" and "tadpole" orbits of the asteroids.

VII. Current Celestial Mechanics

A. Modern Methods

In recent years, there has been a tremendous resurgence of interest in celestial mechanics. This is due mostly to the launching of man-made satellites, but it has also been aided by the availability of modern computing methods. The interest is no longer limited to astronomers and mathematicians, but now extends to engineers.

Modern methods being brought to bear on the problems of celestial mechanics include formalistic, quantitative, and qualitative approaches. The formalistic approach involves mathematical derivation of approximate analytical solutions by transformation of variables and series solution techniques. Great investigators from Leonard Euler (1707–1783) through those of the present day have applied these methods to produce elegant solutions to several difficult problems in celestial mechanics. With the availability of modern computer languages for formula development and manipulation, new analytical treatments are appearing for problems that heretofore were considered too complicated for analytical investigation. Use of the computer allows series expansions to be performed to very high order with unprecedented algebraic reliability and with great speed.

Quantitative methods involve numerical solution of the perturbation equations by many repetitive and tedious calculations that would be possible only with high-speed computers. Such methods do not result in general solutions to problems, but rather produce a specific numerical result corresponding to a specific set of initial conditions. Such methods are very popular in engineering applications where a limited, rather than general, analysis is sought. Some of the most impressive results to date are from a program in which satellites were used to survey the ocean. Actual positions of the spacecraft were modeled within meters by employing very high precision numerical methods. In order to achieve such accuracy, it was necessary not only to include the major effects of earth, moon, and sun gravitation and earth atmospheric drag, but also to model effects of the motion of the earth tides and solar radiation pressure. Such an exhaustive and accurate treatment could never be accomplished with a formalistic approach.

Qualitative methods are employed to obtain general characterizations of motion rather than numerical descriptions of specific motion. Although the complete solution may remain unknown, such characterizations may contribute to a better understanding of the behavior of the system. For example, for the motion of any number of bodies acting only under their mutual gravitational attraction, known as the *n*-body problem, the total energy of the system is constant. Since the total energy is the sum of the velocity-dependent kinetic energy and the position-dependent potential energy, the constancy of total energy can be used to write an equation relating the positions and velocities of the n bodies. Although such an equation does not represent an explicit solution of the problem, it does provide some insight into possible and permissible types of motion of the system. Many important qualitative results have been contributed by such masters as Henri Poincaré (1854–1912), G. D. Birkhoff (1884–1944), and A. N. Kolmogorov (1903–   ). Although such results have no immediate practical or observational application, they nevertheless provide important theoretical inputs to some of the most challenging and complicated problems in celestial mechanics.

B. State of the Art

Developments in the formalistic, qualitative, and quantitative approaches today continue to reach new pinnacles of achievement. Since the pioneering work of Brouwer in 1959, there has been steady progress in the development of analytical solutions for the motion of artificial satellites. Although the earliest treatments included only a few of the major gravitational anomalies, later works have extended the gravitational treatment and have included atmospheric drag effects. Drag modeling has proceeded from primitive exponential or power function density representations all the way to recent formalistic approaches that utilize sophisticated dynamic atmosphere models formerly used only in numerical treatments. For those earth satellites with orbits extending far enough into space to be significantly perturbed by the moon and sun, analytical solutions have been obtained that include the perturbative effects of these bodies. For those orbits that are in a near resonance situation, some work has been done to treat the effects of geopotential resonance analytically.

A relatively new hybrid approach to solving celestial mechanics problems called semianalytical methods has been made possible by modern computer resources. A problem is first approached analytically, and some transformation of variables is performed. The transformed equations are now somewhat simpler and more easily solved numerically than were the original equations. The final solution produced is thus part analytical and part numerical, that is, semianalytical. These methods have been used to produce solutions of problems involving quite extensive gravitational models, dynamic atmo-

sphere models, and perturbations of the moon and sun. Quite impressive results have been obtained that are comparable in accuracy to those obtained by totally numerical techniques.

Before the existence of artificial earth satellites, one of the most difficult problems in celestial mechanics was the analytical construction of a complete lunar theory. To meet the desired accuracy, such a theory must include not only the effects of the earth, sun, and planets, but also the perturbations due to the nonsymmetric shapes of the earth and moon. Many different approaches have been used, from the direct application of the laws of motion by Newton to the exhaustive treatment by E. W. Brown (1866–1938). Brown's theory contains more than 1500 terms and is still used today to compute lunar positions.

The most famous problem in celestial mechanics is the three-body problem, which remains unsolved. Three bodies treated as point masses move under their mutual gravitational attractions. They are given some initial motion, and the problem is to describe their subsequent motion. Describing the motion of the earth, moon, and sun system is a classic example. A multitude of papers have been published on the subject, including contributions from some of the greatest mathematicians the world has known. Yet an exact solution eludes all. An important special case of the three-body problem of practical importance is the restricted three-body problem. In this case, one of the three masses is assumed to be so small that it does not affect the motion of the other two bodies. Thus, the problem is reduced to a two-body problem whose solution is known and a perturbation problem in which the motion of the small body is perturbed as it moves within the two-body system. An example of the restricted problem is the determination of the motion of a spacecraft as it moves under the influence of the earth and moon. Even in the restricted form, the three-body problem has not been solved.

Other questions of a somewhat philosophical nature are of great interest today. Has the solar system existed in its present state ever since the planets were formed? Is it possible for planets or moons to escape their orbits or be captured by one another? These questions involve subtle concepts such as stability and require approaches other than conventional analysis techniques. Existing formalistic solutions for planetary motion are adequate only for a few hundred years because they involve truncated series expansions. Even numerical treatments eventually fail since the longer the prediction, the greater is the number of computations involved and consequently the greater is the accumulation of roundoff errors and the lower is the accuracy.

With little hope for success with the formalistic or quantitative approaches to the problem of planetary motion just cited, most efforts have been directed toward the qualitative approach. New mathematical methods developed in the 1960s have made it possible to make characterizations of the motion that are valid for all time. For instance, the Russian mathematician V. I. Arnold proved that if the sun is sufficiently massive compared with its planets, and if the planetary orbits are sufficiently circular at inclinations sufficiently small, then such a configuration can exist for all time. Although it is difficult to assign a quantitative meaning to the word *sufficiently,* this is still considered an important step toward unraveling the dynamic mysteries of our universe. As we reach ever farther into space with both spacecraft and telescope, new questions unfold daily providing the celestial mechanics practitioner with an endless arena for exploration and analysis.

BIBLIOGRAPHY

Brouwer, D., and Clemence, G. M. (1961). "Methods of Celestial Mechanics." Academic Press, New York.
Danby, J. M. A. (1962). "Fundamentals of Celestial Mechanics." Macmillan, New York.
Fitzpatrick, P. M. (1970). "Principles of Celestial Mechanics." Academic Press, New York.
Liu, J. (1980). *J. Astronaut. Sci.* **3** (4), 304–311.
McCuskey, S. W. (1963). "Introduction to Celestial Mechanics." Addison-Wesley, Reading, Massachusetts.
Pollard, H. (1976). "Celestial Mechanics." The Mathematical Association of America, Washington, D.C.
Roy, A. E. (1982). "Orbital Motion." Adam Hilger Ltd., Bristol.
Szebehely, V. (1967). "Theory of Orbits: The Restricted Problem of Three Bodies." Academic Press, New York.

COSMIC INFLATION

Edward P. Tryon *Hunter College and The City University of New York*

GLOSSARY

Comoving: Moving in unison with the overall expansion of the universe.

Cosmic scale factor: Any two comoving points are separated by a constant fraction of the cosmic scale factor $a(t)$, which governs the scale of cosmic distances and grows as the universe expands.

Cosmological constant: Physical constant Λ that governs the speculative cosmological term in the field equations of general relativity, corresponding to an effective force that increases in proportion to distance and is repulsive for positive Λ.

Critical density: Curvature of space is positive, zero, or negative according to whether the average density of mass-energy is greater than, equal to, or less than a critical density determined by the Hubble parameter and cosmological constant.

Curvature constant: A constant k that is $+1, 0,$ or -1 for positive, zero, or negative spatial curvature, respectively.

Doppler effect: Change in observed frequency of any type of wave due to motion of the source relative to the observer: motion of the source away from an observer decreases the observed frequency by a factor determined by the relative velocity.

False vacuum: Metastable or unstable state predicted by grand unified theories, in which space has pressure P and mass-energy density ρ related by $P = -c^2\rho$; the gravitational effects mimic a positive cosmological constant.

Field: Type of variable representing quantities that may, in principle, be measured (or theoretically analyzed) at any point of space–time; hence, fields possess values (perhaps zero in some regions) at every point of space–time.

Higgs fields: In unified theories, field quantities that, when all are zero, have positive energy density and negative pressure giving rise to false vacuum. At least one Higgs field must be nonzero in the state of minimum energy (true vacuum), which breaks an underlying symmetry of the theory.

Horizon distance: Continually increasing distance beyond which neither light nor any causal influence could yet have traveled since the universe began.

Hubble parameter: Ratio H of the recessional velocity of a comoving object to its distance from a comoving observer; by Hubble's law, this ratio has the same value for all such objects and observers. Its present value is called *Hubble's constant*.

Robertson–Walker (RW) metric: In homogeneous, isotropic, cosmological models, describes space–time and its evolution in terms of a curvature constant and (evolving) cosmic scale factor.

Second-rank tensor: A 4×4 array (i.e., matrix) of quantities whose values depend, in a characteristic way, on the reference frame in which they are evaluated.

Stress-energy tensor: Second-rank tensor whose components include the energy and momentum densities, pressure, and other stresses.

Work-energy theorem: Any change in energy of a system equals the work done on it; as a corollary, the energy of an isolated system is constant.

Cosmic inflation refers to a conjectured period of accelerated, exponential growth during which the universe is believed to have increased in linear dimensions by a factor exceeding 10^{29}. The inflationary period ended when the universe was $\sim 10^{-30}$ seconds old, after which cosmic evolution proceeded in accord with the standard big bang model. Inflation is based on the general theory of relativity together with grand unified theories of the elementary particles: the latter predict a peculiar state called the *false vacuum,* which rendered gravitation repulsive and thereby caused inflation. Several cosmic puzzles are resolved by the conjecture.

I. Einstein's Theory of Gravitation

In Newton's theory of gravitation, mass is the source of gravity and the resulting force is universally attractive. The empirical successes of Newton's theory are numerous and familiar. All observations of gravitational phenomena prior to this century were consistent with Newtonian theory, with one exception: a minor feature of Mercury's orbit. Precise observations of the orbit dating from 1765 revealed an advance of the perihelion amounting to 43 arc sec per century in excess of that which could be explained by the perturbing effects of other planets. Efforts were made to explain this discrepancy by postulating the existence of small, perturbing masses in close orbit around the sun, and this type of explanation retained some plausibility into the early part of the twentieth century. Increasingly refined observations, however, have ruled out such a possibility: there is not sufficient mass in close orbit around the sun to explain the discrepancy.

In 1907, Einstein published the first in a series of papers in which he sought to develop a relativistic theory of gravitation. From 1913 onward, he identified the gravitational field with the metric tensor of Riemannian space–time. These efforts culminated in his general theory of relativity (GTR), which was summarized and published in its essentially final form in 1916. The scope and power of the theory made possible the construction of detailed models for the entire universe, and Einstein published the first such model in 1917. This early model was subsequently retracted, but the science of mathematical cosmology had arrived. [*See* COSMOLOGY; RELATIVITY, GENERAL.]

GTR reproduces the successful predictions of Newtonian theory and furthermore explains the perihelion advance of Mercury's orbit. Measurement of the bending of starlight in close passage by the sun during the solar eclipse of 1919 resulted in further, dramatic confirmation of GTR, which has since been accepted as the standard theory of gravitation. Additional successes of GTR include correct predictions of gravitational time dilation and, beginning in 1967, the time required for radar echoes to return to Earth from Mercury and artificial satellites.

There are no macroscopic phenomena for which the predictions of GTR are known (or suspected) to fail. Efforts to quantize GTR (i.e., to wed GTR with the principles of quantum theory) have met with continuing frustration, primarily because the customary perturbative method for constructing solutions to quantum equations yields divergent integrals whose infinities cannot be canceled against one another (the quantized version of GTR is not renormalizable). Hence there is widespread belief that microscopic gravitational phenomena involve one or more features differing from those of GTR in a quantized form. In the *macroscopic* realm, however, any acceptable theory of gravitation will have to reproduce the successful predictions of GTR. It is reasonable (though not strictly necessary) to suppose that all the macroscopic predictions of GTR, including those that remain untested, will be reproduced by any theory that may eventually supersede it.

For gravitational phenomena, the obvious borderline between macroscopic and microscopic domains is provided by the *Planck length* (L_P):

$$L_P \equiv (Gh/c^3)^{1/2} = 4.0 \times 10^{-33} \text{ cm} \qquad (1)$$

where G is Newton's constant of gravitation, h is Planck's constant, and c is the speed of light. Quantum effects of gravity are expected to be significant over distances comparable to or less than L_P. According to the wave–particle duality of quantum theory, such short distances are probed only by particles whose momenta exceed Planck's constant divided by L_P. Such particles are highly relativistic, so their momenta equal their kinetic energies divided by c. Setting their energies equal to a thermal value of order kT (where k is Boltzmann's constant and T the absolute temperature), the *Planck temperature* (T_P) is obtained:

$$T_P \equiv ch/kL_P = 3.6 \times 10^{32} \text{ K} \qquad (2)$$

where "K" denotes kelvins (degrees on the absolute scale). For temperatures comparable to or greater than T_P, effects of quantum gravity may be important.

The fundamental constants of quantum gravity can also be combined to form the *Planck mass* (M_P):

$$M_P \equiv (hc/G)^{1/2} = 5.5 \times 10^{-5} \text{ g} \qquad (3)$$

The Planck mass and length give rise to the *Planck density* ρ_P:

$$\rho_P \equiv M_P/L_P^3 = 8.2 \times 10^{92} \text{ g/cm}^3 \qquad (4)$$

For mass densities comparable to or greater than ρ_P, the effects of quantum gravity are expected to be significant. (Definitions of L_P, T_P, M_P, and ρ_P sometimes differ from those given here by factors of order unity, but only the orders of magnitude are important in most applications.)

According to the standard big bang model, the very early universe was filled with an extremely hot, dense gas of elementary particles and radiation in (or near) thermal equilibrium. As the universe expanded, the temperature and density both decreased. Calculations indicate that the temperature fell below T_P when the universe was $\sim 10^{-45}$ seconds old, and the density fell below ρ_P at $\sim 10^{-44}$ sec. Hence currently unknown effects of quantum gravity may have been important during the first $\sim 10^{-44}$ sec or so of cosmic evolution. For all later times, however, the temperature and density were low enough that GTR should provide an adequate description of gravitation, and our theories of elementary particles together with thermodynamics and other established branches of physics should be applicable in a straightforward way. We shall assume this to be the case.

As remarked earlier, GTR identifies the gravitational field with a second-rank tensor, the metric tensor of curved space–time. The source of gravitation in GTR is therefore not simply mass (nor mass-energy), but another second-rank tensor: the stress-energy tensor. Hence stresses contribute to the source of gravitation in GTR, the simplest example of stress being pressure.

It is commonly said that gravitation is a universally attractive force, but such a sweeping statement is not supported by the field equations of GTR. It is true that the kinds of objects known (or believed) to constitute the present universe attract each other gravitationally, but GTR contains another possibility: granted an extraordinary medium wherein pressure is negative and sufficiently large, the resulting gravitational field would be repulsive. This feature of GTR lies at the heart of cosmic inflation: it is conjectured that during a very early and brief stage of cosmic evolution, space was characterized by a state called a *false vacuum* with a large, negative pressure. A false vacuum would mimic the effects of a positive cosmological term in the field equations, as will be described in some detail in Section V,D. The resulting gravitational field would have been repulsive, and cosmic dimensions would have grown exponentially in time during this brief period called the *inflationary era*.

The concept of negative pressure warrants a brief explanation. A medium characterized by positive pressure, such as a confined gas, has an innate tendency to expand. In contrast, a medium characterized by negative pressure has an innate tendency to contract. For example, consider a solid rubber ball whose surface has been pulled outward in all directions, stretching the rubber within. More generally, for a medium characterized by pressure P, the internal energy U contained within a variable volume V is governed by $\Delta U = -P \, \Delta V$ for adiabatic expansion or contraction (a result of the work-energy theorem, where ΔU and ΔV denote small changes in U and V, respectively). If U increases when V increases, then P is negative.

The possibility of an early false vacuum with resulting negative pressure and cosmic inflation is suggested by grand unified theories (GUTs) of the elementary particles and their interactions, whose relevant features will be described in Section VI,C. The resulting inflation would have increased cosmic linear dimensions by a factor of $\sim 10^{30}$ over a period of 10^{-32} sec and by $\sim 10^{300}$ if it persisted for 10^{-31} sec. Even greater inflation is plausible.

It should be noted that GUTs constitute a class of speculative theories. At present it is not known which of them is correct, or whether any of them is correct. Hence an era of cosmic inflation is distinctly speculative. We shall see, however, that inflation would explain several observed features of the universe that have heretofore defied explanation. Cosmic inflation also makes one quantitative prediction, which is potentially subject to empirical confirmation or refutation: namely, that the average curvature of space is indistinguishable from zero within the observable universe. As a corollary, the average density of mass-energy is indistinguishable from a *critical density* that arises in the cosmological models of GTR.

We shall next enumerate and briefly describe

some heretofore unexplained features of the universe that may be understood in terms of cosmic inflation, and then we will proceed to a detailed presentation of the theory.

II. Cosmological Puzzles

A. THE HORIZON PROBLEM

Both the special and general theories of relativity preclude the transmission of any causal agent at a speed faster than light. Accepting the evidence that our universe has a finite age, each observer is surrounded by a *particle horizon*: no particle beyond this horizon could have yet made its presence known to us in any way, for there has not yet been sufficient time for light (or anything else) to have traveled the intervening distance. Regions of the universe outside each others' horizons are causally disconnected, having had no opportunity to influence each other. The distance to the horizon naturally increases with the passage of time. In the standard big bang model, an order-of-magnitude estimate for the horizon distance is simply the speed of light multiplied by the time that has elapsed since the cosmos began. (Modifications to this simple estimate arise because the universe is expanding and because space–time is curved, but such details will not be expounded here.)

Astronomical observations entail the reception and interpretation of electromagnetic radiation, that is, visible light and, in recent decades, radiation from nonvisible parts of the spectrum such as radio waves and x rays. We shall refer to that portion of the cosmos from which we are now receiving electromagnetic radiation as the *observable universe*; it is that fraction of the whole which may be studied by contemporary astronomers. With the observable universe so defined, it contains a definite (though not precisely known) number of material particles forming gas, dust, stars, galaxies, etc. The cosmos has been expanding since its earliest moments, so the matter contained within the observable universe spanned a smaller volume of space in the past than at present; the radius of the observable universe has increased in unison with the cosmic expansion.

The universe was virtually opaque to electromagnetic radiation for times earlier than ~700,000 yr (as will be explained). Therefore, the present radius of the observable universe is less than the present horizon distance— ~700,000 light-years less in the standard big bang model, or very much less if cosmic inflation occurred at the very early time characteristic of inflationary models. The cosmos is currently some 10 to 20 billion years old, so the present radius of the observable universe is on the order of 10 to 20 billion light-years. (The precise value depends somewhat on the rate of cosmic expansion and on the curvature of space–time, but these details will not be pursued here.) Approximately 10^{10} galaxies are believed to lie within the observable universe, of which ~10^9 lie within range of existing telescopes.

If the observable universe is hypothetically divided into equal volumes of sufficiently large size (with linear dimensions on the order of 10^8 to 10^9 light-years), approximately equal numbers of galaxies are observed within almost all such volumes yet examined. The observable universe contains a large number (~10^5) of such volumes, so it is meaningful to say that the distribution of matter is approximately homogeneous and isotropic when averaged over sufficiently large distances.

More dramatic and precise evidence for isotropy is provided by the cosmic background radiation, whose existence and properties were anticipated in work on the early universe by George Gamow and collaborators in the late 1940s. These investigators assumed that the early universe was very hot, in which case a calculable fraction of the primordial protons and neutrons would have fused together to form several light species of nuclei (^{2}D, ^{3}He, ^{4}He, and ^{7}Li) during the first few minutes after the big bang. (Heavier nuclei are believed to have formed later, in the cores of stars and during stellar explosions.) The abundances predicted for these light nuclei agreed with observations, which confirmed the theory of their formation. The success of this theory constituted the first evidence that the early universe *was* hot, and it enabled investigators to infer the early temperature.

After this early period of nucleosynthesis, the universe was filled with a hot plasma (ionized gas) of light nuclei and electrons, together with a thermal distribution of photons. When the universe had expanded and cooled for ~700,000 yr, the temperature dropped below 3000 K, which enabled the nuclei and electrons to combine and form electrically neutral atoms that were essentially transparent to radiation. (At earlier times the plasma was virtually opaque to electromagnetic radiation, because the photons could travel only short distances before being scattered by charged particles: the observable universe extends to the distance where we now see it as it was at the age of 700,000 yr.) The thermal photons that existed when the atoms formed traveled freely through the resulting gas, and later through the vast reaches of empty space between the stars and galaxies that eventually formed out of that gas. These photons have survived to the present, and they bathe every observer who looks toward the heavens. (If a television set is tuned to a channel with no local station and the brightness control is turned to its lowest setting, ~1% of the flecks of ''snow'' on the screen are a result of this cosmic background radiation.) [See COSMIC RADIATION.]

The photons now reaching earth have been traveling for ~15 billion yr and hence were emitted by distant parts of the universe receding from us with speeds near that of light. The Doppler effect has redshifted these photons, but preserved the blackbody distribution; hence the radiation remains characterized by a temperature.

Refining the earlier work of Gamow and collaborators, R. A. Alpher and R. C. Herman predicted in 1950 that the present temperature of the residual, cosmic background radiation should be ~5 K. In 1964, Arno A. Penzias and Robert W. Wilson accidently discovered this radiation (at a single wavelength) with a large horn antenna built to observe the Echo telecommunications satellite. Subsequent observations by numerous workers established that the spectrum is indeed blackbody, with a temperature of 2.7 K. The present consensus in favor of a hot big bang model resulted from this discovery of the relic radiation.

Remarkably, the cosmic background radiation is found to be isotropic to a precision of about one part in 10^4 (after the earth's velocity relative to the distant sources of radiation has been factored out; a slight deviation from isotropy at a level of roughly one part in 10^4 is suggested by some data, but such a deviation would be near the limit of observational precision and is not regarded as established.) This high degree of isotropy becomes all the more striking when one realizes that the radiation now reaching earth from opposite directions in the sky was emitted from regions of the cosmos that were 90 times farther apart than the horizon distance at the time of emission, assuming the horizon distance was that implied by the standard big bang model. How could different regions of the cosmos that had never been in causal contact have been so precisely similar?

Of course the theoretical description of any dynamical system requires a stipulation of initial conditions; and one might simply postulate that the universe came into being with perfect isotropy as an initial condition. An initial condition so special and precise begs for an explanation, however; this is called the *horizon problem*.

B. THE FLATNESS PROBLEM

In a sweeping extrapolation beyond the range of observations, it is often assumed that the entire cosmos is homogeneous and isotropic. This assumption is sometimes called the *Cosmological Principle*, a name that is perhaps misleading in suggesting greater sanctity for the assumption than is warranted either by evidence or by logic. (Indeed, current scenarios for cosmic inflation imply that the universe is distinctly different at some great distance beyond the horizon, as will be explained in Section VII,B.) The existence of a horizon implies, however, that the observable universe has evolved in a manner independent of very distant regions. Hence the evolution of the observable universe should be explicable in terms of a model that assumes the universe to be the same everywhere. In light of this fact and the relative simplicity of models satisfying the Cosmological Principle, such models clearly warrant study.

In 1922, Alexandre Friedmann discovered three classes of solutions to the field equations of GTR that fulfill the Cosmological Principle. These three classes exhaust the possibilities (within the stated assumptions), as was shown in the 1930s by H. P. Robertson and A. G. Walker. The three classes of Friedmann models (also called Friedmann–Robertson–Walker or Friedmann–Lemaître models) are distinguished from one another by the curvature of space, which may be positive, zero, or negative. The corresponding spatial geometries are *spherical, flat,* and *hyperbolic,* respectively. Einstein's field equations correlate the spatial geometry with the average density ρ of mass-energy in the universe. There is a critical density ρ_c such that spherical, flat, and hyperbolic geometries correspond to $\rho > \rho_c$, $\rho = \rho_c$, and $\rho < \rho_c$, respectively. A spherical universe has finite volume (space curves back on itself), and is said to be ''closed''; flat and hyperbolic universes have infinite volumes and are said to be ''open.''

The critical density depends on the rate at which the universe is expanding and, if the field equations contain a cosmological term, on the value of the cosmological constant Λ as well. The cosmological term is often assumed to be absent (i.e., $\Lambda = 0$), and we shall adopt this assumption while describing the flatness problem in this section. (In Sections IV and V, some consequences of a cosmological term will be considered, but a nonzero value is speculative and furthermore complicates the flatness problem beyond the scope of this article.)

With $\Lambda = 0$, a closed universe would expand to a maximum size in a finite time and subsequently contract (the "big bang" would be followed by a "big crunch"). The present expansion would continue forever in both of the open cases, but with a difference. In the flat case, the recessional velocity of each galaxy would tend to zero in the limit of infinite time, whereas each such velocity would tend toward a nonzero limit in the hyperbolic case.

Astronomical observations have revealed no departures from Euclidean geometry (on a cosmic scale). While less precise than desired, they are consistent with a flat universe. In addition, the mass-energy density that has been observed, or inferred from the dynamics of galaxies and galactic clusters, is not very different from the critical density. Denoting the ratio of actual to critical densities (with $\Lambda = 0$) by Ω (i.e., $\Omega \equiv \rho/\rho_c$), it is fairly certain that Ω lies between 0.1 and 2.0 at present.

That Ω is near unity is somewhat striking, and becomes much more so in light of a theoretical observation made by Robert H. Dicke and P. James E. Peebles in 1979. If Ω were not precisely unity when the universe came into being, any deviation from unity would have grown rapidly with the passage of time. In order for Ω now to lie between 0.1 and 2.0, some 10 to 20 billion yr after the big bang, Ω could not have differed from unity by more than one part in 10^{15} when the universe was 1 sec old nor by more than one part in 10^{58} at a time of 10^{-44} sec. Of course Ω may be assumed to have had any value whatever when the universe came into being, as an initial condition, but such a special value begs for explanation: this is called the *flatness problem*.

The flatness problem may also be expressed in terms more accessible to intuition. If Ω is precisely unity, the question is simply, Why is the universe flat? If the universe is closed, that is, if $\Omega > 1$, it may readily be shown that the duration (expansion–contraction time) and maximum size (at midlife) are governed by Ω. There is no lower limit to the duration and size of a closed universe; both may be arbitrarily small. For a closed universe to be long-lived and become

large, as ours surely has, it is necessary (and sufficient) for Ω to have been extremely close to unity at early times. Hence if the universe is closed, the flatness problem is equivalent to the question, Why is the universe so large and long-lived?

For a hyperbolic universe, the density of matter can be arbitrarily small. Both the actual density and critical density decrease monotonically with time, because of the expansion; for less obvious reasons, their ratio Ω also tends to zero. If the universe is hyperbolic, why is it not more nearly empty at present (relative to the critical density), especially since some 10 to 20 billion yr have passed since the big bang?

C. THE SMOOTHNESS PROBLEM

Although the observable universe is homogeneous on a large scale, it is definitely lumpy on smaller scales: the matter is concentrated into stars, galaxies, and clusters of galaxies. The horizon problem concerns the establishment of large-scale homogeneity, whereas the smoothness problem concerns the formation of galaxies and the chain of events that resulted in the observed sizes of galaxies. That a problem exists may be seen in the following way.

A homogeneous distribution of matter is rendered unstable to clumping by gravitational attraction. Any chance concentration of matter will attract surrounding matter to it in a process that feeds on itself. This phenomenon was first analyzed by Sir James Jeans, and is called the *Jeans instability*. If the universe has evolved in accord with the standard big bang model since the earliest moment that we presume to understand (i.e., since the universe was 10^{-44} seconds old), then the inevitable thermal fluctuations in density at that early time would have grown much too large by now to be compatible with the observed sizes of galaxies and galactic clusters. Hence even if the universe had been as homogeneous as possible at the age of 10^{-44} sec, the thermal fluctuations and the Jeans instability would have resulted in greater clumping of matter than has in fact occurred, assuming the standard model for cosmic evolution has been valid throughout the intervening time. This observation arises from work by Peebles in 1968, and is called the *smoothness problem*.

D. OTHER PROBLEMS

In addition to the questions or problems thus far described, there are others of a more speculative nature that may be resolved by cosmic inflation. We shall enumerate and briefly explain them here. The first is a problem that arises in the context of GUTs.

GUTs predict that a large number of magnetic monopoles (isolated north or south magnetic poles) would have been produced at a very early time (when the universe was $\sim 10^{-35}$ seconds old). If the universe had expanded in the simple way envisioned by the standard big bang model, then the present abundance of monopoles would be vastly greater than is consistent with observations; this is called the *monopole problem* (see Sections VI,C and VII,A).

The remaining questions and problems that we shall mention concern the earliest moment(s) of the cosmos. In the standard big-bang model, one assumes that the universe began in an extremely dense, hot state that was expanding rapidly. In the absence of quantum effects, the density, temperature, and pressure would have been infinite at time zero, a seemingly incomprehensible situation sometimes referred to as the *singularity problem*. Currently unknown effects of quantum gravity might somehow have rendered finite these otherwise infinite quantities at time zero. It would still be natural to wonder, however, why the universe came into being in a state of rapid expansion, and why it was intensely hot.

The early expansion was not a result of the high temperature and resulting pressure, for it can be shown that a universe born stationary but with arbitrarily high (positive) pressure would have begun immediately to contract under the force of gravity (see Section IV,E). We know that any initial temperature would decrease as the universe expands, but the equations of GTR are consistent with an expanding universe having no local random motion: the temperature could have been absolute zero for all time. Why was the early universe hot, and why was it expanding?

A final question lies at the very borderline of science, but has recently become a subject of scientific speculation and even detailed model-building: How and why did the big bang occur? Is it possible to understand, in scientific terms, the creation of a universe *ex nihilo* (from nothing)?

The theory of cosmic inflation plays a significant role in contemporary discussions of all the aforementioned questions and/or problems, appearing to solve some while holding promise for resolving the others. Some readers may wish to understand the underlying description of cosmic

geometry and evolution provided by GTR. Sections III, IV, and V develop the relevant features of GTR in considerable detail, utilizing algebra and, occasionally, some basic elements of calculus. Most of the equations are explained in prose in the accompanying text, so that a reader with a limited background in mathematics should nevertheless be able to grasp the main ideas. Alternatively, one might choose to skip these sections (especially in a first reading) and proceed directly to Section VI on unified theories and Higgs fields, without loss of continuity or broad understanding of cosmic inflation (the glossary may be consulted for the few terms introduced in the omitted sections).

III. Geometry of the Cosmos

A. METRIC TENSORS

General relativity presumes that special relativity is valid to arbitrarily great precision over sufficiently small regions of space–time. GTR differs from special relativity, however, in allowing for space–time to be *curved* over extended regions. (Curved space–time is called *Riemannian,* in honor of the mathematician G. F. B. Riemann who, in 1854, made a seminal contribution to the theory of curved spaces.) Another difference is that while gravitation had been regarded as merely one of several forces in the context of special relativity, it is conceptually not a force at all in GTR. Instead, all effects of gravitation are imbedded in, and expressed by, the curvature of space–time.

In special relativity, there is an *invariant interval* s_{12} separating events at t_1, $\mathbf{r}_1$ and t_2, $\mathbf{r}_2$ given by

$$
\begin{aligned}
(s_{12})^2 = &-c^2(t_1 - t_2)^2 + (x_1 - x_2)^2 \\
&+ (y_1 - y_2)^2 + (z_1 - z_2)^2
\end{aligned}
\tag{5}
$$

where x, y, and z denote Cartesian components of the position vector $\mathbf{r}$. [Some scientists define the overall sign of $(s_{12})^2$ to be opposite from that chosen here, but the physics is not affected so long as consistency is maintained.] The Euclidean nature of the space of special relativity is expressed by the fact that the value of s_{12} does not change when the coordinates x, y, and z are evaluated in a reference frame that is rotated and/or displaced (but stationary) relative to the initial frame.

The remarkable relationship between space and time in special relativity is expressed by the fact that the value of s_{12} does not change when t, x, y, and z are evaluated in a reference frame that is moving relative to the initial frame. Hence the invariance of the interval given by Eq. (5) is equivalent to a specification of the geometric properties of the space–time of special relativity (called *Minkowski space–time*, after Hermann Minkowski).

The most familiar example of a curved space is the surface of a sphere. It is well-known that the surface of a sphere cannot be covered with a coordinate grid wherein all intersections between grid lines occur at right angles (the lines of longitude on earth are instructive). Similarly, a curved space–time of arbitrary (nonzero) curvature cannot be spanned by any rectangular set of coordinates such as those appearing in Eq. (5). One can, however, cover any small portion of a spherical surface with approximately rectangular coordinates, and the rectangular approximation can be made arbitrarily precise by choosing the region covered to be sufficiently small. In a like manner, any small region of a curved space–time can be spanned by rectangular coordinates, with the origin placed at the center of the region in question. The use of such coordinates (called *locally flat* or *locally inertial*) enables one to evaluate any infinitesimal interval ds between nearby space–time points by application of Eq. (5).

To span an extended region of a curved space–time with a single set of coordinates, it is necessary that the coordinate grid be curved. Since one has no *a priori* knowledge of how space–time is curved, the equations of GTR are typically expressed in terms of arbitrary coordinates x^μ ($\mu = 0, 1, 2,$ and 3). The central object of study becomes the *metric tensor* $g_{\mu\nu}$, defined implicitly by

$$(ds)^2 = g_{\mu\nu}dx^\mu dx^\nu \qquad (6)$$

together with the assumption that space–time is *locally* Minkowskian (and subject to an axiomatic proviso that $g_{\mu\nu}$ is symmetric, i.e., invariant under interchange of its two indices).

In principle, one can determine $g_{\mu\nu}$ at any given point by using Eq. (5) with locally flat coordinates to determine infinitesimal intervals about that point, and then requiring the components of $g_{\mu\nu}$ to be such that Eq. (6) yields the same results in terms of infinitesimal changes dx^μ in the arbitrary coordinates x^μ. The values so determined for the components of $g_{\mu\nu}$ will clearly depend on the choice of coordinates x^μ, but it can be shown that the variation of $g_{\mu\nu}$ over an extended region also determines (or is deter-

mined by) the curvature of space–time. It is customary to express models for the geometry of the cosmos as models for the metric tensor, and we shall do so here.

B. THE ROBERTSON–WALKER METRIC

Assuming the universe to be homogeneous and isotropic, coordinates can be chosen such that

$$(ds)^2 = -c^2(dt)^2 + a^2(t)\left\{\frac{(dr)^2}{1 - kr^2}\right.$$
$$\left. + r^2[(d\theta)^2 + \sin^2\theta(d\varphi)^2]\right\} \qquad (7)$$

as was shown in 1935 by H. P. Robertson and, independently, by A. G. Walker. The ds of Eq. (7) is called the *Robertson–Walker* (RW) *line element,* and the corresponding $g_{\mu\nu}$ (with coordinates x^μ identified as ct, r, θ, and φ) is called the *Robertson–Walker metric*. The cosmological solutions to GTR discovered by Friedmann in 1922 are homogeneous and isotropic, and may therefore be described in terms of the RW metric.

The spatial coordinates appearing in the RW line element are *comoving*; that is, r, θ, and φ have constant values for any galaxy or other object moving in unison with the overall cosmic expansion (or contraction). Thus t corresponds to proper (ordinary clock) time for comoving objects. The radial coordinate r and the constant k are customarily chosen to be dimensionless, in which case the *cosmic scale factor* $a(t)$ has the dimension of length. The coordinates θ and φ are the usual polar and azimuthal angles of spherical coordinates. Cosmic evolution in time is governed by the behavior of $a(t)$. [Some authors denote the cosmic scale factor by $R(t)$, but we shall reserve the symbol R for the curvature scalar appearing in Section IV,A.]

The constant k may be any real number. If $k = 0$, then three-dimensional space for any single value of t is Euclidean, in which case neither $a(t)$ nor r is uniquely defined: their product, however, equals the familiar radial coordinate of flat-space spherical coordinates. If $k \neq 0$, we may define

$$\bar{r} \equiv |k|^{1/2}r \quad \text{and} \quad \bar{a}(t) \equiv a(t)/|k|$$

in terms of which the RW line element becomes

$$(ds)^2 = -c^2(dt)^2 + \bar{a}^2(t)\left\{\frac{(d\bar{r})^2}{1 \mp \bar{r}^2}\right.$$
$$\left. + \bar{r}^2[(d\theta)^2 + \sin^2\theta(d\varphi)^2]\right\} \qquad (8)$$

where the denominator with ambiguous sign corresponds to $1 - (k/|k|)\bar{r}^2$. Note that Eq. (8) is precisely Eq. (7) with $a = \bar{a}$, $r = \bar{r}$, $k = \pm 1$. Hence there is no loss of generality in restricting k to the three values 1, 0, and -1 in Eq. (7), and we shall henceforth do so (as is customary). These values for k correspond to positive, zero, and negative spatial curvature, respectively.

C. The Curvature of Space

If $k = \pm 1$, then the spatial part of the RW line element differs appreciably from the Euclidean case for objects with r near unity, so that the kr^2 term cannot be neglected. The proper (comoving meter-stick) distance to objects with r near unity is of order $a(t)$. Hence $a(t)$ is roughly the distance over which the curvature of space becomes important. This point may be illustrated with the following examples.

The proper distance $r_p(t)$ to a comoving object with radial coordinate r is

$$r_p(t) = a(t) \int_0^r \frac{dr'}{(1 - kr'^2)^{1/2}} \tag{9}$$

The integral appearing in Eq. (9) is elementary, being $\sin^{-1} r$ for $k = +1$, r for $k = 0$, and $\sinh^{-1} r$ for $k = -1$. Thus the radial coordinate is related to proper distance and the scale factor by $r = \sin(r_p/a)$ for $k = +1$, r_p/a for $k = 0$, and $\sinh(r_p/a)$ for $k = -1$.

Two objects with the same radial coordinate but separated from each other by a small angle $\Delta\theta$ are separated by a proper distance given by Eq. (7) as

$$\Delta s = ar \, \Delta\theta \tag{10}$$

For $k = 0$, $r = r_p/a$ and we obtain the Euclidean result $\Delta s = r_p \, \Delta\theta$. For $k = +1$ (or -1), Δs differs from the Euclidean result by the extent to which $\sin(r_p/a)$ [or $\sinh(r_p/a)$] differs from r_p/a. Series expansions yield

$$\Delta s = r_p \, \Delta\theta \left\{ 1 - \frac{k}{6} (r_p/a)^2 \right.$$
$$\left. + \mathcal{O}[(r_p/a)^4] \right\} \tag{11}$$

where "$\mathcal{O}[x^4]$" denotes additional terms containing 4th and higher powers of "x" that are negligible when x is small compared to unity. Equation (11) displays the leading-order departure from Euclidean geometry for $k = \pm 1$, and it also confirms that this departure becomes important only when r_p is an appreciable fraction of the cosmic scale factor. Cosmic inflation predicts

that the scale factor is enormously greater than the radius of the observable universe, in which case no curvature of space would be detectable even if $k \neq 0$.

Equation (7) implies that the proper area A of a spherical surface with radial coordinate r is

$$A = 4\pi a^2 r^2 \tag{12}$$

and that the proper volume V contained within such a sphere is

$$V = 4\pi a^3 \int_0^r dr' \frac{r'^2}{(1 - kr'^2)^{1/2}} \tag{13}$$

The integral appearing in Eq. (13) is elementary for all three values of k, but will not be recorded here.

When $k = 0$, Eqs. (9), (12), and (13) reproduce the familiar Euclidean relations between A, V, and the proper radius r_p. For $k = \pm 1$, series expansions of A and V in terms of r_p yield

$$A = 4\pi r_p^2 \left\{ 1 - \frac{k}{3} (r_p/a)^2 \right.$$
$$\left. + \mathcal{O}[(r_p/a)^4] \right\} \tag{14}$$

$$V = \frac{4\pi}{3} r_p^3 \left\{ 1 - \frac{k}{5} (r_p/a)^2 \right.$$
$$\left. + \mathcal{O}[(r_p/a)^4] \right\} \tag{15}$$

For any given r_p, the surface area and volume of a sphere are smaller than their Euclidean values when $k = +1$, and larger than Euclidean when $k = -1$.

The number of galaxies within any large volume of space is a measure of its proper volume (provided the distribution of galaxies is truly uniform). By counting galaxies within spheres of differing proper radii (which must also be measured), it is possible in principle to determine whether proper volume increases less rapidly than r_p^3, at the same rate, or more rapidly; hence whether $k = +1$, 0, or -1. If $k \neq 0$, detailed comparison of observations with Eq. (15) would enable one to determine the present value of the cosmic scale factor.

Other schemes also exist for comparing observations with theory to determine the geometry and, if $k \neq 0$, the scale factor of the cosmos. Unfortunately, observational and interpretational uncertainties have frustrated the first step in this program, namely, the determination of k. Either space is flat or, if $k \neq 0$, the cosmic scale factor is sufficiently greater than the distance thus far explored that we have not yet observed the curvature in a decisive way. Thus at present we do not know the value of k, nor of $a(t)$, nor (therefore) even the RW radial coordinate r for any object observed (except earth, where $r = 0$ by definition). The machinery of relativistic cosmology is impressive, but its contact with observation remains tenuous.

Further insight into geometry may be gained by noting that if $k = 1$, then space corresponds to the surface of a hypersphere of radius $a(t)$ imbedded in four-dimensional Euclidean space. Such a universe is said to be *spherical* and *closed*; in this case $a(t)$ is called the *radius of the universe*. Going from $r = 0$ to $r = 1$ takes one halfway to the antipodal point. Though lacking any boundary, a spherical universe is finite. The proper circumference is $2\pi a$ and the proper volume is $2\pi^2 a^3$, as may be seen from Eqs. (10) and (13) combined with knowledge that $r = 1$ corresponds to the "equator" when $r = 0$ is taken at a "pole."

Within a curved space, the concept of a "straight" line may be defined as a line that is as straight as possible, that is, everywhere straight relative to the locally flat region surrounding the segment in question. (The formal name for such a line is *geodesic*.) On the surface of the earth, great circles correspond to straight lines, for example, lines of longitude. Note that adjacent lines of longitude are parallel at the equator, but approach each other as one follows them toward either pole, where they intersect. In a spherical universe, straight lines (as defined, for example, by beams of light) that are parallel nearby approach each other as one follows them away, and they intersect at a proper distance equal to one-fourth the circumference of the universe. If one follows a straight line for a distance equal to the circumference, one returns to the starting point.

For $k = 0$ and $k = -1$, space is infinite in linear extent and volume; such universes are said to be *open*. As remarked earlier, a universe with $k = 0$ has zero spatial curvature for any single value of t; such a universe is said to be *Euclidean* or *flat*. The RW metric with $k = 0$ does not, however, correspond to the flat space–time of Minkowski. The RW coordinates are comoving with a spherically symmetric expansion (or contraction), so that the RW time variable is measured by clocks that are moving relative to one another. Special relativity indicates that such clocks measure time differently from the standard clocks of Minkowski space–time. In particular, a set of comoving clocks does not

run synchronously with any set of clocks at rest relative to one another. Furthermore, the measure of distance given by the RW metric is that of comoving meter sticks, and it is known from special relativity that moving meter sticks do not give the same results for distance as stationary ones.

Since both time and distance are measured differently in comoving coordinates, the question arises whether the differences compensate for each other to reproduce Minkowski space–time. For $k = 0$, the answer is negative: such a space–time can be shown to be curved, provided the scale factor is changing in time so that comoving instruments actually are moving.

If $k = -1$, space has negative curvature: straight lines that are parallel nearby diverge from each other as one follows them away. An imperfect analogy is the surface of a saddle; two "straight" lines that are locally parallel in the seat of the saddle will diverge from each other as one follows them toward the front or back of the saddle. (The analogy is imperfect because the curvature of space is presumed to be the same everywhere, unlike the curvature of a saddle, which varies from one region to another.) A universe with $k = -1$ is called *hyperbolic*.

D. HUBBLE'S LAW

It is evident from Eq. (9) that the proper distance r_p to a comoving object depends on time only through a multiplicative factor of $a(t)$. It follows that

$$\dot{r}_p / r_p = \dot{a}/a \equiv H(t) \tag{16}$$

$$\dot{r}_p = H r_p \tag{17}$$

where dots over the symbols denote differentiation with respect to time, for example, $\dot{f} \equiv df/dt$. Thus $\dot{r}_p$ equals the velocity of a comoving object relative to the earth (assuming that earth itself is a comoving object), being positive for recession or negative for approach. An important proviso, however, arises from the fact that the concept of velocity relative to earth becomes problematic if the object is so distant that curvature of the intervening space–time is significant. Also, r_p is (hypothetically) defined in terms of comoving meter sticks laid end to end. All of these comoving meter sticks, except the closest one, are moving relative to earth, and are therefore perceived by us to be shortened because of the familiar length contraction of special relativity. Thus r_p deviates from distance in the familiar sense for objects receding with velocities near that of light. For these reasons, $\dot{r}_p$ only corresponds to velocity in a familiar sense if r_p is less than the distance over which space–time curves appreciably and $\dot{r}_p$ is appreciably less than c.

A noteworthy feature of Eq. (17) is that $\dot{r}_p$ exceeds c for objects with $r_p > c/H$. There is no violation of the basic principle that c is a limiting velocity, however. As just noted, $\dot{r}_p$ only corresponds to velocity when small compared to c. Furthermore, a correct statement of the principle of limiting velocity is that no object may *pass by* another object with a speed greater than c. In Minkowski space–time the principle may be extended to the relative velocity between distant objects, but no such extension to distant objects is possible in a curved space–time.

Equations (16) and (17) state that if $\dot{a} \neq 0$, then all comoving objects are receding from (or approaching) earth with velocities that are proportional to their distances from us, subject to the provisos stated above. One of the most direct points of contact between theoretical formalism and observation resides in the fact that, aside from local random motions of limited magnitude, the galaxies constituting the visible matter of our universe are receding from us with velocities proportional to their distances. This fact was first reported by Edwin Hubble in 1929, based on determinations of distance (later revised) made with the 100-in. telescope that was completed in 1918 on Mt. Wilson in southwest California. Velocities of the galaxies in Hubble's sample had been determined by V. M. Slipher, utilizing the redshift of spectral lines caused by the Doppler effect.

The function $H(t)$ defined by Eq. (16) is called the Hubble parameter. Its present value, which we shall denote by H_0, is traditionally called *Hubble's constant* (a misnomer, since the "present" value changes with time, albeit very slowly by human standards). Equation (17), governing recessional velocities, is known as *Hubble's law*.

The value of H_0 is central to a quantitative understanding of our universe. Assuming that redshifts of spectral lines result entirely from the Doppler effect, it is straightforward to determine the recessional velocities of distant galaxies. A determination of distance, however, is necessarily indirect, and furthermore is problematic for galaxies beyond the limited range where triangulation is feasible. Different observers favor different intrinsic properties of galaxies (and galactic clusters) as criteria for estimating distances, leading to systematic disagreements over the

value of H_0. Additional uncertainty arises from the possibility of galactic evolution. The light now reaching earth from distant galaxies was emitted long ago; if the intrinsic properties of galaxies have changed appreciably over the intervening time, estimates of distance based on these properties will be misleading.

Many calculations in cosmology depend on the value of Hubble's constant. It is useful to deal with the uncertainty in its value by parametrizing H_0 in terms of a dimensionless number h_0, of order unity, defined by

$$H_0 \equiv h_0 \times 50 \text{ km sec}^{-1} \text{ Mpc}^{-1} \quad (18)$$

where Mpc denotes 1 Megaparsec = 3.26×10^6 light-years = 3.09×10^{19} km. The value for h_0 enjoying widest favor is 1.00, reported by A. R. Sandage and G. A. Tammann with a statistical uncertainty of ± 0.14 (1 SD). Using different criteria for estimating distance, however, Gerard de Vaucouleurs and others have obtained values for h_0 in the range from 1.6 to 2.0. Hence the empirical value of h_0 remains unsettled, but seems likely to fall within the range

$$0.8 \leq h_0 \leq 2.0 \quad (19)$$

The value of H_0 is of great importance to our understanding of the universe; it is hoped that a reliable determination will emerge in the near future. Collection of data from satellites above the atmosphere should facilitate such a determination.

Before leaving this section on cosmic geometry, we note that a space–time may contain one or more finite regions that are internally homogeneous and isotropic while differing from the exterior. Within such a homogeneous region, space–time can be described by an RW metric (which, however, tells one nothing about the exterior region). This type of space–time is in fact predicted by theories of cosmic inflation; the homogeneity and isotropy of the observable universe are not believed characteristic of the whole.

IV. Dynamics of the Cosmos

A. EINSTEIN'S FIELD EQUATIONS

A theory of cosmic evolution requires dynamical equations, taken here to be the field equations of GTR:

$$R_{\mu\nu} - \tfrac{1}{2}g_{\mu\nu}R + \Lambda g_{\mu\nu} = 8\pi G c^{-4} T_{\mu\nu} \quad (20)$$

where $R_{\mu\nu}$ denotes the Ricci tensor (contracted Riemann curvature tensor), R denotes the curvature scalar (contracted Ricci tensor), Λ denotes the cosmological constant (introduced and later abandoned by Einstein, but still a possibility warranting study), and $T_{\mu\nu}$ denotes the stress-energy tensor of all forms of matter and energy excluding gravity. The Ricci tensor and curvature scalar are determined by $g_{\mu\nu}$ together with its first and second partial derivatives, so that Eq. (20) is a set of second-order (nonlinear) partial differential equations for $g_{\mu\nu}$. As with all differential equations, initial and/or boundary conditions must be assumed in order to specify a unique solution. (A definition of the Riemann curvature tensor lies beyond the scope of this article. We remark, however, that some scientists define it with an opposite sign convention, in which case $R_{\mu\nu}$ and R change signs and appear in the field equations with opposite signs from those above.)

All tensors appearing in the field equations are symmetric, so there are 10 independent equations in the absence of simplifying constraints. We are assuming space to be homogeneous and isotropic, however, which assures the validity of the RW line element. The RW metric contains only one unknown function, namely the cosmic scale factor $a(t)$, so we expect the number of independent field equations to be considerably reduced for this case.

Since the field equations link the geometry of space–time with the stress-energy tensor, consistency requires that we approximate $T_{\mu\nu}$ by a homogeneous, isotropic form. Thus we imagine all matter and energy to be smoothed out into a homogeneous, isotropic fluid (or gas) of dust and radiation that moves in unison with the cosmic expansion. Denoting the proper mass-energy density by ρ (in units of mass/volume) and the pressure by P, the resulting stress-energy tensor has the form characteristic of a "perfect fluid":

$$T_{\mu\nu} = (\rho + P/c^2)U_\mu U_\nu + P g_{\mu\nu} \qquad (21)$$

where U_μ denotes the covariant four-velocity of a comoving point: $U_\mu = g_{\mu\nu} dx^\nu/d\tau$, with τ denoting proper time ($d\tau = |ds|/c$).

Under the simplifying assumptions stated above, only two of the 10 field equations are independent. They are differential equations for the cosmic scale factor and may be stated as

$$\left(\frac{\dot{a}}{a}\right)^2 = \frac{8\pi G}{3}\rho + \frac{c^2\Lambda}{3} - k\left(\frac{c}{a}\right)^2 \qquad (22)$$

$$\frac{\ddot{a}}{a} = -\frac{4\pi G}{3}\left(\rho + \frac{3P}{c^2}\right) + \frac{c^2\Lambda}{3} \qquad (23)$$

where $\dot{a}$ and $\ddot{a}$ denote first and second derivatives, respectively, of $a(t)$ with respect to time.

Note that the left side of Eq. (22) is precisely H^2, the square of the Hubble parameter.

B. LOCAL ENERGY CONSERVATION

A remarkable feature of the field equations is that the *covariant divergence* (the curved–space-time generalization of Minkowski four-divergence) of the left side is identically zero (*Bianchi identities*). Hence the field equations imply that over a region of space–time small enough to be flat, the Minkowski four-divergence of the stress-energy tensor is zero. A standard mathematical argument then leads to the conclusion that energy and momentum are conserved (locally, which encompasses the domain of experiments known to manifest such conservation). Hence within the context of GTR, energy and momentum conservation are not separate principles to be adduced, but rather are consequences of the field equations to the full extent that conservation is indicated by experiment or theory.

If a physical system is spatially localized in a space–time that is flat (i.e., Minkowskian) at spatial infinity, then the total energy and momentum for the system can be defined and their constancy in time proven. The condition of localization in asymptotically flat space–time is not met, however, by any homogeneous universe, for such a universe lacks a boundary and cannot be regarded as a localized system. The net energy and momentum of an unbounded universe defy meaningful definition, which precludes any statement that they are conserved. (From another point of view, energy is defined as the ability to do work, but an unbounded universe lacks any external system upon which the amount of work might be defined.)

In the present context, local momentum conservation is assured by the presumed homogeneity and isotropy of the cosmic expansion. Local energy conservation, however, implies a relation between ρ and P, which may be stated as

$$\frac{d}{dt}(\rho a^3) = -\frac{P}{c^2}\frac{d}{dt}(a^3) \qquad (24)$$

[That Eq. (24) follows from the field equations may be seen by solving Eq. (22) for ρ and differentiating with respect to time. The result involves $\ddot{a}$, which may be eliminated by using Eq. (23) to obtain a relation equivalent to Eq. (24).]

To see the connection between Eq. (24) and energy conservation, consider a changing spherical volume V bounded by a comoving surface. Equation (13) indicates that V is the product of a^3 and a coefficient that is independent of time. Multiplying both sides of Eq. (24) by that coefficient and c^2 yields

$$\frac{d}{dt}(\rho c^2 V) = -P\frac{dV}{dt} \qquad (25)$$

The Einstein mass-energy relation implies that ρc^2 equals the energy per unit volume, so $\rho c^2 V$ equals the energy U internal to the volume V. Hence Eq. (25) is simply the work-energy theorem as applied to the adiabatic expansion (or contraction) of a gas or fluid: $\Delta U = -P\,\Delta V$. Since energy changes only to the extent that work is done, (local) energy conservation is implicit in this result.

By inverting the reasoning that led to Eq. (24), it can be shown that the second of the field equations follows from the first and from energy conservation; that is, Eq. (23) follows from Eqs. (22) and (24). [To reason from Eqs. (23) and (24) to (22) is also possible; k appears as a constant of integration.] Hence evolution of the cosmic scale factor is governed by Eqs. (22) and (24), together with an equation of state that specifies the relation between ρ and P.

C. EQUATION FOR THE HUBBLE PARAMETER

The present expansion of the universe means that $H(t)$ is currently positive, and the expansion will continue for however long H remains positive. Since $H \equiv \dot{a}/a$, Eq. (22) may be written as

$$H^2 = \frac{8\pi G}{3}\rho + \frac{c^2\Lambda}{3} - k\left(\frac{c}{a}\right)^2 \qquad (26)$$

so the expansion will continue forever unless the right side of Eq. (26) shrinks to zero at some finite time in the future. We next consider that possibility.

The mass-energy density ρ is positive, and has been dominated by matter under negligible pressure (i.e., $P \ll \rho c^2$) since the early hot stage of cosmic evolution. For zero pressure, Eq. (24) implies that ρ varies inversely with the cube of $a(t)$, which simply corresponds to a given amount of matter becoming diluted by the increasing volume of the universe. Neglecting pressure, we can therefore write

$$\rho(a) = \rho_0(a_0/a)^3 \qquad (27)$$

where ρ_0 and a_0 denote present values. Equation (26) then becomes

$$H^2 = \frac{8\pi G}{3}\rho_0\left(\frac{a_0}{a}\right)^3 + \frac{c^2\Lambda}{3} - k\left(\frac{c}{a}\right)^2 \qquad (28)$$

The scale factor $a(t)$ increases as the universe expands, but remains finite for all finite times.

Hence the right side of Eq. (28) remains positive unless a cancellation occurs between positive and negative terms. Assuming (as seems likely) that Λ is zero or positive, no such cancellation is possible if $k = 0$ or -1. Hence an open universe, either flat or hyperbolic, will expand forever. If $\Lambda = 0$, it can furthermore be shown that the recessional velocity (as defined by $\dot{r}_p$) of each comoving object will tend to zero in the limit of long time if $k = 0$, but will approach a nonzero limiting value if $k = -1$. If Λ is positive, then H approaches a constant value of $c(\Lambda/3)^{1/2}$ in the limit of long time for $k = 0$ or -1, in which case the $\dot{r}_p$ of each comoving object increases without limit as its r_p increases without limit.

If $k = +1$, the universe may (or may not) reach a maximum size and then contract, depending on the relation between ρ_0, a_0, and Λ. If Λ is zero, then H will shrink to zero when the scale factor reaches a maximum size given by

$$a_{max} = \frac{8\pi G}{3c^2}\rho_0 a_0^3 = \frac{4G}{3\pi c^2}M_u \qquad (29)$$

where M_u denotes the proper mass of the universe:

$$M_u = \rho V_u = \rho(2\pi^2 a^3) \qquad (30)$$

which is constant in time if the pressure is negligible, as assumed here. It may be shown that the universe reaches its maximum radius in a finite time; under the simplifying assumption of negligible pressure from the very beginning, the proper time required for the radius to grow from an initial value of zero to a_{max} is

$$t(a_{max}) = \frac{2G}{3c^3}M_u \qquad (31)$$

It is remarkable that both a_{max} and the time required to reach it are determined solely by M_u; the dynamics and geometry are so interwoven that the rate of expansion does not enter as an additional parameter. If, for example, a_{max} were reached in 20 billion yr, then the corresponding universe would have a mass of 4×10^{53} kg (and conversely).

After reaching a_{max}, the universe will contract ($H < 0$) at the same rate as it had expanded. Equation (23) ensures that the scale factor will not remain at a_{max}, and Eq. (28) implies that the magnitude of H is determined by the value of the scale factor.

If the universe is closed and Λ differs from zero, then a more careful analysis is required. Assuming that our universe has in fact emerged from an initial state of very high density and small scale factor (i.e., a big bang model), a detailed study of Eq. (28) reveals that the present expansion will continue forever if

$$\Lambda \geq (4\pi c G \rho_0)^{-2} (c/a_0)^6 \qquad (32)$$

If the *inequality* in Eq. (32) holds, the size of the universe will increase without limit. If the equality holds (and $k = +1$), then the universe will expand forever but approach a maximum, limiting size as time tends to infinity. If the relation (32) is violated (and $k = +1$), then the universe will reach a maximum size at some finite time in the future, after which it will contract under the force of gravity. The latter case obviously holds if $\Lambda = 0$, as remarked earlier: in the absence of the cosmological term, a closed universe is destined eventually to collapse.

D. THE CRITICAL DENSITY

The field equation (26) also implies a relation between ρ and the geometry of space. To display this relation, we solve Eq. (26) for ρ and then set $k = 0$, thereby defining a critical density corresponding to the intermediate case of a flat universe:

$$\rho_c(\Lambda) \equiv \frac{3H^2 - c^2\Lambda}{8\pi G} \qquad (33)$$

The term *critical density* has often been used by scientists who presumed the cosmological constant to be zero, which corresponds to $\rho_c(0)$ in our present definition. Neither observation nor theory warrants such a stringent assumption about Λ, however. Furthermore, it may prove difficult to reconcile cosmic inflation with observation unless Λ has a small positive value. Hence we shall retain the possible effects of Λ throughout most of our discussion. [Some investigators entertain the possibility of a nonzero Λ but still refer to $\rho_c(0) = 3H^2/8\pi G$ as the critical density. We prefer the Definition (33) because there is nothing critical about a density of $3H^2/8\pi G$ if $\Lambda \neq 0$.]

Replacing the terms involving H and Λ in Eq. (26) by $\rho_c(\Lambda)$ and solving for ρ, we obtain

$$\rho = \rho_c(\Lambda) + \frac{3k}{8\pi G} \left(\frac{c}{a}\right)^2 \qquad (34)$$

From this equation it is apparent that the three possibilities $\rho > \rho_c$, $\rho = \rho_c$, and $\rho < \rho_c$ correspond to $k > 0$, $k = 0$, and $k < 0$, respectively. Hence within the assumptions stated, the geometry of the cosmos is determined by the relation between ρ and $\rho_c(\Lambda)$.

We note in passing a quantitative prediction of

cosmic inflation, namely that the cosmic scale factor is presently so large ($a_0 \gg c/H_0 \approx 20$ billion light-years) as to render negligible the term involving k on the right side of Eq. (34). Hence inflation predicts that the present value of ρ is virtually indistinguishable from $\rho_c(\Lambda)$. Observations have only revealed, either directly or indirectly, about one-eighth as much mass density as $\rho_c(0)$, and they furthermore place an upper limit of about one-seventh $\rho_c(0)$ on the amount of ordinary (baryonic) matter. This apparent (albeit tentative) discrepancy between ρ and $\rho_c(0)$ is a substantial reason to take seriously the possibility that $\Lambda > 0$.

The critical density obviously provides an important standard of comparison for the actual mass density. Since Λ has often been assumed to be zero (and, if not zero, remains unknown), it is customary to parametrize ρ as a fraction Ω of the critical density for the case $\Lambda = 0$:

$$\rho \equiv \Omega \times \rho_c(0) = \Omega \times (3H^2/8\pi G) \qquad (35)$$

Thus if $\Lambda = 0$, the universe is spherical, flat, or hyperbolic according to whether Ω is greater than, equal to, or less than 1, respectively. Denoting the present value of ρ_c by ρ_{c0}, it may be expressed as

$$\rho_{c0}(0) = h_0^2 \times (4.7 \times 10^{-30} \text{ g/cm}^3) \qquad (36)$$

where use has been made of the parametrization of H_0 defined by Eq. (18).

If $\Lambda = 0$ but $k \neq 0$, then Ω must have been very near unity at early times, as was mentioned in Section II,B. To illustrate this point, we note that Eqs. (27), (28), and (33) may be used to express Ω as

$$\Omega = \left\{ 1 - a(t) \left[\frac{3kc^2}{8\pi G \rho_0 a_0^3} \right] \right\}^{-1} \qquad (37)$$

which clearly approaches unity as one goes backward in time and $a(t)$ tends toward zero. Equation (27) has only been valid since the universe became cool enough for ρ to be primarily nonrelativistic matter ($\sim 10^6$ yr after the big bang). For relativistic matter, however, and/or radiation, ρ is proportional to a^{-4}, in which case Ω approaches unity even more rapidly as one goes back into the early relativistic era with $a(t)$ shrinking toward zero.

It is convenient to parametrize Λ as a fraction λ of the value that would make the critical density vanish, that is,

$$\Lambda \equiv \lambda \times (3H^2/c^2)$$
$$= \lambda_0 h_0^2 \times (8.7 \times 10^{-53} \text{ m}^{-2}) \qquad (38)$$

where λ_0 denotes the present value of λ (Λ is constant; but the evolution of H implies that λ, if not zero, will also evolve). The relation between geometry and mass-energy density may now be expressed as follows: the universe is spherical, flat, or hyperbolic according to whether $(\Omega + \lambda)$ is greater than, equal to, or less than unity, respectively. If one accepts at face value the observational indications that Ω_0 (the present value of Ω) is ~ 0.12, then inflation would imply that λ_0 is ~ 0.88. As will be explained in Section V,B, the corresponding value for Λ is small enough to be consistent with all observations.

E. COSMIC DECELERATION OR ACCELERATION

Although the second field equation contains no information beyond that implied by the first together with local energy conservation, it makes a direct and transparent statement about cosmic evolution that warrants consideration. Recalling again that the proper distance $r_p(t)$ to a comoving object depends on time only through a factor of $a(t)$, we see that Eq. (23) is equivalent to a statement about the acceleration $\ddot{r}_p$ of comoving objects (where again the dots refer to time derivatives):

$$\ddot{r}_p = \left\{ -\frac{4\pi G}{3}\left(\rho + \frac{3P}{c^2}\right) + \frac{c^2\Lambda}{3}\right\} r_p \quad (39)$$

Note that $\ddot{r}_p$ is proportional to r_p, which is essential to the preservation of Hubble's law over time. A positive $\ddot{r}_p$ corresponds to acceleration *away* from us, whereas a negative $\ddot{r}_p$ corresponds to acceleration *toward* us (e.g., deceleration of the present expansion). The same provisions that conditioned our interpretation of $\bar{r}_p$ as recessional velocity apply here; $\dot{r}_p$ equals acceleration in the familiar sense only for objects whose distance is appreciably less than the scale factor (for $k \neq 0$) and whose $\dot{r}_p$ is a small fraction of c.

As a first step in understanding the dynamics contained in Eq. (39), let us consider what Newton's laws would predict for a homogeneous, isotropic universe filling an infinite space described by Euclidean geometry. In particular, let us calculate the acceleration $\ddot{r}_p$ (relative to a comoving observer) of a point mass at a distance r_p away from the observer. (The distinction of "proper" distance is superfluous in Newtonian physics, but the symbol r_p is being used to maintain consistency of notation.)

In Newton's theory of gravitation, the force F between two point masses m and M varies inversely as the square of the distance d between them: $F = GmM/d^2$. If more that two point masses are present, a principle of superposition states that the net force on any one mass is the vector sum of the individual forces resulting from the other masses. While it is not obvious (indeed, Newton invented calculus to prove it), the inverse–square dependence on distance together with superposition implies that the force exerted by a uniform sphere of matter on any exterior object is the same as if the sphere's mass were concentrated at its center. It may also be shown that if a spherically symmetric mass distribution has a hollow core, then the mass exterior to the core exerts no net force on any object inside the core. (These theorems may be familiar from electrostatics, where they also hold because Coulomb's law has the same mathematical form as Newton's law of gravitation.)

The above features of Newtonian gravitation may be exploited in this study of cosmology by noting that in the observer's frame of reference, the net force on an object of mass m (e.g., a galaxy) at a distance r_p from the observer results entirely from the net mass M contained within the (hypothetical) sphere of radius r_p centered on the observer:

$$M = (\tfrac{4}{3}\pi r_p^3)\rho \quad (40)$$

The force resulting from M is the same as if it were concentrated at the sphere's center, and there is no net force from the rest of the universe exterior to the sphere. Combining Newton's law of gravitation with his second law of motion (that is, acceleration equals F/m) yields

$$\ddot{r}_p = -GM/r_p^2 = -\frac{4\pi G}{3}\rho r_p \quad (41)$$

where Eq. (40) has been used to eliminate M, and the minus sign results from the attractive nature of the force.

Note that Eq. (41) agrees precisely with the contribution of ρ to $\ddot{r}_p$ in the Einstein field equation (39); this part of relativistic cosmology has a familiar, Newtonian analog. In GTR, however, stresses also contribute to gravitation, as exemplified by the pressure term in Eq. (39). If $\Lambda \neq 0$, then a cosmological term contributes as well.

Although Eq. (39) is directly applicable only to a homogeneous, isotropic system (the universe being the only known candidate), it is instructive to compare the relative magnitudes of the mass and pressure terms for a familiar system such as the earth, where again both mass and pressure contribute to gravitation. Earth has a mean mass density of 5.5×10^3 kg/m^3. For an extreme example, let us consider the earth's pressure at its center, where P has the impressively large value of 3.6×10^{11} N/m^2. Before comparing pressure with mass density as a source of gravitation, however, the field equations instruct us to divide P by c^2. For the extreme case of Earth's central pressure, $P/c^2 = 4.0 \times 10^{-6}$ kg/m^3, which is nine orders of magnitude smaller than the mean mass density. Hence for terrestrial purposes (and throughout the solar system), pressure is quite negligible as a source of gravitation (which explains why so few nonspecialists are aware that pressure contributes to gravity).

As far as is known, the mass-energy density ρ appearing in the field equations is positive, so that its contribution to $\ddot{r}_p$ in Eq. (39) is negative: mass-energy leads to gravitational attraction. Positive pressure also makes a negative contribution to $\ddot{r}_p$, but pressure may be either positive or negative. Negative pressure contributes a positive term to $\ddot{r}_p$, and hence corresponds to gravitational repulsion. This fact lies at the heart of the inflationary scenario. The cosmological term, if nonzero, contributes to $\ddot{r}_p$ with the same sign as Λ.

The *name* of the big bang model is somewhat misleading, for it suggests that the expansion of the cosmos is a result of the early high temperature and pressure, rather like the explosion of a gigantic firecracker. As just explained, however, positive pressure leads not to acceleration "outward" of the cosmos, but rather to deceleration of any expansion that may be occurring. Had the universe been born stationary, positive pressure would have contributed to immediate collapse. Since early high pressure did not contribute to the expansion, early high temperature could not have done so either. (It is perhaps no coincidence that the phrase *big bang* was initially coined as a term of derision, by Fred Hoyle—an early opponent of the model. The model has proven successful, however, and the graphic name remains with us.)

F. The Growth of Space

It is intuitively useful to speak of ρ, P, and Λ as leading to gravitational attraction or repulsion, but the geometrical role of gravitation in GTR should not be forgotten. The geometrical significance of gravitation is most apparent for a closed universe, whose total proper volume is finite and equal to $2\pi^2 a^3(t)$. The field equations

describe the change in time of $a(t)$, hence of the size of the universe. From a naive point of view, Hubble's law describes cosmic expansion in terms of recessional velocities of galaxies; but more fundamentally, galaxies are moving apart because the space between them is expanding. How else an increasing volume for the universe?

The field equation for $\ddot{a}(t)$ implies an equation for $\ddot{r}_p$, which has a naive interpretation in terms of acceleration and causative force. More fundamentally, however, $\ddot{a}$ and $\ddot{r}_p$ represent the derivative of the rate at which space is expanding. Our intuitions are so attuned to Newton's laws of motion and causative forces that we shall often speak of velocities and accelerations and of gravitational attraction and repulsion, but the geometrical nature of GTR should never be far from mind.

The interpretation of gravity in terms of a growth or shrinkage of space resolves a paradox noted above, namely, that positive pressure tends to make the universe collapse. Cosmic pressure is the same everywhere, and uniform pressure pushes equally on all sides of an object leading to no net force or acceleration. Gravity, however, affects the rate at which space is growing or shrinking, and thereby results in relative accelerations of distant objects. One has no *a priori* intuition as to whether positive pressure should lead to a growth or shrinkage of space; GTR provides the answer described above.

G. The Age of the Universe

The most direct method for estimating the age of the universe is to study the evolution in time of the cosmic scale factor, and thereby determine how long the distant galaxies have been receding from us. The simplest case is that in which gravitational braking (or acceleration) of the expansion has been negligible. From Eq. (23) we see that this corresponds to zero values for ρ, P, and Λ; since $\rho < \rho_c$, the corresponding universe is hyperbolic. If the recessional velocity of a galaxy has been constant, then the time t_H required for it to have reached its present distance from us is simply $r_p/\dot{r}_p$, which, according to Hubble's law, is $1/H_0$ (the same for every galaxy):

$$t_H \equiv 1/H_0 = h_0^{-1} \times (19.6 \times 10^9 \text{ yr}) \quad (42)$$

which is called the *Hubble age* or *Hubble time*.

We note in passing that despite the name *hyperbolic*, a completely empty (i.e., $\rho = P = 0$) universe with $\Lambda = 0$ has precisely the flat space–time of Minkowski. Empty Minkowski space is homogeneous and isotropic, and hence must correspond to an RW metric. It is also a solution to Einstein's field equations with $T_{\mu\nu}$ and Λ set equal to zero, which singles out the empty hyperbolic universe with $\Lambda = 0$. The difference in appearance between the Minkowski metric and the RW metric in question results from a fact noted earlier: that RW coordinates are comoving. Thus the clocks and meter sticks corresponding to the RW metric are moving relative to one another, and they yield results for time and distance that differ from those ordinarily used in a description of Minkowski space–time. In *this* case, the different measures of time and distance compensate for one another: the space–time described is actually the flat one of special relativity.

To make explicit the identification between empty hyperbolic space–time (with $\Lambda = 0$) and Minkowski space–time, we note that the field equations imply $a(t) = ct$ for the hyperbolic case in question. It follows that the relation between coordinates is $r_M = rct$, $t_M = t(r^2 + 1)^{1/2}$, where r_M and t_M denote the usual (stationary) radial coordinate and time of Minkowski space–time, respectively.

If gravitational braking has occurred, then the present age t_0 is less than t_H. Braking means that galaxies were receding more rapidly in the past than is indicated by H_0, so less time was required for them to have reached their present distances from us. Conversely, if negative pressure and/or positive Λ had led to an accelerating expansion, then the present age would be greater than t_H.

To determine the effect of braking more precisely, one needs to know how ρ and P have evolved in time. It is believed that pressure has been negligible for over 99% of the time since the big bang, in which case the time elapsed can be obtained to excellent approximation by assuming $P = 0$ and $\rho = \rho_0(a_0/a)^3$. The amount of braking that has occurred is an increasing function of Ω_0, so larger values of Ω_0 correspond to younger ages for any given H_0. If $\Lambda = 0$, it can be shown that the age decreases from t_H for $\Omega_0 = 0$ to $2t_H/3$ for $\Omega_0 = 1$, and to zero for $\Omega_0 = \infty$. The age for critical density with $\Lambda = 0$ provides an important standard for comparison with observations, and we shall denote it by $t_c(0)$:

$$t_c(0) = 2t_H/3 = h_0^{-1} \times (13.1 \times 10^9 \text{ yr}) \quad (43)$$

A positive Λ would increase the age over the values indicated above.

If inflation is correct in predicting that ρ is close to the critical density, and if $\Lambda = 0$, then the age is $t_c(0)$. Observations indicate that h_0 lies between 0.8 and 2.0, so that $t_c(0)$ lies between 7 and 16 times 10^9 yr; the favored value for h_0 is unity, corresponding to an age of 13×10^9 yr.

An alternative method for estimating the age of the universe is to identify the oldest constituents whose age can be determined. The primary candidates are stars forming globular clusters (spherical conglomerations of $\sim 10^6$ stars found in the haloes of the Milky Way and other galaxies). From the spectral properties of these stars and the theory of stellar evolution, it is believed that globular clusters are $(16 \pm 2) \times 10^9$ years old. It is furthermore believed that the universe was already 2×10^9 years old when the globular clusters formed, indicating an age of $(18 \pm 3) \times 10^9$ years for the universe. Note that this value for the age is distinctly greater than $t_c(0)$ if h_0 is unity and is only marginally consistent when known uncertainties are taken into account.

As mentioned in Section IV,D, efforts to determine the cosmic mass density suggest a value for Ω_0 of roughly 0.12. This amount of matter would have given rise to very little gravitational braking: if $\Lambda = 0$, the predicted age would be almost t_H, which would be compatible with the age indicated by globular clusters if h_0 should prove to be near unity. If $\Omega_0 \approx 0.12$ and the prediction of inflation that $(\Omega_0 + \lambda_0) \approx 1$ is accepted, then $\lambda_0 \approx 0.88$, and the age can be shown to be $1.24t_H$. This would be compatible with the globular clusters for h_0 near 1.3. If it is assumed that $h_0 \approx 1$, that $(\Omega_0 + \lambda_0) \approx 1$, and that the age of the cosmos is 18×10^9 yr, then it would follow that $\Omega_0 \approx 0.4$ and $\lambda_0 \approx 0.6$ (the corresponding age would be $0.89t_H$).

Although the evidence is far from conclusive, neither direct determinations of mass density nor estimates of the age of globular clusters support the prediction of inflation that $\rho \approx \rho_c$, unless $\Lambda > 0$. Hence we take the latter possibility seriously, and proceed now to a detailed consideration of the cosmological term. Even if Λ should prove to be zero, a study of its possible effects will shed light on the dynamics of cosmic inflation.

V. The Cosmological Term

A. HISTORICAL BACKGROUND

For historical and philosophical reasons beyond the scope of this article, Einstein believed during the second decade of this century that the universe was static. He realized, however, that a static universe was dynamically impossible unless there were some repulsive force to balance the familiar attractive force of gravitation between the galaxies. The mathematical framework of GTR remains consistent when the cosmological term $\Lambda g_{\mu\nu}$ is included in the field equations (20), and Einstein added such a term to stabilize the universe in his initial paper on cosmology in 1917.

Although the existence of other galaxies outside the Milky Way was actually not established until 1924 (by Hubble, using the 100-in. telescope on Mt. Wilson), Einstein (in his typically prescient manner) had incorporated the assumptions of homogeneity and isotropy in his cosmological model of 1917. With these assumptions, the field equations reduce to (22) and (23). The universe is static if and only if $\dot{a}(t) = 0$, which requires (over any extended period of time) that $\ddot{a}(t) = 0$ as well. Applying these conditions to Eqs. (22) and (23), it is readily seen that the universe is static if and only if

$$\Lambda = \frac{4\pi G}{c^2}\left(\rho + \frac{3P}{c^2}\right) \tag{44}$$

$$k\left(\frac{c}{a}\right)^2 = 4\pi G\left(\rho + \frac{P}{c^2}\right) \tag{45}$$

Since our universe is characterized by positive ρ and positive (albeit negligible) P, the model predicts positive values for Λ and k. Hence the universe is closed and finite, with the scale factor (radius of the universe) given by Eq. (45) with $k = 1$. We have noted that ρ lies within an order of magnitude of 10^{-30} g/cm^3, which corresponds to a cosmic radius on the order of 10^{23} km: a respectable-sized universe.

The static model of Einstein has of course been abandoned, for two reasons. In 1929, Edwin Hubble reported that distant galaxies are in fact receding in a systematic way, with velocities that are proportional to their distances from us. On the theoretical side, it was remarked by A. S. Eddington in 1930 that Einstein's static model was intrinsically unstable against fluctuations in the cosmic scale factor: if $a(t)$ were momentarily to decrease, ρ would increase and the universe would collapse at an accelerating rate under the newly dominating, attractive force of ordinary gravity; if $a(t)$ were to increase, ρ would decrease and the cosmological term would become dominant, thereby causing the universe to expand at an accelerating rate. For this combination of reasons, Einstein recom-

mended in 1931 that the cosmological term be abandoned (i.e., set $\Lambda = 0$); he later referred to the cosmological term as "the biggest blunder of my life." From a contemporary perspective, however, only the static model was ill-conceived. The cosmological term remains a viable possibility, and it is to Einstein's credit that he proposed consideration of such a novel term.

B. Physical Interpretation

We have seen that Einstein was motivated by a desire to describe the cosmos when he added the $\Lambda g_{\mu\nu}$ term to the field equations, but there is another, more fundamental reason why this addendum is called the cosmological term. Let us recall the structure of Newtonian physics, whose many successful predictions are reproduced by GTR. Newton's theory of gravity predicts that the gravitational force on any object is proportional to its mass, while his second law of motion predicts that the resulting acceleration is inversely proportional to the object's mass. It follows that an object's mass cancels out of the acceleration caused by gravity. Hence the earth (or any other source of gravity) may be regarded as generating an *acceleration field* in the surrounding space. This field causes all objects to accelerate at the same rate, and may be regarded as characteristic of the region of space.

GTR regards gravitational accelerations as more fundamental than gravitational forces, and focuses attention on the acceleration field in the form of $g_{\mu\nu}$ as the fundamental object of study. The acceleration field caused by a mass M is determined by the field equations (20), wherein M contributes (in a way depending on its position and velocity) to the stress-energy tensor $T_{\mu\nu}$ appearing on the right side. The solution for $g_{\mu\nu}$ then determines the resulting acceleration of any other mass that might be present, mimicking the effect of Newtonian gravitational force. The term $\Lambda g_{\mu\nu}$ in Eq. (20), however, does not depend on the position or even the existence of any matter. Hence the influence of $\Lambda g_{\mu\nu}$ on the solution for $g_{\mu\nu}$ cannot be interpreted in terms of forces between objects.

To understand the physical significance of the cosmological term, let us consider a special case of the field equations, namely their application to the case at hand: a homogeneous, isotropic universe. As we have seen, the field equations are then equivalent to Eqs. (26) and (39). From the appearance of Λ on the right sides, it is clear that Λ contributes to the recessional velocities of distant galaxies (through H), and to the acceleration of that recession. This part of the acceleration cannot be interpreted in terms of forces acting among the galaxies for, as just remarked, the cosmological term does not presume the existence of galaxies.

The proper interpretation of the cosmological term resides in another remark made earlier: that the recession of distant galaxies should be understood as resulting from growth of the space between them. We conclude that the cosmological term describes a growth of space that is somehow intrinsic to the nature of space and its evolution in time. Such a term clearly has no analog in the physics predating GTR, which is a major reason why attitudes toward it have been so diverse. In the absence of any compelling theoretical argument for or against the cosmological term, the prudent course is to regard its existence as an empirical question, which currently remains open.

Of course the observational successes of Newtonian gravitation, and of GTR without the cosmological term, set upper limits on the possible value of Λ. In linear approximation, the contribution $\Delta\ddot{r}_p$ of Λ to any acceleration may be inferred from Eq. (39) to be

$$\Delta\ddot{r}_p = \frac{c^2\Lambda}{3}\, r_p \qquad (46)$$

where r_p now denotes the proper distance between the accelerating object and the observer or second object relative to which the acceleration is defined. Note that the relative acceleration caused by Λ is proportional to the distance between the two objects in question: the greater the distance, the greater the equivalent force between them. Conversely, the equivalent force shrinks to zero in the limit of zero separation between two objects. This counterintuitive dependence of relative acceleration on distance reflects the fact that the acceleration does not result from any actual force between the objects in question, but rather from a growth or shrinkage of the space between them.

As an instructive example, let us consider the contribution of Λ to the earth's centripetal acceleration in its orbit around the sun. Assuming for the sake of definiteness that $\lambda_0 = h_0 = 1$, Eqs. (38) and (46) indicate that $\Delta\ddot{r}_p$ would be 22 orders of magnitude smaller than the observed acceleration (caused by the sun's Newtonian gravitation). For Pluto, the outermost planet, the acceleration caused by such a value for Λ would be 18 orders of magnitude smaller than that caused by

the sun's Newtonian gravity. Hence any value for Λ that might be implied by a value for Ω_0 less than unity in an inflationary model would be much too small to have observable consequences within the solar system (or even within the galaxy). In yet another sense we are contemplating a cosmological term: for the magnitude of Λ contemplated here, its effects would only be discernible over cosmic distances.

An observational limit on the magnitude of Λ has been obtained from studies of the dynamics of galactic clusters. No evidence for a nonzero Λ has been observed, which implies that $|\Lambda|$ could not be appreciably greater than 10^{-52} m^{-2}. From Eq. (38), we see that this limit on $|\Lambda|$ is consistent with (and close to) the magnitude of Λ that might be implied by the inflationary prediction that $(\Omega_0 + \lambda_0) \approx 1$.

C. Exponential Growth of Scale Factor

Two further aspects of the cosmological term warrant mention, partly for their own sake and partly because they are intimately related to the possibility of cosmic inflation. The first concerns the behavior of a universe wherein the right side of the field equation (22) is dominated by a cosmological constant. Since the terms involving ρ and k in Eq. (22) both tend to zero as $a(t)$ increases without limit, a positive Λ would necessarily dominate the right side in the limit of long time if $k = 0$ or -1, and might become dominant for $k = +1$.

Neglecting all terms on the right side of Eq. (22) except that involving Λ, and taking the positive square root to describe an expanding universe, we obtain

$$\dot{a} = \Gamma a, \quad \text{where} \quad \Gamma \equiv c(\Lambda/3)^{1/2} \qquad (47)$$

The solution to Eq. (47) is

$$a(t) = a(t_0)\exp[\Gamma(t - t_0)] \qquad (48)$$

where t_0 denotes some initial time after which Eq. (47) is a valid approximation. The solution (48) implies an exponential or geometric growth in the cosmic scale factor: $a(t)$ doubles repeatedly over equal intervals of time, with the doubling time t_d given by

$$t_d = (\log_e 2)/\Gamma \qquad (49)$$

With Λ parametrized as in Eq. (38), the doubling time may be expressed as

$$t_d = \lambda_0^{-1/2} h_0^{-1} \times (1.4 \times 10^{10}\ \text{yr}) \qquad (50)$$

which, for λ_0 (and h_0) of order unity, would be an appropriately cosmic time scale. The doubling time is inversely proportional to the square root of Λ, however; in principle, t_d could be arbitrarily brief for sufficiently large Λ.

D. EQUIVALENCE TO VACUUM ENERGY DENSITY

The final aspect of the cosmological term to be examined here is that it perfectly mimics a certain type of stress-energy tensor, and vice versa. To see the equivalence, recall Eq. (21) for the $T_{\mu\nu}$ of a perfect fluid. If the mass-energy density ρ were characterized by a negative pressure P such that $P = -c^2\rho$, then $T_{\mu\nu}$ would simplify to

$$T_{\mu\nu} = Pg_{\mu\nu} = -c^2\rho g_{\mu\nu} \qquad (51)$$

If ρ were furthermore uniform in space and constant in time, then $T_{\mu\nu}$ would make precisely the same contribution to the field equations (20) as would a cosmological constant Λ_ρ given by

$$\Lambda_\rho = 8\pi Gc^{-2}\rho \qquad (52)$$

Is there any reason to suppose that $T_{\mu\nu}$ might contain a term such as that described above, equivalent to a cosmological constant? Indeed there is. In relativistic quantum field theory (QFT), which describes the creation, annihilation, and interaction of elementary particles, the vacuum state is not perfectly empty and void of activity, because such a perfectly ordered state would be incompatible with the uncertainty principles of quantum theory. The vacuum is defined as the state of minimum energy, but it is characterized by spontaneous creation and annihilation of particle–antiparticle pairs that sometimes interact before disappearing. Straightforward efforts to calculate the energy density of the vacuum state lead to divergent integrals suggesting an infinite result.

For most purposes, only changes in energy are physically significant; this means that the energy of a system is arbitrary up to an overall additive constant. With this rationale, the energy density of the vacuum state is often regarded as zero by definition. In the context of GTR, however, a major question arises: Does the vacuum state of QFT really contribute nothing to the $T_{\mu\nu}$ appearing in Einstein's field equations, or do the inevitable quantum fluctuations in the vacuum state contribute to $T_{\mu\nu}$ and thereby affect the metric $g_{\mu\nu}$ of space–time?

Let us consider the possibility that quantum fluctuations in the vacuum state do lead to an energy density that acts as a source in the field equations. These fluctuations occur throughout space, in a way governed almost entirely by the local physics of special relativity. Hence the energy density of the vacuum would not be expected to change appreciably as the universe expands. Denoting the corresponding mass density by ρ_v (obtained via division by c^2), the corresponding pressure P_v exerted by the vacuum state may be deduced from Eq. (24). Assuming ρ_v remains constant as the universe expands, Eq. (24) would imply that

$$P_v = -c^2\rho_v \qquad (53)$$

which is precisely the relation between P and ρ that mimics a cosmological term in the field equations. The corresponding value for an equivalent cosmological constant would be given by Eq. (52) with $\rho = \rho_v$.

Thus we see a plausible physical mechanism for generating an effective cosmological term: inevitable, spontaneous quantum fluctuations in the vacuum state. Whether such fluctuations actually affect the evolution of space-time in the manner outlined above remains an open question, from both the observational and theoretical points of view.

Although elementary particle physics (via QFT) does suggest the possibility of an effective cosmological term, one can only guess at the magnitude of a Λ_v that might arise in this way. A common attitude of particle theorists is that empirically, Λ is so small by the standards of particle physics (as will be explained in Section VI,C) that it is probably zero for some reason not yet known. It has in fact been speculated that Λ might necessarily vanish in "supergravity" and/or "superstring" theories of the elementary particles and their interactions. This has not been demonstrated conclusively, however, nor has the relevance of either theory to the real world.

We have seen that a vacuum energy density would mimic the effects of a cosmological term. The conjecture of cosmic inflation rests on the assumption that during a brief, early period of cosmic history, space was characterized by an unstable state called the "false vacuum" with a very large energy density mimicking a large Λ. During this brief period the scale factor would have experienced exponential, geometric growth, with a doubling time on the order of 10^{-34} sec. It seems plausible that doubling may have occurred hundreds or thousands of times, enormously expanding the cosmos during the first 10^{-30} sec or so of its existence.

VI. Unified Theories and Higgs Fields

A. QUANTUM THEORY OF FORCES

In quantum field theory (QFT), forces between particles result from the exchange of *virtual quanta,* i.e., transient, particlelike manifestations of some field that gives rise to the force. Electromagnetic forces arise from the exchange of virtual photons, where *photon* is the name given to a quantum or energy packet of the electromagnetic field. The weak nuclear force results from the exchange of *weak vector bosons,* of which there are three kinds called W^+, W^-, and Z. The strong nuclear force results, at the most fundamental level, from the exchange of eight *colored gluons* (a whimsical, potentially misleading name in that such particles are without visible qualities; *color* refers here to a technical property). From the viewpoint of QFT, gravitation may be regarded as resulting from the exchange of virtual *gravitons*. The four types of force just enumerated encompass all of the fundamental interactions currently known between elementary particles. (A few investigators have recently cited evidence suggesting a fifth, very weak force involving a quantity called *hypercharge,* but the current evidence is not conclusive.)

It is well-known that a major goal of Einstein's later work was to construct a unified field theory in which two of the existing categories of force would be understood as different aspects of a single, more fundamental force. Einstein failed in this attempt; in retrospect, the experimental knowledge of elementary particles and their interactions was too fragmentary at the time of his death (1955) for him to have had any real chance of success. The intellectual heirs of Einstein have kept his goal of unification high on their agendas, however. With the benefit of continual, sometimes dramatic experimental advances, and mathematical insights from a substantial number of theorists, considerable progress in unification has now been achieved. A wholly unexpected by-product of contemporary unified theories is the possibility of cosmic inflation. To understand how this came about, let us briefly examine what is meant by a unification of forces.

Each of the four categories of force enumerated above has several characteristics that, taken together, serve to distinguish it from the other types of force. One obvious characteristic is the way in which the force between two particles depends on the distance between them. For example, electrostatic and (Newtonian) gravitational forces reach out to arbitrarily large distances, while decreasing in strength like the inverse square of the distance. Forces of this nature (there seem to be only the two just named) are said to be *long range.* The weak force, in contrast, is effectively zero beyond a very short distance ($\sim 2.5 \times 10^{-16}$ cm). The strong force between protons and neutrons is effectively zero beyond a short range of $\sim 1.4 \times 10^{-13}$ cm, but is known to be a residual effect of the more fundamental force between the quarks (elementary, subnuclear particles) of which they are made. It is the force between quarks that is mediated by colored gluons (in a complicated way with the remarkable property that, under special conditions, the force between quarks is virtually independent of the distance between them).

The range of a force is intimately related to the mass of the virtual particle whose exchange gives rise to the force. Let us consider the interaction of particles A and B through exchange of virtual particle C. For the sake of definiteness, suppose that C is emitted by A and later absorbed by B, with momentum conveyed from A to B in the process. The emission of C appears to violate energy conservation, for C did not exist before it was emitted. This process is permitted, however, by the time-energy uncertainty principle of quantum theory: energy conservation may be violated by an amount ΔE for a duration Δt, provided that $\Delta E \times \Delta t \approx \hbar$ (where $\hbar$ denotes Planck's constant divided by 2π). If particle C has a rest-mass m, then $\Delta E \geq mc^2$, so that $\Delta t \leq \hbar/mc^2$. The farthest that C can travel during its lifetime Δt is $c\,\Delta t$, which limits the range of the resulting force to a distance of roughly $\hbar/mc$ called the *Compton wavelength* of particle C.

The preceding discussion is somewhat heuristic, but is essentially correct. If a particle being exchanged has $m > 0$, then the resulting force drops rapidly to zero for distances exceeding its Compton wavelength; conversely, long-range forces (i.e., electromagnetism and gravitation) can only result from exchange of "massless" particles. (Photons and gravitons are said to be massless, meaning zero rest-mass. Calling them "massless" is a linguistic convention, because such particles always travel at the speed of light and hence defy any measurement of mass when at rest. They may have arbitrarily little energy, however, and they behave as theory would lead one to expect of a particle with zero rest-mass.)

There are other distinguishing features of the four types of force, such as the characteristic strengths with which they act, the types of particles on which they act, and the force direction. We shall not pursue these details further here, but simply remark that two (or more) types of force would be regarded as unified if a single equation (or set of equations) were found, based on some unifying principle, that described all the forces in question. Electricity and magnetism were once regarded as separate types of force, but were united into a single theory by James Clerk Maxwell in 1864 through his equations which show their interdependence. From the viewpoint of QFT, electromagnetism is a single theory because both electrical and magnetic forces result from the exchange of photons; forces are currently identified in terms of the virtual quanta or particles that give rise to them.

B. Electroweak Unification via the Higgs Mechanism

When seeking to unify forces that were previously regarded as being of different types, it is natural to exploit any features that the forces may have in common. Einstein hoped to unify electromagnetism and gravitation, in considerable measure because both forces shared the striking feature of being long-range. As we have noted, Einstein's efforts were not successful. In the 1960s, however, substantial thought was given to the possibility of unifying the electromagnetic and *weak* interactions.

The possibility of unifying the weak and electromagnetic interactions was suggested by the fact that both interactions affected all electrically charged particles (albeit in distinctly different ways), and by the suspicion that weak interactions resulted from exchange of weak vector bosons which had not yet been observed but (if they existed) would have the same intrinsic angular momentum or "spin" as the photon. (Photons and all other particles whose spin is unity times $\hbar$ are called *vector bosons*.) The electromagnetic and weak forces would be regarded as unified if an intimate relationship could be found between the photon and weak vector bosons, and between the strengths of the respective interactions. (More specifically, in regrettably technical terms, it was hoped that the photon and an uncharged weak vector boson might

transform into one another under some symmetry group of the fundamental interactions, and that coupling strengths might be governed by eigenvalues of the group generators. The symmetry group would then constitute a unifying principle.)

The effort to unify these interactions faced at least two obstacles, however. The most obvious was that photons are massless, whereas the weak interaction was known to be of very short range. Thus weak vector bosons were necessarily very massive, which seemed to preclude the intimate type of relationship with photons necessary for a unified theory. A second potential difficulty was that while experiments had suggested (without proving) the existence of electrically charged weak bosons, now known as the W^+ and W^-, unification required a third, electrically neutral weak boson as a candidate for (hypothetical) transformations with the neutral photon. There were, however, no experimental indications that such a neutral weak boson (the Z) existed.

Building on the work of several previous investigators, Steven Weinberg (in 1967) and Abdus Salam (independently in 1968) proposed an ingenious unification of the electromagnetic and weak interactions. The feature of principle interest to us here was the remarkable method for unifying the massless photon with massive weak bosons. The obstacle of different masses was overcome by positing that, at the most fundamental level, there was no mass difference. Like the photon, the weak vector bosons were assumed to be massless at the most basic level of the theory. How, then, could the theory account for the short range of the weak force and corresponding large effective masses of the weak bosons? The theory did so by utilizing an ingenious possibility discovered by Peter Higgs in 1964, called the *Higgs mechanism* for (effective) mass generation.

A QFT is partially defined by its *Lagrangian*, a function of the fields (and their derivatives) that specifies the kinetic and potential energies of the theory and also how the different fields and their corresponding quanta (or particles) interact with one another. The Langrangian contains two kinds of physical constants that are basic parameters of the theory: the masses of those fields that represent massive particles, and *coupling constants* that determine the strengths of interactions.

A theory wherein the Higgs mechanism is operative has a scalar field variable $\Phi(\mathbf{r}, t)$ (a *Higgs field*) that appears in the Langrangian in those places where a mass would appear for certain particles if those particles had a nonzero mass. The potential energy function of the Higgs field is then defined in a way that assures that the state of minimum energy corresponds to a non-zero, constant value v for the Higgs field. The theory's state of minimum energy is identified with the vacuum, and may be regarded as a stage upon which ordinary phenomena are played out. Hence the theory is constructed so that its vacuum has $\Phi = v$, which mimics the effect of a nonzero mass (equal to v, up to a numerical factor that we shall ignore) for certain particles. The mimicking of mass occurs by virtue of Φ appearing in the Lagrangian in those places where a mass would appear if there were a fundamental, intrinsic mass.

The Higgs mechanism may sound like a very artificial procedure for explaining so basic a property as the mass of a particle. It was in fact regarded by many as a mathematical curiosity of no physical significance when first proposed. The Higgs mechanism plays a central role in the Weinberg–Salam electroweak theory, however, for the weak bosons are assumed to acquire their large effective masses in precisely this way. Furthermore, the Weinberg–Salam theory has been experimentally confirmed in striking detail (including the existence of the neutral weak Z boson, with just the mass and other properties predicted by the theory). Verification of the theory led to Weinberg and Salam sharing the 1979 Nobel Prize in Physics with Sheldon Lee Glashow, who had laid some of the ground-work for the theory.

Only a single Higgs field has been mentioned thus far, but the Higgs mechanism actually re-quires at least two: one that takes on a nonzero value in the vacuum, plus one for each vector boson that acquires an effective mass through the Higgs mechanism. (There are four real [or two complex] Higgs fields in the Weinberg–Sa-lam theory, since the W^+, W^-, and Z weak bos-ons all acquire effective masses in this way.) Furthermore, there are many different states that share the same minimum energy, and hence many possible vacuum states for the theory.

As a simple example, consider a theory with two Higgs fields Φ_1 and Φ_2 and a potential en-ergy–density function V given by

$$V(\Phi_1, \Phi_2) = b[(\Phi_1^2 + \Phi_2^2) - v^2]^2 \quad (54)$$

where b and v are positive constants. The mini-mum value of V is zero, which occurs when

$$\Phi_1 = v \cos \psi \quad \text{and} \quad \Phi_2 = v \sin \psi \quad (55)$$

for any value of the "angle" ψ. Given that Φ_1 and Φ_2 satisfy Eq. (55), a more convenient pair of Higgs fields Φ_1' and Φ_2' may be defined by

$$\begin{aligned} \Phi_1' &\equiv \Phi_1 \sin \psi - \Phi_2 \cos \psi \\ \Phi_2' &\equiv \Phi_1 \cos \psi + \Phi_2 \sin \psi \end{aligned} \quad (56)$$

which have the properties that $\Phi_1' = 0$ while $\Phi_2' = v$ in the vacuum state. Hence one may say without loss of generality that "the vacuum" is characterized by $\Phi_2' = v$, so long as one bears in mind that some particular vacuum correspond-ing to a particular value of ψ has been selected by Nature from a continuous set of possible vacuums.

Theories containing Higgs fields are always constructed so that the Higgs fields transform among one another under the action of a symme-try group, i.e., a group of transformations that leaves the Langrangian unchanged. [For the simple example just described, with V given by Eq. (54), the symmetry group is that of rotations in a plane with perpendicular axes correspond-ing to Φ_1 and Φ_2: the group is called $U(1)$.]

Higgs fields are always accompanied in a the-ory by (fundamentally) massless vector bosons that transform among one another under the same symmetry group. The process whereby Nature chooses some particular vacuum (corre-sponding in our simple example to a particular value for ψ) is essentially random, and is called *spontaneous symmetry breaking*. Once the choice has been made, the original symmetry among all the Higgs fields is broken by the con-tingent fact that the actual vacuum realized in Nature has a nonzero value for some particular Higgs field but not for the others. At the same time, the symmetry among the massless vector bosons is broken by the generation of effective mass for one (or more) of them, which in turn shortens the range of the force resulting from exchange of one (or more) of the vector bosons. Theories involving the Higgs mechanism are sometimes said to possess a *hidden symmetry*, because the underlying theory has a symmetry that is not apparent in the phenomena observed in laboratories after some particular vacuum has been selected by Nature.

It seems possible, even likely, that in other regions of space–time far removed from our own the choice of vacuum has been made differ-ently. In particular, there is no reason for the selection to have been made in the same way in regions of space–time that were outside each others' horizons at the time when selection was made. The fundamental laws of physics are believed to be the same everywhere, but the spontaneous breaking of symmetry by random selections of the vacuum state should have occurred differently in different "domains" of the universe.

We have remarked that the state of minimum energy has a nonzero value for one of the Higgs fields. If we adopt the convention that this state (the vacuum) has zero energy, then it follows that a region of space where all Higgs fields are zero must have positive mass-energy ρ_H per unit volume of space. Our example with the Higgs potential given by Eq. (54) would have

$$\rho_H = bv^4/c^2 \qquad (57)$$

Work would be required to expand any such region wherein all Higgs fields vanish, for doing so would increase the net energy. Such a region is therefore characterized by negative pressure P_H that, by the work-energy theorem, is related to ρ_H by $P_H = -c^2\rho_H$ [as may be seen from Eq. (25) for constant ρ and P].

Of all the fields known or contemplated by particle theorists, only Higgs fields have positive energy when the fields themselves vanish. This feature of Higgs fields runs strongly against intuition, but the empirical successes of the Weinberg–Salam theory force one to take the possibility seriously. It is safe to say that our present vacuum has zero (or very nearly zero) energy, for a nonzero vacuum energy density would mimic a cosmological constant, and the latter is known to be very small by the standards of particle physics (see Sections V,D and VI,C). Hence the vacuum energy density is surely positive when all Higgs fields vanish. If all Higgs fields were zero throughout the cosmos, the resulting ρ_H and negative P_H would perfectly mimic a positive cosmological constant, leading to an accelerating expansion of the cosmos. That such was the case during some fraction of the first second after creation is the essence of the inflationary hypothesis. The Higgs fields of central importance for inflation were those of a *grand unified theory*, which we shall next describe.

C. Grand Unified Theories

As evidence was mounting for the Weinberg–Salam theory in the early 1970s, a consensus was also emerging that the strong force between quarks is a result of the exchange of eight colored gluons, as described by a theory called *quantum chromodynamics*. Like the photon and the weak bosons, gluons are vector bosons, which makes feasible a unification of all three forces. (Gravitons are not vector bosons [they have a spin of twice $\hbar$], so that an ultimate unification with gravity would require a different type of unifying principle.) Furthermore, theoretical analyses indicated that the strengths of all three forces become equal at a collision energy of $\sim 10^{15}$ GeV, suggesting that the three forces might indeed be unified by a symmetry that is spontaneously broken at lower energies. Such a unifying symmetry would also explain the equality of magnitude between electron and proton electrical charge.

Guided by these considerations and by insights gained from the earlier electroweak unification, Howard M. Georgi and Sheldon Glashow proposed the first grand unified theory (GUT) of all three forces in 1974 [a group called SU(5) was assumed to describe the underlying symmetry]. The Higgs mechanism was adduced to generate effective masses for several particles in the theory: not only the weak vector bosons, but also supermassive X vector bosons whose existence is required to achieve the grand unification.

Following the lead of Georgi and Glashow, numerous other investigators have proposed competing GUTs, differing in the underlying symmetry group and/or other details. There are now many such theories that reproduce the successful predictions of electroweak theory and quantum chromodynamics at the energies currently accessible in laboratories, while making different predictions for very high energies where we have little or no data of any kind. Thus we do not know which (or indeed whether any) of these GUTs is correct. They typically share three features of significance for cosmology, however, which we shall enumerate and regard as general features of GUTs.

GUTs involve the Higgs mechanism, and the unifying symmetry is spontaneously broken for all energies below $\sim 10^{14}$ GeV. The energy at which symmetry breaking occurs determines the gross features of the Higgs potential and thereby ρ_H, the mass-energy density for the case where all Higgs fields vanish. It is found that

$$\rho_H \approx (10^{14} \text{ GeV})^4/(\hbar^3 c^5) \approx 10^{74} \text{ g/cm}^3 \quad (58)$$

If all Higgs fields were zero throughout the cosmos, then the effective cosmological constant Λ_ρ [according to Eq. (52)] would be

$$\Lambda_\rho \approx 10^{51} \text{ m}^{-2} \qquad (59)$$

If growth of the cosmic scale factor (which governs the linear dimensions of the universe) were dominated by such a value for Λ_ρ, then the corresponding doubling time would be given by Eq. (49) as

$$t_d \approx 10^{-34} \text{ sec} \qquad (60)$$

At this rate of growth, 100 doublings of the cosmic scale factor would occur in 10^{-32} sec, which is the extraordinary kind of growth conjectured for an early era of cosmic inflation.

As noted in Section V,B, the observed dynamics of galactic clusters imply an upper bound on $|\Lambda|$ of $\sim 10^{-52}$ m^{-2}. This empirical bound on $|\Lambda|$ is smaller than the effective Λ_ρ of a GUT false vacuum by a factor of 10^{-103}, which explains our earlier statement that Λ is known to be very small by the standards of particle physics.

A second feature of GUTs is that *baryon number* is not strictly conserved: the total number of protons and neutrons need not remain constant. (Protons and neutrons have baryon number +1, whereas antiprotons and antineutrons have baryon number -1; nonconservation of baryon number means that matter and antimatter need not be created [or destroyed] in equal amounts.) The violation of baryon conservation is intrinsically a high-energy process, however, for it results from emission or absorption of X bosons, whose mc^2 energy is $\sim 10^{14}$ GeV. Laboratories cannot provide this great an energy, so baryon nonconservation (if it occurs in this way) would have escaped detection in collision experiments.

When the universe was sufficiently young, the temperature was high enough for X bosons to have been abundant. Hence one need not assume that the universe was created with more matter than antimatter, as we find it now. Beginning with work by Motohiko Yoshimura in 1978, many scientists have utilized GUTs to show how a universe created with equal amounts of matter and antimatter could have become strongly dominated by matter at a very early time, when thermal energies were dropping below 10^{14} GeV. Such calculations seem capable of explaining not only the dominance of matter since that early time, but also the fact that there are $\sim 10^9$ times as many photons in the cosmic background radiation as there are baryons in the observable universe.

The apparent success (within substantial uncertainties) of such calculations may be regarded as indirect evidence supporting GUTs. More direct and potentially attainable evidence

would consist of observing proton decay, which is predicted by various GUTs to occur with a half-life on the order of 10^{29}–10^{35} yr (depending on the particular GUT and on the values of parameters that can only be estimated). No proton decay has yet been observed with certainty, however, and ongoing experiments indicate that the proton half-life exceeds 10^{32} yr.

A third feature of GUTs results from the fact that there is no single state of lowest energy but rather a *multiplicity* of possible vacuum states. The choice of vacuum was made by Nature as the universe cooled below thermal energies of 10^{14} GeV at an age of $\sim 10^{-35}$ sec. Since different parts of the universe were outside each others' horizons and had not yet had time to interact in any way, the choice of vacuum should have occurred differently in domains that were not yet in causal contact. These mismatches of the vacuum in different domains would correspond to surfacelike defects called *domain walls* and surprisingly, as was shown in 1974 by G. t' Hooft and (independently) by A. M. Polyakov, also to pointlike defects that are magnetic monopoles.

If the universe has evolved in accord with the standard big bang model (i.e., without inflation), then such domain walls and magnetic monopoles should be sufficiently common to be dramatically apparent. None have yet been observed with certainty, however, which may either be regarded as evidence against GUTs or in favor of cosmic inflation: inflation would have spread them so far apart as to render them quite scarce today.

D. PHASE TRANSITIONS

The notion of phase is most familiar in its application to the three states of ordinary matter: solid, liquid, and gas. As heat is added to a solid, the temperature gradually rises to the melting point. Addition of further energy equal to the latent heat of fusion results in a phase transition to the liquid state, still at the melting point. The liquid may then be heated to the boiling point, where addition of the latent heat of vaporization results in a transition to the gaseous phase. The potential energy resulting from intermolecular forces is a minimum in the solid state, distinctly greater in the liquid state, and greater still for a gas. The latent heats of fusion and vaporization correspond to the changes in potential energy that occur as the substance undergoes transition from one phase to another.

In QFT, the state of minimum energy is the

vacuum, with no particles present. If the theory contains Higgs fields, then one of them is non-zero in the vacuum. The vacuum (if perfect) has a temperature of absolute zero, for any greater temperature would entail the presence of a thermal (blackbody) distribution of photons. A temperature of absolute zero is consistent, of course, with the presence of any number of massive particles.

If heat is added to some region of space it will result in thermal motion for any existing particles, and also in a thermal distribution of photons. If the temperature is raised to a sufficiently high level, then particle–antiparticle pairs will be created as a result of thermal collisions: the mc^2 energies of the new particles will be drawn from the kinetic energies of preexisting particles. The Higgs field that was a nonzero constant in the vacuum retains that value over a wide range of temperatures, however, and this condition may be regarded as defining a phase: it implies that certain vector bosons have effective masses, so that the underlying symmetry of the theory is broken in a particular way.

If heat continues to be added, then the temperature and thermal energy will eventually rise to a point where the Higgs fields no longer remain in (or near) their state of lowest energy: they will undergo random thermal fluctuations about a central value of zero. When this happens, the underlying symmetry of the theory is restored: those vector bosons that previously had effective masses will behave like the massless particles they fundamentally are. This clearly constitutes a different phase of the theory, and we conclude that Nature undergoes a phase transition as the temperature is raised above a critical level sufficient to render all Higgs fields zero.

Roughly speaking, this symmetry-restoring phase transition occurs at the temperature where the mass-energy per unit volume of blackbody radiation equals the ρ_H corresponding to the vanishing of all Higgs fields; an equipartition of energy effects the phase transition at this temperature when latent heat equal to $\rho_H c^2$ is added. For the ρ_H given by Eq. (58), the phase transition occurs at a temperature near 10^{27} K. Like all phase transitions, it can proceed in either direction, depending on whether the temperature is rising or falling as it passes through the critical value.

The relation between phase transitions and symmetry breaking may be elucidated by noting that ordinary solids break a symmetry of Nature that is restored when the solid melts. The laws of physics have rotational symmetry, as do liquids: there is no preferred direction in the arrangement of molecules in a liquid. When a liquid is frozen, however, the rotational symmetry is broken because solids are crystals. The axes of a crystal may be aligned in any direction; but once a crystal has actually formed, its particular axes single out particular directions in space.

It is also worth noting that rapid freezing of a liquid typically results in a moderately disordered situation where different regions have their crystal axes aligned in different directions: domains are formed wherein the rotational symmetry is broken in different ways. We may similarly expect that domains of broken-symmetry phase were formed as the universe cooled below 10^{27} K, with the symmetry broken in different ways in neighboring domains. In this case the horizon distance would be expected to govern the domain size.

There is one further aspect of phase transitions that is of central importance for cosmic inflation, namely, the phenomenon of *supercooling*. For a familiar example, consider water, which normally freezes at 0°C. If a sample of very pure (distilled) water is cooled in an environment free of vibrations, however, it is possible to cool the sample more than 20 C° below the normal freezing point while still in the liquid phase. Liquid water below 0°C is obviously in an unstable state: any vibration is likely to trigger the onset of crystal formation (i.e., freezing), with consequent release of the latent heat of fusion from whatever fraction of the sample has frozen.

When supercooling of water is finally terminated by the formation of ice crystals, *reheating* occurs: release of the crystals' heat of fusion warms the partially solid, partially liquid sample (to 0°C, at which temperature the remaining liquid can be frozen by removal of its heat of fusion). Such reheating of a sample by released heat of fusion is to be expected whenever supercooling is ended by a phase transition (though not all systems will have their temperature rise as high as the normal freezing point).

Some degree of supercooling should be possible in any system that undergoes phase transitions, including a GUT with Higgs fields. A conjecture that supercooling (and reheating) occurred with respect to a GUT phase transition in the early universe plays a crucial role in theories of cosmic inflation, which we are now prepared to describe in detail.

VII. Scenarios for Cosmic Inflation

A. THE ORIGINAL MODEL

We now assume that some GUT with the features described earlier (in Section VI,C) provides a correct description of particle physics, and consider the potential consequences for the evolution of the cosmos. According to the standard big bang model, the temperature exceeded 10^{27} K for times earlier than 10^{-35} sec. Hence the operative GUT would initially have been in its symmetric phase wherein all Higgs fields were undergoing thermal fluctuations about a common value of zero.

As the temperature dropped below the critical value of 10^{27} K, there are two distinct possibilities (with gradations between). One possibility is that the GUT phase transition occurred promptly, in which case the expansion, cooling, and decrease in mass-energy density of the universe would have progressed in virtually the same way as in the standard model. The GUT symmetry would have been broken in different ways in domains outside each others' horizons, and one can estimate how many domain walls and magnetic monopoles would have resulted. A standard (albeit numerical) solution to the field equations of GTR tells one how much the universe has expanded since then, so it is a straightforward matter to estimate their present abundances.

No domain walls or magnetic monopoles have yet been observed with certainty, and the abundances predicted by the preceding line of reasoning vastly exceed the levels consistent with observation. For example, the magnetic monopoles of GUTs are calculated to be about 10^{16} times as massive as a proton. If the GUT phase transition had occurred promptly, then the contribution of monopoles to the present cosmic mass density would be roughly 10^{12} times the critical density (with $\Lambda = 0$), and one can show that the cosmos would have evolved to its present stage after a brief period of only 30,000 yr. Such a young age for the universe is grossly inconsistent with an overwhelming amount of evidence from many branches of science. The actual abundance of monopoles is very much less than is predicted by GUTs with a prompt phase transition; this is the monopole problem.

The monopole problem was discovered by elementary particle theorists soon after the first GUT was introduced, and appeared for several years to constitute evidence against GUTs. In 1980, however, Alan H. Guth proposed a remarkable scenario based on GUTs that held promise for resolving the horizon and flatness problems, and perhaps the monopole and other problems as well. Guth conjectured that as the universe expanded and cooled below the GUT critical temperature of 10^{27} K, supercooling occurred with respect to the GUT phase transition. All Higgs fields retained values near zero for a period of time after it became thermodynamically favorable for one (or a linear combination) of them to assume a nonzero value in the symmetry-breaking phase.

Supercooling would have commenced when the universe was $\sim 10^{-35}$ seconds old. The mass-energy density of all forms of matter and radiation other than Higgs fields would have continued to decrease as the universe expanded, while that of the Higgs fields retained the constant value ρ_H corresponding to zero values for all Higgs fields. Within 10^{-34} sec, ρ_H and the corresponding negative pressure $P_H = -\rho_H c^2$ would have dominated the stress-energy tensor, which would then have mimicked a large, positive cosmological constant. The cosmic scale factor would have experienced a period of exponential growth, with a doubling time of $\sim 10^{-34}$ sec. Guth proposed that all this had happened, and called the period of exponential growth the *inflationary era*.

The Higgs fields were in a metastable or unstable state during the inflationary era, so the era was destined to end. Knowledge of its duration is obviously important: How many doublings of the scale factor occurred during this anomalous period of cosmic growth? The answer to this question depends on the precise form of the Higgs potential, which varies from one GUT to another and unfortunately depends, in any GUT, on the values of several parameters that can only be estimated. Guth originally assumed that the Higgs potential had the form suggested by Fig. 1a. For such a potential, a zero Higgs field corresponds to a local minimum of the energy density, and the corresponding state is called a *false vacuum*. (The *true vacuum* is of course the state wherein the Higgs potential is an absolute minimum; that is the broken-symmetry phase where one of the Higgs fields has a nonzero constant value.)

In the classical (i.e., nonquantum) form of the theory, a false vacuum would persist forever. The physics is analogous to that of a ball resting in the crater of a volcano. It would be energetically favorable for the ball to be at a lower eleva-

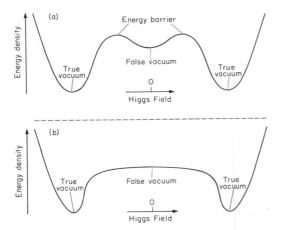

FIG. 1. Higgs potential energy density for (a) original inflationary model, and (b) new inflationary model. For a theory with two Higgs fields, imagine each curve to be rotated about its vertical axis of symmetry, forming a two-dimensional surface. A true vacuum would then correspond to any point on a circle in the horizontal plane.

tion, hence outside the volcano at its base. In classical physics there is no way for the ball to rise spontaneously over the rim and reach the base, so the ball would remain in the crater for all time.

In quantum physics, the story changes: a ball in the crater is represented by a wave packet, which inevitably develops a small extension or "tail" outside the crater. This feature of quantum mechanics implies a finite probability per unit time that the ball will "tunnel through" the potential energy barrier posed by the crater's rim and materialize outside at a lower elevation, with kinetic energy equal to the decrease in potential energy. Such quantum-mechanical tunneling through energy barriers explains the decay of certain species of heavy nuclei via the spontaneous emission of alpha particles, and is a well-established phenomenon. If there exists a region of lower potential energy, for either alpha particles or the value of a Higgs field, then "decay" of the initial state via tunneling to that region is bound to occur sooner or later. The timing of individual decays is governed by rules of probability, in accord with the statistical nature of quantum-mechanical predictions.

The decay of the false vacuum was first studied and described by Sidney R. Coleman. Initially all Higgs fields are zero, but one or another of them tunnels through the energy barrier in a small region of space and acquires the nonzero

value corresponding to the true vacuum, thereby creating a "bubble" of true vacuum in the broken symmetry phase. Once a bubble of true vacuum has formed, it grows at a speed that rapidly approaches the speed of light, converting the surrounding region of false vacuum into true vacuum as the bubble expands. As the phase transition proceeds, energy conservation implies that the mass-energy density ρ_H of the preexisting false vacuum is converted into an equal density of other forms of mass-energy, which Guth hoped would become hot matter and radiation that evolved into our present universe.

A low rate of bubble formation corresponds to a high probability that any particular region of false vacuum will undergo many doublings in size before decay of the false vacuum ends inflation for that region. Guth noted that if the rate of bubble formation were low enough to make it likely that the cosmic scale factor had inflated by a factor of 10^{28} or more in any given region, then the observable portion of the universe would probably have emerged from such a small region that all parts would have been in causal contact with each other before inflation began. Such causal contact would have resulted in a smoothing of any earlier irregularities via processes leading toward thermal equilibrium, thereby explaining (one hopes) the homogeneity and isotropy now observed: the horizon problem would be solved (and perhaps the smoothness problem as well).

It was furthermore noted by Guth that comparable growth of the cosmic scale factor might solve the flatness problem. If the scale factor had grown by a factor of 10^{29} or more during the period of inflation, Ω need not have been very close to unity before inflation. The present similarity between ρ and the critical density would result from the fact that the term involving the curvature constant k in Eq. (34) is rendered negligible by an enormous scale factor (squared) in the denominator. The absence of any observational evidence for curvature of space would also be explained by a scale factor appreciably larger than the radius of the observable universe, since curvature only becomes significant over distances comparable to the scale factor. [It remains unexplained, however, why Λ is either zero or small enough that $\rho_c(\Lambda)$, the critical density as a function of Λ, is not very different from $\rho_c(0)$. It should also be noted that this resolution of the flatness problem is contingent on the observable universe having remained sufficiently homogeneous and isotropic from the beginning of inflation to the present for the RW metric to have been approximately valid throughout the intervening time. We shall see that such persistence of homogeneity cannot be taken for granted.]

The rate at which bubbles form depends on the particular GUT and, within any GUT, on parameters including some whose values are unknown. Guth made the plausible assumption that the inflationary era lasted for at least 10^{-32} sec. With a doubling time of 10^{-34} sec, 100 or more doublings of the scale factor would have occurred during this brief period of cosmic evolution. One hundred doublings is equivalent to multiplication by a factor of 10^{30}; 1000 doublings, requiring only 10^{-31} sec, would result in multiplication of the cosmic scale factor by 10^{300}. It is even plausible that the universe inflated by a factor of 10 raised to the power of 10^{10}.

The range of plausible values for GUT parameters is such that if the universe inflated enough to resolve the horizon and flatness problems, then it is very likely to have inflated by a great deal more. That such is the case is typically regarded as a prediction of the inflationary model. The resulting value for the scale factor would be so large that terms involving k in a description of the observable portion of the universe would be negligible. The observable geometry should correspond very closely to the flat model, and the average density of mass-energy should be virtually indistinguishable from the critical density $\rho_c(\Lambda)$ (again provided that approximate homogeneity has persisted from the onset of inflation to the present, so that an RW metric has been approximately valid throughout the intervening time).

Whether cosmic inflation solves the monopole problem depends on whether the observable universe evolved from a region spanned by sufficiently few bubbles. If so there would be few mismatches between different domains of broken-symmetry phase, hence few domain walls and monopoles. Guth sought to understand whether a region now the size of the observable universe was more likely to have evolved from few or many bubbles, and he encountered a major shortcoming of the model. The role played by bubbles in the phase transition leads to a bizarre distribution of mass-energy, quite unlike what astronomers observe. The problem may be described as follows.

We have remarked that the ρ_H of the false vacuum is converted into an equal amount of mass-energy in other forms as the phase transition proceeds. It can be shown, however, that initially the mass-energy is concentrated in the expanding walls of the bubbles of true vacuum, in the form of moving "kinks" in the values of the Higgs fields. Conversion of this energy into a homogeneous gas of particles and radiation would require a large number of collisions between bubbles of comparable size. Bubbles are forming at a constant rate, however, and grow at nearly the speed of light. If enough time has passed for there to be many bubbles, there will inevitably be a wide range in the ages and therefore sizes of the bubbles: collisions among them will convert only a small fraction of the energy in their walls into particles and radiation. In fact it can be shown that the bubbles would form finite clusters dominated by a single largest bubble, whose energy would remain concentrated in its walls.

One could explain the homogeneity of the observable universe if it were contained within a single large bubble. There remains, however, the problem that the mass-energy would be concentrated in the bubble's walls: the interior of a single large bubble would be too empty (and too cold at early times) to resemble our universe. Guth therefore closed his original publication with an invitation to other workers to find a plausible modification of the inflationary model, one that preserved its strengths while overcoming its initial failure to produce a homogeneous, hot gas of particles and radiation that might have evolved into our present universe.

B. THE NEW INFLATIONARY MODEL

A satisfactory modification of Guth's original scenario was developed in 1981 by A. D. Linde and, independently, by Andreas Albrecht with Paul J. Steinhardt. Success hinges on obtaining a "graceful exit" from the inflationary era, i.e., an end to inflation that preserves homogeneity and isotropy and furthermore results in the efficient conversion of ρ_H into the mass-energy of hot particles and radiation. The new inflationary model achieves a graceful exit by postulating a modified form for the potential energy function of Higgs fields: the form depicted in Fig. 1b, which was first studied by Sidney Coleman in collaboration with Erick J. Weinberg. A Coleman–Weinberg potential has no energy barrier separating the false vacuum from true vacuum; instead of resembling the crater of a volcano, the false vacuum corresponds to the center of a rather flat plateau. (Though not a local minimum of energy, the plateau's center is nevertheless called a false vacuum for historical reasons.)

Supercooling of a Coleman–Weinberg false vacuum results in the formation of contiguous, causally connected domains throughout each of which the Higgs fields gradually evolve in a simultaneous, uniform way toward a single phase of true vacuum. The initial sizes of these domains are comparable to the horizon distance at the onset of supercooling. Within each domain, the evolution of a Higgs field away from its initial value of zero is similar to the horizontal motion of a ball that, given a slight nudge, rolls down the initially flat but gradually increasing slope of such a plateau. The equations describing evolution of the Higgs fields have a term corresponding to a frictional drag force for a rolling ball; in addition to the initial flatness of the plateau, this "drag term" serves to retard the evolution of a Higgs field away from zero toward its true-vacuum, broken-symmetry value. The result is called a "slow-rollover" transition to the true vacuum.

As always, there are different combinations of values of Higgs fields that have zero energy and qualify as true vacua; different domains undergo transitions to different true vacua, corresponding to the possibility of a ball rolling off a plateau in any one of many different directions.

As a Higgs field (or linear combination thereof) "rolls along" the top of the potential energy plateau, the mass-energy density remains almost constant despite any change in the volume of space. The corresponding pressure is large and negative, so inflation occurs. The doubling time for the cosmic scale factor is again expected to be $\sim 10^{-34}$ sec. The period of time required for a Higgs field to reach the edge of the plateau (thereby terminating inflation) can only be estimated, but again inflation by a factor of 10^{30}, or by very much more, is quite plausible. As in the "old inflationary model" (Guth's original scenario), it is typically assumed that the actual inflation greatly exceeded the amount required to resolve the horizon and flatness problems.

When a Higgs field "rolls off the edge" of the energy plateau, it enters a brief period of rapid oscillations about its true-vacuum value. In accord with the field-particle duality of QFT, these rapid oscillations of the field correspond to a dense distribution of Higgs particles. The Higgs particles are intrinsically unstable, and their decay results in a rapid conversion of their mass-energy into a wide spectrum of less massive particles, antiparticles, and radiation. The energy released in this process results in a high temperature that is less than the GUT critical temperature of 10^{27} K only by a modest factor ranging between 2 and 10, depending on details of the model. This is called the "reheating" of the universe.

The reheating temperature is high enough for GUT baryon-number–violating processes to be common, which leads to matter becoming dominant over antimatter. Reheating occurs when the universe is very young, probably less than 10^{-30} seconds old, depending on precisely how long the inflationary era lasts. From this stage onward, the observable portion evolves in the same way as in the standard big bang model. (A slow-rollover transition is slow relative to the rate of cooling at that early time, so that supercooling occurs, and slow relative to the doubling time of 10^{-34} sec; but it is extremely rapid by any macroscopic standard.)

The building blocks of the new inflationary model have now been displayed. As with the old inflationary model, there is no need to assume that any portion of the universe was initially homogeneous and isotropic. It is sufficient to assume that at least one small region was initially expanding, with a temperature above the GUT critical temperature.

So long as the temperature exceeded 10^{27} K, all Higgs fields would have undergone thermal fluctuations about a common value of zero, and the full symmetry of the GUT would have been manifest. The temperature fell below 10^{27} K when the universe was about 10^{-35} seconds old, at which time supercooling began and domains formed with initial sizes comparable to a horizon distance of about 10^{-24} cm. Since domains were causally connected, dynamical processes should have resulted in approximate homogeneity, isotropy, and thermal equilibrium within each domain. Hence the space–time geometry of each domain could be approximated by an RW metric, and the field equations of GTR took the simple form presented in Sections IV,C and IV,E. Homogeneity and thermal equilibrium caused inflation to begin simultaneously throughout a domain, and the process of inflation can be shown to have enhanced any preexisting homogeneity.

As supercooling proceeded, the continuing expansion caused dilution and further cooling of any preexisting particles and radiation, and the stress-energy tensor became dominated at a time of roughly 10^{-34} sec by the virtually constant contribution of Higgs fields in the false vacuum state. With the dominance of false-vacuum energy and negative pressure came inflation: the expansion entered a period of exponential growth, with a doubling time of $\sim 10^{-34}$ sec. The onset of inflation accelerated the rate of cooling; the temperature quickly dropped to $\sim 10^{22}$ K (the *Hawking temperature*), where it remained throughout the latter part of the inflationary era because of quantum effects that arise in the context of GTR.

For the sake of definiteness, we shall assume that domains inflated by a factor of 10^{50}, which would have occurred in $\sim 2 \times 10^{-32}$ sec. Hence a typical domain with an initial size of 10^{-24} cm at the onset of inflation would have grown to span a distance of 10^{26} cm when inflation ended. The observable portion of the universe had a radius of only ~ 10 cm when inflation ended, and it therefore (in all likelihood) lies deep inside a single domain of broken-symmetry phase. Thus neither domain walls nor magnetic monopoles are to be expected within the observable universe, except for a very few that might have resulted from extreme thermal fluctuations after inflation ended.

As the period of inflation neared its end, the density (i.e., number per unit volume) of any preexisting particles had been diluted virtually to zero. It follows that for all times later than $\sim 10^{-30}$ sec, *virtually all of the particles and radiation in the observable universe have been decay products of the mass-energy density ρ_H of the false vacuum.*

The slow-rollover transition ended almost simultaneously throughout each domain with a rapid and chaotic conversion of ρ_H into a hot ($T \geq 10^{26}$ K) gas of particles, antiparticles, and radiation. Note that if baryon number were strictly conserved, inflation would be followed for all time by equal amounts of matter and antimatter (which would have annihilated each other almost completely, with conversion of their mass-energy into radiation). The observable universe, however, is strongly dominated by matter. Hence inflation requires a GUT (or a theory sharing key features with GUTs) not only to explain an accelerating expansion, but also to explain how the symmetry between matter and antimatter was broken: through baryon-number–violating processes at the high temperature following reheating.

Some characteristic features of the new inflationary model are portrayed in Fig. 2. The hori-

zon problem is solved by placing the observable universe deep inside a homogeneous, isotropic domain that evolved from a region once so small that it was (and therefore remains) causally connected. Other regions of the universe may be very different: the GUT symmetry is expected to have been broken in different ways in different domains. Furthermore, there may be regions of the universe that never experienced inflation. Some such regions may have collapsed at an early time, and others may be still expanding but virtually empty. While providing a causal explanation for the homogeneity and isotropy of the region that we observe, inflation also removes any basis for believing that the *entire* universe is similar.

The interior of a domain may be approximated by an RW metric. The cosmic scale factor describing this homogeneous region should be comparable to the domain size, which is expected to be enormously larger than the radius of the observable universe. Hence the observable universe should have a virtually flat geometry and a mass-energy density virtually equal to the critical density $\rho_c(\Lambda)$. The flatness problem is solved [given that Λ is either zero or comparable to ρ in importance at present in Eq. (26)]. Since the observable universe lies inside a single domain of broken-symmetry phase, the monopole problem is solved as well.

The smoothness problem is alleviated by cosmic inflation, for inflation spread out and thereby smoothed any preexisting inhomogeneities of the domain in which we find ourselves. Had the phase transition proceeded and ended in a precisely simultaneous and uniform way throughout the domain, then the present clumping of matter into galaxies, clusters of galaxies, and superclusters would require explanation in terms of thermal fluctuations in density from the time of reheating onward as the seeds of gravitational instability (with turbulence and possibly shock waves also playing roles under active study). Detailed analyses indicate, however, that a nonthermal spectrum of early inhomogeneities is required to explain the present distribution of matter.

In the early 1970s, Edward R. Harrison and, independently, Yakov B. Zel'dovich proposed that a scale-invariant spectrum of early inhomogeneities is both necessary and sufficient to account for the observed clumping of matter into galaxies and clusters of galaxies. By *scale-invariant,* one means that departures from strict homogeneity were of equal magnitude over all

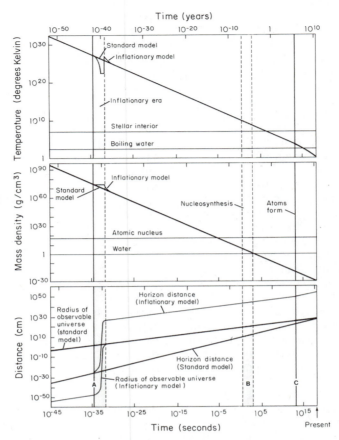

FIG. 2. Comparison of the new inflationary model with the standard big bang model of the universe. (top) Evolution of temperature; (middle) evolution of mass-energy density; and (bottom) evolution of horizon distance and of radius of the currently observable portion of universe. Note that both horizontal and vertical scales grow by powers of 10. [From Guth, A. H. and Steinhardt, P. J. (1984). "The Inflationary Universe." Copyright © 1984 by Scientific American, Inc. All rights reserved.]

distance scales of astrophysical interest. Remarkably, the new inflationary model predicts just such a spectrum of inhomogeneities. Throughout the period of inflation, the Higgs fields were subject to local, microscopic quantum fluctuations. Inflation expanded the spatial extent of these fluctuations so much that they span astronomical distances today. Furthermore, early fluctuations were inflated more than later ones, in such a way as to produce a scale-invariant spectrum of inhomogeneities when inflation ended.

The magnitude of these slight early departures from homogeneity is calculable and is sensitive to details of the underlying particle theory. The original, simplest GUT [minimal SU(5)] can in fact be rejected on the grounds that the predicted magnitude, albeit very small, would have been far too large to be consistent with the observed isotropy of the cosmic background radiation. (This GUT can also be rejected on the grounds that it predicts a proton lifetime of $\sim 10^{30}$ yr, in conflict with recent experiments that imply a lifetime exceeding 10^{32} yr.) The predicted inhomogeneities are nevertheless much smaller than those implied by the standard big bang model, wherein thermal fluctuations as early as 10^{-44} sec after the big bang would have been amplified by gravitational instabilities continually up to the present.

With the proper GUT, inflation might solve the smoothness problem completely by predicting not only a scale-invariant spectrum but also the correct magnitude of early density fluctuations required to explain the present distribution of matter. It is remarkable that given inflation, the distribution of matter on a cosmic scale allows us to distinguish between GUTs whose characteristic differences ordinarily become apparent only over distances of $\sim 10^{-29}$ cm (corresponding to collision energies of 10^{15} GeV).

We have seen that the new inflationary model shares all the successes of the old inflationary model but none of its failures: a suitable slow-rollover transition cures all ills. It should be noted, however, that a slow-rollover transition requires a Higgs potential of the Coleman–Weinberg type (or one close to it), which occurs in most GUTs only for special choices of certain parameters. A "fine tuning" of parameters in order to obtain the new inflationary scenario begs for justification, and undermines somewhat the plausibility of the model. In this connection it is worth noting that many GUTs incorporating *supersymmetry* (which relates particles with integer spin to those of half-integer spin and may incorporate gravity in the unification) lead to slow-rollover transitions without a fine-tuning of parameters. The successes of the new inflationary model are so impressive that most particle theorists regard the need for a slow-rollover transition as a basic criterion in their search for the ultimate theory of elementary particles and their interactions.

Two variations of the new inflationary model warrant mention. A minor modification of the Coleman–Weinberg potential, such as a small crater at the center of the energy-density plateau, could result in tunneling to bubbles still on the plateau followed by a slow-rollover continuation of the phase transition within each bubble. Thus a single bubble might grow large enough to contain the observable universe and then experience a graceful exit from inflation throughout the interior to produce the cosmos we observe. If the rate of bubble formation were low, collisions between bubbles could be rare enough to allow growth to cosmic sizes without collision. Detailed analysis reveals that the fraction of space filled by bubbles would approach unity as time proceeds, but the exponential growth of

space would imply a continual increase in absolute volume of the region occupied by false vacuum. Then bubbles would continue to form forever, and we would have no way of knowing how much time had passed before creation of the bubble in which we find ourselves.

The second variation we shall mention would not alter our understanding of past inflation, but suggests an interesting (albeit remote) possibility for the future. It was conjectured by Frank Wilczek and Michael S. Turner in 1982 that the observable universe might not be in the true vacuum phase, but instead be in a second, metastable false vacuum separated from the true state of minimum energy by a high-energy barrier. The ultimate phase transition would then occur eventually through tunneling to bubbles of true vacuum, which would expand at nearly the speed of light with the latent energy of the phase transition concentrated in their walls. In this case the wall of such a bubble would eventually sweep through the region of the universe that we inhabit. The consequences would be dramatic, but (in addition to being unlikely) would occur so quickly and completely as to render pointless any prior concern.

C. Creation Ex Nihilo

We have seen that the observable universe may have emerged from an extremely tiny region that experienced inflation and then populated the resulting cosmos with particles and radiation created from the mass-energy of the false vacuum. An ancient question emerges in a new context: How did that tiny region come into being, from which the observable universe emerged? Is it possible to understand the creation of a universe *ex nihilo?*

Scientific speculation about the ultimate origin of the universe appears to have begun in 1973, with a proposal by Edward P. Tryon that the universe was created from nothing as a spontaneous quantum fluctuation of some preexisting vacuum or state of nothingness. Central to the conjecture was a hypothesis that the universe has zero net values for all conserved quantities. Accepting the conventional wisdom of that time, Tryon believed that baryon number was strictly conserved, hence that a universe created from nothing would contain equal amounts of matter and antimatter. He therefore predicted equal numbers of matter and antimatter galaxies, which was then marginally consistent with observations simply because ground-based data remained inconclusive about the constitution of distant galaxies.

Speculation qualifies as scientific if it is vulnerable to refutation by observations. The conjecture of creation *ex nihilo* surely qualifies, for within 2 years it appeared to have been refuted: data collected with instruments borne by rockets above the earth's atmosphere led to a consensus that the universe is strongly dominated by matter (no evidence for antimatter galaxies or stars has ever been found). The basic prediction of the conjecture, however, is simply that the universe has zero net values for all conserved quantities. In the face of empirical evidence that matter is dominant, the conjecture may (with hindsight) be construed as predicting that baryon number is not conserved. For completely independent reasons, GUTs also predict the nonconservation of baryon number, so the conjecture of creation *ex nihilo* remains viable. (It should be borne in mind, however, that nonconservation of baryon number has not yet been confirmed by experiment.)

The other conserved quantities of possible concern are electric charge, angular momentum, linear momentum, and energy. Although cosmic net values cannot be measured with precision, observations are eminently consistent with zero net values for electric charge and angular momentum. Linear momentum is not germane, for its value depends on the reference frame of the observer: if nonzero for one observer, it will vanish for a second observer moving relative to the first.

Finally, we come to energy. As explained in Section IV,B, GTR does not permit any rigorous definition of the net energy of a universe. Energy conservation is a strictly local principle in GTR, and is therefore not applicable to the creation of an entire cosmos. It is nevertheless of some interest to note that gravitational potential energy has traditionally been regarded as negative. As was noted by Tryon, a simple and naive calculation suggests that the negative gravitational energy of the cosmos might completely cancel the positive mass-energy, provided that the average density of matter approximates the critical density (with $\Lambda = 0$). In part for this reason, Tryon predicted the universe to be closed (reasoning also that physical processes tend to result in finite systems and that only a closed universe is finite).

Although quantum fluctuations are typically microscopic in scale, no principle limits their potential size and duration, provided that conservation laws are respected. Hence given sufficient time, it seems inevitable that a universe with the size and duration of ours would spontaneously appear as a quantum fluctuation. Large (and long-lived) universes intuitively seem much less likely than smaller ones, however. Before the advent of inflationary theory, it was problematic to understand why our universe is very much larger than required for us to have evolved within it and pose these questions.

A mathematically detailed scenario for creation *ex nihilo* was published in 1978 by R. Brout, F. Englert, and E. Gunzig. This model assumed a large, negative pressure for the primordial state of matter, giving rise to exponential growth that converted an initial microscopic quantum fluctuation into an open universe. This model was a significant precursor to inflationary scenarios based on GUTs, as well as being a model for the origin of the universe.

Other detailed models for creation *ex nihilo* include those by David Atkatz with Heinz R. Pagels (1982) and by Alexander Vilenkin (1982). These scientists favor the creation of finite universes and, like Brout *et al*, utilize negative pressure in the early universe to explain the attainment of cosmic dimensions. The scenario of Vilenkin is the easiest to describe (and as plausible as any). In this scenario, a spontaneous quantum fluctuation gave rise to a microcosm that contained at least one region wherein all Higgs fields were initially zero. That region inflated and gave rise to the observable universe through the same sequence of processes as in the new inflationary model.

All the aforementioned theories of creation *ex nihilo* overcome the singularity problem of infinite quantities at time zero: creation is intrinsically a quantum process, so classical ideas do not extend back to zero time. The initial microcosm came into being with a finite size, and the greatest density contemplated is typically the Planck density. Expansion of the cosmos is understood to be a result of early inflation, which resulted from the large, negative pressure of an early false vacuum. There is no need to assume the universe was hot *before* inflation; if the initial quantum fluctuation yielded a region wherein all Higgs field were zero, then inflation would have ensued just as in the new inflationary model. The high temperature resulting from decay of the false vacuum may have been the *first* instance of high temperature. With creation *ex nihilo,* the "hot big bang" may have been preceded by a "cold big whoosh."

It is obvious that inflation greatly enhances the plausibility of creation *ex nihilo.* There remain, however, profound questions about which one can only speculate: On what stage did the primordial quantum fluctuation occur? What is meant by a vacuum or state of nothingness prior to our universe? What is meant by laws of physics predating the universe? These and other questions lack compelling answers, and may well defy resolution. It is nevertheless interesting that quantum uncertainties suggest the instability of nothingness, in which case inflation might have converted a spontaneous, microscopic quantum fluctuation into our cosmos.

BIBLIOGRAPHY

Davies, P. (1984). "Superforce." Simon and Schuster, New York.

Gibbons, G. W., Hawking, S. W., and Siklos, S. T. C. (Eds.). (1983). "The Very Early Universe." Cambridge Univ. Press, London and New York.

Gribbin, J. (1986). "In Search of the Big Bang." Bantam, New York.

Pagels, H. R. (1985). "Perfect Symmetry." Simon and Schuster, New York.

Trefil, J. S. (1983). "The Moment of Creation." Scribners, New York.

Wald, R. M. (1984). "General Relativity." Univ. of Chicago Press, Chicago, Illinois.

COSMIC RADIATION

Peter Meyer   *University of Chicago*

GLOSSARY

Energy spectrum: Intensity of particles as a function of their energy.

Extensive airshower: Large assembly of secondary particles produced by a high-energy cosmic ray near the top of the atmosphere, simultaneously propagating through the atmosphere.

Heliosphere: Volume of space around the sun that is dominated by the solar wind.

Interstellar gas: Atoms that fill the space between stars in the galaxy.

Mesons: Elementary particles that decay within a short lifetime.

Nuclear fragments: Atomic nuclei resulting from the breakup of a heavy nucleus (spallation).

Nucleosynthesis: Fusion of protons and neutrons to form nuclei of heavy chemical elements in the hot interior of stars.

Solar modulation: Modification of the galactic particle flux in the inner solar system due to the emission of gas by the sun.

Solar wind: Flow of ionized gas outward from the sun.

Spallation products: Fragments of a heavy nucleus that breaks up after collision with another nucleus.

Supernova: Exploding star that has exhausted its nuclear fuel and undergoes gravitational collapse, releasing a large amount of energy.

Synchrotron radiation: Electromagnetic radiation produced when high-energy charged particles orbit a magnetic field.

Universal blackbody radiation: Relic of the thermal radiation present after the "big bang," cooled to 3° absolute temperature due to the expansion of the universe.

Cosmic rays are high-energy nuclear particles and electrons that are accelerated in our galaxy as well as in other galaxies. Their composition and energy distribution point to the nature of their sources. Their interactions in the interstellar medium and in interplanetary space help to unravel the electromagnetic conditions as well as the distribution of magnetic fields and matter in both. Cosmic ray particles are the only matter from outside the solar system that can be directly investigated, and they are the only astronomical radiation we know of that carries mass.

I. Overview and Brief History

Cosmic radiation was discovered in 1912 by the Austrian scientist Victor Hess (Nobel Prize 1936). Hess's discovery was stimulated by the observation that detectors designed to record the radiation from radioactive substances showed the presence of a residual radiation even in the absence of such substances. In several manned balloon flights, Hess demonstrated that the intensity of the unknown radiation increased with increasing altitude and concluded that the radiation either was produced in the upper atmosphere or reached the earth from outer space. Today it is known that the latter conclusion is correct and that cosmic radiation consists of nuclear particles and electrons of high energy that impinge on the earth.

Experiments carried out on high-altitude balloons and on spacecraft established the fact that most cosmic rays are protons (i.e., nuclei of the hydrogen atom) but that they also contain the nuclei of all other known chemical elements, as well as electrons and positrons. The energy of these particles covers a wide range, from those encountered in nuclear physics of $\sim 10^6$ electron volts (eV) to the amazingly high value of 10^{20} eV. Some individual cosmic ray particles thus carry the largest quanta of energy observed in any radiation from the universe.

Being nuclear, strongly interacting particles, cosmic rays collide and interact with nuclei in the uppermost layers of the earth's atmosphere. A wealth of secondary particles is produced in these high-energy interactions: nuclear fragments, as well as many unstable "elementary" particles, including π^+, π^-, π^0 mesons, K mesons, and others. Before the development of high-energy accelerators, cosmic rays were the only nuclear projectiles available for the observation of high-energy phenomena. Many elementary particles were therefore first discovered by cosmic ray experiments (e.g., positrons, π mesons, μ mesons, and K mesons). Except for experiments that investigate phenomena at energies that cannot yet be reached by modern particle accelerators, high-energy physics is no longer part of cosmic ray physics, but is pursued mainly with accelerators.

The interaction of a very high energy cosmic ray nucleus ($E > 10^{15}$ eV) in the upper parts of the earth's atmosphere leads to a spectacular phenomenon: the extensive airshower. Hundreds or thousands of secondary particles can simultaneously be observed to reach the earth's surface as a result of a cascade process in the atmosphere in which secondary particles produce tertiaries and so on, distributing the enormous energy of the primary into a flood of secondaries. Extensive airshowers are observed with arrays of particle detectors spread over many square kilometers, registering the number, distribution, and time of the nearly simultaneously arriving secondary particles. In fact, through careful time measurements, the inclination of the front of particles with respect to the horizontal can be determined and thus the direction of incidence of the primary particle. Since particles at highest energy are so rare, observations of extensive airshowers are presently the only means of observing the energy spectrum and the flux of cosmic rays with energies in excess of $\sim 10^{16}$ eV. As a numerical example one may note that at energy greater than 10^{19} eV only one particle falls on an area of 1 km^2 in 1 yr.

II. Nature of the Primary Radiation

A. Nuclei

The radiation that penetrates the atmosphere bears no resemblance to the primary radiation. The undistorted primary cosmic radiation must be observed at very high altitude in the atmosphere (balloons) or outside the earth's atmosphere (spacecraft). The development of reliable stratospheric balloons in the 1950s and of earth satellites and deep-space probes in the 1960s provided vehicles with which to investigate the undistorted cosmic ray composition and led to measurements of the abundance distribution of the elements in cosmic rays. This distribution, from helium to nickel, is shown in Fig. 1. For comparison, the abundance of the elements in the Solar System is also displayed. There is an astonishing similarity between the cosmic ray abundances and the solar system abundances, but there are also groups of elements where the two abundances are strikingly different. The similarities led to the conclusion that cosmic ray nuclei must originate in nucleosynthesis processes in the interiors of stars before their acceleration. The differences between solar system and cosmic ray abundances can be understood as being due to collisions of the energetic particles with the interstellar gas, with subsequent breakup that produces secondary nuclei. This explains the relatively high flux of the elements lithium, beryllium, and boron, created mostly by spallation of carbon and oxygen nuclei, and also the filling of the abundance valley below iron by iron spallation products. It is now possible to measure with some accuracy the abundances of the very rare elements with atomic number larger than nickel, which are called ultraheavy nuclei in cosmic ray studies. The most recent measurement of this distribution, for the nuclei with even atomic number, is shown in Fig. 2 and again compared with solar system abundances. This comparison leads to the conclusion that the composition in this regime of nuclear charges is not greatly different from that of the solar system. To obtain the composition of cosmic rays at their source, all interactions that take place between source and observer must be carefully taken into account. This is a difficult task since

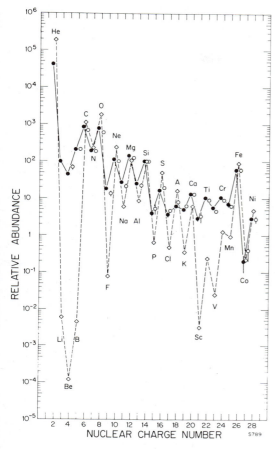

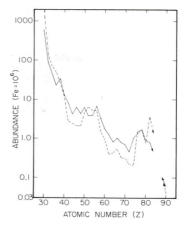

FIG. 2.   Relative abundance distribution of the elements in the cosmic radiation (normalized to Fe = 10^6) with atomic number $Z \geq 30$. For $Z \leq 42$, abundances for elements with even Z only. For $Z \geq 44$, the sum of the element Z and $Z + 1$. Dashed lines represent Solar System abundance distribution. (Courtesy of M. H. Israel, E. C. Stone, and C. J. Waddington.)

FIG. 1.   Relative abundance distribution of the elements in the cosmic radiation (normalized to Si = 100) from He through Ni (●, 70–280 MeV per nucleon; ○, 1000–2000 MeV per nucleon; ◇, ---, Solar System abundance distribution). [Reproduced, with permission, from Simpson, J. A. (1983). *Annu. Rev. Nucl. Part. Sci.* **33**; © 1983 by Annual Reviews Inc.]

many of the cross sections that lead to the production of secondary particles during propagation have not been measured and must therefore be deduced by means of theoretical methods. The resulting cosmic ray source composition resembles in its main features that of the solar system and the interstellar medium but with significant exceptions. (1) Atomic effects appear to influence the selection of the accelerated particles. Observations indicate that the lower the first ionization potential of an element, the higher is its relative abundance compared with solar system material. (2) More important, techniques have been developed for measuring the

isotopic abundances of individual elements. The isotopic composition of cosmic rays should not be influenced by atomic physics. It is found that for some elements the isotopic composition is distinctly different from that of the solar system. For example, neon, magnesium, and silicon all show an enhancement of their neutron-rich isotopes, an enhancement that is as much as $3:1$ for the $^{22}Ne/^{20}Ne$ abundance ratio. The significance of these observations lies in their potential to identify the nature of sources of the cosmic radiation. Attempts to measure isotopic composition with better accuracy and for many elements are therefore at the forefront of cosmic ray research.

Figure 3 shows the energy spectra of several elements over the range of energies where measurements of individual species have been possible. Except for the lowest energies these spectra can be well described by power laws, relating the differential intensity dI in each energy interval dE through the expression $dI/dE \sim E^{-\gamma}$, where γ lies in the range from 2.5 to 3, somewhat different for the different components. At low energies the spectra bend under the influence of modulation in the solar system (see Section V). It should be noted that energy spectra of individual cosmic ray components have been directly measured only to energies much below the region of extensive airshower production. At

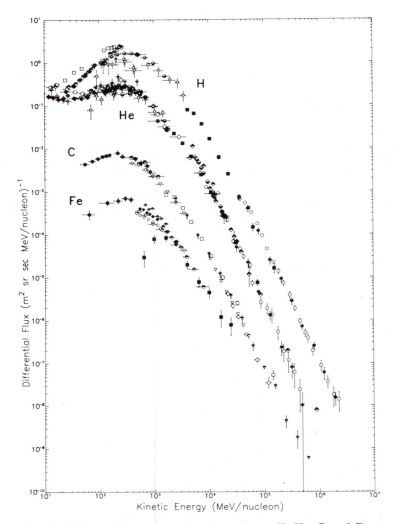

FIG. 3. Differential energy spectra of cosmic ray H, He, C, and Fe as measured near earth. [Reproduced, with permission, from Simpson, J. A. (1983). *Annu. Rev. Nucl. Part. Sci.* **33**; © 1983 by Annual Reviews Inc.]

higher energies, above $\sim 10^{11}$ eV per nucleon, only the spectrum of all components combined is known, because measurements of the shower structure cannot uniquely determine the identity of the primary particle. The all-particle spectrum is shown in Fig. 4.

B. ELECTRONS

A component of high-energy electrons and positrons was discovered in the early 1960s. It had been known before then that the galactic radio background could be understood only in terms of a component of relativistic electrons in the galaxy, producing synchrotron radiation as

they spiral around galactic magnetic fields. The question of whether all electrons (and positrons) are of secondary origin, being produced via the $\pi \rightarrow \mu \rightarrow e$ decay chain that follows the production of π mesons in high-energy collisions, or whether the electrons are directly accelerated, could soon be answered through a separate measurement of the positron and negative electron flux. Secondary negative electrons and positrons would be produced in about equal amounts. However, negative electrons were found to be ~ 10 times more abundant than positrons. This proves that there must be sources of electrons in the galaxy.

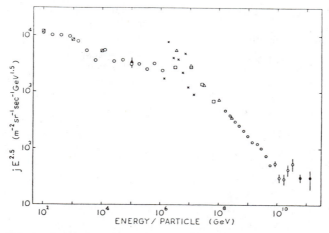

FIG. 4. Differential energy spectrum of all cosmic rays as measured near earth (multiplied by $E^{2.5}$). [After Linsley, J. (1983). "Proceedings of the 18th International Conference on Cosmic Rays, Bangalore," (P. V. Ramana Murthy, ed.) Vol. 12, p. 135. Tata Institute of Fundamental Research, Bombay, India.]

III. Origin of the Primary Radiation

The origin of the cosmic radiation is not yet fully understood. There are good reasons to believe that the bulk of the radiation is produced within our galaxy. Hence, one requires phenomena that are sufficiently energetic to produce cosmic rays at a rate that can compensate for their loss and that have features leading to particle acceleration. Such a large production of energy is observed in explosions of supernovas. These have therefore been candidates for cosmic ray sources for a long time. In recent years much has been learned about the mechanisms of supernova explosions and about the acceleration of particles in shock waves. Two simple and extreme points of view are that (1) cosmic rays are ejecta of supernova material and (2) cosmic rays are interstellar material, accelerated by shock waves that originate in supernovas. In principle one should be able to distinguish between the two alternatives since ejected supernova material has undergone nucleosynthesis processes during the explosion that are different from those occurring in the quiet burning of stars. The elemental and particularly the isotopic composition should exhibit characteristics due to the rapid nucleosynthesis that occurs in the explosion. Some of these characteristics are observed in isotopic distributions. On the other hand, work on ultraheavy nuclei, in particular, indicates a mixture of elements synthesized in slow nuclear burning processes and elements

synthesized in rapid nuclear processes in supernovas. The mixture is similar to that found in the solar system and thus is similar to that expected in the interstellar medium. Advances in the physics of particle acceleration in shock processes made it possible to understand how interstellar material is accelerated to cosmic ray energies. The final answer to the question of cosmic ray acceleration is still unanswered. It is quite possible that several mechanisms are simultaneously at play.

In this context it is of interest to estimate the energy input required to maintain the cosmic ray flux in the galaxy where it has existed for the past 10^9 yr. Radiochemical studies of cosmic-ray-produced isotopes in meteorites have shown that the cosmic ray intensity has not varied by more than a factor of ~ 2 for the past billion years, that is, over the age of the solar system. Individual cosmic ray particles remain in the galaxy for a much shorter time, 10^7 yr, before undergoing a catastrophic collision or being lost to extragalactic space. In the vicinity of the solar system the energy density of cosmic rays is ~ 1 eV/cm^3, comparable to that of starlight. In other words, the flux in energy carried by cosmic rays and falling on the earth is about the same as that of starlight. Since the galactic disk has a volume of $\sim 10^{67}$ cm^3, 10^{67} eV must be created in the form of cosmic rays every 10^7 yr under the assumption that the distribution of cosmic rays is uniform throughout the galactic disk. This means a production rate of 10^{60} eV/yr, or 10^{48}

ergs/yr, which must be compared with the average energy released in supernova explosions. Adopting the best estimates for the average energy released in a supernova explosion to be $\sim 10^{51}$ ergs and for the rate of occurrence of such explosions in our galaxy to be about two per century yields an energy release of 10^{49} ergs/yr by supernovas. Almost 10% of that is needed to accelerate cosmic ray particles. [See SUPERNOVAE.]

Although most of the cosmic rays that one observes near earth are believed to have their origin within our galaxy, the cosmic ray phenomenon is more universal. Synchrotron radiation from other galaxies provides evidence for the presence of high-energy electrons and magnetic fields in these objects. It does not require much speculation to assume that nuclear particles are accelerated in those galaxies as well.

IV. Propagation and Storage of Cosmic Rays in the Galaxy

The interstellar medium consists of a tenuous gas, dust, magnetic fields, starlight, and photons from the universal blackbody radiation. The relativistic, electrically charged cosmic ray particles interact with all of these constituents. Cosmic ray nuclei collide with nuclei in the gas and dust and produce fast secondary nuclei of lower atomic number. In such collisions unstable particles are produced as well: π^0 mesons decay rapidly into two γ rays and are the source of much of the high-energy galactic γ-ray background. The charged π mesons, π^+ and π^-, decay into μ^+ and μ^- mesons, respectively, and these in turn decay into e^+ (positrons) and e^- (negative electrons), forming a fraction of the galactic electron component. The same collisions, if occurring at sufficiently high energy, are the source of the small flux of antiprotons observed in the cosmic radiation.

The galactic magnetic fields deflect the moving electrically charged particles from a straight-line path and force them on spiraling trajectories around the magnetic lines of force. Whenever the direction of the field changes, the direction of the particle changes also. The resulting random walk of the particles in the interstellar medium is the cause of the apparent isotropy of their direction of incidence and also the reason that on the average the particles spend as many as 10^7 yr in the galaxy before escaping into extragalactic space. The amount of matter that cosmic rays traverse during their life in the galaxy can be measured by comparing the flux of secondary nuclei produced by spallation with the flux of primary nuclei. This comparison yields an average thickness of ~ 7 g/cm^2. The galactic magnetic fields do not cause any energy loss to nuclear cosmic rays, unless the particle energy becomes extremely high ($E > 10^{19}$ eV), nor do the encounters with photons. This is different for the electron–positron component, where the particles have a large charge-to-mass ratio. Electrons and positrons lose energy while spiraling around magnetic fields through the emission of synchrotron radiation (this is the source of the galactic radio background). They also lose energy in elastic collisions with the photons of starlight and with the photons of the universal blackbody radiation. These energy losses manifest in the energy spectrum of the electrons, where the power law becomes steeper toward high energy due to the strong losses. The shape of the spectrum provides a clue to the time cosmic rays are contained in the galaxy. This containment time has also been measured directly through a determination of the abundance of the radioactive isotope ^{10}Be. Like the beryllium isotopes with mass 7 and 9, ^{10}Be is entirely a product of collisions, mostly by carbon and oxygen with interstellar gas. It has a half-life of 1.6 million years against radioactive decay, and it is easy to determine its expected abundance if it were stable. Its near absence in cosmic rays means that most of it has decayed before observation—hence, a containment time of about 10 million years.

At energies above $\sim 10^{17}$ eV cosmic ray particles have become so energetic that the galactic magnetic fields cannot bend their trajectories to a large degree. If their sources lie in the galactic disk one would therefore expect a maximum of their flux from directions that point into the galactic plane. Such anisotropy has not been observed, and one therefore concludes that the cosmic rays of highest energy are likely to be of extragalactic origin.

V. Propagation of Cosmic Rays in the Solar System

The heliosphere, a volume around the sun with a radius of about 50 to 100 astronomical units (AU; 1 AU = 1.49×10^{13} cm, or the mean sun–earth distance), is dominated by the solar wind, a stream of plasma from the expanding solar corona. This plasma flows outward from the sun at varying speed, averaging 300–400 km/sec, and carries solar magnetic fields. Due to the rotation of the sun the field has the general pattern of an Archimedean spiral centered on the sun, with many irregularities superimposed on this pattern. As in interstellar space, cosmic ray particles are scattered by field irregularities and proceed by "diffusion" in the solar wind. Three main effects govern the motion of cosmic rays in the heliosphere and influence their intensity in the vicinity of earth: (1) the above-mentioned diffusion, (2) the outward convection by the moving solar plasma, and (3) an adiabatic energy loss due to the fact that the solar wind expands into an increasing volume. Taking into account these three effects one can understand the relation between the particle fluxes and energy spectra observed near earth and those in the nearby interstellar medium. The mechanisms by which the sun modifies the interstellar flux of particles is called solar modulation. Interstellar cosmic rays below ~100 MeV cannot penetrate to the earth's orbit and hence cannot be observed with present detectors. Figure 3 shows the bending of the energy spectra at low energies due to the effect of solar modulation. The higher the energy of the particles, the less does solar modulation influence them. Above ~10 GeV per nucleon the magnetic rigidity of the particles is sufficiently high that the interplanetary fields cease to modify them in a major way. [*See* SOLAR PHYSICS.]

The level of solar modulation changes with time. The most notable change is correlated with the 11-yr solar activity cycle. Figure 5 shows the cosmic ray intensity and its variations as ob-

served with a ground-based monitor. Years of high solar activity as indicated by the average number of sunspots coincide with those of low cosmic ray intensity and vice versa. Short-term and nonperiodic intensity changes, also correlated with solar activity, are observed as well. Studies of the behavior of energetic particles in the solar system have revealed many details of the physics of the interplanetary medium. Deep-space probes (*Pioneer 10 and 11, Voyager 1 and 2*) have already shown that, in the ecliptic plane, the heliosphere probably extends to 100 AU. The International Solar Polar Mission (Ulysses) will for the first time explore the heliosphere far out of the ecliptic plane, over the solar poles.

VI. Particle Acceleration in the Solar System

The most conspicuous particle producer in the Solar System is the sun itself. In association with chromospheric outbursts—solar flares—the sun frequently emits high-energy particles. Protons, electrons, and heavier nuclei are present in those flare emissions, and more recently neutrons have also been identified. Normally the particle energy reaches up to several hundred mega-electron-volts, but in a few flare events particles with up to 20 GeV could be observed. Not all details of the solar acceleration mechanism are known, but coordinated observations of nuclei, electrons, X rays, γ rays, radio emission, and neutrons are now contributing to progress in this area. Through the study of the energetic solar particle composition some elemental and isotopic solar abundances could be measured that are otherwise inaccessible. However, the sun frequently emits very anomalous compositions—for example, large enhancements of the isotope ^{3}He.

The second known point source of particles in the solar system is the planet Jupiter. The Jovian magnetosphere emits electrons in the 1- to 20-MeV energy range with an intensity variation of 10 hr, Jupiter's period of rotation. These electrons are observed not only in interplanetary space but also near Earth. There they are more intense at times when Earth and Jupiter are well connected by the overall spiral interplanetary magnetic field. Due to the orbital periods of the two planets this occurs every 13 months, and the low-energy electron component is observed to vary in intensity with this period. Particles from the sun and from Jupiter are used to probe the

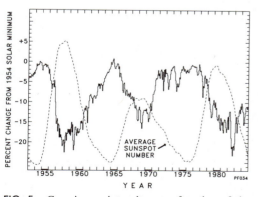

FIG. 5. Cosmic ray intensity as a function of time: University of Chicago climax neutron monitor intensity (27-day averages). The anticorrelation between the 11-yr sunspot activity and the cosmic ray intensity can be immediately noted. (Courtesy J. A. Simpson.)

propagation in the solar system, and hence its electromagnetic conditions. Historically, the observation of the time–intensity dependence of solar flare particles provided the first evidence that charged particles do not travel along straight lines, but diffuse in the solar system. Shocks originating on the sun after a flare not infrequently accelerate particles in interplanetary space. Such shock accelerations may play a larger role in the acceleration of interplanetary particles than is presently recognized.

BIBLIOGRAPHY

Cesarsky, C. J. (1980). *Annu. Rev. Astron. Astrophys.* **18**, 289.

Setti, G., Spada, G., and Wolfendale, A. W., eds. (1981). "Origin of Cosmic Rays." Reidel Dordrecht, Holland.

Shapiro, M. M., ed. (1983). "Composition and Origin of Cosmic Rays." Reidel Dordrecht, Holland.

Simpson, J. A. (1983). *Annu. Rev. Nucl. Part. Sci.* **33**, 323.

COSMOLOGY

John J. Dykla   *Loyola University of Chicago and Theoretical Astrophysics Group, Fermilab*

GLOSSARY

Background radiation: Field quanta, either photons or neutrinos, presumably created in an early high-temperature phase of the universe. The photons last interacted strongly with matter before the temperature fell below 3000 K and electrons combined with nuclei to form neutral, transparent hydrogen and helium atoms. Although shifted to the radio spectrum by the expansion of the universe, this "light" has been moving independently since this decoupling (in the standard model, ~700,000 years after the "big bang") providing direct evidence of how different the universe was at this early time from its appearance now. The measured homogeneity and isotropy of the photon background are very difficult to account for in models, such as the "steady-state" theories, which avoid a "big bang."

Black hole: Region of space-time with so large a curvature (gravitational field) that, classically, it prevents any matter or radiation from leaving. The "one-way" surface, which allows passage into the black hole but not out, is called an event horizon. Black holes are completely characterized by their mass, charge, and angular momentum.

Cosmological principle: The assumption that at each moment of proper time an observer at any place in the universe would find the large-scale structure of surrounding matter and space the same as any other observer (homogeneity) and appearing the same in all directions (isotropy). Combined with the general theory of relativity, this forms the basis of the standard model in cosmology. Observational evidence for the expansion of the universe, combined with the principle of conservation of mass-energy, indicates that the large-scale average density, though everywhere the same at a given time, is decreasing as the universe evolves. The perfect cosmological principle seeks to avoid a "big bang" by asserting that the homogeneous and isotropic density of matter at all points in space is also unchanging in time. Such an assumption could be reconciled with an expanding universe only by postulating the continuous creation of matter and appears untenable in light of the microwave background radiation.

Doppler shift: Change in frequency (or, inversely because of invariant speed of propagation, wavelength) of radiation due to the relative motion of source and observer. For motion purely along the line of sight, recession implies increased wavelength (redshift) and approach implies decreased wavelength (blueshift), complicated by nonlinear relativistic effects when the relative velocity of source and observer is an appreciable fraction of the velocity of light. Purely transverse motion is associated with a relativistic and intrinsically nonlinear redshift.

Event: Point in space at a moment of time. The generalization, in the four-dimensional space–time continuum of relativity theory, of the concept of a point in a space as a geometric object of zero "size."

General theory of relativity: Understanding of gravitation invented by Albert Einstein as the curvature of space-time, described mathematically as a four-dimensional geometric manifold. His field equations postulate the way in which the distribution of mass-energy tells space-time how to curve and imply the equations by which curved space-time tells mass-energy how to move. With the cosmological principle, it forms the basis of the standard model in cosmology.

Missing-mass problem: The discrepancy by approximately an order of magnitude between the ordinary matter that is directly visible through astronomical observations and the mass inferred from dynamic theory applied to the observed motions of galaxies and clusters of galaxies. The calculated average density is roughly the critical value for a flat universe, which is also strongly favored by "inflationary" particle physics scenarios. Thus, various exotic forms of astronomical objects and elementary particles have been proposed to account for the dark matter assumed present.

Nebula: Latin for "cloud," this old term in observational astronomy refers to a confusing variety of extended objects easily distinguished from the pointlike stars. Some are now known to be dust or gas clouds associated with the births or deaths of stars in the Milky Way, but others are galaxies far outside the Milky Way.

Perfect fluid: Idealization of matter as continuous and completely characterized by three functions of space-time: mass-energy density, pressure, and temperature. At each event, these functions are related through a thermodynamic equation of state.

Proper time: Measure of time by a standard clock in the reference frame of a hypothetical observer. In describing periods in the evolution of the entire universe, the reference frame is chosen as one that at each event is comoving with the local expansion. The proper time since the "big bang" singularity is called the age of the universe.

Singularity: Event or set of events at which some physical quantities that are generally measurable, such as density and curvature (gravitational field), are calculated to have infinite values. It represents a limit to the validity of a field theory.

Cosmology as a scientific endeavor is the attempt to construct a comprehensive model of the principal features of material composition, geometric structure, and temporal evolution of the entire physically observable universe. The primary tools of the modern cosmologist are those of observational astronomy and theoretical physics. Signals are gathered over a very broad range of wavelengths of the electromagnetic spectrum, from radio waves collected by the 300-m-diameter dish at Arecibo, Puerto Rico, through visible quanta recorded by charge-coupled devices attached to the 5-m Hale telescope atop Mt. Palomar, California, to γ rays detected by instruments orbiting above the earth's atmosphere. In recent years, experiments in particle physics have become increasingly relevant to cosmological questions. Examples include measurements of the solar neutrino flux, which challenge our understanding of energy generation in the stars, and attempts to detect the superpartners, axions, strings, and other exotic objects predicted by various models in high-energy physics and possible solutions to the missing-mass problem.

The standard theoretical model in cosmology for the past few decades has been based on Einstein's general theory of relativity, supplemented by assumptions about the homogeneity and isotropy of space-time, and data from spectroscopy, consistent with understanding gained from nuclear physics, about the distribution of ordinary matter among various species. Recently, progress in understanding the early universe has been made by fusing this with the standard $SU(3)_C \times SU(2)_L \times U(1)_Y$, model in particle physics, which seems to describe very accurately the interactions of quarks and leptons at energies up to $\sim 10^3$ GeV, the current limit of terrestrial accelerators. Traditional analytical techniques of the theorist are now often supplemented by the raw-number-crunching power of increasingly rapid and sophisticated computers, especially in applications to otherwise intractable calculations involving real or speculative space-times or quantum fields. Further synthesis is clearly dependent on continued advances in observational astronomy, high-energy experimentation, and theoretical physics.

Research in cosmology is subject to one obvious but fundamental limitation that no other field of scientific inquiry shares. By definition, there is only one entire physical reality that we can observe and attempt to understand. Since we cannot acquire data from outside all this, it is impossible in principle to perform a controlled experiment to test a truly comprehensive theory of the cosmos. The best we can hope for are ever more refined models, constrained by data from increasingly diverse observational sources, which become more broadly successful as we continue to measure and to think.

I. Issues from Prescientific Inquiries

A. Philosophical Speculations and Myths

Cosmological thought recorded in diverse cultures during past millennia displays a continued fascination with a relatively small number of fundamental questions about the physical universe and our place within it. Many attempts to provide satisfying answers have been made. It is not appropriate to present here a comparative survey of the myriad world views that have been proposed, but there is value in understanding those issues raised by early thought that are still at the core of modern scientific cosmology.

Logically, the first issue is that of order versus chaos. Indeed, the word *cosmology* implies an affirmative answer to the question of whether the totality of physical experience can be comprehended in a meaningful pattern. We should be aware, however, that adherents of the view that nature is fundamentally inscrutable or capricious have presented their arguments in many cultures throughout recorded history.

Is matter everywhere in the universe of the same type as that which is familiar to us on earth, or are there some distinctively "celestial" substances? Speculations favoring one view or the other have been stated with assurance as long as humans have engaged in intellectual debate.

Is the space of the universe finite or infinite, bounded or unbounded? To this also, pure thought unfettered by empirical evidence has, in many places and times, offered dogmatic pronouncements on either side. The logical independence between the issues of finiteness and boundedness was not fully clarified until the invention of non-Euclidean geometries in the nineteenth century. Thus, we now see that the opening question of this paragraph actually comprises two distinct but related questions. Of the issues discussed in this section, this is perhaps the only one in which progress in pure thought has raised questions not addressed by ancient philosophers and myth makers.

Is the universe immutable in its total structure, or does it evolve with the passage of cosmic time? If the latter, did it have a creation in time, or has it always existed? Similarly, will a changing cosmos continue forever, or will there be an end to time that we can experience? Such

questions have sometimes been distinguished from the structural questions of cosmology and said to be of a different order called cosmogony. We shall see that in scientific cosmology such a separation is artificial and unproductive. Indeed, all the questions posed in this section are, in all current scientific models, inextricably interrelated.

B. Value of Some Early Questions

All modern science is predicated on the philosophical assumption that its subject is comprehensible. Although even Albert Einstein remarked that this is the most incomprehensible thing about the universe, it is difficult to see how this can be a question for scientific debate. The presumption of discoverable regularities, not meaningless chaos, is a necessary underpinning of any scientific endeavor. Of course, the advent of quantum mechanics in the twentieth century has compelled reappraisal of the deterministic paradigm of earlier science. Still, the activities of physicists are grounded in a belief in the existence of objective laws that correlate our observations of natural phenomena and allow at least some limited measure of successful prediction.

Philosophical speculations in earlier ages about the substance of the cosmos often assumed an insurmountable qualitative distinction between the "stuff" of terrestrial experience and the matter of the celestial spheres. The beginnings of modern science included an explicit rejection of this view in favor of one in which all the matter of the universe is of fundamentally the same type, subject to laws that hold in all places and at all times. It is interesting that some recently proposed solutions to the missing-mass problem in astrophysics suggest that the vast majority of the matter of the universe may be in one or more exotic forms that are predicted by elementary particle theory but for which we do not yet have experimental evidence. This is not a return to prescientific notions. It is still assumed that, no matter what the substance of the universe, all physical reality is subject to one set of laws. Most important, it is assumed that such laws can be discovered with the aid of data obtained from terrestrial experiments as well as astronomical observations.

Although most early cosmologists imagined a finite and bounded universe, some thinkers (e.g., the Greek philosopher Democritus in the fifth century B.C.) argued eloquently for an infinite and unbounded expanse of space. Until the latter part of the nineteenth century, however, a finite space was presumed synonymous with a bounded one. The non-Euclidean geometries of Riemann, Gauss, and Lobachevsky first presented the separation of the questions of extension and boundedness in the form of logically possible examples amenable to mathematical study. Originally, generalizations of the theorem of Pythagoras to curved spaces assumed that the square of the distance between two distinct points is greater than zero, described by a positive-definite metric tensor. In relativity theory, time is treated as an additional "geometric" dimension, distinguished by the fact that the square of the interval between two events may be a quantity of either sign (depending on whether there is a frame of reference in which they are at the same place at different times or a frame of reference in which they are at different places at the same time) or zero (if they can be related only by the passage of a signal at the speed of light). Gravitation is understood to be the relation between the curvature of space-time and the mass-energy contained and moving within it. Modern research on the asymptotic structure of space-times is based on the analysis of four-dimensional geometries of indefinite metric within the framework of relativistic gravitational theory. All viable models consistent with minimal assumptions about our lack of privileged position describe spaces that are without a boundary surface at each moment of time, although some are finite in volume.

Myths of creation, whether based on purely philosophical speculations or carrying the authority of religious conviction, are commonplace in almost all cultures of which we have any knowledge. Some suppose the birth of a static universe in a past instant, but most describe a process of formation or evolution of the cosmos into its present appearance. In many eras some individuals or groups have championed an eternal universe. Most of these models of the universe are static or cyclic, but occasionally they represent nature with infinite change, never repeating but with no beginning. In earlier times, a beginning in time was almost always taken to imply finiteness in a bounded space. The new geometries mentioned earlier have made possible serious consideration of a universe that began at a definite moment in the past, is not bounded in space, and may be finite or infinite in extent. The question of whether there will be an end to time or whether there will be eternal evolution is intimately related to the question of fi-

nite or infinite spatial extent in these models. At present, the empirical evidence about whether time will end is inconclusive. The "inflationary scenarios" that are currently in vogue favor a universe remarkably close to the marginally infinite and thus eternally unwinding "asymptotically flat" model, but allow the possibility of a miniscule difference from flatness of either sign, resulting in ultimate closure or eternal expansion. Although the observed density of visible matter appears much too small to halt the present expansion, there are numerous hypothetical candidates for the missing mass.

II. Origins of Modern Cosmology in Science

A. COPERNICAN SOLAR SYSTEM

Centuries before the Christian era, several Greek philosophers had proposed heliocentric models to coordinate the observed motions of the sun, moon, and planets in relation to the "fixed" stars. However, the computational method of Alexandrian astronomer Claudius Ptolemy, which he developed during the second century A.D. using a structure of equants and epicycles based on the earth as the only immovable point, also made successful predictions of apparent motions. Although additional epicycles had to be added from time to time to accommodate the increasingly precise observations of later generations, this system became dominant in the Western world. Its assumption of a dichotomy between terrestrial experience and the laws of the "celestial spheres" stifled fundamental progress until the middle of the sixteenth century. In 1543, the Polish cleric Mikolaj Kopernik, better known by the Latin form of his name, Nicholas Copernicus, cleared the way for modern astronomy with the publication of *De Revolutionibus Orbium Coelestium* (*Concerning the Revolutions of the Celestial Worlds*). Appearing in his last year, this work summarized the observations of a lifetime and presented logical arguments for the simplicity to be gained by an analysis based on the sun as a fixed point.

More important, this successful return to the heliocentric view encouraged the attitude that a universal set of laws governs the earth and the sky. In the generations following Copernicus, the formulation of physical laws and their empirical testing were undertaken in a manner unprecedented in history: Modern science had begun.

The Dane Tycho Brahe observed planetary orbits to an angular precision of 1 arc min. The German Johannes Kepler used Brahe's voluminous and accurate data to formulate three simple but fundamental laws of planetary motion. The Italian Galileo Galilei performed pioneering experiments in mechanics and introduced the use of optical telescopes in astronomy. From these beginnings, a scientific approach to the cosmological questions that humanity had been asking throughout its history developed.

B. Newtonian Dynamics and Gravitation

The first great conceptual synthesis in modern science was the creation of a system of mechanics and a law of gravitation by the English physicist Isaac Newton, published in his *Principia* (*Mathematical Principles of Natural Philosophy*) of 1687. His "system of the world" was based on a universal attraction between any two point objects described by a force on each, along the line joining them, directly proportional to the product of their masses and inversely proportional to the square of the distance between them. He also presented the differential and integral calculus, mathematical tools that became indispensable to theoretical science.

On this foundation, Newton derived and generalized the laws of Kepler, showing that orbits could be conic sections other than ellipses. Nonperiodic comets are well-known examples of objects with parabolic or hyperbolic orbits. The constant in Kepler's third law, relating the squares of orbital periods to the cubes of semimajor axes, was found to depend on the sum of the masses of the bodies attracting one another. This was basic to determining the masses of numerous stars, in units of the mass of the sun, through observations of binaries. As Cavendish balance experiments provided an independent numerical value for the constant in the law of gravitation, stellar masses in kilograms could be calculated. Later, analysis of observed perturbations in planetary motions led to the prediction of previously unseen planets in our solar system. Studies of the stability of three bodies moving under their mutual gravitational influence led to the discovery of two clusters of asteroids sharing the orbit of Jupiter around the sun. These successes fostered confidence in the view of one boundless Euclidean space, the preferred inertial frame of reference, as the best arena for the description of all physical activity. Lacking direct evidence to the contrary, theoretical cosmologists at first assumed that the space of the universe was filled, on the largest scales, with matter distributed uniformly and of unchanging density. Analysis soon disclosed that Newtonian mechanics implied instability to gravitational collapse into clumps for an initially homogeneous and static universe. This stimulated the observational quest for knowledge of the present structure of the cosmos outside the solar system.

C. Technology of Observational Astronomy

Galileo's first telescopes had revealed many previously unseen stars, particularly in the Milky Way, where they could resolve numerous individual images. Based on elementary observations and careful reasoning, the ancient concept of a sphere of fixed stars centered on the earth had been supplanted by a potentially infinite universe with the earth at a position of no particular distinction. Further progress in understanding the distribution of matter in space was dependent on actual measurements of the distances to the stars. The concept of stellar parallax, variation in the apparent angular positions of nearby stars in relation to more distant stars as seen from earth at various points in its orbit around the sun, became useful when the observational precision of stellar location surpassed the level of 1 arc sec. An important contribution was the development of the micrometer, first used by William Gascoigne, later independently by Christiaan Huygens, and systematically appearing in the telescope sights of Jean Picard and others after the latter part of the seventeenth century. The first parallax measurements were reported, by Bessel in Germany, Henderson in South Africa, and Struve in Russia, in rapid succession in 1838 and 1839. By 1890 distances were known for nearly 100 stars in the immediate neighborhood of the sun, and the limits of the trigonometric parallax approach (~100 parsec, based on limits to the useful size of an optical telescope looking through the earth's atmosphere) were being approached. [*See* OPTICAL TELESCOPES.]

As long as sufficient light can be gathered from a star to be dispersed for the study of its spectrum, much can be learned about it even if one is not certain how far away it is. Spectroscopic studies of stars other than the sun were first undertaken by William Huggins and Angelo Secchi in the early 1860s. In 1868, Huggins announced the measurement of a Doppler redshift in the lines of Sirius indicating a recession velocity of 47 km/sec, and Secchi published a catalog of 4000 stars divided into four classes according to the appearance of their spectra. The introduction of the objective prism by E. C. Pickering in 1885 tremendously increased the rate at which spectra could be obtained. The publication of the Draper catalogs in the 1890s, based on the spectral classification developed by Pickering and Antonia Maury, provided the data that led to the rise of stellar astrophysics in the twentieth century.

D. VIEWING THE ISLAND UNIVERSE

Some hazy patches among the fixed stars, such as the Andromeda nebula or the Magellanic Clouds, are clearly visible to the unaided human eye. When Galileo observed the band called the Milky Way through his telescope and was surprised to find it resolved into a huge number of faint stars, he concluded that most nebulosities were composed of stars. Early in the eighteenth century, the Swedish philosopher Emanuel Swedenborg described the Milky Way as a rotating spherical assembly of stars and suggested that the universe was filled with such spheres. The English mathematician and instrument maker Thomas Wright also considered the Milky Way to be one among many but supposed its shape to be a vast disk containing concentric rings of stars. By 1855, Immanuel Kant had further developed the disk model of the Milky Way by applying Newtonian mechanics, explaining its shape through rotation. He assumed that the nebulae are similar "island universes."

Early in the nineteenth century, however, William Herschel discovered the planetary nebulae, stars in association with true nebulosities. This reinforced a possible interpretation of nebulae as planetary systems in the making, in agreement with the theory of the origin of the solar system developed by Pierre Simon de Laplace. John Herschel's observations of 2500 additional nebulae, published in 1847, emphasized that they were mostly distributed away from the galactic plane. The "zone of avoidance" was used by some to argue for the physical association of the nebulae with the Milky Way. (It is now known that dust and gas in the plane of the galaxy diminish the intensity of light passing through it, whether from sources inside it or beyond.) In 1864, Huggins studied the Orion nebula and found it to display a bright line spectrum

similar to that of a hot gas. Also, photographs of Orion and the Crab nebula did not show resolution into individual stars. As the twentieth century began, the question of the nebulae was intricately linked with that of the structure of the Milky Way. The size of nebulae was still quite uncertain in the absence of any reliable distance indicators, but most astronomers believed the evidence favored considering the nebulae part of the Milky Way, thus reducing the universe to this one island of stars.

III. Revolutions of the Twentieth Century

A. NEW THEORETICAL MODELS

The two great conceptual advances of theoretical physics in the twentieth century, relativity and quantum mechanics, have had profound implications for cosmology. The general theory of relativity provides the basis for the evolving space-time of the now standard model, and fundamental particle identities and interactions are keys to understanding the composition of matter in the universe. [See RELATIVITY, GENERAL; RELATIVITY, SPECIAL.]

In 1905, Albert Einstein established the foundations of the special theory of relativity, which connects measurements of space and time for observers in all possible inertial frames of reference. He devoted the next decade to developing a natural way of including observers in accelerated frames of reference, using as a fundamental principle the fact that all test masses undergo the same acceleration in a given gravitational field. The result was Einstein's theory of space-time and gravitation, the general theory of relativity (GTR), completed in 1916. It was immediately apparent that the GTR would have a substantial impact on cosmological questions. Although many model space-times have been studied in the GTR, in 1922 Alexander Friedmann showed that the assumptions of spatial homogeneity and isotropy can be embodied in only three. They are presented here in the modern notation of the Robertson–Walker metric with the choice of units commonly used in theoretical research. The intimate relationship of space and time in relativistic theories is recognized through units of measure in which the speed of light is numerically 1, eliminating the letter c representing its value in units such as those of SI. We may think of the units as geometrized (e.g., time measured in meters) or chronometrized (e.g., length measured in seconds), where 1 sec = 2.99793×10^8 m.

The invariant element of separation ds between two neighboring events in a homogeneous and isotropic universe can be expressed by

$$ds^2 = -dt^2 + R^2[dr^2/(1 - kr^2)$$
$$+ r^2(d\theta^2 + \sin^2\theta \, d\phi^2)] \qquad (1)$$

where $R(t)$ is the scale factor that evolves with cosmic time and k is the curvature constant, which may be positive, negative, or zero. The invariant separation depends on k only through the pure number kr^2, where r is a comoving space coordinate, for which matter is locally at rest. Without loss in generality, we may choose k (and hence also r) dimensionless and set the scale of r so that k is either +1, −1, or 0. Although R carries the units of spatial length, it is not possible to interpret it as a "radius of the universe" if k is not positive, since the volume of a model with $k = 0$ or $k = -1$ is infinite.

To ensure that the physics is described in a manner independent of arbitrary choices, it is useful to introduce the scale change rate, also known as the Hubble parameter, defined as $H = (dR/dt)/R$. Since H has dimensions of velocity divided by length, $1/H$ is a characteristic time for evolution of the model. For an expanding empty universe, $1/H$ would be the time since the beginning of the expansion. Assuming that the distribution of matter on a large scale can be described in terms of a perfect fluid of total density ρ and pressure p, the coupled evolution of space-time and matter in the GTR are determined by the Friedmann equations,

$$H^2 = 8\pi G\rho/3 - k/R^2 \quad \text{and} \quad d(\rho R^3)$$
$$= -p \, d(R^3) \qquad (2)$$

where G is the Newtonian gravitational constant. (Note that the term *density* can be taken as either mass per volume or energy per volume in units where $c = 1$.)

Although the GTR implies that a homogeneous and isotropic space generally expands or contracts, at the time the theory was formulated the common presumption held that the real universe was static. This led Einstein to modify his original simple theory by introducing a "cosmological term" into the field equations. The "cosmological constant" in this term was chosen to ensure a stable static solution. When the evidence for an expanding universe became apparent in the next decade, Einstein dropped consideration of this additional term, calling its introduction the biggest blunder of his career. We shall use the wisdom of hindsight to leave it out of the present presentation. Thus, there are only three possible homogeneous and isotropic universes in the GTR, the Friedmann models. In view of Eq. (2), which determines the evolution of the Hubble parameter, the curvature constant can be positive only if the density exceeds the value $\rho_{crit} = 3H^2/8\pi G$. Thus, an expanding universe in which $k = +1$ must halt its growth and begin to collapse before its mean density drops below this critical value. Such a space-time is finite in volume at each moment but has no boundary. While this model is expanding, the Robertson–Walker coordinate time since the expansion began is less than $2/3H$, depending on how much the density exceeds the critical value. The classical GTR implies that a finite proper time passes between its beginning in a singularity of zero volume and its return to such a singularity. (However, Charles Misner has argued that a logarithmic time scale based on the volume of the universe is more appropriate as a measure of possible change that may occur, giving even this closed cosmology a potentially "infinite" future and past.) If the density is less than the critical value, space-time has negative curvature and will have a positive velocity of expansion into the infinite future of proper time, when H approaches zero. The time since expansion began is greater than $2/3H$ depending on how much the density is less than the critical value, but never exceeds $1/H$. Although such an open universe still had a beginning in proper time at a singularity, its volume is always infinite and it contains an infinite mass–energy. If the density precisely equals the critical value, then the curvature vanishes. Such a flat universe has been expanding for a time $2/3H$, it has infinite mass–energy in an infinite volume described by Euclidean geometry at each moment, and its expansion velocity as well as Hubble parameter will approach zero asymptotically as proper time in comoving coordinates continues toward the infinite future. Which of these three models best describes the universe we inhabit is an empirical as well as theoretical question that remains central to current research in cosmology. [See COSMIC INFLATION.]

B. New Observational Evidence

Beginning in the late 1920s, Vesto Slipher, Edwin Hubble, and Milton Humason used the Hooker telescope at Mt. Wilson, California, then the largest optical instrument in the world, to measure Doppler shifts in the spectra of nebulae that they realized were outside the Milky Way. Convincing reports of a relationship between recession velocities v deduced from Doppler redshifts and estimates of the distances d to these galaxies were published in 1929 and 1931. Hubble pointed out that the velocities, in kilometers per second, were directly proportional to the distances, in megaparsecs (1 Mpc = 3.08×10^{19} km). The observational relation is "smooth" only when the distances considered are at least of the order of a few megaparsecs, allowing the averaging of the distribution of galaxies to produce an approximately homogeneous and isotropic density. Since the ratio v/d is clearly the current value of the scale change rate H in a Friedmann universe, this was the first observational evidence in support of GTR cosmology.

The parameter H was at first called the Hubble constant but is now recognized as a misnomer, because the early data did not extend far enough into space to correspond to looking back over a substantial fraction of the time since the expansion began. Hubble's initial value of H was so large that the time scale for expansion derived from it was substantially less than geological estimates of the age of the earth, $\sim 4.7 \times 10^9$ years. Subsequently, the numerical value of the Hubble parameter has undergone several revisions due to reevaluations of the cosmic distance scale, yielding a universe older than originally supposed and thus a much larger volume from which signals at the speed of light can be observed at our position. Allowing for present uncertainties, H is now believed to be between 50 and 100 (km/sec)/Mpc. Thus, the time scale of the cosmic expansion, $1/H$, is between approximately 1 and 2×10^{10} years.

When early large values of the Hubble parameter were in conflict with geological and astrophysical estimates of the age of the solar system, it was suggested that the simple GTR cosmologies might not be viable. One solution was to reinstate the cosmological constant, not to halt the universal expansion but merely to slow it down in a "coasting" phase. Models studied by George Lemaitre and others after 1927 often had unusual features, such as a closed space of positive curvature that could continue expanding for an infinite proper time. A more radical idea, which had other motivations as well, was to abandon conservation of energy in favor of a "steady-state" cosmology based on the continu-ous creation of matter throughout an infinite past. In such a theory, introduced by Hermann Bondi, Thomas Gold, and Fred Hoyle in 1948, there is no initial singularity, or "big bang," to mark the beginning of the expansion. In 1956, George Gamow predicted that a residual electromagnetic radiation at a temperature of only a few kelvins should fill "empty" space as a relic of the high temperatures at which primordial nucleosynthesis occurred soon after the initial singularity in a GTR cosmology. In 1965, engineers Arno Penzias and Robert Wilson detected a microwave background coming from all directions in space into a communications antenna they had designed and built for Bell Laboratories in Holmdel, New Jersey. Cosmologists Robert Dicke and P. J. E. Peebles at nearby Princeton quickly explained the significance of this 2.7 K blackbody spectrum to them and thus dealt a severe blow to the viability of any cosmological model that avoids a "hot big bang."

A direct determination of whether the universe is spatially closed or open would necessitate precisely measuring density over volumes of many cubic megaparsecs, detecting the sign of departures from Euclidean geometry in an accurate galactic census over distances exceeding several tens of megaparsecs, or finding the change in the Hubble parameter associated with looking back toward the beginning over distances exceeding several hundred megaparsecs. Unfortunately, these conceptually simple observations appear to be somewhat beyond the scope of our present technology. However, there are various indirect ways to estimate, or at least place bounds on, the present mean density of the cosmos. To eliminate the influence of uncertainties in the Hubble parameter on the precise value of the critical density, it is now common to describe the resulting estimates or bounds in terms of the dimensionless quantity $\Omega = \rho/\rho_{crit}$. Clearly, $\Omega > 1$ corresponds to a closed, finite universe and $\Omega \leq 1$ corresponds to an infinite universe. On the basis of the amount of ordinary visible matter observed, as stars and clouds in galaxies, we conclude that $\Omega > 0.01$. Requiring the present density to be low enough so that the age of the universe (which depends inversely on H and, through a monotonically decreasing function, on Ω) is at least as great as that of the oldest stars observed, estimated to be 10^{10} years, yields an upper bound $\Omega < 3.2$. Of the infinite range of conceivable values, it is quite remarkable that the universe can so easily be shown to be nearly "flat." Further evidence and theoretical insight suggest that near coincidence (within one or two powers of 10) of the density and the critical density is not an accident.

The most severe constraints on the contribution of ordinary matter to Ω presently come from demanding that primordial synthesis of nuclei during the "hot big bang" of a standard Friedmann model produces abundance ratios in agreement with those deduced from observation. In nuclear astrophysics, it is customary to specify temperatures in energy units, which corresponds to setting the Boltzmann constant equal to 1. Thus, the relation between the megaelectronvolt of energy and the kelvin of absolute temperature is

$$1 \text{ MeV} = 1.1605 \times 10^{10} \text{ K}$$

Primordial nucleosynthesis models are based on the assumption that the temperature of the universe was once higher than 10 MeV, so that complex nuclei initially could not exist as stable structures, but were formed in, and survived from, a brief interval as the universe expanded and cooled to temperatures of less than 0.1 MeV, below which nucleosynthesis does not occur. Computing all relevant nuclear reactions throughout this temperature range to determine the final products is a formidable undertaking. Of the various programs written since Gamow suggested the idea of primordial nucleosynthesis in 1946, the one published by Robert Wagoner in 1973 has become the accepted standard. With updates of reaction rates by several groups since then, the numerical accuracy of the predicted abundances is now believed to be ~1%. Since the weakly bound deuteron is difficult to produce in stars and easily destroyed there, its abundance, 1×10^{-5} relative to protons as determined in solar system studies and from ultraviolet absorption measurements in the local interstellar medium, is generally accepted as providing a lower bound to its primordial abundance and hence an upper bound of 0.19 to the contribution of baryons to Ω. Analogous arguments may be applied to establishing a concordance between predicted and observed abundances of ^{3}He, ^{4}He, and ^{7}Li, the only other isotopes calculated to be produced in significant amounts during this primordial epoch. The resulting constraint on the contribution of baryons to the critical density, $0.014 \leq \Omega_B \leq 0.14$, shows clearly that baryons alone cannot close the universe. Of course, this does not eliminate the possibility that the universe may have positive curvature due to the presence of less conventional forms of as yet unseen matter.

IV. Unsolved Classical Problems

A. FINITE OR INFINITE SPACE

Although concordance between the standard model and observed nuclear abundances limits baryon density to well below the critical value needed for closure of a Friedmann universe, the question of finite or infinite space remains observationally undecided owing to other complications. In principle, it should be possible to determine the curvature constant k by direct measurement of the deviation from Hubble's simple linear relationship between velocity and distance. Unfortunately, substantial deviation is not expected until sources at distances of the order of $1/H$ are studied, and galaxies are too faint to have their spectra measured adequately at such distances by present technology. Furthermore, observing galaxies at such distances implies seeing them at earlier times, and estimates of their distance could be subject to systematic errors due to unknown evolution of galactic luminosity.

Since the discovery of quasars by Maarten Schmidt at the Palomar Observatory in 1963, cosmologists have hoped that these most distant observed objects could be used to extend the Hubble relationship to the nonlinear regime and decide the sign of the curvature. Quasars are now known with spectral features at nearly 4.8 times their terrestrial wavelengths, corresponding to Doppler recession velocities exceeding 90% of the speed of light, placing them at substantial fractions of the distance from us to the horizon of the observable universe. However, uncertainties in estimating very large distances in the universe due to insufficient understanding of the evolution of galaxies, not to mention the structure of quasars, have prevented unambiguous determination of the sign of the curvature. In fact, Hubble's law is still used to estimate distances to the quasars, rather than they being used to determine both distance and redshift and thus test their relationship. The question of whether space is finite or infinite remains unresolved by observation at this time.

B. ETERNAL EXPANSION OR AN END TO TIME

In the Friedmann cosmologies of the GTR, a finite space implies an end to proper time in the future, but this is not required in some nonstandard models. Assuming the Robertson–Walker form of the metric for a homogeneous and isotropic space-time, it is convenient to discuss the future evolution of any such expanding universe in terms of a dimensionless deceleration parameter, defined as

$$q = -R(d^2R/dt^2)/(dR/dt)^2$$

Throughout most of its history, the dynamics of the universe have been dominated by matter in which the average energy density is very much greater than the pressure. Neglecting the pressure of nongravitational fields, space will reverse its expansion and collapse in finite proper time if and only if $q > \frac{1}{2}$. From the Friedmann equation for the Hubble parameter, it is easy to show that $q = \Omega/2$ in a space-time described by the GTR with zero cosmological constant. In nonstandard models, the deceleration parameter depends on the density, the cosmological constant, and the Hubble parameter in more complicated ways. For some choices of cosmological constant, it is even possible to have an accelerating universe ($q < 0$) with a positive density. However, a cosmological constant whose magnitude substantially exceeded the critical energy density would produce detectable local effects that are not observed. Thus, we can conclude that a sufficiently large density must imply an end to time.

Applications of the virial theorem of Newtonian mechanics to galactic rotation and the dynamics of clusters indicate that masses often exceed those inferred from visible light by about an order of magnitude. Such results push at the upper bound of the baryon density inferred from nucleosynthesis but are still far short of supporting $q > \frac{1}{2}$. However, if only one-tenth of the ordinary matter in galaxies and clusters may be visible to us, is it not possible that there exists mass-energy, in as yet undetected forms or places, of sufficient quantity to produce deceleration exceeding the critical value? Primordial black holes, formed before the temperature of the universe had dropped to 10 MeV, would not interfere with nucleosynthesis in the standard model but could substantially increase the value of Ω. (Notice that black holes due to the collapse of stars are made of matter that contributed to Ω_B during the time of primordial nucleosynthesis and thus are limited by the nuclear abundance data.) Calculations of the primordial black hole mass spectrum, such as those by Bernard Carr, have demonstrated that present observa-

tional data are insufficient to decide whether the contribution of black holes to the density will reverse expansion. Some speculative ideas in high-energy particle physics present other exotic candidates for dark matter that may dominate the gravitational dynamics of the universe as a whole. Empirically, whether there will be an end to time remains an unresolved question.

C. SIGNIFICANCE OF LARGE-SCALE STRUCTURE

The measured homogeneity and isotropy of the cosmic microwave background radiation temperature ($\Delta T/T < 10^{-3}$) is strong evidence that the observable universe is rather precisely homogeneous and isotropic on the largest scale ($1/H$ is ~3000 Mpc). However, it is well known that on only slightly smaller scales, up to 100 Mpc, the universe today is very inhomogeneous, consisting of stars, galaxies, and clusters of galaxies. For example, the variation in density divided by the average density of the universe, $\delta\rho/\rho$, is of the order of 10^5 for galaxies. The density in a large "void" recently discovered is less than average by about one order of magnitude. Since gravitational instability tends to enhance any inhomogeneity as time goes on, the difficulty is not in creating inhomogeneity but rather in deviating from perfect homogeneity at early times in just the way that can account for the structural length scales, mass spectra, and inferred presence of dark matter that are so obvious in the universe today. Opinions about the structure of the universe at early times have run the gamut from Misner's chaotic "mixmaster" to Peebles's quite precisely homogeneous and isotropic space-times. The issue of the origin of structure in the universe on the largest scales remains unresolved, because conflicting scenarios that adequately account for galaxies and clusters can involve so many tunable parameters that it is difficult to distinguish among competing models observationally. Some have even suggested filamentary structures or superclusters of galaxies with lengths of greater than 100 Mpc, but there is controversy about the statistical evidence for the reality of gravitationally bound hierarchies on such a large scale.

D. VIABILITY OF NONSTANDARD MODELS

The assumption that the universe is homogeneous and isotropic on a sufficiently large scale has been called the cosmological principle. This principle, applied to the Riemannian geometry of space-time using the methods of group theory, leads to the Robertson–Walker form of the invariant separation between events. This mathematically elegant foundation for theoretical cosmology is independent of a particular choice of gravitational field equations. However, the evolution of the scale factor and the relation of the curvature constant to the matter distribution are, of course, intimately related to the structure of the gravitational theory that is assumed.

Previous mention was made of the logical possibility of complicating the equations of the GTR by introducing a cosmological constant. There is considerable empirical support for the view that this constant must be small compared with the critical density. Although most theoreticians favor its vanishing, it would be much more satisfying to derive its value from some other theory rather than simply to assert it to be zero.

Numerous alternatives to the GTR consistent with the special theory of relativity have been proposed. Some may be eliminated from further consideration by noncosmological tests of gravitational theory. For example, theories based on a space–time metric that is conformally flat, such as that published by Gunnar Nordstrom in 1912, are untenable because they fail to predict the deflection of light rays in the gravitational field of the sun, first measured by Dyson, Eddington, and Davidson during a solar eclipse in 1919. Others, such as the scalar–tensor theory published by Carl Brans and Robert Dicke in 1961, may be made consistent with current observational data by suitable adjustment of a parameter. Their cosmological consequences present distinct challenges to the standard model for some times during the evolution of the universe. However, sufficiently close to singularities, the predictions of most such theories become indistinguishable from those of the GTR. For example, in 1971 Dykla, Thorne, and Hawking showed that gravitational collapse in Brans–Dicke theory inevitably leads to the "black hole" solutions of the GTR. Hence, cosmological models based on these theories make predictions very much like those of the GTR for the strong fields near the "big bang" and the nearly flat space-time of today. If they are very different in some intermediate regime, perhaps one important to the formation of structures such as galaxies, there presently appear to be no crucial tests that could cleanly decide in their favor. Thus, by an application of Occam's razor, most contemporary models are constructed assuming the validity of the GTR.

An influential exception to the dominance of GTR models was the "steady-state" cosmology of Bondi, Gold, and Hoyle. The philosophical foundation of this work was the extension of the cosmological principle to the "perfect cosmological principle," which asserted that there should be uniformity in time as well as spatial homogeneity and isotropy. As remarked earlier, observations of the microwave background presently render this model untenable. It is also unable to account for the abundances of various light nuclei synthesized when the universe was much hotter and denser than it is now. The apparent overthrow of the perfect cosmological principle encouraged questioning of the assumption of spatial homogeneity and isotropy. Since the empirical evidence in favor of spatial uniformity on very large scales is quite strong, most cosmologists today would like to deduce the cosmological principle rather than assume it. That is, we would like to demonstrate that, starting with arbitrary initial conditions, inhomogeneities and anisotropies are smoothed out by physical processes in a small time compared with the age of the universe. A serious difficulty in attempts to derive the cosmological principle was first emphasized by Misner in 1969. Relativistic space-times of finite age have particle horizons, so that at any moment signals can reach a given point only from limited regions, and parts of the universe beyond a certain distance from one another have not yet had any possibility of communicating. The correlation in conditions at distant regions that is asserted by the cosmological principle can be derived only if chaotic initial conditions smooth themselves out through an infinite number of processes, such as the expansions and contractions along different axes in "mixmaster" universes. In the standard model, the cosmological principle is regarded as an unexplained initial condition.

V. Interaction of Quantum Physics and Cosmology

A. Answers from Particle Physics

The interaction between elementary particle physics and cosmology has increased greatly since 1980, to the benefit of both disciplines. Several initial conditions of classical cosmology are given a tentative explanation in "inflationary universe" scenarios, proposed by Alan Guth in 1981 and modified by Linde, Albrecht, and

Steinhardt in 1982. These models assume a time when the energy density of a "false vacuum" in a grand unified theory (GUT) dominated the dynamics of the universe. Since the density was essentially constant throughout this period, the Robertson–Walker scale factor grew exponentially in time, allowing an initially tiny causally connected region (even smaller than the small value of $1/H$ at the start of inflation) to grow until it included all of the space that was to become the presently observable universe. The original version (1981) assumed that this occurred while the universe remained trapped in the false vacuum.

Unfortunately, such a universe that inflated sufficiently never made a smooth transition to a radiation-dominated early Friedmann cosmology. In the "new inflationary" models (1982), the vacuum energy density dominates while the relevant region of the universe inflates and evolves toward the true vacuum through the spontaneous breaking of the GUT symmetry by nonzero vacuum expectation values of the Higgs scalar. The true vacuum is reached in a rapid and chaotic "phase transition" when the universe is of the order of 10^{-35} sec old, resulting in the production of a large number and variety of particles (and antiparticles) at a temperature of the order of 10^{14} GeV. It is supposed that the universe evolves according to the standard model after this early time.

Since the entire observable universe evolves from a single causally connected region of the quantum vacuum, inflationary models obviously avoid horizon problems. The homogeneity and isotropy of the present observable universe are a consequence of the dynamic equilibrium in the tiny region. Since the density term in the Friedmann equation for the evolution of the Hubble parameter remains essentially constant throughout the inflationary era, while the curvature term is exponentially suppressed, these scenarios also offer a natural explanation for the present approximate flatness of the universe. In fact, plausible suppression of the curvature greatly exceeds that required by any astrophysical observations. Only by artificially contrived choices of parameters in an inflationary universe could we avoid the conclusion that any difference between the current value of Ω and unity is many orders of magnitude less than 1. If this result and the bounds on Ω_B from primordial nucleosynthesis are both true, we must conclude that dark matter of as yet undetermined form dominates the dynamics of the universe. Since GUTs predict the nonconservation of B (baryon number), C (charge conjugation), and CP (product of charge conjugation and parity), the decay of very heavy bosons far from thermodynamic equilibrium offers a way of dynamically generating the predominance of matter over antimatter rather than merely asserting it as an initial condition. In the absence of observation of the decay of the proton or accelerators capable of attaining energies at which GUTs predict convergence of coupling "constants," the apparent baryon asymmetry of the universe is perhaps the best empirical support for some sort of unified theory of quarks and leptons.

B. Constraints from Cosmology

Even as elementary particle theory solves some problems of cosmology, it is subject to limitations derived from cosmological data involving energies far beyond the 10^3-GeV limit of existing terrestrial accelerators. An important example involves the production of magnetic monopoles in the early universe. In 1931, P. A. M. Dirac showed that assuming the existence of magnetic monopoles led to a derivation of the quantization of magnetic and electric charge and a relation between them implying that magnetic charges would have to be very large. However, other properties of the hypothetical monopoles, such as mass and spin, were undetermined in his theory. In 1974, Gerhard t'Hooft and Alexander Polyakov showed that monopoles must be produced in gauge theory as topological defects whenever a semisimple group breaks down to a product that contains a $U(1)$ factor, for example,

$$SU(5) \rightarrow SU(3) \times SU(2) \times U(1)$$

All proposed GUTs, which attempt to unify the strong and the electroweak interactions, are examples of gauge theories in which monopoles are required and have masses of the order of the vacuum expectation value of the Higgs field responsible for the spontaneous symmetry breaking. The present experimental lower limit, of the order of 10^{31} years, for the mean life of the proton implies a lower limit for this mass of the order of 10^{16} GeV. Only within a time of at most 10^{-34} sec after the "big bang" was any place in the universe hot enough to produce particles of so great a mass, either as topological "knots" or as pairs of monopoles and antimonopoles formed in the energetic collisions of ordinary particles.

In addition to their enormous masses and relatively large magnetic charges, monopoles are predicted by GUTs to serve as effective catalysts for nucleon decay. Thus, if present in any appreciable abundance in the universe today, monopoles should make their presence obvious by doing some conspicuously interesting things. If monopoles were made in about the same abundance as baryons, their density alone would exceed the critical value for a closed universe by a factor of $\sim 10^{11}$. Monopoles would use up the potential energy of stationary magnetic fields, such as that of the Milky Way, by converting it to increases in their own kinetic energy. Collecting in a star throughout its history, they would render its collapsed "final state" short-lived by catalyzing nucleon decay. The observational upper bound on Ω, lower limits on the galactic magnetic field, and the life spans of neutron stars place severe constraints on the flux of monopoles at present and hence on their rate of production during the spontaneous symmetry-breaking era. In fact, unacceptably large magnetic monopole production in the simplest GUTs was one of the primary motivations for the development of the new inflationary universe models, which solve the problem through an exponential dilution of monopole density that leaves very roughly one monopole in the entire observable universe at present. While such a scenario appears to make the experimental search for monopoles essentially futile, it is important not to ascribe too much quantitative significance to this result, because the predicted number is exponentially sensitive to the ratio of the monopole mass to the highest temperature reached in the phase transition at the end of spontaneous symmetry breaking. Thus, an uncertainty of a mere factor of 10 (theoretical uncertainties are at least this large) in this ratio changes the predicted number of monopoles by a factor of the order of 10^8.

On the experimental front, Blas Cabrera claimed the detection of a magnetic monopole on February 14, 1982, after 150 days of searching with a superconducting quantum interferometer device (SQUID). If this observation is correct and even approximately corresponds to the typical distribution of monopoles in space, then neither the excessive production of a naive GUT model nor the extreme scarcity of a new inflationary model is to be credited. Confirmation of this monopole detection would leave present theory totally at a loss to explain the monopole abundance, but neither Cabrera nor other observers have yet claimed another detection.

The apparent smallness of the cosmological constant is a fact that has not yet been explained in any viable theory of particle physics or gravitation. Below some critical temperature in electroweak theory or GUTs, the effective potential function of the Higgs fields behaves like a cosmological constant in contributing a term equal to this potential function times the space–time metric to the stress–energy–momentum tensor of the universe. Empirical bounds on the vacuum energy density today imply that this potential at the spontaneous symmetry-breaking minimum was already less than 10^{-102} times the effective potential of the false vacuum. There is no derivation of this extremely small dimensionless number within the framework of GUTs. In fact, the assumption that the cosmological constant is negligible today is an unexplained empirical constraint on the otherwise undetermined scale of the effective potential in a gauge theory.

Another possible success of inflationary models is the natural development of nearly scale-independent density inhomogeneities from the quantum fluctuations in the Higgs field of GUTs during inflation. Inhomogeneities should later evolve by gravitational clumping into galaxies and clusters of galaxies. This opens the possibility of calculation from first principles of the spectrum of later structural hierarchies. Comparison of the results of such calculations with the observed large-scale structure of the universe may provide the most stringent constraints on new inflationary models.

C. SINGULARITIES AND QUANTUM GRAVITY

This survey has looked back nearly 10^{18} sec from the present to the "big bang" with which the standard cosmological model claims the universe began. The attempt to understand its evolution reveals a number of significant eras. Let us review them from the present to the initial singularity in reverse chronological order, which is generally the order of decreasing direct experimental evidence and thus increasing tentativeness of conclusions. At a time of the order of 10^{13} sec after the universe began, when the temperature was ~ 0.3 eV, the photons that now compose the cosmic microwave radiation background last appreciably interacted with matter, which then "recombined" into transparent neutral atoms of hydrogen, helium, and lithium and began to form the large-scale structures familiar to us: stars, galaxies, and clusters of galaxies. At a time of the order of 10^{-2} sec after the beginning, when the temperature was ~10 MeV, the free neutrons and some of the free protons underwent the primordial synthesis that formed the nuclei of these atoms, and the cosmic background neutrinos ceased having significant interactions with matter. At a time of the order of 10^{-5} sec after the beginning and a temperature ~300 MeV, quarks became "confined" to form the hadrons as we now know them. At a time of the order of 10^{-12} sec after the beginning, the temperature was ~10^3 GeV, the present limit of terrestrial accelerators. The distinction between electromagnetic and weak interactions was not significant before then. The reconstruction of earlier history is of necessity much more tentative. The spontaneous symmetry breaking of the grand unification of electroweak and strong interactions is thought to have occurred at a time of roughly 10^{-34} sec and a temperature of the order of 10^{14} GeV. During the "inflation" preceding this epoch the baryon asymmetry of the universe may have been generated by fluctuations from thermal equilibrium in GUTs, and magnetic monopoles may have been produced by symmetry breaking. Any attempt to analyze events at substantially earlier times must address the unfinished program of constructing a quantum theory that unifies gravitation with the strong and electroweak interactions.

In the absence of complete understanding, the time and temperature scales of quantum gravitational effects can be estimated by dimensional analysis applied to the fundamental constants that must appear in any such theory. These are the quantum of action, the Newtonian gravitational constant, the speed of light, and the Boltzmann constant. The results, a time of the order of 10^{-44} sec and a temperature of the order of 10^{19} GeV, delineate conditions so near the classically predicted singularity of infinite density and curvature at the "big bang" that the very concept of a deterministic geometry of space-time breaks down. Violent fluctuations of space-time should generate particles in a manner analogous to that which occurs in the vicinity of a collapsing black hole, as first studied by Stephen Hawking in 1974. Such processes could conceivably be the source of all existing matter, and the possible removal of an infinite-density singularity at time zero would make it scientifically meaningful to ask what the universe was doing before the "big bang." [See BLACK HOLES (ASTRONOMY).]

Not only the existence and structure of matter but even the topology and dimensionality of space-time become properties to be derived rather than postulated in the quantum gravity era. In 1957, John Wheeler suggested that space-time need not necessarily be simply connected at the Planck scale (of the order of 10^{-35} m) but could have a violently fluctuating topology. If so, its description in terms of a smooth continuum would not be appropriate and would have to be replaced by some other mathematical model. The first viable unification of electromagnetism and gravitation, proposed by Theodor Kaluza in 1921 and independently by Oskar Klein in 1926, used a five-dimensional space-time with an additional "compact" spatial dimension subject to constraints that reduced their model to a sterile fusion of Maxwell's and Einstein's field equations. Removing these constraints allows Kaluza–Klein spaces to be used for alternate formulations of gauge field theories. Models with a total of 11 dimensions are currently being actively explored in relation to supersymmetry theories, which seek to provide a unification of bosons and fermions and all interactions among them. Another approach, called string field theory, involves replacing the pointlike particles of conventional quantum field theory with fundamental objects with extent in one spatial dimension. The full implications of such studies for particle physics and for cosmology are not yet clear.

VI. Thoughts on Sources of Future Progress

A. EMERGING OBSERVATIONAL TECHNOLOGIES

Since the pioneering research of Galileo, advances in telescope capabilities have been the source of more and richer data to constrain cosmological speculation. The present generation of astronomers has seen photographic techniques increasingly augmented by electronic image intensifiers. Improvements in photon detectors, such as charge-coupled devices, are beginning to approach their limits, so that achieving substantial gains will involve increasing the aperture in the next generation of optical and infrared instruments. Two 7.6-m telescopes, a monolithic thin meniscus in west Texas and a multiple mirror design in Mauna Kea, Hawaii, are to be completed by approximately the end of the 1980s. A number of other telescopes from 10- through 15-m aperture, possibly as large as 25-m aperture, are being planned or discussed. At least some of these may be operational before the end of the twentieth century. No list of forthcoming instruments would be complete without mention of the much-awaited Hubble space telescope. The instrument is ready, but its deployment into orbit has been delayed by the halt in the space shuttle program following the tragic loss of the Challenger orbiter. When shuttles are again flying, perhaps in 1988 with redesigned boosters, this telescope will be among the highest priority payloads scheduled for launch. Although not quite as large as the greatest ground-based telescopes, the space telescope's unique vantage point completely above the atmosphere should allow previously unattainable performance at optical and infrared wavelengths. It is expected to provide detailed information about the structure of objects at distances corresponding to the formation of galaxies and possibly the birth of quasars.

As more powerful telescopes looking farther into outer space observe signals from earlier in the history of the universe, higher energy accelerators probing farther into inner space measure particle behavior under conditions simulating earlier times during the "big bang." The installation of superconducting magnets in the tevatron at the Fermi National Accelerator Laboratory at Batavia, Illinois, enabled it to produce protons with an energy of 10^3 GeV in 1984. In 1986, it made available a total energy of 2×10^3 GeV by accommodating countercirculating beams of protons and antiprotons that are made to collide. The Russian laboratory at Serpukhov is building a 4×10^2 GeV accelerator, to be completed in 1988 and upgraded in energy to 3×10^3 GeV in the 1990s. Eventually, a second 3×10^3 GeV beam will be made to collide with the first. The U.S. Department of Energy is supporting research and development toward a machine to be built in the 1990s called the superconducting supercollider, which would accelerate protons in opposite directions around a circumference of the order of 100 km and bring them together with total energies of 2×10^4 GeV. There are theoretical indications that this energy might be sufficient to make possible the actual production of the Higgs particle associated with electroweak

unification. Even if this particle is not produced, there is no doubt that the increase in energy to these ranges will lead to important new insights into the structure of matter on a scale of less than 10^{-19} m.

B. CONCEPTS AND MATHEMATICAL TOOLS

The search for a viable extension of GUTs to a totally unified theory (TUT), which would include quantum gravity as well as the strong and electroweak interactions, is an active area of particle theory research. The energies at which such unification was achieved in nature are presumably even higher than those for GUTs. Thus, no data from accelerators, even in the most optimistic projections of foreseeable future technology, can serve to constrain speculation as well as does information from cosmology. The currently fashionable attempts to derive a TUT are based on "supersymmetry," which is the idea that the fundamental Lagrangian contains equal numbers of Bose and Fermi fields and that they can be transformed into each other by a supersymmetry. This immediately doubles the particle spectrum, associating with each particle thus far observed (or predicted by GUTs) a "superpartner" of opposite quantum statistics. There is at present no experimental evidence for the existence of any of these superpartners, inviting doubt as to the necessity of the supersymmetry assumption. However, supersymmetric theories have the potential to address one otherwise unanswered issue of cosmology: Why is the cosmological constant so small, perhaps precisely zero? Supersymmetric theories are the only known quantum field theories that are sensitive to the vacuum energy level. This appears to imply that a derivation of the cosmological constant from first principles should be possible within a supersymmetric theory, but the problem remains unsolved.

As the number of degrees of freedom being considered in field theories increases, increased computing speed and power become more important on working out the consequences of various proposed models. Some new hardware architectures such as concurrent processing appear to be a means of achieving performance beyond the limits of any existing machines but require the further development of software exploiting their distinctive features to achieve their full potential. Progress in discrete mathematics, such as development of the theory of symplectic nets, is of benefit to both computer science and pure mathematics. In the past, fundamental insights have often been derived from mathematical analysis without the benefit of "number crunching," and there is no reason to expect that this process has come to an end. It is, of course, impossible to predict what new closed-form solution of a recalcitrant problem may be discovered tomorrow, or what impact such a discovery may have.

In the last analysis, any attempt to predict the direction of progress in cosmology farther than the very near future seems futile. By the nature of the questions that cosmology seeks to answer, the scope of potentially relevant concepts and information is limitless. It is entirely possible that within a decade carefully reasoned thought, outrageous unexpected data, or some combination of the two may overthrow some of today's cherished "knowledge." Aware of the questions still unanswered and of the possibility that some of the right questions have not yet been asked, we can only hope that future discoveries, anticipated or unforeseen, will result in ever greater insights into the structure, history, and destiny of the universe.

BIBLIOGRAPHY

Barrow, J., and Silk, J. (1983). "The Left Hand of Creation," Basic Books, New York.

Evans, D., ed. (1984). *Texas Symp. Relativistic Astrophys., Ann. N.Y. Acad. Sci.* 422.

Kolb, E., Turner, M., Lindley, D., Olive, K., and Seckel, D., eds. (1986). "Inner Space/Outer Space." Univ. of Chicago Press, Chicago.

Misner, C., Thorne, K., and Wheeler, J. (1973). "Gravitation." Freeman, New York.

Pagels, H. R. (1985). "Perfect Symmetry." Simon and Schuster, New York.

Peebles, P. J. E. (1980). "The Large-Scale Structure of the Universe." Princeton Univ. Press, Princeton, New Jersey.

Isham, C., Penrose, R., and Sciama, D., eds. (1981). "Quantum Gravity II." Oxford Univ. Press, London.

Rowan-Robinson, M. (1985). "The Cosmological Distance Ladder." Freeman, New York.

Weinberg, S. (1977). "The First Three Minutes." Basic Books, New York.

DARK MATTER IN THE UNIVERSE

Steven N. Shore   *New Mexico Institute of Mining and Technology and DEMIRM, Observatoire de Meudon*

GLOSSARY

Baryons: Protons and neutrons that are the basic constituents of luminous objects and which take part in nuclear reactions. These are strongly interacting particles, which feel the electromagnetic and nuclear, or strong, forces.

Cosmic background radiation (CBR):   Relic radiation from the Big Bang, currently having a temperature of 3 K. It separated out from the matter at the epoch at which the opacity to scattering of the universe, because of cooling and expansion, fell to small enough values that the probability of interaction between the matter and the primordial photons became small compared with unity. It is essentially an isotropic background and its temperature (intensity) fluctuations serve to place limits on the scale of the density fluctuations in the universe at a redshift of about $z1000$. The redshift is defined as $\lambda/\lambda_0 = 1 + z$, where λ is the currently observed wavelength of a photon and λ_0 is the wavelength at which it was emitted from a distant object.

Deceleration parameter: Constant that measures both the density of the universe, compared with the critical density necessary for a flat space–time metric [called $\Omega = \rho/\rho_c$], and the departure of the velocity–redshift relation from linearity (q_0).

Faber–Jackson relation: Relation between the bolometric luminosity of a galaxy and its core velocity dispersion. This permits the determination of the intrinsic luminosity of elliptical galaxies without the need to assume a population model for the galaxy. It is used in the measurement of distances to galaxies independent of their Hubble types.

Hubble constant (H_0):   Constant of proportionality for the rate of velocity of recession of galaxies as a function of redshift. Its current value is 75 $\pm$ 25 km sec^{-1} Mpc^{-1}. Often, in order to scale results to this empirically determined constant, it is quoted as $h = H_0/(100$ km sec^{-1} Mpc$^{-1})$. The inverse of the Hubble constant is a measure of the age of the universe, called the *Hubble time*, and is about 1.5×10^{10} yr.

Leptons: Lightest particles, especially the electron, the muon, and tau, and their associated neutrinos. These particles interact via the weak and electromagnetic forces (electroweak).

Tully–Fisher relation: Relation between the maximum orbital velocity observed for a galaxy's 21-cm neutral hydrogen (the line width) and the bolometric luminosity of the galaxy.

Units: Solar mass ($M_\odot$), 2×10^{33} g; solar luminosity ($L_\odot$), 4×10^{33} erg sec^{-1}; parsec (pc), 3×10^{18} cm (3.26 light years).

Dark matter is the subluminous matter currently required to explain the mass defect observed from visible material on scales ranging from galaxies to superclusters of galaxies. The evidence shows that the need for some form of invisible, but gravitationally influential, matter is present on length scales larger than the distance between stars in a galaxy. This article examines the methods used for determining galaxies and clusters of galaxies, and discusses the cosmological implications of various explanations of dark matter.

I. Historical Prolog

The existence of a species of matter that cannot be seen but merely serves some mechanical function in the universe can easily be traced to Aristotelian physics, and was clearly supported throughout the development of physical models prior to relativity. The concept of the *aether,* at first something apart from normal material in being imponderable, eventually metamorphosed into that of a fluid medium, capable of supporting gravitation and electromagnetic radiation. However, since it was the medium that was responsible both for generating and transmitting such forces, its nature was intimately tied to the overall structure of the universe and could not be separated from it. It had to be assumed, and could not easily be studied. Nineteenth century attempts, using high-precision measurements, to observe the anisotropy of the propagation of light because of the motion of the earth, the Michelson–Morley experiment being the best known, showed that the aether's mechanical properties could not conform to those normally associated with terrestrial fluids.

A version of the search for some cosmic form of dark matter began in the late seventeenth century, with Halley's question of the origin of the darkness of the night sky. Later enunciated in the nineteenth century as *Olber's paradox,* it required an understanding of the flux law for luminous matter and was argued as follows: in an infinite universe, with an infinite number of luminous bodies, the night sky should be at least as bright as the surface of a star, if not infinitely bright. Much as in current work on dark matter, the argument was based on the assumption of the premise of a cosmological model and of the dynamical (or in this case phenomenological) character of a particular observable.

In order to circumvent the ramifications of the paradox without questioning its basic premises, F. Struve, in the 1840s, argued for the existence of a new kind of matter in the cosmos, one capable of extinguishing the light of the stars as a function of their distance, hence path length through the medium. The discovery of dark nebulae by Barnard and Wolf at the turn of the century, of stellar extinction by Kapteyn about a decade later, and of the reddening of distant globular clusters by Trumpler in the 1930s served to support the contention that this new class of matter was an effective solution to the *missing light* problem. However, as J. Herschel realized

very soon after Struve's original suggestion of interstellar dust, this cannot be a viable solution for the paradox. In an infinite universe, the initially cold, dark matter would eventually come into thermal equilibrium. The resulting glow would eventually reach, if not the intensity of the light of a stellar surface, an intensity still substantially above the levels required to reproduce the darkness of the sky between the stars.

An additional thread in this history is the explanation of the distribution of the nebulae, those objects we now know to be extragalactic. In detailed statistical studies, Charlier, Lundmark, and later Hubble, among others, noted a *zone of avoidance,* which was located within some twenty degrees of the galactic plane. Several new forces were postulated to explain this behavior, all of which were removed by the discovery of the particulate nature of the dust of the interstellar medium and the expansion of the universe. However, the explanatory power of the concepts of dark matter and unknown forces of nature serves as a prototype for many of the current questions, although the circumstances and nature of the evidence have dramatically altered in time.

The origin of the current problem of dark matter (hereafter called *DM*) really starts in the late 1930s with the discovery of galaxy clusters by Shapley. The fact that the universe should be filled with galaxies that are riding on the expanding space–time was not seen as a serious problem for early cosmology, but the fact that these objects cluster seems quite difficult to understand. Their distribution appears to be very anisotropic, and was seen at the time to be the evidence that the initial conditions of the expansion may be drastically altered by later gravitational developments of the universe.

About the same time, Zwicky, analyzing the distribution of velocities in clusters of galaxies, argued that the structures on scales of megaparsecs (Mpc) cannot be bound without the assumption of substantial amounts of nonluminous material being bound along with the galaxies. His was also the first attempt to apply the virial theorem to the analysis of the mass of such structures. Oort, in his analysis of the gravitational acceleration perpendicular to the galactic plane, also concluded that only about $\frac{1}{2}$ of the total mass was in the form of visible stars.

In the past decade, with the advent of large-scale surveys of the velocities of galaxies in distant clusters, the problem of DM has become central to cosmology. It is the purpose of this article to examine some of the issues connected with current cosmological requirements for some form of nonluminous, gravitationally interacting matter, and the evidence of its presence on scales from galactic to super-cluster.

II. Kinematic Studies of the Galaxy

The Galaxy is a spiral, consisting of a central spheroid and a disk of stars extending to several tens of kiloparsecs. A halo is also present and appears to envelope the entire system extending to distances of over 50 kpc from the center. It is this halo that is assumed to be the seat of most of the DM. Since it is essentially spherical, it contributes to the mass that lies interior to any radius and also to the vertical acceleration. The stars of which it is composed have high-velocity dispersions, of order 150 km sec^{-1}, and are therefore distributed over a larger volume while still bound to the Milky Way galaxy (the Galaxy).

Galactic studies of dark matter come in two varieties: local, meaning the solar neighborhood (a region of several hundred parsecs radius) and large-scale, on distances of many kiloparsecs. The study of the kinematics of stars in the Galaxy produced the first evidence for "missing mass," and there is more detailed information available for the Milky Way than for any other system. Further, many of the methods that have been applied, or will be applied in future years, to extragalactic systems have been developed for the Galaxy. So that the bases of the evidence can be better understood, this section will be a more detailed discussion of some of the methods used for the galactic mass determination. [See GALACTIC STRUCTURE AND EVOLUTION.]

A. LOCAL EVIDENCE

The local evidence for DM in the Galaxy comes from the study of the vertical acceleration relative to the galactic plane, the so-called "Oort criterion." This uses the fact that the gravitational acceleration perpendicular to the plane of the Galaxy structures the stellar distribution in that direction and is dependent on the surface density of material in the plane. The basic assumption is that the stars, which are collisionless, form a sort of "gas," whose vertical extent is dependent on the velocity dispersion and the gravitational acceleration. For instance, taking the pressure of the stars to be $\rho\sigma^2$, where ρ is the stellar density and σ is their velocity dispersion in the z direction, then vertical hydrostatic equilibrium in the presence of an acceleration K_z gives the scale height of an "isothermal" distribution:

$$z_0 \approx \sigma^2/K_z \qquad (1)$$

This represents the mean height that stars, in their oscillations through the plane, will reach. It is sometimes called the *scale height* of the stellar distribution.

The gravitational potential is determined by the Poisson equation:

$$\nabla^2\Phi = -4\pi G\rho = \nabla \cdot \mathbf{K} \qquad (2)$$

where G is the gravitational constant and $\mathbf{K}$ is the total gravitational acceleration. Separation of this equation into radial and vertical components, assuming that the system is planar and axisymmetric, permits the determination of the vertical acceleration. It can be assumed that the stars in the plane are orbiting the central portions of the galaxy, and that

$$K_r = \frac{\Theta^2(r)}{r} \qquad (3)$$

Here, Θ is the circular velocity at distance r from the galactic center. Using this assumption, the Poisson equation becomes

$$\frac{\partial}{\partial z} K_z = -4\pi G\rho + \frac{2\Theta}{r}\frac{d\Theta}{dr} \qquad (4)$$

where the radial gradient of the circular velocity is determined from observation. From the observation of the radial gradient in the rotation curve, it is possible to determine the vertical acceleration from dynamics of stars above the plane, and thus determine the space mass density in the solar neighborhood.

The argument proceeds using the fact that the vertical motion of a star through the plane is that of a harmonic oscillator. The maximum z velocity occurs when the star passes through the plane. From the vertical gradient in the z component of the motion, V_z, one obtains the acceleration, and the mass density follows from this. Assuming that stars start from free-fall through the plane, their total energy is given by

$$\Phi_0 + \tfrac{1}{2}V_{z,0}^2 = \tfrac{1}{2}V_z^2 \qquad (5)$$

so that assuming that $V_{z,0} = 0$, in other words that the star falls from a large distance having started at rest, it follows that the maximum velocity through the plane of a galaxy is determined by the potential, thus the acceleration, at the extremum of the orbit (apogalactic distance). The vertical acceleration is a function of the surface mass density. Thus, star counts can be used to constrain, from luminous material, the mass in the solar neighborhood. Note that $K_z \sim -4\pi G\Sigma$, where Σ is the surface density, given by

$$\Sigma = \int_0^z \rho(r, z')\, dz' \qquad (6)$$

It should therefore be possible to determine the gravitational acceleration from observations of the vertical distribution of stars, obtain the required surface density, and compare this with the observations of stars in the plane.

The mass required to explain K_z observed in the solar neighborhood, about $0.15 M_\odot \, pc^{-3}$, exceeds the observed mass by a factor of about 2. This is not likely made up by the presence of numerous black holes or more exotic objects in the plane, since (for instance) the stability of the solar system is such as to preclude the numerous encounters with such gravitational objects in its lifetime. It appears that there must be a considerable amount of uniformly distributed, low-mass objects or some other explanation for the mass.

The velocity dispersion of stars is given by the gravitational potential in the absence of collisions. Thus, from the determination of the mass density in the solar neighborhood with distance off the plane, the vertical component of the gravitational acceleration can be found, leading to a comparison with the velocity dispersion of stars in the z direction. The resulting comparison of the dynamic and static distributions determines the amount of mass required for the acceleration versus the amount of mass observed.

Most determinations of K_z find that only $\frac{1}{3}$ to $\frac{1}{2}$ of the mass required for the dynamics is observed in the form of luminous matter. In spite of several attempts to account for this mass in extremely low luminosity M dwarf stars, the lowest end of the mass distribution of stellar masses, the problem remains. The mass of the observed halo is not sufficient to explain the vertical acceleration, and the mass function does not appear to extend to sufficiently low masses to solve this problem.

Another local determination of the mass of the Galaxy, this time in the plane, can be made from the observation of the maximum velocity, relative to the so-called local standard of rest, of stars orbiting the galactic center. The argument is best illustrated for a point mass. In the case of a circular orbit about a point, the orbital velocity, Θ, is given by centrifugal acceleration being balanced by the gravitational attraction of the central mass:

$$\Theta(r) = \left(\frac{GM}{r} \right)^{1/2} \qquad (7)$$

and it is easily seen that the escape velocity $v_{esc}(r) = \sqrt{2}\Theta(r)$ for all radii. The coefficient is slightly lower in the case of an extended mass distribution, but the argument still follows the same basic form. The local standard of rest (LSR) is found from the velocity of the sun relative to the most distant objects, galaxies and quasi-stellar objects (QSOs), and the globular cluster system of our galaxy. The maximum orbital velocity observed in the solar neighborhood should then be about $0.4 v_{LSR}$, or in the case of our system about 70 km sec^{-1} in the direction of solar motion. For escape from the system, at a distance of 8.5 kpc with a $\Theta_0 = 250$ km sec^{-1}, the escape velocity from the solar circle should be only about 300 km sec^{-1}. The mass of the Galaxy can be determined from the knowledge of the distance of the sun from the galactic center (determined from mass distributions and from the brightness of variable stars in globular clusters) and from the shape of the rotation curve as a function of distance from the galactic center.

B. LARGE-SCALE DETERMINATIONS

From the stellar orbits, one knows the rate of galactic differential rotation in the vicinity of the sun, a region about 3 kpc in radius, and located about 8.5 kpc from the galactic center. The measurement of galactic rotational motion can be greatly extended through the use of millimeter molecular line observations, such as OH masers, and the 21-cm line of neutral hydrogen. Most of the disk is accessible, but at the expense of a new assumption. It is assumed that the gas motions are good tracers of the gravitational field, and that random motions are small and unimportant in comparison with the rotational motions. It is also assumed that the gas is coplanar with the stellar distribution, and that the motion is largely circular, with no large-scale, noncircular flows being superimposed. These assumptions are not obviously violated in the Galaxy, but may be problematic in many extragalactic systems, especially barred spiral galaxies. The maximum radial velocity along lines of sight interior to the solar radius occurs at the *tangent points* to the orbits. This method of tangent points presumes that the differential rotation of the Galaxy produces a gradient in the radial velocity along any line of sight through the disk for gas and stars viewed from the sun, and that the maximum velocity occurs at a point at which the line of sight is tangent to some orbit. The distance can be determined from knowledge of the distance of the sun from the galactic center, about 8 kpc. Using Θ_{HI} gives a measure of the mass interior to the point r, and thus the cumulative mass of the Galaxy. Recent work has extended this to include large molecular clouds, which also orbit the galactic center.

The determination of the orbits and distances of stars and gas clouds outside of the solar orbit is difficult, since the tangent point method cannot be applied for extrasolar orbits, but from the study of molecular clouds and 21-cm absorption, and stellar velocities of standard stars, it appears that the rotation curve outside of the solar circle is still flat, or perhaps rising. The argument that this implies a considerable amount of extended, dark mass follows from an extension of the argument for the orbital velocity about a point mass, $\Theta^2(r) \sim M(r)/r$. Since the observations support $\Theta = $ constant, it appears that $M(r) \sim r$. The scale length for the luminous matter is small, about 5 kpc, and this is substantially smaller than the radial distances where the rotation curve has been observed to still be flat (of order 15 kpc). The same behavior has been observed in external galaxies, as we will discuss shortly. Thus, there is a substantial need for the Galaxy to have a large amount of dark matter.

As mentioned, mass measurements of the Galaxy from stellar and gas rotation curves beyond the distance of the sun from the center are subject to several serious problems, which serve as warnings for extragalactic studies. One is that there appear to be substantial departures of the distribution from planarity. That is, the outer parts of the disk appear to be warped. This means that much of the motion may not be circular, that there may be sizable vertical motions, which mean that the motions are not strictly indicative of the mass. For the inner galactic plane, there is some evidence of noncircular motion, perhaps caused by barlike structures in the mass distribution, thus affecting the mass studies.

An extension of the Oort method for the vertical acceleration, now taken to the halo stars, can be used to measure the total mass of the Galaxy. One looks at progressively more distant objects. These give handles on several quantities. For instance, the maximum velocity of stars falling through the plane can be compared with the maximum distances to which stars are observed to be bound to the galactic system. Halo stars of known intrinsic luminosity, such as RR Lyrae variable stars, can be studied with some certainty in their intrinsic brightness. From the observed brightness, the distance can be calculated. Thus, the distance within the halo can be found. From an observation of the vertical component of the velocity, one can obtain the total mass of the galactic system lying interior to the star's orbit. The same can be done for the globular clusters, the most distant of which should be nearly radially infalling toward the galactic center. These methods give a wide range of masses, from about $4 \times 10^{11} \ M_\odot$ to as high as $10^{12} \ M_\odot$ to distances of 50 to 100 kpc.

The phenomenological solution to the problem of subluminous mass increasing in fraction of the galactic population with increasing length scale measured assumes that the mass-to-light ratio, M/L, is a function of distance from the galactic center. For low-mass stars, this number is of order 1 to 5 in solar units ($M_\odot/L_\odot$), but for the required velocity distribution in the galaxy, this must be as high as 100. There are few objects, except Jupiter-sized or smaller masses, which have this property. The reason is simple—nuclear reactions that are responsible for the luminosity of massive objects, greater than about $0.08 M_\odot$, cannot be so inefficient as to produce this high value. Even the fact that the flux distribution can be redistributed in wavelength because of the effects of opacity in the atmospheres of stars of very low mass will still leave them bright enough to observe in the infrared, if not the visible, and their total (bolometric) luminosities are still high enough to provide M/L ratios that are less than about 10. Black holes, neutron stars, or cold white dwarfs would appear good candidates within the Galaxy. But limits can be set on the space density of these objects from X-ray surveys, since they would accrete material from the interstellar medium and emit X-rays as a result of the infall. The observed X-ray background and the known rate of production of such relics of stellar evolution are both too low to allow these objects to serve as the sole solution to the disk DM problem. Some other explanation is required.

III. Large-Scale Structure of the Galaxy

A. MULTIWAVELENGTH OBSERVATIONS

1. Radio and Far-Infrared (FIR) Observations

Neutral and molecular hydrogen (H I and H_2), form only a small component of the total mass of the galactic system. Both 21-cm and CO (millimeter) observations can only account for about $\frac{1}{3}$ of the total galactic mass being in the form of diffuse or cloud matter. The IRAS satellite, which performed an all-sky survey between 12 and 100 μm, did not find a significant population of optically unseen but FIR bright point sources in the galaxy. Instead, it showed that the emission from dust in the diffuse interstellar medium is consistent with the amount of neutral gas present in the plane, and that there is not a very sizable component of the galaxy at high-galactic latitude. This severely constrains conventional explanations for the DM, since any object would come into equilibrium at temperatures of the same order as an interstellar grain, and would likely be seen by the IRAS satellite observations.

2. X-Ray Observations

These show the interstellar medium to possess a very hot, supernova-heated phase with kinetic temperatures of order $10^6 - 10^7$ K. But here again, it is only a small fraction of the total mass of luminous matter. The spatial extent of this gas is greater than that revealed by the IR and radio searches, but it is still consistent with the galactic potential derived from the optical studies. As mentioned, X-ray observations also constrain the space density of collapsed but subluminous objects, such as black holes, through limits on the background produced by such objects accreting interstellar gas and dust.

3. Ultraviolet Observations

The stars that produce the UV radiation are the most massive or most evolved members of the disk, being either young stars or the central stars of planetary nebulae. They have very low M/L ratios, are easily seen in the UV, and would not be likely candidates for DM. There is no compelling evidence from other wavelength regions that the missing matter is explained by relatively ordinary matter. Baryonic matter at temperatures from 10^7 K to below the microwave background of 3 K can be ruled out by the currently available observations.

B. The Components of the Galactic Disk and Halo

Star counts as a function of magnitude in specific directions (that is, as a function of l_{II} and b_{II}, the galactic longitude and latitude, respectively) provide the best information on the mass of the visible matter and the structure of the galaxy. This has been discussed most completely by Bahcall and Soniera (1981). It depends on the luminosity function for stars, their distribution in spectral type, and their evolution. However, in the end, it provides more of a constraint on the evolution of the stellar population than on the question of whether the system is supported by a substantial fraction of dark matter.

Although such studies concentrate on the luminous matter, the reckoning of scale heights for the different stellar populations provides an estimating method for the vertical and radial components of the acceleration—direct mass modeling. Such studies are also sensitive to the details of reddening from dust both in and off of the galactic plane, metallicity effects as a function of distance from the galactic center, and evolutionary effects connected with the processes of star formation in different parts of the galaxy. Supplementing the space density studies with space motions (proper motion, tangential to the line of sight, can be found for some classes of high-velocity stars and can be added to the radial velocities to give an overall picture of the stellar kinematics) permits both a kinematic and photometric determination of the galactic mass. These studies have extended the determination of the vertical acceleration to distances of more than 50 kpc into the halo.

Another constraint on the mass of the Galaxy comes from the tidal radii of the Magellanic Clouds, and from the masses and mass distributions of globular clusters. The Magellanic Clouds (the Clouds) are fairly large compared with their distances from the galactic center. They therefore feel a differential gravitational acceleration that counteracts their intrinsic self-gravity and tends to rip them apart as they move around the Galaxy. The maximum distance that a star can be from the Clouds before it is more bound to the Galaxy than the Clouds, or conversely the minimum distance to which the Clouds can approach the Galaxy without suffering tidal disruption, provides a measure of the total mass of the Galaxy. The distribution of the globular clusters does not provide any strong support for a DM halo. In addition, the clusters appear to be bound, even though they are the highest mass separate components of the galactic system, without invoking any nonluminous mass. In order to account for the dynamics, and to be consistent with the known ages of the clusters, the typical M/L ratio is about 3, characteristic of low-mass stars. So it appears that objects of order $10^6 - 10^8 M_\odot$, and with sizes of up to 100 pc, are *not* composed of large quantities of dark mass.

C. Theoretical Studies

The spiral structure of disk galaxies has been a long-standing problem since the discovery of the extragalactic nature of these objects. The first suggestion that the patterns seen in spirals might be due to some form of intrinsic collective mode of a disk gravitational instability was from Lindblad, and elaborated by Lin and Shu (1965) as the "quasi-stationary density wave" model. The picture requires a collective mode, in which the stars in the disk behave like a self-gravitating but collisionless fluid (there is no real viscous dissipation), which clumps in a spatially periodic wave. This wave of density feeds back into the gravitational potential, which then serves as the means for supporting the disk wave structure.

It was soon realized that these waves are not

stationary. Further, because of angular momentum transport by tidal coupling of the inner and outer parts of the disk, the waves will wind up and dissipate in the absence of a continuing forcing. Halos have been shown to help stabilize the disk, thus arguing that perhaps the very existence of spiral galaxies is an indication of some DM being distributed over a large volume of the Galaxy. Ostriker and Peebles (1973), by analysis of the stability of disks against collapse into barlike configurations, came up with an independent argument supporting the need for an extensive halo having at least $\frac{2}{3}$ the total mass of a disk galaxy. Their criterion was established by the stability analysis of a self-gravitating ellipsoid, showing that a disk is unstable to the formation of a bar if the ratio of the kinetic energy to the gravitational potential is large, that is, if $T/W > 0.14$. Here, T is the kinetic energy of the stars and W is the gravitational energy. This is perhaps the first paper to argue for the existence of a hot halo on the basis of the stability of a simple model system. Further modeling, primarily by Sellwood, Carlberg, and Anathassoula, supports this conclusion, which now forms one of the cornerstones of galactic structure models: if the halo is not massive enough to stabilize the disk of a spiral galaxy, the system will collapse to a bar embedded in a more extended spheroid of stars.

The implication is that, since there are many disks that do not possess large-scale, barlike structures, there may be a halo associated with many of these systems. In addition, the fact that a disk is observed is indication enough that there should be a considerable quantity of mass associated with it. Thus, the search was stimulated for rotation curves that remain flat or non-Keplerian (not point masslike) at large distances from the galactic center.

IV. Dark Matter in the Local Group

A. MAGELLANIC CLOUDS

The total mass of the Galaxy can be determined from the stability of the Magellanic Clouds. These two Irr galaxies orbit the Milky Way, being slowly tidally disrupted by the interaction with the disk of the Galaxy. The tidal radius of the Clouds and the fact that they have been stable and bound to each other for far longer than a single orbit (these satellites of the Galaxy appear as old as the Milky Way) provide an upper limit to the mass of the Galaxy and Clouds to a distance of about 60 kpc. This is not far from that obtained from the study of halo stars and clusters, and provides an upper limit of about $10^{12}M_\odot$. The required M/L is thus about 30. The total mass of the Magellanic Clouds may be individually underestimated, but this appears to be a good limit on the total mass of the system and is in qualitative agreement with the mass required to explain the dynamics of the Local Group (the galaxy cluster to which the Galaxy belongs) as well.

B. M 31 AND THE ROTATION CURVES OF EXTERNAL GALAXIES

The study of the rotation curve of the Andromeda Galaxy, M 31, the nearest spiral galaxy to ours and one which forms a sort of binary system with the Milky Way, shows that the rotation curve is flat to the distance at which the stars are too faint to determine the rotation curve. The M/L ratio is close to 30, about the same as that found for the Galaxy. In order to place this result in context, it is necessary to describe the process whereby the rotation curves are determined for extragalactic systems.

First, one can assume that the disk is radially supported by the revolution of collisionless stars about the mass lying interior to their radial distance from the galactic center. This assumption makes the mass within a distance r, $M = M(r)$, a function of radius, and then allows the determination of the orbital velocity assuming that the mean eccentricity of the orbits is small or zero. A slit is placed along, and at various angles to, the symmetry axes of the galaxy, and the radial velocity of the stars is determined. The disk is assumed to be circular, so the inclination of the galaxy can be determined from the ellipticity of the projected disk in the line of sight. There are several methods for fitting the mass model to this rotation curve, most of which assume a power law form for the rotation curve with distance and fit the coefficients of the density distribution required to produce the observed velocities by the solution of the dynamical equations in the radial direction. This density profile is then integrated to give the mass interior to a given radius.

Several assumptions go into this method, most of which appear theoretically well justified. The first is that the system is collisionless. There is little evidence that the stars in a galaxy undergo close encounters—they are simply too small in comparison with the distances that separate them. However, there are very massive objects present in the disk, the Giant Molecular Clouds, which may have masses as great as $10^7 M_\odot$ and which appear to control the velocity dispersion of the constituent stars. These clouds are also the sites for star formation, and as such they form the basis for the interaction of the gas and star components of the Galaxy. They may also transfer angular momentum through the Galaxy, making the disk appear viscous. Such interactions act much like encounters to change the velocity distribution to something capable of supporting a flat rotation curve. However, at the moment, this remains speculation.

The observation of the rotation curve using neutral hydrogen line emission is a better way of determining the mass of a galaxy at a large distance. First, it appears that the H I distribution is considerably more extended than that of the bright stars; H I maps extend, at times, to four or five times the optical radius of the disk. Even at these large distances, studies show that the rotation curves remain flat! This is a very difficult observation to understand if the sole means of support for the rotation is the mechanism just described. Rather, it would seem that some form of extended mass distribution of very high M/L objects is required. The radial extent of such masses may be as great as 100 kpc, where the optical radii of these galaxies is smaller than 30 kpc.

The measurement of the rotation curve for the Local Group galaxies is complicated by the fact that the systems are quite close, and therefore require low resolution observations to determine the global rotation curves. This is also an aid in that the small-scale structure can be studied in detail in order to determine the effects of departures from circular orbits on the final mass determination.

The study of radial variations of velocity curves is essentially the same for external galaxies as for ours. One assumes that there is a density distribution of the form $\rho_*(r, z)$ of the stars, and that this is the only contributor to the rotation curve for the Galaxy. One then obtains $\Theta(r)$, from which one can solve the equation of motion for radial support alone:

$$\Theta^2 = 4\pi(1 - e^2)^{1/2} \int_0^r \frac{\rho(\xi)\xi^2 \, d\xi}{(r^2 - \xi^2 e^2)^{1/2}} \qquad (8)$$

where e is the eccentricity of the spheroid assumed for the mass distribution. One can then expand ρ in a power series of the form

$$\rho(r) = \frac{a_{-2}}{r^2} + \frac{a_{-1}}{r} + a_0 + \cdots \qquad (9)$$

and perform a similar expansion for the velocity at a given distance. The solution is then mapped term by term. Once finished, one can then integrate the entire mass of the disk by assuming that the spheroid is homogeneous so that

$$M(r) = 4\pi(1 - e^2)^{1/2} \int_0^r \rho(\xi)\xi^2 \, d\xi \qquad (10)$$

Such models are, of course, extremely simplified (the method was introduced by Schmidt in the 1950s and is still useful for exposition), but they can illustrate the problem of the various assumptions one must make in obtaining quantitative estimates of the mass from the rotation curve. Since the rotation curves remain flat, one can also assume that the M/L ratio is a function of distance from the galactic center. This is folded into the final result after the mass to a given radius has been determined.

Unlike our system, where one cannot be sure what is or is not a halo star, one can determine with some ease the velocity dispersions for face-on galaxies perpendicular to the galactic plane with the distance from the center of the system. In a procedure much like the Oort method described for the Galaxy, it is possible to place constraints on the amount of missing mass in the disk by using the determination of surface densities from the dynamical information. External galaxies appear to have values similar to our system for the dark matter in the plane, about $\frac{1}{2}$ of the matter being observed in stars.

Many disk systems show warps in their peripheries; M 31, the Galaxy, and other systems show distortions of both their stellar and H I distributions in a bending of the plane on opposite parts of the disk. This argues that the halos of these galaxies cannot be too extensive, beyond the observable light, or one would not be able to observe such warps. Calculations show that these will damp, from dynamical friction (dissipation) and also viscous effects, if the halo is too extensive. However, in the case of some of the current models for the distribution of DM in clusters, it is not yet clear whether this can also constrain a more evenly distributed DM component (one less tied to the galaxies but more evenly distributed in the clusters than the halos).

V. Binary Galaxies

Binary galaxies may also be used to determine the M/L of the DM. The process is much like that employed for a binary star, with the exception that the galaxies do not actually execute an orbit within the time scale of our observations. Instead, one must perform the determination statistically. For a range of separations and velocities, one can determine a statistical distribution of the M/L values, which can then be compared with the mean separations and masses obtained from the rotation curves of the galaxies involved. Again, there is strong evidence for a large amount of unseen matter, often reaching M/L values greater than 100. The method has also been extended to small galactic groups, with similar results.

Using large samples of binary galaxies, so that the effects of orbital eccentricity, inclination, and mass transfer can be accounted, recent studies have shown that about a factor of 3 times the observed mass is required to explain binary galaxy dynamics. Summarized, the estimates of M/L range from about 10 to 50. The wide range reflects uncertainties in both the Hubble constant, which is required for the specification of the separation in physical units, and the projection of the velocities of the galaxies in the line of sight. Unlike clusters, binary galaxies are rather sensitive to the projection factors, since one is dealing only with two objects at a time.

Measurement of the masses of the component galaxies can also add information to this picture, and there have been recent efforts to obtain the rotation curves for the galaxies in binaries and small groups. The interpretation of the velocity differences, the quantity which, along with the projected separation, gives the mass of the system, requires some idea of the potential in which the galaxies move. If there is very extended DM around each of the component galaxies, the use of point mass or simple spheroidal models will fail to give meaningful masses. Also, since some of the binary galaxies may be barred spirals, and others may show subtle and not yet observed tidal effects, it is likely that the mass estimates from these systems should be taken with considerable caution.

Small groups of galaxies, the so-called Hickson groups, can be used to provide an example of dynamical interaction which lies between the clusters and the binaries. Here, however, the picture is also complex. Interactions between the constituent galaxies can alter their mass distributions and affect their orbits, in a manner that vitiates the determination of the orbital properties for the member galaxies. If these groups are bound and stable, the M/L must be of order 100. However, Rubin has provided evidence, from the study of individual galaxies, that the rotation curves do not show the flat behavior of isolated galaxies. It is possible that the halos of the galaxies have been altered in the groups. Deep photographs of the groups provide some handle on this. Complicated mass distributions of the luminous matter are observed, indicating that the galaxies have been strongly interacting with one another. Mergers, collisions, and tidal interactions at a distance are all indicated in the images. Clearly, more detailed study is required of these objects before they can be used as demonstrations of the need for DM.

VI. The Local Supercluster

A. VIRGO AND THE VIRGOCENTRIC FLOW

The Local Group sits on the periphery of the Virgo supercluster, containing several hundred bright galaxies and the Virgo cluster centered on the giant elliptical galaxy M 87, which contains as much as $10^{14}M_\odot$. Because of its mass and proximity, its composition and dynamics have been well studied in the past decade. Virial analyses indicate that the total cluster is bound by DM, which amounts to about ten times the observed mass (a figure that is typical of clusters and which will be discussed in more detail later). Because of its closeness, however, the Virgo supercluster has an even more dramatic effect on motion of the Local Group, which can be used to independently estimate its total mass.

From the motion of the Local Group with respect to the Virgo cluster, it is possible to estimate the total mass of the cluster. The argument is a bit like that of Newton's apple analogy. In the free expansion of the universe, local regions may be massive enough to retard the flow of the galaxies away from each other. This is responsible for the stability of groups of galaxies in the first place. They are locally self-gravitating. In the case of the motion of the Galaxy toward Virgo, there are several ways of measuring this velocity.

One of the most elegant is the dipole anisotropy of the cosmic background radiation (CBR). The idea of using the CBR as a fixed light source relative to which the motion of the observer can be obtained is sometimes called the "new aether drift." The motion of the observer produces a variation in the effective temperature of the background that is directly proportional to the redshift of the observer relative to the local Hubble flow. For motions of the order of 300 km sec^{-1}, for example, the variation in the temperature of the CBR should be about 1 mK, with a dipole (that is cosine) dependence on angle. The CBR shows a dipole temperature distribution, about $\frac{1}{3}$ of which can be explained by the motion of the Local Group toward Virgo, about 250–300 km sec^{-1}, called the *virgocentric velocity*. Since the Local Group is part of the supercluster system that contains Virgo, it is not surprising that we are at least partially bound to the cluster. The total mass implied by this motion is consistent with the virial mass determined for the cluster, which implies a considerable amount of dark matter (about ten times the observable mass). The dipole anisotropy is only partially, however, explained by the virgocentric flow.

There must be some other larger-scale motion, which is about a factor of 2 larger, that is responsible for most of the variation in the CBR. This seems to result from the bulk motion of the local supercluster toward the Hydra-Centaurus supercluster ($l_{II} \approx 270°$, $b_{II} \approx 30°$. This brings up the question of large-scale deviations, on the scale of superclusters, of the expansion from the Hubble law.

B. The Large-Scale Streaming in the Local Galaxies: Dark Matter and the Hubble Flow

Distance determination can be a problem for galaxies outside of the Local Group and Virgo cluster. This has to do with our inability to resolve individual stars and H II regions. In the past decade, several new methods have evolved that allow for determination of the intrinsic brightness of galaxies independent of their Hubble flow velocities. The *Tully–Fisher relation* correlates the absolute magnitude of a galaxy, either in the blue or, recently, in the infrared, with the width of the 21-cm line. Although this depends on the inclination of the galaxy, this can be taken into account from optical imaging, and it provides a calibration for the intrinsic luminosity of the galaxy from which, using the apparent magnitude, the distance can be obtained directly, without cosmological assumptions. Another calibrator, especially useful for elliptical and gas-poor galaxies, is the *Faber–Jackson relation*. This uses the velocity dispersion observed for the core of elliptical galaxies and spheroids to determine the intrinsic luminosity of the parent galaxy. It is representative of a wide class of objects, indicative of dissipative formation of the systems, and is

$$L \sim \sigma^n \qquad (11)$$

where σ is the velocity dispersion of the nucleus and $n \approx 4$ from most studies.

Using these methods, it is possible to obtain the distance to a galaxy independent of the Hubble velocity; one can then look for systematic deviations from the isotropy of the expansion of the galaxies in the vicinity of the Local Group. In the past two years, accumulated observations have shown that there is a large-scale deviation of galaxy motion in the vicinity of the Galaxy. Deviations at large displacement to the virgo-centric velocity of order 600–1000 km sec^{-1}, have shown that there is a large departure from the Hubble law relative to the uniform expansion that is usually assumed. The characteristic scale associated with this deviation is about 50–100 Mpc, much larger than the size usually associated with clusters of galaxies but on a scale of superclusters. No mass

can clearly be identified with this gravitationally perturbing concentration, but it has been argued that it may be a group of galaxies of order $10^{14} M_\odot$, about the size associated with a very large cluster or small supercluster of galaxies. The mystery is its low luminosity, but it is located near the galactic plane, which would account for the difficulty in observing its constituent galaxies. This may be indicative of other large-scale deviations on a larger scale of Gpc.

VII. Clusters of Galaxies

From the first recognition of galaxies as stellar systems like the Milky Way, it has been clear that their distribution is markedly inhomogeneous. Early studies by Shapley indicated large (factor of two) density fluctuations in their distribution, while later work has expanded the complexity of this distribution to larger density contrasts. A variety of large-scale structures is observed in the galactic clustering, which gives rise to the idea of a hierarchy. First, there are clusters, Mpc-scale concentrations of luminous objects with total masses typically of 10^{13}–$10^{14} M_\odot$. These contrast with voids, the most famous of which is the Bootes Void, in which the density of galaxies is about $\frac{1}{10}$ that of the average. Clusters of galaxies cluster themselves, into structures called superclusters, and it appears that even these may have some clustering characteristics, although they are sufficiently rare that the statistical information is shaky at best. Clusters can be dynamically and morphologically separated from the background of galaxies, and these can be studied more or less in isolation from one another.

Several catalogs are currently available for galaxy clusters, generally identified with the names Abell and Zwicky. In addition, large-scale sky counts of galaxy frequency per square degree have been made by Shane and Wirtanen (the Lick survey), Zwicky, and the Jagellonian surveys. These do not contain any dynamical information; they are number counts only. But from information about the spatial correlation on the two-dimensional surface of the sky, the three-dimensional properties of galaxies can be crudely determined.

Recently, this situation has been greatly improved with the Center for Astrophysics redshift survey, a complete radial velocity study for galaxies within 15,000 km sec^{-1} of the Local Group (covering scales $< 0.1c$). With velocity information available, it is possible to place the galaxies in question in three-dimensional space, and to look at the detailed distribution of both the galaxies and clusters on the scale of 100 Mpc. This has given the first evidence for the nature of the large-scale structure of the universe and the need to consider dark matter in the context of both clusters and superclusters of galaxies. First, we examine the basis for the determination of mass within the clusters and the arguments on the reality of clustering. Then, we will discuss the ways in which this information can be used to address the distribution of, and need for, the DM on scales comparable with the size of the visible universe.

A. THE VIRIAL THEOREM

One of the first suggestions that there is a compelling need to include unseen matter in the universe comes from the original determination of velocity dispersions in clusters of galaxies. The argument is quite similar to that used for a gas confined in a gravitational field. Galaxies in a cluster may collide, a problem to which we will later return, but for the moment we will assume that the galaxies are in independent orbits about the center of mass of the clusters. The velocity dispersion is the result of the distribution of orbits of the galaxies about the cluster center, and the radial extent of the galaxies in space results from the fact that they are bound to the cluster. A dynamical system in hydrostatic equilibrium obeys the virial theorem, which states that

$$2T + W = 0 \qquad (12)$$

where T is the kinetic energy and W is the gravitational potential energy. The total energy of the system must be *at most zero,* since the system is assumed to be bound. Therefore, the total energy is given by $E = T + W = \frac{1}{2}W < 0$, and the velocity dispersion is given approximately by

$$\sigma^2 = \frac{GM_{\text{virial}}}{\langle R \rangle} \qquad (13)$$

The mass determined from the statistical distribution of orbits depends on the size of the system, and this requires some estimate of the density profile of the cluster. The rms value of the group radius usually suffices for the component of the radial velocity. It is assumed that the orbits are randomly distributed and that they have not been seriously altered since the formation of the cluster. Also, an assumption is embedded in this equation for the dispersion: the mass distribution has remained unaltered on a time scale comparable with the orbital time scale for the galaxies about the center of mass.

The argument that clusters must be bound and stable comes from the distribution of galaxy velocities within the clusters. The dynamical time scale, also called the crossing time, is the period of a galaxy orbit through the center of the cluster potential. The characteristic time scales are of order 10^9 yr, much shorter than the Hubble time (the expansion time scale for the universe, taken as the inverse of the Hubble constant).

There have been numerous simulations of the evolution of galaxies in clusters, most of which show evidence for encounters of the cluster members. There is also strong evidence for interaction in the numerous disturbed galaxies in these clusters, and the presence of cD galaxies, giant ellipticals with extended halos, in the cluster centers. The latter are assumed to grow by some form of accretion, also called *cannibalism,* of the neighboring galactic systems.

B. INTERACTIONS IN GROUPS OF GALAXIES

Galaxies in clusters collide. This simple fact is responsible for many of the problems in interpreting the virial theorem, because it assumes that the dynamical system is intrinsically dissipationless. Collisions redistribute angular momentum and mass and perturb orbits in stochastic and time-dependent ways. None of these effects can be included in the virial theorem formulation, but instead require a full Fokker–Planck treatment of the dynamics.

Recent observations of small groups of galaxies, the Hickson groups, reveal a complex array of interactions—shells, bridges, and tails are found for many of the members of these small groups. The extent to which this alters the mass distribution of the groups is not presently known nor is it known how this affects the determination of the M/L ratio for the group.

For clusters of galaxies, the situation is even more complicated. In the past five years, many cD galaxies have been found to be embedded in shell systems, which are believed to arise from accretion of low-mass disk galaxies by the giant ellipticals. The shells have been used to argue for the mass distribution of the cDs, although the detailed mass distribution cannot be determined from the use of the shells. Instead, they serve as a potential warning about using simple collisionless arguments for the evolution of the cluster galaxy distribution.

C. THE DISTRIBUTION OF CLUSTERS AND SUPERCLUSTERING

The two-point correlation function, $\xi(r)$, is the measure of the probability that there will be two galaxies within a fixed distance r of each other. The probability of finding a galaxy in volume element dV_1 within a distance r of another in volume dV_2 is given by

$$dP(1, 2) = n^2[1 + \xi(r_{1, 2})] \, dV_1 \, dV_2 \quad (14)$$

where n is the mean number density of galaxies. Defined in this way, it is a good measurement of the trend of galaxies to cluster. If there is a clustering, the correlation is asymptotically vanishing at large distance, very positive at small distance, and negative for some interval. It is only a requirement that the cumulant be positive definite (normalizable). The two-point function is not, however, the sole discriminant of clustering. The problem with the equations of motion is that, in an expanding background and under the influence of gravity, the mass points evolve in a highly nonlinear way. The equations of motion do not close at any order, and there is a hierarchy to the distribution of the correlations of different orders.

The correlation function seems to have a nearly universal dependence on separation, that is $\xi(r) \sim r^{-1.8}$, whether for galaxy–galaxy or cluster–cluster correlations. The coefficient is different, which indicates that the hierarchy is not quite exact and that there may be further differences lurking on the scale of superclusters. It is not likely, however, that this will be easily studied, considering the size of the samples and the extreme difficulty in determining the extent to which these largest scale structures can be separated from the clusters.

VIII. Cosmological Constraints and Explanations of the Dark Matter

Several candidates have been suggested for the particular constituents of DM. Since this is a field of enormous variety and unusual richness of ideas, many of which change on short lifetimes, only a generic summary will be provided. This should be sufficient to direct the interested reader to the literature. [See COSMOLOGY.]

A. COSMOLOGICAL PRELIMINARIES

In attempting to find a likely candidate for the DM, one must first look at some of the constraints placed on models by the cosmological expansion. First, the particles must survive the process of annihilation and creation that dominates their statistical population fluctuations during the early phases of the expansion. The cosmology is given by the Friedmann–Robertson–Walker metric for a homogeneous, isotropically expanding universe with radial coordinate r and angular measures θ and ϕ:

$$ds^2 = dt^2 - \frac{R^2(t) \, dr^2}{1 - kr^2} \\ - R^2(t)(d\theta^2 + \sin^2 \theta \, d\phi^2) \quad (15)$$

where $R(t)$ is the scale factor, the rate of the expansion being given by

$$\frac{d}{dt}(\rho V) + p \frac{dV}{dt} = 0 \quad (16)$$

which is the entropy equation, with the volume $V \sim R^3$ and ρ and p being the energy density and pressure, respectively, and

$$H^2(t) = \left(\frac{\dot{R}}{R}\right)^2 = \frac{8\pi G\rho}{3} - \frac{k}{R^2} \quad (17)$$

which is essentially the Hubble law. The value of k is critical here to the argument, since it is the factor that determines the curvature of the universe. The critical solution, with $\Omega = 1$, has a deceleration parameter $q_0 = 0$. Here, $\Omega = \rho/\rho_c$ where ρ_c is the density needed to give a flat spacetime, $\rho_c = 3H_0^2/8\pi G \approx 2 \times 10^{-29}h^2$ g cm^{-3}, where H_0 is the present value of the Hubble constant and $h = H_0/(100 \text{ km sec}^{-1} \text{ Mpc}^{-1})$. This solution is favored by inflationary cosmologies. During the radiation dominated era, the energy density varies like R^{-4} and the pressure is given by $\frac{1}{3}\rho$, so that during the earliest epoch, $\rho \sim t^{-2} \sim T^{-2}$.

The horizon is given by

$$l_H = \int_{t_0}^{t} \frac{dt'}{R(t')} \quad (18)$$

Perturbations in the expanding universe have a chance of becoming important when they grow to the scale of the horizon. Others will damp, because they cannot be causally connected and will simply locally evolve and die away. The expansion then produces fluctuations in the background radiation on the scale of the horizon at the epoch at which regions begin to coalesce.

The entropy of the expanding universe is constant, so that the number density of particles at the time of their formation is related to the temperature of the background $n_i T^{-3} = $ constant from which we can define the ratio of the photon to baryon number density as a measure of the entropy $S = n_\gamma/n_b = 10^9$, where $n_\gamma \sim T^3$. For particle creation, the particles will freeze out of equilibrium when the energy in the background is small compared with their rest mass energy, or when $T \leq m_i c^2/k_B$, where k_B is the Boltzmann constant. Thus, for particles that are very massive, during the radiation dominated phase, the temperature can be very high indeed. Put another way, the equilibrium temperature is defined as $T_{eq} = 10^{11}m_i(\text{GeV})$ K. Since the temperature during the radiation dominated era varies like R^{-1}, this implies that $R/R_0 \sim 10^{-11}m_i$, where R_0 is the present radius. The point is that the particles which may constitute nonbaryonic explanations for the DM are created in a very early phase of the universe, one which presented a very different physical environment than anything we currently know directly.

B. GLOBAL CONSTRAINTS ON THE PRESENCE AND NATURE OF DARK MATTER

1. Nucleosynthesis

There are few direct lines of evidence for the total mass density of the universe. One of the most direct comes from the analysis of the abundance of the light elements. Specifically, in the early moments of the Big Bang, within the first few minutes, the isotopes of hydrogen and helium froze out. While ^{4}He can be generated in stars by nuclear reactions, deuterium cannot. In the Big Bang, however, it can be produced by several different reactions,

$$p + n \rightarrow {}^2D + \gamma, \qquad p + p + e \rightarrow {}^2D + \gamma,$$
$$^3He + \gamma \rightarrow {}^2D + n$$

and the deuterium can be destroyed very efficiently by

$$^2D + n \rightarrow {}^3H, \qquad {}^2D + p \rightarrow {}^3He + \gamma$$

The critical feature of all of these reactions is that they depend on the rate of expansion of the universe during the nucleosynthetic epoch. The drop in both the density and temperature serves to throttle the reactions, producing several useful indicators of the density.

The argument continues: if the rate of expansion is very rapid, that is if the density is much lower than closure, then the deuterium can be rapidly produced but not effectively destroyed because of the drop in the reaction rates for the subsequent parts of the nuclear network. However, if the density is high, the rate of expansion will be slow enough for the chain of reactions to go to equilibrium abundances of ^{4}He because of the consumption of the ^{2}D. Thus, the D/H ratio and the mass fraction of ^{4}He, and of the ^{3}He/^{4}He ratio, can be used to constrain the value of $\Omega = \rho/\rho_c$, where ρ_c is the critical density required for a flat universe ($2 \times 10^{-29} h^2$ g cm^{-3}). The larger Ω, the smaller the D/H ratio. Current observations of the primordial abundances of helium isotopes and of deuterium provide $\Omega \leq 0.1$.

This appears to provide a number significantly lower than that obtained from the virial masses for clusters and from clustering arguments. Since the baryons in the early universe determined the abundances of the elements that emerged from the Big Bang, the abundances of the light isotopes provide a strong constraint on the fraction of DM that can be in baryonic form, luminous or not.

2. Isotropy of the CBR

In order to form clusters of galaxies, matter has to be clumpy. Since the observed variation in density between clusters and voids is very large, much greater than unity, the development of any density perturbation by the epoch of galaxy formation must have been completely nonlinear. However, sensitive searches for fluctuations in the temperature of the microwave CBR show that $\delta T/T \leq 10^{-5}$, which is the same order as the density fluctuations for adiabatic perturbations in the decoupling epoch. Thus, whatever the distribution may be for galaxies at present, the matter from which they emerged must have been uniformly distributed at the time of formation of the CBR. This constrains the nature of the DM at least in the early universe.

3. The Formation of Galaxies and Clusters of Galaxies

Recent work on the distribution of galaxies has centered on the idea that the visible matter does not represent the distribution of mass overall. The picture that is developing along these lines, called "biased galaxy formation," takes as its starting point the assumption that galaxies are unusual phenomena in the mass spectrum. One usually assumes that the galaxies are the result of perturbations in the expanding universe at some epoch during the radiation dominated period, and therefore representative of the matter distribution. However, biasing argues that the galaxies are the product of unusually large density fluctuations and that the subsequent development of these perturbations is dominated by the smoother background distribution of dark matter—that which never coalesces into galactic mass objects.

Several mechanisms have been suggested, all of which may be reasonable in some part of the evolution of the early universe. One picture uses explosions, or bubbles, formed from the first generation of stars and protogalaxies to redistribute mass, alter the structure of the microwave background, and erase the initial perturbation spectrum.

Gravitational clustering alone is insufficient to explain the dramatic density contrasts observed between clusters and voids, unless one invokes large perturbations at the epoch of decoupling of matter and radiation. The current density contrast is of order 2–10, which implies that at the recombination era the perturbations in the density must be of order 10^{-3}. Either the fluctuations in the density at this pe-

riod are isothermal, in which case the density variation is not reflected in the temperature fluctuations
in the cosmic background radiation, or there must
be some other mechanism for the formation of the
galaxies. In an adiabatic variation, $\delta T/T = (1/3) \times \delta\rho/\rho$, which is clearly ruled out by observations
on scales from arcseconds to degrees to a level of
10^{-5}. If the density fluctuations are smaller than this
value, there is not enough time for them to grow
to the sizes implied by the large-scale structure by
the present epoch. Further, quasars also appear to
show clustering and to be members of clusters, at
$z = 1 - 4$, so this pushes the growth of nonlinear
perturbations to even earlier periods in the history of
the universe. If the density of matter is really the cosmological critical value, that is if $\Omega = 1$, then there
cannot be a simple explanation of the distribution of
luminous matter simply by the effects of primordial
fluctuations.

Biasing mechanisms can be combined with the
DM in assuming that the massive particles are the
ones which show the large-scale perturbations. Here,
the difference between hot and cold DM scenarios
shows up most clearly. If the matter is formed hot,
and stays hot, it will damp out all of the small-scale
perturbations. If formed cold, these will be the ones
that will grow most rapidly, with gravitational effects
later sorting the smallest fragments of the Big Bang
into the larger scale hierarchical structures of clusters
and superclusters. The topologies of the resultant
universes differ between these two extremes, with
the cold matter showing the larger contrast between
voids and clusters and an appearance that is best
characterized by filaments rather than lumps. It
should be added that biasing can act on both scenarios and that the final appearance of the simulations
can depend on the statistical method chosen to determine the biasing.

C. THE PARTICLES FROM
THE BEGINNING OF TIME

The initial scale at which all of the particle theories
start is with the Planck mass, the scale at which
quantum perturbations are on the same scale for
gravity as the event horizon. This gives

$$m_{\text{Planck}} = \left(\frac{hc}{2\pi G}\right)^{1/2} \approx 1.7 \times 10^{19} \text{ GeV} \quad (19)$$

This is the scale at which the ultimate unification of
forces is achieved, independent of the particle theories. It depends only on the gravitational coupling
constant. It is then a question of where the next
grand unification scale occurs (the GUT transition).
This is usually placed at $10^{12}–10^{15}$ GeV, consider-

ably later and cooler for the background. The particles of which the background is assumed to be
composed are usually assumed to arise subsequent to
this stage in the expansion, and will therefore have
masses well above those of the baryons, in general,
but considerably below that of the Planck scale.

1. Inflationary Universe and Soon After

The most recent interest by theorists in the need
for DM comes from the *inflationary universe* model.
In this picture, originated by Guth and Linde, the
universal expansion begins in a vacuum state that undergoes a rapid expansion leading to an initially flat
space–time, which then suffers a phase transition at
the time when the temperature falls below the Planck
mass. At this point, the universe reinflates, now containing matter and photons, and it is this state which
characterizes the current universe. One of the compelling questions that this scenario addresses is why
the universe is so nearly flat. That is, the universe,
were it significantly different from a critical solution
at the initial instant, would depart by large factors
from this state by the present epoch. The same is
true for the isotropy. The universe at the time of
appearance of matter must have been completely
isotropized, something which cannot be understood
within the framework of noninflationary pictures of
the Big Bang. [*See* COSMIC INFLATION.]

The fact that, in these models, $\Omega = 1$ is a requirement, while all other determinations of the density
parameter yield far smaller values (open universes),
fuels the search for some form of nonbaryonic DM.
The ratio $\Omega/\Omega_{\text{observed}}$ is approximately 10 to 100,
of the same order as that required from clusters
of galaxies.

During the expansion, particles can be created and
annihilated through interactions, thus,

$$\frac{dn}{dt} = -D(T)n^2 - 3\frac{\dot{R}}{R}n + C(t) \quad (20)$$

where the first term is the annihilation, which is temperature-dependent; the second is the dilution of the
number density because of the expansion; and the
third is the creation term, which depends on time and
(implicitly) temperature. Let us examine what happens if the rate of creation depends on a particle
whose rate of creation is in equilibrium with its destruction. Then, if a two-body interaction is responsible for the creation of new particles, and they have
a mass m, the Saha equation provides the number in
equilibrium:

$$n_{\text{eq}} \sim (mT)^{3/2} \exp(-m/T) \quad (21)$$

and the rate of particle production becomes a function of the particle mass. Thus, the more massive particles can be destroyed effectively in the early universe, but will survive subsequently because of the rapid expansion of the background and dilution of the number density. Any DM candidate will therefore have a sensitive dependence on its mass for the epoch at which the separation from radiation occurs, as well as the strength of the particle's interaction with matter and radiation.

Following their freeze-out phase, the particles will be coupled to the expanding radiation and matter until their path length becomes comparable with the horizon scale. This phase, called *decoupling,* also depends on the strength of the interaction with the radiation and matter. After this epoch, the particles can freely move in the universe, which is now "optically thin" to them. It is this free-streaming that determines the scale over which the particles can allow for perturbations to grow. For instance, in the case of the massive cold particles, those created with very low velocity dispersions, the scale over which they can freely move without interacting gravitationally with the baryonic matter is determined by the mass of the baryon clumps. For cold DM, this means that the slow particles, those with energies of less than 10 eV, can be trapped within galactic potentials. Those with energies of 100 eV would still effectively be trapped by clusters. If the particles are hot, they cannot be trapped by these clumps and can, in fact, erase the mass perturbations that would lead to the formation of these smaller structures.

As mentioned in discussing galaxy formation in general, simulations of the DM show that for the hot particles, the mass scale that survives the primordial perturbations and damping is the size of superclusters, about 10^{15}–$10^{16} M_\odot$. The superclusters thus form first, then these fragment and separate from the background and form clusters and their constituent galaxies from the "top down." If the particles are cold, they cannot erase the smallest scales of the perturbations, on the order of galaxy size (10^{12}–$10^{13} M_\odot$), and clusters and superclusters form gravitationally from the "bottom up." The problem is that there is little observational evidence, at present, to distinguish which of these is the most likely explanation. In effect, it is the simulations of the galactic interactions, placed in an expanding cosmology and with very restrictive assumptions often being made about the interactions between galaxies within the clusters thus formed, which are used as argumentation for one or the other scenario. In fact, it is for this reason that the models must be called *scenarios,* since there cannot be detailed analytic solution of the

complex problem, and the solutions are only sufficient within the limitations of the precise assumptions that have been applied in the calculations.

The critical mass for gravitational instability and galaxy formation is the *Jeans mass,* the length (and therefore mass) scale at which material perturbations become self-gravitating and separate out in the expanding background. For species in equilibrium with radiation, it depends on both the number density of the dominant particle and its mass, as well as on the temperature of the background radiation. In the case of the expanding universe, since during the radiation dominated era the mass density and temperature are intimately and inseparably linked, the Jeans mass depends only on the mass of the particle that is assumed to be the dominant matter species:

$$M_J \sim \rho_i \left(\frac{a_s^2}{G\rho_i}\right)^{3/2} = \left(\frac{T^3}{Gm_i^2 n_i}\right)^{1/2}$$

$$\approx 3 \times 10^9{}_{m_{i,\,GeV}^{-1}} M_\odot \qquad (22)$$

where a_s is the sound speed and n_i is the number density of the dominant species. The smaller the mass of the particle, the larger the initially unstable mass. This collapse must compete against the fact that viscosity from the interaction with background photons tends to damp out the fluctuations in the expanding medium. The critical mass for stability against dissipation within the radiation background is given by the *Silk mass,* below which the photons, by scattering and the effective viscosity that it represents, damp perturbations. This gives a mass of

$$M_S = 3 \times 10^{13} (\Omega h^2)^{-5/4} M_\odot \qquad (23)$$

Below this mass, which is of the same order as a galactic mass, perturbations are damped out during the early stages of the expansion. This provides a minimum scale on which galaxies can be conceived to be formed. Other particles, however, can also serve to damp out the perturbations at an even earlier epoch if they are hot enough.

2. The Particle Zoo

a. Baryons.   Baryons, the protons and neutrons, are the basic building blocks of ordinary matter. They interact via the strong and electromagnetic forces, and constitute the material out of which all luminous material is composed.

The basic form of baryonic matter, hot or cold gas, can be ruled out on several counts. One has been discussed in the section on nucleosynthesis in the Big Bang; the baryonic density is constrained by isotopic ratios to be small. Another constraint is provided by the absorption lines formed from cold neutral gas through the intergalactic medium in lines of sight to distant quasars. There are not sufficiently high column densities observed along any of the lines of sight to explain the missing mass. The absorption from Lyman α, the ground-state transition of neutral hydrogen, is sufficiently strong and well enough observed, that were the gas in neutral form it would certainly show this transition viewed against distant light sources. While such narrow absorption lines are seen in QSO spectra, they are not sufficiently strong to account alone for the dark matter.

In addition, there is hot gas observed in clusters of galaxies, in the form of diffuse X-ray halos surrounding the galaxies and cospatial with the extent of their distribution. Here again, the densities required to explain the observations are not sufficient to bind the clusters or to account for the larger scale missing mass since the X-ray emission seems to be confined only to the clusters and not to pervade the universe at large scale.

It is possible that the matter could be in the form of black holes, formed from ordinary material in a collapsed state, but the required number, along with the difficulties associated with their formation, makes this alternative tantalizing but not very useful at present. Of course, football- to planet-sized objects could be present in very large numbers, forming the lowest mass end of the mass spectrum that characterizes stars. Aside from the usual problem of not being able to theoretically form these objects in the required numbers, there are also the constraints of the isotropy of the microwave background and of the upper limit to the contribution of such masses to the infrared background. Since these objects would behave like ordinary matter, they should come into thermal equilibrium with the background radiation, and hence be observable as fluctuations in the background radiation on the scale of clusters of galaxies and smaller. Such a variation in the temperature of the CBR is not presently observed. In addition, it would have been necessary for the baryons to have participated in the nucleosynthetic events of the first few minutes, and this also constrains the rate of expansion to rule out these as the primary components of the universe. If inflation did occur, and the universe is really an $\Omega = 1$ world, then the critical missing mass cannot be made up by baryons without doing significant damage to the understanding of the nuclear processing of the baryonic matter.

b. Neutrinos. The electron neutrino (ν_e) is the lightest lepton, and in fact the lightest fermion, making it an interesting particle for explaining DM. It cannot decay into lower mass species, if it is only available in one helicity, and thus would survive from the epoch at which it decoupled from matter and radiation during the radiation-dominated era of the expansion.

It is well known that neutrino processes, being moderated by the weak force, permit the particles to escape detection by many of the classical tests. That these are weak particles means that they decouple from the expansion sooner than the photons and can freely stream on scales larger than those of the photons, also remaining hotter than the photon gas and therefore more extended in their distribution if they accrete onto galaxies or in clusters of galaxies. Should the neutrino have a large enough mass, about 30 eV, the predicted abundance of the 3 known species is sufficient to close the universe and possibly account for the gross features of the missing mass. Current limits (1988) place this maximum mass for the electron neutrino species as ≤ 10 eV. Other problems with this explanation include the rates of nucleosynthesis in the early universe, particularly the neutron to proton ratio, which is fixed by the abundance of light leptons and the number of neutrino flavors.

The argument that the ν_e mass must be nonvanishing is important. If the neutrino is massive, it may be nonrelativistic. This implies that it decoupled from the microwave background at some time before the baryons, but later than the massless particles. Since the particle decoupling depends on the temperature of the radiation, this determines the mean free-streaming velocity. The capture efficiency of baryons to this background of particles is dependent on their mass and velocity. The escape velocity from a galaxy is several hundred km sec^{-1}, while for galaxy clusters it is much higher. Thus, for galactic halo DM to be explained by ν_e, it is necessary that the particles be cold—that is, they must have free-streaming velocities less than 100 km sec^{-1}. The restriction of the particle mass to less than 10 eV by the observations of the width of the burst of neutrinos from Supernova 1987A in the Large Magellanic Cloud likely adds fuel to the argument that the neutrino is an unlikely candidate for DM. [See NEUTRINO ASTRONOMY; SUPERNOVA 1987A.]

If ν_e is the dominant component of DM, an analog of the Silk mass is possible, below which fluctuations are damped because of the weak interaction of matter with the relic neutrinos. The density parameter is now given by $\Omega_\nu = 0.01 h^{-2} m_\nu$ (eV) and

$$M_\nu = 5 \times 10^{14} (\Omega_\nu h^2)^{-2} M_\odot \qquad (24)$$

which is closer to the scale of clusters than galaxies.

With current limits, it looks as if v_e can neither provide $\Omega_v = 1$ nor seriously alter the mass scale for galaxy formation, but may nonetheless be important in nucleosynthesis.

c. Alternative Particles.

After exhausting classical explanations of DM, one must appeal to far more exotic candidates. It is this feature of cosmology that has recently become so important for particle theorists. Denied laboratory access to the states of matter that were present in the early stages of the Big Bang, they have attempted to use the universe as a probe of the conditions of energy and length that dominate the smallest scales of particle interactions at a fundamental level. Above the scale of energy represented by proton decay experiments, the only probable site for the study of these extreme states of matter is in the cosmos.

Many grand unified theories, or GUTs, contain broken symmetries that must be mediated by the addition of new fields. In order to carry these interactions, particles have to be added to the known spectrum from electroweak and quantum chromodynamics (QCD) theory. One such particle, the axion, is a moderator of the charge-parity (CP) violation observed for hadrons. It is a light particle, but not massless, which is weakly interacting with matter and should therefore decouple early in its history from the background radiation. Being of light mass and stable, it will survive to the epoch of galaxy formation and may provide the particles needed for the formation of the gravitational potentials that collect matter to form the galaxy-scale perturbations.

It is perhaps easiest to state that the axion has the attractive feature that its properties can largely be inferred from models of DM behavior, rather than the other way around, and that the formation of these particles, or something much like them, is a requirement in most GUTs. Simulations of galaxy clustering make use of these and other particles, with very generalized properties, in order to limit the classes of possible explanations for the DM of galaxies and clusters of galaxies.

Supersymmetry, often called SUSY, is the ultimate exploitation of particle symmetries—the unification of fermions and bosons. Since there appears to be a characteristic scale of mass at which this can be effected, and since supersymmetry assigns to each particle of one type a partner of the opposite but of different mass, it is possible that the different particles decouple from the primordial radiation at different times (different temperature scales at which they are no longer in thermal equilibrium) and subsequently freely move through the universe at large. This pro-

vides a natural explanation for the mass scales observed in the expanding hierarchy of the universe.

The photon is the lightest boson, massless and with integer spin. It is a stable particle. Its supersymmetric (SUSY) partner is the *photino,* the lowest mass of the SUSY zoo. The photino freezes out of the background at very high temperatures, of order $T_{photino} \approx 4(m_{SUSY,f}/100 \text{ GeV})^{4/3}$ MeV where $m_{SUSY,f}$ is the mass of the SUSY particle (a fermion) by which the photino interacts with the background fermions. If the mass is very small, less than 4 MeV, the decoupling occurs early in the expansion, leading to relativistic particles and free streaming on a large scale; above this, the particle can come into equilibrium with the background through interactions that will reduce its streaming velocity, and annihilation processes, like those of monopoles, will dominate its equilibrium abundance. The final abundance of the photino is dominated by the strength of the SUSY interactions at temperatures higher than $T_{photino}$.

If the gravitino, the partner of the graviton, exists, then it is possible for the photino to decay into this SUSY fermion and a photon. Limits on the half-life can be as short as seconds or longer than days, suggesting that massive photinos may be responsible for mediating some of the nucleosynthesis in the late epoch of the expansion. If the photinos decay into photons, this raises the temperature of the background at precisely the time at which it may alter the formation of the light isotopes, photodisintegrating ^{4}He and changing the final abundances from nucleosynthesis. Recent attempts to place limits on the processes have shown that the photinos, in order to be viable candidates for DM, must be stable, or have a large mass (greater than a few GeV) if they are nonrelativistic, or have a very low mass (< 100 eV). There are problems with large numbers of low-mass particles for the same reason there are problems with v_e, since these would tend to wash out much of the small-scale structure while not being well bound to galaxy halos.

Strings are massive objects that are the product of GUTs. They are linear, one-dimensional objects that behave like particles. Arguments about their structure and behavior have shown that, as gravitating objects, they can serve as sites for promoting galaxy formation, and may be responsible for the biasing of the DM to form potential wells that accrete the baryons out of which galaxies are composed. There is no experimental support for their existence, but there is considerable interest in them from the point of view of particle theories. Strings have the property that they can either come in infinite linear or closed

TABLE I.   Summary of Evidence for Dark Matter in Different Environments

Evidence	$\langle M/L \rangle$	Ω
Galactic stars and clusters	1	0.001
Vertical galactic gravitational acceleration	2–3	0.002
Disk galaxy dynamics	10	0.01
Irregular and dwarf galaxies	10–100	0.101–0.1
Binary galaxies and small groups	10–100	0.01–0.1
Rich clusters and superclusters	100–300	0.2 ± 0.1
Large-scale structure	1000	0.5–1.0
Baryosynthesis	10–100	≤ 0.1
Inflationary universe	$1000h$	1.0

forms. The closed forms serve as the best candidates for inclusion in the DM scenarios.

IX. Future Prospects

The current status of DM is very confusing, especially in light of the wealth of possible models for its explanation. None of the particles presently known can explain the behavior of matter on scales above the galactic, while there are a number of hypothetical particles that can do so for larger scales. The masses of galaxies are such that, in order to explain their halos, the particles of which they are composed must be nonrelativistic; that is, they must have velocities less than a few hundred km sec^{-1}. Thus, cold DM appears the best candidate for the constituents of galactic halos. On the scale of clusters and superclusters, however, it is still not clear whether this is a firm limit. Biased galaxy formation seems a viable explanation of the distribution of luminous matter, but here again there is a wealth of explanation and not much data with which to test it. A summary of the evidence for DM is given in Table I.

The best candidate for testing some of the models is space observation. Two satellites should be launched in the next decade that will greatly extend our ability to test these speculations. The Hubble Space Telescope is a high-resolution optical and ultraviolet spectroscopic and imaging instrument that should be capable of reaching 30th magnitude and of performing high-resolution observations of galaxies to at least $z = 4$. Surveys of redshift distributions in clusters of galaxies with very high velocity resolution should aid greatly in the details of virial mass calculations, while the imaging should delineate the extent to which interactions between cluster galaxies have played a role in the formation of the observed galactic distributions. The Cosmic Background Explorer (COBE) is designed to look at the isotropy of the

microwave background, near the peak of the CBR spectrum. The COBE should place strict limits on the scales on which the background shows temperature variations, thereby delimiting the scale of adiabatic perturbations in the expanding universe at the recombination epoch.

Laboratory tests of particle theories should also contribute to the DM problem. The plans for the Superconducting Supercollider show that it should be able to detect the Higgs boson, responsible for the mass of particles in grand unified theories, and this should also feed the modeling of supersymmetric interactions. As the knowledge of the behavior of the electroweak bosons increases, we should also have a clearer picture of the role played by SUSY in the early universe and in the generation of the particles which are responsible for (at least) some of the DM.

On a final note, it may be said, possibly, we have this all wrong. The study of matter that cannot be seen but only felt on very large scales is, obviously, one driven by models and calculations. These have assumptions built into them that may or may not be justified in the context of the particular studies. Therefore, it may be the case that DM is more an expression of our ignorance of the details of the universe at large distances and on the cosmological scale than we currently believe. However, the need for invoking DM is very widespread in astrophysics: it is required in many explanations, models for it come in many varieties, its presence is indicated by different methods of mass determination, and it is a phenomenon that involves only classical dynamics. Most simple alternatives proposed so far have been very specifically tailored to the individual problems and are often too *ad hoc* to serve as fundamental explanations of all of the phenomena that require the presence of some form of DM. This is a field rich in speculation, but it is also a field rich in quantitative results. As more data is accumulated on the dynamics of galaxies and clusters of galaxies, we will clearly be able to distinguish between the "everything we know is wrong" school and that which detects the fingerprint of the early universe and its processes in the current world.

Appendix: The Discrete Virial Theorem

Consider the motion of a particle about the center of mass of a cluster. The gravitational potential is the result of the interaction with all particles j, not the same as i, for the ith particle:

$$\Phi \equiv -G \sum_{i \neq j} \frac{m_j}{R_{ij}} \qquad (25)$$

where the masses are allowed to be different. The equation of motion for this particle is

$$m_i \mathbf{v}_i = -m_i \nabla \Phi \qquad (26)$$

Taking the product of this equation with the position of the ith particle (the scalar product) is

$$\sum_i \mathbf{r}_i m_i \ddot{\mathbf{r}} = -2T + \frac{1}{2} \ddot{I} \qquad (27)$$

where I is the moment of inertia of the cluster and T is the kinetic energy of the particles. For discrete particles, this is easily calculated. Now assume that R_{ij} is the scalar distance between the ith and jth particle. Then the equation of motion yields the discrete vitrial theorem:

$$\ddot{I} = \sum_i m_i v_i^2 - G \sum_i \sum_{i \neq j} \frac{m_i m_j}{R_{ij}} = 2T + W \qquad (28)$$

Here W is the gravitational energy of the cluster, summed over all masses. This is the usual form of the virial theorem, with the additional term for the secular variation of the moment of inertia included. It is usually assumed that the system has been started in equilibrium, so that the geometry of the configuration is constant. This implies that one can ignore the variation in I. Now, assume that all particles have the same mass, m. This gives

$$-G \sum_{i \neq j} m_i^2 \langle R_{ij} \rangle^{-1} + \sum_i m_i \langle v_i^2 \rangle = 0 \qquad (29)$$

We obtain only a two-dimensional spatial picture and a one-dimensional velocity picture of a three-dimensional distribution of galaxies in a cluster. The virial theorem applies to the full-phase space of the constituent masses and is three dimensional, of course, in spite of the fact that we can only see the line of sight velocities. Therefore, in order to obtain mass estimates from the virial constraint, we must

make some assumption about the isotropy of the velocity distribution. The simplest and conventional assumption is that the radial velocity is related to the mean square velocity by $\langle v_i^2 \rangle = 3 \langle V_{\text{rad}, i}^2 \rangle$ when averaged over the various orientations of the cluster. The *virial mass estimator* is then given by

$$M_{\text{VT}} = \frac{3\pi N}{2G} \left(\frac{\sum_i V_{\text{rad}, i}^2}{\sum_{i<j} \rho_{ij}^{-1}} \right) \qquad (30)$$

where now N is the number of (identical) particles and ρ_{ij} is the projected separation of the objects i and j on the sky. The coefficient results from assuming that these are randomly oriented and that $\langle R_{ij}^{-1} \rangle = (2/\pi) \rho_{ij}^{-1}$.

There are several assumptions that have been built into this derivation, among which are the assumptions of isotropy of the orbits of stars about the center of the cluster and the stability of the shape of the cluster. Notice that there will be an additional term in the equation of motion if there is an oscillation of the cluster with time, and that this will reduce the estimated virial mass (although it will in general be a small term since it is inversely proportional to the square of the Hubble time). An additional potential is contributed by the DM, but it may not enter into the virial argument in the same way as this discrete particle picture. In the preceding discussion, the constituent moving galaxies were assumed to be the only cause of the gravitational field of the clusters. Now if there is a large, rigid mass distribution that forms the potential in which the galaxies move, the mass estimates from the virial theorem are incorrect. There will always be an additional term, W_{DM}, which adds to the gravitational energy of the visible galaxies but does not contribute to the mass, and this can alter the determination of the mass responsible for the binding of the visible tracers of the dynamics. One should exercise caution before accepting uncritically the statements of the virial arguments by carefully examining whether they are based on valid, perhaps case-by-case, assumptions.

BIBLIOGRAPHY

Aaronson, M., Bothun, G., Mould, J., Huchra, J., Schommer, R. A., and Cornell, M. E. (1986). A distance scale from the infrared magnitude/HI velocity–width relation. V. distance moduli to 10 galaxy clusters, and positive detection of bulk supercluster motion toward the microwave anisotropy, *Astrophys. J.* **302**, 536.

Athanassoula, E., Bosma, A., and Papaioannou, S. (1987). Halo parameters of spiral galaxies, *Astron. Astrophys.* **179**, 23.

Blumenthal, G. R., Faber, S. M., Primack, J. R., and Rees,

M. J. (1984). Formation of galaxies and large-scale structure with cold dark matter, *Nature* (London) **311**, 517.

Boesgaard, A. M., and Steigman, G. (1985). Big bang nucleosynthesis: theories and observations, *Annu. Rev. Astron. Astrophys.* **23**, 319.

Dekel, A., and Rees, M. J. (1987). Physical mechanisms for biased galaxy formation, *Nature* (London) **326**, 455.

Dressler, A. (1984). The evolution of galaxies in clusters, *Ann. Rev. Astron. Astrophys.* **22**, 185.

Faber, S. M., ed. (1987). "Nearly Normal Galaxies: From the Planck Time to the Present." Springer-Verlag, Berlin and New York.

Faber, S. M., and Gallagher, J. S. (1979). Masses and mass-to-light ratios of galaxies, *Annu. Rev. Astron. Astrophys.* **17**, 135.

Gibbons, G. W., Hawking, S. W., and Siklos, S. T. C., eds. (1986). "The Very Early Universe." Cambridge Univ. Press, London and New York.

Heisler, J., Tremaine, S., and Bahcall, J. N. (1985). Estimating the masses of galaxy groups: alternatives to the virial theorem, *Astrophys. J.* **298**, 8.

Kaiser, N., and Silk, J. (1986). Cosmic microwave background anisotropy, *Nature* (London) **324**, 529.

Knapp, G. R., and Kormendy, J. F., eds. (1986). "Dark Matter in the Universe: IAU Symp. 117." Reidel, Dordrecht, Netherlands.

Peebles, J. (1980). "Large-Scale Structure of the Universe." Princeton Univ. Press, Princeton, New Jersey.

Rubin, V. (1983). Dark matter in spiral galaxies, *Sci. Amer.* **248** (6), 96.

Trimble, V. (1987). Existence and nature of dark matter in the universe, *Annu. Rev. Astron. Astrophys.* **25**, 425.

van Moorsel, G. A. (1987). Dark matter associated with binary galaxies, *Astron. Astrophys.* **176**, 13.

Vilenkin, A. (1985). Cosmic strings and domain walls, *Phys. Rep.* **121**, 264.

White, S. D. M., Frenk, C. S., Davis, M., and Efstathiou, G. (1987). Clusters, filaments, and voids in a universe dominated by cold dark matter, *Astrophys. J.* **313**, 505.

Zeld'ovich, Ya. B. (1984). Structure of the universe, *Sov. Sci. Rev. E: Astrophys. Space Phys.* **3**, 1.

GALACTIC STRUCTURE AND EVOLUTION

John P. Huchra   *Harvard–Smithsonian Center for Astrophysics*

GLOSSARY

Globular Cluster: Dense, symmetrical cluster of 10^5 to 10^6 stars generally found in the halo of the galaxy. Globular clusters are thought to represent remnants of the formation of galaxies.

Halo: Extended, generally low surface brightness, spherical outer region of a galaxy, usually populated by globular clusters and population II stars. Sometimes containing very hot gas.

H II Region: Region of ionized hydrogen surrounding hot, usually young, stars. These regions are distinguished spectroscopically by their very strong emission lines.

Hubble Constant: Constant in the linear relation between velocity and distance in simple cosmological models. $D = V/H_0$, where H_0 is given in km sec^{-1} Mpc^{-1}. Distance and velocity are measured in megaparsecs and kilometers per second, respectively. Current values for H_0 range between 50 and 100 in those units.

Magnitude: Logarithmic unit of relative brightness or luminosity. The magnitude scale is defined as -2.5 log(flux) + Constant, so that brighter objects have smaller magnitudes. Apparent magnitude is defined relative to the brightness of the A0 star Vega (given magnitude = 0) and absolute magnitude is defined as the magnitude of an object if it were placed at a distance of 10 pc.

Metallicity: Average abundance of elements heavier than H and He in astronomical objects. This is usually measured relative to the metal abundance of the Sun and quoted as "Fe/H," since iron is a major source of line in optical spectra.

Missing Mass: Excess mass found from dynamical studies of galaxies and systems of galaxies over and above the mass calculated for these systems from their stellar content.

Parsec, kiloparsec, Megaparsec: 1 pc is the distance at which one Astronomical Unit (the Earth orbit radius) subtends one arcsecond. 1 pc = 3.086×10^{18} cm = 3.26 light years.

Population I, II: Stars in the galaxy are classified into two general categories. Population II stars were formed at the time of the galaxy's formation, are of low metallicity, and are usually found in the halo or in globular clusters. Population I stars, like the sun, are younger, more metal rich and form the disk of the galaxy.

Solar Mass, Luminosity: $L_\odot = 3.826 \times 10^{33}$ ergs/sec^{-1}; $M_\odot = 1.989 \times 10^{33}$ g

Surface Brightness: Luminosity per unit area on the sky, usually given in magnitudes per square arcsecond. In Euclidean space, surface brightness is distance independent, since both the apparent luminosity and area decrease as the square of the distance.

The object of the study of galaxies as individual objects is twofold, that is (1) to understand the present morphological appearances of galaxies and their internal dynamics, and (2) to understand the integrated luminosity and energy distributions of galaxies both in the framework and timescale of a cosmological model and evolution. The morphology of a galaxy is determined by its formation and dynamical evolution, mass, angular momentum content, and the pattern of star formation that has occurred in it. The brightness and spectrum of a galaxy is determined by its present stellar population, which is in turn a function of the galaxy's star formation history and the evolution of those stars. Significant advances have been made in detailed quantitative modeling of galaxy morphology and internal dynamics. Galaxy form and luminosity evolution is much less well understood despite its great importance for the use of galaxies as cosmological probes.

I. Introduction

The study of galactic structure and evolution began only in the early twentieth century with the advent of large reflectors and new photographic and spectroscopic techniques that allowed astronomers to determine crude distances to external galaxies, and place them in the model universes being developed by Einstein, de Sitter, Friedmann, Lemaître and others. The key discoveries were Harlow Shapley's of a relation between the period and luminosity of Cepheid variable stars, and Edwin Hubble's use of that relation to determine distances to the nearest bright galaxies. Hubble then was able to calibrate other distance indicators (such as the brightest stars in galaxies, which are 10–100 times brighter than Cepheids) to estimate distances to further galaxies. This led to his discovery in 1929 of the redshift-distance relation, which not only established the expansion of the universe as predicted by Einstein, but also established that the age or timescale of the universe was much greater than 100 million years. Current observational estimates for the age of the universe range between 10 and 20 billion years in the basic Hot Big Bang model, the Friedmann–Lemaître cosmological model, which has been favored for the last few decades. [See COSMOLOGY.]

Extragalactic astronomy is still an observationally dominated science, and, as such, tends to be descriptive rather than quantitative.

II. Galaxy Morphology— The Hubble Sequence

The initial steps in the study of galactic structure were the developement of classification schemes that could be tied to physical properties, such as angular momentum content, gas content, mass and age. Another of Hubble's major contributions to extragalactic astronomy was the introduction of such a scheme based only on

the galaxy's visual (or, more correctly, blue light) appearance. This morphological classification scheme, known as the Hubble Sequence, has been modified and expanded by Allan Sandage, Gerard de Vaucouleurs and Sidney van den Bergh. The Hubble Sequence now forms the basis for the study of galactic structure. There are other classification schemes for galaxies based on such properties as the appearance of their spectra or the existence of star like nuclei or diffuse halos; these are of more specialized use. [See STELLAR SPECTROSCOPY; STELLAR STRUCTURE AND EVOLUTION.]

Hubble's basic scheme can be described as a "tuning fork" with elliptical galaxies along the handle, the two families of spirals, normal and barred, along the tines, and irregular galaxies at the end (see Fig. 1). Elliptical galaxies are very regular and smooth in appearance and contain little or no dust or young stars. They are subclassified by ellipticity, which is a measure of their apparent flattening. Ellipticity is computed as $e = 10(a - b)/a$, where a and b are the major and minor axis diameters of the galaxy. A typical elliptical galaxy is shown in Fig. 2. A galaxy with a circular appearance has $e = 0$ and is thus classified E0. These are at the tip of the tuning fork's handle. The flattest elliptical galaxies that have been found have $e \approx 7$, and are designated E7. Both normal and barred spiral galaxies range in form from "early" type, designated Sa, through type Sb, to "late" type Sc. Spiral classification is based on three criteria: (1) the ratio of luminosity in the central bulge to that in the disk, (2) the winding of the spiral arms, and (3) the contrast or degree of resolution of the arms into stars and H II regions. Early type galaxies have tightly wound arms of low contrast and large bulge-to-disk ratios. A typical spiral galaxy is shown in Fig. 3. The important transition re-

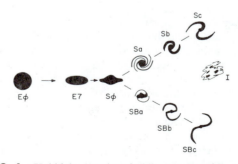

FIG. 1. Hubble's "tuning fork" diagram (adapted from the Realm of the Nebulae, E. Hubble, 1936.)

FIG. 2. The EO galaxy NGC 3379 and its comparison SBO galaxy, NGC 3384. This image was obtained with a Charge Coupled Device (CCD) camera at the F. L. Whipple Observatory by S. Kent. Slight defects in the image are due to cosmetic defects in the CCD.

gion between elliptical and spiral galaxies is occupied by lenticular galaxies that are designated S0. It was originally hypothesized that spiral galaxies evolved into S0 galaxies and then ellipti-

cals by winding up their arms and using up their available supply of gas in star formation. This is now known to be false.

Irregular galaxies are split into two types. Type I or Magellanic irregulars are characterized by an almost complete lack of symmetry, no nucleus, and are usually resolved into stars and H II regions. The Magellanic Clouds, the nearest galaxies to our own, are examples of this type. Type II irregular galaxies are objects that are not easily classified—galaxies that have undergone violent dynamical interactions, star formation events, "explosions", or have features uncharacteristic of their underlying class such as a strong dust absorption lane in an elliptical galaxy.

The modifications and extensions to Hubble's scheme added by Sandage and de Vaucouleurs include the addition of later classes for spirals, Sd and Sm, between classes Sc and Irr I, the further subclassification of spirals into intermediate types such as S0/a, Sab and Scd, and the inclusion of information about inner and outer ring structures in spirals. S0 galaxies have also been subdivided into classes based either on the evidence for dust absorption in their disks, or, for SBO's, the intensity of their bar components. Van den Bergh discovered that the con-

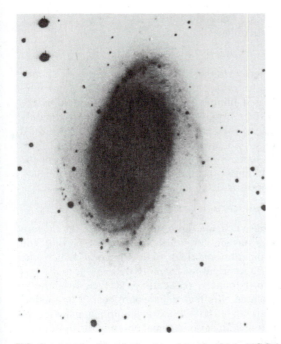

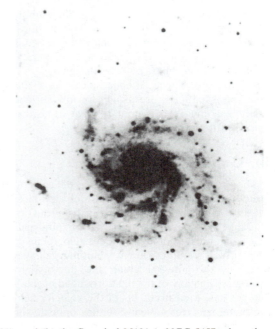

FIG. 3. (a) The Sb spiral galaxy Messier 81 (=NGC 3031), and (b) the Sc spiral M101 (=NGC 5457 a.k.a. the Pinwheel). The CCD images courtesy of S. Kent.

TABLE I.   De Vaucouleurs' T Classification

T Type	Description
-6	Compact elliptical
-5	Elliptical, dwarf elliptical E, dE
-4	Elliptical E
-3	Lenticular L$-$, SO$-$
-2	Lenticular L, SO
-1	Lenticular L$+$, SO$+$
0	SO/a, SO-a
1	Sa
2	Sab
3	Sb
4	Sbc
5	Sc
6	Scd
7	Sd
8	Sdm
9	Sm, Magellanic Spiral
10	Im, Irr I, Magellanic Irregular, Dwarf Irregular
11	Compact Irregular, Extragalactic HII Region

trast and development of spiral arms were correlated with the galaxy's luminosity. Spirals and irregulars can be broken into nine luminosity classes (I, I-II, II, ... V), but there is considerable scatter (≈ 0.6–1.0 magnitude) and thus considerable overlap in the luminosities associated with each class. De Vaucouleurs also devised a numerical scaling for morphological type, call the T type and illustrated in Table I, for his *Second Reference Catalog of Bright Galaxies*.

The most recent 'modification' to the Hubble sequence is van den Bergh's recognition of an additional sequence between lenticular galaxies and normal spirals. This sequence of spirals, dubbed *Anemic*, has objects designated Aa, ABa, Ab, etc. These Anemic spirals have diffuse spiral features, are usually of low surface brightness and are gas poor relative to normal spirals of the same form.

III. Galactic Structure

The quantitative understanding of galactic structure is based on only two observables. These are the surface brightness distribution (including the structure of the arms in spiral galaxies) and the line-of-sight (radial) velocity field. The surface brightness at any point in a galaxy's image is the integral along the line-of-sight through the galaxy of the light produced by stars and hot gas. Measurements of the velocity field are made spectroscopically (either optically or in the radio at the 21-cm emission line of neutral H), and represent the integral along the line-of-

sight of the velocities of individual objects (stars, gas clouds) times their luminosity. Dust along the line-of-sight in a galaxy obscures the light from objects behind it; in dusty, edge-on spiral galaxies only the near side of the disk can be observed in visible light. Extinction by dust is a scattering process and thus is a function of wavelength; all galaxies are optically thin at radio wavelengths. Elliptical galaxies are considered optically thin at all wavelengths. [*See* SPACE, INTERSTELLAR MATTER.]

A. ELLIPTICAL GALAXIES

Most galaxies can be readily decomposed into two main structural components, disk and bulge. Elliptical galaxies are all bulge. The radial brightness profile of the bulge component is usually parameterized by one of three laws. The earliest and simplest is an empirical relation called the Hubble law,

$$\mu(r) = \mu_0(1 + r/r_0)^{-2}$$

and is parametrized in terms of a central surface brightness, μ_0, and a scale length, r_0, at which the brightness falls to half its central value. At large radius, the profiles are falling as $1/r^2$. At radii less than the scale length, the profile flattens to μ_0. Giant elliptical galaxies typically have $\mu_0 \approx 16$ mag/arc-second^2. A better relation is the empirical $r^{1/4}$ law proposed by de Vaucouleurs

$$\mu(r) = \mu_e \exp[-7.67((r/r_e)^{1/4} - 1)]$$

where r_e is the effective radius and corresponds to the radius that encloses $\frac{1}{2}$ of the total integrated luminosity of the galaxy, and μ_e is the surface brightness at that radius, approximately $\frac{1}{2000}$ of the central surface brightness. The third relation is semiempirical and was derived from dynamical models calculated by King to fit the brightness profiles of globular star clusters. These models can be parametrized as

$$\mu(r) = \mu_K[(1 + r^2/r_c^2)^{-1/2} - (1 + r_t^2/r_c^2)^{-1/2}]^2$$

where r_c again represents the core radius where the surface brightness falls to ≈ 0.5, r_t is the truncation or tidal radius beyond which the surface brightness rapidly decreases, and μ_K is approximately the central surface brightness. Isolated elliptical galaxies are best fit by models with $r_t/r_c \approx 100$–200. Small elliptical galaxies residing in the gravitational potential wells of more luminous galaxies (like the dwarf neighbors of our own galaxy) are tidally stripped and have

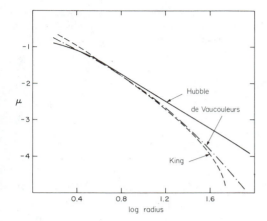

FIG. 4. The three commonly fit surface brightness profiles for elliptical galaxies and the bulge component of spiral galaxies. The Hubble law (solid line) has core radius $r_0 = 1.0$ unit, the de Vaucouleurs law (dash-dot line) has an effective radius $r_e = 5.0$ units, and the King model (dashed line) has a core radius $r_c = 1.0$, and a tidal radius $r_t = 80.0$ units. All three profiles have been scaled vertically in surface brightness to approximately agree at $r = 5.0$ units ($\log r = 0.7$).

$r_t/r_c \approx 10$. Figure 4 shows examples of the Hubble and de Vaucouleurs laws and the King models.

Dwarf elliptical galaxies, designated dE, are low luminosity, very low surface brightness objects. Because of their low surface brightness, they are difficult to identify against the brightness of the night sky (airglow). The nearest dwarf elliptical galaxies, Ursa Minor, Draco, Sculptor and Fornax, are satellites of our own galaxy, and are resolved into individual stars with large telescopes. Although dE galaxies do not contribute significantly to the total luminosity of our own Local Group of galaxies, they dominate its numbers (Table II). As mentioned previously dwarf galaxies are often tidally truncated by the gravitational field of neighboring massive galaxies. If M is the mass of the large galaxy, m is the mass of the dwarf, and R is their separation, then the tidal radius, r_t, is given by

$$r_t = R(m/3M)^{1/3}$$

Frequently the central brightest galaxy in a galaxy cluster exhibits a visibly extended halo to large radii. These objects were first noted by W. W. Morgan and are designated "cD" galaxies in his classification scheme. Unlike ordinary giant E galaxies, whose brightness profiles exhibit the truncation characteristic of de

Vaucouleurs or King profiles at radii of 50 to 100 kpc, cD galaxies have profiles, which fall as $1/r^2$ or shallower to radii in excess of 100 kpc. "D" galaxies are slightly less luminous and have weaker halos; D galaxies are found at maxima in the density distribution of galaxies. The extended halos of these objects are considered to be the result of dynamical processes that occur either during the formation or subsequent evolution of galaxies in dense regions.

The internal dynamics of spheroidal systems (E galaxies and the bulges of spirals) is understood in terms of a self gravitating, essentially collisionless gas of stars. These systems appear dynamically relaxed; they are basically in thermodynamic equilibrium as isothermal spheres with Maxwellian velocity distributions. Faber and Jackson noted in 1976 that the luminosities of E galaxies were well correlated with their internal velocity dispersions.

$$L \approx \sigma^4$$

Here, the velocity dispersion, σ, is a measure of the random velocities of stars along the line-of-sight.

The collisional or two-body relaxation time, t_r, for stars is approximately

$$t_r \sim 2 \times 10^8 \, (V^3/M^2\rho) \text{ years}$$

TABLE II. Presently Known Local Group Members

Name	Type	$B_T{}^a$	Distance (kpc)	Luminosity ($10^9 \, L_\odot$)
Andromeda	Sb	4.38	730	14.7
Milky Way	Sbc	—	—	$(10.0)^b$
M33	Scd	6.26	900	3.9
LMC	SBm	0.63	50	2.2
SMC	Im	2.79	60	0.4
NGC 205	E5	8.83	730	0.24
M32	E2	9.01	730	0.20
IC 1613	Im	10.00	740	0.09
NGC 6822	Im	9.35	520	0.08
NGC 185	dE3	10.13	730	0.07
NGC 147	dE5	10.37	730	0.06
Fornax	dE	9.1	130	0.006
And I	dE	13.5	730	0.003
And II	dE	13.5	730	0.003
And III	dE	13.5	730	0.003
Leo I	dE	11.27	230	0.0026
Sculptor	dE	10.5	85	0.007
Leo II	dE	12.85	230	0.0006
Draco	dE	12.0	80	0.00016
LGS 3	?	$(17.5)^b$	730	0.00008
Ursa Minor	dE	13.2	75	0.00005
Carina	dE	—	170	—

^{a} B_T is the total, integrated blue apparent magnitude.
^{b} Quantities in parentheses are estimated.

where V is the mean velocity in km sec^{-1}, ρ the density of stars per cubic parsec and M is the mean stellar mass in $M_\odot$. This relaxation time in galaxies is 10^{14} to 10^{18} yr—much longer than the age of the universe. Two-body relaxation is generally not important in the internal dynamics of galaxies, but it is significant in globular clusters.

The relaxed appearance of galactic spheroids is probably due to the process called violent relaxation. This is a statistical mechanical process described by Lynden–Bell, where individual stars primarily feel the mean gravitational potential of the system. If this potential fluctuates rapidly with time, as in the initial collapse of a galaxy, then the energy of individual stars is not conserved. The results of numerical experiments are similar to galaxy spheroids.

Between 1970 and 1986, the determination and modeling of the true shapes of elliptical galaxies has been a major problem in galaxian dynamics. Most elliptical galaxies are somewhat flattened. Early investigators assumed that this flattening was rotationally supported as in disk galaxies. Bertola and others observed, however, that the rotational velocites of E galaxies are insufficient to support their shapes (Fig. 5). The velocity dispersion is a measure of the random kinetic energy in the system.

To resolve this problem, Binney, Schwarzchild, and others suggested that E galaxies might be prolate (cigar shaped) or even triaxial systems instead of oblate (disklike) spheroids. Current work favors the view that most flattened elliptical galaxies are triaxial and have internal stellar velocity distributions which are anisotropic. A small fraction do rotate fast enough to support their shapes.

B. S0 GALAXIES

Spiral and S0 galaxies can be decomposed into two surface brightness components, the bulge and disk. Disks have brightness profiles that fall exponentially

$$\mu(r) = \mu_0 \exp(-r/r_s)$$

Bulges have been described previously. Disks are rotating; their rotational velocity at any radius is presumed to balance the gravitational attraction of the material inside. A typical rotation curve (velocity of rotation as a function of radius) is shown in Fig. 6.

Disks are not infinitely thin. The thickness of the disk depends on the balance between the surface mass density in the disk (gravitational potential) and the kinetic energy in motion perpendicular to the disk. This can be a function of both the initial formation of the system and its later interaction with other galaxies. Tidal interaction with other galaxies will "puff up" a galactic disk of stars. Disks of S0 galaxies are composed of old stars and do not exhibit any indications of recent star formation and associated gas or dust, which is part of the definition of an S0.

Bulges of S0 galaxes and spirals *do* rotate and appear to be simply rotationally flattened oblate spheroids. We will return to this point in section V,A when we discuss galaxy formation.

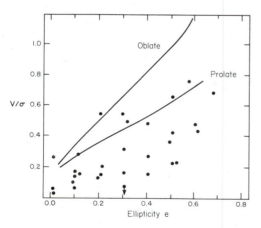

FIG. 5.   The ratio of rotation velocity to central velocity dispersion *versus* the ellipticity, e, for ellipitcal galaxies. Solid lines are oblate and prolate spheroid models with isotropic velocity distributions from Binney. Data points are from Bertola, Cappacioli, Illingworth, Kormendy and others.

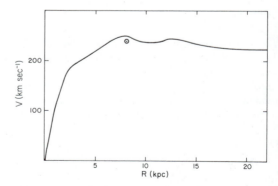

FIG. 6.   A Typical rotation curve for a moderately bright spiral galaxy like our own (adapted from the work of V. Rubin *et al.* and M. Roberts). For reference, the rotational velocity of our own galaxy at the approximate radial distance of the sun, 220 km sec^{-1} at 8 pc, is marked with an $\odot$.

C. Spiral Galaxies

Unlike the smooth and uncomplicated disks of SO galaxies, the disks of spiral galaxies generally exhibit significant amounts of interstellar gas and dust and distinct spiral patterns. The disk rotates differentially (see Fig. 6) and the spiral pattern traces the distribution of recent star formation in the galaxy. Regions of recent star formation have a higher surface brightness than the background disk. Although the spiral is a result of the rotation of the material in the disk, the pattern need not (and generally does not) rotate with the same speed as the material.

The rotation of a spiral galaxy can be described in terms of a velocity rotation curve, $v(r)$, or the angular rotation rate, $\Omega(r)$, where $v = r\Omega$. In 1927, Oort first measured the local differential rotation of our galaxy by studying the motions of nearby stars. He described that rotation in terms of two constants, now known as Oort's constants, which are measures of the local shear (A) and vorticity (B):

$$A = -\frac{r}{2}\frac{d\Omega}{dr}, \qquad B = -\frac{1}{2r}\frac{d}{dr}(r^2\Omega)$$

The values adopted by the International Astronomical Union (IAU) in 1964 are

$$A = 15 \text{ km sec}^{-1} \text{ kpc}^{-1}$$

and

$$B = -10 \text{ km sec}^{-1} \text{ kpc}^{-1}$$

with the sun at a distance of $r_0 = 10$ kpc from the center of the galaxy, and the rotation rate at the sun $V_\odot = 250$ km sec^{-1}. Current estimates support a smaller sun-galactic-center distance (~ 8 kpc) and a smaller rotation velocity (~ 230 km sec^{-1}) as well as slightly different values for A and B.

Studies of rotation in spiral galaxies are undertaken by long-slit spectroscopy at different position angles in the optical or by either integrated (total intensity, big beam) measurements or interferometric mapping in the 21-cm line of neutral hydrogen. The neutral hydrogen (HI) in spiral galaxies is primarily found outside their central regions, reaching a maximum in surface density several kiloparsecs from the center. The gas at the center is mostly in the form of molecular hydrogen, H_2, as deduced from carbon monoxide (CO) maps. From such detailed studies by Rubin, Roberts, and others we know that the rotation curves for spirals generally rise very steeply within a few kiloparsecs of their centers then flatten and stay at an almost constant velocity as far as they can be measured. This result is rather startling because the luminosity in galaxies is falling rapidly at large radii. If the light and mass were distributed similarly, then the rotational velocity should fall off as $1/R$ at large radii as predicted by Kepler's laws. Only a small number of galaxies show the expected Keplerian falloff, leading to the conclusion that the mass in spiral galaxies is *not* distributed as the light.

As in elliptical galaxies, Fisher and Tully noted in 1976 that the luminosities of spiral galaxies are correlated with their internal motions, in particular rotational velocity. The best form of this relation is seen in the near infrared where the effects of internal extinction by dust on the luminosities are minimized (Fig. 7). There the relation is approximately

$$L = (\Delta V)^4$$

where ΔV is the full width of the HI profile measured at either 20% or 50% of the peak. As the HI distribution peaks outside the region where the rotational velocity has flattened, the HI profile is sharp sided and is double peaked for galaxies inclined to the line of sight.

The regularity of the spiral structure seen in these galaxies is exceptional. If the spiral pattern was merely tied to the matter distribution, differential rotation would 'erase' it in a few rotation periods. A typical rotation time is a few hundred million years, or $\approx 1/100$th the age of the universe. To explain the persistence of spiral structure, Lin and Shu introduced the Density Wave theory in 1964. In this model, the spiral pattern is the star formation produced in a shock

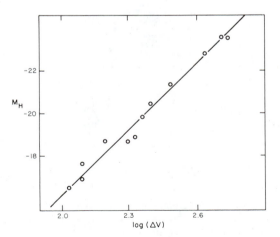

FIG. 7. The relation between the maximum rotation velocity of spiral galaxies and their absolute infrared (H band = 1.6μ) magnitude [Adapted from Aaronson M., Mould J., and Huchra J. P. (1980). *Astrophys. J.* **237**, 655.]

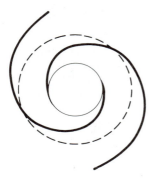

FIG. 8. Typical two-armed spiral density wave pattern. Solid spiral represents the shock front in the gas. This follows closely behind the gravitational potential perturbation. The outer dashed circle is the corotation radius, where the matter in the disk is rotating at the same speed as the spiral pattern. The inner, thin circle represents the inner Lindblad resonance.

wave induced by a density wave propagating in the galaxy disk. The spiral pattern is in solid body rotation with a pattern speed Ω_p. The main features of such a density wave are the corotation radius, where the pattern and rotation angular speeds are the same, and the inner and outer

Lindblad resonances, where

$$\Omega_p = \Omega \pm \kappa/m$$

where m is the mode of oscillation, and κ the epicyclic frequency in the disk, such that

$$\kappa^2 = r^{-3} \, d(r^4\Omega^2)/dr$$

or

$$\kappa = 2\Omega/(1 - A/B)^{1/2}$$

Figure 8 depicts the features of a density wave. The spiral pattern only exists between the Lindblad resonances. Galaxies in which a single mode dominates are called ''Grand Design'' spirals.

Two alternative models have been proposed to account for spiral patterns, the Stochastic Star Formation (SSF) model of Seiden and Gerola, and the tidal encounter model. In the first model, regions of star formation induce star formation in neighboring regions. With proper adjustment of the rotation and propagation timescales, reasonable spiral patterns result. In the second model, spiral structure is the result of galaxy interactions. A possible example of this process is shown in Fig. 9. It is likely that all

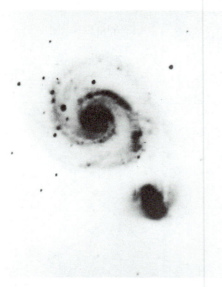

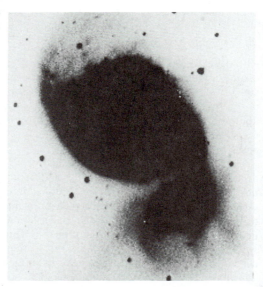

FIG. 9. The interacting galaxy pair NGC 5194 + 5195, also known as M51 or the Whirlpool Nebula. This galaxy pair is also known as Arp 85, VV 1, and Ho 526 from the catalogs of peculiar, interacting and binary galaxies of Arp, Vorontsov–Velyaminov and Holmberg. The bright spiral, NGC 5194, is classified Sbc; its companion, NGC 5195, is classified SB0 P. (a) is a low contrast display of the image to show the interior structure of the galaxies; (b) is a high contrast display of the same image to show the effects of interaction on the outer parts of the galaxies. (CCD photo courtesy of S. Kent.)

three processes operate in nature, with density waves producing the most regular spirals, SSF producing 'flocculent' spirals, and tidal interactions producing systems like the Whirlpool. A significant fraction (~10%) of all galaxies show some form of interaction with their neighbors. However, not all such systems contain spiral galaxies, however.

D. IRREGULAR GALAXIES

Galaxies are classified as Irregular for several different reasons. In the morphological progression of the Hubble sequence the true Irregulars are the Magellanic Irregulars, the Irr I's, which are galaxies with no developed spiral structure that are usually dominated by large numbers of star forming regions. Irregular II's and other peculiar objects have been extensively cataloged by Arp and his coworkers, by Vorontsov-Velyaminov and by Zwicky. These objects are usually given labels to describe their peculiar properties such as 'compact,' which usually indicates abnormally high surface brightness or a very steep brightness profile, 'post-eruptive,' which usually indicates the existents of jets or filaments of material near the galaxy, 'interacting,' or 'patchy.'

There are also galaxies in the form of rings, which are thought to be produced by a slow, head-on collision of two galaxies, one of which must be a gas rich spiral. The collision removes the nucleus of the spiral, leaving a nearly round ripple of star formation similar to the ripples produced when a rock is dropped into a lake. Such ring galaxies almost always have compact companion galaxies, which are the likely culprits.

The Magellanic irregulars are almost always dwarf galaxies (low luminosity), are very rich in neutral hydrogen and have relatively young stellar populations. Their internal kinematics may show evidence for regular structure or may be chaotic in nature. These galaxies are generally of low mass; the largest such systems have internal velocity dispersions (usually measured by the width of their 21-cm Hydrogen line) less than 100 km sec^{-1}. Their detailed internal dynamics have been poorly studied up until now. There are some indications that star formation proceeds in these galaxies as in the SSF theory mentioned previously, however evidence also exists for H II regions aligned with possible shock fronts. Although they do not contribute

significantly to the total luminosity density of the universe, these galaxies and the dwarf ellipticals dominate the total number of galaxies.

IV. Integrated Properties of Galaxies

A. LUMINOSITY FUNCTION

The luminosity function or space density of galaxies, $\phi(L)$, is the number of galaxies in a given luminosity range per unit volume. This function is usually calculated from magnituded limited samples of galaxies with distance information. Distances to all but the nearest galaxies are determined from their radial velocities and the Hubble constant. (Note that in the Local Supercluster, where the velocity field is disturbed by the gravity of large mass concentrations, it is necessary to apply additional corrections to distances measured this way). Figure 10 is the differential luminosity function for field galaxies derived from a recent large survey of galaxy redshifts. The luminosity function is nearly flat at faint magnitudes and falls exponentially at the bright end. A useful parametrization of the $\phi(L)$, derived by Schechter, is

$$\phi(L) = \phi_0 L^{-1} (L/L^*)^\alpha \exp(-L/L^*)$$

L^* is the characteristic luminosity near the knee, ϕ_0 is the normalization, and α is the slope at the faint end. For a Hubble constant of 100 km sec^{-1} Mpc^{-1} and with blue (B) magnitudes from the

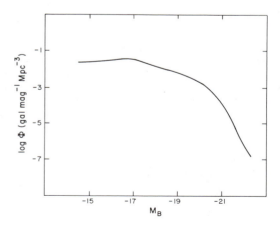

FIG. 10. The differential galaxy luminosity function, $\phi(L)$, in units of galaxies per magnitude interval per cubic megaparsec, derived from the Center for Astrophysics Redshift Survey.

Zwicky Catalog of Galaxies and of Clusters of Galaxies

$$L_B^* \approx 8.6 \times 10^9 \, L_\odot, \quad M_B^* \approx -19.37,$$

$$\phi_0 \approx 0.015 \text{ gal Mpc}^{-3}$$

and

$$\alpha \approx -1.25$$

M_B^* is the characteristic blue absolute magnitude. The Schechter function has the interesting property that the integral luminosity density, useful in cosmology, is just given by

$$L_{\text{int}} = \phi_0 L^* \Gamma(\alpha + 2)$$

where Γ is the incomplete gamma function.

B. Spectral Energy Distributions

The observed integrated spectra of normal galaxies are functions of stellar population, star formation rate, mean metal content, gas content, and dust content. These properties correlate with, and in some cases are, causally related to galaxy morphology. In our own galaxy there are two relatively distinct populations of stars as discovered by Baade in 1944. Population II stars are the old metal poor stars that form the halo of the galaxy. Globular clusters are population II objects. Population I stars are younger and more metal rich. They form the disk of the galaxy. The sun is an intermediate age population I star. [See SOLAR PHYSICS.]

Figure 11a is an example of the optical spectral energy distribution of an old, gasless stellar population. This type of population is typical of the bulge and old disk populations in galaxies. It is dominated by the light of G and K giant stars. The strongest spectral features are absorption lines of (originating in the atmospheres of the giant stars) calcium, iron, magnesium and sodium, usually in low ionization states, as well as molecular bands of cyanogen, magnesium hydride and, in the red, titanium oxide. The very strong, factor of 2, break in the apparent continuum shortwards of 4000Å is the primary distinguishing characteristic of high redshift normal galaxies. The strengths of the absorption features in old bulge populations is primarily a function of metallicity. In populations where star formation has taken place recently (on timescales of $\leq 10^9$ years), most absorption features in the integrated spectra will appear weaker due to the contribution to the continuum emission from hotter, weaker lined A and F

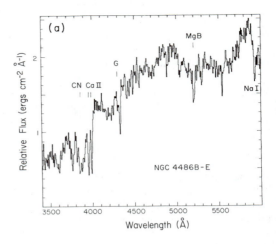

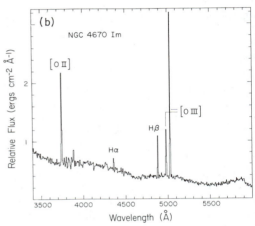

FIG. 11. Galaxy Spectral energy distributions. (a) NGC 4486B, typical of an old, evolved stellar population. The prominent absorption features of Ca, Na, Mg, the CN molecule and the G band blend of Fe and Cr are marked. (b) NGC 4670, typical of a strong emission line galaxy or extragalactic H II region. The O and H emission lines are marked.

stars. The Balmer lines of Hydrogen, which are strong in A and F stars, increase in strength. Table III lists some of the common and strong absorption and emission lines seen in normal galaxy spectra.

Figure 11b is the spectrum of a galaxy who's light is dominated by very young stars and hot gas. The spectrum resembles that of an H II region. This is an irregular galaxy that is undergoing an intense "burst" of star formation. Its optical spectrum is dominated by strong, sharp emission lines of H, He and the light elements N, O and S from the photoionized gas. The con-

TABLE III. Common Spectral Features in Galaxies

λ_0	Element	Comments[a]
Absorption lines		
3810+	CN	(molecular band)
3933.68	Ca II	K
3968.49, 3970.08	Ca II + He	H
4101.75	Hδ	
4165+	CN	(molecular band)
4226.73	Ca I	
4305.5	Fe + Cr	G band
4340.48	Hδ	
4383.55	Fe I	
4861.34	Hβ	F
5167.3, 5172.7, 5183.6	Mg I	b
5208.0	MgH	(molecular band)
5270.28	Fe I	
5889.98, 5895.94	Na I	D
8542.0	Ca II	
Emission lines[b]		
3726.0	[O II]	
3729.0	[O II]	
3868.74	[Ne III]	
3889.05	Hζ	
3970.07	Hε	
4101.74	Hδ	
4340.47	Hγ	
4363.21	[O III]	
4861.34	Hβ	F
4958.91	[O III]	
5006.84	[O III]	
5875.65	He I	
6548.10	[N II]	
6562.82	Hα	C
6583.60	[N II]	
6717.10	[S II]	
6731.30	[S II]	

[a] Letters in comments refer to Fraunhoffer's original designation for lines in the solar spectrum.
[b] Brackets around species indicate that the transition is forbidden.

tinuum is primarily from hot O and B stars with a small contribution from the free–free, two-photon, and Paschen and Balmer continuum emission from the gas.

Population synthesis is the attempt to reproduce the observed spectral energy distributions of galaxies by the summation of spectra of well observed galactic stars, model stars and models for the emission from the gas photoionized by hot stars in the synthesis. An important parameter in such studies is the Initial Mass Function (IMF), which is the differential distribution of stars as a function of mass in star forming regions. The simplest approximation for the IMF is a powerlaw in the mass, M,

$$N(M)\ dM = (M/M_\odot)^{-\alpha}$$

The value of α found by Salpeter for the solar neighborhood is ≈ 2.35. The real IMF is slightly steeper at the high mass end ($M > 10\ M_\odot$), and slightly flatter at the low mass end ($M < 1\ M_\odot$). Results from population synthesis indicate that the mix of stars in most luminous galaxies is very similar to our own in age and metallicity.

C. Galaxy Masses

The masses of galaxies are measured by a variety of techniques based on measurements of system sizes and relative motion, and the assumption that the system under study is gravitationally bound. Masses are sometimes inferred from studies of stellar populations, but these are extremely uncertain as the lower mass limit of the IMF is essentially unknown. It is customary to quote the mass-to-light ratios for galaxies rather than masses. This is both because galaxy masses span several orders of magnitude so it is easier to compare M/L, since mass and luminosity are reasonably well correlated for individual morphological classes, and because the average value of M/L is key in the determination of the mean matter density in the universe. In specifying masses or mass-to-light ratios for galaxies, it is necessary to specify the Hubble constant, scale lengths vary as the distance, while luminosities vary inversely as the square of the distance. It is also useful to specify the magnitude system used for determining luminosity both because different systems measure light to different radii and because galaxies have a wide range of colors and are rarely the same color as the sun.

Masses of individual spiral galaxies are determined from their rotation curves. Spiral galaxies are circularly symmetric, so the apparent radial velocities relative to their centers can be corrected for inclination. Then the velocity of the outermost point of the rotation curve is a measure of the enclosed mass. If the mass were distributed spherically, the problem would reduce to the classical circular orbit of radius, R, around point mass, M,

$$1/2mV^2 = GmM/R$$

where m is the test particle mass, and V is its orbital velocity, implying

$$M = \frac{1}{2}\frac{V^2R}{G}$$

Because the actual distribution is flattened, a small correction must be applied. If the mass distributions of spiral galaxies fall with radius as their optical light, then rotation curves would turnover to a Keplerian falloff, $V \propto R^{-1/2}$. Observations of neutral hydrogen rotation curves made at 21-cm now extend beyond the optical diameters of many galaxies and are still flat. This implies that the masses of spirals are increasing linearly with radius, which in turn, implies that the local mass-to-light ratio in the outer parts of spirals is increasing exponentially.

Masses for individual E galaxies are determined from their velocity dispersions and measures of their core radii or effective radii. For a system in equilibrium, the Virial Theorem states that, averaged over time,

$$\Omega + 2T = 0$$

where T is the system kinetic energy and Ω is the system gravitational potential energy. For a simple spherical system and assuming that the line-of-sight velocity dispersion, σ, is a measure of the mass weighted velocities of stars relative to the center of mass

$$M\sigma^2 = G \int_0^R \frac{M(r)\,dM}{r}$$

If the galaxy is well approximated by a de Vaucouleurs Law, then $\Omega = -0.33G/M^2r_e$. More accurate mass-to-light ratios can be determined for the cores of elliptical galaxies alone by comparison with models.

Masses for binary galaxies are also derived from simple orbital calculations, but, unlike rotation curve masses, are subject to two significant uncertainties. The first is the lack of knowledge of the orbital eccentricity (orbital angular momentum). The second is the lack of information about the projection angle on the sky. The combination of these two problems make the determination of the mass for an individual binary impossible. The formula $M = (V^2R)/(2G)$ produces a minimum mass, the "true" mass is

$$M_T = \frac{V^2R}{2G} \frac{1}{\cos^3 i \, \cos^2 \phi}$$

where i is the angle between the galaxies and the plane of the sky, and ϕ is the angle between the true orbital velocity and the plane defined by the two galaxies and the observer. The correct determination of the masses of galaxies in binary systems requires a large statistical sample to average over projections as well as a model for the selection effects that define the sample (e.g.,

very wide binaries are missed; the effect of missing wide pairs depends on the distribution of orbital eccentricities).

Mass-to-light ratios for galaxies in groups or clusters are also determined from the dimensions of the system and its velocity dispersion via the Virial Theorem or a variant called the Projected Mass method. The Virial theorem mass for a cluster of N galaxies with measured velocities is

$$M_{VT} = \frac{3\pi N}{2G} \frac{\sum_i^N V_i^2}{\sum_{i<j} 1/r_{ij}}$$

where r_{ij} is the separation between the ith and jth galaxy, and V_i is the velocity difference between the ith galaxy and the mean cluster velocity. The projected mass is

$$M_P = \frac{f_p}{GN} \left(\sum_i^N V_i^2 \, r_i \right)$$

where r_i is the separation of the ith galaxy from the centroid. The quantity f_p depends on the distribution of orbital eccentricities for the galaxies and is equal to $32/\pi$ for radial orbits and $16/\pi$ for isotropic orbits.

Table IV summarizes the current state of mass-to-light determinations for individual galaxies and galaxies in systems via the techniques discussed earlier. These are scaled to a Hubble Constant of 100 km sec^{-1}, and the Zwicky catalog magnitude system used earlier for the Luminosity Function.

In almost all galaxies, the expected mass-to-light ratio from population synthesis is very small, on the order of one ore less in solar units, because the light of the galaxy is dominated by either old giant stars, with luminosities several hundred suns and masses less than the sun, or by young, hot main sequence stars with even lower mass-to-light ratios. The large mass-to-light ratios that result from the dynamical studies have given rise to what is called the "missing

TABLE IV. Mass-to-Light Ratios for Galaxies[a]

Type	Method	M/L
Spiral	Rotation Curves	12
Elliptical	Dispersions	20
All	Binaries	100
All	Galaxy Groups	350
All	Galaxy Clusters	400

[a] In solar units, $M_\odot/L_\odot$.

mass'' problem. Astronomers as yet have not been able to determine what constitutes the mass that binds clusters and groups of galaxies and forms the halos of large galaxies. Possibilities include extreme red dwarf (low luminosity) stars, massive stellar remnants (black holes), exotic elementary particles (axions, neutrinos, etc.), and the possibility that either the dynamical state of clusters is much more complex than the existing simple models.

D. Gas Content

Neutral hydrogen (HI) was first detected in galaxies in 1953 by Kerr and coworkers in the 21-cm (1420.40575 MHz) radio emission line. Since then it has been found that almost every spiral and irregular galaxy contains considerable neutral gas. In spiral galaxies, the HI emission distribution usually shows a central minumum and peaks in a ring, which covers the prominent spiral arms. The fractional gas mass ranges from a few percent for early type spirals (Sa galaxies) to more than 50% for some magellanic irregulars. Neutral hydrogen is only rarely detected in elliptical an SO galaxies. The gas mass fraction in these objects is usually less than 0.1%. In spiral galaxies, H I can usually be detected at 21 cm out to two or three times the radius of the galaxy's optical image on the Palomar Schmidt Sky Survey plates.

Ionized gas is found in galaxies in H II regions (H II refers to singly ionized hydrogen), in nuclear emission regions, and in the diffuse interstellar medium. The H II regions are regions of photo-ionized gas around hot, usually young stars. The gas temperature is of order 10^4 K, and depends on the surface temperatures of the ionizing stars and cooling processes in the gas, which depend strongly on its element abundances. Low metallicity H II regions are hotter than those with high metallicity. Ionized gas is often seen in the nuclei of galaxies and is either the result of star formation (as in H II regions) or photoionization by a central nonthermal energy source. In our galaxy, there is a diffuse interstellar medium composed of gas and dust. Much of the volume (although not much mass) of the galaxy is in this state with the gas ionized by the diffuse stellar radiation field. Because its density is so low, ionized gas in this state takes a long time to recombine. (To a first approximation, the recombination time for diffuse, ionized hydrogen is

$$\tau \sim (10^5/n_e)$$

where n_e is the electron density in $1/cm^3$, and τ is in years.) Ionized gas is also found in supernova remnants.

Recently, carbon monoxide (CO), which is found in cool molecular clouds in the galaxy, has been detected in several other nearby spiral and irregular galaxies. The CO clouds are associated with regions of both massive and low mass star formation. In sprial galaxies, the CO emission often fills in the central H I minimum. There is an excellent correlation between the gas content of a galaxy and its current star formation rate as measured by integrated colors or the strength of emission from H II regions.

E. Radio Emission

All galaxies that are actively forming stars emit radio radiation. This emission is from a combination of free-free emission from hot, ionized gas in H II regions, and from supernova remnants produced when massive young stars reach the endpoint of their evolution. In addition, certain galaxies, usually ellipticals, have strong sources of nonthermal (ie., not from stars, hot gas or dust) emission in their nuclei. These galaxies, called radio galaxies, output the greatest part of their total emission at radio and far-infrared wavelengths. The radio emission is usually synchrotron radiation and is characterized by large polarizations and self absorption at long wavelengths. A more detailed treatment of central energy sources can be found in Quasars.

F. X-Ray Emission

As in Section IV,E, all galaxies emit X-ray radiation from their stellar components—X-ray binaries, stellar chromospheres, young supernova remnants, neutron stars, etc. More massive objects, particulary elliptical galaxies, have recently been found by Forman and Jones with the Einstein X-ray Observatory to have X-ray halos, probably of hot gas. A small class of the most massive elliptical galaxies, which usually reside at the centers of rich clusters of galaxies also appear to be accreting gas from the surrounding galaxy cluster. This has been seen as a cooler X-ray emission centered on the brightest cluster galaxy, which sits in the middle of the hot cluster gas. This phenomenon is called a 'cooling flow,' and results when the hot cluster gas collapses on a central massive object and becomes dense enough to cool efficiently. This process is evidenced by strong optical emission lines as well as radio emission. Cooling flows may be sites of low mass star formation at the centers of galaxy clusters.

Active galactic nuclei, such as Seyfert 1 and 2 galaxies (discovered by C. Seyfert in 1943), and quasars are also usually strong X-ray emitters, although the majority are *not* strong radio sources. The X-ray emission in these galaxies is also nonthermal and is probably either direct synchrotron emission or synchrotron–self-Compton emission.

V. Galaxy Formation and Evolution

A. GALAXY FORMATION

The problem of galaxy formation is one that remains as yet unsolved. The fundamental observation is that galaxies exist and take many forms. The interiors of most galaxies are many orders of magnitude more dense than the surrounding intergalactic medium.

In the hot Big Bang model, the simplest description of galaxy formation is the gravitational collapse of density fluctuations (perturbations) that are large enough to be bound after the matter in the universe recombines. At early times, the atoms in the universe—mostly hydrogen and helium—are still ionized so electron scattering is important and radiation pressure inhibits the growth or formation of perturbations. Before recombination, the universe is said to be in the Radiation Era, because of the dominance of radiation pressure. The period of recombination, that is, hydrogen and helium atoms are formed from protons, α particles and electrons, is called the Decoupling Era, because the universe becomes essentially transparent to radiation. After that the universe is Matter Dominated as gravitational forces dominate the formation of structure. The cosmic microwave background radiation, postulated by Gamow and colleagues in the 1950s and detected by Penzias and Wilson in 1965, is the relic radiation field from the primeval fireball and represents a snapshot of the universe at decoupling.

In this simple picture, matter is distributed homogeneously and uniformly before decoupling because the radiation field will tend to smooth out perturbations in the matter. After decoupling statistical fluctuations will form in either the matter density or velocity field (turbulence). The amplitude of fluctuation, which are large enough, will grow, and the fluctuations can fragment and, if gravitationally bound, collapse to form globular clusters, galaxies, or larger structures. A simple criterion for the growth of fluctuations in a gaseous medium was derived by Jeans in 1928. The Jeans wavelength, λ_J, is given by

$$\lambda_J = c_s \, (\pi/G\rho)^{1/2}$$

where c_2 is the sound speed in the medium, and ρ is its density. The sound speed is

$$c_s \sim c/\sqrt{3}$$

in the radiation dominated era, and

$$c_s = (5kT/3m_p)^{1/2}$$

after recombination. If a fluctuation is larger than the Jeans length, gravitational forces can overcome internal pressure. The mass enclosed in such a perturbation, the "Jeans mass," is just

$$M_J \approx \rho\lambda_J^3 \approx (c_s^3/G^{3/2}\rho^{1/2})$$

Before recombination, this mass is a few times $10^{15} \, M_\odot$, which is comparable to the mass of a cluster of galaxies. After recombination, the Jeans mass plunges to $\approx 10^6 \, M_\odot$, or about the mass of a globular star cluster. The amplitude $(\delta\rho/\rho)$ required for fluctuations to become gravitationally bound and collapse out of the expanding universe is dependent on the mean mass density of the universe. The ratio of the actual mass density to the density required to close the universe is usually denoted Ω.

$$\Omega = (8\pi G\rho/3H_0^2)$$

where H_0 is the Hubble Constant. If Ω is large, that is, near unity, small perturbations can collapse.

In the 1970s, work on a variety of problems dealing with the existence and form of large scale structures in the universe made it clear that the formation of galaxies and larger structures had to be considered together. Galaxies cluster on very large scales. Peebles and collaborators introduced the description of clustering in terms of low-order correlation functions. The two-point correlation function, $\xi(r)$, is defined in terms of

$$\delta P = N[1 + \xi(r)]\delta V$$

where δP is the excess probability of finding a galaxy in volume δV, at radius r from a galaxy. The variable N is the mean number density of galaxies. Current best measurements of the galaxy two-point correlation function indicate that it can be approximated as a power law of form

$$\xi(r) = (r/r_0)^{-\gamma}$$

with and index and correlation length (amplitude) of

$$\gamma = 1.8, \quad \text{and} \quad r_0 = 5h^{-1} \text{ Mpc},$$

(where h is the Hubble Constant in units of 100 km sec^{-1} Mpc^{-1}). Power has been found in galaxy clustering on the largest scales observed to date ($\sim$100 Mpc), and the amplitude of clustering of clusters of galaxies is 5 to 10 times that of individual galaxies. In addition, work in the last decade has shown that there are large, almost empty regions of space, called Voids, and that most galaxies are usually found in large extended structures that appear filamentary or shell like, with the remainder in denser clusters. [See STAR CLUSTERS.]

There are currently two major theories for the formation of galaxies and large scale structures in the universe. The oldest is the gravitational instability plus heirarchical clustering picture primarily championed by Peebles and coworkers. This is a 'bottom up' model, where the smallest structures, galaxies, form first by gravitational collapse and then are clustered by gravity. This model has difficulty in explaining the largest structures we see, essentially because gravity does not have time in the age of the universe to significantly agect structure on very large scales. A more recent theory, often called the pancake theory, is based on the assumption that the initial perturbations grow adiabatically, as opposed to isothermally, so that smaller, galaxy sized perturbations are initially damped. In this model the larger structures form first with galaxies fragmenting out later. If dissipationless material is present, such as significant amounts of cold or hot dark matter (eg., massive neutrinos), collapse will usually be in one direction first, which produces flattened or pancakelike systems. Models of this kind are somewhat better matches to the spatial observations of structure, but the hot dark matter models fail to produce the observed relative velocities of galaxies. The Hubble flow or general expansion of the universe, is fairly cool, with galaxies outside of rich, collapsed clusters, moving only slowly ($\sigma < 350$ km sec^{-1}) with respect to the flow.

There are alternative models of galaxy formation, for example, the explosive hypothesis of Ostriker and Cowie. In this model, a generation of extremely massive, pregalactic stars is formed and goes supernova, producing spherical shocks that sweep and compress material soon after recombination. These shells then fragment and the fragments collapse to form galaxies.

There is some recent evidence that favors this model. [See SUPERNOVAE.]

A major problem for all of the above models are the observations of small scale fluctuations in the microwave background radiation. Perturbations produced after recombination should appear as perturbations in the microwave background on scales of a few arc minutes. To date, no fluctuations have been observed to very low limits, $\Delta T/T < 10^{-5}$. This is two orders of magnitude below the level expected in the simple gravitational instability picture in a baryon dominated universe. The existence of cold or hot dark matter in the quantities necessary to bind the universe can abet this problem slightly.

Recently an amplification of the Big Bang model called Inflation has been introduced. In inflationary models, the dynamics of the early universe is dominated by processes described in the new Grand Unified Theories (GUTS). In these models, the universe undergoes a period of tremendous inflation (expansion) at a time near 10^{-35} second after the big bang. If inflation is correct, galaxy formation becomes easier because fluctuations in the very early universe are inflated to scales, which are not damped out and can exist before decoupling. Almost all inflationary models make the strong prediction that Ω is exactly equal to 1. [See COSMIC INFLATION.]

B. POPULATION EVOLUTION

The appearance of a galaxy changes with time (evolves) because its stellar population changes. This occurs because the characteristics of individual stars change with time, and because new stars are being formed in most galaxies. A galaxy's appearance thus reflects its integrated star formation history and the evolution of its gaseous content. The population evolution of galaxies is relatively well understood, although detailed models exist for our galaxy and only a few others.

After the initial "formation' of the galaxy, the more massive of the stars in the first generation evolve more rapidly than the low mass stars. For example, the evolutionary timescale for a 100 $M_\odot$ star is only a few million years, while that for a 1 $M_\odot$ star is nearly 10 billion years. Elements heavier than hydrogen and helium are produced in the cores of these stars and are then ejected into the interstellar medium, either by stellar mass loss or supernova explosions, thus enriching the metal content of the gas. Stars formed later and later have increasing metallicities—the oldest and most metal poor stars in our own galaxy have heavy element abundances only 10^{-3} to 10^{-4} solar. The youngest stars have abundances a few times solar. It is possible for the average metallicity of a galaxy's gas to decrease with time if it accretes primordial low metallicity gas from the intergalactic medium faster than its high mass stars eject material.

Elliptical galaxies are objects in which most of the gas was turned into stars in the first few percent of the age of the universe. Only a few exceptional ellipticals for example, galaxies with cooling flows, show any sign of current star formation. These are objects in which the initial star formation episode was extremely efficient, with almost all of the galaxy's gas being turned into stars. At present, elliptical galaxies probably getting very slightly fainter and redder as a function of time. Their light is dominated by red giant stars that have just evolved off the main sequence, and the number of these stars is a slowly decreasing function of time for an initial mass function with the Salpeter slope. These stars are between 0.5 and 1 $M_\odot$. A competing process, main sequence brightening, can occur in systems where stars still on the main sequence contribute significantly to the light, that is, for systems with steep initial mass functions. Stars on the main sequence evolve up it slightly, becoming brighter and hotter, just before evolving into red giants.

Spiral and irregular galaxies have integrated star formation rates that are more nearly constant as a function of time. In these objects, the gas is used up slowly or replenished by infall. It is probable that in many of these galaxies, star formation is episodic, occurring in bursts caused either by passage of spiral density waves through dense regions, significant infall of additional gas, or interaction with another galaxy. The photometric properties of these galaxies are usually determined by the ratio of the amount of current star formation to the integrated star formation history. The optical light in objects with as little as 1% of their mass involved in recent, $<$ 10^8 years old, star formation will be dominated by newly formed stars. Galaxies with constant star formation rates get brighter as a function of time for any reasonable assumption about the initial mass function.

It is common to model the evolutionary history of galaxies by assuming an unchanging initial mass function and parameterizing the star formation rate in terms of the available gas mass or density, or in terms of a simple functional form, usually and exponential. Examples of the possible color and luminosity evolution of an elliptical and a spiral galaxy are shown in Figure 12. These models assume an exponentially decreasing star formation rate,

$$\Psi(t) \sim Ae^{-\beta t}$$

where β is the inverse of the decay time ($\beta = 0$ is a constant star formation rate). Properties of galaxies along the Hubble sequence are well approximated by a continuous distribution of exponentials, from constant star formation rates (Sd and Im galaxies) to initial bursts with little subsequent star formation (E and S0 galaxies).

C. Dynamical Evolution

There are several processes responsible for the dynamical evolution of galaxies, tidal encounters and collisions, mergers, and dynamical friction. If galaxies were uniformly distributed in space, the probability, P_i, that a galaxy would have undergone a close interaction or merger in

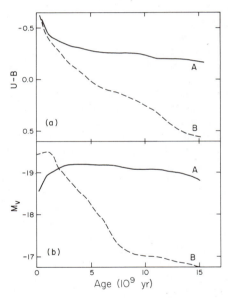

FIG. 12. Luminosity and color evolution for two model galaxies. (a) The color U–B represents the logarithmic difference (in magnitudes) between two broad bandpasses approximately 800 Å wide centered on 3600 Å(U) and 4400 Å(B). More negative U–B's are bluer. (b) The variable M_V is the absolute visual magnitude. Model A has a star formation rate that is nearly constant as a function of time. This might be typical of a spiral galaxy. Model B has an exponentially decreasing star formation rate—15 e-folds in 15 billion years. This would be typical of an E or S0 galaxy with a small amount of present day star formation.

time t is

$$P_i = \pi R^2 \langle v_{rel} \rangle Nt$$

where R is the size of a galaxy, $\langle v_{rel} \rangle$ is the mean relative velocity, and N is the number density of galaxies. For the average bright galaxy R is about 10 kpc (for a Hubble Constant of 100 km sec^{-1} Mpc^{-1}, $\langle v_{rel} \rangle$ is about 300 km sec^{-1} and Pi is thus less than 10^{-4} in a Hubble time. Galaxies are clustered, however, and this significantly increases the probability that any individual object has undergone an interaction. As stated earlier, of order 10% of all galaxies show some evidence of dynamical interaction.

Tidal encounters and collisions that produce observable results are still relatively rare events in the field. Some examples of such events were discussed earlier, eg., the Whirlpool and ring galaxies. Detailed models of individual events have been made by Toomre and others and the models compare well with observed structures and velocity fields. Encounters at large relative velocity usually produce small effects because the interaction time is short and the stellar components of galaxies can easily pass through one another. Encounters have large effects when the relative velocities are comparable or smaller than the internal velocity fields of the galaxies involved. The effects of encounters between spiral galaxies are also enhanced if the spin and orbital angular momenta of the galaxies are aligned. Fast collisions however, may be possibly responsible for sweeping gas from early type galaxies in galaxy clusters, although other mechanisms (ablation by the hot intracluster medium, for example) probably dominate. Tidal encounters between protogalaxies were originally thought to produce most of the observed angular momentum in the universe, however, numerical simulations fail to produce the observed amount. The origin of angular momentum remains a mystery.

It was similary proposed that mergers of spiral galaxies might produce elliptical galaxies, a process that could explain the larger fraction of early type galaxies seen in rich clusters, where interactions are more likely to take place. Unfortunately, the photometric properties of elliptical galaxies as well as their relatively large globular cluster populations strongly argue against this hypothesis. As in tidal encounters, the 'efficiency' of mergers depends on the relative encounter velocity; encounters at low relative velocities are much more likely to produce

mergers than fast encounters. In particular, the line-of-sight velocity dispersion of galaxies in rich clusters is of order 1000 km sec^{-1}, which is much higher than the stellar dispersions internal to the individual galaxies. Dynamical friction is a specialized case of 'encounter' whereby a satellite object slowly loses its orbital energy when orbiting inside the halo of a more massive galaxy. An object moving in the halo of a galaxy produces a gravitational wake which itself can exert drag on the moving object. This effect was first described by Chandrasekhar in 1960. An object of mass M moving through a uniform halo of stars of volume density n with velocity v will suffer a drag of

$$dv/dt = -4\pi G^2 Mnv^{-2}[\phi(x) - x\phi'(x)]\ln \Lambda \,,$$

where ϕ is the error function, $x = 2^{-1/2}v/\sigma$, and Λ is the ratio of the maximum and minimum impact parameters considered. The variable, σ is the velocity dispersion of the stars in the halo.

Mergers via dynamical friction as well as direct mergers of more massive galaxies almost certainly occur at the very centers of rich clusters of galaxies where the central cD galaxy is often accompanied by a host of satellite galaxies. Such processes probably account for the cD galaxy's extended halo, depressed central surface brightness, and excess luminosity relative to unperturbed bright ellipticals.

BIBLIOGRAPHY

Bok, Bart J., and Bok, Priscilla (1974). "The Milky Way." Harvard University Press. Cambridge Massachusettes.

Ferris, Timothy (1982). "Galaxies." Stewart, Tabori and Chang. New York.

Hodge, P. (1986). "Galaxies." Harvard University Press. Cambridge Massachusetts.

Mihalas, Dimitri, and Binney, James. "Galactic Astronomy." Freeman and Co. San Francisco California.

Peebles, P. J. E. (1971). "Physical Cosmology." Princeton University Press. Princeton, New Jersey.

Sandage, Allan (1961). "The Hubble Atlas of Galaxies." Carnegie Institution of Washington. Washington, D.C.

Sandage, Allan, Sandage, Mary and Kristian, Jerome (eds) (1975). "Galaxies and the Universe." University of Chicago Press, Chicago Illinois.

Shu, Frank (1982). "The Physical Universe." University Science Books. Mill Valley, California.

Silk, J. (1980). "The Big Bang." Freeman and Co. San Francisco California.

GAMMA-RAY ASTRONOMY

R. Stephen White *University of California, Riverside*

GLOSSARY

Black holes: Objects so dense that no particles or radiation can escape.

Celestial gamma-ray sources: Sources of gamma rays originating outside the earth, in our galaxy or beyond.

Cerenkov radiation: Radiation produced by a charged particle whose velocity is greater than the velocity of light in the medium.

COS B satellite: European satellite launched in 1975 with a detector that is sensitive to gamma rays of about 50 MeV to 3000 GeV.

Diffuse galactic gamma rays: Gamma rays originating from extended regions (not point sources) in the galactic plane.

Gamma rays: Photons of the electromagnetic spectrum with energies greater than 0.3 MeV.

Gamma-ray bursts: Gamma-ray events with rise times of a fraction of a second and durations of a fraction to a few seconds.

Gamma-ray lines: Discrete energy levels from a gamma-ray source.

Inverse Compton scattering: Scattering of a photon by a high-velocity electron in which the photon gains energy from the interaction.

Neutron star: A star with the density of the nucleus of an atom; core of a star left after a supernova explosion.

Phase diagram (light curve): Plot of the gamma-ray count rate versus fraction of a period of the gamma-ray source.

SAS 2 satellite: NASA satellite launched in 1972 with a detector sensitive to gamma rays from about 35 to 500 MeV.

Pulsars: Rotating neutron stars.

Quasar: Quasistellar source.

Gamma-ray astronomy involves observation, description, and understanding of celestial objects by means of the gamma rays they emit. We define gamma rays here as photons with energies greater than 0.30 MeV, the most energetic photons in the electromagnetic spectrum. Gamma rays are emitted in energetic interactions that take place inside, on the surface, and in the vicinity of stars and galaxies. They are produced by acceleration of electrons, transitions between nuclear levels, decays of certain elementary particles, and particle–antiparticle annihilations. The directions of the gamma rays give information about the identity of the sources; energies, fluxes, and time distributions of the gamma rays give information about the properties of the sources. Gamma-ray lines identify nuclear reactions that take place, their temperatures, and motions. Composition of bombarding particles and target nuclei can be determined. Gamma-ray astronomy complements radio, infrared, optical, ultraviolet, and X-ray astronomy at the lower energies of the electromagnetic spectrum. Gamma rays from the sun are important to gamma-ray astronomy. [*See* SOLAR PHYSICS.]

I. Objectives

The major objective of gamma-ray astronomy, like other astronomies, is to understand the origin and evolution of the universe, its governing laws of physics, and the birth and evolution of the stars, planets, and life. When and how did our universe begin? Current understanding is that it started with the Big Bang about 10 billion yr ago. Will the universe continue to expand or is the expansion slowing down so that contraction will someday begin? Is there enough mass in the universe to cause the deceleration and eventual contraction? The current inventory of observed mass is not sufficient for contraction. But is there additional dark mass that has not yet been observed? How and when did the galaxies form? The most distant and exotic galaxies are the quasars, some rushing away from us close to the speed of light. How do they produce the enormous energies that make them visible at distances near the edge of the observable universe?

We would like to understand the birth, growth, and death of a star. Our understanding of internal structures and atmospheres of stars has developed rapidly over the last 50 yr. Infrared astronomy from ground observatories and, more recently, from satellites in space has identified hot, dense regions in galactic arms and at the centers of galaxies that could be birthplaces of stars. We know stars spend much of their lives on the main sequence only to leave as red giants; they finally die as compact objects: white dwarfs, neutron stars, or black holes. But what occurs in a star's life between the red-giant phase and a supernova explosion? How many compact objects are in our galaxy, the Milky Way, and where are they located? What populates its halo? [See ASTROPHYSICS.]

II. Introduction

Prior to 1600 A.D. all observations of the stars and planets were made with the naked eye. Galileo's telescope, made possible by the discovery of glass lenses and mirrors, revolutionized astronomy. The discovery of the satellites of Jupiter, the phases of Venus, and sunspots changed astronomers' concepts of the solar system. In the next 300 yr we learned much about our universe. So it is surprising that as late as the early 1920s astronomers were still debating whether galaxies other than our own exist. Optical observations in the visible part of the electromagnetic spectrum furnished almost all information about our universe until after World War II.

Scientists used their wartime expertise in radar to develop the radio telescopes of the fifties and the large arrays of today. Radio galaxies were discovered and, in 1960, radio sources associated with optical sources having large red-

shifts. These quasistellar objects were identified as the fast-receding, high energy-emitting quasars.

An important confirmation of the Big Bang came with the discovery by Arno Penzias and Robert Wilson of Bell Labs in 1965 of the background radiation at a temperature of 2.7 K, near the value predicted by George Gamov of the University of Colorado in 1946. The initial radiation of tens of billions of degrees that existed a fraction of a second after the Big Bang was reduced in temperature as the universe expanded to its present radius of 10^{10} light years (l.y.).

With better time resolution, Anthony Hewish and Jocelyn Bell of Cambridge University found unexpected pulsed radio waves with periods of a second and shorter. They quickly dispensed with passing trucks and men from Mars. Within a few months pulsars were explained as rotating neutron stars. The magnetic field axis of the pulsar was oriented at an angle to the spin axis; so each time a magnetic pole pointed toward the earth, a signal was detected, much like the rotating beam from a lighthouse. The pulsar remained as the core of a star, initially with mass between 1.4 and 3 solar masses ($M_\odot$); that had exploded as a supernova. [See PULSARS; SUPERNOVAE.]

X-ray and ultraviolet observations must be carried out above the earth's atmosphere on rockets or satellites, because these photons are absorbed by a small fraction of the earth's atmosphere. Since the discovery of the first X-ray source from a rocket in the early 1960s, thousands of X-ray stars have been cataloged, first by the UHURU satellite, then by the Einstein satellite with its focusing X-ray telescope. Many of the early strong sources were from matter heated by accretion from the second member of a binary pair. X-ray astronomy, much like optical, radio, infrared, and ultraviolet astronomy, has matured rapidly. [See ASTRONOMY, INFRARED; ASTRONOMY, ULTRAVIOLET SPACE; X-RAY ASTRONOMY.]

Gamma-ray astronomy, on the other hand, is still in its pioneer stage. Gamma rays cannot be focused like light, radio, infrared, ultraviolet, or X-ray photons. Therefore, the signals from the gamma-ray sources are usually weak and the backgrounds high. Gamma rays can penetrate only a few tens of gm cm^{-2} of material, so observations must be taken from near the top of the atmosphere. Satellites and high-flying balloons carry gamma-ray detectors and telescopes. Exceptions are the very-high-energy (VHE) gamma rays above about 10^{11} eV. They may be observed from the ground with large dishes that collect the light from Cerenkov radiation produced by the electron secondaries of the gamma-ray interactions high in the atmosphere. Ultrahigh-energy (UHE) gamma rays, with energies above 10^{15} eV, are detected by scintillators spread on the surface of the earth. They measure the energy loss of electrons and muons in extensive air showers (EAS).

From gamma rays we hope to learn about the high-energy interactions that take place on the stars. Gamma rays from pulsars tell us about the acceleration of electrons and the collisions of protons with hydrogen and other nuclei in the vicinity of neutron stars. We can study the surface composition of neutron stars and their atmospheres, the configuration and magnitudes of pulsar magnetic fields, and the variations of these with time.

Black holes are produced as the cores of supernova explosions by stars with masses greater than about 3 $M_\odot$. Neither the degenerate forces of white dwarfs nor the nuclear forces of neutron stars are sufficient to withstand the crushing force of gravity. So the black hole core contracts indefinitely to a singularity at its center. Of course, we have no information from inside the black hole's event horizon. The gamma rays from a black hole are emitted from heated matter in its vicinity—from particles accelerated to high energies that interact outside the event horizon. Cygnus X-1 has been proposed as a black hole candidate. It has been suggested that the active galaxies, including quasars, may be fueled by black holes in this fashion. And the active regions at the centers of spiral galaxies like our own Milky Way may contain gargantuan black holes that feed on other stars. We want to learn as much as possible about these exotic objects, and gamma-ray astronomy will contribute significantly toward this goal. [See BLACK HOLES (ASTRONOMY).]

Active galactic nuclei (AGN), which include quasars and Seyfert galaxies, appear to radiate a sizable fraction of their energy in gamma rays and thus are targets for gamma-ray astronomy. The extremely high power produced by these objects and the associated explosive jetting suggest that these are some of the most interesting objects in the universe. Contributions toward their understanding by gamma-ray astronomy could greatly benefit our knowledge of the universe.

Historically, the structure of the Milky Way has been studied by optical radiation, which is limited because of absorption by dust in the direction of the galactic center. The 21-cm radio waves from hydrogen atoms are more penetrating and have been successful in mapping the galactic arms. Since gamma rays are more penetrating than other parts of the electromagnetic spectrum, they should be especially useful in exploring the dense regions of the galaxy, particularly the galactic center region.

The energy interval from 1 to 20 MeV has promise of rich rewards from nuclear gamma-ray lines. In many cases, the lines are broadened and shifted by the energetic interactions that produce the lines. The information obtained from the gamma-ray lines could be as important to high-energy astrophysics as light from atomic lines has been to classical astronomy.

III. Gamma-Ray Detectors and Telescopes

Gamma rays, unlike visible light or other parts of the electromagnetic spectrum, cannot be focused. They are not reflected by mirrors and are not converged by lenses. The gamma rays travel in straight lines from the source and diverge according to the inverse square law. Only those gamma rays that penetrate into and interact in the detector are counted. For this reason and because the interaction mean-free paths are long (from a few to a few tens of g cm^{-2}), backgrounds from the atmosphere and material surrounding the detector are high, and the fluxes from the celestial sources are low. Thus progress in gamma-ray astronomy has been slow compared with astronomies at other energies.

A. SODIUM IODIDE (THALLIUM) DETECTOR

A favorite detector from 0.3 to a few MeV has been a scintillation detector such as NaI(Tl). Gamma rays are absorbed by the photoelectric effect, giving up all their energies to the scintillator, part of which is converted to light that is collected and amplified by a photomultiplier; or else the gamma rays are Compton scattered by electrons in atoms in the detector, giving up parts of their energies to the scintillator. The NaI(Tl) detector shown in Fig. 1 was used by the Rice Group in 1972 to discover the electron–positron annihilation radiation in the direction of the galactic center. The detector was surrounded with an active NaI(Tl) collimator with an aperture of 24 deg full-width half-maximum

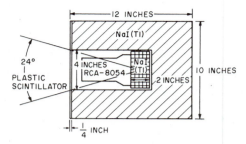

FIG. 1. Schematic diagram of the Rice University NaI(Tl) scintillation detector system. The well-shaped guard crystal rejects charged particles and suppresses Compton scatters by active anticoincidences. The plastic scintillator rejects charged particles from the forward direction but is transparent to γ-rays. [From Johnson, W. N., and Haymes, R. C. (1973). *Astrophys. J.* **184**, 103–125.]

(FWHM). This detector was flown in a gondola suspended from a balloon flying above all but 3 g cm^{-2} of the atmosphere. The detector could be oriented toward any particular right ascension and declination with the gondola system shown in Fig. 2. This type of detector has made many useful measurements but is limited in angular resolution to about 10 deg, energy resolution to about 6%, and is bothered by backgrounds produced by interactions of cosmic-ray protons and their secondaries in the material of the collimator.

B. GERMANIUM DETECTOR

It is possible to obtain considerably better energy resolutions, about 1 keV at 550 keV, with germanium detectors. These are usually cooled to liquid nitrogen temperatures to reduce noise. An example of a 130-cm^3 high-purity Ge detector is shown in Fig. 3. This detector was flown on a balloon by the Bell Labs–Sandia Group to measure accurately the energy of the positron–electron annihilation radiation from the galactic center direction. A NaI active collimator was used. A cold finger extended from a liquid nitrogen tank to the Ge crystal. To date, Ge detectors have been very expensive, so only a few hundred cm^3 have been used in a detector. They have measured very accurate positions and widths of strong narrow lines, but because of their small volumes, their sensitivities are low for wide lines and continuous spectra. Backgrounds of gamma-ray lines produced by cosmic ray–proton interactions in the NaI and Ge can also be a problem.

C. DOUBLE COMPTON SCATTER TELESCOPE

To increase the sensitivity for identifying weak sources from 0.3 to 30 MeV, measure broad lines, and reduce background from cosmic ray–proton interactions with materials

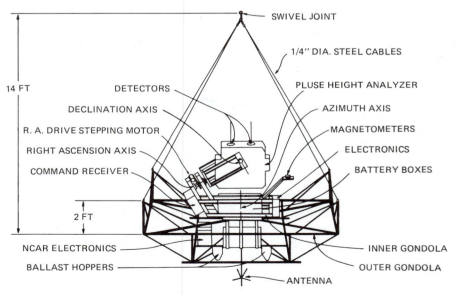

FIG. 2. Rice University balloon-borne telescope orientation system showing the two detectors mounted parallel in an equatorial mount. The swivel joint, connected to the balloon via the recovery parachute, reduces the coupling of the telescope-to-balloon rotation. [From Johnson, W. N., and Haymes, R. C. (1973). *Astrophys. J.* **184**, 103–125.]

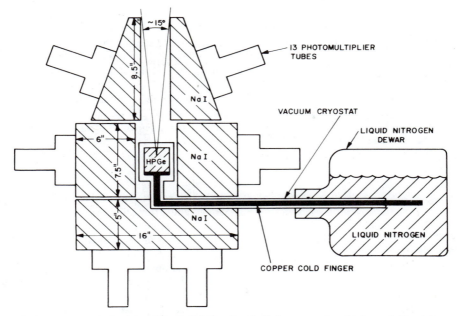

FIG. 3.   Schematic diagram of the Bell Labs–Sandia Labs germanium high-resolution tele-scope. The central detector is a 130-cm³ block of high-purity germanium. [From Levanthal, M., and MacCallum, C. J. (1980). *Ann. N.Y. Acad. Sci.* **336,** 248.]

around the detector, the double Compton scatter telescope is useful. The University of California, Riverside group used the telescope shown in Fig. 4 to measure the gamma rays from the Vela and Crab Pulsars. A gamma ray scatters by the Compton interaction from an electron in the top liquid scintillator tank, continues on, and then Compton scatters from an electron or is absorbed by the photoelectric effect in the second liquid scintillator tank. Each tank is divided into 28 cells, each viewed by a separate photomultiplier. The energy deposits of the electrons are measured in each scintillator. The two cells where the interactions take place are identified to give the scattered gamma-ray direction, and the time of flight of the gamma ray between interactions is measured. Plastic scintillators completely surround each scintillator tank to veto charged particles. Neutrons and upward-moving gamma rays are vetoed by time of flight measured to 0.5 nsec. Since each event is accurately timed to 10 μsec, phase diagrams can be measured even for the shortest period (1.5 msec) pulsar yet discovered. Sufficient information is obtained by this telescope to measure the energy of a gamma ray with an uncertainty of 10 to 20%, depending on the angle of scatter. Its direction is uncertain to a circle on the sky. The position of the gamma-ray source is determined

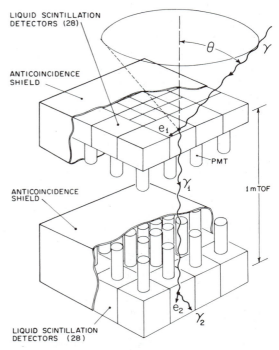

FIG. 4.   Schematic drawing of the UCR double Compton scatter gamma-ray telescope.

from the overlap of circles. The telescope has a wide aperture for sky surveys. And because of its large effective area and accurate timing, it is especially useful for pulsar measurements. Detectors with lots of hydrogen, such as the double Compton scatter telescope, are subject to a background of 2.2-MeV gamma rays. Cosmic-ray interactions in the atmosphere produce neutrons that are slowed by elastic collisions with hydrogen in the detector. When the slow neutrons are captured by protons, the 2.2-MeV gamma rays are emitted.

D. PAIR SPARK CHAMBER

Above a few tens of MeV for low atomic number elements and a few MeV for high, the dominant gamma-ray interaction is electron–positron pair production in the electric field of the nucleus of an atom. The threshold for the reaction is 1.02 MeV, the sum of the electron and positron rest masses. From conservation of energy, the photon energy must equal the pair rest-mass energies plus their kinetic energies. The positron direction is opposite the electron direction in the center of mass system. When transformed into the laboratory system, the pair is emitted in the photon's forward direction; for photon energies $E_\gamma \gg m_0 c^2$, the pair angle is about $m_0 c^2 E^{-1}$. This has two useful consequences. The direction of the incident gamma ray can be determined from the mean of the electron and positron directions (or from the momenta of the electron and positron), and the opening angle is a measure of the gamma-ray energy. The latter is complicated by the scattering of the electron and positron in the high atomic number interaction material. So the scattering itself is also used to measure the energies of the electron and positron.

A schematic diagram of the SAS 2 spark chamber telescope is given in Fig. 5. It is the first satellite spark chamber telescope, a pioneer in celestial high-energy gamma-ray observations, and set the standards for subsequent measurements. The assembly includes 16 wire-grid spark chamber modules above the four central plastic scintillators and 16 below. Each module has two planes of 200 parallel, evenly spaced wires strung across the 25×25 cm opening. Tungsten plates, 0.10 cm thick, separate the spark chamber modules. The spark chamber assembly is triggered by a charged particle that passes through one of the four plastic scintillators and the directional lucite Cerenkov counters immediately below, if there is no coincident pulse in the surrounding plastic scintillator dome. High voltage is then applied to the spark chamber, and a memory scan of the ferrite cores at the end of each wire, set by the spark-induced current, begins. The effective area of the detector is 540 cm^2 and the sensitive aperture about $\frac{1}{4}$ sr. The efficiency for detecting gamma rays is about 0.25, and the timing accuracy is better than 2 msec. The uncertainty in the gamma-ray direction is about 1.5 deg.

IV. Gamma-Ray Observatories

A. BALLOONS

Gamma rays below about 100 GeV cannot be detected from the surface of the earth. Because the earth's atmosphere is about 1000 g cm^{-2} thick and gamma-ray mean free paths vary from a few to tens of g cm^{-2}, celestial gamma rays cannot penetrate to a detector on the ground. Since the 1960s, large zero-pressure balloons

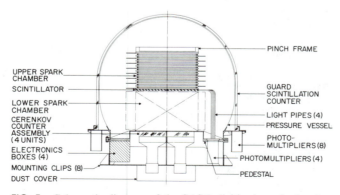

FIG. 5. Schematic diagram of the SAS-2 digitized spark chamber gamma-ray telescope. [From Derdeyn, S. M. *et al.* (1972). *Nucl. Instrum. Methods* **98**, 557–566.]

have been used to carry gamma-ray detectors to within a few g cm^{-2} of the top of the atmosphere. They are typically made of thin, 1-mil, polyethylene filled with 1 million m^3 helium gas at NTP. Detectors that weigh 1000 kg or more are carried for one to two days. Data are telemetered to the launch site at rates up to 50,000 bits sec^{-1}. When the balloon drifts out of telemetry range, the flight is terminated and the payload descends on a parachute. A crush pad under the payload absorbs the shock on landing.

A major limitation of balloon flights has been the short observation times. The longest flights, within telemetry range, are possible at the time of the upper level wind turnaround, at a few g cm^{-2} residual atmosphere, for a couple of weeks in the spring and the fall. Then the high-level winds are at a minimum and variable. At other times, when the winds are more than 100 km hr^{-1}, the balloon speeds out of telemetry range in a few hours. Transcontinental and transatlantic flights have been tried. Down-range telemetry receivers and on-board storage of data are useful and have been successful.

Figure 6 shows a sketch of a typical zero-pressure balloon launch. The balloon and parachute are laid on the ground with the payload suspended from the launch vehicle. The balloon is filled from tubes, not shown, while rolling through the spool held in place by the spool vehicle. At launch the balloon is released from the spool and carried by the ground wind over the launch vehicle. The balloon then lifts the payload from the launch vehicle and climbs to drift altitude. After termination the payload is recovered and trucked back to the launch site.

B. SATELLITES

Satellite observations—which include the discovery of the gamma-ray galactic plane by OSO 3; the Vela and Crab pulsars, other point sources, and the contour of the galactic disk by

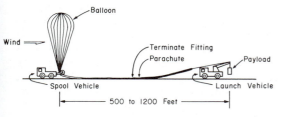

FIG. 6. Schematic drawing of a zero-pressure scientific balloon on the flight line ready for launch. [From National Scientific Balloon Facility. Flight train configuration,'' NSBF Flight Manual.'']

SAS 2; and the detailed measurements of 25 galactic and extragalactic sources by COS B during seven years of observation—have made the major contributions to our knowledge of celestial gamma-ray sources. The satellite contributions have been especially important for the high energies of 35 to 3000 MeV. The successes can be attributed to the long observation times, low backgrounds of the spark chambers, and location of sources to about a degree. Additional information has been obtained by the NaI scintillators on HEAO 1 at gamma-ray energies below about 1 MeV and by the high-resolution Ge detector for gamma-ray lines on HEAO 3. The NaI detector on OSO 7 discovered gamma-ray lines from solar flares, and the one on SMM has studied gamma rays from many flares over several years.

An artist's drawing of the COS B satellite is shown in Fig. 7 with cut-aways to reveal some of the inside components. An incident gamma ray penetrates the thermal blanket, 7, from above and interacts in the spark chamber, 2. The electron and positron produced pass through the trigger scintillators, 3, and deposit their energies in the calorimeter, 4. A few of the important parts of the satellite such as the solar cells, 13, nitrogen attitude control tank, 11, neon flush tank, 12, and spin thruster, 9, along with others are also identified. The mass of the satellite is 300 kg, its diameter 140 cm, and its height 113 cm. The orbit perigee was 350 km, apogee 100,000 km, inclination 90°, and period 37 hr. Measurements were taken outside the intense region of the earth's radiation belt.

C. CERENKOV LIGHT COLLECTORS

At energies above about 10^{11} eV, gamma rays produce electron–positron pairs that grow into showers of electrons and positrons with velocities sufficiently high to produce Cerenkov radiation in the upper atmosphere. The cylinder of Cerenkov light, about 100 m in diameter, reaches the ground; so it may be sampled by dishes on the surface of the earth that focus the light onto one or more photomultipliers. This method was pioneered by the Crimean Russian group in the 1960s. The angular resolution of a gamma-ray point source is about 1 deg. With observation times of about 25 hr on source and comparable times off source, cosmic-ray-induced shower Cerenkov radiation background may be overcome, and statistically significant observations reported from celestial sources.

Improvements have been made by increasing

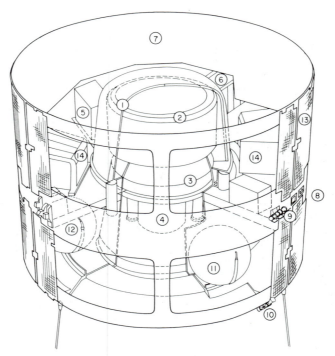

FIG. 7.   Schematic drawing of the COS-B satellite.

sensitivities using larger Cerenkov light collectors, decreasing the cosmic-ray-induced background light, and increasing the observation times. To decrease the cosmic-ray background, the differences in the depth and width of the gamma-ray- and cosmic-ray-induced light pools may be used. The shape of the Cerenkov light pulse may also be exploited to reduce the cosmic-ray background.

Figure 8 is a photograph of two 11-m light collectors located at Sandia Corporation in Albuquerque, New Mexico. They were moved recently to this new location from Edwards Air Force Base, California. These dishes of the United States Department of Energy are used during the day for solar engine tests but are available at night for gamma-ray astronomy. Their large areas and thus high sensitivities permit a reduction in the gamma-ray threshold energy with the resulting increase in photon to cosmic ray Cerenkov light to reduce background. They are located about 100 m apart and may be used in coincidence or in alternate signal–background modes. The ground collectors have the usual advantages of optical telescope observatories: long observation times, accessability, ease of control, convenient data retrieval, and relatively low cost; they have the disadvantage of

poor visibility from clouds or inclement weather, which reduce observation times.

D. EXTENDED AIR SHOWERS

When gamma-ray energies reach about 10^{15} eV some of the charged particles in the shower, mostly electrons, have sufficient energy to reach sea level. Mixed in with the electrons are a few muons. Scintillators, spread over large areas, detect the charged particles in the extended air showers (EAS) by their ionization. Muons interact only weakly with matter, about 10^{-7} that of electrons, so they easily penetrate the 10^3 g cm^{-2} of the atmosphere. Electrons go through many generations before reaching the ground. Cosmic-ray proton or nucleus-induced showers produce a large background. As high energy nucleon-nucleus collisions give large numbers of π° mesons that decay into muons while electrons primarily produce showers with secondary electrons and gamma rays with a few muons, only, the ultra high energy proton showers are expected to be muon rich and the ultra high energy gamma ray showers muon poor.

The direction of the primary gamma ray is determined from the differences in the arrival times of the shower front at the different detec-

FIG. 8.   Cerenkov light collectors located at Sandia Corporation, Albuquerque, New Mexico. They are used for very high-energy gamma-ray observations.

tors. Accuracies of a degree of arc can be obtained for these ultra high energy showers. The first observation of a celestial gamma ray source, Cyg X-3, was reported by the Kiel, West Germany Group for measurements taken from 1976 to 1980. These EAS detectors also have the usual advantages and disadvantages of the ground based optical telescopes.

V. Possible Celestial Gamma-Ray Sources

Celestial gamma-ray astronomy dates back to the discovery in 1967 of gamma rays from the galactic center direction, by George Clark, George Garmire, and William Kraushaar of the Massachusetts Institute of Technology. They used a wide-angle scintillator telescope that detected gamma rays with energies >100 MeV by electron–positron pair production on the satellite OSO 3. During 16 months of observation they detected 621 cosmic gamma rays with ener-

gies above 50 MeV. The gamma rays came primarily from the galactic plane, with a major concentration in the direction of the galactic center and a minor one from the anticenter direction. No point sources were identified, and cosmic-ray interactions with the hydrogen gas in the galactic plane provided a satisfactory explanation of the gamma rays. The detector for this pioneer work was necessarily small with an efficiency of only 2.5 cm^2 sr and angular resolution of about 15 deg half-width at half-maximum (HWHM). Balloon flights followed, with large spark chambers but with short flights of a day, that confirmed the galactic gamma rays with energies $E > 50$ MeV.

The next major contribution to gamma-ray astronomy came from the larger higher sensitivity spark chamber flown on the second small astronomy satellite SAS 2, from November 15, 1972 to June 1973 by Carl Fichtel, Don Kniffen, David Thompson, Robert Hartman, and others from NASA Goddard Space Flight Center. This detector, sensitive to gamma rays from about 35 to 200 MeV, had the advantages of large area (640 cm^2); better angular resolution (2 deg for individual photons and 0.5 deg for local sources), and lower background. Detailed observations of the celestial sphere, concentrated in the region of the galactic plane, found that most of the gamma-ray sources, excluding point sources, were confined to the galactic plane a few degrees thick with a broad maximum in the direction of the galactic center. Four strong gamma-ray sources—the Vela pulsar PSR 0833-45; the Crab pulsar PSR 0833+21 and the Crab Nebula; 195+5 called Geminga; and Cyg X-3, an X-ray source in the Cygnus region—were discovered along with some weaker sources. Unfortunately, this productive flight was terminated prematurely when a power system failed.

Much of the gamma-ray astronomy information to date for $E > 50$ MeV came from seven years of observations with a spark chamber on the European satellite COS B launched by NASA on August 9, 1975. The collaboration included scientists from the Max Planck Institut für Extraterrestrische Physik, Garching, West Germany; Cosmic Ray Working Group, Leiden, Netherlands: Istituto di Scienze Fisiche, Universita di Milano with the Istituto Fisica, Universita di Palermo, Italy; Service d'Electronique Physique Centre d'Etudes Nucleaires de Saclay, Gif-sur-Yvette, France and the Space Science Department, ESTEC, Noordwijk, Netherlands. Their spark chamber was similar to that of SAS 2, but with the addition of a calorimeter for recording the energy of the electron–positron pairs to estimate the energy of the incident photon. The spark chamber had a maximum area–efficiency product of 50 cm^2 that peaked a 500 MeV and was sensitive from 70 MeV to 5 GeV. The angular resolution varied from 7 deg FWHM at 70 MeV to 2 deg FWHM at 5 GeV. The fractional energy uncertainty varied from 0.5 at 70 MeV to 1.0 at 5 GeV, with a minimum of 0.4 at 150 MeV. The background trigger rate in orbit was about 0.2 counts/sec. The strongest sources gave count rates of 1 photon/hr and the weakest about 1 photon/d.

The gamma-ray flux at energies of 70 GeV is given in Fig. 9. The lower graph shows the on-axis counts sec^{-1} sr^{-1} versus galactic longitude. The data points are shown with the error bars, and the solid line is the section through a fitted surface along the galactic equator. The positions of the two gamma-ray pulsars, PSR 0531+21 and PSR 0833-45, and two radio pulsars are identified by the arrows. The shaded area at the bottom of the graph represents the background. The middle figure is a two-dimensional gamma-ray count rate map of the galactic plane. The counts sec^{-1} sr^{-1} are presented as a gray scale where the white region has the highest count rate. The highest intensity regions are identified as the galactic center, Vela, and the anticenter region with the Crab and Geminga. The upper curves are the counts sec^{-1} sr^{-1} versus galactic latitude for 12 cuts through the galactic plane. The longitude integration ranges are shown at the top of the graph.

The picture of the galactic plane that emerges is a narrow source of gamma rays, a few degrees wide in latitude, that is present at all longitudes. There is a broad maximum in longitude at the galactic center that extends from about 60 to 300 deg longitude. Structure is resolved both in the SAS-2 and the COS-B profiles. This has been attributed to the cosmic-ray protons, nuclei, and electrons interacting with the matter in the galactic arms and in molecular clouds. These gamma rays come from the π° meson decay to two gamma rays, each of 70 MeV in the π° meson rest frame, Doppler-shifted to higher and lower energies, and from electrons that emit gamma rays through Bremsstrahlung. The gamma rays are tracers of the matter in the galaxy and of the cosmic rays contained by the magnetic fields of the galaxy.

In addition to the broad galactic features associated with the galactic arms and molecular clouds are the sharp peaks of the point sources. The most intense gamma-ray source observed for $E > 50$ MeV by SAS 2 and COS B is the Vela pulsar PSR 0833-45. Two strong sources, the Crab pulsar PSR 0531+21 and Geminga, 2CG 195+04, are located near the galactic anticenter direction. There is also a prominent peak in the Cygnus region.

Possible gamma-ray sources reported in the literature are summarized in Table I. The first 25 entries are the pointlike gamma-ray sources identified by COS B in their 2C catalog. Of these, PSR 0531+21, PSR 0833-45, CG 195+5, CG 312-1, and CG333+0 were first discovered as gamma-ray sources by the SAS 2 group. The remainder were discovered by COS B. The additional entries include reported gamma-ray pulsars alone and in binaries, Seyfert galaxies, radio galaxies, a normal spiral galaxy, and a black hole. Extended regions such as SN nebula, interstellar clouds and belts, galactic center gamma-ray lines, galactic lines, gamma-ray

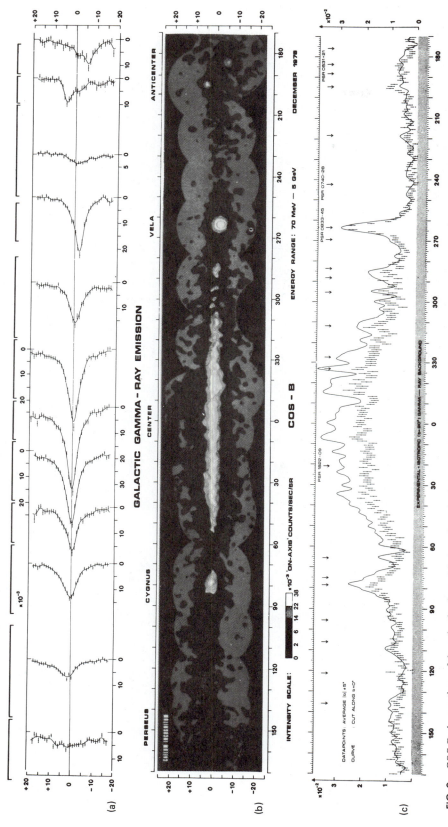

FIG. 9. COS-B observations of the galactic plane: (a) gives the galactic latitude distributions of intensities of gamma rays at different galactic longitudes, (b) gives the contours of intensities, and (c) gives the longitude distribution of intensities. [From Mayer-Hasselwander, H. A. *et al.* (1980). *Ann. N.Y. Acad. Sci.* **336**, 211.]

TABLE I. Astronomical Gamma-Ray Sources

#	Source	R.A. hr min	δ (deg)	ℓ (deg)	b (deg)	Error Radius (deg)	Flux E>100 MeV (photon cm^{-2}s^{-1})	Period	Approx Dist. (Kpc)	Radio	Visible	1-20 KeV	20-300 KeV	0.3-1 MeV	1-30 MeV	0.03-3 GeV	≥10^{12}eV	>10^{15}eV	Verified for E>1 MeV	compelling γ-ray	compelling Identification	Identification	Source	Comments
1	2CG006-00	1800	-23.4	6.7	-0.5	1.0	2.4x10^{-6}			o	o	o	o	o	o	x	o	o	o	x	o			1
2	2CG210-51	1617	-2.8	10.5	-31.5	1.5	1.2			o	o	o	o	o	o	x	o	o	o	x	o			2
3	2CG013+00	1810	-16.8	13.7	+0.6	1.0	1.0			o	o	o	o	o	o	x	o	o	o	x	o			3
4	2CG036+01	1850	+3.8	36.5	+1.5	1.0	1.9			o	o	o	o	o	o	x	o	o	o	x	o			4
5	2CG054+01	1923	+19.6	54.2	+1.7	1.0	1.3			o	o	o	o	o	o	x	o	o	o	x	o			5
6	2CG065+00	1953	+29	65.7	0.0	0.8	1.2	6.13ms 120 d	3.5	x	o	o	o	o	o	x	x	o	o	x	o	PSR 1953+29	magnetic n star In binary	6.1ms radio pulsar 6
7	2CG075+00	2019	+36	75.0	0.0	1.0	1.3			o	o	o	o	o	o	x	o	o	o	x	o			7
8	2CG078+01	2024	+41	78.0	+1.5	1.0	2.5			o	o	o	o	o	o	x	o	o	o	x	o			8
9	2CG095+04	2115	+55	95.5	+4.2	1.5	1.1			o	o	o	o	o	o	x	o	o	o	x	o			9
10	2CG121+04	0025	+70	121.0	+4.0	1.0	1.0			o	o	o	o	o	o	x	o	o	o	x	o			10
11	2CG135+01	0238	+62	135.0	+1.5	1.0	1.0			x	x	x	o	o	o	x	o	o	o	x	x	1E0236+610 possibly		not QSO 11
12	2CG184+05	0531	+21	184.5	-5.8	0.4	3.7	33.5ms	2.0	x	x	x	x	x	x	x	x	o	x SAS 2	x	x	PSR 0531+21 Crab Pulsar	magnetic n star	12
13	2CG195+04	0633	+19	195.1	+4.5	0.4	4.8	59s		x	x	x	x	x	x	x	x	o	x SAS 2	x	x	1E0630+178		Geminga 13
14	2CG218-00	0657	-5.1	218.5	-0.5	1.3	1.0			o	o	o	o	o	o	x	o	o	o	x	o			14
15	2CG235-01	0729	-20.4	235.5	-1.0	1.5	1.0			o	o	o	o	o	o	x	o	o	o	x	o			15

Source	R.A. hr min	δ (deg)	ℓ (deg)	b (deg)	Error Radius (deg)	Flux E>100 MeV (photon cm^{-2}s^{-1})	Period	Approx Dist. (Kpc)	Radio	Visible	1-20 KeV (x-ray)	20-300 KeV	0.3-1 MeV	1-30 MeV (γ-ray)	0.03-3 GeV	>10^{12}eV	>10^{15}eV	Verified for E>1 MeV	γ-ray compelling	Identification compelling	Identification	Source	Comments
2CG263-02	0833	-45	263.6	-2.5	0.3	13.2	89.2 ms	0.5	X	X	X	O	X	X	X	X	X O	X SAS 2	X	X	PSR 0833-45n Vela Pulsar	Magnetic star	16
2CG284-00	1022	-58	284.3	-0.5	1.0	2.7			O	O	O	O	O	O	X	O	O	O	X	O			17
2CG288-00	1048	-60	288.3	-0.7	1.3	1.6			O	O	O	O	O	O	X	O	O	O	X	O			18
2CG289+64	1226	+02	289.3	+64.6	0.8	0.6		8.6×10^5	O	X	X	O	O	O	X	O	O	X	X	X	3C273 Quasar 4U1226+02		19
2CG311+01	1404	-63	311.5	-1.3	1.0	2.1			O	O	X	O	O	O	X	O	O	X SAS 2	X	O			20
2CG333+01	1617	-50	333.5	+1.0	1.0	3.8			O	O	X	O	O	O	X	O	O	X SAS 2	X	O			21
2CG342-02	1705	-44.5	342.9	-2.5	1.0	2.0			O	O	O	O	O	O	X	O	O	O	X	O			22
2CG353+16	1627	-24.7	353.3	+16.0	1.5	1.1		5	X	X	O	O	O	O	O	O	O	O	X	X	ρ Oph cloud	Interstellar cloud. C.R. Interactions	23
2CG356+00	1733	-31.7	356.5	+0.3	1.0	2.6		5	O	O	O	O	O	O	O	O	X	O	O	O		Seen once by COS B. Could be variable	24
2CG359-00	1744	-29.7	359.5	-0.7	1.0	1.8		5	O	O	O	O	O	O	O	O	O	O	O	O			25
PSR 1937+21	1937	+21	57.1	-0.5			1.56ms		X	O	O	O	O	O	O	O	O	O	O	O			26
Her X-1	1656	+35	57.6	37.5			1.24s 1.7d 135d	5	O	X	X	O	O	O	X	X	O	O	O	O	Hz Hercules companion	Magnetic star in binary	27
4U 0115+63	0115	+63					3.61s 24.3 d	>11.4	O	X	X	O	O	O	X	X	O	O	O	O		magnetic n star in binary	Sporadic x-ray binary 28
Cyg X-3	2031	+41	80.1	0.7			4.8 hr 12.6 ms		O	X	X	O	O	O	X SAS2 O COSB	X SAS2 O COSB	X O	X SAS 2 O COSB	O	O		Compact object In binary	29
Vela X-1	0900	-40	262.8	4.1			283 s 8.96 d	1.4	X	O	X	O	O	O	O	O	X	O	O	O		Compact object In binary	30

(continues)

TABLE I *(Continued)*

	Source	R.A. hr min	δ (deg)	ℓ (deg)	b (deg)	Error Radius (deg)	Flux E>100 MeV (photon cm^{-2}s^{-1})	Period	Approx dist. (kpc)	Radio	Visible	1–20 KeV x-ray	20–300 KeV x-ray	0.3–1 MeV	1–30 MeV γ-ray	0.03–3 GeV	<10^{12}eV γ-ray	<10^{15}eV	Verified for E≥1 MeV	γ-ray compelling	Identification compelling	Identification	Source	Comments	
		3	4	5	6	7	8	9	10	11	12	13	14	15	16	17 18	19		20	21	22	23	24	25	
31	Cyg X-1	1956	+35						2.5	X	X	X	X	X	X	O O	O		O	O	O	HDE 226868 optical	Black Hole	Black Hole	31
32	Galactic Center	1706	−29	0.0	0.0					X	O	X	X	X	X	X O	O		X	X	X	Galactic Center	Black Hole	0.511 MeV e$^+$ e$^-$ line 0.5 yr variability	32
33	Galactic line 26Aℓ 1.809 MeV									−	−	−	−	−	X	− −	−		X	X	X	Mostly gal. plane	SN, Novae, M.S. stars	Apparently SN source is factor 10 to low	33
34	SS 433	1921	+05	39.7	−2.2			1.70 d / 13 d / 164 d		X	X	X	X	X	O	O O	O		O	O	O	V1343 Aql companion	Relativ. jets. Binary star		34
35	GBS0526-66	0526	−66	276.0	−33.3					O	O	O	X	X	X	X O	O		X	O	O	N49 LMC Direction	Neutron Star	γ-ray burst, repeated bursts 0.4 MeV line	35
36	Crab Nebula	0531	+21	184.5	−5.8				2.0	X	X	X	X	X	X	X X	X		X	X	X		SN Remnant		36
37	Orion Cloud								0.45	O	O	O	O	O	X	X O	O		O	O	X				37
38	Gould Belt									X	X	O	O	O	O	O O	O		X	X	X				38
39	Dollidze System									X	X	O	O	O	O	O O	X		O	O	O				39
40	NGC 4151	1300	+40	115.5	77.2				2×10^4	X	X	X	X	X	O	O O	O		O	O	O		Seyfert Galaxy		40
41	MCG 8-11-11	0551	+47	165.2	10.7				1.2×10^5	X	X	X	X	X	X	X O	O		O	O	O	1M0600+46	Seyfert Galaxy		41
42	NGC 5128	1324	−42	310.0	20.1				4.4×10^3	X	X	X	X	X O	O	X O	O		O	O	O	Centaurus A	Radio Galaxy		42
43	LMC X-4	0532	−66	275.9	−32.7			13.5 s / 1.41 d	50	O	X	X	X	O	O	O O	X		O	O	O	07111-V star companion Ion to n star	Compact Object In Binary		43
44	M 31 Andromeda	0041	+41	121.3	−21.6				670	X	X	X	X	O	O	O O	O		O	O	O	Sb spiral galaxy	Normal galaxy		44

transient lines, and gamma-ray bursts are also included. Many of these sources have not been verified, and the reliabilities of several are not high. It is not possible to give a complete evaluation of the validity of each source here, but an attempt is made to give the reader a feel for the reality of each. For the first 25 entries of Table I the source name is given in column 2. The 2CG designates the second COS B catalog of gamma-ray sources. The first three digits give the galactic longitude of the source in degrees and the last two digits the galactic latitude in degrees (either + or −). The right ascension (RA) in hours and minutes and the declination (δ) in degrees are listed in columns 3 and 4; columns 5 and 6 give the galactic longitude and latitude in degrees. In column 7, the errors in source locations are given in degrees; and in column 8 the fluxes for energies $E > 100$ MeV are given. In column 9, pulsar rotation and/or binary orbit periods are listed. In column 10, distances to the sources are estimated. In columns 11–19, the electromagnetic spectrum is divided from radio to ultrahigh energy (UHE) gamma rays. An X indicates the source has been seen at that energy band and a 0 means it has not. Columns 20, 21, and 22 evaluate the reliability of the observation. An x in column 20 indicates that two or more convincing observations at energies >1 MeV agree and in columns 21 and 22 indicates whether the gamma-ray source existence is compelling and whether its identification with a source at other energies is compelling. When identified with another source, its name is given in column 23 and the type of source in column 24. Comments are given in column 25. Entries are made in the table only if information is available. The reader is cautioned that many of the

sources and entries for the sources are likely to change as more and better observations with superior telescopes and instruments are carried out.

The COS-B sources are plotted in Fig. 10 on a projection of the galaxy with both the latitude and longitude in degrees. The solid circles signify sources with fluxes >1.3×10^{-6} photon cm^{-2} sec^{-1}, while open circles specify those with less. Twenty-two of the 25 sources lie within 6 deg of the galactic plane. The three sources off the plane are among the weakest observed. One is the quasar 3C273, the second a molecular cloud ρ Oph, and the third 2CG010−31, has not been identified.

A. Galactic Sources

1. Pulsars

The gamma-ray sources most easily identified because of their pulsed period signatures, are the pulsars. Their periods are so precise and variations in time so predictable there can be no mistake in their identification. The pulsars were discovered with radio waves, for which a significant intensity could be built up in light curves of intensity versus phase in a few hundred periods. Phase is measured in fractions of a period. For gamma rays with energies of about 1 MeV to about 1 GeV, it has been difficult to discover pulsars with periods of fractions of seconds. Therefore, searches for gamma-ray pulsars have used previously identified radio or X-ray pulsars whose periods and period derivatives were well established. Searches have been made at $E > 50$ MeV by SAS 2 and COS B and at energies of 1 to 30 MeV by the Max Planck Institut für Extra-

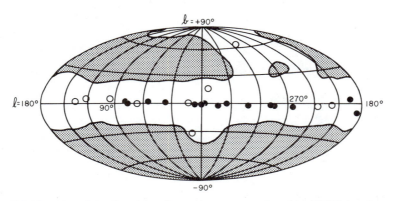

FIG. 10.   Plot of the directions of gamma-ray sources found by COS B in galactic coordinates. No observations were taken in the stippled region. [From Bignami, G. F., and Hermsen, W. (1983). *Annu. Rev. Astron. Astrophys.* **21**, 67.]

terrestrische Physik, Garching, West Germany. However, to date only two gamma-ray pulsars have been confirmed. These are the Crab Pulsar PSR 0531+21 and the Vela Pulsar PSR 0833−45.

A surprising development during 1984–1985 were the reports of Cerenkov detector observations of pulsed gamma rays of $E > 10^{12}$ eV from the Crab and Vela gamma-ray pulsars, from the very fast radio pulsars PSR 1937+21 and PSR 1953+29, from the X-ray pulsars Her X-1 and 4U0115+63, and from Cygnus X-3, a new pulsar not previously seen at radio, X-ray, or any other energy. Verification is required by additional observations before these results are considered compelling.

a. The Crab Pulsar PSR0531+21 (2CG184+05). The Crab pulsar is one of the most studied objects in the sky. It has been observed throughout the electromagnetic spectrum from radio waves of 100 MHz to gamma rays of $E > 10^{12}$ eV, covering more than 10^{17} orders of magnitude in photon energies. The light curves for 430-MHz radio waves, optical, 1.5–10-keV soft X-rays, 18–163-keV medium X-rays, 100–400-keV hard X-rays, 1–20-MeV gamma rays, and >50-MeV gamma rays are shown in Fig. 11. In many respects, the light curves are similar at all energies. The first pulse peaks are aligned at

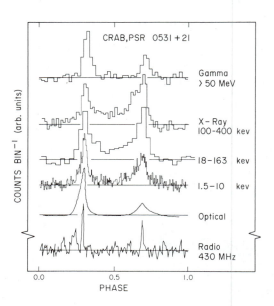

FIG. 11. Phase diagram of the pulsed radiation from the Crab pulsar 0531+21 at six different energies from radio waves to gamma rays. [From Bignami, G. F., and Hermsen, W. (1983). *Annu. Rev. Astron. Astrophys.* **21**, 67.]

the same phase as the radio waves, and the second pulse peaks all follow by the same 0.42 in phase. However, the shapes of the pulse peaks, their relative heights, and the region between peaks vary somewhat with photon energy. The inner pulse region (between the peaks) appears highest at energies of tens and hundreds of keV.

The Crab pulsar photon flux variation with energy has the form $6.17 \times 10^{-4} E^{-2.2}$ photons cm^{-2} sec^{-1} MeV^{-1} over at least the energy range from 50 keV to 5 GeV, five decades of energy. To change from photons to energy, we multiply the photon flux by E to obtain the power flux in units of energy cm^{-2} sec^{-1} MeV^{-1}. To obtain the right-hand ordinate of Fig. 12, we multiply by E again to limit the vertical range of the graph. The ordinate on the left-hand side is in frequency units. The ordinate range in energy (or frequency) is 16 decades. In the energy region where this plot is reasonably flat, from about 5 keV to about 5 GeV, the power flux at the earth is about the same over equal intervals of log E; (i.e., the power arriving from 10 to 100 keV is about equal to that arriving from 10 to 100 GeV). If the photon flux had varied as $E^{-2.0}$, this would, of course, be exactly true. The power flux in the pulsed radiation is particularly low at radio frequencies of 10^{9} Hz and climbs into the optical region but is still considerably below the value at 10 MeV.

The luminosity of the Crab pulsar is calculated in any photon energy range by integrating the power flux over the energy interval and multiplying by $4\pi r^2 f$, where r is the distance to the pulsar and f the fraction of the 4π solid angle fed by pulsar beaming. If the pulsar emits its radiation into a solid angle of 1 sr at each of its magnetic pulses, as shown in Fig. 13, the luminosity of the Crab pulsar is about 2×10^{35} erg sec^{-1} in each two decades of energy from 5 keV to 5 GeV, with the maximum in the region of 10 MeV. This is about 0.05% of total rotational kinetic energy loss, the spin-down energy loss, of the Crab pulsar.

Since COS-B observations extended over seven years, it was possible to measure changes in the gamma-ray light curve for $E > 50$ MeV over that time period. The six light curves from August–September 1975 to February–April 1982 along with one from a balloon flight at energies of 1 to 30 MeV are shown in Fig. 14. Visual observation suggests that the second pulse decreases relative to the first to a low in 1979 then increases again to 1982. Statistical tests verify this variation.

Large dishes on the earth are used to collect

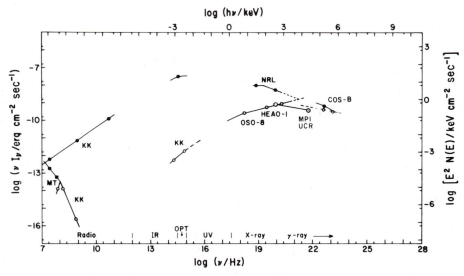

FIG. 12. Log–log plot of the energy distribution of gamma rays per decade of energy versus energy of the gamma ray for the Crab on the right ordinate and upper absissa. Frequency units are used on the left ordinate and bottom absissa. Solid circles represent observations of the Crab nebula, open circles the Crab pulsar and solid squares the sum of the nebula and pulsar. (Groups reporting the observations are: COS B, satellite, 1980; HEAO 1, satellite, 1981; NRL, balloon, 1981; OSO 8, NASA Goddard, satellite, 1981; KK, radio summarized by Kundt and Krotscheck, 1980; MT, radio, Manchester and Taylor, 1977; MPI, balloon, 1982; UCR, balloon, 1984.) [Adapted from Knight, F. K. (1981). PhD Dissertation, University of California, San Diego, UCSD-SP-81-13.]

Cerenkov light from electrons in showers produced by the interactions of VHE gamma-rays with $E > 10^{12}$ eV at the top of the atmosphere. The light arrives at the collectors as a pancake about 200 m in diameter and 5 (nsec) thick. The light is collected by photomultipliers at the focus of the collectors. Sources can be located to about 1 deg. Groups that have reported pulsed radiation from PSR 0531+21 include Dugway, Utah, USA; Ooty, India; and Edwards Air Force Base, California. One group reported that the signal persisted for three months while others saw the signals only occasionally; Whipple Observatory found no pulsed signal although they reported an unpulsed flux from the Crab Nebula. Tien Shan also reported unpulsed VHE gamma rays.

Extensive air showers detected by scintilla-

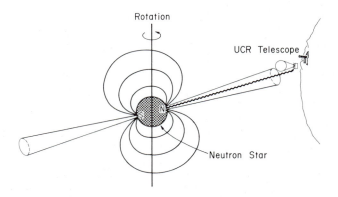

FIG. 13. Schematic drawing of a pulsar. As the neutron star rotates about its axis, the gamma rays are beamed along the magnetic pole direction.

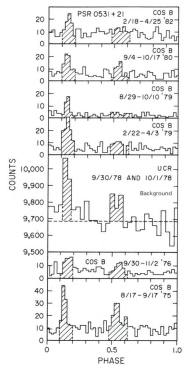

FIG. 14.   Phase plots of the Crab pulsar PSR 0513+21 at various times from 1975 to 1982. The first and second pulses are indicated by the shaded regions. [From White, R. S. *et al.* (1985). *Astrophys. J.* **299**, L23–L27.]

tors placed on the ground over extended areas are used to detect the UHE gamma rays, with $E > 10^{15}$ eV. Pulsed gamma rays have not been reported by any group using this technique; however, observations from Tien Shan gave a steady flux. Only upper limits were reported by Havarah Park, England. The VHE and UHE observations are not yet compelling.

b. The Vela Pulsar, PSR 0833−45 (2CG 263−02). The Vela pulsar is the benchmark of gamma-ray astronomy. As the strongest of the gamma-ray sources, it was used by COS B for onboard calibration. Its light curves, shown in Fig. 15, differ in phase in different parts of the electromagnetic spectrum. The radio signal at 2295 MHz has only a single pulse. The first pulse of gamma rays is delayed by 0.12 and the optical by 0.23 in phase from the radiopulse. The gamma-ray phase separation between the first and second pulses is 0.42, the same as for the Crab pulsar, but is only 0.25 for the optical pulses. Upper limits only exist from measurements of the soft and hard X-rays. Recent results from the University of California, River-

side (UCR) double scatter telescope flown on a balloon at an altitude of 40 km have shown that the light curve from gamma rays of 1 to 30 MeV from double scatters is similar in phase to the one for 50 to 5000 MeV. Using single scatters, the UCR group found no differences down to 0.3 MeV. The Vela pulsar gamma-ray light curve, as for the Crab pulsar, gives significant interpulse signal, especially above 50 MeV. There is also a slight tail following the second pulse. The light curves for $E > 50$ MeV measured by COS B over 6 yr, in contrast to the Crab pulsar, have not changed (Fig. 16). The curves from October 1975 to November 1979 are identical within statistics. The ratio of the second to the first pulse is constant in time.

The upper limits in the X-ray region guarantee that the photon flux distribution must turn over

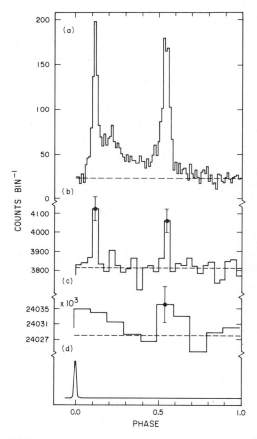

FIG. 15.   Phase diagrams for gamma rays of (a) 50–6000 MeV, COS B; (b) 1–30 MeV, double scatter, UCR; (c) 0.3–1.5 MeV, single scatter, UCR. All were timed absolutely relative to (d), the 2295-MHz radio light curve. [Reprinted with permission from Tumer, O. T. *et al.* (1984). *Nature* **310**, 214–216. Copyright 1984 Macmillan Journals.]

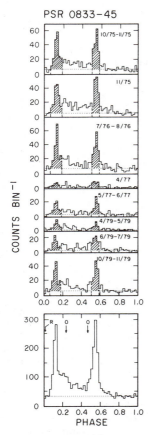

PSR 0833−45

COUNTS BIN⁻¹

10/75−11/75

11/75

7/76−8/76

4/77

5/77−6/77

4/79−5/79

6/79−7/79

10/79−11/79

PHASE

FIG. 16. Gamma-ray phase diagram of the Vela pulsar PSR 0833-45 in the energy interval of 50–3000 MeV at the epochs indicated. The shaded areas mark the first and second pulses, and the horizontal broken lines show the background levels. The bottom graph is the phase diagram summed over all times. The arrows indicate the phases of the radio (R) and optical (O) peaks. [From Hermsen, W. (1982). *Space Sci. Rev.* **36**, 61.]

gamma rays of energy 1 GeV. The luminosity in gamma rays between 10 MeV and 1 GeV, two decades, using the same method and assumptions as for the Crab, is about 2×10^{34} erg/sec. This is a surprisingly large 0.3% of the spin-down energy loss in the rotational energy of the pulsar.

The Narrabri (Australia) Observatory group carried out observations of PSR 0833−45 with Cerenkov light collectors at $E > 3 \times 10^{11}$ eV during 1972 and 1973. One phase pulse was seen at the 3σ significance level in 1972 but not in 1973, unless only the highest pulse height events were selected. The authors consider their results upper limits only. The Tata Institute (India) group with Cerenkov collectors at comparable energies reported the first and second pulses at significance levels of 4.4 and 2.5 σ with a phase separation of 0.42 as seen at lower energies. Their observations were carried out for 35 hr in February–March 1979. Neither group measured the absolute phases. Additional measurements are needed to establish PSR 0833−45 as a source of VHE and UHE gamma rays.

c. The 1.6-msec Pulsar PSR 1937+21. The millisecond pulsar with a period of $0.0015578064488724 \pm 21$ sec is the shortest period pulsar revealed to date. It was discovered in a search at 1400-MHz radio waves with the 305-m antenna at the Arecibo Observatory on December 25, 1982. Its time rate of change of period $\dot{P}$ is $1.05110 \pm 0.00008 \times 10^{-19}$ sec sec⁻¹. Its distance is estimated to be 5 kpc and its magnetic field 5×10^8 G. The low values of magnetic field and period suggest that it is a very old pulsar that has been spun up by accretion from a former companion. The phase curve has two phase maxima separated by about 0.5 in phase, similar to the Crab and Vela pulsars. The pulsed radiation has not been detected in the optical region nor any other part of the electromagnetic spectrum except for the recent report of pulsed VHE gamma rays by the Durham, England group.

d. Hercules X-1. In 1973, Her X-1 was found, by the UHURU satellite X-ray group, to be an X-ray pulsar with a period of 1.24 sec in an eclipsing binary with an orbital period of 1.7 d. Evidence exists for a 35-d amplitude variation of unknown origin. A hard X-ray line was found with a balloon observation using NaI detectors by the Garching Max Planck Institut and Tubingen X-Ray Group at 58 keV. It was attributed to electron cyclotron emission in the strong

from the $6.70 \times 10^{-4} E^{-1.89}$ photons cm⁻² sec⁻¹ MeV⁻¹ found at energies above 50 MeV. Indeed, the UCR group found $3.6 \times 10^{-4} E^{-1.6}$ photons cm⁻² sec⁻¹ MeV⁻¹ at energies of 1 to 30 MeV and a possible sharp decrease in slope with the single scatters down to 0.3 MeV. This information is displayed in Fig. 17 where the power flux for the Vela pulsar multiplied by E is plotted against energy as for Fig. 12. The contrast between the Crab and Vela pulsars is dramatically exhibited in the comparison.

At optical energies the Vela pulsar is the weakest source ever detected and is a factor of 1000 fainter than the Crab pulsar. And the Vela pulsar reaches maximum power at high-energy

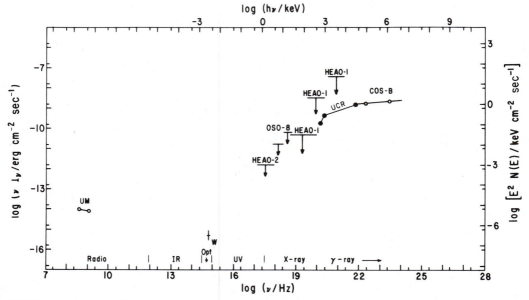

FIG. 17.   Log-log plot of the energy distribution of gamma rays per decade versus energy of the gamma ray for the Vela pulsar on the right ordinate and upper absissa; frequency units are used on the left ordinate and bottom abscissa. (Groups reporting observations are: UM, U. Massachusetts Pulsar List; COS B, satellite, 1980; W, Anglo-Australian Telescope, Siding Spring, Australia; HEAO 2, satellite, 1979; OSO 8, NASA Goddard, satellite, 1981; HEAO 1, satellite, 1981; UCR, balloon, 1984.) [Adapted from Knight, F. K. (1981). PhD Dissertation, University of California, San Diego. UCSD-SP-81-13.]

magnetic field of 5×10^{12} G of a rotating neutron star. The linewidth was less than 12 keV, and the flux was 3×10^{-3} photons cm^{-2} sec^{-1}. Later balloon observations suggested either an emission line at about 52 keV or absorption line at about 38 keV. Upper limits only were given for a possible second harmonic line. The line has been confirmed by the UCSD–MIT hard X-ray experiment on HEAO 1, the NASA–GSFC group with a high-resolution germanium detector, and by several other hard X-ray groups. The companion, HZ Her, has been identified at optical wavelengths. Her X-1 has not been detected as a radio or optical pulsar, a gamma-ray pulsar, or a steady source between 100 keV and 10^{11} eV.

Her X-1, like the binary pulsars PSR 1953+29 and 4U0115+63, was reported as a source of pulsed gamma rays above 1.5×10^{11} eV by the Whipple Observatory group, above 10^{12} eV by the Durham (Dugway) group, and above 3×10^{14} eV by the Utah Fly's Eye group. Each group observed short-time pulsed gamma rays near the X-ray pulsar frequency of 1.2378 sec. The Durham group reported a 3-min interval on April 17, 1983; the Fly's Eye agoup a 40-min interval on July 11, 1983 which was observed at the same time by the Durham Group with a negative result; and a 28-min interval on April 4, 1984 and

one night's observation on May 5, 1984 by the Whipple Observatory group. While the data are suggestive of pulsar gamma rays, the lack of a positive signal over most of the observing time by all groups and the lack of confirmation of pulsed gamma rays during the 40-min interval on July 11, 1983 do not give the required confidence in the results; additional observations are necessary to establish conclusively Her X-1 as a VHE gamma-ray source.

e. The 6-msec Pulsar, PSR 1953+29 (2CG065+00).   The 6.13317-msec pulsar was discovered in radio waves of 430 MHz at Arecibo by Boriakoff, Buccheri, and Fauci and was identified with the COS-B gamma-ray source 2CG065+00. It is in a binary system with an orbit period of 120 ± 4 d. Its pulsed radiation was not detected by COS B or by observers in any other part of the electromagnetic spectrum until the recent report of the Durham group with Cerenkov detection at Dugway Utah at $E > 10^{12}$ eV. One pulse only was seen in the phase curve in the radio and VHE gamma rays. The absolute phase of the VHE gamma rays could not be compared with the radio absolute phase. Additional observations are needed to confirm the pulsed gamma rays from this source.

f. 4U0115+63. The pulsating binary X-ray source discovered by UHURU observations has a 3.61-sec pulsar period, a 24.3-d orbital period, and a 15th magnitude B-type optical flux. Its phase diagram has one broad pulse in soft and hard X-rays. In 1978, the UCSD–MIT hard X-ray group found an absorption line at 20.1 ± 0.5 keV with a width of 3.1 ± 0.5 keV. This implies a magnetic field at the pulsar of 2×10^{12} G. A gravitational redshift of 5–40% might be expected. No pulsed radiation at any energy other than soft and hard X-rays had been observed until the 1985 report of the Durham group of pulsed VHE gamma rays with $E > 10^{12}$ eV. By scanning over a period between 3.613–3.616 sec, they found a flux of $7 \pm 1.4 \times 10^{-11}$ photons cm^{-2} sec^{-1} at a period of 3.61457 ± 0.00001 sec. They reported no detectable variability in pulsed flux over a time of 9 d, September 21–29, 1984. The absolute phase was not measured. Certification of 4U0115+63 as a VHE gamma-ray source awaits confirmation by other observations.

g. Cygnus X-3. Cygnus X-3 was discovered with soft X-rays in a rocket flight in 1966. It was the third most intense X-ray source found in Cygnus and is located close to the cross in the constellation. Its 4.8-hr period was measured by detectors on the UHURU satellite beginning in 1970. In 1972 a large radio burst was observed, followed by a second in 1982 and a third, exceptionally large, in October 1983.

The SAS-2 gamma-ray group first reported gamma rays with the 4.8-hr period in 1977. Their energies were >35 MeV. COS B first confirmed then rejected the period; but now with 7-yr data from seven different observation times, they are firmly convinced no significant periodic signal or steady signal from Cygnus X-3 is seen. Nor has the periodic signal been observed at radio or optical energies or gamma rays below 10^{11} eV.

The Crimean Cerenkov group, Russia, reported gamma rays with $E > 10^{12}$ eV with the 4.8-hr period and a maximum in the phase diagram with respect to X-rays at 0.2 phase. A 5-σ significance was claimed for observations summed from 1972–1980. The Whipple Observatory group confirmed the period, but the maximum moved to 0.75. The Edwards Air Force Base observations confirmed but at a phase of 0.6. The Durham group at Dugway found the maximum at a phase of 0.62. In 1983 the Kiel group, using observations of EAS at $E > 10^{15}$ eV from 1976 to 1980, reported a periodic signal of 4.8 hr but with the phase maximum at 0.35. The

maximum is slightly lower, 0.2–0.3, when the Van der Klis ephemeris is used. Their results were confirmed by the Leeds group, with the phase maximum also at 0.35. Since 1980, both the Cerenkov and EAS observations have reported variations in the results; sometimes they saw the signal, but more often not.

In 1985 the Soudan group in a mine a mile underground in Minnesota, while looking for the proton decay, claimed a muon signal from the direction of Cyg X-3. This could not be caused by charged particles from Cyg X-3 interacting in the earth's atmosphere because the galactic magnetic field would change the particle's direction; so the primary had to be a neutral particle. Ruled out were neutrons because of their lifetimes, neutrinos because the signal was observed only at angles to the zenith <90 deg, and gamma rays because the fluxes measured by the VHE and UHE gamma rays were too low. The authors suggested that a new particle or new interaction producing muons was possible. The results of other underground groups are mixed. The NUSEX group supports the results while others, including the FREJUS and Kamioka groups, are negative.

At the 19th International Cosmic Ray Conference in August 1985, at La Jolla, California, the Durham group announced the discovery of a pulsar in Cyg X-3. It had a period of 12.5908 sec. Its discovery is amazing because it not only is the first pulsar discovered in gamma rays but was discovered with VHE gamma rays of $E > 10^{12}$ eV and in a time interval of just 7 min. It has not been confirmed at any electromagnetic energy, including radio and X-ray.

Because the measurement of a 4.8-hr periodic signal from Cyg X-3 varies from observer group to observer group and from time to time within a group, the position of the maximum in the phase curve varies with group and with gamma-ray energy, because the periodic gamma rays were not seen by COS B over a period of 7 yr, and because the statistical significances of the observations do not seem improved although techniques have improved over the last 10 yr, we conclude that the evidence for observation of gamma rays from Cyg X-3 is not compelling.

h. Vela X-1. Vela X-1, like Her X-1, is an eclipsing X-ray binary pulsar with a pulsar period of 283 sec and orbital period of 8.96 d. It has been extensively studied in soft and hard X-rays up to about 80 keV. The light curve shows two pulse maxima separated by 0.50 in phase. The

NASA Goddard group with high-purity germanium detectors observed Vela X-1 for 5 hr during a balloon flight from Alice Springs, Australia on December 5, 1984. A strong pulsed signal, with the normal two-maxima phase plot, was observed from 20–80 keV, but no lines as strong as 3 σ were observed. No cyclotron lines were seen contrary to the one established for Her X-1. The resulting energy distribution was exponential.

Gamma rays below 35 MeV have not been seen and above 35 MeV were not detected by either the SAS-2 or COS-B groups. The Adelaide, South Australia EAS group reported VHE gamma rays of 3×10^{15} eV from observations taken during 1979–1981. They found a single pulse in the phase diagram for the orbital period that was less than 1-bin wide (1/50 in phase). No evidence was presented for the 283-sec pulsar period seen in X-rays. Before claiming Vela X-1 as a gamma-ray source, confirmation of the Adelaide observations is required.

2. Other Point Sources

a. Geminga CG195 + 04. In Milanese slang, Geminga means "the source that is not there." It has been appropriately named. This source in the COS-B catalog was discovered by the SAS-2 satellite group in 1974 at gamma-ray energies $\geq$35 MeV and confirmed by the COS-B satellite group at similar energies. It was found to be the strongest gamma-ray source in the sky after the Vela pulsar. The SAS-2 group reported a period of 59.0074 sec for December 1972 and for April and January 1973. The phase curve had only one pulse.

The COS-B group gave a preliminary confirmation of the 59-sec periodic signal in 1977 but reversed after a number of additional sightings through 1980 gave negative results. The 59-sec period question was reopened in 1984 when a study of just a part of the COS-B group suggested a 59-sec period in the Einstein X-ray data and also a large P. The gamma-ray data could be interpreted to give the 59-sec period changing with a P that made a discrete jump in 1979 from 2.4×10^{-9} to 4.68×10^{-9} sec sec^{-1}. The latest report of COS B, including the splinter group mentioned above, concluded that the five COS-B observations over 6.7 yr do not confirm the 59-sec period in the gamma-ray flux with the signal characteristics reported by SAS 2.

The 59-sec period has not been detected at either radio or optical energies. Geminga was identified with the bright X-ray source 1E0630+178 with a questioned 59-sec period. Although there are more than 10 radio sources in the gamma-ray location box, there are none at the X-ray position and nothing particularly out of the ordinary in any of those radio sources. Optical candidates at the 21st magnitude, G, at the 24.5 magnitude, G', and at the 25.5 magnitude, G'', have been located in the X-ray error box. For the latter candidates, the ratio of X-ray to optical luminosity is so high that the sources would be at improbable distances of $\geq$200 kpc away. G' and G'' may have high proper motions that would place them quite close, only a few tens of parsecs away. Perhaps Geminga is a nearby neutron star, either alone or in a binary system.

In 1983 Geminga again created a flurry of interest when it was claimed that COS-B observed a 160-min periodicity in gamma rays synchronous with the 160-min oscillations on the sun. Presumably, gravitational waves from Geminga were driving those on the sun. The oscillations were denied by the COS-B group in February 1984 as not being statistically significant, and a rash of theoretical papers appeared pointing out the difficulties with the gravitational wave suggestion. This idea met a quick death.

The 59-sec period turned up again in Cerenkov observations at energies $>2 \times 10^{11}$ eV by the Tien Shan, Russian group. Observations of 216 min were carried out in January–February 1979 and February 1981. A period of 59.28 sec was reported for 1981 with a phase curve with three pulse maxima. In 1983, observations of 144 min gave a period of 59.28 sec. The Gulmarg India group at energies $>10^{12}$ eV found a period of 60.25 sec. The statistical reliability was low for both results. On the other hand, the Whipple Observatory group with 28 ON–OFF scan pairs during November 1983–February 1984 and November 1984–February 1985, with each ON scan 28-min long, found no significant steady or pulsed gamma rays. Three-σ upper limits of 5.5×10^{-11} and 2.0×10^{-11} photons cm^{-2} sec^{-1} were found for the steady and pulsed gamma rays respectively.

The steady source is "present" but the 59 sec pulse may not "be there."

b. Cygnus X-1. Cygnus X-1 is well known as the first and best candidate for a black hole. Since its discovery in 1965, it has been studied intensely in X-rays. Random fluctuations with variations in milliseconds are normal. On the scale of months and years it seems to have several states. The low state (LS) is characterized by low, soft X-ray fluxes and high, hard X-ray fluxes. In the high state (HS) the fluxes of soft and hard X-rays are reversed. More complex behavior was observed during 1979–80 when a new superlow state (SLS) was observed, with low fluxes of both soft and hard X-rays.

A large flux of gamma rays with energies of 1 to 10 MeV was reported by the Southampton, England gamma-ray group from a balloon flight in September 1972. The Toulouse, France group reported a considerably lower flux at energies up to 3 MeV from a balloon flight in June 1976. Upper limits only (above 1 MeV) were obtained by the MISO group, which are not in conflict with the Toulouse results. However, the lower upper limits of the UCR group are at the lower edge of the band of Toulouse fluxes to 3 MeV. The UCR group also obtained upper limits only, up to 20 MeV. Gamma rays above 35 MeV from Cygnus X-1 were not seen by SAS 2 or COS B, or at UHE by Cerenkov detectors, or at VHE by EAS detectors. There have been no gamma rays observed from Cyg X-1 above 10 MeV, and the evidence from 1 to 10 MeV is contradictory, unless there are time variations as large as a factor of 50.

3. Galactic Lines

a. Galactic Center 0.511-MeV Positron Annihilation Line. The positron annihilation line was discovered by the Rice University group from a balloon flight carried out in November 1970. The energy of the line, measured with a NaI detector that had an aperture of 24 deg FWHM pointed in the direction of the galactic center, was reported as 473 ± 30 keV. The next flight in November 1971 confirmed the result, and the combined data gave an energy of 476 ± 24 keV for the line and a flux of $(1.8 \pm 0.5) \times 10^{-3}$ photons cm^{-2} sec^{-1}. The third flight on April 1, 1974, launched from Argentina like the other two, gave a line energy of 530 ± 11 keV. All three were consistent with the 511-keV positron annihilation line.

The Rice observations were confirmed by a precision measurement of the Bell Labs–Sandia group who made a series of flights from Alice Springs, Australia in November 1977, April 1979, November 1981, and November 1984. Using a germanium crystal of high purity with a volume of 130 cm^3 and a resolution of 3.2 keV FWHM, they measured a flux of $(1.22 \pm 0.22) \times 10^{-3}$ photons cm^{-2} sec^{-1} at 511 keV. Accurate

in-flight energy calibrations used positron annihilation in the atmosphere.

While the para state of positronium at rest splits into two monochromatic gamma rays of 511 keV each, the ortho state decays into three gamma rays with a distribution of energies below 511 keV. The measured flux of the ortho state implied that about 90% of the positrons annihilated through orthopositronium, although 0% could not be ruled out. Both the Rice and the Bell Labs–Sandia groups measured continuum radiation from the galactic center. The latter group found a flux of 1.86×10^{-4} $(E/100)^{-2.31}$ photon cm^{-2} sec^{-1} keV^{-1} on their flight in April 1979, while their 511-keV line flux was $(2.35 \pm 0.71) \times 10^{-3}$ photons cm^{-2} sec^{-1}. On the November 1981 and 1984 flights, upper limits only of 7 and 9×10^{-4} photons cm^{-2} sec^{-1} were found.

The JPL HEAO 3 satellite group using high-resolution germanium detectors also measured the turn-off of the galactic center. An observation from September–October 1979 gave a flux of $(1.85 \pm 0.21) \times 10^{-3}$ photons cm^{-2} sec^{-1} and the one from March–April 1980, $(0.65 \pm 0.25) \times 10^{-3}$ photons cm^{-2} sec^{-1}. The variation in flux of the 0.51-MeV line is summarized in Fig. 18.

A combination of the above results gives the source direction as the galactic center ± 4 deg. The annihilation rate is 2×10^{43} positrons sec^{-1} and the luminosity 3×10^{37} erg sec^{-1}. The decrease in flux by a factor of three in 6 months

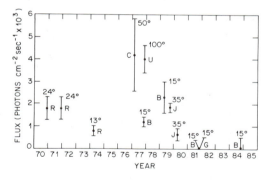

FIG. 18. Measured intensity of the 511-keV annihilation line from the galactic center over times from 1971 to 1985. The field of view (FWHM) of each instrument is indicated. (Groups making the observations are: B, Bell Labs-Sandia Corporation; C, CESR, Toulouse, France; G, Goddard Space Flight Center NASA and CENS, Gif sur Yvette, France; J, JPL HEAO 3 Satellite; R, Rice University: U, University of New Hampshire. All observations except for J were carried out in balloons.] [From Leventhal, M. *et al.* (1986). *Astrophys. J.* **302**, 459–461.]

suggests that both the positron source and the annihilation regions have dimensions less than the velocity of light times 1.5×10^7 sec (6 months) or 10^{18} cm. R. W. Bissard, R. Ramaty, and R. J. Drachmann of NASA Goddard Space Flight Center pointed out that the narrow line width, >2.5 keV, requires that the gas temperature in the annihilation region be less than 5×10^4 K, and the 6-month time variation requires that the gas density be $>10^5$ cm^{-3} for the positrons to slow down and annihilate in that time. A number of sources for the production of the positrons and their annihilation have been suggested, which include a black hole at the center of the galaxy.

b. Diffuse Galactic 1.809-MeV Line from ^{26}Al. The JPL HEAO 3 satellite group, from observations in 1979 and 1980 with their narrow-line Ge detector, discovered ^{26}Al from the galactic center. The energy of the decay was 1808.49 ± 0.41 keV and the flux from the vicinity of the galactic center $(4.8 \pm 1.0) \times 10^{-4}$ photons cm^{-2} sec^{-1} rad^{-1}. The width of the line was 3.0 keV FWHM. The line energy and flux were verified with the SMM satellite gamma-ray group every year from 1980 through 1985 with a NaI detector that had excellent stability but less sharp energy resolution.

The intensity of the line varied with galactic longitude to less than 40% of the galactic center value in the antigalactic center direction. The observed flux gave a ^{26}Al mass of 3 M$_\odot$ in the interstellar medium, an order of magnitude greater than predicted from supernova production. However, production of ^{26}Al in the novae, red giants, and main-sequence stars could probably supply the required amount. This is the first well-verified example of galactic nucleosynthesis. As ^{26}Al has a long lifetime, 1.04×10^6 yr, compared with the time between novae explosions in the galaxy of 1/30 yr, the measurement samples the mixture of debris from millions of novae and thus gives a value for the galactic steady state.

c. The 1.5- and 1.2-MeV Lines from SS433. The binary star SS433 is located in the center of the supernova remnant W50. Optical and radio observations have determined an orbital period of 13 d and a longer period from other causes of about 164 d. SS433 is one of two galactic sources with a pair of oppositely directed beams and the only one where the beams have been imaged in both radio and X-ray energies.

The JPL HEAO 3 gamma-ray group reported a gamma-ray line at 1.5 MeV of significance >5 σ from SS443 during observations in October 1979 and a weaker line at 1.2 MeV. The peak response was located at scan angle 26 ± 3 deg, close to the SS443 angle of 27.2 deg. Additional measurements in 1979 and 1980 gave time averages of $(1.47 \pm 0.28) \times 10^{-3}$ photons cm^{-2} sec^{-1} at 1.5 MeV and $(1.12 \pm 0.18) \times 10^{-3}$ at 1.2 MeV. They suggested these were blue- and red-shifted lines from the decay of ^{26}Mg moving at 0.26 c in the directed beams excited by collisions with protons at rest. The unshifted line is at 1.369 MeV. This model also requires a rest energy 6.176-MeV line, later only marginally observed if at all by the JPL HEAO 3 group. The energy flow of the 800-MeV ^{26}Mg nuclei is 10^{42} erg sec^{-1}. The authors pointed out this was considerably higher than theoretical estimates. Furthermore, very strong lines of ^{12}C and ^{16}O six to seven times the intensities of the ^{26}Mg lines were expected. These were not seen, nor were two strong lines at 1.634 and 1.639 MeV that were expected from spallation reactions.

Several theoretical ways to interpret the results included lines from the thermonuclear reaction ^{14}N(p, γ)^{15}O occurring in the beam and nuclear deexcitations in ^{26}Mg imbedded in grains in the jet inelastically scattered by protons at rest. On the experimental side, the Bell Labs–Sandia group flew their Ge detector on a balloon in September 1980 to search for the JPL HEAO gamma-ray lines. The 1.2, 1.5, and 6.7 MeV lines should have been observed at the 2, 3, and 1 σ levels. None were seen. The SMM satellite group also searched for these lines during 1980 to 1985. The 360-d time-averaged upper limit for each of the shifted 1.37 and 6.1 MeV lines, when SS433 was in the field of view, was 2×10^{-4} photons cm^{-2} sec^{-1} at the 99% confidence level. This is a full order of magnitude lower than the flux reported by the HEAO-3 group. Unusually high variability of SS433 is required to explain the differences in results. The observation of gamma-ray lines from SS433 is certainly not compelling.

4. Extended Sources

a. Crab Nebula. The Crab nebula, a supernova remnant in Taurus, is an expanding mass of gas, spectacular in the visible. The supernova explosion, observed by the Chinese in 1054, was so bright it could be seen in the day. Doppler shifts of the gases require speeds up to 1500 km sec^{-1}. This velocity and the angular extent of the nebula give a distance to the Crab of about 2 kpc.

Extensive measurements have been made in low-energy gamma rays, 0.3–1 MeV; medium energy, 1–30 MeV; high-energy, 30–3000 MeV; and >10^{12} eV. The total gamma rays from the Crab nebula include both pulsed and steady. The total gamma-ray energy distribution may be fitted by the power law $3 \times 10^{-3} E^{-2.3}$ photons cm^{-2} sec^{-1} MeV^{-1} from 1 keV to 3 GeV. At X-ray energies, most of the flux is steady while at gamma ray energies from 30 to 3000 MeV most of the flux is pulsed. The cross-over occurs in the medium energy region from 1 to 30 MeV. The steady radio waves and perhaps more energetic photons are caused by the synchrotron radiation of electrons in the strong magnetic fields of the nebula. The pulsed radiation is associated with the Crab pulsar at the center of the nebula.

b. The ρ Oph Cloud (2CG353+16). The COS-B source 2CG353+16 has been definitely identified with the ρ Oph cloud. The agreement of the COS-B gamma ray map, for energies >100 MeV, with the ρ Oph region in the atlas of dark clouds is striking. The calculated gamma-ray flux, using the cosmic-ray flux in the solar neighborhood and the mass of ρ Oph estimated from OH measurements, is within a factor of 2 of the measured value.

c. The Orion Cloud Complex. The COS-B group did not include the Orion cloud complex in their 2C catalog because it was a resolved extended source covering a few hundred deg^2 in the sky. The flux contours in the COS-B gamma-ray map generally are in good agreement with maps obtained from other tracers of the Orion cloud complex such as CO, OH, and optical absorption. The identification by COS-B is considered certain. The flux from the complex at energies >100 MeV is $(2.0 \pm 0.5) \times 10^{-6}$ photons cm^{-2} sec^{-1}. Using the local cosmic-ray flux and a distance of 450 pc, they obtained a mass of $(1.2 \pm 0.4) \times 10^5$ M$_\odot$ in good agreement with the value of about 1.5×10^5 M$_\odot$ from radio astronomy.

d. The Gould Belt. In 1977, the SAS-2 group suggested that an extended source of gamma rays with energies >100 MeV could be identified with the Gould belt. In the antigalactic direction, the gamma-ray source was tilted toward negative latitudes. From the local cosmic-ray flux and a gas column density of about $2 \times$ 10^{21} atoms cm^{-2}, estimated from reddening or neutral hydrogen measurements, the calculated gamma-ray flux is 4×10^{-5} photons cm^{-2} sec^{-1} sr^{-1} for gamma-ray energies >100 MeV. This calculated value was in good agreement with the observed value of 4.5×10^{-5} photons cm^{-2} sec^{-1} sr^{-1}. The COS-B results confirm the SAS-2 comparisons. Although there appears to be agreement between the measured and expected fluxes on the large scale, the sensitivity of the SAS-2 detector was too low to observe correlations of other extensive sources on smaller scales.

e. Dolidze-Belt. The COS-B group located an extended gamma-ray source that stretches from 117 to 297 deg in galactic longitude and is inclined about 16 deg to the galactic latitude 0 deg plane. They suggest this is the gamma-ray image of the recently discovered Dolidze belt. It is made up of young stars, OB-, T-, and R-associations, diffuse nebulae, and matter. The addition of the Dolidze belt explained some of the discrepancies that existed between the Gould belt alone and COS-B fluxes in the regions 100 < 1 < 140 deg and 10 < b < 20 deg, and 270 < 1 < 330 deg and $-20 < b < 10$ deg. The gamma-ray fluxes are consistent with a constant cosmic-ray intensity throughout the belt.

B. EXTRAGALACTIC SOURCES

1. 2CG135+1

The source 2CG135+1 was discovered by the COS-B group at gamma-ray energies >50 MeV. It was early associated with the nearby quasar QSO 0241+622; now the two appear to be different objects. At lower gamma-ray energies the MISO group from Milan and Bologna, Italy and Southampton, England reported a 5-σ excess of gamma rays from about 20 keV to 1 MeV. They used a double scatter detector with a lead-slat collimator that required arriving gamma-rays to be closer than 3 deg FWHM to the direction of 2CG135+1. Five drift scans were made from a 4-hr balloon observation on October 8, 1978. Upper limits of the MISO group and of the UCR group from a balloon flight in 1979 suggest that the flux must drop off with a power law steeper than about $E^{-2.5}$ above 1 MeV to agree with the COS-B results. The Einstein coverage of the COS-B error box seems to have eliminated all five possible candidate X-ray sources. The variable radio star 1E0236+610 remains a possible candidate for the 2CG135+01 counterpart.

2. The Quasar 3C273 (2CG289+64)

During observations in May and June, 1986 at energies of 50 to 800 MeV, the COS-B source 2CG289+64 has been identified as the quasar 3C273, the brightest discovered to date. It has been observed at radio, infrared, optical, ultraviolet, X-ray, and gamma-ray energies. Observations in hard X-rays extend up to 300 keV; upper limits only exist for gamma rays from 300 keV to 50 MeV. The gamma-ray results were confirmed by a second set of observations in June and July 1978 and again in June and July 1980. The energy distribution follows a steep power law with an exponent of −2.5. No variation in flux more than 50% was found over the observation periods. At the higher energies of Cerenkov detectors $>10^{12}$ eV and EAS $> 10^{15}$ eV, upper limits to the flux, only, have been reported.

3. Seyfert Galaxy NGC 4151

The Seyfert galaxy NGC 4151 has been observed at energies from radio to low-energy gamma-rays. At radio energies it is rather weak and constant. In the infrared it appears to be variable at 2.2 and 3.4 μm but constant at 10 and 33.5 μm. There may be small variations in the visible and ultraviolet. In soft X-rays UHURU satellite measurements gave a 3.8-σ signal. A hard spectrum of $E^{-1.1}$ was observed up to 100 keV, and hard X-rays to 300 keV have been detected. Upper limits only were found from observations of the SAS-2 and COS-B groups at energies of 35 MeV to 3 GeV.

Gamma rays up to 10 MeV were reported by the MISO group from a 3-hr balloon flight launched at Palestine, Texas on May 22, 1977. The detector was the same used for observing 2CG135+1. A second observation was carried out from a balloon on September 30, 1979 with five drift scans over about 2 hr. The measured fluxes were about a factor of two lower than the 1977 values over the energy range of 0.20 to 10 MeV. Two other groups during this time found upper limits only. The Rice group on a balloon flight October 4, 1977 with their NaI detector obtained upper limits comparable to the MISO values, while the UCR group with their double Compton scatter telescope on a flight September 29, 1978 obtained upper limits a factor of five lower than the MISO second flight at 2 MeV and a factor of two lower at 10 MeV. The lack of confirmation by the Rice and UCR groups at energies of 1 to 10 MeV and the negative results

of SAS 2 and COS B at energies above 35 MeV can be explained only by large variations in the fluxes of NGC 4151 of a factor of five at MeV energies and a significantly increased slope in the energy distribution at energies >10 MeV. Confirmation of the MISO results is needed to certify NGC 4151 as a source of gamma rays with energies >1 MeV.

4. Seyfert Galaxy MCG 8-11-11

The Seyfert galaxy MCG 8-11-11 is the third gamma-ray source reported by the MISO group and the second Seyfert galaxy. It was observed on a balloon flight on September 30, 1979 from Palestine, Texas. Like NGC 4151, it was not seen by COS B or SAS 2 at energies >35 MeV. It was first observed in hard X-rays on a transatlantic balloon flight by the Bologna, Italy group with NaI detectors on July 30 and 31, 1976.

The MISO group measured an energy distribution from MCG 8-11-11 with six points from 0.2 to 20 MeV and suggested that a break occurred in the distribution at 3 MeV. The best-fitted slopes below and above 3 MeV were 1.0 and 3.8, taking into account also the upper limits of SAS 2. While a straight-line fit from the MISO 100-keV point to upper limits of SAS 2 is a poor fit to the data, likewise breaking at 3 MeV with only two points from 1 to 20 MeV, both with large statistical errors, stretches the credibility of the measurements. Confirmation of the MISO results are required before MCG 8-11-11 can be considered a compelling gamma–ray source.

5. NGC 5128 (Centaurus A)

Radiation from the radio galaxy Centaurus A has been reported at all energies from radio through a few MeV and at $>3 \times 10^{11}$ eV. It is the brightest extragalactic source of X-rays above 40 keV. Hard X-rays vary by a factor of two over days and a factor of ten over years. The X-ray energy distribution is a flat power law similar to that from other active galactic nuclei.

The Rice Gamma-Ray group measured a continuum from Cen A on a balloon flight from Rio Cuarto, Argentina in April 1974. They obtained a power law energy distribution from 33 keV to 1 MeV, $(0.86 \pm 0.08) E^{-1.9 \pm 0.1}$ photon cm^{-2} sec^{-1} keV^{-1}. Lines at 1.6 and 4.5 MeV with 3.8- and 3.3-σ uncertainties above background were also reported. No indication of the 0.511 positron annihilation line was found. HEAO 1 observations during January and July 1978, also with a NaI detector, confirmed the continuum up to about 1

MeV but found no evidence for lines at 0.511 or 1.6 MeV. The NASA Goddard group, with good resolution Ge detectors, confirmed the continuum within a factor of about two but obtained upper limits only for the 0.511, 1.6, and 4.5 MeV lines. The 1.6-MeV limit was a factor of six lower than the Rice results. The most recent results, from the Max Planck Institut Garching group with a double Compton scatter telescope launched from Uberaba, Brazil in October 1982, confirmed the continuum from 0.7 to 2 MeV. They extended the energy range to 20 MeV and imaged Cen A at energies of 0.7 to 20 MeV. No gamma-ray lines were found. From the above results the evidence for the continuum is strong but for the lines weak. For $E_\gamma > 50$ MeV, COS-B and SAS 2 found upper limits only.

The first evidence for gamma rays from Cen A actually came from the Harvard–Sydney group who used two Cerenkov collectors at energies $>10^{11}$ MeV for observations during April–July 1972, April–June 1973, and March–April 1974. A 4.5-σ excess was obtained from the direction of Cen A. Possible variability over the three years was indicated. If verified, Cen A would be the first extragalactic source of VHE gamma rays detected.

6. LMC X-4

The X-ray source LMC X-4 is thought to be a massive binary system with an O7III-V or O8III-V star of mass about 17 $M_\odot$ and a neutron star of about 1.6 $M_\odot$. The neutron star is a pulsar with a period of 13.5 sec that was discovered during a 40-min X-ray flare. LMC X-4 has an orbital period of 1.408 d. The X-ray source has high and low states with a 30.5-d period attributed to precession of its accretion disk. LMC X-4 has been observed at visible, ultraviolet, and X-ray energies. The only report of gamma rays is at energies $>10^{16}$ eV by the Adelaide, Australia group with measurements of EAS at Buckland Park during 1979, 1980, and 1981. They observed an orbital phase maximum at 0.90 to 0.95 and estimate the probability that this is a fluctuation in background is about 0.01. Additional gamma-ray measurements are necessary to confirm LMC X-4 as a gamma-ray source.

7. M31 (Andromeda Galaxy)

M31 is considered a typical Sb spiral galaxy, in many respects like our own Milky Way. It has been seen at radio, optical, and soft X-ray energies. Upper limits only were obtained with gamma rays from M31 by COS B and SAS 2, both for energies >50 MeV. The only report of gamma rays from M31 is at energies above 10^{12} eV, by the Durham Group. From 15 hr of observations in September and November 1983 using the Dugway Cerenkov collectors, they estimated a flux of $(2.2 \pm 0.7) \times 10^{10}$ photon cm^{-2} sec^{-1}. They found no evidence for variations. Contradictory results were obtained by the Whipple Observatory group from observations in 1972 and more recently in October and November 1984. The group also used Cerenkov collectors but at somewhat lower energies of $>1.5 \times 10^{11}$ eV in 1972 and $>4 \times 10^{11}$ eV in 1984. Their fluxes extrapolated to 10^{12} eV are factors of 10 to 100 below the Durham values. Further measurements are necessary to confirm M31 as a gamma ray source.

C. DIFFUSE GALACTIC RADIATION

The sky in gamma rays is dominated by the galactic plane as can be seen in Figs. 9 and 10. The gamma rays are the sum of the point and extended sources previously discussed and the diffuse radiation. Possible sources of the diffuse radiation are cosmic-ray proton and heavier nuclei interactions with interstellar hydrogen atoms and molecules to give π° mesons that decay into two gamma rays with an energy distribution having a maximum at about 70 MeV; cosmic-ray electrons that create gamma rays by bremsstrahlung; cosmic-ray electrons that interact by the inverse Compton reaction with blackbody radiation and optical and infrared interstellar photons to produce gamma rays; and less importantly, cosmic-ray electrons accelerated in galactic magnetic fields to generate synchrotron radiation.

The diffuse gamma rays have been seen at all energies from the lowest at 0.3 MeV to the highest at 3 GeV observed by COS B. The latitude and longitude distributions of gamma rays observed by COS B are given in Fig. 9. The latitude distributions of COS B and SAS 2 as well, at all energies from 30 MeV to 3 GeV, show sharp maxima a few degrees wide, especially in the galactic center direction. The galactic longitude distribution given in Fig. 9 shows a broad maximum in the direction of the galactic center between 60 and 300 deg that has three times the flux observed in the anti-galactic center direction. A few prominent point sources such as the

Vela, Crab, Cygnus X-3 and Geminga ride on top of the curve.

The energy distribution of the galactic gamma rays falls off with a power law as expected from electron bremsstrahlung from about 0.1 to about 70 MeV and then continues to 3 GeV with a slight offset and drop as might be expected from the addition of π° meson decay gamma rays.

The cosmic-ray electrons undergo bremsstrahlung, and protons and nuclei interactions with hydrogen in the diffuse matter of the galaxy. The ions and dust are minor constituents and can be ignored. Atomic and molecular hydrogen make up about $\frac{2}{3}$, and helium and heavy nuclei about $\frac{1}{3}$, of the matter. The bulk of the target matter is confined to the galactic disk with a scale height of a few hundred parsecs. The atomic hydrogen is mapped by the 21-cm radio waves that originate from the spin flip of the electron spin relative to the proton spin. The molecular hydrogen is more difficult to trace, with the best estimate obtained from the 2.6-mm radio line from carbon monoxide. Much of the CO is found in massive cloud complexes. The galactic material is thought to be more dense by a factor of two in the spiral arms of the galaxy than in between, just as for the HII regions, infrared emission, supernova remnants, and pulsars.

The SAS-2 group assume that the cosmic-ray density in the galactic plane is proportional to the matter density and has a Gaussian distribution with a scale height of 600 pc perpendicular to the plane. This is plausible from the argument that the cosmic ray, magnetic field, and kinetic motion matter densities are about equal and as large as can be contained by the local galactic matter. The bremsstrahlung gamma rays are produced from an electron energy spectrum of $3.9 \times 10^3\ E^{-2.8}$ and $3.2 \times 10^2\ E^{-2.5}$ electrons cm^{-2} sec^{-1} MeV^{-1} sr^{-1} at energies above and below 4 GeV, respectively. An increase in slope at energies below about 50 MeV is required for agreement with gamma-ray measurements at the lower energies.

Reasonable agreement is obtained with the latitude and longitude distributions and with the energy distribution using only cosmic-ray electron bremsstrahlung and nucleon (nuclei) collisions with galactic hydrogen atoms and molecules. Inverse Compton collisions account for less than 20% of the gamma rays, and synchrotron radiation is negligible. The reader is cautioned that the molecular hydrogen normaliza-

tion is usually treated as an adjustable parameter and could fit within a wide range of values. As a consequence, point sources that vary in position like molecular hydrogen would be ignored by the fit. The present feeling is that point sources, other than those already identified, do not make a major contribution to the diffuse galactic gamma rays.

D. DIFFUSE EXTRAGALACTIC RADIATION

When the gamma ray diffuse radiation of the galactic plane, only a few degrees wide in galactic latitude at energies >35 MeV, is subtracted from the total diffuse radiation, an isotropic gamma ray flux remains. This residual gamma ray flux is called the diffuse extragalactic radiation. Early X-ray observations, for example, from OSO 3, established that the diffuse soft X-ray radiation below 40 keV is isotropic to about 2%. So it appeared to be extragalactic in origin. At least 20% of all Type I Seyfert galaxies were found by UHURU and Ariel 5 to emit 2–10 keV X-rays and nearly all were observed as sources by the more sensitive Einstein spacecraft detectors. Therefore, Seyferts were considered leading candidates for the isotropic diffuse radiation. In addition, the Seyfert galaxies had relatively flat energy spectra, so they were expected to be copious gamma ray emitters. The same was expected of the similar BL Lac objects and quasars.

The isotropic diffuse radiation also appears to dominate the galactic diffuse radiation at hard X-ray energies up to hundreds of keV. At energies of 1 to 10 MeV an extragalactic origin is indicated for most of the diffuse radiation. The energy distribution of the diffuse radiation may have a bump at a few MeV that has been attributed to contributions from Seyfert galaxies. The SAS 2 energy distribution of diffuse radiation from 20 to 300 MeV, when extrapolated downward, connects to the MeV diffuse distribution. The energy distribution is interpreted as composed of both galactic and extragalactic (isotropic) diffuse radiation. The extragalactic part connects to the distribution at lower energies and dominates up to about 100 MeV to the crossover, where the galactic diffuse radiation becomes larger.

The SAS 2 group's search for gamma ray emission from the X–ray emitting active galaxies found upper limits only. See, for example, the earlier discussion of NGC 4151. The upper limits were considerably below the values extrapolated from the X-ray energy spectra. This steepening of the spectra at energies between X–rays and gamma rays seems to be a general property of Seyferts and other emission line galaxies. Also X-rays have been observed from a number of quasars, but 3C273 is the only one identified by COS B as a source of gamma rays with energies >50 MeV. See the earlier discussion of 3C273. It also requires a steepening of the spectrum between soft X-rays and 100 MeV gamma rays.

The spectrum, intensity, and isotropy place constraints on possible explanations of the diffuse extragalactic radiation. The SAS 2 group concluded that the summed contribution of Seyfert galaxies to the isotropic diffuse gamma ray background is very significant from 1 to 150 MeV and may account for most if not all of the diffuse radiation at those energies.

E. Gamma-Ray Bursts

Gamma-ray bursts were discovered by Ray Klebesadel, Ian Strong, and Roy Olson of the Los Alamos National Laboratory Vela satellite gamma-ray group. While monitoring nuclear weapons tests in the atmosphere from the Vela satellites, they observed gamma-ray bursts with rise times from milliseconds to a second and time durations of 0.1 to tens of sec. These were seen simultaneously on two or more satellites and so could not have been detector malfunctions. From the time of arrival at the different satellites it was possible to determine the direction of the source of the gamma rays.

The energies of the gamma rays in the burst varied from tens of keV to tens of MeV. The fluences ranged from 10^{-4} to 10^{-6} erg cm^{-2} as measured by the small satellite detectors and down to 10^{-8} erg cm^{-2} by the larger more sensitive balloon detectors. About 100 bursts per year are seen with fluences above 10^{-6} erg cm^{-2} and about 10 per year above 10^{-4} erg cm^{-2}. The satellite detectors normally have observation times of a year or more while the balloon detectors have been limited to a day or two. The Soviet Venera satellites 11 and 12, and more recently 13 and 14, carried larger, more sensitive detectors and therefore have measured and catalogued a large fraction of the observed bursts. The directions are measured to within a few degrees, and the bursts appear to be isotropic in direction. Information on bursts observed on these and other spacecraft are collected in the Konus Catalog of the Ioffe Institute in Leningrad and the first and second catalogs of the Interplanetary Network.

If the bursts originate in the disk of the galaxy, this should be evident from a plot of Ln $N(>S)$ versus Ln S, where S is the fluence. It should give a straight line with a slope $S^{-1/2}$. If the source is isotropic the slope should be $S^{-3/2}$. Early analysis of the Venera data along with that from other satellites and balloons gave a Ln $N(>S)$ versus Ln S curve that tended to bend over at low fluences and was interpreted as favoring a galactic plane origin. However, recent analyses attribute the bending to detector biases caused by energy and time thresholds. So both this plot and the isotropic distribution of direction suggest that the bursts originate within a galactic scale height, about 300 pc, of the earth.

Considerable effort has been spent in identifying gamma-ray bursts with other celestial objects. Arrival times of bursts at spacecraft separated by distances of astronomical units make it possible to obtain accuracies of a few seconds of arc in one direction and a few minutes in the other. Because the detectors on the widely spaced satellites are small, with low sensitivities, only the largest bursts are detected, and less than a dozen have been measured with fine accuracy. Of these only one has been identified with a possible source; the March 5, 1979 burst occurred in the direction of the supernova remnant N49 in the Large Magellanic Cloud (LMC) at a distance of 55 kpc. This remarkable burst had an extremely fast rise time of about 10^{-4} sec and the main peak a width of about 0.15 sec. Its fluence was about 10^{-3} erg cm^{-2}, the largest of any burst seen to date. The first sharp peak was followed by oscillations with a period of about 8 sec for a time of about 100 sec. Since that original burst, about 15 smaller bursts have been seen from the same direction and are thought to be from the same source. If located at the LMC, the original burst source had a luminosity integrated over time of about 10^{44} erg, 10^5 times a typical burst source at a few hundred parsecs. A chance alignment with N49 cannot be ruled out; in that case, the burst could be much closer.

The search in archival plates from optical telescopes for possible optical bursts in the direction of observed gamma-ray bursts has uncovered two appearances of images not on the previous or following plates. These could have been past optical bursts from the gamma-ray burst source. Recently, three optical bursts were seen during the five months in which the March 5, 1979 gamma-ray burst direction was watched. Simultaneous gamma-ray bursts were not observed, but the optical light curves were typical of those of gamma-ray bursts. Arguments concerning the production of light by the gamma rays on an accretion disk around a neutron star, or on a stellar companion, place limits on the maximum luminosity and suggest that the source is much closer than the LMC.

Absorption and/or emission lines in the energy distributions of gamma-ray bursts from the Venera measurements are possible electron cyclotron lines in the strong magnetic fields at the surfaces of neutron stars. These occur in about 20% of the spectra from Venera bursts at energies of about 30 to 60 keV. The resulting magnetic fields are a few times 10^{12} gauss. Because the lines are narrow, the magnetic field over the gamma-ray production or absorption region must be rather constant. Additional emission lines at energies of about 0.4 MeV are suggestive of red-shifted positron–electron annihilation in the strong gravitational fields of the neutron stars.

VI. Closing Comments

Gamma ray astronomy has significantly enriched our knowledge of high energy astrophysics and our understanding of the universe. We now have a much better understanding of the evolution of the stars. The discovery of the ^{26}Al in the galactic plane is verification of explosive galactic nucleosynthesis. So much is observed that additional sources such as novae, red giants, and 0 and Wolf-Rayet stars may be needed. The cores left behind in supernova explosions like the Crab and Vela are the pulsars that generate gamma rays by high energy electron interactions in their very high magnetic fields. Values of magnetic fields of 10^{12} G are obtained from absorption or emission cyclotron lines at tens of keV.

The measurements of the properties of the 0.51 MeV positron annihilation gamma ray line from the galactic center direction place dramatic constraints on the source. The apparent decrease in the flux by a factor of 3 in 6 months excludes multiple or extended sources, supernovae, or primordial black holes previously proposed. It seems to require a single compact object no more than a light year across near the center of the galaxy.

The results of the Cerenkov and EAS observations of Cygnus X-3, if true, would have far reaching implications for astrophysics. Only a few such sources could produce all the very high energy cosmic rays. The Cygnus X-3 accelerator is so strong that the popular Fermi acceleration by scattering on magnetic shocks or other customary acceleration mechanisms would no longer be needed. The implications for high energy fundamental physics are equally far reaching. The standard model would be in deep trouble. The known particles would need different interaction properties than are currently accepted, or new particles would be required. Such new particles might also have an important impact on the amount of dark matter in the universe and therefore on cosmology, whether we have an open or closed universe.

The observations of diffuse galactic gamma rays have contributed to a better understanding of the matter, cosmic rays, and magnetic fields in our galaxy. Gamma rays with energies below about 100 MeV are produced primarily by bremsstrahlung of cosmic ray electrons on galactic hydrogen and other matter. Those above 100 MeV are created mostly by cosmic ray proton and nuclei interactions with galactic hydrogen to produce π° mesons that decay into two gamma rays.

The diffuse extragalactic gamma rays give us information about active galaxies such as Seyferts, BL Lacs, radio galaxies, and quasars. These galaxies are thought to be the unresolved sources of the diffuse extragalactic gamma rays. Accretion onto black holes in these objects may produce the gigantic energies required.

VII. Future

Only the strongest gamma-ray sources have been discovered to date. To locate and measure the properties of weaker sources, gamma-ray telescopes and detectors must have larger areas to collect more photons from the source. These areas can be increased to the limitations imposed by the balloon and satellite carriers. Backgrounds will be reduced by more accurate timing of the detector pulses and analysis of their shapes. New sophisticated detector properties and analysis methods will be used. Directions to celestial sources will be improved dramatically by new techniques such as coded masks with multiple detectors, Anger cameras, and slender scintillator bars with fast timing.

Past experiences with SAS 2 and COS B have

demonstrated the extreme importance of long observation times. Progress will be greatly accelerated with more satellite flights such as the gamma-ray observatory scheduled to fly in the late 1980s. With more flights a greater variety of detectors with different capabilities can be accommodated. In the future, transcontinental, transatlantic, and transpacific balloon flights and combinations will be used. Perhaps the easiest way to obtain long observation times is to use balloon flights that circle the earth. Flights launched and recovered in Australia promise observation times of 30 to 60 days and could make balloon flights competitive with satellite carriers at a fraction of their cost.

With higher sensitivities, good direction measurements, and long observation times, we may expect the discovery of many new gamma-ray sources and their identification with celestial sources known at other energies. A large number of pulsars should be found as counterparts to those discovered at radio and X-ray energies. Sources associated with quasars, Seyfert galaxies, and other active galaxies are expected. Additional sources such as Geminga, not associated with known sources at other energies, will be found. An explanation of the gamma-ray sources should lead to a deeper understanding of high-energy astrophysics and of our universe.

BIBLIOGRAPHY

Bignami, G. F., and Hermsen, W. (1983). *Ann. Rev. Astron. Astrophys.* **21,** 67.

Fichtel, C. E., and Kniffen, D. A. (1984). *Astron. Astrophys.* **134,** 13–23.

Hermsen, W. (1982). *Space Science Reviews* **36,** 61.

Mayer-Hasselwander, H. A., *et al.* (1980). *Ann. NY Acad. Sci.* **336,** 211.

Ramaty, R., and Lingenfelter, R. E. (1982). *Ann. Rev. Nucl. Part. Sci.* **32,** 235.

Ramaty, R., and Lingenfelter, R. E. (1986). *Ann. NY Acad. Sci.* **470,** 215.

Schönfelder, V. (1983). *Adv. Space Res.* **3,** 59.

Swanenburg, B. N., *et al.* (1981). *Ap. J. (Lett.)* **243,** L69.

Weekes, T. C. (1983). "A Review of Very High Energy Gamma Ray Astronomy circa 1982." Proc. Workshop on Very High Energy Cosmic Rays, University of Utah, Salt Lake City, January 10–14.

LUNAR ROCKS

Arden L. Albee   *California Institute of Technology*

GLOSSARY

Craters: Circular depressions excavated by the impact of meteoroids from space. Large craters (>200 km) are called basins.

Extrusive rocks, volcanic rocks: Fine-grained rocks formed by rapid cooling of magma erupted to the surface.

Igneous rocks: Crystallized molten rock (magma).

Impact breccia: Fragmented and molten rock ejected outward from the crater forming widespread ejecta blankets or rays.

Incompatible elements: Elements that concentrate in the residual melt, not being included in the crystallizing minerals.

Intrusive rocks, plutonic rocks: Coarse-grained rocks, formed by slow cooling of magma below the surface.

Magma: Consists of silicate melt, and may include various silicate and oxide mineral phases, volatile-filled bubbles, and molten globules of metal and sulfide. The chemical elements differ in their affinity for these phases and partition themselves between them in a systematic manner.

Magmatic differentiation: Processes by which phases can be separated to produce rocks of differing composition. Crystals of light minerals may float or heavy minerals may sink, in the silicate melt to produce layered cumulate rocks.

Lunar rock studies provide the cornerstone for the scientific findings from lunar exploration by spacecraft since 1959. These findings have completely changed our understanding of moon and its evolution as well as that of earth and the other inner planets. We now understand that the lunar surface features are predominantly the result of impact by numerous huge projectiles during the first half billion years of lunar history and that most of the younger and smaller craters were also formed by impacts, not by volcanism. The role of volcanism is restricted predominantly to the filling, between three and four billion years ago, of the mare basins that resulted from the impacts. The moon did not form by slow aggregation of cold particles that slowly heated up. Instead, it was covered in its early life by molten rock from which a Ca–Al–Si-rich crust formed by accumulation of the mineral plagioclase. As the outer part of moon became rigid, the sources of volcanic lava migrated downward to depths below 500 km. Since three billion years ago, volcanic activity, has been infrequent and localized. Geochemical similarities show that the moon and the earth must have formed in the same general region of the solar system with a relationship not yet understood.

I. Lunar Exploration

A. PRE-APOLLO KNOWLEDGE

Between 1959 and 1976 the United States and the Soviet Union dispatched more than 40 spacecraft missions to explore the moon. These missions (both unmanned and manned) photographed and sensed the moon remotely from orbit, landed and established stations and instruments on the surface, roved across the surface in wheeled and tracked vehicles, drilled holes to collect samples and make measurements, and collected and returned almost 400 kg of rock and soil for detailed study in terrestrial laboratories.

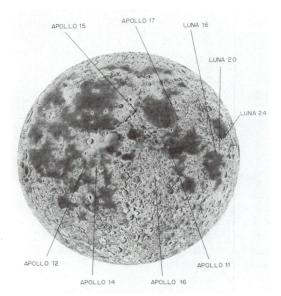

FIG. 1. Landing sites of the Apollo and Luna missions shown on a relief map of moon. The dark colored smooth mare basins contrast sharply with the lighter colored irregular highlands even as seen with the naked eye. [From Burnett (1977). *Rev. Geophys. Space Phys.* **13**(3), 13–34. Copyright © by the American Geophysical Union.]

Figure 1 shows the locations on the surface of moon of the Apollo and Luna missions.

At the start of this period of exploration, the knowledge of our nearest neighbor was based on continuation of Galileo's pioneering use of the telescope in 1610. Galileo provided the first topographic description of the moon. He noted the great dark smooth areas, which he called maria or seas, and the areas pitted with circular depressions, which he called craters. These intensely cratered areas, which appear white in contrast to the dark maria and also rise above them, are now known as terrae or highlands. Figure 1 illustrates this contrast between the dark flat maria and the cratered highlands. Bright streaks extend as radial rays for hundreds of kilometers from some youthful-appearing, well-defined, bowl-shaped craters. [*See* MOON (ASTRONOMY).]

Prior to the return of data by spacecraft most observers argued for a common origin, either impact or volcanic, for all features. Early in the 1960's researchers initiated systematic study of the moon on a stratigraphic basis, recognizing that impacts did not merely excavate basins and raise circular ridges or rings, but also injected rock material to be deposited around the crater

or basin in discrete, three-dimensional strata, or ejecta sheets, of fragmented and melted rock. The lunar stratigraphy is based on the time sequence of such strata. These strata overlap one another in the order that they were deposited, just as do beds deposited in the oceans on the earth. Moreover, the oldest exposed surfaces have a higher abundance of craters, resulting from a continual flux of impacting projectiles over longer duration. Surfaces can be correlated in age and their relative age determined by the abundance and character of craters upon them. As on the earth, volcanism also forms beds of lava, which flows out upon the surface from fissures, and beds of pyroclastic material, which is erupted from vents. The maria contain ponds of lava that fill very large depressions formed by earlier impacts. This stratified character now seems obvious, but it became only gradually accepted as telescopic and spaceflight studies advanced during the 1960's. However, this stratigraphic approach provided the scientific basis for targeting the Apollo landings to insure that they would sample rock units of differing relative ages from both the maria and the highlands.

B. ANALYSIS OF LUNAR ROCK SAMPLES

Hundreds of scientists from many countries and many disciplines have been studying the samples, photographs, and instrumental data returned from the moon by the Apollo and Luna programs. Samples of rock and soil were collected and documented so as to maximize the scientific understanding gained at each landing site. These studies have placed significant limits on chemical and physical parameters, on the timing of many events, and on the rate of many processes, and have given insight into the natural processes that formed the moon and shaped its surface. Determination of the mineral and rock characteristics, along with a detailed chemical analysis in terms of major, minor, and trace elements, can be used to deduce processes that produced the various rock types and formed the major rock units. Isotopic techniques can be used to date major events in the history of a rock unit.

The study of the lunar samples required the coordinated efforts of a variety of scientific disciplines using many kinds of instrumentation. Studies of the mineralogy, mineral chemistry, texture, and bulk chemical composition of rocks are used to determine physical and chemical histories. Mineral composition is particularly important, since mineral assemblages reflect both major element chemistry and conditions of formation. Trace element chemistry is used to determine signatures of specific geochemical processes. Precise isotopic analyses, are used to study a wide variety of chronologic and geochemical problems. Long-lived radioactive species and their decay products (U–Th–Pb, K–Ar, Rb–Sr, Nd–Sm) are used to measure isotopic ages for rocks and to establish an absolute chronology. The stable isotopes of O, Si, C, S, N, and H provide geochemical tracers that are used in conjunction with chronologic data. Isotopic anomalies left by the decay of extinct short-lived radioactive isotopes provide evidence for very early conditions and time scales. Analysis of the rare gases (He, Ne, Ar, Kr, Xe) and their isotopes provide information on the interaction of the surface with the sun's radiation and the space environment. The samples were examined for evidence of magnetization, a clue to the history of the moon's magnetic field, and for a variety of physical properties, such as density, porosity, thermal conductivity, and seismic wave velocity.

These studies required the development of new precision for the analysis of small samples. They heavily utilized the techniques of optical and scanning electron microscopy, the electron microprobe analyzer, X-ray fluorescence, neutron activation analysis, gas mass spectrometry, and isotope-dilution, solid-source mass spectrometry. The lunar rocks are certainly the most intensively and extensively studied materials in the history of science—with the possible exception of the Allende meteorite, which has subsequently been studied with the same techniques. [*See* METEORITES, COSMIC RAY RECORD.]

II. The Lunar Surface

A major scientific task has been to understand the nature of the major surface features of the moon and to recognize the processes that have affected the evolution of the surface. Orbital photographs are available for most of the lunar surface, and higher resolution stereophotographs and remotely sensed data are available for the equatorial area. The lunar samples provide an understanding of the surface in terms of the rocks and their composition and chronology.

A. MAJOR SURFACE FEATURES

Even the naked eye, as illustrated on Fig. 1, detects marked differences in reflectivity over moon's surface. The dark-colored, smooth maria are topographically low basins within the light-colored highlands, which stand more than a kilometer above the maria and have an irregular surface due to the large number of craters. Mountain ranges occur in concentric rings around huge circular basins. Both external and internal processes have shaped this surface. Impacts of projectiles from space have formed cra-

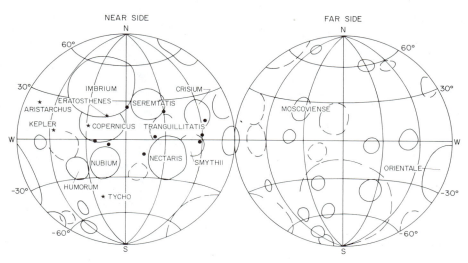

FIG. 2.   Distribution and names of major lunar basins and prominent rayed craters; (★) rayed craters; (●) Apollo/Luna landing sites.

ters of all sizes, including huge basins, and the ejecta thrown out from the impacts have produced secondary craters and extensive mantles of fragmented and melted rock. Basins dominate the highlands—more than 40 basins larger than 300 km in diameter have been recognized (Fig. 2) and referred to 15 age groups that can be dated by the abundance of craters on their ejecta blankets. The younger and larger of these basins (e.g., Orientale, Imbrium, and Crisium) have associated ejecta blankets that can be mapped over much of the moon, as shown in Fig. 3, and help to provide the basis for the stratigraphy (i.e., the time sequence of units, shown in Fig. 4.)

Such basins occur in roughly equal numbers on both the near and far side of the moon; basins on the near side are partially filled by dark-colored lava flows, whereas those on the far side, in general, are not (Fig. 5). The lava-filled maria have surface features such as flow fronts, channels, domes, depressions, and scarps, which are commonly observed on the earth to form during the flow of lava or during its cooling. Locally, very dark-colored thin deposits represent pyroclastic material erupted from volcanic vents. Craters are less abundant on the smooth mare plains, and these younger craters are better defined and bowl-shaped. The youngest craters (e.g., Tycho and Aristarchus) have thin, bright-colored radial rays of ejecta.

B. IMPACT PROCESSES

Bombardment of the lunar surface produced craters ranging in diameter from $<10^{-6}$ to $\sim 10^6$ m. The larger impact events produced major basins, breaking up the outer parts of the lunar crust, transporting material great distances across the lunar surface, and forming layered deposits of impact breccia (i.e., fragmented and recompacted rock). Some of these deposits have been extensively transformed by heat derived from the impact. Clastic fragments are sintered, partially melted, or intimately mixed with impact-derived melt. Some volcanic-textured rocks probably crystallized from impact-melted liquids. The lunar highlands are everywhere broken up, probably to a depth of tens of kilometers, by this process, but the younger mare plains are less heavily and less deeply cratered.

Repetitive bombardment by all sizes of meteorites over long periods of time has produced a generally fine-grained, stratified layer of debris, called regolith or lunar soil, which now blankets nearly all parts of the lunar surface. The regolith differs in thickness from place to place, but is typically a few meters thick on the maria. The debris is composed mostly of the rock types that immediately underlie the regolith, though these rock fragments may themselves be breccias from an impact blanket. Part of the rock fragments in the debris have been derived from a

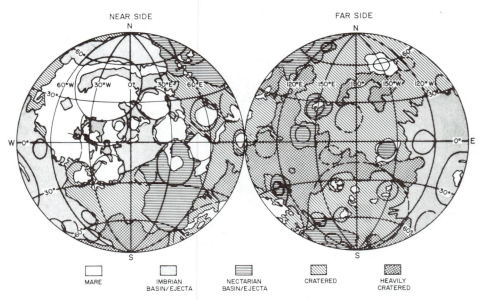

FIG. 3. Map of major geologic provinces of the moon. Ejecta blankets from the Imbrian and Nectarian basins can be seen to cover much of the surface.

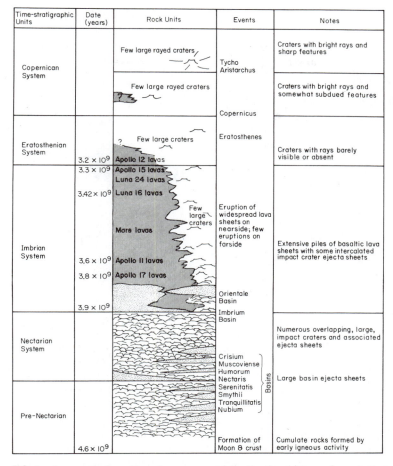

FIG. 4. Lunar stratigraphic column summarizing the time history of events that have shaped the moon's surface. The positions in the column of the Imbrian and Nectarian ejecta blankets provide a basis for widespread correlation, as do the mare lavas that have been dated by isotopic techniques.

considerable distance, having been dispersed across the lunar surface by impacts (and thereby increasing our knowledge of the lunar surface from the few sample sites). The regolith also contains abundant impact-produced glass and a few percent of meteoritic material from the impacting bodies. The impact of micrometeorites upon the regolith produces tiny puddles of silicate melt that cool rapidly to agglutinates of glass intimately mixed with rock and mineral grains. The abundance of such impact-produced glass agglutinates in a layer reflects the length of time that the layer was exposed to micrometeorite bombardment at the surface before being covered by a new layer thrown out from a nearby impact.

C. Highland Breccias and Ancient Rocks

The repeated impact episodes are reflected in the complexities of the fragmental rock (breccia) samples returned from the lunar highlands. The shock of the impact results in intense fragmentation and melting. The molten material may quench to glass during the ejection, or the hot mixture may remelt, sinter, and recrystallize during deposition and cooling in a thick ejecta blanket. The breccia samples range from friable aggregates to hard, sintered material with spherical vesicles that were bubbles filled by a gas phase prior to solidification. Many samples show multiple generations of impacts; fractured fragments of ancient rocks are within irregular

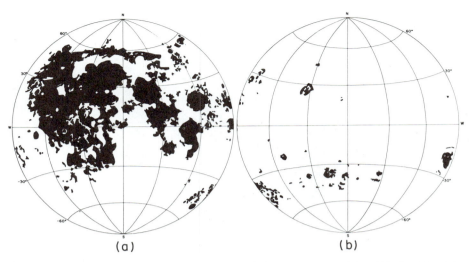

FIG. 5.   Distribution of basaltic flows in the mare basins on the (a) near side and (b) far side of the moon.

fragments of breccia that are themselves contained in a mixture of fragments and melt rock.

The major minerals within the highland breccias are anorthite-rich plagioclase ($CaAl_2Si_2O_8$), orthopyroxene ($[Mg,Fe]SiO_3$), and olivine ($[Mg,Fe]_2SiO_4$); these occur both as mineral fragments and as plutonic rocks made up predominantly of these minerals. The high content of anorthitic plagioclase and the low abundance of iron and titanium oxide minerals is responsible for the light color and for the characteristically high calcium and aluminum composition of the lunar highlands. The words anorthosite, norite, and troctolite are used in various combinations as adjectives or nouns to describe coarse-grained rocks made up of various combinations of these three minerals. Hence the acronym ANT is commonly used to describe this suite of rocks. Such rocks are found on the earth in layered igneous bodies that have crystallized from a silicate melt or magma very slowly deep beneath the surface. The term magma includes not only the complex silicate melt, but the various crystallizing minerals, and may include bubbles of volatiles and globules of sulfide or metal melt. Plagioclase-rich rocks such as the ANT suite do not form by simple crystallization of magma, but represent accumulation of early crystallizing minerals by floating or settling, as evidenced by terrestrial examples of cumulate rocks.

Despite the complex history, a number of fragments of ANT rocks collected from the

breccia have yielded isotopic ages greater than 4.4 billion years (Gy), indicative of crustal formation dating back almost to the origin of the solar system. The existence of an early crust was also inferred from geochemical evidence. The rare earth element europium, unlike the other rare earth elements, is highly concentrated in plagioclase during crystallization of a silicate melt. This element has a relatively high abundance in the highlands rocks and is relatively underabundant in the lunar basalts. These complementary anomalies are ascribed to extensive early differentiation of the primitive lunar material into a plagioclase-rich crustal cumulate of crystals and a more mafic melt, which eventually became the source of the lunar basalts. Hence, it is inferred that much of the outer part of the moon was molten that is, a magma ocean during the early part of lunar history.

This early differentiation seems also to have been responsible for another compositional class of material rich in K, rare earth elements, and P (KREEP). These elements are representative of the "incompatible elements" (which also include Ba, U, Th, and Rb) that do not enter the crystal structure of the major lunar rock-forming minerals and hence become concentrated in the residual liquid during final crystallization of a magma. The abundance of KREEP ranges greatly in the samples of highland breccias and regolith, occurring as both small rock fragments and glass. However, the uniformity of abundance pattern and the isotopic systematics, albeit partially disturbed in some cases, suggest a rather homogeneous source, one that was enriched in the incompatible elements at about 4.4 Gy. Orbital measurements of gamma rays have shown that material rich in K, Th, and U is concentrated in the region of Mare Imbrium and Oceanus Procellarum. The KREEP-rich material may have been distributed from these regions into the regolith by impact scattering.

D. MARE VOLCANISM

Four of the Apollo missions landed on the mare plains and returned samples of mare volcanic rocks. The lunar mare rocks are basalts similar in texture and chemical composition to volcanic basaltic rocks on the earth. The lunar basalts consist chiefly of the silicates—clinopyroxene ($[Ca,Fe,Mg]SiO_3$), anorthitic plagioclase ($CaAl_2Si_2O_8$), olivine ($[Mg,Fe]_2SiO_4$)—and the iron–titanium oxides, ilmenite and spinel. The mineralogy is generally similar to that on earth except for the very low content of sodium in the plagioclase, the high abundance of ilmenite, and the total absence of hydrous alteration minerals. Textures range from fine-grained and partially glassy, such as might be expected in a basalt that chilled quickly at the surface of a flow, to coarser-grained, interlocking textures that result from slower cooling within a flow.

Terrestrial basalts are formed by partial melting in the olivine and pyroxene-rich rocks of earth's mantle. As melting occurs, the early formed silicate melt has the composition of basalt, and it separates and rises to the surface. The detailed composition of a basalt provides a probe into the temperature, depth, and composition of the source; basalt can be readily dated, and the textures provide information on the cooling and crystallization history. Mare basalts differ from terrestrial basalts in some detailed elemental abundances. Lunar basalts (1) contain no detectable H_2O; (2) are low in alkalis, especially Na; (3) are high in TiO_2; (4) are low in Al_2O_3 and SiO_2; (5) are high in FeO and MgO; and (6) are extremely reduced. Lunar basalts contain no trivalent Fe, and the reduced ions (Fe metal, trivalent Ti, and divalent Cr) may be present.

The petrologic and chemical variation within the lunar basalts of the various missions is summarized in Fig. 6; their isotopic ages fall between 3.15 and 3.96 Gy. The samples can be divided into two broad groups: an older, high-titanium group (ages 3.55–3.85 Gy; TiO_2, 9–14%) and a younger, low-titanium group (ages

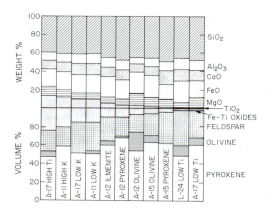

FIG. 6.   Mineral and chemical compositions of the basalt groups identified at the various landing sites. All consist predominantly of pyroxene and plagioclase feldspar, but Fe–Ti oxides (and TiO_2 content) and olivine differ in abundance.

3.15–3.45 Gy; TiO_2 1–5%). Laboratory experiments indicate that the high-titanium mare basalts could be derived from partially-melted titanium-rich cumulates at depths of about 100 km and that low titanium basalts could be produced by partial melting of olivine and pyroxene-rich rock at depths of 200–400 km.

E. LUNAR SURFACE HISTORY

Isotopic analyses of samples from its surface show that the moon, as a body, was formed at 4.6 Gy. It is similar in age to the earth and to meteorites derived from other bodies in the solar system. Isotopic ages have been determined for several events on the lunar geological time scale. Most of the recognizable major geologic events, formation of the basins and infilling of the mare basins, occurred during the first 1.5 Gy of lunar history.

During the first half billion years of lunar history intense meteorite bombardment pulverized and mixed the upper zones to great depths, culminating in the sequential excavation of the great multiringed basins with their widespread deposits of impact breccia (Figs. 3 and 4). Many feldspathic impact breccias have recrystallization ages greater than 3.9 Gy, and fragments of coarse-grained rock from within the breccias have ages of over 4.4 Gy, dating back to the formation of the lunar crust. The bombardment rate declined rapidly after 4 Gy, approaching the current rate.

Commencing about 3.9 Gy vast floods of basaltic lava poured out on the lunar surface within the basins, particularly those on the near side

(Fig. 5). These basaltic eruptions occurred significantly later in time than the excavation of the basins and filled in the topographically low areas of the basins. Although these basalts cover about 17% of the moon's surface, they constitute less than 1% by volume of the crust. The measured isotopic ages range from 3.9 to 3.2 Gy, but photographic and remote sensing studies indicate that dark-colored basalts on the western near side are somewhat younger and were not sampled by the missions.

Except for minor lava flows and a relatively small number of large impact craters, the face of the moon has remained largely unchanged for the last 3 Gy. A few of the very young impact craters have been dated by use of returned samples.

III. The Lunar Interior

Our knowledge of the lunar interior consists of inferences based on geophysical data and the detailed studies of the surface samples. Especially important have been the moonquake and artificial-impact recordings by the landed seismometers. The major results (the existence of a crust and the low level of dynamic activity of the interior) have major implications for our understanding of the moon.

A. Chemical Composition of The Interior

To a first approximation lunar chemistry is dominated by the crustal formation process, and the interior composition can only be inferred. Neither the mare basalts nor the plagioclase-rich rocks of the highlands have compositions that are consistent with the mean density of the moon. Laboratory experiments show that both these rock types would undergo phase changes to higher density materials at fairly shallow depths, which would yield too high a lunar mean density. This evidence, along with the high radioactivity of the surface rocks, the heat flow measurements, and isotopic data, demonstrate that the moon has a crust with a composition different from its bulk composition. Consideration of the mean density indicates that the moon must have less metallic iron than the earth, and that any metallic core must be very small.

The mare basalts were derived by partial melting of the upper mantle and provide a sensitive probe to the composition of the lunar interior and, by inference, the bulk moon. The bulk composition of moon is not primitive, that is, it does not have the composition of the sun or of carbonaceous chondrite meteorites, that are thought to be very primitive because of their similarity to the solar composition. Apart from the obvious paucity of gaseous elements, such as hydrogen and helium, several differences are known.

1. Volatiles

Volatiles are those elements that condense from the solar nebula below about 800°C: The total lack of hydrated minerals in lunar rocks indicates that water has been in very low abundance throughout lunar history. This low water content is also indicated by the very low attenuation of seismic waves in the lunar crust. The abundance of indigenous carbon is also strikingly low; crystalline rocks contain up to a few tens of parts per million by weight of carbon. Regolith samples contain up to an order of magnitude more, but this is largely trapped from the solar wind as it impinges on the surface. Nitrogen is equally rare. The moon is, and probably always has been, an exceedingly inhospitable environment for organic chemicals, prebiotic or biotic. Lunar rocks are also consistently deficient in alkali metals (Na, K, Rb) by up to a factor of ten relative to their terrestrial counterparts. The elements of higher volatility such as Pb, Bi, Tl, Br, etc., are even lower in relative abundance (by factors of 10–100). Most, if not all, of these low abundances probably reflect initial deficiency, not subsequent loss.

2. Refractories

Refractories are those elements with condensation temperatures above 1200°C. Lunar surface rocks, in general, contain the refractory elements (Al, Ca, Sr, Ti, Zr, Th, U, and rare earths) in such high abundances (up to several hundred times chondritic abundances) that it appears certain, especially in conjunction with the heat flow evidence in Section III,B that the bulk moon is enriched in these elements relative to chondritic meteorites. Such an enrichment is consistent with the apparent initial deficiency of volatile elements.

3. Siderophiles

Siderophiles are those elements that tend to enter iron metal rather than oxides, sulfides, or silicates. Metals such as Au, Ir, Re, Ni, etc., are depleted in lunar surface rocks 10^{-4} to 10^{-5} times their solar abundance. They are also much more depleted than on the earth; gold, for example, being more depleted on the moon by two orders of magnitude. This difference suggests that the moon lost its noble metals under physical and chemical conditions quite different from those of the earth. The low abundance of siderophile elements, coupled with the low density of the moon, suggests the separation of a metal phase from the lunar material prior to the formation of the moon as a body.

4. Lunar Oxygen Isotopic Compositions

These compositions are similar to those for terrestrial rocks and basaltic meteorites, but differ significantly from those for chondritic meteorites.

B. INTERNAL TEMPERATURE AND RADIOACTIVITY

Indirect evidence of temperatures in the moon's interior comes from the relative attenuation of seismic waves at various depths, the depth of present seismic activity, the electrical conductivity profile deduced from the interaction of the moon with abrupt changes in the interplanetary magnetic field, and measurements of the near-surface heat flow.

The high attenuation of seismic waves penetrating to depths greater than 800 km and the occurrence of all sizable moonquakes near that depth suggests that partial melts may exist below that depth in the present moon. If so, temperatures must be about 1500°C at that depth. Electrical conductivity profiles of the interior are interpreted as indicating interior temperatures of 1300–1500°C.

Two measurements of the surface heat flow yielded values that are about one-third that of the average heat flow on the earth, but are higher than would be predicted by thermal history calculations for a moon with a bulk uranium and thorium content like that of chondritic meteorites. This greater abundance of uranium supports the indication from the surface samples that the whole moon is enriched in refractory elements relative to the chondrites and the earth. Such a high abundance of long-lived radioactive elements, would lead to total melting of the interior, unless they are mainly concentrated near the surface; this provides further evidence that a substantial portion of the moon's interior is radially differentiated.

C. SEISMICITY

A large number of very small moonquakes have been detected by the Apollo seismic network. The total seismic energy release within the moon appears to be about 80 times less than that in the earth. The moonquakes are concentrated at great depth—between 600 km and 1000 km, which is deeper than earthquakes. The difference in distribution and magnitude of seismic energy release on the earth and the moon is probably related to fundamental differences in interior dynamics. Correlation of deep moonquakes with lunar tides indicates that tidal stresses must play an important role in triggering them.

D. GRAVITY AND SHAPE OF MOON

The gravity field and the shape of the moon have been determined from orbital measurements. Broad-scale gravity anomalies indicate regional variations in the moon's internal density. Furthermore, the center of mass is about 2 km closer to the earth than the center of figure; this has been attributed to greater crustal thickness on the far side.

Striking positive gravity anomalies, the mascons, coincide with major ringed basins. The filled multiringed basins on the near side are mascons, whereas the unfilled multiringed basins on the farside are negative anomalies. Thus, the mascons are associated with the filling rather than the excavation of the basins. The excess mass, in conjunction with the fact that the surface of the maria lies below that of the surroundings, can be understood by the replacement of lower density crustal rocks by higher density basalt. The mascons are associated with topographic basins that have persisted for more than 3 Gy. To support these mascons, requires that the outer few hundred kilometers of the moon have great strength, and seismic data support this interpretation.

E. MAGNETIC FIELD

The moon has a very small magnetic field at the present time. However, the returned lunar samples contain natural remanent magnetism of lunar origin, which is probably due to cooling of fine particles of metallic iron in an ambient magnetic field. Local remanent magnetic fields measured at Apollo landing sides and over large regions scanned by the satellites indicate that magnetic fields were present at the lunar surface during a period of igneous activity. The localized surface fields with scales of 1–100 km suggest that a crust, initially magnetized in a field of global scale, was broken up by impacts. Fields 50–500 times less than earth's fields, but 20–200 times greater than the field in the solar wind, are required to produce such fields. One interpretation is that the moon had a liquid core, rich in metallic iron, which behaved as a dynamo during the time when the mare basalts were crystallizing. The only alternative explanation would be that there existed an external field of unknown origin.

F. Internal Structure

A variety of evidence indicates that the moon has been radially differentiated and is a layered body with a crust, mantle, and possibly core as shown in Fig. 7. It is covered by an extremely dry, highly heterogeneous layer to a depth of about 25 km. This layer is related to the extensive fracturing of the crust by cratering processes, but layered sequences due to impact blankets and lava flows contribute to the general complexity of this zone. Seismic velocities are nearly constant from 25 to 65-km deep with a value consistent with a plagioclase–pyroxene rock. A large velocity increase at about 65 km marks the base of the crust. The 600–900-km thick upper mantle has low attenuation of seismic waves and nearly constant velocities. These and other properties are consistent with a pyroxene-rich composition with olivine, spinel, and possibly plagioclase in small amounts. Below about 700 km the shear-wave velocity decreases, attenuation increases, and deep moonquakes occur; this may represent a partially molten zone. The moment of inertia and overall density of the moon and other data do not rule out the existence of a small iron-rich core, but do place an upper limit of about 500 km on its radius.

IV. The Moon–Space Interface

Unlike the earth, which is partially protected by its magnetic field and atmosphere, the flux of particles and radiation from the sun and of meteorites upon the lunar surface has imprinted evidence of the history of the solar system upon the surface materials. The complex conditions at this interface with space play a major role in giving the surface of the moon its appearance, and in giving the lunar regolith its many novel characteristics.

A. Record of Lunar Surface Environment

Lunar soil properties cannot be explained strictly by broken-up local rock. Distant impacts throw in exotic material from other parts of the moon. About 1% of the soil appears to be of meteoritic origin. Vertical mixing by impacts is important; essentially all material sampled from lunar cores shows evidence of having resided at the surface.

The layers of the regolith preserve a record of meteorite, solar particle, and cosmic-ray bombardment. The surface layers exposed to space contain a chemical record of implanted solar material (rare gases, H, and other elements transported from the sun in the solar wind). During the impact processes, lunar surface material became heated, and lead and other volatile and gaseous elements were mobilized and then trapped in the regolith, enriching it in such elements. Radiation damage effects dominate the physical properties of the surface layers. The surfaces of rock samples record the flux of micrometeorites by the presence of numerous small impact craters, many of them microscopic in size.

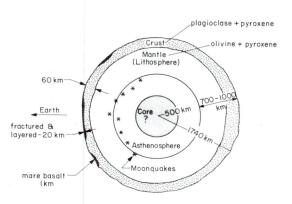

FIG. 7.   Diagrammatic section of the lunar interior.

The abundance of atomic nuclei produced by cosmic-ray bombardment provides a measure of the time duration that rock material resided in the upper few meters of the regolith before being turned under or covered by impact debris. Individual rock fragments found on the surface have been shown to have resided with a few meters of the surface for times (exposure ages) varying from a few million to 500 million years. A few parts of the regolith are known to have remained undisturbed to meter depths for several hundred million years. The ages of several craters have been determined by measuring exposure ages on blocks thrown out from the crater.

B. Solar Record

The isotopic composition of the heavy rare gas component of the solar wind has been measured in soils and lunar samples. Amorphous surface films, produced by solar wind bombardment, are observed on many lunar grains. Comparison with artificial irradiations has set limits on fluctuations in the energy of the ancient solar wind.

The concentrations of hydrocarbons (mainly methane and ethane) correlate with the solar wind irradiation of different samples of lunar soil regolith. It is believed that these compounds formed by trapping and reaction in the surficial damaged layers. Since interstellar space contains both dust clouds and sources of energetic particles, this may be an important model for organic synthesis in the galaxy. Large isotopic fractionation effects for oxygen, silicon, sulfur, and potassium are present in the surface layer.

Dramatic progress has been made in the study of solar flare particles. The energy spectrum of heavy solar particles has been established. It has been shown the lowest energy solar cosmic rays are highly enriched in very heavy nuclei compared to normal solar material, the first demonstration of preferential heavy ion acceleration by a natural particle accelerator.

Information has been obtained on solar flare activity on the sun over geologic time by measuring the induced radioactivities and nuclear particle tracks produced in the outer layers of surface material. The average solar flare activity has not changed appreciably over the past few million years, suggesting that major climatic changes during this time are not related to large-scale changes in solar activity, as had previously been postulated.

V. The Evolution of Moon

A major result of Apollo lunar science has been the understanding of the evolution of another planetary body for comparison with that of the earth. The moon and probably the earth went through a period of major igneous activity and a high rate of impact during their first 1.5 Gy. However, since then the moon has been thermally and mechanically quiescent; if our exploration of the moon had taken place 2.5 Gy earlier, few aspects of the moon would have been different. The importance of understanding this very early history lies mainly in its implication for the earth and the rest of the solar system. Nowhere else have we had access to as detailed an ancient record of planetary evolution. Our knowledge of this evolution is summarized in the cartoon in Fig. 8.

A. Chemical and Isotopic Composition

The bulk composition of the moon is not primitive; it does not have the composition of the sun or of the primitive chondritic meteorites. The moon seems to contain more than its cosmic share of refractory elements and much less than its cosmic share of volatile elements. Yet, the composition of the moon cannot be explained simply by cessation of equilibration with the vapor of the solar nebula at some high temperature. The amount of oxidized iron (in silicates and oxides) and the abundances of volatile metals suggest equilibration of part of the condensate at 200°C. Thus the moon seems to be a mixture of materials that separated from the nebula over a wide range of temperature, like the chondrites and presumably many other bodies in the solar system.

Measurements of oxygen isotopic compositions in meteorites have provided evidence of compositional inhomogeneities in the solar nebula. The earth, moon, and the basaltic meteorites could have formed from a common oxygen isotope reservoir, separate and distinct of that of the chondritic meteorites.

B. Crustal Formation and Early Differentiation

The moon has differentiated into layers. Seismic and gravity measurements indicate that a crust (~60-km thick) of low density (~3.0 g/cm^3) material overlies a higher density (~3.35 g/cm^3) mantle. Isotopic, chemical, and petrologic

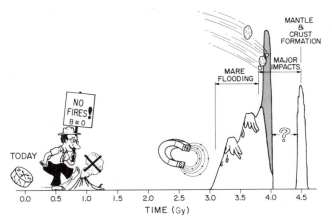

FIG. 8. A cartoon showing the chronology of major events in lunar history. [From Wasserburg *et al.* (1977). *Philos. Trans. R. Soc. London A* **285.** 7–22.]

data indicate that the crust originated by magmatic differentiation very early in lunar history. There was no knowledge of a lunar crust before the Apollo missions; indeed, the moment of inertia suggested that the moon was nearly uniform, and hence undifferentiated. One widely held view at this time was that the moon had accreted cold and had not been heated sufficiently by internal radioactivity to melt.

No samples of "primitive" material were returned by Apollo. However, a variety of igneous cumulate rocks (the ANT suite of rocks) were returned. These probably formed in the early differentiation of the lunar crust, since isotopic ages of greater than 4.4 Gy have been measured on a number of such samples.

Petrologic, geochronologic, geochemical, and geophysical constraints support the idea that accretional melting involved at least the outer few hundred kilometers. Crystallization could have formed a chilled surface layer underlain by (1) an anorthiterich cumulate layer; (2) a pyroxene–olivine–spinel cumulate layer; and (3) trapped residual melts, rich in Ba, U, K, REE, P, Th, and highly concentrated in an intermediate level. The layering would have been disrupted by early impacts, but such magmatic differentiation produced a body systematically zoned in both major and minor elements.

C. IMPACT HISTORY

From 4.6 to 3.9 Gy lunar petrologic history was dominated by impact brecciation and heating to great depths. The large mare basins represent only the last stages of the impact of large bodies on the lunar surface. The profound cratering on all scales of the highlands provides a record of this period of impact and cratering. The Apollo age dates indicate that the Imbrium basin, the second youngest of the large mare basins, was formed only 0.5 Gy after the moon's accretion. During the first half billion years, the rates of impact may have been so great at times that most of the large craters that formed early were totally obliterated by later impacts. Similar catastrophic events probably affected the earth and the other inner planets.

The craters on the mare surface are a record of impacts during the last 3 Gy. The average cratering rate has been much lower than during the first half billion years. Only a limited number of medium-sized (20–100 km) craters such as Copernicus and Tycho formed during this period. Study of particles from the regolith at the Apollo 12 site suggests that Copernicus may have formed 850 million years ago.

D. MARE FILLING AND VOLCANISM

Many large multiringed basins, especially those on the near side, were filled with basaltic flows in the interval 3.9–3.2 Gy. These eruptions occurred significantly later than the formation of the basins. The individual flows are thin and very extensive, reflecting the low viscosity of the lava; they differ slightly in composition from one another. Plagioclase-rich basalts and KREEP-rich basalts with ages greater than about 3.9 Gy may have been derived either by partial melting at shallow depths or from impact-produced melts. From 3.8 to 3.6 Gy titanium-rich mare basalts were extruded, and experimental studies suggest that these were derived from partially-melted titanium-rich cumulates at depths of about 100 km. Low-titanium mare basalts were extruded at 3.4–3.2 Gy, produced by partially melted olivine pyroxenite at depths of 200–400 km. High-titanium flows, somewhat younger than 3.2 Gy, have been identified by photogeologic methods in the western near side of the moon, but were not sampled. It seems likely that the regional lunar volcanic activity stopped by about 2.3 Gy.

E. THERMAL HISTORY

Lunar heat flow, electrical conductivity, seismic velocities and the degree of attenuation at various depths, lunar viscosity, seismicity, the presence of mascons, and tectonism provide evidence on the present-day temperatures and are manifestations of the internal structure. Magnetized lunar rocks, the early differentiated lunar crust, the abundance of heat-producing elements, and the chronology of lunar magmatic activity indicate the thermal state earlier in lunar history. The outer 700 km is relatively cool, strong, and inert—and has been so for about the last 3 Gy. However, the outer part of the moon, at the very least, melted early in lunar history during the closing stages of accretion by impact. This melting permitted extensive magmatic differentiation with upward concentration of radioactive heat sources into the lunar crust. Below 700 km, the moon may still be partially molten with temperatures between 1000 and 1600°C.

VI. The Origin of Moon

To understand the early history of the solar system, it is necessary to know the bulk chemical composition of the planets and their satellites, and also to know the chemical rules that govern the assembly of planetary bodies from a cosmic cloud of dust and gas. The analysis of the lunar rocks is the first step in the chemical and isotopic mapping of the solar system beyond the earth by direct study of returned samples. Similar analyses performed on chondritic meteorites have provided considerable insight into the behavior of the elements during the condensation of the solar nebula.

Prior to lunar exploration it was commonly assumed that the bulk compositions of the inner planets were closely similar to the composition of chondritic meteorites, which are near solar abundance in composition except for the most volatile elements. The sample studies have shown dramatically that the moon is not chondritic in composition, a fact which seriously challenges our earlier assumption about the makeup of the inner planets.

It now appears that the moon and the earth accumulated from material quite changed from solar abundances. These changes consist of enrichment in elements, which are believed to have condensed from the solar nebula at high temperatures and depletion of many less refractory elements as well as the very volatile gases. The moon is apparently enriched in high-temperature condensates, possibly similar to samples preserved as inclusions in the Allende meteorite and similar carbonaceous chondrites. However, meteoritic inclusions have oxygen isotope compositions so different that they must have formed in a different part of the nebula.

Despite the detailed studies, the problem of lunar origin remains. How is it possible for a planet to become enriched in the refractory condensates? What physical processes operate in the solar nebula to concentrate refractory-rich dust from volatile-rich dust? Much more work is obviously required in order to understand this complicated chemical fractionation.

However, our conception of the process that generated the planets has been altered. We now consider that a variety of planetary objects were formed, which ranged from accumulations of material enriched in refractory elements (the terrestrial planets) to relatively unfractionated bodies, represented by the outer planets and all the low-density objects in the solar system. The earth no longer appears peculiar, but seems to belong to a compositional class of planets. The compositions of the planets provide a basis for reconstructing the accumulation processes that governed planet formation. It now appears possible to determine the duration of the accumulation processes and the time scale required for chemical segregation within the planets. The moon preserves the early stages of internal segregation and also shows that the actual accretion process, and major collisions, extended for a much longer period that was considered before. Whatever the origin of the impacting objects (local or distant), the fact that catastrophic impacts occurred as recently as 3.9 Gy on the moon suggests the possibility that the earth suffered similar bombardment, which may have strongly affected its early geological record.

The lunar studies of the past few decades have not produced conclusive evidence against any of the theories of the moon's origin. However, various coaccretion and impact fission models account for both the chemical differences and similarities between the earth and the moon and seem to require the least number of assumptions.

BIBLIOGRAPHY

Basaltic Volcanism Study Project (1981). "Basaltic Volcanism on the Terrestrial Planets." Lunar and Planetary Institute. Houston, Texas.

Burnett, D. S. (1977). Lunar Science: The Apollo legacy, *Rev. Geophys. Space Phys.* **13**(3), 13–34.

Carr, M. H., Saunders, R. S., Strom, R. G., and Wilhelms, D. E. (1984). The geology of the terrestrial planets, *NASA* [*Spec. Publ.*] **SP-469**.

Hartmann, W. K., Phillips, R. J., and Taylor, G. J. (eds.) (1986). "Origin of the Moon." Lunar and Planetary Institute, Houston, Texas.

Murray, B., Malin, M. C., and Greeley, R. (1981). "Earthlike Planets," Freeman, San Francisco, California.

Taylor, S. R. (1982). Planetary Science: A Lunar Perspective: Lunar and Planetary Institute. Houston, Texas.

Wasserburg, G. J., Papanastassiou, D. A., Tera, F., and Huneke, J. C. (1977). I. The accumulation and bulk composition of the moon: Outline of a lunar chronology, *Philos. Trans. R. Soc. London A.* **285,** 7–22.

Wilhelms, D. E. (1986). Geologic history of the moon, *U. S. Geol. Surv. Prof. Pap.,* no. 1348.

METEORITES, COSMIC RAY RECORD

Robert C. Reedy   *Los Alamos National Laboratory*

GLOSSARY

Complex history: Case where a meteoroid was exposed to appreciable intensities of cosmic rays both before and after it suffered a collision in space that changed its geometry.

Cosmogenic: Product of cosmic ray interactions.

Electron volt: Unit of energy, whose symbol is eV, that is necessary to raise one electron through a potential difference of 1 volt. It is equal to 1.602×10^{-19} joules. Nuclei in the cosmic rays typically have energies ranging from about a mega-electron-volt (MeV) (10^6 eV) or less to many giga-electron-volts (GeV) (10^9 eV) per nucleon.

Exposure age: Length of time that a meteorite has been irradiated by a sufficient intensity of cosmic ray particles to produce an observable record.

Meteoroid: Object in space that, after it survives passage through the earth's atmosphere and hits the surface, is called a meteorite.

Noble gases: Elements helium, neon, argon, krypton, and xenon, which are gases at normal temperature and pressure, almost never form compounds, and are usually very rare in meteorites. They are sometimes called rare gases or inert gases.

Radionuclide: Atomic nucleus that is unstable and decays to a different nucleus by the emission of radiation, such as ^{10}Be, which emits an electron and becomes a ^{10}B nucleus. Each radionuclide decays with a specific half-life, the time required for half of an initial population to have decayed away.

Shielding: Location of a sample inside a meteoroid and the size and shape of the meteoroid, which are important in determining the rates that cosmogenic products are produced in that sample.

Simple history: Case where a meteoroid, having previously been deeply buried in its parent object far from the cosmic rays, received its entire cosmic ray record without experiencing a collision that radically changed its geometry.

Spallation: Type of nuclear reaction induced by an energetic particle in which one or more nuclear fragments (nucleons, α particles, and so on) are removed from a nucleus, such as the production of ^{53}Mn from ^{56}Fe. A product of such a reaction is called a spallogenic nuclide.

Thermoluminescence (TL): Property of certain minerals that, after electrons are excited into traps by ionizing radiation, they emit light when moderately heated.

Track: Path of radiation damage made by heavily ionizing particles, such as iron nuclei, in a dielectric medium, such as olivine, or other meteoritic minerals. These paths normally are observable only by electron microscopes but can be dissolved and enlarged by certain chemicals that preferentially etch along the damage zone.

Very heavy (VH) or very very heavy (VVH): Label applied to two groups of cosmic ray nuclei with VH (very heavy) nuclei having $20 \leq Z \leq 28$ and VVH (very very heavy) nuclei having atomic numbers of 30 and greater.

The energetic particles in the cosmic rays interacted with meteorites in space and produced a variety of products. Many of these products can be measured and identified as having been made by the cosmic rays. Most cosmogenic products in meteorites are made by the high-energy galactic cosmic rays, although some products made by energetic particles emitted from the sun are observed. These cosmogenic products include radionuclides with a wide range of half-lives, stable isotopes rare in meteorites (such as noble gases), radiation-damage tracks of heavy cosmic ray nuclei, and trapped electrons observable by thermoluminescence. These cosmogenic products are made by known processes, and their concentrations and distributions in meteorites can be fairly well predicted. The cosmic ray record in meteorites show that the cosmic rays in the past are not much different than those today, although the cosmic ray intensities have varied over time periods of ~10 to ~10^8 yr. Meteorites have been exposed to cosmic rays for periods ranging from about 10^4 to over 10^9 yr, and some fell to the earth as long as 10^6 yr ago. The cosmic ray exposure ages for some classes of meteorites cluster in groups, which suggest that those meteorites were all produced by an event at that time. The exposure histories of some meteorites are complex. These exposure records give information on the recent evolution of meteorites in space.

I. Introduction

A. GENERAL

The solar system is filled with more than just the planets and other large objects like the asteroids and comets. It contains many small stony, iron, or stony-iron objects called meteoroids, which are referred to as meteorites if they survive passage through the atmosphere and reach the earth's surface. Also moving through the solar system is a variety of energetic particles, the cosmic rays. Most of these cosmic ray particles possess sufficient energy (above about 10^6 eV) that they can penetrate and interact with matter. These interactions produce atomic displacements and ionization in solid media; they also can induce nuclear reactions. Many of these cosmic-ray-produced effects persist for long periods of time and can be observed in meteorites and identified as being the products of cosmic ray interactions. The collective ensemble of these cosmic-ray-produced effects in a meteorite forms a record that provides us with information about the cosmic ray interactions and about

the histories of both the cosmic rays and the meteorites. [*See* COSMIC RADIATION.]

There are wide ranges for the mineralogies and chemistries of meteorites. Within each of the three major classes of meteorites (stones, which are silicate rich; irons, which are metal rich; and stony irons, which are half silicates and half metal), there are a number of distinct types. There are two main categories for stones: chondrites, which are fairly primitive and named after the spherical or ellipsoidal chondrules that they contain, and achondrites, which have chemically evolved relative to chondrites. A meteorite is considered a "fall" if it was observed to hit the earth; otherwise it is called a "find," and the length of time that it was on the earth's surface, its "terrestrial age," is unknown. Stony meteorites are normally hard to detect among terrestrial rocks and are fairly rapidly destroyed by weathering (in ~10^4 yr as determined from cosmogenic radionuclides), so most stones are falls (except for those found in unusual regions of the earth). The converse is true for iron meteorites; most are finds that fell a long time ago. Since 1970, a large number of meteorites of many types have been found on several ice fields in Antarctica. During its passage through the earth's atmosphere, a meteorite is subjected to high pressures and its outermost layers are substantially heated. The pressures often cause a meteorite to break into fragments. The heated outer layers are removed by a process called ablation. Both processes affect the geometry of the meteorite and make the study of the cosmic ray record of a meteorite more difficult. Meteorites with low-ablation regions or that fell as a single or mainly in one piece are rare but are valuable in studies of the interactions of cosmic rays with meteoroids.

Here, we shall consider only two types of energetic particles that irradiate meteoroids in space: the galactic cosmic rays (GCR), which come from outside the solar system, and solar energetic particles or solar cosmic rays (SCR), which are emitted irregularly by major flares on the sun. We shall not discuss the irradiation of meteorites by the low-energy particles in the solar wind, although some meteorites do contain a record of solar wind particles. Both the GCR and SCR particles are mainly protons, about 10% α particles (^{4}He nuclei), and 1% heavier nuclei. The nuclei heavier than about calcium ($Z = 20$) produce tracks in certain meteoritic minerals. The protons and α particles induce a variety of nuclear reactions with atomic nuclei in meteorites, many of which produce nuclides, like radionuclides and noble gas isotopes, that are not normally present in a meteorite. All primary and most secondary cosmic ray particles are ionizing radiations that also produce effects observable by thermoluminescence. The cosmic ray particles penetrate at most a few meters into a meteoroid before they are stopped by ionization energy losses or removed by nuclear reactions. Thus meteoritic material that was shielded by more than a few meters of solid matter contain essentially no record of the cosmic rays, and a cosmic ray record only tells us about the most recent history of a meteorite.

A wide variety of techniques are used to study the cosmic ray record of a meteorite. A number of models predict the concentrations of these cosmogenic products expected in meteorites and are used to interpret the record. Meteorites with simple histories in space can be used to study the nature of the cosmic rays in the past. Temporal variations in the intensities of the GCR particles have been observed for periods ranging from over an 11-yr solar cycle to over the last 10^9 yr. The cosmic ray record can be used to infer the size and shape of the preatmospheric body of a meteorite. It also can tell us how long a meteorite was exposed to the cosmic rays. Often the cosmogenic products were made both before and after a meteoroid was altered in space by a collision with another object.

B. HISTORY OF THE FIELD

Although the cosmic radiation was first detected in 1911 by V. Hess from balloon flights with ionization chambers, its exact nature was not known for several decades. By the late 1930s, it was known that the cosmic rays are atomic nuclei moving at high energies. In the late 1940s, W. Libby and co-workers established the use of cosmogenic radiocarbon, ^{14}C, to date terrestrial samples. The activities of ^{14}C measured in objects of known ages agreed well with predicted values, showing that the intensities of the cosmic rays had been fairly constant over the last several thousand years. Also in the 1940s, helium was measured in a number of iron meteorites, but it was assumed that all of the helium was made by the radioactive decay of uranium and thorium. C. Bauer argued in a short note published in 1947 that most of the helium was produced by the cosmic radiation. Soon F. Paneth and others measured that about 20% of

the helium in iron meteorites was ^3He, thus confirming the origin of the helium as the product of nuclear reactions between the energetic particles in the cosmic rays and the iron. At this time, accelerators were starting to produce protons with energies of a few giga-electron-volts (GeV), and nuclear scientists were systematically studying spallation reactions similar to those that make helium in iron meteorites.

In the decade after Paneth's initial measurements of cosmogenic helium in iron meteorites, the study of cosmogenic nuclides in meteorites advanced rapidly. Cosmogenic noble gas isotopes other than helium were measured in several stony and iron meteorites. Then a number of cosmogenic radionuclides were observed in meteorites, starting with the detection of 12.3-yr ^3H by low-level gas counting techniques. Shortly thereafter, other cosmogenic radionuclides, such as ^{26}Al, ^{10}Be, and ^{60}Co, were observed in meteorites, and the cosmogenic pair of $^3H/^3He$ was used to determine the exposure ages of meteorites. Around 1960, a number of systematic studies of cosmogenic nuclides in meteorites were done. Noble gas isotopes were measured in many pieces from slabs of several iron meteorites, such as Carbo and Grant. Many radionuclides were measured in several freshly fallen meteorites, such as the iron Yardymly (then called Aroos) and the stone Bruderheim. Several research groups also measured noble gas isotopes in a suite of meteorites. New techniques to measure cosmogenic radionuclides, such as nondestructive γ-ray spectroscopy, were developed.

In parallel with the rapid growth in cosmogenic nuclide measurements, a number of theoretical models were developed. Simple models in which the primary cosmic ray particles are exponentially attenuated and secondary particles are produced and removed were used by several investigators, such as P. Signer and A. Nier, who applied such a model to their noble gas data for the Grant iron meteorite. Numerous experiments at accelerators in which thin or thick targets were bombarded with high-energy protons established production ratios for many cosmogenic nuclides. J. Arnold, M. Honda, and D. Lal estimated the energy spectrum of primary and secondary cosmic ray particles in iron meteorites and calculated production rates of cosmogenic nuclei using cross sections for many reactions.

These measurements and models for cosmogenic nuclides in meteorites were applied to a number of studies, such as the constancy of the cosmic rays over time. The ratio of the measured radionuclide activities to the calculated values showed no systematic trends for half-lives less than a million years. H. Voshage and H. Hintenberger found that the $^{40}K/^{41}K$ ratios measured in iron meteorites were inconsistent with the ratios for other radioactive/stable pairs of cosmogenic nuclides, implying that the fluxes of cosmic rays have been higher over the last 10^6 yr than over the last 10^9 yr. E. Fireman, R. Davis, O. Schaeffer, and co-workers used the measured $^{37}Ar/^{39}Ar$ ratios in studies of the spatial variation of cosmic rays between 1 and about 3 astronomical units (AU) from the sun (the region of space in which most meteorites probably traveled). A variety of pairs of radioactive and stable nuclides, such as $^3H/^3He$ and $^{39}Ar/^{38}Ar$, were used to determine the lengths of times that meteorites were exposed to cosmic rays.

During the 1960s, additional measurements were made and the ideas used to interpret the observations were refined. Measurement techniques for the tracks produced in certain minerals by heavy nuclei were developed and applied to meteorites. Some meteorites, those with high concentrations of trapped gases and tracks, were realized to have been exposed to energetic solar particles on the surface of some parent object. The orbits of three stony meteorites, Pribram, Lost City, and Innisfree, were accurately determined by several photographic networks and all had aphelia in the asteroid belt and perihelia near 1 AU. Bombardments were done at accelerators to simulate the cosmic irradiation of meteorites and used to predict the profiles for the production of nuclides in meteorites. In the early 1970s, the studies of cosmogenic nuclides in meteorites declined as most investigators were studying lunar samples. Some new methods for measuring cosmogenic nuclides were perfected using lunar samples. The lunar-sample studies confirmed meteoritic results about the galactic cosmic rays and gave us our first detailed knowledge of cosmogenic nuclides produced by the solar cosmic rays. In the late 1970s, interest in the studies of meteorites increased, especially for stones. Cosmogenic nuclides were being measured in samples much smaller than previously possible because of the use of improved or new measurement techniques (like accelerator mass spectrometry). Measurements of the distributions of the cosmic rays in the solar system by various satellites helped in interpreting the meteoritic cosmic ray record. By the mid-1980s, the study of the cosmic ray records of meteorites was a mature field with gradual but steady advances in all of its aspects.

II. Cosmic Rays and Their Interactions with Meteorites

A. NATURE OF COSMIC RAYS

In the solar system move nuclei with a variety of energies and compositions. Energies range from slow moving atoms and molecules to highly relativistic nuclei in the galactic cosmic rays. Here we are only considering the cosmic rays, not the lower-energy particles such as those in the solar wind. Table I summarizes the typical mean fluxes and energies of the particles in the two types of cosmic rays, the galactic cosmic rays (GCR) and the solar cosmic rays (SCR), which are often called solar energetic particles. The nuclei in both types of cosmic rays are mainly protons and α particles (with a proton/α particle ratio of about 10 to 20), with about 1% heavier nuclei ($Z \geq 3$). The elemental

TABLE I. Typical Energies, Fluxes, and Interaction Depths of Cosmic Ray Particles

Radiation	Energies (MeV/nucleon)	Mean flux (particles/cm² sec)	Depth (cm)
GCR protons and α particles	100–3000	3	0–100
GCR nuclei, $Z > 20$	100–3000	0.03	0–10
SCR protons and α particles	5–100	~100	0–2
SCR nuclei, $Z > 20$	1–50	~1	0–0.1

distribution of these heavier nuclei tends to cluster in a number of groups, such as one for carbon, nitrogen, and oxygen, and one around iron.

The temporal and spatial distribution in the inner solar system (out to ~5 AU), where most meteoroids orbit, of both types of cosmic ray particles are strongly influenced by interplanetary magnetic fields that originate at the sun. Earth-based observations established the role of the sun in controlling the variations of the cosmic rays. The sun's activity generally varies with an 11-yr period. During periods of maximum solar activity, sunspots on the solar surface are common and many particles and fields are emitted from the sun. The sun is relatively quiet for the other half of an 11-yr solar cycle. About once every 200 yr, the sun goes through a period of several decades or more where its activity level remains either very high or very low. From studies of sunspots and of the terrestrial record of ^{14}C, we know that the last such solar activity anomaly was a quiet period from about 1645 to 1715, the Maunder minimum. Recently, experiments on many spacecraft have provided important results on the nature of the cosmic rays and their distribution near the ecliptic plane of the solar system. We know much less about the cosmic ray particles in the large region of space outside the ecliptic plane; however, most meteorites probably come from orbits near the ecliptic plane.

The sun is an important source of energetic particles in the inner solar system. These particles are emitted by large solar flares on the surface of the sun. Since 1942, more than 100 particle-accelerating flares have been observed near the earth. Most particle-emitting flares occur when the activity level of the sun is high. During the parts of a typical 11-yr solar cycle when solar activity is low, very few SCR particles are produced. Near the earth, the fluences of protons with $E > 10$ MeV during SCR-producing flare events range from below 10^5 to over 10^{10} protons/cm^2. Most SCR particles have energies below 100 MeV/nucleon, although some solar flares emit many nuclei with energies above a GeV/nucleon. From measurements of SCR-produced radionuclides in the surface layers of lunar rocks, the average fluxes of solar protons with energies above 10 and above 100 MeV over the last few million years are known to be about 70 and 3 protons/cm^2 sec, respectively. Although solar particle events are fairly rare, the average fluxes at 1 AU of SCR particles with energies between 10 and 100 MeV/nucleon

greatly exceed those for GCR particles with the same energies. Only above about 500 MeV/nucleon does the GCR dominate. The average fluence of SCR particles varies with distance from the sun according to the inverse-square law, so the average flux of SCR particles to which most meteorites are exposed at 2 to 3 AU is less than that observed near the Earth. [See SOLAR PHYSICS.]

The GCR particles originate far from the solar system. The sources of the energy that accelerated most of them to high velocities probably involve supernovas, although the exact mechanisms for the acceleration are not known. As the GCR particles diffuse or are transported to the solar system, additional acceleration or other interactions can occur. The transit times of most GCR particles from their sources to the solar system are ~10^7 yr and during that time they past through ~5 g/cm^2 of matter. As these particles enter the solar system, their spectrum is modulated by the interplanetary magnetic fields, which originate at the sun. Solar modulation is the dominant source of GCR variability to which meteorites and the earth are exposed. Over a typical 11-yr solar cycle, the flux of lower-energy GCR particles (E ~100 MeV/nucleon) changes by about an order of magnitude, and the integral flux of GCR particles above 1 GeV/nucleon varies by about a factor of 2. Even during a typical solar minimum, the intensities of GCR particles are reduced compared to what they are in interstellar space near the solar system. The ^{14}C record on earth showed that the GCR-particle fluxes were quite high during the Maunder minimum. Measurements by charged-particle detectors on the Pioneer and Voyager spacecraft that flew past Jupiter showed that the fluxes of GCR protons with $E \geq 80$ MeV increased with distance from the sun by about 2 to 3% per AU. Thus in the ecliptic plane, orbital variations affect the production rates of cosmogenic products much less than the cosmic ray changes over an 11-yr solar cycle.

B. INTERACTION PROCESSES AND PRODUCTS

The energy, charge, and mass of a cosmic ray particle and the mineralogy and chemistry of the meteorite determine which interaction processes are important and which cosmogenic products are formed. Energetic nuclear particles interact with matter mainly in two ways: ionization energy losses and reactions with the nuclei. All charged particles continuously lose energy by ionizing the matter through which they pass. Some of the radiation damage produced during ionization by cosmic ray particles is accumulated in meteorites and can be detected as thermoluminescence (TL). In a number of meteoritic minerals, like olivine, the paths traveled by individual nuclei with atomic number (Z) above about 20 and with energies of the order of 0.1 to 1 MeV/nucleon contain so much radiation damage that they can be etched by certain chemicals and made visible as tracks. The nuclear reaction between an incident particle and a target nucleus generally involves the formation of new, secondary particles (such as neutrons, protons, pions, and γ rays) and of a residual nucleus that is usually different from the initial one.

Low-energy and high-Z nuclei are rapidly slowed to rest by ionization energy losses. High-energy, low Z particles lose energy more slowly and usually induce a nuclear reaction before they are stopped. A 1-GeV proton has a range of ~400 g/cm^2 and a nuclear-reaction mean free path of ~100 g/cm^2, so only a few percent go their entire range without inducing a nuclear reaction. Few GCR particles penetrate deeper than about 1000 g/cm^2 (the thickness of the earth's atmosphere or about 3 m in a stony meteorite) because they are removed by nuclear reactions or stopped by ionization energy losses. Only a few particles, such as muons, reach such depths, and it is hard to detect the few interaction products made there. Only in the few meters near the surface of a meteorite is a cosmic ray record readily determined. Depths are often expressed in units of grams per square centimeter (g/cm^2), which is the product of the depth in centimeters times the density in g/cm^3, because it is the areal density of nuclei that is important for the main nuclear interactions that occur in meteorites. [Densities (in g/cm^3) of meteorites range widely, from about 2.2 for CI carbonaceous chondrites, 3.1–3.4 for most achondrites, 3.4–3.8 for ordinary chondrites, about 5 for stony irons, and near 7.9 for iron meteorites.]

Because of the variety of the cosmic ray particles and of their modes of interactions, the effective depths of the interactions and their products vary considerably, as noted in Table I. This diversity in the types of products and their depths is very useful in studying the cosmic ray record of a meteorite. The meteorite that is found on the earth's surface is not the same object that was exposed to cosmic rays in space. Meteoroids can be suddenly and drastically changed in space by collisions, which remove parts of the old surface and produce new surfaces that usually have not been exposed to the cosmic rays. Micrometeoroids and solar wind ions gradually erode the surface of a meteorite. In passing through the earth's atmosphere, the outer layers of a meteoroid are strongly heated and removed, a process called ablation. The meteoroid usually fragments into several pieces. The cosmic ray record of a meteorite is often used to determine its preatmospheric history.

1. Thermoluminescence

Thermoluminescence (TL) in meteoritic minerals is a consequence of the filling of electron traps when electrons are excited into the conduction band by ionizing radiation and some become trapped in metastable energy states. The concentration of trapped electrons is a function of the ionization rate, the nature of the minerals involved, and their temperature record. Thermoluminescence gets its name because light is emitted when a sample is heated. The light is produced when the electrons are thermally released from the traps and combine with luminescence centers. The TL from ordinary chondrites is mainly in the wavelengths near blue (~470 nm). The natural radioelements, uranium, thorium, and potassium, expose ordinary chondrites to doses of about 10 mrad per year. (Natural radioactivity is the source of radiation that is used in determining when certain terrestrial samples were last heated to a high temperature.) The GCR is the dominant source of ionizing radiation in almost all meteoritic minerals, about 10 rad/year. The natural TL in a meteoritic mineral, like the activity of a cosmogenic radionuclide, is usually at an equilibrium value where the rate for trapping electrons is equal to the rate for thermal draining.

The TL sensitivity of meteoritic samples can vary greatly and depends on bulk composition, the abundance and composition of glasses, and metamorphism. (D. Sears has used the wide ranges of TL sensitivity in primitive types of ordinary chondrites to determine the amount of their metamorphic alterations.) Thus it is useful to normalize the natural TL observed in a sample to the TL induced by an artificial dose of radiation. Comparisons of the TL observed in natural samples to that produced artificially in the same sample show that electrons in traps that can be released by temperatures below about 200°C are relatively rare in meteorites.

Below about 200 to 300°C, the retention of TL varies widely among meteorites. The high-temperature ($T \sim 400°C$) TL observed in most meteorites is usually the same as that from an equivalent dose of 10^5 rad, which is the level acquired from the GCR in about 10^4 yr.

2. Tracks

Unlike nuclear reactions, which occur in all of the phases of meteorites, observations of the solid-state damage due to charged-particle irradiation is almost always limited to the crystalline dielectric phases of the minerals present. This solid-state damage can be produced by energetic heavy nuclei from a variety of sources, such as fission, the solar wind, and the cosmic rays. Here, we are primarily concerned with the damage paths produced in certain meteoritic minerals by cosmic ray nuclei with $Z \geq 20$. These nuclei are divided into two groups: $Z = 20$ to 28, termed the iron or the VH group, and $Z \geq 30$, called the VVH group. This classification is based primarily on the observed abundances of these nuclei in the cosmic rays. In the cosmic rays, the VVH group of nuclei (mainly $Z = 30$ to 40) is less abundant than the VH group by a factor of ~700 for energies above 500 MeV/nucleon. This ratio is similar to abundance ratios for these elements in the sun and meteorites. Nuclei with $Z > 40$ are rare in the cosmic rays, about 5×10^{-5} per iron nucleus.

In dielectrics, when the ionization along the path traveled by the particle exceeds a certain critical value, the radiation damage has altered the material sufficiently that the trails can be dissolved and enlarged by suitable chemical treatment. These chemically developed holes are cylindrical or conical, can be seen with an optical microscope, and are called tracks. In meteoritic and other natural minerals, this ionization threshold is exceeded only by nuclei of calcium and heavier elements ($Z \geq 20$). As Z becomes higher, the range of energies that have ionization rates above the critical value is greater. The VH and VVH groups of nuclei form chemically etchable tracks near the end of their range, where their ionization rates are the largest. Iron nuclei ($Z = 26$) with energies between about 0.1 and 1 MeV/nucleon are capable of producing tracks in meteoritic minerals; the corresponding energy range for krypton nuclei ($Z = 36$) is about 0.1 to 6 MeV/nucleon.

The length of the etchable tracks depends on the mineral. Cosmic ray tracks are mainly studied in three common silicate minerals: olivines, $(Mg, Fe)_2SiO_4$; pyroxenes, $(Mg, Fe)SiO_3$; and feldspars, solid solutions of $CaAl_2Si_2O_8$ and $NaAlSi_3O_8$. The maximum etchable lengths of fresh iron and zinc tracks in olivine are about 13 and 60 μm, respectively. Fossil tracks of these nuclei in meteorites are somewhat shorter because of partial annealing of the radiation damage along the track. However, tracks of VH nuclei do not appear to fade appreciably over long periods of time, so they can be used to determine exposure ages of meteorites. As the length of a track varies rapidly with the atomic number of the nucleus, the Z of the nucleus forming a track usually can be determined to within about ± 2 charge units, and tracks made by VVH-group nuclei are clearly distinguished from those of VH nuclei.

Track density-versus-depth profiles have been measured in a number of materials exposed in space, including meteorites, lunar samples, and a glass filter from the Surveyor III camera returned by the Apollo 12 astronauts. These measurements and theoretical calculations have been used to obtain curves for the production rates of tracks as a function of meteorite radii and sample depth. The track production profile as a function of depth in a large extraterrestrial object is shown in Fig. 1. The tracks between about 10^{-3} and 0.1 cm are made mainly by heavy SCR nuclei that have energies below about 10 MeV/nucleon; the deeper tracks are made by GCR nuclei with $E > 100$ MeV/nucleon after they have been slowed to energies of about 1 MeV/nucleon. The very surface layers of meteorites, which contain the tracks of heavy SCR nuclei, are almost always removed by ablation when passing through the earth's atmosphere. The density of cosmic ray tracks drops rapidly with depth, which make tracks excellent indicators of the preatmospheric depth of a sample.

3. Nuclides

Cosmic ray particles can induce a wide variety of nuclear reactions with any nucleus in a meteoroid. In a nuclear reaction, a particle such as a proton, neutron, pion, or an α particle, interacts with a target nucleus. A large variety of nuclear interactions is possible. The simplest one is elastic scattering, where the incident particle and the target nucleus are unchanged after the interaction, and only the direction and the energy of the particle is changed. Elastic scattering is the only way that low-energy (below ~0.5

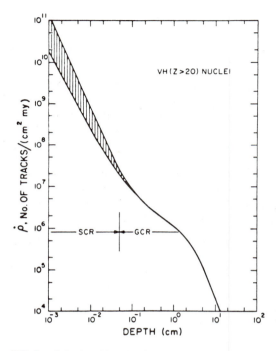

FIG. 1. Calculated production rates of heavy-nuclei tracks as a function of depth in a large stony meteorite. The shaded region represents the uncertainties in the fluxes of VH nuclei in the solar cosmic rays.

MeV) neutrons can be slowed, as neutrons have no charge and hence cannot be slowed by ionization energy losses. Inelastic scattering is like elastic scattering, except that the residual nucleus is left in an excited state. Elastic and inelastic scattering are important in the transport of particles inside a meteorite, but themselves seldom leave a record. The other types of nuclear reactions contribute to the cosmic ray record only when they produce a product nucleus that can be distinguished from the nuclei normally present in the meteorite. Cosmogenic nuclei in meteorites are very scarce: over the $\sim 10^8$-yr exposure age of a meteorite, only about one in every 10^8 nuclei will undergo a nuclear transformation. These cosmogenic nuclides are usually ones that are radioactive or those that are very rare in a meteorite, such as the isotopes of the noble gases.

In a piece of meteorite exposed to cosmic ray particles, the radius- and depth-dependent production rate of a cosmogenic nuclide, $P(R, d)$, is

$$P(R, d) = \sum_j N_j \sum_i \int \sigma_{ij}(E) F_i(E, R, d)\, dE$$

$$(1)$$

where N_j is the abundance of target element j, i indicate one of the primary or secondary particles that can induce nuclear reactions, $\sigma_{ij}(E)$ is the cross section as a function of energy for particle i making the product from element j, and $F_i(E, R, d)$ is the flux of particle i as a function of energy E, meteoroid radius R, and sample depth d. The particles that can induce nuclear reactions are those primary ones in the cosmic rays, such as protons and α particles, or the secondary ones made by nuclear reactions inside a meteorite, such as neutrons or pions. The most important parts of Eq. (1) are the expressions for the fluxes of cosmic ray particles, $F_i(E, R, d)$, that induce the nuclear reactions, and the cross sections, $\sigma_{ij}(E)$, for making the cosmogenic nuclide. The basic shapes for $F_i(E, R, d)$ are fairly well known, especially for the primary cosmic ray particles at higher energies. The numbers of secondary particles and variety of residual nuclei made by nuclear reactions depend on the energy of the incident particle.

Most of the cosmogenic nuclides that have been studied in meteorites are given in Table II. They include radioactive nuclides with half-lives

TABLE II. Cosmogenic Nuclides Frequently Measured in Meteorites

Nuclide	Half-life[a] (yr)	Main targets
^{3}H	12.3	O, Mg, Si, Fe
^{3}He, ^{4}He	S	O, Mg, Si, Fe
^{10}Be	1.6×10^6	O, Mg, Si, Fe
^{14}C	5730	O, Mg, Si, Fe
^{20}Ne, ^{21}Ne, ^{22}Ne	S	Mg, Al, Si, Fe
^{22}Na	2.6	Mg, Al, Si, Fe
^{26}Al	7.1×10^5	Si, Al, Fe
^{36}Cl	3.0×10^5	Fe, Ca, K, Cl
^{36}Ar, ^{38}Ar	S	Fe, Ca, K
^{37}Ar	35 days	Fe, Ca, K
^{39}Ar	269	Fe, Ca, K
^{40}K	1.3×10^9	Fe
^{39}K, ^{41}K	S	Fe
^{41}Ca	1.0×10^5	Ca, Fe
^{46}Sc	84 days	Fe
^{48}V	16 days	Fe
^{53}Mn	3.7×10^6	Fe
^{54}Mn	312 days	Fe
^{55}Fe	2.7	Fe
^{59}Ni	7.6×10^4	Ni
^{60}Co	5.27	Co, Ni
^{81}Kr	2.1×10^5	Rb, Sr, Zr
^{78}Kr, ^{80}Kr, ^{82}Kr, ^{83}Kr	S	Rb, Sr, Zr
^{129}I	1.6×10^7	Te, Ba, La, Ce
$^{124-132}$Xe	S	Te, Ba, La, Ce, (I)

[a] S denotes that the nuclide is stable.

ranging from a few days (^{48}V and ^{37}Ar) to millions of years (^{53}Mn and ^{40}K) and isotopes, especially the minor ones like ^{3}He and ^{21}Ne, of the noble gases. The most commonly measured cosmogenic nuclides include ^{3}He, ^{10}Be, ^{21}Ne, ^{22}Na, ^{26}Al, ^{36}Cl, ^{38}Ar, and ^{53}Mn. In iron meteorites, many stable nuclei lighter than iron, such as ^{45}Sc and the potassium isotopes, can be identified as cosmogenic, usually because they have a different isotopic distribution than normally present in nature. In most pieces of meteorites freshly exposed to the cosmic rays by a collision, the concentrations of cosmogenic nuclides are very low. As the cosmogenic–nuclide production rates at a given position in a meteoroid are fairly constant, the concentrations of cosmogenic nuclides steady build up. After several half-lives, the activity of a cosmogenic radionuclide will approach an equilibrium value where its rate of decay equals its production rate. Stable isotopes, like ^{3}He and ^{21}Ne, continue to accumulate and can be used to determine the total length of time that a meteorite was exposed to cosmic rays, if the production rates of these isotopes are known. Certain processes, however, such as shock or other heating events, can liberate the lighter noble gases, like ^{3}He. When a meteorite falls to the earth's surface, it is exposed to a very low flux of cosmic ray particles, and the activities of the radionuclides start to decrease at rates proportional to their half-lives. The terrestrial age of a meteorite, the length of time it has been on earth, can often be inferred from cosmogenic radionuclides.

The distribution of cosmogenic nuclides made by SCR particles are much different from those made by the galactic cosmic rays. The relatively low-energy solar protons and α particles are usually stopped by ionization energy losses near the surface of a meteoroid. The SCR particles that induce nuclear reactions usually produce few secondary particles and the product nucleus is close in mass to the target nucleus. An example of a SCR-induced reaction is ^{56}Fe$(p, n)^{56}$Co, the production of ^{56}Co when a proton enters a ^{56}Fe nucleus and a neutron is emitted. The fluxes of SCR particles as a function of depth can be calculated accurately from ionization-energy-loss relations, so the production rates of a nuclide can be predicted well if the cross sections for its formation are known. Like densities of heavy-nuclei tracks, the activities of SCR-produced nuclides decrease rapidly with depth, most being made within a few centimeters of the surface of the meteoroid. However, ablation seldom leaves a significant amount of SCR-produced nuclides in a meteorite.

The GCR particles producing nuclear reactions can roughly be divided into four components: high-energy ($E > 1$ GeV) primary particles, medium-energy ($\sim$0.1–1 GeV) particles produced partially from the first component, a low-energy group ($E < 100$ MeV) consisting mainly of energetic secondary neutrons, and slow neutrons with energies below about a kiloelectron volt. The fluxes of the high-energy primary GCR particles decrease exponentially with depth as they are removed by nuclear reactions. The numbers of secondary particles as a function of depth build up near the surface of a meteoroid, where most of them are made, but eventually decrease roughly exponentially. In a very big meteoroid or on the moon, there are about 13 neutrons produced per second per square centimeter of surface area (the solar-cycle-averaged rate). This neutron production rate compares to an average omnidirectional flux of about 3 particles/cm^2 sec for the GCR primaries in space, showing the importance of the large cascade of nuclear reactions in a meteoroid that produce numerous secondary particles. The total fluxes of GCR primary and secondary particles for several depths in spherical stony meteoroids of different radii are shown in Fig. 2. The energy spectrum of primary GCR particles, labeled (0, 0), is shown in Fig. 2. Even in a meteoroid of radius 40 g/cm^2 ($\sim$11 cm), there is a significant flux of low-energy particles, mainly secondary neutrons. This dominance of neutrons at lower energies is shown in Fig. 3, which gives the fractions of neutrons, protons, and pions deep inside a piece of extraterrestrial matter.

Neutrons slowed to energies of kiloelectron volts or electron volts can produce nuclides by neutron-capture reactions, in which only one or more γ rays are emitted after a nucleus captures a neutron, such as ^{59}Co$(n,\gamma)^{60}$Co. However, neutrons can be moderated to such low energies by many scattering reactions only if a meteoroid has a radius of more than about 20 cm. Nuclides made by neutron-capture reactions in meteorites include ^{60}Co, ^{59}Ni, and ^{36}Cl (which decays into ^{36}Ar). Most GCR-induced reactions involve incident particles with energies of about 1 MeV or more, emit one or more nucleons either individually or in clusters like α particles, and are called spallation reactions. Such products are often referred to as ''spallogenic'' nuclides. Sometimes the spallation reactions are divided

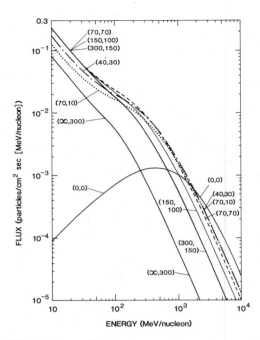

FIG. 2.   Fluxes of both primary and secondary cosmic ray particles as a function of particle energy for certain depths in stony meteoroids of various radii. Numbers in parentheses are the radius and depth (in g/cm²). The curve labeled (0, 0) is that for the primary cosmic rays averaged over an 11-yr solar cycle.

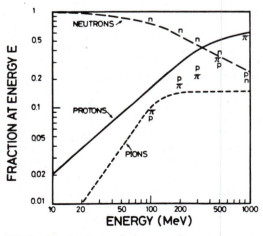

FIG. 3.   The fractions of neutrons, pions, and protons deep (~100 g/cm²) inside meteorites as a function of energy. The lines are based on lunar Monte Carlo calculations by T. Armstrong and R. Alsmiller; the symbols are from meteoritic estimates by J. Arnold, M. Honda, and D. Lal.

into low-energy and high-energy groups. High-energy spallation reactions involve particles with energies above about 100 MeV, produce numerous secondaries, and can make, in relatively low yields, many different product nuclides. Examples of high-energy spallation reactions are $^{16}O(p, X)^{3}$He or ^{24}Mg$(p, X)^{10}$Be, where X can be any one of a large number of possible outgoing particle combinations. Low-energy spallation reactions usually involve particles with energies below 100 MeV and can produce certain nuclides in high yields because both the fluxes of particles and the cross sections [see Eq.(1)] are relatively large. The reaction ^{24}Mg$(n, \alpha)^{21}$Ne is such a low-energy reaction, and is the major source of ^{21}Ne in most stony meteorites.

Because the distributions of low-energy and high-energy GCR particles are not the same for all locations inside meteoroids, the production-rate-versus-depth profiles for different cosmogenic nuclides can vary. A high-energy product like ^{3}He (see Fig. 4) has a profile that is fairly flat near the surface, whereas a low-energy product, like ^{21}Ne (shown in Fig. 5), builds up in concentration considerably with increasing depth near the surface. The size and shape of a meteoroid are important in determining the production profiles of cosmogenic nuclides. When the radius of a meteoroid is less than the interaction length of GCR particles, about 100 g/cm², particles entering anywhere can reach most of the meteoroid, and production rates do not decrease much near the center. In much larger meteoroids, many GCR particles are removed before they get near the center, and production rates for increasing depths decrease from their peak values near the

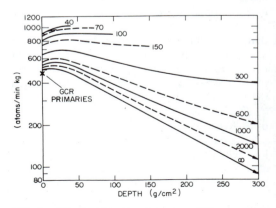

FIG. 4.   Calculated production rates of ^{3}He as a function of meteoroid radius (in g/cm²) and sample depth in the achondrite Shergotty.

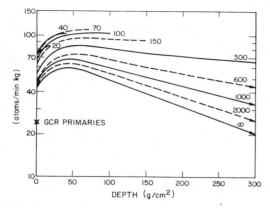

FIG. 5. Calculated production rates of ^{21}Ne as a function of meteoroid radius (in g/cm^2) and sample depth in the achondrite Shergotty.

surface. The production profiles for cosmogenic nuclides made by GCR particles vary with depth much less rapidly than do profiles for tracks or SCR-produced nuclides; however, ratios of a high-energy product to a low-energy one, like ^{3}He/^{21}Ne in Fig. 6 or ^{22}Ne/^{21}Ne, are often used to estimate how much a sample was shielded from the cosmic rays.

As noted in Table II, many cosmogenic nuclides can be made from more than one of the elements that are common in meteorites, so the chemical composition of a sample is often needed to interpret a cosmogenic–nuclide measurement. This diversity of target elements is usually not a problem if one is only studying a single class of meteorites that are chemically homogeneous, such as H chondrites. However,

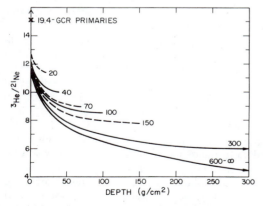

FIG. 6. Calculated ^{3}He/^{21}Ne production ratios as a function of meteoroid radius (in g/cm^2) and sample depth in the achondrite Shergotty.

many cosmogenic nuclides are now studied in very small samples much less than a gram in mass, so chemical variability on that size scale is a concern. Also, when one wants to extend results from one class to another chemical class, elemental production rates for certain cosmogenic nuclides often are needed. Measurements of several cosmogenic nuclides in a suite of chemically different meteorites or in mineral separates from one meteorite have been used to infer elemental production rates. Experimental cross sections with pure target elements can also be used to get relative elemental production rates. One phase of meteorites that is fairly simple chemically is the metal phase, which is almost pure iron with a little nickel. Metallic iron–nickel pieces occur in many chondrites; these pieces are often separated from the bulk of a chondrite, and cosmogenic nuclides, like ^{37}Ar and ^{39}Ar, measured in them.

III. Calculated Production Rates

A. TRACKS

The densities of the tracks made by heavy cosmic ray nuclei have been measured in a number of extraterrestrial objects, such as meteorites, lunar samples, and artificial materials exposed in space. These observations, plus a knowledge of the particle fluxes and processes involved in track formation, have resulted in several operational curves for the production rates of tracks as a function of depth in meteoroids of various radii, such as the profile shown for a large-radius object in Fig. 1. The relations for the ranges in meteoritic material of heavy nuclei as a function of their energy are well known. The energy spectrum of heavy ($Z > 20$) nuclei in the galactic cosmic rays have been measured by detectors on satellites and inferred from track profiles in meteorites and lunar samples with simple exposure histories. Thus the production-rate profiles for tracks of GCR nuclei are predicted fairly well.

As the solar cosmic rays are only emitted irregularly from the sun, there have been only a few chances to directly observe their spectra in space. Another problem is that the low energies of the SCR heavy nuclei make it hard to detect them by experiments on satellites. Each flare has its own energy distribution for the SCR particles, so it is hard to determine the long-term-averaged spectrum of heavy SCR nuclei. Lunar rocks with simple exposure histories of short durations have been used by several research groups to infer the spectrum of SCR nuclei with $Z > 20$ averaged over various time periods. The problem here is that the very surface layers in which the SCR nuclei produce tracks are continuously removed by micrometeoroids and solar wind ions. Typical erosion rates for lunar rocks are of the order of a millimeter per million years. As the tracks from SCR nuclei mainly occur in the top millimeter, the observed track densities very near the surface depend critically on the erosion rate. The range shown in Fig. 1 for the production rates of tracks at depths shallower than 0.1 cm reflects these uncertainties in interpreting such profiles in lunar rocks or in measuring the spectrum of heavy SCR nuclei in space. We do know, however, that the tracks from heavy nuclei in the SCR dominate over those from GCR nuclei for depths less than about 0.1 cm. While such very surface layers of meteorites are removed by ablation, we are still very much interested in tracks of SCR nuclei in meteorites because they occur in grains now inside a meteorite that were once exposed at the very surface of the parent body of the meteorite. A steep gradient of track densities over a distance of much less than 0.1 cm is a sign that the tracks were made by SCR nuclei.

B. NUCLIDES

A number of approaches have been used to predict the rates, ratios, or profiles for the production of cosmogenic nuclides in meteorites. The profiles for the production of trapped electrons observed in thermoluminescence are similar to those for spallogenic nuclides. Experiments at high-energy ($E \sim 1–6$ GeV) proton accelerators have been useful in simulating the cosmic ray irradiation conditions in meteoroids. Some cosmogenic nuclides are only made by cosmic rays with energies above about 1 GeV where secondary particles are not important, examples being ^{36}Cl, ^{37}Ar, and ^{39}Ar from iron. Cross sections or isotope ratios measured with high-energy protons can be used to predict the production ratios for such nuclides in metal phases of meteorites. The distributions of a number of spallation products have been measured in irradiated targets. When the product nuclide has a mass that is more than about 5 units less than that of the target nucleus (this mass difference is called ΔA), the trends for the product yields behave as a power law with respect to the product mass. This relation for the cumulative yield of all isotopes with mass A, $Y(A)$, is usually expressed as

$$Y(A) = c(\Delta A)^{-k}, \tag{2}$$

where k typically varies between about 2 and 3 and is smallest for small meteorites. Measurements of a number of stable cosmogenic nuclides in iron meteorites follow this trend very well. This trend usually fails for very small values of ΔA, such as nuclides made by (p, pn) reactions, and for product nuclides with A less than about 10 (which, like α particles, can be emitted from the excited target nucleus as a fragment by itself). As only a part of the yield for a given A is to a specific isotope, independent yields of radionuclides with mass A is less than $Y(A)$ by a fraction that is related to how close the radionuclide is to the most stable isotope with that mass.

A variety of stationary thick targets have been irradiated with beams of high-energy protons. The distributions of the product nuclides inside such targets varied considerably with depth and with the lateral distance from the beam. High-energy products, like ^{22}Na from iron, decrease exponentially with depth from very near the surface of the target and show little lateral spread. A low-energy product, like ^{55}Fe from iron, shows a buildup in activity near the surface, reaching its peak at depths of 50 g/cm^2 or more, and then shows an exponential drop in its activity with depth. The lateral spread of low-energy products is very large, reflecting the emission at all angles of the secondary particles that mainly induce such reactions. (In studies of low-energy products, care must be taken in planning such thick-target experiments that the target is sufficiently wide to contain all the secondary particles inside the target.)

A thick iron target was bombarded by a collimated beam of 3-GeV protons and the distributions of product nuclides measured by M. Honda. He then transformed the results into that for a hemisphere irradiated by an isotropic flux of particles, an approach valid for large iron meteoroids. In 1967, T. Kohman and M. Bender published a procedure to translate thick-target data to the distribution of nuclides expected inside an isotropically irradiated sphere and applied it to data from several bombardments of thick iron targets. Later B. Trivedi and P. Goel applied the Kohman–Bender model to data for ^{22}Na and ^3H in thick silicate targets. These production profiles based on thick-target bombardments have been very useful in studies of cosmogenic nuclides, although they are usually limited to the nuclides measured in the thick target or those made by similar nuclear reactions. Such transformations from thick-target measurements are not valid to very small meteoroids because the cascade of secondary particles in such small objects is not fully developed. Very recently, R. Michel and co-workers isotropically irradiated several stone spheres at an accelerator using a machine that translated and rotated the spheres in several directions. Even in a 5-cm-radius sphere, there was a noticeable buildup in the activities of low-energy products.

Several research groups have applied a semiempirical model to measured concentrations of cosmogenic nuclides in meteorites. This approach considers the exponential decrease in the flux of primary cosmic ray particles and the buildup and decrease in a flux of secondary particles that can produce a specific nuclide. The basic expression of this model for the production rate P of a nuclide as a function of distance x from the surface is

$$P = A_i[\exp(-\mu_p x) - B_i \exp(-\mu_s x)] \quad (3)$$

where p and s refer to primary and secondary particles, respectively, and $\mu_j = N\rho\sigma_j/A$, j being p or s. Here N is Avogrado's number, ρ the density, A the average atomic weight of the meteorite, and μ_j the interaction cross sections for primaries or secondaries (with $\mu_s > \mu_p$). This expression can be integrated for several simple geometries. Values of μ_p and μ_s can be estimated from nuclear systematics or inferred from meteoritic measurements; A_i and B_i are determined from measured profiles. In 1960, P. Signer and O. Nier applied this model to their ^{3}He, ^{4}He, ^{21}Ne, and ^{38}Ar concentrations measured for many locations in a slab of the Grant iron meteorite and used the model to predict spallogenic noble gas profiles in spherical iron meteorites of any radius. This model has worked quite well in describing the production profiles of noble gas nuclides in meteorites, in spite of several limitations with the model. It does not allow for energy losses or wide angle scattering of the cosmic ray particles, processes that do occur in meteorites. It also is limited to two parameters that represent the effective cross sections, μ_p and μ_s. These limitations are less important for high-energy products, thus the model works fairly well for such nuclides in iron meteorites.

Another series of models for the production profiles of cosmogenic nuclides are based on Eq. (1). The most important parts of these models are the expressions for the fluxes of cosmic ray particles as a function of particle energy, meteoroid radius R, and sample depth d, $F_i(E, R, d)$. In the major versions of this model, all particle types (i) are combined and fluxes are not given for individual particle types. This combination of particle types can be done for most cases as the fractions of various particles usually are not much different from those shown in Fig. 3, being mainly neutrons at lower energies. At higher energies, cross sections for nuclear reactions are less sensitive to the nature of the incident particle. Expressions for the fluxes of cosmic ray particles can be derived in many ways. For SCR particles, ionization-energy-loss relations can be used to calculate them. For GCR particles at fairly high energies, the primary spectrum measured in space, $dF/dE = cE^{-2.5}$, is used. In 1961, J. Arnold, M. Honda, and D. Lal used the expression $dF/dE = c(\alpha + E)^{-2.5}$ for $E > 100$ MeV, where α varied with location in the meteoroid. The same authors used a spectrum measured in the earth's atmosphere for $E < 100$ MeV. R. Reedy and J. Arnold derived spectra for $E < 100$ MeV in the moon that varied with depth. Cross sections used for the production of a nuclide are measured ones, if possible, or else they can be estimated from nuclear models or other systematics, such as spallation formulas that are extensions of Eq. (2). In most cases where there are no experimental cross sections for neutron-induced reactions, results from proton irradiations are used.

J. Arnold and co-workers first used such a model to calculate cosmogenic–nuclide production rates in iron meteorites. In 1972, Reedy and Arnold extended this model to the moon. In 1985, R. Reedy applied the lunar flux expressions to stony meteorites, using measured profiles of cosmogenic nuclides to infer the spectral shapes of cosmic ray particles inside several spherical meteorites. The lunar version of this model has reproduced measurements to depths of about 350 g/cm². For meteorites, there were only a few profiles measured for cosmogenic nuclides. The curves shown in Figs. 4–6 were calculated with this model. The profiles calculated for meteoroid radii less than about 40 g/cm² or more than about 200 g/cm² are less certain than those for intermediate sizes. While the production rates calculated for many cosmogenic nuclides agree well with measured values, there are some serious disagreements, such as for

[53]Mn. Such disagreements could be caused by incorrect cross sections, especially for reactions induced by medium-energy neutrons.

In a few cases, production profiles have been calculated from theoretical expressions for nuclear interactions. T. Armstrong and R. Alsmiller used a Monte Carlo code for intranuclear cascades to calculate the distribution of cosmic ray particles and product nuclides in the moon; few of such calculations have been published for meteorites. Rates for neutron-capture reactions have been calculated for the moon and stony meteorites using several neutron-transport codes. The major neutron-capture-produced nuclides observed in meteorites are ^{60}Co, ^{59}Ni, and ^{36}Cl (and its decay product ^{36}Ar). (For most meteoritic samples, however, ^{36}Cl is made mainly by spallation reactions.) Given an initial distribution of secondary neutrons, which are made with energies of the order of 1 MeV, these calculations transport the neutrons through the object and follow the scattering reactions that slow the neutrons to thermal energies ($E < 1$ eV). P. Eberhardt, J. Geiss, and H. Lutz first did such calculations for meteorites in the early 1960s. They showed that stones with radii of less than about 30 cm are too small to have an appreciable density of thermal neutrons, as most neutrons escape from such small meteoroids before they can be thermalized. Maximum rates for neutron-capture reactions in meteoroids occur in the centers of stony meteorites with radii of about 80 cm ($\sim$300 g/cm^2). The production rates for neutron-capture reactions vary much more with meteoroid radius and sample depth than do those for spallation reactions (like those in Figs. 4 and 5.) This large variability with radius and depth makes neutron-capture products, like track densities, useful in determining sample locations in meteorites.

Production rates, ratios, and profiles of cosmogenic nuclides calculated with all of these various models have been used in studies of meteorites. Each model has its advantages, but each also has some limitations. Most models agree on the basic production profiles that they calculate, giving results that are not much different from those shown in Figs. 4 and 5. These calculated production profiles generally agree well with those measured in meteorites. The model of Signer and Nier is good for cosmogenic noble gases in iron meteorites. There have only been a few profiles for cosmogenic nuclides measured in stony meteorites, such as cores from the Keyes and St. Severin chondrites and

samples from documented locations on the main mass of the Jilin chondrite. Measurements in lunar cores can be used to test models for the production of nuclides in very big objects, although the lunar cores only extend to depths of about 400 g/cm^2 and have been disturbed by meteoroid impacts on various time and depth scales. The models based on thick-target irradiations often do not reproduce well the profiles measured in the moon or large stones like Jilin. Additional production profiles measured in meteorites with a wide range of preatmospheric sizes and shapes would be useful in testing and developing models. The models all work fairly well for typical-sized meteorites (radii between $\sim$40 and 150 g/cm^2), which is also the region where production rates usually do not change much with radius or sample depth. To a first approximation, a single value for the production rate of a cosmogenic nuclide in any type of meteorite can be used, as often was done in meteoritic studies. A better approach has been to use a measured value to estimate the corrections for production-rate variations because of the size and shape of the meteoroid and the location of the sample. The isotope ratio ^{22}Ne/^{21}Ne has often been used for such shielding corrections.

Almost all cosmogenic nuclide production rates used in the studies of meteorites are based on measurements. The absolute rates calculated by these models are sometimes not very reliable, so the calculated production rates are usually normalized to those measured in meteorites. As the activity of a radionuclide is usually in equilibrium with its production rate, measured activities can directly be used as production rates. Sometimes, as with the thin-target cross sections, only production ratios are determined. A measured activity of a cosmogenic radionuclide and a production ratio can be used to infer the production rate for a stable nuclide, especially if the pair of nuclides are made by similar reactions or if the stable nuclide is the decay product of the radionuclide. Radioactive/stable pairs often used with production ratios include ^{3}H/^{3}He (although ^{3}H measurements show much scatter in meteorites), ^{22}Na/^{22}Ne (with care to correct for variations in ^{22}Na activity over an 11-yr solar cycle), ^{36}Cl/^{36}Ar, ^{39}Ar/^{38}Ar, ^{40}K/^{41}K, and ^{81}Kr/^{83}Kr. These radioactive/stable pairs have been frequently used to get exposure ages of meteorites.

Production rates of stable isotopes can also be determined from measured concentrations if the exposure age of the meteorite is known. In meteorites with short exposure ages, the activity of a long-lived radionuclide has often not reached its saturation or equilibrium value. If the activity of a radionuclide is significantly below its equilibrium value and its half-life is known, then the exposure age of the meteorite can be readily calculated. Production rates of several noble gas isotopes, like ^{21}Ne in chondrites, have been calculated from measurements for chondrites with exposure ages that were based on the radionuclides ^{10}Be, ^{22}Na, ^{26}Al, ^{53}Mn, or ^{81}Kr. Recently, these calculated production rates have used the measured ^{22}Ne/^{21}Ne ratio for shielding corrections, and the nominal production rate refers to a specific ^{22}Ne/^{21}Ne ratio. The rates determined from undersaturation of ^{26}Al are higher by about 50% than those based on the other radionuclides, for reasons that are not understood. It should be noted that, until fairly recently, the ^{26}Al-determined production rates were routinely used for ^{21}Ne and other noble gases. These earlier production rates and the exposure ages inferred with them are probably incorrect for stony meteorites, although the relative ages are correct.

IV. Measurement Techniques

A. THERMOLUMINESCENCE

The thermoluminescence of a meteoritic sample is the intensity of light emitted as it is heated. The sample is crushed to a fine powder, magnetic particles are removed, about 4 mg placed in a pan, and heated in a special chamber. The chamber is filled with an oxygen-free gas. The heating strip usually consists of a nichrome plate, and the temperature is monitored with a thermocouple. The temperature is increased from room temperature to about 550°C at a rate of about 2 to 7°C/sec. (Above ~500°C, the blackbody radiation becomes too strong, although a blue filter has be used to transmit the blue TL from chondrites while absorbing most of the longer wavelength blackbody radiation.) The emitted light is detected with a photomultiplier tube that is shielded from magnetic and electrical fields. The plot of the TL light output as a function of the temperature is often called a glow curve.

The sample usually is then exposed to an artificial source of radiation, such as a ^{90}Sr β source, for a test dose of ~10^5 rad (below the saturation part of the TL versus dose curve). The glow curve for this test dose is a measure of the TL sensitivity of the sample. For each temperature, the product of the test dose (in rads) and the ratio of the natural TL to the artificial TL is called the equivalent dose. Below and above a temperature of about 300 to 350°C, the character of the TL output from meteoritic samples varies. At higher temperatures, the TL is more stable. At lower temperatures, the intensity of the TL can cover a much larger range of values. The ratio of low-temperature TL to the high-temperature TL is sometimes used in studies of meteorites.

B. TRACKS

The paths of strong radiation damage that are produced by heavy ($Z > 20$) nuclei in various meteoritic minerals can sometimes be viewed directly by electron microscopy, but are usually observed after chemical etching. The use of simple chemicals to attack these radiation-damage paths enlarges the tracks so that they can be viewed with an optical microscope. Usually a fairly thin piece of meteorite or a series of selected mineral grains are mounted on a suitable support, and the top surface is ground and polished. Polished thin sections with thicknesses of a few tens of micrometers are often used for petrological studies of the minerals. For studies of tracks, the thickness must be ~200–300 μm, and the sample mounted on a medium (like epoxy resin) that can withstand high temperatures and harsh chemicals. The polished thick section is then etched with the appropriate chemicals for a specific length of time to reveal the tracks.

Much work has been done by many researchers in selecting the chemicals and etch times to be used for various materials. Care is needed to etch mainly the radiation damage along the tracks and not the whole surface of the mineral grain. For example, the common meteoritic mineral olivine is often etched using H_3PO_4, oxalic acid, EDTA, water, and NaOH (to adjust the pH) for 2 to 3 hr at 125°C or 6 hr at 90°C. The time and temperature are selected so that the tracks are suitably enlarged, but that a minimum of material beyond the radiation damage region is etched. The etched tracks in most meteoritic minerals usually are cones with a narrow angle (a few degrees) at the apex, as the chemical works in from the surface and enlarges the part of the track near the surface more than that at the bottom. The lengths of the etched tracks vary considerably with the energy and atomic number of the heavy nuclei that produced the track, with how much the radiation-damage paths had annealed prior to chemical treatment, and how much of the track was removed when the grain was polished. Measurements of the complete lengths of etched tracks can be done for tracks inside a mounted thick section, with the chemicals reaching these inner tracks through mineral cleavages or other tracks. The maximum recordable length for tracks of VH-group nuclei are ~15 μm; tracks of VVH nuclei are much longer.

The etched tracks are usually observed with optical microscopes, although electron microscopes are used to observe surfaces with very high track densities ($>10^8$/cm^2). Known magnifications are used to determine the area of the sample scanned for tracks. Several techniques are used to enhance the contrast of the tracks so that they are easier to observe. Often a thin layer of silver or other opaque substance is deposited in the tracks. Occasionally a replica of the holes is made by coating the surface with a material that will fill the holes and removing the coating with the holes now visible in the replica as fingers sticking out from the surface. In scan-

ning the surface being examined, background holes, such as those made by the etching of defects, should not be counted. Backgrounds are usually not a problem for track densities above about 10^4/cm^2. Enough tracks must be counted over a known area to assure a statistically valid number for the track density. In counting the tracks of a sample, some subjectivity is involved as different individuals can vary in what they consider real or background tracks. As with backgrounds, this is usually not a problem when the track densities are fairly high. Special techniques are often used to identify the densities of the long tracks due to VVH-group nuclei, such as grinding about 10 μm from the surface (to remove tracks of VH nuclei) after a heavy etch and silver plating. Tracks from VH and VVH nuclei in grains for the Patwar meteorite are shown in Fig. 7.

C. NUCLIDES

A large number of techniques is needed to measure the cosmogenic nuclides found in meteorites. The nuclides range from the stable isotopes of the noble gases to radionuclides with a

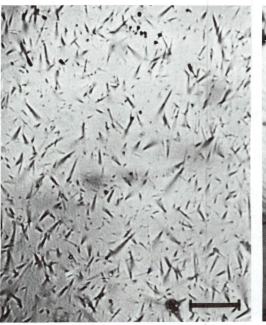 

FIG. 7.   Photomicrographs of tracks due to VH (left) and VVH (right) nuclei in pyroxene grains from the Patwar meteorite (a mesosiderite). The low density of long VVH tracks were revealed by overetching the surface and removing the top layer. The lengths of the bars are 10 μm. (Photographs courtesy of Professor D. Lal.)

great variety of half-lives. Stable isotopes are usually measured by counting the atoms with a mass spectrometer. Radionuclides are usually observed by detecting the radiation that they emit when they decay, although now several more sensitive techniques are used for the longer-lived ones. Preparation of the sample can vary from just almost nothing to detailed chemical separation procedures. Sample sizes have ranged from milligrams to kilograms.

The first measurements of cosmogenic nuclides in meteorites were of noble gas isotopes and were done using mass spectrometers. In a typical noble gas analysis, the gases are released from the sample into an evacuated system by heating, and then purified, ionized, and mass analyzed in the magnetic field of a mass spectrometer. Samples typically have masses of the order of a gram. The whole extraction system is heated under a vacuum to remove gases. Often the sample is first warmed to ~100°C to remove gases adsorbed from the atmosphere. The gases can be collected in a single heating or several fractions can be separated while the sample is heated to successively higher temperatures in a number of steps. Stepwise heating is useful if there are gases of diverse origins, such as trapped solar wind gases, that are released at different temperatures. Gases not released at lower temperatures are released by completely melting a sample (at ~1600°C). The gases are purified, usually by using charcoal at liquid nitrogen temperature or special getters of metals or alloys of titanium or zirconium. The noble gases often are separated by trapping and releasing from charcoal at different temperatures.

The purified noble gases are ionized, and the ions accelerated with an electric field and separated while passing through a magnetic field. The ions of the desired mass-to-charge ratio are detected, often with an electron multiplier. The mass peaks of interest are scanned a number of times. The backgrounds for the various mass peaks are frequently determined by duplicating the whole process without a sample in the heating crucible. Backgrounds usually result from molecules, like hydrocarbons, HD at mass 3, or doubly charged ions of argon or CO_2. The mass spectrometer is calibrated with sources of known isotopic composition (such as atmospheric gases) to get corrections for mass discrimination. Isotopic ratios for a given element can be determined very precisely. Absolute amounts can be measured using standards to calibrate the spectrometer or by isotopic dilu-

tion, the adding a known amount of the same element with a well-known, different isotopic ratio prior to a mass analysis.

Solid isotopes of elements like potassium and calcium have been analyzed mass spectrometrically. In these cases, the sample is usually chemically purified and placed on a special piece of wire that can be heated to very high temperatures. The atoms or molecules to be measured are thermally released and ionized. The ionized species are mass analyzed the same way as described above for gases.

Each radionuclide emits one or more characteristic radiations that can be detected by suitable counting techniques. Some radionuclides, like ^{22}Na and ^{26}Al, emit γ radiations that are sufficiently penetrating and unique that they can be detected nondestructively in a sample. Often a number of cosmogenic or naturally occurring radionuclides emit similar radiations, so the element of interest must be separated from the sample prior to counting. Known masses of the element (called carriers) are usually added to trace the sample through the separation and to determine the final chemical yield. To be certain that there are no contaminations, a sample sometimes is repurified and recounted several times until the ratio of measured radioactivity to the sample mass remains constant. Energy analysis of the emitted radiation can often discriminate against similar radiations from other sources. Standards of known activities are needed to calibrate the counters and determine how efficiently the counter detects the emitted radiation. The efficiencies of the counters must be known to convert the measured counts per minute to disintegrations per minute (dpm). Activities of radionuclides are usually reported as dpm per kilogram (dpm/kg) of the initial sample. In a few cases, the units are dpm per kilogram of the major target element.

The counters are often actively or passively shielded from the cosmic rays and from naturally occurring radiations to reduce the backgrounds from such sources. In active shielding, a special counter covers the main counter, and, if a background radiation (such as a cosmic ray particle) passes through it and gives a signal, then special electronics are used to cancel any signal from the main counter. Passive shielding use large masses of nonradioactive material, such as pure lead, to stop background radiations from reaching the counter. In some cases, the radionuclide emits two or more radiations that can separately be detected with a pair of counters, and backgrounds can be greatly reduced by using a coincidence technique where both radiations must be detected within a very short period of time. Counting with low backgrounds due to shielding or coincidences is called "low level."

Several volatile radionuclides, such as ^{3}H, ^{14}C, ^{37}Ar, and ^{39}Ar, are usually counted as gases inside specialized counters. Known amounts of carrier gases are added before the sample is heated. The gases released from the sample are purified and a known volume is added with counting gases and placed inside a counter designed to detect the radiation from the isotope of interest. Such internal counting is fairly efficient as most radiations are detected.

Cosmogenic radionuclides of nonvolatile elements, such as ^{10}Be and ^{53}Mn, must be counted as a solid and the radiation detected external to the sample. Weak radiations, such as X rays or low-energy β particles, are not very penetrating, and both the sample and the window through which the radiation enters the counter must be as thin as possible. Geiger or proportional counters filled with specialized gases are generally used for X-ray or β counting. Corrections must be made for self-absorption of weak radiations in the sample, so the thickness and composition of the counting sample must be known. Such counting techniques often require fairly large samples and the errors are frequently large because the counting rates are fairly low. γ Rays can be detected using thallium-doped NaI-scintillation crystals and, more recently, with high-resolution solid-state detectors made of germanium. γ-Ray spectra can be measured with chemically purified samples or large, unaltered pieces of a meteorite.

Coincidences between a γ ray and a β particle, X ray, or other γ ray is often used for the low-level counting of radionuclides like ^{22}Na and ^{54}Mn. Two counters are needed, and special electronics determine if both radiations occurred within a very short period of time. The γ ray spectrum when there is a coincidence has a much lower background than one obtained without requiring a coincidence. For coincidences with β particles or X rays, the sample usually is chemically separated and prepared in a thin counting geometry. γ-γ Coincidence counting can be done nondestructively on pieces of meteorites, usually with two large NaI(Tl) crystals positioned on opposite sides of the sample. Radionuclides that emit two or more γ rays (^{60}Co and ^{46}Sc), a positron (a positive electron that

produces two 511-keV γ rays when it annihilates with an electron), or a positron and a γ ray (^{22}Na and ^{26}Al) can be measured very effectively with γ-γ coincidence counting.

Several radioactive or stable cosmogenic nuclides can be measured by neutron-activation techniques. The nuclide is exposed to a high flux of thermal neutrons in a reactor, which produces a radionuclide that can be readily counted. Neutron-activation analysis of stable cosmogenic nuclides have been done for isotopes like ^{45}Sc that are produced in iron meteorites. Several long-lived radionuclides, ^{53}Mn and ^{129}I, can be measured by converting them to the short-lived radionuclides ^{54}Mn and ^{130}I. This activation method is routinely used for ^{53}Mn using samples as small as a few milligrams, and can transform a sample that has only a few decays of ^{53}Mn (which captures an electron and emits a weak X ray) per minute into one that has thousands of energetic ^{54}Mn γ rays emitted per minute. Before being irradiated with thermal neutrons, the sample most have a low activity of ^{54}Mn, and iron, which can make ^{54}Mn by the (n, p) reaction with ^{54}Fe, must be chemically removed. The neutrons that irradiate the purified sample must not have many neutrons with energies above 10 MeV, or else ^{54}Mn will be produced from stable ^{55}Mn by the $(n, 2n)$ reaction. Careful monitoring of interfering reactions is done to correct for the ^{54}Mn not made from ^{53}Mn.

In the late 1970s and early 1980s, a new technique for counting long-lived radionuclides was perfected, accelerator mass spectrometry (AMS). The atoms themselves are counted (as in a regular mass spectrometer) at an accelerator, not the radiation that they emit. A sample with a low level of radioactivity of a long-lived radionuclide like ^{10}Be actually contains many atoms of ^{10}Be. (Only about one in 10^6 atoms of ^{10}Be will decay in a given year.) A regular mass spectrometer normally cannot detect the relatively few atoms of ^{10}Be because it is very hard to remove the atoms of the stable isobar ^{10}B prior to the mass spectrometry or to resolve them with the mass spectrometer. If the nuclei can be accelerated to high energies, however, nuclear-particle-detection techniques can be used to distinguish ^{10}Be from ^{10}B. The energies required are at least a few mega-electron-volts (MeV), which means that large accelerators, usually tandem Van de Graaffs, are needed. The cosmogenic radionuclides that have been frequently done so far with AMS are ^{10}Be and ^{14}C. A number of analyses of ^{26}Al, ^{36}Cl, and ^{129}I have also been done with accelerators, and the list of isotopes measured this way keeps growing. The technique of accelerator mass spectrometry is very sensitive, and sample sizes of the order of milligrams are now used. Such small samples are big improvements over the gram-sized or larger samples that were required for counting the decays of radionuclides like ^{10}Be.

Since about 1970, the new techniques of neutron activation and accelerator mass spectrometry, plus improved γ–γ coincidence counters and noble gas spectrometers, have resulted in many more analyses of cosmogenic nuclides in meteorites. These measurements are now done with much smaller samples, which allows a number of different nuclides to be studied in fairly small pieces. For example, now the light noble gases (helium, neon, and argon) and the radionuclides ^{10}Be, ^{26}Al, and ^{53}Mn are frequently measured in a sample. It is now easier to do such comprehensive sets of analyses on several samples from a given meteorite, such as from cores drilled in several meteorites or from the surfaces or slabs of large pieces. Unfortunately, sensitive measurement techniques have not been developed for certain cosmogenic radionuclides, such as ^{41}Ca and ^{59}Ni, that are very hard to measure by conventional counting methods. Radionuclides like ^{41}Ca and ^{59}Ni would be very useful in the studies of cosmic ray records of meteorites because of their half-lives ($\sim10^5$ years) and because they are often made by neutron-capture reactions and, except for short-lived ^{60}Co, such products are seldom measured in meteorites. It is hoped that the list of cosmogenic nuclides that can be measured with very sensitive techniques will continue to grow. Some will probably be done by accelerator mass spectrometry. The use of lasers to selectively ionize specific elements or isotopes has some promise, but has yet to been applied to studies of cosmogenic nuclides.

V. Histories of Cosmic Rays

A. Heavy Nuclei

The fossil tracks of heavy ($Z > 20$) nuclei in extraterrestrial minerals are the only way to study their record in the past and have provided much of early characterization of heavy nuclei in the cosmic rays. These studies have provided long-term average values of the fluxes, energy spectra, and elemental ratios of cosmic ray nuclei for atomic numbers (Z) from 20 to above 90

and in the energy range $\sim$1–2000 MeV/nucleon. Many of these results depend on knowing additional data that are often model dependent, such as exposure ages, rates for the erosion of meteorites, or their fragmentation by collisions. Studies of large suites of different samples have eliminated many of these possible complications in the studies of tracks. The results from fossil tracks complement, and sometimes extend, what we know about heavy nuclei in the contemporary cosmic rays.

Much is known about the VH or iron group of nuclei in the ancient galactic cosmic rays. Tracks in various samples were formed over the last 10^9 yr. The flux can be determined from the track density if the exposure age can be independently ascertained. The depth-versus-density profile can be used to get the energy spectrum. The flux and energy spectrum of VH-group nuclei in the past are similar to those measured by satellites. For the VVH group ($Z \geq 30$), the main result is the VVH/VH abundance ratio in the energy region 100–1000 MeV/nucleon, which was 1.5 (±0.2) $\times 10^{-3}$. This ratio and relative cosmic ray abundance ratios for the charge groups 52–62, 63–75, 76–83, and 90–96 are in fairly good agreement with the corresponding elemental ratios for the sun or in primitive meteorites.

Most fossil track studies of the VH- and VVH-group nuclei in the solar cosmic rays have been done with lunar samples because the very surface layers of meteorites with tracks of SCR nuclei are usually removed by ablation. The lunar-sample work has shown that the VVH/VH ratio for $E < 40$ MeV/nucleon is much higher, by about an order of magnitude, than that observed for GCR nuclei with $E > 100$ MeV/nucleon. Grains in the lunar regolith contain SCR tracks made a long time ago and since deeply buried. The high density and the steep gradient of tracks in a lunar grain are shown in Fig. 8. Lunar breccias also contain grains with tracks made in the distant past. Their track records show that the energy spectra of ancient heavy nuclei in the SCR have not been very different than the recent ones. The determination of the flux of heavy nuclei in the SCR is difficult because it is hard to get an independent estimate of the exposure ages of these lunar soil grains and because erosion rapidly removed the SCR tracks.

The tracks of SCR heavy nuclei have been observed in mineral grains that are inside certain meteorites called "gas-rich" because they contain high concentrations of solar-wind-implanted noble gases. These tracks are present in very high densities and show pronounced density gradients over distances of a few micrometers, similar to the tracks from SCR nuclei seen inside lunar grains (see Fig. 8). These meteoritic grains were irradiated a long time ago while in a regolith similar to that on the surface of the moon. Certain carbonaceous chondrites also have grains with tracks made during the early history of the solar system. These meteoritic grains are more isotropically irradiated than lunar grains, suggesting a more effective turning process, possibly because of the lower gravity field on the asteroid-sized parent body of these gas-rich meteorites. There is no way to estimate how long these grains were exposed to the solar wind or heavy SCR nuclei. The spectrum of VH nuclei in these gas-rich meteorites is not different from those observed elsewhere. The ancient VVH/VH ratios have varied, but not by more than a factor of 2 or 3, similar to variations in this ratio seen among contemporary flares. These results for solar wind gases and SCR tracks in modern and ancient grains indicate that the mechanisms that accelerate nuclei from the sun has not changed much.

B. SPATIAL VARIATIONS

The intensities of the GCR particles have long been expected to vary with distance from the sun in the inner solar system. Temporal variations over the 11-yr solar cycle had been observed on earth. Spatial variations in the inner

FIG. 8.   An electron micrograph of tracks due to VH nuclei in an Apollo 12 feldspar grain, obtained using the replication technique. Note the high track density and the appreciable track density gradient from the edge. The length of the bar is 1 μm. (Photograph graciously provided by Professor D. Lal.)

solar system were anticipated because the galactic cosmic rays are modulated by the sun and because the solar fields that interact with the cosmic rays become weaker as they move away from the sun. Before satellites traveled far from the earth's orbit at 1 AU, the magnitude of the variation of the GCR flux with distance from the sun was not known, and first estimates were made using cosmogenic nuclides in meteorites. Photographs by several cameras of the fireball trajectories of three recovered meteorites and of many bright meteors showed that most were in orbits with low inclinations, perihelions near 1 AU, and aphelions in the asteroid belt. These orbits are fairly eccentric, and the meteorites spend most of their time near aphelion 2–4 AU from the sun, where most of the longer-lived cosmogenic radionuclides, like 269-year ^{39}Ar, are made. In recent meteorite falls, only the very short-lived radionuclides, such as 35-day ^{37}Ar, are made mainly near the earth. A number of studies used the measured activity ratios of short-lived to long-lived radionuclides in meteorites to study GCR spatial variations.

Most studies of the spatial variations of the GCR used ^{37}Ar/^{39}Ar activity ratios, as this ratio can be determined well using the same or similar counters. Usually the metal phases of meteorites were used because of their chemical simplicities and because production ratios could be inferred directly from high-energy accelerator bombardments of iron. Some measurements of this ratio were also made of the stony material. The ^{37}Ar/^{39}Ar activity ratios measured in various meteorite falls varied widely, although they were usually lower than that expected for no gradient, implying that the cosmic ray intensity increased rapidly with distance from the sun. For energies around 1 GeV, positive gradients as large as ~60%/AU were reported for individual meteoritic analyses. The inferred spatial gradients varied widely, however, depending on the sample that was analyzed.

With additional analyses of ^{37}Ar, ^{39}Ar, and other radionuclides in many meteorites, these early results on spatial gradients were examined more critically. The activities of long-lived radionuclides, such as ^{39}Ar, showed very little spread in meteorites that probably had a range of aphelion, suggesting that the spatial gradients of the GCR were not large. These meteorites fell during different phases of the solar cycle, so the correlation of the ^{37}Ar activity with other measures of the cosmic ray flux, such as count rates of cosmogenic neutrons in terrestrial neutron monitors, were examined. The correlation between the ^{37}Ar activity and the neutron-monitor count rates was weak, although the trend was for higher activities when the sun was less active. Other mechanisms to explain the variations in the ^{37}Ar/^{39}Ar activity ratios have been proposed, such as the perihelions of these meteorites or whether the meteorite was moving toward or away from the sun when it hit the Earth. The generally accepted conclusion now is that cosmogenic radionuclides do not show any evidence of large gradients of the GCR with distance from the sun. Over the last decade, experiments on the Pioneer and Voyager satellites have measured the GCR intensities with distance from the sun, and the GCR spatial gradients are low, only 2–3%/AU.

Unusual activities or activity ratios of cosmogenic radionuclides are sometimes interpreted as reflecting variations of the cosmic ray intensities in the solar system. Orbits determined from visual observations of the meteor trails indicate that some meteorites may have had fairly high inclinations with respect to the ecliptic, such as about 20° and 28° for Allende and Dhajala, respectively. No satellites have yet to travel very far from the ecliptic, although Ulysses (the International Solar Polar Mission) is scheduled to explore such regions in the 1990s, so there are no direct measurements of the cosmic ray fluxes there. In Dhajala, the short-lived radionuclides have activities similar to or slightly higher than those in other meteorites that fell at about the same time, while the long-lived ones tend to be less radioactive than normal. These ratios of short-lived to long-lived radionuclides in Dhajala are higher by about 30 to 50% than usually observed in meteorites. These high ratios could be caused by a variation in the cosmic ray intensity with heliographic latitude, although they could reflect shielding changes in the past. The Malakal and Innisfree chondrites have very high activities of ^{26}Al. Malakal has a very low thermoluminescence level that suggests it was heated in an orbit with a small perihelion (~ 0.5–0.6 AU). While the unusual ^{26}Al activity in Malakal could be a consequence of its orbit, data for other cosmogenic nuclides suggest that Malakal (and Innisfree) possibly had complex exposure histories. Thus there are no clear cases of the cosmic ray record of a meteorite showing unusual spatial variations in the intensities of the cosmic rays.

C. TEMPORAL VARIATIONS

Unlike changes in the fluxes or spectra of heavy nuclei in the past or spatial variations of the cosmic rays, significant temporal variations of the cosmic rays have been seen from the concentrations of cosmogenic nuclides in meteorites. These variations in the intensities of the GCR protons and α particles cover periods that range from an 11-yr solar cycle to $\sim 10^9$ yr. Other time intervals studied with cosmogenic radionuclides include the Maunder minimum, which occurred about 300 years ago, and 10^5 to 5×10^6 years ago. Temporal variations have been based primarily on radionuclides with a variety of half-lives that extend from 16-day ^{48}V to 1.26×10^9-yr ^{40}K. As can be seen from the list of cosmogenic radionuclides in Table II, there are many gaps in the half-lives of such radionuclides studied in meteorites. For example, except for a very few ^{129}I measurements, no radionuclides with half-lives between 3.7×10^6-yr ^{53}Mn and ^{40}K have been measured. Several cosmogenic radionuclides in terrestrial samples, like ^{14}C and ^{10}Be, represent very short time periods because these nuclides are rapidly removed from the atmosphere where they were made and stored in places, such as plants or ice layers, with very little subsequent alterations. In meteorites, radionuclides usually have been produced over their last few half-lives, so the time period examined is determined by the half-life of the radionuclide.

While the variations in the activities of ^{37}Ar meteorites were not strongly correlated with solar activity, such variations have been seen more clearly for other short-lived species. Several research groups have nondestructively measured γ-ray-emitting radionuclides, especially short-lived ^{46}Sc, ^{54}Mn, and ^{22}Na, in a number of recent falls. Although the variations in activity with phase of the solar cycle is visible in the raw radioactivities, it is more visible when corrections are made for the shielding of the cosmic rays by the geometry of the meteoroid and the depth of the sample. Ratios of activities of radionuclides made by similar reactions, such as ^{22}Na/^{26}Al or ^{54}Mn/^{22}Na, eliminate most effects due to shielding and show well the variation in radioactivity with solar cycle. The use of shielding corrections based on the measured ^{22}Ne/^{21}Ne ratios also showed the recent temporal variations in radioactivities more clearly. The shielding-corrected variations in activity is largest for the shorter-lived radionuclides ^{46}Sc and ^{54}Mn, and the inferred production-rate variations are a factor of 2.5 to 3 over the 11-yr solar cycle. This magnitude in the production-rate variation is consistent with the measured fluxes and spectral variations of GCR particles. The GCR-proton fluxes over a solar cycle vary considerably for $E < 1$ GeV and by a factor of 2 for all energies above a giga-electron-volt.

Although the ^{37}Ar/^{39}Ar ratios measured in the metal phases from many meteorites showed much variation, they provided some evidence for the Maunder minimum in solar activity. The 269-yr half-life of ^{39}Ar is ideal for studies of this period from 1645 to 1715 when there were extremely few sunspots or auroras and enhanced production of ^{14}C in the earth's atmosphere. A similar minimum in solar activity, the Sporer minimum, occurred about 200 years before the Maunder minimum. As noted above under spatial variations, the activities of ^{37}Ar were usually lower than those measured for ^{39}Ar, whereas the production rates of these radionuclides from iron are essentially equal. M. Forman and O. Schaeffer noted that the mean activity of ^{37}Ar is below that for ^{39}Ar, even if the observed spatial gradient in the GCR ($\sim$3%/AU) is considered, and interpreted this excess ^{39}Ar ($\sim$18%) as having been made by enhanced fluxes of GCR particles during the Maunder and Sporer minima. Measurements of ^{39}Ar in the few meteorites that fell just after the Maunder minimum could show the effects of the Maunder minimum more clearly than using recent falls.

The radionuclide pair ^{39}Ar and 3×10^5-yr ^{36}Cl are made from iron in nearly equal yields. The measured ratios for the activities of these two radionuclides in iron meteorites are also about unity, so the average fluxes of GCR over the last 300 yr is not very different ($<$10% variation) than over the last 3×10^5 years. As the solar-cycle-averaged ^{37}Ar activities are slightly lower than those of ^{39}Ar and ^{36}Cl (although only by about a standard deviation), there is a hint that the GCR fluxes during the last few decades could be slightly lower than in the last 500–10^5 years, possibly because there have not been any recent periods of unusually low levels of solar activity, like the Maunder minimum.

Several studies have looked for possible GCR variations over the last 5×10^6 years. The measured activities of a number of radionuclides in iron meteorites have been compared with calculated production rates to search for such cosmic ray variability. The ratios of the observed to calculated activities vary by factors of 2 (due probably both to calculational and measurement uncertainties), but show no systematic trends with half-lives. Several research groups have inferred production rates for noble gas isotopes, like ^{21}Ne, from meteorites with known exposure ages. As radionuclide activities were used to get these exposure ages (from either the undersaturation of activity or from a radioactive/stable pair), variations in the inferred production rate of ^{21}Ne could be a consequence of GCR intensity variations in the past. Undersaturation of 7×10^5-yr ^{26}Al, 1.6×10^6-yr ^{10}Be, and 3.7×10^6-yr ^{53}Mn and the pairs ^{22}Na/^{22}Ne and ^{81}Kr/^{83}Kr were used by various groups in inferring ^{21}Ne production rates. Results from all radionuclides except ^{26}Al gave similar results; however, the ^{26}Al-inferred ^{21}Ne production rates were considerably higher, by about 50%. Several simple possible explanations for this anomaly, such as an incorrect half-life for ^{26}Al, have been eliminated, although slight ($\sim$10%) adjustments in several half-lives would reduce the magnitude of the discrepancy. The shielding corrections used in these calculations were reexamined, but the discrepancy still persisted. Temporal variations in the cosmic ray intensities over the last few million years could account for some of this variation in inferred ^{21}Ne production rates but not all of it because the half-life of ^{26}Al is intermediate and not that different from those of 2.1×10^5-yr ^{81}Kr and ^{10}Be. Other possible explanations for the high ^{21}Ne production rates determined from meteorites that are undersaturated in ^{26}Al are some ^{21}Ne from previous irradiations, difficulties in making shielding corrections, or unusual cosmic ray records for the few meteorites that have very low ^{26}Al activities. These meteoritic results for over the last 5×10^6 yr are consistent with measurements for cosmogenic nuclides in lunar and terrestrial samples, which imply that cosmic-ray-intensity variations of $>$30% for periods of $\sim$10^5 yr have been unlikely.

Most stony meteorites have cosmic ray exposure ages considerably below 5×10^7 yr, so studies of cosmic ray variations for longer time periods have used iron meteorites, which have much longer exposure ages. The radioactive/stable pairs that generally have been used to determine exposure ages of iron meteorites are ^{39}Ar/^{38}Ar, ^{36}Cl/^{36}Ar, ^{26}Al/^{21}Ne, ^{10}Be/^{36}Ar, and ^{40}K/^{41}K. The first four radionuclides have half-lives that range from 269 yr to 1.6×10^6 yr, and the exposure ages of various iron meteorites determined with those pairs generally agree well within the experimental uncertainties, and, like the results discussed above, imply no major variations in the fluxes of cosmic rays over the last few million years. The exposure ages determined with 1.26×10^9-yr ^{40}K, however, tend to be 45% greater than the ages inferred with the shorter-lived radionuclides.

These higher exposure ages based on ^{40}K were first discovered around 1960 by H. Voshage and H. Hintenberger. Much work has been done since then that confirms the earlier results. (Voshage cautions, however, that the key parameter in the ^{40}K work, the ^{40}K/^{41}K production ratio, could possibly be incorrect as it was not directly determined from experimental bombardments as were the other production ratios, although he believes that it is known much better than a factor of 1.5.) The question has been whether the higher exposure ages based on ^{40}K represents something in the history of the cosmic rays or of the iron meteorites. For example, complex exposure histories for iron meteorites are fairly common. This explanation is unlikely because at least eight iron meteorites have exposure-age ratios of 1.45 ± 0.10, while other ratios (for irons with complex histories) range from 3 to 8. Another explanation is that these iron meteorites, which have exposure ages of 3×10^8 to 1×10^9 years, have been slowly and steadily eroded by solar wind particles and micrometeoroids in space. An erosion rate of $\sim 2 \times 10^{-8}$ cm/yr could produce a sufficient shielding change to account for the observed differences in the exposure ages. O. Schaeffer, H. Fechtig, and co-workers recently use laboratory simulations to estimate that the erosion rates of iron meteorites in space to be 2.2×10^{-9} cm/yr (30 times slower than for stony meteorites), too low to account for these exposure-age ratios. The generally accepted conclusion is that the high exposure ages determined with ^{40}K were caused by a flux of cosmic rays that has been $\sim$50% higher over the last $\sim 10^6$ years than over the last 10^8–10^9 yr.

As discussed above, the cosmic ray records of meteorites clearly show variations in the intensities of cosmic rays in the inner solar system over an 11-yr solar cycle, hint at an increased flux during the Maunder minimum during 1645–1715, set limits of less than $\sim$30% variations for $\sim 10^5$-yr periods over the last 5×10^6 yr, and strongly imply lower averaged intensities $\sim 10^8$–10^9 yr ago. These cosmic-ray-intensity variations are consistent with what we know about the sources of the cosmic rays, their transport to the solar system, and their modulation by the sun. The effects of the 11-yr solar cycle and the Maunder minimum are clearly seen in other records. The transit times of cosmic ray particles from their sources to the solar system, $\sim 10^7$ yr, the 6.6×10^7-yr period of the sun and solar system in its vertical motion relative to the galactic plane, and the distributions of supernovas in our galaxy (the probable energy source for most cosmic rays) are consistent with the meteorite results for longer time periods.

VI. Histories of Meteorites

A. PREATMOSPHERIC GEOMETRIES

The cosmic ray record of a meteorite can be used to reconstruct its geometry in space while it was being irradiated by the cosmic rays. This reconstruction is needed because the meteorite recovered on earth was ablated and possibly fragmented during its passage through the atmosphere. We can determine the preatmospheric geometry of a meteorite and the location of a given sample within the meteoroid because, as already discussed, the production rates of cosmogenic products vary with geometry and depth. Usually the products that have production profiles that vary the most with shielding are best for such studies. The density of heavy-nuclei tracks and concentrations of neutron-capture-produced nuclides are well suited for such studies. Ratios of the concentrations of certain cosmogenic products, like the ^{3}He/^{21}Ne ratio shown in Fig. 6 or the ^{22}Ne/^{21}Ne ratio, are often used in estimating the preatmospheric radii of meteoritic samples. Evidence for SCR-produced products in meteoritic samples also is a sensitive indicator that a sample had very little preatmospheric shielding. The results of such shielding studies are often needed in the interpretations of other cosmogenic products.

The production rates of heavy-nuclei tracks vary rapidly with depth in a meteoroid (see Fig. 1), so, if the exposure age of a meteorite can be independently determined, the track density of a sample is a very good indicator of how far that mineral grain was from the nearest preatmospheric surface. (Care must be taken in track studies of small meteorites to account for the tracks made by nuclei arriving from all directions.) The usual methods involving radioactive or stable cosmogenic nuclides give exposure ages that are more than adequate for determining the rates that the tracks were produced in a meteoritic sample. The experimental rate is compared with curves like that in Fig. 1 to get the preatmospheric depth of the sample. As the production curve is fairly steep, small errors in the track density or the exposure age do not affect much the determination of the depth. An uncertainty of ±100% in the radius only affects the inferred depth by about ±10%. These preatmospheric depths are very useful in interpreting the record of cosmogenic nuclides. To get the complete preatmospheric geometry of the meteorite requires measurements for samples that were at different locations. For several meteorites, such as St. Severin, cores were drilled through the main mass and complete track profiles measured. For meteorites that fell as a shower of many fragments it is harder to reconstruct the original geometry.

Several studies of preatmospheric depths used both tracks and cosmogenic nuclides. The ^{22}Ne/^{21}Ne ratio is a rough measure of the shielding conditions of a sample. The concentrations of ^{21}Ne (and sometimes ^{3}He) give the exposure age. The cosmogenic nuclide data help to estimate the preatmospheric radius and narrow the ranges of depths consistent with the measured track production rate. Comparisons of the measured track production rates with a cosmogenic nuclide measurement, such as the ^{22}Ne/^{21}Ne ratio or the activity of a radionuclide like ^{53}Mn, can be used to identify meteorites with complex exposure histories and to limit the sets of radius and depth combinations that are possible for that sample. For example, in Fig. 9, the production rates for tracks and ^{53}Mn measured in the St. Severin core samples are compared with calculated profiles. The results in Fig. 9 show that this core had a preatmospheric radius of between 20 and 25 cm, in good agreement with the length of the core plus the ablation losses determined from the track data. Similar plots of track production rates versus ^{22}Ne/^{21}Ne ratios have frequently been used in such studies. The results for a single sample can give a fairly good estimate of the preatmospheric radius in many cases.

For stony meteorites, a major study by N. Bhandari and co-workers showed that the fraction of the preatmospheric mass lost by ablation ranged from about 27 to 99.9% with a median value of about 85%. Preatmospheric masses typically ranged from 10 to 1000 kg with most meteorites having been near the lower end of this range. The probability of a preatmospheric mass

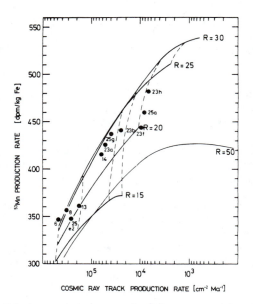

FIG. 9. Production rates for ^{53}Mn versus track production rates in stony meteorites of various radii (in cm) are shown with measured data from core samples (dots with numbers) of the St. Severin LL chondrite using an exposure age of 13×10^6 yr. The figure and the data for ^{53}Mn are from P. Englert; track densities from D. Lal and co-workers or Y. Cantelaube and co-workers; theoretical ^{53}Mn production rates from T. Kohman and M. Bender; calculated track production rates from S. Bhattacharya and co-workers.

in this range varied inversely with the mass. The amount of ablation varies over the surface of a given meteorite, and there are often regions of relatively low (<7 cm) ablation on heavily ablated meteorites. Track results for ablation have shown that several meteorites, such as Bansur and Rosebud, were very lightly ablated in certain places and heavily ablated elsewhere. Studies that related ablation with the velocity of the meteoroid before it reached the earth have shown that the mass fraction removed is a steep function of velocity, but that many parameters besides entry speed, such as mass, shape, entry orientation, and meteorite type, also affect the amount of ablation. Meteoroids approaching the earth with high velocities (>25 km/sec) must have very large masses if they are not to be completely burned during entry, thus meteorites with large aphelia (>3 AU) are not properly represented in our collections.

The concentration of a neutron-capture-produced nuclide, such as ^{60}Co, can frequently be used to determine or set limits on preatmospheric sizes and sample depths. An extensive study of the large Jilin chondrite by G. Heusser and co-workers used activities of ^{60}Co from documented locations of the main mass to establish that Jilin was spherical with a radius of about 85 cm just before it fell in 1976. Usually a ^{60}Co activity about 1 dpm/kg or less implies that the preatmospheric size of the meteorite was too small to thermalize neutrons (i.e., it had a radius of less than ~25 cm). The production of ^{36}Cl, ^{80}Kr, ^{82}Kr, ^{128}Xe, and ^{131}Xe in meteorites by neutron-capture reactions with chlorine, bromine, iodine, and barium have also been used to set lower or upper limits on the preatmospheric radii of meteorites.

In several meteorites, unusual profiles or concentrations of cosmogenic products have been observed that appear to have been produced by solar-cosmic-ray particles. For example, the Antarctic meteorite ALHA77003 has a ^{53}Mn profile that decreases with distance from the surface and implies that ALHA77003 was probably a very small object in space. As the amount of SCR production usually is fairly small and roughly compensates for the decreased GCR production near the surface, it is often hard to detect that SCR production has occurred. The concentrations of ^{21}Ne measured in the outermost 1 cm of a core from St. Severin was higher than in deeper samples, a clear sign of SCR production. A high activity of SCR-produced ^{56}Co was measured in the Salem meteorite, which means that Salem was very small in space. Meteorites that were very small objects in space seem to be very rare.

The ablation of iron meteorites have been estimated by mapping the concentrations of various cosmogenic noble gas isotopes measured from slabs. Contours of equal concentrations for several cosmogenic nuclides that were measured by Signer and Nier in the Grant iron meteorite were slightly ellipsoidal, and comparisons with calculations showed that about 75% of the preatmospheric mass was lost by ablation. Noble gas measurements for a single sample and the exposure age of an iron meteorite can be used with several models to infer the preatmospheric depth of the sample and the radius of the meteoroid.

While measurements for a single sample of a meteorite is adequate to get an estimate of its preatmospheric radius, a suite of samples are needed to get the preatmospheric shape of a meteorite. The few meteorites that have had their shapes in space determined tend to be nonspherical, although Jilin's geometry just prior to

hitting the earth appears to have been very close to a sphere. Cosmogenic products have often been used to help show that two or more meteorites of the same type were part of the same parent body just prior to entry into the earth's atmosphere. If the shielding-corrected exposure ages are similar and the cosmogenic products fit along the profile for a given radius (such as those in Fig. 9), then there is strong evidence that they came from a common parent object. A number of stony meteorites from Antarctica or iron meteorites have been paired with the aid of cosmogenic products. It appears that many of the thousands of meteorites found in Antarctica came from a much smaller set of meteoroids hitting the earth.

B. Terrestrial Ages

The lengths of time that meteorite finds have been on the earth's surface can often be determined from their cosmic ray records. The amount of weathering is a qualitative indication of terrestrial age but varies quite widely with location of falls and meteorite type. As the activities of radionuclides and the amount of thermoluminescence (TL) decays with time, these cosmogenic products can be used to determine or set limits to the terrestrial age of a meteorite. Terrestrial ages are sometimes needed to correct the activities of various radionuclides for decay while the meteorite was on the earth's surface. They also can be used to identify whether several meteorites are pairs, fragmented from a common object during passage through the earth's atmosphere. Falls and finds also have been used to see if the distribution of meteorite types hitting the Earth has varied in the past. As terrestrial ages are usually much shorter than exposure ages, however, it seems unlikely that the present meteorites in our collections will be able to show much difference in the frequencies of meteorite types striking the earth in the past. Meteorites from the moon, as found in Antarctica, are exceptions because of the short time that they are in space, and their terrestrial ages probably will date when they were ejected from the lunar surface.

The decay of TL depends on the temperature record of a meteorite since it fell, but meteorites usually have been stored in museums for most of the time. The decay of TL also varies with the temperature at which the glow was measured, as the lower temperature TL decays faster than that at higher temperatures. The decay of TL is not a pure exponential, so calibration with samples of known terrestrial ages are needed. For the Antarctic meteorites, the mean life for the decay of TL has been measured to be ~10^5 years, much longer than for meteorites that fell in other locations.

Many meteorites from Antarctica have had their terrestrial ages determined from the decay of ^{14}C, ^{81}Kr, ^{36}Cl, ^{26}Al, and ^{10}Be. The distribution of the number of cases versus terrestrial age is concentrated below about 3×10^5 yr and decreases rapidly for higher ages. The longest terrestrial age for an Antarctic meteorite is ~10^6 yr. Such results are important in determining how the Antarctic meteorites were concentrated at the few ice fields on which they are found. The terrestrial ages of the Antarctic meteorites found by the Japanese near the Yamato Mountains generally are shorter than those recovered by American expeditions next to the Allan Hills. These meteorites found in Antarctica are believed to have been carried inside the ice to areas where the ice flow is stopped by an obstacle, such as a mountain range. Thus terrestrial ages of these meteorites can be used to examine ice movement in Antarctica.

The cosmogenic radionuclide ^{14}C has been used to determine terrestrial ages of stony meteorites from areas other than Antarctica. Not many of such stones had terrestrial ages greater than 10^4 yr, although the Potter L6 chondrite, which was found in a dry climate, has a terrestrial age that is greater than 2×10^4 yr. Iron meteorites can have much longer terrestrial ages than stones. The longest terrestrial age is about 3×10^6 yr for the IIIA iron meteorite Tamarugal, found in the desert of northern Chile. The terrestrial ages for many of the 12 hexahedrites (type IIA) found in northern Chile helped to show that these iron meteorites represent at least six different falls. Many of the iron meteorites found with large terrestrial ages were found in mountain areas, probably because alluvial processes make it hard to find old meteorites in the lowlands.

C. COMPLEX HISTORIES

There are a number of indications that certain meteorites have received their cosmogenic products both before and after a major change in their geometry. Such a history for meteorites is called a complex one, as opposed to a simple or one-stage history where all the cosmogenic products were produced without any shielding changes. For a number of meteorites, the exposure ages determined by different methods gave a wide range of ages. For example, some stones have low exposure ages based on ^{26}Al activities and higher ages from their ^{21}Ne concentrations. In interpreting the exposure-age records of meteorites, samples of meteorites with complex histories need to be identified because the age determined assuming a simple history would be incorrect. The fraction of meteorites with complex histories relative to those with simple histories can be used to infer how meteoroids evolved prior to hitting the earth. The rapid erosion of meteoroids by dust and radiation in space, which could create histories that appear complex, has been shown by various arguments and experiments to be unimportant, and thus collisions among larger objects are the sources of geometry changes in space.

A number of methods have been used to detect whether a meteorite had a complex history. Neutron-capture-produced nuclides or a low ^{22}Ne/^{21}Ne ratio (below ~1.07 for chondrites) are indications of production in a large object, and their presence in a meteorite that does not appear to have been very large is a sign to check for a possible complex history. Often the concentration of a stable cosmogenic product, such as ^{21}Ne or tracks, is higher than that predicted from the undersaturation of a radionuclide. The activities of the radionuclides ^{26}Al and ^{53}Mn have been used with ^{21}Ne concentrations to search for complex histories. The experimentally determined trends for the production rate of a radionuclide versus the track production rate (as in Fig. 9) or versus the ^{22}Ne/^{21}Ne ratio can identify complex histories if the data plot outside the allowable range. In some comparisons using radionuclide activities and noble gas concentrations, the exposure age determined from the noble gas is lower than that determined from the activity of the radionuclide, implying loss of the noble gas. It is possible that the excess amounts of a noble gas from an earlier exposure could be later lost, thus creating a record that seems to have been a simple one. The use of the ^{22}Ne/^{21}Ne ratio versus the inferred track production rate can be used to search for complex histories, as there is a range of values prohibited for simple histories.

A problem with these correlation methods is that a sample with a complex history could be shifted from one part to another part of the allowable range. This is especially true if only one sample from a meteorite has been analyzed, and there probably are some meteorites that have been incorrectly identified as having simple histories from such correlation trends. Measurements of a suite of cosmogenic products from a number of different locations can help to distinguish complex versus simple histories. In the Jilin chondrite, the ratios of ^{22}Na to ^{26}Al varied widely from sample to sample, showing that it had a complex history with a large geometry change fairly recently. Extensive analyses of Jilin samples have shown that it was first exposed to cosmic rays for $\sim 9 \times 10^6$ yr as part of a very large object and then was changed to an 85-cm-radius sphere about 4×10^5 yr ago. Such detailed unfolding of a meteorite's complex history, however, is rare.

In iron meteorites, cases of complex histories have been discovered when exposure ages determined with several radioactive/stable pairs have disagreed. For example, exposure ages determined with ^{39}Ar, ^{36}Cl, or ^{10}Be often disagree with those determined with ^{40}K after the factor of 1.45 for the variations in the cosmic ray fluxes are considered. The exposure ages determined with the shorter-lived radionuclides are much smaller than the ^{40}K/^{41}K ages in $\sim 30\%$ of the cases, indicating enhanced production of cosmogenic nuclides fairly recently in the history of the meteorite because of a shielding change. The isotopic ratios in these cases usually indicate that the sample was heavily shielded for most of its history. For very large iron meteorites (such as Odessa, Canyon Diablo, or Sikhote-Alin), different fragments can have different exposure ages, probably as a consequence of fragmentation of only one part of the meteoroid.

A special case of a complex history is when a part of a meteorite received a cosmic ray irradiation prior to its incorporation into the body in which it was found. Gas-rich grains in meteorites received an irradiation by solar wind ions and energetic SCR heavy nuclei early in their history. Cosmogenic nuclei also have been observed in several track-rich grains from gas-rich meteorites. Xenoliths, or foreign inclusions, separated from the H4 chondrite Weston and the LL6 chondrite St. Mesmin had concentrations and ratios of cosmogenic noble gases that differed significantly from the trends for other inclusions in the same meteorite. These xenoliths appear to have been exposed to cosmic rays prior to the compaction of these brecciated meteorites.

There are not very good statistics on the numbers of meteorites with well-documented complex or simple histories, partially because it is hard to be certain which type of history a meteorite had. It appears that iron meteorites are more likely to have complex histories than stones, which would be expected because of their longer exposure ages. The types of histories that meteorites have had probably cover a continuum that ranges from simple to ones with significant production in two very different geometries. As already mentioned, one explanation for the high ^{21}Ne production rates inferred from undersaturation of ^{26}Al could be small amounts of ^{21}Ne made prior to the present geometry. A small excess of cosmogenic ^{21}Ne would normally be hard to detect, and many meteorites might have small amounts of cosmogenic products made prior to their most recent exposure geometry.

Complex histories are useful in examining another stage further back into the evolution of a meteorite than is studied with exposure ages of meteorites with simple histories. G. Wetherill has predicted that complex histories should be more common (of the order of half the cases) than observed if the collisional model for the origins of meteorites is correct and wonders whether this discrepancy is a problem with the model or with interpreting the cosmic ray records of meteorites. R. Greenberg and C. Chapman have argued that many types of meteorites could be delivered to the earth directly by cratering on large main-belt asteroids. Such meteorites are less likely to have had complex histories. Complex histories for the very mechanically weak carbonaceous chondrites could be the result of irradiation in the surface layers of a comet prior to the release of the meteorite after the ice around it has been evaporated or sublimed. Because the real histories of meteorites have not been well determined in most cases, it is hard to use complex histories in studies of the evolution of meteorites. Many more measurements, especially of products with a wide range of production profiles (including tracks and neutron-capture nuclides) in a number of different samples from each meteorite are needed.

D. EXPOSURE AGES

In meteorites, the concentrations of certain cosmogenic products can be used to determine the length of time that the meteoroid was exposed to cosmic rays, the exposure age of the meteorite. An exposure age is the interval of time from when the meteorite was removed from a very heavily shielded location many meters deep in a parent body to when it hit the earth. The cosmic ray exposure ages for most meteorites are orders of magnitude shorter than the radiometric ages, usually 4.55×10^9 yr, for their formation. Exposure ages for meteorites are important in studies of the sources of meteoroids and the mechanisms that caused them to hit the earth. Exposure ages are the youngest of the possible ages for a meteorite. Other ages used in studies of the evolution of meteorites since their initial formation include collisional shock ages, typically 3×10^7 to 7×10^8 yr as inferred from losses of radiogenic gases or resetting of radiometric clocks, and ages for the brecciation of certain meteorites from the regoliths of parent bodies about 1.4×10^9 to 4.4×10^9 yr ago.

The results from studies of some meteorites can be used in understanding how their parent bodies evolved early in the solar system, and it would be nice to know where in the solar system the meteorites formed. For example, it is possible that the Shergottites, Nakhlites, and Chassigny formed on Mars, and, if true, these meteorites can be used to learn much about the evolution of Mars in the same way that lunar samples are used to study the moon. The orbits of the objects from which meteoroids were produced provide an understanding about the fluxes of interplanetary meteoroids and asteroids. Such objects in earth-crossing orbits can produce craters, some of which are very large and which probably caused extinctions on earth in the past, and there are considerable uncertainties in their fluxes.

An exposure age is usually determined from the concentration of a stable cosmogenic product, such as ^{21}Ne, and a production rate. Often corrections to the production rate because of the shielding of the sample are used, such as those based on the ^{22}Ne/^{21}Ne ratio. Many radioactive/stable pairs, such as ^{39}Ar/^{38}Ar, do not require shielding corrections when used to determine exposure ages. Sometimes the activity of a radionuclide, such as ^{26}Al, below its equilibrium value can be used to calculate an exposure age. In calculating exposure ages, it is assumed that the production rate has remained constant, not having varied in the past. As already discussed, this assumption is generally valid. The geometry of the meteoroid in space is assumed not to have changed during its exposure to the cosmic rays. Complex exposure histories, involving major changes in the geometry of the meteoroid with respect to the cosmic rays due to collisions, are known to have happened to some meteorites. Thus it is possible that some exposure ages are incorrect because the exposure history of the meteorite was assumed simple, but actually was complex.

For stony meteorites, concentrations of ^{21}Ne and shielding corrections based on ^{22}Ne/^{21}Ne ratios have been used to determine exposure ages for hundreds of meteorites. As the production rates for ^{21}Ne has recently been revised downward, the exposure ages given here are often higher than those originally reported by the various authors. The stony meteorites have exposure ages that range from ~50,000 yr for the L5 chondrite Farmington to about 8×10^7 yr for the aubrite Norton County. The distributions of exposure ages for the two largest types of meteorites, the H and L chondrites, are shown in Fig. 10. The H chondrites have a major peak near 6 to 7×10^6 yr that contains almost half of this type of meteorite. Its width (~$\pm 20\%$) probably represents the error expected in determining exposure ages due to measurement uncertainties, poor shielding corrections, diffusion losses of ^{21}Ne from some samples, and possibly small amounts of ^{21}Ne made prior to the present exposure geometry. This peak contains all petrologic types of H chondrites, although type H5 is somewhat more frequent and H6 less common in this peak than on the average. The distribution of exposure ages in Fig. 10 for L chondrites shows no statistically significant peaks.

There have been fewer exposure ages measured for stony meteorites of other types, so often the statistics are too poor to see distinct peaks in their exposure-age distributions. The LL chondrites have a very broad cluster of exposure ages near 1.5×10^7 yr. The CM2 type of carbonaceous chondrites are very young with exposure-age groups near ~5×10^5 and ~1.5×10^6 yr. Eight diogenites (all those measured except Manegaon) cluster at exposure ages of ~2.0×10^7 or ~3.5×10^7 yr. The howardites and eucrites, the commonest members of the calcium-rich achondrite family, have similar exposure-age distributions with several clusters between 1×10^7 and 4×10^7 yr. Most of the aubrites have exposure ages near 5×10^7 yr. The exposure ages for three of the four Shergottites are near 2.5×10^6 yr, and those of the three Nakhlites and of Chassigny are about 1.1×10^7 yr. These relatively short exposure ages of the Shergottites (the fourth Shergottite's exposure

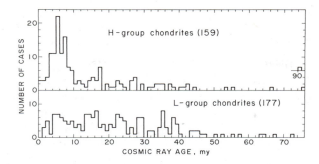

FIG. 10. Cosmic ray exposure ages (in 10^6 yr) of the two major types of stony meteorites as calculated from measured ^{21}Ne concentrations and $^{22}Ne/^{21}Ne$ ratios by J. Smith and K. Marti. [From Reedy *et al.* (1983). Reprinted with permission from *Science* **219,** 127–135. Copyright 1983.]

age is $\sim 1 \times 10^6$ yr), the Nakhlites, and Chassigny, which are all collectively referred to as the "SNC" meteorites, are consistent with a hypothesis, based on a variety of other data, that they came from Mars.

The carbonaceous chondrites tend to have shorter exposure ages than the other types of chondrites, with ages above 2×10^7 yr being fairly rare. The ages generally are shorter for the lower petrological types of the carbonaceous chondrites. Achondrites are much more likely to have exposure ages above $\sim 3 \times 10^7$ yr than are the chondrites. These trends, plus the fact that stony irons and the iron meteorites have much longer exposure ages than stones, indicates that a greater physical strength of a meteorite helps it to survive in space longer.

While about half of the H chondrites seemed to have been produced over a very short period of time, probably by a single large event on a parent body, other meteorite types (such as the L chondrites) have essentially a continuum for their exposure-age distributions. One question is whether the L chondrites were really produced continuously or by a number of individual events. Given the width of the H-chondrite peak, it has been estimated that possibly as few as six events could account for most of the distribution shown for the L chondrites in Fig. 10. The presence of all petrological types and also of gas-rich meteorites in the 6×10^6-yr peak for the H chondrites has been interpreted as implying that the parent body on which this event occurred was one that had earlier been nearly completely fragmented and then gravitationally reassembled into a megabreccia. The relatively large fraction (about 14%) of gas-rich H chondrites in this peak suggests that this event mainly removed objects from near the surface of the parent body.

The exposure ages of iron meteorites are much greater than those of the stony meteorites and range from 4×10^6 yr for the IB Pitts to 2.3×10^9 yr for the anomalous ataxite Deep Springs. Very few iron meteorites have exposure ages greater than 1.0×10^9 yr. The exposure-age frequencies of several types of iron meteorites are shown in Fig. 11. These ages are based on the $^{40}K/^{41}K$ ages as determined by H. Voshage using $^4He/^{21}Ne$ ratios for shielding corrections. (In Fig. 11, the scarcity of exposure ages less than $\sim 2 \times 10^8$ yr is partially because of the difficulty of detecting the relatively small amount of ^{40}K. Other measurements indicate that roughly 10% of iron meteorites have exposure ages below 10^8 yr.) The most prominent peak is near 6.75×10^8 yr for the IIIAB iron meteorites (medium octahedrites) and has a width of about $\pm 1.0 \times 10^8$ yr. The IVA irons (hexahedrites) have exposure ages that cluster near 4.5×10^8 yr. The IIA irons have relatively low ages (most $< 3 \times 10^8$ yr). The exposure-age distributions vary widely among iron-meteorite types. Exposure ages for stony irons are intermediate between those for stones and irons. Stony irons with exposure ages greater than 2×10^8 yr are fairly rare.

Noble gas measurements have shown that major losses of the helium and argon made by the decay of the natural radioisotopes of uranium, thorium, and potassium have occurred occasionally in the past. For example, about a third

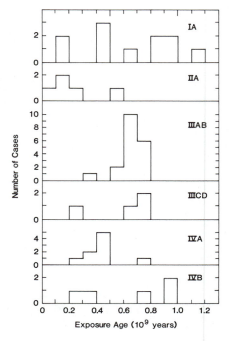

FIG. 11.   Cosmic ray exposure ages of six types of iron meteorites as calculated from measured ratios of $^{40}K/$ ^{41}K and $^4He/^{21}Ne$ by H. Voshage.

In addition to dating the collision events that produced these meteorites, the exposure ages also reflect the probability that these objects eventually hit the earth before being destroyed in space. The long exposure ages for Norton County and the other relatively mechanically weak aubrites could indicate that they had orbits with low probabilities for collisions with other objects in space. While the much longer exposure ages for iron meteorites relative to the stones probably is due to their mechanical strength, it also could be partially caused by their places of origin. The short exposure ages of the lower petrological types of carbonaceous chondrites (CI and CM) have been interpreted as being possibly due to their origin from comets, which have orbital lifetimes of 10^6 to 10^7 yr. Meteoroids in earth-crossing orbits have been calculated to have lifetimes of 10^7 to 10^8 yr, similar to exposure ages of stony meteorites. The exposure ages of the iron meteorites, 10^8–10^9 yr, are similar to the lifetimes for Mars-crossing asteroids. Thus exposure ages of meteorites are consistent with the generally accepted hypothesis that most meteorites come from asteroids.

BIBLIOGRAPHY

Cressy, P. J., Jr., and Bogard, D. D. (1976). *Geochim. Cosmochim. Acta* **40**, 749.
Fleischer, R. L., Price, P. B., and Walker, R. M. (1975). "Nuclear Tracks in Solids." Univ. of California Press, Berkeley.
Honda, M., and Arnold, J. R. (1964). *Science* **143**, 203.
Kirsten, T. (1978). Time and the solar system, *in* "The Origin of the Solar System" (S. F. Dermott, ed.), pp. 267–346. Wiley, New York.
Kirsten, T. A., and Schaeffer, O. A. (1971). High-energy interactions in space, *in* "Elementary Particles: Science, Technology, and Society" (Luke C. L. Yuan, ed.), pp. 75–157. Academic Press, New York.
Lal, D. (1972). *Space Sci. Rev.* **14**, 3.
Millman, P. M. (ed.) (1969). "Meteorite Research." Reidel, Dordrecht, Holland.
Moore, C. B. (ed.) (1962). "Researches on Meteorites." Wiley, New York.
Reedy, R. C., Arnold, J. R., and Lal, D. (1983). *Science* **219**, 127; *Annu. Rev. Nucl. Part. Sci.* **33**, 505.
Voshage, H. (1984). *Earth Planet. Sci. Lett.* **71**, 181.

of the L chondrites lost most of their noble gases in a major collision of their parent body ~5 × 10^8 yr ago. The exposure ages of these L chondrites with low gas-retention ages cover the whole range shown in Fig. 10. The more recent events dated by the exposure ages usually did not cause much loss of such radiogenic gases. The general lack of shock features, such as gas losses, associated with the events that initiated the cosmic ray exposures of most meteorites indicates that most meteorites were probably ejected from their parent bodies with low velocities (~1 km/sec or less). Recently, several mechanisms that can eject meteorite-sized fragments to planetary escape velocities with minimal shock effects (such as spalling or jets of volatiles) have been proposed to account for the lunar meteorites and for possible martian meteorites, which require escape velocities of 2.4 and 5 km/sec, respectively.

MOON (ASTRONOMY)

Bonnie J. Buratti *JPL, California Institute of Technology*

GLOSSARY

Apogee: Farthest distance from the earth in the moon's orbit.

Barycenter: Center of mass of two bodies, around which they revolve.

Bond albedo: Fraction of the total incident radiation reflected by a celestial body.

Geometric albedo: Ratio of the brightness at full moon compared with a diffuse, perfectly reflecting disk.

Librations: Wobbles in the moon's orbit, which cause more than half of it to be visible from earth.

Mare: (pl. maria) Low albedo plains of the moon.

Nodes: Two points in the moon's orbit, at which it intersects the ecliptic.

Perigee: Closest distance to the earth in the moon's orbit.

Planetesimals: Bodies up to kilometers in size, which populated the solar system before the planets accreted.

Primary: Celestial body around which a satellite, or secondary, orbits.

Regolith: Layer of debris and dust, which covers the moon.

Saros cycle: Period during which the lunar and solar eclipses repeat (18.03 years).

Sidereal month: Period of revolution of the moon about the earth with respect to the stars (27.32 days).

Synchronous rotation: Dynamical state caused by tidal interactions in which the moon presents the same face toward the earth.

Synodic month: Period of revolution of the moon about the earth with respect to the sun (29.53 days).

Uplands: Heavily cratered, higher albedo portions of the lunar surface.

The moon is the planet earth's only natural satellite. It orbits the earth approximately once every month as the two bodies journey together about the sun. The moon keeps the same face turned toward the earth because it is tidally evolved. The quest to understand the phases of the moon, the cycle of lunar and solar eclipses, the ocean tides, and the motion of the moon in the sky served as a basis for early scientific investigation. The moon has also figured prominently in the legends and lore of the world's peoples.

I. Introduction

Table I summarizes the gross characteristics of the moon. The earth-moon system is somewhat unusual in that the mass of the moon in comparison to its primary is large (0.012). Charon, the only known companion of Pluto, is the only satellite with a larger relative mass (about 0.1).

In 1610 Galileo first observed the moon through a telescope. He perceived the bright, rough lunar highlands, which were named after terrestrial mountain chains, and the darker lunar plains, or maria (Latin for seas), which were given fanciful names such as the Sea of Tranquility, Sea of Humors, and Sea of Clouds. Lunar craters are generally named after famous scientists. Figure 1 in the article within this encyclopedia, "Planetary Satellites, Natural" shows the near side of the moon as seen through a large telescope.

TABLE I.   Gross Characteristics of the Moon

Mean distance from earth (semi-major axis)	Rotation period	Revolution period	Radius	Mass
384.4×10^3 km	27.32 days	27.32 days	1738 km	7.35×10^{22} kg 0.012 of earth's

Density	Visual magnitude	Visual geometric albedo	Bond albedo
3.34 gm/cc	−12.5	0.11	0.12

Although the moon is the only celestial body to have been visited by human beings, many questions, including that of its origin, remain unanswered. [*See* LUNAR ROCKS.]

II. Properties of the Moon

A. SIZE, SHAPE AND DISTANCE

Aristarchus of Samos, a Greek philosopher of the third century B.C., who ascribed to the heliocentric view of the universe, attempted to measure the relative distances between the earth and the sun and moon by noting the angular distance between the latter two bodies when the moon as seen from the earth was half illuminated, that is, during quarter phase. Although his estimate was grossly inaccurate due to the difficulty of measuring this angular separation, he correctly deduced that the sun was much farther away from the earth than the moon. In the next century, Hipparchus made estimates accurate to within a few percent of both the distance between the earth and moon and the moon's diameter based on a measurement of the moon's parallax.

The experimental techniques developed by the Greeks were lost in the course of the dark ages and the rise of the more abstract Aristotelean school during the Middle Ages (although Arab and Chinese scientists continued to conduct astronomical observations). It was not until the emergence of modern astronomy and the heliocentric world view in the sixteenth and seventeenth centuries that significant progress was made in understanding the nature of the cosmos. The distance of the moon from the earth and its size continued to be estimated by measuring its angular size and parallax until the twentieth century.

The development of radar during World War II led to accurate measurements of the distances to the moon and nearby planets. By measuring the delay required for a radio signal to be returned to earth, scientists calculated the average distance between the earth and the moon to be 384,403 km. It is possible for present laser ranging techniques to measure distances to the moon to within an error of 3 cm. Radar ranging measurements give topographic profiles of lunar craters and mountains with typical errors of tens of meters. Accurate measurements of the angular size of the moon yield a value of 3476 km for the lunar diameter.

The mass of the moon relative to the earth is inversely proportional to the ratio of their distances from the barycenter of the system (the barycenter is the center of mass around which the two bodies revolve). More accurate determinations of the lunar mass can be obtained by measuring the perturbations the moon exerts on spacecraft. The moon is about $\frac{1}{18}$ as massive as the earth, or 7.35×10^{25} gm. The mean density of the moon is 3.34 gm/cm^3, less than the earth's value of 5.52. The lower value for the moon is due to the absence of a large metallic core, and is consistent with a predominantly rocky composition.

The moon is not a perfect sphere. Its earth-facing radius is 1.08 km larger than its polar radius, and 0.2 km larger than its radius in the direction of motion.

B. THE MOTION OF THE MOON IN THE SKY

The moon executes a combination of complex motions, both real and apparent, as it orbits the earth. The period of revolution of the moon about the earth with respect to the stars is 27.32 days and is known as a sidereal month. Because the earth–moon system moves $\frac{1}{13}$ of the way around the sun in one sidereal month the moon must travel another 27° in its orbit about the earth to be in the same position in the sky with respect to the sun. This period is known as the synodic month and takes 29.53 days to complete.

The moon, like other celestial objects, rises in the east and sets in the west. But each night, it is seen to move approximately 13° to the east with respect to the background stars. This apparent motion is due to the moon's revolution about the earth once every lunar month. The revolution of the earth–moon system about the sun at the rate of about 1° per day causes the path of the moon to be displaced only 12° eastward with respect to the sun. This eastward motion of the moon causes it to cross the local celestial meridian an average of 50 min later each day. Because the velocity of the moon in its orbit about the earth varies in accordance with Kepler's second law (which states that a celestial body will sweep out equal areas of its elliptical orbit in equal times) this successive delay ranges from 38 to 66 min. However, for moonrise and moonset, this delay varies (depending on the geographical latitude of the observer) from only several minutes to over an hour. This difference is due to the fact that the path of the moon is in general not perpendicular to the horizon. Thus, on successive nights the moon does not have as far to travel in its apparent motion from below to above the horizon. In mid or far northern latitudes, the extreme example of this phenomenon is the harvest moon, which occurs near the vernal equinox, when the angle subtended by the path of the moon with the horizon is a minimum. On the several nights occurring near the full moon closest to the vernal equinox, the brightly lit moon continues to rise shortly after sunset to provide extra time to farmers reaping their fall crops.

The height of the moon above the horizon varies markedly throughout the year. Like the sun, the moon is higher in the sky during the summer than in the winter, due to the 23.5° inclination of the earth's equator to the plane of its orbit about the sun (the ecliptic). In addition, the moon's orbit about the earth is inclined 5° to the ecliptic. The moon's height also depends on the geographical latitude of the observer. At extreme northern or southern latitudes, the moon appears lower in the sky than near the earth's equator.

Although the moon keeps the same face turned toward the earth, 59% of the moon's surface is visible from the earth because of librations, or wobbles, in the lunar orbit. Geometric librations are caused by three factors in the orbital relations between the earth and moon. A libration in latitude is a consequence of the inclination of 6.7° between the lunar spin axis and orbital plane. A libration of longitude, which causes an additional 7.6° to be seen, is due to the fact the law of inertia constrains the rotational velocity of the moon to be uniform, whereas its orbital velocity varies in accordance with Kepler's second law. The third geometrical libration, known as the diurnal libration, is attributable to the fact the radius of the earth subtends nearly a degree of arc at the distance of the moon. Therefore, the daily rotation of the earth on its axis causes an observer to view a slightly different aspect of the lunar surface. Additional smaller librations, known as physical librations, are caused by pendulum like oscillations induced by the earth in the moon's motions due to the latter's nonspherical shape.

C. Orbital Motions

The major characteristics of the moon's orbit are summarized in Table II. The theory that describes the complete motion of the moon is complex: only the basics are outlined here. The earth and moon revolve about their center of mass, or barycenter. As a first approximation, the path of the moon about the earth is an ellipse with the earth at one of the foci. The eccentricity of the moon's orbit averages 0.0549, but varies from 0.044 to 0.067 due to perturbations by the earth and sun. The closest approach of the moon to the earth (perigee) is 363,000 km and the maximum distance (apogee) is 405,500 km. Tidal forces exerted by the sun cause the semi-major axis, or the line of apsides, to rotate eastward every 8.85 yr. [See CELESTIAL MECHANICS.]

The plane defined by the moon's orbit inter-

TABLE II. Characteristics of the Lunar Orbit

Apogee	Perigee	Mean eccentricity	Inclination to earth's equator	Obliquity
405,500 km	363,300 km	0.055	18 to 29°	6.7 degrees

sects the ecliptic at two points known as the nodes. The ascending node is defined by the moon's motion northward as it rises above the ecliptic as seen by a northern observer, and the descending node occurs when the moon moves below the ecliptic. Perturbations of the moon by the earth and sun cause the line of nodes, defined by the intersection of the orbits of the earth and moon, to move westward along the ecliptic every 18.6 yr in a motion called the regression of the nodes. This 18.6 cycle is known as the nutation period. Two other cycles are the draconitic month, which is the time required for the moon to return to the same node, and the anomalistic month, or the period between two perigee passages.

The regression of the nodes causes the earth's axis to wobble as much as 9 sec. Additional perturbations cause the inclination of the moon's orbit to the ecliptic to vary between 4° 57 min and 5° 20 min.

D. Phases of the Moon

In the course of the lunar month, the moon presents a different fraction of its illuminated face to an observer on earth. Figure 1 is a schematic diagram of the resulting phases of the moon and their names. The moon is said to be waxing when the illuminated face is growing, and waning when it is progressing toward new moon. The quest for an understanding of the moon's phases and eclipses was perhaps the primary impetus for the development of ancient astronomy. By the fourth century B.C., Aristotle presented a clear explanation of these phenomena, although Anaxagoras a century earlier was probably the first western astronomer to offer correct explanations.

The various phases of the moon rise and set at specific times of the night (or day). The full moon always rises shortly after sunset. A waning gibbous moon rises later and later in the evening and the moon at last quarter rises around midnight. The new moon rises at dawn and sets shortly after sunset. The first quarter rises in midday and the waxing gibbous moon rises later and later in the afternoon until it is finally full.

E. Eclipses

Solar and lunar eclipses were the root of much lore and superstition in the ancient world. Both ancient eastern and western astronomers realized that eclipses occurred in cycles and accurately calculated their dates.

When the earth passes between the sun and the full moon and all three objects are in a straight line, the Earth's shadow is cast on the face of the moon to cause a lunar eclipse (Fig. 2). The earth's shadow has both a full zone (the umbra) and a partial zone (the penumbra). If the moon is totally immersed in the full terrestrial shadow, a total lunar eclipse results. If the moon is in the penumbra, the eclipse is said to be penumbral, and is in fact barely visible. If the moon is partly illuminated, the eclipse is said to be partial. A lunar eclipse can last as long as 3 hr, 40 min with the duration of totality as much as 1 hr 40 min. A lunar eclipse is visible to an observer on any point of the earth's nightside portion. During an eclipse, the moon is bathed in a reddish glow, which is due to light refracted by the earth's atmosphere into the earth's shadow.

The more spectacular solar eclipse occurs when the new moon passes in front of the sun

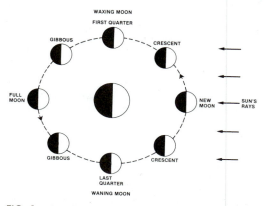

FIG. 1.   A schematic diagram of the phases of the moon as seen from earth (center).

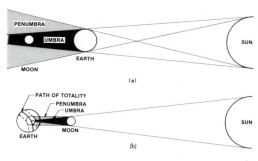

FIG. 2.   (a). A lunar eclipse occurs when the earth's shadow covers the face of the full moon. (b). A solar eclipse occurs during new moon, when the moon's shadow is cast on the earth.

(Fig. 2). Because the angular sizes of the moon and sun are both about a half of a degree, the lunar disk barely obscures the sun to reveal the spectacular sight of the solar corona. The moon's full shadow on the earth is at most 267 km wide, and even the area of partial eclipse, where the observer is in the moon's penumbra, is about 6500 km wide. The duration of totality is always less than 7.5 min. Lunar eclipses are thus more frequently observed than solar eclipses at a specific terrestrial location. If the moon is close enough to the earth so that its angular size is less than that of the sun, the result is an annular eclipse, in which a ring of the sun's disk encircles the moon.

Eclipses occur in cycles of 18.03 yr because they are possible only when both the moon and sun are close to the nodes of the lunar orbit. One eclipse period, also known as the Saros cycle, consists of a whole number of both synodic months and eclipse years (the time required for the sun to return to a node in the moon's orbit, or 346.62 days).

F. THE TIDES

Gravitational interactions between the earth and the moon cause the well-known phenomenon of the oceanic tides (Fig. 3). The gravitational forces experienced by the side of the earth facing the moon is greater than that between the moon and the center of the earth, which in turn is greater than that exerted on the far side of the earth. Thus the differential gravitational forces of the moon on the earth act to flatten the earth with the long axis pointing toward the moon. The liquid seas are influenced more by tidal forces than the solid earth. The sun also exerts

tidal forces on the earth (although they are not as great because they depend on the inverse cube of the distance between the two bodies). At full and new moon, when all three bodies are in a line, spring, or maximum tides result. Similarly, during first and last quarter, the minimum neap tides occur (see Fig. 3).

G. THE TEMPERATURE OF THE MOON

The surface of the moon undergoes greater extremes of temperature than the earth, because it has no atmosphere or ocean to temper the effects of the sun, and because it rotates on its axis only once every month. The temperatures on the sun range from a daytime high of 380K to nighttime lows of 100K.

H. PHOTOMETRIC PROPERTIES

The moon is the second brightest celestial object in the sky, yet it is only little more than 2 millionths of the brightness of the sun. The fraction of visible light that the full moon reflects back to an observer (the geometric albedo) is about 0.1. The intensity of the moon varies markedly over its surface: the lunar maria reflect only 6.5–9% of visible light, whereas the lunar uplands and crater bottoms reflect about 11% and the bright ray craters 13–17%. The Bond albedo (the disk-integrated amount of total radiation emitted compared with that received) is 0.12. The dark portion of the moon is illuminated slightly by earthshine, which is light reflected from the surface of the earth back to the moon.

The moon is associated with a number of striking optical phenomena. On clear dry days in the spring, a ring is often seen around the moon appearing 22° from its disk. The ring is due to the refraction of moonlight by small hexagonal ice crystals high in the earth's atmosphere. Popular belief states that the ring foretells rainy weather: there is a modicum of truth in this belief because cirro-stratus clouds, which are composed partly of ice crystals, are associated with low atmospheric pressure. The deep orange color of the moon when it is near the horizon is caused by the greater degree of scattering of blue light by the atmosphere, which has a much higher optical depth when the observer looks toward the horizon. The rare blue moon is a contrast phenomenon caused by the position of the moon next to pink clouds.

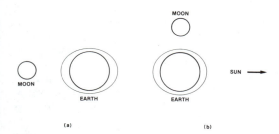

FIG. 3.   The tidal bulges induced on the earth by the moon and sun. Spring tides (a) occur when all three bodies are in a straight line. The minimum or neap tide (b) is seen during the quarter phases of the moon. The earth's rotation causes high tides to appear twice each day.

III. Summary of Observations

A. EARTH-BASED

Because of its proximity and its influence on the earth, the moon is perhaps the most studied celestial body. Detailed studies of the moon's motion, its appearance in the sky, its role in causing the tides and eclipses, has occupied the time of astronomers since ancient times.

In 1610, Galileo was the first to observe the moon through a telescope: he discerned the delineation of the moon into the bright uplands (which he called terrae, or land) and the darker plains (which he called maria or seas). The moon was first photographed in 1840. However, most of the detailed mapping of lunar features from the earth has been done visually, because instabilities in the earth's atmosphere constrain the resolution in photographic exposures to be only a fourth of that possible with the human eye. From the earth, a resolution of about 2 km on the moon's surface is possible with the best telescopes and ideal seeing conditions.

Observations from the earth of the visible and near infrared spectrum of the moon suggested its surface is composed of basaltic minerals. Analysis of the lunar phase curve, which is the variation in disk-integrated brightness with changing phase, suggested the existence of mutual shadowing among the particles in the optically active surface, which is consistent with a fluffy soil. Infrared measurements of how fast the surface loses heat as it comes out of eclipse also pointed to a porous surface. Infrared measurements demonstrated that the moon does not have an internal heat source.

B. SPACECRAFT RECONNAISANCE

A series of reconnaisance craft to the moon sent by US and the USSR began in 1959, when the Soviet Luna 3 photographed the far side of the moon (see Table 2 in the article within this encyclopedia "Planetary Satellites, Natural"). The Ranger program, which was a series of impact probes sent to the moon by the US between 1961–65, returned many thousands of live television images of the moon's surface, some with 1m resolution. Five Lunar Orbiter missions launched by the US between 1966 and 1968 successfully returned high resolution images of 95% of the lunar surface. These spacecraft also mapped gravity anomalies below the moon's surface.

The first spacecraft to land successfully on the moon was the Soviet Luna 9 in 1966. The US successfully landed 5 spacecraft on the moon as part of its Surveyor program. They sent back over 87,000 images, and studied the composition of the lunar soil with an alpha-particle spectrometer. Between 1970 and 1976, three of the Soviet Luna spacecraft sent back to the earth small samples of lunar rock.

C. MANNED EXPLORATION

The United States successfully launched 6 manned Apollo missions to the moon between 1969 and 1972. The major scientific goals of the Apollo mission were to return rock samples for geochemical and morphological analysis, to obtain photographic coverage of the moon for geologic study, to place seismometers, magnetometers, heat flow instruments, gravimeters, and other experiments on the surface, and to employ an array of spectrometers of compositional studies.

Extensive photographic coverage and manned exploration of the moon has transformed it into a tangible geologic world. Although the vast amount of data has shed light on the basic problems of lunar morphology, minerology, and geophysical evolution, some questions, such as the structure of the moon's interior and the origin of the moon, remain unanswered.

IV. Theories on the Moon's Origin

A. INTRODUCTION

The important observational constraints that must be accounted for by a model for the origin of the earth–moon system include: an anomalously large ratio of satellite to primary mass; a lower mean density for the moon than the earth, which indicates the absence of a significant iron–nickel core for the moon; important chemical differences in the lunar crust such as enrichment of refractory elements (especially aluminum, calcium, and titanium); and depletion of volatile materials such as water, sulfur, sodium, and potassium.

The three standard theories for the origin of the moon are: (1) capture, in which the moon forms elsewhere in the solar system and is gravitationally captured by the earth; (2) fission, which asserts that the moon broke off from the earth early in its formation; and (3) coaccretion, which asserts that the earth and moon formed independently but nearly simultaneously near their present locations.

B. CAPTURE

The capture hypothesis explains the chemical differences between the earth and moon by having the moon form in an area of the solar system where refractory elements were rich and volatile elements rare (presumably nearer the sun). However, the likelihood of gravitationally binding such a body in a closed orbit about the earth is small, particularly if the body is not in the earth's immediate neighborhood. In a more recent variation of the capture theory, planetesimals were captured by the earth and later accreted to form the moon.

C. FISSION

In the fission model, the proto-earth had so much angular momentum after the iron core was formed that it was able to fling off a mass of material that became the moon. The model was first developed by George Darwin in 1898, who discovered that the moon was once closer to the earth and has been gradually receding due to tidal interaction (see Section VII A). The fission model explains the chemical differences between the two bodies by having fissioned material consist entirely of the primordial earth's mantle. The problems with the model include explaining how the earth acquired enough angular momentum to fling off a mass as large as the moon, and why the moon revolves in a plane inclined to the earth's own spin equator. In one variation of the fission model, which overcomes these difficulties, a large body, perhaps $\frac{1}{10}$ as massive as the earth, impacted the planet at a grazing angle and blew material into earth's orbit, which then accreted to form the moon.

D. COACCRETION

In the coaccretion model, the moon formed separately from the earth, most likely from debris, which was in orbit around the proto-earth. The large relative mass of the moon is obtained by having a body being captured and subsequently sweeping up material. The major problem with this model is that it is difficult to explain the differences in chemical compositions between the two bodies if they were formed from the same association of materials. One way of overcoming this difficulty is to have heavier iron rich debris preferentially accreted onto the earth.

V. The Lunar Surface

A. SURFACE FEATURES AND MORPHOLOGY

1. The Lunar Uplands

The basic fact of lunar surface morphology is that the moon is divided into two major terrains: the older, brighter heavily cratered lunar uplands and the darker, younger maria. This dichotomy is responsible for the appearance of the "man in the moon" (other cultures have seen a woman in the moon, or a hare). The uplands cover about 80% of the lunar surface as a whole and nearly 100% of the far side.

Figure 4 depicts the rugged, cratered appearance typical of the uplands. Some areas are saturated with craters. The oldest rocks date from the period of the formation and differentiation of the moon (4.3–4.6 billion years ago). Most of the uplands consists of anorthositic gabbro, (a coarse grained, sodium–calcium rich silicate) which has been pulverized by impacts and fused together by shock metamorphism into a rock

FIG. 4.   A Lunar Orbiter picture of the southwest portion of the moon's far side. Heavily cratered highlands dominate this hemisphere. The crater Tsiolkovsky near the center of the picture is filled with dark mare material.

known as breccia. When compared to the maria, the uplands are richer in so-called KREEP basalt, which is higher in potassium (K), rare-earth elements (REE) and phosphorus (P). The highlands are also enriched in aluminum and depleted in magnesium and titanium.

A large area of the lunar uplands consists of the Cayley formation, a smooth higher albedo unit interspersed among the rugged cratered terrain. Although the Cayley formation shows some evidence of having been created by vulcanism, the highly brecciated samples returned by Apollo 16b suggest that it was formed by sheets of material ejected from a large impact event such as the formation of the Orientale Basin.

2. Maria and Basins

The lunar maria (or plains), which were formed between 3.1 and 3.9 billion years ago, are the youngest geologic units on the lunar surface, except for more recent impact craters. The release of heat from large impacts caused extensive melting and extrusion of basaltic lavas on the moon. In some cases the extrusions may have occurred in two stages: first from the impact melt itself and later from eruptions caused by subsequent heating. Several maria are large filled-in impact features, or basins, with clearly circular delineations. Some basins, such as Mare Orientale, are multiring structures.

Large mountain chains at the edges of the large basins were uplifted at the time of impact. Wrinkle ridges, which are believed to be compressional features formed near the end of the period of vulcanism, are seen in many of the maria.

The lunar maria are found primarily on the earth-side of the moon. One possible physical explanation for the unequal distribution is that the maria formed in accordance with hydrostatic equilibrium: the moon's center of mass is closer to the earth's side and the crust on the farside is thicker.

3. Craters

Even when observed with a pair of binoculars or a small telescope, the moon appears covered with craters:about 30,000 are visible from earth. According to a tradition began by Riccioli in 1651 craters are named after well-known scientists. Lunar craters range in size from small millimeter sized pits caused by micrometeorites to

large structures hundreds of kilometers in diameter (see Fig. 5).

Before the period of lunar exploration in the 1960's, one of the major questions of lunar science was whether the craters were of volcanic origin or caused by impacts. Based on morphological evidence, most geologists have determined that craters were caused by impacts. This evidence includes the existence of ejected material (known as ejecta blankets) around crater rims, the creation of secondary impacts in the vicinity of large craters, and the existence of central peaks, which are due to the rebound of material after a large impact.

Throughout its history, the moon has been bombarded by meteoroids ranging in size from a few microns to several kilometers. Most craters, particularly the large ones, date from the period of heavy bombardment, which ended its intense phase about 3.9 billion years ago and tapered off until 3.1 billion years ago. Since then, the moon has experienced only an occasional impact. Because geological processes on the moon do not include wind or water erosion or plate tectonics, which have obliterated nearly all of earth's craters, the moon's face bears the scars of its past history. Extending from the most recent craters,

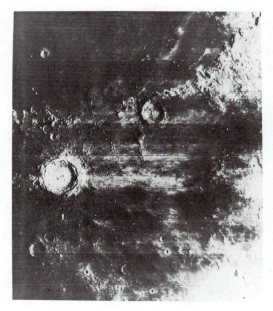

FIG. 5. The rayed cratered Copernicus, showing an extensive ejecta blanket, secondary impacts, and central peak. The crater Erastosthenes is to the upper right of center. Several older craters in the picture have been filled in by lava.

such as Tycho (0.27 billion years old) are bright rays caused by the disturbance and subsequent exposure of fresher, brighter subsurface material.

Geologists use crater counting methods to date lunar surfaces. Although the techniques in practice are complex and involve many assumptions, the principle behind them is simple: if we assume a specific flux (not necessarily constant) of impacting objects throughout the moon's history, the age of a surface (i.e., when it solidified) is proportional to the number of craters on it.

Meter-sized lunar craters are often simple bowl-shaped depressions. As the size of the crater increases, a number of complex phenomena are seen. These include raised crater rims, the formation of central peaks, secondary impact craters caused by large ejected chunks, slumping of material down the crater walls, and raised blankets of ejected material encircling the crater (Fig. 5). In general, larger craters have a smaller depth to diameter ratio than small craters. The largest impacts form the basins, some of which have been filled in by lava flows.

One type of intriguing lunar feature is the dark halo crater, which is a small low-rimmed structure with areas of darker material extending from the rim. Evidence that the dark material may have been deposited during a eruption is given by their tendency to be associated with fissures.

4. Volcanic Features

A number of geologic formations on the moon have volcanic origins. One common feature is the rille, which is a collapsed lava tube or channel (Fig. 6). The Hadley Rille is a sinuous formation 125 km in length. Crater chains are often associated with rilles or fissures, and may be tens of kilometers long.

Domes or even cones are often seen in maria areas and are probably extinct volcanoes.

B. The Regolith

The upper few meters of the moon consists of dust and rubble, which is known collectively as the lunar regolith. This material is the debris from eons of impact events, which have "gardened" the surface. The regolith is fluffy—about 60–80% of its volume is void. It is rich in glassy spherules, which were formed by melting during impacts by micrometeorites.

Another erosional process occurring on the

FIG. 6.   The Alpine Valley (150 km long and 8 km wide) cut by a sinuous rille. At the top is Mare Imbrium.

lunar surface is due to interactions by the solar wind, which tends to darken the upper regolith.

VI. The Lunar Interior

Knowledge of the lunar interior comes primarily from analysis of seismic waves detected by instruments placed on the lunar surface by Apollo astronauts. Interpretations of the data must in addition be consistent with the parameters defining the gross structure of the moon such as its density (3.34 g/cc) and its coefficient of moment of inertia (0.395). The moon's density is lower than that of the earth (5.52), which implies a deficiency of metallic components and a smaller, less dense core, and its moment of inertia is consistent with the differentiation of the moon into crust, mantle, and core. Lunar samples returned by the Apollo astronauts were indeed depleted in metallic elements and enriched in refractory materials.

Because the moon is smaller than the earth and thus cooled more rapidly during its formation, its crust is 60 km—about 4 times as thick as that of the earth. The crust consists largely of anorthositic gabbro and mare basalts. The mantle is composed of ultramafic rocks (minerals richer in heavy elements and silica poor). Seismic evidence indicates partial melting in the lower mantle. Moonquakes detected during the Apollo mission generally originated in the middle mantle (800–1000 km deep). In the core, which has a diameter between 300 and 700 km, attenuation of seismic s-waves, which do not propogate through a liquid, suggest that this part of the moon is molten. The existence of iron and perhaps nickel would give the core its required higher density.

During the Lunar Orbiter mission, the spacecraft experienced a number of anomalous accelerations in its orbit. These perturbations, which were observed only on the near side, were attributed to the implacement of higher density lava in large (>200 km) filled craters and maria. Geologists called these anomalies mascons (mass concentrations).

Instruments placed on the surface of the moon by Apollo 15 and 17 measured the heat flow from the lunar interior, which originates primarily from the decay of radioactive isotopes. The measurements indicate that the moon has a heat flow of only $\frac{1}{5}$ that of earth.

B. Magnetic Field

Planetary magnetic fields are believed to be caused by strong convective currents in planetary cores. These currents require for their production complete melting over of a large region of the planet's interior and rapid rotation of the planet. Because the moon has neither a large core or rotates rapidly, it has no appreciable magnetic field. However, analysis of remnant magnetism in lunar rock indicates that the moon had a weak magnetic field ($\frac{1}{20}$ that of the earth's current field) around the time of the formation of the maria (3.9–3.1 billion years ago).

VII. Evolution of the Moon

A. Dynamical History

The moon is slowly receding from the earth. The cause for this recession is tidal interactions in the earth–moon system. Because the earth is spinning on its axis, the earth's tidal bulge closest to the moon (see Fig. 3) leads the line connecting the two bodies' center of masses. The moon exerts a greater force on this bulge, with the result that the spin rate of the earth, and thus its angular momentum, decreases. Because angular momentum in the system must be conserved, the moon is moving further away from the earth. Evidence that the lunar month and

terrestrial day were shorter in the past is found in the fossilized shells of some species of sea corals, which grow in cycles following the lunar tides.

B. Geologic Evolution

1. Accretion and Early Bombardment

Radiometric dating of lunar samples indicates that the moon solidified about 4.6 billion years ago, at the same time the solar system formed. Accretional heating due to the release of gravitational energy during formation caused the moon to melt and differentiate into core, mantle and crust. Toward the end of the accretional phase, the solid moon was bombarded by remnant planetesimals. The largest of these formed the large ringed impact basins. The heavily cratered portions of the lunar uplands date from the tail end of the bombardment phase.

2. Formation of Maria

Mechanical energy released from the impacting objects during the bombardment phase was sufficient to partially melt the moon's upper mantle. Between 3.1 and 3.9 billion years ago, the large impact basins were flooded with mare basalts. Some amount of localized mare vulcanism perhaps occurred as late as 2.5 billion years ago.

3. Later Evolution

The major geologic processes occurring on the moon today are infrequent cratering events and "gardening" of the regolith by micrometeorites and the solar wind (section V.B.). Although the present cratering rate on the moon is a small fraction of what it was in the past, several well-known craters are of recent origin, such as Tycho (0.27 billion years old), and Copernicus (0.9 billion years old).

The small seismic events (1–2 on the Richter scale) recorded by the Apollo instruments are due to tidal stresses exerted on the moon by the earth or to continued contraction of the moon as it cools.

Although the moon is essentially geologically dead, there have been documented observations of possible small scale current activity. In 1178, five Canterbury monks reported seeing a brightening near the upper horn of the crescent moon. This event may have been the impact that caused the formation of the crater Giordano

Bruno. More recently (1958) a Soviet astronomer obtained an emission spectrum of gas from the crater Alphonsus. In 1963, astronomers in Arizona reported seeing glowing red spots near the crater Aristarchus. The latter two events, which are the best documented of the so-called lunar transient phenomena, may be due to outgassing caused by tidal stresses or other causes.

BIBLIOGRAPHY

Abell, G. (1969). Exploration of the Universe, 2nd ed. Holt, Rinehart and Winston, New York.

Beaty, J. K., B. O'Leary, and A. Chaikin (eds.) (1981). The New Solar System, 2nd ed. Sky Publishing Corp., Cambridge, Massachusetts.

Carr, M., R. S. Saunders, R. G. Strom, and D. E. Wilhelms (1984). The Geology of the Terrestrial Planets. NASA SP–469, U.S. Gov't. Printing Office.

Glass, B. P. (1982). Introduction to Planetary Geology, Cambridge University Press, Cambridge, Massachusetts.

Hartmann, W. K., Moons and Planets, 2nd ed. (1983). Wadsworth, Belmont, California.

Murray, B., M. Malin, R. Greeley (1981). Earthlike Planets, W. H. Freeman and Company, San Francisco, California.

MOON ROCKS—*SEE* LUNAR ROCKS

NEUTRINO ASTRONOMY

Raymond Davis, Jr. *University of Pennsylvania*

GLOSSARY

Carbon–nitrogen cycle: Sequentially ordered set of fusion reactions of hydrogen with carbon and nitrogen isotopes, serving to combine four hydrogen atoms into helium.

Čerenkov light: Light radiated by a charged particle traversing a medium. Čerenkov radiation occurs when the particle has a velocity greater than the velocity of light in that medium.

Leptons: General term for the light elementary particles—electrons, muons, tauons, and their associated neutrinos—that interact with nuclei and with each other by the weak force.

Main sequence: Narrow region on a plot of the luminosity versus surface temperature of stars (Hertzsprung–Russell diagram) where the hydrogen-burning stars are located.

Muon: Unstable, weak-interacting, charged particle ($+$ or $-$) with a mass 207 times that of an electron. It decays to an electron and two neutrinos, $\mu^- \rightarrow e^- + \bar{\nu}_e + \nu_\mu$, mean lifetime 2.2×10^{-6} sec.

Neutrino oscillation: Hypothetical process that would allow different neutrino types, or flavors, to change into one another.

Proton–proton chain: Set of nuclear fusion reactions that converts four hydrogen atoms into helium.

Solar model: Theoretical calculation of the internal structure of the sun that follows the changes in structure, temperature, and composition as the sun ages.

Weak interaction or weak force: One of the fundamental forces in nature that governs the interaction of leptons.

The neutrino is a neutral elementary particle that has the unique capability of easily penetrating matter on the scale of stellar dimensions and densities. Neutrinos are produced abundantly in nuclear processes in the interior of stars or planets, and by a variety of processes in the cosmos. The study of the neutrino radiation from stars is a direct means of observing the energy production processes occurring in their interiors. It is generally believed that intermediate mass stars at the end of their nuclear life suffer a catastrophic collapse (super nova) and emit an intense pulse of neutrinos. Neutrinos are undoubtedly a component in cosmic radiation and in cosmic blackbody radiation, and they conceivably contribute to the mass of the universe. The application of neutrino physics to these problems is a subject of great interest in astronomy and cosmology. Unfortunately, neutrinos are extremely difficult to observe directly. Techniques presently available require that the detector be massive, on the order of 100–3000 tons, to record a neutrino event rate of 0.1–1/day. Because of this constraint, studies have been limited to observing the neutrino radiation from the sun. These studies were carried out to verify and understand the hydrogen fusions mechanisms that are the source of the Sun's energy. We may look forward to the future development of large-scale neutrino detectors that will be capable of observing other sources of neutrinos in the cosmos.

I. Detecting Neutrinos

A. CLASSIFICATION OF NEUTRINOS AS ELEMENTARY PARTICLES

Neutrinos are neutral particles belonging to three separate families of elementary particles called leptons (light-weight particles). Leptons are distinguished from other elementary particles by a unique property. Leptons are coupled to all other particles by the weak force, which results in extremely feeble interactions of leptons with the nuclei of atoms, with other elementary particles, and with themselves. Charged leptons, of course, may interact with other charged particles and fields through the electromagnetic force. As elementary particles, leptons carry an angular momentum of $\frac{1}{2}$ unit of spin, and are therefore regarded as fermions (particles with $\frac{1}{2}$ unit of spin). The three lepton families are designated by their charged members: the electron, e^- and e^+; the muon, μ^- and μ^+; and the tauon, τ^- and τ^+. Each charged set has a negative and a positive member, referred to as particle and antiparticle, respectively. Associated with each of these three sets of charged members is a corresponding set of neutral particles. These are the neutrinos and antineutrinos: ν_e, $\bar{\nu}_e$; ν_μ, $\bar{\nu}_\mu$; ν_τ, $\bar{\nu}_\tau$. The lepton families are listed in Table I.

The masses of neutrinos are much smaller than the masses of their charged partners. The electron, muon, and tauon have masses of 0.511, 105.7, and 1784 MeV, respectively. Many experiments have been performed in an effort to measure the masses of the three neutrinos. The most sensitive experimental measurement published to date reports an electron neutrino mass of 25 ± 10 eV, or 1/20,000th the mass of the electron, but more recent measurements suggest masses less than 5 eV. New experiments to resolve this controversy are underway. Only crude upper limits have been set on the masses of the muon and tauon neutrinos; 0.2 and 250 MeV, respectively. It is generally presumed that the masses of all neutrinos are small, perhaps far too small to be measured directly. Neutrinos play an important role in astronomy and cosmology, and it is of great interest to know the masses of neutrinos. This matter will be discussed later. If the neutrino is a massless particle, as was thought for many years, it would travel with essentially the velocity of light. It would follow that the neutrino would be polarized; that is, its spin axis would be directed along, or against, its direction of motion. A neutrino emitted from a source at the center of our galaxy would arrive only half a second behind a photon! [See NEUTRINO OBSERVATIONAL ASTRONOMY.]

Neutrinos are created in nuclear processes and in various elementary-particle interactions. The most familiar process is nuclear beta decay. In this process, an unstable nucleus simultaneously emits an electron (beta particle) and a neutrino. This process may be visualized as an unstable nucleus radiating its energy by creating a pair of leptons; a neutrino and an electron. It is referred to as beta-minus decay when an electron (e^-) is emitted with an antineutrino ($\bar{\nu}_e$), or beta-plus decay when a positron (e^+) is emitted with a neutrino (ν_e).

In another beta decay process, called electron capture, one of the orbital electrons is absorbed in the nucleus and a neutrino is emitted. Examples of these processes are

$$^{14}\text{C} \rightarrow {}^{14}\text{N} + e^- + \bar{\nu}_e, \qquad \text{beta-minus decay} \qquad (1)$$

$$^{11}\text{C} \rightarrow {}^{11}\text{B} + e^+ + \nu_e, \qquad \text{beta-plus decay} \qquad (2)$$

$$e^- + {}^{37}\text{Ar} \rightarrow {}^{37}\text{Cl} + \nu_e, \qquad \text{electron capture} \qquad (3)$$

In these examples, paired leptons are emitted, one a particle and the other an antiparticle. In electron capture, a particle e^- disappears and a particle ν_e is emitted. The guiding principle is called the principle of lepton conservation. In any process, the total number of leptons does not change; the number before and after is conserved. In Table I are listed lepton numbers for each lepton family. The lepton number is positive for a lepton and negative for an antilepton. By applying the lepton numbers for the electron family, the principle of lepton conservation can be noted for the three examples given.

The principle of lepton conservation applies to each of the three families, and, in addition, the leptons of each family appear to be separately conserved. It has been shown, for exam-

TABLE I.   The Lepton Families

	Electron family	Muon family	Tauon family
Particle	e^-, $l_e = +1$	μ^-, $l_\mu = +1$	τ^-, $l_\tau = +1$
Antiparticle	e^+, $l_e = -1$	μ^+, $l_\mu = -1$	τ^+, $l_\tau = -1$
Particle	ν_e, $l_e = +1$	ν_μ, $l_\mu = +1$	ν_τ, $l_\tau = +1$
Antiparticle	$\bar{\nu}_e$, $l_e = -1$	$\bar{\nu}_\mu$, $l_\mu = -1$	$\bar{\nu}_\tau$, $l_\tau = -1$

ple, that when a neutrino of the muon type (ν_μ) is absorbed in a nucleus, a muon is emitted, never an electron.

$$\nu_\mu + Fe \rightarrow Co + \mu^-$$

It was by this process that the muon neutrino was discovered and shown to be a different particle from the electron neutrino. The principle of lepton conservation for each lepton type (or flavor) has been tested experimentally in many different ways, and this principle appears to be valid. It is possible that lepton conservation is not strictly valid, and additional tests of these principles are needed. The best, and perhaps the only, way of testing for these unique lepton properties is by studying the neutrino spectrum from the Sun. This topic will be discussed further in Section II.

B. DETECTION OF NEUTRINOS

Neutrinos have the unique property of being able to pass through enormous amounts of ordinary matter. This property follows from the basic nature of neutrinos: they are neutral and are coupled to electrons and nuclei in matter only by the weak-interaction force. This force has both a charged and a neutral component, arising from the exchange of massive intermediating particles. The existence of these particles, the $W^\pm$ for charged currents, and the Z^0 for neutral currents, has only recently been demonstrated. The theory of the weak-interaction processes is now well understood, and allows one to calculate the interactions of neutrinos with nuclei and electrons. Using this theoretical foundation, it is possible to determine the sensitivity of detectors to fluxes of neutrinos as a function of their energy. The interaction probability is expressed as the target area of a nucleus, or electron, or other particle that is presented to a neutrino, or any other particle. It is referred to as the cross section (σ) and is usually expressed in units of square centimeters per atom (cm^2/atom).

Neutrinos may be absorbed in nuclei with the emission of an electron, a muon, or a tauon, depending on the neutrino type. These processes are called inverse-beta processes because they are, in the case of electron neutrinos, the inverse of normal radioactive beta decay. Neutrino (ν_e) absorption by an inverse beta reaction can be illustrated by two examples, one for antineutrinos and one for neutrinos:

$$\bar{\nu}_e + p \rightarrow n + e^+ \qquad (4)$$

$$\nu_e + {}^{37}Cl \rightarrow {}^{37}Ar + e^- \qquad (5)$$

Reaction (4) was used by Reines and Cowan in the first experiment to detect antineutrinos emitted by a nuclear reactor. The antineutrinos in this case originate from the beta-minus decay of fission products. This reaction is also used in antineutrino detectors designed for observing antineutrinos from collapsing stars. Reaction (4) is the inverse of the beta decay of the neutron. The positron moves to the other side of the reaction and becomes an e^-. The neutron has a half-life of 12 min, and decays by emitting an electron and an antineutrino, as

$$n \rightarrow p + e^- + \bar{\nu}$$

Reaction (4) requires that the neutrino have sufficient energy to provide the mass difference between the neutron and the proton and to create the positron, a total energy of at least 1.804 MeV. The minimum energy needed to carry out the reaction is called the threshold energy, E(thresh). Given the half-life of the neutron, the threshold energy, the spin change (none in this case), and the weak-interaction coupling constant, the theory of weak interactions may be used to calculate the cross-section as a function of the neutron energy. For example, a 5-MeV antineutrino would have a cross section of 2.4×10^{-43} cm^2/H atom. Using this cross-section value, we can calculate the distance that an antineutrino will penetrate through ordinary water. A beam of neutrinos would be reduced in intensity by 50% in traveling through 45 light years of water!

Reaction (5) is the one used for detecting neutrinos from the sun. It is the inverse of the electron capture decay of ^{37}Ar [see reaction (3)], a decay process with a half-life of 35 days. Again, the ν_e capture cross section can be calculated, knowing the half-life, the threshold energy (0.814 MeV), the spins of the ^{37}Cl and ^{37}Ar nuclei, the weak interaction coupling constant, and the density of various orbital electrons in the nucleus. In the case of complex nuclei such as ^{37}Cl, the neutrino absorption also produces ^{37}Ar in various excited states. The excited states decay rapidly to the ground state by emitting gamma rays. It is very important to include all of these transitions in evaluating the cross section. In the ^{37}Cl–^{37}Ar case, the transition probabilities to all relevant excited states can be evaluated, so that the cross section for capture can be calculated accurately over the range of neutrino energies expected from the sun. One particular excited state in ^{37}Ar, the analog state with an excitation energy of 4.98 MeV, is of great importance. Neutrinos with an energy greater than 5.79 MeV will feed this state, and they have a higher ν_e capture cross section. As we shall see, this fact is crucial for interpreting the results of the chlorine solar neutrino experiment.

These two examples of neutrino absorption reactions illustrate the procedures used in calculating the neutrino capture cross sections. The neutrino capture cross section for reaction (4) is the only one that has been measured experimentally with good accuracy ($\pm 5\%$). An approximate measurement of the cross-section of the reaction

$$\nu_e + d \rightarrow p + p + e^-$$

has also been performed ($\pm 30\%$). This reaction is of potential importance in observing neutrinos from the sun. For all other neutrino capture reactions considered for neutrino detectors, one must rely on a theoretical calculation. There is a particular difficulty in calculating the capture cross section for complex nuclei to produce the product nucleus in an excited state. In these cases, one must resort to nuclear reaction studies and theoretical nuclear models to estimate the cross section.

When a ν_e absorption reaction occurs, the electron that is emitted carries away the energy in excess of that needed to carry out the reaction,

$$E_e = E_\nu - E_\nu(\text{thresh})$$

Therefore, by measuring the electron energy, one can determine the energy of the neutrino that was captured. This feature of a capture reaction can be used to determine the spectrum of neutrinos from a source. In the future, neutrino detectors will be designed to measure the electron energy, and in this way determine the energy spectrum from the sun. One can also obtain some information on the neutrino direction. In general, the emitted electron is in the same direction as the ν_e, but the correlation in their relative directions is not very favorable. These considerations are also of importance in measuring the cosmic-ray neutrino spectrum and neutrino direction. In this case, a high energy ν_e or ν_μ (500 MeV to 100 GeV) is observed by noting the direction of the e^- or μ^- produced. For these high-energy processes, the neutrino direction can be well determined.

Neutrinos of all types, ν_e, ν_μ, and ν_τ, can scatter elastically from an electron. This process is a result of the weak-interaction coupling between the neutrino and the electron, a coupling that has a charged and a neutral current component. The cross section for scattering from electrons depends on the neutrino type, sometimes referred to as neutrino flavor, ν_e, ν_μ, ν_τ, and whether it is a neutrino or an antineutrino. Only the electron neutrino is coupled to the electron by the charged current, whereas all neutrino types are coupled by neutral currents. These processes may be represented by the following equations:

$$\nu_e + e^- \rightarrow \nu_e' + e^-(\text{recoil}), \qquad \text{charged and}$$
$$\text{neutral current}$$

$$(6)$$

$$\nu_\mu + e^- \rightarrow \nu_\mu' + e^-(\text{recoil}), \qquad \text{neutral current}$$

$$(7)$$

$$\nu_\tau + e^- \rightarrow \nu_\tau' + e^-(\text{recoil}), \qquad \text{neutral current}$$

$$(8)$$

The primes represent the neutrino scattered in a new direction and with a reduced energy. These processes are analogous to the Compton scattering of gamma rays. Compton scattering occurs through the electromagnetic coupling, and has a much larger cross section. There is an identical set of equations for the antineutrino, but the recoil electron spectrum is different for antineutrinos. Monoenergetic electron neutrinos, when scattered from electrons, produce a flat recoil energy spectrum. On the other hand, monoenergetic electron antineutrinos produce a spectrum of electron recoils that decreases inversely with the energy. The maximum energy of the recoil electron corresponds approximately to the energy of the neutrino, disregarding the initial binding energy of the struck electron.

The cross section for ν_e elastic scattering from electrons (ν_e–e^- scattering) increases in approximate proportion to the neutrino energy. In general, these cross sections are small for low-energy neutrinos (1–5 MeV) in the range 10^{-46}–10^{-44} cm^2/electron.

The cross section for elastic scattering by neutral current interaction is the same for ν_e, ν_μ, and ν_τ and is smaller than that for ν_e scattering from electrons. A very important characteristic of the ν–electron scattering process is that the recoil electron is in approximately the same direction as the incoming neutrino. Therefore, the neutrino–electron scattering process can be used to determine the direction of the neutrino. However, measuring neutrino energies and directions is difficult because of background processes. One must reduce the flux of external gamma radiation or energetic neutrons to very low levels. This can be accomplished by massive shielding and the use of particle detectors to observe and discount charged particles that enter the active region of the detector. In addition, radioactive elements that produce energetic electrons by beta decay must be removed from the target material and the construction material.

The method used to distinguish the direction of the electron is to observe the Čerenkov light emitted by the fast electron. When an electron (or any other charged particle) passes through a medium faster than light travels in that medium, a cone of light is emitted around the direction of the electron. The cone of Čerenkov light is observed by a large number of sensitive light detectors (photmultiplier tubes) located on the walls of the detector. This technique is now used in very large detectors, filled with 3000 tons of water, to search for the decay of the proton. At the present time, one of these detectors is being used to search for neutrinos from the sun.

II. Neutrinos from the Sun

A. Theoretical Solar Models

Our sun is classified as a dwarf B-type star that is generating energy primarily by the fusion of hydrogen into helium. The overall hydrogen fusion process can be represented as follows:

$$4H \rightarrow {}^4He + 2e^+ + 2\nu_e + \text{gamma radiation}$$
$$+ \text{kinetic energy}$$

The fusion of hydrogen into helium can be accomplished by two separate reaction sequences, known respectively as the proton–proton chain of reactions and the carbon–nitrogen cycle (Table II). It can be observed from Table II that the proton–proton chain has three competing branches, designated for reference as PP-I, PP-II, and PP-III. All stars generate energy by the proton–proton chain in the early stage of their evolution. The sun is a dwarf star, with an age of 4.5 billion years. It is still generating energy chiefly by this mechanism, and will continue to do so for a few billion years. As a star ages, its internal temperature increases, and the carbon–nitrogen cycle begins to play a more prominent role in energy production. All of the reactions shown are used in the solar model calculations.

TABLE II. The Proton–Proton Chain and Carbon–Nitrogen Cycle

	Reaction	Neutrino energy (MeV)	Neutrino flux (cm^{-2} sec^{-1})
	The proton–proton chain		
PP-I	H + H → D + e$^+$ + ν_e (99.75%)	0–0.420 spectrum	6.1 × 10^{10}
	or		
	H + H + e$^-$ → D + ν_e (0.75%)	1.44 line	1.5 × 10^8
	D + H → ^{3}He + γ		
	^{3}He + ^{3}He → 2H + ^{4}He (87%)		
PP-II	^{3}He + ^{4}He → ^{7}Be + γ (13%)	0.861 (90%) line}	
	^{7}Be + e$^-$ → ^{7}Li + ν_e	0.383 (10%) line}	4.3 × 10^9
	^{7}Li + H → γ + ^{8}Be → 2^{4}He		
PP-III	^{7}Be + H → ^{8}B + γ (0.017%)		
	^{8}B → ^{9}Be* + e$^+$ + ν_e	0–14.1 spectrum	5.6 × 10^6
	↳ 2^{4}He		
	The carbon–nitrogen cycle		
	H + ^{12}C → ^{13}N + γ		
	^{13}N → ^{13}C + e$^+$ + ν_e	0–1.20 spectrum	5.0 × 10^8
	H + ^{13}C → ^{14}N + γ		
	H + ^{14}N → ^{15}O + γ		
	^{15}O → ^{15}N + e$^+$ + ν_e	0–1.73 spectrum	4.0 × 10^8
	H + ^{15}N → ^{12}C + ^{4}He		

These are the only reactions considered. [*See* SOLAR PHYSICS; STELLAR STRUCTURE AND EVOLUTION.]

The nuclear reactions shown in Table II have been studied extensively in the laboratory. From our understanding of nuclear reactions of light elements, it has been established that these are the only reactions that are important in the hydrogen fusion processes. The primary reaction that initiates the proton–proton chain,

$$H + H \rightarrow D + e^+ + \nu_e$$

has too low a cross section to be measured in the laboratory. However, it is possible to calculate accurately its cross section at thermal energies. All these reactions are exothermic, producing the energies listed in Table II. Six of the reactions produce neutrinos; two emit monoenergetic neutrinos; and the other four emit a spectrum of neutrinos from near-zero energy to the maximum energy noted.

In order to determine the rates of these various reactions in the sun, detailed calculations must be made of the temperatures, particle densities, and chemical composition of the solar interior. Knowing the rates of the neutrino-producing reactions, one can calculate the energy spectrum of the neutrinos emitted by the sun. In 1964, there was great interest in calculating the solar neutrino spectrum, because it was realized at the time that the neutrino radiation from the sun could be measured and thereby test the theory of solar energy production. Understanding the energy processes in the sun is not only of great interest to us on earth, but is also important to a theoretical understanding of the energy generation and evolution of all stars.

The internal conditions in the sun are derived by a complex solar model calculation, which follows the evolution of the sun from its initial formation to its final state as a cooling white dwarf star, devoid of fuel. A very brief outline will be given of the basic assumptions and information used in the calculations. The sun is the only star for which mass, radius, luminosity, surface temperature, and age are well known. These parameters are used in the model and provide boundary conditions. It is assumed, with good physical justification, that the primitive sun was well mixed and heated by the release of gravitational energy. For this reason the initial elemental composition throughout its mass was identical with that now observed in its photosphere. The elemental composition is introduced in the solar model as the ratio of the abundance of each element heavier than helium to the abundance of

hydrogen. This ratio is a directly measured quantity for each element. The elements heavier than helium, in order of their weight percent, are O, C, Fe, N, Si, Mg, Ne, S, etc. The total percentage of these elements is 1.8% by mass. The helium composition is not directly used in the model, but is calculated from the model. The initial helium composition calculated for the standard model is around 25% by mass. As the sun evolves, the helium content in the core gradually increases as a result of the fusion reactions.

The chemical composition of the sun is an important factor in determining the internal temperatures, because an increase in the composition of the heavier elements affects the rate of energy transport. The transmission of energy depends on a number of processes in which the thermal radiation interacts with unbound electrons, and with atoms in various states of ionization and excitation. These processes must be calculated from simplified atomic models and photon-electron scattering theory. These very complex calculations are provided by scientists from Los Alamos and Livermore National Laboratories. Their results are extensively employed in astronomical calculations.

The hydrodynamic structure of the sun is expressed by an equation of hydrodynamic equilibrium. That is, at every radius shell, the downward gravitational forces are balanced by the outward kinetic and radiation pressures. The calculations follow the initial collapse of the gravitating mass of the sun to the main sequence. The only additional energy is that introduced by the nuclear fusion reactions. When a star, such as the sun, is on the main sequence, where it remains for over 6 billion years, all of the energy is provided by the hydrogen fusion reactions. The nuclear reaction rates are derived from the kinetic energy of the reactants, assuming a Maxwellian velocity distribution and a nuclear barrier penetration factor. Experimental nuclear reaction cross sections are used. They are usually measured in the laboratory at energies above 100 keV, and extrapolated to the energies corresponding to the temperatures in the sun's interior (less than 1 keV). The theoretically calculated value of the primary p–p reaction is used.

The theoretical forecasts of the neutrino fluxes that are usually used to compare with experimental observations are those from the standard model. This model uses the usual values of the solar mass, radius, luminosity, and age (4.7 billion years), and the best-considered values of

TABLE III. Solar Neutrino Fluxes[a] and Cross Sections[b] for the Reaction
$\nu_e + {}^{37}Cl \rightarrow {}^{37}Ar + e^-$

Neutrino source	Flux on earth (cm^{-2} sec^{-1})	Capture cross section (cm^2)	Capture rate (SNU)
$H + H \rightarrow D + e^+ + \nu_e$	6.1×10^{10}	0	0
$H + H + e^- \rightarrow D + \nu_e$	1.5×10^8	1.56×10^{-45}	0.24
^7Be decay	4.3×10^9	2.38×10^{-46}	0.95
^8B decay	5.6×10^6	1.08×10^{-42}	4.3
^{13}N decay	5.0×10^8	1.66×10^{-46}	0.08
^{15}O decay	4.0×10^8	6.61×10^{-46}	0.24
		Total	5.8

[a] From Bahcall, J. N., Huebner, W. F., Lebow, S. H., Parker P. D., and Ulrich, R. K. (1982). *Rev. Mod. Phys.* **54**, 767.
[b] From Bahcall, J. N. (1978). *Rev. Mod. Phys.* **50**, 881.

the nuclear reaction cross sections. In addition, there are some special assumptions that are made. It is presumed that the sun is not rotating, or differentially rotating, rapidly enough in its interior to affect its internal structure or dynamics. Processes that could mix the solar interior, such as diffusion or periodic hydrodynamic oscillation, are not taken into account. Magnetic fields are not regarded as sufficiently intense to affect the sun in any way. A simplified convective theory is used for determining the transport of energy in the convective zone, the outer dynamic region of the sun. The proponents of the standard model consider these processes unimportant for the purpose of forecasting the solar neutrino spectrum.

Table III shows the neutrino flux at the earth for each neutrino source in the proton–proton cycle and the carbon–nitrogen cycle, obtained by standard solar model calculations. It may be noticed that the total neutrino flux at the earth is 6.6×10^{10} neutrinos/cm^2 sec, and 92% of the flux can be attributed to the low-energy neutrinos from the proton–proton reaction.

B. THE CHLORINE SOLAR NEUTRINO EXPERIMENT

In the early 1960s, it was clear that the neutrinos from the sun could be observed. This conclusion was reached after it was realized that the PP-II and PP-III branches were an important part of the energy process of the sun. The processes were sufficiently energetic to drive the neutrino capture reaction

$$\nu_e + {}^{37}Cl \underset{\text{decay(half-life = 35 days)}}{\overset{\text{capture}}{\rightleftharpoons}} {}^{37}Ar + e^-,$$

$$E_\nu(\text{thresh}) = 0.816 \quad \text{MeV}$$

This reaction has too large a threshold energy for use in observing the abundant neutrinos from the p–p reaction but is suitable for measuring the neutrino flux from the PP-II and PP-III branches, and from the carbon–nitrogen cycle. A radiochemical method is used to observe neutrinos by this neutrino capture reaction. This technique employs a very large volume of a suitable chemical compound of chlorine. The radioactive product is removed, purified, and placed in a small detector for measuring the decay of ^{37}Ar back to ^{37}Cl. There are several reasons for choosing the Cl–Ar method for solar neutrino detection: chlorine compounds are inexpensive; the ^{37}Ar can be recovered by a simple efficient chemical procedure; ^{37}Ar has a convenient half-life of 35 days; the decay of ^{37}Ar is easily measured. A radiochemical detector does not observe and characterize a neutrino capture event. It is only capable of measuring a radioactive product that has accumulated over a period of time. In designing the detector, it is therefore important to consider all possible nuclear processes that could produce the radioactive product, and to be sure that they are small compared to the production expected from solar neutrinos. In general, these background processes are small for radiochemical detectors. It is necessary, however, to carry out the measurement deep underground, to reduce the ^{37}Ar production by cosmic ray muons to a level that is low, compared to that expected from solar neutrino capture.

During the period 1965–1967, Brookhaven National Laboratory built a chlorine solar neutrino detector in the Homestake Gold Mine at Lead, South Dakota, to make a quantitative test of the theory of solar energy generation. The detector uses 380,000 liters (615 tons) of tetrachloroethylene (C_2Cl_4) as the target material. This liquid is an inexpensive commercial dry-cleaning fluid, known as perchloroethylene. The experiment was installed 1480 m (4850 ft) underground in a specially designed chamber. Figure 1 shows the experimental arrangement.

In building the detector, great care was taken to keep all background processes small, so that any ^{37}Ar observed could be attributed to neutrinos at the level expected from solar neutrinos. The ^{37}Ar is removed from the liquid by purging with helium gas. This is accomplished by two pumps and a set of 40 nozzles that disperse the helium gas throughout the large volume of liquid. The helium gas from the tank of C_2Cl_4 is circulated through a set of condensation and absorber traps to remove the vapors of tetrachloroethylene. The helium gas then passes through a charcoal trap at $-196°C$ to collect the argon gas and is then returned to the tank. By circulating the helium for about 20 hr, the ^{37}Ar produced in the tank can be recovered with an efficiency of 95%. The argon recovered by this simple procedure is purified and placed in a small proportional counter for measuring the ^{37}Ar radioactivity. Experiments are repeated five to six times a year. During the time between experiments, the

liquid in the tank remains undisturbed but is, of course, exposed to the neutrino flux from the sun. During the exposure period, the number of ^{37}Ar atoms increases, following the normal equations for the growth and decay of a radioactive product. After an exposure interval of 2 months or so, the sample of Ar is collected, purified, and counted. Since the ^{37}Ar production rate is very low, the experiment must be repeated many times.

It was learned, in the first few years of operation, that the ^{37}Ar production rate in the detector was lower than expected from the standard theoretical model of the sun and, in fact, was comparable to the expected background role in the detector. It was necessary to improve the sensitivity of the detection, which was accomplished by improving the counting technique so that ^{37}Ar decay events could be distinguished clearly from background events. Following these improvements, a long series of measurements from 1972 to 1985 produced a better result. A radiochemical detector observes the sum of the production rate from all solar neutrino sources having an energy above the threshold energy. Table III lists the theoretical neutrino flux and neutrino capture cross-section of ^{37}Cl for each neutrino

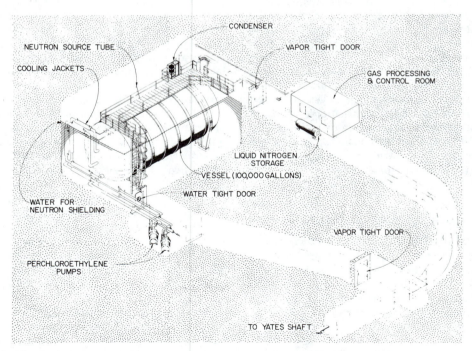

FIG. 1. The chlorine solar neutrino experiment in the Homestake Gold Mine, Lead, South Dakota.

reaction that occurs in the sun. The chlorine experiment would respond to the sum of flux × cross section for each of these sources. The expected neutrino capture rate in the chlorine experiment, derived from the most recent theoretical calculation, is 5.8×10^{-36} captures per second per ^{37}Cl atom. This rate is usually expressed in solar neutrino units (SNU), an SNU being defined as 10^{-36} captures per second per atom. The theoretical rate of 5.8 SNU is to be compared to the experimental rate of 2.1 ± 0.3 SNU. The error limits indicate that there is a 67% probability that the result is between the limits given. It is difficult to express quantitatively the uncertainties in the solar model prediction, since the model may not correctly represent the structure and dynamics of the sun. If one uses the combined errors in the astronomical and nuclear reaction data employed in the calculation, an uncertainty of about 1 SNU is obtained. There is therefore a clear disagreement between the results of the chlorine solar neutrino experiment and the solar model calculations based on the generally accepted theory of the structure and evolution of stars.

C. Interpretation of the Chlorine Solar Neutrino Experiment

The chlorine solar neutrino experiment is the first, and so far the only, direct observational test of the energy production of the sun. Because of the lack of reasonable agreement between theory and experiment, the results have been taken rather seriously. The first matter of importance was to test thoroughly the reliability of the experimental results. The method used to obtain these results is an extreme example of a radiochemical technique: fifteen or so ^{37}Ar atoms were removed from 615 tons of liquid, and measured. After many tests, the experimental results were accepted as valid. A primary consideration is that argon is an unreactive element and does not form stable compounds. Also, a small measured amount of argon is added in each experiment as a carrier gas to measure the chemical recovery yield. An excellent means of testing the chlorine detector would be to expose it to a man-made source of neutrinos with a known intensity. If a source of 1 million curies (Ci) of ^{65}Zn could be prepared in a reactor, a modest test could be performed, but preparing a source of this intensity is beyond our present capability.

In interpreting the chlorine experiment, it must be noted that the expected ^{37}Ar production rate is quite sensitive to the internal temperatures of the sun. This follows from the fact that neutrinos with an energy above 6 MeV have sufficient energy to produce ^{37}Ar in the isobaric analog state. The cross section for the absorption of neutrinos in Cl to form this state in ^{37}Ar is high. Consequently, the chlorine experiment is unusually sensitive to the comparatively low flux of neutrinos from ^{8}B decay; 74% of the expected neutrino capture rate for the chlorine detector would be attributable to neutrinos from ^{8}B decay in the sun. The production of ^{8}B by the combined effect of the PP-II and PP-III branches is very sensitive to the temperatures in the solar core.

With these considerations in mind, a number of solar models have been devised that would lead to lower internal temperatures and thus reduce the ^{8}B flux.

One means of reducing the sun's internal temperatures is to assume a continual modest mixing of the sun in the course of its evolution. This is an artificial process, contradicting the view that the sun becomes more stable as it ages. Mixing could, however, occur in a natural way by several reasonable mechanisms. If the primitive sun were rotating rapidly, as is generally believed, it could gradually slow down through releasing its rotational momentum by mass loss from its surface in the form of solar wind. Such a process could result in a differential rotation in the interior, and thereby introduce a degree of mixing. It is also possible that periodic mixing could be accomplished by oscillating standing gravity waves (g-mode) in the sun's interior. Waves of this nature do undoubtedly exist in the sun. Pressure-mode oscillations that penetrate the outer convective zone of the sun have been observed optically. These observations have provided useful information on the depth, structure, and rotation of the sun's outer layers. Perhaps, in the near future, g-mode oscillations will be observed, thereby adding to our information on the solar core. E. Schatzman and A. E. Maeder (Observatoire de Nice) suggested and carried out calculations in which diffusion processes were considered. Diffusion would effectively mix the solar interior. Solar models that introduce mixing in one form or another result in a lower predicted rate of ^{8}B production and yield a total neutrino capture rate in ^{37}Cl in the range 1.5–2.5 SNU.

Another concept that has been suggested is that the sun has a heavy-element concentration lower than the accepted value of 1.8%. In that case, the solar material would be more transparent to thermal radiation. A model that introduces an initial heavy-element concentration of 0.5% also reduces the solar neutrino capture rate in ^{37}Cl to the observed value.

This brief qualitative discussion of solar models leads us to the conclusion that the standard solar model calculations are probably oversimplified, and do not properly model a complex dynamic star like the sun. Some solar phenomena to which the chlorine experiment could be very sensitive are not adequately treated by the standard solar model. Further progress in understanding the energy-generating process in the sun can be achieved only by new experiments, capable of providing additional information on the solar neutrino spectrum.

We also must keep in mind that, when exploring the interior of the sun by observing its neutrino radiation, we are assuming that we understand the physics of neutrinos and their interaction with matter. When the chlorine experiment began in 1965, physicists were confident that the neutrino behaved as a massless particle and that leptons were conserved, as explained above. So far in this discussion, we have presumed that this view is correct. In exploring the sun using neutrinos as probes, we are extending our knowledge of neutrino physics into a domain of great distances and high matter densities, and regions of intense magnetic fields. Small effects, not observable in terrestrially based experiments, may be important. In this endeavor we should be prepared for surprises. In the opposite sense, we may use astronomical observations to learn more about neutrinos.

Let us consider some properties of neutrinos that could alter the flux of solar neutrinos by the time they reach the earth. One early suggestion was that the neutrino could decay into another neutrinolike particle and a boson (photon?). This process now seems unlikely, since one cannot conceive of any particle into which the neutrino could decay. If the principle of lepton conservation among lepton types, or flavors, is not strictly valid, then a neutrino of one type could change into a neutrino of some other type. We may represent this process by an equilibrium, as

$$\nu_{\mu} \rightleftarrows \nu_{\tau}$$
$$\nwarrow \quad \nearrow$$
$$\nu_e$$

This phenomenon, called neutrino oscillation, was suggested by B. Pontecorvo (Dubna,

U.S.S.R.) in 1968. This process could occur during the 8 min required for the neutrino to travel 1.5×10^8 km from the sun to the earth. If this process occurs, the oscillation length depends on the neutrino energy and the squared differences in the neutrino masses, that is, $\Delta m^2 = m^2_{\nu_e} - m^2_{\nu_{\mu}}$. The extent of mixing among neutrinos depends upon the square of the sine of a mixing angle, $\sin^2 2\theta$. About 10 years later, L. Wolfenstein (Carnegie-Mellon University) pointed out a similar process that could occur in their passage through matter, as a result of the difference in scattering cross sections for electron-neutrinos and other neutrino types. Matter oscillations do not require that the neutrino have a mass. Until very recently, it was believed that neither vacuum oscillations nor matter oscillations could affect the solar neutrino flux because the mixing angle θ was probably very small.

This view, however, was changed in 1984, when S. P. Mikheyev and Yu. Smirnov (Institute of Nuclear Research, Moscow) pointed out that there is a resonance effect in the relative scattering processes of electron neutrinos and other neutrino types in passing through the dense matter (130–160 g/cm^3) of the sun's core. As a consequence of this resonance, the neutrino does not oscillate between neutrino types, but would be converted to neutrinos of some other type, thereby greatly reducing the flux of electron-neutrinos. This effect would occur for small mixing angles and small differences in neutrino masses. Because we do not know the value of the mixing angle or the neutrino mass differences, it is not clear whether low- or high-energy neutrinos would be affected by this resonance conversion process. If the differences in neutrino masses were on the order of 0.01 eV and the mixing angle were small, neutrinos above 7 MeV would be converted to another type of neutrino, ν_{μ} or ν_{τ}, and would not be observed by the chlorine experiment. This process depends on separate lepton number violation, and also assumes that the electron-neutrino has the lowest mass of all neutrinos. If this is so, resonance conversion of solar neutrinos will occur, and will thereby effect the results of solar neutrino experiments.

It is clear that more solar neutrino experiments, capable of determining the neutrino energy spectrum, are needed to resolve this complex question. The most satisfactory test for resonance conversion of electron-type neutrinos would be to observe a modification in the solar neutrino spectrum beyond that which could be explained by solar processes. A crucial observation would be to measure the flux of neutrinos from the proton–proton reaction, since the PP-I branch must be operating at the expected rate to provide the sun's energy. Are these low-energy neutrinos affected by resonance conversion?

D. New Solar Neutrino Experiments

Many new experimental methods have been suggested for observing neutrinos from the sun, and in some cases extensive development work has been carried out. This discussion, however, will be limited to experiments that are already funded and will definitely be performed.

In recent years, the emphasis has been on the following scientific questions: Are the neutrinos coming from the sun? Do they have the expected energy spectrum? Is the basic proton–proton reaction operating in the sun? The question of the neutrino direction can best be answered by an electronic detector that can observe neutrino direction by neutrino–electron scattering. Only the neutrinos from ^{8}B decay have sufficient energy to enable their direction to be determined. If the sun is generating energy predominantly by the PP-I branch of the proton–proton cycle, as we believe, the primary proton–proton reaction must be operating. The production of these low-energy neutrinos is high, and is directly connected, in a known way, to the solar luminosity. Do the neutrinos from this source reach the earth, or are they affected in some way during their passage to the earth? In the current state of development of neutrino detectors, these low-energy neutrinos are best detected by a radiochemical method.

Such a method has been developed and tested on a pilot scale. Full-scale experiments are in preparation. Neutrino detection is based on the neutrino capture reaction

$$\nu_e + {}^{71}\text{Ga} \underset{\text{decay(half-life = 11.4 days)}}{\overset{\text{capture}}{\rightleftarrows}} {}^{71}\text{Ge} + e^-$$

$$E_\nu(\text{thresh}) = 0.233 \quad \text{MeV}$$

In this method, germanium is extracted from many tons of gallium in some suitable chemical form. The isolated germanium is purified, chemically converted to germane gas, and placed in a miniature proportional counter for observing the radioactive decay of ^{71}Ge. Two separate gallium experiments are underway. One is being prepared at the Institute of Nuclear Research, Moscow, and will be installed in the institute's underground laboratory in Baksan Valley, near Mt. Elbrus, in the Caucasus. In this experiment, 60 tons of gallium metal (melting point 30°C) will be used. The germanium will be extracted by an oxidizing hydrochloric acid solution. The Soviet experimental apparatus will be completed in 1987, and measurements will begin in 1988.

The second gallium experiment is being prepared by the GALLEX collaboration of Western Europe (Max-Planck Institut für Kernphysik, KFZ Karlsruhe, TU München, CEN Saclay, Collège de France, University of Nice, University of Milano, University of Rome, WIS Rehovot). This experiment will use 30 tons of gallium in the form of a concentrated gallium chloride–hydrochloric acid solution. Germanium is easily removed from this solution as germanium tetrachloride by purging with helium gas. This extraction technique was developed on a pilot basis at Brookhaven National Laboratory. The GALLEX collaboration will receive its 30 tons of gallium in 1989, and will begin observations in 1990. The Western European experiment will use the Gran Sasso underground laboratory in Italy.

A gallium experiment responds to the entire spectrum of solar neutrinos; it is important to determine the neutrino capture cross section in ^{71}Ga to produce ^{71}Ge in various excited states. The ground state is easily calculated from the known half-life of ^{71}Ge, but the contribution to excited states is more difficult to determine. By employing nuclear reaction studies and nuclear model calculations, however, the cross-section values were satisfactorily determined for all neutrino sources in the sun. The total neutrino capture rate expected from the standard model calculation is 122 SNU; the proton–proton reaction alone would contribute 65 SNU. The corresponding total neutrino capture rate in an experiment using 30 tons of gallium is expected to be only one per day.

The gallium experiment alone will not be sufficient to determine the solar neutrino spectrum, but will provide essential information. Additional radiochemical solar neutrino experiments will be needed. A radiochemical experiment has been developed, using bromine as a target element. A bromine detector would be particularly sensitive to the ^{7}Be decay neutrinos from the sun, and would thereby measure the rate of the PP-II branch of the proton–proton chain. The three radiochemical experiments (gallium, bromine and chlorine) would together be a very good means of determining the solar neutrino spectrum.

In recent years, very large water Čerenkov detectors detectors were developed to observe proton decay, a process that was forecasted by theories uniting the electromagnetic, strong-interaction, and weak-interaction forces into one grand unified theory. Proton decay has not been observed so far. One of these detectors, the Kamiokande detector at Kamioka, Japan, was modified to improve resolution and sensitivity to a variety of proton decay modes. In its present form, this detector may be able to observe the neutrinos from ^{8}B decay in the sun. The neutrinos will be detected by elastic scattering from electrons, a process that allows the neutrino direction to be determined. Recoil electrons having energies above 7 MeV will emit Čerenkov light, to be observed by 50-cm-diameter photomultiplier tubes distributed over the walls of a tank of water 16 m in diameter and 16 m high. To achieve a high sensitivity to Čerenkov light, 20% of the area of the tank is covered with photomultiplier tubes. The detector is now operating in its modified form, and we can look forward to results in a year or so.

III. The Earth, Cosmic Rays, and Supernovas

A. Neutrinos from the Earth

The earth is a source of antineutrinos, produced by beta decays of long-lived radioisotopes in the earth. The decay of these isotopes helps keep the interior of the earth hot. Most of these radioactive isotopes are thought to occur in the lithosphere, the earth's outer layer, which is only a few hundred kilometers thick. It would be of great interest to know the total concentration and distribution of these heat sources in the earth. They are associated with regions of intense volcanic activity, and are related to continental drift. The radioactive isotopes primarily responsible are ^{40}K, ^{232}Th, and ^{238}U. Most of the heat is produced by the alpha decay of uranium and thorium and the series of radioactive decays that follow. This series includes beta-decaying isotopes that emit a spectrum of antineutrinos ($\bar{\nu}_e$). The total flux of these antineutrinos is approximately 2×10^7 cm^{-2} sec^{-1}, and they have energies in the range 0.5–3 MeV.

Unfortunately, these antineutrinos are extremely difficult to observe because of their low energy and low flux. At present, the best approach to their detection is by a radiochemical method employing a massive detector and based on a reaction with the lowest possible threshold energy. One must also recognize that the absorption of an antineutrino produces a positron, requiring an energy of 1.022 MeV. Perhaps a detector based on the reaction $\bar{\nu}_e + {}^{35}\text{Cl} \rightarrow {}^{35}\text{S} + e^-$, with a threshold of 1.19 MeV, is the best prospect.

B. Cosmic-Ray Neutrinos

Cosmic rays entering the earth's atmosphere produce a detectable flux of neutrinos of all types. These cosmic-ray particles, mostly protons and helium nuclei, have extremely high energies. Upon interacting with the earth's atmosphere, they produce a cascade of high-energy particles. Among the most abundant products are pi-mesons (π^+ and π^-) and K-mesons (K$^+$ and K$^-$), which decay to muons and muon-neutrinos, as

$$\pi^+ \rightarrow \mu^+ + \nu_\mu \quad \text{and} \quad \pi^- \rightarrow \mu^- + \bar{\nu}_\mu$$
$$\text{K}^+ \rightarrow \mu^+ + \nu_\mu \quad \text{and} \quad \text{K}^- \rightarrow \mu^- + \bar{\nu}_\mu$$

These secondary particles and their associated muon-neutrinos have a steeply falling energy spectrum, extending from a few mega-electron-volts to many giga-electron-volts. The muons are highly penetrating and have been studied in detail at great depths underground. It is of interest to observe these cosmic-ray-produced neutrinos, and to use them to test for neutrino oscillations and other interesting phenomena. A number of large-area detectors now operating underground are capable of observing the direction of these neutrinos and, in some cases, their energies. This is done by observing muons produced by muon-neutrino absorption in one of the elemental constituents of the surrounding rock, for example, by the reaction

$$\nu_\mu + {}^{16}\text{O} \rightarrow {}^{16}\text{F} + \mu^-$$

If the ν_μ is energetic, the muon produced will penetrate the rock layer and can be observed by a charged-particle detector. If the neutrino arrives from the opposite hemisphere of the earth, it will produce a muon traveling in an upward direction. Upward-traveling muons can be easily distinguished from the abundant flux of downward-traveling cosmic-ray muons if the detector is deep enough underground. Two experimental methods have been used: timing the muon's passage through many layers of detector elements, and observing the cone of Čerenkov light resulting from the muon's passage through water. Six detectors have been operating for several years. The total number of ν_μ events observed worldwide is about 800. [See COSMIC RADIATION.]

The primary cosmic-ray energy spectrum and the angular distribution at the earth are sufficiently well known to permit the calculation of the muon-neutrino energy spectrum and angular distribution. Observations are in satisfactory agreement with these calculations. From this result one can draw some rather qualitative conclusions. In the first place, there is no evidence for neutrino oscillations. If muon-neutrinos changed into electron-neutrinos in their passage through the earth, the electrons produced by their absorption would not escape as easily from the rock, and a reduced intensity of muons would be observed. However, neither the calculations nor the observations are accurate enough to draw firm conclusions. Furthermore, since the observed neutrino rate can be explained by known cosmic-ray sources, one may conclude that there are no intense extraterrestrial sources of high-energy muon-neutrinos.

To make real advances in high-energy neutrino astronomy, larger detectors are needed, with better energy and directional resolution, and they must be operated for many years. The existing detectors built for proton decay have many of the characteristics needed for a neutrino telescope. Following the great advances of recent years, perhaps a new generation of detectors, dedicated to neutrino astronomy, will be built.

For many years there has been interest in developing an underwater detector, perhaps as large as a cubic kilometer. Two active projects are working on this concept. Their goal is the detection of neutrinos with energies in the trillion-electron-volt range, by observing the Čerenkov light from a shower of particles produced by these high-energy neutrinos. One project is based at the University of Hawaii (DUNAND), and the other is at Lake Baikal in Siberia.

C. SUPERNOVAS

Stars have been seen to explode, radiating an enormously bright pulse of light. The light from a single supernova event exceeds the total amount of light radiated by an entire galaxy! Supernova events are separated into two classes, depending on their spectra, light curves, and stellar associations. Type I supernovas arise from older stars of low mass, while type II supernovas appear to result from the collapse of young massive stars. One of the major puzzles in stellar evolution is the physical mechanism by which a star explodes. The current concept is that a type I event results from a companion star dumping matter on a white dwarf, followed by explosive ignition of the in-falling matter. A type II event arises from a star whose core is generating energy by helium fusion. while the outer layers are heated by hydrogen fusion. The delicate balance between inward gravitational forces and outward pressure from radiation is upset, and the star collapses. A set of nuclear synthetic reactions follows, forming a mixture of all the elements heavier than helium. It is generally believed that this event produces a pulse of neutrinos, a few hundredths of a second in duration, and a neutron star is formed. The neutrino pulse is thought to occur by the electron capture process $e^- + p \rightarrow n + \nu_e$ as a result of intense pressures. Following the event, the neutron star cools by radiating neutrino–antineutrino pairs, a phase that, according to the theory, lasts only a few seconds! [See SUPERNOVAE.]

In a galaxy, one supernova event may be expected every 30 years or so. If an event occurred in our galaxy, the light intensity would overwhelm astronomical instruments at observatories normally used to study the spectrum and intensity of the light emitted. Neutrino detectors would not be overwhelmed, however, and could obtain important information. The neutrino energies are expected to be in the range 10–20 MeV. If an event occurred 1 kiloparsec (kpc) (3200 light years) from the earth, some 500 ^{37}Ar atoms would be produced in the Homestake chlorine detector. The number of ^{37}Ar atoms would decrease with the square of the distance from the event. An event at the center of our galaxy, 8.5 kpc away, would produce only seven ^{37}Ar atoms, an amount indistinguishable from the solar neutrino rate. The Kamiokande water Čerenkov detector would have a similar sensitivity in terms of the number of neutrino-induced events, but a higher sensitivity for antineutrino events. The Kamiokande detector could, of course, give much more information. Of great importance would be the time distribution of the sharp neutrino pulse and the longer pulse of neutrinos and antineutrinos from the cooling neutron star.

The interpretation of the neutrino signal from a supernova event may be extremely complicated because of the possibility of neutrino oscillations. Our views on neutrino physics and the mechanism of supernova explosions may be very different by the time such an event occurs!

BIBLIOGRAPHY

Bahcall, J. N., and Davis, R., Jr. (1976). *Science* **191,** 264.

Barnes, C. A., Clayton, D. D., and Schramm, D. N. (eds.) (1982). "Essays in Nuclear Astrophysics." Cambridge Univ. Press, London and New York.

Cherry, M. L., Fowler, W. A., and Lande, K., (eds.) (1985). Solar Neutrinos and Neutrino Astronomy. *Am. Inst. Phys. Conf. Proc.* No. 126, American Institute of Physics, New York.

Clayton, D. D. (1968). "Principles of Stellar Evolution and Nucleosynthesis." McGraw-Hill, New York.

Lande, K. (1979). Experimental neutrino astrophysics. *Annu. Rev. Nucl. Particle Sci.* **29,** 395–410.

Unsold, A., and Baschek, R. B. (1983). "The New Cosmos," 3rd ed. Springer-Verlag, Berlin and New York.

NEUTRINO OBSERVATIONAL ASTRONOMY

Alfred K. Mann and Raymond Davis, Jr. *University of Pennsylvania*

GLOSSARY

Binding energy: Energy difference between the assemblage (system) of free particles and the assemblage of the bound system of those particles. Binding energy is released in the transition from the free particle state to the bound particle state and must be supplied to the bound particle system to reverse the transition.

Čerenkov radiation: Electromagnetic radiation emitted by a charged particle in traversing a medium in which the velocity of the charged particle is greater than the velocity of light in that medium. Čerenkov radiation is a very small fraction of the total energy loss of a charged particle in its interaction with matter.

Neutrino types (or flavors): Neutrinos are elementary particles of zero (or very small) mass and zero electric charge, which interact only very weakly with matter. At least three types or flavors are recognized in nature: electron type, ν_e; muon type ν_μ; tauon type, ν_τ. The ν_e and e form a lepton pair, as do ν_μ and μ, and ν_τ and τ. These pairs and their respective antiparticles appear to be separately conserved in all processes studied thus far.

Neutron star: "Dead" star of roughly one solar mass and extremely small radius of 10 km, consisting largely of neutrons, which results from the core collapse in the first stage of supernovas of type II.

Particle reaction: There are many elementary particle and nuclear reactions just as there are many chemical reactions. The reaction $a + b \rightarrow a + b$, where a may be a particle and b a particle or a nucleus, is elastic. The reaction $a + b \rightarrow c + d$, where c may be a particle and d a nucleus is inelastic and may involve the internal excitation or transmutation of the nuclei involved. Such reactions must satisfy the conservation of energy, momentum, and electric charge, as well as certain other particle type conservation laws.

Photomultiplier tube (PMT): Vacuum tube with a photosensitive cathode from which one or more electrons may be released by light falling on the cathode. The internal structure of the PMT is such as to multiply the initial number of released (photo) electrons to give a measurable electric current at the output of the tube. The current is proportional to the incident light intensity falling on the cathode.

In 1987, there occurred two observations of special interest in neutrino astronomy: (1) the observation on Earth of a burst of neutrinos emitted by the supernova SN 1987A, and (2) the second, independent observation of a deficit in the number of neutrinos reaching Earth from the sun relative to the number predicted by the standard solar model. These observations, taken in conjunction with the earlier measurement of neutrinos from the sun, are a sign of the "coming of age" of observational neutrino astronomy. [*See* NEUTRINO ASTRONOMY; SUPERNOVAE.]

I. Neutrinos from Supernova 1987A

On February 23, 1987, at 7:35 UT (universal time), the core of a massive blue giant star, later identified as Sk-69° 202, collapsed in the first stage of its transformation into a type II supernova. Sk-69° 202 was located in the Large Magellanic Cloud (LMC), a satellite galaxy of our galaxy, at a distance of approximately 55 kpc from Earth. The resulting supernova, SN 1987A, is only the eighth supernova in the past 2000 yr bright enough to be seen with the naked eye and the first since the supernova of 1604, reported then by Kepler. [*See* SUPERNOVA 1987A.]

Supernovas of type II are thought to involve stars of mass greater than about 8 solar masses wherein the central core of roughly 1.5 solar mass, having exhausted its nuclear fuel, can no longer resist the pressure exerted by its internal gravitational force and collapses to a neutron star of small dimension, perhaps 10 km in radius, and nuclear density (10^{17} kg m^{-3}). The remainder of the star is blown off and, excited by the collapse, radiates energy over most of the electromagnetic spectrum.

It was recognized that the release of the binding energy of the neutron star, amounting to 0.1 to 0.2 solar mass, or approximately 10^{53} ergs, would not be possible by the emission of electromagnetic radiation, but would be possible through the emission of roughly 10^{57} neutrinos of average energy in the vicinity of 10 MeV. The extremely weak interaction of neutrinos with matter allows them to pass relatively freely from the neutron star in a few seconds while photons would be indefinitely confined. Neutrinos would be created in the formation of the neutron star in the capture of electrons by protons to yield neutrons and electron type neutrinos (ν_e). This produces a pulse of ν_e lasting only a few tens of milliseconds. Subsequently, in the rapid cooling down of the neutron star from an initial temperature, kT, in the vicinity of 4 MeV, neutrino–antineutrino pairs of all types (ν_e, $\bar{\nu}_e$; ν_μ, $\bar{\nu}_\mu$; ν_τ, $\bar{\nu}_\tau$, . . .) would be emitted and would constitute the dominant (by a factor of 10^2) energy release from the supernova. This neutrino–antineutrino pair emission lasts 10 to 20 sec. The total electromagnetic radiation from the remainder of the progenitor would amount to approximately 1% of the energy released in the form of neutrino–antineutrino pairs.

This description of type II supernovas, based on elementary particle physical and astrophysical principles and computer modeling, had been current for roughly two decades without experimental confirmation. The direct, essentially simultaneous observation of a burst of neutrinos from SN 1987A in two massive, imaging water Čerenkov detectors on Earth provided that confirmation. The number of neutrino interactions observed in the detectors, their characteristics, and the time interval and total neutrino energy of the burst are all in excellent agreement with the general, quantitative aspects of the model of type II supernovas outlined above.

Imaging water Čerenkov detectors make use of the energy, mainly blue light, radiated by a fast charged particle traversing a medium in which the velocity of the particle is greater than the velocity of light divided by the index of refraction of the medium (Čerenkov radiation). The radiated light is emitted along the entire particle path at a fixed angle with respect to the path direction and may be detected with suitable photosensitive devices. The observed ring pattern, or image, of the struck photosensitive elements (photomultiplier tubes), which follows from the fixed angle of the light emission, is distinctive and allows the energy and direction of a low-energy charged particle to be calculated.

The detectors that observed the neutrino burst from SN 1987A were constructed primarily to search for the possible decay of protons and bound neutrons (nucleon decay) and secondarily to study cosmic ray neutrinos by their interactions in the detectors. For both purposes, it was desirable to have detectors of very large mass and of an inexpensive inert material transparent to Čerenkov radiation, which accounts for the use of water. The abundance of the charged particle component of cosmic rays, particularly penetrating muons, at Earth's surface requires that the detectors be located deep underground. For simplicity, we describe some of the salient properties of only one of those detectors; the essential properties of both are similar.

The detector known as Kamiokande-II (an acronym for the Kamioka nucleon decay detector in its second version) is a research project that has involved the collaboration of physicists from the universities of Tokyo and Niigata, the National Laboratory for High Energy Physics in Japan, the University of Pennsylvania, and the California Institute of Technology. The detector is located in the Kamioka Mine in Gifu prefecture in western Japan. The overhead shielding at the detector site is a minimum of 800 m of rock or 2400 m of water equivalent. A schematic view of the detector and its associated equipment is shown in Fig. 1. The useful mass of water is 2140 metric tons, which is viewed by 948 photosensitive elements (photomultiplier tubes), each 0.5 m in diameter, uniformly placed facing inward on a 1 m grid on the entire inner surface of the cylindrical steel tank containing the water. The total photosensitive area is approximately 20% of the total surface area of the useful volume. Completely surrounding the main detector is an anticounter, which serves to shield passively against gamma rays and neutrons from the mine rock as well as to record the entrance of unwanted charged particles entering the detector.

The detector records the passage of electrons or positrons with total energies as low as about 6 MeV and allows the position and direction of the particle

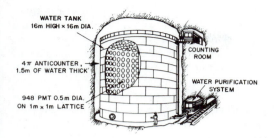

FIG. 1. Schematic view of the Kamiokande-II detector and associated equipment.

to be reconstructed, as well as its energy. The Monte Carlo calculated standard deviation error on each of the Cartesian coordinates of the reconstructed vertex position of 10 MeV electrons is 1.0 m, and the mean energy and angular resolutions for 10 MeV electrons are 22% and 28°, respectively. The electron energy scale is calibrated with an uncertainty of less than 5%.

In Fig. 2, scatter plots of event energy against event time for three 17-min intervals on February 23, 1987 are shown. The first interval is approximately

5 hr prior to the observed neutrino burst, the second interval includes the neutrino burst, and the last interval is 4.5 hr after the burst. The first and third intervals are typical of the usual count rate in the detector. In these plots, each datum point represents a single recorded event with a given value of energy that occurred at the time indicated. Most of the observed events in Fig. 2 have energies less than 7.5 MeV and are background events due to ^{214}Bi decay, itself a product of the decay of ^{222}Rn, which is present in the mine and dissolved in the water of the detector. Events with higher energy with an average occurrence rate $< 10^{-2}$ Hz (counts per second) are consistent with higher energy products of radioactivity at or outside the detector container.

The neutrino burst at 7:35:35 UT in Fig. 2(b) is evident. The occurrence times, energies, and angles relative to the LMC of the 12 events in the burst are given in Table I. The observed pattern of hit photomultiplier tubes for event 9 is shown as an example of a supernova induced event in Fig. 3. The data have also been intensively searched for evidence of other event bursts in the 10 hr period from 2:27 UT to 12:27 UT, and in a larger sample of 42.9 days from January 9 to February 25, 1987. No statistically sig-

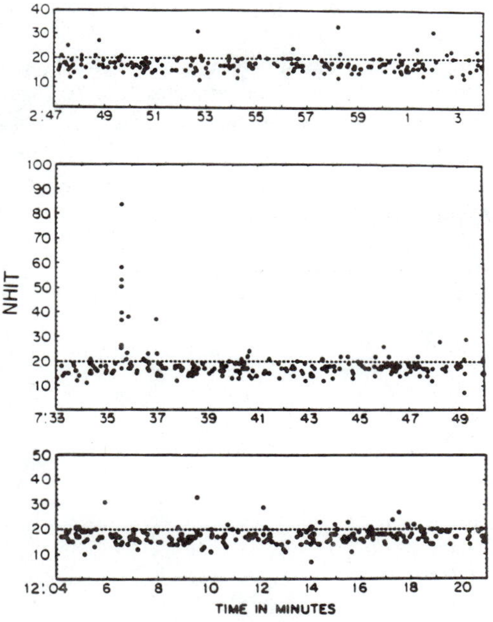

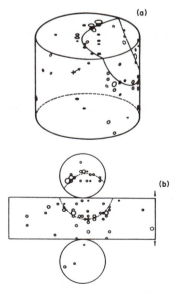

FIG. 3. (a) Pattern of struck photomultiplier tubes in the ninth event of the neutrino burst in a simulated three-dimensional view of the Kamiokande-II detector. (b) The same pattern superimposed on the "exploded" view of the detector in which the cylindrical wall has been "unrolled" and the top and bottom of the tank pressed into the same plane as the unrolled cylinder.

FIG. 2. Scatter plots of event energy versus universal time (UT) for three 17-min intervals on February 23, 1987. Each of the points represents an event observed in the Kamiokande-II detector. The height of a point above the time axis is proportional to the event energy.

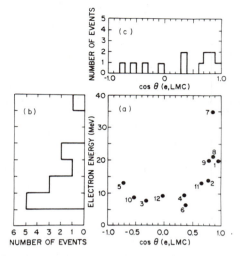

FIG. 4. (a) Scatter plot of the electron energy, E_e, versus the cosine of the angle of the electron track with respect to the direction of LMC for the 12 events observed in the Kamiokande-II detector. (b) Projection of (a) giving the energy distribution of the 12 events. (c) Projection of (a) showing the angular distribution of the 12 events.

nificant evidence for any excess of neutrino-induced events was found, excepting only the burst at 7:35:35 UT.

A scatter plot of electron energy E_e against the cosine of the angle of the electron track with respect to the direction of the LMC [cos $\theta(e, \text{LMC})$] is shown in Fig. 4(a), and projections on each axis are

shown in Figs. 4(b) and 4(c). The angular distribution in Fig. 4(c) is consistent with the distribution expected for the particle reaction $\bar{\nu}_e + p \rightarrow e^+ + n$, where $\bar{\nu}_e$ is from SN 1987A, p is one of the protons in the water of the detector, e^+ is the detected positron, and n is an undetected neutron. This reaction is known to be the most probable, by a factor of approximately 10, of all possible reactions of ν_e, $\bar{\nu}_e$, ν_μ, $\bar{\nu}_\mu$, etc., in the detector water. From the positron energies in Table I, and the assumption of the particle reaction, the resultant integrated flux of $\bar{\nu}_e$ in the observed burst is 1.1×10^{10} cm^{-2} for $\bar{\nu}_e$ with energies above 8.8 MeV, which in turn leads to the value of 9×10^{52} ergs of $\bar{\nu}_e$ output from SN 1987A for an average $\bar{\nu}_e$ energy of 15 MeV. This corresponds to a total energy carried away from SN 1987A by all neutrino species of 5×10^{53} ergs in a time interval of approximately 13 sec.

The neutrino burst observed in the IMB detector, a collaboration among the University of California at Irvine, the University of Michigan, and the Brookhaven National Laboratory, consisted of 8 events beginning at 7:35:41 UT. The energies of observed events are consistent with those determined in the Kamiokande-II detector (see Fig. 5); and the resultant total $\bar{\nu}_e$ energy release and total neutrino energy from SN 1987A obtained from the IMB data are in good agreement with the observation of the Kamiokande-II detector.

The internal time structure and energies of the events in each of the observed bursts have been addressed in studies that attempt to extract the initial

TABLE I.   Measured Properties of the Twelve Electron Events Detected in the SN 1987A Neutrino Burst

Event number	Event time (sec)	Number of PMT (NHIT)	Electron energy[a] (MeV)	Electron angle[a,b] (degrees)
1	0	58	20.0 ± 2.9	18 ± 18
2	0.107	36	13.5 ± 3.2	40 ± 27
3	0.303	25	7.5 ± 2.0	108 ± 32
4	0.324	26	9.2 ± 2.7	70 ± 30
5	0.507	39	12.8 ± 2.9	135 ± 23
6	0.686	16	6.3 ± 1.7	68 ± 77
7	1.541	83	35.4 ± 8.0	32 ± 16
8	1.728	54	21.0 ± 4.2	30 ± 18
9	1.915	51	19.8 ± 3.2	38 ± 22
10	9.219	21	8.6 ± 2.7	122 ± 30
11	10.433	37	13.0 ± 2.6	49 ± 26
12	12.439	24	8.9 ± 1.9	91 ± 39

[a] The errors on electron energies and angles are one-standard-deviation Gaussian errors.
[b] The angle is relative to the direction of the LMC.

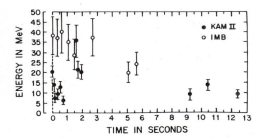

FIG. 5.   Scatter plot of the energies of the 12 events observed in the Kamiokande-II detector and the 8 events observed in the IMB detector versus time. The first event recorded by each detector is arbitrarily set to zero time in the plot.

state properties and time evolution of SN 1987A. A detailed, precise comparison of the burst data from the Kamiokande-II and IMB detectors is not possible because the absolute time of the beginning of the neutrino burst in Kamiokande-II is given with an error of ± 1 min, the uncertainty arising from the absence of an absolute time calibration source in the Kamiokande-II equipment. If it is assumed, arbitrarily but not unreasonably, that the earliest events observed by the two detectors coincided in time, the plot in Fig. 5 is obtained. Figure 5 suggests that the two observations agree on a cluster of 14 events within the first 2 sec and indicates a tailing off of the remaining 6 events to 12.44 sec.

The properties of the neutrino burst coincide extremely well with the current model of type II supernovas and neutron star formation. The observed energies of the neutrinos, their number, and type of interaction, in conjunction with the time duration of the burst, are consistent with the free-fall collapse of the core of a massive star, and the evaporation within a few seconds of all flavors of neutrino–antineutrino pairs with total energy amounting to ~ 3 to 4×10^{53} ergs from the newly born neutron star at temperature $kT \approx 4$ MeV.

There are two principal conclusions of significance in elementary particle physics that may be drawn from the neutrino burst data. First, the lifetime of ν_e and $\bar{\nu}_e$ must be greater than $\approx 1.7 \times 10^5\, m(\nu_e)/E(\nu_e)$ yr, taking the distance to the LMC to be 55 kpc, where $m(\nu_e)$ and $E(\nu_e)$ are the mass and total energy of the electron type neutrino and antineutrino. Second, an upper limit on the mass of ν_e and $\bar{\nu}_e$ may be obtained from the burst data subject to simplifying assumptions. Attempts to do so have led to upper limit estimates on $m(\nu_e)$ ranging from a few eV to 24 eV.

II. Neutrinos from the Sun

The sun is an intense source of neutrinos that are the products of the nuclear processes that provide the solar energy. Our knowledge of the specific nuclear processes is based on laboratory measurements and theoretical models of the internal structure of the sun. Solar model calculations provide a basis for designing experiments to observe the detailed processes in the sun and for interpreting the results. Of primary interest is the theoretical prediction of the entire neutrino energy spectrum, which is the sum of the contributions of eight specific neutrino sources in the sun.

There are two principal scientific objectives that might be achieved from the study of neutrinos from the sun. One is to understand as completely as possible the nuclear energy processes that power the sun and all main sequence stars, and thereby to understand the interior of the sun. The second is to discover intrinsic neutrino properties that will perhaps be revealed by their interactions with matter and by the great distances that neutrinos travel in passing from the core of the sun to a detector on Earth. We summarize the new experimental results from the two solar neutrino detectors operating at present and also discuss briefly forthcoming experiments, as well as certain theoretical speculations stimulated by solar neutrino experiments.

A. The Chlorine Experiment

The chlorine solar neutrino experiment in the Homestake Gold Mine has been taking data since 1970. The experiment depends upon the neutrino capture process $\nu_e + {}^{37}Cl \rightarrow {}^{37}Ar + e^-$, where the radioactive ${}^{37}Ar$ decays with a 35-day half-life. The radioactive argon is removed from the 615-ton tetrachloroethylene detector approximately 5 to 6 times per year, and the activity is measured in a miniature proportional counter. The average ${}^{37}Ar$ production rate measured over the period 1970–1987 is 2.2 ± 0.3 SNU (1-σ error), where SNU is a solar neutrino unit: 10^{-36} neutrino captures per second per target atom. The experimental result may be compared to the rate of 7.9 ± 2.6 SNU (3 σ) derived from solar model calculations. This is shown in the plot of Fig. 6.

The discrepancy between the solar model calculations and the experimental observation has been known for 15 yr. The chlorine experiment is principally sensitive to the flux of energetic neutrinos from ${}^8B$ decay in the sun. The ${}^8B$ results from the temperature-sensitive reactions ${}^3He + {}^4He \rightarrow {}^7Be + \gamma$ and $p + {}^7Be \rightarrow {}^8B + \gamma$.

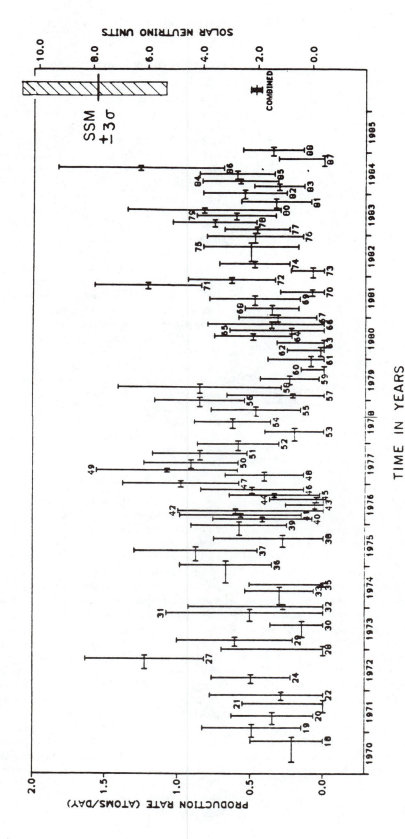

FIG. 6. Results from the radiochemical chlorine solar neutrino experiment that is based upon the reaction $\nu_e + {}^{37}Cl \rightarrow {}^{37}Ar + e^-$. Shown are the ${}^{37}Ar$ production rates in 615 tons of C_2Cl_4 from 1970–1985. The average value and the prediction from the standard model in solar neutrino units are indicated.

B. THE WATER ČERENKOV EXPERIMENT

Kamiokande-II is sensitive to electrons with energies as low as approximately 6 MeV, and is therefore capable of observing low-energy neutrinos from the sun through the neutrino scattering reaction $\nu_e + e^- \rightarrow \nu_e + e^-$. Here, again, the incident neutrino is produced in the sun through the beta decay of the radioisotope ^{8}B, which emits a continuous spectrum of neutrino energies up to 14 MeV, and the target electrons are the electrons in the water molecules in the detector. The recoiling electron in the scattering reactions is observed in the detector if its energy is greater than the detector threshold energy and if the level of the residual background radioactivity in the detector is sufficiently low. Great care is taken to remove radioactive elements from the water.

An important property of the particle reaction $\nu_e + e \rightarrow \nu_e + e$ is that the path of the recoil electron is essentially directed along a continuation of the path of the incident ν_e. Hence, the electron signal, if observed, can demonstrate that the incident ν_e are in fact coming from the sun. In addition, the energy spectrum of the observed electrons would yield information on the energy spectrum of the incident ν_e and on the real time of each event.

During 220 days of useful data from the detector, the average rate of low-energy event triggers was 0.33 Hz, while the rate predicted by the standard solar model was 0.3 electron events per day with energies $\geq$ 10 MeV in a 680-ton fiducial region. To extract a meaningful signal required, therefore, excellent discrimination against the background. This was accomplished by event selection criteria, which reduced the total observed event rate in the fiducial region to 2 events/day with energies $\geq$ 10 MeV.

In the last step in the data analysis, the directional correlation of the observed events relative to the sun was used. This is shown in Fig. 7, which plots the

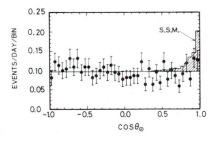

FIG. 7.   Distribution in the cosine of the angle between the observed track of an electron and the direction to the sun of the event sample obtained in 220 complete days of searching for solar neutrinos in the Kamiokande-II detector.

events as a function of the cosine of the angle between each observed electron path and the sun and indicates the angular region predicted to be populated by the standard solar model. No excess of events is seen in the region $\cos \theta_\odot \approx 1$. A quantitative analysis of the angular distribution in Fig. 7 yields the 90% confidence upper limit on the flux of ^{8}B solar neutrinos as $3.2 \times 10^6 \ \nu_e$ per (cm^2 sec). The prediction from the standard solar model is $(5.8 \pm 2.0) \times 10^6 \ \nu_e$ per (cm^2 sec).

C. CONCLUSION FROM THE CHLORINE AND WATER ČERENKOV EXPERIMENTS

The preliminary result of the Kamiokande-II experiment, that the ^{8}B solar neutrino flux is less than approximately one-half the value predicted by the standard solar model, qualitatively confirms the result of the earlier radiochemical chlorine experiment. The latter experiment found a positive signal at approximately one-third the standard solar model value. Accordingly, the discrepancy between the experiments and the model persists. Possible explanations, some of high potential interest, which are mentioned below, will be explored with more data and in experiments planned for the near future.

III. Solar Neutrino Detectors for the Future

We turn now to a description of two planned solar neutrino detectors, which should be ready to take data in two to three years, and then very briefly mention ideas and plans for detectors, the realization of which is somewhat further in the future.

A. THE GALLIUM DETECTORS

The proton–proton chain of reactions is believed to be the main source of solar energy. This chain is initiated by the fusion of two protons to form a deuteron, $p + p \rightarrow D + e^- + \nu_e$. If the sun is indeed producing energy by the proton–proton chain, the neutrinos from this single reaction must be produced at the expected rate, which yields a flux of 6.1×10^{10} neutrinos per (cm^2 sec) at Earth, exceeding all other solar neutrino sources. The energy spectrum of neutrinos from this reaction is continuous from 0 to 0.420 MeV, which is too low in energy to be detected by existing detectors.

Because of the interest in observing the neutrinos from $p + p$ fusion, significant effort has been devoted to developing a radiochemical gallium detector in which the neutrino capture process, $\nu_e + {}^{71}$Ga $\rightarrow$ ^{71}Ge $+ e^-$ will occur. The advantages of this reaction are that it has a low-energy threshold, 0.23 MeV, and is therefore capable of detecting neutrinos from the proton–proton reaction, and that Gallium 71 has a favorable neutrino capture cross section for low-energy neutrinos. According to the standard solar model, a total neutrino capture rate of approximately 134 SNU is expected, and half of that rate is provided by the p + p reaction neutrinos.

The technology for extracting a few atoms of radioactive ^{71}Ge from tons of gallium was developed in the United States and the Soviet Union. The ^{71}Ge radioactivity is easily measured in a miniature proportional counter, as has been the ^{37}Ar activity in the chlorine experiment. There are now two separate radiochemical gallium detectors being built: one by a collaboration of European scientists and the other by the Institute of Nuclear Research in the Soviet Union. The European group will use 30 tons of gallium in the form of an acid gallium chloride solution. The Soviet scientists will use 60 tons of metallic gallium in their detector. The European experiment will be built in the Gran Sasso Underground Laboratory in Italy. The Soviet experiment will be located in the Baksan Underground Laboratory in the north Caucasus near Mt. Elbrus. Both experimental groups have performed successful large-scale pilot experiments to test the chemical recovery of a few atoms of ^{71}Ge from a few tons of target material. Full-scale experiments should be operating by 1990, and results should follow soon thereafter.

B. IDEAS AND PLANS FOR OTHER SOLAR NEUTRINO EXPERIMENTS

There has been increasing interest in devising new solar neutrino detectors capable of measuring neutrino energy and direction, and perhaps even capable of observing neutrinos from the sun that have changed their type to ν_μ or ν_τ after originating, for example, as ν_e from decay of ^{8}B in the sun (see Section IV). This information is probably needed to understand in detail the solar interior and to resolve questions concerning the physics of the neutrino. It is likely that no single experiment will give all the information needed to resolve the discrepancy described previously.

An experiment based on the reaction ν_e + D → p + p + e$^-$, where D stands for the isotope of atomic mass 2 of hydrogen, known as deuterium, has been proposed by a group of Canadian and United States physicists. It would use 1000 tons of heavy water (D$_2$O) on loan from the Canadian Atomic Energy Commission and would be built in a deep mine in Sudbury, Canada. The detector concept is similar

to that of Kamiokande-II. By using deuterium, a large neutrino reaction rate will be obtained, which should allow a more accurate measurement of the neutrino flux from ^{8}B. It is also possible that ν_μ and ν_τ might be detected in the heavy water detector through the weak neutral current processes, for example, ν_μ + D → p + n + ν_μ. In this case, only the neutron released would be observed by its capture in hydrogen or oxygen nuclei.

A boron loaded scintillation detector to acquire essentially the same information as the heavy water detector has also been suggested. In this detector, ^{11}B would be the target of the incident neutrinos, and possibly permit detection of ν_μ and ν_τ as well as ν_e.

An interesting new idea in solar neutrino detectors would use several tons of liquid helium at 0.020 K. At this low temperature, neutrinos excite rotons in the liquid helium by electron scattering, which can be detected dynamically. This method would detect solar neutrinos of all energies.

All of the detectors mentioned require large quantities of local shielding, for example, several meters of water, surrounding them, as well as location in a deep mine to reduce the cosmic-ray muon rate traversing them. All require that they be prepared relatively free of radioactive contaminants, which are in general the most serious source of background counting rate. If these conditions can be satisfied, it may also be possible to use, for example, 50 tons of liquid scintillator without additional loading to measure the flux of solar neutrinos with energy 0.86 MeV from the reaction e$^-$ + ^{7}Be → ^{7}Li + ν_e.

IV. The Solar Neutrino Puzzle: Possible Implications for Solar and Neutrino Physics

The discrepancy described previously between the results of the two operating solar neutrino detectors and the prediction of the standard solar model is the stimulus of efforts to study the solar neutrino spectrum in more detail by means of different neutrino detectors. The deeper motivation to carry out experiments of such difficulty lies in the possibility that an overlooked basic property of the sun or of the neutrinos may be responsible for the discrepancy.

For example, physical processes in the sun not properly accounted for by the standard solar model, for example, turbulent mixing, diffusion, or internal rotational degrees of freedom, might reduce the production of ^{8}B and explain the observed low value of the solar neutrino flux. But also, in predicting the outcome of experiments, the theory presumes that neutrinos created in the core of the sun travel to the detector without any new, as yet unknown interaction or change in properties. It is possible, however, that neutrinos of different types, that is, different lepton number (for example, ν_e, ν_μ, and ν_τ) have different masses and are able to mix, that is, to violate the conservation of different (separate) lepton number. This mixing could give rise to a diminution of ν_e at a detector on Earth because neutrinos of one type, for instance, ν_e, might be transformed or oscillate into neutrinos of another type, such as ν_μ or ν_τ, which the detector would fail to detect for certain well-understood fundamental reasons in particle physics. It is expected that such transformations or oscillations might be enhanced by the changing density of the matter traversed by the neutrinos in moving from the core to the surface of the sun (matter oscillations). Alternatively, neutrinos may have as yet undetected electromagnetic properties that might cause them to interact with the internal magnetic field of the sun, and thereby fail to be detected on Earth.

It is the possibility that one of these fundamental phenomena, or one not as yet proposed, may be revealed as the source of the solar neutrino puzzle, which is what attracts physicists and astronomers and leads them to expend great effort to find the solution to the puzzle.

BIBLIOGRAPHY

Bethe, H. A., and Brown, G. (1985). How a supernova explodes, *Sci. Am.* **252,** 60.

Bionta, R. M. *et al.* (1987). *Phys. Rev. Lett.* **58,** 1494.

Burrows, A. (1987). The birth of neutron stars and black holes, *Physics Today* **40**(9), 28.

Cherry, M. L., Fowler, W. A., and Lande, K., eds. (1985). Solar Neutrino Conference Proceedings, Am. Inst. of Physics, Vol. 126.

Helfand, D. (1987). Bang: the supernova of 1987, *Physics Today* **40**(8), 24.

Hirata, K. *et al.* (1987). *Phys. Rev. Lett.* **58,** 1490.

Woosley, S. E., and Phillips, M. M. (1988). *Science* **240,** 750.

Woosley, S. E., and Weaver, T. A. (1986). The physics of supernova explosions, *Annu. Rev. Astron. Astrophys.* **24,** 205.

NEUTRON STARS

Steven N. Shore   *New Mexico Institute of Mining and Technology and DEMIRM, Observatoire de Meudon*

GLOSSARY

Glitch: Discontinuous change in the rotational frequency of a pulsar.

Millisecond pulsars: Weakly magnetized neutron stars, either isolated or in binary systems in which there is no mass transfer between components, which have periods near the rotational stability limit.

Pulsar: Radio pulsars are isolated or nonaccreting binary neutron stars with strong ($> 10^9$ G) oblique magnetic fields. X-ray pulsars are strongly magnetized neutron stars imbedded in accretion disks, which are fed from mass-losing companions in close binary systems.

Units: Solar mass ($M_\odot$), 2×10^{33} g; solar radius ($R_\odot$), 7×10^{10} cm.

URCA process: Neutrino energy loss mechanism whereby electron capture on nuclei produces neutrinos, followed by subsequent beta decay of the resultant nucleus with an additional neutrino emitted. The name derives from the Casino de Urca, at which gamblers lost their money little by little but without seeming to be aware of it. The process was first described by Gamow and Schoenberg.

Neutron stars are the product of supernovae (SN), gravitational collapse events in which the core of a massive star reaches nuclear densities and stabilizes against further collapse. They occur in a limited mass range, from about 0.5 to $4M_\odot$, being the product of stars between about 5 and $30M_\odot$. If they are rapidly rotating and magnetized, their radio emission can be observed as pulsars. Some occur in binary systems, where mass accretion from a companion produces X-ray and gamma-ray emission over a wide range of accretion rates. If the rate is in the proper range, periodic nuclear reactions can be initiated, which create episodic mass ejection events called bursters. Magnetic neutron stars in binary systems produce pulsed X-ray emission without strong radio emission. Neutron stars are unifications of nuclear and relativistic physics, requiring gravitation theory to explain their structure and serving as useful laboratories for constraining the nuclear equation of state. [*See* SUPERNOVAE.]

I. Historical and Conceptual Introduction

The first work on dense equations of state, following the advent of quantum theory in the 1920s, showed that it is possible for the pressure of a gas to depend solely on the density. This implies that it is possible to construct stable cold configurations of matter, independent of the previous thermal history of the matter. The first application of this idea to stellar evolution was by Landau who, in a prescient 1931 paper, discussed the possibility of the end state of evolution being composed of a sphere in which the pressure support is entirely provided by electrons in this extreme state of matter. Referred to as degenerate, the condition results from the fact that, in a quantum gas, the electrons are constrained to have only one particle per "box" of phase space. Landau pointed out that it might be possible for the configuration to depend only on the mass of the particle, which determines the momentum at a given energy, and thus to have objects in which the particle is neutral and heavy, but Fermi-like.

Chandrasekhar, in the early 1930s, derived an upper limit to the mass of such a configuration, above which the pressure exerted by the electrons is always insufficient to overcome the pull of gravity. The ultimate fate of a more massive object would be collapse. While he did not follow the evolution of such a state, it was clear from this work that the upper mass for such stars, called white dwarfs, is of order the mass of the sun. He argued that more massive stars cannot end their lives in hydrostatic equilibrium but must gravitationally collapse.

The discovery of the neutron and Yukawa's meson theory clarified the basic properties of nuclear matter. Application of quantum statistics demonstrated that the neutron is a fermion, a spin of $\frac{1}{2}$. It is consequently subject to the Pauli exclusion principle, like the electron, but since its mass is considerably heavier than the electron, by about a factor of 2000, degenerate neutron configurations are considerably denser than white dwarfs (WD). If a neutron sphere (NS) were formed, it would be sufficiently compact at a given mass that the Newtonian calculations previously used for stellar structure would not suffice for its description. Following the work of Tolman, who in 1939 derived the interior structure metric for a homogeneous nonrotating mass, Oppenheimer and Volkoff, in the same year, derived the equation for relativistic hydrostatic equilibrium. They showed that there is a maximum mass above which no stable configuration for a neutron star is possible for a degenerate neutron gas. While Chandrasekhar had presented a limiting mass, which is an asymptotic upper limit for infinite central density, Oppenheimer and Volkoff showed that the masses above about 1 $M_{\odot}$ must collapse. A subsequent paper by Oppenheimer and Snyder showed that the ultimate fate of more massive objects was to rapidly collapse past their Schwarzschild radii and become completely relativistic, singular objects, later called black holes (BH) by Wheeler. By the end of the 1930s, it was understood that the end state of stellar evolution was either a stable degenerate configuration of electrons (WD), neutrons (NS), or catastrophic gravitational collapse (BH). This picture has not been significantly altered with time.

Following these calculations, Zwicky and Baade nominated supernovae as the sites for neutron star formation. Having first shown that there is a class of explosive stellar events with energies significantly above the nova outburst, by about three orders of magnitude, which they dubbed "supernovae," they showed that the magnitude of energy observed could be provided by the formation of a neutron star. They argued that collapse of the core of the star from about the radius of the sun to the order of 10 km would be sufficient to eject the envelope of the star (by a process at the time unknown) with the right magnitude of energy observed in the SN events. The only object that could thus be formed is a neutron star.

The first detailed calculations for the structure of such objects awaited the advent of computers in the early 1960s and the improvement in the equations of state for nuclear matter. The fact that degenerate equations give an upper limit of about $0.7 M_{\odot}$ is in part an artifact of the softness of the equation of state at lower than nuclear densities. However, from the energy level structure in nuclei, it is possible to specify more exact forms for the nucleon–nucleon (N–N) interaction potential, which can be used to calculate the equation of state. The nuclear, or strong, force depends on the exchange of integer spin particles, π mesons. It is thus possible to have an attractive potential at large separations, which becomes progressively more repulsive at short distances. The central densities, which result in such models, are somewhat higher near the maximum mass, but they do not change the fact that there is a maximum in the stable mass of such a star. The inclusion of exotic particles as the outcome of N–N scattering changes the details of the equation of state at the highest densities, but is still incapable of removing this maximum entirely.

Neutron stars remained theoretical entities until November 28, 1967, when, using a radio telescope designed for the study of scintillation of radio sources at a wavelength of 3.7 m, Bell and Hewish discovered the first pulsar CP 1919 + 21. They observed a rapidly varying, periodic, pulsed, point radio source with pulse frequency of about 1 Hz. In rapid succession, several more of these objects were discovered. By the end of the year, Gold had suggested that a pulsar is a rapidly rotating magnetized neutron star. The pulses, he argued, result from the magnetic field being inclined to the rotation axis; it is from the magnetic polar regions that the signal is radiated. The short duration of the pulses compared with the interpulse period supports this view. Pacini showed that such stars must spin down via the emission of low-frequency, rotationally generated, electromagnetic waves, and such a spindown was indeed observed by Radakrishnan in the Vela pulsar in 1969. The polarization and frequency characteristics of the radio pulses left no doubt as to their nonthermal origin, and implied the presence of extremely strong magnetic fields, of order 10^{12} G (the field of the earth is about 1 G). The formation of such strong fields had previously been suggested by Ginzberg in 1963 as the consequence of flux freezing during stellar gravitational collapse.

The association of PSR $0532 + 27$ with the Crab Nebula, the remnant of the supernova of 1054, and its optical identification with a faint blue star of featureless continuum and pulsed optical output near the center of symmetry of the nebula, showed that the initial guess for the origin of neutron stars is likely correct. However, with the exception of the Vela pulsar and one or two others (the matter is still in doubt), pulsars are generally not found imbedded in supernova remnants.

The discovery of binary X-ray sources came with the determination of the period of Cen X-3, in the early 1970s following the launch of the UHURU satellite. The spectrum of several other sources, and the mass determinations, notably Her X-1 = HZ Her, have been used for the determination of both the magnetic and mass properties of neutron stars. These sources are regular X-ray pulsars, formed by the structuring of the mass accretion region into a column that corotates with the magnetic field of the star and has a period of order a few seconds. The observation of cyclotron lines in the X-ray from these stars shows that the magnetic fields are $> 10^{10}$ G as expected for neutron stars. None has a mass above $4M_\odot$ regardless of the mass of the companion.

The first binary pulsar, PSR $1913 + 16$, was discovered by Manchester and Taylor in 1974. With its superb timing stability relative to the errors associated with velocity determinations for mass-accreting X-ray pulsars, it helped to set, for the first time, accurate mass limits for neutron stars. Only upper limits are possible, given uncertainties in the mass of the companion. About a half dozen more of these have since been found. The discovery of PSR $1937 + 214$, a pulsar with a period of about 1 msec, in 1982 was soon followed by several others, including several members of binary systems. These pulsars have extremely weak magnetic fields and appear to result from the spin-up of neutron stars by accretion during phases of rapid mass transfer in a close binary system.

II. Formation of Neutron Stars

Neutron stars are formed by gravitational collapse of intermediate mass stars. Nucleosynthesis in such stars can proceed exothermally to build elements up to, but not past, ^{56}Fe because of the progressive increase in the binding energy of nuclei until this point. Following Fe synthesis, the primary loss mechanisms

for the stellar core become electron captures onto seed nuclei via the URCA process:

$$e + (Z, A) \rightarrow (Z - 1, A) + \nu_e \quad (1)$$

where Z is the atomic number and A the atomic mass, followed by the subsequent decay:

$$(Z - 1, A) \rightarrow (Z, A) + e + \bar{\nu}_e \quad (2)$$

which represents a bulk rather than surface cooling. Because the neutrino opacity of normal stellar envelopes is small, these particles can freely escape from the core in rapid order, producing rapid cooling that is very density and temperature sensitive. The core is thus forced to contract, increasing the reaction rates and producing accelerating losses. Finally, with increasing *neutronization*, the core reaches a critical density at which the free-fall time becomes short compared with the sound speed, and the subsequent gravitational collapse cannot be avoided.

The free protons, which result from the destruction of nuclei as the density passes $\approx 10^{11}$ g cm^{-3} and electron capture that converts them to neutrons, alter the equation of state. Should the core mass be small enough, less than the maximum mass allowable for a neutron star, the collapse will halt for the core, and the subsequent release of gravitational potential energy, which is essentially the gravitational binding energy of the final configuration (the collapse takes the core from about 10^6 km to about 10 km) amounts to $\approx 10^{53}$ erg available for the lifting of the envelope.

Most of this energy is radiated away in a neutrino burst that accompanies the formation of the stable core, while a fraction, of order a few percent, is available to power the shock wave that expels the envelope. The visible supernova event that is the product of the collapse contains but a small portion of the energy liberated by the core formation. While the details depend on the evolutionary history of the progenator star, and the visibility of the shock depends as much on the environment of the collapsed star as the energetic processes powering the ejection, it is generally believed that the supernova event that heralds the end of the collapse should be visible. The recent discovery of neutrino bursts from SN 1987A in the Large Magellanic Cloud, with about the right order of magnitude and spectrum, as well as duration (a few seconds), provides strong support for the general picture as outlined here.

There are two critical physical problems connected with the formation of the stable configuration that will become a neutron star: how does the N–N interaction halt the collapse of the core; and how does the final configuration look subsequent to the ejection of the envelope? To answer the first, one must examine the equations of state, as they are the products of the nucleon–nucleon interaction. Then, one must consider the conditions for equilibrium, both mechanical and thermal, and what constraints they provide on the interior structure.

III. Equations of State: Neutron Stars as Nuclei

Nuclear physics meets the cosmos nowhere as dramatically as in the context of the interior structure of neutron stars. Since these stars are in essence nuclei, albeit several kilometers across, delimiting their density profiles is an excellent check on our understanding of the structure of nuclear matter.

The simplest equation of state, the relation between pressure, density, and temperature, comes from application of the Pauli exclusion principle to neutrons. Since neutrons are spin-$\frac{1}{2}$ particles, their phase space is strictly limited to single occupation. To show how a simple equation of state can be derived, consider a gas of quantum mechanical, chargeless particles that interact only via the Pauli principle. That is, the number of particles in a volume of phase space, h^3, where h is Planck's constant, is limited by

$$p_F^3 n^{-1} = h^3 \qquad (3)$$

which defines the *fermi momentum*, p_F, for nucleon density n. The distance between the particles is essentially the de Broglie wavelength. These particles exert a mutual pressure, resulting from the quantum condition, so that $P \approx n(p_F^2/2m) \approx h^2 n^{5/3}/m$. Notice that this is independent of temperature, a condition which is called *degenerate*. The pressure comes primarily from the unavailability of phase space "bins" into which particles can be placed. The degeneracy of a gas is measured by

$$\alpha = \frac{p_F^2}{2mk_BT} \qquad (4)$$

where k_B is the Boltzmann constant, T is the temperature, and m is the mass of the particle in question. If α is large, the quantum effects dominate the available phase space volume for the gas and the temperature plays no role in determining the pressure. The pressure is really the average of this over the total distribution function for

$$P \approx \int_0^{E_F} \frac{E^{3/2} \, dE}{\exp[(E/kT) - \alpha] + 1} \qquad (5)$$

Therefore, the pressure is strongly dependent on the density, and not temperature, unless the total thermal energy significantly exceeds the Fermi energy.

White dwarfs and neutron stars are similar, in this picture, because both are composed of spin-$\frac{1}{2}$ particles. Their differences arise chiefly from the mass of the supporting particles, electrons being the primary supporters for white dwarfs. There is, however, a more profound difference between these two stars—neutrons, being baryons, interact with each other via the strong force. Thus, the details of nuclear potentials play critical roles in the equilibrium of such extreme states of matter.

In order to remain bound in nuclei, neutrons must interact attractively for some separations; the stability of matter, however, demands that the core turn repulsive (or at worst have vanishing potential) at very small distances. The finite sizes of nuclei, of order a few fermi (about 10^{-13} cm), comes from the large mass of the π meson, the particle principally responsible for the transmission of the strong force. Degeneracy alone yields a *soft* equation of state, that is, one which asymptotically vanishes at infinite density. The maximum mass such a gas can support against collapse must be much smaller than one which is due to a strongly repulsive interaction at small separation (high density limit). The star that results from a soft pressure law will consequently be smaller in radius and have a higher central density than for a strongly repulsive short-range potential, called a *hard* equation of state. However, it is only once the neutrons and protons become free that such considerations must be included in structure calculations. Below the density at which this occurs, the nuclei are formed into an ionic lattice, through which a degenerate electron gas flows. This portion of the neutron star, the crust, is a solid ranging from densities of about 10^8 to 10^{11} g cm^{-3}. The electrons at these very high densities are not only degenerate but form pairs called BCS superconductors.

The neutrons are not bound in nuclei in neutron star cores. That is because the free energy is greater for the bound than the "unbound" state, and the neutrons simply diffuse, or "drip" out of the nuclei. This occurs well above the neutronization density, at about 4×10^{11} g cm^{-3}. Its significance for neutron star structure is that until this density is reached, the system can be thought of as a normal ionic gas, or indeed solid, but at higher densities no such state is possible. Pairing becomes important in the reactions of the nucleons, again at densities of the order of the neutron drip state, and a superfluid can result. This involves the formation of N–N pairs, with their subsequent behavior as a bosonic system. The analogous situation occurs much earlier in density for the electrons, which form pairs as in a superconductor and also behave as a superfluid within the ionic lattice that can be set up from the nuclei.

This superfluid is one of the most important components in the interior of the neutron star, not because of its effect on pressure so much as its other dynamical properties. For instance, the heat capacity of a superfluid is essentially zero. The nucleons are completely degenerate and therefore do not have large thermal agitation. It has the property of rotating as a quantum liquid, which means certain properties such as the vorticity are quantized, and therefore exercises a strong dynamical effect on the coupling between the outer and core regions of the neutron star. In effect, the superfluid represents a collective dynamical mode for the neutrons, similar to that observed in nuclei when many-particle states are included in the calculation of nuclear structure. On the role it plays in the rotational history of the star, we shall say more later.

The core region dominates the structure of the neutron star. It is here that the stiffness of the nuclear matter equation of state is most strongly felt. Since the density is about $1-3$ particles fm^{-3}, about 10^{15} g cm^{-3}, it is essential that the small distance part of the nuclear potential be properly evaluated. The pressure is given by the gradient, with separation, of the N–N potential; in fact, the maximum mass of the star is essentially determined by the gradient in the N–N potential at distances less than 1 fm. The steeper the gradient in the potential, the stiffer the equation of state; at the same pressure, a lower density is required. The star can be more distended and have a higher moment of inertia, a less centrally condensed structure, a lower density throughout the envelope, and a higher maximum mass than for a softer potential.

The interior structure can, in fact, be observationally studied. The rotational properties of a neutron star depend on its equation of state through its moment of inertia. If the magnetic field can be determined independently, the spindown rate for pulsars gives an estimate of this quantity. The maximum mass is dependent on the N–N potential. Various models provide substantially different results. While the softest equations of state give a lower limit of about $0.7-1$ $M_\odot$ for the maximum mass, the stiffest give $4-5M_\odot$.

IV. Equations of Structure

In order to go from the equation of state to the structure of a neutron star, it is essential that general relativity play a role. Electrons, being lower mass than neutrons, produce degenerate configurations that are large, with radii about the size of the earth (about $0.01R_\odot$). Nucleons, however, are about 2000 times more massive, so that for degeneracy to be

reached and for the N–N potential to play a part in their support, the sinal configurations must be very much more compact. Neutron stars are so close to the stability limit for a compact configuration and their gravitational binding energies are so close to its rest mass, that relativistic rather than Newtonian equations are required to describe their structure.

To begin with, the mass equation must take into account the fact that the rest mass and the energy density are both involved with the total density. The metric, for simplicity, can be assumed to be that of an isolated, nonrotating (or slowly rotating) point mass outside of the star, which has an increasing mass with the distance from the center of the star (the so-called Schwarzschild interior solution). Take the mass to be scaled in terms of *geometrical* units, so that $G = c = 1$. Then the equation for the mass interior to some point is given by integration on spherical shells:

$$\frac{dm}{dr} = 4\pi r^2 \rho \tag{6}$$

The pressure gradient is given by the reaction of the matter to compression, which depends on the equation of state. Thus,

$$\frac{dP}{dr} = -\frac{(\rho + P)(m + 4\pi Pr^3)}{r^2(1 - 2m/r)} \tag{7}$$

The gravitational potential, actually the term in the metric that provides the redshift, is given by

$$\frac{d\phi}{dr} - \frac{1}{\rho + P}\frac{dP}{dr} \tag{8}$$

This last term is the closest link to the actual metric, since $ds^2 = -e^{2\phi} dt^2 + 1/[1 - (2m/r)]\, dr^2$. For a point mass, the vacuum field has a singularity at $r = 2m$, which in physical units is $r^* = 2GM/c^2$, the Schwarzschild radius.

For a star with a surface gravity of 10^{14} cm sec^{-2}, for instance, the gravitational redshift is about 0.2. This redshift reduces the intensity of the radiation seen by a distant observer in a given energy band by a factor of $(1 + z)^3$ and affects the temperature determined from the flux. Thus, in all of the calculations of neutron star properties, it should be borne in mind that the observational quantities are those detected at infinity, that is by a distant observer, but that the interior values are calculated in the *rest frame* of the neutron star.

These equations, originally based on Tolman's interior solution for the metric of a relativistic nonrotating mass, form the basis of the Oppenheimer–Volkov solution. As one moves progressively farther from the stellar core, the increasing interior mass causes an increasing curvature of space time, which in turn affects the dependence of the pressure on the distance. The gravitational mass, the amount of mass that is felt by the overlying matter outside of a radius r, increases more slowly than would be expected in the Newtonian model, and the pressure therefore changes appropriately.

The equation of state is inserted into these equations, for which the baryon density n is the independent "coordinate," with the initial conditions that the density at the center, ρ_c, is a free variable, with $M(0) = 0$. The mass of the configuration is determined by the boundary condition that $\rho(R) = 0$ and $P(R) = 0$, while $M(R) = M$. Most of the mass of the star is taken up in binding energy, which becomes progressively larger as the mass is increased. Such behavior is already understood from white dwarf models, since degenerate equations of state behave similarly. As mass is added to the star, the star shrinks. The limiting mass is when the radius becomes small enough for a black hole to develop, past which no stable configuration is possible. This is why neutron stars lead necessarily to collapse if they are made massive enough. There is no stable state between them and catastrophe should they begin to collapse.

The maximum mass is a product of general relativity. While for a Newtonian star an infinite central density is asymptotically approached leading to a progressively smaller star (which becomes asymptotically small), there is a lower limit to the size of a relativistic object, the Schwarzschild radius. Degenerate matter produces an approximate mass–radius relation $MR^3 = $ constant, so that for $R = 2GM/c^2$ there is a fixed mass. Above this mass, a black hole is the inevitable result.

The upper mass limit for neutron stars can be set by the observation that, for at least the X-ray pulsars, the mass of the degenerate object must exceed that of a white dwarf and is always lower than 4 or $5M_\odot$. For Cyg X-1 and LMC X-3, the masses appear inconsistent with this value and likely represent cases of black holes as the sources of the X-ray emission.

V. Pulsars

The first observations of radio pulsars served to confirm the existence of neutron stars—over 400 are now known. The detailed modeling of these objects has centered on the mechanisms that are capable of producing the emission. It is clear that they are rapidly rotating collapsed stars, with periods as short as a few milliseconds. In order to discuss how observations of pulsars help to elucidate the interior properties of neutron stars, it is useful to look at some of the observational constraints they can provide.

First, the fact that they shine means that pulsars possess strong magnetic fields; that they pulse is the result of their rotation and the fact that the magnetic field is not axisymmetric. A rotating, magnetized neutron star loses rotational energy by emission of magnetic dipole radiation:

$$\frac{dE_{\text{rot}}}{dt} = \frac{2\mu^2}{3}\omega^4 \sin^2 \beta \qquad (9)$$

where μ is the magnetic moment ($\mu = B_0 R_0^3$, R_0 is the stellar radius, and B_0 is the surface magnetic field) and β is the angle between the magnetic moment and the rotational axis. The change in the rotation period, P_{rot}, is given by

$$P_{\text{rot}}\dot{P}_{\text{rot}} \sim \frac{B^2 R_0^6}{I} \qquad (10)$$

where I is the moment of inertia. Secular variations of the magnetic field, such as decay due to the finite conductivity of the interior matter, produce a change in the deceleration rate. For instance, if the field is exponentially decreasing with time,

$$\dot{\omega} \sim B_0^2 e^{-2t/\tau}\omega^3 \qquad (11)$$

where τ is the decay time for the field. Should the alignment angle of the field also depend on time, the pulse profile and the rate of spindown will also change. Unfortunately, the time scales implied by the models are very long, and while these provide some probes of the electromagnetic properties of neutron star interiors, they are not well understood presently.

In a few other ways, pulsars provide us with windows into the interiors of neutron stars, most dramatically in the form of glitches. These will be discussed later in more detail. Glitches are discontinuous changes in the rate of rotation of the star, by $\Delta\omega/\omega \approx 10^{-7} - 10^{-6}$ superimposed on the secular spindown of isolated pulsars. In addition, however, there is *timing noise*, which is a fluctuation in the rotation rate of the star manifested by changes in the arrival time of pulses not connected with the properties of the interstellar medium. This may be caused by small thermal effects, within the stellar superfluid, which produce random walking between pinning sites of vortices.

Pulsar emission may also reveal properties of neutron star crust, although now only within the context of specific models. The Ruderman–Sutherland model uses the fact that as a magnetic star rotates, it generates a latitude-dependent electric field in the surrounding space. Crust electrons experience an acceleration, initially flowing freely away from the pulsar in the magnetic polar cone, but being trapped closer to the magnetic equator. The pulsar environment fills with charge, which alters the electric field at the stellar surface. Electrons are being ripped out of the crust faster than they can be resupplied from charge migration within the star so that a potential drop builds up over the polar cap. When the potential drop at the poles exceeds that of the rest mass energy for pair creation, $2m_ec^2$, positron–electron pairs, which produce cascades at the polar caps, are created. These, in turn, are responsible for the observed emission. The emission is confined to the polar region, and it is therefore the obliquity of the field that is responsible for the visibility of the pulsar.

There is no evidence for any thermal emission from any known pulsar that is not in a mass-accreting binary system. The neutron star cooling is sufficiently fast and uniform over the stellar surface that there is no evidence for pulsed thermal X-rays. Instead, any X-ray emission can be attributed either to small synchrotron nebulas in the vicinity of the neutron star or the nonthermal high-energy emission from the polar cascade.

The spindown rate of a pulsar, all other things being equal, is a measure of the age of the star. Provided the magnetic field does not change with time, the age can be estimated from $\Pi/\dot{\Pi}$. Usually, this gives ages of about 10^3–10^6 yr. There is one class of pulsar, however, for which this estimate is completely unreliable, the millisecond pulsars. For these, the rate of spindown is extremely long, over 10^8 yr, although their periods are incredibly small. The answer is to be found, in part, from the very low values of their magnetic fields and in part is due to their rotation being the artifact of their histories. These neutron stars have been, or still are, members of close binary systems in which accretion has torqued the neutron star up to rotational frequencies near the limit of stability. We shall return to this point in Section VIII.

VI. Glitches: Evidence for Superfluids in Neutron Star Interiors

At densities in excess of the neutron drip density, the free nucleons feel an attractive potential. This favors the formation of pairs, which behave like bosons and form a superfluid. Pairing is favored at densities between 10^{11} and 10^{14} g cm^{-3}. This fluid feels no viscosity, except by collisions with the normal fluid, which serves to drag on the particles and couple them to the rotation of the star as a whole. A rotating superfluid quantizes vorticity. These vortices have the property of pinning to irregularities in the crustal substructure by threading through nuclei.

Should there be sufficient thermal agitation of the vorticies, they will creep from one nuclear site to another.

The surface is continually slowing down by the emission of low-frequency (the rotation frequency of the star) electromagnetic waves causing the angular velocity gradient to become large across the superfluid mantle. This in turn generates a shear which will produce a drift of the pinning sites. Major depinning events result in glitches; the glitching rate depends on the temperature of the normal fluid and serves as a probe of the internal temperature of the neutron star mantle. Post-glitch relaxation of the rotational frequency depends critically on scattering properties in the superfluid for the phonons generated by the event. In effect, the star rings for a while as it settles down to a new rotationally stable state. The detailed behavior of this system has yet to be completely understood, although phenomenological models have been successful in predicting that the glitch behavior should be a property of only young, that is, still relatively hot, neutron stars.

The reader might think this contradicts the fact that the equation of state for the system is degenerate. However, that does not mean that there are no random motions for the particles; it simply implies that the equation of state does not have a temperature dependence. In addition, neutrons in the superfluid are no longer constrained by Fermi statistics, and so have a different behavior than the electrons or free nucleons.

VII. Cooling Processes

Neutron star matter is completely optically thick to photons. Therefore, only the surfaces can cool by the emission of such radiation. If this were the only mode for the decay in the temperature of such objects, one would expect them to be easily detected by X-ray observations for quite a long time after their formation, of order 10^6 yr, since their formation surface temperatures are in excess of 10^{10} K. Other processes than photon emission, however, are far more important in the cooling of such objects, specifically those associated with the emission of neutrinos. For very dense matter, the cross sections for the production of electron and muon neutrinos by collision processes among nucleons are high. The URCA processes, in which neutrinos are the sole emitted particles, have been shown to be important. Specifically, processes of the form

$$n + n \rightarrow n + p + e + \nu_e \qquad (12)$$

and

$$n + p + e \rightarrow n + n + \nu_e \qquad (13)$$

which are modified URCA processes in which the neutron bystander absorbs some of the momentum necessary for the emission, have been shown to be important. Bremsstrahlung processes, in which the exit channels are essentially the same as the incoming particles with the addition of the radiation of the excess momentum acquired during the collision, can be of the form:

$$n + p \rightarrow n + p + \nu_e + \bar{\nu}_e \qquad (14)$$

$$n + n \rightarrow n + n + \nu_e + \bar{\nu}_e \qquad (15)$$

The URCA processes may occur in the crust, where entire nuclei are present:

$$e + (Z, A) \rightarrow e + (Z, A) + \nu_e + \bar{\nu}_e \qquad (16)$$

One of the earliest suggested processes, by Bahcall and Wolf, is

$$\pi^- + n \rightarrow n + e + \bar{\nu}_e \qquad (17)$$

a weak interaction, like the nn or pn interactions, which for high enough interaction energy is replaced by the (μ, ν_μ) pair. While the rates for $\pi^- n$ and URCA processes have modest temperature dependence of T^6, the nn and np bremsstrahlung processes are more temperature sensitive, varying like T^8.

Superfluids play a role in altering the rates of the bremsstrahlung processes. This is because of pairing of the baryons, which introduces a gap energy for the free particles so that the rates are suppressed by a factor of $\exp(-\Delta_j/kT)$ where Δ_j is the BCS gap for either protons or neutrons ($j = p, n$, respectively). Since this gap energy depends on the density, it alters the predicted rates of cooling in a way that can probe the interior structure of the star.

It must be kept in mind that, since neutron stars are relativistic objects, the gravitational metric plays a role in the cooling. Neutrinos and photons escaping from the interior suffer a redshift simply because of the gravitational field. This reduces their luminosity at the stellar surface and affects the cooling rate. Their observed surface temperature and luminosity are related to their "actual" values by $T_s e^{-\phi}$ and $L_s e^{2\phi}$, respectively. Because the structure of the star is virtually independent of temperature, however, the redshift can be determined from the interior model and then included in the calculation of the luminosity after the fact.

No young pulsars, which are isolated neutron stars or binary neutron stars that are not accreting matter from companions, show observable thermal X-ray emission, including the Crab pulsar. The cooling time estimates from all current calculations indicate that the stars should be detectable for at least

10^3–10^4 yr, yet the Crab pulsar is less than 1000 yr old and is still unobservable in soft X-rays; its inferred surface temperature is less than 2×10^6 K. This is at odds with most of the current theories for neutrino formation, and the properties of the cooling function for neutron stars serve as a useful test of ideas in neutrino and intermediate energy nuclear physics. One possible explanation is that rapid cooling takes place via charged pion condensation because of the rapid increase in the neutrino emission. This occurs at densities of order 3×10^{14} g cm^{-3}.

Recent calculations have included the effects of an atmosphere, although the calculations of the opacities at the probable densities and temperatures associated with neutron stars in the cooling phase are difficult to obtain. The gravities are of order 10^{16} cm^2 sec^{-1}, in comparison with those of a normal star (about 10^4) and a white dwarf (10^8). These show, however, that most of the emission is typical of a blackbody, and that flux redistribution because of the opacity edges at ionization limits alters the shape of the flux profile but cannot explain the lack of detection of thermal emission from young pulsars. Magnetic fields can act to alter the cooling rates as well, but this tends to increase the visibility of the star by suppressing many of the cooling and conduction processes. The apparently rapid cooling associated with neutron stars remains a serious problem.

VIII. Neutron Stars in Binary Systems

The number of pulsars known to be in binary systems is steadily growing, including several of the most rapidly rotating objects known—the millisecond pulsars. The presence of neutron stars in binary systems was signaled by the discovery of Cen X-3 and Her X-1, both of which display rapid X-ray pulsation. Accreting material from a companion, which is losing mass because of tidal interactions with the neutron star, is funneled into a column at the magnetic poles. The accretion columns are tied to the stellar surface and corotate with the star. The subsequent discoveries of the gamma burst sources, the rapid burster, and the X-ray burst sources have added to the picture of binaries.

If the neutron star does not possess a strong magnetic field, matter accreted from a companion will pile up on the surface until it undergoes nuclear reactions. Since these occur with underlying degenerate matter, the heat conductivity away from the site is very poor. The consequence is a thermonuclear runaway, resulting in the matter blowing off the surface and producing a characteristic burst profile. The maximum luminosity of the burst is at the Eddington limit, the luminosity at which the radiative acceleration exceeds that of gravity. The burst signature is a very rapid rise to maximum light (of order 10 msec) and a slower ($<$ 1 sec) decay time. Many of the burst sources have been observed to flare on numerous occasions. Some show quasi-periodic variations in the X-ray, as seen by EXOSAT. These presumably come from accretion onto the magnetosphere of rapidly rotating neutron stars, with the development of a two-stream and Rayleigh–Taylor instability at the boundary layer. This results in the occasional excitation of X-ray variations as the matter rains onto the stellar surface.

Since neutron star formation requires a supernova event, one might think it unlikely that the binary could survive such a cataclysm. Yet clearly many have. This leads to the suggestion that the amount of mass lost from the system must be small compared with the masses of the stars remaining, and that the events must systematically be like type I SN events. This implies that the collapsed star had begun as either a white dwarf or helium star that was induced to collapse, perhaps by the excess accretion of matter. No such events have yet been observed, so the model remains untested except by statistics on the binaries. Many are circular orbits (a few notable exceptions exist, such as Cir X-1), and this requires efficient tidal dissipation on time scales of 10^6 yr to circularize the orbit. The binary X-ray sources have been found in globular clusters, which also contain burst sources, so that the formation of neutron stars is clearly not limited only to the young population of the galaxy. The sources observed for globular clusters may be the result of tidal capture of the neutron star.

IX. Magnetic Field Generation and Decay

While there are some strongly magnetic main sequence stars with fields of several kilogauss, most normal stars show upper limits of about 100 G. If the flux of such a star is somehow conserved in the collapse process, the resultant field would scale as R_*^{-2}, so that the expected field of a neutron star should be of order 10^{12} G. While the details of the magnetic field freezing process have yet to be understood, the fact that this is the right order of magnitude for many neutron star fields is very interesting and indicates that some form of flux freezing must play a role in the stars.

A central problem for neutron star studies is the generation of the wide range observed for pulsar magnetic fields. The millisecond pulsars are all objects that have intrinsically low fields, of order 10^9 G or so, and have been spun up by accretion during their earlier histories. It is important to note that the subsequent decay of the rotation is dependent on the magnetic field strength and configuration, so that the weaker the field, the longer the time over which the star will be a rapid rotator.

A recent model generates strong magnetic fields after the collapse using the currents generated in the crust due to the thermal gradients present immediately after the supernova event and while the neutron star is cooling. It is assumed that a small magnetic field (by pulsar standards) may be present initially after the collapse. Since in a solid, heat is transported by electrons moving within the crystal lattice, a crustal temperature gradient produces a net electron current, the *Nernst effect*. The field saturates when the ohmic dissipation overwhelms the amplifications at about the value observed for pulsars, about 10^{12} G. The fields reach peak strength within the crust, not at the surface, where they become essentially quantum limited, at about 10^{14} G. The problem is that this is a local, not global, effect, and it is not clear how the field manages to organize the large-scale dipolar structure thus far observed for these systems. Further, there is the problem of the obliquity. No obvious alternative models have yet been proposed. What one knows is that many neutron stars do possess strong fields, the weakest of which almost coincide in strength with those of the strongest white dwarf stars, and that many of these fields are present even in stars that have been around for a while. The decay times for the fields appear to be long, but here again there are not sufficient data to judge the matter.

The fields of many neutron stars are highly inclined to the rotation axis. Some of this is due to the discovery procedure—one looks for periodically variable emission lines, pulsed X-ray continuum, or pulsed radio emission, so one systematically discovers the highly oblique fields. Highly oblique fields are known to occur in main sequence stars and in white dwarfs and have even recently been discovered for planets (e.g., Uranus), so they may not be a special feature of magnetic field formation in neutron stars. It remains, however, an unexplained phenomenon for pulsars.

Several important departures from normal matter result from the strength of the surface magnetic fields. One is that the electrons in the degenerate gas at the surface quantize in their orbits about the magnetic field lines. The result is that each of the particles, depending on its momentum (and recall that since there is complete degeneracy in the electron gas this ensures that there will be only one particle per magnetic level, or Landau orbital), occupies an energy level. Thus, the electron gas has a strongly polarized response and also highly anisotropic properties. There are several important consequences of this. For one, heat conductivity is strongly direction dependent. Even with the gas being degenerate, there will still be far more efficient heat transfer parallel than perpendicular to the magnetic field lines. This in turn affects the cooling of the neutron star and alters the details of the temperature gradient. In addition, there is also a strong perturbation felt by the electrons still tied to the ions in the crust. This perturbation causes the electron clouds to align with the magnetic field, producing one-dimensional crystals as a dominant crustal structure. There is also an effect on the coupling of the crust and core via the current represented by the electron gas in the mantle of the neutron star.

X. Future Prospects

Several major problems clearly remain, some of which cut to the core of our understanding of the formation and subsequent evolution of neutron stars. No extragalactic supernovas have yet been observed to produce pulsars, and, with the exception of the Crab and Vela pulsars and RCW 103, supernova remnants do not appear to have associated pulsars. Additionally, no point X-ray sources are associated with Cas A (believed to have exploded in the 1670s) or the remnants of SN 1604, SN 1572, or SN 1006. The neutrino burst from SN 1987A in the Large Magellanic Cloud indicated the formation of a neutron star, but as of this writing (early 1988), it has yet to be seen through the remnant.

Supernova models have still to explain the cutoff between stars that will yield stable final cinders and those that continue to collapse to form black holes. The effects of mass loss in the progenitor structure are still unclear.

The origin and evolution of pulsar magnetic fields, while qualitatively explained by several models, have yet to produce a reason for the large obliquities inferred for neutron star magnetic fields. The interaction of these oblique fields with accreting matter in close, mass-exchanging binary systems is still poorly understood. In particular, much work is still required to pull the observations of the different regions of the accretion disk together, a task that involves correlating observations at X-ray, ultraviolet, optical, and radio wavelengths. The discovery of quasi-periodic X-ray sources has begun the probe of neutron star magnetospheric structures, but much is left to do.

The details of the internal properties of neutron stars, especially the equation of state for the core, the solidification of nuclear matter, whether or not pion condensates or other exotic forms of matter occur, and the effects of superstrong magnetic fields on the equation of state, are among the important problems that will need to be understood before a complete model for neutron stars can be determined.

At this juncture, more than twenty years after the discovery of pulsars, one can at least say that neutron stars exist. They are likely to remain the best available physical laboratory for the study of most extreme stable states of matter for some time to come.

Bibliography

Alpar, M. A., and Pines, D. (1985). Superfluidity in neutron stars, *Nature* (London) **316,** 27.

Alpar, M. A., Anderson, P. W., Pines, D., and Shaham, J. (1984). *Astrophys. J.* **276,** 325.

Baym, G., and Petchek, C. (1979). Physics of neutron stars, *Annu. Rev. Astron. Astrophys.* **17,** 415.

Bethe, H. A. (1971). *Annu. Rev. Nuclear Sci.* **21,** 93.

Blandford, R. D., Applegate, J. H., and Hernquist, L. (1983). Thermal origin of neutron star magnetic fields, *M.N.R.A.S.* **204,** 1025.

Canuto, V. (1974). Equations of state at ultra-high densities I, II, *Annu. Rev. Astron. Astrophys.* **12,** 167.

Canuto, V. (1974). Equations of state at ultra-high densities II, *Annu. Rev. Astron. Astrophys.* **13,** 335.

Drechsel, H., Kondo, Y., and Rahe, J., eds. (1987). "Cataclysmic Variables: Recent Multi-frequency Observations and Theoretical Developments. Reidel, Dordrecht, Netherlands.

Joss, P. C., and Rappaport, S. A. (1984). Neutron stars in interacting binary systems, *Annu. Rev. Astron. Astrophys.* **22,** 537.

Manchester, R. N., and Taylor, J. H. (1977). "Pulsars." Freeman, San Francisco, California.

Michel, C. (1982). Theory of pulsar magnetospheres, *Rev. Mod. Phys.* **54,** 1.

Oppenheimer, J. R., and Volkoff, G. M. (1939). On massive neutron cores, *Phys. Rev.* **55,** 374.

Reynolds, S. P., and Stinebring, D. R., eds. (1984). "Green Bank Workshop on Millisecond Pulsars." NRAO, Charlottesville.

Shapiro, S., and Teukolsky, S. (1983). "Black Holes, White Dwarfs and Neutron Stars: The Physics of Compact Objects." Wiley (Interscience), New York.

Taylor, J. H., and Stinebring, D. R. (1986). Recent progress in the understanding of pulsars, *Annu. Rev. Astron. Astrophys.* **24,** 285.

Tsuruta, S. (1986). Neutron stars: current cooling theories and observational results, *Comments Astrophys.* **11,** 151.

Tsuruta, S. (1979). Thermal properties and detectability of neutron stars—I. Cooling and heating of neuron stars, *Phys. Rep.* **56,** 237.

Woosley, S., and Weaver, T. A. (1986). Physics of supernova explosions, *Annu. Rev. Astron. Astrophys.* **24,** 205.

OPTICAL TELESCOPES

L. D. Barr   *National Optical Astronomy Observatories**

GLOSSARY

Airy disc: Central portion of the diffracted image formed by a circular aperture. Contains 84% of the total energy in the diffracted image formed by an unobstructed aperture. Angular diameter = 2.44 λ/D, where λ is wavelength and D the unobstructed aperture diameter. First determined by G. B. Airy in 1835.

Aperture stop: Physical element, usually circular, that limits the light bundle or cone of radiation that an optical system will accept on-axis from the object.

Coherency: Condition existing between two beams of light when their fluctuations are closely correlated.

Diamond turning: Precision-machining process used to shape surfaces in a manner similar to lathe turning. Material is removed from the surface with a shaped diamond tool, hence the name. Accuracies to one microinch ($\frac{1}{40}$th μm) are achievable. Size is limited by the machine, currently about 2 m.

Diffraction-limited: Term applied to a telescope when the size of the Airy disc formed by the telescope exceeds the limit of seeing imposed by the atmosphere or the apparent size of the object itself.

* Operated by the Association of Universities for Research in Astronomy, Inc., under contract with the National Science Foundation.

Diameter-to-thickness ratio: Diameter of the mirror divided by its thickness. Term is generally used to denote the relative stiffness of a mirror blank: 6 : 1 is considered stiff, and greater than 15 : 1 is regarded as flexible.

Effective focal length: Product of the aperture diameter and the focal ratio of the converging light beam at the focal position. For a single optic, the effective focal length (EFL) and the focal length are the same.

Focal ratio (f/ratio):   In a converging light beam, the reciprocal of the convergence angle expressed in radians. The focal length of the focusing optic divided by its aperture size, usually its diameter.

Field of view: Widest angular span measured on the sky that can be imaged distinctly by the optics.

Image quality: Apparent central core size of the observed image, often expressed as an angular image diameter that contains a given percentage of the available energy. Sometimes taken to be the full width at half maximum (FWHM) value of the intensity versus angular radius function. A complete definition of image quality would include measures of all image distortions present, not just its size, but this is frequently difficult to do, hence the approximations.

Infrared: For purposes of this article, wavelength region from about 0.8 to 40 μm.

Optical path distance: Distance traveled by light passing through an optical system between two points along the optical path.

Seeing: Measure of disturbance in the image seen through the atmosphere. Ordinarily expressed as the angular size, in arc seconds, of a point source (a distant star) seen through the atmosphere, that is, the angular size of the blurred source.

Ultraviolet: For purposes of ground-based telescopes, the wavelength region from about 3000 to 4000 Å.

Optical telescopes were devised by European spectacle makers around the year 1608. Within two years, Galileo's prominent usage of the telescope marked the beginning of a new era for astronomy and a proliferation of increasingly powerful telescopes that continues unabated today. Because it extends what the human eye can see, the optical telescope in its most restricted sense is an artificial eye. However, telescopes are not subject to the size limitation, wavelength sensitivities, or storage capabilities of the human eye and have been extended vastly beyond what even the most sensitive eye can accomplish. Properties of astronomical telescopes operating on the ground in the optical/IR spectral wavelength range from 3000 to about 40,000 Å (0.3–40 μm) will be considered. The atmosphere transmits radiation throughout much of this range. Telescopes designed for shorter wavelengths are either UV or X-ray telescopes and for longer wavelengths are in the radio-telescope category. Emphasis is placed on technical aspects of present-day telescopes rather than history.

I. Telescope Size Considerations and Light-Gathering Power

An astronomical telescope works by capturing a sample of light emitted or reflected from a distant source and then converging that light by means of optical elements into an image resembling the original source, but appropriately sized to fit onto a light-sensitive detector (e.g., the human eye, a photographic plate, or a phototube). Figure 1 illustrates the basic telescope elements. It is customary to assume that light from a distant object on the optical axis arrives as a beam of parallel rays sufficiently large to fill the telescope entrance, as shown.

The primary light collector can be a lens, as in Fig. 1, or a curved mirror, in which case the light would be shown arriving from the opposite direction and converging after reflection. The auxiliary optics may take the form of eyepieces or additional lenses and mirrors designed to correct the image or modify the light beam. The nature and arrangement of the optical elements set limits on how efficiently the light is preserved and how faithfully the image resembles the source, both being issues of prime concern for telescope designers.

The sampled light may have traveled at light speed for a short time or for billions of years after leaving the source, which makes the telescope a unique tool for studying how the universe was in both the recent and the distant past. Images may be studied to reveal what the light source looked like, its chemistry, location, relative motion, temperature, mass, and other properties. Collecting light and forming images is usually regarded as a telescope function. Analyzing the images is then done by various instruments designed for that purpose and attached to the telescope. Detectors are normally part of the instrumentation. The following discussion deals with telescopes.

A. TELESCOPE SIZE AND ITS EFFECT ON IMAGES

The size of a telescope ordinarily refers to the diameter, or its approximate equivalent, for the area of the first (primary) image-forming optical element surface illuminated by the source. Thus, a 4M telescope usually signifies one with a 4-m diameter primary optic. This diameter sets a maximum limit on the instantaneous photon flux passing through the image-forming optical train. Some telescopes use flat mirrors to direct light into the telescope (e.g., solar heliostats); however, it is the size of the illuminated portion of the primary imaging optic that sets the size.

The size of a telescope determines its ability to resolve small objects. The Airy disc diameter, generally taken to be the resolution limit for images produced by a telescope, varies inversely with size. The Airy disc also increases linearly with wavelength, which means that one must use larger-sized telescopes to obtain equivalent imaging resolution at longer wavelengths. This is a concern for astronomers wishing to observe objects at infrared (IR) wavelengths and also explains in part why radio telescopes, operating at

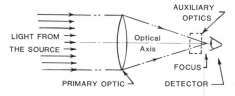

FIG. 1.   Basic telescope elements. Refractive lens could be replaced with a curved reflective mirror.

even longer wavelengths, are so much larger than optical telescopes. (Radio telescopes are more easily built larger because radio wavelengths are much longer and tolerances on the "optics" are easier to meet.)

Telescopes may be used in an interferometric mode to form interference fringes from different portions of the incoming light beam. Considerable information about the source can be derived from these fringes. The separation between portions of the primary optic forming the image (fringes) is referred to as the baseline and sets a limit on fringe resolution. For a telescope with a single, round primary optic, size and maximum baseline are the same. For two telescopes directing their beams together to form a coherent image, the maximum baseline is equal to the maximum distance between light-collecting areas on the two primaries. More commonly, the center-to-center distance would be defined as the baseline, but the distance between any two image-forming areas is also a baseline. Thus, multiple-aperture systems have many baselines.

As telescope size D increases so does the physical size of the image, unless the final focal ratio F_f in the converging beam can be reduced proportionately, that is,

$$\text{final image size} = F_f \cdot D \cdot \theta$$

where θ is the angular size of the source measured on the sky in radians. With large telescopes this can be a matter of importance when trying to match the image to a particular detector or instrument. Even for small optics, achieving focal ratios below about f/1.0 is difficult, which sets a practical limit on image size reduction for a given situation.

Another size-related effect is that larger telescopes look through wider patches of the atmosphere which usually contain light-perturbing turbulent regions that effectively set limits on seeing. Scintillation (twinkling) and image motion are caused by the turbulence. However, within a turbulent region, slowly varying isotropic subregions (also called isoplanatic patches) exist that affect the light more or less uniformly. When a telescope is sized about the same as, or smaller than, a subregion and looks through such a subregion, the instantaneous image improves because it is not affected by turbulence outside the subregion. As the subregions sweep through the telescope's field of view, the image changes in shape and position. Larger telescopes looking through many subregions integrate or combine the effects, which enlarges the

combined image and effectively worsens the seeing. However, these effects diminish with increasing wavelength, which means that larger telescopes observing at IR wavelengths may have better seeing than smaller ones observing in the visible region.

Studies of atmospheric turbulence effects have given rise to the development of special devices to make optical corrections. These are sometimes called rubber mirrors or adaptive optics. An image formed from incoming light is sensed and analyzed for its apparent distortion. That information is used to control an optical element (usually a mirror) that produces an offsetting image distortion in the image-forming optical train. By controlling on a star in the isoplanatic patch with the object to be observed (so that both experience similar turbulence effects), one can, in principle, form corrected images with a large ground-based telescope that are limited only by the telescope, not the atmosphere. In practice, low light levels from stars and the relatively small isoplanatic patch sizes (typically a few arc seconds across) have hampered usage of adaptive optics on stellar telescopes.

B. TELESCOPE CHARACTERISTICS RELATED TO SIZE

At least three general, overlapping categories related to telescope size may be defined:

1. Telescopes small enough to be portable. Sizes usually less than 1 m.
2. Mounted telescopes with monolithic primary optics. Sizes presently range up to 6 m. Virtually all ground-based telescopes used by professional astronomers are in this category.
3. Very large telescopes with multielement primary optics. Proposed sizes range up to 25 m for ground-based telescopes. Only a few multielement telescopes have actually been built.

A fourth category could include telescopes small enough to be launched into earth orbit, but the possibility of an in-space assembly of components makes this distinction unimportant.

One cannot, in a short space, describe all of the telescope styles and features. Nevertheless, as one considers larger and larger telescopes, differences become apparent; and a few generalizations can be postulated.

In the category of small telescopes, less than 1 m, one finds an almost unlimited variety of telescope configurations. There are few major size limitations on materials for optics. Polishing of optics can often be done manually or with the

aid of simple machinery. Mechanical requirements for strength or stiffness are easily met.
Adjustments and pointing can be manually performed or motorized. Weights are modest. Opportunities for uniqueness abound and are often
highly prized. Single-focus operation is typical.
Most the telescopes used by amateur and professional astronomers are in this size range. Figure 2 illustrates a 40-cm telescope used by professional astronomers.

In the 1–2 m size range, a number of differences and limitations arise. Obtaining high-quality refractive optics is expensive in this range
and not practical beyond. Simple three-point
mechanical supports no longer suffice for the optics. The greater resolving-power potential demands higher quality optics and good star-tracking precision. Telescope components are
typically produced on large machine tools. Instrumentation is likely to be used at more than
one focus position. Because of cost, the domain
of the professional astronomer has been
reached.

As size goes above 2 m, new issues arise. The
need to compensate for self-weight deflections
of the telescope becomes increasingly important
to maintaining optical alignment. Flexure in the
structure may affect the bearings and drive
gears. Bearing journals become large enough to

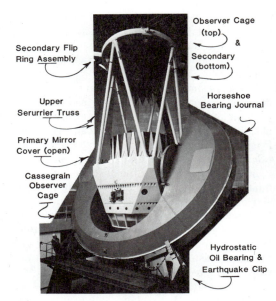

FIG. 3.   Mayall 4-m telescope, with equatorial horseshoe yoke mountings. [Courtesy National Optical Astronomy Observatories, Kitt Peak.]

require special bearing designs, often of the hydrostatic oil variety. The observer may now be
supported by the telescope instead of the other
way around. Support of the primary optics is
more complex, and obtaining primary mirror
blanks becomes a special, expensive task. Automated operation is typical at several focal positions. Star-tracking automatic guiders may be
used to control the telescope drives, augmented
by computer-based pointing correction tables.
Figure 3 illustrates a 4-m telescope with all of
these features.

At 5 m, the Hale Telescope on Mount Palomar
is regarded as near the practical limit for equato-
rial-style mountings (see Section V). Altitude–
azimuth (alt–az) mountings are better suited for
bearing heavy rotating loads and are more compact. With computers the variable drive speeds
required with an alt–az telescope can be managed. Mounting size and the length of the telescope are basic factors in setting the size of the
enclosing building. For technical reasons and
lower cost, the present trend in large telescopes
is toward shorter primary focal lengths and alt–
az mounts. This trend is evident from Table I,
which lists the telescopes 3 m in size or larger
that have been built since about 1950. Also listed
are the major large telescopes proposed for construction in the late 1980s and the 1990s, which
will be discussed in the next section. The largest

FIG. 2.   40-cm telescope on an off-axis equatorial
mount. [Courtesy National Optical Astronomy Observatories, Kitt Peak.]

TABLE I. Telescopes 3 Meters or Larger Built Since 1950

Date completed	Telescope and/or institution	Primary mirror size (m)	Primary focal ratio	Mounting style
1950	Hale telescope, Palomar Observatory	5	3.3	Equatorial horseshoe yoke
1959	Lick Observatory	3	5.0	Equatorial fork
1973	Mayall telescope, Kitt Peak National Observatory	4.0	2.7	Equatorial horseshoe yoke
1974	Cerro Tololo	4.0	2.7	Equatorial horseshoe yoke
1975	Anglo-Australian telescope	3.9	3.3	Equatorial horseshoe yoke
1976	European Southern Observatory (ESO)	3.6	3.0	Equatorial horseshoe yoke
1976	Soviet Special Astrophysical Observatory	6.0	4.0	Alt–az
1979	Infrared Telescope Facility (IRTF)	3.0	2.5	Equatorial English yoke
1979	Canada-France-Hawaii Telescope (CFHT), Hawaii	3.6	3.8	Equatorial horseshoe yoke
1979	Infrared telescope (UKIRT), United Kingdom	3.8	2.5	Equatorial English yoke
1979	Multiple mirror telescope (MMT) Observatory, Mt. Hopkins	4.5[a]	Six 1.8-m[b]	Alt–az at f/2.7
1983	German–Spanish Astronomical Center, Calar Alto	3.5	3.5	Equatorial horseshoe fork
(1986)	Wm. Herschel telescope, La Palma	4.2	2.5	Alt–az
(1987)	European Southern Observatory	3.5	2.2	Alt–az
Proposed	University of Washington, Chicago, Princeton	3.5	1.75	Alt–az
Proposed	University of Texas	7.6	1.8	Alt–az
Proposed	Japanese National Telescope	7.6	2.0	Alt–az
Proposed	Array of separate telescopes, European Southern Observatory	16[a]	Four 8.0-m[b] at f/2	Alt–az
(1991)	Segmented parabolic primary with 36 hexagons, Keck Observatory	10[a]	1.75	Alt–az
Proposed	National New Technology Telescope (NNTT), MMT style	15[a]	Four 7.5-m[b] at f/1.8	Alt–az
Proposed	MMT style, Royal Greenwich Observatory	18[a]	Six 8.0-m[b] at f/1.8	Alt–az
Proposed	Segmented spherical primary with 400 hexagons, USSR	25[a]	2.7	Alt–az

[a] Equivalent circular mirror diameter with equal area.
[b] Number, size, and f/ratio of individual primary mirror.

optical telescope in operation today is the Soviet 6 m, which incorporates a solid, relatively thick (650 mm) primary mirror that had to be made three times in borosilicate glass and finally in a low-expansion material before it was successful. Such difficulty indicates that 6 m may be a practical limit for that style of mirror. New approaches are needed to go beyond.

C. The New Giant Telescopes

The desire for greater light-gathering power and image resolution, especially at IR wavelengths, continues to press astronomers to build telescopes with larger effective apertures. Costs for a given telescope style and imaging performance have historically risen nearly as the primary aperture diameter to the 2.5 power. These factors have given impetus to a number of new technology telescope designs (see Proposed Projects in Table I) that are based on one or more of the approaches discussed in the following. Computer technology plays a strong part in all of these approaches.

1. Extending the Techniques for Making Lightweight Monolithic Mirror Blanks

Sizes up to about 8 m are considered feasible, although the Soviet 6 m is the largest telescope mirror produced before 1985. Further discussion on blank fabrication methods is provided in Section III. Supporting such large mirrors to form good images will be difficult without some active control of the surface figure and thermal conditions in the mirror blank.

Several American universities and Japan are planning telescopes of this variety.

2. Making a Large Mirror from Smaller Segments

Also known as segmented mirror telescope, or SMT. In principle, no limit exists for the size of a mosaic of mirror segments that functions optically as a close approximation to a monolithic mirror. For coherency each segment must be precisely and continuously positioned with respect to its neighbors by means of position sensors and actuators built into its support. The segments may be hexagonal, wedge shaped, or other to avoid large gaps between segments. Practical limits arise from support structure resonances and cumulative errors of the segment positioning system. Manufacturing and testing the segments require special methods since each is likely to be a different off-axis optic that lacks a local axis of symmetry but must have a common focus with all the other segments.

The University of California and the California Institute of Technology have adopted this approach for their Keck Observatory ten-meter telescope, expected to be completed in 1991, which will look similar to the SMT in Fig. 4. The USSR also has announced plans for a 25-m

FIG. 4. Segmented mirror telescope (SMT) concept adopted for the Ten-Meter Telescope (TMT) at Keck Observatory (Mauna Kea, Hawaii) and considered for the 15-m National New Technology Telescope (NNTT). The NNTT model is shown with a model of the Mayall 4-m telescope to the same scale. [Courtesy of the National Optical Astronomy Observatories, Kitt Peak.]

SMT utilizing a spherical primary to avoid the problems of making aspheric segments.

3. Combining the Light from an Array of Telescopes

Several methods may be considered:

1. Electronic combination after the light has been received by detectors at separate telescopes. Image properties will be those due to the separate telescopes, and coherent combining is not presently possible. Strictly speaking, this is an instrumental technique and will not be considered further.

2. Optical combination at a single, final focus of light received at separately mounted telescopes. To maintain coherency between separate light beams, one must equalize the optical path distance (OPD) between the source and the final focus for all telescopes, a difficult condition to meet if telescopes are widely separated.

3. Placing the array of telescopes on a common mounting with a means for optically combining the separate light beams. All OPDs can be equal (theoretically), thus requiring only modest error correction to obtain coherency between telescopes. This approach is known as the multiple mirror telescope (MMT).

The simplest array of separately mounted individual telescopes is an arrangement of two on a northsouth (NS) baseline with an adjustable, combined focus between them (the OPD changes occur slowly with this arrangement when observing at or near the meridian). Labeyrie pioneered this design in the 1970s at Centre D'Etudes et de Recherches Géodynamiques et Astronomiques (CERGA) in France, where he used two 25-cm telescopes on a NS variable baseline of up to 35 m to measure successfully numerous stellar diameters and binary star separations, thereby showing that coherent beam combination and the angular resolution corresponding to a long telescope baseline could be obtained. Other schemes for using arrays of separate telescopes on different baselines all require movable optics in the optical path between the telescopes to satisfy the coherency conditions, and so far none has been successfully built. However, the European Southern Observatory is considering construction of an in-line array of four 8-m telescopes with an "optical trombone" arrangement for equalizing OPDs.

The MMT configuration was first used by the Smithsonian Astrophysical Observatory (SAO) and the University of Arizona (UA). The SAO/ UA MMT on Mount Hopkins uses six 1.8-m image-forming telescopes arrayed in a circle around a central axis. Six images are brought to a central combined focus on the central axis, where they may be incoherently stacked, coherently combined, or used separately. The effective baseline for angular resolution (i.e., the maximum separation between reflecting areas) is 6.9 m, and the combined light-gathering power is equivalent to a single 4.5-m diameter mirror. Figure 5 shows a version of the MMT that has been adopted for the 15-m National New Technology Telescope (NNTT), where it is planned to use four 7.5-m mirrors with an angular resolution baseline of about 21 m. The NNTT design includes exchangeable secondary optics modules that enable switching from the combined beam mode to a mode allowing usage of the four telescopes for individual instrumentation.

II. Optical Configurations

Light entering the telescope is redirected at each optical element surface until it reaches the focal region where the images are most distinct. The light-sensitive detector is customarily located in an instrument mounted at the focal region. By interchanging optics, one can create more than one focus condition; this is commonly done in large telescopes to provide places to mount additional instruments or to produce different image scales. The arrangement of optics and focal positions largely determines the required mechanical support configuration and how the telescope will be used.

The early telescopes depended solely upon the refractive power of curved transparent glass lenses to redirect the light. In general, these telescopes were plagued by chromatic aberration (rainbow images) until the invention in 1752 of achromatic lenses, which are still used today in improved forms. Curved reflective surfaces (i.e., mirrors) were developed after refractors but were not as useful until highly reflective metal coatings could be applied onto glass substrates. Today, mirrors are more widely used than lenses and can generally be used to produce the same optical effects; they can be made in larger sizes, and they are without chromatic aberration. These are still the only two means used to form images in optical telescopes. Accordingly, telescopes may be refractive, reflective, or catadioptric, which is the combination of both.

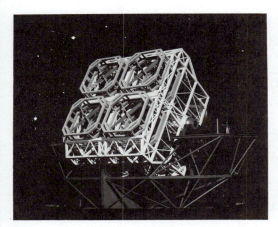

FIG. 5. Planned 15-m National New Technology Telescope based on the multiple mirror telescope (MMT) concept. Four 7.5-m image-forming telescopes will be operated separately or as a square coherent array with a 21-m diagonal baseline and light-collecting area equivalent to a 15-m diameter mirror. Primary mirrors will be lightweight honeycombs, and the top end optics are in modules that can be rapidly interchanged. The combined beam configuration is shown. [Courtesy of the National Optical Astronomy Observatories, Kitt Peak.]

A. BASIC OPTICAL CONFIGURATIONS: SINGLE AND MULTIELEMENT

The telescope designer must specify the type, number, and location of the optical elements needed to form the desired image. The basic choices involve material selections and the shapes of the optical element surfaces. Commonly used surfaces are flats, spheres, paraboloids, ellipsoids, hyperboloids, and toroidal figures of revolution.

The optical axis is the imaginary axis around which the optical figures of revolution are rotated. Light entering the telescope parallel to

TYPE	PRIMARY OPTIC	SECONDARY OPTIC	CONFIGURATION 1-PRIMARY 2-SECONDARY 3-EYEPIECES/CORRECTORS 4-FOCUS
KEPLERIAN GALILEAN (if refractive)	SPHERE or PARABOLA	NONE	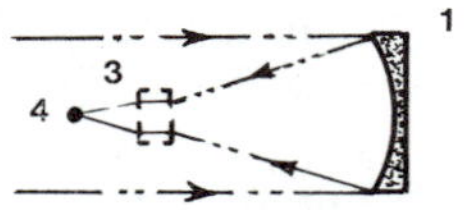
HERSCHELIAN	OFF-AXIS PARABOLA	NONE	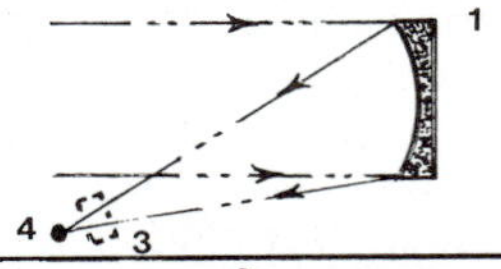
NEWTONIAN	PARABOLA	DIAGONAL FLAT	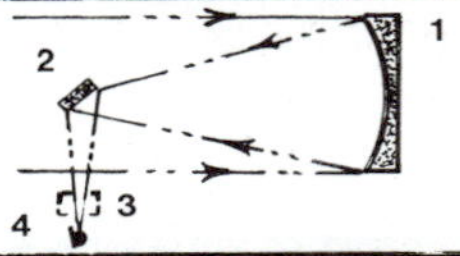
GREGORIAN	PARABOLA	ELLIPSE	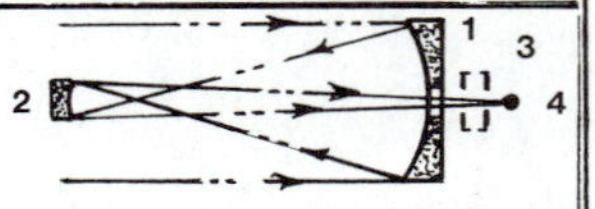
MERSENNE	PARABOLA	PARABOLA	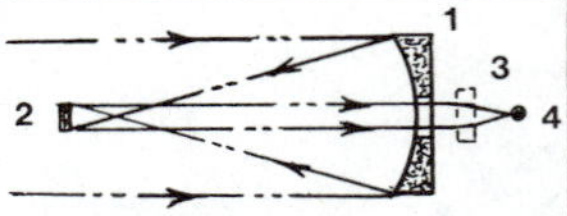
CASSEGRAIN	PARABOLA	HYPERBOLA	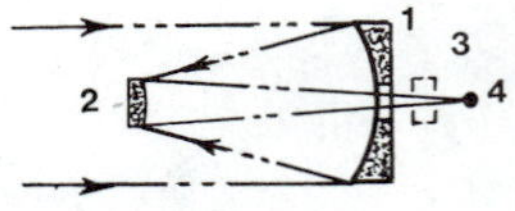
RITCHEY-CHRÉTIEN	MODIFIED PARABOLA	MODIFIED HYPERBOLA	
DALL-KIRKHAM	ELLIPSE	SPHERE	
SCHMIDT	ASPHERIC REFRACTOR	SPHERE	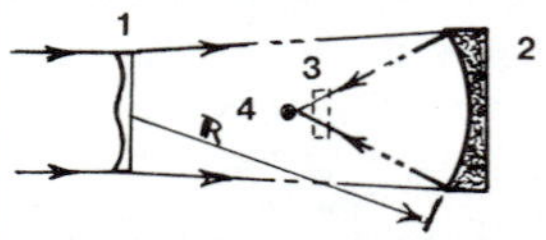
BOUWERS-MAKSUTOV	REFRACTIVE MENISCUS	SPHERE	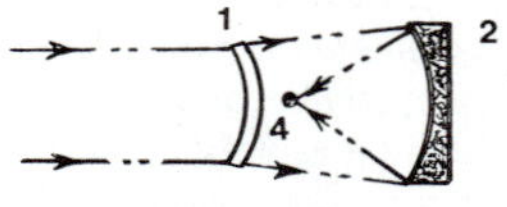

FIG. 6.   Basic optical configurations for telescopes.

this axis forms the on-axis (or zero-field) image directly on the optical axis at the focal region. The field of view (FOV) for the telescope is the widest angular span measured on the sky that can be imaged distinctly by the optics.

In principle, a telescope can operate with just one image-forming optic (i.e., at prime focus), but without additional corrector optics, the FOV is quite restricted. If the telescope is a one-element reflector, the prime focus and hence the instrument/observer are in the line of sight. For large telescopes (i.e., >3 m) this may be used to advantage, but more commonly the light beam is diverted to one side (Newtonian) or is reflected back along the line of sight by means of a secondary optic to a more convenient focus position. Figure 6 illustrates the optical configurations most commonly used in reflector telescopes. In principle, the reflectors shown in Fig. 6 could be replaced with refractors to produce the same optical effects. However, the physical arrangement of optics would have to be changed.

In practice, one tries to make the large optics as simple as possible and to form good images with the fewest elements. Other factors influencing the configuration include the following:

1. Simplifying optical fabrication. Spherical surfaces are generally the easiest to make and test. Nonsymmetric aspherics are the opposite extreme.

2. Element-to-element position control, which the telescope structure must provide. Tolerances become tighter as the focal ratio goes down.

3. Access to the focal region for viewing or mounting instrumentation. Trapped foci (e.g., the Schmidt) are more difficult to reach.

4. Compactness, which generally aids mechanical stiffness.

5. Reducing the number of surfaces to minimize light absorption and scattering losses.

Analyzing telescope optical systems requires a choice of method. The geometric optics method treats the incoming light as a bundle of rays that pass through the system while being governed by the laws of refraction and reflection. Ray-tracing methods based on geometric optics are commonly used to generate spot diagrams of the ray positions in the final image, as illustrated in Fig. 7a. More rigorous analysis based on diffraction theory is done by treating the incoming light as a continuous wave and examining its interaction with the optical system.

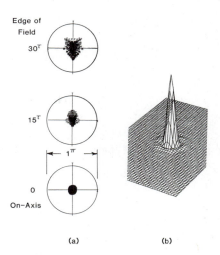

FIG. 7. Examples of image analysis methods. (a) Typical spot diagrams of well-corrected images in the focal plane of a 1 deg field. Each image represents the ray bundle at that field location. (b) Computer-generated 3-dimensional plot of intensity in a perfect image that has been diffracted by a circular aperture.

Figure 7b is a computer-generated plot of light intensity in a diffracted image formed by a circular aperture. The central peak represents the Airy disc. For further detail on the use of these methods the reader is referred to texts on optics design.

B. Wide Field Considerations

Modern ground-based telescopes are usually designed to resolve images in the 0.25–1.0-arcsec range and to have FOV from a few arc minutes to about one degree. In general, distortions or aberrations due to the telescope optics exist in the images and are worse for larger field angles. Table II lists the basic types. Space-based telescopes (e.g., the Hubble telescope) can be built to resolve images in the 0.1 arcsec region because atmospheric seeing effects are absent, but compensation must still be made for optical aberrations. Aberrations can also arise from nonideal placement of the optical surfaces (i.e., position errors) and from nonuniform conditions in the line of sight.

To correct distortions produced by optics, the designer frequently tries to cancel aberrations produced at one surface by those produced at another. Extra optical surfaces may be introduced for just this purpose. Ingenious optical corrector designs involving both refractors and reflectors have resulted from this practice.

TABLE II. Basic Image Aberrations Occurring in Telescopes

Type	Condition
Spherical aberration[a]	Light focuses at different places along the optical axis as a function of radial position in the aperture.
Coma[a]	Image size (magnification) varies with radial position in the focal region. Off-axis flaring.
Field curvature[a]	Off-axis images are not focused on the ideal surface, usually a plane.
Astigmatism[a]	Light focuses at different places along the optical axis as a function of angular position in the aperture.
Distortion[a]	Focused off-axis image is closer to or further from the optical axis than intended.
Chromatic aberration	Shift in the focused image position as a function of wavelength.

[a] Also known as Seidel aberrations.

Many of these designs are described in texts on optics under the originators' names (e.g., Ross, Baker, Wynne, Shulte, and Meinel). It is possible, however, to cancel some field aberrations by modifying the principal optical surfaces or by using basic shapes in special combination. For example, the Ritchey–Crétian telescope design for a wide FOV reduces coma by modifying the primary and secondary surfaces of a Cassegrain telescope. The Mersenne telescope cancels aberrations from the primary (a parabola) with the secondary (another parabola).

A spherical mirror with the aperture stop set at its center of curvature has no specific optical axis and forms equally good images everywhere in the field. Images of distant objects have spherical aberrations, however, and the focal region is curved. The Schmidt telescope compensates for most of the spherical aberration by means of an aspheric refractor at the center of curvature. The Maksutov telescope introduces an offsetting spherical aberration by means of a spherical meniscus refractor. Numerous variations of this approach have been devised yielding well-corrected images in field sizes of 10 deg and more. However, large fields may have other problems.

During a long observing period, images formed by a telescope with a very large FOV (i.e., ~1 deg or more) are affected differently across the field by differential refraction effects caused by the atmosphere. Color dispersion effects (i.e., chromatic blurring) effectively enlarge the images as the telescope looks through an increasing amount of atmosphere. Differential image motion also occurs, varying as a function of position in the FOV, length of observation, and telescope pointing angle. Partial chromatic correction can be made by inserting a pair of separately rotatable prisms, called Risley prisms, ahead of the focal position. Even with chromatic correction, however, images are noticeably elongated (>0.5 arcsec) at the edge of a 5-deg field compared with the on-axis image after a continuous observing period of a few hours.

C. INSTRUMENTAL CONSIDERATIONS: DETECTOR MATCHING AND BAFFLING

At any focal position, the distance measured along the optical axis within which the images remain acceptably defined is referred to as the depth of focus. The detector should be adjusted to the most sharply focused position in this region; however, it is more common to adjust the focus (e.g., by moving the secondary) to sharpen the images at the detector. Certain instruments containing reimaging optics (e.g., spectrographs) require only that the image be formed at the entrance to the instrument (e.g., at a slit or aperture plate).

To view all or part of the focused FOV, it is common to insert a field mirror into the beam at 45 deg to divert the desired portion of the field out to an eyepiece or a TV monitoring camera. The undiverted light to be observed passes on to the instrument. The diverted light can be used for guiding purposes or to make different observations.

The angular size of the focused image (i.e., the image scale) and its sensitivity to defocusing changes is governed by the effective focal length (EFL) of the optical system that formed the image:

$$\text{image scale} = \frac{1}{\text{EFL}}$$
$$= \frac{\text{radians on the sky}}{\text{length in the focal plane}}$$

Large EFL values produce comfortable focal depths, but the image sizes are relatively large. The physical size of the detector may place a limit on the FOV that can be accommodated for a given scale, hence a potential need for a different focal position or reimaging optics.

Light baffles and aperture stops are important but frequently neglected aspects of telescope design. It is common to place a light baffle just ahead of the primary to block out unwanted edge effects, in which case the baffle may become the aperture stop. In some cases, the primary surface may be reimaged further along the optical path where a light baffle can be located. This may be done either to reduce the size of the required baffle or to locate it advantageously in a controlled environment (e.g., at cryogenic temperature to reduce thermal radiation effects). In other cases, the aperture may be set by undersizing one of the optical elements that is further along the optical path. An example is an IR-optimized telescope, where the secondary is made undersized to ensure that the detector cannot see past the edge of the primary mirror. The effective light-gathering power of the telescope can be significantly reduced under these conditions.

Obstructions in the light path further reduce light-gathering power. The most common obstructions are the secondary and auxiliary optics along with the mechanical struts that support them. It is common for 10–20% of the aperture to be obstructed in this manner. Depending upon the telescope style and the instrument location, it may be necessary to use a light baffle to prevent the detector from seeing unwanted radiation. For example, a detector at the Cassegrain focus of a two-mirror telescope can see unfocused light from stars directly past the perimeter of the secondary mirror unless obstructing baffles are provided. The size and location of these baffles are generally determined empirically by ray-tracing, and it is a fact that larger FOV requires larger baffles that mean more obstruction. Thus, one cannot truly determine telescope light-gathering power until the usage is considered.

III. Telescope Optics

Telescope performance depends fundamentally upon the quality of optics, especially the surface figures. That quality in most telescopes is a compromise between what the optical designer has specified, what the glassmakers and opticians can make, and how well the mechanical supporting structures perform. Especially for larger sizes, one should specify the optics and their supports at the same time. How the optics are to be tested should also be considered since the final figure corrections are almost always guided by test results. Optical figuring methods today range from simple manual lapping processes to sophisticated computer-controlled polishers (CCP) and direct machining using diamond tools. The remarks to follow are only a summary of a complex technical field.

A. Refractive Optics

Light passing through the surface of a refractive optic is changed in direction according to Snell's law of refraction:

$$\eta_1 \sin \theta_1 = \eta_2 \sin \theta_2$$

where η_1 and η_2 are the refractive indices of the materials on either side of the surface and θ_1 and θ_2 the angles with respect to the surface normal of incidence and refraction. Producing a satisfactory refractor is, therefore, done by obtaining transparent glass or another transmitting material with acceptable physical uniformity and then accurately shaping the surfaces through which light passes. Sometimes the optician can alter the surface to compensate for nonuniform refractive properties. Losses ordinarily occur at the surface due to scattering and also to the change in refractive index. Antireflection coatings can be applied to reduce surface transmission losses, but at the cost of restricting transmission to a specific wavelength range. Further losses occur internally by additional scattering and absorption.

Manufacturing methods for refractive optics are similar to those for reflectors.

B. Reflective Optics

Light reflecting off a surface is governed by the law of reflection:

$$\theta_1 = -\theta_2$$

where θ_1 and θ_2 are the angles of incidence and reflection, respectively. Controlling the slope at all points on the reflecting surface is, therefore, the means for controlling where the light is directed. Furthermore, any part of the surface that is out of its proper position, measured along the light path, introduces a change at that part of the reflected wavefront (i.e., a phase change) equal to twice the magnitude of the position error.

Surface accuracy is thus the prime consideration in making reflective optics. Achieving high reflectivity after that is usually done by applying a reflective coating.

The most common manufacturing method is to rough-machine or grind the mirror blank surface and then progressively refine the surface with abrasive laps. Certain kinds of mirrors, especially metal ones, may be diamond turned. In this case, the accuracy of the optical surface may be governed by the turning machine, whereas ordinarily the accuracy limits are imposed principally by the optical test methods and the skill of the optician. High-quality telescope mirrors are typically polished so that most of the reflected light (>80%) is concentrated in an image that is equivalent to the Airy disc or the seeing limit, whichever is larger.

C. Materials for Optics

Essential material requirements for all optical elements are related to their surfaces. One must be able to polish or machine the surfaces accurately, and afterward the blank should not distort uncontrollably. Residual stresses and unstable alloys are sources of dimensional instability to be avoided. Stresses also cause birefringence in refractors.

1. Refractors

Refractors should transmit light efficiently and uniformly throughout the operating wavelength region. However, there is no single material that transmits efficiently from 0.3 to 40 μm. One typically chooses different materials for the UV, near-IR (to $\simeq 2$ μm), and far-IR regions. Glassmakers can control the index of refraction to about one part in 10^6, but only in relatively small blanks (<50-cm diameter). In larger sizes, index variations and inclusions limit the availability of good-quality refractor blanks to sizes less than 2-m diameter.

2. Reflectors

The working part of a reflector optic is usually the thin metallic layer, 1000–2000 Å thick, that reflects the light. An evaporated layer of aluminum, silver, or gold is most commonly used for this purpose, and it obviously must be uniform and adhere well to the substrate. Most of the work in making a reflector, however, is in producing the uncoated substrate or mirror blank. The choice of material and the substrate configuration are critically important to the ultimate reflector performance.

Reflector blanks, especially large ones, require special measures to maintain dimensional (surface) stability. The blank must be adequately supported to retain its shape under varying gravitational loads. It must also be stable during normal temperature cycles, and it should not heat the air in front of the mirror because that causes thermal turbulence and worsens the seeing. A mirror-to-air temperature difference less than about 0.5°C is usually acceptable. The support problem is a mechanical design consideration. Thermal stability may be approached in any of several ways:

1. Use materials with low coefficients of thermal expansion (CTE).
2. Lightweight the blank to reduce the mass and enhance its ability to reach thermal equilibrium quickly (i.e., by using thin sections, pocketed blanks, etc.). Machineability or formability of the material is important for this purpose.
3. Use materials with high thermal conductivity (e.g., aluminum, copper, or steel).
4. Use active controls for temperature or to correct for thermally induced distortion. Elastic materials with repeatable flexure characteristics are desirable. Provisions in the blank for good ventilation may be necessary.

Materials with low CTE values include borosilicate glass, fused silica, quartz, and ceramic composites. Multiple-phase (also called binary) materials have been developed that exhibit near-zero CTE values, obtained by offsetting the positive CTE contribution of one phase with the negative CTE contribution from another. Zerodur® made by Schott Glaswerke, ULE® by Corning Glass Works, and epoxy–carbon fiber composites are examples of multiple-phase materials having near-zero CTE over some range of operating temperature. Fiber composites usually require a fiber-free overlayer that can be polished satisfactorily.

One usually considers a lightweight mirror blank to reduce costs or to improve thermal control. This is particularly true for large telescopes. Reducing mirror weight often produces a net savings in overall telescope cost even if the lightweighted mirror blank is more expensive than a corresponding solid blank. The important initial step is to make the reflecting substrate or faceplate as thin as possible, allowing for polishing tool pressures and other external forces. One then has three hypothetical design options:

1. Devise a way to support just the thin monolithic faceplate. This approach works best

for small blanks. Large, thin blanks require complex supports and, possibly, a means to monitor the surface shape for active control purposes.

2. Divide the faceplate into small, relatively rigid segments and devise a support for each segment. Segment position sensing and control is required: a sophisticated technical task.

3. Reinforce the faceplate with a gridwork of ribs or struts, possibly connected to a backplate, to create a sandwichlike structure. One may create this kind of structure by fusing or bonding smaller pieces, by casting into a mold, or by machining away material from a solid block.

All of these approaches have been used to make lightweight reflector blanks. Making thin glass faceplates up to about 8 m is considered feasible by fusing together smaller pieces. Titanium silicate, fused silica, quartz, and borosilicate glass are candidate materials. Castings of borosilicate glass up to 6 m (e.g., the Palomar 5-m mirror) have been produced; 8 m is considered feasible. Structured (i.e., ribbed) fused silica and titanium silicate mirrors up to 2.3 m have been produced (e.g., the Hubble Space Telescope mirror); up to 4 m is considered feasible.

Metals may also be used for lightweight mirrors but have not been widely used for large telescopes because of long-term dimensional (surface) changes. The advent of active surface control technology may alter this situation in the future. Most metals polish poorly, but this can be overcome by depositing a nickel layer on the surface to be polished. Most nonferrous metals can also be figured on diamond-turning machines; however, size is limited presently to about 2 m.

IV. Spectral Region Optimization: Ground-Based Telescopes

The optical/IR window of the atmosphere from 0.3 to 35 μm is sufficiently broad that special telescope features are needed for good performance in certain wavelength regions. Notably, these are needed for the UV region, less than about 0.4 μm and the thermal IR region centered around 10 μm. In the UV, it is difficult to maintain high efficiency because of absorption losses in the optics. When observing in the IR region, one must cope with blackbody radiation emitted at IR wavelengths by parts of the telescope in the light path, as well as by the atmosphere. Distinguishing a faint distant IR source from this nearby unwanted background

radiation requires special techniques. Using such techniques, it is common for astronomers to use ground-based telescopes to observe IR sources that are more than a million times fainter than the IR emission of the atmosphere through which the source must be discerned.

A. UV REGION OPTIMIZATION

The obvious optimizing step for the UV region is to put the telescope into space. If the telescope is ground-based, however, one good defense against UV light losses is to use freshly coated aluminum mirror surfaces. Reflectivity values in excess of 90% can be obtained from freshly coated aluminum, but this value rapidly diminishes as the surface oxide layer develops. Protective coatings such as sapphire (Al_2O_3) or magnesium fluoride (MgF_2) can be used to inhibit oxidations. Also, multilayer coatings can be applied to the surfaces of all optics to maximize UV throughput. However, these coatings greatly diminish throughput at longer wavelengths, which leads to the tactic of mounting two or more sets of optics on turrets, each set being coated for a particular wavelength region. The desired set is rotated into place when needed. This tactic obviously works best for small optics, not the primary.

Many refractive optics materials absorb strongly in the UV. Fused silica is good low-absorbing material. If optics are cemented together, the spectral transmission of the cement should be tested. Balsam cements are to be avoided. [See ASTRONOMY, ULTRAVIOLET SPACE.]

B. IR REGION OPTIMIZATIONS

One cardinal rule for IR optimization is to minimize the number and sizes of emitting sources that can be seen by the detector. This includes mechanical hardware such as secondary support structures and baffles, as well as seemingly empty spaces such as the central hole in the primary mirror. All of these emit black body radiation corresponding to their temperatures. [See ASTRONOMY, INFRARED.]

Since the detector obviously must see the optical surfaces, another cardinal rule is to reduce the emissivity of these surfaces with a highly reflective coating. If a coating is 98% reflective, it emits only 2% of the blackbody radiation that would otherwise occur if the surface were totally nonreflective. If possible, the detector should see only reflective surfaces, and these should be receiving radiation only from the sky or other reflective surfaces in the optical train. Objects that must remain in the line of sight can also be advantageously reflective provided that they are not looking at other IR-emitting objects that could send the reflected radiation into the main beam.

Achieving these goals may require one or more of the following special telescope features:

1. Using an exchangeable secondary support structure. This enables elimination of oversize secondaries and baffles that might be needed for other kinds of observation.

2. Using the secondary mirror as the aperture stop and making it sufficiently undersized that it cannot see past the rim of the primary mirror.

3. Putting all of the secondary support (except the struts) behind the mirror so that none of the hardware is visible to the detector.

4. Placing a specially shaped (e.g., conical) reflective plug at the center of the secondary to disperse radiation emitted from the central hole region of the primary mirror.

A basic technique for ground-based IR observing is that of background subtraction. This involves alternating the pointing direction of the telescope between the object (thus generating an object-plus-background signal from the detector) and a nearby patch of sky that has no object (thus generating a background-only signal). Signal subtraction then eliminates the background signal common to both sky regions. Methods for alternating between positions include (1) driving the telescope between the two positions, usually at rates below 0.1 Hz, (2) wobbling the secondary mirror at rates between 10 and 50 Hz (called chopping), and (3) using focal plane modulators such as rotating aperture plates (called focal plane chopping).

V. Mechanical Configurations

The mechanical portion of a mounted, ground-based telescope must support the optics to the required precision, point to and track the object being observed, and support the instrumentation in accessible positions. It is customary to distinguish between the telescope (or tube), which usually supports the imaging optics, and the mounting, which points the telescope and includes the drives and bearings. Tracking motion is usually accomplished by the mounting.

Most telescope mountings incorporate two, and occasionally three, axes of rotation to en-

able pointing the telescope at the object to be observed and then tracking it to keep it centered steadily in the FOV. In general, the rotating mass is carefully balanced around each axis to minimize driving forces and the location of each rotation axis is chosen to minimize the need for extra counterweights.

Figure 8 illustrates the basic mounting styles that are discussed in the next section.

A. Mounting Designs

A hand-held telescope is supported and pointed by the user. The user is the mounting in this case. The mechanical mounting for a telescope performs essentially the same function, except that a mechanical mounting can support heavier loads and track the object more smoothly. As a rule of thumb, short-term tracking errors in high-quality telescopes are less than 10% of the smallest resolved object that can be observed with the telescope. Smoothness of rotation is important for long-term observations (i.e., no sudden movements) which mandates the use of high-quality bearings. Pressurized oil-film bearings (hydrostatics) are used in large telescopes for this reason.

Telescope drives range widely in style. The chief requirements are smoothness, accuracy, and the ability to move the telescope rapidly for pointing purposes (i.e., slewing) or slowly for

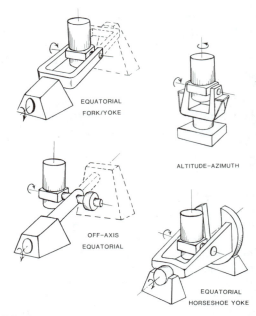

FIG. 8. Basic telescope mounting styles in popular use. Numerous variants on each style are in existence.

tracking (i.e., at one revolution/day or less). Electric motor driven traction rollers, worm gears, or variants on spur gears are most commonly used. Position measuring devices (encoders) are often used to sense telescope pointing and to provide input data for automatic drive controls. Adjustments in tracking rates or pointing are accomplished either by manual control from the observer or, possibly, by star-tracking automatic guiding devices. Pointing corrections may also be based on data stored in a computer from mounting flexure and driving-error calibrations done at an earlier time. Telescope pointing accuracies to about 1 arcsec are currently possible with such corrections. Once the object is located in the FOV, the ability to track accurately is the most important consideration.

1. Equatorial Mounts

Astronomical telescopes ordinarily are used to observe stars and other objects at such great distances that they would appear stationary during an observation period if the earth did not rotate. Accordingly, the simplest telescope tracking motion is one that offsets the earth's rotation with respect to "fixed" stars (i.e., sidereal rate) and is done about a single axis parallel to the earth's north–south (N–S) polar axis. Equatorial mountings are those that have one axis of rotation (i.e., the polar or right ascension axis) set parallel to the earth's N–S axis. This axis is tilted toward the local horizontal plane (i.e., the ground) at an angle equal to local latitude. A rotatable cross-axis (also called declination axis) is needed for initial pointing and guiding corrections, but the telescope does not rotate continuously around this axis while tracking.

The varieties of equatorial mountings are limited only by the designer's imagination. The basic varieties, however, are the following:

1. Those that mount the tube to one side of the polar axle and use a counterweight on the opposite side to maintain balance. For an example, see Fig. 2. These are sometimes called off-axis or asymmetric mounts. It is also possible to mount a second telescope in place of the counterweight.

2. Those that support the tube on two sides in a balanced way to eliminate the need for a heavy counterweight. Yokes and forks are most commonly used, especially for larger telescopes. These are sometimes called symmetric mounts. For an example, see Fig. 3 which shows a horseshoe yoke mount.

2. Other Mounting Styles

The alt–az mounting is configured around a vertical (azimuth) axis of rotation and a horizontal cross-axis (the altitude or elevation axis). The altitude–altitude (alt–alt) mounting, not widely used, operates around a horizontal axis and a cross-axis that is horizontal when the telescope points at the meridian and is tilted otherwise (similar to an English yoke with its polar axis made horizontal). With either of these styles, because neither axis is parallel to the earth's rotation, it is necessary to drive both axes at variable rates to track a distant object. Furthermore, the FOV appears to rotate at the focal region, which often necessitates a derotating instrument mounting mechanism, also moving at a variable rate. These factors inhibited the use of these mountings, except for manually guided telescopes, until the advent of computers on telescopes. Computers enable second-to-second calculation of the drive rates, which is required for accurate tracking. The alt–az configuration cannot track an object passing through the local zenith because the azimuth drive rate theoretically becomes infinite at that point. In practice, alt–az telescopes are operated to within about 1 deg of zenith.

The famous Herschel 20-ft telescope, built in England in 1783, was the first large alt–az telescope. Very few were built after that, but the trend today is toward alt–az mounts (see Section I,B and Table I). The ability to support the main azimuth bearing with a solid horizontal foundation is advantageous, as is the fact that the altitude axis bearings do not change in gravity orientation. These are important considerations when bearing loads of hundreds of tons must be accommodated. The alt–alt mounting is not as suitable for carrying heavy loads because the cross-axis is usually tilted with respect to gravity.

B. TELESCOPE TUBES AND INSTRUMENT CONSIDERATIONS

Design of the tube begins with the optical configuration. Tube structures are designed to maintain the optics in alignment, either by being stiff enough to prevent excessive deflections or by deflecting in ways that maintain the optics in the correct relative position. The well-known Serrurier truss first used on the Palomar 5-m telescope is a much-copied example of the latter (see Fig. 9). The tube structure is normally used to support the instruments at the focal positions, sometimes along with automatic guiders, field

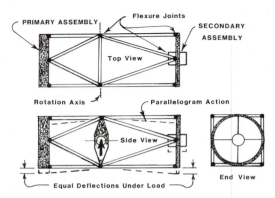

FIG. 9.   Serurrier truss used to maintain primary-to-secondary alignment as the tube rotates. Equal deflections and parallelogram action at both ends keep the optics parallel and equidistant from the original optical axis. Similar flexure is designed into most large telescope tube structures.

viewing TV monitors, calibration devices, and field de-rotators.

Focal positions (i.e., instrument mount locations) on the tube obviously move as the telescope points and tracks, which can be a problem for instruments at those locations that work poorly in a varying gravity environment. In those cases, one can divert the optical beam out of the telescope tube along the cross-axis to a position on the mounting or even outside the mounting. Flat mirrors are normally used for this purpose. To reach a constant-gravity position with an equatorial mounting, one must use several mirrors to bring the converging beam out: first along the declination axis, then the polar axis, and finally to a focus off the mounting. This is known as the coudé focus and is commonly used to bring light to spectrographs that are too large to mount on the telescope tube.

One can reach a constant-gravity focus (instrument location) on an alt–az telescope by simply diverting the beam along the altitude cross-axis to the mounting structure that supports the tube. This is called the Nasmyth focus after its Scottish inventor. The instrument rides the mounting as it rotates in azimuth but does not experience a change in gravity direction.

VI.   Considerations of Usage and Location

Considering the precision built into most optical telescopes, one would expect them to be sheltered carefully. In practice, most telescopes must operate on high mountains, in the dark,

and in unheated enclosures opened wide to the night sky and the prevailing wind. Under these conditions, it is not unusual to find dust or dew on the optics, a certain amount of wind-induced telescope oscillation, and insects crawling into the equipment. Certain insects flying through the light path can produce a noticeable amount of IR radiation. Observer comforts at the telescope at minimal.

In designing a telescope, one should consider its usage and its environment. A few general remarks in this direction are provided in the following sections.

A. SEEING CONDITIONS

The seeing allowed by the atmosphere above the telescope is beyond ordinary control. Compensation may be possible as discussed in Section I,A. but the choice of site largely determines how good the imaging is. Seeing conditions in the region of 0.25 arcsec or less have been measured at certain locations, but more typically, good seeing is in the 0.5–1.0 arcsec range. Beyond 2–3 arcsec, seeing is considered poor. To the extent possible, one should build the telescope to produce images equal to or better than the best anticipated seeing conditions.

Locating the telescope at high altitudes usually reduces the amount of atmosphere and water vapor that is in the line of sight (important for IR astronomy), however, the number of clear nights and the locally produced thermal turbulence should also be considered. In many locations, a cool air layer forms at night near the ground which can be disturbed by the wind and blown through the line of sight. In other cases, warm air from nearby sources can be blown through the line of sight. In either case, telescope seeing is worsened.

Other seeing disturbances can originate inside the telescope enclosure. Any source of heat (including observers) is a potential seeing disturbance. Also, any surface that looks at the night sky, and hence is cooled by radiative exchange, may be a source of cooled air that can disturb seeing if it falls through the line of sight. If possible, it is desirable to allow the telescope enclosure to be flushed out by the wind to eliminate layers and pockets of air of different temperatures. Some telescope buildings have been equipped with air blowers to aid in the process, but dumping the air well away from the building has not always been possible even though it should be done.

The study of atmospheric seeing has become a relatively advanced science, and the telescope builder is well advised to consult the experts in choosing a site or designing an enclosure. Having chosen a site, one may be guided by the truism that seeing seldom improves by disturbing Mother Nature.

B. NIGHTTIME VERSUS DAYTIME USAGE

With the advent of IR astronomy, optical telescopes began to be used both day and night because the sky radiation background is only slightly worse at IR wavelengths during the day compared with night. During the day it is much harder to find guide stars, and the telescope must often point blindly (and hence, more accurately) at the objects to be observed; but much useful data can be obtained. Some problems arise from this practice, however.

A major purpose of the telescope enclosure, other than windscreening, is to keep the telescope as close as possible to the nighttime temperature during the day so that it can equalize more rapidly to the nighttime temperature at the outset of the next night's observing. Obviously, this cannot be done if the telescope enclosure has been open during the day for observation. The condition is worsened if sunlight has been allowed to fall on the telescope during the day. Accordingly, optical/IR telescopes should be designed for rapid thermal adjustment.

Some of the design options in thermal control are (1) to insulate heavy masses that cannot equalize quickly, (2) to reduce weights and masses, (3) to provide good ventilation (i.e., avoiding dead air spaces that act as insulators), (4) to make surfaces reflective so that radiative coupling to the cold night sky is minimized, and (5) to isolate or eliminate heat sources. One should also consider using parts made from materials with low thermal expansion, but these have limited value if their heating effects are allowed to spoil the telescope seeing.

C. REMOTE OBSERVING

The traditional stereotype of an astronomer is a person perched on a high stool or platform, peering through the eyepiece and carefully guiding the telescope. The modern reality is likely to be quite different. Sophisticated electronic detectors replace the eye. Automatic star-tracking guiders take over the guidance chore. The astronomer sits in a control room sometimes far away from the telescope. A TV monitor shows the FOV or, at least, that part of the field not falling on the detector. A computer logs the data and telescope conditions. The telescope is not even seen by the astronomer: It can be in the next room or even a continent away if the communication link is properly established.

The advent of space-based astronomy clearly marked the time when the astronomer and the telescope were separated. The same separation is taking place in ground-based astronomy, albeit less dramatically. Numerous demonstrations have occurred during the 1970s and 1980s in which astronomers conducted observing runs on telescopes located at distant sites. In one case, the astronomer was in Edinburgh, Scotland and the telescope was on Hawaii. The connection was through a communications satellite. This trend is likely to accelerate as the cost for such connections reduces and the data transmission rates increase.

The future stereotype astronomer is likely to be perched at a computer terminal, not a telescope. For some, the romance of astronomy will be gone; for most, the gain in capability will far outweigh the loss. Remotely located telescopes, attended by highly skilled operators, with automated instrumentation linked by computer to the astronomer, are seen by this author as the forefront astronomical equipment of the future. Galileo would be pleased.

BIBLIOGRAPHY

Bell, L. (1981). "The Telescope." Dover, New York.
Burbidge, G., and Hewitt, A., eds. (1981). "Telescopes for the 1980s." Annual Reviews, Palo Alto, CA.
Driscoll, W. G., and Vaughan, W., eds. (1978). "Handbook of Optics." McGraw-Hill, New York.
King, H. C. (1979). "The History of The Telescope." Dover, New York.
Kingslake, R. (1983). "Optical System Design." Academic, Orlando, Florida.
Kuiper, G., and Middlehurst, B., eds. (1960). "Telescopes," Stars and Stellar Systems, Vol. 1. University of Chicago Press, Chicago.
Learner, R. (1981). "Astronomy Through the Telescope." Van Nostrand Reinhold, New York.
Marx, S., and Pfau, W. (1982). "Observatories of the World." Van Nostrand Reinhold, New York.

PLANETARY ATMOSPHERES

Joel S. Levine *NASA Langley Research Center*

GLOSSARY

Atmospheric pressure: Weight of the atmosphere in a vertical column, 1 cm^2 in cross section, above the surface of a planet. On earth, the average value of atmospheric pressure at sea level is 1.013×10^6 dyne cm^{-2}, or 1013 mbar, which is equivalent to a pressure of 1 atmosphere.

Cosmic abundance of the elements: Relative proportion of the elements in the cosmos based on abundances deduced from astronomical spectroscopy of the sun, the stars, and interstellar gas clouds and chemical analyses of meteorites, rocks, and minerals.

Gravitational escape: Loss of atmospheric gases from a planetary atmosphere to space. If an upward-moving atmospheric atom or molecule is to escape the gravitational field of a planet, its kinetic energy must exceed its gravitational potential energy. The two lightest atmospheric gases, hydrogen and helium, usually possess enough kinetic energy to escape from the atmospheres of the terrestrial planets. In photochemical escape, some heavier atmospheric species, such as atomic nitrogen and atomic oxygen, are imparted with sufficient kinetic energy from certain photochemical and chemical reactions to escape from planetary gravitational fields. Over geological time, gravitational escape has been an important process in the evolution of the atmospheres of the terrestrial planets.

Greenhouse effect: Increase in the infrared opacity of an atmosphere which leads to an increase in the lower atmospheric and surface temperature. For example, water vapor and carbon dioxide, the two most abundant outgassed volatiles, increase the infrared opacity of an atmosphere by absorbing outgoing infrared radiation emitted by the surface and lower atmosphere. The absorbed infrared radiation is then re-emitted by the absorbing molecule. The downward directed component of the re-emitted radiation heats the surface and lower atmosphere.

Magnetosphere: Region in upper atmosphere of planet possessing magnetic field where ions and electrons are contained by magnetic lines of force. The earth and Jupiter are surrounded by magnetospheres.

Mantle: One of the three major subdivisions of the earth's interior (the core and the crust being the other two). The mantle contains about 70% of the mass of the earth and is iron-deficient. The mantle surrounds the core, which is believed to consist mainly of iron. Surrounding the mantle is the relatively thin-layered crust. The core, mantle, and crust are composed of refractory elements and their compounds.

Mixing ratio: Ratio of the number of atoms or molecules of a particular species per cm^3 to the total number of atmospheric atoms or molecules per cm^3. At the earth's surface, at standard temperature and pressure, there are about 2.55×10^{19} molecules per cm^{-3}. The mixing ratio is a dimensionless quantity, usually expressed in parts per million by volume (ppmv = 10^{-6}), parts per billion by volume (ppbv = 10^{-9}), or parts per trillion by volume (pptv = 10^{-12}).

Photodissociation: Absorption of incoming solar radiation, usually radiation of visible wavelengths or shorter, that leads to the dissociation of atmospheric molecules to their constituent molecules, atoms, or radicals. For example, the photodissociation of water vapor leads to the formation atomic hydrogen (H) and the hydroxyl radical (OH).

Primary or primordial atmosphere: Atmosphere resulting from capture of the gaseous material in the primordial solar nebula from which the solar system condensed about 4.6 billion years ago. The atmospheres of Jupiter, Saturn, Uranus, and Neptune are believed to be remnants of the primordial solar nebula and, hence, contain atoms of hydrogen, helium, nitrogen, oxygen, carbon, and so on in the same elemental proportion as the sun. The atmospheres of these planets are composed of molecular hydrogen and helium, with smaller amounts of methane, ammonia, and water vapor, and their photodissociation products.

Primordial solar nebula: Interstellar cloud of gas, dust, and ice of a few solar masses, at a temperature of about 10 K, that collapsed under its own gravitational attraction to form the sun, the planets, and the rest of the solar system about 4.6 billion years ago. Compression caused the temperature of the contracting cloud to increase to several thousand degrees, vaporizing all but the most refractory compounds, while conservation of angular momentum flattened the cloud into a disk. The refractory elements in the equatorial plane began to accumulate into large bodies, eventually forming the planets by accretion and coalescence. The bulk of the mass of the primordial solar nebula, composed primarily of hydrogen and helium, formed the sun.

Refractory elements: Elements or their compounds that volatilize only at very high temperatures, such as silicon, magnesium, and aluminum. Refractory elements and their compounds formed the terrestrial planets through the processes of accretion and coalescence.

Secondary atmosphere: Atmosphere resulting from the outgassing of trapped volatiles, that is, the atmospheres surrounding earth, Venus, and Mars.

Troposphere: Lowest region of the earth's atmosphere, which extends from the surface to about 15 km in the tropics and to 10 km at high latitudes. About 80% of the total mass of the atmosphere is found in the troposphere (the rest of the total mass of the atmosphere is found in the stratosphere, which extends to about 50 km, with only a fraction of a percent of the total mass of the atmosphere found in the atmospheric regions above the troposphere and stratosphere: the mesosphere, thermosphere, exosphere, ionosphere, and magnetosphere).

Volatile elements: Elements that are either gaseous or form gaseous compounds at relatively low temperatures.

Volatile outgassing: Release of volatiles trapped in the solid earth during the planetary formation process. The release of the trapped volatiles led to the formation of the atmosphere and ocean.

Gravitationally bound to the planets are atmospheres, gaseous envelopes of widely differing masses and chemical compositions. The origin of the atmospheres of the planets is directly related to the origin of the planets some 4.6 billion years ago. Much of our knowledge and understanding of the origin, evolution, structure, composition, and meteorology of planetary atmospheres has resulted from the exploration of the planets and their atmospheres by a series of planetary fly-bys, orbiters, and landers. The atmospheres of the terrestrial planets (earth, Venus, and Mars) most probably resulted from the release of gases originally trapped in the solid planet during the planetary formation process. Water vapor, carbon dioxide, and molecular nitrogen outgassed from the terrestrial planets to form their atmospheres. By contrast, it is generally thought that the very dense hydrogen and helium atmospheres of the outer planets (Jupiter, Saturn, Uranus, and Neptune) are the gaseous remnants of the primordial solar nebula that condensed to form the sun and the planets. Of all of the planets in the solar system, the atmosphere of the earth has probably changed the most over geological time in response to both the geochemical cycling of a geologically active planet and the biochemical cycling of a biologically active planet.

I. Formation of the Planets and Their Atmospheres

The sun, earth, and the other planets condensed out of the primordial solar nebula, an interstellar cloud of gas and dust, some 4.6 billion years ago (orbital information and the physical characteristics of the planets are summarized in Table I). The chemical composition of the primordial solar nebula most probably reflected the cosmic abundance of the elements (see Table II). Volatiles, elements that were either gaseous or that formed gaseous compounds at the relatively low temperature of the solar nebula, were the major constituents. The overwhelmingly prevalent volatile element was hydrogen, followed by helium, oxygen, nitrogen, and carbon (see Table II). Considerably less abundant in the solar nebula, but key elements in the formation of the solid planets, were the nonvolatile refractory elements, such as silicon, iron, magnesium, nickel, and aluminum, which formed solid elements and compounds at the relatively low temperature of the solar nebula. The terrestrial planets (Mercury, Venus, earth, and Mars) formed through the processes of coalescence and accretion of the refractory elements and their compounds, beginning with grains the size of dust, to boulder-sized "planetesimals", to planetary-sized bodies. The terrestrial planets may have grown to their full size and mass in as little as 10 million years. Volatiles incorporated in a late-accreting, low-temperature condensate may have formed as a veneer surrounding the newly formed terrestrial planets. The chemical composition of this volatile-rich veneer resembled that of carbonaceous chondritic meteorites, which contain relatively large amounts of water (H_2O) and other volatiles. The collisional impact of the refractory material during the coalescence and accretion phase caused widespread heating in the forming planets. The heating was accompanied by the release of the trapped volatiles through a process termed volatile outgassing. The oxidation state and, hence, the chemical composition of the outgassed volatiles depended on the structure and composition of the solid planet and, in particular, on the presence or absence of free iron in the upper layers of the solid planet. If the terrestrial planets formed as geologically differentiated bodies, i.e., with free iron having already migrated to the core (as a result of the heating and high temperature accompanying planetary accretion), surrounded by an iron-free mantle of silicates, the outgassed volatiles would have been composed of water vapor, carbon dioxide (CO_2), and molecular nitrogen (N_2), not unlike the chemical composition of present-day volcanic emissions. Current theories of planetary formation suggest that the earth, Venus, and Mars formed as geologically differentiated objects. Some volatile outgassing may have also been associated with the impact heating during the final stages of planetary formation. This outgassing would have resulted in an almost instantaneous formation of the atmosphere, coincident with the final stages of planetary formation. As a result of planetary accretion and volatile outgassing, the terrestrial planets are characterized by iron–silicate interiors with atmospheres composed primarily of carbon dioxide (Venus and Mars) or molecular nitrogen (earth), with surface pressures that

TABLE I.   The Planets: Orbital Information and Physical Characteristics

	Mercury	Venus	Earth
Mean distance from sun (millions of km)	57.9	108.2	149.6
Period of revolution	88 days	224.7 days	365.26 days
Rotation period	59 days	−243 days Retrograde	23 hr 56 min 4 sec
Inclination of axis	2°	3°	23°27′
Inclination of orbit to ecliptic	7°	3.4°	0°
Eccentricity of orbit	0.206	0.007	0.017
Equatorial diameter (km)	4880	12,104	12,756
Atmosphere (main components)	Virtually none	Carbon dioxide	Nitrogen Oxygen
Known satellites	0	0	1
Rings	—	—	—

range from about 1/200 atm (Mars) to about 90 atm (Venus) (the surface pressure of the earth's atmosphere is 1 atmosphere).

Since Mercury does not possess an appreciable atmosphere, it is not discussed in any detail in this article, which concentrates on the chemical composition of planetary atmospheres. Measurements obtained by Mariner 10, which encountered Mercury three times in 1974–75 after a 1974 Venus fly-by, indicated that the surface pressure of the atmosphere of Mercury is less than a thousandth of a trillionth of the earth's, with helium resulting from radiogenic decay and subsequent outgassing as a possible constituent. Mercury was found to possess an internal magnetic field, similar to but weaker than the earth's. Mariner 10 photographs indicated that the surface of Mercury is very heavily cratered, resembling the highlands on the moon (Fig. 1). A large impact basin (Caloris), about 1300 km in diameter, was discovered. Long scarps of cliffs, apparently produced by crustal compression, were also found.

In direct contrast to the terrestrial planets, the outer planets (Jupiter, Saturn, Uranus, and Neptune) are more massive (15–318 earth masses), larger (4–11 earth radii), and possess multiple satellites and ring systems (see Table I). The atmospheres of the outer planets are very dense and contain thick clouds and haze layers. These atmospheres are composed primarily (85–95% by volume) of molecular hydrogen (H_2) and helium (He) (5–15%) with smaller amounts of compounds of carbon, nitrogen, and oxygen, primarily present in the form of saturated hydrides [methane (CH_4), ammonia (NH_3), and water vapor] at approximately the solar ratio of carbon, nitrogen, and oxygen. The composition of the atmospheres of the outer planets suggests that they are captured remnants of the primordial solar nebula that condensed to form the solar system, as opposed to having formed as a result of the outgassing of volatiles trapped in the interior, as did the atmospheres of the terrestrial planets. It has been suggested that a thick atmosphere of molecular hydrogen and helium, the overwhelming constituents of the primordial solar nebula, may have surrounded the terrestrial planets very early in their history (during the final stages of planetary accretion). However, such a primordial solar nebula remnant atmosphere surrounding the terrestrial planets would have dissipiated very quickly, due to the low mass of these planets and, hence, their weak gravitational attraction, coupled with the rapid gravitational escape of hydrogen and helium, the two lightest gases, from the "warm" terrestrial planets. Therefore, an early atmosphere composed of hydrogen and helium surrounding the terrestrial planets would have been extremely short-lived, if it ever existed at all. The large masses of the outer planets and their great distances from the sun (and colder temperatures) have enabled them to gravitationally retain their primordial solar nebula remnant atmospheres. The colder temperatures resulted in a "freezing

Mars	Jupiter	Saturn	Uranus	Neptune	Pluto
227.9	778.3	1427	2869	4496	5900
687 days	11.86 yr	29.46 yr	84.01 yr	164.1	247.7 yr
24 hr	9 hr	10 hr	17.24 hr	22 hr	−6 days
37 min	55 min	39 min		or less	9 hr
23 sec	30 sec	20 sec			18 min
					Retrograde
25°12′	3°5′	26°44′	97°55′	28°48′	60°?
1.9°	1.3°	2.5°	.8°	1.8°	17.2°
0.093	0.048	0.056	0.047	0.009	0.25
6787	142,800	120,400	51,800	49,500	3500
Carbon dioxide	Hydrogen	Hydrogen	Hydrogen	Hydrogen	Methane (?)
	Helium	Helium	Helium	Helium	
	Methane	Methane	Methane	Methane	
2	16	21	15	2	1
—	Yes	Yes	Yes	—	—

TABLE II.   Cosmic Abundance of the Elements[a]

Element	Abundance[b]	Element	Abundance[b]
$_1$H	2.6×10^{10}	$_{44}$Ru	1.6
$_2$He	2.1×10^9	$_{45}$Rh	0.33
$_3$Li	45	$_{46}$Pd	1.5
$_4$Be	0.69	$_{47}$Ag	0.5
$_5$B	6.2	$_{48}$Cd	2.12
$_6$C	1.35×10^7	$_{49}$In	2.217
$_7$N	2.44×10^6	$_{50}$Sn	4.22
$_8$O	2.36×10^7	$_{51}$Sb	0.381
$_9$F	3630	$_{52}$Te	6.76
$_{10}$Ne	2.36×10^6	$_{53}$I	1.41
$_{11}$Na	6.32×10^4	$_{54}$Xe	7.10
$_{12}$Mg	1.050×10^6	$_{55}$Cs	0.367
$_{13}$Al	8.51×10^4	$_{56}$Ba	4.7
$_{14}$Si	1.00×10^6	$_{57}$La	0.36
$_{15}$P	1.27×10^4	$_{58}$Ce	1.17
$_{16}$S	5.06×10^5	$_{59}$Pr	0.17
$_{17}$Cl	1970	$_{60}$Nd	0.77
$_{18}$Ar	2.28×10^5	$_{62}$Sm	0.23
$_{19}$K	3240	$_{63}$Eu	0.091
$_{20}$Ca	7.36×10^4	$_{64}$Gd	0.34
$_{21}$Sc	33	$_{65}$Tb	0.052
$_{22}$Ti	2300	$_{66}$Dy	0.36
$_{23}$V	900	$_{67}$Ho	0.090
$_{24}$Cr	1.24×10^4	$_{68}$Er	0.22
$_{25}$Mn	8800	$_{69}$Tm	0.035
$_{26}$Fe	8.90×10^5	$_{70}$Yb	0.21
$_{27}$Co	2300	$_{71}$Lu	0.035
$_{28}$Ni	4.57×10^4	$_{72}$Hf	0.16
$_{29}$Cu	919	$_{73}$Ta	0.022
$_{30}$Zn	1500	$_{74}$W	0.16
$_{31}$Ga	45.5	$_{75}$Re	0.055
$_{32}$Ge	126	$_{76}$Os	0.71
$_{33}$As	7.2	$_{77}$Ir	0.43
$_{34}$Se	70.1	$_{78}$Pt	1.13
$_{35}$Br	20.6	$_{79}$Au	0.20
$_{36}$Kr	64.4	$_{80}$Hg	0.75
$_{37}$Rb	5.95	$_{81}$Tl	0.182
$_{38}$Sr	58.4	$_{82}$Pb	2.90
$_{39}$Y	4.6	$_{83}$Bi	0.164
$_{40}$Zr	30	$_{90}$Th	0.034
$_{41}$Nb	1.15	$_{92}$U	0.0234
$_{42}$Mo	2.52		

[a] From Cameron, A. G. W. (1968). In "Origin and Distribution of the Elements." L. H. Ahrens, ed. Pergamon, New York. Copyright 1968 Pergamon Press.
[b] Abundance normalized to silicon (Si) = 1.00×10^6.

out" or condensation of several atmospheric gases, such as water vapor, ammonia, and methane forming cloud and haze layers in the atmospheres of the outer planets.

The most distant planet in the solar system, Pluto, has a very eccentric orbit, which at times brings it closer to the sun than Neptune's orbit.

By virtue of its great orbital eccentricity and small mass, it is suspected that Pluto may have originally been a satellite of another planet. Methane has been detected on Pluto. Very little is known about Pluto. Most of what we know about Pluto is summarized in Table I. Of the numerous smaller bodies in the solar system, including satellites and asteroids, only Saturn's satellite, Titan, has an appreciable atmosphere (surface pressure about 1.5 atm), composed of molecular nitrogen and a small amount of methane. [*See* PRIMITIVE SOLAR SYSTEM OBJECTS: ASTEROIDS AND COMETS.]

II. Earth

The atmospheres of the earth, Venus, and Mars resulted from the outgassing of volatiles originally trapped in their interiors. The chemical composition of the outgassed volatiles was not unlike that of present-day volcanic emissions: water vapor = 79.31% by volume; carbon dioxide = 11.61%; sulfur dioxide (SO_2) = 6.48%; and molecular nitrogen = 1.29%. On earth, the bulk of the outgassed water vapor condensed out of the atmosphere, forming the earth's vast ocean. Only small amounts of water vapor remained in the atmosphere, with almost all of it confined to the troposphere. Some atmospheric water is in the condensed state, found in the form of cloud droplets. Water clouds cover about 50% of the earth's surface at any given time and are a regular feature of the atmosphere (Fig. 2). Near the ground, the water vapor concentration is variable, ranging from a fraction of a percent to a maximum of several percent by volume. Once the ocean formed, outgassed carbon dioxide, the second most abundant volatile, which is very water soluble, dissolved into the ocean. Once dissolved in the ocean, carbon dioxide chemically reacted with ions of calcium and magnesium, also in the ocean, and precipitated out in the form of sedimentary carbonate rocks such as calcite ($CaCO_3$), and dolomite [$CaMg(CO_3)_2$].

The concentration of carbon dioxide in the atmosphere is about 0.034% by volume, which is equivalent to 340 parts per million by volume (ppmv). It has been estimated that the preindustrial (ca. 1860) level of atmospheric carbon dioxide was about 280 ppmv, with the increase to the present level attributable to the burning of fossil fuels, notably coal. For each carbon dioxide molecule in the present-day atmosphere, there

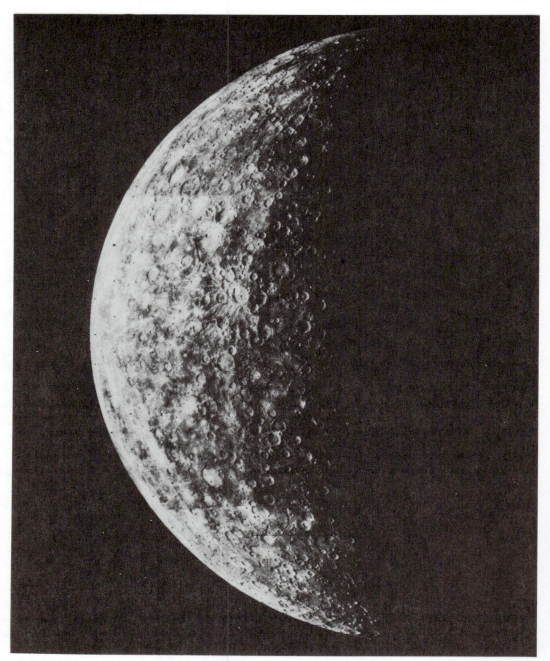

FIG. 1.   Mosaic of Mariner 10 photographs of Mercury. With no appreciable atmosphere, we can see right down
to the cratered surface of Mercury, which is similar to the cratered highlands on the moon. The largest craters in
the photograph are about 200 km in diameter.

FIG. 2.   Planet earth as photographed by Apollo 17 astronauts on their journey to the moon. Scattered clouds, which cover only about 50% of the earth at any given time, permit viewing the surface of the earth. Earth is a unique planet in many respects, including the presence of life, the existence of liquid water on the surface, and the presence of large amounts of oxygen in the atmosphere.

are approximately 50 carbon dioxide molecules physically dissolved in the ocean and almost 30,000 carbon dioxide molecules incorporated in sedimentary carbonate rocks. All of the carbon dioxide presently incorporated in carbonate rocks originally outgassed from the interior and was at one time in the atmosphere. Hence, the early atmosphere may have contained 100 to 1000 times or more carbon dioxide than it presently contains. Sulfur dioxide, the third most abundant gas in volcanic emissions, is chemically unstable in the atmosphere. Sulfur dioxide is rapidly chemically transformed to water-soluble sulfuric acid (H_2SO_4), which readily rains out of the atmosphere. Hence, the atmospheric lifetime of sulfur dioxide is very short.

Molecular nitrogen is the fourth most abundant gas in volcanic emissions. Nitrogen does not condense out of the atmosphere (as does water vapor), is not water soluble (as is carbon dioxide), and is not chemically active (as is sulfur dioxide). As a result, the bulk of the outgassed molecular nitrogen accumulated in the atmosphere, and over geological time became the most abundant atmospheric constituent (about 78% by volume). Molecular oxygen (O_2), produced as a by-product of photosynthetic activity, built up in the atmosphere to become the second most abundant species (about 21% by volume). Argon (isotope 40), the third most abundant atmospheric species (about 1% by volume), is a chemically inert gas resulting from the radiogenic decay of potassium (isotope 40) in the crust. Hence, the bulk chemical composition of the earth's atmosphere can be explained in terms of volatile outgassing and the ultimate sinks of the outgassed volatiles, including condensation/precipitation, dissolution and carbonate formation in the ocean, biogenic activity, and radiogenic decay. These and other processes, including photochemical and chemical reactions, have resulted in a myriad of other trace atmospheric gases in the atmosphere, which are listed in Table III.

TABLE III.   Composition of the Earth's Atmosphere[a]

	Surface concentration[b]	Source
Major and minor Gases		
Nitrogen (N_2)	78.08%	Volcanic, biogenic
Oxygen (O_2)	20.95%	Biogenic
Argon (Ar)	0.93%	Radiogenic
Water vapor (H_2O)	Variable, up to 4%	Volcanic, evaporation
Carbon dioxide (CO_2)	0.034%	Volcanic, biogenic, anthropogenic
Trace gases		
Oxygen species		
Ozone (O_3)	10–100 ppbv	Photochemical
Atomic oxygen (O) (ground state)	10^3 cm^{-3}	Photochemical
Atomic oxygen [O(^{1}D)] (excited state)	10^{-2} cm^{-3}	Photochemical
Hydrogen species		
Hydrogen (H_2)	0.5 ppmv	Photochemical, biogenic
Hydrogen peroxide (H_2O_2)	10^9 cm^{-3}	Photochemical
Hydroperoxyl radical (HO_2)	10^8 cm^{-3}	Photochemical
Hydroxyl radical (OH)	10^6 cm^{-3}	Photochemical
Atomic hydrogen (H)	1 cm^{-3}	Photochemical
Nitrogen species		
Nitrous oxide (N_2O)	330 ppbv	Biogenic, anthropogenic
Ammonia (NH_3)	0.1–1 ppbv	Biogenic, anthropogenic
Nitric acid (HNO_3)	50–1000 pptv	Photochemical
Hydrogen cyanide (HCN)	~200 pptv	Anthropogenic(?)
Nitrogen dioxide (NO_2)	10–300 pptv	Photochemical
Nitric oxide (NO)	5–100 pptv	Anthropogenic, biogenic, lightning, photochemical
Nitrogen trioxide (NO_3)	100 pptv	Photochemical
Peroxyacetylnitrate ($CH_3CO_3NO_2$)	50 pptv	Photochemical
Dinitrogen pentoxide (N_2O_5)	1 pptv	Photochemical
Pernitric acid (HO_2NO_2)	0.5 pptv	Photochemical
Nitrous acid (HNO_3)	0.1 pptv	Photochemical

TABLE III.   (*Continued*)

	Surface concentration[b]	Source
Nitrogen aerosols		
Ammonium nitrate (NH_4NO_3)	~100 pptv	Photochemical
Ammonium chloride (NH_4Cl)	~0.1 pptv	Photochemical
Ammonium sulfate [$(NH_4)_2SO_4$]	~0.1 pptv(?)	Photochemical
Carbon species		
Methane (CH_4)	1.7 ppmv	Biogenic, anthropogenic
Carbon monoxide (CO)	70–200 ppbv (N hemis.) 40–60 ppbv (S hemis.)	Anthropogenic, biogenic, photochemical
Formaldehyde (H_2CO)	0.1 ppbv	Photochemical
Methylhydroperoxide (CH_3OOH)	10^{11} cm^{-3}	Photochemical
Methylperoxyl radical (CH_3O_2)	10^8 cm^{-3}	Photochemical
Methyl radical (CH_3)	10^{-1} cm^{-3}	Photochemical
Sulfur species		
Carbonyl sulfide (COS)	0.5 ppbv	Volcanic, anthropogenic
Dimethyl sulfide [$(CH_3)_2S$]	0.4 ppbv	Biogenic
Hydrogen sulfide (H_2S)	0.2 ppbv	Biogenic, anthropogenic
Sulfur dioxide (SO_2)	0.2 ppbv	Volcanic, anthropogenic, photochemical
Dimethyl disulfide [$(CH_3)_2S_2$]	100 pptv	Biogenic
Carbon disulfide (CS_2)	50 pptv	Volcanic, anthropogenic
Sulfuric acid (H_2SO_4)	20 pptv	Photochemical
Sulfurous acid (H_2SO_3)	20 pptv	Photochemical
Sulfoxyl radical (SO)	10^3 cm^{-3}	Photochemical
Thiohydroxyl radical (HS)	1 cm^{-3}	Photochemical
Sulfur trioxide (SO_3)	10^{-2} cm^{-3}	Photochemical
Halogen species		
Hydrogen chloride (HCl)	1 ppbv	Sea salt, volcanic
Methyl chloride (CH_3Cl)	0.5 ppbv	Biogenic, anthropogenic
Methyl bromide (CH_3Br)	10 pptv	Biogenic, anthropogenic
Methyl iodide (CH_3I)	1 pptv	Biogenic, anthropogenic
Noble gases (chemically inert)		
Neon (Ne)	18 ppmv	Volcanic
Helium (He)	5.2 ppmv	Radiogenic
Krypton (Kr)	1 ppmv	Radiogenic
Xenon (Xe)	90 ppbv	Radiogenic

[a] From Levine, J. S. (1985). In "The Photochemistry of Atmospheres: Earth, the Other Planets, and Comets," J. S. Levine, ed. Academic, Orlando, FL.

[b] Species concentrations are given in percentage by volume, in terms of surface mixing ratio, parts per million by volume (ppmv $\equiv 10^{-6}$), parts per billion by volume (ppbv $\equiv 10^{-9}$), parts per trillion by volume (pptv $\equiv 10^{-12}$), or in terms of surface number density (cm^{-3}). The species mixing ratio is defined as the ratio of the number density of the species to the total atmospheric number density (2.55×10^{19} molec cm^{-3}). There is some uncertainty in the concentrations of species at the ppbv level or less. The species concentrations given in molec cm^{-3} are generally based on photochemical calculations, and species concentrations in mixing ratios are generally based on measurements.

III. Venus

Venus has been described as the earth's twin because of its similar mass (0.81 earth masses), radius (0.95 earth radii), mean density (95% that of earth), and gravity (90% that of earth) (see Table I). However, in terms of atmospheric structure and chemical composition, Venus is anything but a twin of earth. The mean planetary surface temperature of Venus is about 750 K, compared with about 300 K for earth; the surface pressure on Venus is about 90 atm, compared with 1 atm for earth; carbon dioxide at 96% by volume, is the overwhelming constituent in the atmosphere of Venus, while it is only a trace constituent in the earth's atmosphere

(0.034% by volume). In addition, Venus does not have an ocean or a biosphere and is completely covered by thick clouds, probably composed of sulfuric acid (Fig. 3). Hence, the atmosphere is inhospitable and very unlike the earth's atmosphere.

Much of our information on the structure and composition of the atmosphere of Venus has been obtained through a series of U.S. and U.S.S.R. Venus flybys, orbiters, and landers, which are summarized in Table IV.

The clouds on Venus are thick and contain no holes, hence, we have never directly observed the surface of Venus from earth. These clouds resemble a stratified low-density haze extending from about 45 to about 65 km. The total extinc-

FIG. 3.   Venus as photographed by the Pioneer Venus Orbiter. The thick, cloud-covered atmosphere continually and completely hides the surface of Venus, which at very high temperature and atmospheric pressure is not a hospitable environment.

TABLE IV.   U.S. and U.S.S.R. Missions to Venus[a]

Name	Designator (U.S.S.R.)	Launch date	Mission remarks
Venera 1	61-Gamma 1	Feb. 12, 1961	Passed Venus at 100,000 km May 19–21, 1961; contact lost Feb. 27, 1961.
Mariner II	—	Aug. 27, 1962	Planetary exploration: First successful interplanetary probe. Found no magnetic field; high surface temperatures of approximately 800°F. Passed Venus Dec. 14, 1962 at 21,600 miles, 109 days after launch.
Zond 1	64-16D	April 2, 1964	Passed Venus at 100,000 km July 19, 1964; communications failed after May 14, 1964.
Venera 2	65-91A	Nov. 12, 1965	Passed Venus at 24,000 km, Feb. 27, 1966; communications failed.
Venera 3	65-92A	Nov. 16, 1965	Struck Venus March 1, 1966; communications failed earlier.
Venera 4	67-58A	June 12, 1967	Probed atmosphere.
Venera 5	69-1A	Jan. 5, 1969	Entered Venus atmosphere May 16, 1969.
Mariner V	—	June 14, 1967	Planetary exploration: All science and engineering subsystems nominal through encounter with Venus; data indicate Venus has a moonlike effect on solar plasma and strong H_2 corona comparable to Earth's, 72 to 87% CO_2 atmosphere with balance probably nitrogen and O_2. Closest approach, 3,900 km on Oct. 19, 1967.
Venera 6	69-2A	Jan. 10, 1969	Entered Venus atmosphere May 17, 1969.
Venera 7	70-60A	Aug. 17, 1970	Soft landed on Venus; signal from surface.
Venera 8	72-21A	Mar. 27, 1972	Soft landed on Venus; sent data from surface.
Mariner 10 Venus/Mercury	—	Nov. 3, 1973	Conducted exploratory investigations of planet Mercury during three flybys by obtaining measurements of its environment, atmosphere, surface, and body characteristics and conducted similar investigations of Venus. Mariner 10 encountered Venus on Feb. 5, 1974 and Mercury on Mar. 29 and Sept. 21, 1974 and Mar. 16, 1975. Resolution of the photographs was 100 m, 7000 times greater than that achieved by earth-based telescopes.
Venera 9-orbiter	75-50A	June 8, 1975	Orbited Venus Oct. 22, 1975. Orbiter and lander launched from single D-class vehicle (Proton), 4650 kg thrust.
Venera 9-lander	75-50D	June, 1975	Soft landed; returned picture.
Venera 10-orbiter	75-54A	June 14, 1975	Orbited Venus Oct. 25, 1975. Orbiter and lander launched from single D-class vehicle (Proton), 4659 kg thrust.
Venera 10-lander	75-54D	June 14, 1975	Soft landed; returned picture.
Pioneer 12 Pioneer 13 Pioneer Venus	—	May 20, 1978 Aug. 8, 1978	Orbiter launched in May studied interaction of atmosphere and solar wind and made radar and gravity maps of the planet. The multiprobe spacecraft launched in August returned information on Venus' wind and circulation patterns as well as atmospheric composition, temperature and pressure readings. Pioneer 12 entered Venus orbit Dec. 4, 1978; Pioneer 13 encountered Venus Dec. 9, 1978.
Venera 11-orbiter	78-84A	Sept. 9, 1978	Passed Venus as 35,000 km Dec. 25, 1978; served as relay station. Orbiter and lander launched from single D-class vehicle (Proton), 4650 kg thrust.
Venera 11-lander	78-84E	Sept. 9, 1978	Soft-landed on Venus.

(continued)

TABLE IV.   (*Continued*)

Name	Designator (U.S.S.R.)	Launch date	Mission remarks
Venera 12-orbiter	78-86A	Sept. 14, 1978	Passed Venus at 35,000 km Dec. 21, 1978; served as relay station. Orbiter and lander launched from single D-class vehicle (Proton), 4650 kg thrust.
Venera 12-lander	78-86E	Sept. 14, 1978	Soft-landed on Venus.
Venera 13-orbiter	1981-106A	Oct. 30, 1981	Both orbiter and lander launched from single D-class vehicle (Proton), 4650 kg thrust.
Venera 13-lander	None	Oct. 30, 1981	Soft-landed on Venus Mar. 3, 1982; returned color picture.
Venera 14-orbiter	1981-110A	Nov. 4, 1981	Both orbiter and lander launched from single D-class vehicle (Proton), 4650 kg thrust.
Venera 14-lander	None	Nov. 4, 1981	Soft-landed on Venus Mar. 5, 1982; returned color picture.

[a] From NASA (1983). "Planetary Exploration Through Year 2000: A Core Program." Solar System Exploration Committee of NASA Advisory Council, Washington, DC.

tion optical depth of the clouds in visible light is about 29. The extinction of visible light is due almost totally to scattering. The lower clouds are found between 45 and 50 km; the middle clouds from 50 to 55 km; and the upper clouds from 55 to 65 km. The tops of the upper clouds, which are the ones visible from earth, appear to be composed of concentrated sulfuric acid droplets (see Fig. 3).

As already noted, carbon dioxide at 96% by volume, is the overwhelming constituent of the atmosphere of Venus. The next most abundant atmospheric gas is molecular nitrogen at 4% by volume. The relative proportion by volume of carbon dioxide and molecular nitrogen in the atmospheres of Venus and Mars is almost identical. The chemical composition of the atmosphere of Venus is summarized in Table V. At the surface of Venus, the partial pressure of carbon dioxide is about 90 bar, molecular nitrogen is about 3.2 bar, and water vapor is only about 0.01 bar (more about water on Venus later). For comparison, if the earth were heated to the surface temperature of Venus (about 750 K), we would have a massive atmosphere composed of water vapor at a surface partial pressure of about 300 bar (resulting from the evaporation of the ocean), a carbon dioxide partial pressure of about 55 bar (resulting from the thermal composition of crustal carbonates), and a molecular nitrogen pressure of about 1 to 3 bar (resulting from the present atmosphere plus the outgassing of crustal nitrogen).

A major puzzle concerning the chemical composition of the atmosphere of Venus (as well as

the atmosphere of Mars) is the stability of carbon dioxide and the very low atmospheric concentrations of carbon monoxide (CO) and oxygen [atomic (O) and molecular], which are the photodissociation products of carbon dioxide. In the daytime upper atmosphere (above 100 km), carbon dioxide is readily photodissociated with a photochemical atmospheric lifetime of only about one week. The recombination of carbon monoxide and atomic oxygen in the pres-

TABLE V.   Composition of the Atmosphere of Venus[a]

Gas	Volume mixing ratio	
	Troposphere (below clouds)	Stratosphere (above clouds)
CO_2	9.6×10^{-1}	9.6×10^{-1}
N_2	4×10^{-2}	4×10^{-2}
H_2O	10^{-4}–10^{-3}	10^{-6}–10^{-5}
CO	$(2–3) \times 10^{-5}$	5×10^{-5}–10^{-3}
HCl	$<10^{-5}$	10^{-6}
HF	?	10^{-8}
SO_2	1.5×10^{-4}	5×10^{-8}–8×10^{-7}
S_3	$\sim 10^{-10b}$	?
H_2S	$(1–3) \times 10^{-6b}$	?
COS	$<2 \times 10^{-6}$	?
O_2	$(2–4) \times 10^{-5b}$	$<10^{-6b}$
H_2	?	2×10^{-5b}
^{4}He	10^{-5}	10^{-5}
^{20,22}Ne	$(5–13) \times 10^{-6}$	$(5–13) \times 10^{-6}$
^{36,38,40}Ar	$(5–12) \times 10^{-5}$	$(5–12) \times 10^{-5}$
^{84}Kr	$<2 \times 10^{-8}$–4×10^{-7}	$<2 \times 10^{-8}$–4×10^{-7}

[a] From Lewis, J. S., and Prinn, R. G. (1986). "Planets and Their Atmospheres." Academic, New York. Copyright 1984 Academic Press.

[b] Single experiment; corroboration required.

ence of a third body to reform carbon dioxide is efficient only at the higher atmospheric pressures occurring at and below 100 km. However, at these lower altitudes, atomic oxygen recombines with itself in the presence of a third body to form molecular oxygen considerably faster than the three-body reaction that leads to the recombination of carbon dioxide. Thus, essentially all of the photolyzed carbon dioxide produces carbon monoxide and molecular oxygen. Yet, the observed upper-limit atmospheric concentration of molecular oxygen above the cloud tops could be produced in only about one day, and the observed abundance of carbon monoxide could be produced in only about three months. Photodissociation could easily convert the entire concentration of carbon dioxide in the atmosphere to carbon monoxide and molecular oxygen in only about 4 million years, geologically a short time period.

This dilemma also applies to carbon dioxide on Mars. Considerable research has centered around the recombination of carbon monoxide and molecular oxygen back to carbon dioxide. It became apparent that the only way to maintain low carbon monoxide and oxygen concentrations and high carbon dioxide concentrations in the 100–150-km region is by the rapid downward transport of carbon monoxide and oxygen, balanced by the upward transport of carbon dioxide. It is believed that carbon dioxide is reformed from carbon monoxide and oxygen at an altitude of about 70 km through various chemical reactions and catalytic cycles involving chemically active compounds of hydrogen and chlorine.

If Venus and the earth contained comparable levels of volatiles and outgassed them at comparable rates, then Venus must have somehow lost about 300 bar of water vapor. This may have been accomplished by a runaway greenhouse. In the runaway greenhouse on Venus, outgassed water vapor and carbon dioxide entered the atmosphere, contributing to steadily increasing atmospheric opacity and thus to increasing surface and atmospheric temperatures via the greenhouse effect. On earth, water vapor condensed out of the atmosphere forming the ocean, and the oceans then removed atmospheric carbon dioxide via dissolution and subsequent incorporation into carbonates. The greater proximity of Venus to the sun and its higher initial surface temperature appear to be the simple explanation for the divergent fates of water vapor and carbon dioxide on Venus and earth. In the runaway greenhouse scenario, the photodissociation of massive amounts of outgassed water vapor in the atmosphere of Venus would have led to the production of large amounts of hydrogen and oxygen. Hydrogen could have gravitationally escaped from Venus, and oxygen could have reacted with crustal material. The runaway greenhouse and the accompanying high surface and atmospheric temperatures, too hot for the condensation of outgassed water on Venus, would explain the present water vapor-deficient and carbon dioxide-rich atmosphere of Venus. An alternative suggestion is that Venus may have originally accreted without the levels of water that the earth contained, resulting in a much drier Venus.

IV. Mars

The atmosphere of Mars is very thin (mean surface pressure only about 6.36 mbar), cold (mean surface temperature about 220 K, with the temperature varying from about 290 K in the southern summer to about 150 K in the polar winter), and cloud-free, making the surface of Mars readily visible from the earth (Fig. 4). Much of our information on the structure and composition of the atmosphere of Mars has been obtained through a series of flybys, orbiters, and landers, which are summarized in Table VI. As already noted, the composition of the atmosphere of Mars is comparable to that of Venus. Carbon dioxide is the overwhelming constituent (95.3% by volume), with smaller amounts of molecular nitrogen (2.7%) and argon (1.6%), and trace amounts of molecular oxygen (0.13%) and carbon monoxide (0.08%), resulting from the photodissociation of carbon dioxide (the composition of the atmosphere of Mars is summarized in Table VII). Water vapor and ozone (O_3) are also present, although their abundances vary with season and latitude. The annual sublimation and precipitation of carbon dioxide out of and into the polar cap produce a planet-wide pressure change of 2.4 mbar, or 37% of the mean atmospheric pressure of 6.36 mbar.

The amount and location of water vapor in the atmosphere of Mars are controlled by the temperature of the surface and the atmosphere. The northern polar cap is a source of water vapor during the northern summer. The surface of Mars is also a source of water vapor, depending on the location and season. The total amount of water vapor in the atmosphere varies seasonally between the equivalent of 1 and 2 km^3 of liquid

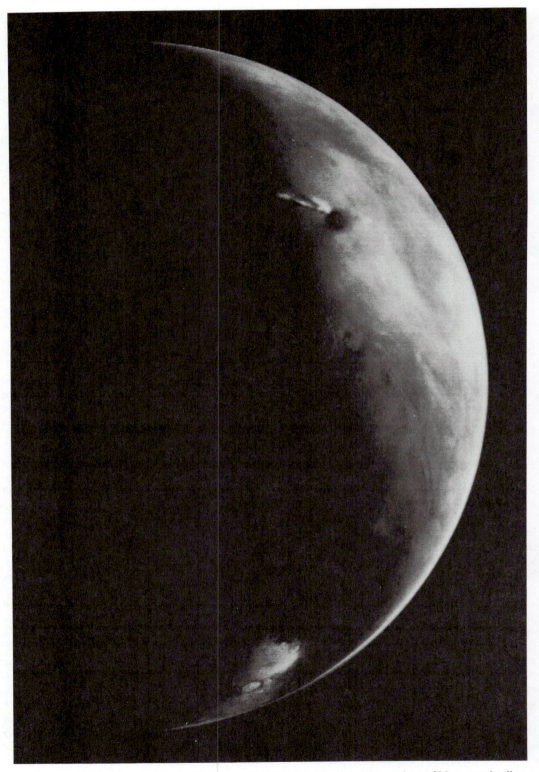

FIG. 4.   Mars as photographed by the Viking 2 orbiter. The thin, cloud-free atmosphere of Mars permits direct observation of the Martian surface from space.

TABLE VI. U.S. and U.S.S.R. Missions to Mars[a]

Name	Designator (U.S.S.R.)	Launch date	Mission remarks
Mars 1	62-Beta Nu 3	Nov. 1, 1962	Passed Mars June 19, 1963 at 193,000 km; communications failed March 21, 1963.
Mariner IV	—	Nov. 28, 1964	Planetary and interplanetary exploration: Encounter occurred July 14, 1965 with closest approach 6100 miles. Twenty-two pictures taken.
Zond 2	64-78C	Nov. 30, 1964	Passed Mars at 1500 km Aug. 6, 1965; communications failed earlier.
Mariner VI	—	Feb. 25, 1969	Planetary exploration: Mid-course correction successfully executed to achieve a Mars flyby within 3330 km on July 31, 1969. Designed to perform investigations of atmospheric structures and compositions and to return TV photos of surface topography.
Mariner VII	—	Mar. 27, 1969	Planetary exploration: Spacecraft identical to Mariner VI. Mid-course correction successful for 3518 km flyby on Aug. 5, 1969.
Kosmos 419	71-42A	May 10, 1971	Failed to separate.
Mars 2-orbiter	71-45A	May 19, 1971	Orbited Mars Nov. 27, 1971. Mars 2 orbiter and lander launched from single D-class vehicle (Proton), 4650 kg thrust.
Mars-lander	71-45E	May 19, 1971	Landed 47°E.
Mars 3-orbiter	71-49A	May 28, 1971	Orbited Mars Dec. 2, 1971. Mars 3 orbiter and lander launched from single D-class vehicle (Proton), 4650 kg thrust.
Mars 3-lander	71-49F	May 28, 1971	Landed 45°S, 158°W.
Mariner IX	—	May 30, 1971	Entered Mars orbit on Nov. 13, 1971. Spacecraft responsed to 38,000 commands and transmitted 6900 pictures of the Martian surface. All scientific instruments operated successfully. Mission terminated on Oct. 27, 1972.
Mars 4	73-47A	July 21, 1973	Passed Mars at 2200 km Feb. 10, 1974, but failed to enter Mars' orbit as planned.
Mars 5	73-49A	July 25, 1973	Orbited Mars Feb. 2, 1974 to gather Mars data and to serve as relay station.
Mars 6-orbiter	73-52A	Aug. 5, 1973	Mars 6 orbiter and lander launched from single D-class vehicle (Proton), 4650 kg thrust.
Mars 6-lander	73-52E	Aug. 5, 1973	Soft landed at 24°S, 25°W; returned atmospheric data during descent.
Mars 7-orbiter	73-53A	Aug. 9, 1973	Mars 7 orbiter and lander launched from single D-class vehicle (Proton), 4650 kg thrust.
Mars 7-lander	73-53E	Aug. 9, 1973	Missed Mars by 1300 km (aimed at 50°S, 28°W).
Viking 1 Lander and orbiter		Aug. 20, 1975	Scientific investigation of Mars. United States' first attempt to soft land a spacecraft on another planet. Successfully soft landed on July 20, 1976. First in situ analysis of surface material on another planet.
Viking 2 Lander and orbiter		Sept. 9, 1975	Scientific investigation of Mars. United States' second attempt to soft land on Mars. Successfully soft landed on Sept. 3, 1976 and returned scientific data. Orbiter from both missions returned over 40,000 high resolution photographs showing surface details as small as 10 m in diameter. Orbiter also collected gravity field data, monitored atmospheric water levels, thermally mapped selected surface sites.

[a] From NASA (1983). ''Planetary Exploration Through Year 2000: A Core Program.'' Solar System Exploration Committee of NASA Advisory Council, Washington, DC.

TABLE VII. Composition of the Atmosphere of Mars[a]

Species	Abundance (mole fraction)
CO_2	0.953
N_2	0.027
^{40}Ar	0.016
O_2	0.13%
CO	0.08%
	0.27%
H_2O	(0.03%)[b]
Ne	2.5 ppm
^{36}Ar	0.5 ppm
Kr	0.3 ppm
Xe	0.08 ppm
O_3	(0.03 ppm)[b]
	(0.003 ppm)[b]

Species	Upper limit (ppm)
H_2S	<400
C_2H_2, HCN, PH_3, etc.	50
N_2O	18
C_2H_4, CS_2, C_2H_6, etc.	6
CH_4	3.7
N_2O_4	3.3
SF_6, SiF_4, etc.	1.0
HCOOH	0.9
CH_2O	0.7
NO	0.7
COS	0.6
SO_2	0.5
C_3O_2	0.4
NH_3	0.4
NO_2	0.2
HCl	0.1
NO_2	0.1

[a] From Lewis, J. S., and Prinn, R. G. (1984). ''Planets and Their Atmospheres.'' Academic, New York. Copyright 1984 Academic Press, New York.
[b] Very variable.

water, with the maximum occurring in the northern summer and the minimum in the northern winter. Ozone is also a highly variable constituent of the atmosphere of Mars. Ozone is present only when the atmosphere is cold and dry.

There is evidence to suggest that significant quantities of outgassed carbon dioxide and water vapor may reside on the surface and in the subsurface of Mars. In addition to the polar caps, which contain large concentrations of frozen carbon dioxide and, in the case of the northern polar cap, of frozen water, there may be considerable quantities of these gases physically adsorbed to the surface and subsurface material. It has been estimated that if the equilibrium temperature of the winter polar cap would increase from its present value of about 150 K to 160 K, sublimation of frozen carbon dioxide would increase the atmospheric pressure to more than 50 mbar. This in turn would cause more water vapor to leave the polar cap and enter the atmosphere. Mariner and Viking photographs indicate the existence of channels widely distributed over the Martian surface. These photographs show runoff channels, tributary networks, and streamlined islands, all very suggestive of widespread fluid erosion. Yet, there is no evidence for the existence of liquid water on the surface of Mars today. In addition, a significant quantity of water vapor may have escaped from Mars in the form of hydrogen and oxygen atoms, resulting from the photolysis of water vapor in the atmosphere of Mars. If the present gravitational escape rate of atoms of hydrogen and oxygen has been operating over the history of Mars, then an amount of liquid water covering the entire planet about 2.5 m high may have escaped from Mars. Viking measurements of argon and neon in the atmosphere of Mars suggest that Mars may have formed with a lower volatile content then either earth or Venus. This is consistent with ideas concerning the capture and incorporation of volatiles in accreting material and how volatile incorporation varies with temperature, which is a function of the distance of the accreting terrestrial planets from the sun.

Unlike the very thick atmosphere of Venus, where the photolysis of carbon dioxide occurs only in the upper atmosphere (above 100 km), on Mars the photodissociation of carbon dioxide occurs throughout the entire atmosphere, right down to the surface. For comparison, the 6.36-mbar surface pressure of the atmosphere of Mars corresponds to an atmospheric pressure at an altitude of about 33 km in the earth's atmosphere. On Mars, carbon dioxide is reformed from its photodissociation products, carbon monoxide and oxygen, by reactions involving atomic hydrogen (H) and the oxides of hydrogen.

Viking photographs indicate that the surface rocks on Mars resemble basalt lava (see Fig. 5). The red color of the surface is probably due to oxidized iron. The soil is fine-grained and cohesive, like firm sand or soil on earth. Viking experiments gave no evidence for organic molecules or for biological activity in the Martian soil, despite unusual chemical reactions pro-

FIG. 5. Surface of Mars as photographed by the Viking 1 lander. Upper photograph shows the Martian landscape with features very similar to those seen in the deserts of earth. Lower photograph, the first ever taken on the surface of Mars, was obtained just minutes after the Viking 1 landed on July 20, 1976.

duced by the soil and measured by the life detection experiments.

V. The Outer Planets

Jupiter, Saturn, Uranus, and Neptune are giant gas planets—great globes of dense gas, mostly molecular hydrogen and helium, with smaller amounts of methane, ammonia, water vapor, and various hydrocarbons produced from the photochemical and chemical reactions of these gases. They formed in the cooler parts of the primordial solar nebula, so gases and ices were preserved. These gas giants have ring systems and numerous satellites orbiting them. As already noted, the outer planets are more massive and larger and have very dense atmospheres that contain thick clouds and layers of aerosol and haze. The solid surfaces of the outer planets have never been observed, and we have only observed the top of the cloud and haze layers.

In many ways, Jupiter and Saturn are a matched pair, as are Uranus and Neptune. Jupiter and Saturn appear to have cores of silicate rocks and other heavy compounds comprising about 25 earth masses, surrounded by thick atmospheres of molecular hydrogen and helium. The total mass of Jupiter and Saturn are 318 and 95 earth masses, respectively. Uranus and Neptune appear to possess much less massive hydrogen/helium atmospheres relative to their cores. The total mass of Uranus and Neptune are only 14.5 and 17 earth masses, respectively.

The large satellites of Jupiter, Saturn, and Neptune are all larger than the earth's moon, with several comparable to the size of Mercury (see Table VIII for a summary of the orbital information and physical characteristics of these

satellites). One of these satellites, Titan, the largest satellite of Saturn, has an appreciable atmosphere. Much of the new information about the atmospheres of Jupiter, Saturn, Uranus, their rings and satellites was obtained by the Voyager encounters of these planets (see Table IX for a summary of the missions to Jupiter and Saturn). [See PLANETARY SATELLITES, NATURAL.]

The Voyager spacecraft obtained high-resolution images of Jupiter, Saturn, and Uranus, their rings and satellites. Voyager instrumentation gathered new information on the chemical composition of their atmospheres. Jupiter, with its colorful banded, turbulent atmosphere, photographed by Voyager 1, is shown in Fig. 6. The Great Red Spot of Jupiter can be clearly seen in this photograph. The helium abundance in the atmosphere of Jupiter was found to be 11% by volume (with molecular hydrogen at 89% by volume), very close to that of the sun. The presence of methane, ammonia, water vapor, ethylene (C_2H_4), ethane (C_2H_6), acetylene (C_2H_2), benzene (C_6H_6), phosphine (PH_3), hydrogen cyanide (HCN), and germanium tetrahydride (GeH_4) in the atmosphere of Jupiter was confirmed (see Table X). The magnetosphere of Jupiter was found to be the largest object in the solar system, about 15 million km across (10 times the diameter of the sun). In addition to hydrogen ions, the magnetosphere was found to contain ions of oxygen and sulfur. A much denser region of ions was found in a torus surrounding the orbit of Jupiter's satellite, Io. The Io torus emits intense ultraviolet radiation and also generates aurora at high latitudes on Jupiter. In addition to a Jovian aurora, huge lightning flashes and meteors were photographed by Voyager on the nightside of Jupiter. A thin ring

TABLE VIII.   The Large Satellites of the Outer Planets: Physical Characteristics[a]

Planet	Satellite	Distance from planet (10^3 km)	Sidereal period (days)	Radius (km)	Density (g cm^{-3})	Surface gravity (cm sec^{-2})
Earth	Moon	384	27.3	1738	3.34	162
Jupiter	Io	422	1.77	1815	3.55	181
	Europa	671	3.55	1569	3.04	132
	Ganymede	1070	7.15	2631	1.93	144
	Callisto	1883	16.7	2400	1.83	125
Saturn	Titan	1222	16	2575	1.89	136
Neptune	Triton	355	6	~2500	~2.1	~150

[a] From Strobel, D. F. (1985). In "The Photochemistry of Atmospheres: Earth, the Other Planets, and Comets," J. S. Levine, ed. Academic, Orlando, FL. Copyright 1985 Academic Press.

TABLE IX.   U.S. and U.S.S.R. Missions to Jupiter and Saturn[a]

Name	Launch date	Mission remarks
Pioneer 10	Mar. 3, 1972	Investigation of the interplanetary medium, the asteroid belt, and the exploration of Jupiter and its environment. Closest approach to Jupiter 130,000 km on Dec. 3, 1973. Exited solar system June 14, 1983; still active.
Pioneer 11 Jupiter/Saturn	Apr. 6, 1973	Obtained scientific information beyond the orbit of Mars with the following emphasis: (a) investigation of the interplanetary medium; (b) investigation of the nature of the asteroid belt; (c) exploration of Jupiter and its environment. Closest approach to Jupiter on Dec. 2, 1974; Saturn encounter: Sept. 1, 1979.
Voyager II Voyager I	Aug. 20, 1977 Sept. 5, 1977	Voyager II encountered Jupiter July 9, 1979, Saturn Aug. 26, 1981, and Uranus Jan. 24, 1986 and is targeted for a Neptune encounter in Sept. 1989. Voyager I encountered Jupiter Mar. 5, 1979 and Saturn Nov. 13, 1980. Both returned a wealth of information about these two giant planets and their satellites including documentation of active volcanism on Io, one of the Galilean satellites.

[a] From NASA (1983). "Planetary Exploration Through Year 2000: A Core Program." Solar System Exploration Committee of NASA Advisory Council, Washington, DC.

surrounding Jupiter, much narrower than Saturn's, was discovered by Voyager. The four large Galilean satellites (Ganymede, Callisto, Europa, and Io) were studied in detail (see Figs. 7–10 for Voyager photographs of these geologi-

TABLE X.   Composition of the Atmosphere of Jupiter[a]

Constituent	Volume mixing ratio[b]
H_2	0.89
He	0.11
CH_4	0.00175
C_2H_2	0.02 ppm
C_2H_4[c]	7 ppb
C_2H_6	5 ppm
CH_3C_2H[c]	2.5 ppb
C_6H_6[c]	2 ppb
CH_3D	0.35 ppm
NH_3[d]	180 ppm
PH_3	0.6 ppm
H_2O[d]	1–30 ppm
GeH_4	0.7 ppb
CO	1–10 ppb
HCN	2 ppb

[a] From Strobel, D. F. (1985). In "The Photochemistry of Atmospheres: Earth, the Other Planets, and Comets," J. S. Levine, ed. Academic, Orlando, FL. Copyright 1985 Academic Press.

[b] ppm ≡ parts per million; ppb ≡ parts per billion.

[c] Tentative identification, polar region.

[d] Value at 1 to 4 bar.

cally varied satellites of Jupiter). Io was found to have at least 10 active volcanos (see Fig. 10). Sulfur resulting from the volcanic emissions on Io is responsible for the orange color of its surface, as well as the presence of sulfur dioxide in its atmosphere (at a partial pressure of only about one ten millionth of a bar). Io's volcanic emissions are also responsible for the ions of oxygen and sulfur in Jupiter's magnetosphere.

After encountering Jupiter and its satellites, both Voyager spacecraft visited Saturn and its satellite system (see Figure 11). The six previously known rings were found to be composed of innumerable, individual ringlets with very few gaps observed anywhere in the ring system. Complex dynamical effects were photographed in the ring system, including spiral density waves similar to those believed to generate spiral structure in galaxies. The helium content of the atmosphere of Saturn was found to be about 6% by volume (with molecular hydrogen at about 94% by volume), compared with about 11% for Jupiter. The trace gases in the atmosphere of Saturn are similar to those in the atmosphere of Jupiter and include methane, acetylene, ethane, phosphine, and propane (C_3H_8) (see Table XI).

Titan, the largest satellite of Saturn, was found to have a diameter slightly smaller than that of Jupiter's largest satellite, Ganymede (see Table VIII). The atmosphere of Titan is covered by clouds and layers of aerosols and haze and has a surface pressure of about 1.5 bar, which makes it about 50% more massive than the

FIG. 6.   Jupiter as photographed by Voyager 1. The Great Red Spot can be seen in the lower center of the photograph. The atmosphere of Jupiter is massive and completely covered with clouds and aerosol haze layers.

earth's atmosphere. The surface temperature of Titan is a cold 100 K. The cloud- and haze-covered Titan is shown in Figs. 12 and 13, obtained by Voyager. Titan's atmosphere is mostly molecular nitrogen, with smaller amounts of methane and trace amounts of carbon monoxide, carbon dioxide, and various hydrocarbons (see Table XII for the chemical composition of Titan in different regions of its atmosphere). The surface of Titan may hold a large accumulation of liquid methane.

After encountering Saturn, Voyager 2 was targeted for Uranus. On January 24, 1986, Voyager 2 had its closest approach to Uranus. Prior to this encounter, very little was known about Uranus, one of the three planets (Neptune and Pluto, the other two) not known to the ancients. Uranus was discovered accidentally by Sir William Herschel in March 13, 1787. Uranus is so far away from the sun (2,869.6 million km) it only receives about 1/400 of the incident solar radiation that the earth receives. Voyager's ra-

TABLE XI.   Composition of the Atmosphere of Saturn[a]

Constituent	Volume mixing ratio
H_2	0.94
He	0.06
CH_4	0.0045
C_2H_2	0.11 ppm
C_2H_6	4.8 ppm
CH_3C_2H[b]	No estimate
C_3H_8[b]	No estimate
CH_3D	0.23 ppm
PH_3	2 ppm

[a] From Strobel, D. F. (1985). In "The Photochemistry of Atmospheres: Earth, the Other Planets, and Comets," J. S. Levine, ed. Academic, Orlando, FL. Copyright 1985 Academic Press.
[b] Tentative identification.

dio signals took 2 hr and 45 min to reach the earth from Uranus.

As Voyager approached Uranus, its cameras indicated that Uranus did not exhibit the colorful and very turbulent cloud structure of Jupiter or the more subdued cloud banding and blending of Saturn. The very low contrast face of Uranus exhibited virtually no detail (see Fig. 14). The atmosphere of Uranus, like those of Jupiter and Saturn, is composed primarily of molecular hydrogen (about 85%) and helium (15 ± 5%). Methane is present in the upper atmosphere and is also frozen out in the form of ice in the cloud layer. The methane in the upper atmosphere selectively absorbs the red portion of the spectrum and gives Uranus its blue-green appearance. The volume percentage of methane may be as much as 2% deep in the atmosphere. Acetylene (C_2H_2) with a mixing ratio of about 2×10^{-7} was also detected in the atmosphere of Uranus. The temperature of the atmosphere was found to drop to a minimum of about 52 K (at the 100 mbar pressure level) before increasing to about 750 K in the extreme upper atmosphere.

Uranus has a ring system (as do Jupiter and Saturn). Two new rings (designated 1986 U1R and 1986 U2R) were discovered in Voyager 2 images of Uranus. The ring system of Uranus consists of 11 distinct rings that range in distance from about 37,000 to 51,000 from the center of Uranus.

Prior to the Uranus encounter, five satellites were known to be orbiting Uranus: Miranda

TABLE XII.   Composition of the Atmosphere of Titan[a]

Constituent		Volume mixing ratio	
N_2		0.76–0.98[b]	
	Surface	Stratosphere	Thermosphere (3900 km)
CH_4	0.02–0.08	≤0.026	0.08 ± 0.03
Ar	<0.16		<0.06
Ne	<0.002		<0.01
CO	60 ppm		<0.05
H_2	0.002 ± 0.001		
C_2H_6		20 ppm	
C_3H_8		1–5 ppm	
C_2H_2		3 ppm	~0.0015 (3400 km)
C_2H_4		0.4 ppm	
HCN		0.2 ppm	<0.0005 (3500 km)
C_2N_2		0.01–0.1 ppm	
HC_3N		0.01–0.1 ppm	
C_4H_2		0.01–0.1 ppm	
CH_3C_2H		0.03 ppm	
CO_2		1–5 ppb	

[a] From Strobel, D. F. (1985). In "The Photochemistry of Atmospheres: Earth, the Other Planets, and Comets," J. S. Levine, ed. Academic, Orlando, FL. Copyright 1985 Academic Press.
[b] Preferred value.

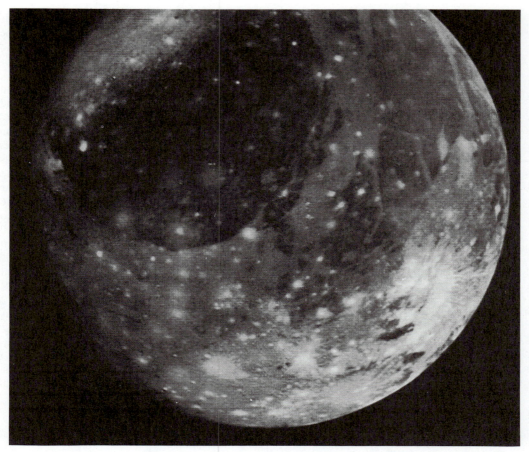

FIG. 7.   Ganymede, the largest satellite of Jupiter, as photographed by Voyager 2. The photograph shows a large, dark circular feature about 3200 km in diameter. The bright spots dotting the surface are relatively recent impact craters, while lighter circular areas may be older impact areas.

(distance = 129,000 km from the center of Uranus; diameter = 484 ± 10 km), Ariel (distance = 190,900 km; diameter = 1160 ± 10 km), Umbriel (distance = 266,000 km; diameter = 1190 ± 20 km), Titania (distance = 436,300 km; diameter = 1610 ± 10 km), and Oberon (distance = 583,400 km; diameter = 1550 ± 20 km). Ten new satellites were discovered on Voyager 2 images. All 10 satellites orbit Uranus within the orbit of Miranda (at distances that range from 49,700 to 86,000 km from the center of Uranus) and have diameters that range from about 40 to 80 km. Two of the newly discovered satellites are located within the ring system of Uranus and "shepherd" one of the rings.

After its encounter with Uranus in January 1986, Voyager 2 was targeted for an encounter with Neptune in September 1989. After its encounter with Neptune, Voyager 2 will join Voyager 1 and Pioneer 10 and 11 and escape the gravitational pull of the sun and head for the stars.

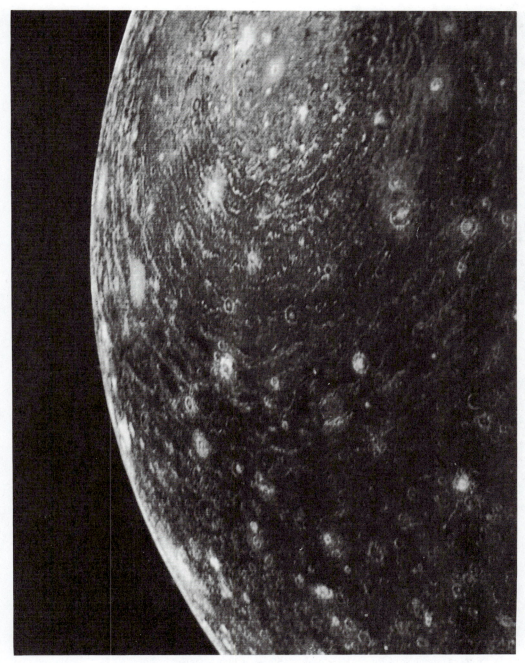

FIG. 8. Callisto, the second largest satellite of Jupiter, as photographed by Voyager 1. Far more craters appear on the surface of Callisto than on the surface of Ganymede, suggesting that Callisto may be the oldest satellite of Jupiter.

FIG. 9. Europa, smallest of Jupiter's Galilean satellites, as photographed by Voyager 2. It is believed that Europa has a reasonable quantity of water in the form of a mantle of ice with interior slush, perhaps 100 km thick.

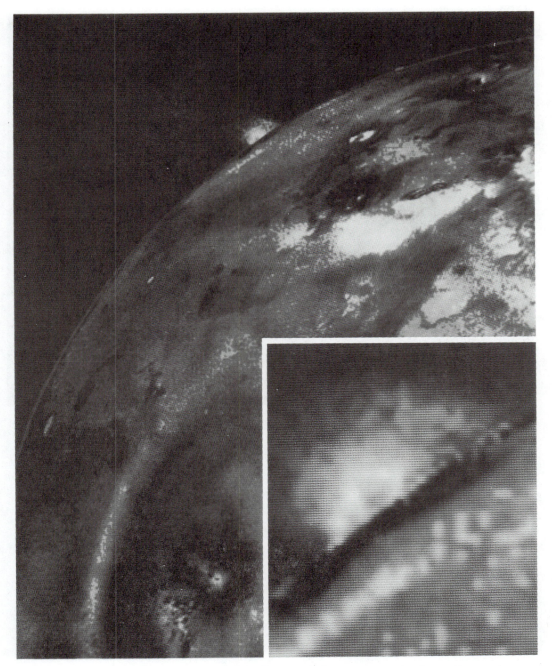

FIG. 10.   Io, satellite of Jupiter photographed by Voyager 1. An enormous volcanic eruption can be seen silhouetted against space over Io's bright limb. Solid volcanic material has been ejected up to an altitude of about 160 km.

FIG. 11.   Saturn photographed by Voyager 1. Like Jupiter, the atmosphere of Saturn is massive and completely covered with clouds and aerosol and haze layers. The projected width of the rings at the center of the disk is 10,000 km, which provides a scale for estimating feature sizes on the image.

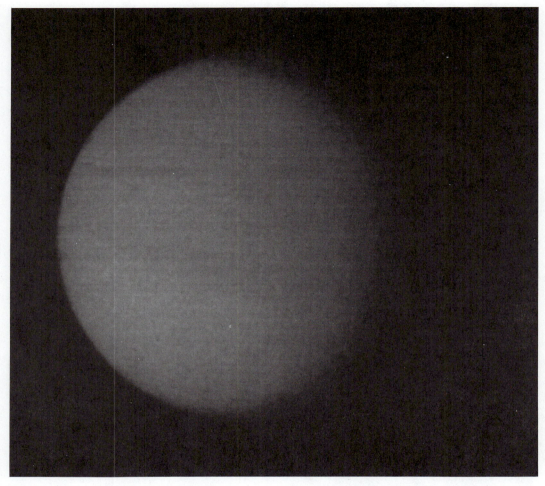

FIG. 12.   Titan, satellite of Saturn photographed by Voyager 1. Titan is the only satellite in the solar system with an appreciable atmosphere. The brownish-orange atmosphere of Titan contains clouds and haze and aerosol layers.

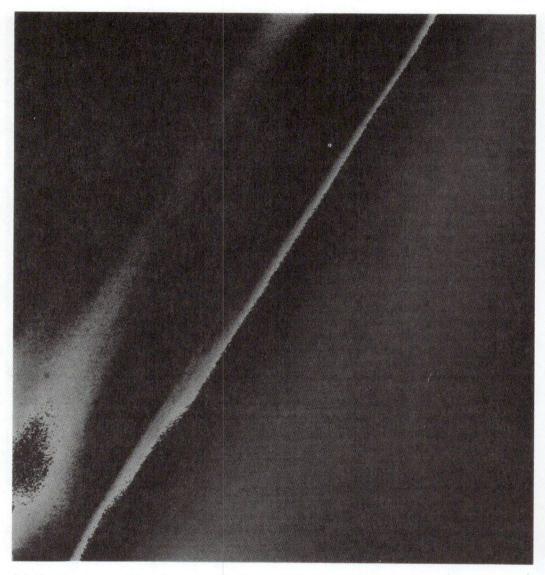

FIG. 13.   Haze layers of Titan photographed by Voyager 1. The upper level of the thick aerosol layer above the satellite's limb appears orange. The divisions in the haze occur at altitudes of 200, 375, and 500 km above the limb.

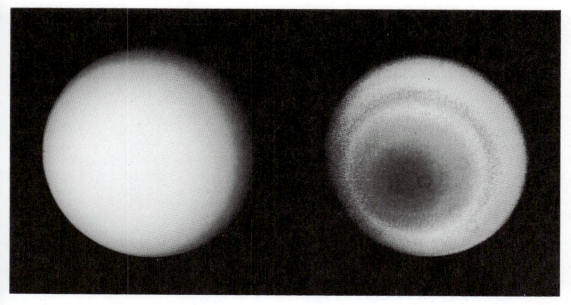

FIG. 14.   Two Voyager 2 views of Uranus—one in true color (left) and the other in false color. The picture on the left shows how Uranus would appear to human eyes from the vantage point of the spacecraft. The picture on the right uses contrast enhancement to bring out subtle details in the polar region of Uranus.

BIBLIOGRAPHY

Barth, C. A. (1985). In "The Photochemistry of Atmospheres: Earth, the Other Planets, and Comets," J. S. Levine, ed. Academic, Orlando, FL, pp. 337–392.

Cameron, A. G. W. (1968). In "Origin and Distribution of the Elements," L. H. Ahrens, ed., Pergamon, New York, pp. 125–143.

Levine, J. S., ed. (1985a). "The Photochemistry of Atmospheres: Earth, the Other Planets, and Comets." Academic, Orlando, FL.

Levine, J. S. (1985b). In "The Photochemistry of Atmospheres: Earth, the Other Planets, and Comets," J. S. Levine, ed. Academic, Orlando, FL, pp. 3–38.

Lewis, J. S., and Prinn, R. G. (1984). "Planets and Their Atmospheres: Origin and Evolution." Academic, New York.

NASA (1983). "Planetary Exploration Through Year 2000: A Core Program." Solar System Exploration Committee of the NASA Advisory Council, Washington, DC.

Prinn, R. G. (1985). In "The Photochemistry of Atmospheres: Earth, the Other Planets, and Comets," J. S. Levine, ed. Academic, Orlando, FL, pp. 281–336.

Strobel, D. F. (1985). In "The Photochemistry of Atmospheres: Earth, the Other Planets, and Comets," J. S. Levine, ed. Academic, Orlando, FL, pp. 393–434.

Walker, J. C. G. (1977). "Evolution of the Atmosphere." Macmillan, New York.

PLANETARY RADAR ASTRONOMY

Steven J. Ostro   *Jet Propulsion Laboratory*[†]

I. Introduction
II. Techniques and Instrumentation
III. Radar Measurements and Target Properties
IV. The Future of Planetary Radar Astronomy

GLOSSARY

Aliasing: Overlapping of echo at different frequencies or at different time delays.

Antenna gain: Ratio of an antenna's sensitivity in the direction it is pointed to its average sensitivity in all directions.

Circular polarization ratio: Ratio of echo power received in the same sense of circular polarization as transmitted (the SC sense) to that received in the opposite (OC) sense.

Doppler shift: Difference between the frequencies of the radar echo and the transmission, caused by the relative velocity of the target with respect to the radar.

Echo bandwidth: Dispersion in Doppler frequency of an echo, that is, the width of the echo power spectrum.

Ephemeris: Table of planetary positions as a function of time (plural: ephemerides).

Klystron: Vacuum-tube amplifier used in planetary radar transmitters.

Radar albedo: Ratio of a target's radar cross section in a specified polarization to its projected area; hence, a measure of the target's radar reflectivity.

Radar cross section: Most common measure of the intensity of a target's echo, equal to the projected area of the perfect metal sphere that would give the same echo power as the target if observed at the target's location.

Scattering law: Function giving the dependence of a surface element's radar cross section on viewing angle.

Synodic rotation period: Rotation period of a target as viewed by a moving observer, to be distinguished from the sidereal rotation period measured with respect to the fixed stars.

Time delay: Time between transmission of a radar signal and reception of the echo.

Planetary radar astronomy is the study of solar system entities (the moon, asteroids, and comets, as well as the major planets and their satellites and ring systems) by transmitting radio signals toward a target and receiving and analyzing the echoes. This field of research has primarily involved observations with earth-based radar telescopes but is now moving increasingly to include certain experiments with the transmitter and/or the receiver on board spacecraft orbiting or passing near planetary objects. Radar studies of the earth's surface, atmosphere, and ionosphere from spacecraft, aircraft, and the ground are not usually considered part of planetary radar astronomy. Radar studies of the sun involve such distinctly individual methodologies and physical considerations that solar radar astronomy is considered a field separate from planetary radar astronomy.

I. Introduction

A. SCIENTIFIC CONTEXT

Planetary radar astronomy is a field of science at the intersection of planetology, radio astronomy, and radar engineering. A radar telescope is essentially a radio telescope equipped with a high-power radio transmitter and specialized

[†] This research was conducted at the Jet Propulsion Laboratory, California Institute of Technology, under contract with the National Aeronautics and Space Administration.

TABLE I.   Radar-Detected Planetary Targets

Year of first detection	Terrestrial objects	Jupiter's satellites	Ring system	Mainbelt asteroids	Near-earth asteroids	Comets
1946	Moon					
1961	Venus					
1962	Mercury					
1963	Mars					
1968					1566 Icarus	
1972					1685 Toro	
1973			Saturn's rings			
1974		Ganymede				
1975		Callisto			433 Eros	
		Europa				
1976		Io			1580 Betulia	
1977				1 Ceres		
1979				4 Vesta		
1980				7 Iris	1862 Apollo	Encke
				16 Psyche		
1981				97 Klotho	1915 Quetzalcoatl	
				8 Flora	2100 Ra–Shalom	
1982				2 Pallas		Grigg–Skjellerup
				12 Victoria		
				19 Fortuna		
				46 Hestia		
1983				5 Astraea	1620 Geographos	IRAS–Araki–Alcock
				139 Juewa	2201 Oljato	Sugano–Saigusa–Fujikawa
				356 Liguria		
				80 Sappho		
				694 Ekard		
1984				9 Metis	2101 Adonis	
				554 Peraga		
				144 Vibilia		
1985				6 Hebe	1627 Ivar	
				41 Daphne	1036 Ganymed	
				21 Lutetia	1866 Sisyphus	
				33 Polyhymnia		
				84 Klio		
				192 Nausikaa		
				230 Athamantis		
				216 Kleopatra		
				18 Melpomene		

In the mid-1970s, echoes from Jupiter's Galilean satellites Europa, Ganymede, and Callisto revealed the manner in which these icy moons backscatter circularly polarized waves to be extraordinarily strange and totally outside the realm of previous radar experience. In 1984, it was discovered that the echo polarization properties of the near-earth asteroid 2101 Adonis resemble those of Callisto more than those of any other observed asteroid or comet. Our theoretical understanding of the anomalous radar behavior is still not complete, but it has improved substantially during the past decade.

By 1985, the list of radar-detected planetary objects included four comets, 13 near-earth asteroids, and 27 mainbelt asteroids (Table I). Comets and asteroids comprise an enormous, diverse population of bodies containing important clues to the origin and evolution of the solar system, and radar observations are now providing new information about these objects' sizes, shapes, spin vectors, surface structure, and composition (e.g., metal concentration). [*See* PRIMITIVE SOLAR SYSTEM OBJECTS: ASTEROIDS AND COMETS.]

II. Techniques and Instrumentation

A. Echo Detectability

How close must a planetary target be for its radar echo to be detectable? For a given transmitted power P_T and antenna gain G, the power flux at distance R from the radar is $P_T G/4\pi R^2$. We define the target's radar cross section σ as 4π times the backscattered power per steradian per unit incident flux at the target. Then, letting λ be the radar wavelength and defining the an-electronic instrumentation designed to link transmitter, receiver, data-acquisition, and telescope-pointing components together in an integrated radar system. The principles underlying operation of this system are not fundamentally very different from those involved in radars used, for example, in marine and aircraft navigation, measurement of automobile speeds, and satellite surveillance. However, planetary radars must detect echoes from targets at interplanetary distances ($\sim 10^5$–10^9 km) and therefore are the largest and most powerful radar systems in existence. [See RADIO ASTRONOMY, PLANETARY.]

The advantages of radar observations in astronomy stem from the high degree of control exercised by the observer on the signal transmitted to illuminate the target. Whereas virtually every other astronomical technique relies on passive measurement of reflected sunlight or naturally emitted radiation, the radar astronomer controls all the properties of the illumination including its intensity, direction, polarization, and time/frequency structure.

The properties of the transmitted waveform are selected to achieve particular scientific objectives. By comparing the properties of the echo to the very well-known properties of the transmission, one can deduce the target's properties. Hence, the observer is intimately involved in an active astronomical observation and, in a very real sense, performs a controlled laboratory experiment on the planetary target.

Computer analysis of the echo enables the target to be resolved spatially in a manner independent of its apparent angular extent, thereby bestowing a considerable advantage on radar over optical techniques in the study of asteroids, which appear as point sources through ground-based optical telescopes. Furthermore, by virtue of the centimeter–meter wavelengths employed, radar is sensitive to scales of surface structure many orders of magnitude larger than those probed in visible or infrared regions of the spectrum. Radar is also unique in its ability to "see through" the dense clouds that enshroud Venus and the glowing gaseous atmosphere or coma that conceals the nucleus of a comet. Because of its unique capabilities radar astronomy has made essential contributions to planetary exploration for a quarter of a century.

B. History

Radar technology was developed rapidly to meet military needs during World War II. In 1946, soon after the war, groups in the United States and Hungary obtained echoes from the moon, giving birth to planetary radar astronomy. These early efforts were motivated primarily by interest in electromagnetic propagation through the ionosphere and the possibility for using the moon as a relay for radio communication.

During the next two decades the development of nuclear weaponry and the need for ballistic missile warning systems prompted enormous improvements in radar capabilities. This period also saw rapid growth in radio astronomy and the construction of huge radio telescopes. In 1957, the Soviet Union launched Sputnik and with it the Space Age; and in 1958, with the formation by the U.S. congress of the National Aeronautics and Space Administration (NASA), a great deal of scientific attention turned to the moon and to planetary exploration in general. During the ensuing years exhaustive radar investigations of the moon were conducted at wavelengths from 0.9 cm to 20 m, and the results generated theories of radar scattering from natural surfaces that still see wide application.

By 1963, improvements in the sensitivity of planetary radars had permitted the initial detections of echoes from the terrestrial planets (Venus, Mercury, and Mars). During this period radar investigations provided the first accurate determinations of the rotations of Venus and Mercury and the earliest indications for the extreme geologic diversity of Mars. Radar images of Venus have revealed small portions of that planet's surface at increasingly fine resolution since the late 1960s; and in 1979, the Pioneer Venus Spacecraft Radar Experiment gave us our first look at Venus's global distributions of topography, radar reflectivity, and surface slopes.

The first echoes from a near-earth asteroid (1566 Icarus) were detected in 1968; it was nearly another decade before the first radar detection of a mainbelt asteroid (1 Ceres in 1977), followed in 1980 by the first detection of echoes from a comet (Encke).

During 1972 and 1973, detection of 13-cm wavelength (λ13 cm) radar echoes from Saturn's rings shattered prevailing notions that typical ring particles were 0.1–1.0 mm in size; the fact that decimeter-scale radio waves are backscattered efficiently requires that a large fraction of the particles be larger than a centimeter. Observations by the Voyager spacecraft have confirmed this fact and further suggest that particle sizes extend to at least 10 m.

tenna's effective aperture as $A_e = G\lambda^2/4\pi$, we find the received power to be

$$P_R = \frac{P_T G A_e \sigma}{(4\pi)^2 R^4} \quad (1)$$

This power might be much less than the receiver noise power $P_N = kT_S \, \Delta f$, where k is Boltzmann's constant, T_S the receiver system temperature, and Δf the frequency resolution of the data. However, the mean level of P_N constitutes a background that can be determined and removed, so P_R is detectable as long as it is at least several times larger than the standard deviation of the random fluctuations in P_N. These fluctuations can be shown to have a distribution that for usual values of Δf and the integration time Δt is nearly Gaussian with standard deviation $\Delta P_N = P_N/(\Delta f \, \Delta t)^{1/2}$. The highest signal-to-noise ratio (SNR), $P_R/\Delta P_N$, is achieved for a frequency resolution equal to the intrinsic bandwidth of the echo. That bandwidth is proportional to $D/\lambda P$, where D is the target's diameter and P the target's rotation period, so let us assume that $\Delta f \sim D/\lambda P$. By writing $\sigma = \hat{\sigma}\pi D^2/4$, where the radar albedo $\hat{\sigma}$ is a measure of the target's radar reflectivity, we arrive at the following expression for the echo's SNR:

$$\text{SNR} \sim (\text{system factor})(\text{target factor}) \, (\Delta t)^{1/2} \quad (2)$$

where

$$\text{system factor} \sim \frac{P_T A_e^2}{\lambda^{3/2} T_S} \quad (3)$$
$$\sim \frac{P_T \, G^2 \lambda^{5/2}}{T_S}$$

and

$$\text{target factor} \sim \frac{\hat{\sigma} D^{3/2} P^{1/2}}{R^4} \quad (4)$$

The inverse-fourth-power dependence of SNR on target distance is a severe limitation in ground-based observations but can be overcome by constructing very powerful radar systems.

B. RADAR SYSTEMS

Two active planetary radar facilities exist: the National Astronomy and Ionosphere Center's Arecibo Observatory in Puerto Rico and the Jet Propulsion Laboratory's Goldstone Solar System Radar in California. The Arecibo telescope (Fig. 1) consists of a 305 m diameter, fixed reflector whose surface is a section of a 265 m radius sphere. Movable line feeds designed to correct for spherical aberration are suspended from a triangular platform $\sim$130 m above the

reflector and can be aimed toward various positions on the reflector, enabling the telescope to point to within about 20° of the overhead direction (declination 18.3°). The Goldstone main antenna is a fully steerable, 65 m parabolic reflector with horn feeds (Fig. 2). Radar wavelengths are 13 and 70 cm for Arecibo and 3.5 and 13 cm for Goldstone; with each instrument, greater sensitivity is achievable with the shorter wavelength. Order-of-magnitude values of optimum system characteristics for each radar are $P_T \approx$ 400 kw, $G \approx 10^{7.1}$, and $T_S \approx 25$ K. Arecibo is about an order of magnitude more sensitive at $\lambda 13$ cm than Goldstone is at $\lambda 3.5$ cm. However, Goldstone has access to the entire sky north of declination $-50°$ and can track targets continuously for longer periods.

Figure 3 is a simplified block diagram of a planetary radar system. A waveguide switch is used to connect the antenna to the transmitter or to the receiver. In most observations one transmits for a duration near the round-trip propagation time to the target (that is, until the echo from the beginning of the transmission is about to arrive) and then receives for a similar duration.

The heart of the planetary radar transmitter is one or more klystron vacuum-tube amplifiers. In these tubes, magnets focus electrons falling through a potential drop of some 60 kV and modulate their velocities, and hence the electron density and energy flux, at radio frequencies (RF). Internal resonant cavities enhance this modulation, and about half of the nearly one megawatt of input dc power is converted to RF power, sent out through a waveguide to the antenna feed, and radiated toward the target. The other half of the input power is waste heat that must be transported away from the klystron by cooling water. The impact of the electrons on the collector anode generates dangerous X-rays that must be contained by heavy metal shielding surrounding the tube, a requirement that further boosts the weight, complexity, and cost of the klystron.

In the receiving system, the maser-amplified echo signal is converted in a superheterodyne mixer from RF frequencies (e.g., $\sim$2380 MHz for Arecibo at $\lambda 13$ cm) down to intermediate frequencies (IF, e.g., $\sim$30 MHz), for which transmission line losses are small, and passed from the proximity of the antenna feed to a remote control room containing additional stages of signal-processing equipment, computers, and digital tape recorders. The signal is converted to very low (baseband) frequencies and is filtered

FIG. 1. (a) Arecibo Observatory's radio/radar telescope in Puerto Rico. The diameter of the spherical reflector is 305 m. (b) Close-up view of the structure suspended above the reflector. Antenna feeds extend from the bottom of the two ''carriage houses'' that contain receiver and transmitter equipment.

FIG. 2.   The 65-m Goldstone Solar System Radar antenna in California.

and amplified, and samples of the signal's voltage are converted from analog to digital form. The nature of the final processing prior to recording data on magnetic tape depends on the nature of the radar experiment and particularly on the time/frequency structure of the transmitted waveform.

C. ECHO TIME-DELAY AND DOPPLER SHIFT

The time between transmission of a radar signal and reception of the echo is called the echo's round-trip time delay τ and is of order $2R/c$, where c is the speed of light. Since planetary targets are not points, even an infinitesimally short transmitted pulse would be dispersed in time delay, and the total extent $\Delta\tau_{TARGET}$ of the distribution $\sigma(\tau)$ of echo power (in units of radar cross section) would be D/c for a sphere of diameter D and in general would depend on the target's size and shape.

The translational motion of the target with respect to the radar introduces a Doppler shift ν in the frequency of the transmission. Both the time delay and the Doppler shift of the echo can be predicted in advance from the target's ephemeris, which is calculated using the geodetic position of the radar and the orbital elements of the earth and the target. The predicted Doppler shift can be removed electronically by continuously tuning the local oscillator used for RF to IF frequency conversion (Fig. 3). The predicted Doppler must be accurate enough to avoid smearing out the echo in frequency, and this requirement places stringent demands on the quality of the observing ephemeris.

Because different parts of the rotating target have different velocities relative to the radar, the echo is dispersed in Doppler frequency as well as in time delay. The basic strategy of any radar experiment always involves measurement of some characteristic(s) of the function $\sigma(\tau, \nu)$, perhaps as a function of time and perhaps using more than one combination of transmitted and received polarizations. Ideally, one would like to obtain $\sigma(\tau, \nu)$ with very fine resolution, sampling that function within intervals whose dimensions $\Delta\tau \times \Delta\nu$ are minute compared with the echo dispersions $\Delta\tau_{TARGET}$ and $\Delta\nu_{TARGET}$. Unfortunately, SNR is proportional to $(\Delta\nu)^{1/2}$, so

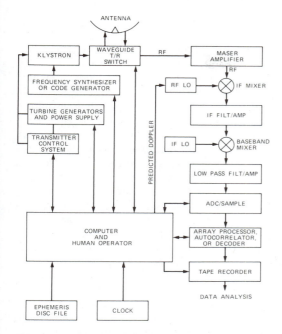

FIG. 3. Block diagram of a planetary radar system. RF LO and IF LO denote radio frequency and intermediate frequency local oscillators, and ADC denotes analog-to-digital converter.

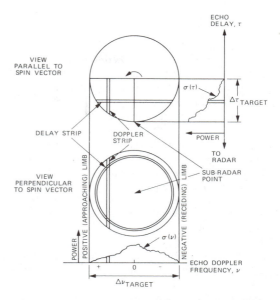

FIG. 4. Time-delay and Doppler-frequency resolution of the radar echo from a rotating spherical target.

one's ability to resolve $\sigma(\tau, \nu)$ is necessarily limited by the available echo strength. Furthermore, an intrinsic upper bound on the product $\Delta\tau\,\Delta\nu$ forces a trade-off between delay resolution and Doppler resolution. Under these constraints, many planetary radar experiments employ waveforms aimed at providing estimates of one of the marginal distributions, $\sigma(\tau)$ or $\sigma(\nu)$. Figure 4 shows the geometry of delay-resolution cells and Doppler-resolution cells for a spherical target and sketches the relation between these cells and $\sigma(\nu)$ and $\sigma(\tau)$. Delay–Doppler measurements are explored further below.

D. RADAR WAVEFORMS

In the simplest radar experiment, the transmitted signal is unmodulated (continuous wave or cw) and highly monochromatic. Analysis of the received signal comprises Fourier transformation of a series of time samples and yields an estimate of the echo power spectrum $\sigma(\nu)$, but it contains no information about the distance to the target or $\sigma(\tau)$. To avoid aliasing, the sampling rate must be at least as large as the bandwidth of the low-pass filter (Fig. 3), and usually

is comparable to or larger than the echo's intrinsic dispersion $\Delta\nu_{\text{TARGET}}$ from Doppler broadening.

Fast-Fourier transform (FFT) algorithms, implemented via software or hard-wired in an array processor, greatly speed the calculation of discrete spectra from time series and are ubiquitous in radar astronomy. In a single FFT operation, a string of N time samples taken at intervals of Δt sec is transformed into a string of N spectral elements with frequency resolution $\Delta f = 1/N\,\Delta t$. Many thousands of FFTs may be needed to reduce the data from a single transmit/receive cycle to echo spectra, and data reduction times can drastically exceed data acquisition times even with the most efficient FFT algorithms. (Needless to say, the high-speed computer is utterly essential in radar astronomy.) Most planetary radar targets are sufficiently narrowband ($\Delta\nu_{\text{TARGET}} \lesssim$ a few tens of kilohertz) for power spectra to be computed and accumulated in an array processor (Fig. 3) and recorded directly on magnetic tape at convenient intervals. In some situations it is desirable to record time samples on tape and Fourier analyze the data later, perhaps using FFTs of different lengths to obtain spectra at a variety of frequency resolutions. For wideband echoes (e.g., $\sim10^6$ Hz for Saturn's rings) the sampling-rate requirements are met most readily by passing the signal through an autocorrelator, recording autocorrelation functions, and then applying FFTs to extract spectra.

To obtain delay resolution one must apply some sort of time modulation to the transmitted waveform. For example, a short-duration pulse of cw signal lasting 1 μsec would provide delay resolution of 150 m. However, the echo would have to compete with the noise power in a bandwidth of order 1 MHz, so the echo power from many consecutive pulses would probably have to be summed to yield a detection. One would not want these pulses to be too close together, however, or there would be more than one pulse incident on the target at once and interpretation of echoes would be insufferably ambiguous. Thus, one arranges the interpulse period t_{IPP} to exceed the target's intrinsic delay dispersion $\Delta\tau_{TARGET}$, ensuring that the echo consists of successive, nonoverlapping replicas of $\sigma(\tau)$ that are separated from each other by t_{IPP}. To generate this pulsed cw waveform, the transmitter is switched on and off while the frequency synthesizer (Fig. 3) maintains phase coherence from pulse to pulse. Then Fourier transformation of time samples, taken at the same position within each of N successive replicas of $\sigma(\tau)$, yields the power spectrum of echo from a certain delay resolution cell on the target. This spectrum has

an unaliased bandwidth of $1/t_{IPP}$ and a frequency resolution of $1/Nt_{IPP}$. Repeating this process for a different position within each replica of $\sigma(\tau)$ yields the power spectrum for echo from a different delay resolution cell, and in this manner one obtains $\sigma(\tau, \nu)$.

In practice, instead of pulsing the transmitter, one usually codes a cw signal with a sequence of 180° phase reversals and cross-correlates the echo with a representation of the code (e.g., using the decoder in Fig. 3), thereby synthesizing a pulse train with the desired values of Δt and t_{IPP}. With this approach, one optimizes SNR by transmitting power continuously and extends the klystron's lifetime by avoiding on–off switching intervals comparable to the tube's internal thermal time constants.

A limitation of coherent-pulsed or phase-coded cw waveforms follows from combining the requirement that there never be more than one echo received from the target at any instant (i.e., that $t_{IPP} > \Delta\tau_{TARGET}$) with the antialiasing frequency requirement that the rate $1/t_{IPP}$ at which echo from a given delay resolution cell is sampled be no less than the target bandwidth $\Delta\nu_{TARGET}$. Therefore, a target must satisfy $\Delta\tau_{TARGET}\Delta\nu_{TARGET} \leq 1$ or it is overspread (Table II) and cannot be investigated completely and simultaneously in delay and Doppler without

TABLE II.   Characteristics of Selected Planetary Radar Targets [a]

Target	Minimum echo delay[b] (min)	Radar cross section, σ_{OC} (km²)	Radar albedo $\hat{\sigma}_{OC}$	Circular polarization ratio, μ_C	Maximum dispersions[c] Delay (ms)	Doppler (Hz)	Product
Moon	0.04	6.6×10^5	0.07	0.1	12	60	0.7
Mercury	9.1	1.1×10^6	0.06	0.1	16	110	2
Venus	4.5	1.3×10^7	0.11	0.1	40	110	4
Mars	6.2	2.9×10^6	0.08	0.3	23	7600	200
1 Ceres	27	3.3×10^4	0.05	0.0	3	3100	9
2 Pallas	25	2.1×10^4	0.09	0.0	2	2000	4
12 Victoria	17	2.1×10^3	0.15	0.1	0.5	590	3
16 Psyche	28	1.4×10^4	0.29	0.1	0.8	2200	2
1685 Toro	2.3	1.7	0.1	0.2	0.02	14	10^{-4}
1862 Apollo	0.9	0.2	0.1	0.4	0.01	16	10^{-3}
2100 Ra–Shalom	3.0	1.0	0.1	0.2	0.01	5	10^{-4}
2201 Adonis	1.5	0.02	?	1.0	?	2	?
Comet IRAS–Araki–Alcock	0.5	2.2	?	0.1	?	300	?
Io	73	1.9×10^6	0.2	?	12	2400	30
Europa	73	7.7×10^6	1.0	1.6	10	1000	10
Ganymede	73	1.3×10^7	0.6	1.6	18	850	15
Callisto	73	5.3×10^6	0.3	1.2	16	330	5
Saturn's rings	134	10^8–10^9	0.7	0.5	1600	6×10^5	10^6

[a] Typical 13-cm-wavelength values. Question marks denote absence of radar data (μ_C) or of prior information about target dimensions.

[b] For asteroids this is the minimum delay for observations to date.

[c] Doppler dispersion for transmitter frequency of 2380 MHz ($\lambda = 12.6$ cm). The product of the dispersions in delay and Doppler is the overspread factor at 2380 MHz.

aliasing, at least with the waveforms discussed so far. Various degrees of aliasing may be acceptable for overspread factors $\lesssim 10$, depending on the precise experimental objectives and the exact properties of the echo. One can deal with extremely overspread targets (e.g., Saturn's rings with an overspread factor $\sim 10^6$ at $\lambda 13$ cm) by using frequency-stepped and frequency-swept waveforms, but only at the expense of more complex data acquisition and reduction procedures. Virtually all modern ground-based radar observations of planetary targets employ cw or phase-coded cw waveforms.

III. Radar Measurements and Target Properties

A. ALBEDO AND POLARIZATION RATIO

A primary goal of the initial radar investigation of any planetary target is estimation of the target's radar cross section σ and its normalized radar cross section or radar albedo $\hat{\sigma} \equiv \sigma/A_p$, where A_p is the target's geometric projected area. Since the radar astronomer selects the transmitted and received polarizations, any estimate of σ or $\hat{\sigma}$ must be identified accordingly. The most common approach is to transmit a circularly polarized wave and to use separate receiving systems for simultaneous reception of the same sense of circular polarization as transmitted (i.e., the SC sense) and the opposite (OC) sense. The handedness of a circularly polarized wave is reversed on normal reflection from a smooth dielectric interface, so the OC sense dominates echoes from targets that look smooth at the radar wavelength. In this context a surface with minimum radius of curvature very much larger than λ would look smooth. SC echo can arise from single scattering from rough surfaces, multiple scattering from smooth surfaces, or certain subsurface refraction effects. The circular polarization ratio $\mu_C \equiv \sigma_{SC}/\sigma_{OC}$ is thus a useful measure of near-surface roughness.

When linear polarizations are used, it is convenient to define the ratio $\mu_L = \sigma_{OL}/\sigma_{SL}$, which would be close to zero for normal reflection from a smooth dielectric interface. For all radar-detected planetary targets, $\mu_L < 1$ and $\mu_L < \mu_C$. Although the OC radar albedo $\hat{\sigma}_{OC}$ is the most widely used gauge of radar reflectivity, some radar measurements are reported in terms of the geometric albedo, equal to $(\hat{\sigma}_{OC} + \hat{\sigma}_{SC})/4 = (\hat{\sigma}_{SL} + \hat{\sigma}_{OL})/4$. A perfectly smooth metallic sphere would have $\hat{\sigma}_{OC} = \hat{\sigma}_{SL} = 1$, a geometric albedo of 0.25, and $\mu_C = \mu_L = 0$.

If μ_C is close to zero (as for the planet Venus and the mainbelt asteroid 2 Pallas), its physical interpretation is unique, because the surface must be smooth at all scales within about an order of magnitude of λ, and there can be no subsurface structure at those scales within several $1/e$ power absorption lengths L of the surface proper. In this special situation, we can interpret the radar albedo as the product $g\rho$ where ρ is the Fresnel power-reflection coefficient at normal incidence, and the backscatter gain g depends on target shape, the distribution of surface slopes with respect to that shape, and target orientation. For most applications to date, g is $\lesssim 10\%$ larger than unity, so the radar albedo provides a reasonable first approximation to ρ. Both ρ and L depend on very interesting characteristics of the surface material, including bulk density, porosity, particle size distribution, and metal abundance.

If μ_C is $\gtrsim 0.3$ (e.g., Mars and some near-earth asteroids), then much of the echo arises from some backscattering mechanism other than single coherent reflections from large, smooth surface elements. Possibilities include multiple scattering from buried rocks or the interiors of concave surface features such as craters, or reflections from very jagged surfaces with radii of curvature much less than a wavelength. Most planetary targets have values of $\mu_C \lesssim 0.2$ at decimeter wavelengths, so their surfaces are dominated by a component that is smooth at centimeter-to-meter scales.

The observables $\hat{\sigma}_{OC}$ and μ_C are disk-integrated quantities derived from integrals of $\sigma(\nu)$ or $\sigma(\tau)$ in specific polarizations. Later, we shall see how their physical interpretation profits from knowledge of the functional forms of $\sigma(\nu)$ and $\sigma(\tau)$.

B. DYNAMICAL PROPERTIES FROM MEASUREMENT OF DELAY AND DOPPLER

Consider radar observation of a point target a distance R from the radar. As noted earlier, the round-trip time delay between transmission of a pulse toward the target and reception of the echo would be $\tau = 2R/c$. It is possible to measure time delays to within 10^{-6} sec. Actual delays range from $2\frac{1}{2}$ sec for the moon to $2\frac{1}{2}$ hr for Saturn's rings. For a typical target distance ~ 1 AU, the time delay is ~ 1000 sec and can be measured within a fractional timing uncertainty of 10^{-9}. Since the fractional error in our knowledge of the speed of light is $\sim 10^{-7}$, we know the light-second equivalent of a radar-determined distance better than the distance in, say, meters. Consequently, results of very precise planetary radar range measurements are generally reported in units of time delay instead of distance, which is a more poorly known, derived quantity.

If the target is in motion and has a line-of-sight component of velocity toward the radar of v_{LOS}, the target sees a frequency that, to first order in v_{LOS}/c, equals $f_{TX} + (v_{LOS}/c)f_{TX}$, where f_{TX} is the transmitter frequency. The target reradiates the Doppler-shifted signal, and the radar receives echo whose frequency is, again to first order, given by

$$f_{TX} + 2(v_{LOS}/c)f_{TX}$$

That is, the total Doppler shift in the received echo is

$$2v_{LOS}f_{TX}/c = v_{LOS}/(\lambda/2)$$

so a 1-Hz Doppler shift corresponds to a velocity of half a wavelength per second (e.g., 6.5 cm sec^{-1} for $\lambda13$ cm). It is not difficult to measure echo frequencies to within 0.01 Hz, so v_{LOS} can be estimated with a precision ~1 mm sec^{-1}. Actual values of v_{LOS} for planetary radar targets can be as large as several tens of kilometers per second, so radar velocity measurements have fractional errors as low as ~10^{-8}. At this level, the second-order (special relativistic) contribution to the Doppler shift becomes measurable; in fact, planetary radar observations have provided the initial experimental verification of the second-order term.

By virtue of their high precision, radar measurements of time delay and Doppler frequency are very useful in refining our knowledge of various dynamical quantities. The first delay-resolved radar observations of Venus during 1961 and 1962 yielded an estimate of the light-second equivalent of the astronomical unit (AU) that was accurate to one part in 10^6, constituting a thousand-fold improvement in the best results achieved with optical observations alone. Subsequent radar observations provided additional refinements of nearly two more orders of magnitude. In addition to determining the scale of the solar system precisely, these observations greatly improved our knowledge of the orbits of earth, Venus, Mercury, and Mars, and were essential for the success of the first interplanetary missions. Radar observations still contribute to maintaining the accuracy of planetary ephemerides for objects in the inner solar system and have played an important role in dynamical studies of Jupiter's Galilean satellites.

Precise interplanetary time-delay measurements have also been used to test Einstein's theory of general relativity. Radar observations verify the theory's prediction that for radar waves passing near the sun, echo time delays are increased because of the distortion of space by the sun's gravity. The extra delay would be ~100 μsec if the angular separation of the target from the sun were several degrees. (The sun's angular diameter is about half a degree.)

Since planets are not point targets, their echoes are dispersed in delay and Doppler, and the refinement of dynamical quantities and the testing of physical theories are tightly coupled to estimation of the mean radii, the topographic relief, and the radar scattering behavior of the targets. The key to this entire process is resolution of the distributions of echo power in delay and Doppler. In the next section we shall consider inferences about a target's dimensions and spin vector from measurements of the dispersions ($\Delta\tau_{TARGET}$, $\Delta\nu_{TARGET}$) of the echo in delay and Doppler. Then we shall examine the physical information contained in the functional forms of the distributions $\sigma(\tau)$, $\sigma(\nu)$, and $\sigma(\tau, \nu)$.

C. DISPERSION OF ECHO POWER IN DELAY AND DOPPLER

Each backscattering element on a target's surface returns echo with a certain time delay and Doppler frequency (Fig. 4). Since parallax effects and the curvature of the incident wave front are negligible for most ground-based observations (but not necessarily for observations with spacecraft), contours of constant delay are intersections of the surface with planes perpendicular to the line of sight. The point on the surface with the shortest echo time delay is called the subradar point; the longest delays generally correspond to echoes from the planetary limbs. As noted already, the difference between these extreme delays is called the dispersion $\Delta\tau_{TARGET}$ in $\sigma(\tau)$, or simply the delay depth of the target.

If the target appears to be rotating, the echo is dispersed in Doppler frequency. For example, if the radar has an equatorial view of a spherical target with diameter D and apparent rotation period P, the difference between the line-of-sight velocities of points on the equator at the approaching and receding limbs is $2\pi D/P$. Thus the dispersion of $\sigma(\nu)$ is $\Delta\nu_{TARGET} = 4\pi D/\lambda P$. This quantity is called the bandwidth B of the echo power spectrum. If the view is not equatorial, the bandwidth is simply $(4\pi D \sin \alpha)/\lambda P$, where the aspect angle α is the acute angle between the spin vector and the line of sight. Thus, a radar bandwidth measurement furnishes a joint constraint on the target's size, rotation period, and pole direction.

In principle, echo bandwidth measurements obtained for a sufficiently wide variety of line-of-sight directions can yield all three scalar coordinates of the target's intrinsic (i.e., sidereal) spin vector ω_s. This capability follows from the fact that the apparent spin vector ω is the vector sum of ω_s and the contribution ($\omega_0 = \dot{e} \times e$, where the unit vector e points from the target to the radar) arising from the changing position of the radar on the celestial sphere as seen from the target's center of mass. Variations in e, $\dot{e}$, and hence ω_0, all of which are known, lead to measurement of different values of $\omega = \omega_s + \omega_0$, permitting unique determination of ω_s.

These principles were applied in the early 1960s to yield the first accurate determination of the rotations of Venus and Mercury (Fig. 5). Venus's rotation is retrograde with a 243-d sidereal period. The period is close to the value (243.16 d) characterizing a resonance with the relative orbits of earth and Venus, wherein Venus would appear from earth to rotate exactly four times between successive inferior conjunctions with the sun. However, two decades of Venus observations yield a refined value (243.01 ± 0.03 d) for the period that seems to establish nonresonance rotation. To date, a satisfactory explanation for Venus's curious spin state is lacking.

For Mercury, long imagined on the basis of optical observations to rotate once per 88-day revolution around the sun, radar bandwidth measurements (Fig. 5) demonstrated direct rotation with a period of 59 days, equal to $\frac{2}{3}$ of the orbital period. This spin–orbit coupling is such that during two Mercury years, the planet rotates three times with respect to the stars but only once with respect to the sun, so a Mercury-bound observer would experience alternating years of daylight and darkness.

What if the target is not a sphere but is irregular and nonconvex? In this situation, which is most applicable to small asteroids and cometary nuclei, the relationship between the echo power spectrum and the derivable information about the target's dimensions is shown in Fig. 6. We must interpret D as the sum of the distances r_+ and r_- from the plane ψ_0 containing the line of sight and the spin vector to the surface elements with the greatest positive (approaching) and negative (receding) line-of-sight velocities. In differ-

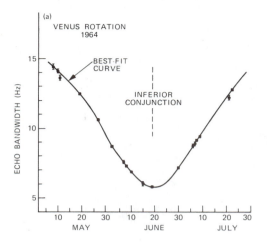

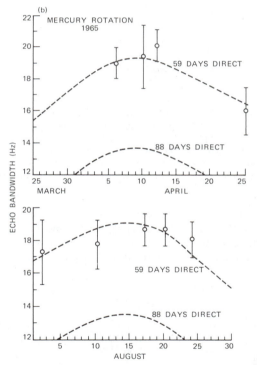

FIG. 5. Measurements of echo bandwidth (i.e., the dispersion of echo power in Doppler frequency) used to determine the rotations of Venus and Mercury.

ent words, if the planes ψ_+ and ψ_- are defined as being parallel to ψ_0 and tangent to the target's approaching and receding limbs, then ψ_+ and ψ_- are at distances r_+ and r_- from ψ_0. Letting f_0, f_+, and f_- be the frequencies of echoes from portions of the target intersecting ψ_0, ψ_+, and ψ_-, we have $B = f_+ - f_-$. Any constant-Doppler contour lies in a plane parallel to ψ_0.

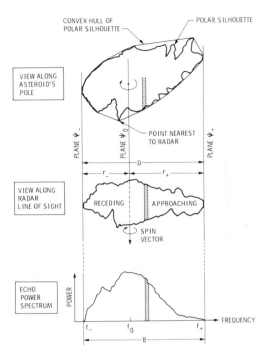

FIG. 6. Geometric relations between an irregular, nonconvex rotating asteroid and its echo power spectrum. The plane ψ_0 contains the asteroid's spin vector and the asteroid–radar line. The cross-hatched strip of power in the spectrum corresponds to echoes from the cross-hatched strip on the asteroid.

It is useful to imagine looking along the target's pole at the target's projected shape (i.e., its polar silhouette). D is simply the width, or breadth, of this silhouette (or, equivalently, of

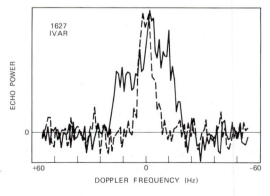

FIG. 7. Two $\lambda 13$-cm, OC radar echo spectra obtained for the near-earth asteroid 1627 Ivar near rotational phases ~90° apart. The asteroid evidently is two to three times longer than it is wide.

the silhouette's convex envelope or hull H_p) measured normal to the line of sight (Fig. 6). In general, r_+ and r_- are periodic functions of rotation phase ϕ and depend on the shape of H_p as well as on the target's mass distribution (i.e., the point about which H_p rotates). If the radar data sample all rotational phases modulo 180°, then in principle one can determine $f_+(\phi)$ and $f_-(\phi)$ completely and can recover H_p. For many small, near-earth asteroids, pronounced variations in $B(\phi)$ reveal highly noncircular polar silhouettes. For example, echo spectra obtained for the ~7-km object 1627 Ivar (Fig. 7) indicate a polar silhouette two to three times longer than it is wide.

D. DELAY AND DOPPLER DISTRIBUTIONS OF ECHO POWER

1. Angular Scattering Law

The functional forms of the distributions $\sigma(\tau)$ and $\sigma(\nu)$ contain information about the radar-scattering process and about the structural characteristics of the target's surface. Suppose the target is a large, smooth, spherical planet. Then echoes from the subradar region (near the center of the visible disk; see Fig. 4), where the surface elements are nearly perpendicular to the line of sight, would be much stronger than those from the limb regions (near the disk's periphery). This effect is seen visually when one shines a flashlight on a smooth, shiny ball: A bright glint appears where the geometry is right for backscattering. If the ball is roughened, the glint is spread out over a wider area, and in the case of extreme roughness, the scattering is described as diffuse rather than specular.

For a specular target $\sigma(\tau)$ would have a steep leading edge followed by a rapid drop. The power spectrum $\sigma(\nu)$ would be sharply peaked at central frequencies, falling off rapidly toward the spectral edges. If, instead, the spectrum were very broad, severe roughness at some scale(s) $\gtrsim \lambda$ would be indicated. In this case, knowledge of the echo's polarization properties would help to ascertain the particular roughness scale(s) responsible for the absence of the sharply peaked spectral signature of specular scattering.

By inverting the delay or Doppler distribution of echo power one can estimate the target's average angular scattering law, $\sigma_0(\theta) \equiv d\sigma/dA$, where dA is an element of surface area and θ the incidence angle between the line of sight and the normal to dA. For the echo's polarized (i.e., OC or SL) components, $\sigma_0(\theta)$ can be related to statistics describing the probability distribution for the slopes of surface elements. Examples of scattering laws applied in planetary radar astronomy are the Hagfors law,

$$\sigma_0(\theta) \sim s_0^{-2}(\cos^4 \theta + s_0^{-2} \sin^2 \theta)^{-3/2} \quad (5)$$

and the Gaussian law,

$$\sigma_0(\theta) \sim \frac{s_0^{-2} \exp(-s_0^{-2} \tan^2 \theta)}{\cos^4 \theta} \quad (6)$$

where s_0 is the rms slope in radians. In the following paragraphs, we shall explore the diversity of radar signatures encountered for solar system targets.

2. Radar Signatures of the Moon and the Inner Planets

Echoes from the moon, Mercury, Venus, and Mars are characterized by sharply peaked OC echo spectra (Fig. 8). Although these objects are collectively referred to as quasi-specular radar targets, their echoes also contain a diffusely scattered component and have circular polariza-

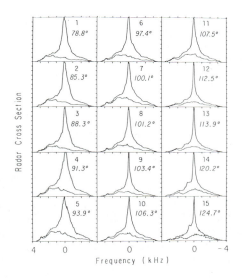

FIG. 8. Mars λ13-cm radar echo spectra for subradar points along 22° north latitude at the indicated west longitudes obtained in the OC (upper curves) and SC (lower curves) polarizations. In each box, spectra are normalized to the peak OC cross section. The echo bandwidth is 7.1 kHz. Very rough regions on the planet are revealed as bumps in the SC spectra that move from positive to negative Doppler frequencies as Mars rotates. (Courtesy of J. K. Harmon, D. B. Campbell and S. J. Ostro.)

tion ratios averaging about 0.07 for the moon, Mercury, and Venus, but ranging from 0.1 to 0.4 for Mars, as discussed below.

Typical rms slopes obtained at decimeter wavelengths for the four quasi-specular targets are around 7°, and consequently these objects' surfaces have been described as gently undulating. As might be expected, values estimated for s_0 increase as the observing wavelength decreases. For instance, for the moon s_0 increases from ~4° at $\lambda \sim 20$ m to ~8° at $\lambda \sim 10$ cm, and to ~33° at $\lambda \sim 1$ cm. At optical wavelengths ($\lambda \sim 0.5$ μm), the moon shows no trace of a central glint, and the scattering is entirely diffuse. This phenomenon arises because the lunar surface (Fig. 9) consists of a regolith (a layer of loose, fine-grained particles) with much intricate structure at the scale of visible wavelengths. Surface structure with radii of curvature near radar wavelengths $\geqslant 1$ cm is relatively rare, so the diffuse component of lunar radar echoes is much weaker than the quasi-specular component, especially at very long wavelengths. At decimeter wavelengths the ratio of diffusely scattered power to quasi-specularly scattered power is ~1/3 for the moon, Mercury, and Venus. This ratio can be determined by assuming that all the SC echo is diffuse and calculating the diffusely scattered fraction (x) of OC echo by fitting to the OC spectrum a model employing a composite scattering law (e.g., $\sigma_0(\theta) = x\sigma_{\text{DIF}}(\theta) + (1 - x)\sigma_{\text{QS}}(\theta)$). Here $\sigma_{\text{QS}}(\theta)$ might be the Hagfors law, and usually $\sigma_{\text{DIF}}(\theta) \sim \cos^m \theta$. Estimated values of m usually fall between one (i.e., geometric scattering, which describes the optical appearance of the full moon) and two (Lambert scattering). Physical interpretations of the diffusely scattered echo employ information about albedo, scattering law, and polarization to constrain the size distributions, spatial densities, and electrical properties of wavelength-scale rocks near the surface.

3. The Radar Heterogeneity of Mars

Diffuse scattering from Mars is much more substantial than that from the other quasi-specular targets and often accounts for most of the echo power. It seems, therefore, that Mars possesses an unusually high concentration of near-surface, centimeter-to-meter-scale rocks. Furthermore, features in this planet's SC spectra reveal the existence of regions of extreme small-scale roughness (Fig. 8). The precise geographic location of the source of a spectral feature can-

FIG. 9.   Structure on the lunar surface near the Apollo 17 landing site. Most of the surface is smooth and gently undulating at scales much larger than a centimeter. This smooth component of the surface is responsible for the predominantly quasi-specular character of the moon's radar echo at $\lambda \geqslant 1$ cm. Wavelength-scale structure produces a diffuse contribution to the echo. Wavelength-sized rocks are much more abundant at $\lambda \sim 4$ cm than at $\lambda \sim 10$ m (the scale of the boulder being inspected by Astronaut H. Schmitt), and hence diffuse echo is more substantial at shorter wavelengths.

not be ascertained from a single spectrum, since echo from anywhere along a constant-Doppler contour contributes to a given spectral element (Fig. 4). However, intersections between the contours corresponding to the feature's Doppler frequency in spectra obtained at different rotational phases can yield the longitude and the absolute value of the latitude of the source region. (The ambiguity in the sign of the latitude is discussed later.) For Mars the sources of SC spectral features evident in Fig. 8 apparently are volcanic regions, and the best terrestrial analog for this extremely rough terrain might be young lava flows (Fig. 10).

Since the motion in longitude of the subradar point on Mars (whose rotation period is only 24.6 hr) is rapid compared with that on the moon, Venus, or Mercury, and since the geometry of Mars' orbit and spin vector permits subradar tracks throughout the Martian tropics,

ground-based investigations of Mars have achieved more global coverage than those of the other terrestrial targets. Bistatic (i.e., two-station) radar observations of Mars, consisting of transmissions from an orbiting Viking spacecraft and ground-based reception of echoes from the Martian surface, have added information about surface properties near polar regions. The existing body of Mars radar data reveals extraordinary diversity in the degree of small-scale roughness as well as in the rms slope of smooth surface elements. For example, Fig. 11 shows the variation in OC echo spectral shape as a function of longitude for a subradar track along $\sim 16°S$ latitude. Surface slopes on Mars have rms values from less than $0.5°$ to more than $10°$. Chryse Planitia, site of the first Viking Lander, has fairly shallow slopes ($s_0 \sim 4$–$5°$), and in fact, radar rms slope estimates were utilized in selection of the Viking Lander sites.

FIG. 10.   This lava flow near Mount Shasta in California is an example of an extremely rough surface at decimeter radar wavelengths, possibly analogous to regions on Mars with very high circular polarization ratios. (Courtesy of D. Evans.)

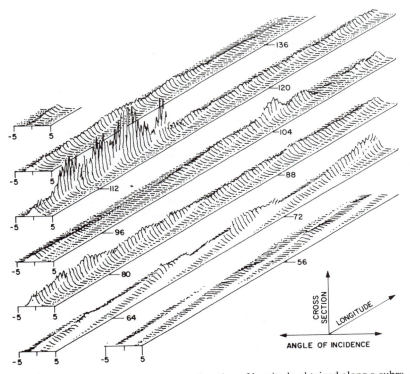

FIG. 11.   Mars echo power spectra as a function of longitude obtained along a subradar track at 16° south latitude. The most sharply peaked spectra correspond to the smoothest regions (i.e., smallest rms slopes). (Courtesy of G. S. Downs, P. E. Reichley, and R. R. Green.)

4. Radar Signatures of Asteroids

Echo spectra for the several dozen radar-detected asteroids show no hint of the sharply peaked signature of quasi-specular scattering seen for the terrestrial planets. However, the circular polarization ratios for mainbelt asteroids average about 0.1, so the echoes arise primarily from single-reflection backscattering from smooth surface elements. The broad spectra indicate that the rms slopes of these elements must be steeper than, say, the typical lunar value of 7°. For example, the nearly spherical, ~540-km-diameter asteroid 2 Pallas (Fig. 12) has $\mu_C = 0.05 \pm 0.02$, and fitting a spectral model based on a Gaussian scattering law indicates that $s_0 \gtrsim 20°$. This result suggests rather dramatic topographic relief, perhaps the outcome of impact cratering.

The much smaller near-earth asteroid 433 Eros (mean diameter ~20 km) has λ3.5-cm and λ13-cm values of μ_C about six times larger than Pallas's value, but not very different from the mean value for Mars. (Of course, very different physical processes are responsible for the small-scale roughness on Mars and Eros.) The Eros OC spectra vary in bandwidth and spectral

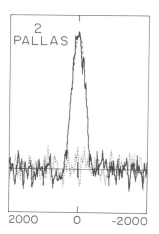

FIG. 12.   Radar echo power spectra obtained at λ13 cm for the asteroid 2 Pallas. Echo power is plotted against Doppler frequency (Hz). The circular polarization ratio μ_C of power received in the same sense of circular polarization as transmitted (i.e., the SC sense; dotted curve) to that in the opposite or OC sense (solid curve) is 0.05 ± 0.02, indicating an extremely smooth surface at decimeter scales. However, the model (dashed curve) fit to the OC spectrum indicates that the surface is very rough at some scale(s) no smaller than several meters and possibly as large as many kilometers.

shape as the asteroid rotates (Fig. 13). Some of the spectra are noticeably asymmetrical, and the mean frequency of the distribution $\sigma(\nu)$, labeled center frequency in Fig. 13, also oscillates with rotational phase. As indicated in that figure, the Eros data can be represented by a triaxial ellipsoid scattering according to a Lambert law, although the postfit residuals reveal significant departures from this model. In general, radar data obtained to date for near-earth asteroids seem very difficult to reconcile with models invoking homogeneous scattering and axisymmetric shapes.

E. TOPOGRAPHIC RELIEF

Topography along the subradar track superimposes a modulation on the echo delay above or below that predicted by ephemerides, which generally are calculated for a sphere with the target's a priori mean radius. There frequently are at least small errors in the radius estimate as well as in the target's predicted orbit. These circumstances generally require that an extended series of measurements of the time delay of the echo's leading edge be folded into a computer program designed to estimate simultaneously parameters describing the target's orbit, mean radius, and topography. The analysis program might also contain parameters from models of wave propagation through the interplanetary medium or the solar corona as well as parameters used to test general relativity, as noted earlier.

Radar has been used to measure topography on the moon and on the inner planets. For example, Fig. 14 shows a three-dimensional reconstruction of topography derived from altimetric profiles obtained for Mars in the vicinity of the giant shield volcano Arsia Mons. The altimetric resolution of the profiles is about 150 m, corresponding to 1-μsec delay resolution, but the surface resolution, or footprint, is very coarse (~75 km). Radar altimetry is best carried out from orbiting spacecraft, and the next generation of space missions to inner-solar-system bodies are likely to employ radar altimeters to make topographic measurements with footprints ~1 km and altimetric resolution as fine as ~10 m.

F. DELAY-DOPPLER RADAR MAPS

As illustrated in Fig. 4, intersections between constant-delay contours and constant-Doppler contours constitute a two-to-one mapping from

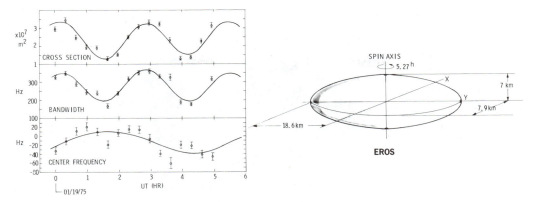

FIG. 13. Measurements of the radar cross section, spectral bandwidth, and apparent center frequency for λ3.5-cm OC echoes from the near-earth asteroid 433 Eros. The solid curves correspond to the sketched ellipsoid. (Courtesy R. F. Jurgens and R. M. Goldstein.)

the target's surface to delay-Doppler space: For any point in the northern hemisphere there is a conjugate point in the southern hemisphere at the same delay and Doppler. Therefore, given $\sigma(\tau, \nu)$, the source of echo in any delay-Doppler resolution cell can be located only to within a two-fold ambiguity. This north–south ambiguity can be avoided if the radar beamwidth is comparable to or smaller than the target's apparent angular radius, as in the case of observations of the moon (angular radius ~15 arcminutes) with the Arecibo λ13-cm system (beamwidth ~2 arcminutes). Similarly, no such ambiguity arises in the case of side-looking radar observations from spacecraft (e.g., the Pioneer Venus Orbiter), for which the geometry of delay-Doppler surface contours differs somewhat from that in Fig. 4. For ground-based observations of Venus and Mercury, whose angular radii never exceed a few tens of arcseconds, the separation of conjugate points is achievable interferometrically, using two receiving antennas, as follows.

The echo waveform received at either antenna from one conjugate point is highly correlated with the echo waveform received at the other antenna from the same conjugate point. However, echo waveforms from the two conjugate points are largely uncorrelated with each other, no matter where they are received. Thus, echoes from two conjugate points can, in principle, be distinguished by cross-correlating echoes received at the two antennas with themselves and with each other and then performing algebraic manipulations on long time averages of the cross product and the two self-products.

The echo waveform from a single conjugate point experiences slightly different delays in reaching the two antennas, so there is a phase difference between the two received signals, and this phase difference depends only on the geometrical positions of the antennas and the target. This geometry changes as the earth rotates, but very slowly and in a predictable manner. The antennas are usually positioned so that contours of constant phase difference on the target disk are as orthogonal as possible to the constant-Doppler contours (which connect conjugate points). Phase difference hence becomes a measure of north–south position, and echoes from conjugate points can be distinguished on the basis of their phase relation.

The total number of fringes, or cycles of phase shift, spanned by the disk of a planet with diameter D and distance R from the radar is approximately $(D/R)/(\lambda/b_{PROJ})$, where b_{PROJ} is the projection of the interferometer baseline normal to the mean line of sight. For example, the Arecibo receiving interferometer links the main antenna to a 30.5-m antenna about 11 km further north. It places about 7 fringes on Venus, quite adequate for separation of the north–south ambiguity. The Goldstone 65-m dish has been linked to smaller antennas to perform three-element as well as two-element interferometry. Tristatic observations permit one to solve so precisely for the north–south location of a given conjugate region that one can obtain the region's elevation relative to the mean planetary radius. Altimetric information can also be extracted from bistatic observations using the time history of the phase information, but only if the variations in the projected baseline vector are very large.

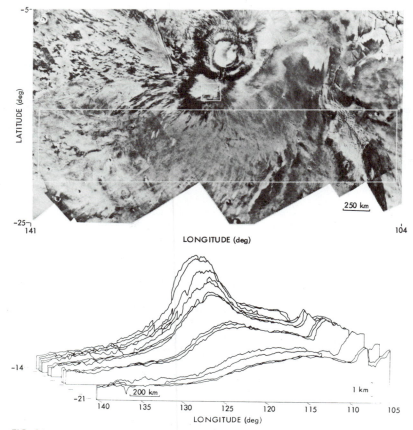

FIG. 14. Topographic contours for the southern flank (large rectangle) of the Martian shield volcano Arsia Mons, obtained from radar altimetry. (Courtesy L. Roth, G. S. Downes, R. S. Saunders. and G. Schubert.)

In constructing a radar map, we transform the unambiguous delay-Doppler distribution of echo power to planetocentric coordinates and fit a model to the data using a maximum-likelihood or weighted-least-squares estimator. The model contains parameters for quasi-specular and diffuse scattering as well as prior information about the target's dimensions and spin vector. For Venus, effects of the dense atmosphere on radar wave propagation must also be modeled. Residuals between the data and the best-fit model constitute a radar reflectivity map of the planet. Figure 15 shows a λ13-cm OC radar map of portions of Venus and Fig. 16 shows a λ3.8-cm SC radar map of a portion of the moon. Figure 17 shows a global map of Venus's λ17-cm SL radar reflectivity obtained by the Pioneer Venus orbiting spacecraft.

G. Physical Interpretation of Radar Reflectivities

Variations in radar reflectivities evident in delay-Doppler maps can be caused by many different physical phenomena, and their proper interpretation demands due attention to the radar wavelength, echo polarization, viewing geometry, prior knowledge about surface properties, and the nature of the target's mean scattering behavior. Similar considerations apply to inferences based on disk-integrated radar albedos.

In Fig. 15a, the alternating bright and dark bands probably result from modulation of the echo strength by preferential orientation of slopes toward and away from the radar, since this area was nearly 70° from the subradar point and even a diffuse ($\cos^m \theta$) scattering law falls

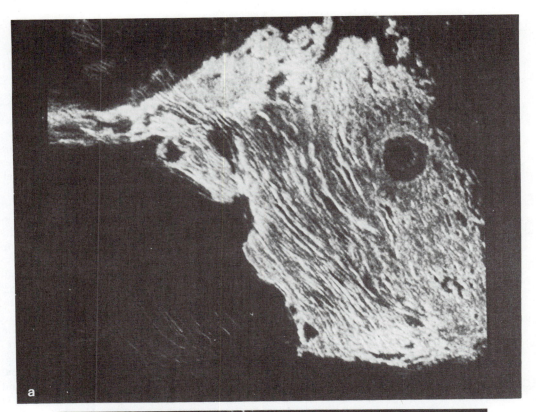

FIG. 15. Arecibo radar maps of portions of Venus. Maxwell Montes (a) is a mountain range produced by tectonic deformation. It is unknown whether the ~100-km-diameter circular feature, named Cleopatra, is of volcanic or impact origin. Beta Regio (b) contains a huge rift zone of linear faults connecting features that appear to be two giant shield volcanoes, Rhea Mons and Theia Mons. All these features are identified in the global map in Fig. 17. (Courtesy D. B. Campbell.)

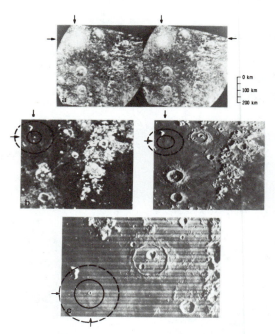

FIG. 16. Lunar crater Piton B (located by arrows) is surrounded by an ejecta blanket that is conspicuous in (a) the λ3.8-cm radar image (displayed in both continuous- and incremental-tone maps) but invisible in (b) earth-based and (c) Lunar Orbiter photographs. The sketched, 50- and 100-km-diameter circles are concentric to the crater and coplanar with the local mean surface. (Courtesy T. W. Thompson.)

off rapidly with angles of incidence that large. However, some of the contrast in Fig. 15a might arise from variations in decimeter-scale roughness.

In the SC lunar radar map in Fig. 16, the enhanced radar reflectivity around the crater Piton B is due to near-surface concentrations of wavelength-scale rocks ejected from the crater during its formation during a hypervelocity impact event. In other lunar radar maps there are reflectivity enhancements thought to be caused by surface chemistry (particularly by iron or titanium concentrations) rather than by structure.

Surface chemistry also appears to be the best available explanation for the approximately five-fold variation in the radar albedos ($\hat{\sigma}_{OC}$) of main-belt asteroids. These objects are thought to be blanketed with thick regoliths, and the variance in radar albedo suggests large variations in regolith porosity or metal abundance. The highest radar albedo estimated for an asteroid (16 Psyche) is consistent with porosities typical of the lunar regolith and a composition nearly en-

tirely metallic, suggesting that this 250-km-diameter object might be the largest piece of "refined" metal in the solar system.

In the Venus reflectivity map (Fig. 17), the brightest regions are more than five times brighter than the darkest regions and have normal-incidence Fresnel reflection coefficients equal to 0.3 or higher. The extremely bright areas might contain substantial amounts of basaltic rock enriched in such titanium- and iron-bearing minerals as ilmenite, magnetite, or pyrite. If this hypothesis is correct, those so-called high-dielectric minerals might be products of recent, if not currently active, volcanism. In any case, Venus's high disk-integrated reflectivity ($\hat{\sigma}_{OC} \sim 0.11$) suggests that Venus is dominated not by a global regolith but rather by extensive exposures of bedrock covered by variable, modest amounts of soil.

As is evident from Fig. 17, there is correlation between the elevation, rms slope, and radar reflectivity of Venus's surface. It has been suggested that some of the roughest, brightest, and most elevated terrain units (e.g., Rhea Mons and Theia Mons in Beta Regio and Maxwell Montes) are monumental volcanic constructs. Ground-based maps and recent maps obtained from Soviet Venera spacecraft show ancient impact craters as well as terrain that appears to be tectonically produced. The Magellan spacecraft, which was to be launched via the Space Shuttle as early as 1989 was expected to provide maps with a resolution (~200 m) much finer than in existing maps. Such maps would clarify the global geologic character of earth's intriguing sister planet.

H. Jupiter's Icy Galilean Satellites

Among all the radar-detected planetary bodies in the solar system, Europa, Ganymede, and Callisto have the most bizarre radar properties. Their reflectivities are enormous compared with those of the moon and inner planets (Table II). Europa is the extreme example (Fig. 18), with a λ13-cm radar albedo (~1.0) that is indistinguishable from that of a metal sphere. Since the radar and optical albedos and estimates of fractional water-frost coverage increase by satellite—in the order Callisto, Ganymede, Europa—the presence of water ice presumably plays a critical role in determining the unusually high reflectivities. For any given porosity, ice is less radar-reflective and less absorbing than silicates.

In spite of the satellites' smooth appearances

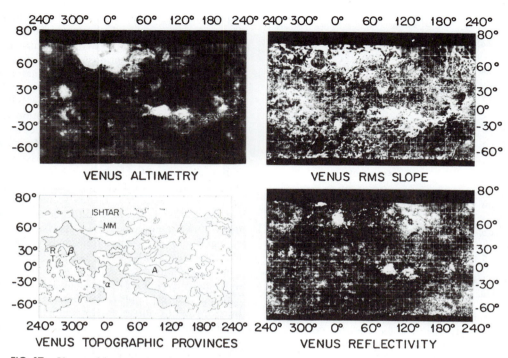

FIG. 17. Venus altimetry, rms slope, and radar reflectivity maps derived from Pioneer Venus radar data. Brighter tones indicate higher, rougher, and more reflective areas, respectively. The map of topographic provinces shows (i) the distribution of highlands (vertical hatching), lowlands (shaded), and rolling plains (blank); and (ii) locations of Ishtar Terra, Aphrodite Terra (A), Alpha Regio (α), Beta Regio (α), Beta Regio (β), Rhea Mons (R), Theia Mons (T), and Maxwell Montes (MM). (See Fig. 15.) (Courtesy G. H. Pettengill.)

at the several-kilometer scales of Voyager high-resolution images, a high degree of near-surface structure at some scale(s) $\geq \lambda 13$ cm is suggested by (i) broad spectral shapes, indicative of a diffuse scattering process, and (ii) fairly large linear polarization ratios ($\mu_L \approx 0.5$).

The precise configurations of the satellite surfaces are constrained most severely by the man-

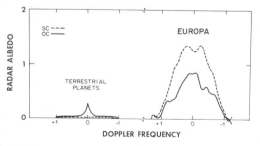

FIG. 18. Typical $\lambda 13$-cm echo spectra for the terrestrial planets are compared with echo spectra for Jupiter's icy moon Europa. The abscissa has units of half the echo bandwidth.

ner in which they backscatter circularly polarized waves. Measured values of the $\lambda 13$-cm circular polarization ratio (μ_C) for Venus, the moon, and Mars average ~ 0.1, and one can easily image a surface so rough (e.g., Fig. 10) that incident waves would be completely unpolarized ($\mu_C \rightarrow 1$). However, μ_C actually exceeds unity for Europa, Ganymede, and Callisto. That is, the scattering largely preserves the transmitted handedness and the circular polarization is "inverted."

Weighted-mean $\lambda 13$-cm values of μ_C for Europa, Ganymede, and Callisto are within 0.2 of 1.6, 1.6, and 1.2, respectively, but the distributions of values of μ_C are wide, as indicated schematically in Fig. 19. Significant polarization and/or albedo features are present in the echo spectra, and in a few cases the feature's source can be identified tentatively in images acquired by Voyager spacecraft. Observations of Ganymede at $\lambda 3.5$ cm yield $\mu_C = 2.0 \pm 0.1$, indicating that the polarization inversion depends on wavelength.

If the radar echo arises from external reflec-

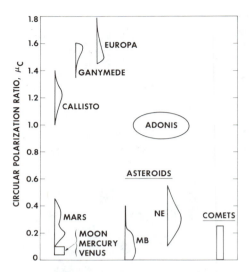

FIG. 19. Circular polarization ratios (μ_C) of radar-detected planetary targets. Approximate distributions of available estimates of μ_C are sketched for the icy Galilean satellites, Mars, mainbelt asteroids, and near-earth asteroids. The curve for near-earth asteroids excludes 2101 Adonis whose polarization ratio 1.0 ± 0.2 is close to values obtained for Callisto.

tions from the surface per se, the surface material must have a refractive index ~20% higher than that (1.8) of solid water ice and must be covered with nearly hemispherical craters. A different model postulates subsurface multiple scattering from randomly oriented planar interfaces between electrically dissimilar components (e.g., ice and void), with the regolith components' refractive indices in a ratio less than 1.4. In the hemispherical crater model, the polarization inversion arises from double-reflection backscattering from crater interiors. In the random facet model, the inversion results from many successive grazing reflections and particularly from total internal reflections in the denser medium.

Perhaps the most promising explanation for the polarization inversion and the huge reflectivities is a process involving, ironically, no reflections at all. Instead, the incident wave enters the surface, is refracted around volumes of material whose density (and hence refractive index) is somewhat higher than that of their surroundings, and reemerges from the surface with its incident polarization (SC) and power largely intact. If this refraction scattering theory is correct, the scale of the putative scattering centers

is probably between a centimeter and a meter. The variation in refractive index must be very smooth and continuous to avoid subsurface reflections, and the uppermost surface must be very tenuous, creating an impedance match between free space and the target. At present, we still lack precise information about the subsurface structure needed to produce refraction scattering and the particular geologic processes responsible for generating the requisite structures. However, one of the more plausible possibilities is that the unusual echoes arise because ejecta within the upper few decameters of the regoliths are thermally annealed. In this scenario, sharp dielectric boundaries between large solid fragments and fine-grained, porous debris have been changed into gentle gradations by exposure to temperature $\gtrsim 130$ K, so the large fragments become refraction scatterers.

As indicated in Fig. 19, the near-earth asteroid 2201 Adonis has polarization properties resembling those of Callisto. Since Adonis's orbit resembles those of some short-period comets, this object might be an extinct cometary nucleus, whose supply of near-surface volatiles (i.e., ices) is so exhausted that sunlight can no longer stimulate cometary (coma/tail) activity. Optical observations reveal no cometary activity for Adonis, and Adonis's μ_C is certainly not cometary. Of course, the same statements apply to Callisto, so it seems plausible that the surface of Adonis might be thermally altered and might resemble that of Callisto.

I. COMETS

Since a cometary coma is nearly transparent at radio wavelengths, radar is much more capable of unambigous detection of a cometary nucleus than are optical and infrared methods, and radar observations of several comets (Table I) have provided useful constraints on nuclear dimensions. The radar signature of one particular comet (IRAS–Araki–Alcock, which came within 0.03 AU of earth in May 1983) revolutionizes our concepts of the physical nature of these intriguing objects. Echoes obtained at both Goldstone (Fig. 20) and Arecibo have a narrowband component from the nucleus as well as a much weaker broadband component from a cloud of particles ejected from the nucleus. Models of the echo suggest that most of the particles, which must be 1–5 cm in size, escaped

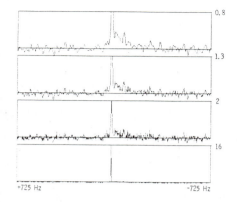

FIG. 20. Echo power spectrum of Comet IRAS–Araki–Alcock, shown here with four different vertical truncations and degrees of smoothing. Echo power is plotted against Doppler frequency. The narrow spike is echo from the comet's ~7-km nucleus. The broadband component, or skirt, is echo from a $\geq 10^3$-km debris swarm consisting of particles no smaller than a centimeter and, for the most part, not gravitationally bound to the nucleus. (Courtesy R. M. Goldstein, R. F. Jurgens, and Z. Sekanina.)

from the ~7-km-diameter nucleus to form a debris swarm whose size is $\geq 10^3$ km. Echo spectra for the nucleus reveal a nonspherical shape and show several features that are consistent with roughness on meter-to-kilometer scales. Thus, one envisions this object as extremely rough with an explosively active surface from which chunks of material are being blasted into space by subsurface sublimation of volatiles.

J. SATURN'S RINGS

Saturn's rings are the most distant radar-detected planetary entity and the only radar-detected ring system; neither distinction is likely to be relinquished in the foreseeable future. The rings are quite unlike other planetary targets in terms of both the experimental techniques employed and the physical considerations involved. For example, the relation between ring-plane location and delay-Doppler coordinates for a system of particles traveling in Keplerian orbits is different from the geometry portrayed in Fig. 4. The rings are grossly overspread (Table II), requiring the use of frequency-stepped waveforms in the more sophisticated radar studies.

Radar determinations of the rings' backscattering properties complement results of the Voy-

ager spacecraft radio occultation experiment (which measured the rings' forward scattering efficiency at identical wavelengths) in constraining the size and spatial distributions of ring particles. The rings' circular polarization ratio is ~1.0 at λ3.5 cm at ~0.5 and λ13 cm, more or less independent of the inclination angle δ between the ring plane and the line of sight. Whereas multiple scattering between particles might cause some of the depolarization, the lack of strong dependence of μ_C on δ suggests that the particles are intrinsically rougher at the scale of the smaller wavelength. The rings' total-power (OC + SC) radar albedo shows only modest dependence on δ, a result that seems to favor many-particle-thick models of the rings over monolayer models. Delay-Doppler resolution of ring echoes indicates that the portions of the ring system that are brightest optically (the A and B rings) are also responsible for most of the radar echoes. The C ring has a very low radar reflectivity, presumably because of either a low particle density in that region or bulk compositions or particle sizes leading to inefficient particle scattering efficiencies.

IV. The Future of Planetary Radar Astronomy

During the remainder of this century new interplanetary spacecraft will carry radar experiments, and the sensitivity of ground-based telescopes will be improved, adding to the momentum of planetary radar astronomy as it pursues several exciting scientific directions.

For Mercury, ground-based delay-Doppler maps will provide our first looks at the hemisphere left unimaged by the Mariner 10 spacecraft a decade ago, and the dual-polarization radar techniques applied so successfully to the other terrestrial objects will characterize the planet's small-scale surface properties.

The next major step in spacecraft exploration of the inner solar system will be taken by the Magellan mission, which will obtain high-resolution images and altimetry for most of Venus. This mission will clarify the detailed nature and evolutionary history of the planet's surface as well as the principal geologic processes influencing the interior and the atmosphere.

Following Magellan, both Mars and the moon are to be studied intensely from spacecraft in polar orbits. Radar altimeters are key components of the instrument packages, which are designed to define the global elemental and mineralogical nature of the surface, the gravitational field, and the topography. These investigations are logical predecessors to Mars rover and sample-return missions and to a second generation of lunar landings by human beings.

The primary emphasis in ground-based radar astronomy will be on asteroids, comets, and the satellites of Mars, Jupiter, and Saturn. The goal of asteroid and comet science is to understand enormous populations; the strategy for reaching that goal is to use a few space missions to selected objects as fiducials to calibrate the interpretation of ground-based observations of many objects. In this context, radar observations will continue to refine our knowledge of the physical properties of individual objects and will play a critical role in the reconnaissance of the asteroid and comet populations. Radar astronomy will also play a role in selection of the target for a Near-Earth Asteroid Rendezvous Mission, analogous to its role in Viking Lander Site selection.

The Martian satellites Phobos and Deimos will be detectable with ground-based radar during several close approaches of Mars beginning in 1988. These objects, which were studied in great depth by the Viking spacecraft, may actually be captured asteroids, and measurement of their radar properties would help guide inferences about asteroid radar signatures. Phobos and Deimos are also very interesting in their own right and are potentially important as convenient, accessible outposts for the human exploration of Mars. One primary objective of radar observations would be to assess the bulk density and near-surface rock populations of the satellites' regoliths.

Ground-based observations of Jupiter's Galilean satellites will devote special attention to volcanically active Io, whose polarization properties have never been measured and whose radar albedo and angular scattering behavior are poorly known. For the icy moons, Europa, Ganymede, and Callisto, a major objective is to identify regions of polarization and albedo anomalies and to locate these regions in high-resolution (~20 m) visual images to be acquired by the Galileo Mission in the 1990s.

One of the most intriguing objects in the solar system is Saturn's largest satellite, Titan. Available constraints on Titan's surface temperature and pressure and on atmospheric chemistry suggest that the solid surface of this satellite might be composed largely of organic compounds and might be partially covered by ethane-rich oceans. However, the actual nature of Titan's surface, which is concealed by clouds, remains unknown. As with Venus, radar techniques provide our only means to discern the global character of the surface. Attempts to detect Titan from Goldstone and Arecibo will be carried out during the coming decade, and the results of these efforts will assist design of a radar instrument for a Titan spacecraft mission.

In summary, there are many exciting prospects for new observations, and the future of planetary radar astronomy is potentially very bright. However, a necessary condition for sustaining the vitality of ground-based efforts is major improvement in the sensitivity of the Arecibo and Goldstone telescopes. Activity involving spacecraft radars certainly should be vigorous and fruitful well into the 21st century.

BIBLIOGRAPHY

Campbell, D. B., Head, J. W., Harmon, J. K., and Hine, A. A. (1984). *Science* **226,** 167–169.
Eshleman, V. R. (1986). *Nature* **319,** 755–757.
Garvin, J. B., Head, J. W., Pettengill, G. H., and Zisk, S. H. (1985). *J. Geophys. Research* **90,** 6859–6871.
Goldstein, R. M., Jurgens, R. F., and Sekanina, Z. (1984). *Astronom. J.* **89,** 1745–1754.
Hagfors, T., Gold, T., and Ierkic, H. M. (1985). *Nature* **315,** 637–640.
Harmon, J. K., and Ostro, S. J. (1985). *Icarus* **62,** 110–128.
Jurgens, R. F. (1982). *Icarus* **49,** 97–108.
Ostro, S. J. (1985). *Publ. Astronom. Soc. Pacific* **97,** 887–884.
Ostro, S. J. (1982). "Satellites of Jupiter," D. Morrison, ed. University of Arizona Press, Tucson, pp. 213–236.
Ostro, S. J., Campbell, D. B., and Shapiro, I. I. (1985). *Science* **224,** 442–446.
Ostro, S. J., and Pettengill, G. H. (1984). In "Planetary Rings," A. Brahic, ed. Cepadues-Editions, Toulouse, France, pp. 49–55.
Pettengill, G. H., Eliason, E., Ford, P. G., Loriot, G. B., Masursky, H., and McGill, G. E. (1980). *J. Geophys. Research* **85,** 8261–8270.
Roth, L. E., Downs, G. S., Saunders, R. S., and Schubert, G. (1980). *Icarus* **42,** 287–316.
Shapiro, I. I., Campbell, D. B., and De Campli, W. M. (1979). *Astrophys. J.* **230,** L123–L126.
Simpson, R. A., and Tyler, G. L. (1982). *IEEE Trans. Antennas and Propagation* **AP-30,** 438–449.
Thompson, T. W., Zisk, S. H., Shorthill, R. W., and Cutts, J. A. (1981). *Icarus* **46,** 201–225.

PLANETARY SATELLITES, NATURAL

Bonnie J. Buratti   *JPL, California Institute of Technology*

GLOSSARY

Bond albedo: Fraction of the total incident radiation reflected by a planet or satellite.

Carbonaceous material: Carbon–silicate material rich in simple organic compounds. It exists on the surfaces of several satellites.

Differentiation: Melting and chemical fractionation of a planet or satellite into a core and mantle.

Geometric albedo: Ratio of the brightness at a phase angle of zero degrees (full illumination) compared with a diffuse, perfectly reflecting disk of the same size.

Greenhouse effect: Heating of the lower atmosphere of a planet or satellite by the transmission of visible radiation and subsequent trapping of reradiated infrared radiation.

Lagrange points: Five equilibrium points in the orbit of a satellite around its primary. Two of them (L4 and L5) are points of stability for a third body.

Magnetosphere: Region around a planet dominated by its magnetic field and associated charged particles.

Opposition effect: Surge in brightness as a satellite becomes fully illuminated to the observer.

Phase angle: Angle between the observer, the satellite, and the sun.

Phase integral: Integrated value of the function which describes the directional scattering properties of a surface.

Primary body: Celestial body (usually a planet) around which a satellite, or secondary, orbits.

Regolith: Surface layer of rocky debris created by meteorite impacts.

Roche's limit: Distance (equal to 2.44 times the radius of the primary) at which the tidal forces exerted by the primary on the satellite equal the internal gravitational forces of the satellite.

Synchronous rotation: Dynamical state caused by tidal interactions in which the satellite presents the same face towards the primary.

A natural planetary satellite is a celestial body in orbit around one of the nine principal planets of the solar system. The central body is known as the primary and the orbiting satellite its moon or secondary. Among the nine planets only Mercury and Venus have no known companions. There are 54 known natural planetary satellites in the solar system; there may exist many more undiscovered satellites, particularly small objects encircling the giant outer planets. The satellites range in size from planet-sized objects such as Ganymede and Titan to tiny, irregular bodies tens of kilometers in diameter (see Table I and Fig. 1).

I. Summary of Characteristics

A. DISCOVERY

The only natural planetary satellite known before the advent of the telescope was the Earth's moon. Phenomena such as the lunar phases and the ocean tides have been studied for centuries. When Galileo turned his telescope to Jupiter in 1610, he discovered the four large satellites in the Jovian system. His observations of their orbital motion around Jupiter in a manner analogous to the motion of the planets around the sun provided critical evidence for the acceptance of the heliocentric (sun-centered) model of the solar system. These four moons—Io, Europa, Ganymede, and Callisto—are sometimes called

TABLE I.   Summary of the Properties of the Natural Planetary Satellites

Satellite	Distance from primary (10^3 km)	Revolution period (days) R = Retrograde	Orbital eccentricity	Orbital inclination (degrees)
Earth				
Moon	384.4	27.3	0.055	18 to 29
Mars				
M1 Phobos	9.38	0.32	0.018	1.0
M2 Diemos	23.50	1.26	0.002	2.8
Jupiter				
J14 Adrastea	128	0.30	0.0	0.0
J16 Metis	128	0.30	0.0	0.0
J5 Amalthea	181	0.49	0.003	0.4
J15 Thebe	221	0.68	0.0	0.0
J1 Io	422	1.77	0.004	0.0
J2 Europa	671	3.55	0.000	0.5
J3 Ganymede	1070	7.16	0.001	0.2
J4 Callisto	1880	16.69	0.010	0.2
J13 Leda	11110	240	0.416	26.7
J6 Himalia	11470	251	0.158	27.6
J10 Lysithea	11710	260	0.130	29.0
J7 Elara	11740	260	0.207	24.8
J12 Ananke	20700	617R	0.17	147
J11 Carme	22350	692R	0.21	164
J8 Pasiphae	23300	735R	0.38	145
J9 Sinope	23700	758R	0.28	153
Saturn				
S17 Atlas	138	0.60	0.002	0.3
S16 1980S27	139	0.61	0.004	0.0
S15 1980S26	142	0.63	0.004	0.1
S10 Janus	151	0.69	0.007	0.14
S11 Epimethus	151	0.69	0.009	0.34
S1 Mimas	186	0.94	0.020	1.5
S2 Enceladus	238	1.37	0.004	0.0
S3 Tethys	295	1.89	0.000	1.1
S14 Calypso	295	1.89	0.0	1?
S13 Telesto	295	1.89	0.0	1?
S4 Dione	377	2.74	0.002	0.0
S12 1980S6	377	2.74	0.005	0.2
S5 Rhea	527	4.52	0.001	0.4
S6 Titan	1220	15.94	0.029	0.3
S7 Hyperion	1480	21.28	0.104	0.4
S8 Iapetus	3560	79.33	0.028	14.7
S9 Phoebe	12950	550.4R	0.163	150
Uranus				
1986U7	49.7	0.33		
1986U8	53.2	0.37		
1986U9	59.2	0.43		
1986U3	61.8	0.46		
1986U6	62.7	0.47		
1986U2	64.6	0.49		
1986U1	66.1	0.51		
1986U4	69.9	0.56		
1986U5	75.3	0.62		
1985U1	86.0	0.76		
U5 Miranda	130	1.41	0.017	3.4
U1 Ariel	191	2.52	0.003	0.0
U2 Umbriel	266	4.14	0.003	0.0
U3 Titania	436	8.71	0.002	0.0
U4 Oberon	583	13.46	0.001	0.0
Neptune				
N1 Triton	355.5	5.89R	0.00	160
N2 Nereid	5567	359.9	0.749	27.7
Pluto				
P1 Charon	19.3	6.3R	0?	120

Radius (km)	Density (gm/cm³)	Visual geometric albedo	Discoverer	Year of discovery
1738	3.34	0.11		
14 × 10	1.9	0.05	Hall	1877
8 × 6	2.1	0.05	Hall	1877
20		<0.1	Jewitt et al.	1979
20		<0.1	Synott	1979/80
135 × 85 × 75		0.05	Barnard	1892
40		<0.1	Synott	1979/80
1815	3.55	0.6	Galileo	1610
1569	3.04	0.6	Galileo	1610
2631	1.93	0.4	Galileo	1610
2400	1.83	0.2	Galileo	1610
10			Kowal	1974
90		0.03	Perrine	1904/5
10			Nicholson	1938
40		0.03	Perrine	1904/5
5			Nicholson	1951
15			Nicholson	1938
20			Melotte	1908
15			Nicholson	1914
20 × ? × 10		0.4	Voyager	1980
70 × 50 × 37		0.6	Voyager	1980
55 × 45 × 33		0.6	Voyager	1980
110 × 95 × 80		0.6	Dollfus	1966
70 × 58 × 50		0.5	Fountain and Larson	1978
197	1.4	0.8	Herschel	1789
251	1.2	1.0	Herschel	1789
530	1.2	0.8	Cassini	1684
12 × 11 × 11		0.6	Space Telescope Tm.	1980
15 × 10 × 8		0.9	Smith et al.	1980
560	1.4	0.55	Cassini	1684
17 × 16 × 15		0.5	Laques and Lecacheux	1980
765	1.3	0.65	Cassini	1672
2575	1.88	0.2	Huygens	1655
205 × 130 × 110		0.3	Bond and Lassell	1848
730	1.2	0.4–0.08	Cassini	1671
110		0.06	Pickering	1898
~25			Voyager 2	1986
~25			Voyager 2	1986
~25			Voyager 2	1986
~30		~0.04	Voyager 2	1986
~30		~0.04	Voyager 2	1986
~40		~0.06	Voyager 2	1986
~40		~0.09	Voyager 2	1986
~60		~0.04	Voyager 2	1986
			Voyager 2	1986
85			Voyager 2	1985
242	1.2	0.22	Kuiper	1948
580	1.6	0.38	Lassell	1851
596	1.8	0.16	Lassell	1851
805	1.6	0.23	Herschel	1787
773	1.6	0.20	Herschel	1787
1750	2.8	0.4	Lassell	1846
300			Kuiper	1949
600	0.8		Christy	1978

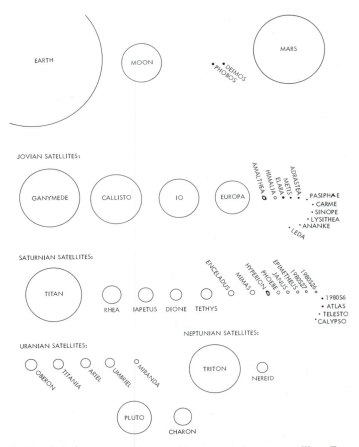

FIG. 1.   Relative shapes and sizes of the natural planetary satellites. For comparison, the relative sizes of smaller planets are shown. Jupiter and Saturn would span about twice the height of the figure and Uranus and Neptune would be about as wide as the figure.

the Galilean satellites. [*See* MOON (ASTRONOMY).]

In 1655 Christian Huygens discovered Titan, the giant satellite of Saturn. Later in the seventeenth century, Giovanni Cassini discovered the four next largest satellites of Saturn. It was not until over one hundred years later that the next satellite discoveries were made: the Uranian satellites Titania and Oberon and two smaller moons of Saturn. As telescopes acquired more resolving power in the nineteenth century, the family of satellites grew (see Table I). The smallest satellites of Jupiter and Saturn were discovered during flybys of the Pioneer and Voyager spacecraft (see Table II).

The natural planetary satellites are generally named after figures in classical Greek and Roman mythology who were associated with the namesakes of their primaries. They are also des-

ignated by the first letter of their primary and an Arabic numeral assigned in order of discovery: Io is J1, Europa, J2, etc. When satellites are first discovered but not yet confirmed or officially named, they are known by the year in which they were discovered, the initial of the primary, and a number assigned consecutively for all solar system discoveries, e.g., 1980J27.

When planetary scientists were able to observe and map geologic formations of the satellites from spacecraft images, they named many of the features after characters or locations from Western and Eastern mythologies.

B. PHYSICAL AND DYNAMIC PROPERTIES

The motion of a satellite around the center of mass of itself and its primary defines an ellipse

TABLE II.   Summary of Major Missions to the Planetary Satellites

Mission name	Object	Encounter dates	Type of mission
Luna 3 (USSR)	Moon	1959	Flyby (far side)
Ranger 7,8,9	Moon	1964–65	Crash landing; image return
Luna 9,13 (USSR)	Moon	1966	Soft landing
Luna 10,12,14 (USSR)	Moon	1966–68	Orbiter
Surveyor 1,3,5,6,7	Moon	1966–68	Soft landing
Lunar Orbiter 1–5	Moon	1966–68	Orbiter
Apollo 7–10	Moon	1968–69	Manned orbiter
Apollo 11,12,14–17	Moon	1969–75	Manned landing
Luna 16,20 (USSR)	Moon	1970–72	Sample return
Luna 17,21,24 (USSR)	Moon	1970–76	Rover
Mariner 9	Deimos	1971	Orbiter
	Phobos		
Pioneer 10	Jovian satellites	1979	Flyby
Pioneer 11	Jovian satellites	1979	Flyby
	Saturnian satellites	1979	Flyby
Viking 1,2	Phobos	1976	Orbiter
	Deimos		
Voyager 1	Jovian satellites	1979	Flyby
	Saturnian satellites	1980	Flyby
Voyager 2	Jovian satellites	1979	Flyby
	Saturnian satellites	1981	Flyby
	Uranian satellites	1986	Flyby

with the primary at one of the foci. The orbit is defined by three primary orbital elements (1) the semimajor axis, (2) the eccentricity, and (3) the angle made by the intersection of the plane of the orbit and the plane of the primary's spin equator (the angle of inclination). The orbits are said to be regular if they are in the same sense of direction (the prograde sense) as that determined by the rotation of the primary. The orbit of a satellite is irregular if its motion is in the opposite (or retrograde) sense of direction, or if it has a high angle of inclination. The majority of satellites move in regular, prograde orbits. Those satellites that do not move in regular, prograde orbits are believed to be captured objects (see Table I). [*See* CELESTIAL MECHANICS.]

Most of the planetary satellites present the same hemisphere toward their primaries, which is the result of tidal evolution. When two celestial bodies orbit each other, the gravitational force exerted on the near side is greater than that exerted on the far side. The result is an elongation of each body to form tidal bulges, which can consist of either solid, liquid, or gaseous (atmospheric) material. The primary will tug on the satellite's tidal bulge to lock its longest axis onto the primary-satellite line. The satellite, which is said to be in a state of synchronous rotation, keeps the same face toward the primary. Since this despun state occurs rapidly (usually within a few million years), most natural satellites are in synchronous rotation.

The natural satellites are unique worlds, each representing a vast panorama of physical processes. The small satellites of Jupiter and Saturn are irregular chunks of ice and rock, perhaps captured asteroids, which have been subjected to intense meteoritic bombardment. Several of the satellites, including Phoebe (which is in orbit around Saturn) and the Martian moon Phobos, are covered with dark carbonaceous material believed to be representative of the primordial, unprocessed material from which the Solar System formed. The medium-sized satellites of Saturn and Uranus are large enough to have undergone internal melting and subsequent differentiation and resurfacing. The Saturnian satellite, Iapetus, presents a particular enigma: one hemisphere is ten times more reflective than the other. Three of the Galilean satellites show evidence of geologically active periods in their history; Io is presently undergoing intense volcanic activity. The Earth's Moon experienced a period of intense meteoritic bombardment and melting soon after its formation.

Saturn's largest satellite, Titan, has a predominantly nitrogen atmosphere thicker than that of the Earth. There is evidence that Triton, the large satellite of Neptune, also has an appreciable atmosphere. Io has a thin, possibly transient sulfur dioxide atmosphere that is thought to be related to outgassing from active volcanoes. None of the other satellites have detectable atmospheres. [*See* PLANETARY ATMOSPHERES.]

II. Formation of Satellites

A. THEORETICAL MODELS

Because the planets and their associated moons condensed from the same cloud of gas and dust at about the same time, the formation of the natural planetary satellites must be addressed within the context of the formation of the planets. The solar system formed 4.6 ± 0.1 billion years ago. This age is derived primarily from radiometric dating of meteorites, which are believed to consist of primordial, unaltered matter. In the radiometric dating technique, the fraction of a radioactive isotope (usually rubidium, argon, or uranium), which has decayed into its daughter isotope, is measured. Since the rate at which these isotopes decay has been measured in the laboratory, it is possible to infer the time elapsed since formation of the meteorites, and thus of the solar system.

The sun and planets formed from a disk-shaped rotating cloud of gas and dust known as the proto-solar nebula. When the temperature in the nebula cooled sufficiently, small grains began to condense. The difference in solidification temperatures of the constituents of the proto-solar nebula accounts for the major compositional differences of the satellites. Since there was a temperature gradient as a function of distance from the center of the nebula, only those materials with high melting temperatures (e.g., silicates, iron, aluminum, titanium, and calcium) solidified in the central (hotter) portion of the nebula. The Earth's Moon consists primarily of these materials. Beyond the orbit of Mars, carbon, in combination with silicates and organic molecules, condensed to form a class of asteroids known as carbonaceous chondrites. Similar carbonaceous material is found on the surfaces of Phobos, several of the Jovian and Saturnian satellites, and perhaps the Uranian satellites. Beyond the outer region of the asteroid belt, formation temperatures were sufficiently cold to allow water ice to condense and remain stable. Thus, the Jovian satellites are primarily ice-silicate admixtures (except for Io, which has apparently outgassed all its water). On Saturn and Uranus, these materials are joined by methane and ammonia. For the satellites of Neptune and Pluto, formation temperatures were probably low enough for other volatiles, such as nitrogen and carbon monoxide, to exist in liquid and solid form. In general, the satellites, which formed in the inner regions of the solar system are denser than the outer planets' satellites, because they retained a lower fraction of volatile materials.

After small grains of material condensed from the proto-solar nebula, electrostatic forces caused them to stick together. Collisions between these larger aggregates caused meter-sized particles, or planetesimals, to be accreted. Finally, gravitational collapse occurred to form larger, kilometer-sized planetesimals. The largest of these bodies swept up much of the remaining material to create the protoplanets and their companion satellite systems. One important concept of planetary satellite formation is that a satellite cannot accrete within Roche's limit, the distance at which the tidal forces of the primary become greater than the internal cohesive forces of the satellite.

The formation of the regular satellite systems of Jupiter, Saturn, and Uranus is sometimes thought to be a smaller scaled version of the formation of the solar system. A density gradient as a function of distance from Jupiter does exist for the Galilean satellites (see Table I). This implies that more volatiles (primarily ice) are included in the bulk composition as the distance increases. However, this simple scenario cannot be applied to Saturn or Uranus because their regular satellites do not follow this pattern.

The retrograde satellites are probably captured asteroids or large planetesimals left over from the major episode of planet formation. Except for Titan and perhaps Triton, the satellites are too small to possess gravitational fields sufficiently strong to retain an appreciable atmosphere against thermal escape. [*See* PRIMITIVE SOLAR SYSTEM OBJECTS: ASTEROIDS AND COMETS.]

B. EVOLUTION

Soon after the satellites accreted, they began to heat up from the release of gravitational potential energy. An additional heat source was provided by the release of mechanical energy during the heavy bombardment of their surfaces by remaining debris. The satellites Phobos, Mimas, and Tethys all have impact craters caused by bodies that were nearly large enough to break them apart; probably such catastrophes did occur. The decay of radioactive elements found in silicate materials provided another major source of heat. The heat produced in the larger satellites was sufficient to cause melting and chemical fractionation; the dense material, such as silicates and iron, went to the center of the satellite to form a core, while ice and other volatiles remained in the crust.

Some satellites, such as the Earth's Moon, Ganymede, and several of the Saturnian satellites underwent periods of melting and active geology within a billion years of their formation and then became quiescent. Others, such as Io and possibly Enceladus and Europa, are currently geologically active. For nearly a billion years after their formation, the satellites all underwent intense bombardment and cratering. The bombardment tapered off to a slower rate and presently continues. By counting the number of craters on a satellite's surface and making certain assumptions about the flux of impacting material, geologists are able to estimate when a specific portion of a satellite's surface was formed. Continual bombardment of satellites causes the pulverization of the surface to form a covering of fine material known as a regolith.

III. Observations of Satellites

A. TELESCOPIC OBSERVATIONS

1. Spectroscopy

Before the development of interplanetary spacecraft, all observations from Earth of objects in the solar system were obtained by telescopes. One particularly useful tool of planetary astronomy is spectroscopy, or the acquisition of spectra from a celestial body.

Each component of the surface or atmosphere of a satellite has a characteristic pattern of absorption and emission bands. Comparison of the astronomical spectrum with laboratory spectra of materials which are possible components of the surface yields information on the composition of the satellite. For example, water ice has a series of absorption features between 1 and 4 microns. The detection of these bands on three of the Galilean satellites and several satellites of Saturn and Uranus demonstrated that water ice

is a major constituent of their surfaces. Other examples are the detections of SO_2 frost on the surface of Io, and methane in the atmosphere of Titan, and the possible detection of liquid nitrogen on Triton.

2. Photometry

Photometry of planetary satellites is the accurate measurement of radiation reflected to an observer from their surfaces or atmospheres. These measurements can be compared to light scattering models that are dependent on physical parameters, such as the porosity of the optically active upper surface layer, the albedo of the material, and the degree of topographic roughness. These models predict brightness variations as a function of solar phase angle (the angle between the observer, the sun, and the satellite). Like the Earth's Moon, the planetary satellites present changing phases to an observer on Earth. As the face of the satellite becomes fully illuminated to the observer, the integrated brightness exhibits a nonlinear surge in brightness that is believed to result from the disappearance of mutual shadowing among surface particles. The magnitude of this surge, known as the "opposition effect," is greater for a more porous surface.

One measure of how much radiation a satellite reflects is the geometric albedo, p, which is the disk-integrated brightness at "full moon" (or a phase angle of zero degrees) compared to a perfectly reflecting, diffuse disk of the same size. The phase integral, q, defines the angular distribution of radiation over the sky:

$$q = 2 \int_0^\pi \Phi(\alpha) \sin \alpha \, d\alpha$$

where $\Phi(\alpha)$ is the disk integrated brightness and α is the phase angle.

The Bond albedo, which is given by $A = p \times q$, is the ratio of the integrated flux reflected by the satellite to the integrated flux received. The geometric albedo and phase integral are wavelength dependent; whereas, a true (or bolometric) Bond albedo is integrated over all wavelengths.

Another ground based photometric measurement, which has yielded important information on the satellites surfaces, is the integrated brightness of a satellite as a function of orbital angle. For a satellite in synchronous rotation with its primary, the subobserver geographical longitude of the satellite is equal to the longitude of the satellite in its orbit. Observations showing significant albedo and color variegations for Io, Europa, Rhea, Dione, and especially Iapetus suggest that diverse geologic terrains coexist on these satellites. This view was confirmed by images obtained by the Voyager spacecraft.

3. Radiometry

Satellite radiometry is the measurement of radiation which is absorbed and re-emitted at thermal wavelengths. The distance of each satellite from the sun determines the mean temperature for the equilibrium condition that the absorbed radiation is equal to the emitted radiation:

$$\pi R^2 (F/r^2)(1 - A) = 4\pi R^2 \varepsilon \sigma T^4$$

or

$$T = \left(\frac{(1 - A)F}{4\sigma\varepsilon r^2} \right)^{1/4}$$

where R is the radius of the satellite, r is the sun-satellite distance, ε is the emissivity, σ is Stefan—Boltzmann's constant, A is the Bond albedo, and F is the incident solar flux (a slowly rotating body would radiate over $2\pi R^2$). Typical mean temperatures in degrees Kelvin for the satellites are: the Earth's Moon, 280; Io, 106; Titan, 97; the Uranian satellites, 60; and the Neptunian satellites, 45. For thermal equilibrium, measurements as a function of wavelength yield a blackbody curve characteristic of T: with the exception of Titan, the temperatures of the satellites closely follow the blackbody emission values. The discrepancy for Titan may be due to a weak greenhouse effect in the satellite's atmosphere.

Another possible use of radiometric techniques, when combined with photometric measurements of the reflected portion of the radiation, is the estimate of the diameter of a satellite. A more accurate method of measuring the diameter of a satellite from Earth involves measuring the light from a star as it occulted by the satellite. The time the starlight is dimmed is proportional to the satellite's diameter.

A third radiometric technique is the measurement of the thermal response of a satellite's surface as it is being eclipsed by its primary. The rapid loss of heat from a satellite's surface indicates a thermal conductivity consistent with a porous surface. Eclipse radiometry of Phobos, Callisto, and Ganymede suggests these objects all lose heat rapidly.

4. Polarimetry

Polarimetry is the measurement of the degree of polarization of radiation reflected from a satellite's surface. The polarization characteristics depend on the shape, size, and optical properties of the surface particles. Generally, the radiation is linearly polarized and is said to be negatively polarized if it lies in the scattering plane, and positively polarized if it is perpendicular to the scattering plane. Polarization measurements as a function of solar phase angle for atmosphereless bodies are negative at low phase angles; comparisons with laboratory measurements indicate this is characteristic of complex, porous surfaces consisting of multi-sized particles. In 1970, ground-based polarimetry of Titan that showed it lacked a region of negative polarization led to the correct conclusion that it has a thick atmosphere.

B. SPACECRAFT EXPLORATION

1. Imaging Observations

Interplanetary missions to the planets and their moons have enabled scientists to increase their understanding of the solar system more in the past 20 years than in the previous total years of scientific history. Analysis of data returned from spacecraft has led to the development of whole new fields of scientific endeavor, such as planetary geology. From the earliest successes of planetary imaging, which included the flight of a Soviet Luna spacecraft to the far side of the Earth's Moon to reveal a surface unlike that of the visible side, devoid of smooth lunar plains, and the crash landing of a United States Ranger spacecraft, which sent back pictures showing that the Earth's Moon was cratered down to meter scales, it was evident that interplanetary imaging experiments had immense capabilities. Table II summarizes the successful spacecraft missions to the planetary satellites.

The return of images from space is very similar to the transmission of television images. A camera records the level of intensity of radiation incident on its detector's surface. A series of scans is made across the detector to create a two-dimensional array of intensities. A computer onboard the spacecraft records these numbers and sends them by means of a radio transmitter to the Earth, where another computer reconstructs the image.

The first spacecraft to send pictures of a moon other then the Earth's was Mariner 9, which began orbiting Mars in 1971 and sent back images of Phobos and Deimos showing that these satellites are heavily cratered, irregular objects. Even more highly resolved images were re-

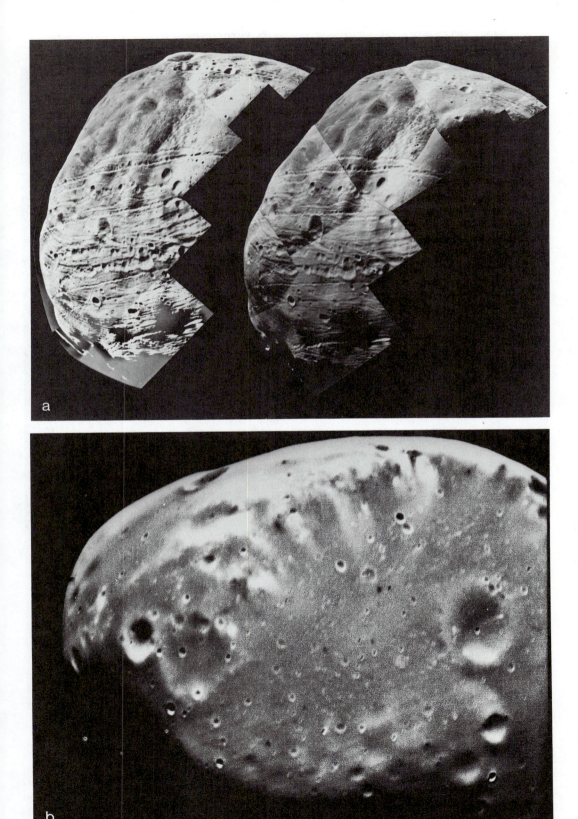

FIG. 2. The two moons of Mars: (a) Phobos and (b) Deimos. Both pictures were obtained by the *Viking* spacecraft.

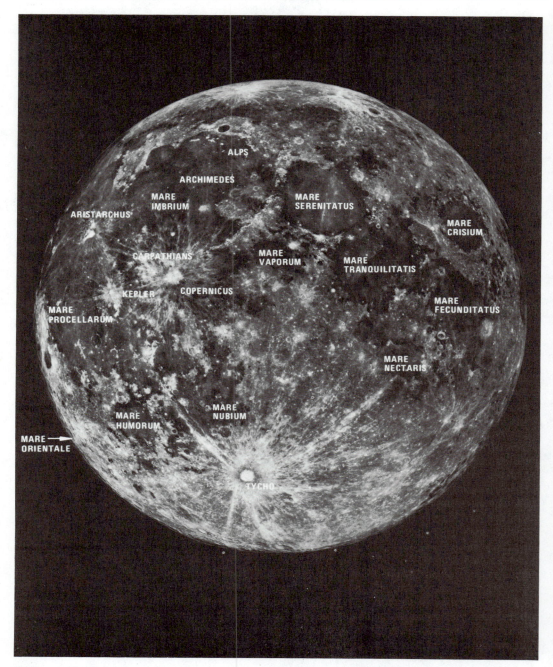

FIG. 3.   A telescopic view of the moon, with the major features marked. (Photograph courtesy of Lick Observatory.)

turned by the Viking orbiters in 1976 (see Fig. 2). The Pioneer spacecraft, which were launched in 1972 and 1973 toward an encounter with Jupiter and Saturn, returned the first disk-resolved images of the Galilean satellites. By far the greatest scientific advancements were made by the Voyager spacecraft, which returned thousands of images of the Jovian and Saturnian satellites, the best of which are shown in Section IV. Color information for the objects was obtained by means of six broad-band filters attached to the camera. The return of large numbers of images with resolution down to a kilometer has enabled geologists to construct geologic maps, to make detailed crater counts, and to develop realistic scenarios for the structure and evolution of the satellites.

2. Other Experiments

Although images are the most spectacular data returned by spacecraft, a whole array of equally valuable experiments are included in each scientific mission. For example, a gamma ray spectrometer aboard the lunar orbiters was able to map the abundance of iron and titanium across the Moon's surface. Seismometers placed on the Moon recorded waves from small moonquakes. Measurements of heat flow from the interior of the Moon enabled scientists to understand something about its composition and present evolutionary state. On board the Voyager spacecraft were several experiments which are valuable for satellite investigations including an infrared spectrometer capable of mapping temperatures; an ultraviolet spectrometer; a photopolarimeter, which simultaneously measured the color, intensity, and polarization of light; and a radio science experiment that was able to measure the pressure of Titan's atmosphere by observing how radio waves passing through it were attenuated.

3. Future Missions

There are two currently scheduled spacecraft missions to the planetary satellites: the Voyager encounter to the satellites of Neptune (in 1989), and the Galileo mission to Jupiter. The Galileo spacecraft is due to be launched by the space shuttle in 1990 toward an encounter in 1996. It will consist of a probe to explore the Jovian atmosphere and an orbiter, which will make several close flybys of each Galilean satellite. The orbiter will contain both visual and infrared imaging devices, an ultraviolet spectrometer, and a photopolarimeter. The visual camera will be capable of obtaining images with 20 m resolution.

IV. Individual Satellites

A. THE EARTH'S MOON

1. Introduction

The Earth's Moon has played a key role in the lore and superstition of the world's peoples.

When Galileo first turned his telescope on the lunar disk in the early 1600s he clearly perceived the most obvious fact of lunar morphology: the demarcation of the Moon's surface into a dark smooth terrain and a brighter more mountainous terrain. This delineation is responsible for the appearance of the 'Man in the Moon'. The dark areas were called maria (Latin for seas; the singular is mare) because of their visual resemblance to oceans and the bright areas were called terrae (Latin for land; the singular is terra), or the lunar uplands. The lunar mountains were named after terrestrial mountain chains and the craters after famous astronomers. The appearance of the Moon through a large telescope is shown in Fig. 3.

The physical characteristics of the Moon are listed in Table 1. Since the Moon has been tidally despun, it is locked in synchronous rotation, and it was not until 1959 that the back side of the Moon was observed by a Soviet spacecraft. Because gravitational perturbations on the Moon cause it to wobble, or librate, in its orbit, about 60% of the surface of the Moon is visible from the Earth. The far hemisphere has fewer maria than the visible side.

According to theories of tidal evolution, the Moon is receding from the Earth and the Earth's spin rate is in turn slowing down. Thus in the past the lunar month and terrestrial day were shorter. Evidence for this is found in the fossilized shells of certain sea corals, which deposit calcareous material in a cycle following the lunar tides.

2. Origin

The three standard theories for the origin of the Moon are (1) Capture, in which the Moon forms elsewhere in the solar system and is gravitationally captured by the Earth; (2) Fission, which asserts that the Moon broke off from the Earth early in its formation; and (3) Coaccretion, which asserts that the Earth and Moon formed independently but nearly simultaneously near their present locations.

3. Early History of the Moon

Soon after the Moon accreted it heated up due to the reasons outlined in Section IIB. The result was the melting and eruption of basaltic lava onto the lunar surface between 3.8 to about 2.8 billion years ago to form the lunar maria. Such lava would be highly fluid under the weaker gravitational field of the Moon and could flow over vast distances.

Before Soviet and American spacecraft explored the Moon, there was considerable debate over whether the craters on the Moon were of impact or of volcanic origin. Morphological features, such as bright rays and ejecta blankets, around large craters show that they were formed by impacts.

The ringlike structures that delineate the maria are the outlines of impact basins, which filled in with lava. The maria are not as heavily cratered as the uplands because the lava flows which created them obliterated pre-existing craters.

4. Nature of the Lunar Surface

The physical properties of the Moon are known more accurately than those of any other satellite because of the extensive reconnaissance and study by spacecraft (see Table II). Rock samples have been returned by both American manned and Soviet unmanned missions.

The astronauts who walked on the Moon found that the upper few centimeters of the lunar surface was covered by fine dust and pulverized rock. This covering, or regolith, is the result of fragmentation of particles from constant bombardment of the Moon's surface during its history and may be structurally similar to the regoliths of other satellites.

Analysis of lunar samples revealed important chemical differences between the Earth and Moon. The lunar uplands consist of a low density, calcium-rich igneous rock known as anorthosite. The younger maria are composed of a dark basalt rich in the minerals olivine and pyroxene. Some of the lunar basalts, known as KREEP, were found to be anonymously rich in potassium (K), rare earth elements (REE), and phosphorous (P). Formations such as lava tubes and vents, which are similar to terrestrial volcanic features, are found in the maria.

The Moon has no water or atmosphere. Surface temperatures range from 100 to 380°K.

B. Phobos and Deimos

Mars has two small satellites, Phobos and Deimos, which were discovered by the American astronomer Asaph Hall in 1877. In Jonathan Swift's moral satire *Gulliver's Travels* (published in 1726), a fanciful but coincidentally accurate prediction of the existence and orbital characteristics of two small Martian satellites was made. These two objects are barely visible in the scattered light from Mars in Earth-based telescopes. Most of what is known about Phobos and Deimos was obtained from the Mariner 9 and the Viking 1 and 2 missions to Mars (see Table II). Their physical and orbital properties are listed in Table I. Both satellites are shaped approximately like ellipsoids and are in synchronous rotation. Phobos, and possibly Deimos, has a regolith of dark material similar to that found on carbonaceous asteroids common in the outer asteroid belt. Thus the satellites may have been asteroids or asteroidal fragments, which were perturbed into a Mars crossing orbit and captured.

Both satellites are heavily cratered, which indicates that their surfaces are at least 3 billion years old (Fig. 2). However, Deimos appears to be covered with a fine, light-colored dust, which gives its surface a smoother appearance. The dust may exist because the surface is more easily pulverized by impacts, or simply because it is easier for similar material to escape from the gravitational field of Phobos, which is closer to Mars. The surface of Phobos is extensively scored by linear grooves that appear to radiate from the huge impact crater Stickney (named after the surname of Asaph Hall's wife, who collaborated with him). The grooves are probably fractures caused by the collision that produced Stickney. There is some evidence that tidal action is bringing Phobos, which is already inside Rôche's limit, closer to Mars. The satellite will either disintegrate (perhaps to form a ring) or crash into Mars in about 100 million years. The suggestion that Phobos' orbit is decaying because it is a hollow extraterrestrial space station has no basis in fact.

C. The Galilean Satellites of Jupiter

1. Introduction and Historical Survey

When Galileo trained his telescope on Jupiter he was amazed to find four points of light which orbited the giant planet. These were the satellites Io, Europa, Ganymede, and Callisto, planet-sized worlds known collectively as the Galilean satellites. Analysis of telescopic observations over the next 350 years revealed certain basic features of their surfaces. There was spectroscopic evidence for water ice on the outer three objects. The unusually orange color of Io was hypothesized to be due to elemental sulfur. Orbital phase variations were significant, particularly in the cases of Io and Europa, which indicated the existence of markedly different terrains on their surfaces. Large opposition effects observed on Io and Callisto suggested their surfaces were porous, whereas the lack of an opposition effect on Europa suggested a smooth surface. The density of the satellites decreases as a function of distance from Jupiter (Table 1).

Theoretical calculations suggested the satellites had differentiated to form silicate cores and (in the case of the outer three) ice crusts. There is the possibility that the mantles of the outer three satellites are liquid water.

The Voyager missions to Jupiter in 1977 and 1979 (Table 2) revealed the Galilean satellites to be four unique geological worlds. Current knowledge of these objects is summarized in Fig. 4.

2. Io

About the size of the Moon, Io is the only body in the Solar System other than the Earth on which active volcanism has been observed. The Voyager spacecraft detected nine currently erupting plumes, scores of calderas (volcanic vents), and extensive lava flows consisting of nearly pure elemental sulfur (Fig. 5). As sulfur cools, it changes from dark brown, to red, to orange, and finally to yellow, which accounts for the range of colors on Io's surface. There are nearly black liquid sulfur lava lakes with floating chunks of solid sulfur. Sulfur dioxide is driven out of the volcanoes to condense or absorb onto the surface as white deposits. A thin, transient atmosphere with a pressure less than one millionth of the Earth's and consisting primarily of sulfur dioxide has been detected. There appears to be a total absence of water on the surface of Io, probably because it has all been degassed from the interior from extensive volcanism and escaped into space. The total lack of impact craters means the entire surface is young and geologically active.

The heat source for melting and subsequent volcanism is the dissipation of tidal energy from

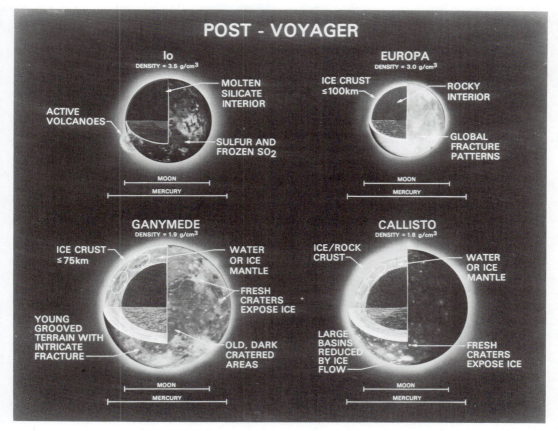

FIG. 4.   A summary of current knowledge of the Galilean satellites.

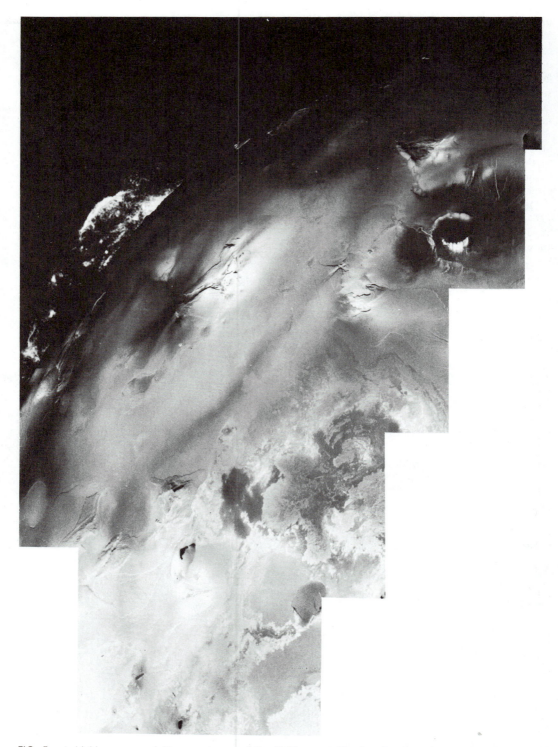

FIG. 5. A highly processed *Voyager* image of the Galilean satellite Io, showing wispy structures in the volcanic plumes (upper left), volcanic vents, and sulfur lava flows.

Jupiter and the other Galilean satellites. As Io moves in its orbit, its distance from Jupiter changes as the other satellites exert different forces depending on their distance from Io. The varying tidal stresses cause Io to flex in and out. This mechanical energy is released as heat, which causes melting in Io's mantle.

A spectacular torus of ionized particles, primarily sulfur, oxygen, and sodium, co-rotates with Jupiter's strong magnetic field at Io's orbital position. High energy ions in the Jovian magnetosphere knock off and ionize surface particles, which are swept up and entrained in the field lines. An additional source of material for the torus is sulfur and sulfur dioxide from the plumes. Aurorae seen on Jupiter are caused by particles from the torus being conducted to the planet's polar regions. Because Io is a conducting body moving in the magnetic field of Jupiter it generates a flux tube of electric current between itself and the planet. Radio emissions from the Jovian atmosphere, which correlate with the orbital position of Io, appear to be triggered by the satellite.

3. Europa

When the Voyagers encountered the second Galilean satellite, Europa, they returned images of bright, icy plains crisscrossed by an extensive network of darker fractures (see Fig. 6). The existence of only a handful of impact craters suggested that geological processes were at work on the satellite until a few hundred million years ago or less. Europa is very smooth: the only evidence for topographical relief is the scalloped ridges with a height of a few hundred meters (see bottom of Fig. 6).

Part of Europa is covered by a darker mottled terrain. Dark features also include hundreds of brown spots of unknown origin, and larger areas, which appear to be the result of silicate laden water erupting onto the surface (bottom left of Fig. 6). The reddish hue of Europa is believed to be due to contamination by sulfur from Io.

The mechanism for the formation of cracks on Europa is probably some form of tidal interaction and subsequent heating, melting, and refreezing. Calculations show that Europa may still have a liquid mantle. Although some scientists have discussed the possibility of a primitive life form teeming in the mantle, there is no evidence that life does indeed exist there.

4. Ganymede

The icy moon Ganymede, which is the largest Galilean satellite, also shows evidence for geologic activity. A dark, heavily cratered terrain is transected by more recent, brighter grooved terrain (see Fig. 7). Although they show much diversity, the grooves are typically 10 km wide and one-third to one-half km high. They were implaced during several episodes between 3.5 and 4 billion years ago. Their formation may have occurred after a melting and refreezing of the core, which caused a slight crustal expansion and subsequent faulting and flooding by subsurface water.

The grooved terrain of Ganymede is brighter because the ice is not as contaminated with rocky material that accumulates over the eons from impacting bodies. The satellite is also covered with relatively fresh bright craters, some of which have extensive ray systems. In the cratered terrain there appear outlines of old, degraded craters, which geologists called palimpsests. The polar caps of Ganymede are brighter than the equatorial regions; this is probably due to the migration of water molecules released by evaporation and impact toward the colder high latitudes.

5. Callisto

Callisto is the only Galilean satellite that does not show evidence for extensive resurfacing at any point in its history. It is covered with a relatively uniform, dark terrain saturated with craters (Fig. 8). There is, however, an absence of craters larger than 150 km. Ice slumps and flows over periods of billions of years and is apparently not able to maintain the structure of a large crater as long as rocky material. One type of feature unique to Callisto is the remnant structures of numerous impacts. The most prominent of these, the Valhalla basin, is a bright spot encircled by as many as 13 fairly regular rings (as Figure 8).

6. The Small Satellites of Jupiter

Jupiter has eleven known small satellites, including three discovered by the Voyager mission. They are all probably irregular in shape (see Table I). Within the orbit of Io are at least three satellites: Amalthea, Adrastea and Metis. Amalthea is a dark, reddish heavily cratered object reflecting less than 5% of the radiation it receives; the red color is probably due to contamination by sulfur particles from Io. Little else

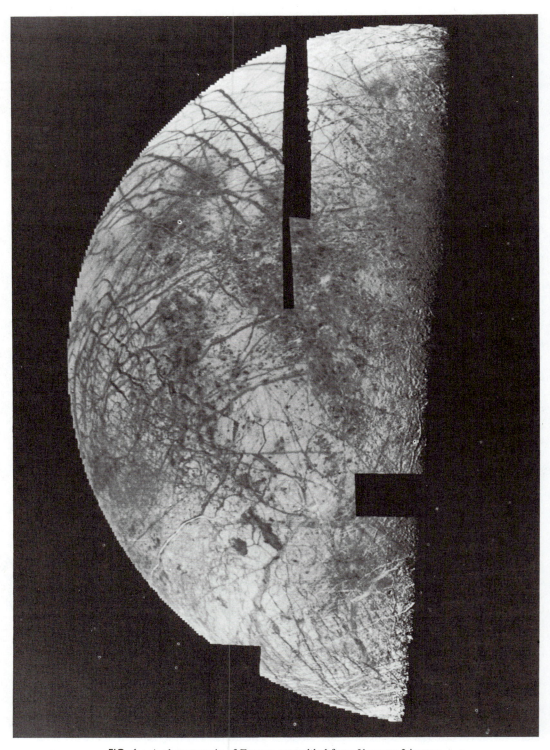

FIG. 6.   A photomosaic of Europa assembled from *Voyager 2* images.

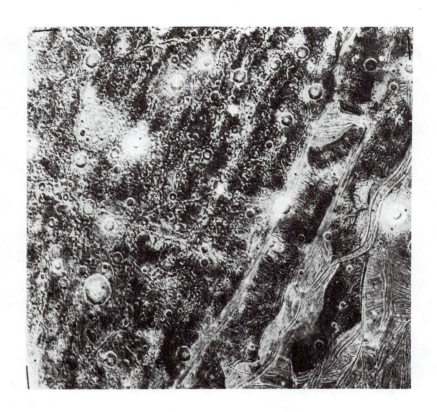

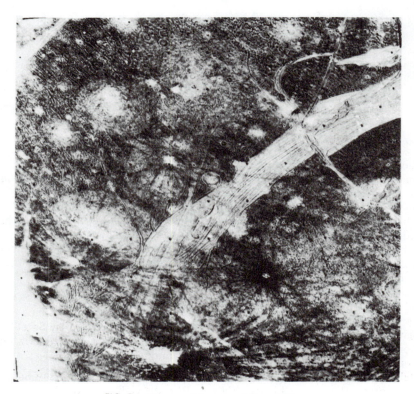

FIG. 7. *Voyager 2* images of Ganymede.

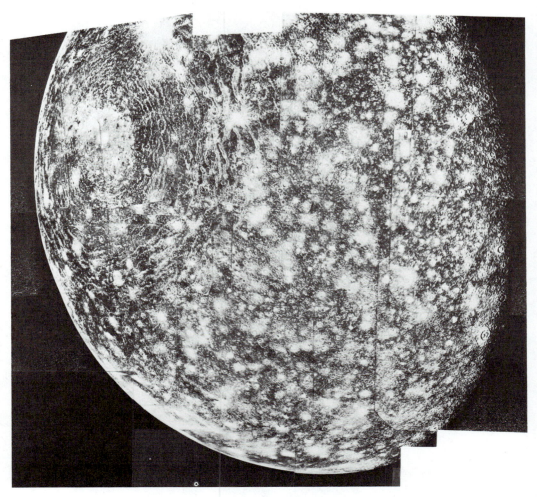

FIG. 8.   A *Voyager 1* photomosaic of Callisto. The Valhalla impact basin, which is 600 km wide, dominates the surface.

is known about its composition except that the dark material may be carbonaceous.

Adrastea and Metis, both discovered by Voyager, are the closest known satellites to Jupiter and move in nearly identical orbits just outside the outer edge of the thin Jovian ring, for which they may be a source of particles. Between Amalthea and Io lies the orbit of Thebe, also discovered by Voyager. Little is known about the composition of these satellites, but they are most likely primarily rock-ice mixtures. The three inner satellites sweep out particles in the Jovian magnetosphere to form voids at their orbital positions.

Moving outward from Jupiter, we find a class of four satellites moving in highly inclined, orbits (Lysithea, Elara, Himalia, and Leda). They

are dark, spectrally neutral objects, reflecting only 2 or 3% of incident radiation and may be similar to carbonaceous asteroids.

Another family of objects is the outermost four satellites, which also have highly inclined orbits, except they move in the retrograde direction around Jupiter. They are Sinope, Pasiphae, Carme, and Ananke, and they may be captured asteroids.

D. THE SATURNIAN SYSTEM

1. The Medium Sized Icy Satellites: Rhea, Dione, Tethys Mimas, Enceladus, and Iapetus

The six largest satellites of Saturn are smaller than the Galilean satellites but still sizable—as such they represent a unique class of icy satellite. Earth-based telescopic measurements showed the spectral signature of ice for Tethys, Rhea, and Iapetus; Mimas and Enceladus are close to Saturn and difficult to observe because of scattered light from the planet. The satellites' low densities and high albedos (Table I) imply that their bulk composition is largely water ice, possibly combined with ammonia. They have smaller amounts of rocky silicates than the Galilean satellites. Resurfacing appears to have occurred on several of the satellites. Most of what is presently known of the Saturnian system was obtained from the Voyager flybys in 1980 and 1981.

The innermost medium-sized satellite Mimas is covered with craters, including one (named Arthur), which is as large as a third of the satellite's diameter (upper left of Fig. 9). The impacting body was probably nearly large enough to

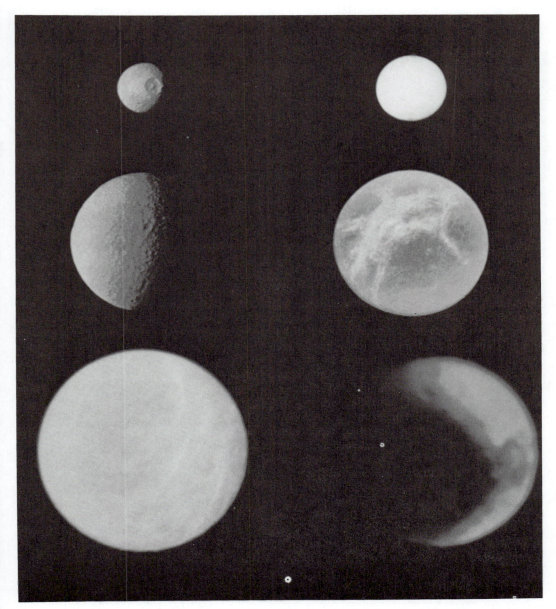

FIG. 9. The six medium-sized icy Saturnian satellites. From the upper left, in order of size: Mimas, Enceladus, Tethys, Dione, Rhea, and Iapetus.

break Mimas apart; such disruptions may have occurred to other objects. There is a suggestion of surficial grooves that may be features caused by the impact. The craters on Mimas tend to be high-rimmed, bowl shaped pits; apparently surface gravity is not sufficient to have caused slumping.

The next satellite outward from Saturn is Enceladus, an object that was known from telescopic measurements to reflect nearly 100% of the visible radiation incident on it (for comparison, the Moon reflects only about 11%). The only likely composition consistent with this observation is almost pure water ice. When Voyager 2 arrived at Enceladus, it transmitted pictures to Earth which showed an object that had been subjected, in the recent geologic past, to extensive resurfacing; grooved formations similar to those on Ganymede were evident (see Fig. 10). The lack of impact craters on this terrain is consistent with an age less than a billion years. It is possible that some form of ice volcanism is presently active on Europa. The heating mechanism is believed to be tidal interactions, perhaps

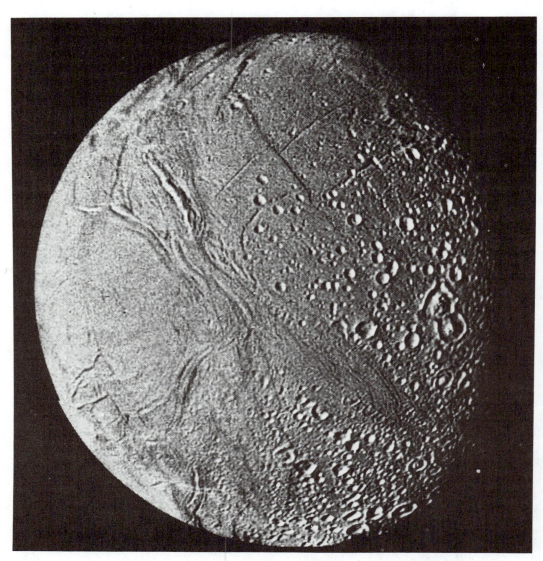

FIG. 10.   A *Voyager 2* photomosaic of Enceladus. Both the heavily cratered terrain and the recently resurfaced areas are visible.

with Dione. About half of the surface observed by Voyager is extensively cratered and dates from nearly 4 billion years ago.

A final element to the enigma of Enceladus is the possibility that it is responsible for the formation of the E-ring of Saturn, a tenuous collection of icy particles that extends from inside the orbit of Enceladus to past the orbit of Dione. The position of maximum thickness of the ring coincides with the orbital position of Enceladus. If some form of volcanism is presently active on the surface, it could provide a source of particles for the ring. An alternative source mechanism is an impact and subsequent escape of particles from the surface.

Tethys is covered with impact craters, including Odysseus, the largest known impact structure in the solar system. The craters tend to be flatter then those on Mimas or the Moon, probably because of relaxation and flow over the eons under Tethys' stronger gravitational field. Evidence for resurfacing episodes is seen in regions that have fewer craters and higher albedos. In addition, there is a huge trench formation, the Ithaca Chasma, which may be a degraded form of the grooves found on Enceladus.

Dione, which is about the same size as Tethys, exhibits a wide diversity of surface morphology. Most of the surface is heavily cratered (Fig. 11), but gradations in crater density indicate that several periods of resurfacing events occurred during the first billion years of its existence. One side of the satellite is about 25% brighter than the other. Wispy streaks (see Figs. 9 and 11), which are about 50% brighter than the surrounding areas, are believed to be the result of internal activity and subsequent implacement of erupting material. Dione modulates the radio emission from Saturn, but the mechanism for this phenomenon is unknown.

Rhea appears to be superficially very similar to Dione (see Fig. 9). Bright wispy streaks cover one hemisphere. However, there is no evidence for any resurfacing events early in its history. There does seem to be a dichotomy between crater sizes—some regions lack large craters while other regions have a preponderance of such impacts. The larger craters may be due to a population of larger debris more prevalent during an earlier episode of collisions.

When Cassini discovered Iapetus in 1672, he noticed that at one point in its orbit around Saturn it was very bright; whereas, on the opposite side of the orbit it nearly altogether disappeared. He correctly deduced that one hemisphere is composed of highly reflective material, while the other side is much darker. Voyager images show that the bright side, which reflects nearly 50% of the incident radiation, is fairly typical of a heavily cratered icy satellite. The other side, which is centered on the direction of motion, is coated with a material with a reflectivity of about 3–4% (see Fig. 9).

Scientists still do not agree on whether the dark material originated from an exogenic source or was endogenically created. One scenario for the exogenic deposit of material entails dark particles being ejected from Phoebe and drifting inward to coat Iapetus. The major criticism of this model is that the dark material on Iapetus is redder than Phoebe, although the material could have undergone chemical changes after its expulsion from Phoebe to make it redder. One observation lending credence to an internal origin is the concentration of material on crater floors, which implies an infilling mechanism. In one model, methane erupts from the interior and is subsequently darkened by ultraviolet radiation.

Other aspects of Iapetus are unusual. It is the only large Saturnian satellite in a highly inclined orbit. It is less dense than objects of similar albedo; this implies a higher fraction of ice or possibly methane or ammonia in its interior.

2. Titan

Titan is a fascinating world that one member of the Voyager imaging team called 'a terrestrial planet in a deep freeze.' It has a thick atmosphere that includes a layer of photochemical haze (Fig. 12) and a surface possibly covered with lakes of methane or ethane. Methane was discovered by G. P. Kuiper in 1944: the Voyager experiments showed that the major atmospheric constituent is nitrogen, the major component of the Earth's atmosphere. Methane (which is easier to detect from Earth because of prominent spectroscopic lines) may comprise only a few percent or less. The atmospheric pressure of Titan is 1.5 times that of the Earth's; however, Titan's atmosphere extends much further from the surface (nearly 100 km) on account of the satellite's lower gravity. The atmosphere is thick enough to obscure the surface entirely.

Titan's density (Table I) implies a bulk composition of 45% ice and 55% silicates. It probably has a differentiated rocky core. Titan was able to retain an appreciable atmosphere, while the similarly sized Ganymede and Callisto were

FIG. 11.   The heavily cratered face of Dione is shown in this *Voyager 1* image. Bright wispy streaks are visible on the limb of the satellite.

not because more methane and ammonia condensed at Titan's lower formation temperature. The methane has remained; whereas, ammonia has been photochemically dissociated into molecular nitrogen and hydrogen, the latter being light enough to escape the gravitational field of Titan. The escaped hydrogen forms a tenuous torus at the orbital position of Titan. Although it has not been directly detected, argon may comprise a few percent of the atmosphere.

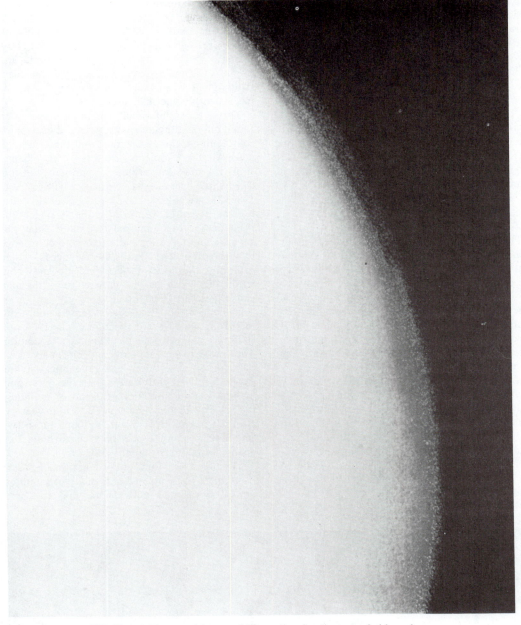

FIG. 12.   A *Voyager 1* image of Titan, showing the extended haze layer.

The infrared spectrometer on voyager detected nearly a dozen organic compounds such as acetylene (C_2H_2), ethane (C_2H_6), and hydrogen cyanide (HCN), which plays an important role in prebiological chemistry. These molecules, which are constantly being formed by the interaction of ultraviolet radiation with nitrogen and methane, constitute the haze layer of aerosol dust in the upper atmosphere. Much of this material, which gives Titan a reddish color, "rains" onto the surface and possibly forms lakes of ethane or methane. The role of methane on the surface and in the atmosphere of Titan may be similar to that of water on Earth; the triple point of methane (where it can coexist as a solid, liquid, or gas) is close to the surface temperature of Titan (93°K). In this scenario, methane in the atmosphere covers methane lakes and ice on the surface.

The northern polar area of Titan is darker than the southern cap. Ground-based observations detected long-term variations in the brightness of Titan. Both observations may be related to the existence of 30-year seasonal cycles on Titan. Even though only a small fraction of the incident solar radiation reaches the surface, Titan is probably subjected to a slight greenhouse effect.

3. The Small Satellites

The Saturnian system has a number of unique small satellites. Telescopic observations showed that the surface of Hyperion, which lies between the orbits of Iapetus and Titan, is covered with ice. Because Hyperion has a visual geometric albedo of 0.30, this ice must be mixed with a significant amount of darker, rocky material. It is darker than the medium-sized inner Saturnian satellites, presumably because resurfacing events have never covered it with fresh ice. Although Hyperion is only slightly smaller than Mimas, it has a highly irregular shape (see Table 1). This suggests, along with the satellite's battered appearance, that it has been subjected to intense bombardment and fragmentation. There is also evidence for Hyperion being in nonsynchronous rotation—perhaps a collision within the last few million years knocked it out of a tidally locked orbit.

Saturn's outermost satellite Phoebe, a dark object (Table I) with a surface similar to that of carbonaceous asteroids, moves in a highly inclined, retrograde orbit. Voyager images show definite variegations consisting of dark and bright (presumably icy) patches on the surface. Although it is smaller than Hyperion, Phoebe has a nearly spherical shape.

Three types of small satellites have been found only in the Saturnian system: the sheparding satellites, the co-orbitals, and the Lagrangians. All these objects are irregularly shaped (Fig. 13) and probably consist primarily of ice. The three shepherds, Atlas, 1980S26, and 1980S27, are believed to play a key role in defining the edges of Saturn's A and F rings. The orbit of Saturn's innermost satellite Atlas lies several hundred kilometers from the outer edge of the A-ring. The other two shepherds, which orbit on either side of the F-ring, not only constrain the width of this narrow ring, but may cause its kinky appearance.

The co-orbital satellites Janus and Epimetheus, which were discovered in 1966 and 1978, exist in an unusual dynamical situation. They move in almost identical orbits at about 2.5 Saturn radii. Every four years the inner satellite (which orbits slightly faster than the outer one) overtakes its companion. Instead of colliding, the satellites exchange orbits. The four-year cycle then begins over again. Perhaps these two satellites were once part of a larger body that disintegrated after a major collision.

The three remaining small satellites of Saturn orbit in the Lagrangian points of larger satellites: one is associated with Dione and two with

FIG. 13. The small satellites of Saturn. They are, clockwise from far left; Atlas, 1980S26, Janus, Calypso, 1980S6, Telesto, Epimetheus, and 1980S27.

Tethys. The Lagrangian points are locations within an object's orbit in which a less massive body can move in an identical, stable orbit. they lie about 60 degrees in front of and in back of the larger body. Although no other known satellites in the solar system are Lagrangians, the Trojan asteroids orbit in two of the Lagrangian points of Jupiter.

E. THE SATELLITES OF URANUS: MIRANDA, ARIEL, UMBRIEL, TITANIA, AND OBERON

The rotational axis of Uranus is inclined 98 degrees to the plane of the solar system; observers on Earth thus see the planet and its system of satellites nearly pole-on. The orbits of Ariel, Umbriel, Titania, and Oberon are regular whereas Miranda's orbit is slightly inclined. Figure 14 is a telescopic image of the satellites. Theoretical models suggest the satellites are composed of water ice (possibly bound with carbon monoxide, nitrogen, and methane), and silicate rock. The higher density of Umbriel implies its bulk composition includes a larger fraction of rocky material. Melting and differentiation have occurred on some of the satellites. Theoretical calculations indicate that tidal interactions may provide an additional heat source in the case of Ariel.

Water ice has been detected spectroscopically on all five satellites. Their relatively dark albedos (Table I) are probably due to surficial contamination by carbonaceous material. Another darkening mechanism that may be important is bombardment of the surface by ultraviolet radiation. The four outer satellites all exhibit large opposition surges, which may indicate that the regoliths of these objects are composed of very porous material.

The Voyager 2 spacecraft encountered Uranus in January 1986 to provide observations indicating that at least some of the major satellites have undergone melting and resurfacing. One feature on Miranda consists of a series of ridges and valleys ranging from 0.5 to 5 km in height (Fig. 14b). Ariel, which is the geologically youngest of the five satellites, and Titania are covered with cratered terrain transected by grabens, which are fault-bounded valleys. Umbriel is heavily cratered and is the darkest of the major satellites, which indicates that its surface is the oldest. Oberon is similarly covered with craters, some of which have very dark deposits on their floors. The satellites are spectrally flat with

visual geometric albedos ranging from 0.2–0.4, which is consistent with a composition of water ice (or methane-water ice) mixed with a dark component such as graphite or carbonaceous chondritic material.

Voyager 2 also discovered 10 new small moons, including two which act as shepherds for the outer (epsilon) ring of Uranus (Table I). These satellites have visual geometric albedos of only 4–9%. They move in orbits that are fairly regularly spaced in radial distance from Uranus, and have low orbital inclinations and eccentricities.

F. THE SATELLITES OF NEPTUNE: TRITON AND NEREID

Neptune has an unusual family of satellites (Table I and Fig. 15). Triton, one of the largest satellites in the Solar System, moves in a highly inclined retrograde orbit. The orbit of Nereid, Neptune's other known satellite, is prograde and highly inclined. This situation implies an anomolous origin, perhaps involving capture of two remnant planetesimals by Neptune. The suggestion that Triton and Pluto were once both satellites of Neptune that experienced a near encounter, causing Triton to go into a retrograde orbit and the expulsion of Pluto from the system, is not plausible on dynamical grounds (see Section G). The detection of a third satellite in 1982 has not been confirmed.

The diameter and mass of Triton have not been well determined (Table I). The low formation temperature of the satellite constrains the composition to be primarily water ice, silicates, methane, and ammonia. The center of the satellite may consist of a differentiated rock-ice core, and a possibly liquid water-ammonia mantle. Methane ice has been detected on the surface and there is some spectral evidence for liquid nitrogen oceans or lakes. Because some of this liquid would evaporate, it can be deduced that Triton probably has a nitrogen atmosphere. If Triton was captured early in its history, it may show evidence of vast resurfacing events from a period of tidal heating.

Nereid is very faint and nothing about its surface composition has been directly observed.

G. THE PLUTO–CHARON SYSTEM

Soon after Pluto was discovered in 1930, scientists conjectured that the planet was an es-

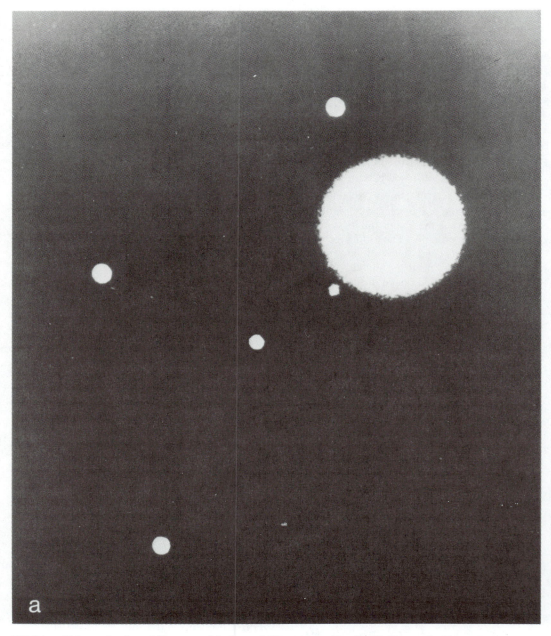

FIG. 14a.   Telescopic view of Uranus and its five satellites obtained by Ch. Veillet on the 154-cm Danish–ESO
telescope. Outward from Uranus they are: Miranda, Ariel, Umbriel, Titania, and Oberon.

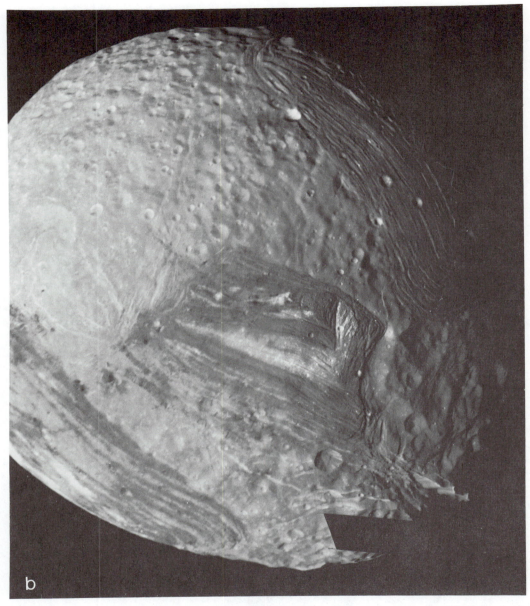

FIG. 14b.   A mosaic of Miranda produced from images taken by the *Voyager 2* spacecraft at 30–40 thousand km from the moon. Resolution is 560 to 740 m. Older, cratered terrain is transected by ridges and valleys indicating more recent geologic activity.

FIG. 15.   The two satellites of Neptune: Triton (near Neptune), and the faint Nereid. (Photograph courtesy of Lick Observatory.)

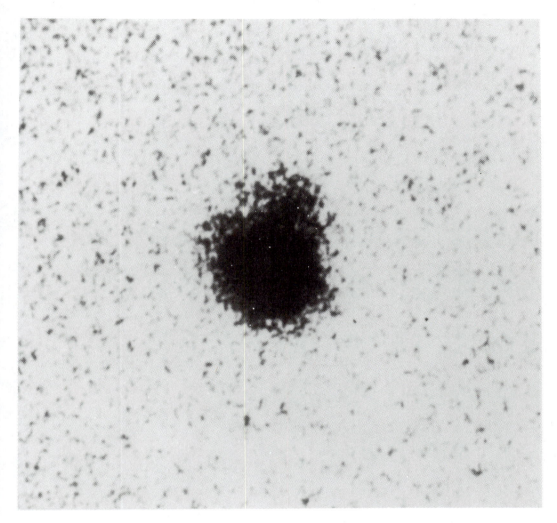

FIG. 16.   A negative image of the Pluto–Charon system obtained by J. W. Christy. Charon is the extended blob to the upper right of Pluto. (Photo courtesy of the U.S. Naval Observatory.)

caped satellite of Neptune. More recent calculations have shown that it is unlikely that Pluto could have acquired its present amount of angular momentum during an ejection event. Pluto is most likely a large planetesimal dating from the formation period of the solar system.

In 1978, Pluto was shown to have a moon of its own, which was named Charon (Fig. 16). Charon appears in specially processed high quality images as a fuzzy mass orbiting close to Pluto. It is more massive in comparison with its primary than any other satellite: its mass is between 5 and 10% that of Pluto. Nothing is known about the surface composition of Charon. Its low formation temperature implies it is probably a conglomerate of ices (including possibly solid nitrogen), and some small fraction of silicates. Between 1985 and 1991, Pluto and Charon will be undergoing a series of mutual eclipses, which will allow a more accurate determination of their radii, masses, and thus density. A more accurate knowledge of Charon's density can be used to constrain its bulk composition.

BIBLIOGRAPHY

Beaty, J. K., B. O'Leary, and A. Chaikin (eds.) (1981). *The New Solar System*, 2nd ed. Sky Publishing Corp., Cambridge Mass.

Burns, J. (ed.) (1977). *Planetary Satellites*, University of Arizona Press, Tucson.

Gehrels, T. (ed.) (1984). *Saturn,* University of Arizona Press, Tucson.

Hartmann, W. K. (1983). *Moons and Planets,* 2nd ed., Wadsworth, Belmont, California.

Morrison, D. (1980). *Voyage to Jupiter,* NASA SP–439. U.S. Government Printing Office, Washington, D.C.

Morrison, D. (1982). *Voyage to Saturn,* NASA SP–451. U.S. Government Printing Office, Washington, D.C.

Morrison, D. (ed.) (1982). *The Satellites of Jupiter,* University of Arizona Press, Tucson, 1982.

Soderblom, L. A. (1980). *The Galilean satellites of Jupiter, Scientific American* **242,** 88–100.

PRIMITIVE SOLAR SYSTEM OBJECTS: ASTEROIDS AND COMETS

Lucy-Ann McFadden *University of Maryland*

GLOSSARY

Albedo: Measurement of the fraction of sunlight that a surface reflects.

Aphelion: Point at which a solar system body is farthest from the sun in its orbit.

Apparition: Time at which an object in the solar system can be viewed from the earth in the night-time sky.

Astronomical unit: 1.496×10^8 km, the mean distance from the earth to the sun.

Eccentricity: Parameter of an elliptical orbit describing how much the orbit deviates from that of a circle, where $e = 0$. When $e = 1$, the orbital motion described would be that of a paraboloid and a body on such an orbit would be gravitationally unbound from the sun.

Ecliptic: Plane defined by the earth and the sun.

Inclination: Orbital element that is the angle measured between the ecliptic and the orbital plane of an asteroid.

Librations: Rotations of a celestial body about its mean position arising from gravitational forces acting on the equatorial region of the body. The result is that the direction of the greatest principal axis of inertia moves around the rotation axis of the body producing rotational oscillations of the body.

Magma: Molten rock.

Maria: (Latin: seas; singular: mare). Regions of the moon that are basaltic rock plains.

Meteorite: Rock of extraterrestrial origin that has survived passage through the earth's atmosphere and reached the surface of the earth.

Orbit: Path that an object takes as it travels about another object.

Orbital elements: Quantities that express the orientation of a celestial body in space.

Perihelion: Point in the orbit of a solar system body at which it is closest to the sun.

Plasma: Electrically neutral gas consisting of charged particles such as ions, electrons, and neutrals. A plasma has a high temperature but its density is very low.

Radioactive decay: The exponential decrease in the amount of material that spontaneously and naturally emits radiation.

Semimajor axis: Distance at which an orbiting body is half-way between its aphelion and perihelion.

Solar nebula: Cloud of gas and dust that surrounded the sun during its early stages of formation. The planets formed from material in this cloud.

Solar wind: Constant, radial flow of energetic, charged particles from the sun.

Asteroids and comets, among the smallest objects in the solar system are thought to be composed of material that has remained relatively unaltered since accretion from the solar nebula. Scientists refer to unaltered material as primitive because it is suspected that this material has not been heated as much as the earth and other planets and thus represents the starting composition of the solar system. The goals of the study of asteroids and comets are to determine how the smallest members formed and to determine the starting composition of the solar system. From the study of asteroids and comets and the study of the larger planets, we expect to piece together the sequence of events that are the history of our solar system. In this article the current state of our knowledge of the asteroids and comets is presented as well as some issues to be resolved in the coming decades.

The asteroids are planetesimals that failed to condense into a full-sized planet. Because of the nonrandom distribution of asteroids between Mars and Jupiter (see Section I,A), it is suspected that Jupiter had a significant role in preventing the formation of the planet between Mars and Jupiter. The number of asteroids with known and catalogued orbits is close to 4000 at the time of this writing. The total mass of these asteroids however is approximately 1.5×10^{24} g, about one billionth the mass of the sun. About 30% of this mass is contained in one asteroid, the largest, named 1 Ceres, using the convention of giving an asteroid a number and a name. The numbers are assigned numerically as the orbits of the asteroids are determined precisely enough for their position to be predicted and verified at a later apparition. The name is chosen by the discoverer. Until that time, asteroids are given a temporary designation that contains the year in which they were discovered and a two-letter code that first references the particular two-week interval and the order in that interval in which they were discovered. Thus an asteroid named 1986 AB was the second asteroid (referring to the B) discovered in the first two weeks (referring to the A) in 1986.

Comets are primarily icy bodies with highly elliptical orbits. With their large orbital periods they are located at great distances from the sun for most of their lifetime. They are observed only when they pass relatively close to the earth. Comets have been observed as far as 12 AU (astronomical unit) from the earth and as close as 0.02 AU (8 times the earth–moon distance). There are records of observations of approximately 1300 comets, but there are an estimated 10^{10} to 10^{12} comets in the solar system. By convention, the comets are divided into three groups, long period comets with periods greater than 200 yr, short period comets with periods less than 200 yr and Jupiter family comets with periods less than 20 yr. Comets making their first perihelion passage are termed new comets; their periods may be hundreds of thousands of years. Old comets are short period comets that have made many perihelion passages. There have been no direct determinations of the mass of a comet, but an estimate of 4×10^{16} g is derived assuming a mean density of 1.3 g/cm³ and a spherical nucleus with a radius of 2 km. Comets

are named after their discoverer and given a preliminary designation including the year and letter designating the order in the year in which they were discovered or in the case of a previously known comet, recovered. A year after the comet has reached perihelion, it is assigned a roman numeral designation indicating its order relative to its date of perihelion. For example, 1985I was the first comet to reach its perihelion point in 1985. The same comet will then have a different designation at each apparition, but its name will not change. Since some people have discovered more than one comet the designations are necessary to distinguish different comets.

I. Asteroids

A. Orbits

Asteroids orbit the sun predominantly between Mars and Jupiter. An orbit is defined by six parameters or orbital elements. Only three, the semimajor axis, a, the eccentricity, e, and the inclination, i, are needed to describe the major aspects of the asteroid population. For most asteroids, the semimajor axis ranges from 2.2 to 4.3 AU, a range used to define the main asteroid belt. Most of these asteroids have eccentricities between about 0.03 and 0.4 and inclinations between 0° and 30°. [See CELESTIAL MECHANICS.]

1. Planet-Crossing Asteroids

A small fraction of the asteroids have orbits that cross the orbits of the inner planets. Other asteroids cross the orbits of Jupiter and Saturn. One asteroid is known to lie between the orbit of Saturn and Uranus. The presence of this asteroid, named 2060 Chiron, suggests that there might be a population of such objects too distant and small to be detected from available earth-based telescopes.

The planet-crossing asteroids that cross the orbits of the inner planets, also called near-earth asteroids, are important to study because they may be the source of the meteorites that occasionally fall to earth. Asteroids with orbits that approach the earth, passing inside the perihelion of Mars, are called Amors. Those that cross the orbit of the earth are called Apollo asteroids. A third group of near-earth asteroids that have orbits crossing that of the earth's but that do not travel further from the sun than the earth's aphelion, are called Aten asteroids, after the name of the first asteroid discovered with this type of orbit.

An estimated 10^2 to 10^3 tons of extraterrestrial material, ranging from submicron-sized dust to boulders weighing many tons, reach the earth's surface every day. Some of this material is derived from the near-earth asteroids. An asteroid that is 1 km in diameter or larger will collide with the earth on the average about once every 1.5 to 2 million yr. Most near-earth asteroids are not on collision course with the earth but may be located near enough to the earth to be used for building and manufacturing in space in the future. It is important to understand what these asteroids are made of in order to study the origin and evolution of the solar system, to evaluate the effects of a collision with the earth and to determine what resources are available for future use in space activities. The destruction of a near-earth asteroid is caused by either a collision with the inner planets or ejection from the solar system by gravitational perturbations due to near misses with either Mars, Earth, or Venus. Calculations of their lifetime indicate that Atens, Apollos, and Amors have been in their present orbits for 10–100 million yr. Since the age of the solar system as determined from radioactive dating of meteorites is at least 4.5 billion yr, there must be a source of these asteroids to keep replenishing that part of the population that is lost. One possibility is that some of the near-earth asteroids are the solid remains of comets that have exhausted all their gases and now orbit the sun in planet-crossing orbits. Another is that collisions between large asteroids in the main asteroid belt occasionally throw out asteroid-sized chunks into planet-crossing orbits. One of the goals of near-earth asteroid studies is to locate the source of these bodies. Other equally important goals are to determine the mechanism that places an asteroid in a planet-crossing orbit, to determine their physical properties, and to assess their importance as economic resources as well as hazards.

The asteroids crossing the orbits of the outer solar system planets are difficult to study because of their great distance from the earth and their size. Nevertheless it is important to search for more of these asteroids and study their physical properties to determine whether or not they are the same composition as the main belt asteroids and to determine their relationship to the composition of the outer planets. Hypotheses have also been presented suggesting that plane-

tesimals in this region of the solar system may be reservoirs for short-period comets (see Section II).

2. Kirkwood Gaps

Asteroid orbits are not distributed totally randomly in the main belt. There are regions in which asteroids are relatively numerous and regions in which they are nearly absent. This structure, shown in Fig. 1, is most apparent when the number of asteroids at a given distance from the sun is plotted as a function of semimajor axis. It is suspected that asteroid orbits were initially randomly distributed throughout the main belt region, but that asteroids in certain orbits were removed by the gravitational forces of Jupiter. These regions of relatively few asteroids are called the Kirkwood gaps, after the nineteenth-century American astronomer who first recognized their existence. The Kirkwood gaps are located at distances from the sun where the orbital period of the asteroid is an integer multiple of the orbital period of Jupiter. When these asteroids and Jupiter line up with respect to the sun, Jupiter's mass exerts a gravitational force on the asteroid. These periodic perturbations alter the orbit of these asteroids creating a gap in the distribution of asteroids in the main belt. Whether or not the gaps formed from collisions subsequent to the formation of the solar system when the gaps were cleared of asteroids,

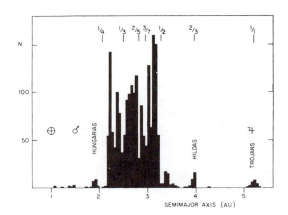

FIG. 1. Distribution over heliocentric distance of the first 1,978 numbered asteroids, in increments of 0.05 AU. Fractions indicate the ratio of orbital periods of the asteroids to that of Jupiter, which form the dynamical resonances called Kirkwood gaps. The orbits of the planets are indicated by symbols at 1.0, 1.5 and 5.2 AU. [Figure complied by B. Zellner and published in NASA-Conference Publication 2053.]

or whether they failed to form in these regions in the first place is not known at present.

Asteroids once located near the Kirkwood gaps have been proposed as sources of near-earth asteroids (this hypothesis assumes that collisions depleted the Kirkwood gaps subsequent to the initial formation of the asteroid belt). Numerical models of the mechanisms by which the asteroids might be perturbed into planet-crossing orbits have been derived and tested through computer simulations. Studies of the physical properties and mineralogical composition of asteroids in both of these regions from ground-based telescopes also contribute to the testing of this hypothesis. With the present classification scheme and techniques available to determine asteroid composition remotely, without directly sampling the material such as was done from the moon during the Apollo missions, some of the asteroids located near the Kirkwood gaps can be classified into the same groups as some of the near-earth asteroids. In addition, the compositions of some of the near-earth asteroids is consistent with those of some of the asteroids near Kirkwood gaps. However, the relationship cannot be proven because our current techniques of classification and compositional analysis do not prove a unique and genetic relationship to these asteroids. (See Sections I,D and I,E.)

Examination of Fig. 1 shows that there are two resonances that are populated by asteroids. These are the Hilda and Trojan regions. The Hildas are in a 3 : 2 resonance with Jupiter, their semimajor axis is at 4.0 AU. Librations with periods of 250 to 300 yr prevent these asteroids from making close approaches to Jupiter, thus they are not perturbed in their orbits and they remain stable. The periods of the Trojans are the same as Jupiter's. They are located at the preceding and following lagrangian points of the Jupiter-Sun system. At these points, the gravitational forces on the asteroids due to Jupiter and the sun are equal to the radial forces of the asteroids' motion and the orbits are stable. Asteroids at these locations will not be ejected out of the solar system by forces due to Jupiter's mass.

3. Hirayama Families

Some asteroids are found to have nearly identical orbital elements. It is thought that these groups of asteroids are remnants of a collision or fragmentation of a large (200-km diameter) asteroid. The early twentieth-century Japanese astronomer, K. Hirayama identified these groups of asteroids. If they are collisional remnants of a larger asteroid, then the study of its fragments will provide the equivalent of a stratigraphic view of the interior of the asteroid. Physical studies of a few asteroid families have shown that the members of some families have similar photometric properties that are different from the surrounding nonfamily population of asteroids. These results support the theory that some families at least resulted from the collisional disruption of a single parent body and that their composition is homogeneous. There are indications from reflectance spectroscopy that some other family parent bodies are compositionally inhomogeneous.

B. SIZE

The size of an asteroid is fundamental information from which the density and albedo of the object can be derived. In addition, the size distribution of asteroids in the main belt is related to the formation mechanism and subsequent evolution of the asteroid population. A range of asteroid sizes is observed from 960 km to less than 0.2 km. There are only three asteroids with diameters larger than 500 km and 27 with diameters greater than 200 km. Asteroids with diameters between 50 and 200 km number more than 300. The number of asteroids of smaller diameter increases exponentially but not uniformly, which would indicate that the population was in equilibrium in terms of collisions. From these observations one must determine what portion of this distribution represents the initial distribution remnant from the formation of the solar system and which portion is the product of subsequent collisional destruction of the asteroid belt. Different processes may represent different size distributions, thus knowledge of this size distribution is important to understanding the history of the main belt. Examining the size distribution of asteroids as a function of their orbital elements or taxonomic type (to be discussed) may also reveal the nature of the dynamical processes presently active in the asteroid belt. Indeed studies have shown that the size distributions differ in different regions of the asteroid belt and for different taxonomic types and families of asteroids in the same region. Thus, it is important to know how many asteroids there are and how big they are.

1. Thermophysical Models

The diameter of hundreds of asteroids have been determined based on a thermophysical model and measurements of the emitted radiation at 10 and 20 μm and the reflected component in the visible or near-infrared. Assuming the surface is in equilibrium with the incident sunlight, the sum of the reflected and emitted radiation must equal the total incident solar flux on the asteroid surface. The reflected component, measured by its visible brightness is proportional to the product of the geometric albedo (p) and cross section of the asteroid. The measured thermal emission is proportional to the product of the absorbed incident sunlight ($1 - A$), where A is the total albedo integrated over all wavelengths [called the Bond albedo, $A = q(\lambda)p(\lambda)$], and the cross section. With measurements of the reflected and emitted radiation, and assumptions concerning the physical properties of the surface: values of the emissivity, its angular distribution, and the relationship between the geometric and total albedo, the size and albedo of an asteroid can be determined with an accuracy of 10–20%. This model is applied to asteroids larger than 30 km in diameter, objects large enough to retain a dusty surface layer. Modifications have been made to this standard model that take into account the thermal properties of an asteroid with a bare rock surface, characteristics that are more physically reasonable for smaller asteroids.

In 1983, the Infrared Astronomical Satellite (IRAS) was launched to make a survey of celestial objects at infrared wavelengths. Contained in this data set are the infrared fluxes of any of the known asteroids. By applying the standard thermal model described above, the diameter and size frequency of these asteroids can be analyzed with the same model.

2. Occultation Diameters

One of the most accurate yet most difficult methods of measuring the diameter of an asteroid is by observing stellar occultations, which occur when an asteroid passes across the line of sight of a background star as seen from the earth. When this occurs, it is possible to map the asteroid's shadow as cast on the earth by the star. This is done by recording the time and magnitude of the decrease in flux from the star at different locations on the earth. Because the angular diameter of all asteroids is small, the path

that the shadow traverses is small as projected on the surface of the earth. The observations are difficult because telescopes and recording equipment have to be placed in the path of the occultation, the time and path of the occultation have to be predicted, and the weather has to be good. With good coverage of the occultation path, both the diameter and projected shape of the asteroid can be determined to an accuracy of a few percent.

Stellar occultations by large asteroids are quite common, but predicting and organizing an observing expedition is costly and time consuming. It is often the case that the position of the asteroids and the coordinates of the stars are not known well enough to know if the shadow will even hit the earth. Consequently, only about 10 diameters have been determined by this method. These results can then be compared to the results obtained by using the thermophysical model to improve our understanding of asteroid thermophysical properties and improve the accuracy of the model calculations for which the data set is much larger.

C. SHAPES AND ROTATION RATES

By monitoring the reflected sunlight from the surface of Ceres as a function of time as the asteroid rotates, we see only a 10% change in brightness called a light curve. This variation indicates one of three things: (1) that there is a change in the cross section or shape of the asteroid as it rotates, (2) that the surface reflectance properties vary across the surface, or (3) that the asteroid has a companion body orbiting around it, eclipsing the primary asteroid and changing the amount of reflected light seen when the companion is at different locations in its orbit. A spherical body with no color variations would produce no light curve. Any rotating, smooth spheroid (that is not perfectly spherical) with a homogeneous surface composition would have a singly periodic light curve (two peaks of equal magnitude). A spherical asteroid with color variations across its surface would also have a singly periodic light curve, its period being equal to the asteroid's rotation rate. In order to estimate the shapes and relative albedos of eclipsing asteroids, at least one of the components has to be elongated and/or have different albedos, otherwise there would be no change in the light curve. Doubly periodic light curves occur in cases of irregular shape and/or variations in albedo across the surface. These types of light curves are the most common. It is believed that most of the asteroid brightness changes are due to their irregular shape because there is little change in both the degree of polarized light and overall color as observed with rotation. However, some of the larger asteroids have been carefully studied for reflectance changes as a function of wavelength and rotation and have been shown to have compositional heterogeneity across their surface. The light curves of the majority of the asteroids are probably due to irregularities of shape however.

Rotation rates range from about 2 hr to several days and reflect the process of formation and the subsequent collisional history of the asteroids. Rotation rates of less than 2 hr result in loss of debris from the surface. Very few asteroids rotate this fast and those that do were products of collisions. Fast-rotating material would tend not to form in the first place, and collisions often result in higher spin rates. The strength of the asteroid material plays an indirect role in determining the rotation rate. As an asteroid made of weak material increases in rotation rate from collisions, it tends to deform from a spherical body. This deformation results in a slower spin rate. The combined effects of collisions resulting in an increase in rotation period, and the slowing down from loss of ejecta and deformation, has probably produced the observed average rotation rate for large (100 km or greater) main belt asteroids of between 8 and 12 hr. The effects due to the strength and internal composition of the material have not been resolved yet due to inadequate statistical sampling of the data. The taxonomic type (see Section I,D) of ~10–15% of the known population has been measured, so the correlation between rotation rate and composition is not statistically meaningful for all types of asteroids.

D. Taxonomy

The asteroids have been classified into eight different types according to their photometric properties: color and albedo. These types are summarized in Table I. This system is based on a cluster analysis of the photometric magnitudes of eight spectral regions (corresponding to different colors) ranging from 0.3 to 1.1 μm and their albedos. The distribution of asteroid types with respect to orbital elements has been studied. As can be seen in Fig. 2, the frequency of different asteroid types varies with distance

TABLE I. Summary of Asteroid Compositional Types[a]

Type	Visual geometric albedo	Spectral reflectivity (0.3 to 1.1. μm)
C	Low (<0.065)	Neutral, slight absorption blueward of 0.4 μm
S	Moderate (0.07–0.23)	Reddened, typically an absorption band at 0.9 to 1.0 μm
M	Moderate (0.07–0.23)	Featureless, sloping up into red
F	Low (<0.065)	Flat
P	Low (<0.065)	Similar to M, hence pseudo-M or P
D	Low (<0.065)	Very red longward of 0.7 μm
R	Very high (>0.23)	Very red, bands deeper than S
E	Very high (>0.23)	Featureless, flat, or sloping up into red
U		Unclassifiable in this system

[a] Information taken from Gradie, J., and Tedesco, E. (1982). Compositional structure of the asteroid belt, *Science* **216**, 1405–1407. Copyright 1982 AAAS.

from the sun. In the inner regions of the main belt, the E, R, and S types predominate. The M and F types are located mostly between 2.5 and 3 AU but are not as abundant as the C types in the middle of the main belt. The outer portion of the main belt is composed of C, P, and D type asteroids. This zonal structure has probably existed since the formation of the solar system. These asteroid types correspond to different

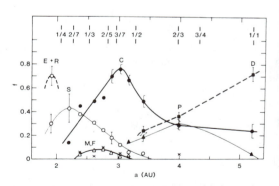

FIG. 2. Distribution of asteroid types within the main asteroid belt based on photometric data for 656 asteroids. Definitions of the designations are in Table I. [Reprinted with permission from Gradie, J., and Tedesco, E. (1982). Compositional structure of the asteroid belt, *Science* **216**, 1405–1407. Copyright 1982 AAAS.]

surface compositions that are inferred from reflectance spectra (see Section I,E). The S-type asteroids are predominantly composed of iron and magnesium silicates and metallic iron. The E-types are iron-poor silicates while the M's may be composed of metallic iron and iron-poor silicate assemblages. The C types, with their low albedo and low ultraviolet reflectance, are optically most similar to carbonaceous chondrites, meteorites composed of hydrous silicates, iron and magnesium-rich silicates, and oxides with varying amounts of elemental carbon. These meteorites are significant because the ratios of their nonvolatile elements are the same as those in the sun. The presence of asteroids of similar composition suggests that they have been unaltered since their initial formation. The P and D type asteroids are probably composed of carbon-rich assemblages, although this interpretation has to be taken with caution since there are no diagnostic absorption bands by which to identify their components (see Section I,E). Nevertheless, the spectra of these types are consistent with carbon-based compounds called kerogens, which are a plausible analogue to outer solar system material based on thermodynamic considerations. The inferred compositions of the asteroids in the observed zones in the asteroid belt are consistent with the observed trend in composition seen among the major planets. Those forming closer to the sun are enriched in iron and silicate minerals that are stable at higher temperatures whereas those found in the outer regions of the solar system are enriched in volatiles and silicate minerals that are stable at lower temperatures.

E. MINERALOGICAL COMPOSITION AND CHEMISTRY

The composition of an asteroid has to be determined without the benefit of direct examination and analytic techniques. A number of different techniques are used to provide information on the mineral composition and texture of asteroids using large telescopes and sensitive instruments from the ground. By measuring the intensity of reflected sunlight from the surface of an asteroid as a function of wavelength (color), certain rock-forming minerals and their chemistry can be determined. With this knowledge we can infer some of the formation processes of the asteroids. For example, Ceres reflects about 6% of the sunlight incident upon it. Also, as shown in Fig. 3, it has a featureless reflectance spectrum

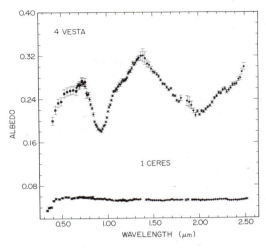

FIG. 3. Reflectance spectra of asteroid 1 Ceres and 4 Vesta from 0.3–2.5 μm plotted as a function of albedo. The low albedo and flat spectrum of Ceres indicate that the surface is probably dominated by claylike minerals. The strong absorptions in the spectrum of 4 Vesta below 0.6 μm and at 0.9 and 1.9 μm indicate the presence of a common mineral found in differentiated bodies called clinopyroxene.

in the visible and near-infrared region of the spectrum and its reflectance is low in the blue and ultraviolet region. It is inferred that Ceres is composed of dark, clay-like material similar to carbonaceous chondrite meteorites. In addition, at 3.0 μm an absorption band due to water in a crystal structure and possibly free water is present, consistent with clay-like mineralogy. Its density, 2.1 g/cm^3 is also consistent with a carbonaceous chondrite-like composition. [See METEORITES, COSMIC RAY RECORD.]

The asteroid 4 Vesta, on the other hand, has a different dominant mineralogy on its surface. Its albedo is about 26% and its spectrum has three strong absorption bands, one with an absorption edge beginning at about 0.8 μm and continuing into the ultraviolet, and two others at 0.93 and 2 μm. A fourth weaker one is at 1.25 μm and is superimposed on the band at 0.93 μm (Fig. 3). The band centers are at wavelengths indicative of a combination of the minerals pyroxene and plagioclase, and both are minerals that commonly form from the cooling of silicate magma such as are prevalent on the moon and volcanic regions of the earth and the other inner planets. It is clear that the history of Ceres and Vesta have been different. In addition, observations of the surface of Vesta have been made as the asteroid rotates, and a compositional map of its surface has been derived based on the variation of the strength and position of the absorption bands. There are large regions of basaltic magma of varying iron compositions, and regions of brighter, more plagioclase-rich rocks.

Probably most of the meteorites that are studied in Earth-based laboratories come from asteroids (some come from the moon and possibly Mars) but we do not know which ones. In determining the composition of an asteroid, one of the first steps in the analysis procedure is to compare the spectrum to that of the meteorite types studied with the same techniques. Ceres is about three times as bright as the carbonaceous chondrite meteorites as measured in reflectance in the laboratory. This difference is presently not explained, but the other aspects of its composition and density are similar to the primitive, carbonaceous chondrite meteorites. Vesta has a composition analogous to a group of meteorites called basaltic achondrites. They have clearly been melted and cooled slowly so that crystals could form. They do not represent the original state of material from the initial condensation of the solar nebula. They are differentiated rocks. Why some asteroids apparently have not been heated during their formation and others have is presently an unsolved question.

Knowledge of the surface composition of a few asteroids is known in terms of the presence of certain but not all of the mineral components. Only the brightest asteroids can be observed over a wide range of wavelengths and the observations are very time consuming. Within the next decade, spacecraft will travel past and fly around more than one asteroid. Our view of these small solar system objects will be greatly expanded and the information obtained will contribute to our understanding of the formation of the rest of the solar system. We might also expect to use material from some of the nearer asteroids for space activities in the upcoming decades.

II. Comets

A. Introduction

Every so often we read about the appearance of a comet in the sky. If we live away from city lights or have binoculars or a telescope, comets can be seen sitting in the nighttime sky as a diffuse glow that sometimes extends out to a tail. About once a decade, on the average, there is a comet that can be seen by just looking in the right part of the sky without the aid of binoculars or telescopes. Even more rarely, a comet is close enough to the earth and the sun at the same time so that it can be seen in daylight. Seeing a comet is an awesome sight especially if it is bright and has a long tail. The opportunity to look at or even search for new comets should not be passed up. When comets were first seen, they were thought to be omens from the gods; Aristotle believed that they formed in the earth's atmosphere in response to gaseous emissions from the earth. Tycho Brahe, tested Aristotle's hypothesis with observations of the comet of 1577. He reasoned that if the comet were an atmospheric phenomenon, there would be a sizable parallax observed. Brahe could measure no such parallax and concluded that the comet was much further from the earth than the moon. After Isaac Newton recorded his theory of gravitation in the *Principia*, it was Edmond Halley who showed that, based on the laws of gravitation, comets orbit the Sun. He hypothesized that the comets seen in 1531, 1607 and 1682 were the same one seen repeatedly as it passed close to the sun. Furthermore, he predicted that it would return again in 1758. He thus successfully applied the basic principles of the scientific method, which include fitting observations to theory and then proposing a test to prove or disprove the theory. He was proven correct when, sixteen years after his death, the comet was again seen and thereafter named comet Halley.

Comet Halley made its 30th recorded appearance in 1986. Scientists from all over the world launched spacecraft and ground-based telescopic observing campaigns to study the physical, chemical, and morphological state of the comet as it traveled through the inner solar system. Undoubtedly, when the data have been analyzed and compared with the results of the various different experiments, concepts of the nature and origin of comets will be drastically different than they are today and there will be many more questions to answer. This portion of this article will present the starting point of existing knowledge of comets before the study of Halley at its 30th recorded apparition in human history.

B. Structure

Comets are seen as fuzzy balls of light sometimes with one or more wispy tails attached to it

FIG. 4.   Comet West Photographed March 9, 1976, by Dr. Elliott P. Moore, Joint Observatory for Cometary Research (JOCR). [JOCR is co-administered by NASA–Goddard Space Flight Center and the New Mexico Institute of Mining and Technology.]

(Fig. 4). The comet shines or gives off light because of two processes. Comets are made of mostly ices and other molecules that are frozen or in the solid state when the comet is far from the sun. As it heats up upon getting closer to the sun, the ices and solids turn to vapor and are thrown away from the comet by the impact of light rays or photons from the sun that push them off the surface. More important may be eruptions on comets and subsequent jetting during which gas pockets are released. When the sunlight interacts with the gases, the electrons within are excited to a higher energy state for a short period of time. Upon returning to their normal state, the molecular fragments give off the excess energy they initially absorbed as light. This process is called fluorescence. The light seen with the naked eye is due mostly to the fluorescence of C_2 molecules. But other molecules such as OH (hydroxyl), H_2O (water), CN (cyanogen), NH (amine), S (sulfur), O (oxygen), H (hydrogen), and CS (carbon–monosulfide) have been detected in regions of the spectrum outside of the sensitivity range of the human

eye. Mixed in with these gases are small particles of dust that reflect the incident sunlight when it hits the dust's surface. Thus the two processes that make the comet shine are fluorescence and reflectance, both of which depend on the close proximity of the comet to the sun and earth.

1. Nucleus

Figure 5 shows a schematic diagram of the anatomy of a comet, which includes the gas-rich cloud surrounding the solid core or nucleus; this region, called the coma is the diffuse glow of the comet. The nucleus cannot be seen through the coma, thus we know very little about it. The premise upon which our knowledge of cometary nuclei is tested is that of a dirty iceball. This model was proposed by F. Whipple in 1950. With the nucleus composed predominantly of ice with dust and other solid material mixed in, Whipple explained the observed phenomena such as the splitting of comets, meteor showers representing only the remaining dust trails from

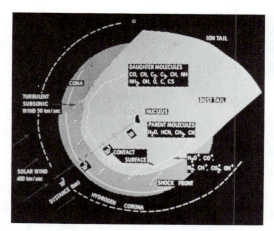

FIG. 5. Schematic diagram of a comet showing the solid nucleus of a few kilometers diameter, inner coma consisting of dust and gas molecules, and the outer coma where ionized particles from the solar wind interact with gases and the ion tail forms. The hydrogen corona is the most extensive structure due to its low mass and great abundance. [Drawing courtesy of Jet Propulsion Laboratory, Pasadena, California.]

comets, and deviations in cometary orbits not attributable to gravitational forces. Observational studies of comets are aimed at testing and elucidating this model of comet nuclei.

A few comets have been studied when they are far from the sun and are not surrounded by gases. Other periodic comets have expelled most of their gas and dust and can be studied through filters that block out the light emitted by the gas. So far we have measured the size and albedo of some of these older comets. The two comets measured in this way are about 10 km in diameter and reflect from 2 to 4% of the sunlight that reaches the surface. The study of the nucleus is important because its size, shape and composition control all cometary phenomena that we observe. Presumably, the composition of the nucleus is controlled by the pressure, temperature, and composition of the material from which the comet formed. Since comets are made up mostly of ices that are not stable in the inner solar system, close to the sun, they must have formed far from the sun where the ices are the stable phases. Thus, knowledge of the composition and structure of the nucleus of comets will tell us about the composition and thermodynamic conditions in the outer solar system.

2. Coma

By studying the composition and chemical processes active in the coma surrounding the nucleus, some of the composition of the nucleus can be inferred. Unfortunately, not all of the comet evaporates to a gas, so the whole complement of cometary material is not known. So far it has been determined that about 80% of a comet is composed of water. An estimated 10% of cometary material is dust composed of the same basic material as rocks found on the earth, that is, its chemistry is based on combinations of silicon and oxygen and presumably other elements commonly found in rocks. The remaining 10% is composed of organic molecules and gases other than water. Some comets have passed very close to the sun during times that we could observe them. At these times we have seen the spectral signature of elements such as calcium, sodium, potassium, iron, magnesium, nickel, copper, and manganese, all of which are suspected to be components of the dust; at least they are common elements in terrestrial rocks and meteorites. However, these elements have been detected when they were so hot that they were torn away from the solid material with which they were initially associated and have not been directly associated with the dust according to studies carried out at this time.

3. Tails

Comet tails are of two types. The large and often curved tails, an example of which is shown in Fig. 4, are type-II tails and composed mostly of dust. The light seen is reflected sunlight from the surface of the dust particles. The tail is curved because when dust particles are ejected from the surface of the comet, their velocity changes and they go into orbits of their own. Because the velocity of the particles changes, the shape of the orbit changes slightly and a curved tail relative to the comet–sun line is seen. The tail of the comet will always point away from the sun because for these dust particles the radiation pressure from sunlight exerted in the direction away from the sun is greater than the force of gravity acting in the direction toward the sun. Therefore, after a comet has passed behind the sun and is in a path taking it toward the outer edges of the solar system, the tail is in front of the comet. The tail fades as the radiation pressure from the sunlight falls off at a rate equal to the square of the distance from the sun.

Some comets also have tails pointing straight back from the comet–sun line such as can be seen in Fig. 4 illustrating Type-I tails. These are composed of molecules that have physically interacted with charged particles (such as electrons and protons) emitted from the sun called the solar wind. Such interactions result in charged molecules, which in the aggregate are electrically neutral (called plasma), that emit light. Plasma tails follow the path of the solar wind (which travels radially away from the sun) because they are not massive enough to go into their own orbit around the sun. Thus the tails form straight behind the comet head. They eventually cease to receive enough light to be seen and dissipate in the solar system. Sometimes these tails form kinks, ropy structures, or break up entirely. This is due to changes in the flux of particles leaving the sun in the direction of the comet or to jetting from the comet itself. Studying the dynamics of these tails tells us about the interaction of the sun with other particles in the solar system.

C. DYNAMICS AND ORIGIN

We have reviewed what is currently known about the structure and composition of comets and why we need to know such things. But where do these comets come from and how do they get here? By observing the motion of a comet its orbital path through the solar system can be plotted. Comet Halley travels out beyond the orbit of Neptune, but many other comets travel so far away from the sun that they are seen only once in tens of thousands of years. Comets are grouped into two classes: the short period comets, which orbit the sun in 200 years or less, and the long period comets with periods greater than 200 years. A subset of the short period comets is the Jupiter family of comets with orbits less than 20 years. Are the short period comets related to the long period comets? We believe they are because their chemistry appears to be the same. Then how do the orbits change? The current theory, awaiting confirmation by some experimental test that has yet to be determined, is that there is a reservoir of trillions of comets at a distance of 20,000 to 50,000 AU from the sun, known as the Oort Cloud, after J. Oort, the Dutch astronomer who proposed its existence in 1950. The occasional passing star exerts a gravitational force on a comet that happens to be in the sphere of influence of that star and pushes the comet toward the sun. The sun's gravity takes over from there and pulls the comet in to the inner solar system where it can be studied. Some of these comets have their orbit changed again when they pass close to the giant planets of Jupiter, Saturn, and

Uranus. Thus the short period comets probably evolve from the Oort cloud. It is possible that there is a reservoir of comets in the region of Saturn and Uranus that has been there since the beginning of the solar system. One planetesimal (asteroid 2060 Chiron) has been discovered orbiting between Saturn and Uranus; there is no reason to suspect that there aren't others remaining to be discovered. If some of the short period comets do originate in this region of the solar system, then the over abundance of carbon observed in cometary spectra relative to the solar composition must be explained. We suspect that this carbon is present in the form of contamination from the interstellar medium, which is rich in carbon compounds. This is further evidence that the comets formed at the very outer edges of the solar system, where interstellar contamination would be greater than closer in toward the sun.

III. Future Directions

A. AREAS OF STUDY

The problems currently pursued in studies of both asteroids and comets will remain the major areas of study for a number of years to come. Among the enigmas of the asteroids is the source and mechanism by which apparently some of the asteroids are heated while others are not. The nature of this heating mechanism and its distribution, which probably occurred shortly after the formation of the asteroids, represents a gap in the time line of solar system history. It is not clear how this event will be elucidated in future explorations of the asteroids. Hopefully, the results from spacecraft missions that are planned for a few asteroids in the upcoming decades will shed some light on this issue.

While we use the meteorites that land on the earth as a low-cost space mission, one that provides samples for study *in situ*, it is of paramount importance that we discover the source region of the most abundant meteorite type, the ordinary chondrites. Other meteorite types that are not currently recognized among the asteroid belt or planet-crossing asteroids are also objects of pursuit. We are currently not sure whether our interpretive techniques are in error or whether we just cannot find what we are looking for. They may be hiding, or they may be too small to see.

The intriguing prospect that some of the planet-crossing asteroids may be the solid rem nants of comets is at the earliest stages of investigation. We can expect to gain insight into this subject in the next few years as spacecraft fly by comets and measure their basic physical parameters, size, shape, spin, and, hopefully, composition. If similar information can be obtained from the candidate asteroids, this hypothesis may be tested.

B. FUTURE OPPORTUNITIES

The world launched a number of expeditions to encounter comet Halley in 1986 with spacecraft approaching within a range of 500 km to millions of kilometers. There were cameras for imaging the comet, spectrometers for determining the mass and chemistry of the coma and dust constituents, plasma wave sensors to study the magnitude and directions of the plasma fields in the vicinity of the comet interacting with the solar wind, and dust counters to determine the abundance of the dust. At the same time there was an energetic effort underway to study the comet from the ground with the world's complement of telescopes and state of the art instruments to determine the chemical abundance of the coma constituents and the time evolution of the composition and structure of the comet's coma and tail. Earth-orbiting instruments also monitored the comet in regions of the spectrum that cannot penetrate the earth's atmosphere. The International Ultraviolet Explorer (IUE) measured the increasing production of the ionization products of water as the comet approached the sun as well as the proportion of dust to gas and its changes with changing heliocentric distance. This rapid increase in the amount of information available on one comet will revolutionize our understanding of the nature and evolution of all comets. Future missions currently in the planning stages include the Comet Rendezvous Asteroid Flyby (CRAF)

mission which would fly by a main belt asteroid and spend several months at a comet measuring its evolution as it travels toward the sun. Scientists anticipate developing an understanding of the dynamics of comets, their composition, and hence mode of origin and relationship to the composition of molecules in interstellar space. The ability to monitor the comet as it approaches the sun will provide a wealth of data beyond that available from just a short flyby, lasting minutes to hours.

A mission to rendezvous with a planet-crossing asteroid is in its early planning stages. Like the CRAF mission, the Near-earth Asteroid Rendezvous (NEAR) mission would spend months at the targeted asteroid, studying it from various ranges with different instruments to determine its size, mass, shape, spin rate, composition, and characteristics of its surrounding environment. The possibility of landing on the asteroid and injecting a probe into the interior of the asteroid has also been proposed. The direction of solar system studies and the availability of funding will determine whether or not this mission flies within the reasonably near future.

BIBLIOGRAPHY

Beatty, J. K., O'Leary, B., and Chaikin, A. (1981). "The New Solar System." Sky Publishing Corp., Cambridge, Massachusetts.
Gehrels, Tom (Ed.) (1979). "Asteroids." University of Arizona Press, Tucson, Arizona.
Hartmann, W. K. (1983). "Moons and Planets," 2nd ed. Wadsworth Publishing Co., Belmont, California.
Littman, M., and Yeomans, D. K. (1985). "Comet Halley Once in a Lifetime." American Chemical Society, Washington, D.C.
Sagan, C., and Druyan, A. (1985). "Comet." Random House, New York.
Wilkening, L. (Ed.) (1982). "Comets." University of Arizona Press, Tucson, Arizona.

PULSARS

F. Curtis Michel   *Rice University*

GLOSSARY

Dispersion measure: Observational measure of the extent to which radio pulses are delayed by propagation through the interstellar medium; used to estimate distances to individual pulsars.

Drifter: Pulsar that exhibits drifting subpulses.

Drifting subpulse: Subpulse that moves systematically across the integrated pulse profile.

Integrated pulse profile: Apparent pulse shape (intensity versus time) of a pulsar as determined by averaging together a large number of otherwise weak and noisy individual pulses.

Neutron star: A star more massive than ~1.4 times the mass of the sun cannot become a white dwarf but instead must collapse (once it has exhausted its supply of hydrogen) to what is in effect a single atomic nucleus ~10 km in diameter.

Nulling: A number of consecutive pulses may be missing in any long train of pulses. A few pulsars exhibit this "nulling."

Subpulse: Persistent feature within the integrated pulse profile, often noticeable in single pulses.

Supernova: Explosion of a star that temporarily produces a "star" as bright as an entire galaxy (~10^{11} ordinary stars). Left behind are an expanding shell of gas and a remnant collapsed object often observed as a pulsar.

White dwarf: Star that has exhausted its internal sources of energy and has therefore shrunk to a dense sphere comparable in size to the earth.

One of the very important discoveries in astronomy of modern times has been that of the so-called pulsars in stellar objects that emit high-intensity pulses of coherent radio emission. These astronomical "clocks" can be extraordinarily stable and provide important probes of interstellar space in addition to being enigmatic objects themselves.

I. Introduction

In astronomy, *pulsar* denotes an object emitting sharp, rapid pulses of radio emission with clocklike periodicity of about a second. The discovery of pulsars by Anthony Hewish and Jocelyn Bell, announced in 1968, earned both a permanent place in the history of astronomy and Hewish won a Nobel prize.

Pulsars are now believed to be neutron stars, stars of mass comparable to that of the sun but collapsed to a sphere of only ~10 km in radius (i.e., nearly 100,000 times smaller than our own sun). These neutron stars are thought to be intensely magnetized and in rapid rotation. The rotation serves two roles. First, the rotation of the highly conducting neutron through its own magnetic field induces strong electric fields, which pull charged particles from the surface; second, this magnetic field is seen from different aspects by a distant observer on the earth owing to the neutron star rotation. Theories generally assume that the radio emission is concentrated into a beam, with electrons being accelerated out of the magnetic polar "caps" and emitting radio waves, which sweep the sky like a lighthouse beam as the star rotates. However, a fully satisfactory theoretical description of the pulsar phenomenon is still not at hand.

A second type of rapidly pulsating object was later discovered to emit X rays in certain binary systems. These objects, called pulsating X-ray sources, are also believed to be rotating magnetized neutron stars, but in this case the neutron star orbits a companion star that is transferring mass from its outer atmosphere onto the neutron star. This mass influx is diverted by the magnetic field and falls primarily on the poles, which are heated to millions of degrees by this bombardment and emit X rays. Again, rotation of the neutron star modulates the X rays seen on earth.

As a class, the two are distinct both observationally (X-ray emission versus radio emission) and theoretically (accretion of mass onto a neutron star versus ejection of charged particles). However, a few pulsars emit both radio waves and X rays, notably the so-called Crab pulsar located in the center of the Crab nebula.

II. Distribution of Pulsars

Radio pulsars are dispersed among the ordinary stars in the galaxy, and 99% of them are single objects not in binary systems, again unlike the pulsating X-ray sources and also unlike ordinary stars, half of which are binary. Nor are they visible as stars even to the best telescopes. Virtually all the detectable energy output is in radio waves, with a spectrum of emission that declines rapidly with increasing frequency. This decline would in itself make radio pulsars too feeble to be seen at the much higher frequencies of the visible spectrum. Most of the stars in the galaxy are concentrated in the disk of the galaxy (the Milky Way), and pulsars are markedly concentrated in the same way. The distances to pulsars can be estimated from their observed dispersion measure. Because radio waves interact with the very few electrons in outer space (only ~30 per liter!), the lower-frequency waves in the spectrum travel more slowly, which causes a sharp pulse to be smeared out; the same thing happens to the sharp pulse created by a lightning stroke, giving the Whistler phenomenon when the waves are detected at large distances. The dispersion measure corresponds to the amount of correction for delay necessary to reconstruct a sharp pulse, which also determines the integrated electron density along the path traveled. The above estimate for the electron density then gives an idea of the distance, the closest being ~80 parsecs (pc) away and some are detected as far away as 55,000 pc.

Most pulsars do not have proper names but are named according to their place in the sky. Thus, the Crab pulsar is also known as PSR 0531+21, meaning a pulsar (PSR) found at a right ascension of 5 hr and 31 min and a declination of 21° north.

Although ~400 pulsars have been discovered, the fastest ones are among the most unusual, and it is instructive to examine the properties of the latter.

III. Exceptional Pulsars

The fastest known pulsar is called the Millisecond pulsar, or PSR 1937+214 (here the extra digit refines the declination to 21.4° north) and has a period of only 1.558 msec, which is quite close to the maximum rotation rate that a neutron star could theoretically have without flying apart owing to the centrifugal forces exceeding gravity. The surface velocity for a typical neutron star according to theory would then be 4×10^9 cm/sec, or 13% the velocity of light. Other than being extremely rapid, the radio emissions from this pulsar are very similar to those of most other pulsars. Indeed, as a class, pulsars are similar to fingerprints; each is readily identifiable as such but is nevertheless distinct on close examination.

The next fastest pulsar is 1953+290, with a period of 6.133 msec. Although slower, this pulsar is remarkable in that it is one of the few pulsars (~1%) to be found in a binary star system. The orbital period is 117 days (about that of Mercury about the Sun), with the pulsar being orbited by an unseen companion about seven times less massive. The companion mass could be higher if we happen to be looking nearly down the orbital axis.

The third fastest pulsar, at 33.1 msec, is the Crab pulsar (PSR 0531+21), which sits centered on what is probably the remnant of a historical supernova observed by the Chinese in 1054 A.D. This remnant, the Crab nebula, is expanding at a measurable rate consistent with such a birthdate. The pulsar is remarkable in also being a source of visible light, which is pulsed at the same 33-msec period as the radio, too fast for the eye to follow! As with typical pulsars the radio emission declines rapidly with frequency, and some researchers believe a separate mechanism may cause the pulsar to become visible again at visible frequencies. [Roughly speaking, radio frequencies are of the order 10^9 cycles/sec (hertz), whereas visible light is ~10^{15} Hz.] Moreover, the high-frequency part of the spectrum extends into X-ray and γ-ray energies (of the order of 10^{21} Hz), again pulsed.

The fourth fastest is PSR 0540-693, with a period of 50 msec and also surrounded by a nebula, resembling quite closely the Crab pulsar except for the distance. This pulsar is in the large Magellanic Cloud some 55 kpc away. Like the Crab, it emits visible pulses.

The fifth fastest pulsar, at 59.0 msec, is again in a binary system and is often called the Hulse–Taylor binary pulsar (PSR 1913+16) after its discoverers. (About 20% of the known pulsars have right ascensions of 19 hr, which is largely a selection effect owing to the fact that the look direction of the giant fixed radio telescope at Arecibo rotates across the Milky Way at this location.) This pulsar was the first binary pulsar to be discovered and is possibly the most important. The orbital period is only 7 hr and 45 min, and general relativistic effects are important. This binary system is the first single-line binary (i.e., only the pulsar is detected) for which all of the orbital elements have been deduced (by means of the general relativity theory), including observation of the advance of perihelion (the same effect explained for Mercury) at a rate entirely consistent with theory. More importantly, the binary pair are spiraling together at a rate consistent with energy loss by gravitational radiation. Unlike the 6.1-msec binary pulsar, the unseen companion also has a mass of 1.4 solar masses, suggesting that it may also be a neutron star.

The sixth fastest pulsar (PSR 0833-45) is the Vela pulsar associated with the Vela supernova remnant, with a period of 89.2 msec. Like the Crab pulsar, it has high-frequency emissions. Unlike any other pulsar, however, it exhibits extremely large "glitches," wherein its period abruptly decreases by a small but readily detectable amount of about one part in a million. These events repeat at an irregular interval of ~3 yr and do not have exactly the same behavior each time.

This listing of pulsars illustrates a number of important observational inferences, detailed in Section IV. Most of the next dozen or so pulsars having periods between 100 and 200 msec tend to be isolated pulsars without striking or unusual properties.

IV. Nature of Pulsars

Pulsars are thought to be formed in supernova events. There are actually only a few pulsar–supernova associations. Many young supernova remnants do not contain detectable pulsars, and most pulsars are not in supernova remnants. The first is partly a selection effect because the remnants themselves are intense radio sources, and only the brighter pulsars can be seen against such a background. Even if the pulsar is bright, its beam might not sweep the earth. The reason for the second is that the remnant expands and fades over a period of $\sim 10^4$ yr, whereas the pulsars live on for perhaps a million years.

The fastest pulsars provide the strongest test of the rotating neutron star hypothesis. Early theories involving white dwarfs as the pulsar object were ruled out by the short time scales involved; such stars are simply too large to behave coherently for periods much less than a second.

All pulsars are observed to be slowing down. Again, this observation is consistent with a rotating object for which the source of free energy is the kinetic energy stored in rotation. The energy output can then be calculated by estimating the moment of inertia of the neutron star (essentially its mass times the square of its radius) and using the observed rate of increase of period to determine how rapidly this energy is declining. The rate of change of the periods spans a large range, but a change of period of $\sim 10^{-15}$ sec/sec is representative. A pulsar with a period of 1 sec (again typical; the pulsars noted above are the fastest) has a stored energy of 10^{46} ergs, and therefore the energy loss rates are $\sim 10^{31}$ ergs/sec. These figures are always much larger by a comfortable factor ($\sim 10^5$) than the energy output in radio waves. Most of the pulsar energy output seems to be invisible and is thought to be in the form of a relativistic magnetized "wind" flowing away from this pulsar. However, the entire Crab nebula supernova remnant radiates power comparable to that lost from the pulsar, suggesting that it is the wind from the Crab pulsar that lights up the nebula.

The pulsating X-ray sources behave quite differently; they have quite large spin-up and spin-down rates, so that the period of a given source may increase over about a year and then decline. The spin-up is thought to be caused by angular momentum carried in by the accreted matter. The spin-down mechanism is less certain but is usually attributed to electromagnetic coupling of the neutron star, assuming it has a strong magnetic field just as the radio pulsars do, to an orbiting disk of matter.

The glitch phenomenon has proved difficult to explain. Most theories concentrate on some change in the physical body of the neutron star itself. A "starquake" that released internal stresses and allowed the neutron star to shrink slightly would explain the spin-up seen in the Vela glitches, but these events happen too frequently. The star cannot endlessly shrink. Present models argue instead for an interior core, which is decoupled from the crust and rotates more rapidly. The glitches might then be abrupt braking phenomena that transfer angular momentum from the core to the crust and spin up the latter. The strong magnetic field, however, would lock the crust and core into corotation, so these models impose special conditions on this field (e.g., it would have to reside entirely in the crust and be excluded from the core). The Crab pulsar also displays these events, but with much smaller amplitude and much more frequently. However, PRS 1509-26 (also associated with a supernova remnant) has an exceptionally large spin-down rate of 1.5 $\times$ 10^{-12} sec/sec and yet has shown no glitches whatsoever. A few slow old pulsars (PSR 1641-45 and PSR 1325-43) have experienced single glitches of amplitude intermediate between those of the Crab and Vela.

With these few exceptions, pulsars have extremely stable slowing-down rates, and if their pulse rates are corrected for spin-down they rival the finest atomic clocks. In particular, the Millisecond pulsar (PSR 1937+214) has such a small slowing-down rate ($\sim 10^{-19}$ sec/sec) that even uncorrected it would serve as an excellent clock. Indeed, its period is known to be 1.5578064488737 (± 6) msec.

The location of a pulsar in the sky can be determined to high accuracy (comparable to that for visible stars) using the above clocklike properties. Owing to the earth's orbital motion about the sun, the pulsar "clock" appears to run fast and slow at different times of the year. This variation can be removed only if the pulsar is located at a very specific location in the sky. In general, there are no stars found at these well-determined locations that seem likely visible counterparts (except for the Crab, Vela, and PSR 0540-693, which display optical emission).

V. The Pulses

Pulsars would not sound like the good clocks that they are if they could be heard. There are large pulse-to-pulse variations, and occasional pulses may even be missing. Several hundred individual bursts of radio emission added together on top of one another form a generally stable integrated pulse profile, but this profile serves more as a "window" through which a variety of puzzling pulse variations are witnessed. In general, the individual pulses bear little resemblance to the integrated profile, often as sharp spikes appearing more or less randomly within the window. In some cases, however, the behavior is quite coherent, with a smaller pulse (subpulse) appearing at one edge of the window and steadily working its way across to disappear at the other edge. These drifting subpulses are found in perhaps 10% of the pulsars, PSR 0809+74 being a classic example. Another pulsar (PSR 0826-34) is a remarkable example in that it has an extremely wide profile within which can be seen four or five subpulses at once. These subpulses are spaced about 30° apart (defining 360° as one pulse period) and move back and forth together. The interpretation is that we are looking nearly along the spin axis of a neutron star that is also magnetized nearly along the spin axis, and therefore we are almost constantly in the pulsar emission beam. Both of these drifters also display nulling (as do some nondrifters), a phenomenon in which the pulsar disappears, becoming entirely undetectable for 10 to even 1000 periods before reappearing abruptly. One pulsar, PSR 0904+77, has never been seen since its first detection. If not a spurious observation, it is an extreme example of nulling.

The phenomenon of nulling shows that pulsar action can cease, and indeed it must if all pulsars slow down. The inevitable consequence should be an accumulation of older and slower pulsars, whereas observationally the average pulsar has a period of ~ 1 sec and yet the slowest pulsar (PSR 1845-19 at present) has a period of only 4.3 sec. The pulsating binary X-ray sources, on the other hand, have periods of 700 sec or longer. Apparently, pulsars as such must rapidly disappear once they slow below a second or so. Pulsars with a 1-sec period have slowing-down rates of $\sim 10^{-14}$ sec/sec, implying that they will live for only $\sim 10^{14}$ sec = 3 $\times$ 10^6 yr, and since most of the time is spent as a slow pulsar, the typical pulsar lifetime is still expected to be of the order of a few million years regardless of how fast the pulsar was rotating at birth.

The slowing down of the fastest pulsars is much less than average, a seeming contradiction. It would take $\sim 4 \times 10^8$ yr for the 1.558-msec pulsar to become even a 3-msec pulsar and extraordinarily long for it to become a 4-sec pulsar. In the magnetized rotating neutron star picture, it is the strong magnetic field that couples the rotating star to its surroundings. Thus, an unusually weak magnetic field would account for the small slowing-down rate. Indeed, it is only when a rapid pulsar has a weak magnetic field that we can ever hope to observe it unless it is seen shortly after birth. The Crab pulsar is a thousand years old and spins at 33 msec. If we scale the lifetime it could have had (with its apparently strong magnetic field of ~4 × 10^{12} G) with a spin rate of 1.5 msec, we find that it would double its period in only a few years. Thus, only in a fresh supernova could one hope to see a strongly magnetized millisecond pulsar. Such events are rare; centuries can pass between known supernova events in our own galaxy, and it is not evident that the surrounding dense cloud of debris would permit the pulsar to shine through early on. Supernovas are more frequently seen in other galaxies (there are so many of them), but associated pulsars have not been detected (possibly they are simply not bright enough to be seen at such huge distances—megaparsecs). We do not know that the millisecond pulsar will actually continue to function after 4 × 10^8 yr. It could disappear, as apparently typical pulsars do after a few million years. Dividing slowing-down rate into the period gives a characteristic "age," but this need not be either the actual age of the object or its likely life expectancy.

VI. Theory of Pulsars

A large amount of circumstantial evidence, as shown in the preceding section, now exists on the behavior of pulsars. Yet the very fact that we can detect them at all is poorly understood. The radio luminosity is ~10^{28} ergs/sec for a typical pulsar, and yet a neutron star is only ~10 km in radius, which would require it to be an extremely efficient antenna. (If much more energy than that in the radio spectrum went into making the system glow in the visible spectrum, some should have been seen as dim stars.) The simplest picture of how so much radiation might be produced in such a small region is the "bunching" hypothesis, which assumes that the energetic electrons producing the radio waves are bunched together. In a simple geometry one might consider a spherical bunch of N particles. At long wavelengths the electrons radiate as if they were fused into a single particle of charge Ne, giving emission N times more intense than

N separate electrons radiating independently. At wavelengths much shorter than the electron spacing, the electrons would radiate more or less independently despite the bunching. Thus, bunching gives a natural qualitative account of why pulsar emission is intense to low frequencies and declines rapidly at high frequencies. However, it has been difficult theoretically to obtain the required bunching, at least in simple models. Early models suggested that electron–positron pairs would be produced by a cascade process. In these models, energetic electrons radiate γ rays, and the γ rays interact with the magnetic field to produce electron–positron pairs, which in turn radiate γ rays, and so on. However, once they are produced as pairs together, it is difficult to separate the electrons from the positrons fast enough that bunches produced in a cascade would be able to radiate coherently. The small scale of a neutron star system together with the high velocities of relativistic particles require that bunches be formed extraordinarily fast. Possibly the bunches are actually waves excited in nonneutral plasma about the neutron star. Some sort of maser action has also been seriously considered.

A considerable theoretical effort has been invested in analyzing the magnetized rotating neutron star model. At first it was thought that the simplest possible model, a dipole magnetic field aligned with the spin axis, would emit particles from the polar caps as a pulsar does and be a valuable test bed for radiation models. Indeed, most radiation models have assumed acceleration of electrons from the polar caps as a starting point (often attempting to include pair production). It was later realized that such particle emission could be only a transient phenomenon, because only electrons would be lost from the system. The resultant electrostatic charging of the neutron star eventually would halt any further loss of electrons. Positive particles that might neutralize this charging should also be emitted, but they would be trapped very close to the star on the closed equatorial magnetic field lines. Although the model turns out to be disappointingly unlike a pulsar, it is nevertheless interesting because the neutron star ends up being surrounded by an equatorial torus of positive charge and two clouds of electrons over the polar caps, with vacuum elsewhere. It is unusual to find a plasma that consists only of electrons because the self-repulsion of the electrons should blow the plasma apart. However, laboratory workers have succeeded in producing just such plasmas by trapping a large number of particles in magnetic Penning traps. These nonneutral plasmas have the remarkable property of behaving like liquids, with a sharp boundary separating a constant density of particles from the vacuum surrounding them.

From a theoretical perspective, it appears that a more complicated model is required. One avenue has been to consider models with a nonzero angle between the magnetic axis and the spin axis. At present it is not known how large this angle must be to avoid the trapping properties of the system. If the angle must be large, alternative interpretations must be found for apparently aligned pulsars such as PSR 0826-34, which have such wide pulses that it would seem one is looking very nearly along the rotation axis.

A completely different model has been put forward in which the neutron star does not rotate in a vacuum but is surrounded by a disk of matter (like the rings surrounding Saturn). Disks, in fact, have been found to play a role in a number of astrophysical objects that were previously difficult to understand as single stars. The action of a disk would be to introduce a conducting element into the highly charged neutron star system, with the consequence that electric currents would have to flow between the disk and star, and dynamic activity would have to ensue. The electrical current pattern envisioned would involve electrons emitted into space from the polar caps with a neutralizing current of electrons to the star from the inner edge of the disk and a corona discharge of positive particles from the distant outer edge of the disk into space. It is uncertain, however, whether disks are generally formed in the pulsar formation event (presumably a supernova).

In all of these models one expects a variation in the apparent (projected) direction of the magnetic field as the emission region swings past the observer. A large number of pulsars indeed show a systematic rotation in the polarization direction of their radiation during each pulse, the rate being fastest at the center of the pulse (which need not be the point of maximum brightness). These observations are broadly consistent with theory assuming a simple tilted dipole magnetic field, and attempts have even been made to deduce the actual angles between spin axis and observer plus magnetic field for some pulsars. Unfortunately, the rotation rate gives only one number, whereas there are two unknown angles to be solved for, so some additional assumption must be made.

The simplest expectation would be that the radio emission is polarized parallel to the magnetic field lines. (Curvature radiation from relativistic electrons followed curved magnetic field lines would have this behavior.) In fact, one observes in many pulsars abrupt 90° changes in polarization at certain points in the pulse, so if the radiation was originally along the field lines, it must suddenly become orthogonal to them, which is not a property of curvature radiation. The attractive idea of relativistic electrons somehow bunching and radiating as they move on curved magnetic field lines is therefore incomplete. It may be that the more of less static nonneutral plasma thought to be trapped near the neutron star complicates the emission mechanism in interesting ways.

Returning to the supernova origin hypothesis, the supernova event rate seems broadly consistent with that necessary to maintain pulsars against their apparent death rate, there being considerable uncertainty in each. (For example, only ~1% of the pulsars in our galaxy have been detected, and the supernova rate is actually an average over supernovas in other galaxies apparently similar to our own.) These rates are thought to be about one per 50 yr. It is not believed that the observed pulsars centered on supernova remnants are chance associations (particularly the Crab pulsar, which seems to be exciting the nebula). Supernova remnants occupy ~1% of the sky in a 10° wide band along the Milky Way. The pulsars are similarly distributed, showing that they are produced mainly by stars in the disk of the Milky Way and that they are seen at distances that are large compared with the thickness of this disk. A few of the approximately 400 observed pulsars could therefore be accidentally in line with a remnant. However, the fact that the pulsars seen in these remnants are also very fast (<100 msec) reduces significantly the chances of accidental association. Presently, three of the six pulsars faster than 100 msec are in supernova remnants.

The supernova association is also complicated by uncertainty in the pulsar beam geometry. Presumably, one observes a one-dimensional slice through the beam cross section it sweeps across the earth, but there is no information about the extent of the beam above and below. If the beam is roughly circular, there must be a large number of unseen pulsars, because the beams are observed to be so narrow (~10° of the latitude if rotating), and statistically only about one in five would be expected to sweep over the earth, about the same ratio as remnants with pulsars to those without.

Extinct pulsars may be resuscitated as γ-ray burst sources long after they fade from view. These sources emit single isolated but intense bursts of γ rays lasting ~1 sec. There is little concensus on what produces these bursts, with models ranging from an astroid falling onto the neutron star to rapid accretion of material from a disk. One such event, the brightest ever observed, which occurred on March 5, 1979, has even been localized near a supernova remnant designated N49 in the Large Magellanic Cloud. This is tantalizing but not necessarily supportive of the idea that the source is an extinct neutron star, because then the remnant should long since have dissipated. Alternatively, the association could be accidental, the γ-ray burst source being actually much closer than average rather than being much farther than average, which would explain the brightness. [*See* GAMMA-RAY ASTRONOMY.]

BIBLIOGRAPHY

Manchester, Richard N., and Taylor, Joseph H. (1977). "Pulsars." Freeman, San Francisco.
Smith, F. G. (1977). "Pulsars." Cambridge University Press, New York.

QUASARS

Steven N. Shore   *New Mexico Institute of Mining and Technology**

GLOSSARY

Active galactic nucleus (AGN): General class of galaxies showing strong, broad emission lines and high (order 10^{46} ergs sec^{-1} or higher) luminosities.

Balmer discontinuity: Ionization edge of hydrogen from the $n = 2$ level; the $n = 1$ edge is called the Lyman discontinuity or Lyman limit.

Blazar (BL Lac object): Optically violent variable galaxy showing featureless continuum with strong ultraviolet emission.

Covering fraction: That portion of the central source of ionizing radiation obscured by absorbing matter.

Eddington limit: Luminosity at which radiation pressure can support an overlying mass of gas.

Friedmann–Robertson–Walker cosmology (FRW): Standard cosmological model characterized by the two parameters of the Hubble constant H_0 and the deceleration parameter q_0; a homogeneous, isotropic, matter-dominated cosmological model.

* Also associated with DEMIRM, Observatoire de Meudon, France.

H II Region of Strömgrem sphere: Region of ionized hydrogen produced by ultraviolet photoionization.

Jansky: Standard unit of flux, 10^{-23} ergs cm^{-2} sec^{-1} Hz^{-1}.

Linear: Low ionization emission line galaxies, which appear to be very low luminosity Sy 1 galaxies, related to the faint end of the quasar luminosity distribution.

Synchrotron emission: Mechanism for the emission of photons due to relativistic particles (electrons) spiraling in magnetic fields. Power law energy spectra produce featureless power law continua.

Very long baseline interferometry (VLBI): Technique of aperture synthesis using continental or larger baselines for radio observations; capable of resolutions on the order of milliarcseconds.

A quasar, also called a quasi-stellar object, or QSO (name coined by H.-Y. Chiu in 1963), is an object with a dominant starlike component, unresolved at the level of 0.001 arcsec (radio) or 0.1 arcsec (ground-based optical), with an emission line spectrum showing a large redshift—up to $z = 3.78$ (limit, 1985)—and extremely broad emission lines (full-width zero intensity up to 20,000 km sec^{-1}) and an absolute blue magnitude $M_B < -23$. Typical luminosities are on the order of 10^{47} ergs sec^{-1}. Quasars represent the most luminous members of the sequence of active galaxies, being intrinsically overlapping in properties with the Seyfert 1 (Sy 1) galaxies. The permitted lines have widths much larger than the forbidden lines. A few show the Lyman continuum edge, but most show relatively featureless power law ($F_\nu \approx \nu^\alpha$) continua with slopes ranging from -0.7 in the X-ray to -1.5 in the visible, with an overall slope of -1. All quasars observed with resolutions better than 1 Å and noise of less than 5% have absorption line systems of heavy elements and/or the Lyman series of neutral hydrogen. About 10% of all quasars are *broad absorption line* (or BAL) systems showing broad, essentially unresolved absorption by hydrogen, C IV, Si IV, and N V. About 10% are radio sources. Most quasars are weak X-ray sources, but the majority ar not strong infrared (IR) sources. Several are associated with gravitational lenses. Many QSOs are variable on timescales of about 1 yr at all wavelengths. Several show timescales on the order of hours (BL Lac objects) to days. Many QSOs have extended weak optical regions, identified with the host galaxy in which they are imbedded. A subgroup of quasars show "superluminal" motion of radio knots (VLBI jets), observed with very long baseline interferometric methods. Many quasars show core-lobe structure in radio and radio jets; a few show optical jets (especially 3C 273, one of the intrinsically brightest members with $M_V = -27.5$). The radio morphology for the extended sources is typically a double lobe structure. Strong core sources are often associated with these objects.

Quasars are the most distant observable galaxies in the universe and appear to be associated with the early history of galaxy evolution. The host galaxies are usually, although not exclusively, spirals, and the QSO therefore appears most often in the nuclei of spiral galaxies. The host galaxies have also been shown to sometimes be members of clusters of galaxies. The probable source of the energy is accretion onto a central black hole, with a mass of 10^8 to 10^{10} solar masses. The superluminal motion is evidence of bulk relativistic ejection of material in the jets emerging from the inner regions of the nuclear accretion disk. The continuum appears to be chiefly produced nonthermally by synchrotron radiation from relativistic electrons in weak magnetic fields, and by thermal emission from a turbulent accretion disk around the central engine. Photoionization by this continuum appears responsible for observed properties of the emission line spectrum.

I. Historical Introduction

The discovery of discrete radio sources in the 1940s and 1950s, and the identification of synchrotron radiation as the mechanism for the radio emission, had permitted the relatively easy identification of the major sources of the Third Cambridge, or 3C, survey by 1960. However, in a few cases of strong sources, there were no obvious optically extended objects within the error boxes of the sources. This was due to the comparatively poor resolution of the instruments available at the beginning of the decade, and to the fact that the fields appeared to generally contain only starlike objects. The real breakthrough occurred in 1960 with the observation by the group at Manchester headed by Palmer, using interfereometric methods with a baseline of about 6×10^4 wavelengths, of the position of the source 3C 48. An additional confirmation was provided by the Owens Valley telescope. This enabled Sandage to obtain an optical identification of the object, which had as its probable counterpart a faint, fuzzy image of about $V = 16$. Optical variations were also noted, seemingly in support of a stellar identification for the counterpart. A spectrum was also obtained, but the emission lines were not those normally observed in galactic stars. The identification of emission lines, and of a blue continuum, yielded no easy explanation, and the object was largely ignored during the next few years. In 1963, by a clever use of lunar occultations observed in the radio with a fast single element system, Hazard and collaborators succeeded in determining the position of 3C 273 and 195. Radio structure was observed in 3C 273, which consisted of two peaks, one of which was associated with a stellar object and the other with the optical extension—the jet. The observation of the spectrum of 3C 273, by Schmidt, quickly yielded the explanation of the emission lines as being the Balmer series of hydrogen redshifted to $z = 0.16$, which at the time was one of the farthest objects known. The redshift was confirmed spectrophotometrically by Oke, who observed the Hα line. At the same time, Greenstein and Mathews explained the spectrum of 3C 48 as being due to the ultraviolet (UV) lines of Mg II and several other species never optically observed, again redshifted by a large value.

Within a few months, Smith and Hoffleit had demonstrated that 3C 273 is variable on timescales of less than 1 yr, with a moderate amplitude, through the use of historical plate collections. The source was also shown to consist of several components, both optically and in radio, and it was shown that there is an extension of the source that has the appearance of a jet of the sort familiar from the Virgo cluster galaxy M 87. The timescales argued for the source lifetime were based on the size of the jet, and yielded lifetimes of at least 10^5 yr. In addition, as was noted in the first optical images of 3C 48, the object was embedded in a fuzzy nebulosity that surrounded the semi-stellar nucleus.

In 1965, it had already become clear that many of the blue stellar objects at high galactic latitude (so out of the extincting dust layer that obscures the stars of the plane of the Galaxy) are quasars. Also, a large number of them were shown to be variable. However, the discovery statistics were poor and based entirely on the follow-up work on the radio surveys.

This situation changed in the 1960s when Markarian and collaborators began to use Schmidt cameras with UV-passing objective prisms to look at large parts of the sky for QSO candidates. This revealed that most quasars and active galaxies are in fact "radio quiet," that is, they are generally not strong radio sources. The observations were followed up by higher resolution spectra, by Khachikian and Weedman, which by 1974 demonstrated that QSOs are a high luminosity extension of the Seyfert galaxies. These were first observed and described by Seyfert in 1943 and are galaxies with extremely broad emission lines and semi-stellar nuclei. The study showed that there are several classes of such galaxies, distinguished on the basis of the line strengths and widths, and it provided the most important piece of evidence then available for the cosmological nature of QSOs.

The advent of space observation came in the late 1960s with the launch of the *UHURU* X-ray survey satellite, which discovered that many of the QSOs were associated with discrete X-ray sources. The observation of a diffuse soft X-ray background also occurred at about the same time, and the argument began over whether the QSOs were the primary source of that background. The launch of the high-resolution imaging and spectroscopic X-ray satellite *EINSTEIN* in the late 1970s led to the serendipitous discovery of a large number of active galaxies by a different method and made it clear that quasars were indeed a major constituent of the earlier universe.

The nature of the redshift and energy source was first argued with the original discoveries of the quasars. The first suggestions, by Woltjer in 1957, of a black hole as the source for the energy were not amplified until the mid-1960s, following the discovery of the redshift of 3C 273 and 3C 48. The arguments for a noncosmological origin of the redshift go back to the original suggestions by Ambartsumian that the QSOs were active objects ejected from galactic nuclei. The statistical controversy that followed, largely due to Arp and collaborators, has finally been settled by the observations of the fuzz of the galaxies in which the quasars are embedded. There was, however, an additional theoretical reason for the controversy. The original observations, which indicated that the quasars are strong radio sources, was near the limit of the energy density that such objects can have (given the size of the emission region) to be stable radio emitters. Burbidge and Hoyle pointed out the problem soon after the observations of the redshift in 3C 273, and the problem remained until the surveys by nonradio methods began to turn up large numbers of radio-quiet QSOs. Gravitational redshifts or more exotic mechanisms are not required for the QSOs.

Perhaps the most remarkable observational point to note concerning quasars is that information about them is available over a wider range of energy than almost any other class of astrophysically interesting objects, with the possible exception of the sun (for 3C 273, there is coverage over 15 orders of magnitude in energy). This is, in part, a product of their high redshifts: they present portions of the spectrum in any wavelength interval that is normally inaccessible by the special methods required to observe in that interval. However, the drawback is their extreme faintness resulting from that redshift.

II. Selecting Quasar Candidates from Surveys

The definitions of the following three types of quasars are essentially the same in all searches. Since the major surveys are still optical, these are defined from ground-based observation in the wavelength region between 3300 and 8000 Å (observer's frame):

1. Photometric: $U - B < -0.40$ (although this may not be the best criterion for the highest redshift objects due to the effects of absorption lines on the colors).

2. Morphological: The object has a starlike appearance on the blue Palomar Sky Survey Plates (the only currently available large photographic sky survey) so that most of the light should come from a region <2 arcsec in diameter.

3. Spectroscopic: Broad emission lines and a substantial redshift. No galactic object has a redshift >0.002; the criterion generally used is

$z > 0.025$. This is the major discriminator among the classes of active galaxies. Sy 1 and QSOs have broad permitted and narrow forbidden emission lines. BL Lac objects show no emission lines and sometimes weak absorption lines in otherwise featureless flat continua, and Sy 2 show narrow permitted and forbidden lines of comparable width.

In the first catalog of QSOs with measured redshifts, in 1971, there were 202 objects. In 1973, a similar list was still dominated by radio discoveries listing 196 radio-loud and only 57 radio-quiet quasars. The latest listing (1985) is 2835 quasars with measured redshifts, and an additional nearly 740 active galaxies (including BL Lacs, and Seyfert galaxies, with 236 being Sy 1). The dramatic increase in the number of QSOs with measured redshifts and spectral properties is the combined result of major improvements in the sensitivity of detectors and the advent of large-scale nonradio search techniques for discovering QSO candidates.

A. PHOTOMETRIC SEARCH PROCEDURES

This article deals with general quasar phenomenology, but will also include the properties of active galaxies as they relate to QSOs. Therefore, for the purposes of this discussion, there will be a distinction drawn between Seyfert galaxies and quasars only where necessary for purposes of identifying the relevant literature. In fact, the data accumulating appears to link the entire sequence of activity in galactic nuclei from Low Ionization Emission Line Galaxies (or *liners*) through BL Lacertae objects (or *blazars*) to Seyfert galaxies and the QSO phenomenon.

The spectrum will be divided into regions on the basis of wavelength. The optical portion of the spectrum will be defined as the interval from 3000 to 10,000 Å, the ultraviolet (UV) as extending below the Lyman limit (912 Å), and the infrared (IR) as the 1 to 1000 μm region. Millimetric observations are discussed along with the radio, since the techniques are similar. The observations of various regions of QSO spectra are complicated by the fact that one object's UV is another's optical, in the observer's frame (due to the redshift distribution), and thus the discussion will be from the reference frame of the quasar.

Quasars may be separated from stars and normal galaxies through the use of the extreme strength of the near UV continuum. The UBV system consists of broad ($\sim$600 Å) filters that are centered on the regions 3600, 4400, and 5500 Å. The observation of most stellar objects shows that the (U-B, B-V) diagram is an excellent discriminator of continuum morphology. The broader band response of the J (blue) and F (red) emulsions of photographic plates has also been employed in searching for the faintest QSO candidates ($J < 24^m$) by automated photometric methods.

Another search method is the use of variability statistics for blue objects. It is now known that the majority of QSOs are variable on timescales on the order of 1 yr. A systematic coverage of a large area of the sky in U permits the selection of optically variable objects, many of which subsequently turn out to be QSOs and BL Lac objects or *optically violently variable* (or OVV) galaxies. Colors alone, however, are not the best determinant of the quasar characteristics, nor is variability. Confirmation requires spectroscopic observation subsequent to any photometric search, at whatever wavelength is used.

B. SPECTROSCOPIC SEARCH PROCEDURES

A frequently used method for the discovery of both QSOs and active galaxies in general has been objective prism methods. In this, the low dispersion spectrum of many objects in the field of view can be obtained simultaneously. The use of wide field cameras, such as Schmidt telescopes, allows for the rapid search of large portions of sky at once. However, the detectors must (at the moment) remain photographic, and therefore one is limited by the characteristics of the emulsions currently available. The lowest dispersion survey, that of Markarian and his collaborators, was the first such systematic search. It is based on the extended UV continuum in these objects and has produced a large fraction of the known systems. Groups at Cerro Tololo, European Southern Observatory, Kitt Peak, and the Anglo-Australian Schmidt have employed this procedure to determine both the UV continuum and some spectral information. The observation of lines in the spectrum allows for the determination of crude redshifts and has also proven to be a better discriminator for QSOs among the objects discovered. A recent development is that of automated quasar detection (AQD) using dedicated scanners for Schmidt plates and pattern recognition by automating the determination of the spectral properties. This procedure, using calibrated plates, promises to permit the searching of large portions of sky in a short time to produce new QSO candidates that might be missed by manual scanning of the plates.

An additional spectral method is the use of grisms, which are diffraction grating–prism combinations on large telescopes. This method has the advantage of going quite faint, but only in a narrow field. However, it permits the use of charge-coupled device (CCD) detectors because of the size of the field and can be consequently used for the determination of spectrophotometric properties of the objects as well as redshift and to much lower brightness limits than the other methods.

C. NONOPTICAL SEARCHES

The advent of a series of X-ray satellites in the 1960s and 1970s (*UHURU, Ariel V, HEAO-1,* and *EINSTEIN*) opened this high-energy regime of the QSOs and permitted X-ray selection of their population. It is now clear that these different search strategies produce complementary samples of objects, and that they are useful cross-checks on the biases in any one technique. Many of the point sources picked up by chance in the *EINSTEIN* fields have since been shown to be low luminosity QSOs; the number of known objects has therefore increased significantly by the so-called serendipitous discoveries from the X-ray satellites.

As the first method by which quasars were discovered, it is clear that the radio surveys produced an initially biased sample of objects. Radio sources have been followed up from ground-based telescopes, usually with imaging but often spectroscopically. These searches have produced many quasar candidates and have also been instrumental in determining the nature of the radio-loud QSOs. The opposite procedure, large-scale radio surveys of optically-selected QSOs, has also yielded basic statistics on the frequency of radio-loud systems and is often employed in tandem with optical search methods.

Infrared surveys have just begun, with the launch of *IRAS,* and follow-up observations are continuing at this time. Several satellites are currently being planned, including *ISO* and *SIRTF,* which will have imaging capabilities in the 1 to 10 μm region. To date, however, it appears that no new candidates can be found using this method. Although many known quasars do show some IR emission, the *IRAS* survey did not turn up new QSO candidates.

To date, there have been no systematic searches in the ultraviolet range. With the launch of the Hubble Space Telescope, it will be possible to perform surveys in the UV of the same sort that have been employed in the optical. Additional X-ray instruments, including *AXAF,* are currently being studied, which will allow for higher spatial resolution and more sensitive detection than *EINSTEIN.*

D. DESIGNATIONS

The confusing wealth of names is mainly the result of the many methods of discovery of quasars. A number of active galactic nuclei (AGNs) occur in bright galaxies, like NGC 4151. Those discovered by radio methods generally have 3C, 4C (Third and Fourth Cambridge surveys), or PKS (Parkes) designations. Objective prism searches have produced the Mk, Ark, and related catalog designations. One of the most extensive lists is that of Green and Schmidt, the PG survey, which contains about 1500 objects with follow-up spectroscopy and redshifts on all. For either the PG or *EINSTEIN* surveys, positional designations are used.

III. Cosmological Notions

It is well to keep in mind that QSOs are objects at high redshift and therefore cannot be interpreted without taking proper account of the cosmological arena. The facts that the curvature of spacetime plays a role in interpreting the luminosity and distance information and that the high redshift alters the widths of lines and their relative strengths must be included in any discussion that attempts to go from observed to intrinsic quantities. Without deriving them here, some basic quantities will be discussed that are useful for interpretation of quasar characteristics, as a function of redshift and cosmological model. Throughout, a standard *Friedmann–Robertson–Walker* (FRW) model will be used, which is the simplest description of an isotropic, homogeneous expansion with matter. The cosmological constant $\Lambda = 0$. The two basic parameters are the Hubble constant H_0, which is the measure of the expansion rate at the present epoch, and the deceleration parameter q_0, which measures the second derivative. The current values are $H_0 = 50$ km sec^{-1} Mpc^{-1} and $q_0 = 0$, and it is these to which the discussion of absolute quantities should be scaled throughout.

The redshift is defined to be positive for displacement of lines to the red. If λ_e is the rest wavelength (the wavelength at which the line was emitted in the frame of the quasar), and λ_o is that at which it is observed, then the redshift is defined by:

$$\lambda_o = (1 + z)\lambda_e \tag{1}$$

which, in the low-velocity limit, yields the usual formula for the doppler effect $z = v/c$ (for $v \ll c$) where v is the velocity and c is the speed of light. For the typical QSO, the values of z range from about 0.1 to the current record-holder at about 3.8. The redshift is related to the distance of the QSO and therefore also to the scale length of the universe. An object whose emitted flux is $f_e(\nu_e)$ will have an observed luminosity per unit frequency of

$$F_o(\nu_o) = \frac{4\pi D_L^2}{1 + z} f_e \left(\frac{\nu_e}{1 + z} \right) \tag{2}$$

The "luminosity" distance, $D_L(z)$, is given by

$$D_L = \frac{c}{H_0} z \left(1 + \frac{1}{2} z \right) \tag{3}$$

if $q_0 = 0$, and

$$D_L = \frac{c}{H_0 q_0^2} (q_0 z + (1 - q_0)[1 - (1 + 2q_0 z)^{1/2}]) \tag{4}$$

if $q_0 > 0$; it is a measure of the curvature of spacetime to an object at redshift z. The absolute magnitude, for a power law continuum with a spectral index α, is given by:

$$M = m - 5 \log D_L$$
$$+ 2.5(1 + \alpha) \log(1 + z) + C \tag{5}$$

where m is the apparent magnitude and C corrects the fluxes to the photometric standard, Vega, in the appropriate bandpass. The factor of 2.5 comes from the definition of magnitude. The factor depending on α is the *K-correction* for a power law continuum which is the change in the apparent magnitude of a galaxy due to wavelength stretching from the cosmic expansion the change in wavelength of the *observed* part of the spectrum. An angular separation on the plane of the sky $\Delta\theta$ between two objects with redshift difference z is equivalent to a physical separation of

$$\Delta l = \frac{c\Delta\theta}{H_0 q_0^2 (1 + z)^2} D_L \tag{6}$$

where $q_0 > 0$. Timescales also have to be corrected so that periods of variation must be cor-

rected for the interval in the frame of the QSO. The equivalent width of an emission or absorption line is defined by

$$EW = \int_{\text{line}} \frac{F_c - F_l}{F_l} d\lambda \tag{7}$$

where F_c is the continuum flux and F_l is the flux in the line, both wavelength dependent. It is independent of foreground reddening. Since this quantity is essentially the same as a wavelength interval, it too must be corrected for the stretching of the wavelength due to the expansion. This gives

$$EW_o = (1 + z)EW_e. \tag{8}$$

The same is true for the observed line widths.

IV. Optical Observations

A. Optical Photometry and Variability

The defining optical morphological characteristic of the quasars is their stellar appearance. The colors of quasars vary considerably from one object to another. For a fiducial value, the original observations of 3C 48 by Sandage gave B-V = 0.38, U-B = −0.61, with V = 16.06, far higher in the U filter than normal stars in the two-color (U-B, B-V) diagram. Several surveys have employed the colors to determine likely QSO candidates. The ultraviolet excess is characteristic.

Recalling the earlier discussion of the K-correction, it is clear that there will be problems with this correction if the redshift is sufficiently large that the strong emission lines from the UV part of the spectrum, such as $Ly\alpha$ and CIV, are shifted into the optical bandpass. This can only be checked on direct inspection of the spectra of candidate objects. It should also be noted that although the colors are an excellent criterion for distinguishing potential QSOs from stars and more mundane objects, the broad absorption lines and strong emission lines, which characterize many of these objects, can alter the colors considerably, so follow-up observations are essential. In other words, a QSO is not verified merely from photometric properties—it must be observed spectroscopically.

The QSOs have, as a subclass, a set of objects that are strong radio sources and that show large amplitude variations on short time scales in the optical. These are the optically violent variables, or OVVs. In general, they are associated with BL Lac objects and are characterized by flat, featureless (no emission lines) spectra from the radio through the optical range. Their polarizations are high, and they show variability on time scales as short as hours. The OVVs form a distinct class and are easily found by their extreme UV excesses. However, because only a few show absorption lines in extended fuzz or from intervening galaxies, they have not been properly placed as a group in their cosmological context. Their radio properties are better known, and these are discussed in that section.

The variability statistics presently available are that for a moderately faint ($B = 19$ magnitude) sample, about $\frac{1}{3}$ of the objects are variable with $\Delta B > 0.3$, while 70% have amplitudes ≥ 0.1. In one survey covering the period from 1950 to 1983, for which photographic plates were available (like the Palomar Sky Survey), about 40 looked at had amplitudes $\Delta B > 0.4$. This indicates that the majority of the QSOs are low level variables and probably only a small number are OVVs, in agreement with the radio statistics.

B. OPTICAL SPECTROSCOPY

For low redshift systems ($z < 0.5$), the visible spectrum (to the atmospheric cutoff at about 3000 Å) does not include the UV. The spectrum is characterized by strong emission lines in almost all quasars of the Balmer series of hydrogen ($n \rightarrow 2$), which have extremely large widths, and the [O III] forbidden lines. The permitted lines are always broader than the forbidden ones, leading to the two-component description of the broad line and narrow line regions for the phenomenology of the line formation. The permitted lines in some systems also include He I 6678 and 5876, although these are usually weak. The lines consist of extremely broad wings with superimposed narrow cores, although these are unresolved in low-dispersion observations.

In their first discussion of the systematics of active galaxy spectra, Khachikian and Weedman identified two distinct types of Seyfert galaxies on the basis of the velocity widths (W) of the emission lines: those with W(permitted) $\gg$ W(forbidden) (Sy 1) and those with W(permitted) $=$ W(forbidden) (Sy 2) with W(Sy 2) $\ll$ W(Sy 1). The quasars fall into the Sy 1 category and are the most luminous of that sequence ($M_V < -23$). The classification has been later refined to include a composite class of profiles (Osterbrock has identified the subclass Sy 1.9 with composite profiles), but this does not change the basic partitioning of the sample of active galaxies.

Recent observations show that the Sy 1 sequence extends to very low luminosity (liners) and that many of the weak emission line sources have very broad wings when observed with high enough signal-to-noise ratio. They also provide the first association of QSO-like spectral phenomenology with elliptical galaxies, thus, it now appears that the quasarlike phenomenon of nuclear activity can be associated with galaxies other than spirals. Their spectroscopic observations can be explained by photoionization, further supporting the extension of the QSO and Sy 1 phenomenon to lower luminosities.

It is possible to define a set of three excitation classes for active galaxies. These augment the Sy 1 and Sy 2 classification. Class A shows Fe II emission, Hβ/[O III] proportional to the absolute luminosity of object (either L_x or L_{opt}) (subclasses A1: Hγ/Hβ larger than A2, [O III]/Hβ correlates well with full width half maximum (FWHM) (Hβ)). Class B shows no Fe II emission, and permitted lines are broader than forbidden. Class C shows no Fe II emission, and permitted lines about the same width as forbidden. For Class A only, the emission profiles tend to be smooth; radio luminosities are usually low, but when large, the source is compact and the spectrum flat; [O III]/Hβ correlates well with Lyα luminosities, F (4000 Å), and nonthermal optical luminosity. [O III]/Hβ is anticorrelated with luminosity both in Hβ and the continuum. For Class B and C, all extended radio sources belong to these classes, class B frequently showing complex line profiles. Class C only has [O II], [O III], [N II] stronger than Balmer lines. Their X-ray emission is frequently weak, and no UV bump is seen in these objects.

QSO emission lines are often asymmetric and variable. The line asymmetry correlates with F(Hγ)/F(Hβ). The time scales for the variations are best studied on the Hβ line, which is often blended with the [O III] lines but which is more variable (the typical time scale is years). This variation manifests itself as structural changes to the profile, with variation of the asymmetry and appearance of additional emission (or absorption) components and overall strength changes. The behavior is very similar to Sy 1 galaxies.

C. POLARIZATION IN THE OPTICAL

To date, polarization has been measured for quasars only in optical and radio wavelengths. Since these appear to be intrinsically linked in the samples so far studied, polarization will be discussed here. The reader is referred to the section on radio properties for comparison information.

Quasars can be separated into two classes on the basis of their polarization properties. A small minority (1% of all QSOs) resemble BL Lac objects with high polarization (3–20%) and similar optical and radio continua. Low polarization QSOs (LPQs) (0–2%, with $< P > = 0.6\%$) have $\Delta P/P \leq 0.16$ and changes in position angle $\Delta\theta \leq 8°$ on a time scale of about 1 yr. For a subsample of LPQ (low polarization quasars), the position angle is constant with wavelength, while the polarization increases toward shorter wavelength. Radio and optically selected LPQs have a similar probability distribution of polarization. The wavelength dependence is approximately power law, with an average value for the spectral index of -0.6. Low polarization QSOs may be either radio loud or radio quiet, show harder optical continua, and only show moderate photometric variability.

In contrast, the high polarization quasars (HPQs) are generally associated with compact radio cores (radio-loud, flat spectrum sources) which exhibit low-frequency variability and superluminal motion, have steep optical nonthermal spectra, exhibit large-amplitude optical (OVVs) photometric variability, and show excess X-ray emission. About 15% of radio-loud QSOs (or about 1% of all known QSOs, since only about 10% are radio sources) are HPQs. There is no correlation with the equivalent width of the emission lines, redshift, or optical luminosity with polarization. The HPQ appear quite similar to BL Lac objects in their basic properties.

The polarization vectors of nearly all QSOs are aligned with the axis of the radio source, as in Sy 1 galaxies. For Sy 2 galaxies, the polarization tends to be perpendicular. Variations in the amount and direction of the optical polarization are common, but not related in a simple way to the flux level. Radio galaxies include a significant excess of parallel objects. This may indicate that electrons scattering optical light in the accretion disk region are responsible for at least some of the observed polarization.

D. QSO FUZZ

The advent of extremely sensitive photometric detectors in the past 5 yr has made it possible to determine the characteristics of many of the host galaxies in which the quasar resides. While this is clearly a field that is progressing rapidly,

several generalizations appear possible. Observationally, it is clear that many QSOs are imbedded in extended regions of faint emission. On detailed spectroscopic examination, many of these regions of "fuzz" display absorption and emission lines. The emission lines are the most characteristic and are those normally found in H II regions: [O II], [O III], [Ne III], H I, and sometimes weak [Fe II]. Their strengths are consistent with photoionization models, and it appears that they are due to a population of young stars. However, there is a second group, comprising about half of the systems that have been studied spectroscopically, which display only Hα in emission, if any is present at all, and a red continuum. This group also shows, for two QSOs (including the infamous 3C 273) the Mg Ib line, which is characteristic of the late type stellar population of a galaxy. It appears then that there are two classes of objects that contain nuclear activity.

An additional, intriguing observation is that of a possible supernova in the host galaxy of QSO 1059 + 730. The QSO is relatively nearby and has a redshift of $z = 0.089$. Although it was not possible to obtain a spectrum for the object, its photometric behavior is consistent with being a supernova.

It is expected that one of the most important products of the Hubble Space Telescope will be the direct imaging, in the ultraviolet as well as the optical range, of the quasars and their environs to magnitude limits on the order of 29 or 30.

V. Ultraviolet Observations

The most important observation, early in the history of QSO research, was that one is able from ground-based observations to study that portion of the spectrum of an active galaxy that is usually invisible due to the high redshifts—the region below the atmospheric cutoff (due to ozone absorption) at about 3000 Å. In fact, this realization provided the key to deciphering the 3C 48 spectrum by Schmidt at the beginnings of quasar research. With a redshift of $z = 2$, the Lyα line is placed at a wavelength of about Hβ, ideal for ground-based observation and making the Lyman continuum limit at 912 Å accessible as well.

A. EMISSION LINES

The importance of UV observations is that, in high-excitation objects in which the highly ionized species of carbon, nitrogen, and oxygen are observed, the resonance lines (ground state transitions) of these elements are readily observed. These are typically very strong lines, and since they arise from the ground states of their respective ions, the observation of their strengths gives an immediate handle on the abundance of the species, more readily than for the highly excited states usually observed in the visible wavelengths. Specifically, the most important permitted lines are Mg II 2800, Al III 1860, C IV 1548, Si IV 1400, O I 1304, N V 1238, and Ly 1216. The He II 1640 line is also very important in that it is connected with the optical He II 4686 line and therefore provides information about the possible role of reddening due to dust mixed into the interstellar medium of the nucleus. There are also numerous lines of Fe II and [Fe II]. In addition, there are a wealth of so-called intercombination lines, lines that involve magnetic dipole and electric quadrupole transitions and that are strongly collisionally damped at high density, like C III] 1910, N III] 1750 and O III] 1667. In fact, most of the plasma diagnostics, which are of the most interest in the determination of the gas properties, are in the UV part of the spectrum.

Finally, because the Lyman series of hydrogen arises from the ground state, it is the strongest set of lines in the interstellar medium of our own and other galaxies. Therefore, in an effort to search for both intergalactic matter and to study the medium of intervening systems, the high redshift quasars provide useful "lighthouses" in the universe between the active nuclei and us.

B. UV EVIDENCE FOR DUST

The interstellar medium of our galaxy and other galaxies is the solid particulate material, also called dust. It has a large absorption cross section in the UV, contributing a broad absorption feature at 2175 Å attributed to a phase of graphite. The observation of this feature is usually very difficult in faint objects due to the high signal-to-noise ratio required to achieve good definition of the continuum morphology, and it usually lies in a portion of the spectrum accessible only from space observation. However, for those systems that show redshifts on the order of 1.5 or higher, enough of the continuum is shifted into the visible to permit study of this feature. There are several problems associated with the interpretation of this feature, however. It is known to vary in strength along different lines of sight in our galaxy and also to vary among galaxies. In addition, the reddening law may not be the same in the UV for all galaxies.

C. THE BALDWIN EFFECT

The equivalent width of the C IV 1550 doublet in high redshift QSOs is anticorrelated with the luminosity of the central source. This has since been confirmed by direct observation of the ultraviolet using the International Ultraviolet Explorer (IUE) satellite. This anticorrelation, known as the "Baldwin effect," has also been observed between the luminosity and Lyα. A weak anticorrelation between the Hβ line width and luminosity has been reported. For Sy 1 galaxies there is a correlation between the equivalent width of Hβ and the FWHM. This does not appear to hold for the Fe II lines, indicating that here may be a collisional damping of the emission line intensity.

D. LYMAN TO BALMER LINE RATIOS

One of the early observations that holds a key to the excitation conditions in the nuclear region is that of the Hβ/Lyα. The upper level of Lyα, n = 2, is the lower level of the Hβ line, so they are directly connected in any cascade. It is possible from radiative recombination to calculate the ratio of the two lines in the same way that the optical line ratios are derived. Of course, the exact value depends both on the local electron density (entering through both the collisional effects on the level and the recombination rate) and on the details of the continuum energy distribution for photoionization; however, the basic result is that the Lyα/Hβ ratio should be large (about 33) if the Lyman continuum is optically thin. It will be smaller if that continuum is thick, but still about 20 (see discussion below in theoretical section). The mean value is about 4, far too low to be accounted for by any of the standard radiative transfer models normally used for galactic H II regions. More recent observations have increased this ratio to as high as 10, but nonetheless, the medium of the broad line region must be quite optically thick.

E. INTRINSIC METALLIC ABUNDANCES

The abundances from emission lines for refractory elements such as Fe, Si, Al, Mg, and C lines observed in QSOs can be determined using the photoionization models for the broad and narrow line regions. Studies show that the abundances cannot be much less than solar in the gas phase and do not match the characteristics observed in either planetary nebulae or the ISM of our galaxy. It appears, therefore, that the metallicity of these extremely young galaxies is considerably elevated above the primordial values and that processing has to have occurred in stars even at the early epochs in which the QSOs are observed.

VI. Observational Constraints on Sizes of Emission Regions

The variability of the strong ultraviolet lines of C IV, C III], and Mg II have been used to study a prototypical Sy 1 galactic nucleus, NGC 4151, during outburst in 1979 May. A sudden increase was observed in the continuum in the ultraviolet range. The C IV line lagged the continuum by about 13 days and C III] by more than 1 yr; Mg II variations are consistent with a length of 60 light days. The observations show that events are first seen in the thermal continuum component, then in the nonthermal part, and last in the emission lines. The mass of the central object can be obtained by using the time delay of the broad line region (BLR) versus the narrow line region (NLR) to provide a measure of the radial scale, and the velocity difference between these two regions to provide the gradient in the orbital motion. Although the lines usually have different widths, when the nonthermal continuum is low, they are relatively narrow (HWZI 4000 km sec^{-1}). The central mass is computed to be $(0.5 - 1) \times 10^8 M_\odot$.

Four years of line and continuum monitoring set a limit of 30 light days to the size of the BLR in the Sy 1 galaxy Ark 120. The BLR is optically thick in the LyC, since the Balmer lines vary with the continuum; the BLR has a crossing time of less than 4 yr. The problem is still that few multiwavelength studies are available for QSOs for extended periods of time. Those few that have been studied (like F9) generally show behavior consistent with the NGC 4151 results. In particular, the changes observed in the structure of 3C 273 and other superluminal sources shows that the central region must be less than 1 light yr across. The jets must originate in a region interior to the minimum size currently set by very long baseline interferometry (VLBI), that is, about 1 light-month, and that collimation must be achieved very near the central engine. This is true for radio galaxies as well. Multiwavelength coverage using IUE, radio, IR, X-ray, and optical observations has been performed for a number of OVVs and BL Lac objects, but long-term projects are still ongoing and dependent on the difficult task of coordinating observations. The advent of Space Telescope, space IR and X-ray observatories and the expansion of the VLBI network, the very long baseline array (VLBA), and millimeter interferometers all promise to make more feasible the task of observing these objects simultaneously.

VII. Absorption Lines in QSOs

In addition to the observations of the emission line spectra, absorption lines were also observed in QSO spectra early in the history of work on these objects. These usually occur at wavelengths shorter than the emission lines they accompany, that is, they occur at lower redshift than the emission. There are, however, several clear cases of absorption at higher reshift than the emission line, but always within a few percent of the emission redshift. In a few QSOs, these take on the appearance of saturated, outflowlike profiles. The majority of the absorption line QSOs show more complicated line distributions and less windlike profiles. The absorption lines occur principally on the UV lines, such as the Lyman series, because they arise from the ground states of their respective ions.

A. FORMATION OF ABSORPTION LINES

If there is no continuum source behind a region of hot gas, or if the intrinsic emissivity of the region in front of the light source is higher than the local photon intensity from the source, an emission line will form. For instance, in planetary nebulae or H II regions in the galaxy, the emissivity of the gas derives from the excitation of the spectrum through collisions as well as excitation of lines by photons followed by cascading de-excitation. In the case of the absorption lines, the local source function (the rate of photon production) is exceeded by the excitation of the atom by absorption of a more intense background source and by the subsequent collisional de-excitation of the level or scattering of the photon. Either way, photons are removed from the line of sight. The line will saturate at some level depending on its atomic parameters, that is, its *oscillator strength*. Therefore, there will be some value of the column density (the amount of absorbing matter per unit area along the line of sight) above which there will no longer be a simple linear relationship between the strength of the line and the number of absorbing atoms. This phenomenon complicates the modeling, especially when there are separate clouds, because the velocity dispersion in the clouds and of the ensemble tends to broaden the line and thus delay the critical density at which it saturates from setting in. The ground state lines of any ion will be the most likely absorbers and the strongest lines.

In a moving medium, the scattering will shift the photons into or out of the line of sight in a specific range of frequency. The scattered photons undergo a doppler shift relative to their emitted frequency because of the motion of the scatterer. In the wind or expanding shell of an object with a strong continuum, the formation of an absorption trough shifted to the blue from the rest wavelength is a consequence of the scattering of radiation out of the line of sight to the central source. There will also be emission from the other extended region gas, since this is scattered radiation into the line of sight (to someone in a different direction, this would appear as an absorption line). Therefore, if the line is formed by pure scattering, the emission is approximately equal in intensity to the absorption trough, neglecting the occultation by the source of the back side of the emission region. In general, both absorption and collisional de-excitation will occur, which tend to decrease the strength of the emission relative to the absorption. The blueward edge of the trough of absorption is at the terminal velocity of the flow, arising in direct line to the continuum source and therefore the largest optical depth. The presence of this essentially featureless broad trough is the best indication of a continuous mass loss by the central object. It suggests that there is a steady-state flow with high terminal velocity and that is optically quite thick; the residual intensity being essentially zero is a measure of the degree of saturation of the profile at a given velocity.

B. Observations of Quasars

All quasars that have been observed to date with high resolution ($\Delta\lambda \leq 1$ Å) and good signal-to-noise ratio (S/N ≥ 20) and that can thus distinguish absorption lines from noise contain absorption systems, either of Lyman α or metals. These fall into two basic categories:

Class (a) includes the Narrow Lines systems. These have two subclasses. In subclass (a1), the "Lyman α Forest," there are no metallic lines, and the absorption lines have a simple structure and dense but resolved distribution of lines with $z_{abs} \leq z_{em}$.

In subclass (a2), the Narrow Metal systems, H I, Mg I, Mg II, Fe II, C IV, Si II–IV, O I, O VI, and occasionally C II* (the excited fine structure line of the UV C II doublet) are present. These show sharp absorption line systems for which $|z_e - z_{abs}|/(1 + z_e) \leq 0.01$ are most commonly seen in the high excitation lines, such as C IV. Some type (a2) quasars show sharp displaced features at higher velocity, but otherwise similar characteristics. Several of these systems display, especially on C IV, absorption at *higher* redshift than the emission line. The metal lines observed in quasar spectra to date tend to have small velocity widths. Recent studies have limited the sample to the range of $20 \leq \Delta v \leq 45$, with the mean being about 25 km sec^{-1}. The lines typically are not saturated; this implies that the column densities cannot be too high to the source of the radiation. The lines arise from a manifold of ions, ranging from Al II and Mg II at the lowest excitation end to N V and O VI.

The matter of determining the distribution of the low redshift systems is fundamentally complicated by the fact that *any* absorber along the line of sight is a possible contributor to the spectrum. Therefore, lines from $z = z_{em}$ down (in principle) to $z = 0$ are possible, including the crossing in wavelength of absorption systems from different lines. For example, the Lyα line is at 1216 Å, and the Lyβ at 1025 Å. Therefore, for a $\Delta z \geq 0.2$, the two systems will cross. In addition, the individual lines can possess complicated structure, making the precise determination of the redshift of the emission line problematic.

The class (b) systems are called broad absorption line (BAL) quasars. These were among the first absorption line quasars to be discovered. The emission lines of C IV, Si IV, N V, and O VI are accompanied by strong absorption troughs extending to relative velocities (compared with the emission peak) of $0.1\ c$. Narrow lines are sometimes seen in association with these and are the same as seen in the class (a) objects. Of moderate to high redshift systems found on objective prism plates 3–10% are BALs. The terminal velocities are typically about 20,000 km sec^{-1}, but some have been seen as high as 65,000 km sec^{-1}. Structure tends to be many essentially unresolved components, or complexes of troughs, or broad nearly featureless troughs. The absorption is often detached from the corresponding emission lines. On average, for BALs the C IV line is weaker, the N V line stronger, and the C III] broader than for non-BAL QSOs. Low ions, such as Mg II and Al III, are rarely observed in these systems. The H II/H I ratio is extremely large, on the order of 10^5, and the ionization in general extends up to O VI. Observations show that this subclass of absorption line systems occurs at redshifts as low as 0.29 (PG 1700 + 518).

The Lyman discontinuity overlaps the two narrow absorption line subclasses (a1 and a2). Most frequently, it occurs along with Narrow Metal systems, but sometimes it is observed in the Lyman α Forest objects.

Subgroup (a1) apparently results from either intergalactic hydrogen clouds, not self gravitating, but being slowly (cosmic time scale) destroyed by the flux from the UV in the intergalactic medium or perhaps to intervening dwarf galaxies, while (a2) appears to be due to absorption from extended halos and disks of more massive intervening galaxies. Group (b) quasars seem to be the result of outflow from the central region of the quasar and are thus intrinsic to the source. For our galaxy, EW(C IV) < 1 Å for all lines of sight. It is found that N(Si IV)/N(C IV) for galactic halo gas is similar to QSOs, where N is the column density. The moderately high temperature (C IV and Si IV) gas in our galaxy also has Ca II and Na I absorption associated with it; this is likely the same as for the QSO systems (for Mg I, for example). The multiplicity observed in the QSO spectra must, however, have a different origin than rotation of or structure in a single galaxy, since the probability of intersection of multiple clouds in a single galaxy at very different velocity along a single line of sight is small. One of the basic results of the absorption line study is to indicate that there is considerable matter at distances of three to four times the optical photometric radius (the Holmberg radius) in the intervening galaxies, indicating the presence of extended nonluminous halos in disk systems.

The distribution function for redshift and equivalent width of the observed QSO absorption lines is uncorrelated. That is, the number density of lines of a given equivalent width at a redshift z, $N(W, z) = n(W)Z(z)$, the two being independent. The number of absorption line sys-

tems, if they arise from a uniformly distributed set of galaxies in a FRW cosmology is

$$dN = n(z)\sigma(z)\, dl$$

$$= \frac{c}{H_0} n_0\sigma_0(1 + z)^{1+\beta}(1 + 2q_0z)^{-1/2}\, dz \quad (9)$$

assuming that the number density of the absorbers goes as $(1 + z)^\beta$. Here $\sigma(z)$ is the geometric cross section of an intervening absorber as a function of redshift. The choice of formal representation for $n(z)$ is related to the problem of determining the proper form for the evolution law for the luminosity function of QSOs. Recent observations suggest that the exponent for the total distribution is on the order of 2, and that there may be two populations of absorbers with the lower excitation (Mg II and Si II) being fundamentally distinct from the high excitation (C IV and Si IV) absorption systems, the latter having smaller equivalent widths for the Lyα lines. Analysis of quasar emission lines indicates that the abundances cannot be very far from solar; the weakness of the metallic line systems may therefore be an indication of the ionization, and not necessarily the abundance, of the absorbing medium.

C. 21-cm Neutral Hydrogen Absorption Lines

Although properly speaking, this is a radio property of the systems, it is important to compare the 21-cm searches with those for Lyα and other lines of neutral hydrogen. The 21-cm line arises from a ground state spin-flip transition in H I, and it is one of the most sensitive probes of the structure of the interstellar medium (ISM) in a galaxy. In the most sensitive search to date, only 2 of 18 QSOs observed with the Arecibo radio telescope show the 21-cm line in absorption. These systems were chosen because they showed low excitation (Mg II) absorption. In general, there are no radio absorption or emission lines of neutral hydrogen at the emission line redshift.

VIII. X-Ray Observations

Two satellites are responsible for the observational surveys that have discovered most of the X-ray emitting QSOs. *HEAO-1* operated at 2 to 100 keV from August 1977 to January 1979. It overlapped with the *HEAO-2* or *EINSTEIN* satellite, an imaging and spectrophotometric observatory, which operated in the range of 0.5 to 4.5 keV between November 1978 and May 1981. At this writing, the operating X-ray satellite, *EX-OSAT,* is working in the range 0.04 to 2 keV (imaging) and 1.5 to 50 keV (spectrophotometry). In addition, long-term gamma-ray observations have been made with the satellite *COS-B,* which operated from August 1979 to April 1982 in the energy range 70 to 5000 MeV.

The QSOs in particular, and AGNs in general, follow a remarkably uniform power law from the X-ray to radio (over six decades of frequency) with an overall spectral index of about -1. For the radio-loud systems, $\langle L_{opt}/L_x \rangle = 0.6$, while for radio-quiet systems, this is about 2. Some flattening of the spectrum seems to occur in X-ray wavelengths. The mean value for the spectral index is -0.68 ± 0.15 for the X-ray part of the spectrum for the QSOs observed to date. An upper limit to the neutral hydrogen column density internal to the nuclear region is $N(H) \leq 2 \times 10^{17}$ cm^{-2}. The indicated covering factors for the nuclear region are consequently quite small, less than 0.5 (and these are upper limits). There is *no* relation between the spectral index and the Einstein solid state spectrometer (SSS) luminosity. The range in SSS luminosity is $43.5 < \log L(SSS) < 47$. Since there is no X-ray excess in the soft end (0.75 keV) of the spectrum, there must be a steepening of the spectrum in the range 500 to 13.6 eV, the Lyman limit.

The cumulative X-ray luminosity of quasars is quite large. In fact, they may be the dominant contributor to the diffuse soft X-ray background. The current luminosity function suggests that unresolved QSOs may represent at least 25% of the observed soft X-ray background flux. They also form a significant contribution to the ultraviolet background.

The majority of QSOs do not show strong variability, that is $\Delta L/L < 0.5$. The current observations indicate that the probability of variability is roughly inversely proportional to the brightness of the AGN, but there is some difficulty with the time scale. Some galaxies are variable on time scales as short as days, however, the majority are variable on a scale of years. The radio loud systems tend to be more variable on time scales greater than 6 months. The X-rays may be produced in part by the process of the relativistic electrons scattering the radio emission up to X-ray wavelengths (inverse compton effect) and in part by the emission from the inner region of the accretion disk, very close

to the Schwarzschild radius where the temperature reaches $10°$ K.

IX. Radio Observations

Although first discovered as radio objects, the advent of optical and X-ray search methods has made it clear that less than 10% of the QSO population can be called radio loud. Sensitive searches have demonstrated that the majority of QSOs (although many show extended radio structures when they have central sources) are not in general strong emitters. Again, it was the method of discovery that was initially biasing the sample.

A. Radio Morphology

Radio sources can be classified, providing a useful distinction among the observed properties of QSOs versus radio galaxies. Narrow, edge-brightened double sources have length to width ratios $(l/w) > 4$ and outer emission lobes that symmetrically bracket the central object (E or QSO). The radio luminous galaxies are usually of this type. Hot spots (regions of high emission) in these sources have typical sizes of 1 kpc. Extended tails are sometimes present. A flat spectrum is usually associated with the parent QSO. The linear sizes increase with increasing luminosity $[l \sim P(178 \text{ MHz})^{1.4}]$ with $\langle l \rangle = 170$ kpc for 10^{32} ergs sec^{-1} Hz^{-1}. Fewer than 5% of the sources exceed sizes of 1 Mpc, about 50% are smaller than 100 kpc. Quasars tend to have pronounced hot spots and asymmetric structure when compared with galaxies. Their cores are systematically brighter than radio galaxies at the same intrinsic power. At the level of the inner jets, all QSOs, such as 3C 273, show only one extended jet structure with flux ratios between the opposite lobes of at least 10. One-sided sources appear to be systematically smaller with typical sizes of tens of kiloparsecs. Narrow, edge-darkened double sources generally have strong jets connected with the nucleus. Narrow-tailed sources have asymmetric structure, with high surface brightness head coincident with the nucleus of the parent object, which are also called head-tail galaxies or narrow-angle tail galaxies. Radio cores are usually found in tailed sources.

Wide-tailed sources are the final category. Although there may be an evolutionary link between QSOs of this and the narrow tailed variety, the possibility remains that it is in part a projection effect and partly and artifact of resolution. Radio observations of "wide tail" sources have provided what is probably the best evidence to date for quasars as members of rich clusters. The motion of a galaxy through the diffuse medium of a cluster produces a bending of the extended jets and lobes, probably due to ram pressure. The presence of this structure in QSOs appears to confirm their presence in clusters massive enough to retain a sizable intracluster medium and also to have a high enough velocity dispersion for the member galaxies to produce the bending. The direct CCD imaging of these QSOs provides the best evidence for the correlation between morphology and environment.

While some radio QSOs show S-bending in their radio tails, suggestive of precession of the central source (such as the galactic object SS 433), most QSOs show little or no extended radio structure at the level of $S/N > 100$. Quasars with the strongest and broadest emission lines tend to show extended structure; narrow line systems are compact. BL Lac objects show some weak extended structure; about half show radio sizes ≥ 1 arcsec. Only about 20% of the BAL systems are detected with fluxes ≥ 1 mJy. There is a notable deficiency of strong radio sources for this class.

B. Spectral Properties

The diffuse emission has a spectral index of $-0.7 \geq \alpha \geq -1.2$, sometimes steepening with distance from the parent galaxy. It is sometimes polarized to the level of 60% at 1 GHz. Hot spots have similar or flatter spectra than the diffuse emission, with 1 GHz polarization of about 20 to 30%. For the cores, ultracompact sources ($\ll 1$ pc) have $\alpha > -0.4$. Sources with sizes of kiloparsecs have steeper spectra with indices smaller than -0.4. The radio-extended QSOs tend to have core luminosities about 20 times those of radio galaxies at the same luminosity, and there is a weak dependence of the luminosity of the extended lobes on the core brightness: $P(\text{core}) \approx P(\text{extended})^{1/2}$, where P is the power. Quasar cores appear to be extremely compact, with sizes well below the resolution of the best VLBI measurements. This places an upper limit of about 0.1 marcsec on their sizes. There is, however, a limit to how small the central emitter can be. If the region becomes too compact for a given luminosity, it will become synchrotron

self-absorbed. This limits the brightness temperature to about 10^{12} K, and gives a cutoff frequency below which the continuum will have a positive power law slope of 2.5:

$$\nu_{\max} = 10^7 S(\text{Jy})^{2/5} \theta_{-3}^{-4/5} B_\mu^{1/5} (1 + z) \text{ GHz} \quad (10)$$

where θ_{-3} is the size of the source in milliarcseconds, B_μ is the magnetic field in microgauss, and $S(\text{Jy})$ is the flux in Janskys. For particles having a power law distribution in energy, $N(E) \, dE = N_0 E^{-\gamma} \, dE$, the optically thin part of the spectrum will have a slope of $\alpha = (1 - \gamma)/2$. Flat spectra appear due to multiple sources in the nucleus, and observationally, their occurrence is well correlated with superluminal and OVV activity.

The best studied QSOs have been attacked with both single antennas and aperture synthesis, and sometimes VLBI. Perhaps the best studied quasar is one of the original sources, 3C 273. It shows extended radio structure, on the scale of the large lobes observed in the sources of the Cygnus A type, and a complex jet. The mapping, which goes from the arcsecond scale of the optical structure to tenths of milliarcseconds for the VLBI (at this distance, on the order of parsecs), shows that the jet is a generally coherent structure, but that it consists of knots that have been observed to move in time. Their velocity, in excess of 5 c is among the best available evidence for superluminal motion. Probably the best case, however, is 3C 345. This is characteristic of some dozen sources that have been well studied and appears to be a common phenomenon among the most active radio galaxies.

C. Superluminal Motion in Quasars

Although the name appears to imply a serious breach of relativistic physics, the superluminal nature of the motion is that which is observed. It is not a result of the cores of quasars actually being tachyonic. They do, however, indicate extremely large intrinsic expansion velocities for the knots in nine sources, and suspected in about another six, and they indicate that the accelerations in the ejection phase must be high enough to produce motions with Lorentz factors in excess of unity.

It has long been known that the best studied superluminal sources show only one-sided ejection. The simplest explanation still appears to be relativistic beaming, which causes the receding jet to have a lower surface brightness. The approaching jet is amplified in surface brightness by the bulk blueshift. It must be kept in mind that the velocities required for this effect are much higher than those indicated by the P Cygni profiles of the BALs. Relativistic motion gives

$$\beta_0 = \beta_{bulk} \sin \theta (1 - \beta_{bulk} \cos \theta)^{-1} \quad (11)$$

for the transverse velocity, if $\beta_{bulk} c$ is the actual bulk velocity and θ is the angle relative to the line of sight. The Doppler boost is given by:

$$\delta = \gamma_{bulk}^{-1}(1 - \beta_{bulk} \cos \theta)^{-1} \quad (12)$$

where γ is the standard Lorentz factor: $(1 - \beta^2)^{-1/2}$. The receding beam will be fainter than the one approaching the observer by a factor of

$$S_{approach} = S_{receed} \left(\frac{1 + \beta_{bulk} \cos \theta}{1 - \beta_{bulk} \cos \theta}\right) \quad (13)$$

The most important factor is that the strongest radio sources should be markedly asymmetric, which appears to be the case with the superluminal sources. The current weight of evidence is that no special physics is required to produce the observed structural behavior. The motion is consistent with the idea that a dominant mode of bulk ejection of relativistic particles from the quasar nuclei is in the form of collimated jets.

X. Infrared Observations

A. GROUND-BASED RESULTS

There is a well-established class of radio sources that have no optical counterpart: the "blank field" sources. These often have infrared objects coincident with the radio position. They are sometimes associated with flat spectrum sources, such as the OVVs and BL Lac objects. Many of them are very highly redshifted QSOs.

To date, no QSOs have shown the dust emission features which are, however, observed in two Sy 1 galaxies: NGC 7469 (3.3 μm, 11.25 μm) and NGC 7582 (3.28 μm, 8.65 μm, 11.25 μm). Infrared spectroscopy of OVV sources shows that their spectra are similar to the few QSOs studied so far. No evidence is seen for significant dust emission, and they display smooth continua from the near IR to millimeter wavelengths. While the lack of IR dust features is in agreement with the lack of the 2175 Å absorption in the UV, there are large variations in the property of dust in the galaxy. The IR observations are therefore essential in assessing the role

played by dust in the radiative transfer in the QSO nuclear regions.

The summary of the near IR properties of quasars is that they present a continuation of the power law behavior seen in other portions of the spectrum. In general, they show no spectroscopic evidence for dust or for IR excesses over the extrapolated radio continua, and only the emission lines which are characteristic of the BLR (that is, emission of the higher series of H I). The use of airborne detectors has yielded measurements at 60 and 10 μm of the blazars OJ 287 and BL Lac. Again, there is no evidence for dust, and the spectrum appears to connect smoothly to the radio and optical continuum. In OJ 287, for instance, most of the energy emerges in the infrared, and the integrated luminosity is about $10^{13} L_\odot$.

A most interesting example of the use of the observer's IR to study a QSO is of the most luminous QSO now known: S5 0014 + 81. The redshift for this system is sufficiently high, $z = 3.41$, that most of its optical is shifted into the 2 and 3 μm windows of the IR. Here, too, a major fraction of the quasar's energy emerges in the IR. However, at present, ground-based observations of QSOs is still limited. One innovation is the scan imaging of the jet in 3C 273. In addition, there is a growing use of near infrared (1–10 μm) imaging detectors, although at this date they are still not sensitive enough for observations of QSOs. This is a next generation project.

B. IRAS RESULTS

The advent of infrared space astronomy came with the launch of the Infrared Astronomical Satellite (*IRAS*) in 1982. In one year of operation, roughly 96% of the sky was observed at 12, 25, 60, and 100 μm. While the analysis of this data is still under way, it is possible to summarize the basic results.

Of 186 Sy galaxies surveyed with *IRAS*, 116 galaxies were detected. The luminosity function for Sy 1 and 2 are similar, with the luminosity for Sy galaxies at 60 μm being $\geq 10^{10} L_\odot$ for about 50% of the sample. In the first survey results, five quasars were observed and found to be similar in their overall properties. The radio-loud systems fit well the extrapolation of the radio continuum, with a power law index of about -1. This suggests that in the radio-loud systems, the continuum is overall nonthermal. A few, however, do appear to show an IR excess at 100 μm, which could be due to "cirrus" (diffuse, patchy

100 μm emission seen throughout the ISM of our galaxy and attributed to dust mixed into the diffuse medium, that is, particulate matter and large molecules not structured into large cloud complexes). As mentioned previously, new QSOs have not turned up from the *IRAS* observations; it appears difficult to distinguish them from other classes of galaxies solely on the basis of their infrared properties.

High luminosity *IRAS* galaxies, associated with ''blank'' fields appear to be associated with extremely reddened galaxies. The situation is much like the radio sources discussed above. The luminosities of these objects are comparable in many cases to those of Sy and QSOs. This is clearly an observational challenge for space infrared imagers. The integrated luminosities appear to be $11.6 \leq \log(L/L_{\odot}) \leq 12.6$, virtually all of which is in the infrared.

XI. Quasars and Gravitational Lenses

The possibility of gravitational lensing of distant objects by galaxies was first discussed by Zwicky in 1937, but the first evidence for such objects was the discovery, by Walsh, Carswell, and Weymann in 1979, of the pair of QSOs 0957 + 561. These have redshifts of 1.41 (A) and 1.39 (B) and are separated by 6 arcsec on the sky. It was quickly shown that there is a lower redshift galaxy in the midst of the images, which are aligned roughly north–south, about 1 arcsec to the north of the B image. Only a handful of these objects are known (0957 + 561, 1115 + 080, 2345 + 007, and 1635 + 267), and probably only the widest pairs have so far been found. It has, however, been suggested that many of the systems would be sufficiently poorly resolved that some of the highest luminosity QSOs may simply be lensed Sy 1 galaxies.

General relativity predicts that the trajectory of photons in gravitational fields, such as those of massive galaxies, will be bent by an angle that is to first order proportional to the mass of the deflector. Lensing also dramatically increases the surface brightness of the image. If the intervening galaxy has an extended halo, different rays may correspond to the same focus. Therefore, depending on the mass distribution in the reflector, it is possible to produce multiple images (in general, and odd number). In the case of the known systems, the simple single image characteristic of a point deflector does not ap-

ply—the intervening galaxies are massive ellipticals that have extended halos. Multiple images have been observed in all well-studied systems. Observations of the time scale for variability between the two images is a measure of the differential pathlength for the two (or more) rays through the lensing galaxy.

The lensing systems have been used to test ideas about the formation of absorption systems in QSOs. The quasar 0957 + 561 has been carefully studied in its two optical images. The quasar has $z_{em} = 1.41$, while the lensing galaxy is at about 0.37. In the two image spectra, the C IV lines are found to be of different equivalent width for the same redshift systems, indicating that there are differences along the path through the same intervening galaxy. It therefore appears that the use of the lenses will for some time provide the best confirmation of the cosmic absorber hypothesis for the narrow emission lines.

XII. Quasar Luminosity Function

The observed distribution of quasar magnitudes is the product both of distance and an intrinsic brightness distribution. For example, if all QSOs have the same intrinsic luminosity, then their relative numbers in a given bin of apparent magnitude represent the relative density of objects at a given redshift (cosmic epoch). On the other hand, any evolution of the luminosity will reflect in a distortion of the distribution, producing too many or too few objects at some redshift, depending on the details.

Independent of the models, there seem to be fewer quasars with $z > 3.5$ than at smaller redshift. It is increasingly difficult to find objects with redshifts in excess if this value, although many may be BALs and therefore not be simple to find on the basis of their colors alone. Some luminosity evolution is required in the sample, although the assumption that the luminosity function has remained formally invariant is uncertain at present. The number of QSOs is best described by the ''success'' rate of discovery. For instance, at $B = 16$, there are about 0.004 QSOs per sq degree; at $B = 18.3$, this is 0.61 per sq degree, and at $B = 19.2$ the number rises to almost 5. Recall that each interval of one magnitude is a factor of about 2.5 in brightness. The distribution appears, however, to level off. At $B = 21$, there are about 80 per sq degree, but this number is uncertain at present by as much as 30%. Finally, a useful limit is provided by the cumulative count. For $B \leq 22.65$, there are between 70 and 140 QSOs per sq degree. The magnitudes at which these various values are given are the result of analysis of the completeness of the sample: estimating how many objects are likely to have been missed at a given redshift and magnitude. These are referred to as "complete" surveys.

A rough idea of what these statistics translate to is given by a comparison of the space densities for active galaxies and QSOs. For Sy 1 with $v_{rad} < 1000$ km sec^{-1}, the space density is $\geq 10^{-4}$ Mpc^{-3}. This is roughly 1000 times the local QSO density at $M_v < -24$, and about ten times the density at $z = 2$. Some evolution appears required; the QSOs have probably not been a constant fraction of all galaxies for all of cosmic time. In other words, the early universe had a larger population of active galaxies at high luminosity than observed today.

Quasar evolution has been discussed in terms of two extreme models. In one, the luminosity function is assumed to vary with time, called *luminosity evolution. Density evolution* is the model that assumes that the number density has varied simply due to the decrease in the space density of galaxies from cosmic expansion and that the intrinsic luminosity function is invariant with time. In the case of density evolution, one assumes that the luminosity function, per comoving volume element of the universe, is given by

$$\Phi(M, z) = \Phi(M, 0)\rho(z)$$

where $\Phi(M, 0)$ is the present function, and $\rho(z)$ is a redshift (cosmic time) dependent density. That is, one assumes that the luminosity function is perfectly stable with time and only the relative density is changing due to the universal expansion. This function is known from cosmological models, and so a complete sampling of the population to some fixed magnitude limit, allowing for the redshift of the object, immediately allows one to determine the luminosity function of the present epoch. In contrast, for luminosity evolution, one assumes that

$$\Phi[M + \delta M(z), z] = \Phi(M, 0)$$

where $\delta M(z)$ is some luminosity evolution law for the object as a function of redshift. All objects are assumed to scale by the same law, so that for any quasar the luminosity is a function of redshift as well as the density. This means that the shape of the luminosity distribution is invariant with time, but the brightness of any single galaxy will change as time passes so that one point may move within the distribution. The observations to date almost certainly require some combination of these, although the form of the law is still a matter of debate.

XIII. Theoretical Interpretations

The models for the structure of the emitting regions of quasars depend on the mechanism chosen for the ionization and, to some extent, on the phenomenon being modeled. For instance, in the case of the emission lines, the primary interest has focused on the line ratios and on the continuum morphology without too much attention being paid directly to the variability of the lines. However, models have been constrained by those variations.

A. Structure of the Emission Line Region

The density of the medium can also be determined from using the fact that in the broad line region, there is a component due to the intercombination lines, such as O III] and C III]. These lines have small enough transition probabilities that their upper states will be collisionally depopulated if there is a substantial density, specifically, higher than

$$A_{ul}/n_e C_{ul} < 1 \qquad (14)$$

where A_{ul} is the Einstein transition probability for the state and C_{ul} is the collision rate. The subscripts u and l indicate the upper and lower states, respectively. This assumes collisional excitation and subsequent radiative de-excitation. Typically, the densities are $n_e \leq 10^{10}$ cm^{-3}. For the forbidden lines, like [O III], there is an even smaller critical electron density for damping, $n_e \leq 10^6$ cm^{-3}. Therefore, the reason for separating out the NLR (associated with the forbidden lines) and the BLR (associated more or less with the permitted lines) is justified on the grounds of the critical densities involved. The densities for the permitted lines, as determined from photoionization models, are typically higher than either of these limits (see below).

B. Central Engine

The size of the emitting region can be determined from the consideration of resolution. In no case does the emitting region BLR appear to be resolved, and the observations in the radio set severe constraints on the angular sizes. The region must be ≪1 pc in size. This means that about 10^{47} ergs sec^{-1} must be emitted in a region about the size of the distance between stars in the disk of the galaxy. To constrain the mass within this region, it is assumed that the line widths observed arise from orbital motion. Therefore, from the fact that the matter is gravitationally bound to the central source, a limit can be obtained on the mass of

$$M = \frac{R}{G} (\Delta v)^2 \qquad (15)$$

where M is the mass of the central object and G the gravitational constant. The model for production of the observed luminosity of the QSOs is accretion onto a compact object, most likely a black hole. The radius of the hole (the Schwarzschild radius, R_*) is given by

$$R_* = \frac{2GM}{c^2} \qquad (16)$$

The luminosity, if the accretion efficiency is ε, is given by

$$L \approx \varepsilon \frac{GM}{R} \dot{M} \qquad (17)$$

where $\dot{M}$ is the mass accretion rate. Models for most gravitational accretion yield efficiencies of about 0.01 to 0.1. There is a maximum rate at which matter can be fed into a gravitational potential in the presence of a radiation field. This is because the increase in the luminosity of the central source on having a higher rate of mass accretion will produce sufficient radiation pressure that the flow will become unstable. This is the Eddington limit for the luminosity, and is given as

$$L_{Edd} = \frac{4Gc}{\sigma_e} M \qquad (18)$$

where σ_e is the opacity due to electron scattering (0.4 cm^2 g^{-1}). Thus, for a given rate of feeding of the central source, the mass must be at least

$$M \geq \varepsilon \dot{M} (\Delta v)^2 \frac{\sigma_e}{4Gc} \qquad (19)$$

It is likely that the hole is surrounded by a turbulent accretion disk of large optical thickness and having a relativistic inner region dominated by compton processes. The observed line variations also make use of this picture in deriving the mass of the central object.

In the current picture, the permitted lines arise in the accretion disk and the denser gas surrounding it. The line widths should therefore be due to orbital motion. The narrow line region is a more tenuous plasma surrounding the entire system, a picture consistent with the observations to date of spectrum variability. The broad absorption line systems appear to display large-scale winds in the nuclear region as well. These may be connected with matter being driven radiatively from the disk of the surrounding gas by radiation pressure. The implication is that super-Eddington accretion rates will produce such effects. Such phenomena have been observed in galactic stars and may occur on the scale of the central region in AGNs.

The radio jets observed in AGNs are similar to those observed in radio galaxies in general. They are highly collimated and appear to be formed very near to the central object. Current models place the formation region in the funnel formed by the inner boundary of the accretion disk, where radiative acceleration appears capable of producing large-scale flows at high velocity and collimation. It is probable that some form of super-Eddington accretion is necessary and that the inner regions of the disks must be thicker than the outer, possibly self-gravitating, portions (where the emission lines are formed).

That the central object is likely a massive black hole does not appear to be a matter for serious debate at present. Also likely is that an accretion disk is responsible for the permitted lines and at least a portion of the continuum. However, the origins of this situation are not well understood presently, nor are the details of the mechanism responsible for the infall of the large amounts of gas required to explain the observed luminosities (on the order of 1 solar mass per yr). Rather, because it is more directly accessible from observation, the structure of the gas surrounding the central source has been more critically studied. Since the information about the abundances in this material and of the rate of supply of the central object all derive from the analysis of the line spectrum, it is important to consider the assumptions that go into the various models for radiative transfer in this medium.

C. CONTINUUM

The power law continua, which are likely of nonthermal origin, are consistent with the synchrotron emission from relativistic electrons in a weak magnetic field (about order 10^{-4} to 10^{-3} G). The ultraviolet properties appear in part due to emission from the accretion disk presumed to exist about the central collapsed object. The luminosity of the disk is generated by the viscous dissipation of orbital energy of the accreting material, much as observed in binary stars that are exchanging matter. The power law dependence of the ultraviolet spectrum is due to the power law dependence of the temperature as a function of radius in the disk. This model can also account for the change in the slope of the continuum in the UV (nearer to thermal than other parts of the spectrum). The radio emission requires some form of re-acceleration of particles to compensate for synchrotron losses, a problem that also holds for radio jets in other types of galaxies.

D. PHOTOIONIZATION MODELS

Irrespective of the ultimate source of the continuum radiation, the appearance of the spectrum of AGNs appears to be a direct result of the intense radiation field on the gas of the surrounding medium. The fact that there appear to be two distinct regions, the BLR and NLR, complicates the models considerably. Ionization of the gas is dominated by the radiation in the Lyman continuum (LyC) above 912 Å. The primary opacity in this part of the continuum is due to hydrogen. The formation of a sphere of ionized gas is studied as follows. One counts the number of ionizing photons arriving per second into the medium,

$$Q(H) = \int_{\nu_{LyC}}^{\infty} \frac{L_\nu}{h\nu} \, d\nu \qquad (20)$$

where L_ν is the luminosity of the illuminating continuum in the LyC; this is balanced, in equilibrium, by the number of recombinations so that the radius of the sphere is given by

$$R_{H\,II} = \left(\frac{3Q(H)}{4\pi n_e^2}\right)^{1/3} \alpha(T)^{-1/3} \qquad (21)$$

where $\alpha(T)$ is the recombination coefficient. It is assumed, in this equation, that every photon in the LyC is absorbed and ionizes the atoms and that the local electron density is entirely due to the ion formation. The medium is called ionization bounded if the total mass to be ionized exceeds the number of available photons; otherwise it is referred to as density bounded.

In normal galactic H II regions (ionized hydrogen regions or emission nebulae), the density bounded is a good approximation. However, in

AGNs, there is considerable matter emitting in the neutral and once-ionized state from atoms whose ionization potentials are considerably lower than that of H. The medium is likely to be ionization bounded. Calculations indicate that the size of the ionized layers in clouds responsible for the production of the emission lines are about 10^{11} cm, a few solar radii, which is only a small fraction of the BLR. For a completely optically thin medium, the cascade of electrons on recombining to the ground state of H produces the observed lines, whose ratios are fixed by the relative branching probabilities for each level. That is, there will be a fixed value for the ratio of the lines of a given series, say the Balmer lines, to each other. This approximation, called case A, gives $H\alpha/H\beta/H\gamma = 3.3:1:0.5$, where the intensity of $H\beta$ is normalized to 1.0 and the $Ly\alpha/H\beta$ ratio is about 33. In case B, the Lyman continuum is assumed to be optically thick and so there is detailed balance among the lines of the Lyman series with a value of $Ly\alpha/H\beta$ between about 20 and 30.

If the $Ly\alpha$ is produced in part by the degradation of LyC photons into lines, then there should be a substantial absorption seen at the LyC edge at 912 Å. This is because essentially all of the photons should be converted eventually into $Ly\alpha$. However, strong absorption edges are generally not observed, and the intensity of the $Ly\alpha$ line is such that it would require that every photon from the central source would eventually be converted to $Ly\alpha$. Therefore, the gas must be optically thick to both $Ly\alpha$ and to LyC to produce the observed line ratios. The original mechanism for accomplishing this, the destruction of $Ly\alpha$ photons by dust absorption, can now be largely ruled out. Dust can be evaluated as the causitive agent for reddening the Lyman lines by looking at the IR emissivity of the AGNs. It is generally found that QSOs are not strong IR sources and that the Sy 1 have a more substantial emissivity at IR wavelengths. Optical depth effects in the gaseous component, rather than some solid phase, must be responsible for the observed spectral characteristics. The escape of $Ly\alpha$ due to velocity gradients plays a role, but trapping of the Ly line photons is the primary agent for the reduction of the Lyman to Balmer series ratios. The explanation for the lack of the Lyman discontinuity is that the clouds are extremely optically thick but small, covering only a small percentage of the central continuum source.

With all of these caveats in mind, the mass of emitting gas can be estimated under the assumption that the lines are formed by recombination. It is found to be, at least

$$M \geq 1.2 \times 10^{-23} L_{Ly\alpha} n^{-1} N(H) \qquad (22)$$

where $N(H)$ is the column density of neutral hydrogen, n is the number density, and $L_{Ly\alpha}$ is the Lyα luminosity.

The covering factor for the nuclear region can also be evaluated using X-ray observations to determine the hydrogen column density. The input spectra from the nuclear source appear to be simple power laws. Thus, a spectral turnover due to hydrogen absorption provides a measure of the fraction of covering of the central source by the line emitting gas. These fractions in general are low, less than 50%, and indicate that the broad line region and the NLR as well are extended structures but with low filling factors.

The ionization parameter, U, is defined to be $Q(H)/4\pi r^2 n_e$. This is dimensionally a velocity, which measures the rate at which a region will have an ionization front moving into it (an ionization velocity). The typical values inferred for galactic nuclei are $\log(U/c) = -2$, compared to the sound speed of about 10^6 cm sec^{-1} for $T = 10^4$ K. The ratio of U to the sound speed, which is like an ionization front Mach number, is the inverse fractional thickness of the ionization zone. Recent work argues that the Baldwin effect is a result of the ionization parameter U being a function of the total luminosity of the central source. Photoionization calculations give $U \approx L^{-1/4}$, although it is the anticorrelation rather than the precise value of the exponent that is most important. Since the strength of the C IV line depends on U, there will be an anticorrelation with the luminosity of the line strength. This model also predicts that the luminosity of the continuum at 1450 Å is essentially linearly correlated with the Lyα intensity.

The use of these models in explaining the observations of luminosity–line strength correlations extends beyond merely describing the physics of the nuclear region. It helps explain why there are several good luminosity indicators in AGN spectra. The emission lines therefore provide a check on the determinations of the luminosity function. This, in turn, makes the active galaxies useful cosmological probes by virtue of this ability to independently determine the absolute magnitude of the object, in addition to the redshift-determined value.

XIV. Concluding Remarks

The QSOs are extremely ancient objects, as indicated by their redshifts, representing stages not very far from the formation of the parent galaxies in which they reside. Material appears to have already been processed in stars by the time it falls into, and is consequently excited by, the nuclear engine. This provides evidence of the extreme efficiency with which stars are formed in galaxies. The observed abundances, however, remain a puzzle. These observations make the quasars important as tools for studying the early history of host galaxies and their constituent stellar populations. That one can also study, through their absorption spectra, the abundances in the halos of intervening galaxies makes them indispensible as probes of cosmological evolution of the galaxian population.

That there is a connection between the Sy 1 and QSO phenomena also argues that there may be some link in time between the two and that whatever feeds the central engine in quasars is no longer operating. The recent observations of the center of the Milky Way, which reveal evidence for a compact, probably relativistic, central body on the order of $10^7 M_\odot$ and a central nonthermal source, show that the formation of black holes may not be unusual for galaxies. There are several others in the vicinity, such as M 81, that have compact central bodies; it appears that the QSOs may thus be the galaxies in which the most massive and most vigorously fed central beasts are found.

Many of the planned observations of the Hubble Space Telescope are connected with quasar and galactic nuclear activity. It is expected that within the next decade regular observations will be possible with large millimeter single antennas and interferometers, as well as the very long baseline array (VLBA) and continued improvements of VLBI methods. Several imaging space infrared satellites are being built or discussed, and the next generation of X-ray satellites is currently being considered. In short, this article has been written at the opening of a new epoch in quasar and active galaxy astrophysics.

The quasars therefore represent among the best physical laboratories available for the study of the physics of the early universe, the early stages of galaxy evolution, and the processes that can occur in the vicinity of compact objects of high mass. As is now clear after almost three decades of research, they represent the most extreme physical environments observationally available.

BIBLIOGRAPHY

Angel, J. R. P., and Stockman, H. S. (1980). Optical and infrared polarization of active extragalactic objects, *Ann. Rev. Astr. Ap.* **18,** 321.

Balick, B., and Heckman, T. M. (1982). Extranuclear clues to the origin and evolution of activity in galaxies, *Ann. Rev. Astr. Ap.* **20,** 431.

Begelman, M., Blandford, R. D., and Rees, M. (1984). Theory of extragalactic radio sources, *Rev. Mod. Phys.* **56,** 255.

Bridle, A. H., and Perley, R. A. (1984). Extragalactic radio jets, *Ann. Rev. Astron. Ap.* **22,** 319.

Burbidge, E. M. (1967). Quasi-stellar object, *Ann. Rev. Astron. Ap.* **5,** 399.

Davidson, K., and Netzer, H. (1979). The emission lines of quasars and similar objects, *Rev. Mod. Phys.* **51,** 715.

de Young, D. S. (1976). Extended extragalactic radio sources, *Ann. Rev. Astr. Ap.* **14,** 447.

Hazard, C., and Mitton, S. (Eds.) (1979). "Active Galactic Nuclei." Cambridge Univ. Press, Cambridge.

Miley, G. K. (1980). The structure of extended extragalactic radio sources, *Ann. Rev. Astron. Ap.* **18,** 165.

Pacholczyk, A. G. (1970). "Radio Astrophysics." W. H. Freeman, San Francisco.

Robinson, I., Schild, A., and Schucking, E. L. (Eds.) (1965). "Quasi-Stellar Sources and Gravitational Collapse." Univ. of Chicago Press, Chicago.

Schmidt, M. (1969). Quasistellar objects, *Ann. Rev. Astron. Ap.* **7,** 527.

Strittmatter, P. A., and Williams, R. E. (1976). The line spectra of quasi-stellar objects, *Ann. Rev. Astr. Ap.* **14,** 307.

Ulrich, M. H. (1981). 3C 273—A survey of recent results, *Space Sci. Rev.* **28,** 89.

Ulrich, M. H. (1984). Line variability in active nuclei and the structure of the broad-line region, *Ann. N. Y. Acad. Sci.* **422,** 291.

Veron-Cetty, M. P., and Veron, P. (1985). "A Catalogue of Quasars and Active Nuclei," 2nd ed. European Southern Observatory, Munich, Germany.

Weedman, D. W. (1977). Seyfert galaxies, *Ann. Rev. Astr. Ap.* **15,** 69.

RADIO ASTRONOMY

Steven N. Shore   *New Mexico Institute of Mining and Technology and DEMIRM, Observatoire de Meudon and Radioastronomie, Ecole Normale Superieure*

GLOSSARY

Brightness temperature: Temperature that a source of thermal radiation would have to obtain to produce, at a given frequency and bandwidth, the same flux as an observed radio source.

Faraday effect: Rotation of the plane of polarization in a dispersive circularly polarized medium.

H I emission (21-cm neutral hydrogen line): Emission from the spin transition in the ground state of neutral hydrogen, occurring at 1420 MHz.

H II regions: Regions of ionized hydrogen, produced by thermal ionization of low-density gas.

Jansky: Unit of radio flux, given in multiples of 10^{-26} W m^{-2} Hz^{-1}.

Recombination lines: High-frequency atomic emission lines from very high principal quantum numbers, produced by the cascade of electrons recombining in ionized regions with ions.

Synchrotron radiation: Radiation emitted by relativistically moving electrons. Nonthermal form of radiation in which the radiation is due to the strength of the magnetic field and the effect it has on electron trajectories. The nonrelativistic version of the emission is called gyrosynchrotron or cyclotron emission.

u–v Plane: The transform of the interferometer baselines projected on the sky. This is the basic data for the Fourier transform used in inverting the radio telescope array measurements.

Very Large Array (VLA): A Y-shaped array of 27 independent raido telescopes located about 50 miles from Socorro, New Mexico, on the plains of San Agustin at an altitude of about 7000 ft.

The array can grow or shrink between 4 standard configurations. Each antenna has a 25-m aperture.

Very long baseline interferometry (VLBI): Intercontinental interferometry using independent radio telescopes. Signals are correlated after data are taken independently at each site. The array form of this is currently under construction by the National Radio Astronomy Observatory (NRAO) and is called the very long baseline array.

The history of radio astronomy can be viewed as a chain of interesting accidents, with only a few predictions. The first observation of cosmic radio waves was made in 1932 by Karl Jansky's accidental detection of galactic radio noise. At the time, Jansky was performing a series of experimental determinations of noise sources for transcontinental radio communication. Using a steerable antenna, he mapped out the galactic plane and determined that the Galaxy itself is a source of low-frequency radiation. With an antenna pattern having a half-power beamwidth of about 30°, he was unable to resolve discrete sources.

The next important step was taken in the late 1930s by Grote Reber, an amateur who built a steerable parabolic antenna with a sensitive receiver and was able, using an alt-azimuth mount, to map the Galaxy at 1 m. He also can be credited with the discovery of the first discrete radio sources, although the resolution of his antenna was decidedly poor.

At the time of these observations, the radio emission was interpreted as thermal radiation from the diffuse interstellar medium. Shklovskii, shortly after the identification of Tau A with the Crab Nebula (M 1 = NGC 1952), the remnant of the supernova of 1054, suggested that the radiation results from high-energy electrons spiraling in weak magnetic fields in the diffuse gas. This emission, called synchrotron radiation after the laboratory accelerators in which it was first observed, is highly polarized and nonthermal in origin. The detection of the diffuse emission in external galaxies, produced by cosmic ray electrons in the interstellar medium, confirms this explanation and serves as a sensitive probe of the conditions of the medium. It is also one of the few ways to study the magnetic fields of external galaxies.

Solar radio emission was discovered accidentally by Hey and several others during the solar activity maxima of 1942–1943. This has demonstrated that magnetic phenomena, normally associated with sunspot activity, account for an appreciable emission from the sun. It is evidence for many detailed plasma processes, which are the likely sources for high-energy particles, such as magnetic field reconnection. The radio emission from active regions is still one of the best ways to study acceleration processes.

Interferometry was invented as a technique of radio measurement during the late 1940s and was exploited heavily in the following decade to measure the positions and structures of discrete radio sources. Aperture synthesis instruments have been the product of the past decade, with the construction of the Westerbork Radio Synthesis Telescope in the Netherlands, the Very Large Array in the United States, and the Australian Array. The extension of interferometric arrays to millimeter wavelengths is still under way, with a few interferometers having been constructed in Japan and the United States and one under construction in France.

More than any other field of observational astronomy, radio astronomy is a technical field where many of the achievements and limitations of the science are driven by the details of the instruments. In order to properly set the stage for many of the discoveries discussed elsewhere in this encyclopedia, this article will concentrate on the methods by which radio observations are made and some of the emission mechanisms responsible for such signals. The discussion of the results of radio observations of specific objects can be found under specific articles on those objects. [See GALACTIC STRUCTURE AND EVOLUTION; QUASARS; SOLAR PHYSICS.]

I. Radio Telescopes

Radio astronomers typically work in the wavelength range of 1 cm to 100 m, limited by the water vapor of the troposphere and stratosphere on the short wavelength, as well as a difference in receivers required for operation, and by the iono-

spheric opacity at the long wavelength end. Millimeter astronomers typically concentrate on 1 to 3 mm, again limited primarily by atmospheric water vapor. Most radio surveys have been conducted between about 400 and 1500 MHz.

A. SINGLE APERTURES

The angular resolving power of a telescope is related to both the wavelength at which the light is observed and the aperture of the instrument (independent of the focal length of the instrument) by

$$\Delta\theta \approx \frac{\lambda}{D} \qquad (1)$$

where D is the aperture of the instrument and λ is the wavelength, the so-called Rayleigh criterion. The size of the telescope must therefore be of order $2 \times 10^5 \lambda$ in order to achieve a resolution of one arc second or better. For optical telescopes, an aperture of 10 cm will suffice; for radio wavelengths of order 10 cm, a single antenna would need to be about 20 km across to obtain the same resolution. It is for this reason that single dish observations generally achieve low resolution and produce a smeared image of the sky. The largest steerable single radio telescope is the 300-m telescope, operated by the Max Planck Institute for Radio Astronomy near Bonn, Federal Republic of Germany.

Several clever artifices have been contrived to obtain high resolution without the constraint of mobility. The Nancy, France, and Ohio State radio telescopes are fixed parabolic reflectors with a tilting secondary reflector that produce elongated vertical beams and use the earth's rotation to carry the sources across the field of view. The Arecibo radio telescope is a vertically pointing spherical bowl, suspended above the caldera of an extinct Puerto Rican volcano, which uses a tilting feed to scan the aperture between tight declination limits. It also uses the earth's rotation for scanning in right ascension. None of these instruments, however, can obtain resolutions better than about 10 arcsec at diffraction limit.

Since the beam of a single antenna is quite large, subtending a solid angle Ω, the measured signal is actually the intensity of the source, I, integrated over the beam of the telescope, or $\int I \, d\Omega$; thus, rather than measuring the intensity, one obtains a flux value for a point source that may not be due to a single object in the beam of the instrument.

Most millimeter observations are still performed with single antennas, the largest of which are the 30-m IRAM telescope in Spain and the NRAO 12-m telescope at Kitt Peak in Arizona. In the past few years, however, arrays of millimeter telescopes have been constructed at Nobyama in Japan and Hat Creek in California. A three-element system of 50-m telescopes is under construction in the Alps as a part of the Institut Radioastronomie Millimetrique (IRAM) project.

All radio telescopes measure signals at long wavelength, where the Rayleigh–Jeans limit of the blackbody distribution can be applied. The intensity of the measured radiation, S_ν is compared with the temperature that a thermal (blackbody) radiator would have to have to radiate the same intensity at frequency ν:

$$S_\nu = \frac{2kT}{c^2} \nu^2 \qquad (2)$$

Consequently, one often speaks of the antenna temperature rather than the power received by the system from a source or of the system temperature for the equivalent noise power produced by all effects in the sky or system electronics other than the celestial source. First of all, the system temperature is given by

$$T_{sys} = T_{3K} + T_{GB} + T_{ground}$$
$$+ T_{sky} + T_{loss} + T_{cal} + T_{rx} \qquad (3)$$

where T_{3K} and T_{GB} are contributed by the 3-K background radiation and the galactic diffuse radio emission, respectively; T_{ground} is the spillover from the ground thermal emission; T_{sky} is the emission from atmospheric and ionospheric sources; T_{loss} is the noise from the losses in the feed and line; T_{cal} is the injected noise from the internal calibration; and T_{rx} is contributed by the receiver. Thus, the power radiated by the system is

$$P_{noise} = GkT_{sys} \Delta\nu \qquad (4)$$

where G is the gain and k is the Boltzmann constant. The bandwidth is $\Delta\nu$. The antenna temperature is given by a similar expression for the power from the source, presumably with the same system gain (this may not be true for the strongest sources like the sun, but it is usually true for generally weak astronomical radiators).

Assuming that the antenna has a geometric area A and an efficiency η, which is determined by the precise measurement of the beam characteristics, the sensitivity of the telescope is given by the minimum flux that can be observed by the antenna and is

$$\Delta S_{RMS} = C \frac{2kT_{sys}}{\lambda^2 (\Delta\nu \, \Delta t)^{1/2}} (\eta AG)^{-1} \qquad (5)$$

The integration time is Δt and the bandwidth of the entire system is $\Delta\nu$. The factor C depends on the number of elements (and is unity for a single filled aperture).

While for many applications, the wide beam of single apertures is not a debilitating limitation, such as in millimeter observations, for longer wavelength, one requires much larger telescopes to achieve reasonable angular resolutions, which forces the change to some form of interferometric detection system.

B. INTERFEROMETERS AND APERTURE SYNTHESIS

Suppose that two antennas observe a source simultaneously. Assume that the source is very distant, so that the incoming electromagnetic signal is a plane wave. Then the two antennas observe both amplitude and phase information. If they are separated by a distance b, the length of the baseline, the antennas will receive the wave at slightly different times. If the signals from the two elements are then combined, the resulting wave will either constructively or destructively interfere, depending on the phase shift. By the observation of fringing, as the source moves over the elements, one can determine the angular size of the source and its direction in the sky. This is the principle of the interferometer, the same idea that is used in optical systems, such as the Michelson stellar interferometer.

For two elements separated by a distance b, the delay in the arrival time of the wave will be

$$\tau = \frac{b}{c}\mathbf{b} \cdot \mathbf{n} \qquad (6)$$

where $\mathbf{n}$ and $\mathbf{b}$ are, respectively, the unit vector along the source direction and along the baseline. As the source crosses the interferometer, the phase shift will be given by $\phi = 2\pi\nu\tau$, where ν is the frequency at which the observations are made.

The fringes move at a rate that is directly related to the length of the baseline. As the source changes position relative to the center of the baseline, one gets a fringe frequency that grows proportionally to $(b/\lambda)\mathbf{b} \cdot \mathbf{s}$. For small interferometers, operating at a wavelength of order 10 cm, this rate can be managed—it will be of order 1 MHz for baselines of order 1 km. The antennas can be steered in order to track a position on the sky, but the different portions of the sky onto which the baseline is projected will change orientation relative to the telescopes. For larger baselines, however, the fringe modulation becomes too fast for most systems to handle. In order to manage the amplification and source tracking better, the incoming radio frequency signal is mixed, via a local oscillator in the feed system of the antenna, to two sidebands, at $|\nu_{rf} \pm \nu_{LO}|$, where ν_{rf} is the radio frequency and ν_{LO} is the frequency of the local oscillator. It is the lower frequency that is most easily amplified, and therefore, one often uses single-sideband receiver systems in radio astronomical observations.

In order to amplify the signal, the incoming radio frequency (rf) is mixed with a local oscillator at a frequency ν_0 to produce two so-called intermediate frequencies (IF) at $\nu + \nu_0$. In single-sideband receivers, a filter is placed in the system after the mixer to choose only one of these IFs, which is then used as the basis for the correlation. In general, the lower sideband is employed since the amplifiers can be made most efficiently at the lower frequency. The fringe rate is consequently also slowed down, increasing the accuracy with which the fringe center can be tracked. This signal is now ready to be combined with other antenna outputs in the correlator, which produces the final datum from the baseline:

$$R(t) = B(\tau(t) - \tau_D)$$
$$\times \exp(i\omega_0[\tau(t) - \tau_D] - \phi(t)) \qquad (7)$$

where $B(x) = \int \alpha(\Delta\omega)e^{ix\Delta\omega} \, d\Delta\omega$ is the fringe washing function, depending on $\alpha(\Delta\omega)$, which is the bandwidth of the frequency response for the sideband. The time delay τ_D is added to the system to compensate for the time delay in transferring the signal from each antenna to the correlator, a quantity that depends on the physical length of the signal path and not simply on the separation of the antennas. The delay $\tau(t)$ is that given by the phase of the wave intercepted by each antenna. As a result of the mixing of the rf with the local oscillator to form the IF signal, the fringe rate is more reasonable and easier to track. Without this intermediate step, the rate of oscillation of the signal would render the source fringes invisible in general.

The basic element of an interferometer is the baseline, not the individual antennas. These baselines, when part of an array, become the equivalent of different portions of a single mirror when seen at the focus of a telescope. The phase shift between different antennas being detected at a correlator is like the reflections from the different portions of a mirror being combined at the focus with all of the phase and amplitude information being preserved. As in Michelson interferometry or speckle interferometry, the coherence time of the signal, the time over which this information is "remembered" by the wave seen at the different antennas, is the longest time scale for integration of the radiation. Generally, this is of order milliseconds for optical photons but can be considerably longer (up to minutes) for radio observations. The signals are recorded for each baseline at

some fixed sampling rate, and then averaged together for longer intervals, which are short compared with the rate of motion of the source across the beam of a single telescope. These intensities, recorded for each baseline, are called the visibilities and are the time correlated signals received by a pair of antennas. For an array of N antennas, there are $N(N - 1)/2$ such combinations. For instance, the Very Large Array, or VLA, operated by NRAO on the plains of San Agustin near Socorro, New Mexico, has 27 antennas, distributed 9 at a time along each of three arms of a north-oriented Y-shaped array; these provide 351 individual baselines or elementary interferometers with which to form visibilities. Note that the density of projected baselines on the sky increases rapidly with the number of antennas used, like N_2, so that the sky coverage can be economically increased by increasing even by a small factor the number of elements in an array. The VLA and other arrays also have the feature that the relative positions of the antennas in the array can be altered by moving the telescopes to different stations, thereby allowing the observer to select the resolution of the instrument and also the density of sky coverage. In effect, both the earth rotation synthesis and the variation of the baselines allow for very good coverage of the sky.

The basic principle of aperture synthesis is that once sufficiently dense coverage of the sky has been achieved, either by virtue of the number of elements in the interferometric array or by the time of observation producing a large enough track on the sky, one can invert the problem to obtain the brightness distribution on the sky. Most of the research in radio astronomical imaging currently centers on this process. Stated briefly, the problem is as follows. Given that one never sees the complete distribution of intensity on the sky, except when using a single element, and given that there may be confusing sources in the field and "holes" in the image because of the array configuration, how does the signal (visibility) observed along a set of baselines correspond to the image of the source in the sky?

A source can be distributed about the center of the field, toward which the array is pointing. This is also the tracking center for an array of steerable antennas. This point is taken to be the center of the observation, whether there is a source located at that point or not. The nominal displacement of any source from this point is taken to be a small value, given the size of the beam of an individual antenna in an array, so that the direction to any source, $\mathbf{s}$, is given by

$$\mathbf{b} \cdot \mathbf{s} = \mathbf{b} \cdot \mathbf{s_0} + ux + vy \qquad (8)$$

where (x, y) are the displacements in right ascension and declination on the sky from the direction of the phase center, and (u, v) are the projected baselines in those directions in wavelength units. The signals coming from these displaced positions on the sky modulate the amplitude of the fringes produced from the center of the field, since they are effectively coming from off-axis sources (perhaps part of the same object that is being viewed). As the source rotates across the array, the fringes produced by the field vary quite rapidly (recall that the rate is dependent on the frequency at which the observation is taken and so will be quite rapid). Therefore, the signals are combined at the central correlator only after corrections for the delays across the array have been normalized to the "phase center" of the image. This is the same as the optical procedure of "fringe tracking," which means that the only modulation in the fringes is due to structural information in the source and not just the overall drift of the source over the antennas. In general, the modulation of the intensity from the source will be slower because of the structure than the variation of the fringe rate from the phase center of the field.

The (u, v) plane maps the projections of the baselines in equatorial coordinates onto the plane of the sky as seen by the array, taking the hour angle (right ascension) and latitude of the array into account. The projected paths are ellipses given by

$$u^2 + \left(\frac{v - l_z \cos \delta_0}{\sin \delta_0} \right)^2 = l_x^2 + l_y^2 \qquad (9)$$

where $l_j = L_j/\lambda$ is the length of the baselines in the equatorial system of celestial coordinates in units of the wavelength of the observation λ. Here, δ_0 is the declination of the phase center for the observation. Thus, for a source on the equator, the tracks in the (u, v) plane are straight lines, while for the pole they are circles. The more telescopes one has in the array, the denser the coverage on the sky of the projected baselines.

The visibility seen along a baseline with projection (u, v) is given by the integral of the celestial sphere seen by the array since each portion of the source contributes a signal, as seen by the array, at slightly different delays. The beam pattern of the array must also be taken into account, so that the fringes are the result of the convolution of the source with the beam of an antenna when averaged over the solid angle subtended by the array. Therefore, the signal seen along any baseline is given by

$$V(u, v) = \iint V'(u - u', v - v')S(u', v')\, du'\, dv'$$
$$= \int_{-\infty}^{\infty}\int_{-\infty}^{\infty} I(x, y)e^{2\pi i(ux+vy)}\, dx\, dy \qquad (10)$$

where $\dot{V}'$ is the visibility without the instrumental profile included, and S is the instrumental beam pattern in the (u, v) plane. In other words, the appearance of the source along the baselines is given by the Fourier transform of the source on the sky. Having $V(u, v)$ for a time-ordered set of observations by the array, the intensity of the source on the sky is recovered by the inverse Fourier transform, now taken over the projected baselines:

$$I(x, y) = \int dv \int V(u, v)e^{-2\pi i(ux+vy)}\, du\, dv \qquad (11)$$

However, one rarely has regular sampling or completely sampled images with all of the antennas seeing everything that fills the holes in the array. Consequently, there will be fringes left in the image that will be the effective lobes of the array on the sky. Worse still, if there are antennas that are not functioning during the observation, there will be additional holes in the "map." Most of the effort directed toward radio astronomical image processing is spent in developing methods to fill these holes and remove the fringes still left in the map.

The detector bandwidth, $\Delta\nu$, affects the resolution of the radio telescope. In the determination of the phase shift across an array, the bandwidth acts like an uncertainty in the baseline, producing a smearing of the image. In effect, coherence can be lost across the bandwidth of the receiver. The trade-off is that, for each antenna, the bandwidth also affects the minimum detectable signal for a fixed integration time, so that broad bandwidths allow for higher velocity ranges and higher sensitivities. This is one of the key design considerations in the construction of interferometers.

C. VERY LONG BASELINE INTERFEROMETRY

Very long baseline interferometry, or VLBI, is the extension of the technique of interferometry to larger baselines, of order the size of a continent. Distant radio telescopes can be linked either by direct satellite access or by using atomic clocks, synchronized at a central location and then used as time markers at the individual sites. The signals are correlated some time after the observations have been made rather than in the real-time mode used in aperture synthesis arrays. The highest resolutions that have been achieved are of order several milliarcseconds. The most serious difficulties associated with VLBI methods arise from the relatively small number of antennas associated with a network performing such observations and the heterogeneity of the antenna and receiver characteristics. These will be partially alleviated by the Very Long Baseline Array, or VLBA, the continental extension of the VLA. With 10 antennas located from the Canadian to Mexican borders, and from the Caribbean to Hawaii, this array, under construction by NRAO, provides the same resolution normally obtained with VLBI methods but with identical receivers and antennas at each location, compatible with those of the VLA (which is the "low-resolution" subarray in this collection).

The small number of antennas in VLBI observations means that image reconstruction is somewhat more model or algorithm dependent. Nevertheless, reliable, repeatable maps can be obtained at the highest possible resolution for bright radio sources.

II. Emission Mechanisms

Broadly speaking, there are two classes of mechanisms that produce emission at longer than infrared wavelengths (although they may contribute to shorter wavelength emission as well). These are thermal processes, which include both line and continuum emissions from molecular, atomic, or ionized media and nonthermal continuum and line emissions.

A. LINE EMISSION

Atomic and molecular emission and absorption is the result of differences between level populations in the atom or molecule in question. Assuming that n_l and n_u are the populations of the lower and upper states, the equation for statistical equilibrium can be used to give the rate of change, with time, of these level popluations:

$$\frac{dn_u}{dt} = \sum_{j<u}^{u-1} (B_{ju}I_{ju} + C_{ju}n_e)n_j$$

$$- \sum_{j=u+1}^{\infty} (B_{uj}I_{uj} + C_{uj}n_e)n_u - \sum_{j=0}^{u-1} A_{uj}NM_u$$

$$+ \sum_{j=u+1}^{\infty} A_{ju}n_j + \sum_{j=u+1}^{\infty} C_{ju}n_j \qquad (12)$$

Here, we have included all stimulated, B_{ij}, spontaneous, A_{ij}, and collisional, C_{ij}, processes that do not ionize the system. It may require a detailed model of the system to be able to determine what the level populations will be in order to compute the emission and absorption coefficients. At radio and millimeter wavelengths, however, one is generally concerned with only two levels at a time. Therefore, the emission coefficient is defined by

$$j_\omega = n_u A_{ul} h\nu \qquad (13)$$

and the absorption coefficient is

$$\kappa_\omega = n_l B_{lu} - n_u B_{ul} \qquad (14)$$

The equation of radiative transfer then gives

$$\frac{dI_\nu}{dl} = -\kappa_\nu I_\nu + j_\nu \qquad (15)$$

as the change of the intensity along the line of sight dl to the source for radiation with a frequency ν. Generally, the effects of stimulated emission can be neglected; but in the radio and millimeter regime, there are many available transitions that can both overpopulate and be overpopulated by the various excitation mechanisms.

$$\kappa_\nu = n_l B_{lu}\left(1 - \frac{g_u B_{ul}}{g_l B_{lu}}\right)$$
$$= \kappa_0(1 - e^{-h\nu/kT}) \qquad (16)$$

The optical depth of a region is determined by the integral of the opacity along a line of sight through the medium:

$$\tau_\nu \equiv -\int \kappa_\nu n \, dl \qquad (17)$$

so that the intensity of a source at the earth is given by

$$I_\nu(\tau_\nu) = \int_0^\infty \frac{j_\nu}{\kappa_\nu} e^{-\tau_\nu(l)} \, dl \qquad (18)$$

For a simple absorption, with no reemission, this translates into

$$I_\nu(\tau_\nu) = I_\nu(0)e^{-\tau_\nu} \qquad (19)$$

and for an emitting region, which is uniform along the line of sight, as

$$I_\nu(\tau_\nu) = S_\nu(1 - e^{-\tau_\nu}) \qquad (20)$$

where S_ν is called the source function and is the ratio of the emission to absorption rates. The approximate rate of emission of an optically thin medium is, therefore, $S_\nu \tau_\nu$.

1. 21 Centimeter Emission of Neutral Hydrogen

Several classes of line emission are observed from celestial sources. The most important, for studying galactic structure, is the line from the ground state of atomic hydrogen, H I. This line occurs at 21 cm, a frequency of 1420 MHz, and was predicted by van de Hulst in 1943. It is due to a spin flip of the electron in the ground state, which has a very low transition probability, $A_{ul} \approx 10^{-15} \text{sec}^{-1}$, so on average the excited state lasts for about 10^6 yr. The path lengths through the interstellar medium are such, however, that in spite of the very small local densities (of order 0.1 to 1 cm^{-3}), the line can be observed using most radio telescopes. Typical column densities, given by $\int n \, ds$, are of order 10^{14} to 10^{22} cm^{-2} for the galactic diffuse interstellar medium. The excitation temperature needed to give rise to emission in this line is only about 100 K, so that it is not observed for most stellar sources.

The 21-cm line arises in the diffuse clouds that permeate the interstellar medium. Its importance to radio astronomy is that it has a known rest wavelength and thus can be directly used to determine the dynamics of the interstellar gas. Since the line arises from the most abundant species in the gas of any galaxy, it can be used to observe dynamics in very distant, faint sources. Because the gas occupies such a substantial fraction of the volume of a galaxy, it can also be used to measure the mass of the galaxy. In particular, the width of this line has been found to correlate with both the mass and luminosity of the parent galaxy, a relation discovered by Tully and Fisher in the late 1970s.

2. Recombination Lines

When an electron falls back onto an ion, the resulting cascade through the atomic energy levels produces a rich spectrum of emission lines. These allow direct access to the dynamics and thermal conditions of ionized gas in stellar envelopes and the interstellar medium. For He and C, one can see individual transitions (labeled by the quantum number of the upper state involved with the cascade). These become progressively more difficult to resolve as the upper state approaches the continuum, and usually only the lines from the upper hundred or so levels are accessible with current instruments. The strengths of these transitions provide a measure of the abundance of the species of interest, although this is model dependent for most ionized regions and may be affected by the details of the line formation process.

3. Molecular Emission

Two types of lines are possible from molecules. In the infrared, the lines come primarily from vibrations of atoms within the molecules, while at radio wavelengths, one usually sees the lower energy rotational transitions. For a simple molecule, having a dipole moment $\mathbf{d}_{ul}$ and a moment of inertia I, the frequency of the emission is given by

$$\nu_{j'j} = \frac{h}{4\pi I}(j'(j' - 1) - j(j - 1)) \quad (21)$$

Here, j and j' are the rotational quantum numbers of the levels. The rate of emission is therefore a function of the moment of inertia of the molecule (the rotational constant), as is the rate of emission, which varies like $A_{ul} \sim \nu^3|\mathbf{d}_{ul}|^2$.

There is little difference between the formation of molecular and atomic lines, other than the details of the collisional excitation processes. One of the main differences arises from the fact that molecules have structure. The exchange of energy between vibrational and rotational modes can occur by internal processes subsequent to excitation, and this contributes substantially to the disequilibrium of rotational level populations. The energies at which most ground and first excited state transitions take place are also typically quite close to the peak of the 3-K background radiation, so that galactic and cosmic backgrounds can produce level populations that are significantly out of equilibrium with those observed in atomic transitions.

The most important molecular species, both by virtue of their line strengths and abundances, are CO, H_2O, H_2CO, CS, and NH_3. Both ground state and excited state transitions have been studied for these species.

4. Masers

As in the laboratory, maser emission results from the overpopulation of an excited state of a molecule by radiative excitation. Briefly stated, the condition that must be fullfilled is that, in the absence of collisions,

$$\kappa_\nu = n_1 B_{1u} - n_u B_{ul} < 0 \quad (22)$$

which is possible if the radiative excitation rate from lower states and the de-excitation of upper states produces a ratio of $n_u/n_1 > 1$.

The signature of "masing" is a very high brightness temperature, highly polarized line, which is exceedingly narrow. Large velocity shifts will desaturate the maser, whose existence depends on strong radiative coupling between different parts of the radiating medium, no matter how physically distant they might be. This high line intensity can be thought of as the result of converting broader band, higher frequency radiation to narrow-band, low-frequency line emission. The OH radio lines at 1665 and 1667 MHz are especially susceptible to this pumping effect in the presence of a strong infrared continuum, such as one finds in some highly evolved red supergiants and giants. The SiO and NH_3 radio lines also show this effect. Many molecules are known to show masing transitions in cosmic sources because of the low collision rates characteristic of such environments as molecular clouds near regions of active star formation and in red giant and supergiant expanding envelopes.

The VLBI observations have been very important in studying the structure of H II regions containing masers. Actual space velocities for the regions and the time development of the maser sources have both been studied to scales of roughly a few astronomical units (AU).

B. CONTINUUM PROCESSES

1. Thermal Emission

An optically thick plasma radiates like a blackbody. This is because in an optically thick medium the photons are all absorbed and reradiated, regardless of their previous histories, according to the local gas temperature. This condition, also called local thermodynamic equilibrium, is important as a measure of the intensity of a distant source. The radio portion of the spectrum, for a hot source (T higher than a few thousand degrees Kelvin), radiates like the long wavelength approximation of the Planck function:

$$S_\nu = \frac{2kT}{c^2}\nu^2 = B_\nu(T) \qquad (23)$$

and can be used to define the brightness temperature. This is the temperature a source, observed at a frequency ν, must have to radiate at an intensity S_ν. The total flux of an object with an angular diameter θ is therefore $\int S_\nu \, d\Omega$.

With increasing frequency, the absorption coefficient of the plasma decreases and the medium becomes progressively more optically thin. At this point, the frequency dependence becomes that expected for thermal bremsstrahlung, also called free–free radiation since the radiating electrons are transiting between a continuum of free trajectories in the ionized gas as a result of collisions with ions.

For an optically thin plasma, the absorption and emission coefficients have almost the same frequency dependencies, with the result that the observed flux is nearly gray, that is, it has little dependence on frequency. The absorption coefficient is given by

$$\kappa_\nu = \frac{8Z^2e^6}{3^{3/2}m_e^2c}\left(\frac{\pi}{2}\right)^{1/2}\left(\frac{m_e}{kT}\right)^{3/2}n_e n_i \nu^{-2} <g_\nu> \qquad (24)$$

where the Gaunt factor, $<g_\nu>$ depends only weakly on frequency and is given by

$$<g_\nu> = \frac{3^{1/2}}{\pi}\ln\left[\left(\frac{2kT}{\gamma m_e}\right)^{3/2}\left(\frac{m_e}{\pi\gamma Z e^2 \nu}\right)\right] \qquad (25)$$

where γ is Euler's constant. Thus, $j_\nu = \kappa_\nu B_\nu$ is approximately frequency independent. Here, n_i and n_e are the number densities of ions and electrons, respectively (for a pure hydrogen plasma that is completely ionized, $n_e \approx n_i$), and e and m_e are the charge

and mass of the electron, respectively. The turnover of the frequency dependence of the spectrum occurs at $\tau_\nu = 1$, so the observation of the detailed spectral intensity distribution is an excellent method for the determination of the density and temperature conditions of a plasma. In the case of a complex structure, or of density and temperature variations within the source, the optically thick part of the spectrum may have a lower exponent than the canonical ν^2; such spectra are typical of stellar winds and some H II regions, where slopes of unity are often observed.

Typical temperatures of H II regions, which are regions of ionized hydrogen, are about 10^4 K, and densities range from about 10–10^6 cm^{-3}; they are typically strong thermal emitters. Because their densities are quite low, they are usually optically thin at frequencies above 1 GHz, being partially optically thick at lower frequency. For higher frequencies, of order $\kappa T/h$, the spectrum passes to the Wien part of the Planck function and declines exponentially with increasing frequency. This part of an optically thin plasma is best observed in X-ray wavelengths for high temperature regions, such as stellar coronae and X-ray emission regions of external galaxies. It is not normally observed in the radio region. For H II regions, this transition wavelength is in the infrared.

Perhaps the best example of a thermal source is the cosmic background radiation (CBR). Discovered by Penzias and Wilson in 1965, this radiation, which has a temperature of about 2.9 K, is a predominantly isotropic background resulting from the cooling of the expanding universe after the initial moments of the Big Bang.

2. Nonthermal Processes—Synchrotron Radiation

Electrons in a magnetic field move along curved trajectories. Since they are consequently being accelerated, they radiate at a frequency ω_L, the Larmor frequency, given by

$$\omega_L = eB/m_e c \qquad (26)$$

where B is the magnetic field strength. Normally, for the kinds of fields encountered in the diffuse interstellar medium and in most radio sources, of order 10^{-5} to 10^{-6} G, this radiation would be unobservable. However, a relativistic particle boosts the frequency because of its motion to $\omega \gamma_2 \omega_L$, where $\gamma = (1 - (v/c)^2)^{-1/2}$ and v is the speed. Thus, for the weak fields observed in the interstellar medium (ISM), cosmic rays with $\gamma \approx 10^4$ are capable of producing observable emission. This radiation has been well observed in terrestrial accelerators and is known as synchrotron radiation.

The shape of the synchrotron spectrum is entirely determined by the energy distribution of the emitting particles. Observations of cosmic rays near the sun show that the energy distribution follows a power law of the form $N(E) \, dE \sim E^{-s} \, dE$, so that the collective emission from these particles has a characteristic frequency dependence of $S \approx \nu^{\alpha}$ where $\alpha < 0$ for nonthermal emission. The frequency dependence is related to the energy spectrum through α, which is given by

$$\alpha = -\tfrac{1}{2}(s - 1). \qquad (27)$$

Thus, nonthermal sources are easily distinguished from thermal sources by the slope of their continua. An optically thick synchrotron source also has a different slope than a thermal plasma, being given by $\nu^{5/2}$ because of the frequency dependence of the absorption coefficient.

The high-frequency end of the spectrum is dominated by the effects of the finite radiative lifetimes of the individual particles. One way of looking at this is that the acceleration mechanism produces high energy for a single particle, but as soon as enough energy is lost, the emission from this electron falls in a wavelength region inaccessible to currently employed ground-based radio telescopes. Thus, the high-energy end of the spectrum reflects the lifetime since the last acceleration of the radiating particles. The rate of emission is given by

$$dE/dt \sim E^2 B^2 \qquad (28)$$

so that the lifetime is given by

$$t_{\text{synch}} \sim E^{-1} B^{-2} \qquad (29)$$

Thus, for the highest energy particles, the rate of emission is greatest and the lifetime shortest. This produces an increased slope that is a larger α than if the high-energy electrons are continually resupplied, a process called reacceleration.

Nonthermal sources are generally highly polarized because of the anisotropy of the emission mechanism. Radiation viewed along the field is circularly polarized, but on transmitting a magnetized medium the left circularly polarized waves (ordinary waves) are more refracted than the right circularly polarized ones (extraordinary). This results in the rotation of the plane of polarization, called the Faraday effect. The angle, in rad m^{-2}, is called the rotation measure (RM) and is given by

$$\text{RM} = \frac{\omega}{2c} \int (n_{\text{ord}} - n_{\text{ex}}) \, dl \qquad (30)$$

where n is the index of refraction and l is the path length through the plasma. The effect is most pronounced when the Larmor frequency is approximately the same as the plasma frequency, $\omega_{\text{p}} = (4\pi e^2 n_{\text{e}}/m_{\text{e}})^{1/2}$, so that

$$\text{RM} = 2.4 \times 10^4 \nu^{-2} \int n_{\text{e}} \, \mathbf{B} \cdot \mathbf{l} \, dl \qquad (31)$$

Here, $\mathbf{B} \cdot \mathbf{l}$ is the projection of the field along the line of sight. For linear polarization measurements, one observes a rotation of the plane of polarization with frequency. Thus, in order to determine the direction of the magnetic field intrinsic to a source, one must obtain the correction to the angle of polarization due to the Faraday effect. Rotation measures of order 10 to 10^3 rad m^{-2} are common in many galaxies.

Additional depolarization results from scattering of the radiation by thermal plasma in the line of sight or mixed in with the source. Both the Faraday effect and depolarization are, however, due to nonrelativistic electrons either between the observer and the source or mixed into the emitting region. An optically thick source is generally unpolarized.

The radiation from a synchrotron source has an enormous brightness temperature, far higher than a thermal source can achieve. It is limited, however, to $T_{\text{b}} < 10^{12}$ K. If a higher emissivity is achieved in a volume, the relativistic electrons scatter the radiation, by the Compton effect, into the X-ray, thereby limiting the rate of radio emission. This is a kind of self-absorption effect, called inverse Compton scattering, which depends only upon the energy density of the radiation.

III. Types of Sources

Detailed entries for most classes of radio emitters will be found under separate entries for the individual objects. This section should serve as a brief summary of the most important types of objects for which radio astronomical observations have contributed to a fundamental understanding. Individual sections are cross-referenced for convenience.

A. Solar System

Solar system sources other than the Sun include Jupiter, Saturn, and the Moon. The lunar emission is strictly thermal, being due to the heating of the surface and the proximity of the radiator. For Jupiter, the primary form of emission is due to nonthermal processes occurring connected with Io and the Jovian magnetosphere. The details of the origin of Saturn's emission are not precisely defined, but appear due in part to interaction of the ring system with the planetary magnetic field. Thermal emission from Jupiter is observed at wavelengths of 2 mm or longer, while at longer wavelengths, especially around 10 m, the bursting nonthermal radiation dominates.

B. RADIO STARS

The sun is a strong radio source, both of thermal and nonthermal emission. Its outer atmosphere, the corona, has a temperature of about 2×10^6 K and is a powerful emitter of free–free radiation. In addition, due to flaring, the sun emits nonthermal gyro-cyclotron and synchrotron radiation at wavelengths of tens of meters to millimeters. Flare radiation is highly circularly polarized, arising from the interaction regions above sunspots, where the magnetic field strengths are several gauss. Therefore, for most of the emission, the particles do not have to be terribly relativistic and γ's of 10 to 100 are probable. Flares are classified on the basis of their time development and duration. The most energetic and highly polarized, the type III bursts, have rise times of milliseconds and decay times of order 1 sec. The type I bursts are longer duration, low-frequency noise storms, which may last up to a week or more.

The sun is a powerful radio source, though, only by virtue of its proximity. In reality, one would not be able to detect solar radio emission with current technology from even the nearest stars. Radio emission is, consequently, an indication of abnormal phenomena occurring in solar-type stars. Two classes of strong radio emitters are known among stars of about the same temperature as the sun or cooler. The dMe stars, also called the UV Ceti or BY Dra stars after their prototypes, show both optical and radio flares that are several orders of magnitude more energetic than the sun. Unlike the solar case, where one strong flare may be observed per rotation, the dMe stars frequently show many such events per day. The strongest flares are circularly polarized, and show the characteristic drift from high to low frequency as the flare progresses. Recent observations with the VLA have permitted the study of flare activity on these stars with the same temporal and frequency resolution obtained for the sun, and it is probable that this area will develop rapidly as phase-array radio telescopes come on line.

Another class of the stellar radio source is the RS Canes Venaticorum stars. These stars, typically evolved red subgiants, are members of binary systems. Several are found with massive, main-sequence primaries or companions that are not radio sources. They are strong sources of nonthermal emission, which appears to arise above active regions. These star spot regions have been observed at other wavelengths, notably in the optical, ultraviolet, and X-ray. The strength of the radio emission is evidence that

flare activity is occurring on a much larger scale in these stars than on the solar-type stars.

Wolf–Rayet stars, massive stars with strong stellar winds, have also been detected as weak radio sources. Typically, they show thermal radiation due to the strengths of their stellar winds and the high emission measures that can be achieved in their envelopes. They share, with some of the luminous stars, the property that some of their wind emission may be nonthermal as well. It is conjectured that this emission arises from sychrotron radiation due to shocks in the stellar envelope accelerating particles in chaotic magnetic fields. The resultant emission may, however, also be due to stellar envelopes that are distorted and therefore not subject to the normal emission characteristics expected for spherical stellar winds.

Novae are the final class of strong, time variable, stellar radio objects. These are sources of both thermal and synchrotron emission because of the expansion of the envelope blown off of the white dwarf during the process of nuclear shell ignition.

C. PULSARS

Rotating, strongly magnetized neutron stars are detected by their radio emission. These objects, called pulsars, constitute the primary evidence for gravitational collapse accompanying the supernova event. [See NEUTRON STARS; PULSARS; SUPERNOVAE.]

D. H II REGIONS AND PLANETARY NEBULAE

Interstellar clouds in the vicinity of hot stars, with surface temperatures in excess of about 3×10^4 K, are ionized by Lyman continuum photons. When this gas recombines, it radiates line emission; when in the ionized state, it radiates like a thermal plasma. Both processes are temperature and density dependent and serve as sensitive probes of the interior conditions of the gas. The H II regions are found near sites of active star formation, where the gas is ionized by the radiation from massive stars whose lifetimes are short. Since the number of ionizing photons determines the rate of radio emission, the determination of integrated flux from such regions is a useful measure of the rate of star formation.

Planetary nebulae are the ionized portions of the outer envelopes of low-mass stars ejected in the strong stellar wind that accompanies the red supergiant end stages of evolution. As the envelope of the star is progressively stripped off, the hot inner core region is revealed. This central star, a white dwarf, cools rapidly on time scales of about 10^5 yr to temperatures below that capable of maintaining the ionization of the envelope, and the nebula disappears from both radio and optical visibility. The primary emission seen in planetaries is thermal free–free from a density stratified medium. Molecular emission from the cold outer parts of the wind has been detected for several planetaries, notably NGC 7293, mainly in CO lines. Recent observations have also detected neutral hydrogen in the envelopes of several planetaries.

E. DIFFUSE NEUTRAL HYDROGEN CLOUDS IN THE GALAXY AND EXTRAGALACTIC SYSTEMS

The major contribution of radio astronomy to the study of galactic structure comes from the determination of masses and rotation curves for the neutral interstellar medium through observations of the 21-cm line. For the Galaxy, the interstellar medium is known to consist of discrete clouds (1 to 100 cm^{-3} and $T \approx 10$ to 100 K) imbedded in a tenuous (0.01 to 1 cm^{-3}), hot ($10^3 - 10^4$ K) medium. The H I emission from these clouds serves as a tracer of large-scale motion of the gas.

For extragalactic systems, two modes of observation are employed. Single-dish radio telescopes generally observe only lines from the integrated disk of spiral galaxies, so that the line profile is all of the information available. Here, the width of the line and its profile as a function of wavelength (or velocity) provide structural information. The linewidth is a measure of the mass of the galaxy, while the profile is crude map of the velocity field of the system. The latter can be thought of in the following fashion.

Every portion of a rotating galaxy contributes emission at a fixed velocity relative to the observer, or radial velocity, which is characteristic of its distance from the galactic center. The inner portions of the galaxy contribute to the highest velocity parts of the profile, for a differentially rotating system with $v(r) \sim r^{-n}$, where $n > 0$, while the flatter the profile, the less differentially rotating the system will be. The total width, v_{max}, has been found to be proportional to the mass of the galaxy and therefore its luminosity, the Tully–Fisher relation. Missing neutral material, either because of voids in the distribution or from ionization, will produce structure in the line profile and can be used as rough maps of the hydrogen distribution.

There are several components to the emission seen from our galactic plane and from other galaxies. The normal disk gas is confined within about 100 pc of the galactic plane and has a low-velocity dispersion of order 10 km sec^{-1}. High-velocity, high-latitude clouds have also been detected in our system and in a few extragalactic objects. Finally, the Magellanic Stream, a trail of H I that surrounds the Galaxy as a result of the tidal disruption of the Large and Small Magellanic Clouds, circulates around the disk at intermediate galactic latitudes at a distance of about 50 kpc.

The H I observations show that the peripheral parts of galactic disks are often warped, a possible indication of interactions among galaxies in clusters. An important point to note is that H I observations often extend far beyond the optical limits of galaxies, sometimes past 40 kpc, and show that even at these large distances the disks display "flat" rotation curves, that is the rotational velocity is constant. This observation is one of the cornerstones of the dark matter argument—that there must be significant mass in disk galaxies of a nonluminous or subluminous variety compared with the stars.

Elliptical galaxies have recently been detected in increasing numbers in H I. While generally very gas poor, these systems appear to possess low-mass disks in their inner regions, possibly as a result of interactions with cluster galaxies.

F. NONTHERMAL EMISSION FROM GALAXIES

Perhaps the largest body of work with modern synthesis telescopes has centered on the imaging of radio galaxies in continuum radiation. Here, the greatest progress has been made by the introduction of large aperture synthesis arrays, such as Westerbork and the VLA. What had previously appeared as distended emission regions flanking a compact core have often turned out to be lobes connected by tenuous jet-like structures to the radio core. The best examples are M 87 and Cygnus A, where scales of several milliarcseconds (about 1 pc) to the many kiloparsec scales are now accessible.

The main understanding that radio astronomy has contributed is that there is an active source in most extended radio sources feeding the radio lobes via particle transport and acceleration in the jets. The details of the structures observed in the lobes and jets, the filamentation mechanisms, and the relation between the dynamics of the extruded matter and the magnetic fields required to produce the observed emission are all problems for future research.

IV. Concluding Remarks

The field of radio astronomy is one of the most rapidly developing areas of observational astrophysics. With its strong dependence on improvements in computational facilities and algorithms, the development of low-noise receivers, and the continuing proliferation of aperature synthesis instruments at a variety of wavelengths, it is probable that the field will have significantly evolved even by the appearance of this article. Radio observations cover the entire range of physical processes and environments observable in the cosmos and serve as a vital link between the thermal and magnetic properties of astrophysical plasmas. On the theoretical side, an understanding of the structures discovered by modern synthesis instruments in extragalactic systems is only beginning to emerge, and it is in this area, too, that there should be much work for future generations of astrophysicists.

BIBLIOGRAPHY

Christiansen, W. N., and Hogbom, J. A. (1985). "Radiotelescopes," 2nd ed. Cambridge Univ. Press, London and New York.

Hey, J. S. (198). "The History of Radio Astronomy," 3rd ed. Gordon & Breach, New York.

Kraus, J. D. (1983). "Radio Astronomy." Cygnus-Quasar Books, Columbus, Ohio.

Lovell, B. (1968). "The Story of Jodrell Bank." Oxford Univ. Press, London and New York.

Meeks, B. (1976). In "Methods of Experimental Physics, 12B; Radio Telescopes." Academic Press, New York.

Napier, P., Thompson, A. R., and Ekers, R. D. (1983). The very large array: design and performance of a modern synthesis radio telescope, *Proc. IEEE* **71,** 1295.

Pearson, S., and Readhead, A. C. S. (1984). Image-formation by self-calibration in radio astronomy. *Annu. Rev. Astron. Astrophys.* **22,** 97.

Perley, R. A., Schwab, F. R., and Bridle, A. H., eds. (1986). "Synthesis Imaging; Course Notes from an NRAO Summer School Held in Socorro, New Mexico, August 5–9, 1985." NRAO, Charlottesville, Virginia.

Rolfs, E. (1986). "Tools of Radio Astronomy." Springer-Verlag, Berlin and New York.

Sullivan, W. T., III (1982). "Classics of Radio Astronomy." Cambridge Univ. Press, London and New York.

Thompson, A. R., Moran, J. M., and Senson, G. W., Jr. (1986). "Interferometry and Aperture Synthesis in Radio Astronomy." Wiley, New York.

Verschuur, G., and Kellermann, K. I, eds. (1988). "Galactic and Extragalactic Radio Astronomy." Springer-Verlag, Berlin and New York.

RADIO ASTRONOMY, PLANETARY

Samuel Gulkis *Jet Propulsion Laboratory*

GLOSSARY

Antenna temperature: Measure of the noise power collected by the antenna and delivered to the radio receiver. Specifically, the temperature at which a resistor, substituted for the antenna, would have to be maintained in order to deliver the same noise power to the receiver in the same frequency bandwidth.

Blackbody: Object that absorbs all electromagnetic radiation that is incident on it. The radiation properties of blackbody radiators are well known. Planetary radio astronomers use the properties of blackbody radiators to describe the radiation from planets.

Brightness temperature: Temperature at which a blackbody would radiate an intensity of electromagnetic radiation identical to that of the planet for a specific frequency, frequency bandwidth, and polarization under consideration.

Effective area: Equivalent cross section of an antenna to an incident radio wave; a measure of a radio telescope's capability to detect weak radio signals.

Effective temperature: Temperature at which a blackbody would radiate over all frequencies an intensity of electromagnetic radiation identical to that radiated from a planet.

Equivalent blackbody disk temperature: Temperature of a blackbody radiator with the same solid angle as the planet that gives the same radiation intensity at the earth as observed from the planet at a specified frequency and bandwidth.

Flux density: Power per unit area and per unit frequency of an electromagnetic wave crossing an imaginary plane surface from one side to the other. In observational radio astronomy, the mks system of units is generally used, and the units of flux density are watts per square meter per hertz.

Flux unit or jansky: Commonly used unit of flux density equal to 1×10^{-26} W m^{-2} Hz^{-1}. The size of the unit is suited to planetary radio emissions, which are very weak.

Nonthermal radio emission: Radio emission produced by processes other than those that produce thermal emission.

Thermal radio emission: Continuous radio emission from an object with a temperature above absolute zero.

Planetary radio astronomy is the study of the physical characteristics of the planets in the solar system by means of the electromagnetic radio radiation emitted by these objects. The term is also used more generally to include the study of planetary ring systems, the moon, asteroids, satellites, and comets in the solar system. Radio astronomy generally refers to the (vacuum) wavelength range from about 1 mm (300 GHz = 300×10^9 Hz) to 30 m (10 MHz = 10×10^6 Hz) and longward. Observations from the ground cannot be carried out below a few megahertz because of the earth's ionosphere, which is opaque to very low frequency radio waves. Radio emissions have been measured from all of the planets with the exception of Pluto, which has thus far gone undetected because of the faintness of its radio emission. The observed emissions from the planets can be broadly classified as quasi-thermal (having the same general shape as a blackbody emitter) and nonthermal (i.e., cyclotron, synchrotron). Planetary radio emissions originate in the solid mantles, atmospheres, and magnetospheres of the planets. Several planetary spacecraft have carried radio astronomical instrumentation.

I. Introduction

A. Brief History

The science of radio astronomy began with the pioneering work of Karl G. Jansky, who discovered radio emission from the Milky Way galaxy while studying the direction of arrival of radio bursts associated with thunderstorms. Ten years later, in 1942, while trying to track down the source of radio interference on a military antiaircraft radar system in England, J. S. Hey established the occurrence of radio emission from the sun.

The first intentional measurement of a solar system object was made in 1945 by R. H. Dicke and R. Beringer working in the United States. They observed radio emission from the moon at a wavelength of 1.25 cm, thus beginning the first scientific studies at radio wavelengths of the planets and satellites of the solar system. Subsequent observations revealed that the microwave emission from the moon varies with lunar phase but the amplitude of these variations is much smaller than that observed at infrared wavelengths. This result was interpreted in terms of emission originating below the surface of the moon, where the temperature variations are smaller than at the surface. Within a few years, it was widely recognized that the long wavelengths provided by radio measurements offered a new and important tool for solar system studies, namely the capability of probing into and beneath cloud layers and surfaces of the planets. However, thermal radio emissions from the planets are exceedingly weak and nearly a decade elapsed before system sensitivity was sufficiently improved to enable the detection of planetary thermal emission.

Meanwhile, another unanticipated discovery was made. In June 1954, when the angular separation between the sun and a supernova remnant known as the Crab Nebula was small, astronomers B. Burke and K. Franklin of the Carnegie Institution were attempting to study the effect of the solar corona on radio waves from the Crab Nebula. Occasionally, they observed bursts of radio interference, which they initially thought were due to the sun. That hypothesis was discarded when it was discovered that the origin of

the interference bursts was nearly fixed with respect to the background stars. Further observations and examinations of the data revealed that the emissions were in fact originating from Jupiter, which happened to be located in the same region of the sky as the sun and the Crab Nebula. Because the intensity of the emissions was much too strong to be of thermal origin, Franklin and Burke concluded that Jupiter was a source of nonthermal radio emission.

The first successful measurements of thermal emission from the planets were made in 1956 at the Naval Research Laboratory in Washington, D.C. C. H. Mayer, T. P. McCullough, and R. M. Sloanaker scanned Venus, Mars, and Jupiter with a 15-m parabolic antenna equipped with a new 3-cm wavelength radio receiver. They detected weak thermal emission from these three planets when each was observed at its closest distance to the earth.

In the intervening years, thermal emission has been measured from all of the planets in the solar system with the exception of Pluto. A few asteroids, satellites, and comets have also been measured. Nonthermal radio emission has been measured from Jupiter, Saturn, and Uranus. In this article we give an overview of the techniques used by planetary radio astronomers and discuss what has been learned from the measurements and what can be done in the future.

B. Measurement Objectives

The primary goal of planetary exploration is to determine the physical characteristics of the planets, satellites, asteroids, and comets in order to obtain an understanding of the origin and evolution of the solar system, including the origin of life on the planet earth. One component objective associated with this goal is to determine the composition and physical characteristics of these bodies and their atmospheres. Studies of the energy budget and redistribution of energy within solid surfaces and atmospheres are part of this work. Another objective is to investigate the magnetic fields and ionized plasmas that surround some of the bodies in the solar system and to understand the interaction of the magnetic fields with the solar wind and the cosmic environment.

Planetary research involves many scientific disciplines and requires a variety of instruments and techniques, including astronomical studies from the earth, planetary spacecraft flybys, orbiters, probes, and eventually manned landings.

Each of the various approaches used has a particular strength that the experimenters try to exploit. Planetary radio astronomy contributes several unique tools to planetary exploration. Thus far, most planetary radio astronomy has been carried out from the ground, but the techniques carry over to spacecraft as well.

First, radio waves can be used to provide information about planetary atmospheres and planetary subsurface materials to much greater depth than other remote sensing techniques. Neutral gases and minerals are generally more transparent to radio waves than higher-frequency waves such as infrared or visible light. Also, the scattering from particulate materials in planetary atmospheres is generally less at radio wavelengths than at shorter wavelengths. The relative transparency of atmospheres, clouds, and surfaces to radio waves allows the planetary radio astronomer to measure thermal profiles of planetary atmospheres beneath the cloud layers in the atmosphere and to measure temperatures beneath the solid surface of the planet. Taking advantage of this property, radio astronomers were the first to measure the very high surface temperature of cloud-covered Venus.

A second unique tool provided by radio waves is the capability of measuring both synchrotron and cyclotron radiation. This radiation originates in the ionized plasmas and magnetic fields that surround some of the planets. Planetary conditions are such that this radiation is confined to the radio spectral region. The existence of Jupiter's strong magnetic field was first deduced from measurements of its polarized radio emission.

Spectroscopy is another tool that has been successfully used by the planetary radio astronomer. While the technique is not unique to the radio spectral range, the radio observations are complementary to those in other wavelength ranges. It is possible to achieve very high frequency resolution in the radio region by translating the frequency of the radio signal under study to a more convenient place in the frequency spectrum where spectrum analysis is easier to achieve with either digital or analog techniques. This process takes place without disturbing the relation of the sidebands to the carrier frequency. The process of frequency translation is referred to by such names as heterodyne, mixing, or frequency conversion. The heterodyne techniques makes it possible to measure absorption and emission line shapes in greater detail than has been possible at shorter wavelengths. Both composition and altitude distribution of an absorbing chemical species (e.g., NH_3, H_2O, CO) can be deduced from spectroscopic measurements. [*See* PLANETARY ATMOSPHERES.]

C. Physical Properties of the Planets

The physical characteristics of the planets are required in order to make qualitative estimates of the radioactive power expected from the planets. Table I presents physical data for the plan-

TABLE I. Physical Data for the Planets

	Mean distance (AU)[a]	Mass (earth = 1)	Radius (equator) (km)	Obl.[b]	Density (g/cm³)	Bond albedo[c]	Diameter (arc sec)[d] Min.	Diameter (arc sec)[d] Max.
Mercury	0.387	0.055	2,440	Small	5.44	0.06	4.7	12.2
Venus	0.723	0.815	6,050	Small	5.269	0.77	9.9	62.2
Earth	1.000	1	6,378	1/298.2	5.517	0.39	—	—
Mars	1.524	0.107	3,397	1/156.6	3.945	0.16	3.5	24.6
Jupiter	5.203	317.9	71,600	1/16.7	1.314	0.45	30.5	49.8
Saturn	9.523	95.2	60,000	1/9.3	0.704	0.61	14.7	20.5
Uranus	19.164	14.6	25,900	1/100	1.21	0.42	3.4	4.2
Neptune	29.987	17.2	24,750	1/38.5	1.66	0.42	2.2	2.4
Pluto	39.37	0.0017	1,300?	Unknown	1.0?	?	?	?

[a] AU, Astronomical unit = 149.6×10^6 km.

[b] Obl., Oblateness of planet = 1 − (polar radius equatorial radius).

[c] Bond albedo, ratio of reflected solar radiation to incident solar radiation. Spectral range should be specified, that is, visual Bond albedo.

[d] Diameter, angular extent of disk when planet–earth distance is greatest (min.) and least (max.).

ets, some of which will be referred to later in this article.

II. Basic Concepts

A. Thermal (Blackbody) Radiation

Any object with a temperature above absolute zero emits a continuous spectrum of electromagnetic radiation at all wavelengths including the radio region. This emission is referred to as thermal emission. The concept of a "blackbody" radiator is frequently used as an idealized standard with which real absorbers and thermal emitters can be compared. A blackbody radiator is defined as an object that absorbs all electromagnetic radiation that falls on it at all frequencies over all angles of incidence. No radiation is reflected from such an object. According to thermodynamic arguments embodied in Kirchhoff's law, a good absorber is also a good emitter. The blackbody radiator emits the maximum amount of thermal radiation possible for an object at a given temperature. The radiative properties of a blackbody radiator have been well studied and verified by experiments.

To be sure, a blackbody radiator is a concept rather than a description of an actual radiator. Only a few surfaces, such as carbon black, carborundum, platinum black, and gold black, approach a blackbody in their ability to absorb incident radiant energy over a broad wavelength range. Many materials are spectrally selective in their ability to absorb and emit radiation, and hence they resemble blackbody radiators over some wavelength ranges and not over others. Engineers use the properties of real materials to control the heat flow into and out of their designs. For example, radio telescopes are sometimes painted with a special kind of white paint that reflects the incident visible radiation from the sun and that radiates well at longer wavelengths. The spectrally selective properties of the white paint serve to keep the telescope at a lower temperature than it might have if no attention were paid to the surface coating.

Over large ranges of the radio and infrared spectrum, planets behave as imperfect blackbodies. Later, we will see how the deviations from the blackbody spectrum contain information about physical and chemical properties of these distant objects.

An important property of a blackbody radiator is that its total radiant energy is a function only of its temperature; that is, the temperature of a

blackbody radiator uniquely determines the amount of energy that is radiated into any frequency band. Planetary radio astronomers make use of this property by expressing the amount of radio energy received from a planet in terms of the temperature of a blackbody of equivalent angular size. This concept is developed more fully in the following paragraphs.

The theory that describes the wavelength dependence of the radiation emitted from a blackbody radiator was first formulated by the German physicist Max Planck in 1901. Planck's theory was revolutionary in its time, requiring assumptions about the quantized nature of radiation. Planck's radiation law states that the brightness of a blackbody radiator at temperature T and frequency ν is expressed by

$$B = (2h\nu^3/c^2)(e^{h\nu/kT} - 1)^{-1} \qquad (1)$$

where B is the brightness, watts per square meter per hertz per radian; h Planck's constant (6.63×10^{-34} J sec); ν the frequency, hertz; $\lambda = c/\nu$ the wavelength, meters; c the velocity of light (3×10^8 m/sec); k Boltzmann's constant (1.38×10^{-23} J/K); and T the temperature, kelvin.

Equation (1) describes how much power a blackbody radiates per unit area of surface, per unit frequency, into a unit solid angle. The three curves in Fig. 1 show the brightness for three blackbody objects at temperatures of 6000, 600, and 60 K. The radiation curve for the undisturbed sun is closely represented by the 6000 K curve over a wide frequency range. The other two curves are representative of the range of thermal temperatures encountered on the planets. It should be noted in Fig. 1 that the brightness curve that represents the sun peaks in the optical wavelength range while representative curves for the planets peak in the infrared. This means that most of the energy received by the planets from the sun is in the visible wavelength range while that emitted by the planets is radiated in the infrared. Radio emissions are expected to play only a small role in the overall energy balance of the planets because the vast majority of the power that enters and leaves the planets is contained within the visible and infrared region of the spectrum.

A useful approximation to the Planck radiation law can be obtained when $h\nu$ is small compared with kT ($h\nu \ll kT$). This condition is generally met over the full range of planetary temperatures and at radio wavelengths. It leads to the Rayleigh–Jeans approximation of the

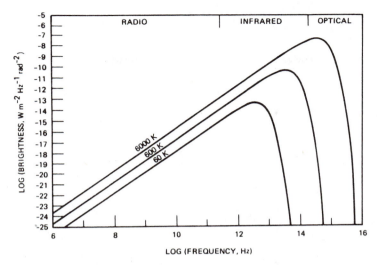

FIG. 1. Blackbody radiation curves at 6000, 600, and 60 K. The 6000 K curve is representative of the solar spectrum.

Planck law, given by

$$B = 2\nu^2 kT/c^2 = 2kT/\lambda^2 \qquad (2)$$

The Rayleigh–Jeans approximation shows explicitly that the brightness decreases as the inverse square of the wavelength. Using the Rayleigh–Jeans approximation overestimates the brightness by $\sim h\nu/2kT$. The error is largest at the shortest wavelengths (highest frequencies) and at the lowest temperatures. At a temperature of 100 K and a wavelength of 1 mm, the error is $\sim 8\%$.

Planetary radio astronomers estimate the radio power emitted by the planets by measuring with a radio telescope the power flux density received at the earth. Figure 2 illustrates the geometry involved in the measurement of power from an ideal blackbody radiator. The spectral power (per unit frequency) emitted by an elemental surface element of the blackbody of area dA into a solid angle $d\Omega$ is given by $B \cos(\theta)\, d\Omega\, dA$, where θ is the angle between the normal to the surface and the direction of the solid angle $d\Omega$. The total power (per unit frequency interval) radiated by a blackbody radiator is obtained by integrating the brightness over the surface area and over the solid angle into which each surface element radiates. The total spectral power density produced by a spherical blackbody radiator of radius r at a distance d from the blackbody is given by

$$S = \tfrac{1}{4}\pi d^2 \iint B \cos(\theta)\, ds\, d\Omega \qquad (3a)$$

$$= 2\pi kT(r/d)^2/\lambda^2 \qquad (3b)$$

The double integral represents integration over the surface area of the emitting body and over the hemisphere into which each surface element radiates. The quantity S is called "flux density." Flux density has units of power per unit area per unit frequency. A common unit of flux density is the flux unit (f.u.) or jansky (Jy), which has the value 10^{-26} W m^{-2} Hz^{-1}.

If a planet radiates like a blackbody and subtends a solid angle of Ω steradians at the observer's distance, then the flux density produced by the planet is given by (using the Rayleigh–Jeans approximation)

$$S = 2kT\Omega/\lambda^2 \qquad (4a)$$

$$= B\Omega \qquad (4b)$$

The convention generally adopted for calculating Ω for a planet is to use the polar (PSD) and equatorial (ESD) semidiameter values in the expression

$$\Omega = \pi \times \text{PSD} \times \text{ESD} \qquad (5)$$

The American Ephemeris and Nautical Almanac (AENA) values for PSD and ESD have generally been used, but improved values derived from spacecraft measurements are becoming available. Differences between AENA and current measured values of several percent are not uncommon. Equations (4) and (5) can be combined to yield the expression

$$S = 5.1 \times 10^{-34}\, T\theta_E\theta_P/\lambda^2 \quad \text{W m}^{-2}\,\text{Hz}^{-1} \qquad (6a)$$

$$= 5.1 \times 10^{-8}\, T\theta_E\theta_P/\lambda^2 \quad \text{Jy} \qquad (6b)$$

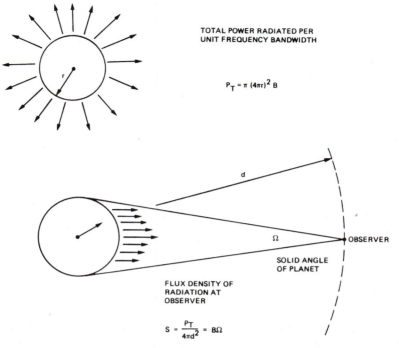

TOTAL POWER RADIATED PER
UNIT FREQUENCY BANDWIDTH

$$P_T = \pi (4\pi r)^2 B$$

d

Ω

OBSERVER

SOLID ANGLE
OF PLANET

FLUX DENSITY OF
RADIATION AT
OBSERVER

$$S = \frac{P_T}{4\pi d^2} = B\Omega$$

FIG. 2. Relationship between the brightness and power radiated by a blackbody spherical radiator of radius r and the flux density at a distance d from the blackbody.

where θ_E and θ_P are the apparent equatorial and polar diameters of the planets in seconds of arc and λ is in meters.

Even though the planets do not radiate like a blackbody, planetary radio astronomers express the observed brightness in terms of the temperature of an equivalent blackbody that would produce the same brightness. This temperature is called the brightness temperature T_B, defined as follows [from Eq. (4)]

$$T_B = B\lambda^2/2k = (S/\Omega)\lambda^2/2k \qquad (7)$$

The brightness temperature for a planet can be calculated once the flux density S and solid angle Ω are known. The brightness temperature approximates the physical temperature the more the planet behaves like a blackbody radiator.

For problems that deal with the energy budget of the planets, it is necessary to know the total amount of power radiated over all frequencies. Once again the concept of the blackbody is useful. The Planck radiation law can be integrated over all frequencies and solid angles to obtain the relationship known as the Stefan–Boltzmann law, given by

$$R = \sigma T^4 \qquad (8)$$

where R is the rate of emission, expressed in units of watts per square meter in the mks system. The constant σ has a numerical value of 5.67×10^{-8} W m^{-2} K^{-4}. Using the blackbody concept, we can now estimate the energy balance of planets that absorb visible light and radiate energy into the infrared.

B. THERMAL EMISSION FROM ATMOSPHERES AND SURFACES

1. Effective Temperatures of the Planets

The amount of thermal radiation expected from a given planet depends in detail on the physical characteristics of the planet's atmosphere and surface. A starting point for understanding the observed flux densities of the planets is to assume that the planets are blackbodies in equilibrium with the energy they receive from the sun and that which is radiated into free space. The radiation energy incident from the sun on a unit area per unit time is 1.39×10^3 W m^{-2} sec^{-1} at the earth. This quantity is called the solar constant. The incident solar flux available to heat a planet is given by

$$(1 - A)S_0 \pi R^2/d^2 \qquad (9)$$

where A is the fraction of the incident solar flux that is not absorbed, R the radius of the planet, S_0 the solar constant at astronomical unit 1 (AU), and d the mean distance of the planet from the sun in astronomical units. The quantity A is known as the Bond albedo or Russell–Bond albedo of the planet. Disregarding any significant internal heat sources, the total flux of absorbed radiation must equal the total flux of outgoing radiation when the planet is in equilibrium. The Stefan–Boltzmann law provides the relationship between effective temperature T_E and the absorbed flux:

$$\int \sigma T_E^4 \, ds = \pi R^2 (1 - A) S_0 / d^2 \qquad (10)$$

If the planet rotates rapidly, equilibrium will be reached between the insolation and the radiation from the entire planetary surface area, $4\pi R^2$. This leads to the estimate

$$T_E = 277(1 - A)^{1/4} d^{-1/2} \quad \text{K} \qquad (11)$$

If a planet did not rotate and its emitted radiation came only from the sunlit hemisphere, the effective temperature of the sunlit hemisphere would increase by the factor $2^{1/4}$ because of the reduction in the emission surface area. The equilibrium temperature for this case would become

$$T_E = 330(1 - A)^{1/4} d^{-1/2} \quad \text{K} \qquad (12)$$

Figure 3 shows the calculated effective temperatures of the planets for the rapidly rotating and nonrotating cases. The albedos and distances used are those given in Table I. Having obtained the effective temperatures, it is possible to predict flux densities for the planets by using the effective temperatures from Eq. (11) or (12) and the angular diameter data for the planets in Table I in Eq. (6a) or (6b).

Thus far, we have discussed the ideal model in which the planets behave like blackbody radiators. This gives planetary astronomers a crude model from which they can estimate flux densities and search for departures. The planets would not be very interesting to study if they behaved like blackbodies since a single parameter, namely the temperature, could be used to define their radiation properties. More important, it is the departures from the simple model that allow radio astronomers to deduce the physical properties of the planets.

The observed temperatures of the planets depart markedly from this ideal model for a number of different reasons. The presence of atmospheres on the planets produces strong perturbations from the ideal model. Atmospheres modify the amount of heat that enters and leaves the planets over the entire electromagnetic spectrum. Strong "greenhouse" effects can raise the temperatures considerably over that calculated from the ideal models. The presence of internal sources of energy within a planet can modify its effective temperature and affect the thermal profile of the atmosphere.

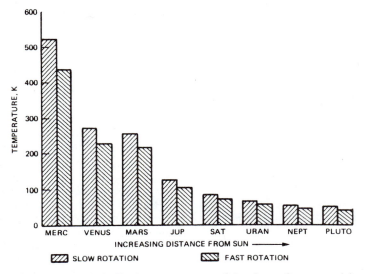

FIG. 3. Theoretical effective temperatures of the planets for two models. The higher-temperature model is for the case where the planet is not rotating [Eq. (12)]. The lower-temperature model is for the case where the planet is rapidly rotating [Eq. (11)].

Surface emissivity effects modify the apparent temperature of the planets. Nonuniform heating of the planets by the sun, due to orbit eccentricity and rotational effects, also produces departures from the ideal model. Nonthermal emission from planetary magnetospheres produces the largest departures from the blackbody model. The role of the planetary astronomer is to sort out the various effects that take place and to measure and infer the actual physical characteristics of the planets.

2. Radiative Transfer in Planetary Atmospheres

The apparent brightness temperature of a deep atmosphere is related to the physical parameters of the atmosphere, such as pressure, temperature, and composition, through the equation of radiative transfer. To a good approximation at radio wavelengths, the equation of radiative transfer for a ray making an angle $\cos^{-1} \mu$ with the vertical in a lossy medium is

$$T_{\mathrm{B}}(\nu, \mu) = \int T(z) \exp \left[- \int \right.$$
$$\left. \times\ \alpha(z', \nu)\mu^{-1}\, dz' \right] \alpha(z, \nu)\mu^{-1}\, dz \tag{13}$$

where $\alpha(z, \nu)$ is the absorption coefficient of the atmosphere at frequency ν and depth z and $T(z)$ the physical temperature along the line of sight. This equation states that the brightness temperature in any given direction is the sum of the radiation emitted at each point along the trajectory, each component being attenuated by the intervening medium. The equation neglects scattering and variations of the index of refraction. These effects are important in certain wavelength ranges and for certain ray trajectories. Sometimes it is necessary to add another term to Eq. (13) to account for the presence of a solid surface.

Measurements with single-dish antennas generally have insufficient angular resolution to determine the brightness distribution across the planetary disk. The mean disk brightness temperature T_{D} can be calculated by integration of Eq.(13) over all angles of incidence. This yields

$$T_{\mathrm{D}}(\nu) = 2 \int_0^1 T_{\mathrm{B}}(\nu, \mu)\mu\, d\mu \tag{14}$$

This equation is usually used to calculate the average disk brightness temperature for a model atmosphere for comparison with observations. Unlike the blackbody radiation model, this model predicts a wavelength dependence of the brightness temperature. The frequency dependence is introduced by the frequency dependence of the absorption coefficient and the thermal gradients in the atmosphere.

The absorption coefficient may either vary slowly with frequency or exhibit an abrupt change over a narrow frequency range. The two kinds of absorption are referred to as nonresonant and resonant absorption, respectively. The standard theory of nonresonant absorption is due to Debye. Resonant absorption is produced by the discrete transitions from one energy level to another in a molecule that cause the molecule to absorb or emit at particular frequencies. The study of resonant absorption lines in planetary atmospheres is referred to as planetary spectroscopy.

A brief discussion of the energy levels in the molecule CO is helpful in understanding resonant absorption and the usefulness of planetary spectroscopy as a tool for studying planetary atmospheres. The rotational energy levels in CO are quantized according to the relation

$$E_{\mathrm{r}} = J(J + 1)B \tag{15}$$

where J ($J = 0, 1, 2, \ldots$) is the rotational quantum number and B the rotational constant. The rotational constant for a diatomic molecule is defined by

$$B = h^2/8\pi^2 I \tag{16}$$

where I is the moment of inertia of the molecule about the axis perpendicular to the line connecting the nuclei and h is Planck's constant. The transitions between the various energy levels are limited by the exclusion principle to $\Delta J = \pm 1$. The energy of a quantum emitted or absorbed during a transition is given by

$$h\nu = J(J + 1)B - (J - 1)(J)B = 2JB \tag{17}$$

When B is expressed in megahertz, the transition frequency ν in megahertz is expressed directly as $2JB$. The value of B for CO is 57,897.5 MBz. Thus, CO has two rotational transitions in the millimeter spectrum, a ground state ($\sim$115 GHz) and the first excited state ($\sim$230 GHz). The ground state transition corresponds to a transition between $J = 0$ and 1; the first excited state corresponds to a transition between $J = 1$ and 2. Both of these transitions have been observed in the atmospheres of Venus and Mars.

A number of factors cause the energy levels in a molecule to vary slightly and cause the molecule to emit or absorb over a range of frequencies. The range of frequencies or "width" of a spectral line is determined primarily by three factors: (1) natural attenuation, (2) pressure broadening, and (3) Doppler broadening. Only pressure broadening and Doppler broadening have proved to be important for planetary radio astronomy. Natural attenuation is interpreted as the disturbance of the molecule by zero-point vibration of electromagnetic fields, which are always present in free space.

Doppler broading is caused by the frequency shift introduced by the molecule's motion. The statistical effect of the simultaneous observation of a large number of molecules moving at various velocities is to spread the frequency of the spectral line over a range of frequencies. The spectral line produced by thermal motions in gas is symmetric and has a full width at half-maximum of

$$\Delta\nu = 7.2 \times 10^{-7}(T/M)^{1/2}\nu \qquad (18)$$

where T is the temperature of the gas, M the molecular weight of the molecule, and ν the frequency of the spectral line without Doppler shift. The Doppler linewidth of the ground state transition ($J = 0$ to 1) of CO at 200 K is ~200 kHz. A measure of the Doppler width can be used to estimate the temperature of the gas.

Pressure broadening can be interpreted as the effect of collisions disrupting the processes of emission and absorption in a molecule. A number of different theoretical line shapes have been derived to explain pressure broadening. The one most widely used is the Van Vleck–Weisskopf line shape. Linewidths due to pressure broadening are proportional to pressure, the proportionality constant depending on the particular molecules involved in the collisions. For CO broadened by CO_2, the linewidth of the ground state transition of CO is approximately

$$\Delta\nu = 3.3p(300/T)^{0.75} \qquad (19)$$

where p is the pressure in millibars and $\Delta\nu$ the linewidth in megahertz. At a pressure of 1 mbar and temperature of 200 K, the pressure-broadened linewidth of CO is ~4.5 MHz. Observations of pressure-broadened spectral lines can be used to determine the altitude distribution of a molecule in a planetary atmosphere.

3. Radiative Transfer in Planetary Subsurfaces

While some planets have deep atmospheres, others such as Mercury and Mars have relatively tenuous atmospheres. For these planets (and the moon and satellites) the atmospheres are nearly transparent at radio wavelengths, except possibly in narrow wavelength ranges, where resonant absorption lines can produce strong absorption. Thermal emission from the surfaces of these planets is easily observed at radio wavelengths; it is possible to interpret the measurements in terms of the physical properties of the near-surface materials.

Observations of the thermal radiation at radio wavelengths provide measurements that are complementary to infrared measurements of the subsurface materials. Thermal emission in the infrared is emitted very close to the surface of the planet because the opacity of most minerals is high at infrared wavelengths. Consequently, infrared thermal emission reflects the physical characteristics of the near-surface materials. The opacity of typical planetary materials is considerably less at radio wavelengths, and the observed thermal emission originates at greater depth.

The interpretation of thermal emission data from a planetary surface begins with an analysis of heat transfer in solids. The solid surface receives heat from the incoming solar radiation and transports it downward, mainly by conduction and radiation. Boundary conditions are set by the heating of the surface by the sun, the nighttime cooling, and the internal sources of heat, if any. The surface temperature responds to the heating and cooling, being controlled by the "thermal inertia" of the near-surface material. Because the planets are heated and cooled periodically due to their spin, a thermal wave is set up in the surface layers. The equation of radiative transfer is used to relate the temperature structure in the subsurface layers to the observed thermal emission.

The formal solutions of the equations of heat transfer in solids and radiative transfer depend on the following properties of the near-surface material: the complex dielectric constant ε, thermal conductivity k (ergs per centimeter per second per kelvin), specific heat c (ergs per gram per kelvin), and density ρ (grams per cubic centimeter). The analytic theory of heat transfer at planetary surfaces begins by assuming that the temperature at any point on the surface can be expanded in a Fourier series in time:

$$T_s(t) = T_0 + \sum_{n=1}^{\infty} T_n \cos[(n\omega t) - \Phi_n] \qquad (20)$$

where ω (radians per second) is the fundamental

heating frequency (i.e., rotation rate of the planet as seen from the sun) and t is time. Assuming that the planet is a semi-infinite homogeneous slab with constant thermal properties, the equilibrium solution for the subsurface temperature distribution is

$$T(x, t) = T_0 + \sum_{n=1}^{\infty} T_n \exp(-x\beta_n) \cos$$

$$\times [(\omega t - \beta_n x - \Phi_n)] \qquad (21)$$

where x is the depth beneath the surface and β_n is given by

$$\beta_n = [n\omega\rho c/2k]^{1/2} = \beta_1 n^{1/2} \qquad (22)$$

Equation (21) represents a series of thermal waves propagating into the surface and attenuating with distance. The higher harmonics are attenuated more rapidly than the lower harmonics since β_n increases as the square root of the harmonic number n. The attenuation and phase of each harmonic depend on the quantity β_1, which is termed the "thermal absorption coefficient" of the planetary material. The reciprocal of β_1 ($L_t = 1/\beta_1$) is termed the "thermal skin depth." At a distance of 3 to 4 thermal skin depths, the fluctuations in subsurface temperature are practically zero.

Given the thermal absorption coefficient and the boundary conditions on the heating, it is possible to determine the constants of temperature (T_0 and T_n) and phase (Φ_n) in Eq. (21). The inverse problem is faced by the radio astronomer, namely to determine the thermal absorption coefficient from measurements of the thermal emission. This is done in the following manner. The temperature distribution given by Eq. (21) is used in the equation of radiative transfer

$$T_B = [1 - R_P(\nu, \theta_0)] \int_0^{\infty} T(x) \exp$$

$$\times [-k_\nu x/\cos(\theta_i)][k_\nu/\cos(\theta_i)] \, dx \quad (23)$$

to compute the radio brightness temperatures at any wavelength for comparison with observations. In this expression k_ν is the power absorption coefficient at frequency ν, θ_i the angle of incidence of radiation just below the surface, and θ_0 the angle of incidence of the observation. The function $R_P(\nu, \theta_0)$ is the Fresnel reflection coefficient of polarization P emerging at angle θ_0.

After substituting (the series) Eq. (21) for $T(x, t)$ in Eq. (23), the integral can be evaluated as a series of integrals. Each integral can be put

in the form of a standard Laplace transform and integrated directly. The resulting brightness temperature at time t at a specified point on the surface is given by

$$T_B(\nu, p, t) = [1 - R_P] \sum_{n=1}^{\infty}$$

$$\frac{T_n \cos[n\omega t - \Phi_n - \Psi_n(\theta_i)]}{[1 + 2\delta_n(\theta_i) + 2\delta_n 2(\theta_i)]^{1/2}} \qquad (24)$$

where

$$\delta_n(\theta_i) = n^{1/2}\beta_1 \cos(\theta_i)/k_\nu \qquad (25)$$

$$\Psi_n = \tan^{-1}\{\delta_n[\theta_i/(1 + \delta_n(\theta_i)]\} \qquad (26)$$

Defining $1/k_\nu$ as the radio absorption length (L_e), δ_1 reduces to the ratio of the radio absorption length to the thermal absorption length at normal incidence. Equations (24)–(26) relate the observational data to the physical parameters of the surface materials. Only the first few terms of Eq. (24) are normally important because the higher-order terms are small. The most important terms determined from the observations are δ_1 and Ψ_1. The δ_1 term defines the reduction in amplitude of the fundamental diurnal wave component from its surface value. It is best determined from observations at several different wavelengths spanning a wavelength range of 2 : 1 or more. The Ψ_1 term defines the phase shift in the diurnal wave at depth. It also can be deduced from measurements at several wavelengths, but usually with somewhat less precision than δ_1.

The parameter δ_1 can be written in terms of the physical characteristics of the surface material as follows:

$$\delta_1 = L_e/L_t = (\Omega\rho c/2k)^{1/2}\lambda[2\pi\sqrt{\varepsilon} \tan(\Delta)]^{-1} \quad (27)$$

where $\tan(\Delta) = 2\sigma/\varepsilon_r\nu$ is the loss tangent, ε_r the real part of the dielectric constant, and σ the electrical conductivity (mhos per meter) of the medium.

Radio observations by themselves do not permit separation of the physical parameters contained in δ_1 and Ψ_1. Nevertheless, the radio data, when combined with infrared data, radar data, and laboratory data for real materials, constrain the material properties and in some cases allow one to exclude certain classes of materials in favor of others.

C. NONTHERMAL RADIO EMISSION

Thermal radio emission in solids and neutral gases arises from the emission of quanta from individual atoms and molecules in thermodynamic equilibrium with each other. Random collisions between ions and electrons in thermal equilibrium with each other in an ionized gas also produce thermal emission. The electrons and ions in this case have a Maxwellian velocity distribution. When energy sources are present that produce particles having a non-Maxwellian velocity distribution, the system is not in thermodynamic equilibrium. Efficient processes can arise under these conditions that produce large amounts of radio energy. Nonequilibrium conditions frequently arise in an ionized gas or plasma typical of those found in space. In a fully ionized gas, nonequilibrium conditions can lead to coherent and incoherent plasma emissions. The radiation that arises from these mechanisms is called nonthermal emission. Cyclotron emission and synchrotron emission are examples of nonthermal emission. In the case of thermal emission, the blackbody radiation laws limit the radiation to an amount corresponding to the temperature of the body. For nonthermal radiation, this limit does not exist. The brightness temperature of nonthermal radiation sometimes exceeds millions of degrees even though the effective temperature of a planet does not exceed several hundred degrees.

The classic sources of nonthermal radio emission within the solar system are Jupiter's magnetosphere and the solar corona. Nonthermal radio emissions have also been observed from the earth's magnetosphere and from Saturn. Cyclotron and coherent plasma emissions account for much of the low-frequency (<10 MHz) nonthermal emission from the earth, Jupiter, Saturn, and the solar corona. This emission is highly variable and the details of the generation processes involved are not clearly understood.

Synchrotron radiation is the dominant source of emission from Jupiter from about 50 MHz to 5 GHz; it also accounts for continuum bursts of type IV from the sun. The theory of synchrotron radiation is well developed. A review article on magnetospheric radio emissions by Carr, Desch, and Alexander, contained in the book by Dessler cited in the bibliography, lists a number of references.

Synchrotron radiation is produced by high-energy electrons moving in a magnetic field. Although the observed synchrotron radiation from a planet or the sun is the integrated emission from many electrons, an understanding of the radiation characteristics of a single electron in a magnetic field is sufficient to understand the qualitative aspects of the observations.

A single charged electron moving in a magnetic field is accelerated unless its velocity is solely in the direction of the magnetic field. This causes the electron to emit electromagnetic waves. The nature of these waves depends on whether the electron is nonrelativistic (velocity $\ll 3 \times 10^{10}$ cm/sec) or relativistic (velocity $\sim 3 \times 10^{10}$ cm/sec). The radiation emitted by nonrelativistic electrons is referred to as cyclotron radiation; radiation emitted by relativistic electrons is referred to as synchrotron emission.

A nonrelativistic electron with mass m and charge e, in the presence of a magnetic field (of magnitude B), moves in a helical path with the sense of rotation of a right-hand screw advancing in the direction of the magnetic field. The frequency of rotation about the magnetic field, sometimes called the electron cyclotron frequency or the gyrofrequency, is given by

$$f_c = Be/2\pi m = 2.8B\text{(gauss)} \quad \text{MHz} \quad (28)$$

Cyclotron radiation is emitted in all directions and has a frequency equal to the gyrofrequency. The radiation is polarized with the polarization depending on the direction of propagation. The polarization is circular when viewed along the direction of the magnetic field and linear when viewed in the plane of the orbit. At intermediate angles the polarization is elliptical.

Relativistic electrons radiate not only at the gyrofrequency but also at the harmonics. The relativistic mass increase with energy causes the harmonic spacing to decrease with increasing energy until the synchrotron spectrum is essentially smeared into a continuum. The radiation from a relativistic electron is highly nonisotropic. The emitted radiation is concentrated within a narrow cone about the instantaneous direction of the velocity vector with an approximate half cone width given by

$$\theta \cong mc^2/E \quad \text{rad} \quad (29)$$

where E is the electron energy. When E is expressed in millions of electron volts (MeV) and θ in degrees, the expression becomes

$$\theta \cong 29/E\text{(MeV)} \quad \text{deg} \quad (30)$$

An observer situated in the plane of the electron orbit would see one pulse per revolution of the electron. These pulses recur at the relativistic gyrofrequency of the electron.

A single electron of energy E radiates synchrotron emission with an intensity spectrum that varies as $\nu^{1/3}$ (ν is frequency) up to a critical frequency $0.29\nu_c$ and decreases exponentially at higher frequencies. The critical frequency is defined as

$$\nu_c = (3e/4\pi mc)(E/mc^2)B$$
$$= 16.08\,B(\text{gauss})\,E^2(\text{MeV})\quad\text{MHz}\quad(31)$$

Thus a 10-MeV electron in a 1-gauss magnetic field will radiate a maximum intensity near 0.29×1608 MHz. The spectral density of the radiation near the frequency of the maximum intensity is

$$I(\nu = 0.29\nu_c) \cong 2.16 \times 10^{-29}B(\text{gauss})\quad\text{W/Hz}$$
$$(32)$$

The total energy radiated per second is given by

$$P = 6 \times 10^{-22}B^2(\text{gauss})E^2(\text{MeV})\,\sin^2\alpha\quad\text{W}$$
$$(33)$$

where α is the pitch angle of the electrons ($\alpha = 90°$ for electrons with no motion in the direction of the magnetic field).

As in the case of nonrelativistic electrons, the polarization is linear when the electron orbit is seen edge-on and elliptical or circular elsewhere. Since the intensity of the radiated power is beamed in the plane of the orbit, synchrotron radiation is predominantly linearly polarized.

It is possible to deduce many properties of Jupiter's magnetic field and of the high-energy particle environment from radio measurements of Jupiter by judicious use of the equations given above. Detailed calculations of synchrotron emission are complex since they involve integrals over the volume of the emitting electrons and over the electron energy spectrum while taking into account the complex geometry of the magnetic field and polarization properties of the radiation.

III. Instrumentation for Planetary Radio Astronomy

A. RADIO TELESCOPES

A radio telescope is a device for receiving and measuring radio noise power from the planets as well as from galactic and extragalactic sources of emission. A simple radio telescope consists of an antenna for collecting the noise power (in a specified polarization and from a limited range of directions) and a sensitive receiver–recorder for detecting and recording the power. The antenna is analogous to the objective lens or pri-

mary mirror of an optical telescope; the receiver is analogous to the recording medium of the optical telescope (i.e., photographic plate, photodetectors, etc.). Figure 4 illustrates the basic components of a simple radio telescope. Single antennas may be connected together electrically to form a radio telescope with multiple antennas. There are many different types of radio telescopes in use today and their capabilities and visual appearances show much diversity.

Three important properties of an antenna are its effective area $A_e(\theta, \phi)$, normalized antenna power pattern $P_N(\theta, \phi)$, and gain G. The effective area is a measure of the wave front area from which the antenna can extract energy from a wave arriving at the antenna from different directions. It can be thought of as the equivalent cross section of the antenna to the incident wave front. For most radio telescopes, the effective area is less than the physical area. A working definition of the effective area is given by

$$P = \tfrac{1}{2}SA_e\,d\nu\qquad(34)$$

where P is the power the antenna can deliver to a matched load in bandwidth $d\nu$ when flux density S is incident on the antenna. The factor $\tfrac{1}{2}$ is introduced because it is assumed that the radiation is unpolarized and that the antenna is responsive to only one polarization component.

The effective area of an antenna is a function of the direction of arrival of the waves. The effective area to a signal from a distant transmitter, as a function of direction, is called the antenna power pattern. The effective area normalized to unity in the direction of the maximum effective area $A_{e-\max}$ is the normalized power pattern of the antenna:

$$P_N(\theta, \phi) = A_e(\theta, \phi)/A_{e-\max}\qquad(35)$$

By reciprocity arguments, the normalized power pattern is the same for both transmitting and receiving.

A general expression for the power delivered to a radio receiver from a planet is obtained by

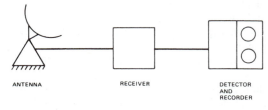

FIG. 4.   Basic components of a simple radio telescope.

integrating the brightness over the solid angle of the planet and over the bandwidth of the receiver as given below:

$$P = \tfrac{1}{2} \iiint B(\theta, \phi, \nu)$$

$$\times A_e(\theta, \phi, \nu) \sin(\theta) \, d\theta \, d\phi \, d\nu \quad (36a)$$

$$P = \tfrac{1}{2} A_{e-\text{max}} \iiint B(\theta, \phi, \nu)$$

$$\times P_N(\theta, \phi, \nu) \sin(\theta) \, d\theta \, d\phi \, d\nu \quad (36b)$$

A typical antenna power pattern consists of a large number of lobes with one or a few lobes being much larger than the others, as shown in Fig. 5. The lobe with the largest maximum is called the main lobe, while the remaining lobes are called side or back lobes.

The half-power beamwidth of the main lobe, Θ, is the angle between the two directions in which the received power is half of that in the direction of maximum power. The half-power beamwidth is a measure of the ability of the antenna to separate objects that are close together in angle. In optical systems, this is known as resolving power. An estimate of the beamwidth in radians of the main lobe is given by $\Theta \sim \lambda/D$, where D is the linear dimension of the antenna in the plane in which the beam is measured and λ the wavelength in the same units as D. The resolving power can be improved by using either a shorter wavelength or a larger-diameter telescope.

Two major classifications of large radio telescopes are (1) filled-aperture telescopes and (2) unfilled-aperture telescopes. Filled-aperture radio telescopes generally consist of a single reflecting element that focuses the received radio waves to a point or, in some cases, along a line. Examples of filled-aperture radio telescopes are the 64-m parabolic reflector antenna at Parkes, Australia, the 100-m parabolic reflector at Bonn, Germany, and the ~300-m spherical reflector antenna at Arecibo, Puerto Rico. Filled-aperture antennas are limited in size by the structural deformations of the antenna. The resolving power of a filled-aperture radio telescope is far less

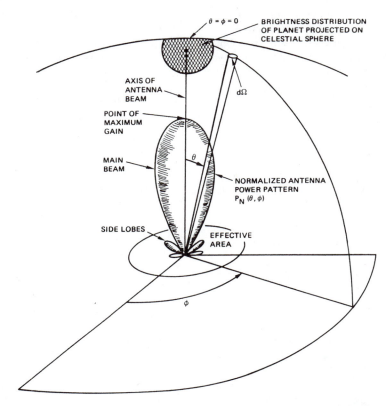

FIG. 5. Antenna power pattern and its relation to a measurement of the flux density from a planet.

than is required for adequate resolution of the disks of the planets. The largest single-dish antennas presently available to radio astronomers are capable of partially resolving the disks of the planets of largest angular diameter, Venus and Jupiter. For example, an antenna with a maximum dimension of 100 m operating at a wavelength of 10 cm would have a main lobe width of approximately 1/1000 rad or 3.44 arc min. The angular diameters (Table I) of Venus and Jupiter are approximately 1 arc min; thus an antenna with an aperture of ~344 is required just to "fill the main lobe." Adequate resolution of even the largest planets (at 10 cm wavelength) requires apertures at least 10 times larger.

Higher resolving power can be achieved by connecting together the outputs of two or more filled-aperture antennas at the input of a common radio receiver. The resolving power of such a system depends on the maximum separation between the individual elements, even though the space between the elements is unfilled. Multiple-element radio telescopes, with spaced separations between the elements to achieve high resolving power, are termed unfilled-aperture radio telescopes. The simplest unfilled-aperture radio telescope is a total power interferometer with two identical elements, as shown in Fig. 6. The output from the two-element interferometer modifies the power pattern of a single element with an angular modulation of scale λ/D superimposed.

The Very Large Array (VLA) radio telescope in Socorro, New Mexico, is an example of a modern unfilled-aperture radio telescope. This instrument is especially important to planetary radio astronomers. It has the equivalent resolving power of the largest ground-based optical telescopes. The VLA currently consists of 27 antennas of 25 m diameter (effective aperture of >7500 m^2 on a Y-shaped baseline, each leg of which is 21 km long. The antennas are mounted on tracks so that they can be moved into a variety of positions along the Y-shaped baseline. This allows the radio telescope to be custom tailored to a particular kind of measurement. Continuum, polarization, and spectral line observations at four separate wavelengths (1.3, 2.0, 6, and 18–20 cm) are now supported. The VLA has sufficient resolving power to measure the brightness distributions across most of the planets at its shortest wavelength of operation, and the resolution allows satellite emission to be separated from the parent planet emission.

The performance of a filled-aperture antenna is sometimes expressed by specifying the gain of the antenna. The antenna gain is a measure of the ability of the antenna to concentrate radiation in a particular direction. Gain is defined as the ratio of the flux density produced in direction (θ, ϕ) by an antenna when transmitting with an input power P to the flux density produced by the same transmitter feeding an antenna that radiated equally in all directions. The antenna

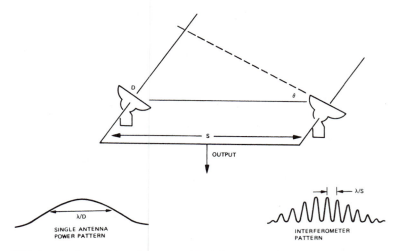

FIG. 6.   Total power interferometer. (Top) Geometry of a two-element interferometer. (Bottom) Antenna response for a single element of the interferometer (left) and response of the interferometer (right) to a completely unresolved planet.

gain, effective area, and beam solid angle (Ω) are interrelated quantities. The relationship between them is given by

$$G = 4\pi A_{e-max}/\lambda^2 = 4\pi/\Omega \qquad (37)$$

Radio astronomers generally specify the received power P delivered to the receiver in the frequency bandwidth dv as the antenna temperature T_a, defined by

$$T_a = P/k\,dv \qquad (38)$$

where k is Boltzmann's constant. This definition follows from the Nyquist theorem, which states that the noise power delivered from a resistor at temperature T into a matched load in bandwidth dv is

$$P_N = kT\,dv \qquad (39)$$

The antenna temperature can be thought of as the physical temperature at which a resistor would have to be maintained to deliver the same power to the receiver as the antenna.

The antenna temperature of an unresolved radio source can be calculated directly from the definition of the effective area and the definition of antenna temperature if its flux density is known. For a planet of known solid angle and temperature, the antenna temperature can be calculated by substituting the Rayleigh–Jeans approximation for the brightness, $B = 2kT_p/\lambda^2$, and the relationship between effective area and beam solid angle

$$A_{e-max} = \lambda^2/\Omega \qquad (40)$$

into Eq. (36). When the planet is unresolved by the main beam of the telescope, the antenna temperature is given by

$$T_a = (\Omega_p/\Omega)T_p \qquad (41)$$

where Ω_p and T_p are the solid angle and temperature of the planet, respectively, and Ω is the beam solid angle of the telescope. The relationship states that the antenna temperature of a planet is proportional to the brightness temperature of the planet with the constant of proportionality being the ratio of the solid angle of the planet to the solid angle of the beam. This approximation holds when $\Omega_p \ll \Omega$; consequently, the antenna temperature is always less than the brightness temperature.

The concept of specifying the received power in terms of an equivalent temperature is useful for estimating the signal-to-noise ratio for a particular measurement since random noise in the receiving equipment is easily expressed as a temperature. A discussion of signal-to-noise ratio follows.

B. RADIO RECEIVERS

Radio astronomy receivers, like radio telescopes, are highly varied, depending on the type of measurement to be performed. The function of the receiver is to detect and measure the radio emission with as much sensitivity as possible. The receiver also defines the frequency range or ranges of the measurement. Most modern receivers consist of a low-noise amplifier to boost the power of the incoming signal (without adding significant noise), followed by a heterodyne mixer and square law detector. The heterodyne mixer transforms the signal frequency to a convenient (often lower) frequency for detection or further processing.

The inherent noise fluctuations of a radio receiver usually determine the weakest signal strength that can be measured with a radio telescope. The statistical nature of noise radiation is such that statistical fluctuations are proportional to the noise power itself. Furthermore, the average of N independent measurements of the noise power is $\sqrt{N}$ times more accurate than a single measurement. Noting that a single independent measurement can be made in the minimum time interval, $1/dv$, the maximum number of independent measurements that can be made in time t is $t\,dv$. Thus, the sensitivity equation for an ideal receiver is given by

$$\Delta T = \text{rms noise power} \approx T_S/(t\,dv)^{1/2} \quad (42)$$

where the "system temperature" T_S is a measure of the noise power from the receiver and t is the integration time of the measurement. The rms noise power is expressed in kelvins and can be directly compared with the antenna temperature to determine the signal-to-noise ratio (SNR) of a particular measurement.

A modern radio telescope may have a system temperature of 20 K or less in the frequency range 1–10 Ghz, where the radiation from the terrestrial atmosphere and galaxy are both low. For measurements of the radio continuum, dv may be chosen to be 10 MHz or larger, depending on the characteristics of the radio receiver being used. If we adopt a value of 100 MHz for dv and 20 K for T_S, then the rms noise power obtained in 1 sec of integration is 0.002 K. For planetary spectroscopy, dv would have to be reduced to 1 MHz or less and the rms noise power would increase to 0.02 K. These noise fluctuations can be further reduced by increasing the integration times; however, systematic effects within the receiving equipment prevent ΔT from being pushed to zero.

C. SPACECRAFT

Earth-based radio observations of the planets suffer from (1) lack of spatial resolution, (2) restrictions on the viewing geometry of the planets, (3) the opaqueness and variability of the terrestrial atmosphere, and (4) the intrinsic faintness of the radio emissions from planetary bodies. The opacity of the terrestrial atmosphere varies with frequency. The atmosphere is opaque at frequencies lower than about 5 MHz due to the terrestrial ionosphere. Attenuation due to the atmospheric gases, water vapor, and oxygen affects the centimeter and millimeter bands but observations are possible from the ground by working in the transparent "windows" in the spectrum. Rain, fog, and clouds occasionally limit the usefulness of the centimeter and millimeter bands.

To overcome these difficulties, a number of spacecraft radio systems have been proposed, and several have flown on U.S. spacecraft. The first planetary radio system on a U.S. spacecraft was a two-channel microwave radiometer that operated at wavelengths of 13.5 and 19.0 mm, flown on the *Mariner II* spacecraft to Venus in 1962. The microwave radiometer system weighed ~10 kg and used an average power of 4 W. This early system was designed to take advantage of the high spatial resolution and sensitivity that could be achieved from a spacecraft. Another radio astronomy experiment was placed on the Voyager spacecraft that was launched in the late 1970s to the outer planets and targeted to fly by Neptune in 1989. This experiment measures the radio spectra of planetary emissions in the range 1.2 kHz to 40.4 MHz. The system was designed to measure planetary spectra below the frequency range that is cut off by the earth's ionosphere and to take advantage of the unique viewing geometry provided by the spacecraft.

In the future, we expect to see many more radio astronomy spacecraft experiments. Spacecraft experiments will allow the submillimeter spectral range to be observed without hindrance from the terrestrial atmosphere. Planetary spectroscopy in the submillimeter spectral range is expected to reveal new information about the upper atmospheres of the planets.

IV. Results—Mercury to Neptune

A. MERCURY

Radio measurements of Mercury show no evidence of an atmosphere. Estimates based on other techniques including spacecraft flyby instruments show the atmosphere on Mercury to be extremely tenuous, probably with a surface pressure of less than 0.1 mbar. For the purpose of interpreting the radio data, it can safely be assumed that Mercury is an airless planet.

The radio emission from Mercury is thermal in character, strongly controlled by the high eccentricity of Mercury's orbit and the synchronism between Mercury's spin period and its period of revolution. Solar tidal effects have caused the period of axial rotation of Mercury to be 58.642 days, precisely two-thirds of its orbital period of 87.97 days. One solar day on Mercury is equal to 3 stellar days or 2 Mercurian years. This period equals 176 mean earth solar days.

Because of the synchronism between spin period and revolution period, the sun takes a curious diurnal path in the sky as seen from the surface of Mercury (Fig. 7). At some longitudes, the sun rises and sets twice a Mercury day. At perhelion the insolation is approximately twice its value at aphelion. The insolation reaches a maximum value of ~14 × 10³ W/m², 10 times the value at earth. The visual albedo of Mercury is similar to that of the moon. The spin–orbit coupling and eccentricity combine to cause the surface of Mercury to be heated very nonuniformly in longitude. A pair of longitudes 180° apart alternatively face the sun at perhelion. These longitudes are preferentially heated because their midday insolation occurs when the sun is nearly stationary on the meridian and the solar distance is smallest. At longitudes 90° away from these hot longitudes, the heating is identical but much less than that received at the hot longitudes. The longitudinal temperature variations on Mercury are indicated schematically by the "hot," "warm," and "cold" regions shown on the left in Fig. 7. This pattern of temperature variations is fixed on Mercury because the spin–orbit coupling causes the heating to be cyclic at each longitude (i.e., the same pair of longitudes faces the sun at perhelion).

The solar heating cycle on Mercury suggests that the two longitudes that see the sun directly overhead at perhelion (receiving more than twice as much energy as the longitudes 90° away) will be hotter than those 90° away. This is

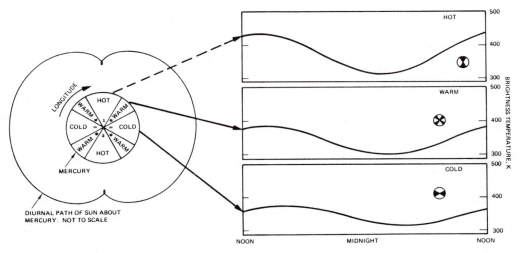

FIG. 7. (Left) Diurnal path of the sun about Mercury and (right) representative brightness temperature curves near 4 cm wavelengths for "hot," "warm," and "cold" regions on Mercury.

borne out by the radio observations. The right-hand portion of Fig. 7 shows a schematic representation of the variation of temperature at "hot," "warm," and "cold" regions on Mercury at a wavelength near 4 cm.

Most existing microwave observations of Mercury do not have sufficient resolving power to resolve the disk of Mercury, so the reported temperatures are averages over the entire visible disk. The measurement results are presented as a brightness temperature T_B as a function of phase angle θ (the sun–Mercury–earth angle) and longitude (on Mercury) L for a particular wavelength. The results are of the form

$$T_B = T_0 + T_1 \cos(\theta + \Phi_1)$$
$$+ T_2 \cos(2L + \Phi_2) \qquad (43)$$

The terms T_0, T_1, T_2, Φ_1, and Φ_2 can be related to the terms δ_1 and Ψ_1 [Eq. (25) and (26)], which in turn are related to the physical characteristics of the surface materials. Measurements exist over the wavelength range from ~1 mm to 11 cm. Morrison (1970) has given a good review of the observations and interpretation of the data. Estimates of the physical parameters of Mercury based on radio, radar, and infrared data and model studies are as follows: $\delta_1/\lambda = 0.9 \pm 0.3$ cm^{-1}, density $\rho = 1.5 \pm 0.4$ g/cm^3, loss tangent $\tan(\Delta) = 0.009 \pm 0.004$, inverse thermal inertia $(k\rho c)^{-1/2} = (15 \pm 6) \times 10^{-6}$ erg^{-1} cm^2 sec$^{1/2}$ K, conductivity $k = (4 \pm 2) \times 10^2$ erg cm^{-1} sec^{-1} K^{-1}, thermal skin depth $L_t = $ ~10 cm, and dielectric constant $\varepsilon = 2.9 \pm 0.5$.

The thermophysical properties of Mercury appear very similar to those of the moon. The surface appears to resemble a loosely packed rock powder. A rather unexpected result is that the mean brightness temperature of Mercury, averaged over a complete diurnal period, increases with wavelength and hence with depth. The explanation for this is believed to be related to a temperature-dependent term in the thermal conductivity. The temperatures on Mercury are sufficiently high during the periods of intense heating that heat transport is enhanced by grain-to-grain thermal radiation. The effective conductivity increases during the daytime, thereby enhancing the heat flow beneath the surface. At night, the temperature drops and the surface material becomes a good thermal insulator, thereby trapping the heat beneath the surface.

B. VENUS

Venus has a very thick and cloudy atmosphere. Early radio astronomy data indicated that the surface temperature of Venus was above 600 K. The mean physical structure of the atmosphere (pressure and temperature profile) is now reasonably well known from data returned by a number of space probes. The surface pressure and temperature are approximately 90 atm and 730 K, respectively. The principal atmospheric constituent is carbon dioxide; its concentration is about 97% at 22 km altitude. Carbon dioxide is a weak radio absorber at low pressures but it produces a broad nonresonant absorption by quadrupole-induced dipole transitions, which are important at high pressures. Other known or suspected microwave absorbers in the atmosphere are H_2O, SO_2, and the sulfuric acid particles in the clouds.

The effective temperature of Venus, deduced from measurements in the infrared, is about 240 ± 8 K, corresponding to an altitude of approximately 60 km. Below this level, the temperature distribution generally follows that for an atmosphere that is in convective equilibrium. Convective equilibrium implies that the temperature gradient in the atmosphere is close to the adiabatic value. The temperature gradient is approximately 8.6 K/km. The high surface temperature is believed to be due to the greenhouse effect. The physical basis for this is that the visible light from the sun is only partially absorbed by the clouds and atmosphere. Some of the light reaches the surface and warms it. The heated surface reradiates in the infrared. The atmosphere is highly absorbing in the infrared spectral region to CO_2 and perhaps H_2O. The atmospheric opacity traps the infrared radiation, thereby raising the surface temperature.

The atmosphere of Venus is opaque at millimeter and short centimeter wavelengths, gradually becoming transparent at longer wavelengths. Atmospheric attenuation is usually expressed by giving the dimensionless quantity "optical depth" along a specified path. A signal that passes through an atmosphere whose optical depth is τ is attenuated by the factor $e^{-\tau}$. The total vertical optical depth of the atmosphere of Venus at a wavelength of 1 cm is estimated to be slightly less than 20 and to vary approximately as λ^{-2} (optical depth = 1 at ~4 cm). A little more than half of the total opacity is due to CO_2, the remaining opacity being produced by the minor constituents in the atmosphere. Near wavelengths of 6 cm and longward, the atmosphere is sufficiently transparent that it is possible to measure the surface temperature of the planet from remote distances. The right side of Fig. 8 shows the continuum spectrum of Venus from a few millimeters wavelength out to approximately 6 cm. The left side of Fig. 8 shows the temperature versus altitude profile of Venus. The brightness temperature is seen to rise from about 225 K and 3 mm to about 700 K near 6 cm. This increase in brightness temperature is due to the decreasing opacity of the atmosphere with increasing wavelength. The decreasing opacity allows radio waves to escape from deeper regions in the atmosphere, where it is warmer due to the adiabatic lapse rate. At wavelengths longer than ~15 cm the brightness temperature decreases to ~600 K. The decrease in brightness temperature at the longest wavelength is still unexplained.

Radio interferometric data, radar data, and spacecraft data have been used to study the surface of Venus, in particular to determine the dielectric constant. Radar reflectivity data place the dielectric constant in the range 4–5. Radio interferometric and polarization radio data suggest a value near 4. These values are considerably greater than the values for Mercury, Mars, and the moon, which range from 2.0 to 2.5. The

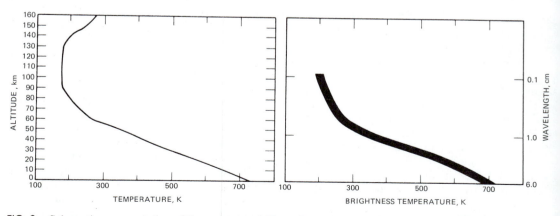

FIG. 8.   Schematic representation of the spectrum of Venus from 1 mm to 6 cm (right) and temperature versus altitude profile (left). The figure illustrates how the atmosphere is probed in altitude by changing the wavelengths of the observations.

higher values are suggestive of a surface composed of dry rock not unlike many rocks of the earth's surface.

Carbon monoxide is an important constituent of the upper atmosphere of Venus. It is formed primarily by the dissociation of carbon dioxide by solar ultraviolet radiation and is removed by chemical and transport processes in the atmosphere. The ground and first excited rotational states of CO (located at very high altitudes in the Venus atmosphere) have been observed to absorb the hot continuous background of the deeper atmosphere. A theoretical line profile is shown in Fig. 9. It has been possible to derive the vertical profile of CO in the upper atmosphere of Venus from these observations. An interesting result noted by planetary radio astronomers is that the concentration of high-altitude CO on Venus varies with solar phase angle. The variability is believed to be the result of large-scale circulation in the upper atmosphere of Venus, although this is by no means proved.

C. MARS

Mars moves in an orbit slightly larger than the earth's, always turning its day side toward the earth as it approaches. Earth-based measurements of the night side are impossible and phase angle coverage is greatly restricted. The axis of rotation is tilted from the perpendicular to the plane of its orbit by 25°, about the same as for

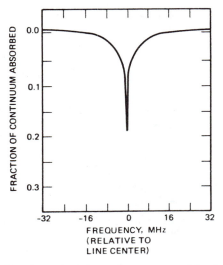

FIG. 9. Theoretical absorption spectrum of $J = 0 \rightarrow 1$ transition of CO in the atmosphere of Venus. Line center frequency is 115 GHz.

the earth. Radio observations of the disk brightness temperature of Mars show it to have a nearly flat spectrum from about 1 mm to 21 cm. The mean disk brightness temperature is about 195 K. As seen from the earth, the average surface disk temperature varies by ±15 K as the subearth point moves from afternoon to morning and from midlatitudes to equatorial latitudes. Thus far, the best observational data used in thermal modeling studies have come from infrared measurements made from spacecraft. The infrared measurements are limited to the near-surface properties; future microwave measurements with the Very Large Array and from spacecraft should add substantial new information about the subsurface properties.

The centimeter radio brightness temperatures of Mars have been found to vary as a function of the central meridian longitude of the planet. Temperature differences as large as 5–10 K are observed over the full range of longitudes. The variations are believed to be due to a nonuniformity in the Martian surface properties; however, no completely satisfactory explanation of the observations exists.

The Martian atmosphere is very tenuous, having a surface pressure some 200 times less than on the earth. The primary constituent of the lower atmosphere is CO_2. Photolysis of CO_2 by solar ultraviolet radiation produces CO and O_2. The diatomic molecule CO plays an important role in determining the millimeter wave spectrum of Mars.

Both the ground state and first excited state transitions of CO have been observed in the Martian atmosphere. The altitude distribution of CO has been inferred by interpreting the observed line shape in terms of pressure broadening. A column abundance of CO equal to 2–5 × 10^{20} molecules/cm^2 has been derived from the measurements. The CO appears to show significant variability, and work to elucidate the physical processes responsible for the variability is continuing.

D. JUPITER

Jupiter is the first planet detected at radio wavelengths. The discovery observations occurred in 1955, at the very low frequency of 22.2 MHz. Prediscovery observations of Jupiter were later traced back to 1950. Subsequent observations of Jupiter revealed that its radio spectrum is exceedingly complex, showing both thermal and nonthermal emission mechanisms.

Thermal emission from the atmosphere dominates the Jovian spectrum shortward of 7 cm. Nonthermal synchrotron emission dominates the spectrum from ~3 m to 7 cm; brightness temperatures exceed 10^5 K for the synchrotron component. Longward of 7.5 m, Jupiter emits strong and sporadic nonthermal radiation. The radiation exhibits complex frequency, time, and polarization structure. The brightness temperature of this component exceeds 10^{17} K, suggesting a coherent source of emission. The schematic appearance of Jupiter's spectrum is shown in Fig. 10.

Observations of Jupiter at high angular resolution with radio interferometers have been used to map the synchrotron radiation from Jupiter's radiation belts and to separate the thermal from the nonthermal synchrotron components. The nonthermal component is easily identifiable with a radio interferometer because it is greatly extended relative to the optical disk of Jupiter and is strongly linearly polarized.

The thermal component originates in the Jovian atmosphere. The observations are consistent with a deep model atmosphere, composed mostly of hydrogen and helium, in convective equilibrium. The principal source of opacity is ammonia (NH_3), which exhibits very strong absorption in the microwave spectral region. The required ammonia abundance is approximately that expected from an atmosphere that contains mixing ratios of elements similar to the solar ratios.

Radio interferometric maps of Jupiter's synchrotron emission have been made at a number of different wavelengths. It has been possible to deduce a great deal of information about Jupiter's magnetosphere from the radio measurements. The radio astronomical measurements provided convincing proof that Jupiter has a strong magnetic field, and this information was used to design the first spacecraft sent to Jupiter. The radio measurements show that the magnetic field is primarily dipolar in shape with the dipole axis tilted about 10° with respect to Jupiter's rotational axis. Using the well-developed theory of synchrotron emission (summarized in Section II,C), it has been possible to estimate energies and densities of the high-energy electrons that are trapped in Jupiter's magnetic field.

At frequencies below 40 MHz, Jupiter is a strong emitter of sporadic nonthermal radiation. A decameter wavelength (DAM) component observable from the ground is characterized by complex, highly organized structure in the frequency–time domain and on the observer's position relative to Jupiter. The satellite Io modulates the DAM emission. The *Voyager* spacecraft added significantly to our knowledge of this low-frequency component when it flew by Jupiter in 1979. A kilometric wavelength (KOM) component was discovered at frequencies below 1 MHz and the observations of the DAM component were significantly improved. The extremely high brightness temperatures ($>10^{17}$ K), narrow-bandwidth emissions, and

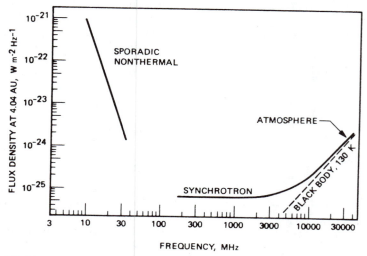

FIG. 10. Schematic representation of the spectrum of Jupiter, showing the frequency ranges for which atmospheric emission dominates, synchrotron emission dominates, and sporadic nonthermal emission dominates.

sporadic nature all suggest that the very low frequency emissions from Jupiter are generated by energetic particles acting coherently and interacting with the plasma that surrounds Jupiter. The details of the emission process are not well understood. This is an active field of research at present.

E. SATURN

Radio emission from Saturn has been observed from the earth over a wavelength range from 1 mm to approximately 70 cm. The emission is thermal throughout this band, arising in both the atmosphere and the rings. The atmospheric emission is similar to that observed from Jupiter. Model studies indicate that Saturn, like Jupiter, has a deep convective atmosphere. Hydrogen and helium form the bulk of the atmosphere, whereas ammonia in trace amounts provides most of the microwave opacity.

Interferometer observations of Saturn at centimeter and millimeter wavelengths have detected thermal emission from the ring particles. The rings have a low brightness, on the order of 10 K. Ring particle sizes larger than a few centimeters are suggested by the radio observations. The observations are consistent with the bulk properties of the ring particles being those of water ice.

The Voyager spacecraft detected two distinct classes of nonthermal emissions from Saturn at frequencies below 1 MHz. These emissions are not observable from the earth because of the nontransparency of the earth's ionosphere at frequencies below a few megacycles. The first class is a relatively narrowband polarized emission, which is called Saturn kilometric radiation. The second class is a broadband, impulsive emission called Saturn electrostatic discharge. The source of these emissions is under investigation.

F. URANUS

Uranus is unique among the planets in having its rotation axis tilted close to the plane of the ecliptic. The north pole of Uranus is inclined ~98° to the ecliptic plane (98° = 8° to the south of the plane), and the seasons on Uranus average 21 terrestrial years in length. The effect of this geometry on the large-scale circulation of the Uranian atmosphere is not yet understood, but it is expected to be significant.

As in the case of both Jupiter and Saturn, the disk brightness temperatures of Uranus significantly exceed the expected equilibrium temperature. The observed temperatures are greater than 100 K at wavelengths greater than a few millimeters and longward, whereas the predicted effective temperature is only about 55 K. The radio emission from Uranus is unpolarized within the measurement uncertainties of a few percent. Interferometer observations of Uranus show the emission to be confined to the solid angle of the visible disk, providing evidence that the excess emission is from the atmosphere and is not synchrotron emission. It is believed that the emission from Uranus is thermal, originating in the atmosphere of the planet.

The radio emissions from Uranus arise from sufficient depths that collision-induced absorption by hydrogen is an important source of opacity at millimeter wavelengths. Ammonia, which appears to exist in solar cosmic abundance in the atmospheres of Jupiter and Saturn, is severely depleted in the Uranus atmosphere. Another interesting aspect of the radio emission from Uranus is that it is variable. The cause of the variability is not understood, although it is believed to be related to the large inclination of Uranus' rotational axis.

G. NEPTUNE

Disk brightness temperature measurements of Neptune are sparse, but existing measurements suggest a similarity to those of Uranus. The disk brightness temperatures exceed the predicted equilibrium temperatures by 50 K or more. Interferometer observations are not yet available, so nonthermal synchrotron emission cannot be ruled out by the data. As in the case of Uranus, model studies suggest that ammonia must be depleted on Neptune to explain the high brightness temperatures that are observed.

V. Prospects for the Future

Until quite recently, planetary observations were limited to measurements of the disk-averaged brightness temperatures, or to high-resolution observations of the planets having the largest angular diameters. Instrumentation that is just starting to become available will eventually allow the planets to be studied in nearly the same detail in which the earth is being studied with combined sensors on weather satellites and ground stations. As angular resolution improves with time, it should be possible to map the vertical and horizontal distribution of certain chemical species within planetary atmospheres, to measure wind speeds, and to search for local variations in subsurface properties. Both ground-based radio telescopes and spacecraft radio receivers will play a role in future observations. Large telescopes on the ground or possibly placed on the moon will have sufficient resolution to examine the global properties of the planets out to Neptune and to study satellites and asteroids. The advent of lighter-weight and lower-power radio receivers will enhance the possibilities of placing radio experiments on spacecraft. Radio telescopes working at submillimeter wavelengths are gradually becoming a reality as a few mountaintop observatories are nearing completion and plans for a submillimeter space telescope are being discussed in a number of countries. The shorter-wavelength region of the spectrum will provide new opportunities to study the upper atmospheres of the planets with spectroscopic techniques that can detect an abundance of heretofore unobserved transitions that occur in the submillimeter and infrared. In addition, the shorter wavelengths will provide high angular resolution with only moderate-size telescopes. In summary, the future prospects for planetary radio astronomy are bright. This optimistic outlook is based on what is possible with technology that is currently available; the evolution of this technology will make the job more affordable.

BIBLIOGRAPHY

Berge, G. L., and Gulkis, S. (1976). Earth-based radio observations of Jupiter: Millimeter to meter wavelengths. *In* "Jupiter" (T. Gehrels, ed.), University of Arizona Press, Tuscon.

Dessler, A. J. (Ed.) (1983). "Physics of the Jovian Magnetosphere." University Press, Cambridge, U.K.

Lewis, J. S., and Prinn, R. G. (1984). "Planets and Their Atmospheres." Academic Press, Orlando, Florida.

Morrison, D. (1970). Thermophysics of the planet Mercury. *Space Sci. Rev.* **11**, 271–307.

Muhleman, D. O., Orton, G. S., and Berge, G. L. (1979). A model of the Venus atmosphere from radio, radar, and occultation observations. *Astrophys. J.* **234**, 733–745.

Sullivan, W. T., III (Ed.) (1984). "The Early Years of Radio Astronomy." University Press, Cambridge, U.K.

RELATIVITY, GENERAL

James L. Anderson   *Stevens Institute of Technology*

GLOSSARY

Binary pulsar: Double star system, one of whose components is a neutron star.

Doppler shift: Fractional change in frequency of light due to relative motion between source and observer.

Electrodynamics: Theory of electric and magnetic fields and of the interactions of the charged particles that produce them.

Hubble constant: Ratio of velocity of recession to distance of galaxies.

Mapping: Association of points of a space–time manifold with other points of this manifold.

Perihelion: Point in the orbit of a planet when it is closest to the sun.

Riemannian geometry: Geometry in which the distance between neighboring points is defined by a metric and is quadratic in the coordinate differences between the points.

The general theory of relativity is first and foremost a theory of gravity. At the same time it is also a description of the geometry of space and time that differs profoundly from all such previous descriptions in that this geometry is identified with the gravitational field. The predictions of the general theory differ both quantitatively and qualitatively from those of the Newtonian theory of gravity. Although the quantitative differences between the two theories are usually small, they have been extensively tested in the solar system and other astro-physical systems. Today the agreement between observation and theory is better than 0.5%. However, the qualitative predictions of the theory are its most exciting and challenging feature. Among others, the theory predicts the existence of gravitational radiation. Although this radiation has not been directly observed, the effects of its emission have been observed in the binary pulsar PSR 1913 + 16 and agree with the predictions of the theory to within 3%. The theory also predicts the phenomenon of gravitational collapse leading to the creation of black holes. There is strong observational evidence that such objects exist in the universe. And finally, the general theory serves as the basis for our best description of the universe as a whole, the so-called hot big-bang cosmology. [*See* BLACK HOLES (ASTRONOMY); COSMOLOGY.]

I. Space–Time Theories of Physics

A. THE SPACE-TIME MANIFOLD

To understand the revolution wrought by the general theory it is useful to set it in a framework that encompasses it as well as the other two major space–time structures of physics, Newtonian mechanics and special relativity. Basic to each of these structures is the notion of the space–time manifold consisting of a four-dimensional continuum of points. It is assumed only that any finite piece of this manifold can be mapped in a one-to-one manner onto a connected region of the four-dimensional Euclidean plane. Otherwise, these points are featureless and indistinguishable from each other, and the manifold as a whole is characterized only by its topological properties. While this manifold is not itself associated with any physical entity, it serves as the basis for the construction of the geometrical structures that are to be associated with such objects. [*See* MANIFOLD GEOMETRY.]

Since the points of the space–time manifold can be mapped onto the four-dimensional Euclidean plane, one can coordinatize the manifold by assigning to each point the coordinates of its image point in the Euclidean plane, x^μ, where the index μ takes on the values 0, 1, 2, 3. Because the points of the manifold are assumed to be indistinguishable, this mapping is to a large extent arbitrary and hence the coordinatization is also arbitrary. Depending on the topological structure of the manifold, it may be necessary to cover it with several overlapping coordinate "patches" to avoid singularities in the coordinates. If, for example, the manifold has the topology of the surface of a ball, it is necessary to employ two such patches to avoid the coordinate singularity one encounters at the pole when using the customary polar coordinates.

If the manifold is coordinatized in two different ways, for example, by using Cartesian or spherical coordinates, the coordinates used for one such coordinatization must be functions of those used for the other and vice versa. This relation is called a coordinate transformation. In order to preserve the continuity and differentiability of the manifold it must be continuous, nonsingular, and differentiable.

B. GEOMETRICAL STRUCTURES

In space–time theories, physical entities are associated with geometrical objects that are constructed on the space–time manifold. These objects can be of many different types. A curve can be associated with the trajectory of a particle and is specified by designating the points of the manifold through which it passes. This can be done by giving the coordinates of these points as functions $x^\mu(\lambda)$ of a monotonically varying parameter λ along the curve. Likewise, a two-dimensional surface could be specified by giving the coordinates of the surface as functions of two monotonic parameters, and similarly for three- and four-dimensional regions.

In addition to collections of points, one can introduce geometrical objects that consist of a set of numbers assigned to a point. These numbers are said to constitute the components of the geometrical object. The components of the velocity of a particle at a particular point in its trajectory would constitute such a collection. If the components are specified along a trajectory, a surface, or any other part of the space–time manifold, they are said to constitute a field. The temperature in a room can, for example, be as-

sociated with a one-component field. Likewise, the electromagnetic field surrounding a moving charge can be associated with a field consisting of six components.

The basic requirement that must be met in order that an object be geometrical is a consequence of the indistinguishability of the points of the space–time manifold. It is that under a coordinate transformation, the transformed components of the object must be functions solely of its original components and the coordinate transformation. This requirement is simply met in the case of curves, surfaces, and so forth; for example, given a curve, the transformed curve can be immediately calculated given the coordinate transformation.

An especially useful group of geometrical objects for associating with physical entities are those whose transformed components are linear, homogeneous functions of the original components. The simplest example is the single-component object called a scalar $\phi(x)$. Under a coordinate transformation, its transformed value is just equal to its original value. The other linear, homogeneous objects constitute the vectors, tensors, and pseudoscalars, pseudovectors, and pseudotensors. Vectors and pseudovectors are four-component objects (they come in two varieties called covectors and contravectors), while tensors and pseudotensors have larger numbers of components. There are also objects whose transformed components are linear but not homogeneous functions of the original components. Finally, there are objects whose transformed components are nonlinear functions of the original components, although such objects have not been used to any great extent. In all cases, however, the nature of the object is characterized by its transformation law.

It should be pointed out that not all objects one can construct are geometrical. The gradient of a scalar is a geometrical object while the gradients of vectors and tensors in general are not.

C. LAWS OF MOTION

The basis for associating geometrical objects with physical entities is purely utilitarian—there is no general procedure for making this association. The numerical values that these objects can assume are taken to correspond to the observed values of the physical entities with which they are associated. Since not all such values can in general be observed, it is necessary to formulate a set of rules, called here laws of motion, that select from the totality of values a given set of geometrical objects can have, a subset that corresponds to possible observed values. Thus, if it is decided to associate a curve with the trajectory of a planet, one would have to discover a system of equations such as those obtained from Newton's laws of motion to select from the totality of all curves the subset that would correspond to actual planetary trajectories. [*See* CELESTIAL MECHANICS.]

One requirement that one would like to be fulfilled by all laws of motion is that of completeness—every set of values allowed by them must, at least in principle, be observable. It is, after all, the purpose of the laws of motion to rule out unobservable sets of values. Nevertheless, there are problems associated with the imposition of such a requirement. There may, for example, be practical limitations on our ability to observe all the values allowed by a given set of laws. It is unlikely that we will ever be able to attain the energies needed to verify some of the predictions of the grand unified theories that are being considered today. However, if one of these theories correctly described all that we can observe about elementary particle interactions, we would not discard it because we could not directly observe its other predictions. More troubling, however, are limitations in principle on what we can observe. When applied to the universe as a whole, the general theory of relativity allows for many different possibilities, yet by its very nature we can observe only the universe in which we live. In the strict sense, then, the general theory should be considered incomplete. Nevertheless, it does correctly describe a vast range of phenomena, and so far there does not exist a more restrictive theory that does so. Therefore, probably, the best we can do is to require that a law not admit values that could be observed if they existed but do not.

D. Principle of General Covariance

However the laws of motion are formulated, they must be such as to be independent of a particular coordinatization of the space–time manifold. This requirement is called the principle of general covariance and was one of the basic principles employed by Einstein when he formulated the general theory. For the principle to hold, the laws of motion for a given set of geometrical objects must be such that all of the transforms of a set of values of these objects that satisfy the laws of motion must also satisfy these laws.

The principle of general covariance is not, as has sometimes been suggested, an empty principle that can be satisfied by any set of physical laws. If, for example, the geometrical object chosen to be associated with a given physical entity is a scalar field $\phi(x)$, then the only generally covariant law that can be formulated involving only this object is the trivial equation

$$\phi(x) = \text{const} \qquad (1)$$

In order to formulate a nontrivial law of motion for ϕ it is necessary to introduce other geometrical objects in addition to the scalar field. One possibility is to introduce a symmetric tensor field $g_{\mu\nu}(x)$ and its inverse $g^{\mu\nu}(x)$, which are related by the equation

$$g^{\mu\rho}g_{\rho\nu} = \delta_\nu^\mu \qquad (2)$$

where δ_ν^μ is the Kronecker delta, with values given by

$$\begin{aligned} \delta_\nu^\mu &= 1 \qquad \mu = \nu \\ &= 0 \qquad \mu \neq \nu \end{aligned} \qquad (3)$$

and where the appearance of a double index such as ρ implies a summation over its range of values. One can then take as the law of motion for ϕ the equation

$$(\sqrt{-g}\, g^{\mu\nu}\phi_{,\mu})_{,\nu} = 0 \qquad (4)$$

where g is the determinant of $g_{\mu\nu}$ and $,\mu := \partial/\partial x^\mu$. The tensor field $g_{\mu\nu}$ cannot itself be given as a function of the coordinates directly since in that case Eq. (4) would not be generally covariant. Rather, it must in turn satisfy a law of motion that is itself generally covariant. If one requires that this law involve no higher than second derivatives of $g_{\mu\nu}$, it can be shown that there are in fact only three essentially different such laws for this object. One can form laws of motion for ϕ other than Eq. (4), but in each case it is necessary to introduce other geometrical objects for the purpose and to formulate laws of motion for them.

E. Absolute and Dynamical Objects

To understand the revolutionary nature of the general theory it is necessary to distinguish between two essentially different types of objects that appear in the various space–time theories. We call them absolute and dynamical, respectively. If the totality of values allowed by the laws of motion for some geometrical object, such as the tensor $g_{\mu\nu}$ introduced above are such that they can all be transformed into each other by coordinate transformations, we say that that object is an absolute object in the theory. This can occur if the law of motion for the object does not involve any of the other objects in the theory. The remaining objects in the theory are called dynamical objects. The electric fields associated with different charge distributions, for example, cannot in general be transformed into one another and hence must be associated with a dynamical object.

Given a theory with absolute objects, it is possible to coordinatize the space–time manifold so that they take on a specific set of values. In the case of the tensor $g_{\mu\nu}$, one of the three possible laws of motion mentioned above is such that every set of values allowed by it can be transformed so that, for every point of the space–time manifold, $g_{\mu\nu} = \mathrm{diag}(1, -1, -1, -1)$. If these values are substituted into the other laws of motion they will no longer be generally covariant, but rather they will be covariant with respect to some subgroup of coordinate transformations. This subgroup will leave invariant the chosen values of the geometrical object (or objects) and will be called the invariance group of the theory. The structure of this group will be independent of which particular set of values allowed by the laws of motion is chosen for the absolute objects. If there are no absolute objects then the invariance group is just the group of all allowed coordinate transformations.

Absolute objects are seen to play a preferred role in a theory—their values are independent of the values of the dynamical objects of the theory while the converse is in general not the case. (If it is, the absolute objects become superfluous and can be ignored.) A theory with absolute objects thus violates a kind of general law of action and reaction. We will see that both Newtonian mechanics and special relativity contain absolute objects while the general theory does not.

II. Newtonian Mechanics

A. ABSOLUTE TIME AND SPACE

In his formulation of the laws of motion, Newton introduced a number of absolute objects, chief of which were his absolute space and absolute time. Absolute time corresponds to the foliation of the space–time manifold by a

one-parameter family of nonintersecting three-dimensional hypersurfaces, which we call planes of absolute simultaneity. All of the points in a given plane are taken to be simultaneous with respect to each other. Furthermore, these planes are such that the curves associated with the trajectories of particles intersect each plane once and only once. The "time" at which such an intersection takes place is characterized by the value of the parameter associated with the plane being intersected. These planes are absolute in that their existence and structure are assumed to be independent of the existence or behavior of any other physical system in the space–time.

In Newtonian mechanics the interaction of particles is assumed to be instantaneous as in Newton's action-at-a-distance theory of gravity. Consequently, such interactions take place between the points on the trajectories that lie in the same plane of absolute simultaneity. As a consequence, these planes can be observed by giving an impulse to one of a number of interacting particles and noting where, on the trajectories of the other particles, the transmitted impulse acts.

Newton's absolute space corresponds to a unique three-parameter congruence of nonintersecting curves that fill the space–time manifold; that is, through each point of the manifold passes one and only one such curve. Furthermore, each curve passes through one and only one point of each plane of absolute simultaneity. The existence of such a congruence would therefore imply that there exists a unique one-to-one relation between the points in any two planes of absolute simultaneity. The "location" of a space–time point would be characterized by the parameters associated with the curve of the congruence passing through it.

The notion of absolute space brings with it the notion of absolute rest: a particle is absolutely at rest if its trajectory can be associated with one of the curves of the congruence. However, unlike the planes of absolute simultaneity that are needed in the formulation of the laws of motion of material particles, these laws do not require the existence of the space–time congruence of curves that constitute Newton's absolute space, nor do they afford any way of detecting a state of absolute rest. This property of the Newtonian laws of motion is known as the principle of Galilean relativity. Furthermore, since the congruence is not needed in the formulation of these laws, we can dispense with it and hence with

Newton's absolute space altogether as an unobservable element of the theory.

B. Free Bodies

In his setting down of the three laws of motion, Newton was careful to give the first law, "Every body continues in its state of rest, or of uniform motion in a right line, unless it is compelled to change that state by forces impressed upon it," as separate and distinct from the second law. He clearly did not consider it, as it is sometimes taken to be, a special case of the second law. In effect, the first law supposes a class of curves, the straight (right) lines, to exist in the space–time manifold. Furthermore, these curves correspond to the trajectories of a class of objects on which no forces act, namely free bodies. As a consequence, these curves are absolute objects of the theory. Furthermore, they, like the planes of absolute simultaneity, are needed to formulate the laws of motion for bodies on which forces act.

C. Galilean Invariance

One can always coordinatize the space–time manifold in such a way that the parameter t, which labels the different planes of absolute simultaneity, is taken to be one of these coordinates. When this is done, the equation that defines these planes is simply

$$t = \text{const} \tag{5}$$

Furthermore, the remaining coordinates can be chosen so that the equations of the curves associated with the trajectories of the free bodies are linear in t; that is, they are of the form

$$x_i = v_i t + x_{0i} \tag{6}$$

where the index i takes on the values 1, 2, 3 and v_i and x_{0i} are constants. The constants v_i are the components of the "velocity" of the free body whose trajectory is associated with this curve and the x_{0i} are its initial positions. When expressed in terms of these coordinates, the laws of motion of Newtonian mechanics take on their usual form.

Since the planes of absolute simultaneity and the straight lines constitute the absolute objects of Newtonian mechanics and enter into the formulation of the laws of motion of all Newtonian systems, the subgroup of coordinate transformations that leave them invariant as a whole constitutes the invariance group of Newtonian mechanics. In addition to the group of spatial rotations and translations and time translations, this group consists of the Galilean transformations given by

$$x_i' = x_i + V_i t \tag{7a}$$

and

$$t' = t \tag{7b}$$

where the V_i are the components of the velocity that characterize a particular transformation of the group.

In terms of the primed coordinates, we see that the equations of a straight line (6) take the form

$$x_i' = v_i' t' + x_{0i} \tag{8}$$

where the transformed velocity components v_i' are given by the Galilean law of addition for velocities:

$$v_i' = v_i + V_i \tag{9}$$

III. Special Relativity

A. Light Cones

The transition from Newtonian mechanics to special relativity in the early part of this century involved the abandonment of the Newtonian planes of absolute simultaneity and their replacement by a new set of absolute objects, the light cones. With the completion of the laws of electrodynamics by Maxwell in the middle of the last century it became evident that electromagnetic interactions between charged particles were not instantaneous but rather were transmitted with a finite velocity, the speed of light. This fact, coupled with the Galilean velocity addition law, made it appear possible that some electromechanical experiment could be devised for the detection of a state of absolute rest and thus reinstate Newton's absolute space. However, all attempts to do so, such as those of Michelson and Morley and Trouton and Noble, proved fruitless. In one way or another, these experiments sought to measure the absolute velocity of the earth with respect to this absolute space. Even though they were sensitive enough to detect a velocity as small as 30 km/sec, which is much less than the known velocity of the earth with respect to the galaxy, no such motion was ever detected. [See RELATIVITY, SPECIAL.]

Einstein realized that if all interactions were

transmitted with a finite velocity there was no way objectively to observe the Newtonian planes of absolute simultaneity and that they, like Newton's absolute space, should be eliminated from the theory. It was his analysis of the meaning of absolute simultaneity and its rejection by him that distinguished his approach to special relativity from those of Lorentz and Poincaré. Since, however, unlike absolute space, the planes of absolute simultaneity were needed in the formulation of the laws of motion for material bodies, it was necessary to replace them by some other structure. The key to doing this lay in Einstein's postulate that the velocity of light is independent of the motion of the source. If this is the case, and to date all experimental evidence supports this postulate, the totality of all light ray trajectories form an invariant structure and can be associated with a corresponding family of three-dimensional surfaces in the space–time manifold, the light cones. Just as in Newtonian mechanics, where through each point there passes a unique plane of absolute simultaneity, in special relativity through each point there passes a light cone that consists of all of the points on the curves passing through this point that correspond to the trajectories of light rays.

In Newtonian mechanics the interaction of particles was assumed to take place between the points on the curves associated with their trajectories that lay in the same plane of absolute simultaneity. In special relativity this interaction is assumed to take place, depending on the type of interaction that exists between the particles, either between points that lie in the same light cone or between one such point and points in the interior of the light cone associated with this point. The electromagnetic interaction, for example, takes place between points lying in the same light cone. Consequently, these light cones can be observed by giving an impulse to one member of a group of charged particles and noting where, on the trajectories of the other particles, the transmitted impulse acts.

B. FREE BODIES

In addition to the absolute light cones, special relativity assumes, like Newtonian mechanics, a family of curves, the straight lines, that are associated with the trajectories of free bodies. There is, however, an important difference between the two theories. In Newtonian mechanics any straight line that intersects all of the planes of absolute simultaneity is assumed to correspond to the trajectory of a free body. In special relativity, on the other hand, only the straight lines that correspond to free bodies with velocities less than or equal to the speed of light are assumed to correspond to observable free bodies. These straight lines are such that, given a point lying on one of them, the other points lying on it are either interior to or lie on the light cone associated with that point.

C. LORENTZ INVARIANCE

Together, the light cones and straight lines constitute the absolute objects of special relativity. It can be shown that one can coordinatize the space–time manifold in such a way that the points with coordinates x^μ lying on a straight line are given by the equations

$$x^\mu = v^\mu \lambda + x_0^\mu \qquad (10)$$

where the v^μ and x_0^μ are constants and λ is a monotone increasing parameter along the line. For the straight lines that correspond to the trajectories of free bodies, the v^μ are constrained by the condition that

$$\eta_{\mu\nu} v^\mu v^\nu \geqq 0 \qquad (11)$$

where $\eta_{\mu\nu} = \text{diag}(1, -1, -1, -1)$. Provided that $\eta_{\mu\nu} v^\mu v^\nu > 0$, it is always possible to choose the parameter λ such that $\eta_{\mu\nu} v^\mu v^\nu = 1$. In this case the v^μ are said to constitute the components of the four-velocity of the particle and $\tau = \lambda$ is the proper time along the line.

In addition to the form (10) for the straight lines, coordinates can be chosen so that the points x^μ lying on the light cone associated with the point x_0^μ satisfy the equation

$$\eta_{\mu\nu}(x^\mu - x_0^\mu)(x^\nu - x_0^\nu) = 0 \qquad (12)$$

For points interior to this light cone the quantity on the left side of this equation is greater than zero, while for points exterior to it it is less than zero. In what follows, coordinates in which the straight lines and light cones are described by Eq. (10) and (12) will be called inertial coordinates.

Since the light cones and straight lines are absolute objects in special relativity, the coordinate transformations that leave these structures invariant constitute the invariance group of special relativity. In an inertial coordinate system, these transformations have the form

$$x'^\mu = \alpha_\nu^\mu x^\nu + b^\mu \qquad (13)$$

where α_ν^μ and b^ν are constants. The b^μ are arbitrary while the α_ν^μ are constrained to satisfy the conditions

$$\eta_{\mu\nu}\alpha_\rho^\mu\alpha_\sigma^\nu = \eta_{\rho\sigma} \qquad (14)$$

These transformations form a group, the inhomogeneous Lorentz group, each member of which is characterized by the 10 arbitrary values one can assign to the α_ν^μ and b^μ. This group contains, as subgroups, the three-dimensional rotation group and the group of spatial and temporal translations. It also includes the group of Lorentz transformations, now called Lorentz boosts. A boost along the x axis takes the form

$$\begin{aligned} x'^0 &= \gamma(x^0 + \beta x^1) \\ x'^1 &= \gamma(\beta x^0 + x^1) \\ x'^2 &= x^2 \\ x'^3 &= x^3 \end{aligned} \qquad (15)$$

where $\gamma = (1 - \beta^2)^{-1/2}$ and β is a parameter that characterizes the boost. These transformation equations take their more familiar form if we set $x^\mu = (ct, x, y, z)$ and similarly for x'^μ and $\beta = v/c$, where c is the velocity of light, in which case v is the velocity associated with the transformation. In special relativity, the Lorentz boosts replace the Galilean transformations of Newtonian mechanics just as the light cones replace the planes of absolute simultaneity. Also, the Galilean law of addition for velocities, Eq. (9), is no longer valid. For a boost in the x direction, the transformed components v_i' of the velocity of a body are related to its original components v_i by the equations

$$v_i' = \delta(v_1 + v) \qquad v_2' = \gamma^{-1}\,\delta v_2 \qquad v_3' = \gamma^{-1}\,\delta v_3 \qquad (16)$$

where $\delta = (1 + v_1 v/c^2)^{-1}$.

D. The Space–Time Metric

Equations (10) and (12) for straight lines and light cones are given in a special coordinate system in which they assume these simple forms. It is possible to write generally covariant equations for these objects by introducing a symmetric second rank tensor $g_{\mu\nu}$ of signature -2 together with its inverse $g^{\mu\nu}$. With its help, the light cones can be characterized by the surfaces $\phi(x) = 0$, where ϕ satisfies the covariant equation

$$g^{\mu\nu}\phi_{,\mu}\phi_{,\nu} = 0 \qquad (17)$$

To construct an equation for a straight line we first introduce the Christoffel symbols $\{^\mu_{\rho\sigma}\}$ defined by

$$\{^\mu_{\rho\sigma}\} = \tfrac{1}{2}g^{\mu\nu}(g_{\rho\nu,\sigma} + g_{\nu\rho,\sigma} - g_{\rho\sigma,\nu}) \qquad (18)$$

These quantities constitute the components of a geometrical object that is linear but not homogeneous. With their help, the equations for the coordinates $x^\mu(\lambda)$ of the points lying on a straight line can now be written as

$$d^2x^\mu/d\lambda^2 + \{^\mu_{\rho\sigma}\}(dx^\rho/d\lambda)(dx^\sigma/d\lambda) = 0 \qquad (19)$$

where again λ is a monotone increasing parameter along the curve. One can choose λ so that $g_{\mu\nu}\,dx^\mu/d\lambda\,dx^\nu/d\lambda = 1$, in which case it is the proper time along the line. One can also use Eq. (19) to characterize the trajectories of light rays if one adds the condition that $g_{\mu\nu}\,dx^\mu/d\lambda\,dx^\nu/d\lambda = 0$. Such rays have the property that they serve as the generators of the light cones. Equation (19) is usually referred to as the geodesic equation since it has the same form as the equation for a geodesic curve, that is, a curve of minimum length connecting two points in a Riemannian space with a metric $g_{\mu\nu}$.

Having introduced the tensor $g_{\mu\nu}$, it now becomes necessary to construct a law of motion for it. In special relativity this law is taken to be

$$R_{\mu\nu\rho\sigma} = 0 \qquad (20)$$

where the tensor $R_{\mu\nu\rho\sigma}$, called the Riemann–Christoffel tensor, is constructed from the tensor $g_{\mu\nu}$ according to

$$\begin{aligned} R_{\mu\nu\rho\sigma} = \tfrac{1}{2}(g_{\mu\rho,\nu\sigma} &+ g_{\nu\sigma,\mu\rho} - g_{\mu\sigma,\nu\rho} - g_{\nu\rho,\mu\sigma}) \\ &+ g_{\alpha\beta}(\{^\alpha_{\mu\rho}\}\{^\beta_{\nu\sigma}\} - \{^\alpha_{\mu\sigma}\}\{^\beta_{\nu\rho}\}) \end{aligned} \qquad (21)$$

It appears in the equation of geodesic deviation that governs the separation between two neighboring, freely falling bodies. When all of the components of $R_{\mu\nu\rho\sigma}$ vanish, this separation remains constant.

It can be shown that every solution of Eq. (20) can be transformed so that $g_{\mu\nu} = \eta_{\mu\nu}$ everywhere on the space–time manifold, in which case $g_{\mu\nu}$ is said to take on its Minkowski values. In a coordinate system in which $g_{\mu\nu}$ has this form, Eq. (17) becomes

$$\eta^{\mu\nu}\phi_{,\mu}\phi_{,\nu} = 0 \qquad (17a)$$

It is seen that the surface defined by Eq. (12) satisfies this equation. Likewise, in this coordinate system, Eq. (19) reduces to

$$d^2x^\mu/d\lambda^2 = 0 \qquad (19a)$$

and has, as its solution, the curves defined by Eq. (16).

Since $g_{\mu\nu}$ can always be transformed to its Minkowski values, it is seen to be an absolute object in the theory and the group of transformations that leave it invariant is again the inhomogeneous Lorentz group. In all respects $g_{\mu\nu}$ is equivalent to the straight lines and light cones of the theory. Furthermore, one can either construct laws of motion that employ inertial coordinates and that are covariant with respect to the inhomogeneous Lorentz group or construct generally covariant laws with the help of the $g_{\mu\nu}$. When $g_{\mu\nu}$ is transformed to take on the values $\eta_{\mu\nu}$, the latter equations reduce to the former.

The Riemann–Christoffel tensor arose in the study of the geometry of manifolds with Riemannian metrics. In such a geometry one defines a distance ds between neighboring points of the manifold with coordinates x^μ and $x^\mu + dx^\mu$ to be

$$ds^2 = g_{\mu\nu}(x)\, dx^\mu\, dx^\nu \qquad (22)$$

where $g_{\mu\nu}$, a symmetric tensor field, is the metric of the manifold. The vanishing of the Riemann–Christoffel tensor can be shown to be the necessary and sufficient condition for the geometry to be flat; that is, the metric can always be transformed to a constant tensor everywhere on the manifold. Since the tensor $g_{\mu\nu}$ introduced above is an absolute object it is sometimes referred to as the metric of the flat space–time of special relativity. Although we will not need to make use of this geometrical interpretation, it will sometimes prove convenient to give an expression for ds as a way of specifying the components of $g_{\mu\nu}$.

IV. General Relativity

A. THE PRINCIPLE OF EQUIVALENCE

After Einstein formulated the special theory of relativity he turned his attention to, among other things, the problem of constructing a Lorentz-invariant theory of gravity. Newton thought of gravity as an action-at-a-distance force between massive bodies and as transmitted instantaneously between them. Since special relativity required a finite speed of transmission, Einstein sought to construct a relativistic field theory of gravity. The simplest object to associate with the gravitational field was a scalar field. However, a difficulty presented itself when he came to construct a source for this field. In the Newtonian theory, the gravitational attraction between bodies was proportional to their masses. In special relativity, however, energy has associated with it an equivalent mass through the relation $E = mc^2$. Consequently, Einstein argued, mass density by itself could not be the sole source of the gravitational field. At the same time, energy density could not be used since it is not, by itself, associated with a geometrical object in special relativity but rather with one component of a tensor field.

While thinking about the problem of gravity, Einstein was struck by a peculiarity of the gravitational interaction between bodies, namely the constancy of the ratio of the inertial to the gravitational mass of all material bodies. In Newtonian mechanics, mass enters in two essentially different ways—as inertial mass in the second law of motion and as gravitational mass in the law of gravitational interaction. Logically, these two masses have nothing to do with one another. Inertial mass measures the resistance of a body to forces imposed on it while gravitational mass determines, in the same way as electric charge determines the strength of the electrical force between charged bodies, the strength of the gravitational force between massive bodies. Galileo was the first to demonstrate this constancy by observing that the acceleration experienced by objects in the earth's gravitational field was independent of their mass. In 1891, Eötvös demonstrated it to an accuracy of one part in 10^8. More recent determinations by groups in Princeton and Moscow have established this constancy to better than one part in 10^{11}.

While there was no explanation for this constancy, Einstein realized that it called into question the existence of one of the absolute objects of both Newtonian mechanics and special relativity, the free bodies. If indeed this ratio was a universal constant, then there could be no such thing as a gravitationally uncharged body since zero gravitational mass would then imply zero inertial mass. Einstein also realized that this constancy meant that it would be impossible to distinguish locally, that is, in a sufficiently small region of space–time, between inertial and gravitational effects through their action on material bodies. An observer in an elevator being accelerated upward with an acceleration equal to that produced by the earth's gravity would see objects fall to the floor of the elevator in exactly the same way that they fall on earth, that is, with an acceleration that is independent of their mass.

After this realization, Einstein made a characteristic leap of imagination. He postulated that it is impossible to distinguish locally between inertial and gravitational effects by any means. One of the consequences of this postulate, called by him the principle of equivalence, is that light should be bent in a gravitational field just as it would appear to be to an observer in an accelerating elevator. But if this is the case, the light cones of special relativity would no longer be absolute objects either, and this in turn would mean that the metric of special relativistic space–time would not be an absolute object.

The principle of equivalence however, implied even more. If inertial and gravitational effects are indistinguishable from each other locally, then one and the same object could be used to characterize both effects. Since it is the metric $g_{\mu\nu}$ that is responsible for the inertial effects one observes in special relativity, $g_{\mu\nu}$ should also be associated with the gravitational field. In effect, geometry and gravity became simply different aspects of the same thing. Actually, one never needs to interpret $g_{\mu\nu}$ as a metric. One can identify it solely with the gravitational field. This identification has, as a consequence, that $g_{\mu\nu}$ must be a dynamical object since the gravitational field clearly must be such. Having recognized this fact, Einstein then turned his attention to the problem of constructing a law of motion for this object.

B. THE PRINCIPLE OF GENERAL INVARIANCE

In his attempts to construct a law of motion for $g_{\mu\nu}$, Einstein proposed that these laws should be generally covariant. However, we have already seen that the laws of motion of special relativity could be cast in generally covariant form with the introduction of a metric satisfying Eq. (20). But such a metric was absolute and Einstein wanted a law of motion for a dynamical $g_{\mu\nu}$. Consequently, what Einstein was really requiring was not general covariance but rather general invariance, that is, that the invariance group of the laws of motion should be the same as their covariance group, namely the group of all arbitrary coordinate transformations. As we have argued above, this can be the case only if there are no absolute objects in the theory. The absence of absolute objects in the theory satisfies a version of Mach's principle which states that there should be no absolute objects in any physical theory.

Although Einstein did not use precisely the reasoning outlined above, it was his recognition of the preferred role played by the inhomogeneous Lorentz group in special relativity that was crucial to the development of the general theory of relativity. And although he formulated his argument in terms of the relativity of motion, it is clear that he was referring to the invariance properties of the laws of motion. His argument that all motion should be relative—hence the term general relativity—was really a requirement, in modern terms, that these laws should be generally invariant. This is not an empty requirement, as some authors have suggested, but rather severely limits the possible laws of motion one can formulate for $g_{\mu\nu}$.

C. LAWS OF MOTION

The search for a generally invariant law of motion for $g_{\mu\nu}$ occupied a considerable portion of Einstein's time prior to the year 1915. At one point he even argued that such a law could not exist. However, he did succeed in that year in finally formulating this law. If one requires that this law contain no higher than second derivatives of the $g_{\mu\nu}$ and furthermore that it be derivable from a variational principle, then there is, in fact, essentially only one law that fills these requirements. This is in marked contrast to the situation in electrodynamics, where there are an infinite number of laws of motion for the vector potential A_μ that satisfy these requirements.

To formulate the law of motion for $g_{\mu\nu}$, we first construct from the Riemann–Christoffel tensor (21) the Ricci tensor $R_{\mu\nu}$, where

$$R_{\mu\nu} = g^{\rho\sigma}R_{\sigma\mu\rho\nu} \qquad (23)$$

and the curvature scalar R, where

$$R = g^{\mu\nu}R_{\mu\nu} \qquad (24)$$

In terms of these quantities this law can be written as

$$R_{\mu\nu} - \tfrac{1}{2}g_{\mu\nu}R + \Lambda g_{\mu\nu} = \kappa T_{\mu\nu} \qquad (25)$$

where Λ and κ are constants and $T_{\mu\nu}$ is the energy–momentum tensor associated with the sources of the gravitational field.

In general, the components of the Riemann–Christoffel tensor will not vanish even when all of the components of the Ricci tensor do. As a consequence, it follows from the equation of geodesic deviation that the separation between neighboring freely falling masses will change with time. Since such changes appear due to tidal forces in many-practicle systems, the Riemann–Christoffel tensor is thus a measure of such forces and vice versa.

The so-called cosmological term $\Lambda g_{\mu\nu}$ was originally not present in the Einstein field equations. It was later added by him to obtain a static cosmological model with matter. When it was later realized that the universe was expanding and that there were solutions of the field equations without the cosmological term that fit the current observations, the motivation for the inclusion of this term disappeared and it is now not usually included in the equations. Also, measurements made on distant galaxies place an upper limit of 10^{-66} cm^{-2} on $|\Lambda|$.

In addition to the law of motion for $g_{\mu\nu}$ it is necessary to formulate generally invariant laws of motion for the other geometrical objects that are to be associated with the physical quantities being observed. One way to do this is simply to take over the generally covariant form of the laws formulated for these objects in special relativity. The law of motion for the electromagnetic field, when this field is associated with a vector A_μ, can, for example, be written in the form

$$(\sqrt{-g}\, g^{\mu\rho}g^{\nu\sigma}F_{\rho\sigma})_{,\mu} = 4\pi j^\nu \qquad (26)$$

where g is the determinant of $g_{\mu\nu}$, j^μ the current density associated with the sources of the electromagnetic field, and

$$F_{\mu\nu} = A_{\nu,\mu} - A_{\mu,\nu} \qquad (27)$$

Likewise, the equation of motion for a body on which no other forces act can be taken to be given by Eq. (19). (In fact, it can be shown that this law of motion is a consequence of the field equations for the gravitational field $g_{\mu\nu}$ and hence need not appear as a separate postulate in the theory.) Such laws of motion are said to involve minimal coupling to the gravitational field. It is also possible to construct laws of motion that do not couple minimally to the gravitational field. In the case of the electromagnetic field, for example, one could include a factor of $1 + R$, where R is the curvature scalar, inside the parentheses in Eq. (26). The only requirement these laws of motion should satisfy is that they reduce to their special relativistic form when the Riemann–Christoffel tensor vanishes.

D. Clocks, Rods, and Coordinates

It has been argued that some kind of postulate concerning the behavior of clocks and measuring rods is required in general relativity. For example, it has been suggested that a class of ob-

jects, ideal clocks, measure proper time along their trajectories, where the proper time along a trajectory is defined to be the integral of the distance ds, given by Eq. (22), along this trajectory. In this view, clocks, and also measuring rods, are assumed to be primitive objects in the theory.

Actually, all such postulates are unnecessary, in both the special and general theories. Clocks, and similarly measuring rods, are, in fact, composite physical systems with laws of motion governing their behavior. Once these laws have been established, there is no need to add additional postulates governing their behavior. It can be shown, for example, that if one takes, as a model for a clock, a classical hydrogen atom, then as long as the forces acting on this clock are small compared to the internal forces acting on its constituents and its dimensions are small compared to the curvature of its trajectory, it will indeed measure approximately the proper time along this trajectory. However, if the forces acting on it are sufficiently strong, the atom will be ionized and cease to measure any kind of time along its trajectory. Thus, the behavior of clocks is seen to be a dynamical question that cannot be decided *a priori* from any kinematic postulate.

In this view, then, clocks and measuring rods, and indeed all measuring devices, are considered to be physical systems with the geometrical objects associated with them obeying their own laws of motion. Furthermore, a physical description would have to be considered incomplete if it did not supply these laws of motion. To avoid the necessity of having to formulate and solve the laws of motion for a particular kind of clock, one may assume that it does satisfy the conditions for measuring proper time, with the proviso that if these conditions are violated it will no longer do so. If this assumption results in inconsistencies it does not mean that a principle of general relativity has been violated but only that these conditions have not been met.

While clocks and rods can be used to measure times and distances, it should be emphasized that these measurements bear no direct relation to the coordinates employed in the formulation of the laws of motion. Since, in all space–time descriptions, these laws are generally covariant, there are no preferred coordinate systems. Consequently, it follows that the predictions of a theory cannot depend on a particular coordinatization. In effect, the coordinates play the same role in space–time theories as do the indices that characterize the various components of a geometrical object and hence, like these indices, are not associated with any physical objects.

It is, however, often convenient to choose a particular coordinatization. Thus in Newtonian mechanics one usually chooses coordinates such that one of them is the parameter that characterizes the planes of simultaneity, and likewise in special relativity one usually employs inertial coordinates. In general relativity one also can employ a coordinatization that is particularly convenient for some purpose. One can, for example, choose coordinates in such a way that one of them corresponds to the time and distance intervals measured by a particular family of clocks and rods. Alternatively, one can choose coordinates so that certain components of the gravitational field have simple values. For example, one can choose coordinates so that $g_{00} = 1$ and $g_{01} = g_{02} = g_{03} = 0$. But in all cases such a choice is arbitrary and devoid of physical content.

V. Gravitational Fields

A. Newtonian Fields

Since Newtonian theory describes, to a high degree of accuracy, the phenomena associated with weak gravitational fields, it is essential that this theory be an approximation to the general theory. Although originally formulated as an action-at-a-distance theory, the Newtonian theory of gravity can also be formulated as a field theory analogous to electrostatics. The gravitational field is characterized, in this version of the theory, by a single scalar field ϕ that satisfies, in suitable coordinates, the field equation

$$\nabla^2 \phi = 4\pi\rho \tag{28}$$

where ρ is the mass density of the sources of the field.

In the general theory we assume that, in the case of weak fields, there exists a coordinate system such that $g_{\mu\nu} = \eta_{\mu\nu} + h_{\mu\nu}$, where $h_{\mu\nu} \ll 1$. We also assume that the velocities of the sources of the gravitational field are all vanishingly small compared to the velocity of light. In this case, the only nonvanishing component of $T_{\mu\nu}$ is $T_{00} = \rho c^2$ and Eq. (25) can be shown to reduce to Eq. (28) if we set $\Lambda = 0$, $\kappa = -8\pi G/c^4$, where G is the Newtonian gravitational constant, and take

$$\phi = (c^2/2)h_{00} \tag{29}$$

Furthermore, one can show that the law of motion (19) reduces to the Newtonian form

$$d^2\mathbf{x}/dt^2 = -\nabla\phi \tag{30}$$

B. The Schwarzschild Field, Event Horizons, and Black Holes

In spite of their enormous complexity, the Einstein field equations (25) possess many exact solutions. One of the first and perhaps still the most important of these solutions was obtained by Schwarzschild in 1916 for the case $\Lambda = 0$ and $T_{\mu\nu} = 0$ by imposing the condition of spherical symmetry on $g_{\mu\nu}$. The nonvanishing components of $g_{\mu\nu}$ are given in spherical coordinates by

$$g_{00} = 1 - 2M/r \qquad g_{11} = -1/(1 - 2M/r)$$
$$g_{22} = -r^2 \qquad g_{33} = -r^2 \sin^2\theta \tag{31}$$

where M is a constant of integration. This solution is seen to be independent of the coordinate x^0 and hence is a static field. The condition that the field be independent of x^0 was originally imposed by Schwarzschild in obtaining his solution of the field equations but has since been shown to be a consequence of the condition of spherical symmetry.

The importance of the Schwarzschild solution lies in the fact that it is the general relativistic analog of the Newtonian field of a point mass. The solution to Eq. (28) in this case is $\phi = Gm/r$, where m is the mass of the point. By making use of Eq. (29) and Eq. (31) for g_{00} we see that $M = Gm/c^2$. The constant $2M$ is referred to as the Schwarzschild radius of the mass m. The Schwarzschild radius of the sun is 2.9 km and of the earth is 0.88 cm. For comparison, the Schwarzschild radius of a proton is 2.4×10^{-52} cm and that of a typical galaxy of mass $\sim 10^{45}$ gm is $\sim 10^{17}$ cm.

The Schwarzschild field has a property that distinguishes it from the corresponding Newtonian field: at $r = 2M$ it becomes singular. Indeed, at this radius g_{11} is infinite! However, this is not a physical singularity, as Eddington first showed in 1924, but rather what is called a coordinate singularity. A final clarification of the structure of the Schwarzschild field came in 1960 with the work of M. Kruskal. He found a coordinate transformation from the Schwarzschild coordinates (x^0, r, θ, ϕ) to the set (u, v, θ, ϕ), where

$$u = a \cosh(x^0/4M) \qquad v = a \sinh(x^0/4M) \tag{32}$$

with $a = [(r/2M) - 1]^{1/2} \exp(r/4M)$, such that the transformed components of the Schwarzschild

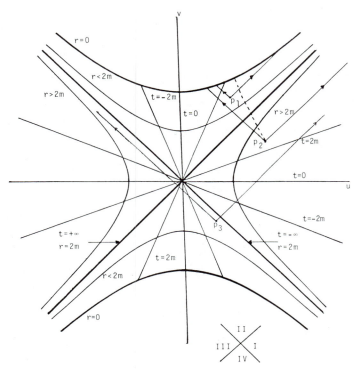

FIG. 1. Kruskal diagram for the Schwarzschild field. The hyperbola $r = 0$ represents a real singularity of the field.

field were given by

$$g_{00} = f \qquad g_{11} = -f \qquad g_{22} = -r^2$$
$$g_{33} = -r^2 \sin^2 \phi \qquad (33)$$

where $f = (8M/r)\exp(-r/2M)$. In these latter expressions, r is a function of u and v obtained by solving Eq. (32) for this quantity. In this form the field is seen to be singular only at $r = 0$, which is a true physical singularity.

Figure 1 depicts a number of the features of the Kruskal transformation in what is called a Kruskal diagram. In this diagram, curves of constant Schwarzschild r correspond to the right hyperbolas $u^2 - v^2 = $ const, while curves of constant x^0 correspond to straight lines passing through the origin. The region I to the right of the two $r = 2M$ lines corresponds to the entire region $r > 2M$, $-\infty < x^0 < \infty$. Hence it follows that the transformation from Kruskal to Schwarzschild coordinates must be a singular transformation, and indeed it is the singular nature of this transformation that is responsible for the singularity of $g_{\mu\nu}$ at $r = 2M$ in Schwarzschild coordinates.

Many of the properties of the Schwarzschild field can be understood by referring to the

Kruskal diagram. In Kruskal coordinates, light rays propagate along straight lines at 45° with respect to the u, v axes. As a consequence, it is seen that all of the rays emitted at a point P_1 in region II above the two lines $r = 2M$ will ultimately reach the singular curve $r = 0$, so no information can be transmitted from this point into region I. Likewise, inward-directed rays emitted at a point P_2 in region I will also reach the $r = 0$ curve while outward-directed rays will continue their outward propagation forever. And finally, some of the rays emitted from a point P_3 in region IV below the two lines $r = 2M$ will ultimately reach points in region I while others will reach the region III to the left of the two lines $r = 2M$.

Because of these features, the surface $r = 2M$ is said to form an event horizon: an observer in region I can never receive information about events taking place in regions II and III of the diagram. In addition, it is seen to be a light cone.

Finally, we note that a material body starting from rest at P_2 will fall into the singularity at $r = 0$ along the dashed path indicated in Fig. 1. Even though it reaches the event horizon at $x^0 = \infty$, the proper time along the curve from P_2 to the point where it reaches the singularity is finite. A local observer falling with the particle would notice nothing peculiar as he passed through the horizon. As long as the particle is outside this horizon it is possible to reverse its motion so that it does not cross it. However, once it does its fate is sealed; its motion can no longer be reversed and it will ultimately reach the $r = 0$ singularity in a finite amount of proper time. For this reason the source of a Schwarzschild field is said to be a black hole—nothing that falls into it can ever get out again and any radiation generated inside the horizon can never be seen from the outside.

C. KERR–NEWMAN FIELDS, NAKED SINGULARITIES

In addition to the Schwarzschild family of solutions, each member of which is characterized by a value of the parameter M, other stationary solutions of the empty-space Einstein field equations have been found by Kerr and Newman. Like the Schwarzschild solution, the Kerr–Newman solutions are asymptotically flat; that is, the value of the Riemann–Christoffel tensor goes to zero as the coordinate r approaches infinity and also the physical singularity in the solution is surrounded by an event horizon. This family of solutions is characterized by three continuously variable parameters, which, because of the asymptotic form of these solutions, can be interpreted as the mass, angular momentum, and electric charge of the black hole.

After the discovery of the Kerr–Newman solutions it was shown by the work of Israel, Carter, Hawking, and Robinson that the Kerr–Newman solutions are the only asymptotically flat, stationary black hole solutions, that is, solutions with event horizons that depend continuously on a finite set of parameters. This result is a version of a conjecture of Wheeler to the effect that "a black hole has no hair." What is still lacking is a proof of uniqueness of the Kerr–Newman solutions, that is, that not more than three parameters is sufficient to characterize completely all stationary, asymptotically flat black hole solutions. To date, the best one can do in this respect is to show that the only such static solutions and the only such neutral solutions belong to the family of Kerr–Newman solutions.

In addition to the Kerr–Newman family of solutions there are solutions of the empty-space field equations that do not have event horizons surrounding the physical singularity in the solution. One such solution is obtained by letting the parameter M in the Schwarzschild solution become negative. Also, a charged Schwarzschild field has no horizon if the charge $Q < M$. In both cases the only singularity occurs at $r = 0$. Because of the absence of a horizon surrounding this point, it is referred to as a naked singularity.

D. FIELDS WITH MATTER—GRAVITATIONAL COLLAPSE

In addition to the empty-space solutions discussed above there are numerous solutions of the Einstein equations with nonvanishing energy–momentum tensors. One can, for example, construct a spherically symmetric, nonsingular interior solution that joins on smoothly to an exterior Schwarzschild solution. The combined solution would then correspond to the field of a normal star, a white dwarf, or a neutron star. What distinguishes these objects is the equation of state for the matter comprising them. It is surprising that no interior solutions that could be joined to a Kerr–Newman field have been found.

Normal stars, of course, are not eternal objects. They are supported against gravitational collapse by pressure forces whose source is the thermonuclear burning that takes place at the center of the star. Once such burning ceases due to the depletion of its nuclear fuel, a star will begin to contract. If its total mass is less than approximately two solar masses it will ultimately become a stable white dwarf or a neutron star, supported by either electron or neutron degeneracy pressure. If, however, its total mass exceeds this limit, the star will continue to contract under its own gravity down to a point, a result first demonstrated by Oppenheimer and Snyder in 1939. Although they treated only cold matter, that is, matter without pressure, their result would still hold once the star has passed a certain critical stage even if the repulsion of the nuclei comprising the star were infinite. This critical stage is reached once the radius of the star becomes less than its Schwarzschild radius. When this happens, the surface $r = 2M$ becomes an event horizon and the matter is trapped inside, forming a black hole. Even an infinite pressure would then be unable to halt the continued contraction to a point because such a pressure would contribute an infinite amount to the energy–momentum tensor of the matter, which in turn would result in an even stronger gravitational attraction. [See STELLAR STRUCTURE AND EVOLUTION.]

Because of the disquieting features of black hole formation (neither Eddington nor Einstein was prepared to accept their existence), theorists looked for ways to avoid their formation. However, theorems of Penrose and Hawking show that collapse to a singularity is inevitable once the gravitational field becomes strong enough to drag back any light emitted by the star, that is, when the escape velocity at the surface of the star exceeds the velocity of light.

What has not been proved to date is what Penrose calls the hypothesis of "cosmic censorship." This hypothesis asserts that matter will never collapse to a naked singularity, but rather that the singularity will always be surrounded by an event horizon and hence not be visible to an external observer. While the hypothesis is supported by both numerical and perturbation calculations, it has so far not been shown to be a rigorous consequence of the laws of motion of the general theory.

The detection of black holes is complicated by the fact that they are black—by themselves they can emit no radiation. If they exist at all then, they can be detected only through the effects of their gravitational field on nearby matter. If a black hole were a member of a double star system, it would become the source of intense X rays when its companion expanded during the later stages of its own evolution. As matter from the companion fell onto the black hole it would become compressed and thus heated to temperatures high enough for it to emit such high-energy radiation. Of course, it is not enough to find an x-ray-emitting binary system in order to prove the existence of a black hole. It is also necessary that the mass of the x-ray-emitting component be larger than the upper limit on the mass of a stable neutron star, that is, larger than two solar masses.

Until quite recently the best candidate for a black hole was the binary system Cygnus X-1. However, the mass of the x-ray source in this system is difficult to determine from orbital measurements, so possibilities other than its being a black hole cannot be entirely ruled out. In 1983 a more convincing candidate for a black hole was found in an intense x-ray source in the Large Magellanic Cloud. Orbital parameters show that this source has a mass of at least seven solar masses. Also, a second object similar to Cygnus X-1 massive enough to be a black hole has been found in this cloud. Finally, evidence is accumulating that the most likely source of energy in a quasar is a massive black hole. The size and luminosity of quasars are consistent with their being black holes of about 10^8 solar masses.

Also, up to 50% of the mass of matter falling into such a hole can be converted into radiation that escapes. If a few solar masses per year were so converted it would be enough to power a typical quasar.

E. GRAVITATIONAL RADIATION, THE QUADRUPOLE FORMULA

Shortly after he formulated the general theory, Einstein showed that, when linearized around the flat field $g_{\mu\nu} = \eta_{\mu\nu}$, the empty-space field equations with zero cosmological constant possessed plane wave solutions similar to the plane wave solutions of electromagnetism. These waves propagate along the light cones defined by $\eta_{\mu\nu}$, that is, at the speed of light. Like their electromagnetic counterparts, they are transverse waves and possess two independent states of polarization. However, unlike the dipole structure of plane electromagnetic waves, gravitational waves have a quadrupole structure, as would be expected from the tensor nature of the gravitational field. Using coordinates such that $g_{\mu\nu} = \eta_{\mu\nu} + h_{\mu\nu}$ and the direction of propagation as the z axis, the two states of polarization are characterized by $h_{xx}^+ = h_{yy}^+$ and $h_{xy}^\times = h_{yx}^\times$ with, in the two cases, the other components having the value zero.

Although the empty-space field equations have been shown to possess exact solutions with wavelike properties, the so-called plane fronted waves, neither they nor the approximate wave solutions found by Einstein are associated with sources. In electromagnetic theory it is possible to find exact solutions of the field equations associated with sources with arbitrary motions. So far, no such solutions have been found for the highly nonlinear equations of the general theory. In order to construct radiative solutions associated with sources it is necessary to employ approximation methods.

In the past, two such methods have been used, the so-called slow- and fast-motion approximations. As its name implies, the slow-motion approximation assumes that the motion of the sources is small compared to the velocity of light and that the gravitational fields produced by these sources are weak. Unfortunately, this method can be shown to lead to inconsistencies in higher orders of approximation. On the other hand, the fast-motion approximation, which assumes only that the fields are weak, while free of these problems is extremely difficult to apply in practice due to the nonlinearities in the field equations. For gravitationally bound systems such as a binary star, the nonlinear terms can be shown to be of the same order of magnitude as the linear terms in the field equations and hence cannot be ignored.

By combining the best features of these two approximations with the method of matched asymptotic expansions, the work of Burke, Kates, Kegeles, Madonna, and Anderson has led to a satisfactory derivation of the so-called quadrupole radiation formula

$$dE/dt = -\tfrac{1}{5}(G/c^5)\langle d^3Q_{ij}/dt^3 \; d^3Q_{ij}/dt^3\rangle \quad (34)$$

where the angle brackets denote an average over a period of oscillation of the source and the indices i and j take on the values 1 through 3. The quantities Q_{ij} are the components of the mass quadrupole moment:

$$Q_{ij} = \int \rho(x)(x_ix_j - \tfrac{1}{3}\delta_{ij}\delta^{kl}x_kx_l) \; dv \quad (35)$$

where $\rho(x)$ is the mass density of the source. Finally, the quantity E appearing on the left side of this equation is the total Newtonian energy, kinetic plus potential, of the source. As a consequence, the quadrupole formula can be interpreted as an expression of energy conservation, in which case the quantity on its right side becomes the energy carried away by gravitational waves.

The quadrupole formula gives the main contribution to the energy loss of a system due to gravitational radiation. Changes in higher multipoles of the mass will also contribute to this loss, although in general their contribution will be much smaller than that of the quadrupole. However, unlike the electromagnetic case, there is no dipole contribution, since both the total momentum and angular momentum of the radiating system are conserved. And in both theories there is no monopole radiation because of conservation of charge in the one case and conservation of mass in the other.

It should be pointed out that the quadrupole formula is an approximation and is valid only for slow-motion sources, that is, for sources whose velocities are all less than the velocity of light. As a consequence it must be applied with caution to emitters of strong gravitational waves.

The amount of energy emitted by slow-motion sources is in most cases very small. A beryllium rod of length 170 m and weighing 6×10^7 kg, spinning on an axis through its center as fast as it can without disintegrating ($\omega \sim 10^3 \; \text{sec}^{-1}$), would radiate energy at the rate of approximately 10^{-7} erg/sec. For the earth–sun system

Eq. (34) yields a rate of energy loss of about 200 W. Only in the case of extremely massive objects moving at high speeds such as in the binary pulsar will the amount of energy radiated be significant.

VI. Observational Tests of General Relativity

A. Gravitational Red Shift

In general relativity the apparent rate at which clocks run is affected by the presence of a gravitational field. Like its counterpart in special relativity, this is a kinematic effect and hence is independent of any direct effect of the gravitational field on the internal dynamics of the clock. Only if gradients of this field result in tidal forces that are comparable to the nongravitational forces responsible for the functioning of the clock will this internal dynamics be altered.

The amount of red shift is most easily calculated in the case of a static gravitational field and for two clocks that are at rest in this field. We suppose that one clock, the emitter, sends out light waves whose frequency ν_{em} is the same as its own frequency. A receiving clock, which is identical in construction to the emitting clock, that is, has the same internal dynamics, is used to measure the frequency ν_{rec} of the received radiation. Then it can be shown that

$$Z := (\nu_{rec} - \nu_{em})/\nu_{em} = (1/c^2)(\phi_{em} - \phi_{rec})$$
$$(36)$$

where ϕ_{em} and ϕ_{rec} are the gravitational potentials at the locations of the emitter and receiver, respectively. If ϕ_{em} is less than ϕ_{rec} then the quantity Z is negative, hence the emitted light appears to be red shifted. This effect can be understood by noting that, in going from the emitter to the receiver, a photon will, in this case, gain potential energy. Since its total energy is conserved, its kinetic energy, which is proportional to its frequency, will decrease and hence so will its frequency. If either the emitter or receiver is moving with respect to the background field then Eq. (36) must be amended to take account of the Doppler shift produced by such motion.

The first attempts to observe the gravitational red shift were made on the spectral lines of the sun and known white dwarfs. For the sun, $Z = -2.12 \times 10^{-6}$ while for white dwarfs it would have values 10–100 times as large. In the case of the sun the shift was masked by the Doppler broadening of the spectral lines due to thermal motion. However, observations near its edge are consistent with a red shift of the magnitude predicted by Eq. (36). In the case of the white dwarfs, it was not possible to measure their masses and radii with sufficient accuracy to determine ϕ_{em}, although again red shifts were observed whose magnitudes were consistent with estimates of these quantities.

The first accurate test of the red shift prediction was carried out in a series of terrestrial experiments by Pound and Rebka in 1960, using the Mössbauer effect. The emitter and detector used by them were separated by a vertical height of 74 ft. In this case the gravitational potential can be taken equal to gz, where g is the local acceleration due to gravity and z is the height above ground level. Equation (36) then yields the value $Z = 2.5 \times 10^{-15}$. In spite of its small value, Pound and Rebka were able to observe a shift equal to 1.05 ± 0.10 times the predicted Z. Later experiments with cesium beam and rubidium clocks on jet aircraft yielded similar results.

The possibility of testing the red shift prediction has improved dramatically with the development of high-precision clocks. In 1976 Vessot and Levine used a rocket to carry a hydrogen-maser clock to an altitude of about 10,000 km. Their result verified the theoretical value to within 2 parts in 10^{-4}.

It has been argued that red shift observations do not bear on the question of the validity of general relativity but rather only on the validity of the principle of equivalence. This is, in fact, only partially true. If one assumes that the gravitational field of the earth is uniform over the 74 ft that separated the emitter and receiver of the Pound–Rebka experiment, then indeed their result can be calculated using only this principle. However, one cannot use it alone to determine the result of the Vessot–Levine experiment. In this case it is necessary to make direct use of Eq. (36). The derivation of Eq. (36), however, makes use of Eq. (19) for a light ray, which, in turn, is a consequence of the field equations (25) of general relativity. Also, although the present red shift measurements are not sufficiently accurate to distinguish between different possible equations for the gravitational field, there is nothing in principle that would preclude such a test.

B. SOLAR SYSTEM TESTS—THE PPN FORMALISM

The first arena used for testing general relativity was the solar system and it remains so to this date. What has changed dramatically over the years, due to the rapid growth of technology, is the degree of accuracy with which the theory can be tested. It is in the solar system that the gravitational field of the sun is of sufficient strength that deviations from the Newtonian theory are observable, but just barely.

In calculating the size of these effects one assumes that the trajectories of planets and light rays obey Eqs. (19). However, rather than take the sun's gravitational field to be the Schwarzschild field in evaluating the Christoffel symbols appearing in these equations, it is useful to use what has become known as the parametrized post-Newtonian (PPN) formalism. In this formalism, first developed by Eddington and extended by Robertson, Schiff, Will, and others, one assumes a more general form for the post-Newtonian corrections to the gravitational field of the sun than those given by general relativity. These corrections are allowed to depend on a number of unknown parameters that one hopes to determine by solar system and other observations. The reason for proceeding in this manner is that it allows one to test the validity of other competing theories of gravity in which these parameters have different values from those they would have if general relativity were valid.

In the most extreme versions of this formalism as many as 10 parameters are employed. However, a number of these parameters can be eliminated if one requires that the equations of motion (19) are a consequence of the field equations for the gravitational field, as they are in general relativity. Also, some of these parameters are known to be small from other experiments. In what follows we will use an abbreviated version of the PPN formalism employed by Hellings in his analysis of the solar system data. In this version the components of the gravitational field have the form

$$g_{00} = 1 - 2U[1 - J_2(R_\odot/r)^2 P_2(\theta)]$$
$$+ 2\beta U^2 + \alpha_1 U(w/c) \qquad (37a)$$

$$g_{0i} = \alpha_1 U w^i/c \qquad (37b)$$

$$g_{ij} = -(1 + 2\gamma U)\delta_{ij} \qquad (37c)$$

where β, γ, and α_1 are PPN parameters $U = GM_\odot/rc^2$ is the Newtonian gravitational potential of the sun, and $R_\odot$ and $M_\odot$ are the radius and mass of the sun. Included in Eq. (37a) is a term proportional to J_2, a dimensionless measure of the quadrupole moment of the sun. In this term P_2 is the Legendre polynomial of order 2 and θ is the angle between the radius vector from the sun's center and the normal to the sun's equator. The so-called preferred frame velocity w^i is taken to be the average of the determinations of the solar system velocity relative to the cosmic blackbody background. In general relativity $\beta = \gamma = 1$ and $\alpha_1 = 0$.

1. Bending of Light

One of the more spectacular confirmations of the general theory came in 1919, when the solar eclipse expedition headed by Eddington announced that they had observed bending of light from stars as they passed near the edge of the sun that was in agreement with the prediction of the theory. Derivations of the bending that use only the principle of equivalence or the corpuscular theory of light predict just half of the bending predicted by the general theory.

The angle of bending for light passing at a distance d from the center of the sun can be computed by using the equation of motion (19) with $g_{\mu\nu} \, dx^\mu/d\lambda \, dx^\nu/d\lambda = 0$. Using the form for the gravitational field given by Eq. (37), the angle of bending is given by

$$\Theta = (1 + \gamma)2GM_\odot/c^2 d \qquad (38)$$

For light that just grazes the edge of the sun $\Theta = 1''.75$ when $\gamma = 1$. Because of this small value it is necessary to observe stars whose light passes very close to the edge of the sun, and this can be done only during a total eclipse. The apparent positions of these stars during the eclipse are then compared to their positions when the sun is no longer in the field of view in order to measure the amount of bending. Unfortunately, such measurements are beset with a number of uncertainties. Thus the measurements made by Eddington and his co-workers had only 30% accuracy. The most recent such measurements were made during the solar eclipse of June 30, 1973 and yielded the value

$$\tfrac{1}{2}(1 + \gamma) = 0.95 \pm 0.11 \qquad (39)$$

The use of long-baseline and very-long-baseline interferometry, which is capable in principle of measuring angular separations and changes in angle as small as 3×10^{-4}, has made possible much more accurate tests of the bending of light. These techniques have been used to observe a number of quasars such as 3C273 that pass very close to the sun in the course of a year. Beginning in 1970, these observations have yielded increasingly accurate determinations, and the most recent, in 1984, agrees with the general relativistic prediction to within 1%.

2. Time Delay

In passing through a strong gravitational field, light not only will be red shifted but also will take longer to traverse a given distance than it would if Newtonian theory were valid. The reason for this delay is that the gravitational field acts like a variable index of refraction, so the velocity of light will vary as it passes through such a field. This effect was first proposed by Shapiro in 1964 as a means of testing general relativity. It can be observed by bouncing a radar signal off a planet or artificial satellite and measuring its round-trip travel time. At superior conjunction, when the planet or satellite is on the far side of the sun from the earth, the effect is a maximum, in which case the amount of delay is given by

$$\delta t = 2(1 + \gamma)(GM_\odot/c^3) \ln(4r_e r_p/d^2) \qquad (40)$$

where r_e, r_p, and d are, respectively, the distance from the sun to the earth, the distance from the sun to the target, and the distance of closest approach of the signal to the center of the sun. Since one does not have a Newtonian value for the round-trip travel time with which to compare the measured time it is necessary to monitor the travel time as the target passes through superior conjunction and look for a logarithmic dependence.

The use of a planet such as Mercury or Venus as a target is complicated by the fact that its topography is largely unknown. As a consequence, a signal could be reflected from a valley or a mountaintop without our being able to detect the difference. Such differences can introduce errors of as much as 5 μsec in the round-trip travel time. Artificial satellites such as Mariners 6 and 7 have been used to overcome this difficulty. Furthermore, since they are active retransmitters of the radar signal they permit an accurate determination of their true range. Unfortunately, fluctuations in the solar wind and solar radiation pressure produce random accelerations that can lead to uncertainties of up to 0.1 μsec in the travel time. Finally, spacecraft such as the Mariner 9 Mars orbiter and the Viking Mars landers and orbiters have been used as targets. Since they are anchored to the planet they will not suffer such accelerations. The most recent measurements by Reasenberg et al. in 1979 have yielded a value

$$(1 + \gamma)/2 = 1.000 \pm 0.001 \quad (41)$$

3. Planetary Motion

Long before the general theory was proposed, it was known that there was an anomalous precession of the perihelion (distance of closest approach to the sun) of the planet Mercury that could not be accounted for on the basis of Newtonian theory by taking into consideration the perturbations on Mercury's orbit due to the other planets. At the end of the last century, Newcomb calculated this residual advance to have a value of $41''.24 \pm 2''.09$ of arc per century.

The field values given by Eq. (37) and the equation of motion (19) together yield an expression for the perihelion advance per period that is given, to an accuracy commensurate with the accuracy of the observations, by

$$\delta\bar{\omega} = (6\pi GM_\odot/c2p)[\tfrac{1}{3}(2 + 2\gamma - \beta)$$
$$+ J_2(R_\odot^2 c^2/12GM_\odot p)] \quad (42)$$

where $p = a(1 - e^2)$ is the semi-latus rectum of the orbit, a its semimajor axis, and e its eccentricity. Using the best current values for the orbital elements and physical constants for Mercury and the sun, one obtains from Eq. (38) a perihelion advance of $42''.95\lambda_p$ of arc per century, where $\lambda_p = [\tfrac{1}{3}(2 + 2\gamma - \beta) + 3 \times 10^3 J_2]$.

The measured value of the perihelion advance of Mercury is known to a precision of about 1% from optical measurements made over the past three centuries and of about 0.5% from radar observations made over the past two decades. If one assumes that J_2 has the value $\sim 1 \times 10^{-7}$, which it would have if it were the consequence of centrifugal flattening due to a uniform rotation of the sun equal to its observed surface rate of rotation, then, using this value, Shapiro gives

$$\tfrac{1}{3}(2 + 2\gamma - \beta) = 1.003 \pm 0.005 \quad (43)$$

which is in excellent agreement with the prediction of general relativity.

This agreement has been called into question by some researchers, notably Dicke and Hill. Observations of the solar oblateness by Dicke and Goldenberg in 1966 led them to conclude that J_2 actually has a value of $(2.47 \pm 0.23) \times 10^{-5}$, leading to a contribution of about $4''$ per century to the overall perihelion advance. If true, this would put the prediction of general

relativity into serious disagreement with the observations. On the other hand, it would agree with the prediction of the Brans–Dicke scalar tensor theory of gravity if an adjustable parameter in that theory were suitably chosen. However, a number of authors have disagreed with the interpretation of their observations by Dicke and Goldenberg. These authors argue that the observations could equally well be explained by assuming a standard solar model with $J_2 \sim 10^{-7}$ and a surface temperature difference of about 1° between the pole and the equator. More recently Hill has given a value of $J_2 = 6 \times 10^{-6}$, based on his measurements of normal mode oscillations of the sun. If true, the general relativistic prediction for Mercury would be inconsistent with the observed value by about two standard deviations. Unfortunately, the present measurements of the orbit of Mercury are not sufficiently accurate to separate the post-Newtonian and quadrupole effects.

A resolution of this difficulty has come from an analysis of the ranging data for the planet Mars. Since the quadrupole contribution to the perihelion advance has a different dependence on the semimajor axis from the gravitational effect, it is in principle possible to separate the two by observing the advance for different planets. In spite of the smallness of these effects on the orbit of Mars, the accuracy of the Viking data from Mars, which are accurate to within 7 km, combined with the radar data from Mercury allows such a determination. Using a solar system model that includes 200 of the largest asteroids, Hellings has found, with $J_2 = 0$, that

$$\beta - 1 = (-0.2 \pm 1.0) \times 10^{-3} \quad (44a)$$
$$\gamma - 1 = (-1.2 \pm 1.6) \times 10^{-3} \quad (44b)$$
$$\alpha_1 = (2.2 \pm 1.8) \times 10^{-4} \quad (44c)$$

When J_2 was allowed to have a finite value, he found that

$$J_2 = (-1.4 \pm 1.5) \times 10^{-6} \quad (45a)$$

and

$$\beta - 1 = (-2.9 \pm 3.1) \times 10^{-3} \quad (45b)$$
$$\gamma - 1 = (-0.7 \pm 1.7) \times 10^{-3} \quad (45c)$$
$$\alpha_1 = (2.1 \pm 1.9) \times 10^{-4} \quad (45d)$$

Hellings also used these data to analyze the nonsymmetric gravitational theory of Moffat, which was consistent with the Mercury data and Hill's value for J_2. The result was that

$$J_2 = (1.7 \pm 2.4) \times 10^{-7} \quad (46)$$

From the above results it appears that the predictions of general relativity are confirmed to about 0.1%. However, by a suitable adjustment of parameters, several competing theories also share this property. What distinguishes general relativity from these other theories is that, aside from the value for the gravitational constant G, it contains no other adjustable parameters.

4. Time Varying G

In addition to the tests discussed above, the solar system data can be used to test the possibility that the gravitational constant varies with time. Such a possibility was first suggested by Dirac in 1937 on the basis of his large number hypothesis. He observed that one could form, from the atomic and cosmological constants, several dimensionless numbers whose values were all of the order of 10^{40}. Rather than being a coincidence that was valid only at the present time, Dirac proposed that the equality of these numbers was the manifestation of some underlying physical principle and that they held at all times. Since one of these numbers involves the present age of the universe through its dependence on the Hubble "constant" and hence decreases as one moves back in time, the other constants must also change with time in order to maintain the equality between the large numbers. One of these numbers, however, involves only atomic constants, being the ratio of the electrical to the gravitational force between an electron and a proton. Hence the Dirac hypothesis requires that one of these atomic constants must be changing on a cosmic time scale. The constant that is usually taken to vary with time in theoretical implementations of the large number hypothesis is the gravitational constant.

There are several ways of constructing a theory with an effective time-varying gravitational constant. In the Brans–Dicke theory, the effective gravitational constant itself varies with time:

$$G_{\text{eff}} = G[1 + (\dot{G}/G)(t - t_0)] \qquad (47)$$

An alternative proposal by Dirac assumed that cosmic effects couple to local atomic physics so that the ratio of atomic to gravitational time is not constant. The rate of change of gravitational time τ_G with respect to atomic time τ_A is then given as

$$d\tau_G/d\tau_A = 1 + \dot{\phi}(t - t_0) \qquad (48)$$

where ϕ is some cosmological field that is supposed to be responsible for the effect. In both cases, the net effect is to produce an anomalous acceleration in the equations of motion for material bodies.

Since the change in atomic constants is tied to cosmic evolution in the large number hypothesis, the expected rate of change in G should be proportional to the inverse Hubble time:

$$\dot{G}/G - H_0 \cong 5 \times 10^{-11} \text{ yr}^{-1} \qquad (49)$$

On the basis of the Viking lander data, Hellings concludes that

$$\dot{G}/G = (0.2 \pm 0.4) \times 10^{-11} \text{ yr}^{-1} \qquad (50a)$$

$$\dot{\phi} = (0.1 \pm 0.8) \times 10^{-11} \text{ yr}^{-1} \qquad (50b)$$

Since these limits are an order of magnitude smaller than what one would expect from simple cosmic scale arguments, they cast serious doubt on the large number hypothesis.

C. THE BINARY PULSAR

A new, and essentially unique, opportunity for testing general relativity came with the discovery of the binary pulsar PSR 1913 + 16 by Hulse and Taylor in 1974. It consists of a pulsar in orbital motion about an unseen companion with a period of 7.75 hr. Its relevance for general relativity is twofold: because $v^2/c^2 \sim 5 \times 10^{-7}$ is a factor 10 larger than for Mercury, relativistic effects are considerably larger than any that have been observed in the solar system. Also, the short period amplifies secular changes in the orbit. Thus the observed periastron advance amounts to $4°.2261 \pm 0.0007$ of arc per year compared to the 43″ of arc per century for Mercury. Furthermore, the pulsar carries its own clock with a period that is accurate to better than one part in 10^{12}. As a consequence, measurements of post-Newtonian effects can be made with unprecedented accuracy. If this were all, the binary pulsar would still be an invaluable tool for testing general relativistic orbit effects. However, it also provides us for the first time with a means for testing an essentially different kind of prediction of general relativity, namely the existence of gravitational radiation.

Considerable effort has gone into identifying the pulsar companion. It was soon found that the pulsar radio signals were never eclipsed by the companion. Also, the dispersion of the pulsed signal showed little change over an orbit, implying the absence of a dense plasma in the system. These two facts together ruled out the possibility of the companion being a main sequence star. Another possibility is that it is a helium star. However, since the pulsar is at a distance of only about 5 kpc from us, such a star would have been seen. In spite of intense efforts, no such star has been observed in the neighborhood of the pulsar. The remaining possibility is that it is a compact object, either a white dwarf, another neutron star, or a black hole.

In the case of conventional spectroscopic binaries, it is usually possible to measure only two parameters of the system, the so-called mass function of the two masses M_1 and M_2 of the components and the product of the semimajor axis a_1 and the sine of the angle i of inclination of the plane of the orbit to the line of sight. However, in the case of the binary pulsar one can use general relativity to determine all four of these parameters from measurements of the periastron advance and the combined second-order Doppler shift and gravitational red shift of the emitted signals. One finds from these combined measurements that $M_1 = M_2 = (1.41 \pm 0.06)M_\odot$. From the fact that the Chandrasekhar limit on the mass of a nonrotating white dwarf is about $1.4 M_\odot$, it appears likely that the unseen companion is either a neutron star or a black hole.

In addition to the measurements discussed above, it was discovered that the orbital period P was decreasing with time. Later measurements gave a value for $\dot{P} = (-2.30 \pm 0.22) \times 10^{-12}$ sec/sec^{-1} or about 7×10^{-5} sec per year. If one computes the period change due to loss of energy by the emission of gravitational radiation using the quadrupole formula (34) one obtains a value for $\dot{P} = -2.40 \times 10^{-12}$ sec/sec^{-1}, in excellent agreement with the observed value.

Of course, there are other effects that could change the orbital period, such as tidal dissipation, mass loss or accretion onto the system, or acceleration relative to the solar system. Furthermore, there could be other contributions to the periastron advance such as rotational or tidal deformation of the companion. Only in the case of a helium-star companion would any of these effects contribute significantly to the calculated or observed period change. Furthermore, it would be truly remarkable if some combination of these effects should conspire to give a value for the period change equal to that predicted by the quadrupole formula. It therefore appears safe to say that for the first time we have evidence of a qualitatively new prediction of general relativity, namely gravitational radiation. Finally, the data from the binary pulsar seem to

rule out a number of competing theories of gravity such as the Rosen bimetric theory. In such theories this system can radiate dipole gravitational waves that result in a period increase. Only a very artificial mechanism could then give rise to the observed period decrease.

D. Gravitational Wave Detection

The first comprehensive attempt to detect gravitational waves impinging on the earth was begun by J. Weber in 1961. His antenna consisted of a large aluminum cylinder ~ 1.5 m long with a resonant frequency of ~ 1660 Hz. In the early 1970s Weber announced the detection of coincident pulses on two of these antennae separated by a distance of ~ 1000 mile. However, attempts to duplicate these results by a number of other groups, using somewhat more sensitive detectors than those used by Weber, proved fruitless, and it is now generally agreed that the events recorded by Weber were not caused by gravitational waves.

Since Weber's pioneering efforts, about 15 different groups from around the world have undertaken the construction of gravitational wave detectors. The sensitivity of a detector can be expressed in terms of the smallest strain $\Delta L/L$, where ΔL is a change in the length L of the detector, that can just be measured. This change in length is produced by tidal forces associated with the incident gravitational wave and hence its measurement leads to a determination of the Riemann–Christoffel tensor of the wave. Since this strain is approximately equal to the dimensionless amplitude h of a gravitational wave incident on the detector, the sensitivity of a detector is usually given as the minimum value of h that can be detected.

The original Weber bars had a sensitivity $h \sim 10^{-16}$. At present one of the main limitations on the sensitivity of Weber bars is thermal noise. As a consequence, second-generation Weber bars are being constructed that will be cooled to liquid helium temperatures. Such bars are estimated to have sensitivities of $h \sim 10^{-19}$. It is technically feasible to construct bars for which $h \sim 10^{-21}$, although that would require cooling to the millidegree level. The latter value appears to be a lower limit to what can be attained with presently available technology.

One of the drawbacks of Weber bar detectors is that they are only sensitive to the Fourier component of the incoming signal whose frequency is equal to the resonant frequency of the bar. Furthermore, most bars have resonant frequencies in the kilohertz range with a smallest reported frequency of 60.2 Hz. Unfortunately, most continuous wave sources such as binary star systems have much lower frequencies. In an attempt to overcome this difficulty and to increase sensitivity, a number of groups have undertaken the construction of laser interferometer detectors. In these devices, a gravitational wave would change the lengths of the interferometer arms and one would measure the resulting fringe shifts. Such detectors can, in principle, record the entire waveform of an incoming wave rather than a single Fourier component. It is possible that sensitivities as low as $h \sim 10^{-22}$ might be achieved. It has also been suggested that gravitational radiation could be detected by the accurate Doppler tracking of spacecraft. Such a scheme would, in principle, be able to detect waves with frequencies in the 1 to 10^{-4} range. Present technology is within one or two orders of magnitude of the sensitivity needed to detect possible signals in this frequency range.

Possible sources of gravitational waves can be divided into two groups, those that emit continuously and those that emit in bursts. Possible continuous wave sources are binary stars and vibrating or rotating stars. In the case of binary stars, the strongest emitter known is μ Scorpii, for which $h = 2.1 \times 10^{-20}$. However, its frequency is 1.6×10^{-5} Hz. The largest binary frequency known is 1.9×10^{-3} Hz. However, for this system $h \sim 5 \times 10^{-22}$. Other possible continuous wave sources have values for h that are this small or smaller. Thus, estimates of h for waves from the Crab and Vela pulsars are of order 10^{-24} to 10^{-27}, at frequencies between 10 and 100 Hz. In spite of these low amplitudes, signal integration over an extended time can effectively increase the sensitivity of a detector by an order of magnitude or more, so the detection of such signals is not totally out of the question.

Bursts of gravitational radiation can be expected to accompany cataclysmic events such as supernova explosions, stellar collapse to form neutron stars or black holes, or coalescence of the neutron stars or black holes in a binary system at the end stage of its evolution. One of the problems in dealing with such systems is the determination of the efficiency with which other forms of energy can be converted into gravitational radiation. Estimates range from a maximum of 0.5 to as low as 0.001. Such events would have characteristic frequencies in the range 10^2 to 10^5 Hz and those occurring in our galaxy would have amplitudes estimated to be in the range $h \sim 10^{-18}$ to 10^{-17}. Here the problem for detection is not so much the frequency or the intensity, as it is in the case of continuous emitters, but rather the scarcity of such events. Thus, the supernova rate in our galaxy has been estimated to be 0.03 per year. If one includes such events in other galaxies the rate increases. For example, at a distance of 10 Mpc the estimated supernova rate is one per year. However, the corresponding amplitude would be $h \approx 3 \times 10^{-21}$ to 3×10^{-20}.

By combining the sensitivity estimates for gravitational wave detectors now under construction and the expected amplitudes and frequencies of possible sources we see that the possibility for detection in the near future is good. Furthermore, as the technology improves, an era of gravitational wave astronomy may soon be possible. Since gravitational waves are not absorbed by intervening matter as is electromagnetic radiation, such an astronomy may allow us to explore regions of the universe, such as the centers of galaxies, that are now blocked to our view.

E. GRAVITATIONAL LENSES

In many of its effects, a gravitational field acts like a medium with a variable index of refraction. Thus, two of the observed effects discussed above, the bending of light and the time delay of signals as they pass through a gravitational field, can be understood on this basis. A further consequence of this notion is that there should exist gravitational lenses with properties similar to those of ordinary optical lenses. The most likely candidates for such lenses are galaxies. If placed between us and a distant point source such as a quasar, a galaxy can, in effect, provide more than one path along which light from the source may reach the observer. As a consequence, one would see multiple images of the source. Applied to a galaxy, the bending formula (38) (with $\gamma = 1$) gives typical bending angles of about 1 arcsec.

As of this date, two "lensed" quasars have been observed. The first of these consists of two quasar images of comparable brightness separated by about 6 arcsec, with identical red shifts of 1.41 within the measuring errors. Both images are also radio sources with a flux ratio that is the same as in the visible. In this case the lens has also been identified as a (probable) elliptic galaxy with a red shift of 0.36. It is in fact a member of a cluster, which must be taken into account in determining the image structure. Shortly after the discovery of the first lensed quasar a second candidate was discovered. It consists of three quasars with identical (to within 100 km/sec) red shifts of 1.722. Although in this case no lensing galaxy has been observed as yet, the configuration can be explained by a nearly edge-on spiral galaxy with a red shift of about 0.8, which would make it hard to detect. In addition to confirming the gravitational lens effect, these observations prove that at least three quasars are at cosmological distances from us.

VII. Gravity and Quantum Mechanics

A. HAWKING RADIATION AND BLACK HOLE THERMODYNAMICS

For most of its history, general relativity has stood apart from quantum mechanics. Early attempts to quantize the gravitational field proved to be largely unsuccessful and quantum theory usually neglected the presence of gravitational fields. For most problems one could justify this neglect. The radius of the first Bohr orbit of a hydrogen atom held together by gravitational rather than electrical forces, for example, would be about 5×10^{30} m, which is almost four orders of magnitude larger than the radius of the visible universe! However, one is not justified in ignoring the effects of strong gravitational fields, such as those that occur near the Schwarzschild radius of a black hole, on the behavior of quantum systems since gravity couples universally to all physical systems. When one takes account of the gravitational field in the quantum description of a system, qualitatively new features emerge. One such feature is the phenomenon of Hawking radiation.

Even in the vacuum, where there are no real quanta, pairs of virtual quanta of the various matter fields observed in nature are being continually created and destroyed in equal numbers. Their presence is manifested in such phenomena as the Lamb shift in hydrogen and the Casimir effect. According to Hawking, a black hole can absorb one member of such a virtual pair, leaving its partner to propagate as a real quantum of the field. The energy needed for this process to occur is supplied by the gravitational

energy of the black hole. Hawking was able to show that, as a black hole formed, such a flux of real quanta should be produced and that it would be equal to the flux produced by a hot body of temperature T given by

$$kT = \hbar g/2\pi c \qquad (51)$$

where k is Boltzmann's constant, $\hbar$ is Planck's constant, and g is the gravitational acceleration at the Schwarzschild radius of the black hole and is equal to $c^4/4GM$, where M is its mass.

For a solar mass black hole this temperature would be 2.5×10^{-6} K. However, for a 10^{12} kg mass black hole it would be 5×10^{12} K. Such a black hole would emit energy at a rate of about 6000 MW, mainly in gamma rays, neutrinos, and electron–positron pairs. Hawking has suggested that ''primordal black holes'' with such masses might have been formed by the collapse of inhomogeneities in the very early stages of the universe and that some of them might have survived to the present day. If so, they probably represent our only hope of observing Hawking radiation. However, measurements of the cosmic-ray background around 100 MeV place an upper limit for black holes with masses around 10^{15} kg of about 200 per cubic light-year.

The fact that a black hole can radiate real quanta might seem to contradict the fact that no radiation can escape from a black hole. However, that restriction is only true classically. One can think of the emitted radiation as having come from inside the event horizon surrounding the black hole by quantum mechanically tunneling through the potential barrier created by its gravitational field. Actually, it is possible for a black hole to emit almost any configuration of quanta, including macroscopic objects. Since we cannot have direct knowledge of the interior of a black hole, all we can determine are the probabilities for the emission of such configurations. The overwhelming probability is that the emitted radiation is thermal with a temperature given by Eq. (50).

That a black hole should have associated with it a temperature fits in with some analogies between black holes and thermodynamics discovered by Bardeen, Carter, Hawking, and Bekenstein. If the energy density of the matter that went to make up the black hole is nonnegative, it can be shown that, classically, the surface area of the event horizon surrounding it can never decrease with time. Moreover, if two black holes coalesce to form a single black hole, the area of its event horizon is greater than the sum of the areas of the event horizons surrounding the two original black holes. These properties are very similar to those of ordinary entropy. Furthermore, when a black hole forms, all information concerning its structure except its mass, charge, and angular momentum is lost, and when ordered energy is absorbed by a black hole it too is forever lost to the outside world. These considerations led Bekenstein to associate with a black hole an entropy S given by

$$S = ckA/4\hbar \qquad (52)$$

where A is the surface area of the event horizon surrounding the hole.

The existence of Hawking radiation raises the question of the ultimate fate of a black hole. As it radiates, its mass decreases. Its temperature therefore increases, and hence so does the rate of emission of radiation. What is left is, at this point, speculation. It might disappear completely, it might cease radiating when its mass reaches some critical value, or it might continue radiating indefinitely, creating a negative-mass naked singularity. While the latter two possibilities seem unlikely, the first one implies that whatever matter went into making the black hole initially would simply cease to exist. In deriving the emission from black holes, the gravitational field was treated classically while the matter fields were treated quantum mechanically. It has been suggested that when the mass of the black hole becomes comparable to the Planck mass, that is, the mass one can form from the constants c, $\hbar$, and G, namely $(\hbar c/G)^{1/2} \sim 5 \times 10^{-15}$ g, the gravitational field can no longer be treated as a classical field. If so, the fate of a black hole will be decided only when we have a consistent theory of quantum gravity.

B. Quantum Gravity

There seems to be little doubt that, in the final analysis, the gravitational field, like all matter fields, must be quantized. The most telling argument in favor of this assertion is that, if the gravitational field were a classical field, it would be possible to determine both the position and the momentum of its sources by measuring all of its components simultaneously with arbitrary accuracy and thus violate the uncertainty principle. The quantization of the gravitational field is, however, beset with many technical difficulties and there is still considerable debate on how best to accomplish this task.

One of the most troubling problems attendant on the quantization of any field theory is the existence of divergent integrals, which arise in various calculations. In all successful quantum field theories such as quantum electrodynamics, these divergent integrals can be dealt with by a renormalization process whereby they are absorbed into the masses and coupling constants that appear in the theories. The "renormalized" values of these quantities are then taken to be their observed values. A conventional quantization of the gravitational field in the absence of other fields does not lead to divergences, at least in the lowest order of approximation. However, once gravity is coupled to matter, divergences arise that cannot be renormalized. A number of attempts to overcome this difficulty are underway.

One such approach is supergravity. It is an extension of supersymmetric quantum field theory in which every boson in the theory has associated with it a fermion and vice versa. Furthermore, the theory is invariant under the interchange of bosons and fermions. Such theories have been used to construct a unified theory of the strong and electroweak interactions between elementary particles. In supergravity, one associates a spin 3/2 particle called the gravitino with the quantum of the gravitational field. This theory has been shown to have no logarithmic divergences in the lowest two orders of perturbation theory even when gravity is coupled to matter. Whether supergravity or one of its extensions proves to be a viable theory is a matter for future investigation.

BIBLIOGRAPHY

Anderson, J. L. (1967). "Principles of Relativity Physics," Academic Press, New York.
Bergmann, P. G. (1968). "The Riddle of Gravitation," Scribner's, New York.
Bertotti, B., de Felice, F., and Pascolini, A. (Eds.). (1984). "General Relativity and Gravitation," Reidel, Dordrecht, Netherlands.
Hawking, S. W., and Israel, W. (Eds.). (1979). "General Relativity," Cambridge Univ. Press, Cambridge, U.K.
Kaufman, W. J., III. (1977). "The Cosmic Frontiers of General Relativity," Little, Brown, Boston.
Misner, C. W., Thorne, K. S., and Wheeler, J. A. (1971). "Gravitation," Freeman, San Francisco.

Rindler, W. (1977). "Essential Relativity," 2nd ed., Springer Verlag, New York.

Sexl, R., and Sexl, H. (1979). "White Dwarfs, Black Holes," Academic Press, New York.

Smarr, L. (Ed.). (1979). "Sources of Gravitational Radiation," Cambridge Univ. Press, Cambridge, U.K.

Wald, R. M. (1984). "General Relativity," Univ. of Chicago Press, Chicago.

Weinberg, S. (1972). "Gravitation and Cosmology," Wiley, New York.

Will, C. (1981). "Theory and Experiment in Gravitational Physics," Cambridge Univ. Press, Cambridge, U.K.

RELATIVITY, SPECIAL

John D. McGervey   *Case Western Reserve University*

GLOSSARY

Electric charge: Basic property of particles of matter such that each particle possessing a charge is surrounded by an electric field.

Electric field: Condition in space that causes a force to be exerted on a charged particle. The force is proportional to the product of the magnitude of the charge and the strength of the field; it acts in the direction of the field on a positive charge and in the opposite direction on a negative charge.

Electron volt (eV): Unit of energy equal to 1.6×10^{-19} J, or the kinetic energy given to an electron that has been accelerated by an electric potential difference of 1 V.

Half-life: Time interval during which there is a probability of one-half that a particular radioactive particle or atomic state will decay.

Inertia: Ability of a body to resist an attempt to accelerate it. The greater the inertia, the smaller will be the acceleration produced by a given force.

Magnetic field: Condition in space that causes a force to be exerted on a moving charged particle. The direction of the force is perpendicular to the direction of the field and to the direction of the velocity of the particle.

Mass: Measure of inertia. Masses may be compared by applying equal forces to two bodies and finding their accelerations.

Vector: A directed line segment, representing a quantity that has a magnitude and a direction, such as velocity or force. Vectors may be added or subtracted graphically like displacements (e.g., a northward displacement of 5 miles added to an eastward displacement of 5 miles yields a northeastward displacement of about 7 miles). The northward and eastward displacements are called *components* of the displacement vector.

The special theory of relativity displays the logical consequences of assuming the nonexistence of absolute space or absolute time. This assumption was made by Albert Einstein on the basis of experimental evidence that the speed of light, relative to an observer, is independent of the state of motion of that observer. The theory that proceeded from this assumption explained all the observations known at that time, and it also predicted other phenomena whose existence had not even been suspected. Among other things, the theory showed that, no matter how much energy one gave to a particle, one could not accelerate it to a speed greater than the speed of light. It also showed that the mass of a particle was a measure of its internal energy, thereby giving a clue to the possibility of obtaining nuclear energy; it provided the basic equations for describing the motion of highly energetic particles; and it led to the prediction of the existence of antimatter.

I. Experimental Background

A. ABERRATION OF STARLIGHT

The first well-verified connection between the motion of an observer and the velocity of light was made by the astronomer James Bradley in 1727. He observed a seasonal change in the apparent position of the star Gamma Draconis, relative to other stars. This effect, now known as stellar aberration, is illustrated in Fig. 1.

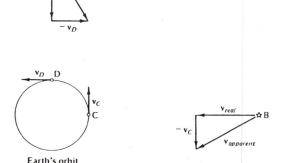

Earth's orbit

FIG. 1.   Aberration of starlight. Stars are located at A and B. Light travels in direction $\mathbf{v}_{real}$ to reach earth. When earth is at C, moving with velocity $\mathbf{v}_C$, earth observers see light from B moving in the direction of $\mathbf{v}_{apparent} = \mathbf{v}_{real} - \mathbf{v}_C$. When earth is at D, moving with velocity $\mathbf{v}_D$, light from A is seen to move in direction $\mathbf{v}_{real} - \mathbf{v}_D$. [Reproduced, with permission, from J. McGervey, (1983). "Introduction to Modern Physics," 2nd ed., p. 29, Academic Press, New York.]

Stars A and B lie in the plane of the earth's orbit. In order to reach the solar system from each star, rays of light must travel in the direction of the vector $\mathbf{v}_{real}$, shown at A and B, respectively. When the earth is at point C, with velocity $\mathbf{v}_C$, the ray from A is traveling in the direction $\mathbf{v}_{real}$, relative to earth, when it arrives. But the ray from B appears to be moving in the direction of the vector labeled $\mathbf{v}_{apparent}$ at B. Thus the measured angle between the two stars as

they appear in the sky is less than 90°. But when the earth is at D, three months later, the same two stars are more than 90° apart in the sky.

Recent observations have confirmed, to a high degree of accuracy, that the angle between $\mathbf{v}_{real}$ and $\mathbf{v}_{apparent}$ equals the ratio of the earth's orbital speed to the speed of light, or about 10^{-4} (in radians). A constant aberration, which might result from the motion of the whole solar system through space, cannot be detected. Only the change in aberration angle is seen, as earth changes direction in its orbit. [See CELESTIAL MECHANICS.]

B. FIZEAU EXPERIMENT

With the development of the wave theory of light, it was believed that all of space was filled with some substance, called ether, that served as a medium to transmit light waves. It was then suggested that the focal length of a lens would depend on the motion of the earth through this ether, because the speed of light relative to earth would vary as the earth moved in different directions relative to the ether. However, when this idea was tested on starlight, it was found that the light was always bent by a lens as it would be if the earth were at rest, and the apparent direction of the light ray could be considered its "true" direction.

To explain why the motion of the earth does not affect the properties of a lens moving with the earth, Fresnel postulated that, in a medium whose index of refraction is n, the ether density is proportional to n^2, and the excess ether in the

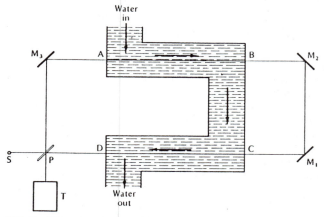

FIG. 2.   Fizeau's experiment. Mirrors M_1, M_2, and M_3 cause light beams to pass through moving water along the rectangle ABCD in both directions. [Reproduced, with permission, from J. McGervey (1983). "Introduction to Modern Physics," 2nd ed., p. 33. Academic Press, New York.]

body is carried with it when it moves. Fresnel showed that his "ether drag" always had the consequence that no observer can ever tell, by any method based on the refraction of light, that he is moving through the ether, as long as the observer's speed is much less than c, the speed of light in vacuum.

Fresnel's postulate was put to a direct test by Fizeau in 1851, using the apparatus illustrated in Fig. 2. Water flows through a bent, transparent tube. Light from the source S shines through the water in two directions after passing through plate P, which partially reflects it and partially transmits it. If the water drags the ether with it, the light ray that goes with the water flow will move faster than the ray that moves against the flow. When the rays come together again at the telescope T, the phase difference between them will vary as the water speed varies. The experiment showed that the changes in phase were in agreement with what one would expect from Fizeau's postulate.

The Fizeau experiment involves the speed of light in water, so it is essentially a refraction experiment. If the entire Fizeau apparatus were to move in any direction whatsoever, it would have no effect on the observed phase dif-ference between the two light rays, as long as the ether is dragged according to Fresnel's postulate.

C. Michelson–Morley Experiment

Attempts to find small effects of motion through the ether culminated in the Michelson–Morley experiment (Fig. 3). Like Fizeau's experiment, the Michelson–Morley experiment is based on the interference between two light rays that combine after traveling on different paths. But no refraction is involved.

Light from source S is partly reflected and partly transmitted by mirror M to give two rays. The transmitted ray (ray 1) travels to mirror M_1 and back to M; the reflected ray (ray 2) travels to mirror M_2 and back to M. The rays then combine and enter telescope T. If the total travel time of ray 1 is the same as the travel time of ray 2 (between S and a spot on the telescope), then the rays produce a bright region at that spot. This will happen if both rays travel at the same speed, and the total lengths of the two paths are equal.

But if light is carried by an ether, the speed of light will be the same for the two rays only if the apparatus is at rest in the ether. If light travels at

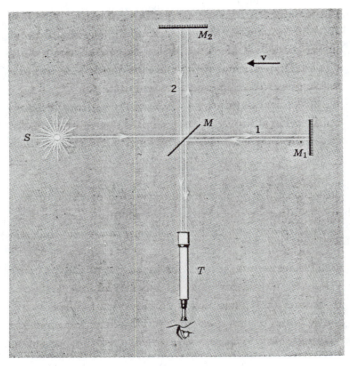

FIG. 3. Simplified diagram of the Michelson–Morley experiment. [Reproduced, with permission, from R. Resnick (1968). "Introduction to Special Relativity," p. 20. John Wiley, New York.]

TABLE I.   Experimental Basis for the Theory of Special Relativity[a][b]

Theory		Light propagation experiments			
	Aberration	Fizeau convection coefficient	Michelson–Morley	Kennedy–Thorndike	Moving sources and mirrors
Ether theories					
Stationary ether, no contraction	A	A	D	D	A
Stationary ether, Lorentz contraction	A	A	A	D	A
Ether attached to ponderable bodies	D	D	A	A	A
Emission theories					
Original source	A	A	A	A	A
Ballistic	A	N	A	A	D
New source	A	N	A	A	D
Special theory of relativity	A	A	A	A	A

[a] From W. K. H. Panofsky and M. Phillips, "Classical Electricity and Magnetism," 2nd ed., © 1962, Addison-Wesley, Reading, Massachusetts. Page 282, Table 15-2. Reprinted with permission.
[b] A, the theory agrees with experimental results; D, the theory disagrees with experimental results; and N, the theory is not applicable to the experiment.

speed c in the ether, and the apparatus moves through the ether, the light will have an apparent speed that differs from c. This apparent speed will depend on the direction in which the light travels. If the apparatus travels through the ether with velocity **v** in the direction shown (from M_1 to P), then ray 1 will take longer than ray 2; the time difference will be approximately Lv^2/c^3, where L is the distance from M to either M_1 or M_2. But if the apparatus is traveling at right angles to this direction, ray 2 will take the longer time. For the length used by Michelson and Morley, with v equal to the orbital speed of the earth, the time difference is about 4×10^{-16} sec.

By rotating the apparatus through a 90° angle, Michelson and Morley were able to reverse the roles of the two rays. If the apparatus were moving through the ether, this should have the effect of changing the locations of bright regions in the field of view of the telescope as the telescope rotated.

The experiment was repeated several times, at different times of the year, to allow for the possibility that the earth happened to be stationary in the ether when the experiment was first performed, but no such effect was ever observed. Michelson had estimated that a speed through the ether of about 2% of the earth's orbital speed should have led to a detectable effect when the apparatus was rotated.

D. PRERELATIVITY EXPLANATIONS

As a result of these and other experiments, it seemed that nature would never permit us to detect motion of the earth through the hypothetical ether. Various explanations were advanced for this situation. It was suggested by Lorentz that bodies moving through the ether were contracted in the direction of motion by just enough to cancel the additional travel time required by ray 1 in the Michelson–Morley experiment. Another possibility was that the earth carried the ether along with it, so the apparatus was always at rest in the ether. It was also suggested that light, unlike other waves, had a speed that depended on the speed of the source. One by one, each theory was tested by additional experiments and found to fail. What was needed was not a series of *ad hoc* theories, but rather, a comprehensive theory based on fundamental principles. Einstein developed that theory—special relativity. It has survived millions of experimental tests.

E. SUMMARY OF EXPERIMENTAL FACTS AND THEORIES

Experiments testing various theories are listed in Table I, showing that each theory, with the sole exception of special relativity, disagrees with one or more experimental results.

Light propagation experiments		Experiments from other fields						
De Sitter spectroscopic binaries	Michelson–Morley using sunlight	Variation of mass with velocity	General mass–energy equivalence	Radiation from moving charges	Meson decay at high velocity	Trouton–Noble	Unipolar induction using permanent magnet
A	D	D	N	A	N	D	D
A	A	A	N	A	N	A	D
A	A	D	N	N	N	A	N
D	D	N	N	D	N	N	N
D	D	N	N	D	N	N	N
D	A	N	N	D	N	N	N
A	A	A	A	A	A	A	A

II. Postulates of Special Relativity

A. THE PRINCIPLE OF RELATIVITY

Galileo referred to the principle of relativity in his "Dialogue on the Great World Systems." The principle may be stated as

No experiment can tell whether one is at rest or moving uniformly

An equivalent way of stating this principle is

The laws of nature are the same in all uniformly moving laboratories

It is understood that uniformly moving laboratories include those considered to be at rest. Another name for a uniformly moving laboratory is inertial frame of reference. The name comes from the fact that Newton's first law of motion, the law of inertia, is valid in such a frame of reference. The law of inertia would not be valid in an accelerated frame of reference.

According to this principle, absolute motion cannot be detected; anyone who is moving uniformly with respect to some object is entitled to consider himself to be at rest and the object to be moving. He can do no experiment to disprove his assumption that he is at rest.

The principle of relativity was considered to be obvious until the concept of the ether was developed. The inability to detect the ether led Einstein to adopt the principle of relativity as the first postulate of his special theory of relativity. This postulate gets rid of the problems associated with the ether by simply stating that there

is no ether. The presence of an ether would allow one to detect absolute motion, if one were to assume the ether to be at rest.

B. THE SPEED OF LIGHT

Light is a wave, and waves travel in a material medium with a characteristic speed relative to that medium. Light waves apparently travel in empty space, where some other rule must govern their speed. Einstein, considering the experimental facts, stated that this rule should be the second postulate of special relativity:

Light in empty space always travels with the same definite speed c with respect to any inertial frame of reference, regardless of the state of motion of the body emitting the light.

This postulate explicitly rules out emission theories, which say that the speed of light depends on the speed of the source. Such theories have been disproven experimentally by observation of radiation from binary star systems (Table I). Einstein's postulate, on the other hand, is in agreement with all known observations of the speed of light, and it has led to many other conclusions that have stood the test of experiment.

The fundamental concern of the postulate is *speed*, not light. The existence of the speed c has consequences that can be stated without reference to the behavior of light. One could rephrase the postulate to say:

There is a characteristic speed in the universe, such that anything traveling at that speed with respect to one inertial frame of reference must also travel at that speed with respect to any other inertial frame.

The experiments on light indicate that light travels at that characteristic speed c.

III. Concepts of Space, Time, and Motion

A. ABSOLUTE TIME

It is natural to think of time in absolute terms, as something that must be the same for all frames of reference. The postulate of an absolute speed conflicts with the notion of absolute time. To appreciate this fact, consider a light source in the middle of a boxcar that is moving uniformly from south to north.

A flash of light emitted in all directions from the source will reach the south end of the car before it reaches the north end, because the south end is moving toward the source and the north end is moving away from it. The light that travels northward (moving at the same speed as the light that travels southward) has a greater distance to travel before it hits its end. But this will be true only in a frame of reference in which the boxcar is moving northward.

An observer inside the boxcar is entitled to assume that the car is at rest. No measurement made inside the boxcar can contradict that assumption. Therefore, for this observer, the same light flash reaches north and south ends simultaneously. In general, events that are simultaneous in one frame and not simultaneous in other frames.

Thus time is not absolute. It is a measure that places all events in a sequence, such that one event may be before, after, or simultaneous with another event. This measure is different in different frames of reference, as we have just seen. The two events, namely, light reaching the north end of the boxcar and light reaching the south end, are simultaneous in only one frame of reference. According to any frame in which the car is moving northward, the same light flash will reach the south end first. There are also frames of reference in which the car is moving southward. Measurements made in any such frame will show that the light will reach the north end first.

B. TRANSFORMATION OF COORDINATES

A frame of reference may be thought of as a set of perfect clocks, which can be placed at any point, plus a set of rigid measuring rods to measure the space coordinates of each point. An event is then specified by using these tools to

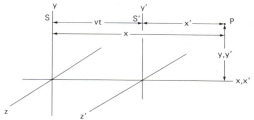

FIG. 4. Two inertial frames with a common x–x' axis and with axis y parallel to y', axis z parallel to z'. Relative to frame S, frame S' is moving in the positive x direction at speed v. Point P gives the location of an event whose space and time coordinates may be determined in each frame of reference. The diagram suggests that the x' coordinate is related to the x coordinate by the equation $x' = x - vt$. The origin of frame S' is chosen so that $x' = x$ at $t = 0$.

obtain the four space–time coordinates x, y, z, and t. The same event in a different (primed) frame of reference would be specified by the four coordinates x', y', z', and t'. If we know x, y, z, and t for a given event, and we know how the primed frame is moving relative to the unprimed frame, we should be able to deduce the values of x', y', z', and t' for that event. The formula for finding these values is called a coordinate transformation.

1. Galilean Transformation

Before relativity, it was simply assumed that time is absolute, so for any event, $t' = t$. We can always choose the directions of our coordinate axes so that the velocity **v** of the primed frame relative to the unprimed frame is parallel to the x axis (Fig. 4). From these initial assumptions we obtain the Galilean transformation

$$x' = x - vt, \qquad y' = y,$$
$$z' = z, \qquad t' = t \tag{1}$$

From Fig. 4, the Galilean transformation seems so obvious that one feels it cannot be wrong. But it conflicts with the second postulate, as we now demonstrate.

A flash of light, starting at $x = 0$ when $t = 0$, will reach coordinate x at time t such that $x/t = c$. The same flash of light will start at $x' = 0$ and $t' = 0$, and its speed in the primed frame will be x'/t'. According to the Galilean transformation, $x'/t' = x'/t = c - v$, the speed of light in the primed frame is not c, and the postulate is violated. Therefore we must rule out the Galilean transformation and find a transformation that gives the same speed of light in both frames of reference.

TABLE II. Coordinates of Events in Two Reference Frames[a]

	Event						
	A	B	C	D	E	F	G
$x =$	0	−3	−1	+1	+3	+5	0
$t =$	0	+5	+5	+5	+5	+5	+8
$x' =$	0	0	+2.5	+5	+7.5	+10	+6
$t' =$	0	+4	+5.5	+7	+8.5	+10	+10

[a] In this table, the speed of light is one unit, and the relative speed of the two reference frames is $\frac{3}{5}$ unit.

2. Lorentz Transformation

To obtain the correct transformation, one must abandon the assumption that $t' = t$, relying instead on the postulate concerning the speed of light. Therefore the coordinates of the light flash described in the preceding paragraph must always satisfy the equations $x/t = c$ and $x'/t' = c$. This objective is met by the Lorentz transformation

$$x' = (x - vt)\gamma, \qquad y' = y, \qquad z' = z$$
$$t' = [t - (vx/c^2)]\gamma \tag{2}$$

where γ is defined by

$$\gamma = [1 - (v/c)^2]^{-1/2}$$

If we divide the expression for x' by the expression for t', and set x/t equal to c, we find that x'/t' also equals c, as required.

We illustrate the Lorentz transformation in Table II, which lists the coordinates of a set of events A, B, C, etc. This table will permit us to compare measurements made in different frames of reference, without repeated substitution into Eqs. (2). The events in the table are described as follows.

1. Event A is chosen as the event for which all coordinates are zero in both frames.

2. Events B–F are simultaneous in the unprimed frame (at $t = 5$), occurring at equally spaced values of x. Their coordinates in the primed frame are obtained from Eqs. (2) with $v/c = -0.6$ and $c = 1$. (Use of the light-second as the unit of length makes c equal to 1 in light-seconds per second.)

3. Event B occurs at the origin of the primed frame, 4 sec after event A, according to a clock fixed at that origin ($x' = 0$).

4. Event G occurs at the origin of the unprimed frame, 8 sec after event A, according to a clock fixed at that origin ($x = 0$).

The reader may verify for each event in Table II that the x' and t' coordinates in the column for that event are correctly obtained from the x and t coordinates by the use of Eqs. (2) with v equal to -0.6.

If the unprimed frame is considered to be at rest, we verify that the velocity of the primed frame along the x axis is $v = -\frac{3}{5}$ by comparing event A with event B. In a time t of 5 sec, point $x' = 0$ moves from $x = 0$ to $x = -3$, so $v = x/t = -\frac{3}{5}$. (The point $x' = 0$ is a fixed point—a mark on a stationary meter stick—in the primed frame. So its speed in the unprimed frame is the same as the speed of the whole primed frame of reference.)

If the primed frame is considered to be at rest, we find the speed of the unprimed frame by comparing events A and G. In a time (t') of 10 sec, the point $x = 0$ moves from $x' = 0$ to $x' = 6$, so the velocity of the unprimed frame along the x' axis is equal to $\frac{6}{10}$, or $\frac{3}{5}$. Thus both observers agree on the relative speed of the two frames.

From events A and F we see that an object moving with speed $u = c = 1$ in the unprimed frame also moves with speed $u = c = 1$ in the primed frame. A body moving at speed 1 in the unprimed frame, starting from event A, reaches $x = 5, t = 5$, which are the coordinates of event F. Event F is described in the primed coordinates by $x' = 10$ and $t' = 10$, so the object moves at speed $x'/t' = \frac{10}{10} = 1$ in that frame also.

IV. Implications of the Lorentz Transformation

A. LENGTH AND TIME MEASUREMENTS IN A MOVING FRAME OF REFERENCE

1. Addition of Velocities

If an object moves at a speed less than c in one frame of reference, then its speed is different in another frame of reference. For example, refer to Table II and consider an object that moves with constant speed from event A to E. (The fact that this object is present at event E must be clear to an observer in any frame of reference. The general principle is that all observers agree on whether an object is present at a given event. Thus, if a second object were also at event E, the two objects would collide, and the fact that a collision had occurred would be indisputable. The collision could leave marks on each object, and the marks would be visible to all observers, no matter what their speed.)

In the unprimed frame, event E has coordinates $x = 3$ and $t = 5$, so the speed of an object moving from event A to E is $\frac{3}{5}$. In the primed frame, event E has coordinates $x' = 7.5$ and $t' = 8.5$, so the speed of the same object is 7.5/8.5, or $\frac{15}{17}$. (If we used the Galilean transformation [Eqs. (1)], x' would be 6, t' would be 5, and the speed would be $\frac{6}{5}$—a speed greater than that of light. We see below that the Lorentz transformations cannot give a speed greater than the speed of light, because when v is greater than c, the factor γ in Eqs. (2) is imaginary.)

A general formula giving the velocity of an object in the primed frame in terms of the velocity in the unprimed frame may be found by simple division, using Eq. (2). The components of the velocity along the x', y', and z' axes are, respectively,

$$u'_x = \frac{x'}{t'} = \frac{(x - vt)\gamma}{[t - vx/c^2]\gamma} = \frac{(x/t) - v}{1 - (vx/c^2t)}$$

$$= \frac{u_x - v}{1 - (u_x v/c^2)}$$

$$u'_y = \frac{y'}{t'} = \frac{y}{(t - (vx/c^2))} = \frac{u_y}{[1 - (u_x v/c^2)]\gamma} \qquad (3)$$

$$u'_z = \frac{z'}{t'} = \frac{z}{(t - (vx/c^2))} = \frac{u_y}{[1 - (u_x v/c^2)]\gamma}$$

where u_x, u_y, and u_z are the components of the velocity in the unprimed frame, and v is the velocity of the primed frame relative to the unprimed frame. (This velocity is directed along the x axis, as mentioned previously.)

Equations (3) show directly that an object moving at speed c in one frame has the same speed in the other frame, regardless of the value of v. For example, if $u_x = c$, $u_y = 0$, and $u_z = 0$, then direct substitution shows that $u'_x = c$ and the other components are zero.

For the example discussed in connection with Table II, with $u_x = \frac{3}{5}$, $c = 1$, and $v = -\frac{3}{5}$, the first part of Eqs. (3) gives

$$u'_x = [\tfrac{3}{5} - (-\tfrac{3}{5})]/[1 - (\tfrac{3}{5})(-\tfrac{3}{5})] = \tfrac{15}{17}$$

the same speed deduced from Table II.

Equations (3) can also be applied to the Fizeau experiment described in Section I. In the (unprimed) frame of reference in which the water is at rest, the speed of light in the water is $u_x = c/n$, where n is the index of refraction of water. If the laboratory is taken to be the primed frame, and the water moves with speed v in this frame, then the primed frame must be moving with speed v relative to the unprimed frame, and the speed of light in the laboratory should be,

according to Eqs. (3),

$$u'_x = [(c/n) - v]/[1 - (v/cn)] \qquad (4)$$

when the light and the water are moving in opposite directions. Fizeau's result agrees with Eq. (4). (However, it must be admitted that the Fresnel ether-drag formula gives almost the same result. The difference between the predictions of the two formulas is too small to have been observed by the Fizeau experiment.) There are now much better grounds for reliance on Eqs. (3).

2. Synchronization of Clocks

When we speak of the times t and t', we refer to an abstraction that can only be realized by reading a clock. The times shown in Table II are known from clock readings, and these readings can be photographed. Observers in all frames of reference will agree that an unprimed clock that reads 5.00 and a primed clock that reads 4.00 are both present at event B. They also agree that an unprimed clock that reads 5.00 and a primed clock that reads 5.5 are both present at event C.

If these readings are to give what we consider to be the actual time, clocks at different places have to be synchronized. One way to synchronize two clocks at different locations is to start each one with a flash of light that originates midway between them. If both clocks are at rest, the light flash will arrive at each clock and start the clocks simultaneously. Being identical, the clocks will remain synchronized as time passes.

An observer O who is at rest in the unprimed frame of reference will conclude that events B and C are simultaneous, because the clocks in his rest frame at those events both read $t = 5.00$. He will say that clocks in the primed frame are not synchronized, because they read different values at the same time t.

The clocks in the primed frame have indeed been synchronized, but the synchronization was done in the primed frame, that is, by an observer O' who was moving along with the clocks in that frame. Observer O says that the procedure of O' did not work, because the light flash that started the primed clocks did not arrive simultaneously at the two clocks. The primed clock at event C was moving toward the origin of the light flash, and the primed clock at event B was moving away from it; therefore, the clock at event C received the signal first, and it started sooner.

On the other hand, observer O' at rest in the primed frame will say that the primed clocks are indeed synchronized, because they are at rest, relative to O'. Observer O' will say that events B and C are therefore not simultaneous, because the primed clocks at those events read $t' = 4$ and $t' = 5.5$, respectively. Assuming observer O' to be at rest, that observer will say that the clocks of observer O are the ones that are moving and are not synchronized. The general principle here is, if all clocks are synchronized in their own rest frames, then

Moving clocks at different locations are not synchronized

That is, at any given time, two moving clocks at different places have different readings, no matter how the clocks are constructed (unless the clocks are faulty). This conclusion also follows directly from the Lorentz transformation, or it can be shown from any other method one might use to synchronize two clocks, given the second postulate.

If a starting signal originates halfway between two clocks, the clock that is moving toward the source of the signal will start first. Thus in Table II, giving readings of clocks that are observed at the same time t, the reading of time t' at event C is greater than the reading of t' at event B, because the primed clock at event C, moving toward the signal, started at an earlier time (t) than did the primed clock at event B. The difference between the readings t' of the two clocks is Lv/c^2, where L is the distance between the two clocks (measured in the frame in which they are at rest) and v is the speed of each clock. Table II shows that, for the clocks present at events B and C, $L = 2.5$, $v = \frac{3}{5}$, and $c = 1$, making the difference between the clock readings equal to 1.5, as shown.

From the point of view of observer O', the unprimed clocks are moving in the $+x$ direction. Thus if two unprimed clocks were observed at the same time t', the clock at the larger value of x would give a smaller reading than the other clock (e.g., see events F and G).

3. Lorentz Contraction

Again refer to Table II. Consider a stick that is 5 units long and is at rest in the primed system. Let one end of the stick be fixed at $x' = 0$, and the other end fixed at $x' = 5$. As the primed frame moves past the unprimed frame, observer O measures the position of both ends of the stick simultaneously (at time $t = 5$). One end is taken at $x = -3$ (event B) and the other end is at $x = +1$ (event D). Thus O would logically say that the stick is 4 units long, because that is the dis-

tance between the two ends at a given time. The general principle is

A moving stick is contracted along the direction in which it moves

The amount of contraction is given by

$$L' = L/\gamma \tag{5}$$

where L is the length of the stick when it is at rest, L' is its length when it is moving, and γ is defined in Eqs. (2).

Again, either O or O' can be considered to be at rest. If a stick is at rest in the unprimed system, O' measures its length by measuring the position of both ends at the same time t'. Assume that the left end of the stick is at $x = 0$ and the right end at $x = 5$. If the measurement by O' is made at time $t' = 10$, O' will find the right end to be at $x' = 10$. At this time ($t' = 10$) we see from event G that the point $x = 0$ has the coordinate x' equal to 6. Thus, when they are measured at the same time t', the ends of the stick are at $x' = 6$ and $x' = 10$. As before, the stick is found to have a length of 4 units when it is moving at a speed of $\frac{3}{5}$. It is contracted to four-fifths of its rest length, in agreement with the factor γ when $v = \frac{3}{5}$.

4. Time Dilation

Notice events A and B in Table II. Both events occur at the same place ($x' = 0$) in the primed frame of reference. Therefore the times t' of these two events are measured by a single clock, which is at rest in the primed frame. Similarly, events A and G occur at the same place ($x = 0$) in the unprimed frame of reference, and the times t of those events can also be measured by a single clock, which is at rest in the unprimed frame of reference.

In each case, the time interval read by the single clock is smaller than the time interval that is measured between the two events in the other frame of reference, in which that clock is moving. This is a further implication of the Lorentz transformation:

Moving clocks run slowly relative to identical clocks at rest

The Lorentz transformation can be used to derive a general formula relating the rates of clocks. Consider two events that occur at different places and a clock that is present at both events. If that clock is moving at speed v in any given frame of reference, and the time interval between the events is t_0 in that frame, then the time interval τ measured by the moving clock is given by

$$\tau = t_0/\gamma \qquad (6)$$

where $\gamma = [1 - (v/c)^2]^{-1/2}$ as in Eq. (2).

For example, the clock located at $x' = 0$ is present at events A and B. In the unprimed frame of reference, this clock is moving, with $v/c = -\frac{3}{5}$, and the time interval t_0 between A and B is 5 sec. Inserting these values in Eq. (6) gives $\tau = 4$, which is just the time interval recorded by the clock at $x' = 0$.

Equation (6) can also be applied to events A and G. According to observer O', these events occur at different places (at $x' = 0$ and $x' = 6$, respectively), and the time interval between these events is 10. Thus in Eq. (6) we can put t_0 equal to 10, and a clock that is present at both events must move at a speed of $\frac{6}{10}$, or $\frac{3}{5}$. If we put v/c equal to $\frac{3}{5}$, we again find $\gamma = \frac{5}{4}$, and Eq. (6) then gives $\tau = 8$. This is just the time interval recorded by the clock located at $x = 0$—the clock that is present at both events A and G.

5. Proper Time

Although observers in different frames of reference may disagree on the time interval between two events, and they may disagree on the speed of the clock that is present at both events, they must agree on the times that are recorded by that clock at each event. They must agree for the simple reason that the clock face can be photographed as it appears at each event. The time interval measured by such a clock is called the proper time interval.

A formula that is equivalent to Eq. (6) gives the proper time interval τ in terms of the space and time coordinates of two events as

$$cτ = \{c^2(t_2 - t_1)^2 - (x_2 - x_1)^2$$
$$- (y_2 - y_1)^2 - (z_2 - z_1)^2\}^{1/2} \qquad (7)$$

where the coordinates x, y, z, and t apply to any frame of reference whatsoever, and the subscripts 1 and 2 refer to their values at event 1 and event 2, respectively. Notice that in the frame of reference for which both events occur at the same place,

$$x_2 = x_1, \qquad y_2 = y_1$$
$$z_2 = z_1, \qquad \tau = t_2 - t_1$$

Thus we define a proper time interval by the statement

The proper time interval between any two events is the time interval recorded by a uniformly moving clock that is present at both events

This statement is equivalent to

The proper time interval between any two events is the time interval measured in the frame of reference in which the two events occur at the same place

It can be shown directly from the Lorentz transformation that τ is a quantity that is the same for all observers. When this quantity is zero, a clock would have to move with the speed of light to be present at both events. Such a clock would register zero time interval between any two events; it would be a clock that does not run at all. We can say that, for something traveling at the speed of light, time simply does not flow.

6. Space–Time Interval between Events

There exist many pairs of events for which the quantity τ^2 is negative, so that the proper time interval τ is imaginary. These events are so far apart that a clock would have to travel with a speed greater than c in order to be present at both events. Thus there is no clock that reads this interval directly.

In such a case we can still define $c\tau$ to be the space–time interval between the events. If τ is real, the interval is timelike, and τ is the proper time interval. If τ is imaginary, the interval is spacelike. We have seen that, when the interval is timelike, there is a frame of reference in which the two events occur at the same place. Similarly, when the interval is spacelike, there is a frame of reference in which the two events occur at the same time. In that frame of reference, the square of the distance between the two events is equal in magnitude to the square of the space–time interval. All observers agree on the space–time interval between two events. We say that this is an invariant quantity.

An invariant quantity is one that has the same value in all inertial frames of reference

Thus, although the distance between two events is different in different frames, the space–time interval, as calculated by Eq. (7), is the same in all frames.

Events F and G of Table II are separated by a spacelike interval. In the unprimed frame, F and G are separated by a time of 3 units and a distance of 5 units, so $c^2\tau^2$ is equal to 9 minus 25, or -16. Thus there is a frame of reference in which these two events occur at the same time; in that frame, the distance between the events is equal to 4. The primed frame of Table II is in fact the frame in which these two events occur at the same time, and in that frame the distance between these events is $10 - 6$, or 4. To be present at both events, a clock would have to travel at a speed of $\frac{2}{3}$ relative to the unprimed frame, or at an infinite speed relative to the primed frame. In both the primed frame and the unprimed frame, the square of the space–time interval as defined by Eq. (7) is equal to -4. All observers agree on the value of $c^2\tau^2$. (Test this for other pairs of events; for example, for C and G the value of $c^2\tau^2$ is 8).

B. THE DOPPLER EFFECT

The Doppler effect is well known in wave phenomena. A sound source has a higher pitch when it is approaching than when it is receding. The Doppler effect for light is similar to the effect for sound, but the formula must be modified for the effect of time dilation—the fact that a moving light source emits light of lower frequency than does an identical source at rest, just as a moving clock runs more slowly than an identical clock at rest. In fact, a light source itself can be used as a standard clock.

For a sound wave, the observed frequency f' may be related to the source's vibration frequency f by the nonrelativistic formula

$$f'/f = (v_o + c_s)/c_s \qquad (8)$$

where v_o is the component of the velocity of the source in the direction toward the observer, c_s is the speed of sound, and the observer is at rest in the medium that transmits the sound.

For light, we can always assume the observer to be at rest, so we can use the same formula, provided we replace c_s by c, and replace f by f/γ. The latter substitution is required by the time dilation effect on the moving source; f is the source's frequency in its rest frame, and the moving source has a lower frequency than it would have if it were at rest. The result is

$$f'/f = (v_o + c)/c\gamma \qquad (9)$$

If the velocity vector $\mathbf{v}$ is directed toward the observer, then v_o becomes equal to the relative speed v, and Eq. (9) can be simplified to

$$f'/f = [(c + v)/(c - v)]^{1/2} \qquad (10)$$

If the source is moving directly away from the observer, Eq. (10) is still valid, but v becomes negative. Finally, if $\mathbf{v}$ is at right angles to the line between source and observer, then v_o equals zero, and Eq. (9) becomes $f' = f/\gamma$. In this case we have what is called the transverse Doppler effect, which results purely from time dilation.

For a sound wave, which travels in a definite medium, the observed frequency f' is given by a different formula when the source is at rest in the medium and the observer is moving. But for light in empty space, it is meaningless to distinguish between moving source and moving observer, so the same formula, Eq. (9), must hold in all situations, with v_o being the relative speed of approach of source and observer.

To illustrate the use of Eq. (10), let us refer again to Table II. Suppose that a source of light is fixed at $x = 0$, and it emits 100 light flashes per second. Further suppose that an observer O', at rest in the primed frame of reference, is located at $x' = 6$. According to Eq. (9), with $c = 1$ and $v = \frac{3}{5}$, O' will see $f' = 2f$, or 200 flashes/sec as the source approaches.

At time $t' = 0$, O' says that the source is 6 light-sec away, and is moving toward him/her at speed $v = \frac{3}{5}$ light-sec/sec. At time $t' = 10$, the source has arrived at O'. (This is event G.) During this 10-sec time interval, O' has received flashes at the rate of 200/sec, for a total of 2000 light flashes. Some of these flashes were already en route to O' at time $t' = 0$, and the rest were emitted by the source during the following 10 sec. Let us try to account for all of these flashes.

Because the source is moving at $v/c = \frac{3}{5}$, it emits light flashes at a frequency of only four-fifths of the frequency seen in its rest frame [according to the time dilation formula of Eq. (6)]. Therefore O' says that the source emits 80 flashes/sec, and that it emits 80×10, or 800 flashes during those 10 sec. The other 1200 flashes must have already been en route to O' at time $t' = 0$. We confirm this by noting that the flashes are arriving at a rate of 200/sec, so the distance between one flash and the next must be 1/200 light-sec. Since the total distance between O' and the source (according to O') was 6 light-sec, there must have been 6×200, or 1200 flashes that were on their way to O' at time $t' = 0$.

C. THE TWIN PARADOX

Time dilation governs all processes in the moving frame of reference. Biological processes proceed at the same rates as measured by the clock in any given frame (otherwise one could tell if one is in a moving frame). Therefore, if the clocks in a moving frame run slowly (as seen by us in our rest frame), then people moving with those clocks will (as seen by us) age more slowly than will people at rest. However, those people have no way of determining that they are living longer than they would if they were at rest; as far as they are concerned, they are at rest.

These facts suggest an apparent paradox for space travelers moving at high speeds. Consider fraternal twins Jay and Kay. On their 20th birthday, Kay leaves on a space trip while her brother Jay remains on earth. Her journey, at a speed of $\frac{3}{5}c$, is to star system X, which is 15 lt-yr from earth. Upon her arrival there, she will very quickly turn around and return to earth at the same speed.

We can analyze this trip by means of three inertial reference frames: frame J is Jay's rest frame; frame K' is Kay's rest frame while the twins are separating; frame K'' is Kay's rest frame while the twins are coming closer together again. The round trip, which takes 50 yr in frame J, takes only 40 yr according to Kay's clocks as she spends 20 yr in frame K' and 20 yr in frame K''. She will reach earth again on her 60th birthday, when Jay is observing his 70th birthday.

This result need not be a surprise to either twin, because they can communicate with each other during the entire trip. Let us assume that each twin, counting time by his or her own clocks, sends the other twin a "happy birthday" message—a light signal—on each anniversary of their separation. Kay will send Jay 40 messages, and Jay will receive all of them (including the last one, which will be delivered in person). Jay will send Kay 50 messages, and she will receive all of these. Both twins will agree that Jay has had 50 birthdays and Kay has had 40 birthdays.

The sequence of events is shown graphically in Fig. 5. Both Jay and Kay move on paths in space–time called "world lines." Figure 5a shows these paths in frame J. In frame J, Jay goes nowhere, remaining at $x = 0$; thus his world line is along the t axis (i.e., the line $x = 0$). (He moves in time and is motionless in space.) In this frame, Kay's path is two straight lines: one from the origin to the point $x = 15$, $t = 25$ and another from that point to $x = 0$, $t = 50$.

The light signals that each sends follow straight lines with a slope of 1 (because their speed is 1 lt-yr/yr). The ones sent by Jay at $t = 0, 5, 10, \ldots$ are shown in Fig. 5a. Signal 10 arrives at Kay's ship just as she reaches planet X. Kay is then 40 years old, receiving the message Jay sent when he was 30-yr old. She has received one signal every two years (in agreement with the Doppler formula (10), which gives $f' = \frac{1}{2}$ when $v = -\frac{3}{5}$ and $f = 1$).

During Kay's 20-yr return trip to earth, she receives Jay's birthday messages at the rate of 2/yr. [The Doppler formula [Eq. (10)], with $v = \frac{3}{5}$, $c = 1$, and $f = 1$, yields $f' = 2$]. She receives 40 messages during the return trip, so she is not surprised to find that her brother is $30 + 40$, or 70 years old when she arrives.

Figure 5b shows the trip in Kay's two inertial frames of reference (K' for the first 20 years, K'' for the next 20 years). Kay receives signal 10 at $t' = 20$, but to deduce how rapidly Jay has been aging, she must make an inference regarding the time when the signal was sent. From the Doppler effect she can infer that Jay has been reced-

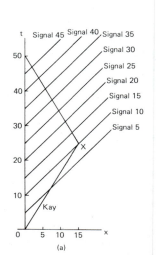

(a)

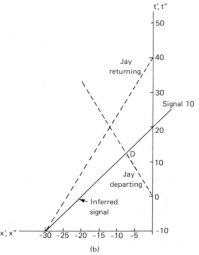

(b)

FIG. 5.   World lines for twins Jay and Kay and for the signals sent from Jay to Kay. (a) Seen in Jay's rest frame and (b) seen in Kay's rest frames.

ing from her at a speed of $\frac{3}{5}$. This tells her that his world line must be the dashed line. The intersection of this line with the world line of the light signal (which has a slope of 1 in all frames) shows that the signal was sent at time $t' = 12.5$, from the event marked D. (To verify this conclusion, note that Jay would reach $x' = -7.5$ at time $t' = 12.5$, and that the signal would take 7.5 yr to reach Kay from that point, making the arrival time t' equal to $12.5 + 7.5$, or 20 yr, as observed.)

Thus each twin can deduce that the other twin is aging less rapidly. Kay sees that Jay's clock has only recorded 10 yr when $t' = 12.5$ (event D in the figure), and Jay can see that Kay's clock has only recorded 20 yr when his time t is equal to 25 yr (event X). The paradox is in the fact that Jay does indeed age 50 yr (and Kay only 40 yr), because the twins are not in equivalent situations. Jay remains in the same inertial frame for the entire time, but Kay switches from one inertial frame to another. At this moment a strange thing happens: Kay's inference regarding the position of her twin must suddenly change!

In Kay's new frame K″, Jay is judged to be approaching rather than receding. Therefore, Jay's inferred world line is different. From Fig. 5b we see that Jay would have had to be 30 lt-yr away when signal 10 was sent. At a speed of $\frac{3}{5}$, it will take Jay 50 yr to reach Kay, but he will send 40 more signals, showing that he will have aged by another 40 yr when he reaches Kay. Again, Jay's clocks are slow, registering 40 yr instead of 50; but he will be 70 yr old when the twins meet.

If Kay's sudden change of perception when she changes reference frames seems unnatural, the reader must realize that the whole situation is unnatural. Any traveler who attempted to change velocity from $\frac{3}{5}c$ to $-\frac{3}{5}c$ in such a short time would be crushed by the enormous force required. In any actual space trip, the traveler would be subjected to acceleration over extended periods of time, and the special theory of relativity would not be able to show what would happen during the periods of acceleration. The general theory, which does deal with acceleration, leads to the same result as the special theory for examples like the one just described. The traveler is indeed younger than her twin when she returns home. This conclusion obviously has not been tested for human beings, but it has been well verified for short-lived elementary particles, to be discussed in Section VII. [See RELATIVITY, GENERAL.]

V. Momentum and Energy

A. Conservation Laws

Among the basic laws of physics are conservation laws stating that certain quantities such as energy and momentum are always conserved in any physical process. An observer in one frame of reference may disagree with an observer in another frame of reference in calculating how much energy or momentum is present, but each observer must find that this calculated quantity does not change as time goes by. That is,

> The laws of conservation of energy and conservation of momentum are valid in all inertial frames of reference.

in agreement with the principle of relativity.

In prerelativity physics, the momentum of an object of mass m is defined as a vector $\mathbf{p}$ whose components are

$$p_x = mu_x, \qquad p_y = mu_y, \qquad p_z = mu_z \quad (11)$$

where u_x, u_y, and u_z are the three components of the object's velocity.

For a collection of objects, the total momentum $\mathbf{P}$ is defined as the vector sum of the momenta of the individual objects. That is, each component of $\mathbf{P}$ is defined by the equations

$$P_x = m_1 u_{x1} + m_2 u_{x2} + \cdots$$
$$P_y = m_1 u_{y1} + m_2 u_{y2} + \cdots \qquad (12)$$
$$P_z = m_1 u_{z1} + m_1 u_{z1} + \cdots$$

where the subscripts 1, 2, etc., refer to the values for objects 1, 2, etc.

When two particles interact, momentum may be transferred from one to the other, but the law of conservation of momentum states that the total momentum vector $\mathbf{P}$ of an isolated collection of particles remains constant, no matter how the particles interact with each other.

1. Effect of Change of Reference Frame

The Galilean transformation equations [Eq. (1)] ensure that $\mathbf{P}$, as defined in Eqs. (12), is conserved in all frames of reference, as long as it is conserved in one frame. This can be illustrated by an interaction (e.g., a collision) between two particles, of mass m and m_2, respectively, and respective speeds u_{x1} and u_{x2} along the x axis. The total momentum before the collision is

$$P_x = m_1 u_{x1} + m_2 u_{x2}$$

In a second frame of reference, moving with velocity v in the $+x$ direction, the respective velocity components are $u_{x1} - v$ and $u_{x2} - v$ according to the Galilean transformation. Then the total momentum in this frame is

$$P'_x = m_1(u_{x1} - v) + m_2(u_{x2} - v)$$

which is easily shown to be

$$P'_x = P_x - (m_1 + m_2)v \qquad (13)$$

Thus if P_x does not change in a collision, and the sum of the masses is unchanged (law of conservation of mass), then P'_x does not change, and the law of conservation of momentum is maintained in both frames by use of the Galilean transformation.

2. THE NEED FOR A RELATIVISTIC DEFINITION OF MOMENTUM

We now know that the Galilean transformation is wrong, the Lorentz transformation is the correct one to use, and Eqs. (3) must be used to transform velocities from one frame to another. Suppose that Eqs. (12) were retained as the definition of momentum, but Lorentz velocity transformation [Eqs. (3)] were used instead of the Galilean transformation to derive an expression for P'_x analogous to Eq. (13). This expression could not be written so simply in terms of P_x, as it is in Eq. (13). The result would be that conservation of P'_x would not require that P_x be conserved. In that case the law of conservation of momentum would violate the principle of relativity.

We cannot escape the need for the Lorentz transformation, and the law of conservation of momentum has proven to be very useful. We can preserve both, without violating the principle of relativity, by adopting a new definition of momentum. The new definition must meet two conditions:

1. It must be conserved in all frames of reference, to satisfy the principle of relativity.
2. It must reduce to the older definition in the limit $u \to 0$.

B. RELATIVISTIC EXPRESSIONS FOR MOMENTUM AND ENERGY

1. Momentum

The clue to the form of the new definition of momentum comes from the definition of proper time. The expression for P'_x in terms of P_x becomes complicated when one uses the old definition [Eq. (12)], because the velocity in each frame is given by the ratio of displacement to time interval, using the time coordinates peculiar to one frame of reference (e.g., $u_x = x/t$). (To simplify the form of the equations, we again assume that the particle starts at $x = y = z = 0$.)

Instead of finding the velocity components by dividing each component of displacement (x, y, and z) by t, let us divide the displacement by the proper time τ (that is, the time in the frame of reference in which the particle is at rest). We then multiply the result by m to obtain each component of momentum. The result is that Eq. (11) is replaced by

$$p_x = m_x/\tau = mu_x\gamma, \qquad p_y = m_y/\tau = mu_y\gamma,$$

$$p_z = m_z/\tau = mu_z\gamma \qquad (14)$$

As $u \to 0$, $\gamma \to 1$. In that limit, Eqs. (14) can be approximated by Eqs. (11), so that requirement 2 is satisfied.

The velocity that results from using proper time instead of one's own time is called the four-velocity, because one can define a fourth component by dividing t, the displacement in time, by the proper time γ. Since $t = \tau\gamma$, this fourth component is simply equal to γ. Multiplying this by the mass gives us a fourth component for the momentum vector:

$$p_t = mt/\tau = m\gamma \qquad (15)$$

It is logical to be concerned with this fourth component, because when the Lorentz transformation transforms the space components of displacement, the result involves the time. Let us find the components of momentum in another frame of reference, by applying the Lorentz transformation to the coordinates x, y, z, and t that appear in the four components of momentum defined here. We assume that the mass m is the same in all frames of reference, and so is the proper time τ, and we obtain

$$p'_x = (p_x - vp_t)\gamma, \qquad p'_y = p_y$$

$$p'_z = p_z, \qquad p'_t = (p_t - vp_x/c^2)\gamma \qquad (16)$$

Equations (16) are identical in form to Eqs. (2); we have simply replaced x by p_x, t by p_t, etc. The four components of momentum are transformed by a Lorentz transformation in exactly the same way as the four components of space (displacement) and time.

2. Energy

It would not be legitimate to have a law that conserved only the first three components of momentum. Therefore, the fourth component must also be conserved. The fourth component of momentum appears to be unrelated to previous theories, but examination of the velocity dependence of this component shows that it is related to the total energy of the particle. The law of conservation of energy, in relativity theory, becomes a part of the law of conservation of momentum.

Let us state the result and then justify it. The energy of an object of mass m and speed u is equal to $c^2 p_t$, or

$$E = mc^2\gamma \qquad (17)$$

where $\gamma = (1 - u^2/c^2)^{-1/2}$. We require that this definition of energy, like the definition of momentum, reduce to the classical definition in the limit $u \to 0$. Let us find this limit by expanding Eq. (17) in a power series in u/c. The result is given by the binomial theorem as

$$E = mc^2(1 + u^2/2c^2 + 3u^4/8c^4 + \cdots) \qquad (18)$$

When $u^2 \ll c^2$, the third term and all succeeding terms in Eq. (18) may be neglected, and the energy becomes

$$E = mc^2 + mu^2/2 \qquad (19)$$

The identification of $c^2 p_t$ with the energy is now justified in the limit $u \to 0$, because the second term in Eq. (19) is the familiar expression for the kinetic energy of an object of mass m and speed u. The first term is the energy possessed by an object when its speed is zero—the rest energy. This term was unknown in classical physics, but since the zero level of energy is arbitrary, its presence does not contradict any previously known law that is valid for small values of u. It does, however, lead to consequences that were quite unexpected when the theory was first proposed.

3. Definition of Kinetic Energy in Relativity Theory

As in classical theory, the kinetic energy of a body is defined as the energy associated with the motion of the body as a whole. The kinetic energy E_k of a body of mass m is thus equal to the total energy minus the rest energy, or

$$E_k = E - mc^2 = mc^2(\gamma - 1) \qquad (20)$$

We see from Eqs. (17) and (18) that the kinetic energy of an object of speed u is always greater than the value given by the classical expression. As u increases, the difference between the classical value and the value given by Eq. (17) increases. If u were to equal the speed of light c, the value of γ would be infinite. This demon-

strates how the speed of light acts as a limiting speed.

4. Definition of Force

In relativity, force is defined as the rate of change of momentum (in agreement with the classical definition). As the speed of a body approaches the speed of light, the momentum approaches infinity, which means that the increase in speed associated with a given change in momentum becomes smaller and smaller.

Classically, with momentum equal to mass times velocity, force may be defined as mass times acceleration. If we used that definition of force, we would find that as the speed of a body increases a greater force is required to provide a given acceleration. This would lead to the conclusion that the mass increases with increasing speed. The mass that increases is the so-called relativistic mass $m_r = m\gamma$, whose introduction is an artifice that permits one to use equations of the same form as classical ones (e.g., $F = m_r a$ and $p = m_r v$).

On the other hand, the definition of force as rate of change of momentum, with momentum given by Eqs. (14), lets us express everything in terms of one mass, the rest mass, which is independent of speed. Because the introduction of the term relativistic mass is unnecessary and often confusing, the word mass in this article will always mean rest mass.

5. Invariance of Rest Mass

By a simple extension of the principle of relativity, we can say that the mass, like proper time, is invariant. This does not mean that it cannot change as the result of an interaction; it means that it has the same value in all frames of reference. If we assume that all objects are made up of identifiable elementary particles whose masses are determined by laws of physics, and the laws are the same in all frames of reference, then the mass of each particle is the same in all frames of reference.

C. CONSERVATION OF ENERGY AND MOMENTUM IN COLLISIONS

1. Inelastic Collisions

An inelastic collision is defined classically as one in which kinetic energy disappears, to be replaced by some other form of energy (e.g., heat). In relativity theory, one can use the same definition, with the understanding that all possible forms of energy are included in the total en-

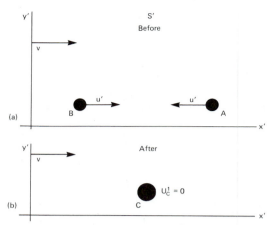

FIG. 6. Inelastic collision of equal-mass objects viewed in frame S'. (a) Before collision and (b) after collision.

ergy E, or the fourth component of the momentum vector. Study of an inelastic collision shows how the conservation of all four components of momentum is applied.

Figures 6 and 7 show, in two different frames of reference, a perfectly inelastic collision between two objects of equal mass M. In a perfectly inelastic collision, the colliding particles stick together, and there is a frame of reference S' in which the composite object formed by the collision is at rest, the kinetic energy having disappeared completely in that frame.

Let us apply the law of conservation of energy to the collision in frame S'. In this frame (Fig. 6), bodies A and B each have speed u', so their energies are equal, and the total energy is twice the energy of either one. Therefore, using Eq. (17) with γ equal to $(1 - u'^2/c^2)^{-1/2}$, we have

$$E' = 2Mc^2(1 - u'^2/c^2)^{-1/2}$$

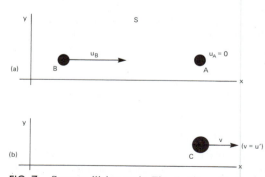

FIG. 7. Same collision as in Fig. 6, viewed in frame S: (a) before collision and (b) after collision. S' has velocity **v** relative to S.

After the collision, the total energy must remain equal to E'. But now there is only one body, and it is at rest, so E' equals the rest energy of that body:

$$E' = M_f c^2$$

where M_f is the mass of the composite body after the collision.

Equating the two expressions for E', we conclude that the mass of that composite body must be

$$M_f = 2M(1 - u'^2/c^2)^{-1/2} \qquad (21)$$

Thus $M_f \neq 2M$, and we have a violation of the law of conservation of mass; the total mass of the particles in the system is changed by the collision. The law of conservation of mass is no law at all. We invoked it to make the Galilean transformation obey the principle of relativity [see the discussion following Eq. (13)], but we have seen that the Galilean transformation must be abandoned. There exists no reason to retain the law of conservation of mass. That law is replaced by the law of conservation of total energy:

The total energy of a collection of particles is unchanged by interactions between the particles

In applying this law, it is understood that the energy of each particle is given by Eq. (17), the value of γ being computed for each particle by using the speed of that particle.

In any inelastic collision, the rest mass must increase, because the kinetic energy is decreased, and the sum of the two is the total energy, which is conserved. In such a collision there is also, in any frame of reference, an increase in the temperature of the colliding objects. The objects have acquired internal energy that may in principle be detected by measuring either the change in temperature or the change in the total rest mass of the objects. If the same change in temperature had been produced by other means, such as by building a fire under these objects, the rest mass of the objects would necessarily increase by the same amount.

When the mass changes in one frame, it must change in all frames. Thus in Fig. 7, which shows the collision of Fig. 6 in a different frame of reference S, the total mass before the collision is $2M$. After the collision, the total mass is $2M(1 - u'^2/c^2)^{-1/2}$, as it was in frame S' [Eq. (21)], with the same value of u'.

The law of conservation of energy is just a

special case of the law of conservation of momentum, which may now be stated in similar fashion:

> The value of any component of the total momentum of a collection of particles is unchanged by interactions between the particles

2. A Numerical Example

Let us suppose that, in the collision shown in Figs. 6 and 7, u' and v are both equal to $0.6c$. In that case, $M_f = 2.5\,M$. In frame S' it is clear that the total momentum is zero at all times, and the total energy is $M_f c^2$. Let us show that momentum, as defined by Eqs. (14), is also conserved in frame S.

We can find the speed of each particle in frame S by using the velocity-addition formula [Eqs. (3)]. Frame S is moving in the $-x$ direction with speed $v = u' = 0.6c$, relative to frame S'. Thus object A has speed zero in this frame, and the formula gives, for the speed of object B,

$$u_B = (0.6 + 0.6)c/(1 + 0.36) = (15/17)c \quad (22)$$

The momentum of object B is then, according to Eq. (14),

$$p_B = Mu_B(1 - u_B^2/c^2)^{-1/2} = \tfrac{15}{17}Mc/\tfrac{8}{17} = \tfrac{15}{8}Mc \quad (23)$$

The momentum of A being zero, we find that the total momentum is $\tfrac{15}{8}Mc$ before the collision. After the collision, the composite object C has a mass of $2.5M$, a speed of $0.6c$, and hence a momentum of

$$P_C = 2.5Mu_C(1 - u_C^2/c^2)^{-1/2} = 1.5Mc/0.8$$
$$= \tfrac{15}{8}Mc \quad (24)$$

We see that momentum is indeed conserved in frame S. We can also verify that energy is conserved in S. Object A, at rest, has an energy of Mc^2. The energy of object B is, according to Eq. (17),

$$E_B = Mc^2(1 - u_B^2/c^2)^{-1/2} = Mc^2/\tfrac{8}{17} = \tfrac{17}{8}Mc^2$$

so the total energy before the collision is

$$E = E_A + E_B = Mc^2 + \tfrac{17}{8}Mc^2 = \tfrac{25}{8}Mc^2 \quad (25)$$

in frame S. After the collision, object C has a mass of $2.5M$ and a speed of $0.6c$. Consequently, its energy is

$$E_C = 2.5Mc^2(1 - u_C^2/c^2)^{-1/2} = 2.5Mc^2/0.8$$
$$= \tfrac{25}{8}Mc^2$$

C is the only object remaining after the collision,

and its energy equals the original energy E. Therefore, we have verified that energy is conserved in this frame of reference.

3. Energy–Momentum Relation

From Eqs. (14) and (17) it follows that

$$mc^2 = [E^2 - c^2(p_x^2 - p_y^2 - p_z^2)]^{1/2}$$

or

$$(mc^2)^2 = E^2 - c^2p^2 \quad (26)$$

The values for object C of Fig. 7 are [Eqs. (24) and (25)]:

$$p_C = \tfrac{15}{8}Mc, \qquad E_C = \tfrac{25}{8}Mc^2,$$
$$M_C = 2.5M$$

Insertion of these values into Eq. (26) yields

$$(2.5Mc^2)^2 = (\tfrac{25}{8}Mc^2)^2 - (\tfrac{15}{8}Mc^2)^2$$

or

$$6.25 = \tfrac{625}{64} - \tfrac{225}{64}$$

which balances.

Equation (26) can also be deduced from Eq. (7), through multiplication of both sides by mc/τ. Just as we can find the proper time interval between two events by measuring the coordinates of the events in any frame of reference, we can find the mass of a particle by measuring its energy and momentum in any frame of reference, because mass, like proper time, is an invariant quantity.

4. Conversion of Rest Mass into Kinetic Energy

The collision shown in Figs. 6 and 7 could be reversed. Particle C could split up into particles A and B. In that case, the total rest mass of the system would decrease as the kinetic energy increased. It is well known that this process occurs in a nuclear reactor or a nuclear weapon. It is not so widely recognized that this conversion of rest mass into kinetic energy occurs in chemical processes as well. Any process that yields useful energy does so at the expense of rest mass.

D. PARTICLES WITH ZERO REST MASS

1. The Photon

Maxwell's classical theory of electromagnetism shows that light is an electromagnetic wave. Such a wave carries momentum as well as energy. The theory shows that the energy density in the wave equals c times the momentum density, and this is verified by independent measurements of energy and momentum. For example, the energy density in solar radiation is well known, and the momentum density can be determined from the effect of the resulting radiation pressure on the orbits of low-density artificial satellites (e.g., balloons).

Einstein's theory of the photoelectric effect states that the energy and momentum carried by a light wave cannot be divided into arbitrarily small amounts. Rather, a light wave is made up of particles called photons, whose energy (and hence momentum) is proportional to the frequency of the wave. This theory is consistent with Maxwell's theory provided that the energy E and momentum p of each photon obey the relation $E = pc$.

This relation is consistent with Eq. (26), provided that $m = 0$. We conclude that

the photon is a particle with a rest mass of zero

Furthermore, if we place $v = c$ in Eqs. (14) and (17), we obtain an impossible result—infinite energy or momentum—unless $m = 0$. Thus the photon must have zero mass if it is to travel with the speed of light.

Placing $m = 0$ and $v = c$, in Eq. (14) or (17), yields zero divided by zero, which of course is indeterminate, so these equations do not give us any further information about an individual photon. But Eq. (26), with $m = 0$, is used routinely in studying interactions involving photons.

2. The Neutrino

The neutrino, a neutral particle produced in radioactive decay or in nuclear reactions, is also believed to have a mass that is zero, or so close to zero that the difference is exceedingly hard to detect. Therefore it, too, travels with the speed of light, and its energy E and momentum p are related by the equation $E = pc$.

In spite of its zero mass, the neutrino has many properties that it does not share with the photon. The existence of the neutrino underscores our discussion of the second postulate of relativity. In the theory of relativity, the significant thing about the speed c is not that it is the speed of light; rather, light is of interest because its speed is c.

E. THE REST MASS OF A COMPOSITE BODY

The energy of a moving body can be divided into two parts: (1) the external energy, known as kinetic energy and (2) the internal energy,

known as rest energy. The external energy is easily observed. Changes in internal energy had been detected before relativity theory by means of temperature measurements, but the relativistic equation [Eq. (17)] showed the connection between internal energy and rest mass.

[The factor c^2 in Eq. (17) is simply a conversion factor between our unit of mass and our customary unit of energy. If we used units such that $c = 1$, it would be easier to realize that mass and energy are basically the same thing. What is often called the conversion of mass into energy could be more accurately called the conversion of internal or rest energy into kinetic energy.]

Internal energy may be visualized by considering the body to be a box containing a gas. When the atoms of the gas move about randomly, the box remains at rest, but the kinetic energy of these atoms may be transferred to the outside. When this happens, the box becomes colder and we say that heat is flowing out of it. The rest mass of the box also decreases, because there is less energy in the box. A sufficiently sensitive measurement would show that the box has less inertia when it is colder; it is easier to accelerate it, because the slow atoms inside can be accelerated more easily than faster atoms could be. (It follows from the momentum–energy relation [Eq. (14)] and the ensuing discussion that a given force produces a smaller change in velocity as the speed increases.)

If a box containing a gas is moving uniformly, its external energy includes some of the kinetic energy of the gas atoms inside, because these atoms must move along with the box while they are bouncing around inside the box. If the box is suddenly stopped in an inelastic collision, the motion of these atoms becomes more disordered, as they no longer have a net motion. Their total kinetic energy may remain the same, but the externally observed energy, resulting from the collective motion of the atoms with the box, disappears, being converted into random motions of the atoms (i.e., heat, or increased rest mass of the box).

The connection between rest mass and temperature was not observed before Einstein because any temperature increase that does not vaporize a body is accompanied by a minuscule increase in mass—perhaps one part in 10^{11}, or 1 μg in 100 kg. But the change in mass always is associated with any kind of change in internal energy. For example, it is known that an energy of 13.6 eV is needed to separate the electron from the proton in a hydrogen atom when this atom is in its normal state. Therefore, the mass of the proton plus the mass of the electron exceeds the mass of the normal hydrogen atom by an amount equivalent to this additional energy of 13.6 eV (often written as a mass of 13.6 eV/c^2, because the energy–mass conversion factor is c^2). The total rest energy of the hydrogen atom is 9.4×10^8 eV; the total mass of its constituents exceeds the mass of the atom by only about one part in 100 million.

Rest mass is changed in any reaction—chemical or nuclear—that causes a change in kinetic energy. For example, consider the chemical reaction

$$2H_2 + O_2 \rightarrow 2H_2O \qquad (27)$$

It is well known that this reaction produces kinetic energy; hydrogen and oxygen form an explosive mixture. Since total energy is always conserved, the total rest mass of the two H_2O molecules that are produced must be less than the total rest mass of the two H_2 molecules and the O_2 molecule.

Now let us consider an explosion-proof, perfectly insulated combustion chamber containing hydrogen and oxygen. Let the total mass M of the system—the chamber and its contents—be determined by a precise measurement. The system is at rest, so that its rest energy is all internal energy, equal to Mc^2. Reaction (27) then takes place, and the mass is determined again. Since the chamber is insulated, no energy could escape; therefore the internal energy has not changed, and the mass of the system remains M. However, the rest mass of the constituents has changed, because some of it has been converted to kinetic energy by reaction (27). This kinetic energy is internal energy in this system, and thus it contributes to the rest mass of the system just as did the molecular rest mass from which it was created.

In this example, the system's temperature has been raised by the reaction, because the temperature is related only to the internal kinetic energy, rather than to the total energy, which is unchanged. If we were to remove the insulation, permitting the chamber to cool down, heat would be lost to the outside. This loss of energy would be reflected in a loss of rest mass by the system. If the entire operation were carried out in a system that was in contact with a constant-temperature bath, heat would flow out continuously and the system would lose mass continuously. The difference between the final mass and the initial mass would be simply equal to the difference between the total rest mass of the reactants (hydrogen and oxygen) and the total rest

mass of the resulting individual H_2O molecules. (If the temperature were fixed below the boiling point of water, some additional rest mass would be lost as the H_2O molecules condensed into water.)

It follows that whenever a body loses energy by radiation, the body's rest mass decreases. The photon that is radiated carries away this mass in the form of energy, even though the photon itself has no rest mass. Now consider a photon that is radiated within a body (e.g., inside an oven). For example, suppose that a photon is emitted from an inside wall of an oven and travels to the opposite wall, where it is absorbed. The energy of that photon, while it exists, is part of the internal energy of the oven, and it contributes to the rest mass of the oven. When the photon is emitted from one wall, the mass of that wall decreases because of the loss of energy; when the photon is absorbed by the other wall, the mass of that wall increases as it gains energy. But the overall rest mass of the oven—the mass m that appears in Eqs. (17)–(20)—remains constant throughout the whole process. In summary:

> The rest mass M of a composite object is NOT equal to the sum of the rest masses of its constituent parts. Rather, M is determined by the equation $E = Mc^2$, where E is the total energy that these constituents have, in the frame of reference in which the object is at rest.

VI. Electromagnetism

A. Moving Charges and Magnetism

When the electric and magnetic forces were first studied, it was thought that they were two distinct entities. In the nineteenth century it was found that an electric current could exert a force on a magnet, and Michael Faraday showed that a changing magnetic field could produce an electric current. The connection between magnetic and electric fields became more strongly established when Maxwell's equations of electromagnetism showed that a changing electric field produces a magnetic field, and vice versa.

Unlike the Newtonian equations of mechanics, the Maxwell equations are completely consistent with the special theory of relativity. That means that Maxwell's equations do not change form when a Lorentz transformation is applied to the coordinate system. This should not be unexpected, because the speed of light may be derived from these equations, and the Lorentz transformation is designed to make speed of light independent of coordinate system.

This is not the place for a discussion of Maxwell's equations. Rather, without considering the form of these equations, we simply show how electric and magnetic fields are affected by a Lorentz transfusion. We begin by considering a coordinate system in which there is no magnetic field, but only an electric field. A purely electric field is produced by stationary (or static) electric charges. In general, if any charges are moving, there will be a magnetic field as well as an electric field.

A charge that is at rest at a single point in space in a given frame of reference produces an electric field like that shown in Fig. 8a. The field is represented by lines that show the direction of the field at each other point in space (i.e., the direction of the force that would be exerted on a different charge located at that field point). The density of the lines in any region of space is proportional to the strength of the field (i.e., the magnitude of the force on a charge of a given size) in that region. One could do a set of experiments to measure the field at each point and thus construct all of these lines.

If the charge were moving with velocity **u**, as in Fig. 8b, the results would be different. However, we could find another frame of reference O in which the charge would be at rest, and in that frame the results would necessarily be the same as shown in Fig. 8a. The difference between Fig. 8b and Fig. 8a can be found from a Lorentz transformation.

B. LORENTZ TRANSFORMATION OF ELECTRIC AND MAGNETIC FIELDS

When the observer in frame O sees the electric field lines of Fig. 8a, the observer in O' sees

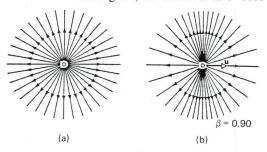

(a)   (b)
$\beta = 0.90$

FIG. 8.   Electric field lines from a point charge: (a) when the charge is at rest; (b) when the charge is moving with velocity **u**, for the case in which u/c is equal to 0.9. [Reproduced, with permission, from R. Resnick (1968). "Introduction to Special Relativity," p. 168. Wiley, New York.]

the entire frame of O contracted in the direction of the velocity **u**. If one takes Fig. 8a and contracts it along the horizontal axis, the result is Fig. 8b. (The effect can be seen by tilting the page sideways and looking from the right or left.)

In any frame of reference in which the charge is moving, there is also a magnetic field. Long before special relativity was developed, it was known, (from the studies of André-Marie Ampère and others) that an electric current (and hence a moving electric charge) produces a magnetic field. In general, no matter what the source of the fields, we can show that, when frame O' is moving relative to O, and there is a purely electric field in O, there will be both electric and magnetic fields in O'.

Without going into the details of electromagnetic fields, we can understand the origin of the magnetic field by considering a simple example. Two particles, each of mass m and charge q, are at rest in frame of reference O—one at the origin, the other directly above it, at $x = 0$, $y = 0$, and $z = d$. According to Coulomb's law, each repels the other with a force of $F = kq^2/d^2$, where k is a constant that depends on the units used for q and d. Let us suppose that this is the only force acting on these charges. Since they are at rest at this time, Newton's law gives the acceleration as $a = F/m = kq^2/d^2m$. This can in principle be confirmed by a measurement of the velocity of each particle as a function of time t, as the particles move apart in the vertical (z) direction.

Next, consider the motion of each charge in reference frame O', which is moving horizontally (in the x direction) with speed v relative to O. In this frame of reference, the acceleration of each charge can be determined in the same way, by measuring the velocity of the charge as a function of the time t'. When the observer in O' measures the vertical component of the velocity, he/she must divide z' by t', just as the observer in O divides z by t. To find the vertical acceleration, he must divide by t' again. According to the Lorentz transformation [Eq. (2)], $z' = z$ and $t' = t\gamma$ (because $x = 0$). Dividing by t' is equivalent to dividing by $t\gamma$, or dividing by t and multiplying by $\sqrt{1 - (v/c)^2}$. When this is done twice, the acceleration a' in O' is found to be the same as in frame O except for a factor of $1 - (v/c)^2$.

An observer in O' who knew nothing of relativity or other frames of reference would therefore say that charges moving in this manner with speed v are subject to a mutual force of

$$F' = ma' = (kq^2/d^2)[1 - (v/c)^2] \quad (28)$$

This reduces to Coulomb's law when $v = 0$. Such an observer would say that the force given by Coulomb's law is augmented by an additional force of $(kq^2/d^2)(v/c)^2$ when the charges are moving parallel to each other with speed v, and that this force is opposite in direction to the electric force. He could then call this additional force the magnetic force, and say that the total force on each charge is given by the vector sum of the electric and magnetic forces, to arrive at the result of Eq. (28).

Magnetic forces were discovered before relativity, but relativity could be used to derive all magnetic forces between moving charges, as we have done in this example. In the general case, with many charges moving in various directions, it is not possible to find a frame of reference like O in the previous example, in which there is no magnetic field. But one can use the Lorentz transformation to find the electric field components E'_x, E'_y, E'_z and the magnetic field components B'_x, B'_y, B'_z in O' in terms of the field components in O. If it is assumed that frame O' is moving in the $+x$ direction with speed v relative to frame O, then the result is

$$E'_x = E_x, \qquad\qquad B'_x = B_x$$
$$E'_y = \gamma[E_y - (v/c)B_z], \quad B'_y = \gamma[B_y + (v/c)E_z] \quad (29)$$
$$E'_z = \gamma[E_z + (v/c)B_y], \quad B'_z = \gamma[B_z - (v/c)E_y]$$

Notice that Fig. 7a shows a special case in which the magnetic field is zero in frame 0. In that case, Eq. (29) show that

$$E'_x = E_x, \qquad E'_y = \gamma E_y,$$
$$E'_z = \gamma E_z$$

Thus the components of the vector **E** in the direction perpendicular to the velocity **v** are simply increased by the factor γ resulting from Lorentz contraction of the frame O, in agreement with the configuration of field lines shown in Fig. 7b.

Notice the symmetry between E and B. The equations are basically the same if we replace E by B and B by E everywhere. The components of the vector **E** and the vector **B**, taken together, are part of a single entity, the electromagnetic field. The essential unity of this field only became apparent in the light of special relativity.

VII. Consequences of Special Relativity

A. Atomic Masses

More than 99.9% of the mass of any atom resides in its nucleus, which is composed of neu-

trons and protons. Atomic nuclei are thus composite bodies. Their masses provide the most well-documented illustrations of the principles governing the rest masses of composite bodies.

Masses of atoms can be determined with great precision by direct observation of their inertia in a mass spectrometer. This device measures the inertia of atoms by deflecting them in a magnetic field. [A neutral atom is not deflected in a magnetic field. Therefore, the atoms must first be ionized by removing (or adding) one or more electrons. Consequently, the direct measurement is that of the mass of an ion, but the very small mass of the missing (or added) electron(s) is accurately known.]

As a result of such measurements, it is known that the mass of any atom (except hydrogen) is almost 1% less than the total mass of its constituents. For example, a helium atom (^{4}He) containing two protons, two neutrons, and two electrons has a mass of (3728.43 ± 0.01) MeV/c^2 (which means that its rest energy is 3728.43 ± 0.01 MeV). The rest energy of an individual proton is 938.280 MeV, that of a neutron is 939.573 MeV, and of an electron 0.511 MeV, so the sum of the rest energies of the six particles in a helium atom is about 3756.73 MeV, or about 28.3 MeV more than the rest energy of the atom.

Thus, to split up a ^{4}He atom completely into its six constituent particles requires that 28.3 MeV be added to the atom. (The energy required to completely split up an atom is called its binding energy. Almost all of this energy is actually the binding energy of the nucleus, because the binding energies of the electrons are minuscule in comparison. For helium, the total energy required to separate the two electrons is only about 80 eV, or 0.0008 MeV.) Binding energies of nuclei or of individual particles in the nucleus (the energy required to separate one particle from the rest) have been measured repeatedly for thousands of atomic species. Measurements have also been made of the energy absorbed or released in nuclear reactions, in which the protons or neutrons are transferred from one atom to another. When the results are compared with mass spectrometer measurements, it is found that the atomic masses are always in exact agreement with the energy measurements and the Einstein formula $E = mc^2$.

B. Nuclear Energy

The fact that energy is equivalent to mass makes it possible to deduce the energy content of an atom by measuring its mass. Unlike chemical reactions, nuclear reactions release an amount of energy that is easily accounted for in a measurement of mass. Thus, even before a given reaction is observed, it can often be recognized from tabulated values of atomic masses that the reaction will be able to release energy—to transform rest energy into kinetic energy.

1. Nuclear Fission

In the 1930s Albert Einstein and others realized the significance of the fact that the uranium atom has about 200 MeV more rest energy than the total mass of lighter bodies into which it can split. One of many possibilities for the splitting of uranium is the induced fission reaction

$$^1n + {}^{235}U \rightarrow {}^{95}Y + {}^{139}I + 2\,{}^1n \qquad (30)$$

in which a neutron (^1n) causes a uranium (^{235}U) nucleus to split. In this example, the total mass of the four resulting particles [an atom of yttrium (^{95}Y), an atom of iodine (^{139}I), and two neutrons] is less than the rest mass of the uranium and the original neutron by an amount equivalent to an energy of 172 MeV. In this and other induced fission reactions involving uranium, the total energy released averages about 0.1% of the original rest mass of the uranium.

Most of the energy released in reaction (30) is given to the two neutrons, which are able to react with other uranium atoms in the same way to release more energy and additional neutrons, thus causing a chain reaction. The kinetic energy of the resulting swarm of neutrons is transferred to the surroundings, raising the temperature to the point where it can produce an explosion (in a bomb) or boil water (in a fission reactor).

2. Nuclear Fusion

Further study of atomic masses reveals that fusion of the lightest elements into heavier ones also leads to a net loss in rest mass, and thus to a gain of kinetic energy (as in the combination of hydrogen and neutrons to form helium, previously discussed). Such fusion processes are the main source of energy for the sun and other stars. [See STELLAR STRUCTURE AND EVOLUTION.]

These processes involve numerous light elements besides hydrogen and helium. For example, three ^4He atoms can fuse to form one atom of carbon (^{12}C), with a decrease of 7.27 MeV in rest energy. The energy thus released is less than 0.1% of the rest energy, but because of the abundance of helium in the sun, it makes a significant contribution to the sun's energy output. The total power radiated by the sun is about 3×10^{26} W. This means that the sun loses a mass of about 3 million tons/sec.

It is tempting to try to build a fusion reactor that would convert the hydrogen in water to helium. The energy that could thus be obtained from a single cup of water would be sufficient to power a thousand all-electric homes for a month. Unfortunately, it has not yet proven possible to produce the combination of high density and high temperature needed to overcome the mutual repulsion of hydrogen atoms and sustain a reaction that would generate useful power.

C. DECAY OF SUBATOMIC PARTICLES

Modern physics has discovered a huge number of subatomic particles with variety of properties. Most of these particles are unstable: they decay, being spontaneously transformed into other particles. For example, the muon is a particle like the electron but with a rest mass that is 207 times that of the electron. The negatively charged muon decays into an electron and two neutrinos, with a half-life of about 1.5 μsec in the frame of reference in which the muon is at rest. In any frame of reference in which the muon is moving, the observed half-life is larger by the time dilation factor γ. Thus, for muons moving at a speed of 0.8 c, $\gamma = \frac{5}{3}$, and the half-life is 2.5 μ sec.

Half-lives of moving particles have been observed millions of times, for every known unstable particle, in laboratories all over the world, with results that are in exact agreement with the time dilation formula [Eq (6)]. The moving particle acts just like a moving clock.

It is significant that the decay of the muon is not the result of an electromagnetic force, yet special relativity still governs this decay. Relativity is a theory about space and time, not specifically about electromagnetism, even though it was developed by considering the speed of electromagnetic waves.

D. ANTIMATTER: PAIR PRODUCTION AND ANNIHILATION

The development of a theory combining special relativity with quantum mechanics led to the prediction, soon verified, of the existence of antimatter. Antimatter behaves in almost all respects like matter. Each particle of matter has an antiparticle with identical mass and opposite charge. An entire universe made up of antimatter would presumably behave like a mirror image of our own universe, with negatively charged atomic nuclei and positively charged electrons (positrons).

In a striking demonstration of the equivalence of rest mass and energy, a particle of matter and its antiparticle can be simultaneously created from the pure energy of a photon. For example, any photon whose energy exceeds twice the rest energy of an electron is capable of converting that energy into rest mass by creating a positron and an electron. Many radioactive materials emit photons of the requisite energy, and this pair production process is routinely observed when these photons pass through matter. (The process cannot occur in empty space, because the momentum of the pair that is created cannot equal that of the photon. Another body must participate in the process, to absorb the remaining momentum as the photon disappears.)

When a particle meets its antiparticle (e.g., when any electron encounters any positron), the pair can be annihilated, producing pure energy—a photon or photons. This process is also governed by the relativistic laws of conservation of energy and momentum. Thus, unless a third particle participates in the process, annihilation of a pair must produce at least two photons. If the total momentum of the pair is zero (as it must be in some frame of reference), creation of a single photon would violate the law of conservation of momentum. A second photon, emitted in a direction exactly opposite to that of the first photon, permits the total momentum of the two photons to be zero.

Many radioactive materials emit positrons, and the resulting pair annihilation process, with the production of two equal-energy photons going in opposite directions, has been observed countless times, always with results in agreement with the equations of special relativity. Observations of this sort are now routinely made by undergraduate students in laboratory physics courses.

E. Red Shifts and Cosmology

Application of relativity formulas is not confined to the domain of subatomic particles. The frequencies of characteristic radiation from elements in distant galaxies can be analyzed with the aid of the relativistic Doppler shift formula [Eq. (10)]. It is found that the radiation from each element in a distant galaxy is shifted toward the red by an amount that can be used, with the aid of Eq. (10), to compute a speed of recession of the galaxy from the earth. [*See* COSMOLOGY.*]

One cannot use this information to verify the correctness of Eq. (10) on this scale, because there is no independent way to verify that the galaxy is actually receding at the calculated speed. However, the distances of the nearer galaxies can be measured independently, because certain stars. (Cepheid variables) can be seen in these galaxies. These stars oscillate in brightness, with a period that is related to the intrinsic average brightness of the star. Knowing the intrinsic brightness of a star, one can calculate its distance from the observed brightness of that star in the sky. Other methods permit estimates of the distances of more remote galaxies. The estimated distances, in conjunction with the redshift measurements, indicate that the speed of recession of a galaxy is directly proportional to its distance from earth.

From this proportionality it has been concluded that the entire universe is expanding. The concept of an expanding universe has led to theories of cosmology that explain numerous details of the known universe. These include the relative abundances of the elements and the presence of low-energy cosmic background radiation that comes uniformly from all directions in space.

Further discussion of this topic would take us from the special theory of relativity to the realm of the general theory, which deals with the relationship between gravitation and space–time.

Bibliography

Einstein, A. (1961). "Relativity: The Special and the General Theory." Crown Publ., New York.
Jackson, J. D. (1975). "Classical Electrodynamics," 2nd ed., Chapter 11. Wiley, New York.
McGervey, J. D. (1983). "Introduction to Modern Physics," 2nd ed., Chapters 2 and 14. Academic Press, New York.
Purcell, E. M. (1965). "Electricity and Magnetism," Chapters 5 and 6. McGraw-Hill, New York.
Resnick, R., and Halliday, D. (1985). "Basic Concepts in Relativity and Early Quantum Theory," 2nd ed. Wiley, New York.

SOLAR PHYSICS

Carol Jo Crannell   *NASA Goddard Space Flight Center*

GLOSSARY

Arcade: Series of arches formed by magnetic fields extending into the outer layers of the solar atmosphere. The closed magnetic loops that form an arcade confine hot gas or plasma.

Astronomical unit (AU): Parameter describing the earth's orbit around the sun, corresponding to the mean distance between the sun and the earth, approximately 1.5×10^8 km. Because the earth's orbit is elliptical, the actual distance is not constant.

Chromosphere: Thin layer of the solar atmosphere that lies above the photosphere and below the corona. Light from this layer is predominantly Hα emission with additional contributions from other strong atomic transitions.

Corona: Outermost and hottest layer of the solar atmosphere. The corona is visible to the naked eye during eclipses when light from the inner layers is blocked by the moon.

Coronal hole: Region of the corona that appears to be dark when observed in the ultraviolet and soft X-ray portion of the electromagnetic spectrum. These features were discovered only recently with instruments carried on space satellites and rockets. They are thought to be the source of high-speed solar wind streams.

Faculae: Regions in the upper photosphere, frequently in the vicinity of sunspots, that are brighter than the surrounding medium due to their higher temperatures and greater densities.

Filament: Dense, massive structure that lies above the chromosphere, generally along a line separating regions of opposite magnetic polarity. Filaments appear as dark, irregular lines when observed on the solar disk.

Flare: Rapid release of energy from a localized region on the sun in the form of electromagnetic radiation and, usually, energetic particles. The spectral range of the radiation may extend from meter waves through high-energy γ rays. The time scales of such events range from fractions of a second to hours.

Footpoint: Lowest visible portion of a magnetic loop, corresponding to the vicinity on the solar disk in which the loop or arch intersects the photosphere.

Granulation: Irregular light-and-dark structures, visible in the lower photosphere, indicating the tops of convection cells. Individual granules appear and disappear on time scales of order 10 min and exhibit a range of sizes, with 2000 km being a typical diameter.

Hα: Light emitted with a wavelength of 656 nm (6563 Å) from an atomic transition in hydrogen, the lowest energy transition in its Balmer series. This wavelength is in the red portion of the visible spectrum and is the dominant emission from the solar chromosphere.

Optical depth: Measure of how far one can ''see'' into a semitransparent medium, such as the solar atmosphere. At the surface of such a medium, none of the photons emitted in the direction of an observer are absorbed, and the optical depth is zero. Beneath the surface, only a fraction of the number of photons that originate within the medium traverse the distance to the surface without being absorbed. That fraction decreases exponentially with increasing optical depth.

Photosphere: Innermost portion of the solar atmosphere. The base of the photosphere is defined to be the lowest level in the sun that can be observed directly in the visible portion of the electromagnetic spectrum.

Plage: Portion of a magnetic active region that appears much brighter in Hα than the surrounding chromosphere. Plages usually are visible before the sunspots with which they are associated and persist after the sunspots disappear.

Plasma: Gas in which some or all of the constituents are partially or fully ionized. The plasma state is generally associated with temperatures in excess of 10^4 K.

Prominence: Filament viewed on the limb of the sun that extends above the chromosphere into the corona. A prominence, like the chromosphere itself, radiates primarily in Hα.

Spicule: Small, filamentary magnetic structure containing material at chromospheric temperatures that extends into the corona. Spicules are dynamic features that rise and then disappear on time scales of minutes.

Sunspot: Region in which strong magnetic fields emerge into the solar atmosphere from below the solar surface. Visible in both Hα and in the white-light continuum, sunspots are cooler than the surrounding medium because the magnetic fields from which they are formed suppress the temperature of the plasma they contain.

Surge: Great eruption of hot material that originates below the chromosphere. A surge may accompany a flare and is considered to be a type of eruptive prominence.

The sun is important to us for many reasons. From our perspective as earth-dwelling mortals, the sun's primary importance is its role as the source of energy that sustains life on earth. From an anthropological perspective, the sun serves as a source of human wonderment, artistic inspiration, and even religious devotion. From the perspective of space science, the sun is both the closest and most accessible star and a laboratory with unique facilities for studies of plasmas and magnetohydrodynamics. The solar laboratory contains matter at temperatures, pressures, and densities attainable nowhere else currently accessible from earth. The information

presented here, a description of the sun, its activity, and the thrust of future investigations, was developed from this latter perspective of space science.

I. Description of the Sun as a Star

A. GENERAL PHYSICAL CHARACTERISTICS

Table I presents numerical parameters characterizing the size and energy output of the sun. These values are the standard yardstick by which other stars are measured. The large number of significant digits tabulated here serve mainly to illustrate the precision to which these parameters are known. Also listed are parameters characterizing the earth's orbit around the sun and the intensity of the sun's radiation at the mean orbital distance.

The appearance of the sun depends critically on how it is observed. When observed from earth through very light cloud cover or haze, it appears to be surrounded by rays of light (Fig. 1a), an effect produced by scattering of the sun's light in the earth's atmosphere. Such a view of the sun has been adopted for the flag of the state of New Mexico, as illustrated in Fig. 1b. Quite another image is seen during a solar eclipse, an event that occurs when the moon passes directly between the sun and an observer. The moon acts as an occulting disk, blocking the light from the most luminous portion of the sun, so that only the outermost portion of the solar atmosphere, known as the solar corona, is visible. The corona presents a wispy appearance similar to flames of a campfire, with large variations in in-

TABLE I. Dimensions, Distances, and Luminosity[a]

Radius of the sun, $R_\odot$	6.9599×10^5 km
Mean distance from sun to earth (astronomical unit, AU)	$1.495979(1) \times 10^8$ km
Range of sun-to-earth distance	
at perihelion	1.4710×10^8 km
at aphelion	1.5210×10^8 km
Mass of the sun, $M_\odot$	$1.989(1) \times 10^{30}$ kg
Luminosity of the sun, $L_\odot$	$3.826(8) \times 10^{26}$ W
Intensity of the sun's radiation at 1 AU	1.360 kW m^{-2}

[a] The numbers in parentheses represent the measurement uncertainty in the last digit of the associated parameter.

FIG. 1. Views of the sun and their stylized representations. (a) Sun observed through earth's atmosphere; (b) sun symbol on New Mexico flag; (c) sun as it appears during total eclipse; and (d) anthropomorphic characterization of the sun from a European brass ornament.

tensity from one portion of the solar limb to another. Figure 1c illustrates this effect, which results from the structure of the solar magnetic field in the corona. These structures can be characterized as arches, both ends of which connect to magnetic fields originating below the surface of the sun, and as streamers formed by magnetic fields that connect at one end to the sun and at

the other end to the magnetic fields of interplanetary space. The sketch in Fig. 1d, taken from a European brass, emphasises the flamelike appearance and may well have been inspired by solar eclipse observations.

There are many ways to observe the sun, both from ground-based observatories and from vantage points in space. Each type of radiation the sun emits carries specific information about the physical processes that determine its dynamical behavior. Techniques for observing the sun's various emissions throughout the electromagnetic spectrum are illustrated in Fig. 2. Optical and most radio emissions are able to penetrate the earth's atmosphere, so observations in these wavelength bands can be carried out with ground-based facilities. The atmosphere and its fluctuations do distort the highest frequency radio emission and, thus, limit the spectral range that can be measured from earth. In the optical domain, fluctuations in the earth's atmosphere limit the angular resolution to approximately 1 arcsec. At the sun, this corresponds to a linear dimension of 700 km, approximately the distance between Washington, D.C., and Cincinnati, Ohio. In the range of wavelengths corresponding to ultraviolet (UV) through soft X rays, the earth's atmosphere is very strongly attenuating.

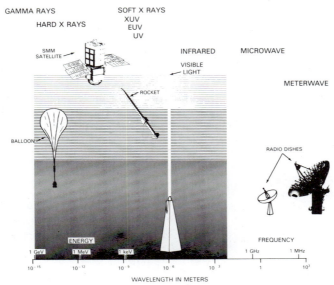

FIG. 2. Spectral range of electromagnetic radiation detected from the sun, illustrating relationships between energy, frequency, and wavelength, together with the types of instrumentation employed for observations. The shading represents the earth's atmosphere, with the open portions designating the spectral intervals for which the atmosphere is transparent.

Thus, at these wavelengths, observations are possible only above the earth's atmosphere. Because solar emission in this range is very intense, useful observations can be obtained in short times with instruments carried onboard rockets, which rise to altitudes of a few hundred kilometers before they fall back to earth. Continuous measurements and observations of transient events require orbital platforms, which carry instrument payloads above the earth's atmosphere for missions that may last many years. Hard X rays and γ rays, the highest energy, shortest wavelength electromagnetic radiations, can penetrate small fractions of the earth's atmosphere. These emissions, therefore, can be observed from balloon-borne platforms at altitudes of 30–40 km. High-altitude scientific balloons are most commonly used for short-duration observations of 1–2 days. The technological capability for long-duration flights of 15–30 days is currently being developed.

B. STRUCTURE AND COMPOSITION

Unlike the earth, the sun is gaseous throughout and so does not have any solid surface. The structure of the sun can be thought of as a series of concentric spherical shells or layers, each characterized by a unique combination of physical processes. At the center of the sun is the nuclear burning core, as illustrated in Fig. 3. Traveling outward, one encounters first the radiative zone, then the convection zone, then the photosphere, the chromosphere, the transition zone, and finally, the corona. All of these regions are powered by the nuclear burning core from which energy is transported outward through successive layers by radiation and con-

vection. The temperature is 15×10^6 K in the core and decreases monotonically outward to a minimum of approximately 4×10^3 K in the chromosphere. The transition from radiative to convective energy propagation occurs in the region in which the temperature drops below 2×10^6 K, whereby convection becomes a more efficient transport mechanism than radiation. This defines the boundary between radiation and convection zones.

The only radiation that carries information out of the sun directly from these inner regions is a flux of neutrinos produced in nuclear burning. The ability of these massless (or almost massless) particles to penetrate so much solar matter is a result of the fact that they interact very weakly with matter. Consequently they are also very difficult to detect. Knowledge of the solar interior regions, therefore, does not come from direct observations but rather from models checked indirectly by neutrino observations and more plentiful observations of the photosphere and outer atmosphere.

The innermost layer of the sun that can be observed in visible light is called the photosphere. The base of the photosphere is defined to be that depth in the solar atmosphere at which a photon of wavelength 500 nanometers (nm) has a 37% probability of escaping without scattering or being absorbed. This condition is referred to as optical-depth unity for a photon of the specified wavelength. From lower depths, even greater quantities of material are encountered along any outward trajectory, so the probability of an emitted photon escaping without interacting is decreased. The photosphere is, as a result, more luminous than the optically thin outer portions of the solar atmosphere and defines the size of the sun as observed in visible

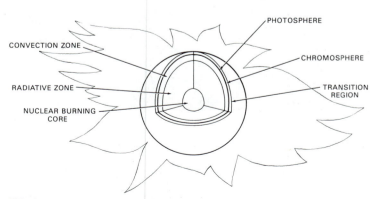

FIG. 3.   Interior layers of the sun, with the solar corona in the background.

TABLE II. Characteristics of the Solar Atmosphere

Region	Height above base of photosphere (km)	Temperature (K)	Density (atoms m^{-3})
Photosphere	0–320	6500–4500	10^{23}–10^{22}
Chromosphere	320–1990	4500–28,000	10^{21}–10^{16}
Transition zone	2000	10^5	10^{16}
Corona	at 7×10^5	1.8×10^6	10^{12}

light. The height, temperature, and density of various layers of the solar atmosphere are given in Table II.

The atmospheric layer that lies just outside the photosphere is the chromosphere. The visible emissions from the chromosphere are overpowered by the full light of the photosphere when observed without special filters and are blocked by the moon during the totality of a solar eclipse. They are, however, visible with the unaided-eye as a flash of red light just before totality. This red light, known as Hα emission, is emission at a wavelength of 656 nm due to an atomic transition in hydrogen. Unlike the simple representation in Fig. 3, the chromosphere is not a smooth spherical shell but, instead, exhibits many large- and small-scale features that may extend well into the corona. Their identity with the chromosphere is based on their temperature which supports Hα emission. Prominences are large-scale chromospheric features with an arch-like structure, visible in Hα at the edge of the solar disk against dark sky. Because these structures absorb light from the underlying photosphere, they appear as dark features when observed on the solar disk. Spicules are fine-scale chromospheric features with an even more flamelike appearance than the corona.

Perhaps the most puzzling feature of the outer solar atmosphere is its temperature structure. Instead of a continued decrease in temperature with distance outward, an increase is observed. The gradual increase observed in the chromosphere becomes a steeply rising increase in the transition zone, so named for the abrupt change in the thermal gradient. The temperature continues to rise well out into the corona. While this temperature increase in the corona appears to violate the elementary thermodynamic principle that a body cannot supply heat to a hotter body without external work being done, the paradox is resolved with the understanding that the pho-

tosphere heats the corona from the nonthermal source of energy stored in its magnetic fields. Two mechanisms are thought to be involved: One is currents, generated by changing magnetic fields; the other is magnetohydrodynamic waves. The relative importance of these two mechanisms is presently the subject of intensive investigation.

In specifying the chemical composition of the sun, one can be at least 98% correct by saying "Hydrogen and helium." In spite of the fact that these are the two lightest elements in the periodic table, this statement is true not only for abundance by number of atoms but also for abundance by weight. The next most abundant elements, in decreasing order, are oxygen and carbon. The rest comprise less than 1% of the sun by weight and less than 0.03% by number. The photospheric abundances are quite similar to those determined for local galactic abundances, but in the corona, lower abundances of carbon and oxygen are reported. Recent γ-ray observations of solar flares show a similar underabundance of carbon and oxygen.

C. The Sun among Stars

Stars can be classified in many different ways. Among the most useful schemes are classifications by luminosity, or intrinsic brightness, and by spectral class, which is a measure of a star's surface temperature. The plot in Fig. 4 illustrates a particularly informative combination of these classification schemes, called a Hertzsprung–Russell (H–R) diagram after its originators. In the H–R diagram surface temperature increases, albeit nonuniformly, from right to left. Normal stars populate the main sequence, the band that runs diagonally across the diagram. These stars are in the early stages of their

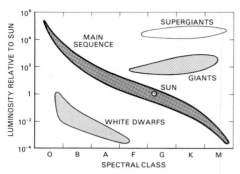

FIG. 4. Plot of luminosity as a function of spectral type for varieties of stars. Such a plot is frequently referred to as a Hertzsprung–Russell diagram. The sun in this diagram, indicated by the symbol ⊙, is on the Main Sequence, as are most stars.

evolution and are known as dwarfs to distinguish them from the much larger, more luminous stars in the upper right-hand portion of the diagram. [*See* STELLAR STRUCTURE AND EVOLUTION.]

The sun is represented in this diagram by its symbol ⊙. As indicated by its location in the diagram, the sun is a G dwarf, an average star both with respect to its luminosity and with respect to its spectral class. Contrary to what one might assume, the lack of superlatives in its description adds to the attractiveness of the sun as an object of astronomical interest. The normalcy enables most of what is learned about the sun to be applied to other stars, all of which are much less accessible. Distance makes spatially resolved observations of stellar dynamics and activity, described for the sun in the following sections, directly observable only on the sun. For the same reason, development of instrumentation for observations of astrophysically significant phenomena on the sun has pioneered the development of instrumentation for observations of similar phenomena outside the solar system.

II. Motion in the Quiet Sun

In this section on the quiet Sun, the focus is on continuous and regular types of activity which characterize the sun's general behavior and condition. One such type of motion can be observed as a slowly changing grainy pattern in the photosphere. When the photosphere is seen with good angular resolution, 1 arcsec or better,

its appearance is similar to that of a simmering pot of grits with individual grits (analogous to granular cells) appearing and disappearing on a timescale of about 10 min. The linear dimensions of the granules are typically 1500 km. They are actually the tops of a pattern established in the convection zone and visible in the photosphere. Their comings correspond to the rise and expansion of hot material, and their goings correspond to the loss of heat by radiation, accompanied by contraction and sinking. Slower and larger convection patterns, called supergranules, are visible on a scale size of 30,000 km and on a time scale of days. Unfortunately these supergranules penetrate only a few percent of the depth of the convection zone, and the small-scale granules penetrate even less. The remainder of the convection zone cannot be probed by observations of this process.

Another even more regular motion of the quiet sun is its oscillation. This rhythmic rise and fall of the solar surface, with a dominant 5-min period, corresponds to the resonant frequency of the convection zone for sound waves. The full acoustic spectrum of solar motions includes the harmonics of this oscillatory period superposed on the stochastic motion of solar granules and modified with two distinct shifts due to the rotation of the sun's surface. These acoustic waves serve as probes of the solar interior, forming the basis for seismological studies of the sun. Observations of the detailed spectrum of solar oscillations have enabled parameters, such as the depth of the convection zone, to be determined from the models of solar structure.

The motion of the sun's surface that has been known for the longest time is solar rotation. Because of its gaseous nature, the sun does not rotate as a solid body. The sun does rotate about an axis, but the regions near the solar equator rotate more rapidly than the regions near the solar poles. This differential rotation, as it is known, is most evident from observations of sunspots.

A sunspot is an outcropping of magnetic field through the visible surface of the sun. These magnetic fields impede the flow of energy into the material they confine. As a result, the plasma in a sunspot is cooler and, hence, darker than the surrounding region. Sunspots and how they appear on the sun are illustrated in Fig. 5. Sunspots and differential rotation play key roles in the solar activity described in Section III.

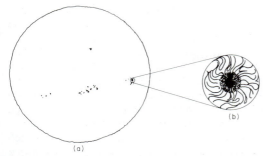

FIG. 5. (a) The sun as seen in visible light, with several active regions, including a complex sunspot group such as was observed in early June 1982. (b) Structure of a sunspot, showing the type of detail visible in Hα.

III. Solar Activity

Solar activity manifests itself in many forms. Numerical parameters characterizing these dynamic aspects of the Sun are presented in Table III.

A. SUNSPOT CYCLE

The most widely known parameter describing solar activity is the approximately 11-yr duration, or period, of the cyclic increase and decrease in the occurrence of sunspots, as illustrated for the most recent past two cycles in Fig. 6. The variation in sunspot number is associated with the drift of sunspot zones toward the solar equator and with increases and decreases in the occurrence of solar flares. This sequence of events is only half the story. The full pageant covers 22 yr, with one half of the full cycle being characterized by a magnetic dipole field that is the reverse of the field that characterizes the other half.

TABLE III. Time Scales Characterizing Solar Activity

Sunspot cycle	11 yr
Full magnetic cycle	22 yr
Solar rotation period	
at solar equator	27 days
at 60° solar latitude	31 days
Lifetime of a solar granule	10 min
Period of solar oscillations	5 min
Duration of a solar flare	
impulsive phase	Few milliseconds to 10 min
impulsive plus gradual	Minutes to hours

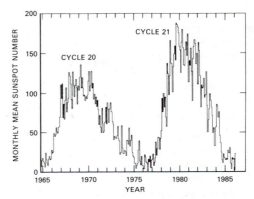

FIG. 6. Plot of the sunspot number for Cycle 20, spanning the time interval from October 1964 through June 1976, and the part of Cycle 21 now completed, from June 1976 until the beginning of 1986. Details of the cycles vary considerably from one cycle to another, but the approximately 11-yr periodicity is known to extend back more than two centuries.

At sunspot minimum, the sun has a weak dipole field aligned with its axis of rotation. The first spots that appear during an 11-yr cycle generally occur at solar latitudes 20–30° above or below the solar equator. As a cycle progresses, both the number of sunspots and the observable magnetic complexity of the groups or active regions that they form increase. During this same time, the sun's dipole field weakens, and the zones of sunspot activity migrate toward the solar equator. At the time of sunspot maximum, the original dipole field vanishes and reappears, but with its north/south magnetic polarity reversed. Also at the maximum, sunspots occur predominately in bands from 5 to 20° north and south of the solar equator. As the cycle approaches a new minimum, the zones of activity shrink towards the equator, eventually forming bands at latitudes from 5 to 10°. When this migration phenomenon is represented as a plot of sunspot latitude versus time of occurrence, the result is commonly referred to as a butterfly diagram, with symmetric wings above and below the solar equator.

Variations in the sunspot number have been observed on much longer time scales, as well. A variation with a 90-yr period, known as the Gleissberg cycle, has been reported for the number observed at sunspot maximum. This periodicity has further significance because the diameter of the sun, as determined from the duration of solar eclipses, has been observed to vary with a 90-yr period, also. The relative phase of these two variations is such that the maximum sunspot number is increasing while the solar diameter is decreasing, and vice versa. Researchers studying these phenomena have suggested that the solar dynamo may cause related changes in the solar luminosity and, consequently, changes in the earth's climate, as well.

B. ROLE OF MAGNETIC FIELDS

The visible structure of sunspots and their periodic migrations are manifestations of magnetic dynamic activity in the outer layers of the sun. This physical process is illustrated in Fig. 7. To understand these diagrams and the significance of the magnetic polarities, in particular, the reader should keep in mind that magnetic fields always form closed loops. By convention, the direction associated with the field lines emerging from the north-seeking pole of a magnet is from

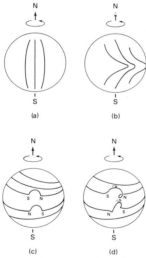

(a)   (b)

(c)   (d)

FIG. 7.   Magnetic fields in the surface of the sun under the influence of a differentially rotating atmosphere. The material that comprises the sun's surface is observed to rotate at a rate that is more rapid near the solar equator and decreases with increasing solar latitude. Field lines that are initially parallel to meridians, as in (a), are stretched and distorted, as in (b), by this differential rotation. The resultant increase in the magnetic field perpendicular to the sun's rotation axis causes the ionized material in the vicinity of the field lines to become less dense and, hence, more buoyant, as in (c). The rising and expanding motion of the plasma bubbles in the rotating reference frame of the sun causes them to twist, an effect known as the Coriolis force, as in (d). This then provides a mechanism for the field lines perpendicular to the sun's rotation axis to delete themselves by reconnecting, leaving behind a magnetic field parallel to the rotation axis but with a reversed north/south polarity.

north to south, while the direction of its continuation closing the loop within the magnet is from south to north. In Fig. 7a, the field lines shown are those connecting the north and south poles of the sun inside the surface. Differential rotation stretches these field lines to the right, as shown in Fig. 7b, giving them a longitudinal component. Wherever these field lines emerge from the surface, they form a new pole with polarity opposite to that of the pole from which they originated. Where they reenter, they form a pole with the same polarity as the original pole, as illustrated in Fig. 7c. These regions of emerging flux are observed as sunspots.

As can be seen in Fig. 7c, a directivity is imparted to the polarity of these sunspot pairs relative to the direction of solar rotation. In each hemisphere of the sun, the polarity of a leading spot is the same as that of the dominant dipole field in that hemisphere. The polarity of a spot trailing relative to the direction of rotation is the opposite. The effect of the sun's rotation on the rising plasma structure causes it to twist so that the newly emerging poles separate laterally. A mechanism is thereby provided by which the longitudinal fields can reform themselves along the north/south direction, but with a polarity opposite to that of the original solar dipole field. This dynamo activity explains both the cyclic 11-yr variations in sunspot number and the fundamental 22-yr cycle that includes reversals of the sun's dipole field.

The process illustrated in Fig. 7 is, of necessity, a simplification, representing only a small part of the magnetic activity on the sun. The whole picture is a concatenation of numerous twists and untwists of the magnetic field. Any one sunspot may connect to any number of others. Some of the field lines reconnect in interplanetary space, so that the motions of bodies outside the sun interact with and perturb the solar field. Magnetic fields play fundamental, complex, and dynamic roles in the physics of the sun.

C. ERUPTIVE BEHAVIOR

The role of magnetic fields in solar flares is less understood, but what is known leaves no doubt that magnetic fields provide the sources of energy and define the settings in which flares occur. Flares are transient events in which energetic particles and electromagnetic radiations, ranging from meter waves to γ rays, are produced. Their rate of occurrence follows the sunspot cycle, with the most important flares, those with the most emission, occurring in the most complex magnetic active regions. The total energy released in a flare is greater the closer the flare site is to a sunspot.

Flares produce both increases in the emissions that can be observed continuously, even from the quiet sun, and great bursts of high-energy emissions that are associated only with flare events. The amount of energy released from the sun in a large flare, the size observed approximately once a month during the three years following sunspot maximum, is 10^{32} erg. This amount is ten million times greater than the energy of a large volcanic eruption on earth but, spread over the typical duration of such a large flare, represents only a few thousandths of a percent increase in the total solar luminosity.

Electrons, protons, and heavier ions are accelerated to relativistic energies in some, and possibly all, flares. Most of these copiously produced energetic particles interact with each other, with quiescent solar material, and with the sun's magnetic fields until all of their excess energy is spent. The others stream out of the sun along magnetic field lines into interplanetary space. The escaping charged particles carry information on their transport from the flare site blended with information on the process that energized them. The electromagnetic radiation and neutrons produced by particle interactions carry information about the flare that is undistorted by transport effects.

The hard X-ray and microwave emissions produced in flares provide the most direct evidence available for investigations of how electrons receive energy in a flare and what role they play in the remainder of the flare dynamics. The spatial, spectral, and temporal aspects of these radiations each yield crucial information for checking and building physical models of the flare process. In Fig. 8, the measured hard X-ray and microwave emissions are shown as functions of time for a particular flare observed on April 27, 1981. This relatively large event illustrates the typically close correlation in time between the impulsive spikes in hard X rays and microwaves. Such close temporal coincidence has been observed all the way into the nuclear γ-ray domain. Because nuclear γ-ray line emission results primarily from interactions of energetic protons with quiescent solar material, while hard X-ray and microwave emissions are due to interactions of energetic electrons, this temporal coincidence suggests that protons and electrons are accelerated simultaneously in flares.

Emission in the ultraviolet through soft X-ray portion of the spectrum evinces heating of the solar atmosphere on time scales ranging from very impulsive, of order seconds, to hours. Meter-wave radio emissions provide evidence for vast flare-associated plasma disturbances traveling through the solar atmosphere, as well as some smaller scale phenomena closely correlated in time with impulsive hard X-ray and microwave events.

Another type of eruptive behavior generally associated with flares is the ejection from the solar surface of whole blobs of ionized material, catapulted into space by the magnetic fields that had supported them. Instances of such phenomena have been observed with ground-based optical telescopes and meter-wave radio interferometers and from manned and unmanned orbiting satellites. These observations have shown prominences suddenly to disconnect and lift off with billions of tons of material accelerated to speeds of many tens of kilometers per second.

Although no theory successfully explains all the processes known to take place in solar flares, many theories or parts of theories have been proposed to explain various aspects. Almost all of these theoretical models involve magnetic arches, the structures formed by magnetic field lines making connections in the solar

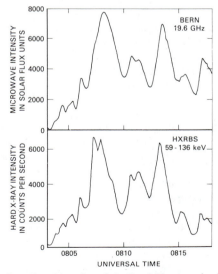

FIG. 8.   Time histories of the hard X-ray and microwave emissions observed from a solar flare on April 27, 1981. The microwave observations were obtained with the solar radio telescopes of the University of Bern, Switzerland, and the hard X-ray data were obtained with the Hard X-Ray Burst Spectrometer (HXRBS) onboard the Solar Maximum Mission (SMM), a NASA satellite.

atmosphere between energing magnetic flux regions on the sun's surface. These are the arch structures that contribute to the corona's distinctive appearance as illustrated in Figs. 1 and 3. The theoretical and observational study of solar flares continues to be actively pursued with special emphasis on investigations of the most energetic processes in solar plasmas.

D. Solar Wind and Coronal Holes

Electrons and ions stream out from the solar corona along trajectories defined by the sun's magnetic field in interplanetary space. The sun's rotation causes the magnetic field lines, and hence the particle streams, to follow a spiral pattern. This flow, known as the solar wind, was first deduced from theoretical considerations and subsequently understood to be the force that orients comet tails. Because the flow is actually an expansion of the solar corona and is continuously moving outward, its effect on comets is to sweep their tails away from the sun. [*See* SOLAR SYSTEM, MAGNETIC AND ELECTRIC FIELDS.]

The extensive coronal studies carried out with the ultraviolet and soft X-ray telescopes on Skylab, including detailed maps of coronal emission, provided the additional information necessary to explain variations that had been observed in the solar wind but not previously understood. When gusts of enhanced solar wind were traced back to their sources on the sun, these sites in the corona were found to be areas that appeared quite dark in the Skylab photographs. These regions are known as coronal holes. Coronal holes were first discovered with instruments on board the sixth satellite in NASA's highly successful Orbiting Solar Observatory (OSO) series. Data from OSO-7 provided the first evidence for the connection between coronal holes and the solar wind. One particularly large and interestingly shaped coronal hole observed from Skylab is illustrated in Fig. 9.

A coronal hole itself is formed by magnetic field lines that stream out of the sun to a connection point in interplanetary space rather than to a magnetic reconnection point on the sun's surface. These features, like the magnetic arches, are often visible during a solar eclipse. Because the magnetic field of the sun is continuously evolving, coronal holes are continuously changing and moving. During sunspot minimum when the solar dipole fields are strongest, coronal holes are invariably found at both of the solar poles. Further advances in the study of the solar

FIG. 9.   Artist's sketch of the coronal hole nicknamed "The Boot of Italy Hole" for its distinctive shape, as observed in soft X rays from Skylab. The same pattern was observed concurrently in ultraviolet light with another of the Skylab instruments.

wind and coronal structure await the flight of detectors out of the plane of the earth's orbit and into the regions above the sun's poles.

IV. Outstanding Questions and the Search for Answers

Solar physics is a vigorously pursued subject that forms vital components of astronomy, physics, and space science. Researchers in universities, private industry, and government laboratories are involved in this effort. Below is a summary of the foremost questions directing the attention of solar physicists and the investigations underway or being planned to address these questions.

A. Neutrinos

Neutrinos are produced in the fundamental nuclear processes that heat the sun's interior. The measured neutrino flux from the sun falls significantly short of predictions based on the best current understanding of nuclear physics and models of stellar structure. How must the theory of the nuclear processes and/or the stellar models be changed to resolve this anomaly?

New ground-based detectors that are more sensitive than earlier detectors and that respond to additional components of the predicted neutron flux are being developed. [*See* NEUTRINO ASTRONOMY.]

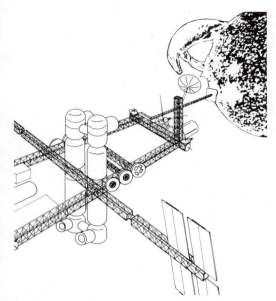

FIG. 10. Artist's concept of the instruments comprising the Advanced Solar Observatory mounted on Space Station. The instruments being studied for this program of the future include hard X-ray and γ-ray imaging devices and spectrometers, high-energy neutron spectrometers, very high-resolution telescopes in the optical through soft X-ray domain, and low-frequency radio spectrographs.

B. DYNAMICS OF THE QUIET SUN

Are there long-term changes in the size of the sun? If so, are they associated with changes in the sun's luminosity? What are the composition, density, and temperature of the sun's interior, and how do they behave dynamically? By what mechanisms is heat supplied to the outer solar atmosphere? These questions are being addressed with continued optical observations using ground-based facilities and will be addressed with optical telescopes being developed for use in space.

C. SOLAR FLARES

What triggers the release of energy in a solar flare? How and where is the energy transferred to electrons, protons, and ions? What role do the energetic particles play in low-energy solar flare emissions?

Currently, only ground-based radio interferometers are available for studies of the source location and geometry of emission from energetic electrons in solar flares. Significant advances are expected both from imaging observations and from spectroscopic studies in the hard X-ray and γ-ray domain with new instrumentation now being developed for space flight. Balloon-borne platforms are being used for initial verification of the instrument concepts.

A solar observatory located in space is illustrated in Fig. 10. This space station of the future provides a stable platform above the distorting effects of the earth's atmosphere, making the earth's unique star and plasma laboratory even more accessible.

BIBLIOGRAPHY

Eddy, J. A. (1979). A New Sun: The Solar Results from Skylab, NASA SP-402 (R. Ise, ed.) National Aeronautics and Space Administration, Washington, D.C.

Giovanelli, R. (1984). "Secrets of the Sun." Cambridge University Press, New York.

Jordan, S. (ed.) (1981). The Sun as a Star, NASA SP-450. National Aeronautics and Space Administration, Washington, D.C.

Kundu, M., and Woodgate, B. E. (eds.) (1986). "Energetic Phenomena on the Sun," Proceedings of the SMM Workshops, NASA CP-2401. National Aeronautics and Space Administration, Washington, D.C.

Noyes, R. W. (1982). "The Sun, Our Star." Harvard University Press, Cambridge, MA.

Sturrock, P. A. (ed.) (1980). "Solar Flares," A Monograph from Skylab Workshop II. Colorado Associated University Press, Boulder, Colorado.

Sturrock, P. A., Holzer, T. E., Mihalas, D. M., and Ulrich, R. K. (eds.) (1985). "Physics of the Sun," Volume I: The Solar Interior; Volume II: The Solar Atmosphere; Volume III: Astrophysics and Solar-Terrestrial Relations. D. Reidel Publishing Company, Hingham, Mass.

Zirker, J. B. (ed.) (1977). "Coronal Holes and High Speed Wind Streams," A Monograph from Skylab Solar Workshop I. Colorado Associated University Press, Boulder.

SOLAR SYSTEM, GENERAL

Stanley F. Dermott *Cornell University*

GLOSSARY

Accretion: Process whereby gases and particles of dust agglomerate to form large bodies such as planets, satellites, comets, and asteroids.

Asteroid: Solar system object, unresolved by ground-based telescopes, without any trace of cometary activity, that is neither a planet nor a satellite of a planet.

Carbonaceous chondrite: High-carbon type of chondrite regarded as the least subject to changes since formation and thus one of the most primitive objects in the solar system.

Chondrite: Stony meteorite characterized by the presence of chondrules, small spherical grains composed predominantly of iron and magnesium silicates.

Comet: Solar system object displaying cometary activity (a coma or a tail) when close to the sun, due to the vaporization of icy volatiles.

Galilean satellites: Four large satellites of Jupiter (Io, Europa, Ganymede, and Callisto) discovered by Galileo in 1610.

Kirkwood gaps: Gaps in the distribution of asteroidal semimajor axes where the ratio of the orbital periods of an asteroid and Jupiter would be close to the ratio of two small integers.

Meteorite: Extraterrestrial material that survives passage through the earth's atmosphere and reaches the earth's surface as a recoverable object.

Oort cloud: Reservoir of, perhaps, 10^{12} comets with aphelia distances in the range 20,000–100,000 AU and orbits that are subject to stellar perturbations with the result that occasionally a comet is deflected into an orbit with a perihelion distance close enough to the sun to allow observation.

Planetesimal: Hypothetical early body of intermediate size (perhaps meters to 100 km) that eventually accreted to a larger body; sometimes referred to as a planetary building block.

Protoplanetary disk: Flattened disk of gas and dust surrounding the early sun out of which solar system objects are believed to have formed.

Resonance: Satellites are in resonance if the geometric configuration of their positions with respect to their orbits is repeated within a short period of time; this requires that the ratio of their orbital periods be close to the ratio of two small integers.

Roche limit: Minimum distance at which the self-gravitation of a satellite is sufficient to overcome the tidal and rotational forces that tend to disrupt satellites orbiting close to a planet.

The major classes of objects in the solar system are stars (this class has only a single representative, the sun), planets, satellites, asteroids, rings, comets, and meteorites. The study of these objects contributes to our understanding of the origin and evolution of the solar system.

I. Overview of the Solar System

The origin of the solar system, the ubiquitousness of other planetary systems, and the exis-

TABLE I.   The Planets[a]

Planet	Mass (10^{24} kg)	Equatorial radius (km)	Density (10^3 kg m^{-3})	Escape velocity (km sec^{-1})
Mercury	0.3303	2,439	5.43	4.25
Venus	4.870	6,051	5.25	10.4
Earth	5.976	6,378	5.518	11.2
Mars	0.6421	3,393	3.95	5.02
Jupiter	1,900	71,398	1.332	59.6
Saturn	568.8	60,330	0.689	35.5
Uranus	86.87	26,200	1.18	21.3
Neptune	102.0	25,225	(1.54)	23.3
Pluto	(0.013)	1,145	1.84	1.3

[a] Values in parentheses are uncertain.

tence of life on other planets are the major problems that challenge planetary science. Stars are formed in galaxies from the collapse of interstellar clouds of gas and dust. During collapse, the clouds, which have masses of thousands of solar masses, fragment to produce thousands of stellar systems. Observations of stellar systems in our own galaxy, the Milky Way, indicate that probably 80% of these fragments must give rise to multiple star systems, that is, double stars, triple stars, or more complex systems. However, anywhere between a few percent and 20% of all stars are not part of multiple systems, and our own star, the sun, is such a star. Theoretical reasoning, guided predominantly by considerations of the conservation of angular momentum, indicates that an intermediate step in star formation is the formation of an extensive disk of gas and dust. This is usually referred to as a protoplanetary disk or an accretion disk. This terminology may be oversuggestive, implying as it does that planet formation may be a normal by-product of star formation in single-star systems. This indeed may be the case, but the question of the existence of planetary systems other than our own should strictly be regarded as open. Nevertheless, it is now widely believed that the various bodies in our solar system were formed from a protoplanetary disk of gas and dust concomitant with the sun.

The nine known planets in the solar system can be divided into two major groups: terrestrial and Jovian. The terrestrial planets (Mercury, Venus, Earth, and Mars) consist largely of iron and ferromagnesium silicate rocks. Jupiter and Saturn consist predominantly of hydrogen and helium and are fluid throughout, except for small cores of rock and ice, which may be solid.

TABLE II.   Planetary Orbits

Planet	Semimajor axis AU[b]	Semimajor axis 10^6 km	Orbital period (yr)	Eccentricity	Inclination[a] (deg)
Mercury	0.387	57.9	0.24085	0.206	7.003
Venus	0.723	108.2	0.61521	0.0068	3.394
Earth	1.000	149.6	1.00004	0.0167	0.000
Mars	1.524	227.9	1.88089	0.0933	1.850
Jupiter	5.203	778.3	11.86223	0.0483	1.309
Saturn	9.539	1427.0	29.45774	0.0559	2.493
Uranus	19.182	2869.6	84.018	0.047	0.772
Neptune	30.058	4496.6	164.78	0.0087	1.779
Pluto	39.44	5900.1	248.4	0.247	17.146

[a] Inclination is measured with respect to the orbital plane of the earth.
[b] An astronomical unit (AU) is the mean distance of the earth from the sun, or 149.60 10^6 km.

TABLE III.   Planetary Spins and Magnetic Fields

Planet	Rotation period	Obliquity of spin axis (deg)[b]	Magnetic moment (G cm³)	Surface field[a] (G)
Mercury	58.65 days	2 ± 3	2.4×10^{22}	0.002
Venus	243.01 days	177.3	$<4 \times 10^{21}$	<0.00002
Earth	23.9345 hr	23.45	7.98×10^{25}	0.3
Mars	24.6299 hr	23.98	2.5×10^{22}	0.0006
Jupiter	9.841 hr (equator) 9.925 hr (interior)	3.12	1.5×10^{30}	4
Saturn	10.233 hr (equator) 10.675 hr (interior)	26.73	4.6×10^{28}	0.2
Uranus	17.24 hr (interior)	97.86	4.1×10^{27}	0.2
Neptune	18.2 ± 0.4 hr	(29.56)	—	—
Pluto	6.387 days	(118.5)	—	—

[a] Measured at the magnetic equator.
[b] Values in parentheses are uncertain.

Uranus and Neptune consist largely of ice-forming material with an elemental composition of carbon, nitrogen, and oxygen, as well as hydrogen and helium. The outermost planet, Pluto, is small and somewhat anomalous, its orbit having more in common with those of the asteroids than with the other planets. The surface composition of Pluto is that of methane ice. The sizes, orbital parameters, rotational periods, magnetic fields, and atmospheric compositions of the planets are given in Tables I–IV.

The internal structure of the earth is compara-

TABLE IV.   Planetary Atmospheres

Planet	Surface pressure (bars)	Average surface temperature (K)	Major gases	Fractional abundance by number[a]
Mercury	$\sim 10^{-14}$	440	Na	0.97
			He	0.03
Venus	90	730	CO_2	0.96
			N_2	0.035
Earth	1	288	N_2	0.77
			O_2	0.21
			H_2O	0.01
Mars	0.007	218	CO_2	0.95
			N_2	0.027
			Ar	0.016
Jupiter	—	—	H_2	0.90
			He	0.10
Saturn	—	—	H_2	0.94
			He	0.06
Uranus	—	—	H_2	0.85
			He	0.15
Neptune	—	—	H_2	0.85?
			He	0.15?
Pluto	0.001	—	CH_4	—
			Ne?	—

[a] This quantity is also referred to as the volume mixing ratio.

TABLE V.   Satellite Atmospheres

Satellite	Surface pressure (bars)	Average surface temperature (K)	Major gases	Fractional abundance by number[a]
Moon	$\sim 2 \times 10^{-14}$	274	Ne	0.4
			Ar	0.4
			He	0.2
Io	$\sim 1 \times 10^{-10}$	~ 110	SO_2	1
Titan	1.6	95	N_2	0.73–0.99[b]
			Ar	0.00–0.28
			CH_4	0.01–0.12
Triton	~ 0.1	~ 57	N_2	—
			CH_4	—

[a] This quantity is also referred to as the volume mixing ratio.
[b] This is the range of uncertainty.

tively well determined. Interpretation of the shape of the earth, in particular the difference between its polar and equatorial radii, and of its external gravitational field yields the moment of inertia of the planet. Interpretation of the amplitudes, phases, and arrival times of seismic waves generated by earthquakes yields a detailed picture of the interior. The earth is divided into five zones: the crust, the upper mantle, and the lower mantle, which consist of ferromagnesium silicate rocks; an outer core of liquid nickel–iron; and an inner core of solid nickel–iron. The internal structures of the other terrestrial planets are not well known, but the fact that these planets have densities that are similar to that of the earth, after allowance has been made for the different degrees of compression, suggests that they have similar compositions. In contrast, there are large differences between their atmospheres and the physical states of their surfaces.

Normal stars can have masses as small as 0.08 that of the sun, but this is still a factor of 80 greater than that of the largest planet. Jupiter, Saturn, and Neptune but not, perhaps, Uranus do have internal heat sources. These planets ra-

diate more heat than they receive from the sun, by factors of 1.7, 1.8, and 2.1, respectively. However, this heat is not derived from thermonuclear reactions in their interiors, but is largely a remnant of the heat derived on formation from the gravitational potential energy of the contracting planetary material.

Unlike the terrestrial planets, the Jovian planets tend to have extensive satellite systems. The inner satellites of Jupiter, Saturn, and Uranus are described as regular in that their orbits are prograde (their orbital motion is in the same sense as that of the spin of the planet), near circular, and lying in or close to the equatorial plane of the planet. The outer, irregular satellites of Jupiter, Saturn, and Neptune have orbits with high inclinations with respect to the planetary equators, and some of these orbits are retrograde. Irregular satellites may be bodies captured by the planet in the later stages of planet formation. The physical properties of the satellites and their orbits are described in Tables V–VII.

The other bodies in the solar system are small and, in terms of their masses, insignificant. However, these bodies may be our best source

TABLE VI.   The Satellites

Planet		Satellite	Mass (10^{20} kg)	Radius[a] (km)	Density (10^3 kg m^{-3})	Surface composition
Mercury	—	None	—	—	—	—
Venus	—	None	—	—	—	—
Earth	—	Moon	734.9	1738	3.34	Rocks?
Mars	MI	Phobos	1.26×10^{-4}	$13.5 \times 10.7 \times 9.6$	2.2	Carbonaceous
	MII	Deimos	1.8×10^{-5}	$7.5 \times 6.0 \times 5.5$	1.7	Carbonaceous

TABLE VI (*Continued*)

Planet	Satellite		Mass (10^{20} kg)	Radius[a] (km)	Density (10^3 kg m^{-3})	Surface composition
Jupiter	JXVI	Metis	—	$? \times 20 \times 20$	—	Rock?
	JXV	Adrastea	—	$12.5 \times 10 \times 7.5$	—	Rock?
	JV	Amalthea	—	$135 \times 82 \times 75$	—	Sulfur/rock
	JXIV	Thebe	—	$? \times 55 \times 45$	—	Rock?
	JI	Io	894	1815	3.57	Sulfur, SO_2
	JII	Europa	480	1569	2.97	Water ice
	JIII	Ganymede	1482.3	2631	1.94	Dirty water ice
	JIV	Callisto	1076.6	2400	1.86	Dirty water ice
	JXIII	Leda	—	~8	—	Carbonaceous?
	JVI	Himalia	—	90	—	Carbonaceous?
	JX	Lysithea	—	~20	—	Carbonaceous?
	JVII	Elara	—	40	—	Carbonaceous?
	JXII	Ananke	—	~15	—	Carbonaceous?
	JXI	Carme	—	~22	—	Carbonaceous?
	JVIII	Pasiphae	—	~35	—	Carbonaceous?
	JIX	Sinope	—	~20	—	Carbonaceous?
Saturn	SXV	Atlas	—	$19 \times ? \times 14$	—	Water ice?
	SXVI	Prometheus	—	$70 \times 50 \times 37$	—	Water ice?
	SXVII	Pandora	—	$55 \times 43 \times 33$	—	Water ice?
	SXI	Epimetheus	—	$70 \times 58 \times 50$	—	Water ice?
	SX	Janus	—	$110 \times 95 \times 80$	—	Water ice?
	SI	Mimas	0.38	$210.2 \times 197.3 \times 192.5$	1.137	Water ice?
	SII	Enceladus	0.8	251	1.2	Pure water ice
	SIII	Tethys	7.6	524	1.26	Water ice
	SXIII	Telesto	—	$? \times 12 \times 11$	—	Water ice?
	SXIV	Calypso	—	$15 \times 13 \times 8$	—	Water ice?
	SIV	Dione	10.5	559	1.44	Water ice
	SXII	Helene	—	$18 \times ? \times <15$	—	Water ice?
	SV	Rhea	24.9	764	1.33	Water ice
	SVI	Titan	1345.7	2575 (solid) 2775 (clouds)	1.882	Atmosphere of N_2, CH_4, Ar
	SVII	Hyperion	—	$175 \times 120 \times 100$	—	Dirty water ice
	SVIII	Iapetus	18.8	718	1.21	Ice/carbonaceous?
	SIX	Phoebe	—	$115 \times 110 \times 105$	—	Carbonaceous?
Uranus	1986U7		—	~20	—	Water ice?
	1986U8		—	~25	—	Water ice?
	1986U9		—	~25	—	Water ice?
	1986U3		—	~30	—	Water ice?
	1986U6	—	—	~30	—	Water ice?
	1986U2	—	—	~40	—	Water ice?
	1986U1	—	—	~40	—	Water ice?
	1986U4	—	—	~30	—	Water ice?
	1986U5	—	—	~30	—	Water ice?
	1985U1	—	—	85	—	Water ice?
	UV	Miranda	0.7	242	1.3	Dirty water ice
	UI	Ariel	13	580	1.6	Dirty water ice
	UII	Umbriel	13	595	1.4	Dirty water ice
	UIII	Titania	35	805	1.6	Dirty water ice
	UIV	Oberon	29	775	1.5	Dirty water ice
Neptune	NI	Triton	1300	1750	(5)[b]	CH_4 ice, N_2?
	NII	Nereid	—	~200	—	—
Pluto	PI	Charon		~640	1.84	Methane ice

[a] For satellites that are markedly nonspherical, we quote the principal axes of the best-fitting triaxial ellipsoid.

TABLE VII.   Satellite Orbits

Planet	Satellite		Semimajor axis		Orbital period[a] (days)	Eccentricity[b]	Inclination[c] (deg)
			(10³ km)	Planetary radii			
Mercury	—	None	—	—	—	—	—
Venus	—	None	—	—	—	—	—
Earth	—	Moon	384.4	60.3	27.3217	0.05490	5.15[d]
Mars	MI	Phobos	9.378	2.76	0.319	0.015	1.02
	MII	Deimos	23.459	6.91	1.263	0.00052	1.82
Jupiter	JXVI	Metis	127.96	1.7922	0.2948	<0.004	~0
	JXV	Adrastea	128.98	1.8065	0.2983	~0	~0
	JV	Amalthea	181.3	2.539	0.4981	0.003	0.40
	JXIV	Thebe	221.90	3.108	0.6745	0.015	0.8
	JI	Io	421.6	5.905	1.769	$(0.0041)-10^{-5}$	0.040
	JII	Europa	670.9	9.397	3.551	$(0.0101)-10^{-4}$	0.470
	JIII	Ganymede	1,070	14.99	7.155	$(0.0006)-0.0015$	0.195
	JIV	Callisto	1,883	26.37	16.689	0.007	0.281
	JXIII	Leda	11,094	155.4	238.72	0.148	27[d]
	JVI	Himalia	11,480	160.8	250.57	0.158	28[d]
	JX	Lysithea	11,720	164.2	259.22	0.107	29[d]
	JVII	Elara	11,737	164.4	259.65	0.207	28[d]
	JXII	Ananke	21,200	296.9	631(R)	0.169	147[d]
	JXI	Carme	22,600	316.5	692(R)	0.207	163[d]
	JVIII	Pasiphae	23,500	329.1	735(R)	0.378	148[d]
	JIX	Sinope	23,700	331.9	758(R)	0.275	153[d]
Saturn	SXV	Atlas	137.64	2.281	0.602	~0	~0
	SXVI	Prometheus	139.35	2.310	0.613	0.0024	~0
	SXVII	Pandora	141.70	2.349	0.629	0.0042	~0
	SXI	Epimetheus	151.422	2.510	0.694	0.009	0.34
	SX	Janus	151.472	2.511	0.695	0.007	0.14
	SI	Mimas	185.52	3.075	0.942	0.0202	1.53
	SII	Enceladus	238.02	3.945	1.370	(0.0045)	0.02
	SIII	Tethys	294.66	4.884	1.888	0.0000	1.09
	SXIII	Telesto	294.66	4.884	1.888	~0	~0
	SXIV	Calypso	294.66	4.884	1.888	~0	~0
	SIV	Dione	377.40	6.256	2.737	(0.0022)	0.02
	SXII	Helene	377.40	6.256	2.737	0.005	0.2
	SV	Rhea	527.04	8.736	4.518	$(0.0010)-0.0003$	0.35
	SVI	Titan	1,221.85	20.25	15.945	0.0292	0.33
	SVII	Hyperion	1,481.1	24.55	21.277	(0.1042)	0.43
	SVIII	Iapetus	3,561.3	59.03	79.331	0.0283	7.52
	SIX	Phoebe	12,952	214.7	550.48(R)	0.163	175.3[d]
Uranus	1986U7	—	49.7	1.90	0.33	—	—
	1986U8	—	53.8	2.05	0.38	—	—
	1986U9	—	59.2	2.26	0.43	—	—
	1986U3	—	61.8	2.36	0.46	—	—
	1986U6	—	62.7	2.39	0.48	—	—
	1986U2	—	64.6	2.47	0.49	—	—
	1986U1	—	66.1	2.52	0.51	—	—
	1986U4	—	69.9	2.67	0.56	—	—
	1986U5	—	75.3	2.87	0.62	—	—
	1985U1	—	86.0	3.28	0.76	—	—
	UV	Miranda	129.783	4.95	1.413	0.0027?	4.22
	UI	Ariel	191.239	7.30	2.520	0.0034?	0.31

TABLE VII   (*Continued*)

| Planet | Satellite | | Semimajor axis | | Orbital period[a] (days) | Eccentricity[b] | Inclination[c] (deg) |
			(10³ km)	Planetary radii			
	UII	Umbriel	265.969	10.15	4.144	0.0050?	0.36
	UIII	Titania	435.844	16.64	8.706	0.0022?	0.14
	UIV	Oberon	582.596	22.24	13.463	0.0008?	0.10
Neptune	NI	Triton	354.3	14.0	5.877(R)	<0.0005	159.0
	NII	Nereid	551.5	219	360.16	0.75	27.6[d]
Pluto	PI	Charon	19.1	16.7	6.387	~0	94.3

[a] R denotes a retrograde orbit.
[b] Eccentricities in parentheses are forced eccentricities due to a resonant interaction with another satellite. All other eccentricities are free eccentricities.
[c] Measured with respect to the equatorial plane of the planet.
[d] Orbits of satellites distant from a planet are strongly perturbed by the sun.

of information about the early history, and even prehistory, of the solar system. Comets, in particular, are thought to be planetesimals, or planetary building blocks, that were flung to the outer reaches of the solar system by close planetary encounters at the time of planetary accretion. Because of their small size and low internal temperatures, the physical and chemical changes that have occurred in comets since their formation, and before their first passage through the inner solar system, are probably negligible.

Asteroids are not the remnants of a disintegrated planet but, again, planetesimals that

failed to accrete fully to a planet (the largest has a diameter of 1000 km), probably due to the gravitational influence of Jupiter. Not all asteroids are primitive bodies. Mutual collisions among small asteroids lead to their disintegration and gradual reduction (or comminution) to smaller and smaller bodies. Meteorites are solar system objects that survive passage through the earth's atmosphere to reach the earth's surface as recoverable objects. Some meteorites are undoubtedly the collision products of asteroids. However, others may have a cometary origin, and we now know that a few

TABLE VIII.   Selected Asteroids

Asteroid	Semimajor axis (AU)	Eccentricity	Inclination (deg)	Diameter (km)	Type[a]	Rotational period (hr)
1 Ceres	2.768	0.077	10.598	1025	C	9.078
2 Pallas	2.773	0.233	34.800	565	C	7.811
3 Juno	2.671	0.255	13.002	244	S	7.211
4 Vesta	2.362	0.090	7.144	533	U	5.342
5 Astraea	2.577	0.189	5.349	122	S	16.812
16 Psyche	2.921	0.138	3.092	249	M	4.303
221 Eos[b]	3.012	0.071	10.02	98	S	10.450
320 Katherina[b]	3.013	0.074	10.08	—	—	—
590 Tomyris[b]	3.001	0.078	10.02	—	—	—
1580 Betulia	2.196	0.490	52.041	6	U	6.130

[a] The major asteroidal classes are C (carbonaceous?), S (stony or silicate?), and M (metal?). A large number of asteroids are designated U (unclassified); these asteroids do not fall into any of the recognized classes.
[b] The asteroids Eos, Katherina, and Tomyris are members of a Hirayama family. For these asteroids we list their proper orbital elements. In all other cases we list the osculating orbital elements.

TABLE IX. Selected Comets[a]

Comet	Name	Semimajor axis (AU)	Orbital period (yr)	Eccentricity	Inclination (deg)	Date of next pericenter passage
1977	IX West	30,000 (exit)	5×10^6 (no return)	0.999815	116.9	—
1937	IV Whipple	16,000 (714)	2×10^6 (2×10^4)	0.999892	41.6	—
1910	I Great Comet	7,400 (1500)	6×10^5 (6×10^4)	0.999983	138.8	—
	P Giacobini–Zinner	3.51	6.59	0.7076	31.88	April 14, 1992
	P Halley	17.9	76.0(R)[b]	0.9673	162.23	July 28, 2061
	P Enke	2.21	3.29	0.8499	11.93	July 17, 1987
	P Tempel 2	3.04	5.29	0.5444	12.43	Sept. 16, 1988

[a] The values quoted are estimates of the orbital parameters the comet had before it reached the inner solar system; the values in parentheses are the orbital parameters the comet will have when it is no longer perturbed by the planets. These comments do not apply to periodic comets.

[b] R denotes a retrograde orbit.

TABLE X. Planetary Rings

Planet	Ring	Boundary radii (planetary radii)	Eccentricity	Inclination (deg)	Ringlet width[a] (km)
Jupiter	Halo	1.41 –1.71	~0	<10	—
	Main ring	1.71 –1.81	~0	0	—
	Gossamer ring	1.81 –1.83	~0	0	—
Saturn	D Ring	1.11 –1.235	0	0	—
	C Ring	1.235–1.525	0	0	—
	B Ring	1.525–1.940	0	0	—
	A Ring	2.025–2.267	0	0	—
	F Ring	2.324	0.00026		16
	G Ring	2.82	0	0	—
	E Ring	3–8	0	0	—
Uranus	1986U2R	1.41 –1.51	—	—	2500
	6	1.5973	0.0010	0.063	1–3
	5	1.6122	0.0019	0.052	2–3
	4	1.6252	0.0011	0.032	2–3
	α	1.7073	0.0008	0.014	7–12
	β	1.7431	0.0004	0.005	7–12
	η	1.8008	0	0.002	0–2
	γ	1.8179	0	0.011	1–4
	δ	1.8439	0	0.004	3–9
	1986U1R	1.9099	—	—	1–2
	ε	1.9527	0.0079	0	22–93
Neptune	Arc[b]?	2.145	—	—	23
	Arc?	2.51	—	—	8
	Arc[b]	2.66	—	—	15

[a] The width of a narrow ring varies from a minimum at pericenter to a maximum at apocenter.

[b] These partial rings have been confirmed by independent observers.

meteorites reach the earth from the surface of the moon. Meteorites are our most important source of extraterrestrial material.

Whether planetary rings are primitive structures is not known. Their importance, however, lies in the fact that some of the dynamic processes operating in the present ring systems may be similar to processes that operated in the early accretion disks. Data on some representative asteroids and comets and their orbits are given in Tables VIII and IX, respectively. Data on planetary rings are given in Table X.

II. Dynamics and Structure

The orbits of most bodies in the solar system are determined, to a very good approximation, by the gravitational influence of a single, central body. The motions of such bodies are well described by the laws of planetary motion published by Johannes Kepler in 1609 and 1619. These state that (1) each planetary orbit is an ellipse with the sun at one focus (1609); (2) a line joining a planet to the sun sweeps out equal areas in equal times (1609); (3) for any pair of planets, the ratio of the squares of their orbital periods is equal to the ratio of the cubes of their semimajor axes (1619). These laws were interpreted by Isaac Newton in terms of the universal law of gravitation, which states that every particle of matter in the universe attracts every other particle of matter with a force directly proportional to the product of their masses and inversely proportional to the square of the distance between them. Hence, the force F between two particles of masses m_1 and m_2 separated by a distance r is given by

$$F = Gm_1m_2/r^2$$

where G is the universal constant of gravitation. Laboratory experiments have determined that $G = 6.668 \times 10^{-11}$ N kg^{-2} m^2.

Kepler's laws tell us nothing about the spacing of the planets, and it was left to Johann Bode in 1772 to popularize a law enunciated by Johann Titius in 1766 and now known as the Titius–Bode law or, more commonly, Bode's law. It is often written as

$$r_n = 0.4 + 0.3(2^n)$$

where r_n is the mean distance of the planet from the sun (in units of the earth's distance), n taking the values $-\infty, 0, 1, 2, 3, \ldots$ for Mercury, Venus, Earth, and so on. The law became popular after the discovery of Uranus by William Herschel in 1781 and played an important role in the discovery of the largest asteroid, Ceres, on January 1, 1801. However, it failed conspicuously in the case of Neptune. The predicted and actual (in parentheses) distances of the planets are as follows: Mercury [$n = -\infty$], 0.4 (0.39); Venus [$n = 0$], 0.7 (0.72); Earth [$n = 1$], 1.0 (1.00); Mars [$n = 2$], 1.6 (1.52); Ceres [$n = 3$], 2.8 (2.77); Jupiter [$n = 4$], 5.2 (5.20); Saturn [$n = 5$], 10.0 (9.54); Uranus [$n = 6$], 19.6 (19.20); Neptune [$n = 7$], 38.8 (30.07); Pluto [$n = 8$], 77.2 (39.46). Even though it is accepted that the distribution of planetary orbits is not entirely random and that similar laws describe the distribution of satellite distances, Bode's law is not given much consideration by modern astronomers.

The structure of the solar system is clearly nonrandom in that there are far more resonances in the solar system than would arise in a system with a random distribution of orbital periods. In celestial mechanics, the orbit of a body is described by the elements a, e, I, Ω, and $\tilde{\omega}$, where a is the semimajor axis, e the eccentricity, I the inclination, Ω the longitude of the ascending node, and $\tilde{\omega}$ the longitude of the pericenter. The position in the orbit is given by the mean longitude,

$$\lambda = \int_0^t n \, dt + \varepsilon \simeq nt + \varepsilon$$

where n, the mean motion of the body, is equal to $2\pi/P$, where P is the body's orbital period; ε is the longitude at epoch. A pair of planets or satellites are in resonance if the geometric configuration of their positions with respect to their orbits is repeated within a short period of time. This requires that the ratio of their orbital periods (or, equivalently, the ratio of their mean motions) be close to two small integers. For example, the ratio of the orbital periods of Neptune and Pluto is $0.6634 \simeq 2:3$. The exact conditions that describe the resonant configuration of these planets is

$$\phi = 2\lambda - 3\lambda' + \tilde{\omega}' = \pi$$

where the primed quantities refer to the orbital elements of Pluto. In fact, ϕ is not exactly π but librates (or oscillates) about π. Differentiating the above equation with respect to time and rearranging, we obtain

$$(n' - \dot{\tilde{\omega}}')/(n - \dot{\tilde{\omega}}') = \frac{2}{3}$$

Thus, the mean motions of the planets relative to the motion of Pluto's pericenter are exactly commensurate. The orbital paths of Neptune and Pluto intersect, leading one to expect that eventually the planets will collide. However, conjunctions of planets occur when $\lambda = \lambda'$, and the above resonance conditions ensure that all conjunctions of Neptune and Pluto occur near Pluto's apocenter, which in turn ensures that these planets never have a close encounter.

Numerous other resonances of this type exist in the solar system, particularly among the satellites of Jupiter and Saturn, and it has been proved that the number of pairs of nearly commensurate mean motions in the solar system is too great to be ascribed to chance. We can conclude from this either that the mechanism of formation of the planets and satellites was such as to favor orbits with commensurate mean motions or that the present distribution of orbits is the result of orbital evolution since the time of formation. It is likely that the resonances in the satellite systems of Jupiter and Saturn are the result of orbital evolution due to tidal friction.

In marked contrast to the planet and satellite systems, asteroids tend to avoid orbits that would be nearly commensurate with Jupiter. These gaps in the distribution of asteroidal mean motions (or semimajor axes) are known as the Kirkwood gaps after their discoverer Daniel Kirkwood. Similar gaps occur in the ring system of Saturn. In particular, particles at the inner edge of Cassini's division (the prominent gap named after J. D. Cassini that separates the outer A ring from the inner B ring; see Fig. 1) are in a 2:1 resonance with the satellite Mimas. Resonances also appear to bound the edges of the Uranian ε ring. Bodies that orbit at the same mean distance from a primary, that is, that have a common orbit, are in 1:1 resonance. Examples include the Trojan asteroids, which coorbit with Jupiter and oscillate about the planet's Lagrangian equilibrium points; some move, on average, 60° ahead of the planet, whereas others trail the planet, on average, by 60°.

Resonances in the solar system can also involve the spin and the orbit of a body. The rotational period of Mercury, for example, is exactly two-thirds of its orbital period. Spin–orbit resonances are not primordial but the result of spin–orbit evolution due to tidal friction. Tidal oscillations always result in the dissipation of mechanical energy. Since angular momentum is conserved, the decrease in the total energy of the system due to tidal interactions results in an exchange of angular momentum between the spin of a planet and the orbit of the tide-raising

FIG. 1. Saturn taken on August 4, 1981, by *Voyager 2*. Three of Saturn's icy satellites are evident at left. They are, in order of distance from the planet, Tethys, Dione, and Rhea. The shadow of Tethys appears on Saturn's southern hemisphere. A fourth satellite, Mimas, is less evident, appearing as a small bright spot between Tethys and the rings. The broad, inner B ring is separated from the outer A ring by Cassini's division. The dark region near the outer edge of the A ring is Encke's gap. Spokes, transient markings produced by electrical phenomena, are evident on the B ring. Note that the planet is markedly oblate with an equatorial radius 8.8% larger than the polar radius. (Courtesy of JPL/NASA.)

satellite. If the orbital period of the satellite is greater than the planet's rotational period, the planet is braked while the orbit of the satellite expands. Observations indicate that the moon may have once orbited within 10 earth radii from the earth, at which time the earth's day may have been less than 15 hr. When resonant states are encountered, resonance capture can occur, with the result that the resonant configuration is maintained despite the continued action of the tidal forces.

III. The Terrestrial Planets

The four terrestrial planets are similar in that (1) they are composed of silicate rock and iron, (2) they are differentiated and probably have nickel–iron cores and solid rock mantles (this has been determined directly only in the case of the earth), and (3) they have anomalous or nonexistent satellite systems. The satellites of Mars, Phobos and Deimos, are extremely small and may be bodies captured in the early stages of planet formation at a time when Mars had an extensive atmosphere. In contrast, the moon is anomalously large. Mercury and Venus do not possess any satellites, although it has been argued that satellites may have existed in the past that have since been lost, due, indirectly, to tides raised on the planets by the sun. The atmospheres, surfaces, and magnetic fields of these planets are remarkable for their lack of similarity. We must not neglect the fact that only one of these planets is known to maintain, at present, any form of life.

The earth has a strong magnetic field, which is generated by convective motions in its liquid nickel–iron core. The earth's "twin" planet, Venus, does not possess any magnetic field. Mercury has a very weak field, as does Mars (in this case the measurements are marginal and controversial). The reasons for these wide variations are not known. In the case of Venus, a likely cause is its extremely low spin rate (the rotational period of Venus is −243 days, with the peculiar, but unimportant consequence that the Venusian "day" is longer than its "year"). The moon does not, at present, possess a permanent magnetic dipole moment. However, analyses of the rocks returned by the Apollo missions and measurements of local magnetic fields on the surface of the moon show that some lunar rocks possess remnant magnetizations. It follows that these rocks were exposed to strong magnetic fields as they cooled through their Curie points (the temperature below which a ferromagnetic rock can retain magnetization) more than 3.6 billion years ago. The origin of these magnetic fields is unknown, but it has been suggested that the moon once possessed an internal, self-generating dynamo that ceased to operate after the moon cooled and the small iron core, in which the putative dynamo operated, solidified.

None of the atmospheres of the terrestrial planets are thought to be primordial. The atmosphere of Mercury is virtually nonexistent. The small amount of sodium that does exist has been released from the surface rocks by sputtering (erosion produced by the impact of energetic charged particles). Surface temperatures on Mercury range from 510°C on the sunlit side to −210°C on the dark side. Venus has a dense, hot atmosphere that consists largely of carbon diox-

ide (96%) and nitrogen (3.5%). The clouds in the upper atmosphere consist of sulfuric acid and circulate with speeds that reach 225 mi/hr. Venus has a surface pressure of 90 atm, that is, a pressure 90 times greater than that at the surface of the earth, and a surface temperature of 482°C, hot enough to melt lead. The high surface temperature on Venus is caused by the "greenhouse" effect: the trapping of incoming thermal radiation by the thick atmosphere.

The atmosphere of the earth consists largely of nitrogen (77%), oxygen (21%), and a small, but important quantity of water (1%). Mars has a thin atmosphere, consisting largely of carbon dioxide (95%), nitrogen (2.7%), and argon (0.93%). The surface pressure of Mars is only 0.007 atm, and the mean surface temperature is as low as −55°C. The Martian air contains only 0.001 the water vapor that exists in the earth's atmosphere. Under the present climatic conditions, water cannot exist in the liquid state on Mars. However, the northern polar cap is made largely of water-ice, with a smaller amount of "dry ice" (frozen carbon dioxide). The Viking missions to Mars returned dramatic images showing the existence of features closely resembling shorelines, gorges, riverbeds, and islands, suggesting that very large quantities of water once flowed on the planet and that the Martian atmosphere was at one time different and more substantial than the present atmosphere (see Fig. 2). Venus is at present devoid of water, but the anomalously high deuterium/hydrogen ratio in the atmosphere strongly suggests that the surface of the planet may have once harbored oceans. These important findings raise important questions about the past existence of life on these planets.

The surfaces of the terrestrial planets are also remarkably different. Mercury, the moon, and, to some extent, Mars are heavily cratered and thus retain memories of their bombardment histories. This includes both the heavy bombardment associated with the tail end of the planetary formation process and the subsequent, continual, but less dramatic pounding produced by the impact of asteroidal and cometary bodies. The earth and Venus must, of course, have had similar bombardment histories. However, on these planets, particularly on the earth, the eroding effects of wind, water, and tectonic activity remove the evidence. Nevertheless, it is now thought that the catastrophic effects of large impacts may have had a profound influence on the evolution of life on the earth. The

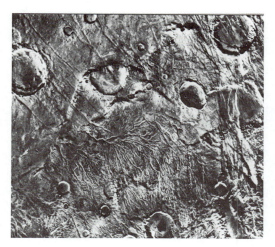

FIG. 2. A network of valleys on the surface of Mars indicates the past existence of liquid water at the surface of the planet. Climatic conditions are now such that water cannot exist in the liquid state. However, the high density of craters on the valleys indicates that the channels are very old and that the warmer, wetter climate that produced the channels may have existed 3 billion (3×10^9) years ago. The scene shown is in Thaumasia Fossae (40°S, 90°W) and is 250 km across. (Courtesy of NASA.)

existence of a large amount of iridium (an element found rarely on the earth but commonly in meteorites) in the thin layer of clay at the boundary separating Cretaceous from younger, Tertiary rocks argues that the impact of a large (diameter ~10 km) extraterrestrial object 65 million years ago may have been responsible for an extinction event that not only extinguished the dinosaurs but also removed 75% of the species existing on the earth at that time.

The earth is remarkable in that its surface shows tectonic activity on a global scale. The earth's surface is divided into "plates" that move around the surface, driven by convection currents in the upper mantle. Rocks with temperatures greater than half their melting point may act as solids to short-time-scale disturbances, the passage of a seismic wave, for example, but on time scales of millions of years hot, "solid" rocks behave more like fluids. Virtually all the large-scale features on the earth, including mountain chains like the Himalayas and the Andes, are a result of the movements and collisions of these tectonic plates. Whether such activity exists on Venus is unknown.

IV. The Moon

Before the exploration of the moon by *Apollo* and *Luna* spacecraft, it was thought that the moon was a cold, primordial object with a composition similar to that of chondritic meteorites. Theories of the moon's origin postulated that it was either captured by or co-accreted with the earth. These views have now been largely abandoned. The rocks returned to the earth revealed that the outer layers of the moon were once molten, possibly to a depth of 300 km; the moon, like the earth, is a differentiated body. Dating of lunar rocks shows that the moon was formed at least 4.4 billion years ago, and the similarity of the oxygen isotope ratios of the earth and moon argues that these bodies formed at a similar radial distance from the sun. Lunar rocks are quite unlike chondritic meteorites but have some similarities with the rocks found in the earth's mantle.

Some half a billion years after formation and after the segregation that produced the lunar highlands, giant impacts produced the mare basins, which were then flooded with lava. Production of lava ceased when the moon was ~1.5 billion years old, and since that time the moon has been largely inactive and unchanged, apart from cratering produced by sporadic bombardment. The surface of the moon is covered with a layer of fine powder and broken rubble. This regolith has a depth of 1 to 100 m and has been formed by continuous meteoritic impact. Seismometers placed on the surface of the moon during the Apollo landings have since detected numerous impacts as well as internal moonquakes. The latter are a factor of 10^{-8} less energetic than corresponding events on the earth; nevertheless, observations of the arrival and travel times of these weak seismic waves have revealed some of the moon's internal structure. The moon has a crust that varies in thickness from 100 km on the lunar farside to 60 km on the side facing the earth. Beneath the crust is a layer of denser rock (the mantle), which extends down to a depth of 800 km, the level at which the deep moonquakes occur. The structure of the deep interior is unknown; it may be partially molten, and the moon may possess a small iron core.

Interpretation of the chemical composition of the moon may reveal its origin. Lunar rocks are devoid of water and, in comparison with undifferentiated planetary material (carbonaceous chondrites), depleted in other volatiles, that is, elements that vaporize at comparatively low temperatures; sodium, potassium, lead, and gold are common examples. This could be accounted for if the formation of the moon was a highly energetic event that resulted in substantial heating of the lunar rocks. The moon is also deficient in iron, even after allowance has been made for a possible small iron core. The moon's mean density (3.344 ± 0.002 g cm^{-3}) is much lower than the uncompressed densities of the terrestrial planets, which range from about 3.7 to 5.4 g cm^{-3}. If the moon has an iron core, it could not exceed ~6% of the lunar mass, whereas the earth's iron core accounts for 32% of its mass. There is a similarity between the abundance patterns of siderophile elements cobalt, nickel, tungsten, phosphorus, sulfur, selenium, and tellurium in the earth and the moon and between their abundance patterns of lithophile elements vanadium, calcium, and manganese. The significance of these similarities is a matter of current dispute, but they suggest that the rocks that comprise the earth's mantle and the moon derived from some common event. It is now thought that the moon may have been formed from material ejected from the earth's mantle as a result of a collision between the earth and a Mars-sized planetesimal during the final stages of planetary accretion.

FIG. 3. Jupiter, its Great Red Spot, and two of its four largest satellites are visible in this photograph taken February 5, 1979, by *Voyager 1*. The innermost large satellite, Io, can be seen against the right-hand side of Jupiter's disk. The satellite Europa is seen to the right of the planet. Jupiter's colorfully banded atmosphere displays complex flow patterns highlighted by the Great Red Spot, a large, circulating atmospheric disturbance seen here in the bottom left portion of the disk. (Courtesy of JPL/NASA.)

V. The Outer Planets

The primordial solar nebula consisted largely of (1) the permanent gases hydrogen, helium, and neon, which never condense except at very low temperatures or very high pressures; (2) the ice-forming elements carbon, nitrogen, and oxygen, which in the presence of hydrogen form methane, ammonia, and water, respectively; and (3) the rock-forming elements iron, magnesium, and silicon. It is probable that the abundances of these elements matched those of the present sun. The outer planets are quite different from the terrestrial planets because they have retained more of the primordial volatile gases and ices. The fact that they are very much more massive than the terrestrial planets (Jupiter has a mass of 318 earth masses) is partly a consequence of the large abundances of the lighter elements, particularly hydrogen and helium.

Unlike the terrestrial planets, the atmospheric compositions of the giant planets (Figs. 1 and 3) are a good guide to their internal structures. The atmospheres are deep, extending to the near centers of the planets, and well mixed. The other sources of information on the internal structures are the mean densities, shapes, and external gravitational fields. The variation of the density of hydrogen with pressure is well understood, and this makes it possible to calculate the mean density of cold hydrogen bodies as a function of increasing mass. The observation that the mean density of Jupiter (1.332 g cm^{-3}) is close to that of a pure hydrogen body of similar mass is a direct indication that the planet must be composed largely of hydrogen. The hydrogen/helium ratio (by volume) in the atmosphere is observed to be 9 : 1 and is similar to that of the sun. Jupiter almost certainly has its full, solar quota (or more) of the other, less abundant elements, but these reside in the deep interior, possibly in a core. Uranus also has a solar complement of hydrogen and helium. The Saturnian atmosphere, however, is markedly deficient in helium. It is possible that helium is condensing, or "raining out," in Saturn's interior and that the associated release of gravitational energy accounts for the excessive thermal output of this planet (in this case, the release of gravitational energy on formation is not enough to account for the present heat output).

The external gravitational field of a nonrotating planet in hydrostatic equilibrium would give no clue as to the nature of its internal structure. However, slow rotation of a planet produces an axially symmetric distortion, and the size of the equatorial bulge is determined, in part, by the internal density distribution. The external gravitational field of an oblate planet has the form

$$V = - \frac{GM}{r} \left[1 - \sum_{n=1}^{\infty} \left(\frac{R_e}{r} \right)^{2n} J_{2n} P_{2n}(\cos \theta) \right]$$

where M is the mass of the planet, R_e the equatorial radius, and θ the angle between the rotation axis and the radial vector $\mathbf{r}$. The P_{2n} are the Legendre polynomials, and the J_{2n} are the gravitational moments. The oblateness f of a planet is defined by

$$f = 1 - R_p/R_e$$

where R_p is the polar radius. After the mean density, the most important constraint on the internal structure of a planet is the moment of inertia factor C (= I/MR_e^2, where I is the moment of inertia). This can be estimated from the Darwin–Radau relation, which can be written

$$C - \frac{15}{8} \left(\frac{2}{5} - C \right)^2 = J_2/f$$

Interpretation of the observed mean densities, shapes, gravitational moments, and atmospheric compositions enables us to construct reliable models for the interiors of the outer planets, particularly Jupiter and Saturn. In fact, more is known, with near certainty, about the interiors of Jupiter and Saturn than about those of our near neighbors, Mercury, Venus, and Mars. The central pressures of Jupiter, Saturn, Uranus, and Neptune are, respectively, 100, 76, ~16, and ~16 Mbar (1 Mbar = 10^6 atm), and the central temperatures are estimated to be 20,000, 10,000, 7,000, and 7,000 K. The enormous pressures ensure that the material in the deep interiors is highly compressed. However, the planets are fluid throughout, except for rocky cores that may be solid.

The outer layers of Jupiter and Saturn are composed of molecular hydrogen and helium.

At pressures greater than 3 or 4 Mbar, which are reached at 0.76 and 0.40 planetary radii in Jupiter and Saturn, respectively, the density of liquid hydrogen exceeds 1 g cm^{-3}, and the electrons are no longer bound to individual protons. The hydrogen is transformed to the metallic state, and electric currents generated in the electron sea by convective motions associated with the transport of thermal energy produce huge magnetic fields of these planets. The magnetic fields external to the planet maintain large magnetospheres of charged particles, and the rapid motions of the latter up and down the magnetic flux tubes give rise to the decimetric and decametric radio signals that emanate from these planets. There are wide variations in the orientations of the magnetic fields with respect to the rotation axes, ranging from nearly zero for Saturn to 60° for Uranus. The Uranian magnetic dipole is also offset from the center of the planet by 0.3 planetary radii. Pressures in the outer layers of Uranus are not high enough for the hydrogen to be metallic, and it is likely that the planet's magnetic field is generated in a weakly ionized fluid layer of water with additions of ammonia and methane. Whether Neptune has a magnetic field is not known at present, but it would be surprising if it did not.

The internal structures of Uranus and Neptune are not as well established as those of Jupiter and Saturn. They probably consist of an outer, fluid layer of molecular hydrogen and helium, a deeper, fluid ice layer, and silicate–iron cores that are probably solid.

VI. Satellites

In the mid-1970s the number of known satellites in the solar system was 32. Since the exploration of the outer solar system by the *Pioneer* and *Voyager* spacecraft, the number has increased to 54 and continues to increase as more of the data are analyzed. We are now able to recognize some of the important processes that either had a part in the formation of these satellites or determined their subsequent evolution. These processes include the role and mechanism of satellite capture, tidal heating due to the damping of orbital eccentricities, and the role of cometary bombardment. Exploration has revealed a variety of worlds that was previously unimagined. Satellites are very much more active, and in some cases volcanic, than we previously supposed; others may have oceans of liquid hydrocarbons or liquid nitrogen.

Satellite capture typically involves a dramatic loss of energy during a single encounter of the planet and satellite, and this can be achieved only in special circumstances. Capture is most directly achieved by a collision with an existing satellite. This mechanism is interesting in that it is not tied to any particular model of planetary formation. If a planet passed through a giant, gaseous protoplanet stage, capture due to gas drag may have been possible. Satellite capture has now been proposed for the origin of the moon (in this case, the composition of the moon is evidence against the hypothesis), the Martian satellites Phobos and Deimos, the outer, irregular Jovian satellites, the Saturnian satellites Phoebe and Iapetus, and the Neptunian satellites Triton and Nereid. A number of features of the two irregular, Jovian satellite groups positively support the gas drag hypothesis. The lack of small satellites is particularly significant, since these would have been most readily removed by gas drag. Although the orbit of the Neptunian satellite Triton is now nearly circular, this may not have been the case in the past. Tidal dissipation in the body of the satellite, due to tides raised on it by the planet, could have circularized an orbit that was initially highly eccentric. The internal heating produced by tidal friction may have been sufficient to melt Triton's interior.

One of the major results of the *Voyager* missions is the realization that cometary bombardment may have repeatedly shattered some of the satellites, the small Saturnian satellites in particular. Spacecraft images showing the surviving small satellites bearing impact craters with diameters comparable to the satellite radii (see Fig. 4) force us to conclude that these satellites have suffered impacts only marginally below the probable disruption level, implying that from time to time more disruptive collisions probably occur.

Another important process in the evolution of some satellites is tidal heating due to tides raised on a satellite by a planet. Examples include the Jovian satellites Io and Europa, the Saturnian satellite Enceladus, and possibly, but this is less certain, the Uranian satellites Miranda and Ariel. Io provides the most dramatic testimony to this process. Io's rotation is such that one face is permanently turned toward Jupiter, and to a first approximation, the tidal bulge raised on Io by Jupiter is ~15 km in height and is stationary on Io. However, there are small periodic variations in the tide height due to Io's orbital eccen-

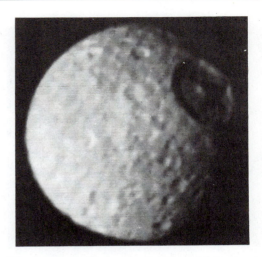

FIG. 4.   Image of the cratered surface of Saturn's satellite Mimas taken by *Voyager 1* on November 12, 1980. The prominent crater, Herschel, with a diameter of 140 km and a depth of 5 km, is the result of an impact that probably came close to shattering the entire satellite. (Courtesy of JPL/NASA.)

tricity, and these amount to 100 m. The heating produced by the repeated flexing of an imperfectly elastic solid can be large. However, tidal forces also act to circularize an eccentric orbit with the result that we would expect the orbital eccentricity and the tidal height variations to decay and the heating event to be short-lived and insignificant. (The case of Triton may be an exception.) Io is special in that its orbital eccentricity is forced to remain at the value 0.0041 by the gravitational influence of Europa, with which it is in a 2 : 1 orbital resonance. Thus, in this case, the heat source does not decay and the thermal input to the satellite from tidal friction is sufficient to drive the observed volcanism (see Fig. 5). The deposition rate of sulfur and sulfur dioxide on Io from both surface flows and volcanic plumes is observed to be ~0.1 cm yr^{-1} and is sufficient to have recirculated a mass equal to the mass of the satellite during the lifetime of the solar system.

The four Galilean satellites of Jupiter have similar sizes but show a striking decrease in mean density with increasing distance from the planet. This is probably a consequence of the strong temperature gradient that existed in the Jovian accretion disk shortly after Jupiter's gravitational collapse. Io probably has a molten silicate interior; the surface consists of sulfur and frozen sulfur dioxide. Europa has a rocky interior with an ice crust, possibly covering a

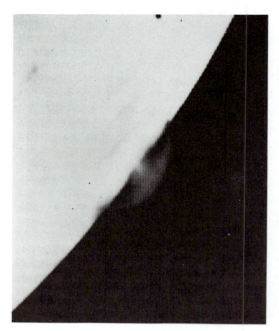

FIG. 5.   Photograph of an active volcanic eruption on Jupiter's satellite Io taken by *Voyager 1* on March 4, 1979. On the limb of the satellite can be seen one of at least four simultaneous volcanic eruptions. The observed volcanism is extremely explosive, with initial velocities of more than 2000 mi/hr (~1 km/sec); the ejecta reach heights of 60 mi or more. Several eruptions have been identified with volcanic structures on the surface of Io, which have also been identified by *Voyager*'s infrared instrument as being abnormally hot—several hundred degrees kelvin warmer than the surrounding terrain. (Courtesy of JPL/NASA.)

global ocean of water. The surface is young with almost no craters but with a global pattern of fractures. Ganymede and Callisto have deep mantles of ice and rocky interiors. The surface of Ganymede is a mixture of old, dark, cratered terrain and young, grooved terrain. Callisto has an old, heavily cratered surface.

The Saturnian and Uranian satellites are mixtures of ice and rock. Recent measurements of the shape of the Saturnian satellite Mimas have shown that it is a triaxial ellipsoid in nearly hydrostatic equilibrium and have revealed some of the internal structure of the satellite. It probably has a rocky core of 0.43 satellite radii and a deep mantle of water ice. Mimas is the only satellite in the solar system, other than the moon, whose internal structure has been determined by direct measurement. Titan has a deep atmosphere of

nitrogen with a small amount of argon and methane. The surface pressure is 1.5 bars, and the surface temperature is 95 K. At that pressure, the surface temperature is between the melting point (90.6 K) and the boiling point (118 K) of methane, and it has been suggested that the surface may be covered with a deep (~1 km) ocean of liquid methane and ethane. Absorption features in the infrared spectrum of the Neptunian satellite Triton suggest that it may have an ocean of liquid nitrogen.

Examples of the dramatic tectonic activity that has occurred on some of the small, icy satellites is evident in the composite image of the Uranian satellite Miranda shown in Fig. 6. Probable compressional (pushed together) fractures are seen in the curvilinear patterns, as are many extensional (pulled apart) faults. The grooved terrain probably developed at the expense of, or replacing, the ancient cratered terrain. Very large scarps, or cliffs, ranging in height from 0.5 to 5 km (0.3 to 3 mi), can be seen at the fracture zones. The highest scarp, approximately 5 km (3 mi) high, is three times higher than the walls of the Grand Canyon. Grooves on the face of this large scarp were probably caused by contact of the fault plates as they rubbed together (leaving what are known as slickensides).

FIG. 6.   A composite image of the Uranian satellite Miranda taken by *Voyager 2* on January 24, 1986. Compressional features are evident near the lower limb. The scarp, or cliff, on the limb at the left is 5 km (3 mi) high. (Courtesy of JPL/NASA.)

VII. Rings

Before 1979, Saturn was the only planet in the solar system known to have a ring system. Saturn's rings are both broad and bright, and if similar systems existed around any of the other planets, they would have been discovered a long time ago. However, we now know that a wide variety of ring systems can exist. All the major planets have rings, but the rings have little in common, apart from the fact that most exist close to the planet, inside Roche's limit.

Roche's limit, the distance from a planet at which the tidal and rotational forces acting on a satellite would tend to disrupt it, depends on the properties of the satellite material and on the satellite's shape. For a satellite in synchronous rotation with no cohesive strength, that is, for a satellite with a rotational period equal to its orbital period that is held together by self-gravitation alone, the shape of the satellite is that of a triaxial ellipsoid and Roche's limit, R_l, is given by

$$R_l = 2.46(\rho_p/\rho_s)^{1/3}R_p$$

where ρ_p and ρ_s are the mean densities of the planet and satellite, respectively, and R_p is the radius of the planet.

The particles of Saturn's rings are primarily icy, but there is evidence of some albedo and therefore some compositional variations within the ring system. Most of the particles are in the 1-cm to 5-m size range, but wave structures detected by the *Voyager* spacecraft are strong evidence that small satellites with radii of ~10 km also exist. The Jovian ring is optically thin and appears to contain little structure. However, because of smear motion, features smaller than ~700 km are difficult to resolve in the *Voyager* images. The Jovian ring particles are micron-sized and have short lifetimes ($<10^4$ yr) limited by erosion due to sputtering and meteoroid impacts. These particles must be replenished, probably by some source within the rings. This fact alone suggests the existence of a number of small, unseen satellites within the rings, in addition, perhaps, to the two small, dark satellites that orbit near its outer edge.

The Uranian rings have a structure that is, to some extent, the opposite of that of the Saturnian ring system. The Saturnian system consists of thousands of ringlets and contains few clear gaps, whereas the Uranian rings are narrow and widely separated. All the Uranian rings, except the innermost ring discovered during the *Voyager* flyby, which is broad and diffuse, are optically thick. Some of the rings are eccentric, inclined to the planet's equator, and nonuniform in width and have very sharp edges. The most prominent ring is the outermost, ε ring. This eccentric ring increases in width from a minimum of 22 km at pericenter (the nearest point to the planet) to a maximum of 93 km at apocenter (the farthest point from the planet). Spectra of the rings in the wavelength range 0.89–3.9 μm show their geometric albedo (a measure of the reflectivity of the particles) to lie between 0.02 and 0.03, ruling out ice-covered particles: The ring particles are black. *Voyager* observations indicate that particles in the ring have diameters of >10 cm.

The theory of shepherding satellites of P. Goldreich and S. Tremaine successfully accounts for the existence of narrow, eccentric rings with sharp edges. This theory postulates that a narrow ring is confined by the tidal torques exerted on it by a pair of nearby satellites, one orbiting interior to the ring and the other exterior to it. The discovery of two small satellites (Fig. 7) bounding the Uranian ring strongly supports this theory. The Neptunian rings are partial arcs and probably consist of

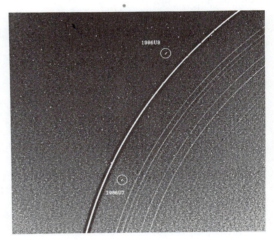

FIG. 7. Image taken by *Voyager 2* on January 21, 1986. This is the first direct observation of the Uranian rings in reflected sunlight. Evident are the two "shepherding" satellites (discovered by *Voyager 2*) that bound and confine the bright, outermost ε ring. Lying inward from the ε ring are the δ, γ, and η rings; then the β and α rings; and the barely visible 4, 5, and 6 rings. *Voyager 2* discovered two other faint rings that are not visible in this image. (Courtesy of JPL/NASA.)

particles trapped in resonances with unseen satellites. Partial arcs also appear to be evident in the *Voyager* Uranian data.

Evidence is accumulating that many, if not all, the observed planetary rings are much younger than the age of the solar system. On the other hand, more rings will probably be created in the future—some by the cometary disruption of small satellites that orbit close to a planet, others by the disruption of small satellites by tidal forces. The Martian satellite Phobos has an orbit inside the planet's synchronous orbit (the orbit for which a satellite's orbital period equals the planet's rotational period), and due to tides raised on the planet by the satellite, the satellite's orbit is decaying. It is anticipated that in 10^8 yr the satellite will be pulled inside Roche's limit, where it may (its cohesive strength is unknown) disintegrate and form a ring system.

VIII. Asteroids, Comets, and Meteorites

Most asteroids have orbits between those of Mars and Jupiter, although some have orbits that cross those of Mars and the earth (the Apollo, Amor, and Aten groups of asteroids). One asteroid, or minor planet, Chiron, discovered in 1979, orbits between Saturn and Uranus. The size–frequency distribution of asteroids can be described by a cumulative power law of the form

$$N(r) = \frac{1}{3(q-1)} \left(\frac{r_0}{r} \right)^{3(q-1)}$$

where $N(r)$ is the number of asteroids with radii $>r$ and r_0 is a constant. The present population of small asteroids (diameters <100 km) is probably the product of catastrophic collisions of larger asteroids. Observational and theoretical considerations indicate that for such a population $q = 1.837$. Since more than 500 asteroids are observed to have diameters of >50 km, it follows that there are probably 10 million kilometer-sized and more than 10^{14} meter-sized asteroids. However, these objects are widely dispersed and are not a threat to interplanetary spacecraft. About half of the known asteroids belong to Hirayama families, groups of asteroids with similar orbital elements that are probably the collision products of a single, large asteroid.

On the basis of various observed surface characteristics, particularly their geometric albedos and UBV colors [a system of stellar magnitudes that consists of measuring an object's apparent magnitude through three color filters: an ultraviolet (U) filter, a blue (B) filter, and a visual (V) filter], the asteroids have been separated into a small number of fairly distinct classes. The two most populous classes are S (moderate albedo and red UBV color) and C (low albedo and neutral UBV color). Other, less populous classes are M, E, D, A, F, and P. Asteroids in the S and C classes are considered to be the most likely sources of, respectively, the abundant stony-iron and the carbonaceous chondritic meteorites. The distribution of asteroidal types is ordered with a tendency for C-type asteroids, which are water rich, to have orbital radii greater than those of S-type asteroids. This distribution is probably original and relates to the temperature and pressure conditions in the primordial solar nebula out of which the asteroids accreted.

Comets are mixtures of ice and dust that formed at low temperatures, probably at the outer reaches of the solar system. The most widely accepted cometary model is the "dirty snowball" model of F. L. Whipple. This term is now known to be slightly misleading. First, unlike the conventional snowball, a typical comet is probably a 50:50 mixture of dust and ice. Second, recent observations have shown that cometary nucleii are black. The diameters of a few cometary nucleii have been measured by radar; sizes range from 1 to 10 km. The recent spacecraft encounter with what must be regarded as the most famous comet, that named after E. Halley revealed a black, potato-shaped object with dimensions of 8, 8, and 15 km (Fig. 8).

Most comets reside in the Oort cloud. This is a spherically symmetric distribution of, perhaps, 10^{12} comets that surrounds our solar system and extends to distances comparable to those of the nearest stars, that is, 20,000–100,000 AU. At these distances, comets are subject to stellar perturbations with the result that occasionally a comet is deflected into the inner solar system, where it is then detected. Comets that reenter the planetary region are subsequently either dynamically ejected into interstellar space on hyperbolic orbits or captured into short-period orbits as a result of perturbations by Jupiter and

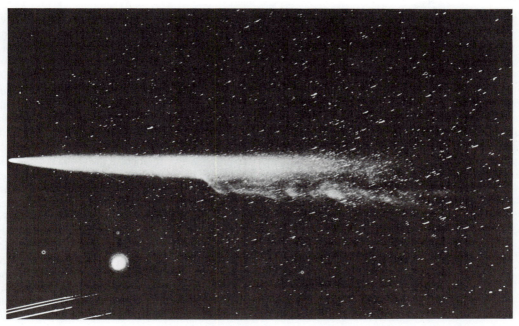

FIG. 8.   Halley's comet as photographed at Lowell Observatory in May 1910. The bright object beneath the comet is the greatly overexposed image of the planet Venus. The streaks at the lower left are the Flagstaff, Arizona, city street lights. The recent flyby of the comet by Soviet, Japanese, and European spacecraft revealed a potato-shaped body, $15 \times 8 \times 8$ km in size, that rotates with a period of 53 ± 3 hr. The surface is black. (Courtesy of Lowell Observatory.)

Saturn. While close to the sun, the icy component of a comet evaporates. Dust particles expelled from the nucleus are blown away, more or less directly outward from the sun, by solar radiation pressure to form an elongated tail seen in scattered sunlight. A second type of cometary tail is also formed—a tail of ionized gases accelerated away from the nucleus by the solar wind's magnetic field. This plasma tail is made visible by the emission of light as the ionized gas relaxes to lower energy states.

Comets trapped in short-period orbits gradually decay to form, in some instances, meteor streams, collections of dust particles with nearly identical orbital elements that intersect the orbit of the earth. These streams give rise to annual meteor showers. The Infrared Astronomical Satellite (IRAS) detected an object, 1983TB, with an orbit identical to that of the Geminid meteor stream, which impacts the earth every December 14. This object is less than 2 km in diameter and is almost certainly a spent comet. Its orbit is also similar to some earth-crossing asteroids, suggesting that some of these asteroids may in fact be defunct comets.

Comets may be icy planetesimals that were formed in the vicinity of Uranus and Neptune and were gravitationally ejected into the Oort cloud as a result of close planetary encounters. Other possible theories, however, favor an origin for comets in orbits outside that of Pluto. According to these theories, comets may be frozen, unaltered samples of the original interstellar material out of which the solar system formed.

Meteorites are rocky objects that survive passage through the atmosphere to land on Earth: "stones from the sky!" Approximately a half-dozen observed falls are recovered each year, and another one or two dozen are found that were not observed to fall. Three meteorites, Pribram, Innisfree, and Lost City (meteorites are named after the location at which they are found) were photographed from more than one point during atmospheric entry. Their orbits have been calculated from the observed meteor trails and traced back to the asteroid belt. It is thought that asteroidal fragments injected into the Kirkwood gaps as a result of collisions are then acted on by Jovian gravitational perturbations (objects in a Kirkwood gap may be in resonance with Jupiter) with the result that from time to time their orbital eccentricities reach values high enough for the fragments to encounter the earth.

A large number of meteorites have been recovered from ice fields in Antarctica. Two of these meteorites have been shown to be lunar rocks. Both appear to be from locations on the moon not visited by the Apollo astronauts or the *Luna* landers; thus, we have the exciting possibility of extending our exploration of the moon by "down to earth" means. These new findings have also revived the hypothesis that the SN [Shergotty (India), Nakhla (Egypt), and Chassigny (France)] meteorites were blasted off the surface of Mars.

IX. Origin of the Solar System

The starting point of most modern theories of the origin of the solar system is the formation and evolution of the primordial solar nebula out of which all the bodies in the solar system, including the sun, accreted. Such a system is expected to develop because interstellar clouds possess angular momentum, which is largely conserved during the process of gravitational collapse. The rotation of the cloud tends to prevent collapse except along those directions parallel to the rotation axis, and thus the formation of an extended flattened disk is a natural consequence of the collapse process. In recent years, these ideas have received dramatic confirmation. The Infrared Astronomical Satellite (IRAS) detected disks of solid particles around several nearby stars, including Vega, Fomalhaut, and Beta Pictoris. The disk around Beta Pictoris has also been detected by ground-based optical telescopes and shown to have a radius of ~400 AU, that is, about 100 times the size of the present solar system.

Frictional processes that arise because of internal shear motions (angular velocity within the disk decreases with increasing distance from the center) result in the redistribution of mass and the developement of a large radial temperature gradient. This much is common to most theories of planet formation, but there are two schools of thought as to how planet formation proceeds.

These mechanisms are not mutually exclusive, and both may have occurred in different parts of the accretion disk. According to the first theory, dust settles to the central plane of the disk, and gravitational instabilities in the dust layer lead to the formation of kilometer-sized planetesimals. Gravitational perturbations of the orbits of the planetesimals result in collisions and growth until only a few bodies, "the planets," remain. In the outer, cooler regions of the disk the solid bodies are able to accrete the more volatile constituents of the nebula, hydrogen, helium, and the ice-forming elements carbon, oxygen, and nitrogen. This scenario is likely if the solar nebula was not particularly massive. In a massive nebula of two solar masses or more, hydrodynamic instabilities may have led to the formation of giant, gaseous protoplanets. According to this theory, the main difference between the inner and the outer planets is that the inner ones had their gaseous envelopes stripped from them, perhaps by a tremendous burst of matter and radiation from the sun.

BIBLIOGRAPHY

Beatty, J. Kelly, O'Leary, Brian, and Chaikin, Andrew, eds. (1982). "The New Solar System," 2nd ed. Cambridge University Press, New York.

Black, David C., and Matthews, Mildred Shapley, eds. (1985). "Protostars and Planets II." University of Arizona Press, Tucson.

Burns, Joseph A., and Matthews, Mildred Shapley, eds. (1986). "Satellites." University of Arizona Press, Tucson.

Dermott, Stanley F., ed. (1978). "Origin of the Solar System." Wiley, New York.

Elliot, James, and Kerr, Richard. (1984). "Rings." MIT Press, Cambridge, Massachusetts.

Gehrels, Tom, ed. (1979). "Asteroids." University of Arizona Press, Tucson.

Gehrels, Tom, and Matthews, Mildred Shapley, eds. (1984). "Saturn." University of Arizona Press, Tucson.

Greenberg, Richard, and Brahic, Andre, eds. (1984). "Planetary Rings." University of Arizona Press, Tucson.

Murray, Bruce, Malin, Michael C., and Greeley, Ronald. "Earthlike Planets." Freeman, San Francisco.

Wilkening, Laurel L., ed. (1982). "Comets." University of Arizona Press, Tucson.

SOLAR SYSTEM, MAGNETIC AND ELECTRIC FIELDS

C. T. Russell   *University of California, Los Angeles*

GLOSSARY

Adiabatic invariants: First, second, and third adiabatic invariants are conserved quantities associated with the three periodic motions (gyro, bounce, and drift) of charged particles trapped in a magnetic mirror configuration, such as the Earth's dipolelike field.

Bow shock: Collisionless shock wave in the solar wind plasma that stands in the flow, slows and heats the flow, and deflects it around all planetary obstacles.

Corotational electric field: Electric field in a planetary magnetosphere associated with the rotation of the plasma at the same angular rate as the planet because of the high electrical conductivity of the plasma along magnetic field lines.

Debye length: Electrical shielding length in a plasma. A test charge in a plasma can not be sensed beyond this distance.

Geomagnetic storm: Period of several days in which currents circling the Earth in the equatorial plane of the magnetosphere become enhanced. The energization of these currents is caused by changes in the solar wind and interplanetary magnetic field.

Gyrofrequency: Number of times per second that a charged particle orbits a magnetic field line. Depends directly on the particle's charge and magnetic field strength and inversely as the mass of the particle.

Gyroradius: Radius of orbit of charged particle in a magnetic field. Depends directly on particle mass and velocity perpendicular to the magnetic field and inversely as the charge and magnetic field strength.

Interplanetary magnetic field: Magnetic field of the solar wind carried out from the Sun by the solar wind flow.

Ionopause: Upper boundary of an ionosphere that interacts directly with the solar wind.

Ionosphere: Ionized part of the atmosphere of a planet.

Magnetohydrodynamics: Physics of magnetized electrically conducting fluids. Often applied to plasmas in situations in which they display fluidlike behavior.

Magnetopause: Outer boundary of a magnetosphere confined by the solar wind.

Magnetosheath: Shocked plasma behind a planetary bow shock flowing around the planetary magnetopause or ionopause.

Magnetosphere: Magnetic cavity formed by the interaction of the solar wind with a planetary obstacle.

Magnetotail: Long cylinder of magnetic field lines dragged out in two oppositely directed tail lobes behind a planetary obstacle.

Plasma: Gas, fully or partially ionized, having equal densities of electrons and ions in which the energy of motion of the particles exceeds that associated with the electric potential of the charged particles. In an unmagnetized plasma, particle motions are nearly straight lines.

Plasma frequency: Natural oscillation frequency of a plasma set into motion by a small separation of the electrons and ions.

Plasma parameter: Number of electrons in a cube whose side is a Debye length. "Collective" plasma behavior occurs when this number is greater than unity. It also is a rough measure of the number of plasma oscillations between inter-particle collisions.

Reconnection: Process in which the magnetic topology of the magnetic field in an element of plasma changes. For example, at the dayside magnetopause, the magnetic field lines in the post-shock solar wind plasma become linked with the terrestrial field lines and accelerate the plasma on the newly joined field lines.

Solar wind: Supersonically expanding upper atmosphere of the Sun that because of its high electrical conductivity carries the solar magnetic field with it.

Substorm: Disturbance in the night-time current systems in the terrestrial magnetosphere usually lasting a couple of hours and accompanied by enhanced auroral activity.

Sunspot cycle: Eleven-year cycle in which cool, dark, magnetized regions seen in the photosphere become first more frequent and then less frequent. Because the magnetic polarity pattern of sunspot groups changes every 11 years also, the cycle is in actuality a 22-year cycle. This period is not exactly constant but varies slightly from cycle to cycle.

Tail lobes: Two regions of a magnetotail of oppositely directed magnetic flux in which the magnetic energy density greatly exceeds the plasma pressure.

The study of solar system electric and magnetic fields includes the investigation of both the magnetic fields generated by electrical currents flowing in the interior of the Sun and the planets, as well as electric and magnetic fields and their associated current systems, flowing in both the solar wind and in the magnetospheres and the ionospheres of the planets. Magnetic fields are generated by electric currents that arise when more particles of one charge flow in a particular direction per unit time than particles of the opposite sign. Such currents can flow in the highly electrically conducting fluid cores of the planets, which are both differentially rotating and convecting. Such currents are also found in the various plasmas of the solar system. These plasmas, or electron–ion gases, are highly electrically conducting. The material inside the Sun, its at-

mosphere, the expanding solar wind and ionized material in the upper atmospheres of all the planets are plasmas. One of the more important of these current systems is the one that generates the solar magnetic field. Finally, there are currents at the atomic level in solid materials. While our usual exposure to magnetic fields of this kind is from man-made magnets, nature also produces magnetized materials. Such remanent magnetization often preserves a record of the magnetic field at the time of the formation of the material and thus the study of magnetized rocks has led to an understanding of the magnetic history of both the Earth and the Moon. Magnetic fields have both magnitude and direction. The most common instrument for sensing this direction is the compass. The scientific instrument for measuring magnetic fields is called a magnetometer. There are several different types of magnetometers in common use. Proton precession magnetometers measure total field strength and are frequently used in terrestrial studies. Fluxgate magnetometers measure components of the magnetic field in a particular direction and are most commonly used on spacecraft. Solar magnetic fields are sensed remotely through the splitting of spectral lines by the Zeeman effect. Magnetic fields are measured in gammas, gauss, and teslas ($1 \text{ T} = 10^4 \text{ G} = 10^9$ gammas). The magnetic field on the surface of the Earth at the equator is about 0.31 G. In the outer reaches of the Earth's magnetosphere it is about 0.001 G or 100 gammas. Outside the magnetosphere in the solar wind it is about 10 gammas. The magnetic field on the surface of the Sun is highly variable but on average it has a magnitude of several Gauss.

Electric fields are generated by the separation of positive and negative charges. Perhaps the most spectacular naturally occurring separation of charges is produced in convecting clouds leading to lightning discharges. Since electrons and ions differ greatly in mass they often react quite differently to plasma effects. The different reactions also lead to charge separations in plasmas. If an electric field is applied to a magnetized plasma in a direction perpendicular to the magnetic field, the plasma will drift in a direction perpendicular to both the applied electric field and the magnetic field. This process is referred to as "**E** cross **B**" drift. Conversely, when a magnetized plasma is observed to be drifting, there must be an electric field perpendicular to the magnetic field. Clearly, if an observer is moving with the plasma, she or he sees no drift and hence in her or his frame of reference there is no electric field. Thus the electric field in a plasma is frame dependent; it depends on the velocity of the observer. Electric fields have both magnitude and direction. They accelerate charged particles in the direction of the electric field. The energy gained by a particle moving in an electric field is the particle's charge times the electric field times the distance travelled. The electric field times the distance is called the potential drop or potential difference between two points. An electric field is most often detected by measuring the electrical potential difference between two points and dividing by the distance between them. Spacecraft designed to measure electric fields in space often carry long (up to 200 m) antennas. The common units of electric fields are volts/meter and statvolts per centimeter. One statvolt/cm equals 3×10^5 V/m. A typical electric field in the equatorial magnetosphere of the Earth is 0.2 mV/m. A typical electric field in the solar wind outside of the Earth's magnetosphere is 2 mV/m.

I. The Physics of Solar System Electric and Magnetic Fields

A. SOURCES OF MAGNETIC FIELDS

There are four basic equations governing the interrelationship of charges, currents, magnetic and electric fields. These four equations are called Maxwell's laws. In an electrical resistor, current is usually directly proportional to the applied electric field. In a magnetized flowing electrical conductor, an additional term arises because of the electrical field associated with the motion of the conductor. The equation describing the relationship between the current, the electric and magnetic fields, and the flow velocity is called the Ohm's law. Maxwell's laws and Ohm's law can be combined to give what is known as the dynamo equation. This equation shows that the change in a magnetic field is the difference between the resistive decay of currents and the regeneration of the field due to fluid motion. If there is no motion in the conducting fluid, the magnetic field will decay with time. If the fluid core of the Earth froze, the terrestrial magnetic field would decay in a few thousand years. The Jovian field would decay over a few hundred million years, but the solar magnetic field decay time would be comparable

to the age of the solar system. Hence, the solar magnetic field could conceivably be in part primordial, present from the time of formation of the Sun due to the compression of prexisting interstellar magnetic fields. Such a primordial field could explain only a steady component. The principal part of the solar field reverses approximately every 11 yr in the solar magnetic cycle, which because of its effect on geomagnetic records, can be shown to have continued for at least 120 yr. The solar cycle has, by inference, probably existed for over 2000 yr because of the associated sunspot cycle that has been recognized in the available, albeit irregular, optical observations, and various proxy data, such as the fraction of the radioactive isotope of carbon, C^{14}, measured in tree rings. This varying solar magnetic field, the magnetic field of Jupiter and that of the Earth must be actively maintained by a generator of magnetic fields (i.e., a dynamo). [See SOLAR PHYSICS.]

One approach to studying dynamos is to seek patterns of motion of the conducting fluid that can generate magnetic fields. This kinematic approach is not self-consistent because it does not generate the requisite velocity field from first principles. Furthermore, the magnetic field so generated usually acts on the conducting fluid, thus altering the motion. Figure 1 shows how a planetary field might be self-regenerative if it has a conducting fluid core. The left-hand panel shows a typical planetary magnetic field with field lines confined to planes that contain the axis of rotation of the planet. Deep in the fluid core, as shown in the middle panel, this field is twisted out of these planes by differential rotation of the core. At different depths the core rotates more rapidly. This causes an azimuthal component around the rotation axis, which may become much larger than the original field sketched in Fig. 1a. We know such a differential motion, or shear, in the fluid motions exists in the Earth because features in the terrestrial magnetic field drift slowly westward. We also see differential motions in the solar photosphere. Sunspots move faster at the equator than at higher latitudes. This differential motion, thought to extend well down into the interior, is called the omega effect. Heating, such as produced by radioactive sources or by the freezing out of a solid inner core in the Earth, or nuclear fusion in the Sun, produces rising convective cells. These rising convective cells carry the azimuthal field upward as shown in the right-hand panel in Fig. 1. The rotation of the planet causes

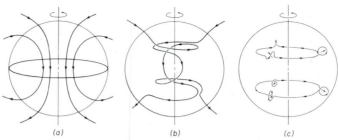

(a)   *(b)*   *(c)*

FIG. 1.   Schematic illustration of a self-regenerative magnetic dynamo. Panel (a) shows the inward extrapolation of the field observed on the surface of the Earth. Panel (b) shows how the field lines sketched in panel (a) became twisted in azimuth by the rotation of the core, which varies with depth. Panel (c) shows the effect of rising convective cells on the azimuthal field of panel (b).

these convective cells to twist and thus creates a component of magnetic field parallel to the initial field. This new component replaces any field lost due to resistive decay. In this way differential rotation and upward convection combine to regenerate the planetary magnetic field. This process is thought to act in the Earth, Mercury, Jupiter, Saturn, the Sun, and many stars. However, the theory of this process is not yet sufficiently developed to provide a prediction of the strength of a planetary dynamo, even if the physical and chemical structure of the interior of the planet were well known.

One final source of planetary magnetic fields is natural remanent magnetization of solid materials. The most common means of acquiring such natural remanence is the cooling in an external magnetic field of magnetic material through one or more "blocking" temperatures below which the material can retain its acquired magnetization. Typical blocking temperatures are several hundred degrees Celsius and typical carriers of remanence are small particles of free iron metal and nickel and the iron oxide, magnetite, Fe_3O_4. It is often possible to determine the direction and magnitude of the ancient magnetizing field from rocks containing natural remanence. Such studies have been crucial on the Earth for demonstrating that continents drift and the ocean floor spreads and that the terrestrial field periodically but irregularly reverses its direction. Lunar rocks also possess natural remanent magnetism leading to the conjecture that the Moon once possessed its own dynamo generated internal magnetic field that ceased operating over 3 billion years ago.

B. Charged Particle Motion in Electric and Magnetic Fields

In order to understand the structure and behavior of solar system magnetic and electric fields, it is necessary to understand how charged particles move in these fields, for, to a large extent, these electric and magnetic fields arise self-consistently from these same particles. As shown in the left-hand panel of Fig. 2 charged particles gyrate around magnetic field lines. The frequency of this motion is proportional to the strength of the magnetic field and the charge on the particle and inversely proportional to the mass. A proton in a 100 gamma or nanotesla field gyrates around the field 1.5 times per second. An electron in a 100 gamma field gyrates 2800 times per second. The direction of rotation of the charged particle depends on the sign of the charge. Positively charged particles gyrate in a left-handed sense, clockwise, viewed with the

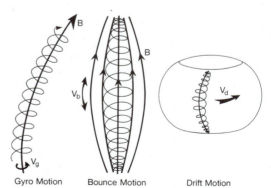

Gyro Motion   Bounce Motion   Drift Motion

FIG. 2.   The three periodic motions of charged particles: gyro-motion, bounce motion, and drift motion.

magnetic field pointing toward the observer. Electrons rotate in a right-handed sense. These circulating charged particles carry a current around the magnetic field in such a direction to reduce the magnetic field strength. The radius of the orbit of a charged particle, its gyroradius, is proportional to the velocity of the particle perpendicular to the field and the mass of the particle, and inversely proportional to its charge and the magnetic field strength. The area enclosed by this orbit times the current due to the gyration of a particle (i.e., its charge times its gyrofrequency) is called the magnetic moment of the particle. It is equal to the energy associated with the motion of the particle perpendicular to the magnetic field divided by the magnetic field strength. If the variation in magnetic field strength is sufficiently slow in space or time the magnetic moment of a particle is conserved. The magnetic moment is also called the first adiabatic invariant.

If magnetic field lines converge as they are shown to do in the middle panel of Fig. 2, the field strength, as seen by the particle, increases as the particle goes from the middle to either end of the field line. Because the magnetic moment of the particle is conserved, the perpendicular energy of the charged particle will increase until it is equal to the total energy of the particle. Because at the point where this occurs the particle has no forward velocity, and because the same forces are still acting on the particle that decelerated its forward motion, it reflects and starts moving away from this "mirror point". A planetary magnetic field generally has two points of convergence, one in the northern hemisphere and one in the south. Thus, particles will bounce back and forth trapped between mirror points. The bounce frequency is determined by how fast the particles are moving along the magnetic field line and by the length of the magnetic field. The parallel momentum of a charged particle is its mass times its velocity measured along the magnetic field. The integral of the parallel momentum over a bounce period (i.e., the momentum summed over each portion of the path) is conserved. This quantity is called the second adiabatic invariant. If mirror points of a particle move closer together, the particle must gain energy along the field due to this conservation principle. This process is called Fermi acceleration.

In nonuniform magnetic fields, charged particles drift perpendicular to the magnetic field. If the magnetic field strength decreases with altitude as it does in the terrestrial magnetic field, the gyroradius of a particle will be larger when it is farther from the Earth in its gyromotion than when it is closer. This leads to a net drift of the particle whose direction depends on the charge of the particle. In the Earth's magnetic field, which in the equatorial regions points northward, positively charged ions such as protons drift westward opposite to the Earth's rotation and electrons drift eastward. This process is referred to as gradient drift. If a particle is bouncing back and forth along a curved field line as shown in the right-hand panel of Fig. 2, it will experience a centrifugal force outward as its motion is deflected by the curved field. This has exactly the same effect as a field gradient and electrons and protons drift in opposite directions. The drift path around a planetary magnetosphere is generally closed for most nonrelativistic particles in most planetary fields if the planetary field approximates that of a dipole. This drift requires from minutes to hours, compared to seconds to minutes for the bounce motion, and milliseconds to seconds for the gyromotion. When a charged particle completes its motion around a planetary magnetosphere the path traced out by the drifting bouncing particle is a roughly spheroidal shape with open ends. This so-called drift path or shell encloses a certain amount of magnetic flux, or equivalently a certain number of magnetic field lines. The magnetic flux enclosed by the drift path, or equivalently the magnetic flux through the open ends, is conserved when the magnetic field changes slowly on the time scale of the drift motion. This conserved amount of magnetic flux is referred to as the third adiabatic invariant. If the magnetic field of the Earth, for example, increased then the radiation belts would move outward to conserve this invariant. The gradient and curvature drift of charged particles around the Earth's equator cause a net current to flow there, called the ring current. The ring current causes a depression in the strength of the magnetic field observed on the surface of the Earth. The size of this depression is linearly related to the energy of the ring current particles. If the particles in the terrestrial trapped radiation belts possessed 4×10^{15} J of kinetic energy, there would be a 100-nT depression with the horizontal component of the Earth's surface field. A geomagnetic storm is such a period of enhanced ring current.

A charged particle is accelerated by an electric field. In the case of an electric field at right angles to a magnetic field, a gyrating particle will be accelerated for one-half of its gyration and decelerated in the other half. When it is moving fastest, its gyroradius is largest and when it is moving slowest its gyroradius is smallest. Thus, it moves farther in one-half of its gyro-rotation than the other. Protons and electrons gyrate in opposite directions about the field and also are accelerated on opposite halves of their gyration. The net effect is that electrons and protons drift in the same direction and at the same velocity perpendicular to both the magnetic and electric fields. In the Earth's magnetic field, which is northward in the equatorial regions, an electric field from dawn to dusk produces drift toward the Sun. Since electrons and ions drift together, there is not current associated with this drift.

C. THE PHYSICS OF PLASMAS

Thus far we have considered only single particle motion. However, a gas of charged particles can exhibit collective behavior. The term plasma is usually restricted to a gas of charged particles in which the potential energy of a particle due to its nearest neighbor is much smaller than its kinetic energy. If we put a test charge in a plasma it gathers a screening cloud of oppositely charged particles around it that tends to cancel the charge. In effect, the membership of a given particle in many screening clouds produces the collective behavior. Beyond some distance, called the Debye length, there is no observable effect of an individual charge. The Debye length is proportional to the square root of the ratio of the plasma temperature to its density. A plasma such as the solar wind with a temperature of 10eV and a density of ten per cubic centimeter has a Debye length of 740 cm. The number of particles in a cube with the dimensions of the Debye length, is called the plasma parameter. The value of this parameter determines whether the ions and electrons can be treated as a plasma exhibiting collective behavior or as an ensemble of particles each exhibiting single particle behavior. We can consider an electron–ion gas to be a plasma when the plasma parameter is much greater than 1. As illustrated in Fig. 3, in the Earth's ionosphere and the Sun's outer atmosphere (the corona) this number is about 10^5. In the Earth's magnetosphere and in the solar wind it is about 10^{10}. The characteristic frequency of a plasma at which

the plasma would oscillate if the ions and electrons were pulled apart and allowed to move back together (a collective plasma effect) is called the plasma frequency. It is equal to 9 kHz times the square root of the number of electrons per cubic centimeter. The maximum plasma frequency in the Earth's ionosphere is somewhat greater than about 10 MHz. The number of collisions per second in a fully ionized plasma is very roughly the plasma frequency divided by the plasma parameter. Thus, particles in plasmas can oscillate many, many times between collisions, and hence, plasma processes are often referred to as collisionless processes. When the plasma parameter is large, charged particles move in almost straight lines. As the plasma parameter decreases the individual interactions between charged particles become more important and large angle deflections become more frequent. Eventually, for small plasma parameters, electrons become trapped in the potential wells of individual ions. This same effect occurs in metals and the interior of the Sun.

The key to understanding the behavior of the electric and magnetic fields in the solar system lies in understanding the behavior of plasmas and the various instabilities that transfer energy from one form to another in a plasma. The collective interactions allow us often to ignore the individual particle nature of a plasma and consider it to be an electrically conducting fluid. The laws governing the behavior of this fluid are known as the Maxwell equations and Ohm's law together with the conservation of mass and momentum. Use of these equations is called the magnetohydrodynamic, or MHD, approximation. Plasmas in the solar system often find themselves in unstable situations in which the

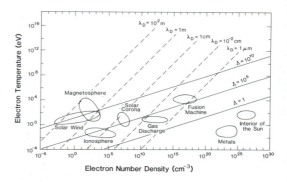

FIG. 3.   The Debye length, λ_D, and the plasma parameter, Λ, for plasmas of geophysical interest.

MHD equations would predict a rapid change of configuration. For example, there is the interchange instability in which entire magnetic flux tubes interchange position because by doing so they acquire a lower energy state. This might occur if a magnetic flux tube lay on top of a lightly loaded magnetic flux tube in a gravitational field. This is analogous to the situation in which a heavy fluid sits on top of a lighter one. Another MHD instability of some importance is the Kelvin–Helmholtz or wind-over-water instability, in which surface waves on a boundary are induced by a shear in the flow velocity at the boundary. This instability is often invoked to explain magnetic pulsations in the Earth's magnetosphere. It is analogous to the mechanism for the formation of ocean waves. Plasmas, because they are collisionless, can have different temperatures along and across magnetic field lines. However, too large a difference can be unstable. The fire hose instability arises when the thermal pressure along a field line exceeds the sum of the pressure across a field line and the magnetic pressure. When this occurs the magnetic field lines wiggle back and forth like a fire hose whose end is not being held. An instability also occurs when the thermal pressure across a field line greatly exceeds the pressure along it. This is known as the mirror instability and it creates pockets of denser plasma along the field lines.

Many of the various plasma instabilities cause fluctuations in the plasma: oscillations in number density, in electric and magnetic field strength and direction, in the velocity of the plasma, etc. Such oscillations can also be stimulated by processes external to the plasma being studied. There are three types of magnetohydrodynamic waves: fast, intermediate, and slow. Fast waves compress both the plasma and magnetic field. Intermediate waves bend the magnetic field lines but do not alter the density or magnetic field strength. Slow waves compress the field when they are decompressing the plasma and vice versa so that the total pressure, thermal plus magnetic, is nearly but not quite constant. Such waves are found throughout the solar system plasma. The velocity of an intermediate wave, also called the Alfven velocity, is proportional to the magnetic field strength and inversely proportional to the square root of the mass density. In a plasma of 9 protons per cubic centimeter and a field strength of 6 gammas, or nT, which is typical of the solar wind near Earth, the Alfven velocity is 44 km/sec. Fast and slow waves travel somewhat faster and

slower than this velocity by a factor that depends on the temperature of the plasma and the direction of propagation relative to the magnetic field.

Fast and slow waves can steepen into shock waves if their amplitudes are sufficient. A shock wave is a thin discontinuity that travels faster than the normal wave velocity and causes irreversible changes in the plasma. Such a collisionless shock is analogous to the shock produced by a supersonic aircraft flying through a collisional gas, air. The process of shock formation is described as steepening because the passage of a fast or slow wave alters the plasma so that a following wave will travel faster and thus catch up with the first wave. Thus the trailing parts of a wave catch up with the leading part and a steep wavefront is formed. This process is similar to the steepening of ocean waves as they approach the shore. The steepening of shock waves in a plasma is limited by collective processes occurring in the plasma over sometimes rather short scale lengths, such as the ion gyroradius. Many of the processes occurring in collisionless shocks generate plasma waves, both electromagnetic waves that transmit energy, and electrostatic waves that do not. These waves, oscillating at and above frequencies in the neighborhood of the gyrofrequencies of the ions and electrons, heat the plasma and equalize the pressures along and across the field. The thickness of collisionless shocks is often close to the ion inertial length, which is the velocity of light divided by the ion plasma frequency (a factor of 43 less than the electron plasma frequency mentioned earlier if the ions are protons). Numerically, this is equal to 228 km divided by the square root of the number of protons per cm^3. In a typical solar wind plasma near the Earth a shock will be about 80-km thick.

Collisionless shocks are found in front of all the planets, forming standing bow waves much like the bow wave in front of a boat. Such a wave occurs because the solar wind must be deflected around each of the planetary obstacles. However, the velocity of the solar wind far exceeds the velocity at which the pressure needed to deflect the flow can propagate in the solar wind. The only means by which a planet can deflect the supersonic solar wind is to form a shock wave that slows down, heats, and deflects the flow. Solar flare initiated blast waves also cause shocks in the solar wind, which are convected out past the planets. When they reach the Earth, such shock waves cause sudden compressions in the Earth's magnetic field that are observed by ground stations. A sudden compression followed by an injection of energy into the Earth's ring current is called a sudden storm commencement or SSC. Otherwise a sudden compression of the Earth's magnetic field is called a sudden impulse or SI. Sometimes shock waves propagating towards the Sun, or reverse shocks, are carried outward by the very supersonic solar wind. These can cause negative sudden impulses in which the Earth's surface magnetic field suddenly decreases. It was the occurrence of the negative and positive sudden impulses in the Earth's magnetic field that originally led to the postulate that collisionless shocks could exist in the solar wind plasma. In ordinary gases, shocks require interparticle collisions to heat the gas across the shock front. In a collisionless plasma this heating occurs both through oscillating magnetic and electric fields and through a steady-state process by which a small fraction of the ions get reflected by the shock and thus attain a high thermal energy relative to the flowing solar wind.

D. Conductivity and Electric Field Sources

As noted above, in a plasma without collisions, an electric field perpendicular to a magnetic field causes a drift of both the electrons and the ions in the same direction and hence, no electrical current. In the dense plasmas of planetary ionospheres and the solar photosphere, collisions either with other charged particles or with neutral particles occur frequently enough that the collisions modify the response of the plasma to an applied electric field. When collisions occur much more rapidly than either the electron or ion gyrofrequencies, the charged particles are no longer controlled by the presence of the magnetic field. The charged particles are then said to be unmagnetized and the electric field drives a current parallel to the electric field as in an ordinary conductor. The ions are unmagnetized at intermediate collision frequencies, because of their lower gyrofrequency. They drift parallel to the electric field, while electrons, because their gyrofrequency is much higher, are magnetized and drift perpendicular to the magnetic field. Thus, there is a component of current carried by electrons perpendicular to the applied electric field and a component of current carried by ions parallel to the electric field. The ratio of proportionality between the current

and the applied electric field is called the conductivity of the plasma. The conductivity perpendicular to the magnetic field and parallel to the electric field is called the Pederson conductivity. The conductivity perpendicular to both the magnetic and electric fields is called the Hall conductivity. The conductivity parallel to the magnetic field is called the direct or longitudinal conductivity. At some altitudes in the terrestrial ionosphere the Hall conductivity greatly exceeds the Pederson conductivity and electric current flows mainly perpendicular to the applied electric field.

Collisions couple the neutral and ionized atmospheres of planets, including the Earth. Motions of either one couple into the other. Neutral winds are driven by solar heating. At low altitudes in the terrestrial ionosphere where both ions and electrons are unmagnetized these collisions simply carry both the ions and electrons along with the neutral wind and cause a drift of the plasma. At high altitudes, the collisions cause the electrons and ions to drift in opposite directions perpendicular to the direction of the wind resulting in a current perpendicular to the wind associated with the differential flow of the ions and electrons. At intermediate altitudes the electrons are magnetized and the ions unmagnetized so that there is an electron current perpendicular to the wind and the magnetic field, and an ion current parallel to the wind. Thus, neutral winds as well as applied electric fields can drive currents. These currents cause variations in magnetic records taken on the surface of the Earth. Variations associated with the normal solar-heating driven convection of the ionosphere have been called S_Q variations because they are in phase with the Sun and are present on quiet days. Daily variations at geomagnetically active times are called S_D variations. Variations caused by the effects of lunar gravitation on atmospheric circulation at quiet times are called L_Q variations. The S_Q variations are caused by both solar gravitational and heating effects whereas L_Q variations are due to gravitational tides only.

The electrons and ions that surround a planet and comprise its ionosphere can be set in motion from above as well as below. The expanding solar atmosphere, or solar wind, flows by the planets at an exceedingly high velocity, about 400 km/sec on the average. Despite the fact that this solar wind is very tenuous, it can couple into a planetary ionosphere or magnetosphere by viscously dragging on the magnetopause which couples to the ionopause, the outer boundary of the ionosphere. In a magnetized ionosphere like the Earth's ionosphere, such motion is equivalent to an electric field as discussed above. Because the parallel conductivity along magnetic field lines is quite high, there is little potential drop along the field lines and the potential drop appears across field lines in the lower ionosphere, driving currents there in the complex manner discussed in the previous paragraph. In short, drag at high altitudes in a planetary magnetosphere can cause circulation in the ionosphere. This, in turn, leads to complex current patterns at low altitudes.

Because the solar wind is magnetized, there is an additional component to the drag on a planetary magnetosphere in addition to the normal particle and wave transfer of momentum across the boundary. When the interplanetary magnetic field has a component antiparallel to a planetary magnetic field where the solar wind and planetary magnetosphere first come into contact, the two fields can become linked in a process called reconnection. This process increases the drag on the magnetized planetary plasma, leads to a long magnetized tail behind the planet, and causes ionospheric flow across the polar cap, which returns at lower latitudes. This process of reconnection is believed to be the primary controlling mechanism for almost all of geomagnetic activity, except for sudden impulses in ground magnetograms that are associated with shocks in the solar wind passing the Earth.

Because the Earth rotates and the terrestrial magnetic field lines are good electrical conductors, the plasma on the field lines tends to rotate with the Earth. The electric field associated with this rotation is called the corotational electric field. The motion of plasma due to the drag of the solar wind, which at high latitudes is directed away from the Sun over the polar caps, is toward the Sun at lower latitudes. The electric field associated with this motion is called the convection electric field. The combination of the two fields produces a plasma circulation pattern in the Earth's equatorial magnetosphere as shown in Fig. 4. In this figure the large arrow labeled E represents the direction of the convection electric field. It is perpendicular to the magnetic field, which is out of the page and to the direction of flow, or convection, of the plasma, which is indicated by the streamlines labeled with smaller arrows. Most of these streamlines are open and carry low-energy plasma from the nightside of the magnetosphere to the dayside and out through its boundary. A subset of the streamlines in the inner magnetosphere is closed. In this region the plasma rotates with the Earth, both being supplied by and losing parti-

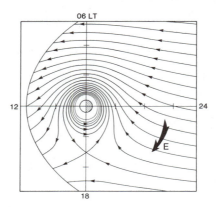

FIG. 4. Drift paths of cold plasma in the Earth's magnetic field as a result of the superposition of a uniform electric field due to the interaction of the solar wind with the Earth and the corotation electric field. Because of the strong coupling of the magnetospheric and solar wind electric fields in this model, it is called the open magnetospheric model. The drift path flow from midnight (24 hours local time, LT), principally past the dawn meridian at 6 LT, and out through the magnetopause near noon, 12 LT. The electric field direction, E, is perpendicular to the streamlines.

cles to the Earth's ionosphere along magnetic field lines.

II. Plasma Interactions with Unmagnetized Atmosphereless Bodies

The simplest interaction of a plasma with a solar system body is that with an unmagnetized atmosphereless body as in the interaction of the solar wind with the Earth's moon. The electrons and ions in the solar wind strike the lunar surface and are absorbed, leaving information about the solar wind energy implanted in the lunar soil but undergoing no esoteric plasma processes. This process leaves a cavity behind the Moon. The solar wind attempts to close behind the Moon to fill in this cavity. As illustrated by Fig. 5, the closure depends on the direction of the interplanetary magnetic field relative to the direction of the solar wind flow. If the magnetic field is aligned with the flow, the cavity behind the Moon is filled with magnetic flux and the cavity closes only slightly, so that pressure balance with the solar wind pressure is maintained. If the magnetic field is perpendicular to the flow, then the solar wind closes slowly behind the Moon as plasma flows along field lines at the thermal speed. The thermal speed is the average random velocity of the particles relative to the drift or bulk velocity of the flow. [*See* MOON (ASTRONOMY).]

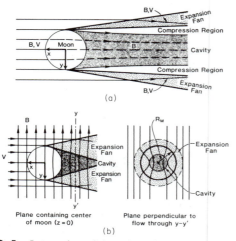

FIG. 5.   Interaction of the solar wind with the Moon for (a) flow, *V*, aligned with magnetic field, *B*, and (b) crossed flow and magnetic field. Plane containing center of Moon ($z = 0$) (left); Plane perpendicular to flow through y–y′ (right).

The Moon is not completely unmagnetized. It possesses remanent magnetization from an earlier epoch. Some regions are coherently magnetized over hundreds of kilometers and sufficiently so that they have enough magnetic pressure to deflect the solar wind when the solar wind flow is tangential to the lunar surface in the region known as the solar wind limbs, or terminator. This deflection causes a disturbance to be launched into the solar wind but it is not strong enough to be called a shock. These disturbances have been called limb compressions.

The Moon has an insulating outer shell but its interior, being much hotter, is quite electrically conducting. The size and conductivity of this highly conducting region is such that it would take at least many days and perhaps hundreds of years for a magnetic field to diffuse from outside the core to the interior of the core. Thus magnetic observations can be used to probe the interior lunar electrical conductivity. One way to do this is to use measurements obtained when the Moon is in the near vacuum conditions of the geomagnetic tail lobes. A satellite, such as one of the Apollo 15 and 16 subsatellites, flying in low-altitude lunar orbit, can measure the distortion of the field caused by its exclusion from the lunar core. A second technique is to measure the frequency spectrum of magnetic fluctuations in the solar wind on a spacecraft and then measure the frequency spectrum of magnetic fluctuations seen on the lunar surface as was done during the Apollo program. The alteration of the frequency spectrum can be used to determine the conductivity profile of the Moon. In particular, this technique can sound the conductivity profile in the cooler outer layers of the Moon. Its sensitivity is limited by the accuracy of the intercalibration of the two instruments used.

Plasmas interact with the moons of Jupiter and Saturn. Except for Titan and to a lesser extent Io, these moons have almost no atmospheres. Energetic charged particles from the radiation belts of these planets impact the surfaces of these Moons and sputter atoms from them. These atoms then become ionized and join the plasma surrounding the planets. Such sputtering also occurs in planetary rings. The collision of energetic particles with moons and rings is a significant loss process for the radiation belts of Saturn and to a lesser extent the radiation belts of Jupiter. [*See* PLANETARY SATELLITES, NATURAL.]

Dust particles, both those in rings and those not in the rings can become electrically charged.

For small dust particles, the charge can significantly alter their motions because the particles will feel the forces of the planetary electric and magnetic fields as well as of gravity. For instance, a planetary electric field is typically such as to enforce corotation of the plasma in the planetary magnetosphere with the rotation of the interior of the planet, whereas the orbital velocity of uncharged particles varies with radial distance. So too will interplanetary and cometary dust be affected by the motional electric field of the solar wind.

Finally, the interaction of the solar wind with an asteroid should resemble the interaction with the Moon except that perhaps the cavity will not be as well-defined because the asteroids are all smaller than the Moon and hence gyroradius effects become more important. If the asteroid outgasses, as an asteroid that is actually a nearly extinct comet might, then its interaction region might be quite large and its interaction processes quite different from those described above. We defer such discussions to Section V.

III. Plasma Interactions with Magnetized Bodies

There are at least five strongly magnetized planets in the solar system, Mercury, the Earth, Jupiter, Saturn, and Uranus. Each of these planets provides us with a different aspect of the interaction of a flowing plasma with a magnetic field. Mercury gives us a small magnetosphere, or magnetic cavity, both in absolute terms and relative to the size of the planet. Mercury also has no atmosphere or ionosphere. The Earth has a sizeable magnetosphere as well as a well-developed atmosphere and ionosphere. Jupiter is a rapidly rotating planet with an immense magnetosphere and a strong source of plasma deep in the magnetosphere. Saturn has a smaller but also rapidly rotating magnetosphere. However, its plasma sources are not as strong as Jupiter's sources. Uranus has a magnetic dipole axis that is at an angle of 60° to its spin axis.

A. Solar Wind Interaction with Mercury

Mercury is the smallest of the terrestrial planets with a radius of 2439 km, intermediate between the Earth's moon and Mars in size. It rotates more slowly than the Moon, rotating with a period of 59 days compared to the Moon's 28 days. It is heavily cratered like the moon but differs from the Moon in its lack of synchronicity of its rotational and orbital periods and in its

density, which is 5.4 gm/cm³ compared to the Moon's density of 3.3 g/cm³. The high density indicates that Mercury has a significant iron core. This core apparently is sufficient to sustain an active dynamo despite the small size and slow rotation of the planet. Mariner 10 passed through the nightside magnetosphere of Mercury twice, in March of 1974 and 1975, and detected a magnetic field arising from a planetary dipole magnetic moment of strength $4 \pm 2 \times 10^{22}$ Gauss-cm³. Figure 6 shows a sketch of the solar wind interaction with Mercury and the field lines in the Mercury magnetosphere. The nose of the magnetopause is only about 0.35 Mercury radii, R_M, above the surface of Mercury. The bow shock is at about 0.9 R_M above the surface at the subsolar point. The magnetic field strength on the surface of Mercury is 300–500 gammas (nT). Its instantaneous value depends on variations in the solar wind pressure relatively more than the magnetic field at the surface of the Earth depends on these variations. Variations observed in the magnetic field during the passages of Mariner 10 past the planet have been interpreted in terms of a dynamic magnetosphere controlled to a large extent by internal processes, but there are few observations on which to judge. The magnetopause and bow shock seem similar in basic properties and structure to these same terrestrial boundaries.

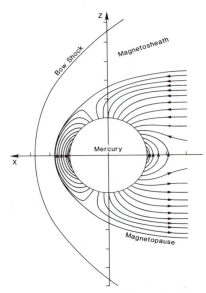

FIG. 6.   The solar wind interaction with Mercury illustrating the large fraction of Mercury's magnetosphere occupied by the planet.

B. Solar Wind Interaction with the Earth

The Earth, like Mercury, presents a magnetic obstacle to the solar wind. The Earth's magnetosphere, though, is about 20 times the size of Mercury's magnetosphere, being about 200,000 km wide at the dawn–dusk or terminator plane. Furthermore, at the feet of the field lines of the terrestrial magnetosphere there is a dynamically significant ionosphere. In front of this magnetosphere stands a bow shock wave much like that of Mercury, both in relative location and in microstructure. The closest approach of the bow shock to the center of the Earth under typical solar wind conditions is about 92,000 km or about 14.5 Earth radii. Under similar conditions the boundary of the magnetosphere (i.e., the magnetopause) lies at about 70,000 km or 10.9 Earth radii. [See SOLAR–TERRESTRIAL PHYSICS.]

The Earth's dipole magnetic moment is 8×10^{25} Gauss-cm^3, which corresponds to a surface field of about 0.31 Gauss at the equator of the Earth and twice this at the poles. There are contributions from higher-order terms such as the quadrupole and octupole terms. The relative importance of these higher-order terms diminishes with altitude so that for most purposes we can consider the field in the inner magnetosphere to be dipolar. The Earth's internal dynamo is not steady. The most noticeable surface field change is a slow westward drift, but on much longer scales, millions of years, the field can actually reverse. These reversals provide a useful clock for geophysical studies. Their existence also suggests that the Earth was not always so well shielded from the solar wind as it is today.

Figure 7 shows a moon-midnight meridian cross section of the terrestrial magnetosphere.

The solar wind, after passing through the bow shock flows through a region known as the magnetosheath and around the magnetopause. The electrons and protons in the magnetosheath near the magnetopause are reflected by the Earth's magnetic field. In this encounter with the Earth's magnetic field they turn 180° about the field line and escape from the magnetopause. The current associated with this gyro-reflection self-consistently generates the proper electric current to bound the Earth's magnetic field under ideal conditions. Thus, plasma cannot directly enter the Earth's magnetic field in the subsolar region. The magnetic field inside the boundary has a pressure equal to the dynamic or ram pressure of the solar wind outside the magnetosphere.

Behind the Earth, a long magnetic tail stretches for perhaps a 1000 Earth radii or more. If there were no viscosity in the solar wind interaction the magnetopause shape would be determined only by normal stresses perpendicular to the magnetopause, and the resulting shape would be a tear drop. Viscosity is presently thought to be supplied in part by the Kelvin–Helmholtz instability discussed above, and in part by the process of reconnection discussed below. As the flow in the magnetosheath passes above and below the polar regions, it passes a weak point in the magnetospheric field that marks the beginning of the tail. The two regions, one in the north and one in the south, are called the polar cusps. In these regions, solar wind plasma can reach all the way down to the ionosphere. Plasma can enter the magnetosphere here and form a boundary layer that has been called the plasma mantle. In a reconnecting magnetosphere the plasma can drift across the tail to a null field point, called the X-point,

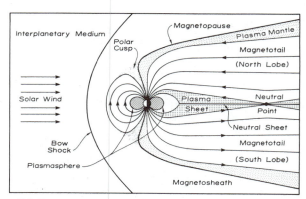

FIG. 7.   The interaction of the solar wind with the Earth.

where it is accelerated. There, some of it enters the closed field region of the magnetosphere. The process of reconnection effectively couples the electric field of the solar wind into the electric field of the magnetosphere. Since motion of magnetized conductors is equivalent to electric fields, we may also view this as the motion of the solar wind causing motion of the terrestrial plasma. However, this coupling of the two magnetized fluids cannot occur without breakdown of the very high conductivity of the fluids at least in some limited region.

The net result of this process is flow in the outer magnetosphere, but the deep interior of the magnetosphere is relatively unaffected by these flows and continues to corotate with the Earth. In this interior region, the ionospheric plasma, which is continually upwelling from the dayside ionosphere, can build up in density to a level at which the upwelling is matched by an equal downward flow. This high-density region of cold ionospheric plasma is called the plasmasphere. In this region the magnetic field is quite dipolar in character and the electric field is simply that necessary to make the plasma rotate uniformly.

The strongest coupling between the solar wind and the magnetosphere is through the process known as reconnection. This process can be steady-state, slowly varying, or highly temporally varying. First we examine the case of steady-state reconnection. Then, we examine how variations in the reconnection rate lead to a dynamic magnetosphere and finally we examine the effects of temporally and spatially patchy reconnection. Figure 8 shows the field lines in an idealized solar wind–magnetosphere interaction. In the top panel, the interplanetary, or solar-wind, field lines are in the direction opposite those in the Earth's magnetosphere. Due to a breakdown in electrical conductivity at the "nose" of the magnetosphere, the interplanetary and planetary magnetic fields join at this point. The flow of the solar wind carries the ends of these joined fields over and under the magnetosphere, pulling the magnetospheric field lines over the poles. These field lines sink in the tail until they meet in the center of the tail at the X-point where they reconnect once more. Here they form a closed field line, which touches the Earth at both ends, and also an interplanetary field line that does not touch the Earth at all. This interplanetary field line flows away from the Earth. The newly closed field line flows toward the Earth, around the Earth out of the

plane of the figure, and then joins up with a new interplanetary magnetic field line.

When the interplanetary magnetic field is northward, parallel to the Earth's magnetic field, as in the bottom panel, reconnection apparently can still occur at high latitudes behind the polar cusps. This reconnection removes flux from the magnetotail and adds it to the dayside magnetosphere if the same field line connects to both the North and South of the magnetotail. Otherwise, the magnetosphere is simply stirred by this process but no net flux transfer occurs.

The processes sketched here are steady-state. If, instead, as occurs in practice, the rates of reconnection and hence flux transfer from one region to another vary and vary differently, flux can build up in one region or another. The sequence that frequently occurs is a sudden increase in the dayside reconnection rate with a southward field in the solar wind. The magnetic flux in the magnetotail increases until the reconnection rate in the tail increases to remove the flux from the tail. This latter rate of reconnection can exceed the rate that built up the flux enhancement in the tail. This generates rapid flows and high electric fields as plasma is accelerated both toward and away from the Earth. The accelerated plasma, in addition to filling the radiation belts and adding energy to the ring current flowing in the magnetospheric equator, also powers the auroral displays. Aurora occur when energetic particles from the magnetosphere collide with neutral atoms and molecules putting

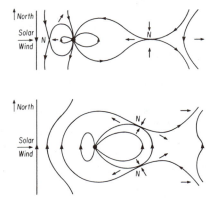

FIG. 8.   The reconnection process between interplanetary magnetic field lines and terrestrial magnetic field lines for situations in which the interplanetary magnetic field is (a) southward and (b) northward. The points labeled N are 'neutral' points in which the magnetic field strength if zero, and across which the magnetic field threading the boundary reverses sign.

their electronic shells into an excited state from which they later decay by giving off light. During these times, flows in the outer magnetosphere can well exceed the flows possible in the ionosphere at low altitudes. This can occur only if the magnetic field line has less than infinite conductivity, and therefore, electric fields can appear parallel to the magnetic field lines. In such situations, particles bouncing on flux tubes become accelerated in the "parallel" electric field, and cause further auroral displays as they excite atmospheric atoms by collision. The overall sequence of events in the Earth's magnetosphere and ionosphere at such times is usually referred to as a magnetospheric substorm.

Even this complex process is an oversimplification. In practice, the reconnection process appears to be very unsteady. This unsteadiness appears to lead to the formation of magnetic islands on both the dayside and the nightside. This unsteadiness also leads to oscillations in the magnetospheric plasma inside the magnetosphere, adding to those oscillations that may be associated with the Kelvin–Helmholtz instability. Yet another source of waves is the region of space, upstream of the Earth's bow shock yet connected to it by field lines. Beams of particles escaping from the bow shock generate waves. These waves are blown back toward the Earth by the supersonic solar wind. If the waves are generated in just the right region, they can be carried to the magnetopause and also add to the oscillations in the magnetosphere. Thus the magnetospheric field lines are almost always vibrating to some degree. Some of these vibrations can occur at resonances of the field line and large standing waves build up. Often the oscillating electric field associated with these waves far exceeds the background or steady electric field value. However, typically the magnetic amplitude is at most a few percent of the background field.

The presence of these oscillating electric and magnetic fields has very significant consequences for the trapped energetic charged particles in the Earth's radiation belts. In a dipolar field such as the Earth's, charged particles have three separable periodic components of motion: gyration around the magnetic field, bounce motion back and forth along the field, and drift around the Earth. As discussed earlier, in Section I,B, these three components of motion are associated with three conserved quantities or adiabatic invariants. These quantities remain constant unless the assumptions about the constancy of the fields in which the particles move

are violated. The lowest frequency oscillations in the magnetosphere, with periods of many minutes to hours, can violate the third adiabatic invariant, that associated with the drift of particles around the Earth causing them to diffuse inward or outward. Fluctuations with periods of seconds to minutes can resonate with the bounce motion causing particles to mirror at different points. Fluctuations at periods of milliseconds to seconds can resonate with the gyro motion of particles causing particles to have different helical paths along the field and to enter the atmosphere rather than bounce back along the magnetic field line.

A very important set of these resonant wave–particle interactions are known as wave–particle instabilities. If waves at the resonant frequency do not exist, they may be spontaneously generated by the plasma if the net effect of the waves on the particles includes a transfer of energy to the waves. For example, if an electron travelling along a magnetic field with some energy across the field and some energy along the field met a right-handed electromagnetic wave head-on, it would resonate with the wave if the particle velocity along the magnetic field caused the wave to appear to the particle to be oscillating at its gyrofrequency. This resonance would cause energy to be transferred from the parallel motion to the perpendicular motion of the electron or vice versa depending on the phase of the interaction. This resonant oscillation also would transfer energy from the wave to the particle in the former case and from the particle to the wave in the latter. If, for a given resonant parallel electron velocity, there are more electrons with energy predominantly perpendicular to the field than along the magnetic field, this random resonance will cause a net diffusion from the perpendicular energy state to the parallel one. Since this liberates energy from the electron, the wave gains energy. This process is but one example of a wave–particle instability. There are many such instabilities in the magnetospheric plasma causing both electromagnetic waves and electrostatic waves. These instabilities help the magnetosphere regulate itself and return to an equilibrium configuration when disturbed.

C. SOLAR WIND INTERACTION WITH JUPITER

The magnetosphere of Jupiter is immense. It could easily contain the Sun and its corona. If one could see it in the night sky, it would appear to be much larger than the full Moon. It also rotates very rapidly, because the planet rotates rapidly and there is good electrical connection between the planet and the magnetosphere. Furthermore, the magnetosphere is filled with heavy ions that appear to come mainly from the satellite Io. These characteristics combine to provide Jupiter with a very interesting and different magnetosphere than that of the Earth.

The reason for the enormity of the Jovian magnetosphere is that the planet's intrinsic magnetic moment of 1.4×10^{30} Gs-cm^3 is over 16,000 times greater than that of Earth, and that the solar wind that confines the magnetospheric cavity is over 25 times weaker. The resulting average distance to the magnetopause at the subsolar point is 70 planetary radii. The bow shock at the subsolar point is on average 85 Jovian radii. The surface magnetic field of Jupiter has a greater contribution from higher-order terms, (quadrupole, octupole, etc.) than the surface field of the Earth. This is at least in part due to the fact that the conducting core of Jupiter in which the dynamo currents flow is relatively closer to the surface of Jupiter than is the Earth's core.

Well inside the Jovian magnetosphere, at 5.9 Jovian radii, orbits the moon Io with its sulfurous volcanoes. These volcanoes release into the Jovian magnetosphere sulfur, oxygen, sodium, and other gases, which in turn are ionized and become part of the trapped plasma of the Jovian magnetosphere. Sputtering from Io, the other satellites, and the Jovian rings also adds atoms and ultimately plasma to the Jovian magnetosphere. Because the Jovian field lines are good electrical conductors, the plasma is forced to corotate with the solid body of the planet. This acceleration is accomplished through an electrical current system closing through the Jovian ionosphere and the magnetospheric plasma joined along field lines. Such a current system also flows through Io itself as Jupiter attempts to force Io into corotation. These processes are thought to generate large potential drops across Io and along the field lines in the vicinity of Io. These electrical potential drops can accelerate charged particles to high energies and may be responsible for some of the radio emissions from Jupiter at decametric (10 m) wavelengths. At decimetric wavelengths (10 cm) radio emissions are due to synchrotron radiation from the intense fluxes of relativistic electrons near the planet. Auroral emissions are also observed at Jupiter and are thought to be caused in much the same way as terrestrial aurora.

The centrifugal force on the plasma in the Jovian magnetosphere far exceeds the gravitational force over much of the magnetosphere. On the front side of the magnetosphere the solar wind opposes this centrifugal force and near static equilibrium is reached. If too much mass is added to flux tubes then they can interchange their positions (via the interchange instability) with lighter flux tubes further out in the magnetosphere, but generally the flow corotates azimuthally with the planet.

One major effect of these rapidly rotating mass-loaded magnetic field lines is that of stretching the magnetospheric cavity from the more common spherical shape to a more disk-like shape. This effect can be seen in observations of the magnetic field and of the location of the bow shock. The streamlined shape of the disk-shaped magnetosphere allows the shock to stand closer to the boundary.

On the nightside there is no barrier to flow away from the planet and a tailward wind is set up. The process of reconnection of the planetary and interplanetary magnetic field occurs at Jupiter but appears to be of minor importance in magnetospheric dynamics. The process of magnetic island formation in the tail also occurs as it does in the case of the Earth, but here, the process is probably associated more with the shedding of plasma down the tail than with removal of excess magnetic flux from the tail lobes.

The size of the Jovian tail is also immense. It is about 200 R_J (14 million km) across or more than 40 times the width of the Earth's tail. It is at least 4 AU long, extending all the way to Saturn.

D. Solar Wind Interaction with Saturn

The magnetic moment of Saturn is a factor of 32 less than that of Jupiter but it is immersed in a solar wind whose pressure is a factor of four less than that at Jupiter. The net result is a Saturn magnetosphere that, while it is only a quarter of the size of that of Jupiter, still dwarfs the magnetosphere of the Earth. Saturn's magnetosphere seems to be less inflated than that of Jupiter. The intrinsic magnetic field of Saturn is highly unusual. The surface magnetic field strength at the equator is 0.21 Gs, not much different than that of the Earth. However, the magnetic moment of Saturn is almost perfectly aligned with the rotation axis, whereas the magnetic dipole moments of the Earth and Jupiter are tilted by about 10° to the rotational axes of the planets. The sizeable tilt of planetary dipole magnetic fields is thought to be essential to the dynamo process. Thus, the alignment of the Saturn magnetic moment is quite puzzling. The contribution to the surface field by the higher-order moments is much less than that at Jupiter indicating that the depth at which the dynamo is acting at Saturn is deeper than at Jupiter.

Saturn has many small and intermediate-sized moons and a well-developed ring system. These bodies are bombarded by the radiation belt particles, mass is sputtered from their surfaces, and mass is added to the magnetic field lines. The Saturn ionosphere, like the Jovian ionosphere, attempts to accelerate this plasma to corotational velocities. However, the mass-loading rates in the inner magnetosphere are not as large as at Jupiter and little distortion of the magnetosphere results. The one large moon of Saturn, Titan, orbits at the outer edge of the magnetosphere at 20 Saturn radii. Its dense atmosphere does strongly interact with the corotating magnetospheric plasma. However, the resulting distortion of the overall shape of the magnetosphere is much less. Again, this is evident from the direction of the magnetic field in the magnetosphere of Saturn and the location of the bow shock relative to the magnetopause.

The radiation belts of Saturn are much more benign than those of Jupiter and the radio emissions are much less intense. The rings absorb the charged particles as they diffuse radially in toward the planet. Because it is the particles in the innermost part of a planetary magnetosphere that are most energetic, the rings play a significant role in reducing the particle and radio flux. Ring particles can become electrically charged. It is thought that some of the exotic behavior of the rings such as the appearance of radial spokes in the rings may be the result of the effects of Saturn's magnetic and electric fields on these charged dust particles.

IV. Plasma Interactions with Ionospheres

A planetary ionosphere is the ionized upper atmosphere of a planet or moon, usually caused by solar ultraviolet radiation but often times impact ionization caused by charged particles plays an important role. If a body has an ionosphere it also has a neutral atmosphere. The plasma also can interact directly with the neutral atmosphere. Thus, it is often difficult to separate those effects due to the interaction of a magnetized plasma with a conductor (the ionosphere) and those due to charged particle–neutral particle interactions. Nevertheless, in the next two sections we attempt just that. In this section, we discuss the interaction of the solar wind with Venus and then with Mars, both of which are probably dominantly, but not exclusively, controlled by ionospheric behavior.

A. SOLAR WIND INTERACTION WITH VENUS

The planet Venus has been visited by many spacecraft including one long-lived orbiter, the Pioneer Venus orbiter, one of whose objectives was to study how the solar wind interacted with the Venus ionosphere. Figure 9 shows schematically the interaction of the solar wind with Venus. Despite the fact that Venus has no detectable intrinsic magnetic field, the solar wind is deflected about the planet and a bow shock formed. The ionospheric pressure balances the

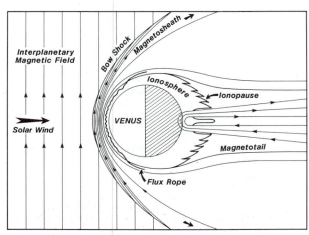

FIG. 9.   The solar wind interaction with Venus.

solar wind pressure to stand off the solar wind. The pressure of the solar wind is applied to the planetary ionosphere through a compression of and pile up of magnetic field lines over the forward hemisphere of the planet. Magnetic field does penetrate the ionospheric barrier in two ways. First, bundles of magnetic field about 20 km across break through the ionopause and slip into the planetary ionosphere. The ionosphere sweeps these flux ropes to the nightside. However, as they move through the ionosphere, these tubes can become highly twisted and form links like a highly twisted rubber band. The magnetic field also can diffuse and convect into the ionosphere.

Usually the rate of diffusion is small enough that any field that enters the planetary ionosphere is swept away by ionospheric flows although, at the subsolar point, downward motion of the ionospheric plasma also can transport magnetic flux from the ionopause to low altitudes. When the solar wind pressure is high, especially when it exceeds the peak ionospheric pressure, diffusion and downward transport of magnetic flux can become fast enough to magnetize the Venus ionosphere with a horizontal field of up to ~150 nT strength. At such times ion-neutral collisions in the atmosphere help support the ionosphere against the solar wind pressure. The magnetic field that enters the Venus ionosphere from the solar wind still has its ends in the solar wind and is dragged antisunward by the solar wind flow. This process contributes to the formation of a magnetic tail behind Venus.

B. Solar Wind Interaction with Mars

Much less is known about the interaction of the solar wind with Mars than with Venus. The magnetic moment of Mars is clearly much less than that of the Earth and is possibly as small as that of Venus. The understanding of the solar wind interaction with Mars takes on added importance because the properties of this interaction have been used as evidence for and against the presence of a significant intrinsic Martian magnetic field. Measurements by the Soviet Mars-3 spacecraft have been variously interpreted as indicating entry into a planetary magnetosphere and as indicating passage only through a magnetosheath like that present at Venus. Mars-5 observations have been interpreted as indicating a magnetotail arising from planetary sources and alternatively from interplanetary sources. The most one can say is that the magnetic dipole moment of Mars is small, perhaps lower than 10^{22} Gs-cm³. The ratio of the strength of the solar wind pressure to the ionospheric pressure at Mars is similar to the ratio that resulted in a magnetized ionosphere at Venus. Thus, even if Mars does not have an intrinsic magnetic field, it should have both a magnetosphere and magnetotail.

V. Plasma Interactions with Neutral Gas

The epitome of the interaction of a plasma with a neutral gas is the formation of a cometary tail in the solar wind. A cometary nucleus evaporates when it is close to the Sun. The expanding cloud of neutral gas is ionized by the solar ultraviolet radiation as well as by charge exchange with the solar wind and by impact ionization. This ionization makes the solar wind heavier and it slows down. However, the ends of the magnetic field lines are not affected and they continue to move at a rapid rate antisunward. This stretches the field lines out in a long magnetic tail behind a comet. Much of our knowledge comes from remote sensing and computer models. In this section we will discuss our present understanding of comets. Then we examine the same processes as they occur at Venus, Io, and Titan. The major difference in the interactions with these three bodies is that their neutral atmospheres are gravitationally bound to them. This restricts the region of mass addition.

A. Solar Wind Interaction with Comets

For purposes of understanding the solar wind interaction with comets, the cometary nucleus can be thought of as a reservoir of frozen gases and dust. The gas may be locked up in a lattice of other material. Such an assemblage is called a clathrate. When the gas is warmed up by the approach of the comet to the Sun, it evaporates, expanding supersonically and carrying dust particles with it. The evaporation can occur from a limited region of the nucleus, and form a jet, or it can occur rather uniformly. The resulting cloud of dust orbits with the comet, although light pressure changes the effective force of gravity on the dust so that the dust follows a slightly different path trailing out behind the comet. This dust cloud forms what is known as a type II cometary tail. The neutral gas expands to great distances up to several million km or more in an

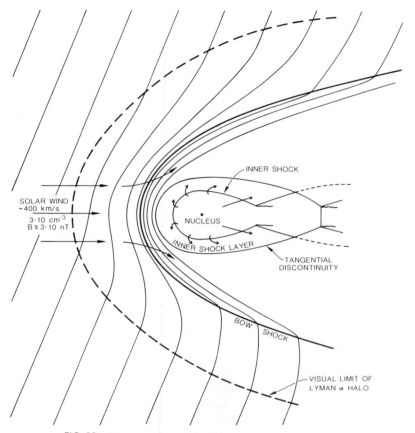

FIG. 10.   The solar wind interaction with an active comet.

active comet, with an emission rate that may exceed 10^{30} molecules per second. The neutral hydrogen in this cloud extends to millions of kilometers as can be determined from observations of the solar Lyman α scattered from the neutral hydrogen and forming a bright ultraviolet halo about the comet as indicated in Fig. 10. The neutral gas becomes ionized by solar ultraviolet radiation, by charge exchange with solar wind ions, and by impact with solar wind electrons. Photoionization simply adds mass to the solar wind, as does impact ionization. Charge-exchange does not add mass to the flow if the charge exchange is symmetric (i.e., between nuclei of the same species). However, a fast ion turns into a fast (400 km/s) neutral in this process, so momentum is removed from the solar wind. If the charge exchange is between a solar wind proton and a heavy neutral atom such as oxygen, the solar wind plasma gains mass. All these processes lead to a slowing down of the solar wind flow, because the electric field of the

solar wind immediately accelerates the newly created ion up to the speed of the flow across the magnetic field. Since momentum is conserved, this acceleration of the ion must be accompanied by a slight deceleration of all the neighboring ions.

The energy of the solar wind flow is converted to energy of gyromotion of the decelerated heavy ions. Ions with energies of the order of 100 keV may be detected millions of kilometers from a comet. Near the comet the energy density in the picked-up ions may significantly exceed the energy density in other constituents in the plasma. The large amount of energy in these heavy ions in turn leads to plasma instabilities and much turbulence downstream of the nucleus.

The interaction with a comet creates a slow spot in the solar wind, which is illustrated in Fig. 10. Well away from this slow spot, the solar wind is moving at an undiminished rate. Since the fast regions and slow regions are linked by the magnetic field lines, the magnetic field gets stretched out, forming a long tail. This long tail is filled with ions, at least in its central region and can often be readily seen in bright comets. This is a type I or ion tail.

If a comet is weak, there is no formation of a bow shock wave. However, if the comet is strong, the solar wind is unable to accommodate the added mass, a shock wave is formed and the incoming flow is deflected around the heaviest mass-loading region. It is this latter situation that is sketched in Fig. 10. The newly added mass may, in fact, have enough pressure to create an ionopause and a magnetic field void in the interior of the interaction region. The physics of this region should be similar to the interaction of the solar wind with an ionosphere as discussed above. Finally, the supersonically expanding plasma has to be stopped and turned around prior to reaching the ionopause. This is accomplished via a shock transition, called the inner shock. Far downstream the tension in the bent magnetic field lines accelerates the decelerated plasma back to the ambient solar wind velocity.

B. Formation of the Venus Magnetotail

The neutral atmosphere of Venus is gravitationally bound to the planet. However, both the hydrogen and oxygen in the Venus atmosphere have hot components that can reach many thousands of kilometers in altitude before returning to the planet. As the solar wind passes through this so-called exosphere, it can pick up mass just as in the cometary interaction described above. This added mass affects the location of the bow shock of Venus, as measured above the terminator plane, since more matter has to flow by the planet. The Venus bow shock is found farthest from the planet at the peak of the solar cycle when the solar ultraviolet radiation is strongest. Thus Venus is most cometlike at solar cycle maximum. Measurements behind Venus reveal a well-developed magnetic tail with two lobes as in the Earth's tail, and a diameter of 4 Venus radii. However, at Venus the location of these two lobes depends on the orientation of the interplanetary magnetic field because they are formed by the draping of this field around the obstacle to the solar wind flow.

C. Plasma Interactions with Io and Titan

Interactions of the corotating magnetospheric plasma occur with both Io at Jupiter and Titan at Saturn, which are very similar to the Venus interaction since these moons both have atmospheres. The major difference is that these moons are connected via the planetary magnetic field to a conducting ionosphere. Bent, or draped magnetic fields, and associated field-aligned currents, have been observed by the Voyager spacecraft on the flux tubes perturbed by the satellites. The Jovian satellite Ganymede also has a magnetic wake disturbance and possible field-aligned currents but the observations were made far behind the satellite and exact details of the interaction remain unknown.

VI. Concluding Remarks

Plasmas in the solar system are complex entities in part because the currents, fields, and particle distributions are all interdependent physical parameters. They are not independent of each other. All are linked through Maxwell's equations and the laws of classical mechanics and gravitation. Oftentimes, intuition based on our observation of neutral gases and fluids fails us so that we find these plasmas behaving in quite unexpected ways. Much of the present understanding of space plasmas is based on a symbiotic relationship between theory and observation and assisted in recent years by large-scale computer simulations. This relationship will have to continue as we explore and attempt to understand the distant reaches of our solar system. The most distant reach of the solar system, as far as electric and magnetic fields are concerned is the heliopause, where the solar wind is stopped. Here the pressure of the galactic plasma, including its cosmic rays and magnetic field, is sufficient to balance the dynamic pressure of the solar wind. Our spacecraft have not reached the heliopause yet but we expect that, if they keep operating, one of the Pioneer or Voyager spacecraft will one day penetrate this boundary, probably before the end of this century.

Bibliography

Battrick, B., and Rolfe, E. (Eds.) (1984). "Achievements of the International Magnetospheric Study (IMS)," European Space Agency, Noordwijk, The Netherlands.

Brandt, J. C., and Chapman, R. D. (1981). "Introduction to Comets," Cambridge University Press, New York.

Carovillano, R. L., and Forbes, J. M. (Eds.) (1983). "Solar Terrestrial Physics," Reidel, Dordrecht, Netherlands.

Dessler, A. J. (Ed.) (1983). "Physics of the Jovian Magnetosphere," Cambridge University Press, London and New York.

Gehrels, Tom and Matthews, M. S. (Eds.) (1984). "Saturn," The University of Arizona Press, Tucson.

Hones, E. W., Jr. (Ed.) (1984). "Magnetic Reconnection in Space and Laboratory Plasmas," Geophysical Monograph 30, American Geophysical Union, Washington, D. C.

Hunten, D. M., Colin, L., Donahue, T. M., and Moroz, V. I. (Eds.) (1983). "Venus," The University of Arizona Press, Tucson.

Lyons, L. R., and Williams, D. J. (1984). "Quantitative Aspects of Magnetospheric Physics," Reidel, Dordrecht, Netherlands.

Merrill, R. T., and McElhinny, M. W. (1983). "The Earth's Magnetic Field: Its History, Origin, and Planetary Perspective," Academic Press, New York.

Potemra, T. A. (Ed.) (1984). "Magnetospheric Currents," Geophysical Monograph 28, American Geophysical Union, Washington, D. C.

Priest, E. R. (1982). "Solar Magnetohydrodynamics," Reidel, Dordrecht, Netherlands.

Russell, C. T. (1980). Planetary magnetism, *Rev. Geophys. Planetary Phys.* **18,** 77–106.

Russell, C. T. (Ed.) (1986). Solar wind interactions, *Adv. Space Res.* **6**(1).

Southwood, D. J., and Russell, C. T. (Eds.) (1982). Special issue on international magnetospheric study, *Rev. Geophys. Planetary Phys.* **20.**

Wilkening, L. L. (Ed.) (1982). "Comets," The University of Arizona Press, Tucson.

SOLAR–TERRESTRIAL PHYSICS

L. J. Lanzerotti   *AT&T Bell Laboratories*

GLOSSARY

Alfvén wave: Basic hydromagnetic wave in a plasma containing a magnetic field; the plasma displacement is transverse to the magnetic field, with propagation directed along the field.

Aurora: Lights in the upper atmosphere (from about 90 to 300 km altitude) produced by the excitation of atmospheric gases by energetic particles; the localized areas in the two polar regions where aurora are typically observed are determined by the detailed topology of the magnetosphere.

Collisionless shock: Type of discontinuity formed in the sunward direction in the solar wind because of the interposed obstacle of the magnetosphere.

Coronal holes: Regions in the solar corona in which the solar magnetic fields are not closed but are open into the interplanetary medium.

Ecliptic plane: Plane of the apparent annual path of the sun on the celestial sphere.

Heliosphere: Region around the sun in our galaxy (the Milky Way galaxy) influenced by the solar wind.

Ionosphere: Ionized region of the upper atmosphere, generally from about 90 km to about 1000 km in altitude; the area beyond this is defined as the magnetosphere.

Magnetopause: Boundary between the flowing solar wind plasma and the magnetosphere.

Magnetosphere: Region of space around the earth in which the terrestrial magnetic and electric fields usually dominate the transport and motions of charged particles.

Magnetosphere cusps: Regions in the northern and southern polar areas separating the magnetic fields that form the dayside magnetopause from those fields that stretch into the magnetotail.

Magnetotail: Region of the magnetosphere in the antisunward direction; if visible, the magnetotail would have characteristics like the tail of a comet.

Photosphere: Visible solar surface with a temperature of about 6400 K.

Plasma: Gas (atomic and/or molecular species) that is ionized. Naturally occurring plasmas usually contain magnetic fields.

Plasma sheet: Sheet of plasma, several earth radii in thickness in the tail of the magnetosphere, that separates magnetic fields of opposite magnetic polarity.

Plasmapause: Outermost boundary of the plasmasphere.

Plasmasphere: Region of plasma in the magnetosphere of ionosphere character.

Radiation belts: Localized regions in the magnetosphere that contain energetic charged particles whose motions are primarily controlled by the magnetic field of Earth.

Solar corona: Region, beginning about 2000 km above the photosphere, with a temperature of over one million degrees kelvin.

Solar flare: Sudden brightening, for several minutes to several hours, of a small area of the solar photosphere that contains a group of sunspots.

Solar wind: Expansion of the solar corona, primarily hydrogen ions, into the interplanetary medium.

Substorm: Interval of one to three hours of auroral, geomagnetic, and magnetospheric activity, usually followed by a several-hour interval of relative quiescence.

Sunspot cycle: Variation with time of the appearance of the number of sunspot groups on the visible solar photosphere.

Sunspots: Darkened areas of the solar photosphere usually occurring in groups and containing intense magnetic fields.

Thermosphere: Region of the earth's outer atmosphere, extending above about 85 km altitude.

Solar–terrestrial physics is the study of the interactions of nonoptical solar emissions with the magnetic field and atmosphere of earth and the results and implications of these interactions for human technologies. These solar emissions include γ rays, X rays, and energetic charged particles produced by solar flares. The major solar emission is the solar wind, a super-Alfvénic flow of charged particles boiled off the top of the sun's hot outer atmosphere, the corona.

I. Introduction

Although unrecognized as such, appearances of the aurora have announced the solar–terrestrial connection for thousands of years. Descriptions of the aurora are common in the oral histories and legends of northern peoples. Although large auroral displays are infrequently seen at very low latitudes, descriptions of phenomena in the sky that can be interpreted as the aurora are found in ancient writings by such authors as Aristotle and Seneca, as well as probably in the first chapter of Ezekiel. Old Chinese texts describe auroral observations in the Orient. Natural science volumes from the middle ages often contain fanciful illustrations of auroral displays over towns of the time. The reality of the solar–terrestrial connection, however, was first identified only in the late 19th century. Not all of the manifestations of this connection are known even to this date.

One of the first modern impacts of the solar–terrestrial connection, unrecognized at the time, was the puzzling report by W. H. Barlow of the spontaneous appearance of anomalous currents measured on electrical telegraph lines in England. Perhaps the first scientific realization of the connection was by the British scientist Richard Carrington. On 1 September 1859, while sketching sun spots in the course of his studies

of solar phenomena, Carrington suddenly observed an intense brightening in the region of one of the spots. This occurrence so excited him that he quickly called his associates to the telescope to witness the event. Within a day, violent fluctuations in the magnetic field and intense auroral displays were observed on earth, with reports of aurora as far south as Honolulu. Carrington was very intrigued about a possible link between his white-light flare and the subsequent aurora. Nevertheless, he urged caution in connecting the two, commenting "one swallow does not a summer make."

During the several days of enhanced auroral displays, telegraph systems throughout Europe and in eastern North America suffered severe impairments in operation or even complete disruptions of service. The new technology of telegraphy had never experienced such widespread impacts on service, occurrences that were attributed by telegraph engineers to currents flowing in the earth, somehow associated with the auroral displays. The strange occurrences of 1859 were a significant spur to action by these engineers. Considerable professional activity in Europe was devoted for many years to the studies of the phenomena of earth currents. While the engineering literature often discussed the relationships of these currents to enhanced fluctuations in the terrestrial magnetic field (magnetic storms), and possibly to disturbances on the sun, natural scientists of the time were less likely to see such a cosmic causal connection. While the British scientist E. Maunder seriously discussed such a connection, the dominant authority of the time, Lord Kelvin, thought otherwise. In discussions of a particular magnetic storm (25 June 1885), he definitively stated that "it ... is absolutely conclusive ... that terrestrial magnetic storms are [not] due to magnetic action of the sun". He further concluded that "the supposed connection between magnetic storms and sunspots is unreal".

The first half of the twentieth century, which saw the implementation of trans-Atlantic radio broadcasts and the discovery of the ionosphere, gave rise to more considered and quantitative discussions of the sun–earth connection. Such considerations were warranted, not only scientifically but for very practical reasons as well. For example, trans-Atlantic telephone traffic via low-frequency radio was often disrupted during magnetic storms, a situation that prompted many scientific and engineering experiments and publications in the 1930s, as well as the laying of the first trans-Atlantic telephone cable in the 1950s.

Nearly 100 years after Carrington's discovery, at the threshold of the space age, the first trans-Atlantic telecommunications cable was severely affected on 11 February 1958 by a magnetic storm that followed closely on the heels of a particularly large solar flare. Toronto suffered an electrical blackout during the geomagnetic disturbances. Geomagnetic disturbances on power distribution systems continue to present an engineering problem of some concern in northern areas such as Canada and Scandinavia. In 1972, the year after the last astronaut had set foot on the moon and safely returned, a series of large solar flares in early August resulted in such extensive and large magnetic disturbances that there was a complete disruption of a transcontinental communications cable in the midwestern part of the United States. The energetic particles produced by these flares, primarily protons and electrons, would have been lethal to an explorer on the surface of the moon or, probably, in a spacecraft in transit there. [See SOLAR SYSTEM, MAGNETIC AND ELECTRIC FIELDS.]

Studies from the earth of the ion tails of comets led the German scientist Ludwig Biermann to suggest in 1951 that the sun emitted invisible gases that controlled the orientations of the tails as the comets traversed the solar system. Finally, the advent of scientific spacecraft, making physical measurements above the atmosphere, demonstrated conclusively that solar–terrestrial connections occur not only through the optical emissions of the sun but also by the "invisible" tenuous mixture of ions, electrons, and magnetic fields that we now call the solar wind. Spacecraft also provided measurements of the connection through the much more energetic particles produced by solar flares. Thus, the space age has provided the means to dispatch unmanned instruments into the environment around earth to measure and study the interplanetary and near-earth regions that can produce both the dramatic, visual spectacle of the aurora and disruptions to technology.

The advances made by the use of spacecraft in understanding the solar–terrestrial environment have been enormous. However, even as the basic morphology of this environment has become better known, it has become clear that there are physical processes occurring in it that yet defy complete understanding. Thus, predictions regarding the state of the environment and its possible effect on the earth remain rather rudimentary. The physics of the processes in the medium is complicated. However, the increasing sophistication of experiments (ground-based, rocket, and spacecraft) and of theoretical and computational capabilities has begun to provide some insights into the fundamentals of the solar–terrestrial system.

II. Solar Processes: The Source for Sun–Earth Couplings

Visible radiation from the sun provides the heat and light that have enabled life to evolve and thrive on our planet. The constant, unvarying nature of the source of the life-sustaining light and heat was an important element in the mythologies of many ancient civilizations, which attributed godlike qualities to the sun. This myth was shattered when Galileo's telescope revealed that the sun was not "perfect": Spots varying with time and with location marred the solar surface. Since Galileo's time the sun has been found to have a number of changeable features, many of which can affect the earth. The existence of a cyclic variation in the number of sunspots was firmly established in the mid-19th century by Heinrich Schwabe, a German druggist and amateur astronomer.

The periodic variation in sunspot numbers is shown in Fig. 1. While the spot cycle is approximately 11 years, the fundamental cycle, based upon the magnetic polarity of the sun, is approximately 22 years. At the beginning of a new cycle the spottedness begins first at latitudes of about 30° on both sides of the solar equator. As the cycle continues to develop and the number of spots increases, the spots migrate slowly towards the equator in both hemispheres until, near the end of the cycle, those nearest the equator begin to disappear. During and following the disappearance, spots again begin to appear at latitudes of ±25–30°. For a significant interval after their discovery, there was a long hiatus when the number of spots was very small, or even zero (Fig. 1). This so-called Maunder minimum period, named after the British scientist who first drew attention to it, is a significant enigma in terms of basic understanding of solar variability. [See SOLAR PHYSICS.]

Sunspots are not just regions of cooler gases on the solar surface. They are also regions where the magnetic fields from the solar interior

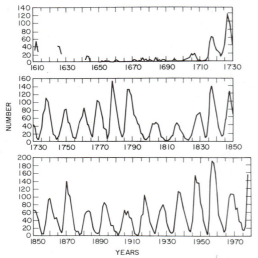

FIG. 1. Sunspot cycle, A.D. 1610–1979, as represented by the annual mean sunspot number for those years. The interval of about seven decades beginning in 1645 when sunspot numbers were unusually low is known as the Maunder minimum, which coincided with a significant drop in global temperatures on earth. The records of sunspots between 1610, when they were first discovered, and 1640 are mostly too scanty to reconstruct with reasonable confidence. [Courtesy of J. A. Eddy, High Altitude Observatory, NCAR.]

penetrate the surface, producing very large, up to several thousand gauss, intensities (compared with the magnetic field intensity at the surface of the earth of approximately one-half gauss). Sunspots occur in groups of two to several, with opposite magnetic polarities, north and south. Magnetic fields arch from one spot to the other. Under conditions that are poorly understood, the magnetic fields of the sunspot structures can become very unstable, ultimately producing a solar flare: Intense brightening of a region on the solar surface occurs abruptly as atomic ions and electrons in the solar atmosphere are suddenly energized. The energy content of a solar storm, 10^{32}–10^{33} ergs, is probably derived from conversion of energy stored in the magnetic fields. Some energized particles impact the solar atmosphere, producing optical and ultraviolet light as well as X rays and γ rays; the remainder are rapidly emitted into interplanetary space and propagate to the orbit of earth. Intense bursts of radio-frequency emissions, easily detectable at earth, accompany the particle energization.

Solar flares are one example of violent solar activity. Another is the episodic injection into the interplanetary medium of large masses of solar coronal material, approximately 10^{12} kg or more. Such coronal transients were identified from data acquired during the Skylab era of solar research in the mid-1970s. While such transients certainly produce disturbed interplanetary conditions, the mode of propagation and dissipation of the ejected mass and energy through the solar system, including their influences on the plasma environment of earth, are largely unknown at present.

III. Interplanetary Medium: The Mechanism for Sun–Earth Plasma Couplings

Measurements of the solar atmosphere by atomic spectroscopy techniques reveal that the outermost region, the solar corona, has a much lower density and a much higher temperature than the underlying surface (the photosphere), where solar flares occur. The precise mechanism(s) for heating this high upper solar atmosphere is (are) unknown. Possible methods could include heating by acoustic waves generated by the convection of heat from the solar interior and heating by the damping of Alfvén waves produced in the solar photosphere that propagate outward from the sun. The plasma structure of the solar corona is complicated and dynamic, often exhibiting long, intense radial streamers and huge arches (Fig. 2).

The end result of heating the corona is that the very hot outer atmosphere continually expands away from the sun, forming the solar wind. The speed of this wind is typically 400–600 km/sec, although much higher velocities have been measured following some large solar flares. At a distance of a few solar radii above the photosphere the velocity of expansion of the corona becomes larger than the Alfvén velocity;

$$V_A = \left(\frac{B^2}{4\pi\rho}\right)^{1/2}, \text{ cm sec}^{-1}$$

where B is the magnetic field intensity in gauss and ρ the ion mass density (g cm^{-3}). That is, the velocity of expansion becomes faster than the velocity with which information can be transferred in the highly conducting (ionized) gas. Beyond this distance the solar wind expands through the solar system, carrying with it the magnetic field from the solar surface. The region around our sun that is influenced by the solar wind is called the heliosphere. It is the solar wind that provides the primary plasma link between the sun and the earth and that is the source of geomagnetic disturbances on earth. Lord Kelvin, in his categorical statements of nearly a century ago, could not have been aware of this invisible, crucial link between the sun and the planets.

The outward expansion of the solar wind, combined with the rotation of the sun, provides an important physical phenomenon in the heliosphere: The solar magnetic field is tied firmly to the sun by the highly conducting solar photosphere. At the same time, the magnetic field is firmly embedded (frozen) in the radially outward-flowing solar plasma. The field in interplanetary space thus forms a spiral pattern similar to that produced by a rotating garden water sprinkler. At the orbit of earth, the interplanetary magnetic field makes an average angle in the ecliptic plane of approximately 45 degrees to the radially outward direction. The angle of the interplanetary field with respect to the magnetic field of earth plays a key role in determining the level of geomagnetic activity.

Sunspot activity and other disturbances on the sun can significantly disrupt the tranquility of the interplanetary medium. In addition to injecting high energy particles, solar flares also emit greatly enhanced solar wind streams. These streams have a higher velocity and a greater particle number density than the normal solar wind. Shock waves are thus formed in the interplanetary medium between the boundaries of the faster and slower moving winds. These shock waves themselves can accelerate interplanetary particles to higher energies and can greatly agitate the magnetosphere.

During solar minimum conditions the winds observed in interplanetary space appear to be emitted primarily from coronal holes, especially in the polar regions of the sun. Such holes are regions of the solar corona in which the solar magnetic fields tend to be more open, extending into interplanetary space. These open field regions can extend, in areas limited in longitudinal width, from the poles to near the solar equator. These coronal hole regions emit high velocity solar wind streams that can significantly disturb the plasma environment on earth, even during the periods when solar flares and sunspots are nearly absent from the solar surface.

At the orbit of earth the solar wind carries an energy density of about 0.1 ergs/cm^2, a factor about 10^7 smaller than the energy in the visible and infrared wavelengths ($\sim 1.4 \times 10^6$ erg/cm^2). Yet, as will be seen, this low energy density, when applied over the magnetosphere, can have profound effects on the terrestrial space environment. At great distances from the sun it is likely that the heliosphere, with the embedded sun and planets, forms a kind of magnetosphere itself in the local interstellar medium. There is likely to be a boundary established between the outward-flowing solar wind and the interstellar plasmas and magnetic fields in the direction in which the sun is moving relative to the nearby stars. The boundary is expected, from present ideas, to occur between 100–150 earth–sun distances. The boundary is likely to be a turbulent region, with perhaps a shock wave established in the interstellar medium, similar to the situation for the earth in the solar wind (see below).

IV. Magnetosphere of Earth: The Extension of Magnetic Fields and Plasmas into Space

A. OVERALL MORPHOLOGY

The earth shows its presence in the solar wind in a manner analogous to that of a supersonic

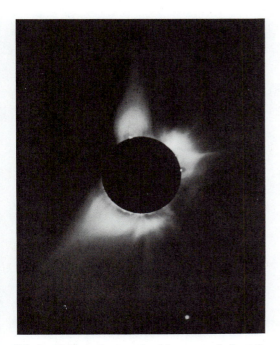

FIG. 2. Eclipse photograph of the sun made from the surface of earth, showing coronal streamers and coronal structure.

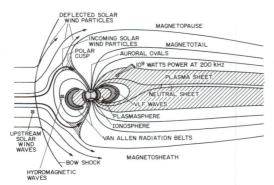

FIG. 3. Schematic illustration of the magnetosphere of earth, the nearby plasma environment formed by the interactions of the solar wind with the geomagnetic field.

airplane traversing the atmosphere. Since the solar wind velocity is faster than the characteristic speed in the ionized gas, a shock wave is established. The average location of this bow shock is some 10–14 earth radii outward on the sunward side (Fig. 3). On the earthward side of the shock wave is a region of more turbulent solar wind flow. Then, at an average altitude above the earth of some nine earth radii, the magnetopause exists as a thin, current-carrying boundary separating regions of solar-wind control of space from regions controlled by the terrestrial magnetic field (Fig. 3). To a reasonable approximation, the solar wind flow around the earth at the magnetopause is similar to the air flow in a wind tunnel past an aerodynamic object. Also to a good approximation, the subsolar magnetopause location can be approximated by equating the dynamic pressure $C\rho V^2$ of the flowing solar wind and the sum of the magnetic field and plasma pressure inside the magnetosphere,

$$C\rho V^2 = \frac{B^2}{8\pi} + P$$

Here C is a constant, V and ρ are the solar wind velocity and density respectively, $B^2/8\pi$ is the pressure due to the magnetic field of the magnetosphere at the boundary, and P (usually negligible in the case of the terrestrial magnetosphere) the gas pressure inside the magnetosphere.

The actual physical processes forming the configuration of the magnetosphere, while approximated by these simple concepts, are in fact much more involved in detail. The formation of the shock wave in front of the magnetopause is more complicated than the simple airplane analogy. For one thing, the magnetic field imbedded

in the solar wind is intimately associated with establishing the detailed characteristics of the shock. Furthermore, the shock wave is a collisionless shock; that is, the solar wind particles do not interact by collisions among themselves. Rather, the presence of the obstacle produces magnetic field and the plasma wave conditions, which cause the plasma distribution to become unstable and form the shock. Waves generated by the formation of the shock can propagate in the solar wind, back toward the sun, making the interplanetary environment near earth quite complicated and very turbulent.

Similarly, the magnetopause is a complicated physical region. While the magnetosphere has a vast extent, the magnetopause boundary itself is a very thin region where electrical currents flow in the plasma. These currents define a thin, cellular boundary separating the two very different plasma regimes. This cellularlike boundary is not completely impenetrable, and its location in space can be quite variable depending upon solar wind conditions. The boundary can move inward, toward the earth, under higher solar wind velocities, and can move outward during periods of lower solar wind speed conditions. Nevertheless, the two plasma regimes separated by the magnetopause boundary can be treated individually in analyzing many problems involving internal processes in each system. The interplanetary magnetic field also plays an important role in the formation of the boundary. A reconnection of the earth's internal magnetic field and the interplanetary field appears to occur sporadically at the boundary, allowing plasma to escape from the magnetosphere and heliosphere plasma to become entrapped in the terrestrial magnetosphere.

The solar wind, by a viscous interaction with the magnetosphere plasma and magnetic field, forms the magnetosphere into a long, cometlike (but invisible) object. A large dawn-to-dusk electric field is created across the magnetosphere by this interaction. The energy transfer rate from the solar wind into the magnetosphere is about 10^{19} ergs/sec. The magnetotail may extent to more than a thousand earth radii (some 6×10^6 km) in the antisunward direction. In the magnetotail a sheet of highly conducting plasma separates the magnetic fields that originate in the southern hemisphere of earth from those that terminate in the northern hemisphere (Fig. 3). A current flows from dawn to dusk through this plasma sheet. The earthward extensions of this plasma sheet protrude along magnetic field lines that connect with the night-side auroral zone (ionosphere currents) in both hemispheres. The power dissipated in these currents is about 10^{18} ergs/sec (10^{11} W). On the front side of the magnetosphere, the separations in each hemisphere between magnetic field lines closing on the dayside and those extending into the magnetotail form a cusplike region through which solar wind plasma can reach the upper atmosphere, forming auroral emissions and currents in the ionosphere. The topology of the auroral zone, shown in Fig. 4 in a picture taken by a U.S. spacecraft during local nighttime over the southern polar region, provides a qualitative, visible measure of the solar–terrestrial connection. The aurora also can be seen under daylight conditions if measured in ultraviolet emissions.

During intervals of geomagnetic substorms, the auroral zone can extend to much lower latitudes, and the currents flowing in the ionosphere can be significantly enhanced in intensity and in latitudinal extent. The energy dissipated during a geomagnetic substorm, which may last 2–3 hr, is about 10^{19} ergs/sec. These magnetic disturbance conditions seem to occur from the energization of the plasma sheet in the magnetotail, often triggered by changed solar wind conditions. The energization is believed to occur from the conversion of magnetic field energy into plasma particle kinetic energy through a process of reconnection of magnetic field lines across the plasma sheet. This enhanced plasma is then transported into the auroral zones. Additional energization of charged particles occurs along magnetic field lines above the aurora. Occasionally the interplanetary medium is disturbed for many hours, even days. At such times the geomagnetic disturbances in the magnetosphere can also last for a day or more, a condition called a magnetic storm.

This overall concept of the terrestrial magnetosphere was rapid in developing once spacecraft were sent above the atmosphere. Soon after the discovery of the radiation belts (Fig. 3) by Van Allen and his students using Explorer 1 in 1958, the magnetosphere boundary and shock were detected. In the last few years, with the detailed measurements of the characteristics of the magnetospheres of Jupiter, Saturn, and Uranus, the minimagnetosphere around Mercury, and the magnetosphere formed by the solar wind interaction with the ionized upper atmosphere of Venus, the concept of a magnetosphere has become quite general in cosmic plasma physics. Indeed, many exotic astrophysical objects, which can only be studied remotely by detection of emitted electromagnetic radiation (e.g., pulsars and some radio galaxies) are now commonly discussed in magnetosphere terms.

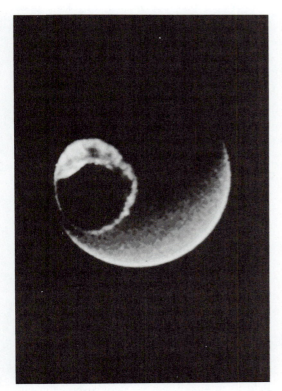

FIG. 4. Southern auroral zone of the earth measured in ultraviolet light, primarily atomic oxygen emission at 1304 Å, under nighttime conditions by the Dynamics Explorer satellite. [Courtesy of L. A. Frank, University of Iowa.]

B. PLASMA PROCESSES

There are many localized and small-scale plasma processes throughout the global magnetosphere system which provide the underlying mechanisms that determine the state of the system. A large variety of plasma waves exist in the magnetosphere. These waves can be electromagnetic, that is, having both electric and magnetic properties, or they can be electrostatic, resulting from the separation of oppositely charged plasmas. Lightning storms in the atmosphere create radio-frequency waves that not only produce static in radios but also propagate, with a whistling tone, along geomagnetic field lines extending into the outer magnetosphere, even to the opposite hemisphere. Other waves, with wavelengths of the order of the length of a magnetic field line (several earth radii or more) are known as Alfvén waves. These waves are generated by the flow of the solar wind past the magnetic field of the earth and by instabilities that can occur in the magnetosphere plasma. Still other radio waves are generated above the auroral regions (within perhaps one earth radius altitude) where electrons and protons are accelerated into and out of the ionosphere.

A most interesting aspect of the magnetosphere, as well as of solar flares, is the possible existence of regions in which magnetic field energy is converted directly to plasma particle energy. Such regions are believed to be those in which magnetic field lines of opposite polarities can suddenly reconnect and, in the process of reconfiguring, release energy to the embedded plasma. As noted above, these regions can exist at the front side of the magnetosphere, possibly in the cusp regions, and in the magnetotail. In the magnetotail, the magnetic field lines in the plasma sheet are always oppositely directed. If the plasma sheet conditions should change in such a manner that the plasma became less conducting, reconnection of the magnetic field lines might occur spontaneously. Alternatively, the magnetic field lines could be pushed together by an external force (such as the solar wind interaction) that would slowly compress the plasma sheet. As the plasma in the center of the region became more dense, collisions among plasma particles or the onset of certain plasma instabilities might alter the conductivity, providing an environment for reconnection to occur. Many theoretical considerations and computer simulations of the reconnection process have been carried out, and it is one of the most active areas of basic theoretical magnetospheric plasma research at present.

The ionosphere of the earth is an intriguing plasma environment, in which both neutral and ionized gases are threaded by the geomagnetic field. At times, the ionization layers become unstable, producing patchy conditions with different ionization densities. This can be particularly prevalent in the currents that flow in the auroral zone. In the equatorial regions of the ionosphere, where the magnetic field lines are parallel to the ionization layers (as well as to the surface of the earth), the ionosphere layers can become unstable; plasma bubbles can form and

rise through the ionosphere to the upper levels. Studies of the basic plasma physics of such bubbles and ionization patches have led to significant new insights into cosmic plasma processes.

The background plasma density in the magnetosphere varies from several thousand particles per cubic centimeter within the first few earth radii altitude (the plasmasphere; Fig. 3) to only a few per cubic centimeter at higher altitudes. There is ordinarily a rather sharp discontinuity between the two plasma regimes, and the boundary is called the plasmapause. The boundary is formed approximately at the location where there is a balance between the electric fields produced by the rotation of the geomagnetic field and the large-scale electric field imposed across the magnetosphere by the solar wind flow. This discontinuity in the plasma distribution can be a source of magnetosphere plasma waves and can significantly affect the propagation of Alfvén waves. The background plasma density inside the plasmapause results from ionosphere plasma diffusing up into the magnetosphere during local daytime conditions. Outside the plasmapause the cold plasma from the ionosphere is swept (convected) out of the magnetosphere by the cross-magnetosphere electric field.

The radiation belts are populated largely by particles accelerated out of the upper atmosphere and ionosphere and by some solar wind ions and electrons. The relative importance attributed to these two sources of radiation belt particles has varied during the history of magnetosphere research, with much emphasis being placed at present on the importance of the ionospheric source. The ions and electrons can be accelerated to their radiation belt energies by internal plasma instabilities and by large-scale compressions and expansions of the magnetosphere under action of the variable solar wind. In the innermost part of the magnetosphere the decay of neutrons, produced by high-energy cosmic rays that strike the upper atmosphere, yields electrons and protons to the trapped radiation belts. The motions of radiation belt ions and electrons are controlled by the terrestrial magnetic field. The inner Van Allen belt of electrons is located earthward of the plasmapause, while the outer belt is outside.

A sketch of major current systems in the magnetosphere–ionosphere system is shown in Fig. 5. The auroral current system is linked to the magnetosphere and the plasma sheet via cur-

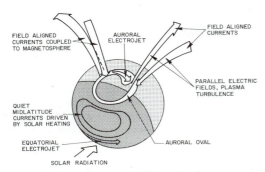

FIG. 5. Simplified picture of ionospheric current systems and the connection of the auroral ionosphere current systems to the magnetosphere via electrical currents flowing along geomagnetic field lines.

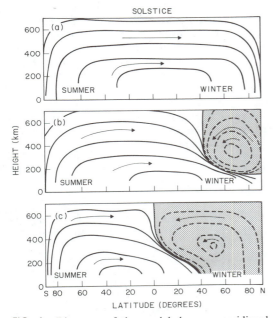

FIG. 6. Diagrams of the modeled mean meridional atmospheric circulation in the thermosphere at the time of the winter solstice in the northern hemisphere, for three levels of auroral (geomagnetic) activity. The arrows indicate the directions of convective heat flow from solar heating of the southern hemisphere and from the auroral heating of the dark, northern hemisphere winter atmosphere. The intensity of auroral heating (cross-hatched regions) of the upper reaches of the atmosphere increases from (a) to (c). As auroral heating of the upper thermosphere increases, the influences of the heating are felt at lower latitudes, inhibiting circulation and heat transport from the summer to the winter hemispheres.

rents that flow along geomagnetic field lines. Solar radiation in the visible and ultraviolet wavelength ranges produces ionospheric currents on the dayside of the earth, including an intense current in the equatorial regions.

The auroral current systems can heat the upper, neutral atmosphere of earth, altering the overall circulation patterns in these regions. Without auroral heating, circulation would tend to flow from the hotter, equatorial regions toward the colder, polar regions. Increased levels of geomagnetic activity (produced by disturbances in the solar wind) produce increased auroral heating. As illustrated in the three panels of Fig. 6, which show the results of theoretical calculations of the effect for the December solstice, increasing the intensity of the auroral currents causes the upper atmosphere circulation patterns to reverse at some latitudes. The latitude of reversal depends upon the intensity of the heating.

V. The Sun–Earth Connection

The magnetic disturbances that followed Carrington's white-light flare had a dramatic impact on the modern technology of the time, the telegraph. Indeed, during some particularly intense auroras the telegraph lines running from Boston to points both north and south could be operated for several hours without benefit of the battery supplies. Even in daylight some telegraph lines were disrupted, indicating to the engineers that the aurora must still be present, although it could not be observed because of sunlight. During a nighttime episode, the extraneous voltage measured on a Boston telegraph line was reported to vary with an approximately 30-sec periodicity, coincident with variations in the auroral display.

Many of the effects of the sun–earth connection on technological activities involve those that use long conductors, such as telephone and telegraph lines, and electric power grids. These phenomena occur because the time-varying geomagnetic field, produced by the variable solar wind, causes currents to flow in the earth. These earth currents enter the long conductors at the points where they are grounded to the earth and can produce deleterious effects on the connected electronics. For example, the building of the Alaskan oil pipeline directly across the auroral zone meant that considerable effort was expended in studying the problem of the induction, by auroral currents, of earth currents in the pipe. Procedures had to be adopted to avoid potential difficulties. While earth currents have been much studied for over a century, their description remains largely empirical for two major reasons. First, it is difficult, if not impossible as yet, to predict the exact spatial scales of the disturbed, time-varying magnetic field. Second, the inhomogeneities in the conductivity structure of the earth are poorly known, so that regions of expected higher current flow for a given geomagnetic field variation cannot easily be predicted.

Prospecting techniques for mineral resource deposits using air and satellite surveys often employ magnetic sensing. Natural geomagnetic field fluctuations can seriously impair the interpretation of such surveys; the spurious fluctuations must be removed by various analytic techniques to resolve geologically important features. It is particularly difficult to perform subtractions of magnetic disturbances in the auroral zones, and ground and aerial survey parties are scheduled, insofar as feasible, to operate during geomagnetically quiet intervals. The ability to predict such intervals is still somewhat rudimentary but is, of course, of considerable economic importance to prospecting companies.

The extension of sophisticated technologies into space has brought a new set of concerns. When communications satellites were originally proposed, it was not realized that the environment in which they would fly would be anything but benign. However, it was quickly recognized that the magnetospheric plasmas (including the energetic particles) can significantly alter the properties of solar cells and spacecraft thermal control blankets. The energetic particles also can damage semiconductor components and produce anomalies in memory devices. Spacecraft have been known to go suddenly into a noncontrolled state because of radiation effects on electronic components.

The space shuttle, flying some 200–300 km above the surface of earth, as well as other spacecraft in similar low-altitude orbits have been found to have a visible glow around them. This glow is produced by the interaction of the vehicle with the space environment at these altitudes, mostly with oxygen ions. However, the precise physical and chemical processes involved are not well understood. Such a glow, if present around earth-sensing and astronomical

telescopes, could seriously affect the sensitivity and operational conditions of these instruments.

The possible influence of the variable features of the sun—solar optical radiation (the solar constant), solar activity, and solar wind—on the lower atmosphere of earth, where weather patterns are established, remains a great enigma. Statistically, certain elements of climate (and weather) in some regions appear to be related to solar variability, particularly the long-interval solar cycles. The several elements of solar variability may all be interrelated in complex ways, which thus far have inhibited understanding of the statistical results. A variability in the solar constant can be incorporated into modern atmosphere circulation models to test for cause and effect relations. On the other hand, relating variabilities in solar activity and the solar wind to weather and climate (if these quantities are uncoupled from the solar constant) is fraught with difficulties (to say nothing of controversies) and presents the significant problem of identifying the physical driving mechanism(s). If such a mechanism (or mechanisms) exists to couple the solar particle and magnetic field flows to the lower atmosphere, it remains to be discovered in the complex and interrelated atmosphere–ionosphere–magnetosphere system.

BIBLIOGRAPHY

Akasofu, S. -I., and S. Chapman, (1972). "Solar-Terrestrial Physics." Oxford University Press, Oxford.

Bray, R. J., and Loughhead, R. E. (1979). "Sunspots." Dover, New York.

Brekke, A., and Egeland, A. (1983). "The Northern Lights." Springer-Verlag, Heidelberg.

Carovillano, R. L., and Forbes, J. M., eds. (1983). "Solar–Terrestrial Physics." D. Reidel Publishing Co., Dordrecht, the Netherlands.

Eather, R. H. (1980). "Majestic Lights." American Geophysical Union, Washington, D.C.

Geophysics Study Committee (1977). "The Upper Atmosphere and Magnetosphere." National Academy of Sciences, Washington, D.C.

Hargreaves, J. K. (1979). "The Upper Atmosphere and Solar–Terrestrial Relations." Van Nostrand Reinhold, New York.

Herman, J. R., and Goldberg, R. A. (1978). "Sun, Weather, and Climate," NASA SP-426. National Aeronautics and Space Administration, Washington, D.C.

Hundhausen, A. J. (1972). "Coronal Expansion and Solar Wind." Springer-Verlag, Heidelberg.

Kennel, C. F., Lanzerotti, L. J., and Parker, E. N., eds. (1979). "Solar System Plasma Physics," 3 vols. North-Holland Publishing Co., Amsterdam.

Lanzerotti, L. J., and Krimigis, S. M. (1985). "Comparative Magnetospheres", in *Physics Today* **38**(11), 24.

Ratcliffe, J. A. (1972). "An Introduction to the Iono-sphere and Magnetosphere." Cambridge University Press, Cambridge.

Volland, H. (1984). "Atmospheric Electrodynamics." Springer-Verlag, Heidelberg.

White, O. R., ed. (1977). "The Solar Output and Its Variation." Colorado Associated University Press, Boulder, CO.

SOLAR WIND—*SEE* SOLAR SYSTEM, MAGNETIC AND ELECTRIC FIELDS

SPACE, INTERSTELLAR MATTER

Donald G. York   *University of Chicago*

GLOSSARY

Absorption: Removal of energy by atoms, molecules, or solids from a beam of radiation, with re-emission at wavelengths other than the absorbing wavelength.

Cosmic rays: Atomic nuclei or electrons accelerated to energies of more than 1 MeV by unknown processes.

Dissociation: Break up of a molecule into smaller molecules or atoms.

Emission: Process whereby excited energy states in atoms and molecules produce radiation as electrons within the particles relax to lower energy states.

Galaxy: Aggregate of stars (numbered 10^6–10^{12}) gravitationally bound in orbits typically 10 kpc in size.

Intercloud medium: Space and material between the interstellar clouds, generally thought of as the largest volume in the interstellar medium.

Interstellar clouds: Condensations in the interstellar medium of various densities and temperatures, typically several parsecs to several tens of parsecs in size.

Interstellar medium: Space between the stars and the dust, gas, and fields that fill it.

Ionization: Removal of electrons from atoms or molecules by any of several processes.

Nucleosynthesis: Formation of heavy elements from lighter elements through fusion.

Parsec: Measure of distance, equal to 3×10^{18} cm or about 3 light-years.

Polarization: Preferential rather than random alignment of the electric vector of incoming radiation.

Recombination: Capture of an electron by a charged atomic nucleus or molecule.

Scattering: Redirection of light without changing its wavelength.

Shock front: Pressure discontinuity within which low-energy particles are accelerated and ionized or dissociated.

As far as we know, galaxies can be regarded as the main building blocks of the Universe. They are the central feature in modern astronomical research, all other fields being aimed at understanding where they came from and what goes on within them. The galaxies we can see are made up of stars, but the largest part of the volume of the galaxies is filled with dust and gas—the most pure vacuum in nature, save only the space between the galaxies themselves. By mass, the interstellar material accounts for about 10% of our galaxy. The space between the stars is referred to as the interstellar medium. According to modern theories, the existence of a visible galaxy as an active, star-forming entity depends upon the passage of atoms made in stars out into this medium. The processes that occur therein are thought to lead to formation of new generations of stars.

The detailed process of removal of atoms from stars and their aggregation into condensing clouds that form new stars in interstellar space is complex. The elaboration of the process is a major thrust in modern astronomical research. We attempt here to describe, from an empirical point of view, what is known about the interstellar medium and the central role played by the gas and dust found there in the unfolding of the larger story of the life, birth, and death of galaxies. Following an overview, the article proceeds to describe the environment of interstellar space, the physical processes thought to be important, and the diagnostic techniques used in the field. The central ideas on the nature of the interstellar medium are then summarized. Near the end of the article, the particular details of the region near the sun are generalized to elaborate on areas of research on interstellar material in other galaxies and to discuss the cosmological importance of interstellar medium research.

I. The Interstellar Medium— An Overview

A. CLOUDS

The interstellar medium in our galaxy contains four readily detectable components, three of which are described in terms of clouds of atoms and molecules. The brightest components are the so called HII regions, regions of ionized hydrogen that emit radiation when the protons (H⁺) and electrons recombine. The regions are constantly re-ionized by radiation from nearby stars. They are readily visible near stars with temperatures above 30,000 K (O stars). The Great Nebula in Orion, around the multiple star θ Orionis, is the best-known example of the HII region because it is visible to the naked eye.

The second component consists of dark clouds that are very cold and emit no visible radiation, but do emit far infrared (100-μm) and millimeter molecular line radiation. Absorption of background star light by dust in these clouds is responsible for the dark bands along the Milky Way in Scorpio and in Cygnus (the Cygnus rift) and for the famous dark spot in the southern sky known as the Coal Sack. [See ASTRONOMY, INFRARED]

A third component, detectable only with sophisticated instrumentation, consists of diffuse clouds. These are lower in mass than either HII regions or dark clouds and are warmer than dark clouds but are not warm enough to emit significant radiation. They contain too little dust to cause discernible extinction to the human eye.

B. THE INTERCLOUD MEDIUM

A fourth component, only recently appreciated, is a very hot phase, with $T \sim 10^6$ K, thought to fill much of the void between the stars. This phase has been detected only indirectly, and its direct detection represents the

most important frontier in the field. Gas with $T \sim 3 \times 10^5$ K (detected in the form of OVI or O⁺⁵) is ubiquitous and is thought to imply the presence of the hotter 10^6-K material. Soft X-rays detected from gas near the Sun may arise in the 10^6-K gas. X rays from supernovae remnants at 10^7 K are easily detectable. As these remnants expand and cool, they should produce large cavities at 10^6 K.

Figure 1 illustrates the location of interstellar gas of various types, in a schematic spiral galaxy, seen from above.

II. Physical Environment in Interstellar Space

A. GALACTIC RADIATION FIELD

Interstellar gas is ionized by radiation from many sources. The dominant source of high-energy photons, capable of ionizing hydrogen ($E > 13.6$ eV) is the O and B stars ($T > 10^4$ K). These stars primarily reside in spiral arms, but they produce a diffuse source of radiation with the spectral shape of a 25,000-K black body and an energy density of about 1 eV/cm³ (equivalent to the radiation detected from a single B1 star at a distance of about 30 light-years). Since the hot-

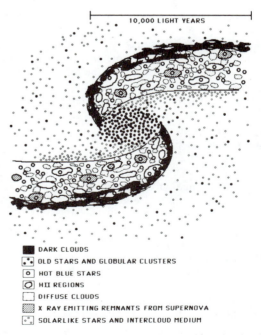

FIG. 1.   Relative location, within the spiral arms of a galaxy, of various types of interstellar material.

test, most massive stars form in groups, there are local maxima in this radiation field, discernible by the copious emission of HII regions. [*See* STAR CLUSTERS; STELLAR STRUCTURE AND EVOLUTION.]

The only place unaffected by this pervasive field is the interior of dark clouds. Hydrogen atoms at the exterior of clouds shield interior atoms from the high-energy photons ($E > 13.6$ eV, the ionization potential of HI), but lower-energy photons may penetrate. However, in the dark clouds, dust at the exterior provides a continuous opacity source, and so even visible light cannot penetrate.

Other sources of radiation include white dwarfs, hot cores of dying stars, and, of course, normal stars like the sun. The first two sources produce high-energy photons, but not in copious amounts. Normal stars produce an important source of photons only at $\lambda > 4000$ Å.

While X-ray sources permeate space, the X rays themselves have little measurable effect on interstellar clouds. X rays probably serve to warm and ionize cloud edges and to produce traces of high-ionization material.

Supernovae from time to time provide radiation bursts of great intensity, but, except perhaps in the distant halo, these have little global effect. (The mechanical impact of the blast wave from the explosion, on the other hand, is very important, as discussed later.) [*See* SUPERNOVAE.]

B. MAGNETIC FIELD

A weak magnetic field exists in space, though its scale length and degree of order are uncertain. The strength is roughly 10^{-6} G. The field was originally discerned when it was discovered that light from stars is linearly polarized ($\sim 1\%$). From star to star the polarization changes only slowly in strength and direction, suggesting that the polarizing source is not localized to the vicinity of the stars, but rather is interstellar in origin. It is thought that the field aligns small solid particles, which in turn lead to polarization of starlight. Since many elements are kept mostly ionized by the radiation field in space, the magnetic field constrains the motion of the gas (ions) and may be responsible for some leakage of gas from the galaxy.

C. COSMIC RAYS

Very high energy particles (10^6–10^{21} eV) penetrate most of interstellar space. In the dark clouds, where ionizing photons do not penetrate, collisions between atoms and cosmic rays provide the only source of ions. Cosmic rays probably provide the basis of much of the chemistry in dark clouds, since, except for H_2, the molecules seen are the result of molecule–ion exchange reactions. Collisions between cosmic rays and atoms are thought to produce most of the observed lithium, beryllium, and boron through spallation. The energy density of cosmic rays is about 1 eV/cm^3, similar to that of stellar photons. Cosmic rays interact with atoms, leading to pion production. These decay to γ rays. The diffuse glow of γ rays throughout the galactic disk is accounted for by this process. [*See* COSMIC RADIATION.]

D. COSMIC BACKGROUND RADIATION

Light from other galaxies, from distant quasars, and even from the primordial radiation has a measurable impact on the interstellar medium. There are about 400 photons/cm^3 (or about 1 eV/cm^3, since $\lambda \sim 1$ mm) in our galaxy from the radiation bath of the Big Bang, now cooled to 2.71 K. At the remote parts of the galaxy, the integrated light of the extragalactic nebulae at 5000 Å, and of the distant quasars at $E > 13.6$ eV ($\lambda < 912$ Å) is comparable to galactic sources of such radiation.

E. MECHANICAL ENERGY INPUT

Stars, in the course of their evolution, propel particles into space, often at thousands of kilometers per second. These may come from small stellar flares; from massive dumps of the envelope of one star onto a binary companion, leading to an explosion (novae or supernovae); from winds driven by, among other things, radiation pressure, especially in the hottest stars; or from detonation of stellar interiors, leading to supernovae. These particle flows generate pressure on the interstellar clouds, which in regions of massive stars and supernovae, substantially heats the gas and reshapes the clouds. The random motion of interstellar clouds is largely the result of the constant stirring caused by supernovae.

III. Physical Processes

A. ABSORPTION AND EMISSION OF RADIATION

Up to a point the physics of interstellar clouds is easy to ascertain. Atoms and simple molecules are easy to detect spectroscopically. Various effects lead to excitation or ionization of atoms and molecules. The extremely low densities (10^{-3}–10^5 particles/cm³), and the resulting great distances between atoms (millimeters to meters), means they seldom interact. When they do, the restoring radiative processes are usually fast enough that the atoms return spontaneously to their original energy states. The net result is quantized emission that is detectable at earth. The strength and wavelength of the emission can be used to discern the density, temperature, abundance, velocity, and physical environment of the emitting region.

The same physics implies that most atoms and molecules are in their ground electronic state, with very few of the higher energy states populated by electrons. Hence, when light from a given star is intercepted by an atom, only a very restricted band of photons corresponds to an energy that the atom can absorb. In these few cases, the absorption is followed immediately by return of the electron, excited by the absorbed photon, to its original state, and subsequent re-emission of an identical photon. There is only a tiny chance that the re-emitted photon will have exactly the same path of travel as the one originally absorbed. To an observer using proper equipment, it appears that a photon has been removed from the beam. Since there are 10^{10}–10^{21} atoms in a 1-cm² column toward typical stars, the many repetitions of the absorption/re-emission process leave a notch in the spectrum, called an "absorption line," because of the appearance of the features as seen in a spectrograph—dark lines across the bright continuous spectrum of a star. The absorption line strengths can be used to measure the physical conditions of the absorbing region. [*See* STELLAR SPECTROSCOPY.]

B. FORMATION AND DESTRUCTION OF MOLECULES

In the densest regions (10–10^5 atoms/cm³), molecules are known to exist in their lowest electronic states. The most abundant molecule, H_2, is thought to be formed on the surfaces of grains, small-sized (<1-μm) particles discussed later. Neutral hydrogen and molecular hydrogen are ionized by cosmic rays (see preceding discussion) as are other species such as neutral oxygen. A complicated chain of reactions (ion–molecular reactions) then ensues, since molecules and ions exchange electronic charge

at rapid rates under the conditions present in interstellar clouds. Neutral molecules are then formed by recombination of ionized molecules and ambient electrons.

Some molecules are too abundant to be formed in the reaction scheme noted (such as ammonia, NH_3). Perhaps formation on grains is important in such cases. While the production of H_2 on grains can be generalized, production of other molecules cannot be predicted without specific knowledge of the makeup of the grains, which we now lack (see later discussion).

Some molecules may be formed in diffuse clouds by shock-triggered processes. Molecules are destroyed by chemical reactions (to make other molecules) or by radiation. In general, the higher the photodestruction cross section of a molecule, the more deeply it will be buried (in an equilibrium between destruction and formation processes) in the denser regions of clouds.

The most ubiquitous molecules are H_2 and CO (10^{-5}–0.5 molecules per H atom). Two- and three-atom molecules are common (e.g., CH, CN, HCN, HCO⁺, with 10^{-13} to 10^{-8} molecules per H atom). The most massive molecule now known is $HC_{11}N$. However, there is speculation that molecules as large as C_{60} may be present in large numbers. Because the nature of grains is poorly understood, it is possible that some observations that require small grains could be accounted for by large molecules (see later discussion).

C. IONIZATION AND RECOMBINATION

Atoms and molecules in clouds are infrequently hit by photons or particles (atoms, molecules, or grains). When sufficient energy exists in the collision, an electron may be removed (ionization), producing a positively charged particle. Coulomb attraction between these particles and free electrons leads to recombination, reinstating the original state of charge. Recombination may be due to an excited state of the neutral species. Subsequent decay of the electron to the ground state, through the quantized energy levels of the neutral atom or molecule, leads to emission of photons. The best-known examples are the Balmer lines of hydrogen seen in recombination from HII regions, the strongest of which is the Hα line at 6563 Å, that leads to the red appearance of some emission nebulae in color photographs. Sometimes the recombination process leads to population of atoms in "forbidden" states not formally permitted to be populated by absorption from the ground state because of quantum mechanical selection rules. These states are normally suppressed by collisional de-excitation in laboratory gases, but at the densities encountered in space ($<10^4$ ions/cm^3), radiative decay occurs before collisional de-excitation can occur. The best-known cases are the oxygen forbidden lines, particularly $\lambda 5007$ of O^{+2}, that gives a greenish appearance to some nebulae.

D. Shock Fronts

Discontinuities in pressure occur in the interstellar medium, caused by explosions of stars, for instance. Propagation of such a pressure shock through the medium, into clouds in particular, leads to observable consequences. Shocks propagating at velocities above 300 km sec^{-1} randomly thermalize the gas impinged upon to temperatures above 10^6 K, where cooling by radiation of a gas with cosmic or solar abundance is inefficient. Such shocks are referred to as adiabatic and lead to growth of large cavities of hot, ionized gas, transparent to radiation of most wavelengths. Eventually, particularly when a dense cloud is hit by a shock, the shock speed is reduced; hence it thermalizes gas to a lower temperature, and cooling is more important. The radiation that cools the gas is just the recombination radiation discussed earlier. Thus, if a supernovae explosion occurs in a low-density region containing higher-density (hence cooler) clouds, the propagating shock wave will lead to observable radiation from cooling just where the clouds are. Supernovae remnants, such as the famous Vela or the Cygnus Loop, are thus largely filamentary to patchy in visible light due to recombination radiation from cloud edges.

E. Dust Scattering

Several observations of stars indicate the presence of dust particles in space. The first and chief indication of their presence, historically, is that stars of the same temperature (based on stellar spectra) have different colors. In general, fainter, more distant stars are redder. Some clouds have so much dust they extinguish all background stars at $\lambda < 2$ μm.

Models that try to account for the detailed features of the reddening suggest the existence of silicate grains, and perhaps some graphite grains. Sizes range from 100 Å to 1000 Å (0.01–0.1 μm) with perhaps a core of silicates and a

mantle or outer shell of amorphous ices. Such grains could produce extinction by direct scattering into directions away from the line of sight of the observer, by pure absorption in the bulk of the grains (with reradiation into longer wavelengths) and by interference of refracted and scattered radiation on the observer's side of the grains. Such grains, if irregular in shape, can be aligned by the interstellar magnetic field, leading to polarization of starlight.

F. Grain Formation and Destruction

Since the exact makeup of the grains is unknown, it is not clear how they form. Specific models for grain makeup allow one to pose questions as to how certain types of grains might form. For instance, cool stars are known to expand and contract periodically (over periods of months). At the maximum expansion, the outer envelopes reach temperatures below 1300 K. At the relevant densities, solid particles can condense out of the gas. The content of the grains depends on the composition of the stellar atmospheres, but the atmospheres seem varied enough to produce, in separate stars, silicate- and carbon-based cores. Such grains may be separated from the infalling, warming gas during the contraction of the atmosphere by radiation pressure, depending upon residual electric charge on the grains.

Subsequently, grains may grow directly in interstellar clouds. In this case, small-grain cores (or perhaps large molecules) must serve as seeds for growth of the remainder of the grain by adhesion of atoms, ions, and molecules with which the seed grains collide. The type of grains formed would then depend on the charge on the grains and the charge on various ions and on the solid-state properties of the grain surface and of monolayers that build up on it.

G. Heating and Cooling of the Gas

Cooling processes in interstellar gas can be directly observed because cooling is mainly through line radiation. Diffuse clouds are cooled mainly by radiation from upper fine structure levels of species such as C^+, which are excited by collisions of atomic or ionic species with electrons, neutral hydrogen, hydrogen ions, or H_2. In dense clouds, this process is important in atoms such as carbon. In hot, ionized regions (HII regions, $T \sim 10^4$ K) recombination radiation of H^0 and O^+ carries away energy from the gas, whereas at temperatures of 10^5 K (shocks in supernova remnants) recombination and subsequent radiation from forbidden levels of O^{+2} and other multiply ionized species dominates.

Heating of the interstellar gas is poorly understood. The heating sources observable (X rays, cosmic rays, exothermic molecular reactions) are inadequate to explain the directly observed cooling rates, given the temperatures observed in diffuse clouds ($T \sim 100$ K). The prime candidate for the primary heating mechanism is photoelectron emission from grains caused by normal starlight at $\lambda \lesssim 3000$ Å striking very small grains.

Heating of dense molecular clouds ($T \sim 10$ K) is also uncertain. Obvious gas phase processes appear to be inadequate. An additional possibility in such clouds is that star formation activity (directly by infrared radiation or indirectly by winds or shocks caused by forming stars) heats the clouds. Localized shocked regions with $T \sim 25$ K have now been directly observed in dense molecular clouds through emission from rotationally and vibrationally excited H_2.

IV. Diagnostic Techniques

A. Neutral Hydrogen Emission

The spin of the nucleus (proton) of hydrogen and the electron can be parallel or antiparallel. This fact splits the ground state of hydrogen into two levels so close together that the populations are normally in equilibrium. In practice this means that the higher energy state is more frequently population and that the resulting emission, in spite of its low probability, yields a detectable emission at $\lambda = 21$ cm. In many cases, when the column density in each cloud (or velocity component) is not too high, the power detected at 21 cm is directly proportional to the number of hydrogen atoms on the line of sight of the main lobe of the radiation pattern of the radio telescope. The total number of atoms is generally expressed as a column density N_{HI} in units of atoms per square centimeter. The detection is not dependent on the *volume* density of the gas (denoted as n_{HI} atoms/cm³) and yields no direct information on the distance of the emitting atoms: all atoms in the velocity range to which the receiver channel is sensitive are counted. Generally, the local expansion of the universe removes atoms not in our galaxy from the receiver frequency by Doppler shifting their 21-cm emission to longer wavelengths.

B. Molecular Emission

Molecules, while generally in the ground electronic state, are excited by collisional or radiative processes to higher rotational and vibration levels. Of the some 40 molecular species known in interstellar clouds, most are detected in the millimeter wavelength region through rotational excitation. Hydrogen has been detected in rotational and vibrational emission and may be detectable in electronic emission (fluorescence).

If the mechanism populating the higher level of a transition is known, the emission strength gives some information about the mechanism. For instance, if emission from several different upper levels of a molecule is detectable, the relative population of the states can be determined. Given molecular constants and the temperature, the density of hydrogen can be determined if excitation is by collisional processes.

There are several unidentified emission features in the near infrared that are apparently related to interstellar material, some of which have been attributed to molecules in the solid phase or to polycyclic aromatic hydrocarbons. Identification of these features is being actively pursued in several laboratories.

C. Atomic Emission

Recombination lines and forbidden emission lines of atoms allow determination of abundances, densities, and temperatures in HII regions. Optical telescopes provide the main data for these lines. Eventually, fine structure emission in the infrared will give us detailed abundance data inside molecular clouds. Such data are currently available only for selected regions.

D. Atomic and Molecular Absorption

Absorption lines, already explained, are used to derive column densities of many species. Because of collisional de-excitation mechanisms, H_2CO is seen in absorption against the microwave cosmic background of only 3 K. Twenty-one–centimeter radiation is absorbed by H atoms in the lowest hyperfine state against background radio continuum sources. Resonance absorption lines of molecules such as C_2, H_2, CN, and CH^+ are seen in the optical and ultraviolet spectral regions. Resonance (ground-state) transitions of most of the first 30 elements in atomic or ionic form are seen as optical or ultraviolet absorption against stellar continuum spectra. In special circumstances, the degree of absorption is related linearly to the number of atoms leading to derived column densities (particles per square centimeter) to be compared with corresponding values of N for hydrogen. The ratio $N(X)/N(HI)$ gives an abundance of the species X, though in practice several ionization states of a species X must be accounted for. Conversely, some ions of heavy elements may be detected when hydrogen is ionized (HI unobservable), and the number of ionized hydrogen atoms must be accounted for (using, for instance, knowledge of the electron density and the intensity of the Balmer recombination radiation).

When ratios $N(X)/N(HI)$ are available for clouds, they are frequently referred to solar abundance ratios. If there are 1/10 as many atoms of a certain kind per H atom as found in the sun, the element is said to be depleted by a factor of 10 in interstellar space.

E. X-Ray Emission and Absorption

X-ray absorption lines have not yet been detected from the hot gas mentioned earlier. While this is one of the most important measurements to be done in the study of interstellar material, it awaits the arrival of a new generation of very large X-ray telescopes. X-ray absorption edges have been detected, but these are contaminated by circumstellar absorption in the X-ray source itself, with little possibility of the velocity distinction possible in resonance absorption lines.

X-ray emission arises from radiative recombination and collisional excitation followed by radiative decay. Broad band X-ray emission from interstellar gas at 0.1 to 1-keV energies has been detected, but no high-quality spectra are yet available of the resolved emission lines from the hot gas. Such measurements are extremely important because the large hydrogen column densities at distances >100–200 pc absorb such soft X-ray photons. Any detections will thus refer only to very nearby gas, which can therefore be studied without confusion from more distant emission. Broadband X-ray emission at higher energies from the diffuse hot gas cannot yet be separated from a possible continuum from distant QSOs and Seyfert Galaxies.

Higher-energy (>1-keV) X-ray lines and continuum have been detected from supernova remnants and from very hot ($T \sim 10^7$ K) gas falling into distant clusters of galaxies.

F. γ-Ray Emission

As noted earlier, cosmic rays interact with interstellar clouds to produce γ rays. A knowledge of the distribution of interstellar clouds and of the observed distribution of diffuse γ radiation may lead to a detailed knowledge of the distribution of cosmic rays in the galaxy. Our current knowledge is based largely on the cosmic rays detected directly at one point in the galaxy (earth).

V. Properties of the Interstellar Clouds

Given the many possibilities for detection of radiation emitted or modified by interstellar gas, astronomers have pieced together a picture of the interstellar medium. In many cases, the details of the physical processes are vague. In most cases, detailed three-dimensional models cannot be constructed or are very model dependent. On the other hand, in some cases, sufficient knowledge exists to learn about other areas of astrophysics from direct observations. The example of the cosmic-ray distribution has already been given. Others are mentioned subsequently.

Interstellar clouds are complex aggregates of gas at certain velocities, typically moving at ±6 to 20 km sec^{-1} with respect to galactic rotation, itself ~250 km sec^{-1} over most of the galaxy. Each cloud is a complex mixture of a volume of gas in a near pressure equilibrium and of isolated regions affected by transient pressure shocks or radiation pulses from star formation or from supernova explosions. Clouds are visible in optical, UV, or X-ray emission (or continuum scattering) when they happen to be close to hot stars and are otherwise detectable in absorption (molecules or atoms) or emission from low-lying excited levels (≥ 0.01 eV) or from thermal emission of the grains in the clouds.

A. Temperatures

Molecular clouds are as cold as 10 K. Diffuse clouds are typically 100 K. HII regions have $T \sim$ 8000 K, depending on abundances of heavy elements that provide the cooling radiation. Low-column-density regions with 10,000 K $< T <$ 400,000 K are seen directly, presumably the result of heating at the cloud edges from shocks, X rays, and thermal conduction. Isolated regions with $T > 10^6$ K are seen near sites of supernova explosions.

B. Densities

Densities, as determined from direct observation of excited states of atoms and molecules, are generally inversely proportional to temperature, implying the existence of a quasi-equilibrium state between the various phases of the medium. The effects of sources of disequilibrium in almost all cases last $\lesssim 10^7$ years, or less than one-tenth of a galactic rotation time, itself one-tenth of the age of the sun. The product nT (cm^{-3} K) is ~3000 to within a factor of 3 where good measurements exist. Thus, the molecular (dark) clouds have $n > 10^2 \ cm^{-3}$, while in diffuse clouds, $n < 10^2 \ cm^{-3}$. Higher densities (up to $10^5 \ cm^{-3}$) occur in disequilibrium situations such as star-forming regions inside dense clouds and in HII regions.

C. Abundances: Gas and Solid Phases

By measuring column densities of various elements with respect to hydrogen, making ionization corrections as necessary, abundances of elements in interstellar diffuse clouds can be determined. Normally, the abundances are compared with those determined in the sun.

Different degrees of depletion are found for different elements. Oxygen, nitrogen, carbon, magnesium, sulfur, argon, and zinc show less than a factor of 2 depletion. Silicon, aluminum, calcium, iron, nickel, manganese, and titanium show varying degrees of depletion. Correlations of depletion with first ionization potential, or with the condensation temperature (the temperature of a gas in thermal equilibrium at which gas phase atoms condense into solid minerals), have been suggested, but none of these scenarios actually fits the data in detail.

The pattern of depletion suggests no connection with nucleosynthetic processes. Those elements that are depleted are presumed to be locked into solid material, called grains. Such particles are required by many other observations attributed to interstellar gas, as discussed earlier. In principle, the unknown makeup of the grains can be determined in detail by noting exactly what is missing in the gas phase. However, since there must be varying sizes and probably types of grains and since the most obviously depleted elements do not constitute enough mass to explain the total extinction per H atom, most of the grains by mass must be in carbon and/or oxygen. Establishing the exact mass of the grains amounts to measuring the depletions of C and O accurately, a task still beyond reach.

High-resolution spectrographs on earth-orbiting UV telescopes should succeed in making the required measurements by 1990.

The grain structure (amorphous or crystalline) is not known. There are unidentified broad absorption features, called diffuse interstellar bands, that have been attributed to impurities in crystalline grains. However, these features may be caused by large molecules. It has been argued that even if grains are formed as crystalline structures, bombardment by cosmic rays would lead to amorphous structures over the life of the galaxy.

Theories of grain formation are uncertain. A general scenario is that they are produced in expanding atmospheres of cool supergiants, perhaps in very small "seed" form. They may then acquire a surface layer, called a mantle, probably in the form of water ice and solid CH_4, NH_3, etc. This growth must occur in cold dense clouds. The detailed process, and the distribution of atoms between minerals and molecules in solid phase, is unknown.

D. Evolution

Interstellar clouds can be large, up to 10^6 solar masses, and are often said to be the most massive entities in the galaxy. In this form, they may have a lifetime of more than 10^8 years. They are presumably dissipated as a result of pressure from stars formed within the clouds. Over the lifetime of the galaxy, interstellar clouds eventually turn into stars, the diffuse clouds being left over from the star formation process. Growth of new molecular clouds from diffuse material is poorly understood. Various processes to compress the clouds have been suggested, including a spiral density wave and supernovae blast waves. No one mechanism seems to dominate and several may be applicable. However, the existence of galaxies with up to 50% of their mass in gas and dust and of others with less than 1% of their mass in interstellar material leads to the inference that diffuse material and molecular clouds are eventually converted into stars.

VI. Properties of the Intercloud Medium

A. Temperatures

As already suggested, the medium between the clouds is at temperatures greater than 10^4 K. The detection of soft X rays from space indicates that temperatures of 10^6 K are common locally. Detection of ubiquitous OVI absorption suggests there are regions at $\sim 3 \times 10^5$ K. Various observations suggest widespread warm neutral and ionized hydrogen. Attempts to explain this 10^4-K gas as an apparent smooth distribution caused by large numbers of small clouds with halos have not been successful. Thus there is evidence for widespread intercloud material at a variety of temperatures, though large volumes of gas near 80,000 K are excluded. The more tenuous diffuse clouds, however formed, may be constantly converted to intercloud material through evaporation into a hotter medium.

B. DENSITIES

In accordance with previous comments, all indications are that approximate pressure equilibrium applies in interstellar space. The above temperatures then imply intercloud densities of 10^{-1} H atoms/cm^3 to 10^{-3} H atoms/cm^3. While direct density measurements in such regions are possible at $T < 30,000$ K, through studies of collisionally excited C$^+$ and N$^+$, direct determinations in hotter gas are spectroscopically difficult. X-ray emission has not yet been resolved into atomic lines and so is of limited diagnostic value. Direct measurements of ions from 10^4 to 10^7 K in absorption over known path lengths must be combined with emission line data to fully derive the filling factor, hence the density, at different temperatures. Since emission lines arise over long path lengths, velocities must be used to guarantee the identity of the absorption and emission lines thus observed. Such data will not be available for several years.

C. ABUNDANCES

The depletion patterns already noted in the discussion of clouds appear qualitatively in all measures of gas at 10^4 K, ionized or neutral. In general, the intercloud gas is less depleted. Perhaps shocks impinging on this diffuse medium lead to spallation of grains and return of some atoms to the gas phase. Data on hotter gas come mainly from absorption lines of OVI, CIV, and NV. Since ionization corrections are not directly determinable and since the total H$^+$ column densities are not known at the corresponding temperatures of 5×10^4 to 5×10^5 K, abundances are not available. However, the ratios C/O and N/O are the same as in the Sun. It is not known whether elements such as iron,

calcium, and aluminum are depleted in this hot gas. Optical parity-forbidden transitions of highly ionized iron or calcium may some day answer this important question.

D. EVOLUTION

The evolution of the intercloud medium depends on the injection of ionization energy through supernova blast waves, UV photons, and stellar winds. A single supernovae may keep a region of 100-pc diameter ionized for 10^6 years because of the small cooling rate of such hot, low-density gas. Ionizing photons from O stars in a region free of dense clouds may ionize a region as large as 30–100 pc in diameter for 10^6 years before all the stellar nuclear fuel is exhausted. Thus in star-forming regions of galaxies with low ambient densities and with supernova rates of 1 per 10^6 years per (100 pc)^3 and/or comparable rates of massive star formation, a nearly continuous string of overlapping regions of 10^4 K to 10^6 K can be maintained. When lower rates of energy input prevail, intercloud regions will cool and coalesce, forming new clouds. In denser regions, comparable energy input may not be enough to ionize the clouds, except perhaps near the edges of the dense region, for periods as long as 10^8 years.

VII. The Interstellar Medium in Other Galaxies

While much is unknown about the actual balance of mechanisms that affect the distribution of gas temperatures and densities in our own galaxy, the facts that are known make interstellar medium observations in other galaxies an important way to determine properties of those galaxies as a whole. A few examples are mentioned here.

A. SUPERNOVA RATES

Supernova are difficult to find because the visible supernova occur only once per 30 to 300 years. Regions with higher rates are often shrouded in dust clouds or extinguished by nearby dust clouds. Thus, little is known about supernovae rates and their global effects on galaxies and on the orgin of elements. Nucleosynthesis in massive stars and in the supernovae explosions, with subsequent distribution to the diffuse interstellar medium, may have occurred at variable rates, perhaps much more frequently in the early stages of galaxy formation than now.

Studies of interstellar media in other galaxies can shed light on these subjects. X-ray emission and atomic line emission can reveal the presence of supernovae remnants, which, since they last up to 10^5 years or more, are easier to find than individual supernovae, which last only a few months. Absorption line measurements and emission line measurements can be used to determine abundances in other galaxies. Absorption line measurements reveal the velocity spread of interstellar gas, the stirring effect caused by supernovae integrated over 10^6 to 10^7 years. (Quasi-stellar objects, clumps of O stars, or the rare supernova can be used for background sources in such absorption line studies.)

By making the above-noted interstellar medium studies on samples of galaxies at different redshifts, the history of galaxy formation can be discerned over a time interval of roughly three-fourths the expansion age of the universe.

B. Cosmic-Ray Fluxes

The origin of cosmic rays is unknown. However, they account for a large amount of the total energy of the galaxy. In situ measurements are only possible near earth. However, cosmic rays provide the only explanation of the observed abundances of boron, beryllium, and lithium ($\sim 10^{-9}$–10^{-10} atoms/H atom). Thus the abundance of any of these elements is a function of the integrated cosmic-ray flux over the lifetime of the galaxy. Variations in the ratio [B/H] would imply different cosmic-ray fluxes. Comparison of [B/H] with other parameters related to galaxy history (mass, radius, total $H\alpha$ flux, and interstellar cloud dispersions) may indicate the history of cosmic rays.

C. History of Element and Grain Formation

As is clear from previous sections, studies of various kinds of clouds in our own galaxy have not led to a clear empirical picture of how the interstellar medium changes with time. The trigger or triggers for star formation are poorly understood, and the history of the clouds themselves is unknown in an empirical sense. Numerous problems exist on the theoretical side as well.

Study of interstellar media in other galaxies should be very important in changing this situation. Absorption and emission measurements that provide data on individual clouds and on ensembles of clouds should allow classifications of the gas phase in galaxies that can be compared with other classifications of galaxies by shape and total luminosity. Because of the rarity of background sources for absorption studies, studies of optical emission lines offer the best chance of tracing several key parameters back through time. Key questions include these: Do elements form gradually over time, or are they created in bursts at the beginning of the life of the galaxy? Do earlier galaxies have the same extinction per hydrogen atom as is found in our own galaxy, or are differences seen perhaps because grains require long growth times to become large enough to produce extinction at optical wavelengths? Do the many unidentified interstellar features (optical absorption, IR emission and absorption) occur at all epochs, or are they more or less present at earlier times? Does the star formation rate depend only on the amount of hydrogen, or are other factors such as metallicity important?

D. Cosmological Implications

For various reasons the interstellar medium has proved to be fruitful ground for determining cosmological quantities. Without detailed comparison with other techniques, a few examples are given. [*See* COSMOLOGY.]

The light elements hydrogen and helium are thought mainly to be of nonstellar origin. They are currently thought to be formed in the Big Bang. Expansion of the early cosmic fireball leads to a small interval of time when the gas has the correct temperatures for fusion of hydrogen to deuterium and of deuterium to helium (^{3}He and ^{4}He). The helium reactions are so rapid that deuterium remains as a trace element, the abundance of which is dependent on the density of matter at the time of nucleosynthesis. By knowing the expansion rate of the universe, the derived density can be extrapolated to current densities. Comparison with other determinations of the current mean density of matter today shows that astronomers are detecting less than 10% of all the matter in protons and neutrons at the time of nucleosynthesis.

The amount of helium present is similarly important in deriving the properties of matter at the time of nucleosynthesis. In principle, the helium abundances today provide tests of fundamental particle physics because they depend on the number of neutrino types (currently thought to be three) and on the validity of general relativity.

Deuterium abundances are currently best determined in UV absorption line experiments in local interstellar matter. Helium abundances are best determined by measurements of optical recombination lines of He^{++} and He^+ from HII regions in dwarf galaxies with low metal abundance.

The preceding comments suggest how interstellar medium studies allow a view of the very earliest stages of the formation of the universe through measurements of relic abundances of deuterium and hydrogen. Earlier comments related the importance of such studies to observing and understanding the formation and evolution of galaxies through studies of abundances and star formation rates. A third example is the study of clustering of galaxies at different redshifts. Since galaxies are very dim at cosmological distances and since they may not have central peaks in their light distributions, they are difficult to detect at $z > 2$. Even at $z = 1$, only the most luminous galaxies are detectable. However, interstellar medium observations of absorption lines depend not on the brightness of the galaxy, but on the brightness of an uncorrelated, more distant object, say a QSO. Thus, galaxies at very high redshift can be studied. Many QSOs side by side will pass through a number of galaxies in the foreground, and with adequate sampling, the clustering of absorption lines reflects the clustering of galaxies on the line of sight.

Depending on the total matter density of the universe, the galaxies at high redshift will be clustered to a comparable or lesser degree than they are today. Thus, changes in clustering between high-z galaxies (interstellar media) and low-z galaxies (direct photographs and redshifts) reflect the mean density of the universe. The material measured in this way includes any particles with mass, whereas the density measurement discussed earlier, using deuterium, counts only neutrons and protons. The difference between the mass density determined from changes in clustering and the density determined from deuterium gives a measure of the mass of the universe in particles that are weakly interacting with matter, such as axions and neutrinos. Current estimates are that 1%–2% of the total mass is in luminous stars, 10%–20% is in dark matter made of baryons and protons, (e.g., planets or black holes), and 80%–90% is in unseen, weakly interacting particles of unknown nature. Thus the clustering experiment described here can be regarded as fundamental to our understanding of the nature of the Universe. [*See* BLACK HOLES (ASTRONOMY).]

BIBLIOGRAPHY

Bally, J. (1986). Interstellar molecular clouds. *Science* **232**, 185.

Boesgaard, A. M., and Steigman, G. (1985). Big Bang necleosynthesis: Theories and observations. *Annu. Rev. Astron. Astrophys.* **23**, 319.

Cowie, L., and Songaila, A. (1986). High-resolution optical and ultraviolet absorption line studies of interstellar gas. *Annu. Rev. Astron. Astrophys.* **24**, 499.

McGray, R., and Snow, T. P. (1979). The violent interstellar medium. *Annu. Rev. Astron. Astrophys.* **17**, 213.

Savage, B. D., and Mathis, J. S. (1979). Observed properties of interstellar dust. *Annu. Rev. Astron. Astrophys.* **17**, 73.

STAR CLUSTERS

Steven N. Shore *New Mexico Institute of Mining and Technology**

GLOSSARY

Association: Loosely gravitationally bound group of coeval stars, including massive (OB) stars and often numerous subgroups of more tightly bound systems. Generally associated with massive molecular clouds and sites of recent star formation.

Hertzsprung–Russell diagram (H–R diagram): Plot of loci of stars by surface temperature or color versus luminosity or absolute magnitude.

Horizontal branch: Stage of helium core burning. In evolved clusters, this appears as a nearly horizontal grouping of stars (i.e., nearly constant luminosity) in the Hertzsprung–Russell diagram.

Main sequence: Stage of a star's evolution when energy is generated via hydrogen processing in the stellar core.

OB Stars: Massive stars, typically more massive than $\sim 10\ M_\odot$, which have short main-sequence lifetimes and evolve into luminous supergiants and, probably, Wolf–Rayet stars. Generally found in associations.

Pre-main-sequence star: Star in the stage of evolution, before hydrogen core ignition, when energy is still being generated by gravitational contraction and mass accretion.

Stellar populations: Differentiation between stars of different ages, determined from their metal abundances and distribution. Population I stars are young stars confined

* Also with DEMIRM, Observatoire de Meudon.

to the disk of the galaxy and of metal abundances near the solar value; Population II stars are the oldest stars in the galaxy and reside primarily in the galactic halo, with metallicities much less than the sun's.

Star clusters are gravitationally bound, presumably coeval groups of stars. Clusters can be morphologically distinguished between open or galactic and globular on the basis of both the overall geometry and the stellar density. Globular clusters, which reside primarily in the halo of the galaxy, are associated with an older population and are typically an order of magnitude more metal poor than the disk. They contain upwards of 10^5 members with typical radii of the order of 1 to 10 parsecs (pc). Open clusters are usually of much lower mass, containing of the order of 100 to 1000 members, and are among the recently formed stars in the galaxy. They span the age and metallicity range of the Population I stars. The OB associations are looser groups of massive stars and contain the most recently formed stars. These are thought to be a possible extension to the lower-mass open-cluster sequence. The populous blue clusters are observed in a number of external galaxies, notably the Magellanic Clouds. The use of clusters for the study of stellar evolution and metals production in galactic history is facilitated by the fact that they display a mass range that is similar to that observed for disk stars and appear to form stars sufficiently rapidly that there is no great spread in their basic chemical properties. These objects provide the basic test of models for stellar evolution and also a fundamental distance calibration.

I. Historical Introduction

The first observations of a cluster can be found in the earliest compilations of the constellations by the Greeks. The Hyades, Pleiades, and Praesepae are all mentioned. These are the most visible northern open clusters, the first two located in Taurus and the last in Cancer. That these stars are in fact of higher density than the rest of the field was initially recognized by Galileo in 1610, in the first use of the astronomical telescope. In the "Siderius Nuncius" he mentions the Pleiades specifically, showing that it has a considerably larger population than can be seen with the naked eye and larger even than the neighboring galactic plane. A more complete collection of cluster data can be found in the first telescopic surveys by Halley in the seventeenth century and Messier about a century later. These revealed the existence of several classes of clusters—specifically the globular and compact open varieties. For a time, it was generally believed that all such objects were nebulous, in agreement with the Kant–Laplace nebular hypothesis, which appeared to predict the preplanetary stage these objects represented. The resolution by Herschel and later observers of many of these nebulae into stars for a time precluded the general acceptance of the idea of star formation from gaseous matter, but the spectroscopic observations of Huggins and Secchi in the late nineteenth century assisted in distinguishing clusters from nebulae.

In the first years of the twentieth century, observations of the globular clusters at Harvard under S. Bailley and later H. Shapley showed that there is a distinct class of variable star associated with the clusters. These are the RR Lyrae stars, a group of horizontal branch stars with periods of $\sim\frac{1}{2}$ day and that display (like many pulsating variables) a distinctive period–luminosity relation. The use of this relation, coupled with the velocity and positional measurements for the clusters, enabled Shapley to determine that the globular clusters form a spherical distribution about the galactic center and that we are displaced from that center. The precise value is still under debate, but is of the order of 8 kpc.

In the 1930s, R. Trumpler discovered that the most distant of the clusters also appeared to be more reddened than the local stars. This led to the discovery of interstellar dust and further served to correct the determination of the distance scale. The discovery of globular clusters in the halos of external galaxies has continued at a rapid pace in recent years, extending their use as distance indicators.

Observations of the colors, motions, and brightnesses of cluster stars had become suffi-

ciently advanced by the mid-1940s to reveal that the stars are in fact coeval. M. Schwarzschild and F. Hoyle, and later A. Sandage, exploited this property of the members to test models for stellar evolution. The argument proceeds as follows. If the stars are formed at the same time and have different masses, then the most massive ones should evolve the most rapidly toward the red giant phase. By using the luminosity and effective temperature of the brightest main-sequence star, one should therefore be able to find a corresponding mass for the most evolved hydrogen-core-burning star from which the age of the cluster follows. E. Salpeter extended this to the determination of the initial mass function (IMF) by determining the relative number of stars of each mass that are needed to synthesize the morphology of the H–R diagram of the cluster. Fitting a power law shows that many more low-mass than higher-mass stars are formed in these clusters. The same is true for the globular clusters, although there is a narrower mass range. [See STELLAR STRUCTURE AND EVOLUTION.]

The globular clusters were shown by W. Baade to have the same population mix as, and similar metallicities to, the stars found in the halo of the galaxy and M31, the Andromeda galaxy. Specifically, when a sample of high-velocity field stars (presumed to be the stars that have the most extended distribution and thus are moving statistically faster) is obtained, their overall properties closely match those of the globular clusters. The distributions in space, around the disk of the galaxy, are also quite similar. From the low metallicities and the extended spatial distributions, Baade and subsequent investigators argued that the globular clusters were formed in a rapid sequence of events in the early history of the galaxy, thus forming a sample of the stages before the formation of the disk stars.

Observations from space began for these objects in the late 1960s with the launch of OAO-2, which performed UV photometric observations in many of the globular clusters and individual stars in open clusters. Additional work by International Ultraviolet Explorer (IUE) has shown that a number of globular clusters possess UV-bright cores. UHURU and EINSTEIN observations have revealed the existence of a class of bright X-ray sources, at first thought to be central block holes and now believed to be low-mass binaries, in $\sim 10\%$ of the globular clusters. None of the open or globular clusters appears to be a γ-ray source. [*See* ASTRONOMY, ULTRA-VIOLET SPACE.]

II. Galactic Open Clusters

A. INTRODUCTION

The young stellar population of the galaxy, Population I, whose metal abundance is about the same as that of the sun, appears to form only loose clusters, consisting of some hundreds of members, which have ages ranging from only a few hundreds of thousands of years to about the age of the oldest globular clusters, of the order of 10^{10} yr. Among the youngest are NGC 2244, 2264, and 6530 (all with ages of less than a few tens of millions of years), and among the oldest are NGC 188, NGC 2506, Mel 66, and M 67, which are as old as the oldest disk stars (about 10 Gyr). The mechanism of formation appears to be the same as that of the associations; that is, the stars are formed from a parent molecular cloud in which many stellar masses are present. If we take the mass of the typical cluster to be several hundred solar masses, this places a lower bound to the mass of the parent cloud. The formation time for the stars in the cluster appears to be quite short, of the order of the contraction time for a star of a few solar masses, though we shall return to this point later.

The stars in the cluster all have a common proper motion and thus were formed in a gravitationally bound system. The mass of the cluster can be estimated as follows. There is a simple relation between the total kinetic and gravitational potential energies of a group of bound objects in a gravitational field, called the virial theorem, which is that

$$2T + \Omega = 0 \qquad (1)$$

where

$$T = \tfrac{1}{2}Nm_*\sigma^2$$

is the mean kinetic energy of the N stars in the cluster, which are all assumed to have about the same mass m_*, and σ is the velocity dispersion, and

$$\Omega = -GN^2m_*^2\langle R\rangle^{-1}$$

is the mean gravitational potential energy, where $\langle R\rangle$ is the mean radius of the stars from the center of the cluster. Assuming that the stars are collisionless, which can be shown to hold to a good approximation for the lifetime of the cluster, it is possible to estimate the mass of the

cluster. The agreement between the calculated mass required to bind the cluster and the observed masses of the stars (inferred from their luminosities and spectral types) shows that the clusters were formed as gravitationally bound systems.[1]

B. ISOCHRONES AND THE INTERPRETATION OF H–R DIAGRAMS FOR CLUSTERS

The fact that the stars were all formed at approximately the same time means that the cluster represents a snapshot of stellar evolution for the masses of the component stars. Assume that we have two stars of different mass, $M_1 > M_2$, which start their life at the same time. The more massive star will evolve faster owing to the higher rate of core nuclear burning and thus will become a red giant in a shorter time than the lower-mass star. If we know that the stars on the main sequence, the hydrogen-core-burning stage of evolution, have no more than some mass, say M_{max}, then we can place a lower limit on both the age and the mass of the evolved star. If we know, for example, that M_2 is a main-sequence star and M_1 is a red giant, then a lower limit on the mass of M_1 can be determined. By fitting stellar evolutionary tracks to the entire ensemble of cluster stars, one can determine the number of stars in a given mass range that were formed at the time of formation of the cluster, the age of the cluster, the initial chemical composition, and the distance to the cluster through the knowledge of the masses and absolute luminosities of the component stars.

In order to employ isochrones, however, the cluster's distance must be determined. The raw data are simply the visual apparent brightness (or V magnitude) of the star plotted versus its color, or B–V (the more negative B–V, the bluer is the star; the lower the V magnitude, the brighter is the star). Figure 1 shows the color–magnitude or H–R diagram for a moderately young open cluster, M11. Note the well-populated main sequence and the scattering of a few red giants to the right of the figure. Figure 2 shows the old disk cluster M67. Here, the sub-

---

[1] The total energy of the cluster is given by

$$E = T + \Omega < 0$$

which shows that the clusters are bound (that gravity wins out). It should also be added that the virial theorem holds for any gravitationally bound cluster, even of galaxies.

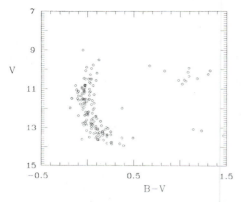

FIG. 1.   Young open cluster, M11. [Courtesy of B. Anthony-Twarog.]

giant branch is well populated, and the turnoff point is at about the age of the sun.

An additional use of isochrones concerns the determination of physical properties of envelope and core convection. In the stars more massive than ~1.5 $M_{\odot}$, the core is dominated by carbon-nitrogen-oxygen (CNO) processing on the main sequence and is consequently completely convective. The envelopes are radiative, so that there is a turbulent interface between the two regions. Overshooting of the turbulent cells from the core can promote mixing of fresh, hydrogen-rich material into the nuclear processing region, with the consequent increase in the mass of the chemically helium-rich core with time. This produces, at the end of the main-sequence stage, a more rapid than expected contraction of the core and a gap in the H–R diagram of the cluster immediately after the main sequence.

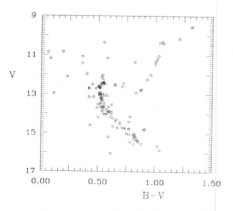

FIG. 2.   Old open cluster, M67. [Courtesy of B. Anthony-Twarog.]

Several clusters, notably M67 and NGC 188, have been used for the study of the interiors models, but the definitive answer must await the development of better interiors models. The post-main-sequence gap, as this feature of the diagrams is known, is one of the few probes we have, using the aggregate population of the clusters, of the detailed properties of the interiors of stars more massive than the sun. This gap can be clearly seen at $V = 13$ in Fig. 2, the color–magnitude diagram for M67.

C. INITIAL MASS FUNCTION

Since all of the stars were formed from the same composition and at the same time in the same place in the galaxy, one can use this important feature to trace the chemical history of the galaxy. The older clusters in the disk have been shown to have a slightly lower metallicity than the sun, whereas the ones just now being formed have a slightly higher or about the same value of metal abundance. This is one of the key observations indicating that the metallicity of the disk of the galaxy has been secularly changing in time. In addition, it appears that for the lifetime of the disk, during which the open clusters have been formed, there has been little change in the so-called initial mass function. This is the population distribution of the stellar masses.

The actually measured quantity for a cluster is the number of stars in a given range of color and visual magnitude. Stellar atmosphere models are used to determine the transformation between color and effective (surface) temperature, and this provides the bolometric (total) luminosity. The stars are then placed on a "theoretician's" H–R diagram (luminosity versus effective temperature) and traced back to the main sequence. Thereafter, their masses (assuming no mass loss has substantially altered their abundances or masses) can be specified, and the number per unit mass can be determined. This is the initial mass function of the cluster.

The observational picture is still only approximately known, but there is strong evidence that the IMF for cluster stars is the same as that for field stars, and that for at least the past 10^{10} yr it has remained stable, with few high-mass and many low-mass stars being formed. Simple power law fits to the distribution show that

$$\Phi(M)\, dM \sim M^{-\alpha}\, dM$$

with α being ~2. There is still considerable uncertainty in the exponent (± 0.5). A more accurate fit for field stars is a log-normal distribution. However, the basic result, that the massive stars are considerably rarer than the low-mass ones like the sun, is unchanged by this detail. The details of the origin of the mass function are, presently, unknown. Qualitatively, it is attributed to a stochastic fragmentation cascade to some minimum stable mass. It is also possible that it entails a kind of fragmentation–coagulation phenomenon.

The lower mass of star formation is also an important parameter for stellar evolution theory. The lower end of the main sequence, representing the minimum mass at which stable nuclear reactions can be initiated in the stellar core, appears to be at ~ 0.1 $M_\odot$. The minimum mass from fragmentation is of the same order, so the census of this population in clusters is a very important task. However, these stars are so faint that we will have to wait for the launch of the Hubble Space Telescope (HST) before they can be observed in any but the nearest clusters.

Another check on stellar evolution theory provided by open clusters is that there are white dwarf stars present in many of the older clusters. These are extremely compact remnants of stellar evolution, formed from stars of about the mass of the sun. They are no longer generating luminosity by nuclear processes, so their presence indicates their cooling time and rate of formation. Again, it is important to be able to observe these in systems of known age, since then both the ages of the white dwarfs and the masses of the progenitor stars can be specified. There are no pulsars associated with any cluster presently known, nor are clusters generally associated with supernova remnants. In general, the massive stars appear to have a connection with associations and not the open clusters, and this may be an additional clue to the formation mechanism for such systems.

In some clusters, notably the Pleiades (among the younger group of open clusters), there is a sequence above the main sequence, in the lower-mass star region of the H–R diagram, that may be an indication of a continuing star formation in the cluster. The blue stragglers may also represent such a group. The bounds on the time scale required for star formation to proceed in these systems is thus of the order of the pre-main-sequence contraction time for a very low-mass star to the main-sequence lifetime of a roughly 3 $M_\odot$ star. This is from about 10^7 to 10^8 yr.

D. ROTATION IN OPEN CLUSTERS

Studies of stellar rotational velocity in clusters have shown that there is also a distribution of angular momentum among young stars, such that the angular momentum J is larger for the more massive stars. Although at present we do not know why this is the case, the study of stars in clusters greatly aids the determination of this important parameter for star formation theory. The lower-mass stars are typically slow rotators, owing to loss of angular momentum during the main-sequence phase through stellar winds, but in some clusters there is an excess of rapid rotation on the lower main sequence.

One of the few relations that appears to hold for rotation has been determined by A. Skumanich. It is that the rate of chromospheric activity appears to decline with age and that, if this is due to the spin-down of the stars owing to angular momentum loss via a stellar wind, then the rotation frequency should vary as $\Omega \sim t^{-1/2}$, where t is the age. For the youngest clusters, like the Pleiades, the low-mass stars also display strong stellar coronas, suggesting that the enhanced dynamo activity presumed to be the origin of the X-ray-emitting region is connected with the statistically higher rotation rate of these stars due to their age.

Recent observations of rapidly rotating lower-main-sequence stars in the Pleiades and possibly other young clusters may be an indication of a long time (in comparison with the age of the upper-main-sequence stars) of formation for the cluster members. In associations, this appears to be short in the subgroups of the systems but in galactic clusters may take as long as 10^8 yr. Much remains to be done in relation to this problem.

E. BINARY FREQUENCY IN OPEN CLUSTERS

There is still considerable uncertainty about both the frequency of occurrence of, and distribution of mass ratio in, binary systems in clusters. The problem is that few long-term surveys of clusters have been undertaken to determine the presence of small-amplitude velocity variations in main-sequence stars. About half of the stars in the field are binaries, and the mass ratio for many of these systems is nearly unity. (It is still uncertain what the true value is.) There appears to be a dearth of binaries among the Pleiades stars, but the claimed excess in IC 4665 has been shown to be incorrect. There may be some clusters for which the binary statistics are unusually high or low compared with this value, but this is a problem for the future.

F. STAR FORMATION IN CLUSTERS

The youngest clusters, like NGC 6530 and NGC 2264, are still imbedded within their parent clouds. Many of the individual stars in these young clusters still have dust shells and are contracting toward their initiation of hydrogen core burning. In a few such young clusters, one can see stars evolving both toward (the lower-mass stars) and away from (the highest-mass ones) the main sequence, thus providing a check on the pre-main-sequence contraction time and the rate of stellar evolution on the main sequence. There is also some evidence, from the fact that the lower-mass stars evolve more slowly throughout their lifetimes, that the star formation in open clusters may take place over a time scale comparable to the age of the cluster. In some clusters, like the Pleiades and NGC 752, there is evidence, albeit weak at the moment, that there are several epochs of star formation. The spread in the rotational and luminosity properties of the lower main sequence is larger than would be expected for a sample of stars formed simultaneously.

Some clusters display a population of stars that lie above the main-sequence turnoff points of the massive stars, the so-called blue stragglers. These may be stars that have been the product of mass transfer between the members of close binary systems, in which one of the stars has been moved above the turnoff point due to mass transfer from a companion that is evolving toward the red giant phase. However, there is little evidence for binarity among these stars. An interesting alternative is that they have somehow become completely chemically mixed and are evolving as homogeneous configurations. The evolution of such stars, rather than being toward the cooler parts of the H–R diagram, is almost along the main sequence toward slightly smaller and hotter configurations as the stars age as a result of the global increase in the mean molecular weight of the gas in the interior. Normally, stars massive enough (~ 1.5 $M_\odot$ or greater) to burn hydrogen via the CNO cycle would have convective, completely mixed cores whose mean composition would be considerably heavier than the stellar envelope (which is radiative). As a result, as the cores contract under the increase in the mass of the constituent species, the released gravitational binding energy is ca-

pable of lifting the envelope. This causes the star to increase, overall, in radius and move toward lower surface temperatures (the mechanism responsible for producing red giants). However, the mechanism whereby the mixing is initiated and maintained remains conjectural. The blue stragglers remain one of the important puzzles in the otherwise largely coherent picture of cluster evolution. They may even be additional evidence for multiple epochs of star formation.

G. ABUNDANCE STUDIES

An important use of galactic clusters is the study of the age dependence of the metallicity of the stars in the galactic disk. There are several elements for which it is especially important to know the absolute, rather than statistical, ages. One such element is lithium. It is well known that lithium is destroyed by nuclear processing in solar-type stars. The element is mixed into the core by convection and totally consumed by proton nuclear processes. Thus, it is a useful constraint on the rate of mixing in stellar envelopes, as well as being a valuable potential age discriminator in field stars. Studies of intermediate-age clusters show that the time scale for this mixing to occur is on the order of 1 Gyr and is also consistent with the change in the rotation rate for solar-type stars due to angular momentum loss via stellar winds.

We do not yet have a good determination of the deuterium abundance in cluster stars, but this would also serve as a useful constraint on both the rate of consumption of deuterium via a mechanism similar to that of lithium destruction and also on the primordial deuterium abundance.

The abundance of heavy metals appears to vary among cluster stars, but this is likely due primarily to processes connected with mixing and diffusion that are intrinsic to the stellar envelope. As yet, there is no clear measure of the intrinsic dispersion in initial abundances among stars in a cluster. The abundances overall follow the trend of the field stars—the youngest clusters are a bit more metal rich than the sun, whereas the oldest clusters are considerably more metal poor.

H. VARIABLE STARS IN CLUSTERS

Several classes of pulsating variable stars are observed in open clusters. The main-sequence population comprises the δ Scuti stars, which are ~1.5 M$_\odot$ and are late A and early F stars. These have periods of about an hour and amplitudes that are typically several hundredths of a magnitude. They are the same type as the field population, but their properties can be best studied, as for all pulsators, when their distances can be independently determined and their absolute magnitudes assigned. Perhaps the most important aspect of clusters, however, for the purposes of both stellar evolution and cosmology is that a few clusters have δ Cepheid or classical Cepheid variables as members. These stars, which are about 3–5 M$_\odot$, are evolved stars that pulsate with periods of several days to several months and have large amplitudes (typically near a magnitude). They have a regular period–luminosity relation from which their absolute brightnesses can be determined given their period of variation, a comparatively easily measured quantity. They form the anchor of the extragalactic distance scale. It is, in part, the fact that the Cepheids occur in clusters that has led to one of the most interesting questions of stellar evolution theory. When one determines the masses of the stars, they typically have evolutionary masses that are greater than indicated by pulsation-theoretical calculations. Although a number have been shown to be members of binary systems, thereby altering their placement in the H–R diagram, not all have been shown to be such. Thus, again, clusters provide one of the most important handles on the absolute properties of these stars.

Flare stars and T Tau stars are associated with the youngest clusters. The flare stars, which are chromospherically active M dwarfs, may be more frequent because of greater dynamo activity in the newly formed systems. These stars are rapidly rotating systems and may not yet have spun down to their more sedate rotation rates due to breaking by stellar wind-driven angular momentum loss. One serious complication in this observational picture is that at present there are few surveys of such activity in clusters.

I. Magnetic Fields in Cluster Stars

Additional important information provided by galactic clusters is the occurrence of magnetic fields in main-sequence stars. Only a few clusters have so far been studied in any detail for direct observation of magnetism, but there is some indirect information provided by the presence of Ap stars. These are stars with anomalously high (compared with the average Population I star and with the sun) abundances of heavy metals—especially the rare earths and silicon—which possess strong magnetic fields. They occur in clusters having ages greater than ~10^7 yr, suggesting that the magnetic fields are present but that there is some lag time after formation for the development of the abundance anomaly. These stars are massive, typically greater than 2 M$_\odot$.

For the lower-mass stars, the statistics for the presence of chromospheric activity and X-ray emission form a useful tool with which to study the development of dynamo-generated magnetic fields in stars. The youngest systems, like the Pleiades, have already well-developed coronal activity indicators, and it seems that most stars in the youngest clusters on the lower main sequence have some X-ray emission associated with them. Indeed, in agreement with the fact that the mean rotational velocity decreases with time, the oldest clusters show weaker dynamo activity among the lower-main-sequence stars than do the young clusters. Again, because the ages of these stars can be determined independently from the H–R diagram for the cluster, the study of the aggregate properties is far more valuable than the detailed work on field stars for which such information is only statistically available. There is only weak evidence at presence for the decay of the strong magnetic fields of the upper-main-sequence stars with age, although the time scale for such a phenomenon should be of the order of 10^9 yr or fewer due to diffusive losses. This remains one of the most important areas of research in open cluster populations.

III. Associations

A. OB Associations

The most massive stars in the galaxy do not appear to form in the tight configurations that open clusters represent. Instead, they form dynamically fragile associations of several thousand solar masses. Some, like the Sco–Cen association, contain open clusters (NGC 6231). Others, like Orion, contain tight multiple-star systems of the Trapezium variety, which contain ~100 stars but which in total have masses about like those of very small open clusters. A few show evidence not only of having formed over a considerable period of time, of the order of 10^7 yr, but also of forming in sequences (e.g.,

Cep OB3 and Ori OB1). Many OB associations, these two groups being prototypical, are still associated with their molecular clouds and show evidence of continuing star formation.

The physics governing the division between cluster and association is not well understood. One suggestion is that the formation of massive stars stops the continued formation of the lower-main-sequence objects and, by altering the thermal and dynamic structure of the environment, brings the system's star-forming activity to a halt. A related point is that only the most massive molecular clouds may have sufficient mass and stability to allow for the formation of OB associations, whereas the average cloud may serve as the nursery for the lower-mass objects. If this is the case, the OB associations should show at the same age as the clusters a systematic preference for the upper-main-sequence stars, a test that has yet to be completed. The problem is that there are few open clusters known from optical observation to be of sufficient youth with which to test this idea. Instead, the infrared astronomy satellite (IRAS) observations of clusters should be used to determine the mass function for the stars forming currently in clouds of different mass.

Another important feature of the associations is that they can evaporate. Due to the gravitational interactions of a bound star in the group with the others in the vicinity, there can occasionally be a large enough kick from a close encounter that the star is sent at escape velocity or higher out of the system. The main reason is that the associations are rather fluffy, from the gravitation point of view—quite distended even considering the masses of the individual stars. The escape of stars from the clusters produces a contraction of the cluster core, increasing the rate of encounters and gravitationally "heating" the stars in the association. The contraction of the core of the association therefore increases, with a further increase in the rate of escape of the stars. This evaporative dissipation of the cluster, first discussed by S. Chandrasekhar and J. Oort in the 1940s and 1950s, is the main mechanism for the dissolution of the associations.

It is likely that the evaporation time scale for the associations is always short, of the order of the main-sequence time for the B stars, and may be important for understanding the formation mechanism for these systems. The Ori OB1 and Cep OB3 associations have among the best determined dynamic ages, and these agree quite

well with the inferred stellar ages from evolutionary models.

B. R and T Associations

If the open cluster or association is still connected with dust, it is sometimes called an R association, for *reflection*. Dust has the property of scattering, as well as absorbing, starlight. The presence of B and A stars in an association, in the absence of massive stars, does not provide sufficient UV radiation to ionize the environment. A result is that the neighborhood dust survives, without the formation of an emission nebula, which then scatters the starlight and forms an extended filamentary diffuse nebula. These R associations are among the youngest clusters in the galaxy, having ages a bit greater than those of the OB associations.

Finally, if the association contains T Tauri stars, which are pre-main-sequence objects, it is sometimes called a T association. The purpose of separating these groups was to call attention to the star-forming activity associated with them. However, in light of the recent observations by IRAS that star formation is present in the molecular clouds connected with many or most of the OB associations in the galaxy, even if it were optically invisible to us, there seems little reason to continue this designation. The T Tau stars are well emerged from their parent cloud material, and the indication of ongoing star formation is better provided by the wealth of point IR sources observed to be associated with many of the OB associations in the galaxy.

IV. Globular Clusters

A. Introduction and Basic Properties

The most striking property of globular clusters is their overall optical appearance. They form nearly spherical, compact systems that are enormous aggregates of stars. It is this feature that was first selected as their distinguishing characteristic. The morphologically distinguishing features of individual clusters are the degree of central concentration, the so-called Shapley classes (on a scale of I to XII, depending on decreasing concentration), and the ellipticity. The latter, though originally looked to as providing evidence of the rotation of the cluster as a whole, actually appears due to both the tidal interaction with the galaxy as a whole and the formation of nonaxisymmetric internal velocity and spatial structures.

The frequently employed model for the overall mass distribution, the so-called King model, assumes that the globular cluster is an isothermal sphere of stars. This assumption means that the velocity dispersion is assumed to be homogeneous and isotropic. The stars behave as if they are particles in the mutually generated gravitational potential well moving around at uniform temperature. The rate of escape of stars from such a system is seen to be lower than the rate of escape from the open clusters because of the relative compactness of the potential and of the mass of the cluster. The central portion of such spheres is called the core radius, a measure of the distance over which the surface brightness of the cluster falls to about half its central value. The Shapley classes are not well correlated with this more quantitative measure of concentration, for which reason the former is preferred as a characterization of the cluster morphology.

B. INTEGRATED PROPERTIES OF THE CLUSTER SYSTEM OF THE GALAXY AND COMPARISON WITH EXTERNAL GALAXIES

An important feature of the globular clusters is that they can be studied as if they were extremely bright stars and, because of their compactness and brightness, can be observed to very large distances. This includes the halos of external galaxies. The cluster system in our galaxy appears to consist of two components, the halo and disk systems, which are distinguished by their metal abundances and spatial distributions. [See GALACTIC STRUCTURE AND EVOLUTION.]

For the halo system, the mean integrated absolute visual magnitudes of the disk and halo clusters are about the same, $\langle M_V \rangle = -6.9$, to within ~20%. For external galaxies, only a few systems have so far been observed. These include most of the galaxies in the Local Group (the cluster to which the galaxy belongs) and NGC 5128 = Cen A, a nearby active and morphologically extremely peculiar galaxy.

There exists in the Large Magellanic Cloud a population of young, globularlike value clusters, the so-called populous blue clusters. These appear to be still in a mode of active star formation. Clusters having ages as young as 0.01 Gyr are observed. Recent observations of several

blue clusters show that they can have ages as great as several gigayears but, again, that they are not as old as the galactic globular clusters. Similarly, these have been observed in the even more metal poor Small Magellanic Cloud. They typically have metallicities at most 10% of the solar value, comparable with the most metal rich galactic clusters. This latter point is likely the most important clue to their origin, since the Magellanic Clouds, being of lower mass than the galaxy, also have an overall lower abundance of metals.

For M31, the clusters appear to have an excess of blue evolved stars for the same metallicities (determined from integrated spectra), but much remains to be done on the populations of clusters in general. Their luminosity distribution is not markedly different from that of our galaxy. For M33, the mean magnitude is about the same as in our galaxy, and the population of the LMC-like blue clusters is lower, more in agreement with our experience locally. The same appears to be true (although this is very poorly determined at present) for the clusters about the active peculiar elliptical galaxy NGC 5128. The cluster system of M87, the central giant elliptical galaxy in the Virgo cluster, has also shown that many of the clusters have a higher metallicity than those in our galaxy and that there is a substantial population of brighter systems than we possess. There appears to be no radial gradient in the abundances. More detailed information on external galactic systems will have to await high-resolution space observations.

C. METALLICITIES AND SPACE DISTRIBUTION

The globular clusters are characteristic members of the old population, Population II, of the galaxy. They are distributed in a halo around the disk, although some are present in the disk. Their metallicities range from the lowest values observed in the oldest open clusters to $\sim 10^{-3}$ the solar metallicity (which is ~0.02 by mass). The usual means of quoting metallicity for these clusters, as with the open galactic clusters, is in terms of the Fe/H ratio. This is normally quoted as a differential measure relative to the sun, as $[Fe/H] = \log(Fe/H) - \log(Fe/H)_\odot$.

An important observational fact, still not well understood, is that there are no galactic globular clusters with the characteristics of the Population I stars and that they do not now appear to form in the disk. Their distribution can be separated into two fairly distinct groups, differentiated on the basis of metallicity. One is distributed approximately spherically around the disk, centered on the galactic center, and concentrated toward the bulge of the galaxy. It was this property that allowed Shapley, around 1920, to determine that the sun is located in the periphery of the disk, since the distribution of the clusters is not centered on the solar position. These are the most metal poor clusters, having values of Z, the metal fractional mass abundance of metals, ranging from $10^{-4} Z_\odot$, where $Z_\odot \approx 0.02$.

The metallicity scale for the globular clusters is not as well established as that for the open systems, since they are intrinsically fainter and only the more evolved stars can in fact be individually measured. These tend to be stars on the red giant branch, which may have undergone internal mixing processes and so may not be representative of the entire cluster abundances. Furthermore, the metallicity determined from integrated cluster spectra, in which the entire system is treated as if it were a single star, are severely affected by the morphology of the cluster H–R diagram. Specifically, the population of the blue end of the horizontal branch can cause spurious abundance results when photometric indices are used. The lines from these stars are extremely strong and the stars are far brighter than the main sequence so that they also wash out the contribution from these fainter, although more numerous, stars. However, in spite of the uncertainty, it appears that even the most metal rich globular clusters, like 47 Tuc, barely if at all reach the current abundances of the disk.

Observations of the distribution of the clusters suggest that the break point in the flattening of the cluster system occurs at [Fe/H] = −1 to −1.5 (depending on the metallicity scale employed). The spatial distribution of the more metal rich globulars appears flattened, with a scale height (distance above the galactic plane) of the order of 0.5 kpc. These are far fewer in number than the more metal poor systems and appear to have the metallicity of the oldest field stars in the disk, the so-called old Population I distribution. They too are concentrated toward the galactic center.

D. H–R DIAGRAMS AND AGGREGATE PROPERTIES

Two important observational features dominate the study of these clusters. One is the fact that the morphology of the H–R diagram appears to depend on the metallicity of the stars. The stage of the helium core burning is represented in clusters by a horizontal branch (HB), which lies about a magnitude or so above the main sequence and stretches across the diagram from red to blue, connecting with the nearly vertical red giant branch on the cool end. The lower the mean metallicity of the cluster, the bluer the horizontal branch is. This is reflected in the $B/(R + B)$ ratio for the HB stars. Since the presence of RR Lyrae stars is a function of the morphology of the HB, the clusters of lower metallicity have a larger population of these variables. In addition, there appears to be a relation between the HB morphology and distance of the cluster from the galactic center, such that the lower-metallicity stars and the bluer HB clusters lie farther from the center. The relation is a weak one, but it nonetheless appears to indicate that the objects of higher metallicity lie closer to the galactic disk.

There is one other metallicity-dependent parameter for globular clusters. The brightness of the tip of the giant branch relative to the HB is also a function of the mean cluster metallicity, although it may depend as well on the initial helium abundance. One serious problem in understanding this is that there is a considerable spread in metal abundance, especially of CNO, among the stars on the giant branches of several clusters. The study is hampered by the fact that only the brightest stars, in the nearest clusters, can be studied presently. Nonetheless, there is a spread of upwards of an order of magnitude in the abundance of CN in ω Cen and in 47 Tuc, the two brightest such clusters.

The integrated spectra of the clusters can also be observed. The most important parameter here is ΔS, the difference between the spectral type provided by the hydrogen and the metallic lines (especially Ca II H and K). This can be derived for both individual stars (e.g., halo high-velocity stars) and for the integrated light of the clusters. It is an excellent indicator of metallicity. The metallicities of globular clusters are summarized in Table I. [*See* STELLAR SPECTROSCOPY.]

E. AGES

The globular clusters are the oldest members of the galactic system that we can study in detail. Their ages are inferred to be about 14 to 17 Gyr, with no large age spread even with the large observed metallicity dispersion. It is clear

TABLE I.   Metallicity of Select Globular Clusters

NGC	Other name	[Fe/H]	$B/(B + R)$	$\log(Z/Z_\odot)$
104	47 Tuc	−1.1	0.0	−0.4
5272	M3	−1.6	0.5	−1.2
5904	M5	−1.1	0.8	−0.8
6205	M13	−1.4	1.0	−1.2
6341	M92	−2.1	1.0	−1.7
7078	M15	−1.8	0.8	−1.4

that the epoch of formation of these objects lasted only for a brief interval, in contrast to the ongoing activity of star formation in the disk of the galaxy. Both the optically observed color–magnitude diagrams and the UV properties of the stellar populations of the clusters give similar ages. It appears that the disk increase in abundance was very rapid indeed, taking no more than a few gigayears.

F. Luminosity Functions

The mass function for globular clusters is heavily weighted toward the low-mass stars because of evolutionary effects. The turnoff points for most of the clusters are well below 1 $M_\odot$, leading to functions that can be easily compared with only the solar neighborhood in the disk of the galaxy. Until the introduction of charge-coupled devices (CCDs), the determination of this population was hampered by the faintness of the stars; however, recently a few clusters have been studied to lower than 0.5 $M_\odot$. One of the most thoroughly studied is M13, the Hercules globular cluster. It has been taken to $M_V = +9.5$. The IMF is steeper above 0.65 $M_\odot$ than for the galactic disk, flattening out for the lower-mass (≤ 0.5 $M_\odot$) objects compared with the field. The intrinsic width of the main sequence is quite narrow, of the same order as observed for the most populous open clusters, less than 0.1 magnitude, and there is no evidence for a sizable population of equal-mass binary stars. This has been seen in a number of open clusters as well. The low spread, even in light of the rotation of stars (see discussion for the open galactic clusters), is in part due to the extreme age of the systems (these stars would have slowed in their rotation by now) and the fact that lower-main-sequence stars are intrinsically slow rotators anyway. There is also no evidence for a second epoch of star formation in the clusters, which

would agree with the narrow spread in their various evolved members across the H–R diagram.

G. Initial Mass Function

The globular clusters now consist of only low-mass stars, but this is likely due to their epoch of observation and formation. If they were formed in the early stages of the evolution of the galaxy and all of the stars within them were born at the same time, then we would not see only stars lower in mass than the sun due to stellar evolution. We therefore cannot use them as direct confirmation of the stability of the IMF with time, since all of the massive stars have disappeared.

H. Variable Stars

The primary variable stars in these clusters are the RR Lyrae type, which have periods of about 0.3–0.6 day. Originally called cluster variables, they have been extensively catalogued in the past several decades. The light curves fall into two types—the *ab*, which have sharp rising portions and slow declines, and the *c* type, which is more nearly sinusoidal. The *ab* types tend to be longer periods and have larger amplitude. First recognized by Bailly in the late nineteenth century, they are apportioned differently in different globular clusters. This is the Oosterhoff dichotomy. The bluer the horizontal branch, the larger is the ratio of *c* to *ab*. There appears to be some single parameter, as yet unknown, that controls both the population of the HB in general and the population of instability strip in particular. Both metallicity and initial helium composition have been implicated. The RR Lyr stars have nearly the same absolute magnitude, about +0.5, and thus allow the distances to the clusters to be determined independent of parallax measurements. Since the light of the cluster is dominated by the HB and the red giants, these colors and synthetic population models can be used to determine the distances to galaxies in which the clusters can be located. This is yet another use of the globulars as distance indicators on the cosmological scale, one that will be increasingly important with the advent of the HST.

I. X-Ray Emission

About 10% of the known globular clusters in the galaxy are X-ray sources. Not all of them, however, show the bursting activity with which the class is usually identified. The burst sources are likely due to accretion onto a neutron star of matter from a low-mass companion, which may have formed a binary by means of capture in the cluster's lifetime. Presently, there is no evidence from X-ray data for the presence of diffuse gas in globular clusters. About 10 clusters show diffuse X-ray emission centered on their cores, but this is probably due to many unresolved stellar sources. [*See* X-RAY ASTRONOMY.]

J. BINARY STARS AND X-RAY SOURCES

Although nothing is known about the occurrence of binary stars in globular clusters from direct observations of the radial velocity variations of individual stars, several clusters do contain X-ray sources. These are probably accreters onto neutron stars and are members of close binary systems. This observation strongly supports the contention that at least some stars manage to survive as close binaries in the high stellar density environment of the clusters. In addition, there are novas observed in clusters, also presumably due to accretion of matter onto a compact component in a close binary.

Since the formation of neutron-star-containing binary systems involves supernova explosions, it is probable that they are formed from the remnants of low-mass stars, which have been pushed past the Chandrasekhar limiting mass for stable white dwarfs rather than from recently formed massive objects. The presence of such stars in the clusters is also strong support for the mechanism of neutron star binary formation for SN type I in the galaxy.

Simulations argue, however, that such systems should be quite rare in globulars. The lifetime for a binary against tidal disruption from a close encounter with another star is so short that there should be no longer period (days or weeks) separations permitted in these systems. The best model for the formation of any close binaries still appears to be tidal capture of the companion, but there are still too many uncertainties in the tidal calculation—especially the dissipation of energy and angular momentum—to inspire confidence in the mechanism.

K. DIFFUSE GAS AND STAR FORMATION

There is no compelling evidence for diffuse interstellar gas in globular clusters. Sensitive searches have failed to turn up extended emission, and there are no known instances of planetary nebulae in any cluster, although in at least one case there is a chance superposition of a foreground nebula with M15. So there is again strong support for the contention that such clusters cannot and do not undergo star formation in the present epoch. Finally, the evidence for blue stars in the cores of these clusters comes from the HB objects, and not from new massive stars. The clusters contain red giants that should have strong stellar winds, of the order of 10^{-7} M$_{\odot}$ yr^{-1}. However, the fact that the material does not accumulate within the cluster also suggests strongly that it is swept out on time scales short compared with the cluster lifetime. Such a mechanism as the ram pressure-induced stripping of the clusters as they pass through the interstellar medium in the plane of the galaxy has been successful in explaining the absence of this extended material. Material may also be blown out by supernovas or removed via a galactic shock due to the tidal potential of the plane. The gravitational binding energy of globulars is such that if the gas is heated by supernovas and other hydrodynamic heating sources, it will simply be blown out of the cores of the clusters and lost, unlike clusters of galaxies in which the matter can be retained to much higher temperatures. Again, considerable work remains to be done on this important problem in the chemical history of the galaxy.

V. Concluding Remarks

At present, from the IRAS observations and related work on the chemical evolution of the galaxy, it appears that stars arise in clouds of considerable mass—from hundreds to millions of solar masses. Very likely, they also arise in association with other stars, perhaps by stimulated processes such as supernova-induced collapse of clouds or spontaneously. It therefore is likely that clusters hold the most important key to an understanding of the formation of stars and ultimately of the chemical input for all of subsequent galactic history. The question of the evolution of the mass function for star formation is best answered with cluster data, since we have the additional certainty that the stars are essentially simultaneously formed (on the time scale of the galactic lifetime). This has direct implications in cosmology, the interpretation of the spectra of distant galaxies at earlier epochs in their histories. [*See* COSMIC INFLATION; COSMOLOGY.]

The comparison of galactic and extragalactic clusters and associations is likely to yield an important test of stellar evolution: the dependence of stellar population properties on the environment. This will have to await high-resolution observations from the HST and related future large telescope projects. The advent of CCD imaging has greatly improved knowledge of the lower main sequence of old galactic and globular clusters, and it is likely that in the next few years age determinations will be routinely performed for globulars by fitting main-sequence turnoff points. The helium abundance determination for evolved clusters can also be greatly improved by such observations, an important cosmological parameter. Finally, the observation of spectra for individual cluster stars in the lower main sequence will also improve as telescope imaging technology and solid-state detectors are improved. The revision of the metallicity scale for stellar populations and of the understanding of the origin of H–R diagram morphology for globular and open clusters remains an intriguing prospect for the next decade.

BIBLIOGRAPHY

Blaauw, A. (1964). The O associations in the solar neighborhood. *Annu. Rev. Astron. Astrophys* **2**, 213.

Freeman, K., and Norris, J. (1981). Physical properties of globular clusters. *Annu. Rev. Astron. Astrophys.* **19**, 319.

Goudis, C. (1982). "The Orion Complex: A Case Study of Interstellar Matter." Reidel, Dordrecht, Holland.

Harris, G. H. (1970). "Atlas of Galactic Open Cluster Color–Magnitude Diagrams, Vol. 4." Publications of the David Dunlap Observatory.

Lewin, W. H. G. (1980). X-ray burst sources in globular clusters and the galactic bulge. *In* "Globular Clusters" (D. Hanes and B. Madore, eds.). Cambridge Univ. Press, New York, p. 315.

Philip, A. G. D., and Hayes, D. S. (1981). "Physical Parameters of Globular Clusters: IAU Colloquium 68." Davis, New York.

Pilachowski, C. A., Sneden, C., and Wallerstein, G. (1983). The chemical composition of stars in globular clusters. *Astrophys. J. Suppl.* **52**, 214.

Sandage, A., and Roques, P. (1984). Main sequence photometry and the age of the metal-rich globular cluster NGC 6171. *Astrophys. J.* **89**, 1166.

Walker, M. F. (1983). Studies of extremely young clusters. VII. Spectroscopic observations of faint stars in the Orion nebula. *Astrophys. J.* **271**, 642.

Zinn, R. (1985). The globular cluster system of the galaxy. IV. The halo and disk subsystems. *Astrophys. J.* **293**, 424.

STELLAR SPECTROSCOPY

John B. Lester   *University of Toronto*

GLOSSARY

Chromosphere: Upper part of the atmosphere of a solar-type star where the temperature starts to rise with increasing height.

Corona: Tenuous outer part of the atmosphere of a solar-type star where the temperature may exceed a million degrees.

Curve of growth: Relationship between the equivalent width and the corresponding total abundance of an element.

Doppler imaging: Use of the Doppler effect to determine the distribution of features on the surface of a star.

Double star: Two stars that are bound together by their mutual gravitational forces. They each orbit about the center of mass of the system. The study of such systems provides direct information about stellar masses.

Effective temperature: Representative temperature of the stellar atmosphere. A perfect radiator at this temperature would emit as much radiation per unit area as does the star. Abbreviated T_{eff}.

Equivalent width: Integrated absorption of a spectral line produced by the gas in the stellar atmosphere.

Microturbulence: Small-scale random motions in the stellar atmosphere that are invoked to obtain consistent abundance determinations from both weak and strong lines.

Radial velocity: Component of motion that is toward or away from the earth.

Spectrum synthesis: Wavelength-by-wavelength computation of the stellar spectrum, including all known sources of absorption, for comparison with the observed spectrum. Abundances can be derived by using this method when the spectrum is too complicated to be studied in any other way.

Stellar atmosphere: Thin layer of gas at the stellar surface from which light can escape.

Surface gravity: Acceleration of gravity at the surface of a star. This quantity sets the pressure in the stellar atmosphere through the requirement of pressure equilibrium.

The light from a star contains many different colors or wavelengths of light, but these wavelengths are mixed together so thoroughly that the overall impression is generally a color close to white; the amount or brightness of the light creates a greater impression than does the underlying mix of colors. It is possible, however, to break the white light into its constituent colors, which then can be displayed in order of increasing or decreasing wavelength; the wavelength provides the quantitative means of arranging this order. Simple devices such as glass prisms or diffraction gratings can accomplish this decomposition. The resulting ordered display of light by wavelength is the spectrum of the light. The measurement and analysis of the stellar spectrum, wavelength by wavelength, is the activity of stellar spectroscopy.

I. Introduction

Stellar spectroscopy has a long and distinguished history, beginning with Isaac Newton, the discoverer of the spectral nature of sunlight. The importance of Newton's discovery was not realized until the nineteenth century when continuing research on the spectral properties of solar and stellar light showed the tremendous information content of the spectrum. The first of these discoveries found that the sun's spectrum is not a continuous, smooth function of wavelength. Instead, the sun's brightness is greatly reduced at certain, precise, narrow places in the spectrum. This discovery was made by examining a small region of the solar spectrum in greater detail. This detailed examination is generally expressed as the resolution with which the observation is made; the greater the resolution, the finer the wavelength detail that can be seen. The sun's spectrum is sometimes referred to as a dark line spectrum because of the presence of these narrow lines at which the intensity is diminished. [*See* SOLAR PHYSICS.]

Another great advance came with the recognition that the spectrum of the sun, with its distinctive pattern of dark lines, is qualitatively the same as the spectra of the stars; the blindingly bright sun is just the nearest star. This discovery provides tremendous insight into the properties of stars because we can be guided by our greater familiarity with the easily observed sun.

A third great discovery of the nineteenth century was the identification of a prominent pair of dark lines in the solar spectrum with the chemical element sodium. Laboratory studies with flames containing sodium found that this element emits light at exactly the same pair of wavelengths. This correspondence showed that the dark lines in the sun's spectrum are produced by familiar chemical elements absorbing light at certain, precise wavelengths. This discovery opened up the way to determining the composition of an astronomical object that we could never hope to visit and sample directly. It also connected the studies being done in the physics laboratories with the conditions in the sun. With the connection between the sun and the stars, there was a continuous bridge between what could be learned by direct laboratory experimentation and what could be observed in the most distant object; astrophysics was born. [*See* ASTROPHYSICS.]

The results of stellar spectroscopy are firmly anchored in the laboratories where the properties of matter and light are studied. We have knowledge of, rather than just speculation on, the physical conditions in astronomical environments because of this link. However, the physical conditions present on a star or in space are frequently very different from the conditions that can be produced in the laboratory. This permits astronomy to test ideas about matter and light under novel conditions. In this way astrophysics is not just an application of what has been learned in the laboratory, but an equal partner in the study of matter and light.

II. Spectroscopic Information

Building on the discoveries of the nineteenth century, astronomers have developed spectroscopy into the primary tool used to determine the physical properties of stars. The spectrum has been found to hold a rich treasure of clues that we have learned to decipher through an interaction with the results of the laboratory work done under known conditions. [*See* STELLAR STRUCTURE AND EVOLUTION.]

A. STELLAR TEMPERATURES

The spectrum of a star contains direct information about the temperature of the surface layers of the star, the only part of the star open to inspection. A qualitative estimate of the surface temperature can be made from the wavelength distribution of the starlight. In analogy with the glow of incandescent objects, a hot star radiates most brightly at the short wavelengths seen as blue or violet colors, whereas cool stars radiate most strongly at longer-wavelength red colors.

A second qualitative clue comes from identifying the source of the absorption lines in the spectrum. In the coolest stars absorptions are caused by molecules and neutral atoms. The molecules are so fragile that they can only survive in a low-temperature environment. The molecular absorption lines fade away in warmer stars as the molecules dissociate, leaving the neutral atoms of elements such as iron to dominate the spectrum. In hotter stars the environment is severe enough to strip away an electron from many of the atoms; the ions that result from this process absorb at wavelengths distinct from the neutral atoms of the same element. The sun's spectrum, for example, has prominent absorption lines of ionized calcium, as well as many weaker lines of ionized iron, nickel, and other elements. In still hotter stars, the absorption lines can be identified with atoms that have been stripped of several electrons and with elements that can only be ionized in extremely hot conditions, such as helium. Studying these clues in the stellar spectrum enables the astronomer to place the star on the continuum between the coolest and the hottest stars. The process, however, does not yield a value for the surface temperature; we only learn the relative placement of the star on the temperature scale. Figure 1 shows the blue spectral region of stars spanning the entire range of normal surface temperatures.

To determine a quantitative value of the stellar surface temperature, it is necessary to mea-

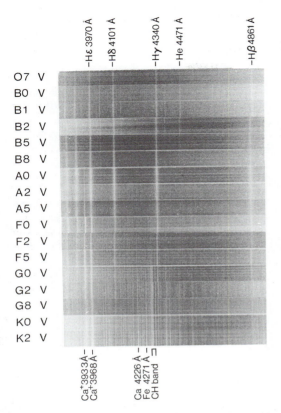

FIG. 1. The blue–violet spectral region of normal, main-sequence stars spanning the temperature range from 40,000 down to 4000 K. The spectra are displayed as negatives, and so the dark parts are where the light is bright and the light lines are where the light is dim. Note that the spectra of the hot stars at the top of the figure extend all the way to the violet on the left-hand side, whereas the cool stars at the bottom of the figure have little violet radiation. Also note that several of the most prominent spectral features, such as those that are identified, display a systematic variation in strength as the temperature changes. [Reproduced with permission from David Dunlap Observatory, University of Toronto.]

sure temperature-sensitive indicators in the spectrum. Our experience with incandescent objects suggests that one good indicator is the amount of blue light relative to the amount of red light. The measurement of the ratio of blue to red light will capture this brightness distribution, the measurement being made either through filters transmitting broad bands of the blue and the red spectral regions or with instruments designed to sample the light monochromatically from widely separated wavelengths. A second temperature indicator is the relative

prominence of absorption lines of a neutral atom and an ion of the same element. With increasing temperature the lines of the ion will increase in prominence while the lines of the neutral atom will decrease because there will be fewer neutral atoms present in the gas on the stellar surface.

The measurements alone only provide a quantitative display of the qualitative variations that result from temperature changes; these do not yield temperatures until they are interpreted. In a very limited number of cases the interpretation can be done by using only the observations; the resulting temperatures are fundamental reference points for other methods. The less direct but more widely applicable methods involve modeling, an extremely important method that is the main means of extracting astrophysical information.

A model is a description of the object under study that is required to behave in a way consistent with our knowledge of physical systems. For example, the surface layers of a star are taken to be in pressure balance; the force of gravity pulling matter toward the center of the star must exactly equal the expansion pressure of the hot gas. The surface layers are also required to conserve energy, with the energy radiated to space being exactly equal to the energy coming up from the deep interior of the star. These requirements are expressed as mathematical equations that can be solved to give the physical conditions present in the model. It is still beyond our computational power, however, to solve the equations in their full generality. For example, the sun is obviously spherical, but the models neglect this curvature by using only the depth dimension in the equations. Similarly, it is common to assume that thermodynamic equilibrium holds for the local value of the temperature at each location, although the propagation of light from a region of one temperature to a region of a different temperature invalidates this assumption in principle if not always in practice. Models that incorporate spherical geometry or that replace the assumption of local thermal equilibrium by a microscopic consideration of absorption, emission, and collision processes have been computed, but they have not yet been widely applied to the interpretation of stellar spectra because of their great computational demands. The conclusions based on the simplified models must always be treated with caution, but they appear to be valid for many stars.

The mathematical model is used to predict the

indicators of temperature that have been measured. If the predicted and observed indicators disagree, the model is changed and recomputed until agreement is achieved. The model that produces the best match to the observations is taken as an accurate representation of the surface layers of the star.

It is found that the light comprising the stellar spectrum originates from a range of depths; this region from which light reaches us is referred to as the stellar atmosphere. The temperature varies throughout the atmosphere, being hottest at the deepest layers visible and dropping to about half the hottest value at the top of the atmosphere; there is no single temperature valid throughout the whole atmosphere. Because of this, the "temperature" of the atmosphere is defined in terms of the total flux, integrated over all wavelengths, passing through the atmosphere to space. For the simple case when the atmosphere can be accurately represented without using spherical geometry, the effective temperature of the atmosphere T_{eff} is defined as the temperature

at which an ideal radiator would emit as much radiation per unit surface area as the star or the model. That is,

$$\sigma T_{eff}^4 = L/4\pi R^2$$

where σ is the Stefan–Boltzmann constant, L the total radiated energy from the atmosphere, and R the stellar radius. In practice, T_{eff} characterizes that level in the stellar atmosphere where the light has about a 50 : 50 chance of escaping to space without being absorbed again. The match of the predicted and observed temperature indicators give the T_{eff} of the star's atmosphere, and the model gives the details of how the temperature changes with depth. Figure 2 compares the observed energy distribution of the bright star Vega with the fluxes from a model of the stellar atmosphere. Vega is one of the few stars for which T_{eff} can be found directly from the observations; the empirical and the model T_{eff}'s agree to within the uncertainty of each method.

Modeling shows that stars span a wide range of effective temperature. Some stars are as hot

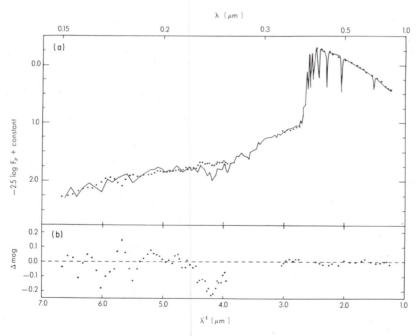

FIG. 2. (a) The observed energy distribution of Vega (α Lyrae), represented by the dots, compared with the flux of the model that best matches the observations. The comparison covers the spectral band from 1500 Å in the far ultraviolet to 8000 Å in the near infrared by using a combination of ground-based and satellite observations. (b) The difference between the observations and the model. There are some localized disagreements, but the general shapes match closely. [Reproduced with permission from M. C. Lane and J. B. Lester (1984). *Astrophys. J.* **281**, 723.]

as 100,000 K, although temperatures this high are only found for stars just becoming white dwarfs. Main-sequence stars have effective temperatures cooler than about 50,000 K and extend down to less than 3000 K. For example, the sun has an T_{eff} of 5770 K, whereas that of Vega is about 9500 K. Stars, therefore, exhibit a variation of more than a factor of 10 in their effective temperatures.

B. Surface Gas Pressure

The stellar spectrum also contains information about the brightness of the stars. Two stars of the same effective temperature will differ in brightness depending on which star has the larger radius. Even though stars of the same temperature radiate the same amount of light from equal-size patches of their surfaces, the larger star has a larger total surface area. Another consequence of a large radius is a lower atmospheric pressure because the gas is spread out over a larger volume. It is the gas pressure that has a detectable effect on the spectrum.

As with the temperature, the atmospheric pressure can be estimated qualitatively as well as measured quantitatively. Both methods require features in the spectrum that are sensitive to gas pressure. A prime indicator for stars hotter than about 9000 K is the strength of the hydrogen absorption lines. The density of charged particles surrounding an atom of hydrogen has a powerful effect on the width and strength of the absorption lines (linear Stark effect); the greater the density of charges (negative electrons and positive protons), the more the hydrogen lines are broadened. The temperature of a star can be estimated from its color or by which elements and ions are visible in its spectrum; the luminosity or pressure can be determined from the breadth of the hydrogen lines. Figure 3 shows stars of one temperature arranged in order of relative luminosity as determined by the hydrogen lines.

Stars cooler than 9000 K require other indicators because there are too few charged particles to have a strong influence on the hydrogen lines. One type of indicator involves a comparison of absorption lines of a neutral atom and an ion of the same element. When a neutral atom has been ionized, it possesses a residual positive charge that is capable of attracting any free electrons in its vicinity; the lifetime of an ion is short. The lifetime, however, depends on the density of free electrons. In an environment with few free electrons, the ion might have to wait a long time before recombining to become neutral again. While waiting for the recombination, the ion is capable of absorbing light at its characteristic wavelengths. On the other hand, when there are many free electrons, the ion can recombine immediately, leaving it little opportunity to absorb light. Because the density of free electrons depends on the pressure of the stellar atmosphere, the relative prominence of the ion of an element relative to the neutral will be an indicator of pressure, size, and luminosity. Of course, the ratio of the strength of an ion to a neutral of the same element also depends on temperature. The temperature and pressure can be found separately either by using several elements that differ in their individual responses to the two parameters or by first using the color to find the temperature.

The quantitative determination of the pressure in the stellar atmosphere again involves the measurement of pressure-sensitive indicators and the use of models to predict the values of those indicators under known conditions. It is here that knowledge gained from laboratory studies becomes of great importance. The hydrogen lines, which are so useful in the hotter stars, have been studied extensively in laboratory plasmas as well as with the theoretical methods of quantum physics. As a result, the response of the lines to various environments is well known. With this knowledge it is possible to compute the appearance of the hydrogen lines for a model of the stellar atmosphere having a certain effective temperature and a variation of the pressure with depth. This predicted appearance must then be compared to the observed line profile.

To observe a line profile it is necessary to measure the brightness variation of the absorption from the center of the line to the point where the absorption is negligible. Each observed point in this measurement must be cleanly separated from adjacent points, but they must be spaced closely enough in wavelength to map the line's shape accurately. For example, the hydrogen absorption line known as Hα, in the red spectral region, has its line core at a wavelength of 6562.8 Å. The absorption extends to as much as ±100 Å from the line center, but it is necessary to measure the profile at points separated by a fraction of an Ångstrom, and these measurements must not be contaminated by the flux at adjacent points in the profile. This represents a very detailed examination of the line.

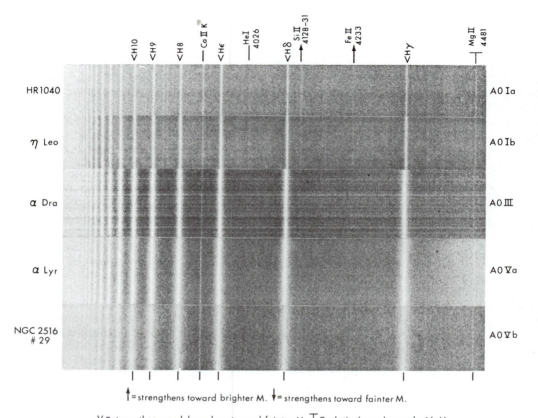

↑ = strengthens toward brighter M.   ↓ = strengthens toward fainter M.

V = strengthens, and broadens, toward fainter M.   ⊤ = relatively unchanged with M.

FIG. 3. The blue–violet spectral region for stars with an effective temperature of about 9500 K but of different luminosities. The spectra are arranged with the most luminous stars at the top of the figure. Note the dramatic change in the width and strength of the hydrogen absorption lines as the luminosity and atmospheric pressure vary. [Reproduced with permission from W. W. Morgan, H. A. Abt, and J. W. Tapscott, (1978). "Revised MK Spectral Atlas for Stars Earlier than the Sun." Yerkes Observatory, The University of Chicago, and Kitt Peak National Observatory.]

If the observed line profile disagrees with the profile predicted for a certain atmospheric pressure, the model of the stellar atmosphere can be altered and the prediction repeated until agreement is achieved. The resulting model is taken as representing the structure of the star's surface layers. As with the temperature, the pressure increases with depth in the stellar atmosphere. To have a convenient way of referring to this range of pressures, it is common to use the acceleration of gravity at the surface of the star,

$$g = GM/R^2$$

where G is the gravitational force constant, M the mass of the star, and R the stellar radius. This quantity is related to the pressure because of the requirement that gas pressure must exactly balance the force of gravity. Because the atmosphere of a star is so thin compared to the stellar radius, the quantity g is a constant. Figure 4 shows the observed and computed Hα profiles for Vega.

The surface gravity and the associated gas pressure span a wide range of values. The sun, a typical star of moderate temperature, has $g = 27{,}540$ cm sec^{-2}. The corresponding density at a representative place in the solar atmosphere is 10^{17} particle cm^{-3}. The earth's atmosphere, for comparison, contains about 2×10^{19} molecules cm^{-3} at sea level. Considering the stars, we find surface gravities as small as $g = 1$. These extremely luminous stars have radii more than

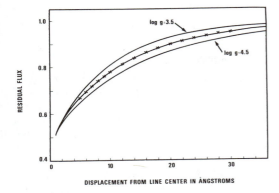

FIG. 4.   A comparison of the observed Hα line profile, represented by the crossmarks, and the computed profiles for a range of surface gravities for the star Vega. Because the profile is symmetrical about its central wavelength, only half of it is shown. The computed profiles are labeled with the log of the surface gravity. The best match can be found by interpolation.

1000 times the solar radius, or many times greater than the distance between the sun and the earth. At the other extreme, compact white dwarf stars no larger than the earth have $g = 10^8$.

C. RADIAL VELOCITIES

In addition to information on the physical conditions in the stellar atmosphere, the spectrum of light also contains information about the motions of the star and the motion of the atmosphere relative to the center of the star.

The length of a wave is related to two of its other properties, its speed and the frequency with which it passes a given point. The relationship is

wavelength × frequency = wave speed

For light waves this equation is simplified somewhat because the wave speed in a vacuum is a constant, 2.9979×10^{10} cm sec^{-1} = c, for all wavelengths. The frequency of the light waves, however, will depend on whether the observer or the source is stationary or moving. For example, if we are moving toward the source of light, we will encounter the light waves more frequently than we would if we were stationary. We would detect a higher frequency of waves, which could also be stated as being a shorter wavelength of light. The same result is produced if the source is moving and we are stationary; it is the relative motion of the source and the ob-

server that determines the apparent frequency and wavelength.

This interaction of relative motion with the wavelength and frequency was first realized by Johann Christian Doppler (1803–1853), who demonstrated the effect by using sound waves. It is the explanation of the familiar change of pitch of an automobile horn or siren as the automobile speeds by us. The Doppler effect also holds for light waves, with the relative motion between the observer and the source of light being known as the relative radial velocity. The frequency and the wavelength are not affected (to first order) by motions perpendicular to the line-of-sight direction.

The Doppler effect is widely used in studying the spectrum of starlight because the absorption lines provide accurately known reference wavelengths. For example, the Hα line of hydrogen described earlier is centered at 6562.817 Å. That is, the wavelength is known to 0.001 Å, or one part in 10^7, and comparable measurements have been made for many other lines. If the measurement of a stellar spectrum shows that all the spectral lines are displaced to either longer or shorter wavelengths, we can conclude that the star and the earth are in relative motion away from or toward each other, respectively. The wavelength displacement is related to the speed of relative radial motion according to

$$v = c \times \Delta\lambda/\lambda$$

where v is the relative radial velocity, $\Delta\lambda$ the wavelength displacement, λ the laboratory value of the line's wavelength, and c the speed of light. This relationship is valid as long as the radial velocities are much smaller than the speed of light, a condition that is true for all stars that can be observed individually.

The smallest detectable radial velocity depends on the smallest measurable wavelength displacement. In turn, this depends on the elimination of all systematic instrumental complications and scrupulous attention to experimental detail. Extremely careful measurements with a precision as small a 10 m sec^{-1} have been reported for stellar sources. Without exercising extraordinary effort, many studies have routinely measured velocities with an accuracy of 1 km sec^{-1} or greater, and velocities as large as several hundred kilometers per second have been measured for individual stars. Even a velocity of 300 km sec^{-1} amounts to a wavelength shift of only 5 Å in the green portion of the spectrum. Such a small shift would have no effect on the perceived color of the star; a red star has its color because of its effective temperature, not its radial velocity.

An analysis of the radial velocity measurements of a large number of stars shows that stellar motions are systematic. The sun, and the neighboring stars, orbit the distant center of our Milky Way Galaxy. The sun's orbital speed is about 250 km sec^{-1}, with the earth, of course, being carried along for the ride. The sun's distance from the galactic center is about 26,000 light-years (1 light-year $\approx$ 9,470,000,000,000 km), and one orbit takes about 220 million years. It is also found that, in the space around the sun, the orbital speed decreases with increasing distance from the galactic center. As a result, the sun is being passed on the inside by the faster stars closer to the galactic center, while, in turn, the sun is overtaking the slower stars at greater distances from the center of the galaxy.

A major application of the Doppler effect is the determination of stellar masses. Most stars in space seem to occur in bound systems of two or more stars instead of as single stars, such as the sun appears to be. The stars in such systems orbit around the center of mass of the system, in addition to moving as a unit through space in orbit about the galactic center, just as a moon orbits a planet while the planet orbits the sun.

The existence of a bound system can be detected by repeatedly measuring the radial velocities of the stars. At one time we observe one star approaching us (wavelengths shifted to smaller values) while the other star is moving away (wavelengths shifted to larger values). At another time the two stars might both be moving perpendicular to our line of sight in opposite directions, and consequently they would have no wavelength shifts. At still another time the star that was originally approaching us will be moving away, while its companion will now be moving toward us. The periodic variation of the radial velocities is the proof that the two stars form a bound system. Figure 5 displays the spectrum of a double star at two different phases.

As the two stars orbit their center of mass, they are held in the system by the gravity of their companion. Therefore, one star becomes a measure of the strength of the gravity, and hence the mass, of the other star. Long experience studying the planets and their satellites in the solar system has shown how to interpret this information. In particular, the size and the period of the orbit give the sum of the masses of

the two objects using

$$M_1 + M_2 = 4\pi^2 a^3 / G p^2$$

where M is the mass of the star; p the orbital period, found from the variation of the radial velocities; a the semimajor axis of the orbit, which will, in general, be elliptical; and G the gravitational force constant. The individual masses are found from observing the orbital speeds of the two stars. The more massive star is closer to the center of mass of the system, just as the heavier child must sit closer to the balance point of a seesaw. The two stars must complete one orbit around the center of mass in the same length of time, so the more massive star moves more slowly because it has a shorter distance to travel. The ratio of masses is found from the inverse ratio of the orbital speeds.

$$M_1 / M_2 = v_2 / v_1$$

The sum of the masses and the ratios of masses make it possible to find the individual mass of each star. These are generally expressed in terms of the sun's mass, 1.99×10^{33} g.

To actually apply the method outlined in the preceding discussion, one additional piece of information is essential; we must know the angle between our line of sight and the orbital axis of the system. This information is needed to convert the observed velocities, which are projections, to the true orbital velocities. The geometrical information required for this conversion is available if the stars periodically occult each other as seen from the earth. The eclipse produced as one star passes in front of its companion enables us to determine the projection angle.

The masses found for the stars range from a low of about 0.1 of a solar mass up to a high of about 100 times the sun's mass. The upper limit has been challenged recently by some astronomers who feel that there is evidence for stars as massive as 1000 solar masses, but this evidence is secondary, not involving the measurement of orbital motions. As the range of masses shows, the sun is a typical star of modest mass, consistent with its effective temperature and brightness.

The Doppler effect also plays a central role in two of the most interesting current research projects in stellar spectroscopy. The planets are prominent and fascinating members of our solar system, but there is only scanty evidence for planets in orbit around other stars. As mentioned earlier, most stars seem to belong to bound systems. In trying to understand this as a

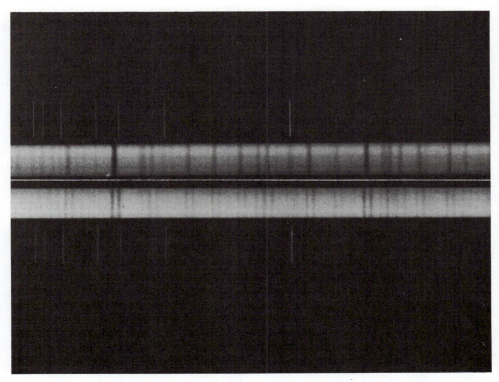

FIG. 5.   Spectra of the double star Mizar at two different phases. The top spectrum shows the system when the two stars are moving perpendicular to our line of sight, and so the spectral lines are coincident. The bottom spectrum shows the system when one star is approaching us, so that its lines are shifted to shorter wavelengths, while its companion is moving away from us, producing a shift toward longer wavelengths. [Reproduced with permission from David Dunlap Observatory, University of Toronto.]

natural product of star formation, theories suggest that planetary systems should also be common, perhaps being present around those stars that appear to be single. The problem is the difficulty of detecting a planet, which is only visible by reflected light, associated with a star that emits vast amounts of light. One possibility that is being actively studied is the detection of the orbital motion of the star about the center of mass of the star–planet system. For example, the sun and Jupiter orbit about the center of their system with a period of 11.86 years. The sun is not the stationary center of Jupiter's orbit, but moves about the center of mass that is located 7.4×10^{10} cm, 1.07 times the solar radius, from the center of the sun. Because of this motion, the sun shows a radial velocity variation with an amplitude of 12.5 m sec^{-1} and a period of 11.86 years. The discovery of a planet with Jupiter's mass in orbit about a nearby star by means of the star's periodic velocity shift would push the ca-

pabilities of stellar spectroscopy to the current limit. There have been no detections to date, but several groups have embarked on long-term projects to monitor stars for low-amplitude periodic radial velocity variations because of the astronomical and social importance such a detection would have.

At the other extreme, another active research area is the detection of massive but compact objects that might be black holes. Satellite measurements have discovered that certain stars are strong X-ray emitters. Many of these have been found to be double stars in which the X rays come from a companion that is not visible in optical light. The companion's powerful gravity is the source of energy that heats the gas surrounding it to temperatures as high as 10^7 K. Gas at this temperature is capable of emitting X rays. Such a strong gravity requires a high mass and an extremely small size, a radius of 10 km or less. Current theories of stellar structure suggest two possibilities: a neutron star or a black hole. A neutron star is a whole star compressed until it consists entirely of neutrons packed together with the density of an atomic nucleus. A black hole has an even greater density, producing a gravity so strong that not even light can escape from it. The X rays that are observed would originate from the hot gas that has not yet been pulled into the black hole. [*See* BLACK HOLES (ASTRONOMY).]

The theory of stellar evolution predicts that stars less than a few times the sun's mass can become neutron stars; stars above this limit can only be black holes. The issue, then, is to determine the masses of the compact companions in the X-ray double stars. Again, this is done by finding the effect of the compact star's gravity on the other star of the system by measuring the periodic variation of the star's radial velocity. The analysis is complicated for various reasons, but a conservative assessment is that at least two of the X-ray stars appear to have compact companions that are too massive to be neutron stars. This finding is of great importance for astronomy because of the information it provides about stellar structure and the development of double star systems. However, it is also of great importance for our understanding of the properties of matter and gravity under extreme conditions that cannot be reproduced in the physics laboratory.

Radial velocity measurements also show that stars experience complex atmospheric motions. Some stars exhibit periodic changes in both radial velocity and brightness. In some cases this behavior can be attributed to eclipsing double stars, but in other cases this explanation can be rejected by considering the relative timing of the two variations. Instead, the stars are swelling and shrinking periodically, with their surfaces moving at speeds up to 50 km sec^{-1}. There are several different types of stars that experience this pulsation, but they are almost all stars that are undergoing internal adjustment as they leave the main-sequence and approach the red giant stage of their lives. One type of pulsating star that has played a particularly large role in astronomy is the Cepheid variable, named after the prototype δ Cephei. These are stars with T_{eff} comparable to the sun, but they are much larger, up to 300 times the solar radius, making them extremely bright stars. In addition, their intrinsic average brightness has been found to correlate well with the period of their pulsation; the larger, brighter stars have longer pulsation peri-

ods. Because of this, the distance to these stars can be determined by comparing their apparent brightness with the intrinsic brightness found from the pulsation period. This method is an important way of finding the distances to nearby galaxies in which these stars can be seen individually because of their great luminosities.

The type of pulsation described, known as radial pulsation because the star moves to increase or decrease its radius, has been known for many decades. Recent studies of stellar spectra have shown that stars also exhibit nonradial pulsation in which a wave motion moves across the surface of the star. Again this motion is detected as a Doppler shift because the wave motion is primarily a vertical displacement.

The sun was the first star to be observed to have wave motions in its atmosphere, but these motions, to be described later, may be fundamentally different from the motions now observed in hot stars. The stellar motions must be organized on a larger scale to be detected in the integrated flux from the stellar surface. The observation of nonradial pulsation requires high spectral resolution and low noise data. Then, as Fig. 6 shows, the profile of a single spectral line is found to vary its shape in a systematic way as the wave moves across the stellar surface. The process generating these waves is still not known, but their detection provides us with some idea of the properties of the stellar surface even though we cannot resolve the disk.

The analysis of radial velocities also provides information about the dynamic state of a stellar atmosphere. In strict equilibrium the gas pressure just balances gravity. Under such a condition the star would not lose any matter. Stars, however, have additional pressures present in their atmospheres, and occasionally these can force matter off the star. The most important of these additional pressures is produced by the light emitted by the star itself. The force of light under typical terrestrial conditions is extremely weak; it can only be demonstrated by sensitive experiments. On the surface of a star, however, the intensity is vastly greater than anything in our experience. So great, indeed, that the force of this intense light is not negligible. This becomes of increasing importance as the effective temperature of the atmosphere increases, because the intensity is proportional to T_{eff}^4. The force of the light is captured by the absorption lines in the stellar atmosphere, with the resulting pressure adding to the gas pressure. Sometimes

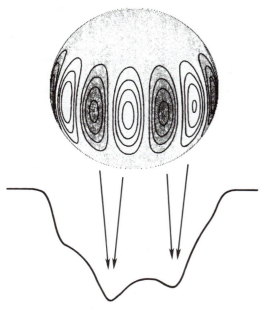

FIG. 6. A depiction of a star experiencing nonradial pulsation and a schematic line profile showing the effect of the nonradial motions on the line shape. The shaded regions on the stellar surface represent motions away from the observer relative to the center of the star, and the contours are drawn at 5-km-sec^{-1} intervals. [Reproduced with permission from S. S. Vogt and G. D. Penrod (1983). *Astrophys. J.* **275**, 661.]

the total pressure exceeds the ability of gravity to balance the force.

The evidence for this process comes from the Doppler shift of the lines. Weak absorption lines tend to form deep in the stellar atmosphere, where the gravity has its strongest hold on the gas. Stronger absorption lines form higher in the atmosphere, where the strength of gravity is diminished. The strongest absorption lines form at the outermost levels of the atmosphere. A comparison of the observed wavelengths and the laboratory values for lines of different strength shows a systematic progression; the displacement toward smaller wavelengths increases directly with the strength of the line. Interpreted as a Doppler shift, this shows that the outer layers of the stellar atmosphere are moving outward toward us compared to the deeper layers. In the case of the strongest lines, the outward velocity can exceed the speed needed to escape from the gravity of the star entirely. This matter will then flow away from the star, decreasing its total mass. Figure 7 shows some of these strong lines for one star.

The rate of observed mass loss spans a wide range. In the most extreme cases stars can lose as much as 10^{-5} of the mass of the sun in a single year. Hot stars losing mass at this rate are probably in the range of 20 to 50 times the sun's mass, so in one year they lose only a small fraction of their material. They keep this up, however, for many years. Such stars are thought to live for more than a million years, during which time they will blow off more than 10 solar masses, a significant fraction of their total mass. Cool stars losing mass at a high rate are much less massive, a few times the sun's mass. The effects of mass loss are felt much more quickly for these stars. At the other extreme of mass loss are cool stars such as the sun, which are losing matter at a rate of only 10^{-14} solar masses a year. This rate is so small that it is not detectable as a shift in the solar spectral lines; we make use of our proximity to the sun to measure the solar wind flowing past the earth. Even after 5 billion years the sun has only managed to shed a small fraction, 0.005%, of its total mass if it has maintained a constant mass loss rate over its lifetime.

The consequences of mass loss may be important for the whole range of values observed. The high mass loss rates of the hot, massive stars cause them to shed a large fraction of their total mass during their lives. A star's mass, however, governs how rapidly it evolves; as a star loses more matter, it will slow its aging process. The result might be to prolong the star's life significantly. The exact details of the consequences are still somewhat uncertain because the mass loss process must be characterized for the calculations in ways that may not be entirely appropriate. The large mass loss rates for the much less massive cool stars may have the opposite effect of accelerating their aging process. The loss of matter is a primary activity of these stars. Therefore, increasing the rate of this activity for a star that is only a few times the mass of the sun hastens the end of the star's life.

The consequences for low rates of mass loss may be no less important, although for an entirely different reason. The matter escaping from the sun has been subjected to a high temperature that has converted the neutral gas to a collection of free electrons and ions. These charged particles attach to the solar magnetic field lines, and as the sun spins, it sweeps around this attached matter. This process, which acts as an energy drain on the sun's rotation, is thought to be the explanation of why the sun, and all other stars

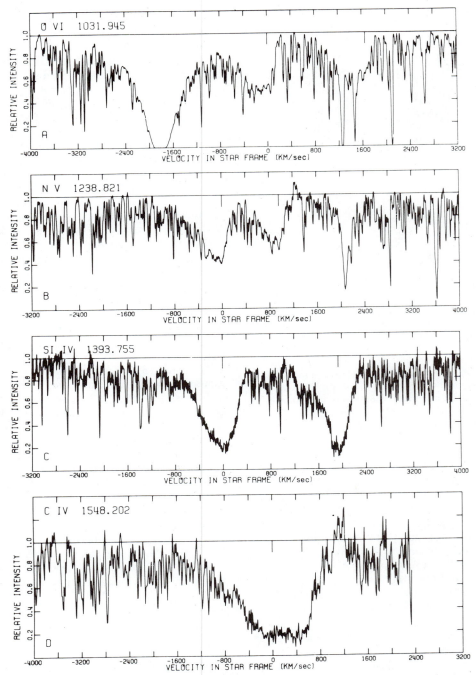

FIG. 7. Ultraviolet spectra of the star τ Scorpii showing the extremely strong absorption lines produced by multiply ionized carbon, nitrogen, oxygen, and silicon. The absorption extends to shorter wavelengths than the line center, indicating that matter is flowing toward us relative to the star. The horizontal axis is expressed in the velocity relative to the line center. [Reproduced with permission from H. J. G. L. M. Lamers and J. B. Rogerson (1978). *Astron. Astrophys.* **66,** 417.]

cooler than about 6500 K, rotate much more slowly than the hotter stars, which may not have the same magnetic field structure.

As described earlier, systematic Doppler shifts can be used to detect double stars, and if the stars are eclipsing, the velocity variations can be used to determine the masses of the stars in the system. The stellar spectrum and the Doppler effect also show that the stars in these systems frequently interact more directly than just through their gravities; gas can flow from one star to the other. The presence of gas in the system can be detected by the emission, instead of the absorption, of light at specific wavelengths. This emission results from the hot gas being projected against black space instead of the bright star. The combined effect of emission from the gas plus absorption lines from the stars can produce either an emission above the surrounding wavelengths or a partial filling in of the absorption line, depending on the amount of gas between the stars. This complex combination changes as the stars rotate about the center of mass, presenting different perspectives to our line of sight.

Although they are complicated, studies of such interacting systems are important for several reasons. First, multiple systems of stars seem to be more common than single stars. This interaction may represent the norm for stars, and this common event should be understood. Second, gas flowing between stars is increasing the mass of one star at the expense of the other, something like the flow of sand between the two halves of an hourglass. The loss of matter may significantly alter one star, and the gain may be equally important for its companion. The gain and loss are not always peaceful. Some interacting systems are known to undergo explosive events that blow some matter into space and raise the total brightness of the system to thousands of times the pre-explosion value. A stellar system undergoing this event is called a nova, from the Latin for *new,* because they suddenly appear where they had been too faint for visual detection.

The solar velocity fields provide the final examples of the kind of information that can be gleaned from the Doppler shift of spectral lines. Because the sun is so near, its surface can be observed point-by-point rather than just in the light from the whole disk. In this way we detect phenomena that must be typical of many stars, but are buried in the integrated light. As Fig. 8 shows, recording the solar spectrum with high

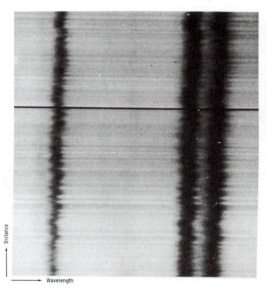

FIG. 8. A small part of the solar spectrum photographed with high spectral and spatial resolution. The spectral lines exhibit a wiggly shape because of the rising and falling motions of the solar atmosphere produced by the convection at the base of the atmosphere. [Reproduced with permission from National Solar Observatory/SP, AURA, Inc.]

spectral and spatial resolution reveals that the absorption lines are wiggly; different places on the solar surface show small wavelength shifts because of vertical motions. These motions are produced by rising and sinking elements in the deep solar atmosphere, called granules, which are interpreted as the top of a region of convective motion at the base of the atmosphere. On a larger scale, there are also supergranulation cells extending over a few tens of granule diameters. The supergranulation pattern is seen as rising gas at the center of the cell, a systematic flow of gas horizontally over most of the cell, and then sinking at the junction of several cells. Because of the predominantly horizontal flow, the supergranular cells are most easily seen toward the solar limbs.

The oscillation of the solar surface is a final example of the complex nature of the sun's velocity field. The best-studied oscillation has a period of about 5 min, although oscillations with longer periods, 20–160 min, have been reported recently. The 5-min oscillations are produced by localized wave motion extending over about 10,000 km of the solar surface. The oscillations begin spontaneously, have a velocity amplitude of about 1 km sec^{-1}, and then die out after four or five cycles. The various oscillations are currently topics of intense study because it is thought that they are produced by sound waves trapped in the solar interior, and so they provide a means of determining the conditions in regions of the sun that had been completely unobservable.

Because of the localized nature of the oscillations, a detection of an analogous phenomenon in the integrated light from a star would be unlikely. Other stars, however, might have wave motions covering more of their surfaces, presenting a motion coherent enough to be seen in the flux spectrum. There is a recent report of 5-min oscillations detected in the spectrum of the solar-type star α Centauri A, which might represent the opening of the study of stellar oscillations.

D. ROTATION

In addition to their radial velocities, orbital motion, pulsation, or complex atmospheric oscillations, all stars exhibit another motion: they rotate. The solar rotation can be detected easily by tracking the positions of sunspots, but the spectrum also shows this rotation. The equator at the east limb of the sun is approaching us at 2 km sec^{-1}, which corresponds to a rotation period of about 27 days. As a result, the spectral lines from the eastern solar hemisphere are systematically shifted toward shorter wavelengths and the lines from the western hemisphere are all shifted toward longer values.

The same systematic wavelength shifts are present in the light from a star, but we only see the light integrated over the whole disk. Instead of shifts, we observe spectral line profiles that are broadened compared to the nonrotating case. Because the shifts are all blended together, it is necessary to use a model to derive the stellar rotation speed.

The model of stellar rotation must take several factors into account. First, the maximum speed occurs at the equator because the matter there must travel the greatest distance during the rotation period; moving away from the equator the rotational speed drops, reaching zero at the poles. Second, we only observe part of the rotational speed of any point on the stellar disk, that component that is moving toward or away from us that can produce a Doppler shift; the material at the limbs produces the largest shifts. A third factor is the fraction of the star's light that has a particular projected velocity. Matter at the equatorial limbs has the largest radial velocity, but this represents a tiny fraction of the stellar disk. On the other hand, matter along the rotation axis represents the largest portion of the disk, but it has no velocity shift because the material is moving perpendicular to our line of sight. A final factor is the brightness decrease from the center of the disk to the limb. This limb darkening, easily observed for the sun, results because our line of sight at the limb enters the atmosphere at a slanting angle. As a result, we only see to shallow depths where the star is cooler and fainter than the depths reached at the center of the disk.

The projection and brightness factors described above are used to combine the profiles of a line from every location on the stellar disk into the total line profile of the star. The result will be a rotationally broadened line profile for some equatorial rotation speed.

The preceding description is for the case in which the rotation axis is perpendicular to our line of sight. This is not true in general, so that our line of sight only gets some unknown projection of the equatorial rotation speed. Despite this uncertainty, the method of computing a rotationally broadened profile is unchanged as long as we understand that the equatorial velocity is a projection. The rotation speed of the star is found by computing the broadened profile for various values of the equatorial rotation speed and interpolating to find the value giving the closest match to the observations. Figure 9 illustrates this process. Recently, another method of comparison has been developed based on the Fourier transform of both the observed and the computed line profiles.

Many studies of stellar rotational velocities have been conducted by using these methods, showing that stars rotate with a wide range of speeds. At the low end of the scale are solar-type stars that have equatorial rotation speeds of the order of a few kilometers per second. At the other extreme, some stars have equatorial speeds of more than 400 km sec^{-1}. These stars, which rotate in less than a day even though they are larger than the sun, cannot rotate any faster; at a higher rotational speed the star would break up.

While the whole range of rotational velocities is found for some types of stars, there is a well-established variation of the average rotation speed with the effective temperature of the star. The hot stars rotate rapidly as a group, with the

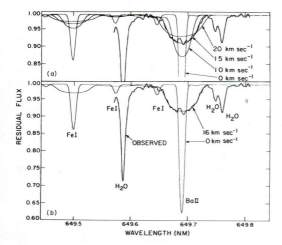

FIG. 9. (a) The observed profile of the absorption line of ionized barium in the spectrum of the bright star Sirius. Also shown is the computed spectrum for four assumed values of the projected equatorial rotational velocity. (b) The computed spectrum with a projected velocity of 16 km sec^{-1} that matches the observations most closely. [Reproduced with permission from R. L. Kurucz, W. A. Traub, N. P. Carleton, and J. B. Lester (1977). *Astrophys. J.* **217**, 771.]

peak values found around T_{eff} = 20,000 K. The average value decreases toward cooler temperatures, dropping to low values cooler than about 6500 K. Upon closer inspection the rotational properties of the various groups have interesting variations. For example, the most rapidly rotating group also has members that show episodes of mass ejection. Presumably these are stars right at the limit of stability. Some disturbance in the stellar atmosphere can then push some matter beyond the gravitational threshold, triggering a mass loss event. Among the stars around 10,000 K there are two groups of stars. The stars with normal, rapid rotation also appear normal in other ways. A second group of stars in the same temperature range have low rotation speeds. These stars also show pronounced spectral peculiarities compared to the rapidly rotating stars. The slow rotation seems to be the condition necessary for the appearance of the abnormalities. Among the cool stars there are also some anomalous rotators, this time more rapid than normal. Many of these stars are in double star systems in which the gravitational interaction of the stars increases the spin at the expense of the orbital motion. These stars radiate X rays and ultraviolet more intensely than the more slowly rotating stars.

Another systematic characteristic of rotation is the variation from one star cluster to another. A star cluster consists of a group of stars with a range of masses that formed at the same time out of a common gas cloud. Instead of clusters showing the same range and distribution of values found among the field stars, there seem to be clusters of rapidly rotating stars and clusters of slowly rotating stars. There differences are not yet well understood, but there is an inverse correlation between the cluster's average rotation speed and the frequency of double stars.

The discussion of rotation to this point, as well as the description of the method by which rotation is calculated, has assumed that a star can be characterized by a single equatorial rotation speed. Observation of the sun shows that this is not true. The sun's surface rotates differentially, with the rotation period increasing from about 27 days at the equator to more than 35 days near the poles. In addition, the sun's rotation appears to vary with time and to exhibit a north–south asymmetry. These facts all show that the solar rotation is a much more complex phenomenon that might be apparent at first. We have no information about such complications in stellar rotation, but they are likely to be present along with other, unknown behavior.

E. TURBULENCE

As might be expected, the gases in stellar atmospheres are not quiescent. In addition to the ordered, large-scale motions of rotation and pulsation, there is disordered motion on a variety of smaller scales that is referred to as turbulence. This is not used in the precise hydrodynamical sense, but just to suggest the existence of quasi-random motions.

Motions such as those observed in the sun are at the large end of the scale. They are called macroturbulence because the moving units are large compared to the average distance traveled by light before it is reabsorbed by the gas of the stellar atmosphere. These motions, of course, are unresolved in the light of a star; their effect can only be seen in the changes to the line profile. Because we see the combined light of many moving elements, the total effect is similar to the Doppler shifts of rotation. For a particular star it is extremely difficult to decide whether a small amount of rotation or a large amount of turbulence is present, although statistical studies provide evidence for macroturbulence. These studies show that among some groups of stars there are no cases of negligibly small rotation. Because some stars should be viewed from directly above the rotation pole, the line width for the stars showing the least amount of broadening is attributed to turbulence.

There is also evidence for small-scale turbulence that cannot be spatially resolved even on the sun (note that the smallest structure visible on the sun is about 500 km). The limiting case of this motion, called microturbulence, is pictured as moving elements that are small compared to the average distance between absorptions of light. The main effect of microturbulence is to broaden the cores of strong lines so that they become more effective at absorbing light. By using microturbulence it is possible to match both weak and strong absorption lines with the same elemental abundance. Even the analysis of a single point on the solar disk requires a small-scale turbulence of 0.5–1.0 km sec^{-1} to achieve consistency. Typical values for other stars are about 2 km sec^{-1}, and sometimes values as large as 10–20 km sec^{-1} are needed. These large values are suspect because they exceed the sonic velocity in the stellar atmosphere. Motions this fast would be energetic enough to alter the populations of the atomic levels. The small values, however, are reasonable given the likely presence of mechanical energy from various sources in the atmosphere.

F. MAGNETIC FIELDS

In addition to temperature, pressure, and motion, the spectrum also contains information about the magnetic environment in the stellar atmosphere. Spectral lines that appear single consist of multiple, overlapping lines. This multiplicity is not apparent except in the presence of a magnetic field external to the atom, a phenomenon known as the Zeeman effect. When the gas is subjected to a magnetic field, the lines split apart into a distinctive pattern, with the amount of the splitting being proportional to the strength of the field. The splitting is very small, however, even for strong magnetic fields. For example, a magnetic field of 1000 G, more than 1000 times the mean field of the earth, only splits lines by about ±0.01 Å at a wavelength of 5000 Å. Such a split is smaller than the detection limit for many stars because of the masking effects of rotation, turbulence, and pressure broadening.

Because of the small amount of splitting, almost all measurements of stellar magnetic fields rely on another property of the Zeeman effect;

the components that split to opposite sides of the line center have opposite senses of circular polarization. Detectors that can separate the two senses of polarization, either in time or in space, enhance the magnetic signature. Initially the detection was done photographically, with the opposite senses of polarization being projected onto different parts of the emulsion. Now the measurements are done by using much more sensitive detectors that either concentrate on a single line or use special masks to measure many lines at once. These techniques have proved successful even for rapidly rotating stars with broad lines and for stars with intrinsically weak magnetic fields.

The sun was the first astronomical source to have its magnetic field measured. The general field of the sun is weak, comparable to the general field of the earth, but the sunspots are regions of powerful, concentrated magnetic fields reaching strengths of thousands of Gauss. As is shown in Fig. 10, the sun's size, brightness, and slow rotation make it possible to resolve easily the Zeeman pattern of a sunspot's magnetic field. [*See* SOLAR SYSTEM, MAGNETIC AND ELECTRIC FIELDS.]

Although the measurement of stellar magnetic fields is much more difficult because of the limitation of observing only the integrated light from the star's disk, detections have been made for

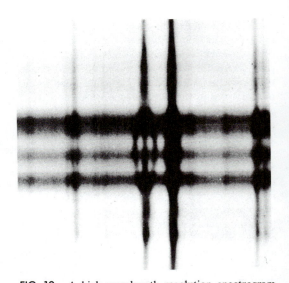

FIG. 10. A high-wavelength resolution spectrogram showing the splitting of the absorption lines produced by the intense magnetic field of a sunspot. [Reproduced with permission from National Optical Astronomy Observatories, AURA, Inc.]

several different kinds of stars. The strongest stellar magnetic fields, with strengths greater than 10^8 G, have been measured in white dwarf stars. The magnetic fields of the main-sequence stars have been observed to be as large as 30,000 G. Of more importance, the strong magnetic fields in these young stars are not constant. Instead, they vary in strength and polarity with regular periods of a few days. This is attributed to the star's magnetic poles, which are offset from the rotation poles, being carried through our line of sight by the star's rotation. This insight has been a powerful aid in understanding several different kinds of stars. The early results were limited to slowly rotating stars, but the use of more sensitive detectors has led to similar insights into the properties of rapidly rotating stars and stars with weaker field strengths. It is now possible to push the limit of detection for some stars down to the level of 10 G, which is small enough to provide an interesting constraint for these stars.

G. Abundances

The determination of stellar abundances is a primary application of spectroscopy. The analysis goes beyond the mere identification of absorption lines in the spectrum to a quantitative measurement of the amounts of the various elements. Abundance studies have been an important type of analysis for many years, but the field continues to advance for several reasons: the quality of the data is constantly improving, our knowledge of atomic structure is becoming increasingly accurate, and the growth of computer power permits more detailed studies.

The basic method of determining abundances is built on the measurement of the strength of an absorption line. The integrated line absorption, known as the equivalent width, is found first by estimating the spectrum's shape in the absence of the absorption line and then by measuring the sum of the absorption relative to that reference. Because the absorption at each wavelength is expressed relative to the reference level, the equivalent width has units of wavelength, and it can be thought of as the width of a completely dark line that would have the same total absorption as the observed line. The strongest lines can have equivalent widths of more than 10 Å. At the other extreme, the weakest detectable line depends on the resolution and the amount of noise in the data, but equivalent widths weaker than 0.01 Å have been measured with high-quality observations.

A model of the stellar atmosphere is needed to interpret the equivalent width. The flow of light through the atmosphere must be computed, taking into account the absorption by the atoms of the element of interest. This requires a knowledge of the atom's intrinsic ability to absorb at this particular wavelength as well as the density of atoms in the appropriate absorbing state. Computing the line strength for different assumed values of the total abundance of the element gives a relation known as the curve of growth. The shape of the curve exhibits three different segments. The equivalent widths of weak lines increase directly as the abundance is increased; lines on this portion of the curve of growth give the most reliable results because they are free of many complications that plague stronger lines. Lines of medium strength are saturated. Their profiles have flat cores from which the sides rise almost vertically to the continuum level. Increasing the abundance has little influence on the equivalent width because the line only broadens its saturated core. As a result, these lines give uncertain results; in addition, they are subject to the presence of microturbulence in the stellar atmosphere. The equivalent widths of strong lines are proportional to the square root of the abundance, but the line profiles also have extensive wings because of the interaction of the gas with the atoms. A major uncertainty in the interpretation of these lines is the need to treat these pressure effects correctly. Figure 11 shows both a typical curve of growth and the kinds of lines that are found on its three main segments.

The use of equivalent widths discards all the shape information present in the line profile: only the integrated strength is used. This is appropriate for observations made with moderate wavelength resolution because the equivalent width is not altered by the resolution, except for possible systematic effects related to the estimation of the continuum reference level, whereas the line shape is dominated by instrumental effects. Measurements made with high-wavelength resolution, however, minimize the instrumental contamination, permitting a direct comparison of the computed and observed line shapes. This kind of comparison avoids some of the sources of error present in the use of equivalent widths.

Both the equivalent width and the line profile

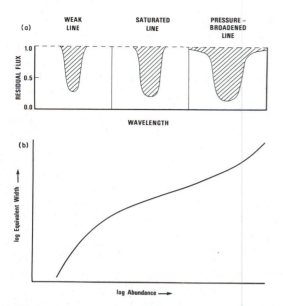

FIG. 11.   (a) The schematic appearance of the absorption lines that are found on the three portions of the curve. (b) The curve of growth that displays the dependence of the equivalent width on the abundance of the element.

approaches assume that the absorption line is isolated in the stellar spectrum. This might be true, especially at green wavelengths and longer, but at shorter wavelengths there is often severe crowding and overlapping of lines. Important absorption lines are frequently badly blended together, and there is no clean spectrum for the measurement of an equivalent width or a line profile. In such cases it is necessary to compute the predicted spectrum including all the overlapping and blended absorption lines, although this pushes our knowledge of atomic physics to the limit. If the synthetic spectrum does not match the observed spectrum, the abundances of the elements are adjusted and the calculation is repeated until agreement is achieved. Figure 12 exhibits a comparison of observed and synthesized spectra, showing the close agreement that is possible.

Methods such as these enable one to determine the chemical composition of the stars. As with most aspects of stellar spectroscopy, the sun is the primary reference and benchmark. The sun, however, does not give the full picture because some elements are difficult to study under the conditions present in the solar atmosphere. A prime example is helium, the second-most-abundant element. Although helium was discovered on, and named after, the sun, the

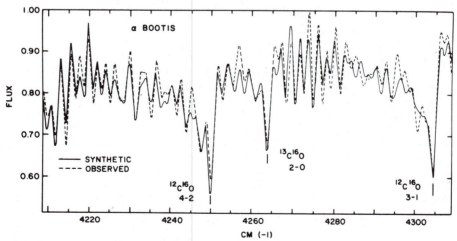

FIG. 12.   High-resolution infrared spectrum of the star α Bootis (dashed line); computed spectrum that best matches the observations (solid line). This portion of the spectrum displays the many lines associated with the CO molecule, with the isotopes ^{12}C and ^{13}C giving distinctly different wavelengths of absorption. The fit to the observations requires a $^{12}C : ^{13}C$ ratio of about 10, indicating that the ^{13}C abundance in the atmosphere has been enhanced by mixing of material from the interior. [Reproduced with permission from S. T. Ridgway (1974). *Astrophys. J.* **190**, 591.]

sun's atmosphere is too cool to exhibit helium absorption lines. The helium lines seen in the solar spectrum come from regions that are much hotter than the atmosphere and that are subject to much greater uncertainties in the analysis. Because of this, we must rely on studies of hotter stars to fill in this important element. The combined results from various sources and using various techniques give the standard solar abundances shown in Table I. This listing is thought to represent the composition of the solar system at the time of its formation.

Even a brief inspection of Table I shows that the standard composition is different from our experience of terrestrial compositions. In every thousand atoms in a typical star there are about 935 atoms of hydrogen, 63 helium atoms, and 2 atoms drawn from all the other elements. On the earth hydrogen is not common in a free, gaseous form, although a large amount is present in water. Helium is so rare on the earth that it was not even discovered until 1895, almost 30 years after it was detected in the sun's emission. The tremendous difference in composition is produced by the inability of the earth to hold on to light elements in the presence of the amount of solar heating we receive. The large masses of the stars, however, enable them to hold on to their light gases even at temperatures more than 10 times greater than the earth's temperature.

More subtle differences are also present in Table I. For example, cosmically the abundance of iron is comparable to silicon, whereas on the earth's surface silicon is much more abundant. This is attributed to the separation of the elements on the earth, with the heavy elements, such as iron, sinking to the core. The ability of elements to separate from each other illustrates

TABLE I.   Abundance of the Elements Normalized to Hydrogen = 10^{12} [a]

Element	log A	Element	log A	Element	log A
1 Hydrogen	12.00	32 Germanium	3.64	63 Europium	0.55
2 Helium	10.84	33 Arsenic	2.37	64 Gadolinium	1.20
3 Lithium	3.36	34 Selenium	3.40	65 Terbium	0.46
4 Beryllium	1.66	35 Bromine	2.54	66 Dysprosium	1.14
5 Boron	2.53	36 Krypton	3.19	67 Holmium	0.54
6 Carbon	8.63	37 Rubidium	2.36	68 Erbium	0.94
7 Nitrogen	7.94	38 Strontium	2.93	69 Thulium	0.12
8 Oxygen	8.84	39 Yttrium	2.26	70 Ytterbium	0.88
9 Fluorine	4.47	40 Zirconium	2.65	71 Lutetium	0.12
10 Neon	7.99	41 Niobium	1.53	72 Hafnium	0.81
11 Sodium	6.36	42 Molybdenum	2.18	73 Tantalum	-0.12
12 Magnesium	7.61	43 Technetium	—	74 Tungsten	1.05
13 Aluminum	6.51	44 Ruthenium	1.85	75 Rhenium	0.28
14 Silicon	7.58	45 Rhodium	1.18	76 Osmium	1.41
15 Phosphorus	5.39	46 Palladium	1.69	77 Iridium	1.43
16 Sulfur	7.28	47 Silver	1.24	78 Platinum	1.72
17 Chlorine	5.26	48 Cadmium	1.77	79 Gold	0.90
18 Argon	6.61	49 Indium	0.85	80 Mercury	0.90
19 Potassium	5.12	50 Tin	2.14	81 Thallium	0.85
20 Calcium	6.38	51 Antimony	1.07	82 Lead	1.99
21 Scandium	3.07	52 Tellurium	2.39	83 Bismuth	0.72
22 Titanium	4.96	53 Iodine	1.68	84 Polonium	—
23 Vanadium	3.98	54 Xenon	2.34	85 Astatine	—
24 Chromium	5.68	55 Cesium	1.17	86 Radon	—
25 Manganese	5.55	56 Barium	2.26	87 Francium	—
26 Iron	7.53	57 Lanthanum	1.14	88 Radium	—
27 Cobalt	4.92	58 Cerium	1.65	89 Actinium	—
28 Nickel	6.26	59 Praseodymium	0.83	90 Thorium	0.23
29 Copper	4.31	60 Neodymium	1.47	91 Protactinium	—
30 Zinc	4.68	61 Promethium	—	92 Uranium	0.01
31 Gallium	3.15	62 Samarium	0.96		

[a] Adapted from H. G. W. Cameron (1982). Elemental and nuclidic abundances in the solar system. *In* "Essays in Nuclear Astrophysics" (C. A. Barnes, D. D. Clayton, and D. N. Schram, eds.), p. 23. Cambridge Univ. Press, New York.

the pitfalls that must be avoided in interpreting the results of analysis. Another result is the saw-tooth variation in cosmic abundances superimposed on the general trend. For example, chromium, iron, nickel, and zinc have greater abundances than their neighboring elements of vanadium, manganese, cobalt, copper, and gallium. This is seen as a signature of the nuclear reactions that formed these elements in the cores of stars.

The focus of this discussion has been on the elemental abundances. Each element, however, can exist as different isotopes. Isotopes of a given element have different masses because they contain different numbers of neutrons in the atomic nucleus. Carbon, for example, can exist in two different stable isotopes, ^{12}C and ^{13}C. The isotopes of an element all absorb at nearly the same wavelengths. The small differences produced by the different atomic masses are normally undetectable spectroscopically. An important exception to this is the detection of isotope shifts in the absorption lines of molecules. Because molecules consist of two or more atoms bound together, the mass of the atoms can have a small but measurable effect on the wavelengths at which the molecular system absorbs. This enables a molecule such as CO to give information about both forms of carbon. Referring again to Fig. 12, we see that the absorptions by ^{12}CO and ^{13}CO are cleanly separated, permitting a determination of the abundance of each isotope. This is important because isotopic abundances show the physical conditions under which the element was fused by nuclear reactions. For example, $^{12}C : {}^{13}C$ has a ratio of 89 on the earth and in certain types of primitive meteorites that appear to have been spared any chemical or isotopic separation. This ratio is indicative of nuclear fusion operating in an equilibrium condition. There are, however, certain red giant stars that are observed to have ratios of $^{12}C : {}^{13}C$ equaling 5 to 10. This increase in the relative amount of the isotope ^{13}C is attributed to mixing between the stellar atmosphere and the deeper, hotter layers. The ratio departs so much from the standard value because the nuclear reactions are not able to reach an equilibrium rate before the gas is mixed out of the interior and back to the atmosphere. The study of isotopic abundances provides direct information about processes that occur in regions of the star beyond our normal observation.

Many of the stars with high levels of ^{13}C are also observed to have technetium in their spectra. This is a sure sign of mixing between the stellar atmosphere and the interior because this element has no stable isotopes. The isotope ^{99}Tc has the longest half-life, 2×10^5 years, but this is shorter than the lifetimes of stars. Its presence in the atmosphere is only possible if some of the stellar material has been exposed to a low flux of neutrons (s-process) deeper within the star and then brought to the surface.

Another characteristic of these stars is a variation of the total amount of carbon, nitrogen, and oxygen. For normal stars the ratios C : N : O are approximately $5 : 1 : 8$. Because oxygen is more abundant than carbon, the formation of the CO molecule, the molecule of oxygen with the strongest binding, controls the availability of carbon to form other molecules such as CN, CH, and C_2. In some red giant stars, however, the ratio of C : O can reverse to be greater than unity. As a result, the CO formation cannot lock up large amounts of carbon, which is then free to form other molecules. We observe stars with spectra dominated by carbon molecules that are not usually prominent. Again, this behavior can be understood by mixing between the surface and the hotter interior where nuclear reactions can occur.

A final, and extreme, example of the compositional change that can occur at the end of a star's life, is the group of stars that no longer have atmospheres dominated by hydrogen. Instead, these stars have largely helium atmospheres, with He : H ratios ranging from 1 to as high as 10,000, instead of the usual ratio near 0.1. Other elements, such as carbon, can also be greatly enhanced.

There are also young stars that exhibit peculiar abundances. A prime case is the group of moderately hot stars that have tremendous enhancements of cosmically rare elements. For example, europium is measured to be up to a million times more abundant than normal in the atmospheres of these stars. Moreover, the strengths of these lines vary with the rotation period of the star. There seem to be regions on the stellar surface where the elements are concentrated to large enhancements. The important key for these stars is that they are those stars that have also been found to have strong, variable magnetic fields. We picture the stars as having magnetically confined regions on their surfaces where the element enhancement can take place, although the quantitative verification of this picture is still in progress. It is necessary to determine the star's magnetic geometry and then to compare the variation in line strengths with that geometry. Both types of studies push the observational methods currently available to their limits.

In addition to the extreme abundances of certain groups of stars, there are also systematic variations of the elements heavier than helium. Iron, for example, ranges from 0.001 to 3 times the solar value. Moreover, this variation is correlated with the age of the star and the way it moves within our galaxy. The oldest stars, those that formed 10–15 billion years ago, have the smallest amounts of the heavy elements. The stars that formed after that time contained greater amounts of the heavy elements, with the sun having formed with its composition about 5 billion years ago. Since then there has been a very slow rate of increase with time. The oldest stars also move in elongated orbits in our galaxy, whereas the stars of the sun's age and younger travel in orbits that are nearly circular. The correlation of age and motion with average composition is combined into a picture of the process by which our galaxy has formed. Early in the history of our galaxy there must have been many short-lived, massive stars that exploded violently to create quickly the heavier elements that were then incorporated into the following generations of stars. However, a continuing problem with this description is the absence of some low-mass stars with no heavy elements that should have survived from that early epoch. [See GALACTIC STRUCTURE AND EVOLUTION.]

There are also compositional variations with location in our galaxy. The quantitative rates of change are difficult to determine, but recent results show a radial variation in the galactic plane such that stars 5000 light-years farther from the galactic center than the sun have only about 85% of the amount of the heavier elements compared to stars at the sun's location. The stars 5000 light-years closer to the galactic center have about 120% of these elements. It is difficult to extend this study to greater distances because of absorption by interstellar dust in the plane of the galaxy, but similar gradients are observed over the entire disks of other galaxies. In the direction perpendicular to the galactic plane, the gradient seems to be even larger, although age effects greatly complicate this determination. At a distance of 5000 light-years above or below the galactic plane, the stars have abundances that are only 10% of the values found in the plane surrounding the sun. Such variations reflect the history of the formation of the galaxy and the nuclear reactions in the stars.

Our galaxy is just one of an uncountable number of galaxies, most of which are so distant that it is far beyond our capability to see individual stars in them. Some of the nearest galaxies, however, are close enough for us to observe the spectra of individual stars with large telescopes and modern instruments. The two closest galaxies are small, irregular-shaped galaxies visible only from the earth's southern hemisphere, the Large and the Small Magellanic Clouds. By observing stars in these other systems, we can determine if the pattern of abundance we find for the stars in the vicinity of the sun represents the compositions in other, totally isolated stellar systems. While this is an area of current research activity, certain results are emerging. The amount of helium in these galaxies seems to be close to the amount found locally. The heavier elements, however, are less abundant than in the solar neighborhood. In the Large Magellanic Cloud the heavy elements are only 60% of the solar values, and in the Small Magellanic Cloud they are down to only 25%. Other studies of this type will be done with increasing frequency as instrumentation continues to improve.

H. SURFACE STRUCTURE

As has been stated repeatedly, an important advantage of studying the sun is the ability to observe its surface point-by-point in addition to using the integrated light from the whole disk. Because of this we know of sunspots, with their concentrated magnetic fields, as well as many other solar phenomena. It is still not possible to observe the disk of any other star, although some angular diameters have been measured. Stellar spectroscopy, however, already provides a means of deriving some information about the surfaces of some stars. The method, called Doppler imaging because of its use of the Doppler effect, has been applied to cool stars thought to have spots analogous to sunspots.

A starspot is an area of decreased atmospheric brightness, probably associated with a localized magnetic field. If a large spot is located on the surface of a star, the star's light will be diminished by an amount that depends on the temperature decrease in the spot and on its area relative to the entire disk. If the star were not rotating, the diminished brightness would always make the same contribution to every measurement of the stellar spectrum. All stars rotate, however, so the decreased brightness will be concentrated at the shifted wavelength appropriate for the projected rotation speed of the spot. As the spot is carried across the stellar disk by rotation, it will have different projected speeds and, therefore, different wavelength shifts. By making repeated observations of the line profiles, we can follow the progress of the spot across the stellar surface. Of course, if several spots are present simultaneously, the interpretation becomes more complicated, but some progress can be made, especially if the spots last for more than one stellar rotation period.

In addition to a dimming, a starspot can also produce a brightening of the star's light in certain circumstances. The magnetic field that produces a dark spot also causes a brightening of the tenuous upper parts of the stellar atmosphere. We can sample these upper layers by observing strong spectral lines that form completely in the lower-density gas. Because of this, a strong line might show a brightening in the presence of a starspot.

An analysis of the combined dimming and brightening shows something of the surface appearance of stars. The observations, however, have demanding requirements: high spectral resolution to isolate different wavelengths shifts cleanly, low noise data so that variations in brightness at a given wavelength in a line profile are true signals, and continuous nightly coverage for times of several days to several weeks.

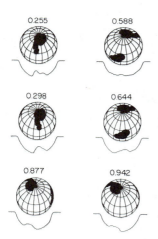

FIG. 13. Schematic representations of the profiles of a single line of the star HR 1099 are shown beneath the models of the surface distribution of starspots that reproduce the profiles. [Reproduced with permission from S. S. Vogt and G. D. Penrod (1983). *Publ. Astronom. Soc. Pacific* **95**, 565.]

To date such observations have only been attempted for a few stars. Figure 13 shows the line profiles and the reconstruction of the surface structure of one well-studied star.

III. Current Developments

Stellar spectroscopy is a mature field of astronomy, but it is also a field that is now experiencing a time of rapid development, a renaissance. This advance is being driven by several factors: the great expansion of the spectral region available for analysis, the development of new instruments and detectors, and the dramatic growth in computing power that has enabled the use of increasingly realistic models to interpret the data. Because the increased availability of computers is well known, we concentrate on the other two factors.

A. EXPANDED SPECTRAL COVERAGE

Throughout most of its history, astronomy has been restricted to the light visible to the eye, which corresponds to wavelengths from about 3900 to nearly 7000 Å. The development of photographic detection altered this somewhat; it was possible to detect some ultraviolet light to which the eye is blind (3000–3900 Å) and also some infrared (7000–10,000 Å), which is again beyond our limit of vision. The primary range of photography, however, is from a wavelength of about 5000 Å down to the limit of atmospheric transparency at 3000 Å. The photographic sensitivity in this spectral region is so low that 99% or more of the light is not used; the efficiency at wavelengths longer than 5000 Å is even lower. Therefore, most astronomy was done in the blue–violet spectral region to make the most of the available light.

This situation has now changed dramatically. An important ingredient in this change has been the development of the space program and the ability to make astronomical observations from above the earth's atmosphere. Our atmosphere absorbs much of the radiation that strikes it from space. The ultraviolet spectral region, wavelengths less than 3000 Å, is prevented from penetrating the atmosphere by ozone, O_3. This protects life on the earth from harmful radiation, but it blocks a large portion of the spectrum. The earth's atmosphere is also opaque to extreme ultraviolet and X-ray radiation, wavelengths less than 1000 Å, because of absorption by N_2 and O_2. Satellite-borne telescopes circumvent this obstruction. The ultraviolet alone more than doubles the amount of spectral coverage compared to the traditional blue–violet band; the inclusion of the X ray expands this even more. This greatly increases the amount of astronomical information available to us. Figure 14 displays the total spectrum of light, showing that the visible radiation is a small fraction of the total. [*See* ASTRONOMY, ULTRAVIOLET SPACE.]

The drive to expand the spectral coverage into the ultraviolet was strongly motivated by solid expectations as well as by the desire to probe the unknown. For example, the wavelength at which a star radiates most strongly decreases as the star's temperature increases. The measurement of a star's output at different wavelengths helps establish its effective temperature, but for hotter stars only a small portion of the radiation is emitted at visible wavelengths; our knowledge of a star's temperature is uncertain unless we have measured the region of its peak output. Even a star with an effective temperature as low as 10,000 K radiates a large portion ot its energy below 3000 Å; it is even more of an advantage to observe hotter stars in the ultraviolet. Another clear motivation for making observations of the ultraviolet spectral region comes from the structure of atoms. Most elements produce their strongest absorption lines—resonance lines—in the ultraviolet. If an element is rare cosmically, these resonance lines may be the only ones we can hope to observe; without them we would have no knowledge of the abundance of that element. Other elements have weak lines in the visible spectral region, but we often lack enough atomic information to analyze these lines accurately. However, we do know enough to analyze the stronger ultraviolet lines. For reasons such as these it was possible to predict confidently that the ultraviolet would be a rich mine of important spectral information.

With this motivation, a high priority of space astronomy has been the construction of astronomical satellites capable of measuring the ultraviolet spectral region. Even before it was possible to launch satellites carrying astronomical instruments, space astronomy went forward by using sounding rockets to snatch a few minutes of data above the atmosphere before dropping back to the earth with their prize. The sun was an early target, but so were hot, bright stars. These early stellar observations were the first to find large mass-loss rates by discovering the large Doppler shifts of the strong ultraviolet lines.

The sun has been the target of a long series of orbiting solar observatories because it occupies a central place in the study of stellar properties. Not only is it bright, allowing more detailed spectral study, and large, enabling astronomers to study surface structures, but its light does not

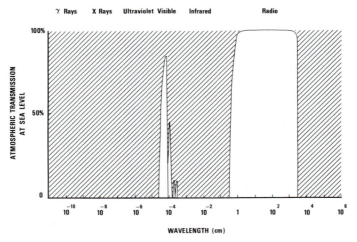

FIG. 14. The transmission of the earth's atmosphere to the complete electromagnetic spectrum. Radiation only reaches the earth's surface in the visible, radio, and narrow bands in the infrared. Additional parts of the infrared can be studied by observing at high altitudes and in extremely dry locations. The expansion of astronomy into the spectral bands that are completely blocked by the atmosphere has led to the rapid advances of recent years.

suffer absorption by interstellar gas. The space between the stars is filled by low-density gas, about one atom per cubic centimeter. Although better than any terrestrial vacuum, this sparse gas can have a tremendous effect on starlight because of the enormous distances between stars. For example, even the nearest star is 4×10^{18} cm from the sun, so its light must pass through about 4×10^{18} atoms. Most of these atoms are hydrogen that can absorb light at all wavelengths shorter than 912 Å. Because of this, interstellar space becomes opaque at wavelengths shorter than this limit, although there may be some holes in this screen because of low gas density in some directions. Only the sun is close enough for us to explore this spectral region thoroughly. We learn from the sun what other stars might also be like, and perhaps we can develop indirect methods of studying this hidden spectral region.

Another advantage of being so close to the sun is the opportunity to observe it from different perspectives. Anchored to the earth, we are restricted to observing every star, including the sun, from just one orientation. For example, we see the sun from almost the plane of its equator. We see the other stars from various orientations, depending on the random positioning of their rotation axes. Some of the observed star-to-star variations have been attributed to viewing stars from different aspects, but this cannot be tested directly. There is now a plan to send a satellite over the solar poles, giving us our first view of the sun, or any star, from different orientations. This will certainly be important for our understanding of the sun, but it will also be a valuable piece of information for stellar spectroscopy.

The first satellites that were launched to observe the stars carried instruments that were designed to measure the spectrum with low resolution. This resolution was inadequate to observe individual line features, but it was appropriate for the measurement of stellar energy distributions. Notable contributions were made in this area by the *Orbiting Astronomical Observatory-2 (OAO-2)*, which carried instruments built by the University of Wisconsin at Madison and by the Smithsonian Astrophysical Observatory. Another highly successful satellite with the same scientific goals was the *European Astronomical Satellite TD-1*, which carried the S2/68 Ultraviolet Sky Survey Telescope.

The next in the OAO series of satellites carried an instrument designed by Princeton University Observatory for high spectral resolution. Nicknamed *Copernicus* because it was launched in 1972, the 500th anniversary of the birth of Nicolas Copernicus, this instrument was intended primarily for the study of absorption lines created by the interstellar gas, but it was also used with good success to study ultraviolet stellar spectral lines. The data displayed in Fig. 7 were obtained with this satellite. Its resolution of 0.05 Å is the highest yet employed in an astronomical satellite. This was achieved, however, by measuring one spectral resolution element at a time, which made for slow observing limited to bright stars. It was also restricted to the spectral range of 912 to 1500 Å.

In 1978 the *International Ultraviolet Explorer (IUE)* satellite was launched. This satellite, a collaborative effort between NASA, the European Space Agency, and the U.K. Science Research Council, carried a 45-cm telescope equipped with instrumentation to observe the spectral region 1150 to 3200 Å at both high and low resolution. This was the first satellite with this dual capacity, making it applicable to a much greater range of astronomical topics. It was also the first astronomical satellite to be placed in a geosynchronous orbit. The earlier satellites in low orbits had periods of about 90 minutes. Because of this, their exposures had to be limited to avoid contamination by earthshine. The *IUE*, on the other hand, has an orbital period of 24 hours, placing it always above one part of the earth. This allows it to continue one exposure for many hours, reaching fainter objects than had ever been studied before in the ultraviolet. With this satellite it has become routine to measure the energy distributions for both bright and faint stars and to observe the detailed absorption line spectra for the brighter objects. Although its best resolution is less than that provided by *Copernicus, IUE* has the significant advantages of covering the whole ultraviolet region and of observing more than 1000 Å in a single exposure.

Not all astronomical observations from space have been made remotely by unmanned satellites. *Skylab*, a space laboratory manned by three teams of astronauts, carried telescopes designed to study both the sun and the stars. An advantage of manned observations is the opportunity to record the data on film that can then be returned to the earth; unmanned satellites must rely on electronic detectors that can transmit their measurements to the ground in a digital form. While such electronic detectors have many important advantages, they have smaller recording areas than film. The *Skylab* telescopes photographed the entire solar disk through filters isolating specific ultraviolet spectral lines. These images show the distribution of the regions emitting this radiation and how these regions change over an extended period of time. Another *Skylab* telescope recorded low-resolution stellar spectra used to assign effective temperatures for many stars. The completeness achieved in this sample complements the more detailed study of individual stars possible with satellites such as *IUE*.

Finally, plans are being made for the launch of the Hubble Space Telescope (HST), a telescope with a 2.4-m-diameter primary mirror. The HST is designed to provide a permanent space observatory with a working life of several decades. It will carry a battery of instruments, two of which will return high- and low-resolution spectra. The size and the quality of the primary mirror, which represents a tremendous advance over existing satellite telescopes, are expected to translate into a comparable increase in the scientific output of the observatory. [*See* OPTICAL TELESCOPES.]

The results from ultraviolet stellar observations have reshaped our understanding of stars. One extremely important result, mentioned earlier, was the discovery that hot, luminous stars are losing large amounts of mass. This was first found from data gathered by sounding rockets. It was explored in a limited way using *Copernicus,* but it has been thoroughly studied using the *IUE* satellite. Mass loss is such an important process for stars that it must now be included in calculations of the way a star evolves. The consequences of mass loss are now also being considered even when it is too small to alter the star's evolution. For example, a small mass loss rate, 10^{-12} solar mass per year, is now thought to be enough to alter the surface layers of a star enough to change the apparent composition derived from its absorption lines. Such a possibility would never have been considered without the ultraviolet observations.

Another finding has been the discovery of anomalously ionized elements in the atmospheres of some hot stars. We can determine values of the effective temperatures by studying the energy distributions and the visible spectra of these stars. Coupled with our knowledge of atomic structure, it is then possible to use a model of the stellar atmosphere to predict the ionization state of each element. For example,

silicon should be an ion that has been stripped of three electrons at an effective temperature of 25,000 K. Instead, we observe that the elements are missing more electrons than would be expected from the star's temperature. Triply ionized silicon is observed in stars as cool as 11,000 K, although these stars are nearly 10,000 K too cool for this process to occur according to our standard ideas. The same anomaly is also observed for ions of nitrogen and oxygen. The explanation of this phenomenon is still not clear, but obviously these hot stars have regions in their atmospheres that are much hotter than expected from the way they radiate light.

Ultraviolet spectroscopy has also made important contributions to the study of chromospheres and coronas. Above the sun's atmosphere the density of gas decreases with height, but the temperature begins to increase after reaching a minimum value of about 4300 K. The temperature rises gradually to a value of about 7000 K, and then rapidly heats up to more than 1 million K. The cooler, inner region (the chromosphere) and the hot, outer region (the corona) are both visible during total solar eclipses. In addition, both regions also produce a spectrum of emission lines as the hot gas is seen in front of a much cooler solar atmosphere or the blackness of space. It is possible to detect some of these emission lines by careful observations made from the earth's surface, but they are not prominent. Ultraviolet observations made from space, however, show the chromospheric emission much more clearly for two reasons. First, the background brightness of the cool solar disk diminishes rapidly at shorter, ultraviolet wavelengths. At wavelengths less than 2000 Å, the spectrum becomes dominated by the chromosphere and corona. Second, the strong lines of the ions present at the higher temperatures of the chromosphere and corona are concentrated in the ultraviolet.

The study of stellar chromospheres and coronas can also be pursued spectroscopically. Emission by singly ionized calcium is the primary indicator in the visible spectral region, although emission can occasionally be seen in the hydrogen lines. However, it is the rich array of ultraviolet emission lines that is most prominent: neutral hydrogen and oxygen and singly ionized magnesium, silicon, and iron. The transition region between the chromosphere and the corona is characterized by temperatures in the range 50,000 to 250,000 K. Because ions that are present at these temperatures do not emit at optical wavelengths, the presence of transition regions in stars was not shown until ultraviolet observations in the 1200- to 2000-Å region were possible. These observations found emission lines of ions of helium, carbon, nitrogen, oxygen, and silicon. The presence of a transition region implies the existence of a corona, too, but unfortunately the ions that exist at temperatures of more than 1 million K emit in the extreme ultraviolet region, wavelengths less than 900 Å. Because of the absorption by interstellar hydrogen, it is impossible to see such emission. Therefore, we are left to infer a corona if we see evidence of the transition region, except for those stars that have been found to emit X rays indicative of temperatures of more than 1 million K.

The *IUE* satellite has provided the means of surveying many cool stars to establish the broad properties of their chromospheres and coronas. The survey has shown that these stars can be divided into two groups. There are stars that have chromospheres, transition regions, and, by implication, coronas. These regions are tenuous, similar to the conditions present in the sun. There is another group of stars, however, that have chromospheres but no transition zones. From this we conclude that these stars also lack coronas. Stars in this group do have the spectral line asymmetries produced by mass loss. This is an excellent example of the interplay between studies of the sun and those of the stars. Our long experience with the sun has educated us to the spectroscopic indicators of the chromosphere, transition region, and corona. With the development of appropriate satellite instrumentation we have turned this experience to stellar observations, where we find that some stars are similar to the sun, but other stars are not because of different effective temperatures or radii or compositions. We certainly learn something about the stars in this way, but we also learn about the sun, because these stars show us how the outer parts of the sun's atmosphere depend on the basic parameters. Figure 15 gives examples of the chromospheric and transition region spectra measured by the *IUE*.

The ultraviolet spectral region has clearly been a major area of development in the past two decades. A more moderate effort has also been made to push stellar spectroscopy into the longer, infrared, wavelengths. While there are some partially transparent bands, atmospheric absorption, primarily due to water vapor, has severely hampered the exploration of this region. Ground-based observations must be made

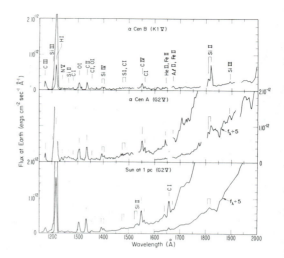

FIG. 15. A comparison of the ultraviolet spectrum of the sun with the spectra of the two nearest stars, α Centauri A and B. The spectra show that these stars have features indicative of chromospheres and transitions regions similar to the sun. [Reproduced with permission from T. R. Ayres and J. L. Linsky (1980). *Astrophys. J.* **235,** 76.]

from the tops of mountains located in arid regions to get above as much of the atmospheric blocking as possible. Observations are also made from aircraft, high-altitude balloons, and now from space with the successful mission of the *Infrared Astronomical Satellite*.

Limited atmospheric transparency is not the only obstacle to observing in the infrared. The detectors pick up signals from any source of heat, including the telescope and its housing. Because of this, the detector, and sometimes the whole telescope, is actively cooled to temperatures approaching 0 K. The coolant is an expendable material that limits the life of a satellite mission. Another complication of infrared spectroscopy is the difficulty in achieving high spectral resolution. Instruments that work well at visible and ultraviolet wavelengths are not appropriate for the infrared. New techniques have been required to work at the longer wavelengths. [*See* ASTRONOMY, INFRARED.]

Despite the difficulties, there is strong motivation to do infrared stellar spectroscopy. The natural targets for infrared observations are the cool stars that radiate most strongly at long wavelengths. Measurement of the infrared energy distribution of a cool star gives a firm grip on the effective temperature. This is particularly true because the H^- ion, the major source of atmospheric opacity in cool stars, displays its greatest variation in the near infrared, reaching a minimum at a wavelength near 1.5 μm. The amount of H^- in a stellar atmosphere changes rapidly with effective temperature, so the infrared observations have a sensitive temperature indicator. Because only the shape of the energy distribution is needed for this type of determination, low-spectral-resolution measurements are adequate.

Higher-resolution observations make it possible to study individual spectral features, which are usually produced by molecules in these cool stars. Molecules are capable of absorbing by several mechanisms. Some of the absorptions appear in other parts of the spectrum, but the combination of rotational and vibrational absorption produces extensive bands in the infrared spectral region for many common molecules. Therefore, there is a convergence of a wealth of spectral information and an abundance of radiation in the infrared for cool stars.

Infrared stellar spectroscopy has already found several important applications. For example, the light of pre–main-sequence stars, which are still shrouded by the gas and dust from which they formed, is so strongly absorbed at visual wavelengths that it is frequently impossible to see them. Infrared light, however, penetrates this obscuring material much more easily. The effective temperatures and possibly the radii of these newly formed stars can be determined from energy distributions measured by low-resolution infrared observations. This information represents an important addition to our knowledge of the properties of stars soon after birth. Without the ability to make infrared observations, we would not be able to see these stars until the dust had dispersed and they were much older.

Low-resolution infrared measurements have also detected excess amounts of radiation from some stars. This is interpreted as a sign of matter surrounding the stars, probably produced by mass loss such as is shown by Doppler-shifted absorption lines in the spectrum. The infrared energy distribution complements the spectral line data, and it is often easier to derive a mass-loss rate from the infrared data because the detailed velocity distribution of the outward flowing gas is not required.

The high-resolution observation of molecular absorption lines also provides a wealth of information. For example, the fundamental bands of CO are located in the infrared at a wavelength of 5 μm. As described earlier, in most stars oxygen is more abundant than carbon. Because CO is the dominant molecule of these two elements, the amount of CO determines the availability of carbon to form other molecules. Therefore, it is necessary to understand the CO formation correctly before it is possible to interpret properly other molecules such as CN, C_2, and CH. Infrared spectroscopy provides this crucial piece of information.

In addition to CO, the infrared is also rich with bands of molecules formed from the other common elements: C_2, CN, H_2O, SiO, OH, NH, and MgH. Because these are the most important elements, the infrared spectrum provides an unprecedented opportunity to derive accurate abundances, both for the total element and for the isotopes of each element. The importance of measuring ^{12}C : ^{13}C was mentioned earlier. Infrared observations provide the data to study other key ratios, such as ^{16}O : ^{17}O : ^{18}O, ^{28}Si : ^{30}Si, and ^{35}Cl : ^{37}Cl. Some of these ratios depart substantially from the ratios that are considered normal from terrestrial studies. The wealth of information present in high-resolution infrared spectra is clearly evident in Fig. 12.

A recent development in infrared spectroscopy has been the discovery of atomic lines of Mg and Al in the solar spectrum at a wavelength near 12 μm. This is an important find for magnetic studies because the Zeeman splitting increases with increasing wavelength more rapidly than does Doppler broadening. Even though these lines have a small intrinsic sensitivity to the magnetic field, the large wavelength makes them the most useful measures of the magnetic field strength known. This discovery might also have application to stellar studies.

Finally, infrared stellar spectroscopy is capable of measuring radial velocities with high precision. This is because of the many individual absorption lines associated with a particular vibration–rotation molecular band. The wavelengths of the individual lines are known with great accuracy, and the large number of lines reduces the accidental errors of measurement. Furthermore, because different molecules can form in different strata of the stellar atmosphere, it is possible to study the relative motions of these levels.

Stellar spectroscopy at still longer wavelengths moves from the infrared into the radio spectral region. The transition between infrared and radio occurs in a region where the earth's atmosphere is opaque, but at wavelengths greater than 1 mm the atmosphere is increasingly transparent, and beyond 1 cm is totally transparent out to the long wavelength cutoff of the ionosphere. Therefore, radio spectroscopy enjoys a significant advantage over infrared and ultraviolet observations in being able to make unobstructed observations from the earth's surface. Radio detectors are also capable of high spectral resolution, although they have been limited to small segments of the spectrum. The main disadvantage to radio observations in general has been the poor angular resolution of the telescopes. A single radio antenna cannot pinpoint an object as small as a star. If the source is in a crowded field, the radio observation is subject to considerable confusion. This problem is circumvented by using two or more telescopes together to simulate the ability of a single, much larger telescope.

The spectral features present at radio wavelengths tend to be molecular lines, and this selects cool stars as the most likely targets of study. Even cool stars, however, emit mostly at infrared wavelengths rather than the radio. Therefore, radio spectroscopy is best suited to studying the cooler gas surrounding stars as the result of mass loss, mass exchange among double stars, or remnants of formation.

The circumstellar gas shells of cool stars have proved to be rich sources of radio molecular emission. Normal, oxygen-rich stars exhibit thermal emission from SiO and CO, as well as maser emission in lines of OH, H_2O, and SiO. This list is even longer for carbon-rich stars because of the formation of complex carbon-based molecules. The study of these features show several things about the stars and their environments. First, they provide information on the composition of the material, both of the total element and of the isotopes. This may confirm the results of spectroscopic studies in the visual and infrared spectral regions. In other cases, the star's visible light is so obscured by the circumstellar matter that the radio data provide the only source of information. Second, the amount of circumstellar matter gives the rate at which mass loss is occurring. Often the derived rates are large, indicating that a significant amount of stellar mass is recycled to the interstellar medium. Third, the circumstellar matter is not always spherically symmetric, as is generally assumed in models of mass loss. The radio observations give some indication of the importance of processes such as rotation, nonradial

pulsation, and magnetic fields that might affect the geometry. Finally, the SiO, H_2O, and OH maser emission, which has been observed in more than 300 stars, gives important information about the energy sources and physical conditions in the circumstellar matter.

B. Instrumentation

The expansion of spectroscopy into new spectral regions has been paralleled by the development of new instruments and detectors. The spectrographs used to make high-resolution measurements of line profiles and equivalent widths have traditionally been large, so large that they are housed in stationary rooms away from the telescope, to which the light must be brought by a chain of mirrors. This design has restricted these powerful instruments to just the largest telescopes at the largest observatories because of the need to focus a large amount of light down the long mirror train and because of the high cost involved. Manufacturing methods can now produce high-quality components for a new form of high-resolution spectrograph using an echelle in place of the ordinary grating. The echelle differs from the conventional grating by having coarse grooves, typically 50 mm^{-1}, which are tilted at a steep angle. The advantage of the echelle is that it can produce high resolution in a compact instrument. This advantage has been realized for a long time, but production techniques now can turn this realization into a reality. Because of the much smaller size of the echelle spectrograph, it is possible to mount it directly on to the telescope, preventing the light loss associated with the train of mirrors. This enables the instrument to observe fainter stars or to observe brighter stars more quickly. Also, because the echelle spectrograph is smaller, it can be used with telescopes of modest size, greatly increasing the number of locations at which high-resolution spectroscopic observations can be made. An echelle spectrograph was used in the *IUE* satellite to fit a high-resolution instrument into the available space.

The Fourier transform spectrometer (FTS) is another recent addition to the array of spectroscopic instruments. It is a Michelson interferometer with a means of varying the path difference between the two arms of the instrument. It has an impressive list of attractive features, including variable resolution from high to low, excellent efficiency, high signal-to-noise properties, the absence of scattered light, and high photometric accuracy. Its only drawback is that it uses a single detecting element. When the random fluctuations in the arriving light dominate the noise, the FTS is slower than instruments that use multiple detecting elements, such as a conventional photograph, although this may be more than offset by its good features. This is the situation in the visible spectral band, where the FTS has been used extensively for solar observations, but not for stars because its slowness becomes a disadvantage for the fainter sources. In the infrared, however, the dominant source of noise is usually in the detector. The use of a single detector becomes an advantage, and the FTS is the instrument of choice for spectroscopic studies of the sun, stars, planets, and other astronomical objects. It is now possible to make spectroscopic observations of bright stars that are comparable to the quality that could only be obtained for the sun a few years ago, and both solar and stellar spectra are of the same quality as laboratory measurements. This is an area of considerable potential that should develop rapidly.

The development of new detectors has paralleled the development of new instruments. As mentioned earlier, the photograph has been the workhorse for astronomical detection for more than a century. Even now it is unrivaled if a detector of large area and information capacity is required; emulsions covering an area of 20 by 25 cm are routinely available, and larger sizes can be manufactured. There are, however, several serious limitations to photographic detection. First, the potentially high efficiency of the photographic process has never been reached. Under the best conditions, less than 5% of the light gathered by the telescope and transmitted by the instrument is used to form the photographic image. With less care and at longer wavelengths only a small fraction of a percent is used. A second deficiency of photography is the need to calibrate the sensitivity of the emulsion to light of different intensity. This calibration, which is crucial to the measurement of line profiles and equivalent widths, depends on the wavelength of the light being studied as well as the way the emulsion is developed, and many errors have been traced to this step in the acquisition of research data.

For these reasons, astronomers have always sought new detectors with improved properties. A significant recent advance has been the development of solid-state silicon photodiode detectors. These devices have the property of producing an electrical signal when exposed to light; the greater the exposure, either from a brighter source or a longer exposure time, the greater the signal. Moreover, there is a direct one-to-one relationship between the exposure and the signal. The signal can then be converted to a digital representation for immediate computer manipulation. In addition to the considerable convenience of digital output, these devices enjoy several other advantages. First, they have efficiencies reaching 80% in the red part of the spectrum. This is an enormous gain compared to the small fraction of a percent for photography at the same wavelengths. Another advantage is a well-defined, stable geometry. Each photodiode, or pixel, has a fixed width of about 0.02 mm. A third advantage is the ability to measure a large range of intensities in one exposure. Not only is it possible to measure deep spectral lines in this way, but the noise can be suppressed to a tiny fraction of the signal; it is possible to make observations where the noise is less than 0.5% of the signal being studied. Such observations show spectral features, such as the signature of starspots shown in Fig. 13, which had been lost in the noise previously. The main disadvantages of these detectors are their need to be cooled to low temperatures (-100 to $-150°C$) to suppress thermal contamination during exposures of several hours' duration, the need to calibrate accurately the pixel-to-pixel variation in sensitivity, and their small size (typically 25 mm) set by the dimensions of single silicon wafers.

The silicon photodiodes currently used for astronomical observations come in different forms. Devices consisting of a single row of pixels have been used successfully for spectroscopy for several years. Two-dimensional arrays of diodes are now being used for direct imaging and for spectroscopic detection with both conventional and echelle spectrographs. Because there is commercial and industrial interest in compact two-dimensional detectors, we can expect continued development of these devices.

The solid-state devices are exposed directly to the light from the instrument for times up to several hours to record a single observation. For faint objects, however, the signal is so weak that noise generated by the process of electrically reading the detector is dominant, making it necessary to amplify the light before its measurement. This approach records the arrival of individual photons of light, and the resulting noise is just determined by the fluctuations in the stellar light. Such counting devices also make it possible to record simultaneously the light coming from the adjacent blank sky and subtract this from the stellar signal. The result is a detector system that can extend the observational limit to extremely faint stars. The disadvantages of these devices are their small detecting areas and the difficulty of maintaining a strict one-to-one correlation between input light and output signal in a large, complex system. Because of these constraints, these devices have been used primarily with low-resolution spectrographs and for projects where line position or gross line strengths are desired instead of high-accuracy intensities. As mentioned earlier, these newer devices are much more sensitive than the photographic emulsions, and this is particularly true in the yellow and red portions of the spectrum. Because of this improvement, much greater attention is now being directed toward these spectral regions even though they have been accessible in a limited way for some time.

With the recent development of these new detectors coupled with more efficient spectrometers and the ability to observe over the entire stellar spectrum from the ultraviolet to the ra-dio, the old field of stellar spectroscopy faces a bright future.

BIBLIOGRAPHY

Chaffee, F. H., Jr., and Schroeder, D. J. (1976). Astronomical application of Echelle spectroscopy. *Annu. Rev. Astron. Astrophys.* **14**, 23.

Garrison, R. F., ed. (1984). "The MK Process and Stellar Classification." David Dunlap Observatory, Toronto.

Jordan, S. D., ed. (1981). "The Sun as a Star." National Aeronautics and Space Administration, Washington, D.C.

Linsky, J. L. (1980). Stellar chromospheres. *Annu. Rev. Astron. Astrophys.* **18**, 439.

Mackay, C. D. (1986). Charge-coupled devices in astronomy. *Annu. Rev. Astron. Astrophys.* **24**, 255.

Merrill, K. M., and Ridgway, S. T. (1979). Infrared spectroscopy of stars. *Annu. Rev. Astron. Astrophys.* **17**, 9.

Ridgway, S. T., and Brault, J. W. (1984). Astronomical Fourier transform spectroscopy revisited. *Annu. Rev. Astron. Astrophys.* **22**, 291.

Wolff, S. C. (1983). "The A-Type Stars." National Aeronautics and Space Administration, Washington, D.C.

STELLAR STRUCTURE AND EVOLUTION

Peter Bodenheimer   *Lick Observatory, University of California, Santa Cruz*

GLOSSARY

Degenerate gas: Gas in which the elementary particles of a given type fill most of their available momentum states as determined by the Pauli exclusion principle. Electron degeneracy occurs in the cores of highly evolved stars and in white dwarfs; neutron degeneracy occurs in neutron stars.

Effective temperature: Surface temperature of a star calculated from its luminosity and radius under the assumption that it radiates as a black body.

Galactic cluster: Group of a few hundred or a few thousand stars found in the disk of the galaxy. In a given cluster all the stars were formed at about the same time, but a wide range of ages is represented among the various clusters.

Globular cluster: Compact group of 10^5 to 10^6 stars, generally found in the halo of a galaxy and formed early in the history of the galaxy.

Hertzsprung–Russell diagram: Plot of a collection of stars, each of which is represented by a point whose ordinate is the luminosity and whose abscissa is the effective temperature (or color or spectral type).

Horizontal branch: Sequence of stars on the Hertzsprung–Russell diagram of a typical globular cluster, above the main sequence. The stars all have approximately the same luminosity and are in the evolutionary phase where helium is burning in the core.

Luminosity: Total rate of radiation of electromagnetic energy from the surface of a star, in all wavelengths and in all directions.

Main sequence: Sequence of stars in the Hertzsprung–Russell diagram, on which a large fraction of all stars fall, running diagonally from upper left to lower right, and associated with the evolutionary phase in which the stars burn hydrogen to helium in their cores.

Neutrino: Subatomic particle with no measurable mass or charge, produced in beta-decay reactions, which travels at the speed of light and interacts only very weakly with matter.

Neutron star: Highly compressed remnant of the evolution of a star of high mass. Its main constituent is free neutrons, the degenerate pressure of which supports the star against gravitational collapse.

Nova: Sudden but temporary brightening of a star by a factor of hundreds to thousands, occurring in a binary system where a white dwarf star is accreting mass from its main-sequence companion. The outbursts are caused either by instability in the accretion disk surrounding the white dwarf or by nuclear reactions in the material recently accreted on to its surface.

Nucleosynthesis: Chemical evolution of the galaxy, as studied by observations of abundances of the elements and comparison with production rates of the elements through nuclear reactions in stars.

Protostar: Star in its earliest stage of evolution, during which it undergoes hydrodynamic collapse and during which it is observable in the infrared part of the spectrum.

Supernova: Sudden increase in luminosity of a star, by a factor of up to 10^{10}, followed by a slower decline over a time of months or years. The event is caused by explosion of the star and dispersal of much of its matter.

Thermonuclear reaction: Nuclear fusion reaction in which the high relative particle velocities required for its operation are provided by the thermal energy of the gas.

White dwarf: Compact star representing the final stage of evolution of a star of low to moderate mass. It is supported against its gravity by the high pressure of the degenerate electrons in its interior.

Zero-age main sequence: Line in the Hertzsprung–Russell diagram corresponding to the points where stars of different masses first arrive on the main sequence. They have completed the phase of gravitational contraction and are deriving their energy from nuclear burning of hydrogen, but their chemical composition has not yet been significantly changed by the burning.

Stellar structure is the study of the internal properties of a star (e.g., the temperature or the energy production mechanism) and their variation from the center to the surface. The structure can be determined through a combination of theoretical calculations based on known physical laws and observational data. Stellar evolution refers to the change in these physical properties with time, again as determined by both theoretical and observational arguments. The three main phases of stellar evolution are (1) the pre–main-sequence phase, during which gravitational contraction provides most of the star's energy, (2) the main sequence phase, in which nuclear fusion of hydrogen to helium in the central regions provides the energy, and (3) the post–main-sequence phase, in which hydrogen burning away from the center, as well as the burning of helium, carbon, or heavier elements, may contribute to energy production. The evolutionary properties are strongly dependent on the initial mass of the star and to some extent on its original chemical composition.

I. Introduction

The study of the structure and evolution of the stars, which constitute the major fraction of the directly observable mass in the universe, is of critical importance for the understanding of the production of the chemical elements heavier than helium, of the evolution of the solar system, of the structure and energetics of the inter-stellar gas, and of the evolution of galaxies as a whole. The structure of a star is determined by the interaction of a number of basic physical processes, including nuclear fusion; the theory of energy transport by radiation, convection, and conduction; atomic physics involving especially the interaction of radiation with matter; the equation of state of a gas; and thermodynamics. These principles, combined with basic equilibrium relations, allow the construction of mathematical models of stars that give the temperature, density, pressure, and chemical composition of the object as a function of distance from the center. A star is not static, however; it must evolve in time, driven by the loss of energy from its surface, primarily in the form of radiation. This energy is provided by two fundamental sources—nuclear energy and gravitational energy—and in the process of providing this energy the star undergoes major changes in its structure. To follow this evolution mathematically requires the solution of a complicated set of equations, a solution that requires the use of high-speed computers to obtain sufficient detail. The goal of the calculations is to obtain a complete evolutionary history of a star, as a function of its initial mass and chemical composition, from its birth in an interstellar cloud to its final state either as a white dwarf, a neutron star, or a black hole or possibly as an object completely disrupted by a supernova explosion.

The heart of the study of stellar structure and evolution comes, however, through the comparison of models and evolutionary tracks with the observations. There are numerous ways in which such comparisons can be made, for example, by use of the Hertzsprung–Russell (H–R) diagrams of star clusters, the mass–luminosity relation on the main sequence, the abundances of the elements at different phases of a star's evolution, and the mass–radius relation for white dwarfs. There are many exotic stars that the theory is not yet able to fully explain, such as pulsars, novae, X-ray binaries, γ-ray bursters, or stars showing rapid mass loss. These systems provide a challenge for the future theorist. However, the general outline of the phases of stellar evolution has by now fallen into place through a complex interplay between theoretical studies, observations of stars, and laboratory experiments, particularly those required to determine nuclear reaction rates. The period of development of ideas concerning the structure of stars extends at least 100 years into the past. Among the noteworthy historical developments were the clarification by Sir Arthur Eddington (1926) of the physics of radiative energy transport, the development by S. Chandrasekhar (1931) of the theory of white dwarf stars and the derivation of their limiting mass, and the work of H. Bethe (1939) and others, which established the precise mechanisms by which fusion of hydrogen to helium provides most of the energy of the stars. However, much of the detailed development of the subject has occurred between 1955 and 1985, spurred by the availability of high-speed computers and by the extension of the observational data base from the optical region of the spectrum into the radio, infrared, ultraviolet, X-ray, and γ-ray regions. Numerous scientists have collaborated to advance our knowledge of the physics of stars in all phases of their evolution. [*See* ASTROPHYSICS.]

II. Observational Information

The critical pieces of observational data include the luminosity, surface temperature, mass, radius, and chemical composition of a star. A further fundamental piece of information that is required to obtain much of this data is the distance to the star, a quantity that is in general difficult to measure because even the nearest star is 2.6×10^5 astronomical units (AU) away, where the astronomical unit, the mean distance from the earth to the sun, is 1.5×10^{13} cm. The astronomical unit, measured accurately by the use of radar reflection experiments off the surface of Venus, is used as the baseline for trigonometric determinations of stellar distance. The apparent shift in the position of a nearby star, against the background defined by more distant stars or galaxies, when viewed from different points in the earth's orbit, allows the distance to be determined. The parsec (pc) is defined as the distance of a star with an apparent shift of 1 sec of arc on a baseline of 1 AU and has the value of 3.08×10^{18} cm. The nearest star is 1.3 pc away, and the method becomes inaccurate once the distance is greater than 200 pc. Beyond this distance indirect methods must be used, based, for example, on properties of the spectrum of the stars and a calibration from stars with similar spectra and known distances.

A. LUMINOSITY

The standard value of luminosity is that of the sun, which is obtained by a direct measurement of the amount of energy $S_\odot$, received per square

centimeter per second, over all wavelengths, outside the earth's atmosphere, and at the mean distance of the earth from the sun. This quantity, known as the solar constant, is then converted into the solar luminosity by using the formula

$$L_\odot = 4\pi d_\odot^2 S_\odot = 3.86 \times 10^{33} \text{ erg/sec,}$$

where $d_\odot = 1$ AU. For other stars, in principle, the stellar flux S, in ergs per square centimeter per second received at the earth, is corrected for the effects of the atmosphere and interstellar absorption and is extended to include all wavelengths of radiation. If the star's distance d is known, its luminosity follows from $L = 4\pi d^2 S$.

B. SURFACE TEMPERATURE

A number of different methods are used to determine the surface (effective) temperature, most of which are based on the assumption that the stars radiate into space with a spectral energy distribution that approximates a black body. For a few stars, such as the sun, whose radius R can be measured directly, the value of T_{eff} is obtained from L and R by use of the black-body relation $L = 4\pi R^2 \sigma T_{\text{eff}}^4$, where σ is the Stefan–Boltzmann constant. Otherwise, the temperature may be estimated by four different methods.

1.   The detailed spectral energy distribution of the star is measured and the temperature of the black-body distribution that best fits it is found.

2.   The wavelength $\lambda_{\max}$ of maximum intensity in the spectrum is measured, and the temperature is found from $\lambda_{\max} T = 2.89 \times 10^7$, if $\lambda_{\max}$ is expressed in angstroms.

3.   The "color" of the star is obtained by measurement of the stellar flux in two different wavelength bands. For example, the color "B–V" is obtained by measurement of the star's flux in a wavelength band centered at 4400 Å and about 1000 Å broad (blue) as compared to that measured in a band of similar width centered at 5500 Å (visual). In principle, the transmission properties of the B and V filters could be used in connection with the black-body curves to determine the temperature. In practice, stars are not perfect black bodies, and the temperature is determined by comparison with a set of standard stars.

4.   From the strength in the stellar spectrum of absorption lines of various chemical elements, the temperature can be determined. This

method does not depend on the black-body assumption. The spectral type is determined by comparison with a set of standard stars. [*See* STELLAR SPECTROSCOPY.]

C. RADIUS

For only a few stars can the radius be measured directly. In the case of the sun, the angular size can be measured and the distance is known. In the case of Sirius and a few other stars, the angular diameter can be measured by use of interferometry. For certain eclipsing binary systems, if the orbital parameters are known and the light variation with time can be accurately measured, the radius of the eclipsing star can be found from the time required for the light to decrease from maximum to minimum at the start or end of the eclipse. In all other cases the radius must be estimated from measurement of L and T_{eff} and the formula $L = 4\pi R^2 \sigma T_{eff}^4$.

D. MASS

A stellar mass can be measured directly only if the star is a member of a binary system, in which case Kepler's third law can be applied. If M_1 and M_2 are the stellar masses, in units of the solar mass $M_\odot$, P the orbital period of the system in years, and a the semimajor axis of the relative orbit in astronomical units, the law states that $(M_1 + M_2)P^2 = a^3$. In the case of the sun the orbital periods and distances of the planets can be used for an accurate determination of $M_\odot$. If both components of a binary system are visible (visual binary) and the angular separation as well as the period can be measured, the sum of the masses follows from Kepler's law as long as the distance is known. The individual masses can be found if the relative distances of the stars from the center of mass can be measured. If the binary system has such a close separation that the components cannot be visually resolved, it may be possible to resolve them spectroscopically. If the Doppler shifts of spectral lines as a function of time can be measured for both components, then the period can be determined as well as the orbital velocity of each star. However, the system must show an eclipse, so that it is known that the orbit is being observed edge on, since the Doppler-shift method measures only the velocity component in the line of sight. In this case the period and velocity combine to give a, so that Kepler's law can be used to obtain $M_1 + M_2$. The individual masses follow from the ratio of the individual velocities. Note that the distance to the system does not enter into the mass determination. However, there are very few systems with the required orbital characteristics to allow reasonably accurate mass determinations.

E. ABUNDANCES

The relative abundances of the elements in the solar system have been compiled by A. G. W. Cameron, based in most cases on measurements of the oldest meteoritic material and in some cases from the strengths of absorption lines in the solar atmosphere. Abundances for selected elements are given in Table I. In stellar evolution theory, the fractional abundance of H by mass is known as X, that of He as Y, and that of all other elements as Z. In stars, the abundances are determined from the strengths of the absorption lines in the spectrum, together with other parameters of the stellar atmosphere, and compared with the solar values. In practically all measured systems, the values of X and Y are very similar to solar values. However, Z can vary, and in the oldest stars and the globular clusters it can fall below the solar value by a factor of up to 100. [*See* STELLAR SPECTROSCOPY.]

F. HERTZSPRUNG–RUSSELL DIAGRAM

The Hertzsprung–Russell (H–R) diagram is a plot of luminosity versus surface temperature for a set of stars. Although the data can be plotted in various forms, the sample H–R diagram

TABLE I. Abundances of the Elements in the Solar System

Element	Abundance by number of atoms (Si = 10⁶)	Fractional abundance by mass
1 H	2.66×10^{10}	0.772
2 He	1.80×10^9	0.209
3 Li	60	1.21×10^{-8}
6 C	1.11×10^7	3.91×10^{-3}
7 N	2.31×10^6	9.42×10^{-4}
8 O	1.84×10^7	8.57×10^{-3}
10 Ne	2.6×10^6	1.52×10^{-3}
12 Mg	1.06×10^6	7.48×10^{-4}
14 Si	1.0×10^6	8.16×10^{-4}
20 Ca	6.25×10^4	7.28×10^{-5}
26 Fe	9.0×10^5	1.46×10^{-3}
28 Ni	4.78×10^4	8.14×10^{-5}

shown here (Fig. 1) includes data converted from observed quantities to L and T_{eff}. Most of the stars fall along the main sequence, which represents the locus of stars during the phase of hydrogen burning in their cores, with increasing T_{eff} corresponding to increasing mass. The stars well below the main sequence are in the white dwarf phase, having exhausted their nuclear fuel. The stars in the upper-right part of the diagram are red giants (e.g., Aldebaran); these are stars that have exhausted their central hydrogen and are now burning hydrogen in a shell region around the exhausted core. At higher luminosities, roughly $10^3 L_\odot$ or above, the red giants are also burning helium. The density of stars in the H–R diagram in any given region is roughly pro-

portional to the evolutionary time spent in that region; thus the main-sequence, or core hydrogen-burning, phase is that in which stars spend most of their lifetime.

G. MASS–LUMINOSITY RELATION

For stars known to be on the main sequence, the observational information on M and L can be combined to produce a reasonably smooth mass–luminosity relation (Fig. 2). Empirically, the luminosity increases as M^3 for stars in the region of $10M_\odot$, as $M^{4.5}$ for stars of solar mass, and more slowly as M^2 for the low-mass main-sequence stars. A theoretical curve of the main-sequence mass–luminosity relation is given in

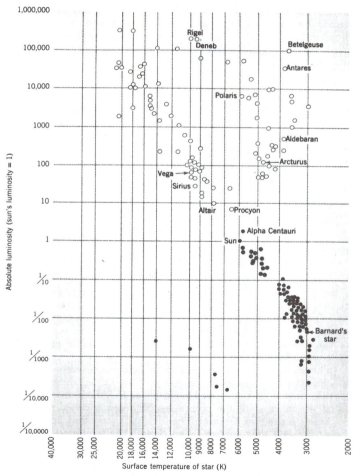

FIG. 1.   H–R diagram for the 100 brightest stars (light circles) and the 90 nearest stars (dark circles). [Reprinted with permission from Jastrow, R., and Thompson, M. H. (1984). "Astronomy: Fundamentals and Frontiers," 4th ed. John Wiley and Sons, New York. © 1984, Robert Jastrow.]

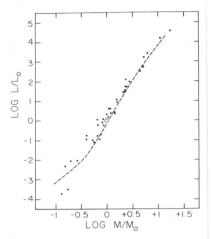

FIG. 2. Observed masses of main-sequence stars, as determined from binary orbits, plotted against their luminosities (dots). The sun is indicated by the symbol ⊙. The dashed line is the theoretical mass–luminosity relation for stars on the zero-age main sequence, with an assumed chemical composition similar to that of the sun.

the figure, based on an assumed composition of $X = 0.71$, $Y = 0.27$, $Z = 0.02$. The theoretical curve exhibits the same trends in its slope as the observed points. Because stars evolve and change their luminosity considerably while retaining the same mass, a mass–luminosity relation cannot be specified for most regions in the H–R diagram, only near the zero-age main sequence. Evidently, the luminous, hot stars near the upper end of the main sequence in the H–R diagram are of high mass, while those near the lower end of the main sequence are of low mass.

H. STELLAR AGES

Although the age of the solar system (and therefore presumably the sun) can be determined fairly accurately from potassium–argon age dating of the oldest moon rocks and meteorites, the ages of other stars cannot be determined directly. Several indirect methods exist, which depend for the most part on the theory of stellar evolution, and generally there is some uncertainty in the derived ages.

1. The H–R diagrams of galactic or globular clusters can be compared with evolutionary calculations. The stars in a cluster are assumed to be all of the same age, to have the same composition, and to all lie at the same distance from earth. The observed cluster diagram is compared with a theoretical line of constant age, or

the isochrone, obtained by calculating the evolution of a set of stars of different masses and connecting points on their evolutionary tracks that correspond to the same elapsed time since formation. This procedure is illustrated in Fig. 3, which shows how an artificial, computed star cluster looks at two different times. All stars were assumed to have been formed at the same time and to have a distribution of masses that approximates that in an actual cluster. Evidently, the main sequence at the earlier time extends to higher L and T_{eff} than that at the later time, for the reason that the time spent by a star on the main sequence decreases with increasing mass. Since stars evolve relatively quickly away from the main sequence after core hydrogen burning, the main-sequence turnoff, defined as the point of maximum blueward extent of the main sequence, is a reliable indicator of age. In this method, the age of the stars is really being determined by the nuclear-burning time scale.

2. The abundance of the light element lithium is an approximate indicator of age in stars of around a solar mass. This element is easily destroyed by nuclear reactions with protons at temperatures of 3×10^6 K or above. The abundance of lithium at a stellar surface will therefore decrease on a time scale comparable to that required to circulate surface material down to

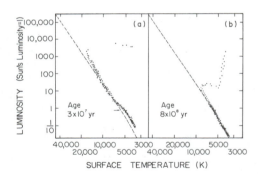

FIG. 3. H–R diagram of a theoretical star cluster, all of whose stars were assumed to have been formed at the same time and to have the same composition. The dashed line represents the zero-age main sequence, with the arrow indicating the zero-age sun. At a time of 3×10^7 yrs since formation (a), stars below $1M_\odot$ have not yet contracted to the main sequence, while stars above $7M_\odot$ have left the main sequence and evolved to the red giant region. At a time of 8×10^8 yrs (b), all stars above $2M_\odot$ have evolved away from the main sequence. Evolutionary stages beyond the onset of helium burning are not shown. [Adapted by permission from Kippenhahn, R., and Weigert, A. (1966). "Sterne und Weltraum." No. 8/9, p. 182.]

layers at that temperature. The youngest stars and the meteorites all have comparable abundances of lithium. However, the present abundance in the sun (age 4.6×10^9 yr) is almost a factor of 100 below this primordial value. Solar-type stars in the Hyades cluster (age 8×10^8 yr) have depleted lithium in their surface layers by a factor of 2, while those in the Pleiades cluster (age 6×10^7 yr) are found to have no depletion. After calibration of lithium abundances in clusters of known age in this manner, age determinations can be extended to stars outside clusters when they have sufficient lithium to be observable and are on the main sequence. The lithium depletion time scale, which so far has been only empirically determined, varies with mass, with the lower masses having a shorter time scale.

3.   The youngest stars, known as T Tauri stars, have ages of 1×10^6 yr or less and are identified by their high lithium abundances, their irregular variability, their association with dark clouds in the galaxy, and their location in the H–R diagram above and to the right of the main sequence with T_{eff} in the range 4000 to 5000 K. Massive stars near the upper end of the main sequence are also known to be young, since their nuclear burning time scale is only a few million years.

4.   The stars in the galaxy have been roughly divided, according to age, into two populations. The Population I stars are associated with the galactic disk, have relatively small space motions with respect to the sun, and have metal abundances comparable to the sun's. Although precise ages cannot be determined, this population as a group is younger than the Population II stars, which are distributed in the galactic halo and in globular clusters and have high space velocities and low metal abundances. These objects were formed early in the history of the galaxy, up to 15×10^9 yr ago, and their low metal abundance supports the point of view that the elements heavier than helium have been synthesized in the interiors of the stars and in part ejected from these stars, mainly in supernova explosions, so that later generations of stars form from interstellar material that has been gradually enriched in the heavy elements.

III. Physics of Stellar Interiors

A. TIME SCALES

The dynamical time scale, the contraction time scale, the cooling time scale, and the nu-

clear time scale are important at different stages of stellar evolution. If the star is not in hydrostatic equilibrium, it will evolve on the hydrodynamic time scale given by $t_{ff} = (3\pi/(32G\rho))^{1/2}$; this is in fact the free-fall time from an initial density ρ. Examples of gravitationally unstable stars that evolve on this time scale are protostars or the evolved cores of massive stars, which undergo photodissociation of the iron nuclei and consequently are forced into gravitational collapse. The resulting supernova outburst, of course, also takes place on a dynamical time scale. Another example is the light variation in Cepheid variables, which is caused by radial oscillations about an equilibrium state; the oscillation period of a star is of the same order as t_{ff}. Characteristic time scales are 10^5 yr for the protostar collapse, 0.1 sec for the collapse of the iron core, and 10 days for the pulsation period of a Cepheid.

For stars in hydrostatic equilibrium but without a substantial nuclear energy source, a slow contraction takes place with the release of gravitational energy. The associated time scale, known as the Kelvin–Helmholtz time scale, is given by the gravitational energy divided by the luminosity or $t_{HK} = GM^2/(RL)$. Stars contracting to the main sequence evolve on this time scale as do, for example, stars at the end of their main-sequence phase when they run out of hydrogen fuel in the core. For the present sun the value of t_{HK} is about 3×10^7 yr, which represents the time required for it to contract to its present size without any contributions from nuclear reactions. Because t_{HK}, as well as the other time scales discussed later, depend on L, they are controlled by the time required for energy to be transported from the interior of the star to the surface.

Stars that derive all their energy from nuclear burning evolve on a nuclear time scale, which can be estimated from the total nuclear energy available divided by the luminosity. For the case of hydrogen burning, four protons, each with a mass of 1.008 atomic mass units, combine to form a helium nucleus which has 4.0027 atomic mass units. The difference in mass of .0073 atomic mass units (amu) per proton is released as energy. The corresponding time scale is $t_{nuc} = 0.007Mc^2/L$, where M is the total mass and c the velocity of light. For the sun, if $X = 0.75$, $L = L_\odot$, and half the hydrogen can actually be burned, this estimate gives $t_{nuc} = 3.75 \times 10^{10}$ yr. In fact the main-sequence lifetime of the sun is only 1×10^{10} yr and much of the burning takes place at later stages when L is higher. For a star of $30M_\odot$ the value of L is about $10^5 L_\odot$ and $t_{nuc} = 1.1 \times 10^7$ yr. Again the actual main sequence lifetime is shorter. For helium burning, where three helium nuclei combine to produce carbon, the energy release per atomic mass unit is a factor 10 smaller than for hydrogen burning, and correspondingly the time spent by a star in the core helium-burning phase is roughly a factor of 10 shorter than its main sequence lifetime.

A final time scale associated with stellar evolution is the cooling time scale. For white dwarf stars that can no longer contract and also have no more nuclear fuel available, the radiated energy must be supplied by cooling of the hot interior. The electrons by this time are highly degenerate (see later discussion), so the available thermal energy is only that of the ions. The time scale can be estimated from

$$t_{cool} = E_{thermal}/L = 1.5 R_g TM/(\mu_A L)$$

where T is the mean internal temperature, R_g the gas constant, and μ_A the mean atomic weight of the ions. For example, the cooling time for a star of $0.7M_\odot$ composed of carbon from the beginning of the white dwarf phase to a luminosity $L = 10^{-3}L_\odot$ is 1.15×10^9 yr. The cooling time scale also applies to substellar objects that have insufficient mass to ignite nuclear reactions once they have contracted to their limiting radius.

B. EQUATION OF STATE OF A GAS

In the pre–main-sequence and main-sequence stages of the evolution of most stars, the ideal gas equation holds. The pressure of the gas is given by $P = NkT$, where k is the Boltzmann constant and N the number of free particles per unit volume. In the stellar interior the gas may be considered to be fully ionized. If X, Y, and Z are the mass fractions of H, He, and heavy elements, respectively, then

$$N = (2X + 0.75Y + 0.5Z)\rho/m_H$$

where m_H is the mass of the hydrogen atom. It has been assumed that the number of particles contributed per nucleus is 2 for H, 3 for He, and $A/2$ for the heavy elements, where A is the atomic weight. In the outer layers of the stars N must be adjusted to take into account partial ionization. The internal energy for an ideal gas is $1.5kT$ per particle or $1.5kTN/\rho = 1.5 P/\rho$ per unit mass. The equation of state can also be written $P = R_g\rho T/\mu$, where μ, the mean atomic weight per free particle, is given by $\mu^{-1} = 2X + 0.75Y + 0.5Z$ for a fully ionized gas.

At high temperatures and low densities the pressure of the radiation must be added. When photons are absorbed by matter, there is momentum transfer between radiation and gas. According to quantum theory a photon has energy $h\nu$, where ν is the frequency and h Planck's constant, and momentum $h\nu/c$. The radiation pressure is the net rate of transfer of momentum per unit area, normal to an arbitrarily oriented surface. Under conditions in stellar interiors where the radiation is nearly isotropic, it can be shown that the radiation pressure is given by $P_R = \frac{1}{3}aT^4$, where a is the radiation density constant, which is equal to 7.56×10^{-15} erg cm^{-3} degree^{-4}.

At high densities an additional physical effect must be considered, the phenomenon of degeneracy. The effect is a consequence of the Pauli exclusion principle, which does not permit more than one particle to occupy one quantum state at the same time; it applies to elementary particles with half-integral spin, such as electrons or neutrons. In the case of electrons in an atom, the principle governs the distribution of electrons in the various energy states. If the lowest energy states are filled, any electrons added subsequently must go into states of higher energy; only a discrete set of states is available. The same principle applies to free electrons, although there are many more states available. In a momentum interval $dp_x\, dp_y\, dp_z$ and in a volume element $dx\, dy\, dz$ the number of states is $(2/h^3)\, dp_x\, dp_y\, dp_z\, dx\, dy\, dz$, where the factor of 2 arises from the two possible directions of the electron spin. Degeneracy occurs when most of these states, up to some limiting momentum p, are occupied by particles. Complete degeneracy is defined as the situation in which all states are occupied up to some limiting momentum p_0 and all higher states are empty. Under this assumption, if the particles are electrons, their pressure can be derived to be $P_e = 9.91 \times 10^{12}(\rho/\mu_E)^{5/3}$, where μ_E, the mean atomic weight per free electron, is $2/(1 + X)$ for a fully ionized gas. This formula is valid if p_0 is low enough that the electron velocities are not relativistic. In the limiting case that all electrons are moving with the velocity of light, the corresponding expression is $P_e = 1.231 \times 10^5(\rho/\mu_E)^{4/3}$. In either case most of the electrons are forced into such high momentum states that their pressure, defined again as the rate of momentum transport per unit area, is much higher than the ideal gas pressure. The electrons in stellar interiors become degenerate when the density reaches 10^3–10^6 g cm^{-3}, depending on T. The protons or neutrons become degenerate only at much higher densities, about 10^{14} g cm^{-3}; these densities are characteristic of neutron stars, and it is the degenerate pressure of the neutrons that supports the star against gravity. The regions in density and temperature where ideal gas, radiation pressure, and electron degeneracy dominate in the equation of state are shown in Fig. 4.

C. ENERGY SOURCES

From the previous discussion of time scales it is clear that only two fundamental energy sources are available to a star. Gravitational energy can be released either on a dynamical or Kelvin–Helmholtz time scale. Nuclear energy in moderate-mass stars is produced by conversion of hydrogen to helium or by conversion of helium to carbon and oxygen. Only in the most massive stars do nuclear reactions proceed further to the synthesis of the heavier elements up to iron and nickel. During cooling phases stars draw on their thermal energy, which, however, has been produced in the past as a consequence of nuclear reactions or gravitational contraction. As the star evolves, the energy is taken up by radiation, heating of the star, expansion, neutrino emission, ionization, and dissociation of

nuclei. Here we discuss in somewhat more detail the nuclear energy source.

The nuclear processes involve reactions between charged particles. At the temperatures characteristic of nuclear burning regions in stars, the gas may safely be assumed to be fully ionized. The Coulomb repulsive force between particles of like charge presents a barrier for reactions between the ions. Particles must come within a distance of 10^{-13} cm of each other before the strong (but short-range) nuclear attractive force overcomes the Coulomb force. The energy that is required for particles to pass through the Coulomb barrier is $E_{coulomb} = Z_1 Z_2 e^2/r$, where r is the required maximum separation of 10^{-13} cm, e is the electronic charge, and Z_1 and Z_2 are, respectively, the charges on the two particles involved, in units of the electronic charge. Evidently, conditions for reactions to occur will occur most readily for particles of small charge, but even for two protons, $E_{coulomb} = 1000$ keV. The only source for this energy is the thermal energy of the particles, which, however, in the temperature range around 10^7 K, is only $\frac{3}{2}kT = 1$ to 2 keV. It would seem that nuclear reactions under these conditions would not be possible.

However, three factors contribute to a small but nonzero probability of reaction: (1) The particles have a Maxwell velocity distribution, so that a small fraction of them have thermal energies much higher than the average. (2) According to the laws of quantum mechanics, there is a small probability that a particle with much less energy than required can actually "tunnel" through the Coulomb barrier. (3) A star has a very large number of particles, so that even if an individual particle has a very small probability of reacting, enough reactions do occur to supply the required energy. It turns out that particles with about 20 times the mean energy satisfy the two requirements of existing in sufficient numbers and having enough energy to have a reasonable chance of passing through the Coulomb barrier. Most reactions occur in a fairly limited energy range about this value.

At temperatures of 1 to 2×10^7 K, which characterize the interiors of main-sequence stars, only reactions between particles of low Z need be considered, usually those involving protons. However, even if the Coulomb barrier is overcome, there is still only a small probability that a nuclear reaction will occur. This nuclear probability varies widely from one reaction to another, and it can be determined theoretically for

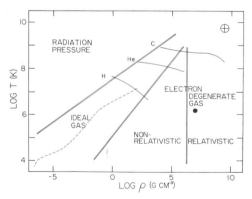

FIG. 4. Density–temperature diagram. The double lines separate regions in which different physical effects dominate in the equation of state. Solid lines, denoted by H, He, and C, respectively, indicate the central conditions in stars that are undergoing burning of hydrogen, helium, and carbon in their cores. The dashed line represents the structure of the present sun, while the solid dot gives typical central conditions in a white dwarf of 0.8 $M_\odot$. The circled cross gives central conditions in a star of 25$M_\odot$ just before the collapse of the iron core.

only the simplest systems; in general these probabilities must be measured for each reaction in laboratory experiments at low energy. An intense effort was made during the period 1960 to 1980 to make measurements of reasonable accuracy of all of the nuclear reactions that are important in stars. The group headed by W. A. Fowler at the California Institute of Technology led in the task of carrying out these difficult measurements.

Two reaction sequences have been identified that result in the conversion of four protons into one helium nucleus, the proton–proton (pp) cycle and the CNO cycle. The reactions of the main branch of the pp cycle are as follows:

$$^1H + {}^1H \rightarrow {}^2D + e^+ + \nu \qquad (1)$$

$$^2D + {}^1H \rightarrow {}^3He + \gamma \qquad (2)$$

$$^3He + {}^3He \rightarrow {}^4He + {}^1H + {}^1H \qquad (3)$$

The symbol e^+ denotes a positron, ν the neutrino, and ^2D the deuterium nucleus, composed of one proton and one neutron. The positron immediately reacts with an electron, with the annihilation of both and the production of energy. The total energy production of the sequence is 26.7 MeV or 4.27×10^{-5} erg. The neutrino immediately escapes from the star, as it interacts only very weakly with matter, and it carries with it a certain fraction of the energy. The remainder of the energy is deposited locally in the star, typically in the form of γ rays.

Reaction (1) turns out to be very improbable, as the conversion of a proton into a neutron requires a beta decay. The reaction rate is so slow that it cannot be measured experimentally and must be calculated theoretically. At the center of the sun a proton has a mean lifetime of almost 10^{10} yr before it reacts with another proton; this reaction controls the rate of energy production, as subsequent reactions are more rapid. The neutrino in reaction (1) carries away an average of 0.25 MeV. Once ^2D is formed, it immediately (within 1 sec) captures another proton to form ^3He. This reaction can be measured in the laboratory down to energies close to those in stars. Any ^2D initially present in the star will be burned by this reaction at temperatures of around 10^6 K. When reactions (1) and (2) have occurred twice, the two resulting nuclei of ^3He combine to form ^4He and two protons. There is perhaps a 10% uncertainty in this reaction rate and in most others in the proton–proton chains because in general measurements cannot be made at the

relevant energies of about 20 keV but must be extrapolated down to that point.

The second branch of the pp chain occurs when the ^3He nucleus produced in reaction (2) reacts with ^4He. The sequence of reactions is then

$$^3He + {}^4He \rightarrow {}^7Be + \gamma \qquad (4)$$

$$^7Be + e^- \rightarrow {}^7Li + \nu + \gamma \qquad (5)$$

$$^7Li + {}^1H \rightarrow {}^4He + {}^4He \qquad (6)$$

Under conditions of the solar interior, reaction (4) proceeds at approximately one-quarter of the rate of reaction (3), and as the temperature increases it becomes relatively more important. Reaction (5) involves an electron capture on the ^7Be and the production of a neutrino of 0.81 MeV. The ^7Li nucleus immediately captures a proton and produces two nuclei of ^4He. This reaction also results in the destruction of any ^7Li that may be present at the time of the star's formation, since it becomes effective at temperatures of 2.8×10^6 K or above. Note that this second branch of the pp chain has a total neutrino loss of 1.06 MeV, but that otherwise the net result is the same: the conversion of four protons to a ^4He. The third branch of the pp chain occurs if a proton, rather than an electron, is captured by ^7Be:

$$^7Be + {}^1H \rightarrow {}^8B + \gamma \qquad (7)$$

$$^8B \rightarrow e^+ + \nu + {}^8Be \rightarrow {}^4He + {}^4He \qquad (8)$$

Less than 1% of pp-chain completions occur through reaction (7) in the sun; however, it is very important because the neutrinos produced in reaction (8) have an average energy of 7.2 MeV and therefore result in most of the detections in the solar neutrino experiment. The total rate of energy generation by the pp chains involves a complicated calculation of the rates of all the reactions given above, which are functions of temperature, density, and concentration of the species involved. A very approximate expression, under solar conditions near $T = 1.5 \times 10^7$ K, for ε, the amount of energy produced per second per gram of material, is

$$\varepsilon = 0.45\rho X^2 (T_6/15)^{3.95} \quad \text{erg} \quad \text{g}^{-1} \quad \text{sec}^{-1} \qquad (9)$$

where X is the fraction of hydrogen by mass, ρ the density, and T_6 the temperature in units of 10^6 K.

Protons can also interact with the CNO nuclei, but because the Coulomb barrier is higher these reactions become more important than the pp chain only at temperatures above 2×10^7 K.

The reactions are the following:

$$^{12}C + {}^1H \rightarrow {}^{13}N + \gamma \qquad (10)$$

$$^{13}N \rightarrow {}^{13}C + e^+ + \nu \qquad (11)$$

$$^{13}C + {}^1H \rightarrow {}^{14}N + \gamma \qquad (12)$$

$$^{14}N + {}^1H \rightarrow {}^{15}O + \gamma \qquad (13)$$

$$^{15}O \rightarrow {}^{15}N + e^+ + \nu \qquad (14)$$

$$^{15}N + {}^1H \rightarrow {}^{12}C + {}^4He \qquad (15)$$

The net effect is the conversion of four protons into a helium nucleus, two positrons (which annihilate) and two neutrinos, which carry away energies of 0.71 and 1.0 MeV, respectively. The CNO nuclei simply act as catalysts; however, their relative abundances change as a result of the operation of the cycle. Note from Table I that ^{12}C is considerably more abundant than ^{14}N in the solar system. However, the rate of reaction (10) under stellar conditions is about 100 times faster than that of reaction (13), which uses up ^{14}N. All the other reactions are much faster. The reaction chain tends to reach an equilibrium, in which each CNO nucleus is produced as fast as it is destroyed. This equilibrium is obtained only when most of the ^{12}C and other participating nuclei are converted to ^{14}N. This change in relative abundances could be observed if the layers in which the reactions occur could later be mixed to the surface of the star.

A secondary branch of the CNO cycle also affects the abundance of ^{16}O. Starting at ^{15}N, the reactions are

$$^{15}N + {}^1H \rightarrow {}^{16}O + \gamma \qquad (16)$$

$$^{16}O + {}^1H \rightarrow {}^{17}F + \gamma \qquad (17)$$

$$^{17}F \rightarrow {}^{17}O + e^+ + \nu \qquad (18)$$

$$^{17}O + {}^1H \rightarrow {}^{14}N + {}^4He \qquad (19)$$

This branch accounts for only 0.1% of the completions, but the ^{16}O is generated less rapidly than it is converted into ^{14}N by reactions (17), (18), and (19), so that when the branch comes into equilibrium the abundance ratio O/N will be reduced. The overall effect of the CNO cycle, apart from its energy generation, is the conversion of 98% of the CNO isotopes into ^{14}N. An approximate expression for ε, valid in the temperature range $T_6 = 22$ to 28, is

$$\varepsilon = 2.16 \cdot 10^4 \rho X X_{CNO} (T_6/25)^{16.7}$$

$$\text{erg} \quad \text{g}^{-1} \quad \text{sec}^{-1} \qquad (20)$$

where X_{CNO} is the total mass fraction of all CNO isotopes.

The synthesis of elements heavier than helium has proved in the past to be a considerable problem, primarily because there is no stable isotope of atomic mass 5 or 8. Thus if two helium nuclei react to produce ^8Be, the nucleus will immediately decay back to He. If a proton reacts with a ^4He, the result is ^5Li, which also is unstable. Helium burning must, in fact, proceed through a three-particle reaction, which can be represented as follows:

$$^4He + {}^4He \leftrightarrow {}^8Be \qquad (21)$$

$$^8Be + {}^4He \leftrightarrow {}^{12}C + \gamma + \gamma \qquad (22)$$

Although the ^8Be is unstable, its decay is not instantaneous. Under suitable conditions, e.g., $T = 10^8$ K and $\rho = 10^5$ g cm^{-3}, it was shown by E. Salpeter that a sufficient equilibrium abundance exists so that a third ^4He nucleus can react with it [reaction (22)]. It was also predicted by F. Hoyle that for this triple-α process to proceed at a significant rate, reaction (22) must be resonant; that is, there must be an excited nuclear state of ^{12}C accessible in the range of stellar energies whose presence greatly enhances the probability that the reaction will occur. This resonance was later found in experiments performed at the Kellogg Radiation Laboratory at the California Institute of Technology. The amount of energy produced per reaction is 7.272 MeV per reaction or 0.606 MeV per atomic mass unit, a factor of 10 less than in hydrogen burning (per atomic mass unit). An approximate expression for the rate of energy production at $T_8 = T/10^8 = 1$ is

$$\varepsilon = 4.4 \cdot 10^{-8} \rho^2 Y^3 T_8^{40} \quad \text{erg} \quad \text{g}^{-1} \quad \text{sec}^{-1}$$

$$(23)$$

where Y is the mass fraction of He. A further helium burning reaction, which occurs at slightly higher temperatures than the triple-α process, is

$$^{12}C + {}^4He \rightarrow {}^{16}O + \gamma \qquad (24)$$

with an energy production of 7.162 MeV per reaction.

D. Energy Transport

The transport of energy outward from the interior of a star to its surface depends in general on the existence of a temperature gradient. Heat will be carried by various processes from hotter regions to cooler regions; the processes that need to be considered include (1) neutrino transport, (2) radiative transport, (3) convective transport, and (4) conductive transport. In each case a relation must be found between the energy flux F_r, defined as the energy flow per square centimeter per second at a distance r from the center of the star, and the temperature gradient dT/dr.

The neutrinos produced by nuclear reactions in stellar interiors have a very tiny probability of absorption by matter. At a density of 1 g cm^{-3} the mean free path before absorption is 10^{20} cm or 30 pc! Only in the very dense cores of evolved stars where $\rho = 10^{10}$ g cm^{-3} or more is the mean free path sufficiently small so that the neutrino can be absorbed. Here one must consider neutrino transport, but in most phases of stellar evolution it is sufficient simply to subtract the neutrino loss locally from the energy production of the nuclear reaction concerned.

Energy transport by radiation depends on the emission of photons in hot regions of a star and absorption of them in slightly cooler layers. We may define the mass absorption coefficient κ as follows: if radiation of intensity I strikes and passes through a layer of material with density ρ and thickness dx, the intensity is reduced by $\kappa\rho I\,dx$. The mean free path of a photon before it is absorbed is then $1/(\kappa\rho)$, which in a typical stellar interior can be only 1 cm or less. Thus a typical photon is absorbed at practically the same temperature as it is emitted. An observer measuring the radiation at a point in a star will ''see'' to a temperature in the outward direction only slightly lower than that in the inward direction. Because the radiation inside the star can be well approximated by a black body, the observer will measure a net flux (outward flux minus inward flux) of

$$\sigma[(T + \delta T)^4 - T^4] \simeq 4\sigma T^3\,\delta T$$

where δT is the inward–outward temperature difference that the observer can measure. The corresponding distance δr is about one photon mean free path. Thus dimensionally, the net flux is given by

$$F_r \simeq -4\sigma T^3(\kappa\rho)^{-1}\,dT/dr$$

A more exact derivation gives the following relation between flux and temperature gradient:

$$F_r = -(4ac)(3\kappa\rho)^{-1}T^3(dT/dr) \qquad (25)$$

where c is the velocity of light and a the radiation density constant. This expression is known

as the diffusion approximation for radiative transfer and is valid when the temperature change is small over the mean free path of a photon. Radiative transfer in stars can be regarded as a diffusion process because a photon is absorbed almost immediately after it is emitted, so that an enormously large number of absorptions, re-emissions, or scatterings must occur before the energy of a photon is transmitted to the surface. The time required for energy to diffuse in this manner from the center of the sun to the surface is estimated to be about 10^5 yr. The quality of the radiation changes during this process. As the photons diffuse to lower temperatures, their energy distribution corresponds closely to the Planck distribution at the local temperature. Thus the γ rays produced by nuclear reactions at the center are gradually transformed into optical photons by the time the energy reaches the surface.

As Eq. (25) indicates, the energy transport in a radiative star is controlled by the opacity κ. Numerous atomic processes contribute to this quantity, and in general the structure of a star can be calculated only with the aid of detailed tables of the opacity, calculated as a function of ρ, T, and chemical composition. Starting at the highest temperatures characteristic of stellar interiors and proceeding to lower temperatures, the main processes are (1) electron scattering, also known as Thomson scattering, in which a photon undergoes a change in direction but no change in frequency upon interaction with a free electron, (2) free–free absorption, in which a photon is absorbed by a free electron in the region of a nucleus, with the result that the electron increases its kinetic energy, (3) bound–free absorption on metals, also known as photoionization, in which the photon is absorbed by an atom of a heavy element (e.g., iron) and one of the bound electrons is removed, (4) bound–bound absorption of a heavy element, in which the photon results in an upward transition of an electron from a lower quantum state to a higher quantum state in the atom, (5) bound–free absorption of H and He, which generally occurs fairly near stellar surfaces where these elements are being ionized, (6) bound–free and free–free absorption by the negative hydrogen ion H$^-$, which forms in stellar atmospheres in layers where H is just beginning to be ionized, (7) bound–bound absorption by molecules, which can occur only in the atmospheres of the coolest stars and in protostars, and (8) absorption by dust grains, which can occur in the early stages of protostellar evolution at temperatures below the evaporation temperature of grains (1400–1800 K).

The Thomson scattering from free electrons is given by $\kappa = 0.2(1 + X)$. For stars this process is important generally only at temperatures above 10^8 K unless the density is very low. The free–free absorptions and bound–free absorptions from heavy elements have the approximate dependence $\kappa \propto \rho T^{-3.5}$. As one moves outward in a star from higher to lower T, the number of bound electrons per atom increases, and correspondingly, κ increases. A maximum in κ occurs at $T = 3$ to 5×10^4 K, where H and He, the most abundant elements, undergo ionization. Below this temperature range, the dependence is no longer valid and the opacity drops considerably. In the temperature range 3000 to 10,000 K, the opacity is approximated by $\kappa \propto \rho T^{10}$, dropping to a minimum of 10^{-2} at $T = 2000$ K. At lower T, where grains exist, the opacities are higher, on the order of 1 cm^2 g^{-1}.

Under certain conditions the temperature gradient calculated according to Eq. (25) can be unstable, leading to the onset of convection. Suppose that a small element of material in a star is displaced upward from its equilibrium position. It may be reasonably assumed that the element remains in pressure equilibrium with its surroundings. If the material has the equation of state of an ideal gas an upwardly displaced element with density less than that of the surroundings will have a temperature greater than that of the surroundings. If, as the element rises, its density decreases more rapidly than that of the surroundings, it will continue to feel a buoyancy force and will continue to rise. In this case the layer is unstable to convection, and heat is transported outward by the moving elements themselves. If, however, the density of the upward-moving element decreases less rapidly than that of the surroundings, the densities soon will be equalized, there will be no buoyancy force, and the layer will be stable.

The condition for occurrence of convection can be expressed in another form. If the actual temperature gradient in the layer is steeper than the adiabatic temperature gradient, then convection will occur:

$$|dT/dr| = 3\kappa\rho F_r(4acT^3)^{-1} > |dT/dr|_{ad} \quad (26)$$

A radiative layer with high opacity is therefore likely to be unstable; this situation can occur near the surface layers of cool stars. A layer with a high rate of nuclear energy generation in a small volume is also likely to be unstable because of high F_r; this situation can occur in the cores of stars more massive than the sun where the temperature-sensitive CNO cycle operates.

Convection involves complicated, turbulent motions with continuous formation and dissolution of elements of all sizes. The existence of convection zones is important in the calculation of stellar structure because the overturning of material is rapid enough that the entire zone can be assumed to be mixed to a uniform chemical composition at all times. It is also important for heat transport, which is accomplished mainly by the largest elements. No satisfactory theory for convective energy transport exists, and so an approximate theory based on the concept of a "mixing length" is used. It is assumed that a convective element is formed, travels one mixing length vertically, and then dissolves and releases its excess energy to the surroundings. The mixing length is approximated by αH, where α is a parameter of order unity and H is the distance over which the pressure drops by a factor e. By use of this theory the average velocity, excess thermal energy, and convective flux can be estimated. In most situations in a stellar interior the estimates show that it is adequate to assume that $dT/dr = (dT/dr)_{ad}$. Although an excess in dT/dr over the adiabatic value is required for convection to exist, convection is efficient enough to carry the required energy flux even if this excess is negligibly small. The uncertainty in stellar structure arising from the use of mixing length theory is limited to the surface layers of stars where in fact the actual temperature gradient in a convection zone is significantly greater than the adiabatic gradient. The parameter α can be calibrated by comparison of calculated mixing lengths with the size of the granular elements in the solar photosphere that represent solar convective elements. Theoretical models of the sun can also be compared with the observed $R_\odot$ and $L_\odot$. In this manner it has been determined that the parameter α should be in the range 1 to 2.

Conduction of heat by the ions and electrons is in general inefficient in stellar interiors because the density is high enough that the mean free path of the particles is small compared with the photon mean free path. In the interiors of white dwarfs and the evolved cores of red giants, however, the electrons are highly degenerate, their mean free paths are very long, and conduction becomes a more efficient mechanism than radiative transfer. The flux is related to the temperature gradient by $F_r = -K \, dT/dr$, where K, the conductivity, is calculated by a complicated theory involving the velocity, collision cross section, mean free path, and energy carried by the electrons. Numerical tables are constructed for use in stellar structure calculations.

E. BASIC EQUILIBRIUM CONDITIONS

During most phases of evolution a star is in hydrostatic equilibrium, which means that the force of gravity on a mass element is exactly balanced by the difference in pressure on its upper and lower surface. In spherical symmetry this condition is written

$$(1/\rho)(dP/dr) = -GM_r/r^2 \qquad (27)$$

where M_r is the mass within radius r, ρ the density, and P the pressure. In certain phases of evolution, such as the collapse of a protostar or of an evolved iron core, this equilibrium is not satisfied and the corresponding equation is the equation of motion

$$Dv/Dt = -GM_r/r^2 - (1/\rho)(dP/dr) \qquad (28)$$

where v is the velocity and D refers to the derivative following the motion. A second equilibrium condition, which applies in all situations, is the mass conservation equation, which simply states that

$$dM_r/dr = 4\pi r^2 \rho \qquad (29)$$

where dM_r is the mass element contained in a spherical shell of thickness dr. The hydrostatic condition can be used to estimate the internal pressures required to support a star against gravity. By saying that dP is the difference between the central pressure and the surface pressure, dr the total radius, ρ the mean density, M_r half the total mass, and r half the total radius, we can estimate to order of magnitude the central pressure $P_c \approx GM^2/R^4 \approx 10^{16}$ dynes cm^{-2} for the case of the sun.

The third condition is that of thermal equilibrium, which refers to the situation where a star is producing enough energy by nuclear processes to balance exactly the loss of energy by radiation at the surface. The condition can be expressed

$$L = \int_0^M \varepsilon \, dM_r \qquad (30)$$

where M is the total mass, and when it is satisfied the evolution time scale is t_{nuc}. In many stages of stellar evolution this condition is not

satisfied, since the star may be expanding, contracting, heating, or cooling. The more general equation of conservation of energy must then be used:

$$dL_r/dM_r = \varepsilon - (DE/Dt) - P(DV/Dt) \qquad (31)$$

where $V = 1/\rho$, L_r is the total amount of energy crossing a sphere at radius r, and E is the internal energy per unit mass. If $\varepsilon = 0$, the star contracts and obtains its energy through the third term on the right-hand side.

A final important equilibrium condition is the virial theorem, which, for a nonrotating, nonmagnetic spherical star in hydrostatic equilibrium, can be written

$$2T_i + W = 0 \qquad (32)$$

Here W is the gravitational energy, qGM^2/R, where R is the total radius and q a constant of order unity that depends on the mass distribution, and T_i is the total internal energy $\int_0^M E \, dM_r$. This expression can be used to estimate stellar internal temperatures for the case of an ideal gas. Then $T_i = 1.5(R_g/\mu)TM$, where T is the mean temperature. Then $T \approx GM\mu/(RR_g) = 6 \times 10^6$ K for the case of the sun. This expression also shows that a star without any nuclear energy sources that is contracting is also heating up, if the equation of state is that of an ideal gas, with T increasing roughly as $1/R$.

To obtain a detailed solution for the structure of a star if hydrostatic and thermal equilibrium hold, one must solve Eqs. (27), (29), and (30). In addition, one must solve the radiative transfer equation (25). However, each point in the star must be tested to see whether the condition for convection is satisfied. If the point is unstable to convection, the radiative temperature gradient is replaced by the adiabatic temperature gradient. At the center the boundary conditions $M_r = 0$ and $L_r = 0$ at $r = 0$ are applied. At the surface the simple boundary conditions $P = 0$ and $T = 0$ will suffice, but in practice, in order to get a reasonable value of T_{eff} a more accurate integration of the atmospheric layers should be used. The four differential equations are supplemented by the equation of state and expressions for ε and κ as functions of ρ, T, and composition. The total mass and the distribution of chemical composition as a function of M_r must also be specified. For the calculation of stellar evolution, in general equation (31) must be used instead of Eq. (30), and in some situations Eq. (28) must be used instead of Eq. (27). Then the calculation is started from an assumed initial model, and a sequence of models is calculated, separated by small intervals of time. The time appears explicitly in Eqs. (28) and (31), and during nuclear burning phases the changes in chemical composition caused by reactions (e.g., conversion of H to He) are also calculated at each layer. The system of equations is solved numerically, and even in spherical symmetry the solution can involve hours of time on a large computer for some of the more complicated stages of stellar evolution.

IV. Stellar Evolution before the Main Sequence

The early phases of stellar evolution can be divided into three stages: star formation, protostellar collapse, and slow contraction. The characteristics of these three stages are summarized in Table II. As the table indicates, there is a vast difference in physical conditions between the time of star formation in an interstellar cloud and the time when a star arrives on the main sequence and begins to burn hydrogen. An increase in mean density by 22 orders of magnitude, and in mean temperature by 6 orders of magnitude, occurs. In the star formation and protostellar collapse stages, it is not sufficient to assume that the object is spherical, and a considerable variety of physical processes must be considered. The problem involves solution of the equations of hydrodynamics, including rotation, magnetic fields, turbulence, molecular chemistry, and radiative transport of energy. Some two- and three-dimensional hydrodynamic calculations have been performed but with a limited range of physical effects and limited spatial resolution for the star formation phase; detailed calculations have been performed in the spherical approximation for the protostellar collapse. A number of important

questions remain to be solved: (1) What is the rate and efficiency of conversion of interstellar matter into stars? (2) What determines the distribution of stars according to mass? (3) What determines whether a star will become a member of a double or multiple star system or a single star with a planetary system?

A. STAR FORMATION

It is clear that star formation is limited to regions of unusual physical conditions compared with those of the interstellar medium on the average. The only long-range attractive force to form condensed objects is the gravitational force. However, a number of effects oppose gravity and prevent contraction, including gas pressure, turbulence, rotation, and the magnetic field. The chemical composition of the gas, the degree of ionization or dissociation, the presence or absence of grains (condensed particles of ice, silicates, or iron with characteristic size 5×10^{-5} cm), and the heating and cooling mechanisms in the interstellar gas all have an important influence on star formation.

We first consider only the effects of thermal gas pressure and gravity in an idealized spherical cloud with uniform density ρ, uniform temperature T, and mass M. The self-gravitational energy is $E_{grav} = -0.6GM^2/R$, where R is the radius. The internal energy is $E_{int} = 1.5R_gTM/\mu$, where μ is the mean atomic weight per particle. Collapse will be possible roughly when $|E_{grav}| > E_{int}$, that is, when the cloud is gravitationally bound, a criterion that has been verified by numerical calculations. Thus, for collapse to occur the radius must be less than $R_J = 0.4GM\mu/(R_gT)$, known as the Jeans length. By eliminating the radius in favor of the density, we obtain the Jeans mass, which is the minimum mass a cloud with density ρ and temperature T must

TABLE II. Major Stages of Early Stellar Evolution[a]

Phase	Size (cm)	Observations	Density (g cm^{-3})	Internal temperature (K)	Time (yr)
Star formation	10^{20}–10^{17}	Radio	10^{-22}–10^{-19}	10	10^7
Protostellar collapse	10^{17}–10^{12}	Infrared	10^{-19}–10^{-3}	10–10^6	10^6
Slow contraction	10^{12}–10^{11}	Optical	10^{-3}–1.0	10^6–10^7	4×10^7 [b]

[a] Reproduced with permission from P. Bodenheimer (1983). Protostar collapse, *Lect. Appl. Math.* **20**, 141. © American Mathematical Society.
[b] For 1 solar mass.

have for collapse to occur:

$$M_J = (2.5R_gT/\mu G)^{3/2}(\tfrac{4}{3}\pi\rho)^{-1/2} \qquad (33)$$

This condition turns out to be quite restrictive. A typical interstellar cloud of neutral hydrogen has $T = 50$ K, $\rho = 1.7 \times 10^{-23}$ g cm^{-3}, and $\mu = 1$, and the corresponding $M_J = 3600 M_\odot$. The actual masses are far less than this value, so the clouds are not gravitationally bound. On the other hand, typical conditions in an observed molecular cloud are $T = 10$ K, $\rho = 1.7 \times 10^{-21}$, and $\mu = 2$, with $M_J = 8 M_\odot$, and so collapse of stellar mass fragments seems to be quite possible, since molecular clouds have masses up to 10^4 to $10^5 M_\odot$. The question of how the clouds of neutral hydrogen accumulate and compress to molecular cloud densities has not been solved, but possible contributing processes include cloud–cloud collisions, the shock wave associated with the spiral arms of the galaxy, shock waves arising from supernova explosions, or Rayleigh–Taylor instability induced by the bending of the magnetic field lines in the galactic plane.

The angular momentum of the interstellar clouds also provides an obstacle to star formation. The problem can be stated in the following way. The minimum angular momentum of a cloud is calculated under the assumption that it rotates at the same angular velocity as that associated with its motion around the center of the galaxy, 10^{-15} radians/sec. The corresponding angular momentum per unit mass is $J/M = 10^{24}$ cm^2 sec^{-1} for a cloud of $10^4 M_\odot$. The rotational velocities of young stars that have just started their pre–main-sequence contraction indicate that their J/M is only 10^{17} cm^2 sec^{-1}. Evidently some fragments of dark cloud material must lose 7 orders of magnitude in J/M before they reach the stellar state. Two processes have been proposed that could provide a considerable reduction in angular momentum. If the matter is closely coupled to the magnetic field, the twisting of the field lines can generate Alfven waves that would transfer angular momentum from inside the cloud to the external medium. It has been estimated that significant reduction of angular momentum could take place in 10^6 to 10^7 yr. Once the density increases to about 10^{-19} g cm^{-3}, the field no longer is strongly coupled to the gas because the density of charged particles becomes very low. Then this mechanism of angular momentum loss becomes less effective. However, during the later stages of star formation another effect could be important. When fragmentation occurs in a rotating cloud, the central region of the cloud breaks up into orbiting fragments. Each fragment retains some angular momentum of spin, but it is small enough that it can begin to collapse on its own. After further collapse, the fragments themselves could break up into subfragments, again converting some spin angular momentum into orbital motion. After several such stages the angular momentum of spin of the smallest fragments could be reduced by several orders of magnitude. This picture is consistent with the fact that most stars are members of binary or multiple systems.

A sequence of fragmentation is in any case a likely event during the star formation process. Equation (33) shows that the mass required, under interstellar conditions, for collapse to start is much larger than stellar masses. However, the same equation shows that once the large mass starts to collapse, fragmentation into smaller masses is possible. As the collapse proceeds, the density increases but the temperature remains constant or even decreases. This behavior occurs because as the cloud tends to heat by gravitational compression, the internal energy is rapidly converted into radiation, which can immediately escape from the cloud if the density is low enough. Therefore M_J decreases. Further fragmentation can, in principle, occur until the density increases to the point where the radiation can no longer escape. This critical point occurs at about 10^{-13} to 10^{-14} g cm^{-3} and the corresponding value of M_J (at $T = 10$ K) is $0.005 M_\odot$. A further increase in density results in heating, and M_J rises. That fragmentation does indeed occur during star formation is supported by a considerable amount of evidence: (1) numerical simulations of rotating collapsing clouds, (2) observations of multiple infrared and radio sources embedded in molecular clouds, (3) the existence of stellar multiple systems with both a long and a short period, suggesting the conversion of spin angular momentum into orbital motion, and (4) the existence of clusters of stars, suggesting that the fragmentation of a large cloud has occurred (see Fig. 5). It is now becoming evident from observations that star formation occurs in the cores of dense clumps embedded in molecular clouds. However, most regions of the clouds are not forming stars; turbulent random motions and magnetic fields may be supporting most of the cloud matter against gravitational collapse.

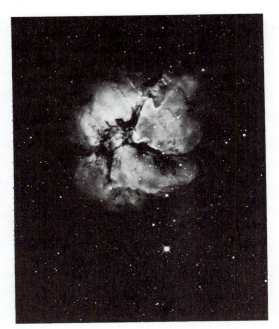

FIG. 5. The Trifid Nebula in Sagittarius (M20, NGC 6514), a region of recent star formation (Shane 120-in. reflector). (Lick Observatory photograph.)

B. PROTOSTAR COLLAPSE

Although rotational, magnetic, and fragmentation effects are probably important during at least the early part of the protostar collapse stage, we describe the evolution under the assumption of spherical symmetry, starting at the point where a fragment of a given mass first is able to evolve as an independent object. The evolution can be divided into an isothermal phase, an adiabatic phase, and an accretion phase. For $T = 10$ K, a fragment of $1 M_\odot$ with solar composition can begin collapse at $\rho = 10^{-19}$ g cm^{-3}. In the initial stages up to $\rho = 10^{-14}$ g cm^{-3}, instability to collapse continues because the cloud is transparent to its own radiation and loses energy rapidly. The pressure cannot increase rapidly enough to maintain hydrostatic equilibrium. The temperature stays close enough to 10 K that an isothermal collapse can be assumed. The equations to be solved are (28) and (29) plus the equation of state $P = K\rho$, where K is a constant. At the outer boundary, either the pressure is set to a constant value or constant volume is assumed, with no mass flow across the boundary. The results of numerical calculations show that even if the cloud initially has uniform density, a density gradient is soon set up because of the rarefaction wave that proceeds inward from the outer boundary; soon the denser central regions are collapsing much faster than the outer regions and the protostar becomes highly condensed toward the center. The infall velocity increases from zero at the center to a maximum some distance out and then drops again toward the surface. Typical maximum velocities are 1 to 2 km sec^{-1}, which are supersonic. The pressure gradient that develops slows the collapse down somewhat from free-fall velocities.

The dense central regions eventually are no longer transparent to their emitted radiation, and some of the released gravitational energy is trapped and begins to heat the cloud. The collapse time is still close to the free-fall time, and as the density increases further, this time becomes short compared to the time required for radiation to diffuse out of the central regions. The collapse then becomes nearly adiabatic. The pressure gradient increases to the point where it exceeds the force of gravity, and the central parts of the cloud are decelerated. A small region forms that is close to hydrostatic equilibrium; however, as it continues to compress, the temperature rises to about 2000 K. At that point the molecular hydrogen dissociates and thereby absorbs a considerable fraction of the released gravitational energy. As a result, the very center of the cloud becomes unstable to further collapse, which continues until most of the molecules have been dissociated at a density of 10^{-2} g cm^{-3} and a temperature of 3×10^4 K. The collapse is again decelerated, and a core forms in hydrostatic equilibrium. The mass of this core, which can be thought of as the nucleus of the star, has initially only 0.1% of the cloud mass, but it grows in time as the infalling material farther out falls onto it.

Once the cloud starts to heat up, the calculation of the evolution involves the solution of Eqs. (25), (28), (29), and (31), as well as the equation of state of an ideal gas. Special numerical treatment is required for the shock front that forms at the outer edge of the stellar core once the accretion phase begins. The low-density material outside the core falls at supersonic speeds, passes through the shock front, and is decelerated rapidly as it comes nearly to rest. Most of the kinetic energy of infall is converted to heat behind the shock front, and this in turn is radiated from behind the front back out into the infalling envelope. The luminosity generated in this way is given by $L = GM_cM/R_c$, where M is the core mass, $\dot{M}$ the mass accretion rate in grams per second, and R_c the core radius. This energy is transmitted through the envelope and emerges at the outer edge of the opaque dust layer, where it is observable in the infrared part of the spectrum. The temperature at this observable surface increases with time, from around 100 K near the beginning of the accretion phase to about 3000 K when most of the envelope has fallen onto the core. The time for the completion of this phase is essentially the free-fall time of the outer layers, which lies in the range 10^5 to 10^6 yr, depending on the initial density ρ_i. Evolutionary tracks for protostars in the H–R diagram are shown in Fig. 6. The radius R_f of the star when it finally comes into equilibrium also depends on ρ_i. For example, for $1M_\odot$, calculations show that $R_f = 2R_\odot$ for $\rho_i = 10^{-19}$ g cm^{-3}, but $R_f = 6R_\odot$ for $\rho_i = 10^{-16}$ g cm^{-3}; the corresponding evolutionary tracks are shown as curves a and b in Fig. 6. The location of the youngest stars in the observed H–R diagram suggests that the actual value is between these two. It is difficult to confirm observations of stars in the collapse phase itself because of the short evolutionary time and because the objects are heavily obscured by dust.

C. Pre–Main-Sequence Contraction

Once internal temperatures are high enough (above 10^5 K) that hydrogen is substantially ionized, the star is able to reach an equilibrium state with the pressure of an ideal gas supporting it against gravity. The star radiates from its surface, and this energy is supplied by gravitational contraction, which can be regarded as a passage through a series of quasi-equilibrium states. The virial theorem shows that half of the released gravitational energy goes into radiation and the other half into heating of the interior as long as the equation of state remains ideal. The calculation of the evolution involves the solution of Eqs. (25), (27), (29), and (31), with the adiabatic gradient substituted for Eq. (25) when the zone is convective.

The solutions are shown in the H–R diagram in Fig. 7. The point where a star first appears on these evolutionary tracks depends on the initial conditions in the preceding protostellar phase; thus the earlier phases shown in Fig. 7 may be bypassed. The results of the calculations show in general that the stars first pass through a convective phase (vertical portions of the tracks) and later a radiative phase (relatively horizontal

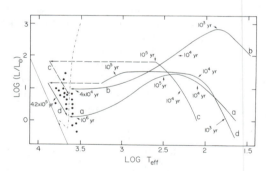

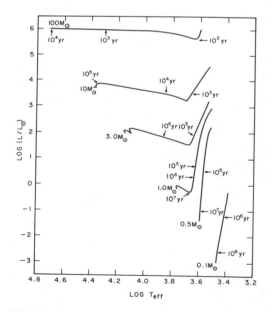

FIG. 6. Evolution of a protostar of 1 solar mass in the H–R diagram, according to R. Larson (a, b), S. Stahler, F. Shu, and R. Taam (c), and K.-H. Winkler and M. Newman (d). The solutions differ in their physical assumptions and initial conditions. Dashed arrows refer to rapid transitions. Schematic positions of the youngest T Tauri stars, which represent the end of the protostar collapse phase, are given by solid dots. The dashed-dotted line shows the evolutionary track for $1M_\odot$ under the assumption that the contraction occurs in hydrostatic equilibrium, while the light solid line represents the zero-age main sequence.

FIG. 7. Evolutionary tracks for stars of various masses during pre–main-sequence contraction. Tracks end at the zero-age main sequence. [Reproduced with permission from Bodenheimer, P. (1972). *Rep. Prog. Phys.* **35**, 1. © The Institute of Physics.]

portions of the tracks). The relative importance of these two phases depends on the stellar mass. During the convective phase, energy transport in the interior is quite efficient, and the rate of energy loss is controlled by the thin radiative layer right at the stellar surface. The opacity is a very strongly increasing function of T in those layers, and the star is able to adjust its surface boundary condition with very slight changes in T_{eff}. As the surface area decreases, L drops and T_{eff} stays between 3000 and 4000 K. As the star contracts, the interior temperatures increase and in most of the star the opacity decreases as a function of T. The star gradually becomes stable against convection, starting at the center. When the radiative region includes about 75% of the mass, the rate of energy release is no longer controlled by the surface layer, but rather by the opacity of the entire radiative region. At this time the tracks make the sharp bend to the left, and the luminosity remains nearly constant or increases gradually as the average interior opacity decreases.

Contraction times to the main sequence for various masses are given in Table III. A complete summary of the evolution of the sun, up to the present time, is given in Table IV; the entries for the protostellar phases are based on curve a in Fig. 6. The high-mass stars have relatively high internal temperature and therefore relatively low internal opacities and are able to radiate rapidly. The contraction times are short, and because the luminosity is relatively constant during the radiative phase, during which they spend most of their time, the contraction time is well approximated by the Kelvin–Helmholtz time $t_{KH} = GM^2/(RL)$. A star of $1M_\odot$ spends

about 10^7 yr on the vertical track, known as the Hayashi track after the Japanese astrophysicist who discovered it. For the next 2×10^7 yr, the star is primarily radiative, but it maintains a thin outer convective envelope all the way to the main sequence. The final 10^7 yr of the contraction phase represents the transition to the main sequence, during which nuclear reactions begin to become important at the center, the contraction slows down, and as the energy source becomes more concentrated toward the center, the luminosity declines slightly. For the lower-mass stars, the evolution is entirely along the Hayashi track. Stars of $0.3M_\odot$ or less remain fully convective all the way to the main sequence. Because the luminosity varies rapidly during the contraction, t_{KH} does not give a good estimate of the total contraction time; it actually takes a star of $0.1M_\odot$ about 10^9 yr to reach the main sequence.

A number of comparisons can be made between the contraction tracks and the observations.

1. The theoretical tracks in Fig. 7 indicate that no stars in hydrostatic equilibrium should be found with T_{eff} below 2500 to 3000 K, in agreement with observations. (Post–main-sequence tracks also do not go below these temperatures.)

2. The lithium abundances of stars that have just reached the main sequence can be compared with calculations of the depletion of lithium in the surface layers during the contraction. Lithium is easily destroyed by reactions with protons starting at $T = 2.8 \times 10^6$ K. The lower-mass stars have convection zones extending

TABLE III.   Evolutionary Times

Mass ($M_\odot$)	Pre–main-sequence contraction time (yr)	Main-sequence lifetime (yr)	End of main sequence to onset of core He burning (yr)	Core He burning lifetime (yr)
30	3×10^4	4.8×10^6	1.0×10^5	5.4×10^5
15	6.2×10^4	1.0×10^7	2.7×10^5	1.6×10^6
9	1.5×10^5	2.1×10^7	8.5×10^5	3.6×10^6
5	5.8×10^5	6.5×10^7	5.4×10^6	1.4×10^7
3	2.5×10^6	2.3×10^8	2.6×10^7	5.6×10^7
2	7×10^6	7×10^8	1.1×10^8	1.8×10^8
1	4×10^7	9.7×10^9	2.5×10^9	—
0.5	2×10^8	1.5×10^{11}	—	—
0.3	4×10^8	4.5×10^{11}	—	—
0.1	1×10^9	2.6×10^{12}	—	—

TABLE IV.   Summary of the Evolution of the Sun

Point in evolution	Time since line above (yr)	T (surface) (K)	$L/L_{\odot}$	T (central) (K)	ρ (central) (g cm^{-3})	$R/R_{\odot}$
Onset of protostellar collapse	0	10	10^{-4}	10	10^{-19}	2×10^6
Formation of stellar core	3×10^5	10	10^{-4}	2×10^4	2×10^{-2}	2×10^6
Core contains half of material	7×10^4	300	26	8×10^5	0.25	2×10^3
Onset of quasi-static contraction	8×10^5	4400	1.6	4×10^6	1.5	2.1
Minimum L on convective track	8×10^6	4400	0.5	6×10^6	11	1.6
Maximum L just before main sequence	1.6×10^7	5900	1.1	1.3×10^7	83	1.0
Main sequence, age zero	10^7	5700	0.7	1.4×10^7	90	0.87
Present sun	4.6×10^9	5800	1.0	1.5×10^7	156	1.0

down to this T during the contraction; the higher-mass stars do not. The low-mass stars would be expected to have on the average less observable lithium when they arrive at the main sequence, in agreement with observations.

3.   The youngest stars, known as the T Tauri stars, have high lithium abundances, are associated with dark clouds where star formation is taking place, display irregular variability in light, as well as mass loss that is much more rapid than that of most main-sequence stars. All these characteristics indicate youth; their location in the H–R diagram falls along the Hayashi tracks for stars in the mass range 0.5 to $2.0M_{\odot}$.

4.   The H–R diagrams of young stellar clusters can be compared with lines of constant age drawn between the tracks of Fig. 7. The results show that there is a considerable scatter of observed points about a single line of constant age,

indicating that star formation probably occurs continuously in a given cloud over a period of about 10^7 yr.

V. The Main Sequence

A. GENERAL PROPERTIES

The zero-age main sequence (ZAMS) is defined as the time when nuclear reactions first provide the entire luminosity radiated by a star. The chemical composition is assumed to be spatially homogeneous at this time; later the nuclear reactions result in a change of composition, with the ratio of helium to hydrogen increasing inward toward the center. The characteristics of a range of stellar masses on the ZAMS are given in Table V; the assumed composition is $X =$

TABLE V.   Zero-Age Main Sequence

$M/M_{\odot}$	Log $L/L_{\odot}$	Log T_{eff} (K)	$R/R_{\odot}$	T_c (10^6 K)	ρ_c (g cm^{-3})	M_{core}/M	M_{env}/M	Energy source
30	5.15	4.64	6.6	36	3.0	0.6	0	CNO
15	4.32	4.51	4.7	34	6.2	0.39	0	CNO
9	3.65	4.41	3.5	31	10.5	0.30	0	CNO
5	2.80	4.29	2.3	27	17.5	0.23	0	CNO
3	2.0	4.14	1.7	24	40.4	0.18	0	CNO
2	1.3	4.01	1.4	21	68	0.12	0	CNO
1	−0.13	3.76	0.9	14	90	0	10^{-2}	PP
0.5	−1.42	3.59	0.44	9	74	0	0.4	PP
0.3	−1.9	3.55	0.3	8	125	0	1	PP
0.1	−3.0	3.51	0.1	5	690	0	1	PP

0.71, $Y = 0.27$, $Z = 0.02$. The following general points can be made.

1.   The equation of state in the interior is close to that of an ideal gas. There are deviations only at the high-mass end, where radiation pressure becomes important, and at the low-mass end, where electron degeneracy begins to be significant.

2.   The radius increases roughly linearly with mass at the low-mass end and roughly as the square root of the mass at the high-mass end.

3.   The theoretical mass–luminosity relation agrees well with the observations (Fig. 2); the location of the ZAMS in the H–R diagram is also in good agreement with observations.

4.   The central temperature increases with mass, as expected from the virial theorem. The higher-mass stars burn hydrogen on the CNO cycle, and because of the steep T dependence of these reactions and the corresponding strong degree of concentration of the energy source to the center, they have convective cores. The fractional mass of these cores increases with total mass. The lower-mass stars run on the pp chain, and because of the more gradual T dependence, they do not have convective cores. However, because they have cooler surface layers and therefore higher opacity near the surface, they have convection zones in their envelopes, which become deeper as the mass decreases. At $0.3M_\odot$ this convection zone extends all the way to the center. The lowest mass stars do not have high enough central temperatures to allow ^{3}He to react with itself; therefore the pp chain proceeds only through reactions (1) and (2).

5.   There is no substantial evidence for stars above about $50M_\odot$ on the main sequence. The upper limit for stellar masses arises either from the fact that radiation pressure from the core can prevent further accretion of envelope material during the protostellar phase when the core reaches some critical mass or from the fact that main-sequence stars above a given mass are pulsationally unstable and as a result tend to lose mass rapidly from their surfaces.

6.   The lower end of the main sequence occurs at about $0.08M_\odot$. Below that mass nuclear reactions cannot provide sufficient energy. Physically, the limit arises from the fact that as low-mass stars contract they approach the regime of electron degeneracy before the temperature becomes high enough to start nuclear burning. As the electrons are forced into higher and higher energy states, the gravitational energy

supply of the star is used up in providing the required energy to the electrons. The ions, whose temperatures determine nuclear reaction rates, reach a maximum temperature and then begin to cool. Their thermal energy is required, along with the gravitational energy, to supply the radiated luminosity. Stars below $0.08M_\odot$ simply contract to a limiting radius and then cool.

B. EVOLUTION ON THE MAIN SEQUENCE

During the burning of hydrogen in their cores, stars move very slowly away from the zero-age main sequence. The main-sequence lifetimes are given in Table III. High-mass stars go through this phase in a few million years, while stars below about 0.8 to $1.0M_\odot$ have not had time since the formation of the galaxy to evolve away from the main sequence. The rate at which energy is lost at the surface, which determines the evolutionary time scale, is controlled by the radiative opacity through much of the interior. The nuclear reaction rates adjust themselves to maintain thermal equilibrium, which means that the energy production rate is exactly matched by the rate at which the energy can be carried away by radiation. If for some reason the nuclear processes were producing energy too rapidly, some of this energy would be deposited as work in the inner layers, these regions would expand and cool, and the strongly temperature-dependent energy production rate would return to the equilibrium value.

The structure of the star changes during main-sequence evolution as a consequence of the change in composition. In the case of upper main-sequence stars the H is depleted uniformly over the entire mass of the convective core, while in the lower-mass stars, which are radiative in the core, the H is depleted most rapidly in the center. The conversion of H to He results in a slight loss of pressure because of the smaller number of free particles; as a result the inner regions contract very slowly to maintain hydrostatic equilibrium. The conversion also reduces the opacity somewhat, which tends to cause a slow increase in L. The outer layers see no appreciable change in opacity; a small amount of the energy received from the core is deposited there in the form of work, and these layers gradually expand. In the case of high-mass stars, when X goes to zero at the center, it does so over the entire mass of the convective core, and the star is suddenly left without fuel. A rapid overall contraction takes place, until the layers of unburned H outside the core reach temperatures high enough to burn. For solar-mass stars, the main-sequence evolution does not end so suddenly, because the hydrogen is depleted only one layer at a time. The evolution of important parameters of the sun is given in Table IV. The value of L has increased slightly since age zero, T_{eff} has increased slightly, R has become slightly larger, the depth of the surface convection zone has decreased slightly, the central density has increased considerably, and the central temperature has increased somewhat. An important calibration for the entire theory of stellar evolution is the match between theory and observation of the sun. The procedure is to choose a composition and to evolve the sun from age zero to an age of 4.7×10^9 yr. The value of Z is fairly well constrained by observations of photospheric abundances; thus if the solar L does not match the model L, the He abundance in the model can be used as a parameter to be adjusted to provide agreement. The derived value of Y, of course, must still be in agreement with the observed Y, within the observational error. Once L of the model is satisfactory, the radius can be adjusted to match the observed R by small adjustments in α, the mixing length parameter that is used in the calculation of the surface layers.

The resulting value must be in reasonable agreement with the sizes of the observed solar granulation elements. All of these requirements can be met for an assumed $X = 0.73$, $Y = 0.25$, $Z = 0.02$, and $\alpha = 1.8$, which are the parameters used in the "standard" solar model, at its present age, shown in Table VI. The value of Y determined in this manner is thought to be accurate to ± 0.01.

C. The Solar Neutrino Problem

In the Homestake Mine in South Dakota, a mile underground, R. Davis, Jr., has set up an experiment to detect the neutrinos that are given off in the interior of the sun as a byproduct of the pp chains. Because of the very small probability that a neutrino will be absorbed by matter, a tank of perchloroethylene (C_2Cl_4) containing 10^5 gallons is used; it is shielded by water to reduce background events. The reaction used to detect the neutrinos is $^{37}\text{Cl} + \nu \rightarrow {}^{37}\text{Ar} + e^-$. The argon is radioactive and its decay can be detected. The reaction will not occur unless the neutrino has energy 0.81 MeV or more. Only a small fraction of solar neutrinos have the required energy, and the ones that are the most detectable are those produced in reaction (8). Although the flux of solar neutrinos at the surface of the earth is $7 \times$

TABLE VI.   Standard Solar Model[a]

$M(r)/M_\odot$	$r/R_\odot$	T (10^6 K)	ρ (g cm^{-3})	$L(r)/L_\odot$	X	κ (cm^2 g^{-1})
0.0	0.00	15.5	156.3	0.00	0.355	1.1
0.0099	0.046	14.8	133.9	0.079	0.417	1.2
0.0385	0.076	13.8	108.1	0.264	0.497	1.3
0.1038	0.113	12.4	78.9	0.555	0.592	1.4
0.1620	0.138	11.4	63.2	0.718	0.641	1.6
0.2100	0.156	10.8	53.6	0.809	0.668	1.7
0.2580	0.173	10.2	45.7	0.874	0.688	1.8
0.3100	0.190	9.60	38.5	0.921	0.702	1.9
0.3900	0.217	8.77	29.4	0.964	0.716	2.1
0.4700	0.245	8.00	22.1	0.986	0.724	2.4
0.5500	0.275	7.27	16.1	0.996	0.728	2.8
0.6900	0.336	6.03	8.03	1.000	0.731	3.5
0.8300	0.430	4.65	2.85	1.000	0.732	4.7
0.9264	0.554	3.40	0.773	1.000	0.732	8.0
0.9602	0.641	2.72	0.338	1.000	0.732	12.2
0.9784	0.718	2.12	0.169	1.000	0.732	16.8
0.9954	0.849	0.95	0.050	1.000	0.732	CONV
1.00	1.00	0.0058	2.8×10^{-7}	1.00	0.732	0.3

[a] Adapted with permission from Bahcall, J. N., et al. (1982). *Rev. Mod. Phys.* **54,** 767. ©1982 The American Physical Society.

10^{10} neutrinos cm^{-2} sec^{-1}, the total number of detections expected in the entire tank is only about one per day. A long series of experimental results has been obtained over a period of about 20 years, with experimental errors improving with time. As of 1984, solar neutrinos had definitely been detected, and the rate of capture was 2.1 ± 0.3 solar neutrino units (SNU), where 1 SNU = 10^{-36} captures per target atom per second. The best theoretical estimate, based on the standard solar model, of the expected rate of capture is 7.6 ± 3.3 SNU. The discrepancy is clearly too large to be accounted for by the theoretical or experimental uncertainties.

The following possibilities have been considered as solutions to the solar neutrino problem.

1. The solar model is inappropriate. The production rate of the neutrinos produced by ^{8}B decay is very sensitive to the temperature near the center of the sun. A decrease in this temperature would bring the predictions more into line with the experiment. Essentially the only parameter in the solar model is the abundance of He. A decrease in its assumed value requires a compensating decrease in Z so that the solar luminosity will be matched by the model at the solar age. Although this procedure does result in a lower central temperature, the value of Z is well enough known that a change in Y sufficient to bring T_c down sufficiently would result in a required value of Z that is outside observational uncertainties. In general, a change in the solar model sufficient to bring about consistency with the solar neutrino experiment will result in inconsistency with at least one other well-observed property of the sun.

2. The nuclear reaction rates, when extrapolated to stellar energies, are incorrect. However, their expected uncertainties have already been included in the theoretical error estimate given earlier.

3. The neutrinos decay before they reach the earth. However, at the moment there is no evidence from particle physics that the required decays occur.

4. The sun undergoes periodic episodes of mixing of the chemical composition in the interior, which would result in temporary expansion and cooling of the central regions. The associated drops in the solar luminosity could possibly be associated with the occurrence of ice ages on earth. It is also possible that the sun is continuously mixed through a diffusive process, but, in either case, the mechanism for mixing is not understood. The solution of the solar neutrino problem may well have to wait until more accurate experimental measurements can be made. For example, a detector made of ^{71}Ga could have an expected detection rate of about 100 SNU and because of a lower threshold could actually detect not only the neutrinos resulting from ^{8}B decay, but also the much more numerous ones arising from the basic reaction (1).

VI. Stellar Evolution beyond the Main Sequence

Following the exhaustion of hydrogen at the center of a star, major structural adjustments occur; the physical processes that result in the transition to a red giant were first made clear by the calculations of F. Hoyle and M. Schwarzschild. The central regions, and in some cases the entire star, contract until the hydrogen-rich layers outside the exhausted core can be heated up to burning temperatures. Hydrogen burning becomes established in a shell source, which becomes narrower and hotter as the star evolves. The central regions, inside the shell, continue to contract, while the outer layers expand rapidly as the star evolves to the red giant region. The value of T_{eff} decreases to 3500 to 4500 K, and a convection zone develops in the outer layers and becomes deeper as the star expands. When the temperature in the core reaches 10^8 K, helium burning begins, resulting in the production of ^{12}C and ^{16}O. Evolutionary time scales up to and during helium burning for masses above $1M_\odot$ are given in Table III. Whether further nuclear burning stages occur that result in the production of still heavier elements depends on the mass of the star. For masses less than $8M_\odot$, interior temperatures never become high enough to burn the carbon or oxygen. The star ejects its outer envelope, goes through the planetary nebula phase, and ends its evolution as a white dwarf. The more massive stars burn C and O in their cores, and nucleosynthesis proceeds to the production of an iron–nickel core. Collapse of the core follows, resulting in a supernova explosion, the ejection of a large fraction of the mass, and the production of a neutron star remnant. The details of the evolution are strongly dependent on mass. The following sections describe the post–main-sequence evolution of stars of Population I with masses of 1, 5, and $25M_\odot$ and of a star of Population II of about $0.8M_\odot$, which represents a typical observable star in a globular cluster. Evolutionary tracks for the Population I stars are shown in Fig. 8.

A. FURTHER EVOLUTION OF $1M_\odot$ STARS

During the main-sequence evolution, the hydrogen burns most rapidly at the center. A distribution of composition is set up with the hydrogen fraction decreasing smoothly inward (see Table VI). As the central hydrogen is exhausted, the star makes a gradual transition to H burning in a shell; during this process the central regions contract slowly but the outer regions continue to expand. The shell-burning region narrows gradually, and as it does the star evolves to the right in the H–R diagram. Although this phase is short compared with the main-sequence lifetime, it is slow enough that stars making the transition to the red giant region should be observable in old clusters. As the convective envelope develops, the evolution changes direction in the H–R diagram, and the luminosity increases considerably, while T_{eff} decreases only slowly. The inner edge of the convection zone moves inward to a point just outside the hydrogen-burning shell; inside the shell is the dense,

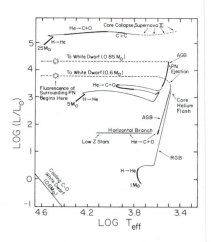

FIG. 8. Post–main-sequence evolutionary tracks in the H–R diagram for stars of $1.0M_\odot$, $5.0M_\odot$, and $25M_\odot$ under the assumption that the metal abundance Z is comparable to that in the present sun. The heavy portions of each curve indicate where major nuclear burning stages occur in the core. The label RGB refers to the red giant branch, AGB to the asymptotic giant branch, and PN to the planetary nebula phase. Helium burning occurs on the horizontal branch only if Z is much less than the solar value. [Adapted with permission from Iben, I., Jr. (1985). *Quart. J. Roy. Astronom. Soc.* **26**, 1. © Royal Astronomical Society.]

burned-out core consisting mainly of helium and increasing in mass with time. Temperatures in the shell increase to the point where the CNO cycle takes over as the principal energy source. The outer convective envelope becomes deep enough so that it reaches layers in which C has been converted to N through the first reaction of the CNO cycle. The modification of the C to N ratio at the surface of the star, which is expected because of convective mixing, has been verified by observations of abundances in red giant stars.

Helium burning finally begins in the core when its mass is about $0.45M_\odot$ and L equals $10^3 L_\odot$. At this time $T = 10^8$ K and $\rho = 10^5$ g cm^{-3} in the core, and the electron gas is degenerate. The helium-burning reaction is very sensitive to temperature, but the electron pressure is practically independent of temperature. When the reaction starts, the local region heats somewhat and the reaction rate increases, resulting in further heating. Under normal circumstances, the increased temperature would result in increased pressure, and the region would expand to the point where the rate of energy generation matched the rate at which the energy could be transported away. Under degenerate conditions the pressure does not respond appreciably to the temperature, expansion does not occur, and the region simply heats, resulting in a runaway growth of the energy generation. This thermal instability is known as the helium flash during which the luminosity can increase to $10^{11} L_\odot$ for a brief period. This enormous luminosity, however, is absorbed in the heating and gradual expansion of the core, and the luminosity at the stellar surface is not affected. The temperature of the core rises to the point where it is no longer degenerate, cooling can occur, and the nuclear reaction once more becomes regulated. Relatively little He is actually burned during the flash, and the star then settles down, at relatively constant L, on the red giant branch, with the energy production coming from both core helium burning and shell hydrogen burning. Although a convection zone develops in the region of the helium flash, it does not link up with the outer convection zone, and so the products of helium burning are not mixed outward to the surface of the star at this time.

The evolution of a solar mass beyond the onset of helium burning has not been calculated in detail, but the events that must occur later on are as follows. When the helium is exhausted in the center of the core, that region begins to contract and heat until helium burning is established in a shell region. The hydrogen-burning shell is also still active; thus the star now has a double-shell source surrounding a core of carbon. The convective envelope still extends inward to a point just outside the hydrogen-burning shell. The star resumes its climb up the giant branch, and the shell sources become narrower and also closer together. The core becomes degenerate, and the increase in its temperature eventually stops because the energy released by contraction must go into lifting the electrons into higher and higher energy states. The structure of the star is divided into three regions: (1) the degenerate carbon core, which grows in mass to about $0.6M_\odot$ and maintains a radius of about 10^9 cm, (2) the hydrogen- and helium-burning shells, which are separated by a very thin layer of helium and which contain a tiny fraction of the total mass, and (3) the extended convective envelope, which still has essentially its original abundances of H and He and a mean density of only 10^{-8} g cm^{-3}; it expands to a size of 6×10^{12} cm. The helium-burning shell is subject to a thermal instability during which the energy generation increases to very high values. This event can repeat periodically during this evolutionary phase, but as in the case of the helium flash, the surface luminosity does not change appreciably. As the star increases in luminosity, it develops an increasingly stronger stellar wind, as a result of which it loses mass gradually. Also, as a consequence of the deep convection zone and the instability in the helium-burning shell, some of the carbon from the core may be mixed outward to the surface. The observed properties of carbon stars, high-luminosity red giants with abnormally strong spectral features of carbon compounds, can be explained in this manner.

As the luminosity increases to near $10^4 L_\odot$, the envelope of the star develops an instability that leads to its ejection. Although the details of the mechanism are not fully understood, it is probable that the star first becomes pulsationally unstable. In fact, most of the stars in this region of the H–R diagram are variable in light with periods of about a year. As the luminosity increases, the amplitude of the oscillations increases and eventually grows to the point where a small amount of mass at the outer edge is brought to escape velocity. As mass begins to be lost, the whole envelope becomes unstable and is ejected, leaving behind the core, the shell sources, and a thin layer of hydrogen-rich mate-

rial on top. The star now enters the planetary nebula phase.

The system now consists of a compact central star and an expanding diffuse envelope, known as the planetary nebula, which is observable as a consequence of its illumination by the star (Fig. 9). The star evolves to the left in the H–R diagram at a nearly constant luminosity of $10^4 L_\odot$, with T_{eff} increasing from about 10^4 K to 2×10^5 K and with the radius decreasing from $33R_\odot$ to $0.08R_\odot$. The time spent in the planetary nebula phase is only a few times 10^4 yr. The star consists mainly of carbon, and the two nuclear burning shells are still active near the outer edge. The hydrogen-rich outer envelope decreases in mass as the fuel is burned, and eventually the star comes to the point where it has no more fuel to burn. The outer layers contract rapidly until the radius approaches a limiting value. Only a very slight amount of further gravitational contraction is possible because practically the entire star is highly degenerate, and the high electron pressure supports it against gravity. Essentially the only energy source left is from the cooling of the hot interior. From this point the star evolves downward and to the right in the H–R diagram, with decreasing L and T_{eff}, and soon enters the white dwarf region. Thus the final state of the

FIG. 9.   The giant planetary nebula NGC 7293 (Shane 120-in. reflector). (Lick Observatory photograph.)

evolution of $1M_\odot$ is a white dwarf of about $0.6M_\odot$.

B. EVOLUTION OF $5M_\odot$ STARS

The evolution of higher-mass stars differs from that of lower-mass stars in several respects. On the main sequence the star of $5M_\odot$ burns hydrogen on the CNO cycle and has a convective core that includes about 23% of the mass at first but which decreases in mass with time. The H is uniformly depleted within that core. With the exhaustion of H in the core, the star undergoes a brief overall contraction that leads to heating and ignition of the shell source; the convection zone has now disappeared. As the shell narrows, the central regions interior to it contract while the outer regions expand. The star rapidly crosses the region between the main sequence and the red giant branch. Because of this rapid crossing, very few stars would be expected to be observable in this region, and for that reason it is known as the Hertzsprung gap. As T_{eff} decreases, a convection zone develops at the surface and advances inward; when it includes about half of the mass, the star turns upward in the H–R diagram. When $\log L/L_\odot = 3.1$ and the hydrogen-exhausted core includes about $0.75M_\odot$, helium burning begins at the center. Because this region is not yet degenerate, the helium flash does not occur and the burning is initiated smoothly. A central convection zone develops because of the strong temperature dependence of the triple-alpha reaction. The structure of the star now consists of (1) the convective helium-burning region, (2) a helium-rich region that is not hot enough to burn helium, (3) a narrow hydrogen-burning shell source, and (4) an extended outer envelope with unmodified hydrogen and helium abundances. This outer convection zone has extended downward into layers where the CNO cycle has previously operated when the star was in its hydrogen-burning phase. The relative depletion of carbon and enhancement of nitrogen could now be observable at the stellar surface.

As helium burning progresses, the central regions expand, resulting in a decrease in temperature at the hydrogen-burning shell and a slight drop in luminosity. The expansion of the outer layers is reversed, and the star evolves to higher T_{eff}. The energy production is divided between the core and the shell, with the former becoming relatively more important as the star evolves.

The total lifetime during core helium burning is about 1.4×10^7 yr, about 20% of the main-sequence lifetime. The star passes through the region of the H–R diagram where the Cepheid variables are observed; these objects are interpreted as stars of $5M_\odot$ to $10M_\odot$ in the phase of core helium burning. The evolutionary calculations show that for solar metallicity only stars in this mass range make the excursion to the left in the H–R diagram during helium burning. Lower-mass stars stay on the giant branch during this phase.

As the helium becomes depleted in the core, that region contracts and heats in order to maintain about the same level of energy production. The contraction results in a slight increase in temperature at the hydrogen-burning shell. The energy production increases there, and a small amount of this excess energy is deposited in the outer envelope in the form of mechanical work, causing expansion. The evolution changes direction in the H–R diagram and heads back toward the red giant region. The He is used up at the center, a helium shell source is established, and the star resumes its interrupted climb up the giant branch. The surface convection zone develops again and reaches deep within the star to layers that have been previously enriched in He. The ratio He/H at the surface is thereby increased. As the star rises to higher L, its structure includes the following regions: (1) a degenerate core composed of a mixture of ^{12}C and ^{16}O, (2) a narrow helium-burning shell source, (3) a thin intershell layer composed mostly of He, (4) a narrow hydrogen-burning shell source, and (5) the extended outer envelope.

The star increases in L up to about $2 \times 10^4 L_\odot$, and the mass of the C/O core increases as the H and He are burned. A series of thermal flashes occurs in the helium-burning shell, as a consequence of which some of the carbon produced in helium-burning reactions can be mixed to the surface of the star. If the core reaches a critical mass of $1.4M_\odot$, its internal temperature becomes high enough to burn the C and O; the products of the reactions include ^{24}Mg, ^{32}S, ^{20}Ne, and ^{28}S. Because the core is highly degenerate at this point, the burning will be unstable and explosive; it has been suggested that the whole star could explode into a supernova. However, it is now generally accepted that ejection of the envelope and formation of a planetary nebula will occur before the critical core mass is reached. In that case the final state of the star would be a white dwarf having a mass equal to that of the core at the time of ejection. For $5M_\odot$ the probable result is the ejection of a planetary nebula at a core mass of $0.85M_\odot$ (see Fig. 8). The same result will occur for stars up to $9M_\odot$, with the maximum core mass being about $1.1M_\odot$.

C. EVOLUTION OF $25M_\odot$ STARS

The high-mass stars evolve similarly to the stars of $5M_\odot$ during main-sequence and immediate post–main-sequence evolution, except that the fraction of the mass contained in the convective core is larger. The star evolves toward the red giant region, but helium burning at the center occurs when T_{eff} is still in the range 10^4 to 2×10^4 K, and carbon burning starts afterward. The star expands to become a red giant with $L/L_\odot = 2 \times 10^5$ and $T_{eff} = 4500$ K. A sequence of several new nuclear burning phases now occurs in the core, rapidly enough that little concurrent change occurs in the surface characteristics. Carbon burns at about 9×10^8 K, neon at 1.75×10^9 K, oxygen at 2.3×10^9 K, and silicon at 4×10^9 K. A central core of about $1.5M_\odot$, composed of iron and nickel, builds up, surrounded by successive layers that are silicon rich, oxygen rich, and helium rich, respectively. Outside these layers is the envelope, still with its original composition. The layers are separated by active shell sources. The temperature at the center reaches 7×10^9 K and the density 3×10^9 g cm^{-3}. However, the sequence of nuclear reactions that has built the elements up to the iron peak group in the core can proceed no further. These elements have been produced with a net release of energy at every step, a total of 8×10^{18} ergs per gram of hydrogen converted to iron. However, to build up to still heavier elements, a net input of energy is required. Furthermore, the Coulomb barrier for the production of these elements by reactions involving charged particles becomes very high. Instead, an entirely different process occurs in the core. The temperature, and along with it the average photon energy, becomes so high that the photons can react with the iron, breaking it up into helium nuclei and neutrons. This process requires a net input of energy, which must ultimately come from the thermal energy of the gas. The pressure therefore does not rise fast enough to compensate for the increasing force of gravity, and the core begins a catastrophic gravitational collapse. On a time scale of seconds, the central density rises to 10^{14} g cm^{-3} and the temperature to 3×10^{10} K. As the density increases, the free electrons are captured by the nuclei, reducing the electron pressure and further contributing to collapse. The point is reached where most of the matter is in the form of free neutrons, and when the density becomes high enough, their degenerate pressure increases rapidly enough to stop the collapse. At that point a good fraction of the original iron core has collapsed to a size of 10^6 cm and has formed a neutron star, nearly in hydrostatic equilibrium, with a shock front on its outer edge through which material from the outer parts of the star is falling and becoming decelerated.

The question of what happens after core collapse is one of the most interesting in astrophysics. Can at least part of the gravitational energy released during the collapse be transferred to the envelope and result in its expansion and blowoff in the form of a supernova? Present indications are that it is possible and that the shock will propagate outward into the envelope, partly aided by the neutrinos, produced during the neutronization of the core, which deposit at least part of their energy and momentum in the very outer parts of the core. Assuming that the shock does propagate outward, it passes through the various shells and results in further nuclear processing, including production of a wide variety of elements up to and including the iron peak. It also accelerates all of the material outside the original iron core outward to escape velocities. When the shock reaches the surface of the red giant star, the outermost material is accelerated to 10,000 km sec^{-1}, and the deeper layers reach comparable but somewhat smaller velocities. Luminosity, velocity, and T_{eff} as a function of time in the calculated supernova outburst agree well with observations of one particular class of supernovae. The enormous luminosity arises, in the earlier stages, from the rapid release of the thermal energy of the envelope. At later times, most of the energy comes from the radioactive decay of the ^{56}Ni that is produced mainly by explosive silicon burning in the supernova shock. Supernova observations are best fit with total explosion energies of about 10^{51} ergs. The Crab Nebula (Fig. 10) is consistent with this energy and an expansion velocity of 10,000 km sec^{-1}. Another good test of the theory of stellar evolution is the calculation of the relative abundances of the elements between oxygen and iron

FIG. 10.   Crab Nebula in Taurus (M1, NGC 1952) in red light, the remnant of the supernova explosion in A.D. 1054 (Shane 120-in. reflector). (Lick Observatory photograph.)

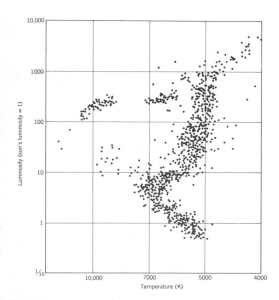

FIG. 11.   H–R diagram of the globular cluster M3. [Reprinted with permission from Jastrow, R., and Thompson, M. H. (1984). "Astronomy: Fundamentals and Frontiers," 4th ed. John Wiley and Sons, New York. © 1984, Robert Jastrow.]

in the ejected supernova envelope; for $25M_\odot$ stars the values agree well with solar system abundance ratios.

D. EVOLUTION OF LOW-MASS STARS WITH LOW METAL ABUNDANCE

The oldest stars in the galaxy are characterized by metal abundances that are 0.1 to 0.01 that of the sun and by a distribution in the galaxy that is roughly spheroidal rather than disklike. These stars can be found in globular clusters or in the general field; in particular the H–R diagrams of globular clusters (Fig. 11) give important information on the age of the galaxy and the helium abundance at the time the first stars were formed. There are observational difficulties in determining the properties of these stars, since most of them are very distant. For example, it has not proved possible to determine observationally the mass–luminosity relation of the main sequence. However, observations of the locations in the H–R diagram of a few nearby low-metal stars show that they fall somewhat below the main sequence defined by stars of solar metal abundance. This information, com-

bined with detailed analysis of the H–R diagrams of globular clusters, indicates that helium abundances in the old stars are not very different from that of the sun: $Y = 0.25 \pm 0.03$.

The age estimates of globular clusters, based on detailed comparisons of observed H–R diagrams with theoretical evolutionary tracks, range from 10 to 17×10^9 yr. Recent work, based on improved observations and improved theoretical fits, favors ages toward the high end of this range. The high-mass stars in these clusters have long ago evolved to final states that are now unobservable. The mass of the stars that are now evolving off the main sequence and becoming relatively luminous red giants is about $0.8M_\odot$. On the main sequence, however, these stars have approximately solar luminosity because of their lower metal content, and hence lower opacity, when compared with stars of normal metal abundance. The evolution during the main-sequence phase and the first ascent of the red giant branch is very similar to that of $1M_\odot$ stars described previously. Hydrogen burning occurs on the proton–proton chain, there is no convection in the core, and when hydrogen is exhausted in the center, the energy source shifts to a shell and the star gradually makes the transition to the red giant region. The core contracts and becomes degenerate, and the envelope expands and becomes convective. A helium flash in the core occurs when the core mass is $0.45M_\odot$ and the total luminosity is $10^3L_\odot$.

When the helium flash is completed and the core is no longer degenerate, the star settles down to a stable state with helium burning in the core and hydrogen burning in a shell. However, the location in the H–R diagram during this phase differs from that of the low-mass stars with solar metals. The star evolves rapidly to a location on the horizontal branch (HB), with T_{eff} considerably higher than that on the giant branch but with L somewhat less than that at the helium flash (see Figs. 8 and 11). Calculations show that as the assumed value of Z is decreased for a given mass, the position of the model on the HB shifts to the left. This behavior is in agreement with observations of globular clusters, which show that the leftward extent of the HB is well correlated with observed Z, in the sense that smaller Z corresponds to an HB extending to higher T_{eff}. In theoretical models, as Z is increased to 10^{-2}, the HB disappears altogether and the helium-burning phase occurs on the red giant branch. This calculation is also in agreement with the observed fact that old clusters with solar Z do not have a HB. The HB itself is not an evolutionary track. The spread of stars along the HB in a cluster of given Z represents a spread in stellar mass, ranging from about $0.6M_\odot$ at the left-hand end to about $0.8M_\odot$ at the right. The implication is that stars lose mass on the red giant branch during the period of their evolution just prior to the helium flash and that the total mass lost varies from star to star. The mass loss probably occurs by a mechanism similar to that which drives the observed solar wind, but much stronger.

The phase of core helium burning on the HB lasts about 10^8 yr, after which the star develops a double shell structure and evolves back toward the red giant branch, which during the following phase of evolution is referred to as the asymptotic giant branch. The evolution from this point on is very similar to that of the star of $1M_\odot$ with $Z = 0.02$: ascent of the asymptotic giant branch, development of a carbon core that becomes increasingly degenerate, development of a pulsational instability in the convective envelope, ejection of the envelope, evolution through the planetary nebula phase, and finally the transition to the white dwarf phase.

VII. Final States of Stars

A. BROWN DWARFS

Stars with mass less than $0.08M_\odot$ never attain internal temperatures high enough that nuclear burning can supply their entire radiated luminosity. These stars contract to the point where the electrons become highly degenerate, and they reach a limiting radius that depends on the mass and composition. At $0.07M_\odot$ the time to contract to this limiting radius is about 10^9 yr; for lower masses it is less. Beyond this point the star simply cools at constant radius, with decreasing L and T_{eff}. For $0.07M_\odot$ the value of T_{eff} during contraction is about 2500 K; for $0.01M_\odot$ it is about 2000 K. A typical brown dwarf after an evolution time of 2 to 3×10^9 yr will have $T_{eff} = 1000$ to 1500 K and $L/L_\odot = 10^{-5}$ to 10^{-6} and is therefore very difficult to observe. These stars are distinguished from white dwarfs first by their very cool surface temperatures and second by the fact that their internal composition is practically unchanged from the time of formation— that is, it is about 75% hydrogen by mass. An example of a brown dwarf is the companion of VB8, observed in 1984 to have an estimated $L/L_\odot = 3 \times 10^{-5}$, $T_{eff} = 1360$ K, and a deduced mass $0.03M_\odot$ to $0.07M_\odot$. The dividing line between brown dwarfs and planets is arbitrary; Jupiter at $0.001M_\odot$ could be considered to be a brown dwarf because it is radiating at the expense of its own internal energy at the rate of $L/L_\odot = 10^{-9}$. The formation mechanism may be the only way to distinguish between high-mass planets and low-mass brown dwarfs.

B. WHITE DWARFS

Observational parameters of the most commonly detected white dwarfs are $T_{eff} = 10,000$ to 20,000 K and $\log L/L_\odot = -2$ to -3. The objects therefore lie well below the main sequence, and the deduced radii fall in the range 10^{-2} $R_\odot$ and the mean densities about 10^6 g cm^{-3}. At these high densities most of the mass of the object falls in the region of complete electron degeneracy; only a thin surface shell is nondegenerate. Three white dwarfs, Sirius B, Procyon B, and 40 Eridani B, occur in well-observed binary systems, and their masses are found to be, respectively, $1.05M_\odot$, $0.63M_\odot$, and $0.43M_\odot$.

The structure of a white dwarf can be determined in a straightforward way under the assumptions of hydrostatic equilibrium and uniform composition, with the pressure supplied entirely by the degenerate electrons. The ion pressure and the effect of the thin surface layer are small corrections. Because the pressure is a function simply of density—for example, $P \propto \rho^{5/3}$ in the nonrelativistic limit—it can be eliminated from Eq. (27). The structure can then be calculated from the solution of Eqs. (27) and (29). The results show that for a given composition there is a uniquely defined relation between mass, radius, and central density. The theoretical mass–radius relation, appropriate for any composition of elements heavier than hydrogen, is shown in Fig. 12, where it is compared with the observations of a few white dwarfs. Note that the radius decreases with increasing mass and vanishes altogether at $1.44M_\odot$. This upper limit to the mass of a white dwarf was first derived by S. Chandrasekhar and corresponds to infinite central density in the formal solutions of the equations. In fact, however, the physical process that limits the mass of a white dwarf is capture of the highly degenerate electrons by the nuclei, predominantly carbon and oxygen. Suppose that a white dwarf increases in mass upward toward the limit. The central density goes up, the degree of electron degeneracy goes up, and a critical density is reached, about 10^{10} g cm^{-3} for carbon, where electron capture starts. As free electrons are removed from the gas, the pressure is reduced and the star can no longer

exist in hydrostatic equilibrium. Collapse starts, more electron capture takes place, and neutron-rich nuclei are formed. Collapse stops when the neutron degeneracy results in a very high pressure, in other words, when a neutron star has been formed. The white dwarf mass limit is reduced by about 10%, depending upon composition, from the value of $1.44M_\odot$ given earlier.

The theoretical mass–radius relation agrees well with observations, as Fig. 12 indicates. The location of the theoretical white dwarfs in the H–R diagram also agrees with observations. From the L and T_{eff} of observed white dwarfs, one can obtain radii and from them the corresponding masses. The typical mass of a white dwarf, determined in this fashion, is $0.6M_\odot$ to $0.8M_\odot$. These objects have presumably evolved from stars of somewhat higher original mass, generally in the range $0.8M_\odot$ to $1.5M_\odot$ or higher. Their composition is therefore that of the evolved core of these stars, mainly carbon and oxygen with only a thin layer of hydrogen on the outside, the remnant of the hydrogen-rich envelope. The internal temperatures are not high enough to burn nuclear fuel, and the only energy source available is the thermal energy of the ions in the interior. The star's luminosity, T_{eff}, and internal temperature all decrease as it cools at constant radius. The energy loss rates are sufficiently low that cooling times are generally several billion years.

C. NEUTRON STARS

These objects, whose existence as condensed remnants left behind by supernova explosions was predicted by F. Zwicky in 1934, were first discovered in 1967 from their pulsed radio radiation. The typical mass is about $1M_\odot$, the radius 10 km, and the mean density above 10^{14} g cm^{-3}, close to the density of the atomic nucleus. At this high density the main constituent is free neutrons, with a small concentration of protons, electrons, and other elementary particles. The surface temperature is very difficult to determine directly. Neutron stars form as collapsing cores of massive stars and reach temperatures of 10^9 to 10^{10} K during formation. However, they then cool very rapidly by neutrino emission; within a month T_{eff} is less than 10^8 K and within 10^5 yr it is less than 10^6 K. Their luminosity soon becomes so low that they would not be detected unless they were very nearby. The fact that neutron stars are observable arises from the remark-

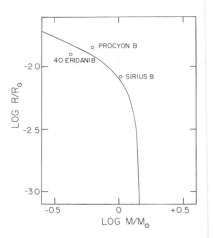

FIG. 12.   Theoretical mass–radius relation derived by S. Chandrasekhar for fully degenerate white dwarfs (solid line) compared with a few observed points (open circles). The radius for Procyon B is very uncertain.

able facts that (1) they are very strongly magnetic, (2) they rotate rapidly, and (3) they emit strong radio radiation in a narrow cone about their magnetic axes. The rotation axis and magnetic axis are not coincident, and so if the star is oriented so that the radio beam sweeps through the observer's instrument, he or she will observe radio pulses separated by equal time intervals. The observed pulsars are thus interpreted as rapidly spinning neutron stars with rotation periods of typically 1 sec but also possibly shorter. For example, the pulsar in the Crab Nebula (Fig. 10) has a period of $\frac{1}{30}$ sec and the shortest observed period is a few milliseconds. The origin of the beamed radiation, which is also observed in the optical and X-ray regions of the spectrum, is not known. The electromagnetic energy is emitted at the expense of the rotational energy, and the rotation periods are gradually becoming longer.

D. BLACK HOLES

Neutron stars also have a limiting mass, but it is not known as well as that for white dwarfs because the physics of the interior, in particular the equation of state, is not fully understood. A reasonable estimate is $2M_\odot$. If the collapsing core of a massive star exceeds this limit, the collapse will not be stopped at neutron star densities because the pressure gradient can never be steep enough to balance gravity. The collapse continues indefinitely, to infinite density, and a black hole is formed. Stars with original masses in the range $10M_\odot$ to $20M_\odot$ form cores of around $1.5M_\odot$; they will become supernovae and leave a neutron star as a compact remnant. Stars above $25M_\odot$ to $30M_\odot$ have larger cores at the time of collapse, probably above the neutron star mass limit, and so they will collapse to form black holes. [See BLACK HOLES (ASTRONOMY).]

E. SUPERNOVAE

A very small fraction of stars ever become supernovae. Two main types are recognized. The supernovae associated with the collapse of the iron core of massive stars are called type II. They tend to be observed in the spiral arms of galaxies, where recent star formation has taken place. The maximum luminosity is about $10^{10}L_\odot$, and the decline of luminosity with time takes various forms with a typical decline time of a few months. The observed temperature

drops from about 15,000 to 5000 K during the first 30 to 60 days, and the expansion velocity decreases from 10,000 to 4000 km sec^{-1} during the same time period. The abundances of the elements are similar to those in the sun, with hydrogen clearly present. The deduced mass of the expanding region is $1.5-2.5M_\odot$, with considerable uncertainty. The type I supernovae, on the other hand, tend to be observed in regions associated with an older stellar population, for example, in elliptical galaxies. The luminosity at maximum is about the same as for type I and the curves of the decline of light with time are all remarkably similar, with a rapid early decline for 30 days and then a more gradual exponential decay in which the luminosity decays by a factor 10 in typically 150 days. Temperatures are about the same as in type II, the expansion velocities are somewhat higher, and the abundance ratios again are near solar but with no hydrogen present. The mass in the expanding region is probably less than $1M_\odot$. The type I supernova is interpreted as the end point of the evolution of a close binary system. Just before the explosion the system consists of a white dwarf near the limiting mass and a companion of about $1M_\odot$ on the main sequence. The evolution of the system results in overflow of the mass of the companion onto the white dwarf, which therefore is accreting matter that is hydrogen rich. The hydrogen burns near the surface, producing a layer of helium that increases in mass. By the time the helium is hot enough to burn, it does so under degenerate conditions. The temperature increases rapidly to the point where the carbon ignites. Carbon burning is so rapid that it produces an explosion, processing much of the material to iron and nickel, and blowing up the entire star. There is probably no remnant. The light emitted is produced primarily from the radioactive decay of ^{56}Ni. [See SUPERNOVAE.]

VIII. Summary: Important Unresolved Problems

The study of stellar structure and evolution has resulted in major achievements in the form of quantitative and qualitative agreement with observed features, in the pre–main-sequence, main-sequence, and post–main-sequence phases. A number of aspects of the physics need to be improved. Considerable uncertainty is caused by the lack of a detailed theory of convection that can be applied to stellar interiors.

On the main sequence this uncertainty could result in relatively minor changes in the structure of the surface layers of stars, mainly in the mass range $0.8–1.1 M_\odot$. However in pre–main-sequence and post–main-sequence stars, which have deep, extended convection zones, a better theory could result in some major changes. The degree of mixing of heavy elements, produced by nuclear reactions in the deep interiors of stars, to the surface could be strongly affected even by small changes in the theory of convection. Improvements in laboratory measurements of nuclear reaction rates, particularly for the proton–proton cycle, could reduce the theoretical error bar in the rate of solar neutrino production. The solar neutrino experiment is of particular importance because it is the only direct measurement of conditions in the deep interior of any star. The gallium detection experiment should provide major insight into the reason for the current discrepancy between theory and observation of solar neutrinos.

The assumption of mass conservation is built into much of the theory discussed in this article. However, it is known from observation that various types of stars are losing mass, including the T Tauri stars in the early pre–main-sequence contraction phase, the O stars on the upper main sequence, and many red giant stars. The physics of the ejection process is not well explained, and the mechanism for sudden ejection of a large amount of mass just before the planetary nebula phase also requires further clarification. It is also likely that during the early phases of their evolution stars accrete mass from the surrounding infalling cloud and from a circumstellar disk. The effects of the accretion of mass and angular momentum upon the evolution remain to be studied. During certain phases of the evolution of close binary systems, the combination of mass loss and mass accretion must be considered. As the more massive star in such a system evolves away from the main sequence and expands, matter can overflow onto the companion. As a consequence of this mass exchange, the evolution of the system is very different from that of single stars. A number of problems remain to be solved in the theory of binary star evolution. For example, the short period systems consisting of a white dwarf and a closeby low-mass main-sequence companion are associated with nova outbursts and, under special circumstances, supernovae of type I. One problem is that the orbital angular momentum of such a binary is much smaller than that of the original

main-sequence binary from which the system must have evolved. A goal of current research is to uncover the processes by which the loss of angular momentum occurs.

The effects of rotation and magnetic fields have not been discussed in this article. However, their interaction in the context of stellar evolution is certain to produce some very interesting modifications to existing theory. During the phase of star formation these processes are central, and consideration of them is crucial for the clarification of the formation of binary systems. During main sequence evolution rotational and magnetic energies are considered to be small compared with gravitational energy, and the overall effect on the structure is therefore small. However, the circulation currents induced by rotation could be important with regard to mixing and the exchange of matter between the deep interior and the surface layers. During post–main-sequence phases, the cores of stars contract to very high densities, and if angular momentum is conserved in them, they would be expected to rotate very rapidly by the time the core becomes degenerate. The effect of rotation on the collapse of the iron core in massive stars and on the generation of the type II supernova could be significant and has been studied only in a preliminary way. However, white dwarfs are not observed to rotate rapidly, suggesting that at some stage angular momentum is transferred, in an as yet unexplained way, out of the cores of evolving stars.

A number of other problems present themselves. The complicated physical processes involved in the generation of supernova explosions require further study. The formation, structure, and evolution of neutron stars involve many unsolved problems. The observational and theoretical study of stellar oscillations can provide substantial information on the structure and evolution of stars. For example, the interior of the sun can be probed through the analysis of its oscillations. For many types of stars, the physical mechanism producing the oscillations is not understood. A particularly interesting problem is that of the ages of stars in globular clusters, which seem to be too high to be consistent with current theories of cosmology, which give an age for the universe of 10 billion years. It is clear that the study of stellar evolution involves a wide range of interacting physical processes and that the findings are of importance in numerous other areas of astronomy and physics. [*See* STAR CLUSTERS.]

BIBLIOGRAPHY

Clayton, D. D. (1983). "Principles of Stellar Evolution and Nucleosynthesis." Univ. of Chicago Press, Chicago, Illinois.

Goldberg, H. S., and Scadron, M. (1981). "Physics of Stellar Evolution and Cosmology." Gordon and Breach, New York.

Kaplan, S. A. (1982). "The Physics of Stars." Wiley, New York.

Kippenhahn, R. (1983). "100 Billion Suns." Basic Books, New York.

Shklovskii, I. S. (1978). "The Stars: Their Birth, Life, and Death." Freeman, San Francisco, California.

SUPERNOVA 1987A

David Branch   *University of Oklahoma*

GLOSSARY

Absolute magnitude: Apparent magnitude a star would have if it were at a distance of 10 pc.

Apparent magnitude: Logarithmic measure of the brightness of a star, usually measured through a blue or visual filter. A difference of 5 magnitudes corresponds to a factor of 100 in observed flux; one magnitude corresponds to a factor of 2.512. Increasing magnitude corresponds to decreasing brightness.

Light curve: Graph of apparent magnitude, absolute magnitude, or luminosity plotted against time.

Luminosity: Rate at which electromagnetic radiation is radiated from the surface of a star, at all wavelengths and in all directions.

Parsec (pc):   Distance at which a star would show an annual trigonometric parallax of 1 arcsec. One parsec equals 3.09×10^{18} cm or 3.26 light years.

Photosphere: Atmospheric layer from which a star's continuous spectrum is emitted; equivalently, the layer at which the atmosphere becomes opaque to an external observer's line of sight.

Stellar wind: Flow of matter from the photosphere of a star into surrounding space.

Supernova (SN) 1987A in the Large Magellanic Cloud, the most spectacular observational event in astronomy since the invention of the telescope, dominated the subject of supernovae in 1987 and 1988. The detection of a burst of neutrinos from SN 1987A has confirmed that the highly evolved cores of massive stars do undergo catastrophic gravitational collapse. For many years into the future, observation and analysis of the continuing photon emission across the electromagnetic spectrum will present an extraordinary opportunity to advance our understanding of stellar explosions. [*See* SUPERNOVAE.]

This article is concerned with the observations of photons from SN 1987A, and with the basic physics of the explosion.   [*See* NEUTRINO ASTRONOMY.]

I. Discovery, Location, and Type

During the predawn hours of February 24, 1987, University of Toronto astronomer Ian Shelton, working at the Los Campanas Observatory in Chile, fortuitously discovered a supernova in the Large Magellanic Cloud (LMC). A routine photograph of the LMC, obtained with a 10-inch astrographic reflector, revealed the presence of a star of the fifth apparent magnitude in the place of a previously inconspicuous twelfth-magnitude star (Fig. 1). Intensive study by ground-based optical and radio observatories in the southern hemisphere was undertaken as the night swept around the globe, to South Africa, Australia, and back to Chile. (At a celestial declination of $-69°$, the LMC is never accessible to northern observatories.) Within 14 hr of the discovery, the earth-orbiting International Ultraviolet Explorer (IUE) was also being used to make observations. As the first supernova of the year, the event was designated SN 1987A.

Astronomers recognized immediately that SN 1987A, as the brightest supernova since 1604, would provide a rare opportunity for comprehensive observations of a stellar explosion. The LMC is a relatively small, irregular satellite galaxy of our Milky Way (the Galaxy); at a distance of only 50 kpc (160,000 light years), the LMC is the nearest galaxy

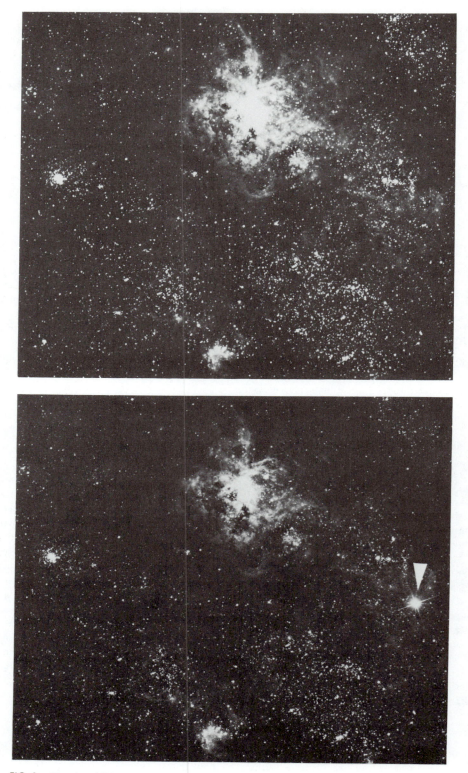

FIG. 1.   The site of SN 1987A, before and after the explosion. The neighboring bright nebulosity is the 30 Doradus nebula. (Courtesy of R. E. Williams, Cerro-Tololo Inter-American Observatory.)

external to our own. No supernova in the Galaxy has been observed since Kepler's, 5 yr before Galileo's first astronomical use of the telescope. Prior to SN 1987A, the brightest supernova of modern times was SN 1885A, a sixth-magnitude event that appeared near the nucleus of Messier 31, the Andromeda galaxy, at a distance of 700 kpc. In the twentieth century, more than 600 supernovae have been discovered in other galaxies, but none were nearer than 3 Mpc and most were beyond 15 Mpc. At such great distances, supernovae become unobservably faint after only a year or two. Being 60 times nearer than the other supernovae that have been accessible to twentieth century telescopes, SN 1987A will be observed for many years to come, with unprecedented angular and spectroscopic resolution.

A supernova is classified as type I or type II according to whether hydrogen lines are absent or present in its optical spectrum. The first spectral observations of SN 1987A established the presence of hydrogen lines, making SN 1987A a type II. From their tendency to appear only in regions of active star formation in spiral and irregular galaxies, type II supernovae are inferred to be the explosions of relatively short-lived ($\lesssim 10^7$ yr) massive stars—those formed with masses in excess of 8 times the mass of the sun ($8 M_\odot$). Supernova 1987A appeared near a conspicuous star-forming region of the LMC, the 30 Doradus (Tarantula) nebula (Fig. 1). Highly evolved massive stars are thought to explode when their iron cores collapse to form neutron stars or black holes.

II. The Progenitor Star

A. IDENTIFICATION AND PROPERTIES

The supernova appeared near the position of a twelfth-magnitude star called Sanduleak (Sk) -69 202 (star number 202 near -69° declination in a catalog of stars published in 1969 by Nicholas Sanduleak of Case Western Reserve University). However, examination of the few high-resolution images of Sk -69 202 obtained before the explosion revealed the additional presence of two faint companion stars (Fig. 2). Following a few weeks of uncertainty regarding which of the three stars had exploded, accurate measurements of the supernova's position established that it was coincident, within the measurement uncertainty of 0.1 arcsec, with the previous position of Sk -69 202, and inconsistent with the positions of the companions. The brighter companion, a main sequence star of apparent visual magnitude 15.3 and spectral type B0 V, is 3 arcsec to the northwest of Sk -69 202, and the fainter companion, a 15.7 magni-

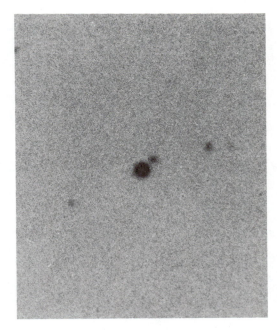

FIG. 2. The supernova progenitor and its companions. The brighter companion is evident to the upper right; the fainter companion is discernible only as a faint extension to the lower left of the Sk -69 202 image. (Courtesy of You-Hua Chu, University of Illinois.)

tude B1.5 V main sequence star, is 1.5 arcsec to the southeast. The identification of the progenitor was confirmed when IUE images, obtained after the ultraviolet brightness of the supernova had faded drastically, showed that Sk -69 202 was missing while the two companions had survived.

Sanduleak -69 202 is the first supernova progenitor whose physical properties can be determined from observations made prior to the explosion. (The progenitor of SN 1961V in the galaxy NGC 1058 was observed before it exploded, but only in a limited way.) Low-resolution photographic spectra of Sk -69 202 showed that it was an intrinsically bright supergiant star of spectral type B3 Ia. The color index, corrected for interstellar reddening in the Milky Way and in the LMC, corresponded to an effective temperature of 16,000 K. The distance to the LMC, the apparent visual magnitude of 12.3, and an allowance of 0.6 magnitudes for interstellar extinction determine an absolute visual magnitude of -6.8. The absolute magnitude and the effective temperature imply that the luminosity was 4.5×10^{38} erg sec^{-1}, which is 1.1×10^5 times the luminosity of the sun ($1.1 \times 10^5 L_\odot$). From the effective temperature and the luminosity, the radius is inferred to have been 3 $\times$

TABLE I.   Properties of SK -69 202

Right ascension (epoch 1950)	5 hr 35 min 50.0 sec (± 0.1 sec)
Declination (epoch 1950)	$-69°$ 17' 58.0'' ($\pm 0.1''$)
Spectral type	B3 Ia
Apparent visual magnitude (m_V)	12.3 (± 0.1)
Absolute visual magnitude (M_V)	-6.8 (± 0.4)
Luminosity	1.1 (± 0.4) $\times 10^5 L_\odot$
Effective temperature	16,000 (± 500) K
Radius	40 (± 15)$R_\odot$
Core mass	6 (± 1)$M_\odot$
Initial mass	19 (± 3)$M_\odot$

10^{12} cm, 40 times the radius of the Sun ($40R_\odot$). In a post main-sequence star, such as Sk -69 202, the luminosity is determined not by the total mass of the star, but by the mass of its core of helium and heavier elements; the Sk -69 202 luminosity implies a core mass of $6M_\odot$. This core mass indicates, in turn, that the initial mass of Sk -69 202 was about $19M_\odot$. The total mass at the time of the explosion must have been more than $6M_\odot$, but less than the original mass of $19M_\odot$, owing to the gradual loss of mass from the surface of the star by means of a stellar wind. The properties of Sk -69 202, with estimates of the associated uncertainties, are summarized in Table I.

B. PREEXPLOSION EVOLUTION

From their characteristic luminosities and the shapes of their light curves, most type II supernovae that had been observed previously were inferred to be the explosions of red supergiants—internally evolved massive stars having low effective temperatures near 4000 K, and very large radii approaching $1000R_\odot$. As explained in Section III, the explosion of Sk -69 202, a much less extended *blue* supergiant, has produced a supernova of relatively low luminosity and a light curve of unusual shape. Why did Sk -69 202 explode as a blue supergiant, rather than a red one? According to numerical computer simulations of the evolution of 19-$M_\odot$ stars, the post main-sequence behavior depends on the initial composition. Stars whose initial composition is like that of the sun are predicted to expand to become red supergiants, and remain red supergiants until they explode. However, stars in the LMC, as in most small irregular galaxies, tend to be deficient in elements heavier than hydrogen and helium, relative to the sun; the typical "metal deficiency" of LMC stars is a factor of 4. Metal-deficient stars, after becoming red supergiants, are predicted to contract to become blue supergiants before exploding. The theoretical evolu-

tion of metal-deficient stars cannot yet be predicted with complete confidence, because it is sensitive to the amount of stellar-wind mass loss and to the precise way that convective energy transport inside the star is simulated in the calculations. The observational fact that the LMC contains numerous red supergiants suggests, however, that Sk -69 202 probably did go through a red supergiant phase. Further evidence for a red supergiant phase is provided by observational signs of an interaction between SN 1987A and its surroundings (Section V).

Why had almost all of the previously observed type II supernovae displayed luminosities and light curves characteristic of red, rather than blue, supergiant progenitors? The answer may be that since the explosions of blue supergiants are less luminous than those of red ones, they are less likely to be detected when they occur in remote galaxies. The LMC is so nearby that even the explosion of a blue supergiant cannot be overlooked. This may account for why type II supernovae had seldom been seen to appear in irregular galaxies despite the numerous massive stars that they contain. The irregular galaxies tend to be metal deficient, so they mainly produce the relatively inconspicuous blue supergiant explosions.

III. Collapse and Explosion

The time scales for the nuclear evolution of a star's core are determined by the star's mass. A 19-$M_\odot$ star, such as Sk -69 202, spends 10 million years as a main sequence star, fusing its innermost $6M_\odot$ of hydrogen into helium, followed by one million years as a supergiant, burning the inner parts of the core to heavier elements. After forming a carbon core, the star has only 1000 yr to live; after forming an iron core it has only hours. Owing to a sudden loss of internal pressure support caused by photodisintegration of nuclei at temperatures above 6×10^9 K (0.6 MeV), the 1.5-$M_\odot$ iron core collapses—from a radius approaching 10^9 cm and a density exceeding 10^9

g cm^{-3} to a radius less than 10^7 cm and a density beyond 10^{14} g cm^{-3}—in a fraction of a second. The collapsed core, a protoneutron star heated to a temperature of 5 MeV, emits a burst of neutrinos, most of which stream freely from the star carrying off almost the entire 3×10^{53} ergs of gravitational potential energy released by the collapse. The foregoing description was a purely theoretical one until the detection of a neutrino burst from SN 1987A on February 23.316 (Universal time or UT), 1987, provided the first direct observation of a stellar core collapse. (The neutrino signal was recognized only after the optical discovery of SN 1987A had been announced.)

Just as the neutrinos testified to the occurrence of a core collapse, the subsequent photon emission shows that the collapse was followed by the ejection of solar masses of matter carrying a kinetic energy on the order of 10^{51} ergs. The mechanism by which a small fraction of the collapse energy is channeled into kinetic energy of ejected matter has been intensively studied by means of elaborate computer simulations, but significant uncertainties persist. When the 1.5-$M_\odot$ iron core of a 19-$M_\odot$ star collapses, the innermost 0.6$M_\odot$ overshoots nuclear density, stiffens, and rebounds on a time scale of milliseconds. The encounter between the bouncing inner core and the infalling outer core creates an outward propagating shock wave that may be able to reverse the infall and eject all but the central 1.4$M_\odot$, which becomes a neutron star. In an iron core as massive as 1.5$M_\odot$, this "prompt" explosion mechanism may fail, owing to the loss of shock energy to photodisintegration of nuclei outside the inner core. If so, the shock may be revived by a "delayed" mechanism, in which a fraction of the neutrinos diffusing out of the core on a time scale of seconds is captured and deposits its energy in the surrounding matter. In numerical simulations, the outcome of the prompt mechanism is found to be sensitive to the uncertain properties of matter at and beyond nuclear densities, and to the treatment of convection. Considering the uncertainties, it is not yet clear whether Sk -69 202 exploded by means of the prompt or the delayed mechanism.

The fundamental property of the ejected matter—the composition as a function of velocity—is not expected to depend strongly on whether the explosion was prompt or delayed. In either case, matter is heated and accelerated as the shock wave sweeps through. The inner 1.6$M_\odot$ of the ejected matter is briefly hot and dense enough to undergo nuclear fusion reactions, but the rest is simply ejected with the composition of the presupernova core. The predicted composition of the ejected matter is shown in Fig. 3. The deepest layers consist primarily of the radioactive isotope ^{56}Ni, which plays a critical role in pro-

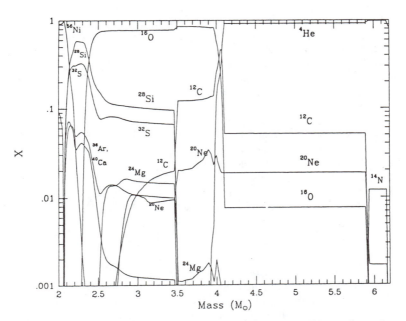

FIG. 3.   Predicted composition of the inner 6$M_\odot$ of ejected matter. Observation established that the mass of ^{56}Ni was 0.07$M_\odot$ rather than 0.4$M_\odot$ predicted here. [Reproduced by permission from Woosley, S. E. (1988). *Astrophys. J.*, **330**, 218.]

ducing the supernova's light curve (Section IV). The amount of ejected ^{56}Ni is difficult to predict, so this quantity must be determined by observation. The shock wave accelerates as it propagates outward through layers of decreasing density, so the initial velocity distribution increases monotonically outward. Each element of matter then coasts at whatever velocity was provided to it by the shock; after a few days, the velocity distribution approaches the simple special case in which, at a fixed time, velocity is proportional to distance from the center of the explosion. A kinetic energy of 10^{51} ergs carried by an ejected mass on the order of $10 M_\odot$ would correspond to a characteristic ejection velocity of 3000 km sec^{-1}; the heavy-element core moves more slowly, and the low-density outer layers attain much higher velocities.

IV. Photon Emission

The emission of electromagnetic radiation from a supernova during its opaque "photospheric" phase is discussed in some detail in the article on supernovae in the Encyclopedia. To summarize briefly, the photosphere emits a thermal continuous spectrum, which to a first approximation is an ultraviolet-deficient blackbody spectrum that reddens with time as the temperature at the photosphere decreases owing to expansion. The luminosity in the continuum is determined by the photosphere's temperature and radius. The radius initially increases as the photosphere is carried outward with the matter, but the radius eventually decreases as expansion causes the photosphere to recede into the deeper layers. Spectral lines, formed above the photosphere primarily by photon scattering, are superimposed on the continuum. The lines have a characteristic "P Cygni" shape, with an emission component at the rest wavelength of the transition (in the frame of the supernova) and an absorption component Doppler shifted to shorter wavelengths by an amount that corresponds to the velocity at the photosphere. For convenience, the discussion here will be divided into two parts, the first dealing with the time dependence of the luminosity, the second with the spectrum.

A. The Light Curve

At the time of Shelton's discovery, February 24.23 (UT), the supernova had brightened to the fifth apparent magnitude. The supernova was soon found to have appeared on even earlier routine photographs of the LMC; the earliest established that the supernova had already reached the sixth apparent visual magni-

tude by February 23.44 (UT), only 3 hr after the core collapse as timed by the neutrino burst. The short interval between the collapse and the initiation of the photon display by the arrival of the shock wave at the surface of the star is consistent with the radius of Sk -69 202 inferred from its luminosity and temperature. The shock wave could have reached the surface of a star of $40 R_\odot$ in less than 3 hr, provided that the mean shock velocity was greater than 2700 km sec^{-1}. Considering that the star needed to have time to increase its surface area to reach the sixth magnitude, the mean shock velocity probably must have been several times greater.

The light curve of SN 1987A is shown in Fig. 4. After reaching a weak local maximum just a few days after the explosion and then fading to a local minimum in early March, the supernova slowly brightened to a peak apparent blue magnitude of 4.5 (and a peak visual magnitude of 2.9) by May 18, 84 days after the explosion. The peak was followed by a decline of one magnitude during the next month, and finally by a linear decay of 0.01 magnitudes per day. The color indices of the supernova, corrected for interstellar reddening, corresponded to those of an ultraviolet-deficient blackbody that cooled quickly from 14,000 K on February 24 to 6000 K only a week later, and then slowly fell below 5000 K in the subsequent months.

As Fig. 4 shows, the light curve was unlike the typical light curves of type II supernovae. To understand the reasons for the difference, it is necessary to review the physical explanation of the typical type II light curve (the discussion here refers to the "plateau" rather than to the less common "linear" type II light curve; the latter bears even less resemblance

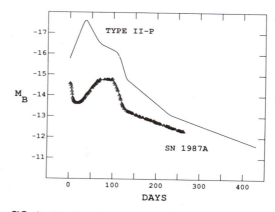

FIG. 4.   The blue light curve of SN 1987A is compared to the mean light curve of "plateau" type II supernovae.

to SN 1987A). The typical type II light curve has two principal components—an early plateau phase (from the period of nearly constant brightness that lasts nearly 90 days), followed by a radioactivity phase. During the plateau phase, the supernova radiates by releasing the internal energy that was deposited in the ejected matter by the shock wave. The brightness of the plateau phase depends strongly on the radius of the progenitor star; because of the nearly adiabatic cooling that accompanies expansion, a small progenitor radius fails to lead to a large, hot photosphere. The bright plateau phases of the previous type II supernovae could only have been produced by extended, red supergiant progenitor stars. After cooling terminates the plateau phase months after the explosion, the radioactivity phase begins. The decay energy of ^{56}Ni (6.1-day half-life) and its radioactive daughter ^{56}Co (77 days), initially carried by positrons and gamma rays, is trapped and thermalized, keeping the matter warm enough to radiate optically.

In SN 1987A, the relatively dim plateau phase produced by the $40R_\odot$ progenitor, combined with the ordinary type II radioactivity phase powered by a more or less normal production of ^{56}Ni, resulted in an unusual, subluminous light curve. The linear decay during the radioactivity phase, which corresponds to an exponential decay of luminosity, had not been observationally well established for previous type II supernovae. The well-established linear decay of SN 1987A corresponds closely to the 77-day half-life of ^{56}Co, thus settling the question of whether radioactivity plays an important role in powering the light curve. The absolute brightness of SN 1987A during the radioactivity phase corresponds to an initial production of $0.07M_\odot$ of ^{56}Ni in the explosion.

Supernova 1909A, which appeared in the outskirts of the spiral galaxy Messier 101 where the metallicity is about as low as in the LMC, had a light curve that resembled that of SN 1987A (Fig. 5). Supernova 1909A was brighter in the absolute sense, implying that it ejected more ^{56}Ni than SN 1987A, presumably because its progenitor was more massive than Sk -69 202.

B. SPECTRA

Analysis of spectra provides information on the velocity and composition of the ejected matter. Shortly after the explosion, the spectrum forms in the high-velocity outermost layers of the ejected matter; later, as the outer layers thin out and the photosphere recedes, the deeper, slower layers are revealed. An early optical spectrum is shown in Fig. 6. The spec-

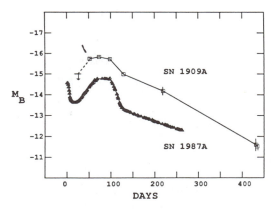

FIG. 5. Comparison of the light curves of SN 1909A in Messier 101 and SN 1987A. [Reproduced by permission from Young, T. R. and Branch, D. (1988). *Nature (London)* **333**, 305.]

tral features have the characteristic P Cygni shape, and can be identified with lines of hydrogen and neutral helium forming in matter moving with velocities between 20,000 and 30,000 km sec^{-1}. Ultraviolet spectra obtained at the same phase show evidence for some matter moving as fast as 40,000 km sec^{-1}, 13% of the velocity of light. By the time of the later spectrum shown in Fig. 7, the underlying continuous spectrum had reddened, and numerous deep, narrow, heavy-element absorption lines had developed. The lines had sharpened because of the reduced Doppler broadening and shifting in the deeper, slower layers

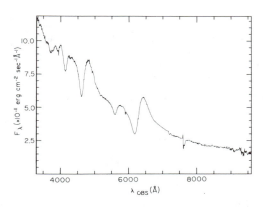

FIG. 6. An optical spectrum of SN 1987A obtained less than two days after the explosion. Spectral lines are produced by hydrogen and neutral helium; for example, the emission near 6500 Å and the associated absorption near 6000 Å are identified with the Balmer-alpha transition in hydrogen. [Reproduced by permission from Phillips, M. M., Heathcote, S. R., Hamuy, M., and Navarrete, M. (1988). *Astron. J.* **95**, 1087.]

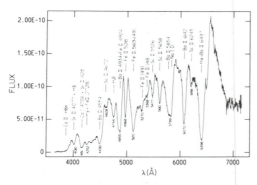

FIG. 7. An optical spectrum of SN 1987A obtained a month after the explosion. [Reproduced with permission from Williams, R. E. (1987). *Astrophys. J. (Lett.)* **320**, L117.]

(3000–5000 km sec^{-1}) in which the spectrum formed. The heavy-element lines had strengthened primarily because of the decreased temperature (also reflected in the reddened continuum) rather than because of a change in composition of the spectrum-forming layers. The deepest layers, enriched in heavy elements, still were hidden from view beneath the photosphere.

An expanding supernova eventually becomes transparent to continuum photons, loses its photosphere, and becomes more like a nebula. The optical, infrared, and ultraviolet spectra then consist of relatively narrow emission lines formed in the deep dense layers. In November 1987, infrared spectra obtained from NASA's Kuiper Airborne Infrared Observatory showed nebular emission lines from singly ionized nickel, cobalt, and argon. Quantitative analysis of the nebular spectra will provide, for the first time, empirical information on the composition of the heavy-element-enriched inner layers of a type II supernova.

Detailed numerical simulations of the hydrodynamics of the explosion give a good fit to the observed light curve and expansion velocities if the ejected mass is taken to be about $13 M_\odot$, and the kinetic energy to be 1.0×10^{51} erg. A summary of the basic properties of SN 1987A is given in Table II.

V. Circumstellar Interaction

In addition to photons received directly from the supernova, others are received from the circumstellar matter that surrounds the supernova as a consequence of stellar winds from the progenitor star. The circumstellar matter is stimulated to radiate by means of direct *hydrodynamical interactions* with the supernova matter, and by means of *light interactions* with the supernova photons; because the light travel time usually plays a role in the light interactions, these often are referred to as *light echoes*.

In the idealized case of a spherically symmetric stellar wind at constant velocity, the density in a circumstellar shell falls off as the inverse square of the distance from the star, by conservation of matter. The density at any particular distance is proportional to the wind mass-loss rate and inversely proportional to the wind velocity. The circumstellar medium surrounding SN 1987A is expected to have a multicomponent structure because of the previous history of the progenitor star. The immediate surroundings should be a medium of low density resulting from a moderate-velocity wind of the blue supergiant phase (~ 500 km sec^{-1}—the wind velocity ordinarily is on the order of the stellar surface escape velocity). If Sk -69 202 was once a red supergiant (Section II), there should be an outer dense shell resulting from a low-velocity (~ 10 km sec^{-1}) wind; the inner edge of the dense shell may have been further compressed by an interaction with the faster blue supergiant wind. Out-

TABLE II.   Properties of SN 1987A

Time of core collapse	February 23.316 (Universal time), 1987
Energy radiated in neutrinos	$3 (\pm 1) \times 10^{53}$ erg
Kinetic energy	$1.0 (\pm 0.5) \times 10^{51}$ erg
Energy radiated in photons	$1.0 (\pm 0.3) \times 10^{49}$ erg
Date of peak luminosity	May 18 (± 3 days)
Peak apparent magnitude (m_V)	2.9 (± 0.1)
Peak absolute magnitude (M_V)	$-15.5 (\pm 0.4)$
Peak absolute magnitude (M_B)	$-14.5 (\pm 0.4)$
Peak luminosity	$1.2 (\pm 0.3) \times 10^{42}$ erg sec^{-1}
Ejected mass	$13 (\pm 4) M_\odot$
Ejected mass of ^{56}Ni	$0.07 (\pm 0.01) M_\odot$

side the dense shell is a low-density "bubble" in the interstellar medium, blown by the high-velocity (~ 1000 km sec^{-1}) wind of Sk -69 202 when it was in its main sequence phase.

The first sign of a circumstellar interaction from SN 1987A was a burst of synchrotron radio photons during the first weeks after the explosion. The radio emission, which amounted to less than a millionth of the total photon output of the supernova, evidently was generated by a hydrodynamical interaction with the expected low-density circumstellar shell immediately surrounding the supernova. The observed properties of the radio emission constrain the density in the shell: assuming that the wind velocity from the blue supergiant phase of Sk -69 202 had been typical of B3 Ia stars, the mass-loss rate is inferred to have been $6 \times 10^{-6} M_\odot$ yr^{-1}.

In July 1987, the ultraviolet spectrum of SN 1987A developed high-excitation emission lines that were too narrow to have originated in the ejected matter. They are likely to have come from a light interaction with the expected dense circumstellar shell. An initial burst of ultraviolet photons from the early hot supernova photosphere ionized the dense shell, and emission from the shell developed as the ions recombined. The inner shell radius appears to be on the order of light years, which implies that the red supergiant phase must have ended thousands of years before the explosion. The relative strengths of the narrow ultraviolet lines suggest that helium and nitrogen are overabundant compared to other elements. This is consistent with a prior red supergiant phase in which the products of hydrogen burning were mixed to the stellar surface by convection and then expelled by a wind. Convective mixing to the surface is less likely to occur in a blue supergiant.

During March and April 1987, the technique of speckle interferometry revealed the temporary presence of an unexpected "mystery spot," 10% as bright as the supernova itself, at a separation of only 0.06 arcsec. A possible interpretation of the spot, but not without quantitative difficulties, is that it was a discrete circumstellar cloud that brightened in response to a hydrodynamical interaction. In any event, the spot served as a reminder that the circumstellar medium may be complex, rather than spherically symmetric and smoothly stratified as usually assumed for simplicity.

Further manifestations of the dense shell are expected to appear in the future. When the leading edge of the ejected matter reaches the shell, on a time scale of decades, the ensuing hydrodynamical interaction may produce strong radio and X-ray emission.

VI. The Remains

During the first months after the explosion, the decay products from the radioactivity were trapped and thermalized inside the ejected matter, but as expansion proceeded, an increasing fraction of the gamma rays were expected to escape directly. In August 1987, emission in the 0.847 and 1.238 MeV lines of ^{56}Co was detected by NASA's orbiting Solar Maximum Mission and by the Soviet space station Mir, at a level on the order of 10^{-3} photons cm^{-2} sec^{-1}. No previous supernova had been detected in gamma rays. The unambiguous detection of the ^{56}Co gamma-ray lines provided the direct, unambiguous confirmation that nucleosynthesis occurs in supernovae. At about the same time, Mir and Japan's Ginga satellite detected X-rays that are believed to have been produced when some of the gamma rays lost energy by Compton scattering in the outer layers of the supernova matter. The gamma- and X-ray emission, although qualitatively consistent with predictions based on hydrodynamical models of SN 1987A, appeared several months earlier than expected. The most likely explanation is that partial radial mixing of matter during the explosion brought some of the ^{56}Co toward the surface, facilitating the early escape of the decay products. Observation and analysis of the gamma- and X-ray data, especially high-spectral-resolution observations of the ^{56}Co emission lines obtained by sensitive balloon-borne spectrometers, will provide tight constraints on the extent of the radial mixing.

The nature of the neutrino signal from SN 1987A implies that the core collapse created a neutron star rather than a black hole. Direct detection of radiation from the hot surface of the neutron star may prove difficult; the best prospect is for thermal X-rays, which might be possible with instruments of the future, such as the Advanced X-Ray Astronomical Facility (AXAF). If the neutron star acts as a pulsar that beams in the direction of the earth, the detection of pulsed radiation at radio, X-ray, or even gamma-ray wavelengths is a possibility. A pulsar is expected to be surrounded by a nonthermal nebulosity that might be detectable at some wavelengths, but separating the spatially unresolved nebulosity from the ejected and circumstellar matter may be difficult. The time at which *direct* detection of the neutron star, pulsar, or associated nebulosity becomes possible depends on whether the ejected matter breaks up into clumps or filaments; smoothly distributed ejected matter tends to veil the events at the center of the supernova from the view of the outside observer. An *indirect* manifestation of a pulsar could be a decrease in the rate of the luminosity decline of SN 1987A, owing to absorption and reemission of pulsar radiation by the ejected matter.

On a longer time scale, the ejected matter together with the accumulated circumstellar matter will sweep up and compress interstellar matter. After a time, on the order of 100 yr, when the mass of the swept interstellar matter becomes comparable with the ejected and circumstellar mass, deceleration will become significant, some of the expansion energy will be converted to radiation, and a *supernova remnant* will form. Just as SN 1987A is providing an extraordinary opportunity to peer into the depths of a stellar explosion, so it will provide an unprecedented opportunity to observe the transition from a supernova to a supernova remnant. Even with present technology, supernova remnants in the LMC at ages of 10,000 yr or more are observed at radio, X-ray, and optical wavelengths. Thus, the aftermath of SN 1987A may be followed observationally for millenia, as long as there are astronomers on the earth.

BIBLIOGRAPHY

Bethe, H. A., and Brown, G. (1985). How a supernova explodes, *Sci. Am.* **252**, 60.
Woosley, S. E., and Weaver, T. A. (1986). The physics of supernova explosions, *Annu. Rev. Astron. Astrophys.* **24**, 205.

SUPERNOVAE

David Branch   *University of Oklahoma*

GLOSSARY

Absolute magnitude: Apparent magnitude a star would have if it were at a distance of 10 pc. The absolute blue magnitude of the sun is $M_B = +5.4$, the brightest stable stars are -10, and supernovae at their peaks are in the range -16 to -20.

Apparent magnitude: Logarithmic measure of the apparent brightness of a star, usually measured through a blue or "visual" filter. A difference of five magnitudes corresponds to a factor of 100 in observed flux; one magnitude corresponds to a factor of 2.512. The apparent blue magnitude of the sun is $m_v = -26.2$, the brightest star, Sirius, is -1.5, the ground-based limit is $+24$, and the Hubble Space Telescope reaches $+28$.

Electron degeneracy: Degeneracy that occurs in high-density matter in which the Pauli exclusion principle prevents free electrons from assuming the Maxwellian distribution of momentum appropriate to the matter temperature. The distribution extends instead to higher values of momentum, and the pressure exerted by the electrons is higher than it would be in the absence of degeneracy.

Light curve: Graph of apparent or absolute magnitude plotted against time.

Luminosity: Rate at which electromagnetic energy is radiated from the surface of a star, at all wavelengths and in all directions.

Neutron star: Star whose internal pressure is provided by degenerate neutrons. The radius of a typical neutron star is on the order of 10 km and the density is comparable to the density of nuclear matter, 10^{14} g/cm^3. The maximum mass is thought to be less than two times the mass of the sun. A neutron star is formed as the final state of a star of high mass.

Parsec: Distance at which a star would show an annual trigonometric parallax of 1 arc-second. One parsec equals 206,625 astronomical units (AU) or 3.09×10^{18} cm. The distances to the nearest stars are conveniently expressed in parsecs, distances across our galaxy are expressed in kiloparsecs, and distances to external galaxies are expressed in megaparsecs.

Planetary nebula: Extended, expanding shell of gas surrounding a compact, hot central star. At the end of its red giant phase a star of low or intermediate mass expels its outer layers at a velocity on the order of 10 km/sec. The expelled mass absorbs ultraviolet radiation from the hot, exposed core and emits an emission-line optical spectrum. The nebula ultimately merges with the interstellar medium and the core cools to become a white dwarf.

White dwarf: Star whose internal pressure is provided by degenerate electrons. The radius of a typical white dwarf is comparable to that of the earth, and the density is on the order of 10^7 g/cm^3. The maximum mass is 1.4 times the mass of the sun. A white dwarf is formed as the final state of a star of low or intermediate mass.

A supernova is a bright, catastrophic explosion of a star. The luminosity of a supernova at its brightest approaches 10^{10} times that of the sun. The explosion throws matter into space at a few percent of the speed of light, carrying a kinetic energy of 10^{51} ergs, the energy equivalent of 10^{28} megatons of TNT. The ejected gas contains nuclei of heavy elements, some created by nuclear fusion reactions during the slow, pre-explosive phases of the star and some created during the explosion itself. Thus, supernovae drive the process of cosmic chemical evolution.

Some supernovae are thought to be caused by thermonuclear explosions in white dwarf stars; others are initiated by the gravitational collapse of the cores of massive stars. Some supernovae may explode completely, but others eject only their outer parts and leave behind compact stellar remnants—white dwarfs, neutron stars, or black holes. The ejected matter sweeps up, compresses, and heats the ambient interstellar gas to form extended supernova remnants. [See STELLAR STRUCTURE AND EVOLUTION.]

Because they are so bright, supernovae can be detected out to distances comparable to the radius of the observable universe, and therefore they can be used to estimate two of the fundamental cosmologic parameters—the universal expansion rate (the Hubble constant) and the rate of its deceleration.

I. Optical Observations

A. DISCOVERY

Seven of the temporary naked-eye stars that have suddenly appeared in the sky during the past 2000 years are now recognized to have been supernova explosions in our galaxy (Table I). Some were bright enough to be noticed in daylight and remained above the nighttime naked-eye limit of the sixth apparent visual magnitude for several years. Radio, X-ray, and optical emission from the extended remnants of these historical galactic supernovae is now detected;

TABLE I. Historical Supernovae in the Galaxy

Year	Constellation	m_v^a	Remnant
185	Centaurus	-8	RCW 86
393	Scorpio	-1	CTB 37
1006	Lupus	-9	PKS 1459-41
1054	Taurus	-4	Crab Nebula
1181	Cassiopeia	0	3C 58
1572	Cassiopeia	-4	Tycho
1604	Ophiuchus	-3	Kepler

^{a} m_v stands for "apparent visual magnitude."

the Crab Nebula in Taurus and the remnants of Tycho Brahe's supernova in Cassiopeia and Johannes Kepler's supernova in Ophiuchus (Fig. 1) are especially conspicuous and well known. More than 100 other, older supernova remnants, which occurred within the past 100,000 years but for which no records of the original explosions exist, are recognized in our galaxy.

No galactic supernova has been seen since the invention of the telescope in the early seventeenth century. The galactic supernova remnant known as Cassiopeia A, the brightest radio source in the sky beyond the solar system, evidently was produced by a supernova that occurred near the year 1680, but the supernova was subluminous and either was not noticed or was dismissed as an ordinary variable star. Until another outburst in our galaxy is seen, studies of the explosive phases of supernovae must be based on observations of much more distant, fainter events in other galaxies.

The study of extragalactic supernovae began in 1885 when a star of the sixth apparent visual magnitude appeared temporarily only 15 arcseconds from the nucleus of the Andromeda Nebula, now known to be the nearest large external galaxy. In 1934, after more than a dozen additional bright events had been noticed in some of the relatively nearby galaxies, Walter Baade and Fritz Zwicky at the California Institute of Technology named and defined supernovae as stellar outbursts that are more luminous than ordinary novae. Novae are known to be caused by less catastrophic thermonuclear explosions near the surfaces of white dwarf stars. Roughly 50 novae per year occur in our galaxy, although absorption of their light by interstellar dust grains in the line of sight prevents us from seeing most of them. A nova explosion ejects only a very small fraction of the star's mass, from 10^{-5} to 10^{-4}, at a velocity on the order of 1000 km/sec. The kinetic energy of ejected matter is only on the order of 10^{44} ergs, and the nova attains a peak luminosity of no more than 10^6 times that of the sun.

After Zwicky began to search systematically for supernovae in 1936, the discovery rate soon increased. The search eventually grew to involve the powerful 48-in. Schmidt telescope on Palomar Mountain and cooperation with observatories in many other countries, notably Italy and Switzerland. In a typical year one or two dozen supernovae have been discovered (Fig. 2), and a convention for designating each event has had to be introduced: SN 1954a, for exam-

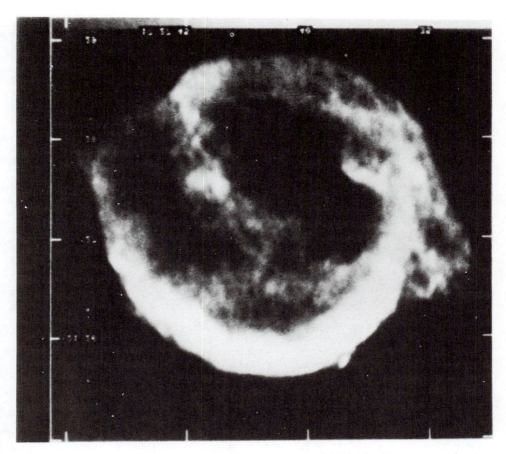

FIG. 1. Radio image of the remnant of Kepler's supernova of 1604. The image was made at a wavelength of 20 cm by using the Very Large Array radiotelescope of the National Radio Astronomy Observatory. [Reproduced with permission from Matsui, Y., Long, K. S., Dickel, J. R., and Greisen, E. W. (1984). *Astrophy. J.* **287**, 295.]

ple, refers to the first supernova to be discovered in 1954, and SN 1979c was the third supernova of 1979. Some extragalactic supernovae of special interest, either because they were very bright or especially well observed, are listed in Table II. More than 600 supernovae had been found by the end of 1985. Almost all were discovered by visual comparison of optical photographs obtained at two different times. Astronomers expect that the discovery rate will increase by a factor of 10 or more when automated supernova searches with telescopes dedicated to the purpose become fully operational in the late twentieth century.

B. TYPES

Supernovae are separated into two major types according to the appearance of their opti-

TABLE II.   Extragalactic Supernovae

Supernova	Galaxy	m_{pg}^a	Type
1885a	Andromeda	7	Ipec
1895b	NGC 5253	8	I
1937c	IC 4182	8	Ia
1940b	NGC 4725	13	II-P
1954a	NGC 4214	10	Ipec
1961v	NGC 1058	12	IIpec
19681	M83	12	II-P
19691	NGC 1058	13	II-P
1970g	M101	12	II-P
1972e	NGC 5253	8	Ia
1979c	M100	12	II-L
1980k	NGC 6946	11	II-L
1981b	NGC 4536	12	Ia
1983n	M83	12	Ib

[a] m_{pg} stands for "apparent photographic magnitude."

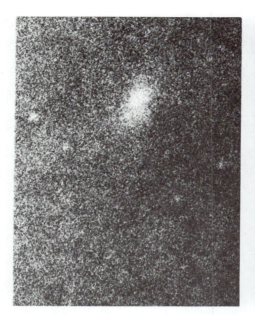 

FIG. 2.   The photograph on the left, taken by Charles Kowal in 1959 with the 48-in. Palomar Schmidt telescope, shows the peculiar galaxy NGC 5253 (as well as several faint foreground stars belonging to our galaxy). The photograph on the right shows the same field (exposed slightly deeper), but includes the eighth-magnitude type I supernova 1972e, discovered by Kowal on May 16, 1972. North is up, east is to the right. Supernova 1972e is one of the two brightest supernovae to have been seen in the twentieth century.

cal spectra. Type I supernovae, so named because the first well-observed events of the 1930s were of this kind, show broad spectral features having characteristic widths corresponding to Doppler broadening at a velocity of 10,000 km/sec, 3% of the velocity of light. The spectra of ordinary type I supernovae evolve with time according to a well-defined pattern (Fig. 3), so that an experienced spectroscopist can estimate the time elapsed since maximum light from the appearance of a spectrum. Until about 1970, type I spectra were said to consist of broad, overlapping, unidentified emission lines. [See STELLAR SPECTROSCOPY.]

The other major kind of supernova, type II, was recognized in 1940. The optical spectra of type II supernovae are nearly continuous at the time of maximum light, but within weeks they develop features that are almost as broad as those of type I. Type II spectra also evolve according to a standard pattern, but they show more individuality than type I. A few of the most conspicuous features in type II spectra easily were identified from the beginning with the Balmer lines of hydrogen, but the other features remained unidentified for 30 years.

Supernovae of both major types can be subclassified according to certain differences in their observed properties. Type I are divided into Ia (ordinary type I) and Ib (less common) on the basis of differences in their optical spectra (Section III,D), and Type II are separated into II-P (plateau) and II-L (linear) according to the shapes of their light curves (Section I,C). A few supernovae that clearly were related to one or the other of the two main types but showed some unique properties are referred to as type Ipec and IIpec, for *peculiar*.

Of the 600 supernovae discovered by 1985, about 150 were classified as type I and 75 as type II; most of the rest were not observed spectroscopically and could not be classified. A few supernovae have at times been assigned to additional types. In 1965 Zwicky defined types III, IV, and V, each based primarily on one supernova. The principal examples of Zwicky's types III (SN 1961i in NGC 4303) and IV (SN 1961f in NGC 3003) were not very well observed; both had hydrogen lines in their spectra and are now regarded to have been more or less peculiar examples of type II. Zwicky's example of a type V (SN 1961v in NGC 1058) also had hydrogen lines

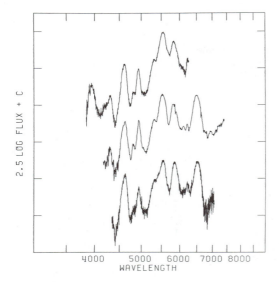

FIG. 3. Optical spectra of the type I supernova 1981b in NGC 4536, 35 days (top spectrum) and 49 days (bottom spectrum) after maximum light, compared to the spectrum of the type I supernova 1983u in NGC 3227, 40 days (middle spectrum) after maximum light. Vertical displacements are arbitrary. The spectra of SN 1981b were obtained by Marshall McCall, Alan Uomoto, and Beverley Wills, who used the 107-in. telescope of the McDonald Observatory at the University of Texas; the spectrum of SN 1983u was obtained by Michael de Robertis and Phillip Pinto, who used the 120-in. telescope of the Lick Observatory at the University of California. Wavelength is in angstroms and flux is in ergs/cm²/sec/angstrom.

in its spectrum and can be classified on that basis as a type II. However, SN 1961v was exceptional in several respects and is suspected to have been the explosion of a very massive star (Section IV,D).

C. LIGHT CURVES

Supernovae of both types reach maximum optical brightness 2 to 3 weeks after their explosions. At their peaks, type I supernovae are brighter than type II by about one magnitude. The peak *absolute* magnitude depends on the extragalactic distance scale, which is known only to a factor of 2. On a "long-distance" scale, corresponding to a Hubble constant of 50 km/sec/Mpc (Section VII,A), type I supernovae reach an absolute blue magnitude of -20.0 and have a peak luminosity of 2×10^{43} ergs/sec, 5×10^9 times the luminosity of the sun. On a "short scale" corresponding to 100 km/sec/Mpc, the

type I absolute magnitude is -18.5 and the luminosity is lower by a factor of 4. Observations made across the electromagnetic spectrum show that type I supernovae emit almost all of their energy in the optical part of the spectrum while type II supernovae emit a significant fraction at ultraviolet wavelengths. Consequently, type II supernovae, although fainter optically, have nearly the same peak luminosity as type I. The total time-integrated emission of electromagnetic energy of both types is on the order of 10^{49} to 10^{50} ergs, much less than the kinetic energy of 10^{51} ergs. Thus, the radiative efficiency of supernova explosions is low.

The average light-curve shapes of supernovae are compared in Fig. 4. The mean type I curve, which is well defined by the observations, consists of an initial rise and fall (the *peak*) lasting until 30 days after maximum light, and a subsequent, slowly fading tail. The tail appears to be nearly linear when magnitudes are plotted against time, but magnitude is a logarithmic measure of brightness and the tail actually corresponds to an exponential decay of brightness with time. The decay rate of the type I tail corresponds to a half-life of 50 days.

Type II supernovae are subdivided into II-P and II-L on the basis of light-curve shape. Two-thirds of observed type II supernovae are II-P, which interrupt the initial declines from their peaks to enter a plateau phase of nearly constant brightness until 80 days after maximum light. A type II-L shows nearly a linear decline from its peak for 80 days after maximum light. The avail-

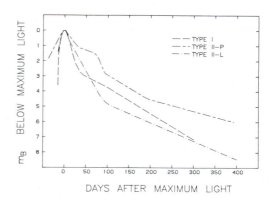

FIG. 4. Comparison of mean blue light curves for type I, type II-L (linear), and type II-P (plateau) supernovae. The curves are normalized at maximum light, and so differences in peak brightness are not shown. [Reproduced with permission from Doggett, J. B., and Branch, D. (1985). *Astron. J.* **90**, 2303.]

able data on the later phases of type II light curves are sparse and the mean curves are not well defined, but both II-P and II-L do appear to have slowly fading, linear tails, with a decay rate corresponding to a half-life of 100 days.

D. Sites, Rates, and Stellar Populations

Galaxies are classified morphologically as spirals, ellipticals, and irregulars. Type II supernovae have been seen only in spirals, almost always in the spiral arms. Type I supernovae are found in all kinds of galaxies, and those in spirals have little or no tendency to be in the arms.

To estimate the relative rates of supernova production in the different kinds of galaxies, some observational selection effects must be taken into account. First, type I supernovae are brighter than type II and are more easily discovered. Second, because most of the light from a spiral galaxy comes from a disk of stellar and interstellar matter that includes the spiral arms, supernovae are more easily discovered in spirals whose disks are oriented face-on to our line of sight than in those whose disks must be observed edge-on. When these and other selection effects are roughly allowed for, elliptical galaxies are found to produce supernovae at a lower rate than spirals by a factor of about 3 per unit galaxian luminosity. In spirals, both the type I and type II rates are correlated with galaxy color. The bluer the galaxy—and by inference the higher the recent rate of star formation—the higher the supernova rate. For a given galaxy type, the supernova rate is roughly proportional to galaxy luminosity.

Absolute supernova rates are more uncertain than relative rates because the relevant selection effects are still more difficult to estimate. The mean interval between supernova explosions in a fairly large spiral galaxy such as our own appears to be 20 yr, to within a factor of 2. A similar rate is derived for our galaxy from the number of historical supernovae that have been seen. The latter estimate involves a large correction for supernovae in the galaxy that are presumed to have been missed owing to our location in the dusty, obscuring disk of the Milky Way. The fact that all of the historical galactic supernovae are within a few kiloparsecs of the sun, while the radius of the galaxy is 15 kpc, confirms that only the nearest galactic supernovae of the past 2000 years have been noticed by observers on earth.

The observation that type II supernovae appear in the arms of spiral galaxies leads unambiguously to the conclusion that type II supernovae are produced by relatively massive stars, more massive than about 8 times the mass of the sun ($8 M_\odot$). The more massive a star, the shorter its nuclear evolutionary lifetime. Only massive stars have lifetimes so short, less than 10^7 yr, that they meet their fates before drifting out of the arms in which they were born.

The observation that type I supernovae occur in elliptical galaxies, but also in spiral galaxies in proportion to their recent rates of star formation, appears at first to be paradoxical. Elliptical galaxies seem to have stopped forming new stars 10^{10} yr ago and now contain only low-mass stars of $1 M_\odot$ or less. On the other hand, the correlation between the type I rate and the recent star-formation rate in spirals implies that most type I supernovae *in spirals* are produced by stars that are fairly short-lived, therefore moderately young and massive. Yet no systematic difference between the properties of type I supernovae in ellipticals and spirals has been established. A possibility that has not been entirely excluded, but is unlikely, is that type I supernovae are produced by explosions of intermediate-mass single (nonbinary) stars in the range $5 M_\odot$ to $8 M_\odot$. This would require that a low rate of continued star formation in at least some elliptical galaxies produces $5 M_\odot$ to $8 M_\odot$ stars in sufficient numbers to account for the type I rate while *not* producing stars more massive than $8 M_\odot$; the latter would lead to supernovae of type II, which are not observed in ellipticals.

A more promising explanation for the type I supernova statistics is that a type I is produced by a white dwarf star in a close binary system; the white dwarf accretes matter from its companion star and eventually is provoked to explode. This model can account for the correlation of the type I rate with the recent star-formation rate in spirals, for the lower rate in ellipticals, and for the similarity of the type I supernovae in the two kinds of galaxies (Section III,C).

II. The Interpretation of Optical Spectra

The spectrum of a supernova carries information on the temperature, velocity, and chemical composition of the ejected matter. However, because the spectral features are so broad and overlapping, the interpretation of supernova spectra is difficult. Information began to be extracted from the spectra only in the early 1970s, and the subject still is in an early stage of development. A supernova spectrum consists of broad emission and absorption lines formed in the outer layers of the ejected matter, superimposed on a thermal continuous spectrum radiated by the deeper, opaque layers. This description applies as long as the ejected matter remains optically thick; that is, it applies to type II spectra emitted within a year of maximum light and to type I supernovae within at least the first months.

A. THE CONTINUOUS SPECTRUM

The continuous spectrum emitted by an ordinary, static star escapes from a thin spherical layer near the star's surface called the photosphere. The photosphere is at the depth at which the star's atmosphere becomes opaque to the line of sight of an external observer. The shape of the continuous spectrum with respect to wavelength is determined primarily by the temperature at the photosphere. The absolute brightness depends also on the radius of the photosphere.

In a supernova the temperature and radius of the photosphere change with time. When a star explodes, its matter is heated and thrown into rapid expansion. The luminosity abruptly increases by orders of magnitude in response to the high temperature at the photosphere, but most of the energy is radiated at X-ray and ultraviolet wavelengths rather than as optical light. As the supernova expands, it cools. For two to three weeks the optical light curve rises as an increasing fraction of the radiation from the expanding, cooling photosphere goes into the optical band. At maximum optical light the temperature at the photosphere has its optimum value for optical emission, on the order of 10,000 K. To be consistent with the absolute brightness and temperature, the radius of the photosphere at maximum light must be 10^{15} cm, that is, 70 AU, almost twice the radius of Pluto's orbit around the sun. The matter density at the photosphere is low, on the order of 10^{-16} g/cm^3.

During the first months after maximum light, further cooling causes the light curve to decline. For a month the radius of the photosphere continues to grow, but owing to the expansion, the photosphere lags behind the expanding matter, and the external observer's line of sight penetrates deeper into the matter distribution. Subse-

quently, the radius of the photosphere begins to decrease, and finally, at a time depending on the total amount of matter ejected, all of the matter becomes optically thin, the photosphere ceases to exist, and the photospheric description of spectrum formation no longer applies.

Optical observations of the shape of the continuous spectra (i.e., the colors) of supernovae of both types show that the temperature at the photosphere falls from above 10,000 K at maximum light to below 6000 K a month later and changes only slowly thereafter.

B. THE SPECTRAL LINES

A schematic model of a supernova that is useful for understanding the formation of its spectral features is shown in Fig. 5a. The cross-hatched region is the photosphere, and the region outside the photosphere is the optically thin atmosphere. The entire supernova is in differential expansion; to a good approximation the expansion velocity is proportional to distance from the center. This simple velocity law is a natural consequence of matter being ejected with a range of velocities and then coasting, without acceleration; each element of matter attains a distance from the center in proportion to its velocity. In this respect the expansion of a supernova is like the expansion of the universe.

As photons in the continuous spectrum travel outward from the photosphere through the ex-

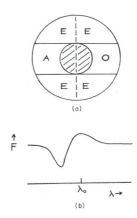

FIG. 5.   (a) Schematic drawing of the atmosphere of a supernova surrounding the photosphere. An observer to the left sees an emission line formed in region E and an absorption line formed in region A; region O is occulted. (b) The characteristic shape, called a P Cygni profile, of a spectral line formed in a supernova or a star having an expanding atmosphere.

panding atmosphere, they undergo a Doppler shift with respect to the matter they are passing through, just as photons do as they move through the expanding universe. Thus, photons can undergo redshift into resonance with electronic transitions in atoms or ions in the atmosphere and be absorbed, their energy going into electronic excitation. In the useful approximation of pure scattering, the absorbed photon is immediately re-emitted, but in a random direction. From the point of view of an external observer, photons scattered by moving atoms may undergo Doppler shift to longer or shorter wavelengths than the rest wavelength of the electronic transition. Photons scattered into the observer's line of sight from the right of the vertical line in Fig. 5 come from atoms having a component of motion away from the observer and are seen to have undergone redshift, while those from the left have undergone blueshift. Therefore, the region labeled E, which wraps around the photosphere in three dimensions, produces a broad, symmetrical emission line superimposed on the continuous spectrum and centered on the rest wavelength of the transition. Region O is occulted by the photosphere from the observer's point of view. In region A, photons in the continuous spectrum originally directed toward the observer can be scattered *out* of the line of sight. Because it has a component of motion toward the observer, matter in region A produces a broad, asymmetrical, blueshifted absorption component to the spectral line profile.

The shape of the profile is in Fig. 5b. This kind of line profile is characteristic of expanding atmospheres and is referred to by astronomers as a P Cygni-type profile, after a bright star whose atmosphere is in rapid but nonexplosive expansion. The precise shape of the profile in a supernova spectrum depends on how the matter density varies with radius and on the expansion velocity. The higher the velocity at the photosphere, the broader the profile and the more the absorption component is blueshifted.

A supernova spectrum consists of the thermal continuous spectrum from the photosphere, with many P Cygni-type line profiles, produced by electronic transitions within the various atomic and ionic species in the atmosphere, superimposed. The expansion velocity is so large that the line profiles overlap. Physically, this overlapping corresponds to multiple scattering of photons in the atmosphere; after a photon is scattered by one electronic transition, it continues to undergo redshift and may come into resonance with other transitions and be scattered again and again before it escapes the atmosphere.

C. Velocity and Composition of Ejected Matter

The spectrum of a type II supernova near maximum light is almost a smooth, featureless continuum. As the light curve begins to fall from its peak, the Balmer lines of hydrogen appear, with widths and blueshifts corresponding to a velocity at the photosphere in the range 7000 to 10,000 km/sec. Within weeks the spectrum assumes a more complicated appearance as many additional lines develop; the strongest are identified with transitions in neutral sodium, singly ionized calcium, and singly ionized iron. The linewidths indicate that the velocity at the photosphere usually decreases within a month to 5000 km/sec or less, reflecting the progressively deeper penetration of the observer's line of sight into the ejected matter. Analysis of the strengths of the spectral lines indicates that the composition in the outler layers of type II supernovae is similar to the composition of the sun and other stars; hydrogen and helium are most abundant and only a small fraction of the matter is in the form of heavier elements.

During the bright, observable phases of a type II explosion, our line of sight evidently does not penetrate the outer, fast-moving, hydrogen-rich matter. A goal for the future is to observe and analyze spectra of a type II explosion years after its outburst, when its hydrogen-rich layers have become optically thin, to probe the composition of its deeper, slow-moving layers, which may be enriched in heavy elements.

Spectra of type I supernovae are more complicated in appearance than those of type II supernovae. Even before maximum light, the spectra are practically covered with strong, overlapping, P Cygni-type line profiles. Therefore, attempts to interpret type I spectra rely on computer simulations of spectrum formation based on models of the supernova atmosphere, rather than on direct identifications of spectra lines on the basis of wavelength coincidence. Figure 6 shows a comparison of an observed maximum-light spectrum of SN 1981b in NGC 4536 with a theoretical spectrum. The strongest features in the synthetic spectrum are produced by lines of neutral oxygen and singly ionized magnesium, silicon, sulfur, and calcium, and the

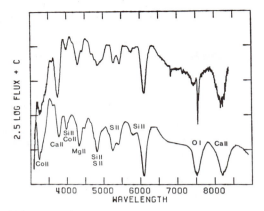

FIG. 6.   A McDonald Observatory optical spectrum of the type I supernova 1981b in NGC 4536 at maximum light (top) compared to a theoretical spectrum (bottom) calculated on the assumptions that the ejected matter contains no hydrogen or helium and that the spectral lines are formed by pure scattering. In the observed spectrum, narrow absorptions near 6800 and 7600 Å are formed in the earth's atmosphere. In the theoretical spectrum, the ions that produce the absorption components of the P Cygni profiles are labeled. Vertical displacements are arbitrary. [Reproduced with permission from Branch, D. (1985). In "Nucleosynthesis: New Challenges and Developments" (Arnett, W. D., and Truran, J., eds.). University of Chicago Press, Chicago.]

velocity at the photosphere is 12,000 km/sec. Spectra observed weeks and months later are found to match synthetic spectra dominated by lines of singly ionized iron moving at lower velocities near 8000 km/sec. The composition of matter ejected from type I supernovae is unlike that of ordinary cosmic matter; hydrogen and helium are deficient in abundance if not absent completely. The matter consists instead of a mixture of heavier elements; in the outer layers, where the maximum-light spectrum forms, the composition is dominated by elements of intermediate mass from oxygen to calcium, and the deeper layers that produce the later spectra consist primarily of iron.

III. The Origin of Type I Supernovae

A. Nickel and Cobalt Radioactivity

When a star explodes, its matter suddenly is heated, but expansion then begins to cool it, nearly adiabatically. Unless the initial radius of the star is extremely large, its outer layers are predicted to cool too quickly to ever attain the large (10^{15}-cm), hot (10^4-K) radiating surface that is required to account for the peak optical brightness of a supernova. Some continuing source of energy must be provided to the expanding gas to partially compensate for the adiabatic cooling. For type I supernovae, the continuing energy source has been thought since 1980 to be the decay of radioactive isotopes of nickel and cobalt. Detailed calculations have shown that the peak luminosity and shape of the type I light curve can be explained by supposing that nuclear reactions during the explosion synthesize from $0.2M_\odot$ to $1.0M_\odot$ of the unstable isotope ^{56}Ni. The mass range reflects primarily the present uncertainty in the extragalactic distance scale and the associated uncertainty in the supernova peak luminosity. Nickel 56 has a half-life of 6.1 days and decays by electron capture with the emission of a γ ray to produce its daughter nucleus, ^{56}Co. The average decay energy is 1.72 MeV. In turn, ^{56}Co decays with a half-life of 79 days to produce stable ^{56}Fe, the most abundant isotope of iron in nature. Cobalt 56 decays primarily by electron capture and γ-ray emission, but sometimes by positron emission. The mean decay energy is 3.58 MeV; 4% is carried by the positrons.

Computer simulations show that during the first month of the explosion, the supernova is optically thick to the γ rays; they are absorbed in the ejected matter and provide the continued source of energy needed to keep it hot. As it expands, the supernova becomes increasingly transparent to the γ rays, and after a month they begin to escape rather than be absorbed. The positrons from the cobalt decay, however, can be trapped by even very small local magnetic fields long enough to deposit their kinetic energy in the gas by means of collisions, before they mutually annihilate with electrons to produce more escaping γ rays. The time-dependent fraction of the γ-ray energy deposited, together with the kinetic energy of the positrons, provides enough energy to account for the tail of the type I light curve. In the tail phase the optical spectrum no longer is a thermal continuum because the supernova no longer is opaque. Detailed computer simulations of spectrum formation in this phase have shown that the fraction of the decay energy that is deposited in the gas maintains the temperature near 5000 K. Collisions with free electrons produce electronic excitation within the first few ionization states of iron and cobalt, and photons subsequently are emitted during spontaneous de-excitation. Synthetic spectra consisting of broad, overlapping emission lines of iron and cobalt ions are in good agreement with type I spectra observed during the late, postphotospheric phases.

B. THE FATE OF STARS OF LOW AND INTERMEDIATE MASS

The decay of radioactive nickel and cobalt isotopes provides an explanation for the nature of the type I light curve, but in what kinds of stars and by what mechanism is a large initial mass of radioactive nickel produced? Stars having initial masses less than $1M_\odot$ have nuclear evolutionary lifetimes longer than the present age of the universe and cannot yet have become highly evolved internally. A star between $1M_\odot$ and $1.4M_\odot$ converts hydrogen into helium by nuclear fusion reactions in its core during the main-sequence phase, at a characteristic core temperature of 10^7 K. When the hydrogen in the core is exhausted, the star expands its envelope to undergo the red giant phase while its core contracts and heats until it is able to fuse the helium to a mixture of carbon and oxygen, at a temperature of 10^8 K. When the helium supply is exhausted, the core contracts again. The carbon and oxygen core would fuse to heavier elements at 10^9 K, but this temperature is not reached because the density in the core becomes sufficiently high, 10^6 g/cm^3, that the inward force of gravity is balanced by the pressure of degenerate electrons—electrons whose momentum distribution is determined by the Pauli exclusion principle. The star forms a stable carbon–oxygen white dwarf that has a radius of only a few percent of the radius of the sun, comparable to the radius of the earth.

The maximum mass of a white dwarf is $1.4M_\odot$. The fate of a star whose initial mass is in the range from $1.4M_\odot$ to $8M_\odot$ depends on the rate at which it loses mass during its red giant phase by means of a stellar "wind" and the formation of a planetary nebula. If mass loss fails to reduce the star below $1.4M_\odot$, the star cannot become a stable white dwarf. Instead, its degenerate carbon–oxygen core will contract until the temperature and density are such that nuclear fusion of carbon begins. Ordinarily, the energy released by the ignition of a nuclear fuel produces a rise in pressure, which leads to expansion and cooling, and the rate of nuclear burning is thermostatically regulated. In a gas supported by the pressure of degenerate electrons, how-

ever, the pressure increase due to nuclear burning is inconsequential and there is no immediate compensating expansion. Thus, the nuclei acquire higher energies and fuse still more furiously, and a thermonuclear runaway takes place. Nuclear reactions quickly incinerate the burning material to a state of nuclear statistical equilibrium, dominated by isotopes as heavy as iron, which are the most tightly bound. Under the circumstances in the cores of intermediate-mass stars, computer simulations predict that the predominant individual isotope to be formed in nuclear statistical equilibrium is ^{56}Ni. The energy released by the fusion of a stellar core from carbon and oxygen to nickel finally would be sufficient to raise the pressure to the point of exploding the star and producing a supernova.

However, mass loss undoubtedly occurs during the red giant phase, and so a star initially more massive than $1.4M_\odot$ may be able to reduce its mass beneath the upper limit for a white dwarf. Observational and theoretical indications are that stars at least as massive as $5M_\odot$ and probably as massive as $8M_\odot$ lose enough mass to become white dwarfs rather than supernovae. Furthermore, it is observationally unlikely that there are enough stars in excess of $5M_\odot$ in elliptical galaxies to account for the type I supernovae that they produce. Thus, it is difficult to explain type I supernovae as the explosions of ordinary single stars.

C. ACCRETING WHITE DWARFS

A more promising explanation for the origin of type I supernovae appeals to the more complicated evolution of stars in close binary systems. A pair of stars of intermediate mass forms, and the more massive one evolves first to form a carbon–oxygen white dwarf. Eventually the other star evolves, and at some phase, probably as a red giant, it transfers matter to the surface of the first white dwarf, thus increasing its mass and ultimately provoking it to explode. The time delay between the formation of the binary system and the supernova explosion is determined by the mass and nuclear evolutionary time scale of the second star. This model can account for the presence of type I explosions in elliptical galaxies (the second star has low mass and a long nuclear time scale, and so binary systems formed long ago are producing type I supernovae now) as well as for the correlation between the type I rate and the recent star-formation rate in spiral galaxies (pairs in which the second star has a larger mass produce type I explosions on a shorter time scale).

Computer simulations of the response of a white dwarf to the accretion of matter lead to a variety of outcomes, depending on the initial mass and composition of the white dwarf and the composition and rate of arrival of the accreted matter. Certain combinations of these parameters do lead to predictions that correspond well to the observed properties of type I supernovae. If the accretion rate of hydrogen-rich matter onto a carbon–oxygen white dwarf is on the order of $10^{-7}M_\odot$ per year, the hydrogen is fused first to helium and then to carbon and oxygen by nuclear reactions near the surface of the star, thus producing an increasingly massive carbon–oxygen white dwarf. As the total mass approaches $1.4M_\odot$, the white dwarf contracts and heats until carbon ignites at its center and incinerates, as described earlier for single stars, to nuclear statistical equilibrium. A subsonic, convectively driven nuclear-flame front, called a deflagration wave, propagates outward from the center at about one-third of the local sound speed, taking a second to reach the surface of the white dwarf. Because the front is subsonic,

pressure waves move out ahead of the front and set the unburned outer parts of the star into expansion before the front arrives, thus decreasing the density of those layers. As the deflagration front moves outward into material of lower density, it loses strength; consequently, only the inner part of the star is converted to nuclear statistical equilibrium. An intermediate part is converted to intermediate-mass elements from oxygen to calcium, and an outermost portion is ejected as unburned carbon and oxygen. The predicted composition for a detailed carbon deflagration model is shown in Fig. 7. The model accounts for the spectrum near maximum light, dominated by lines of oxygen through calcium, and for the later spectra dominated by lines of iron.

In the carbon-deflagration model for a type I supernova, the energy produced by nuclear fusion is more than 10^{51} ergs. A fraction of the fusion energy is used to overcome the negative gravitational energy of the white dwarf, that is, to unbind it, and the remainder goes into the kinetic energy of the explosion. However, for reasons given earlier, the explosion becomes optically bright only because of the smaller amount

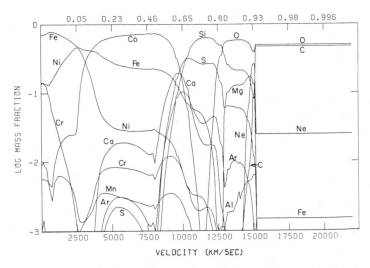

FIG. 7. The predicted composition of a model of a white dwarf exploded by a carbon deflagration, displayed as mass fraction versus ejection velocity. The fraction of the star's mass interior to the point in question is given at the top. This figure applies to 32 days after the explosion. The composition in the inner part of the star, moving slower than 10,000 km/sec, changes with time owing to beta decays; most of the cobalt and nickel ultimately decay to iron. [Reproduced with permission from Branch, D., Doggett, J. B., Nomoto, K., and Thielemann, F. K. (1985). *Astrophys. J.* **294**, 619.]

of energy released on a slow time scale by the radioactive decay of nickel and cobalt. In short, the energy of nuclear fusion makes the star explode, but the energy of radioactive decay makes the explosion shine. To produce nearly $1M_\odot$ of ^{56}Ni, as required to explain the peak luminosity on the "long" extragalactic distance scale, the white dwarf must also produce enough fusion energy to disrupt itself completely. On the "short" distance scale, only a few tenths of a solar mass of ^{56}Ni are required to account for the luminosity, the associated fusion energy is lower, and it is possible that only a portion of the star is ejected, with a less massive white dwarf or a neutron star left behind. The computer simulations of accreting white dwarfs that best match the observed characteristics of ordinary type I supernovae correspond to complete disruptions. Simulations that lead to the formation of compact stellar remnants may correspond to uncommon or subluminous kinds of type I supernovae.

D. Unusual Type I Supernovae

A small fraction of observed supernovae show characteristics that clearly are related to those of ordinary type I supernovae, but differ in some respects. A few whose characteristics are unique are referred to as type Ipec. Others appear to form a distinct, homogeneous subclass of type I and are referred to as type Ib (type Ia then refers to the ordinary kind discussed in the preceding sections). The recognition of the existence of the type Ib subclass came in 1983, with the first well-observed member of the class, SN 1983n in M83. At maximum light, the spectra of type Ib supernovae resemble the spectra of type Ia supernovae at later phases in their evolution; thus type Ib supernovae can be said to be "born old" spectroscopically. At maximum light type Ib supernovae are observed to be fainter than type Ia by more than a magnitude. Whether this is caused by an intrinsic subluminosity of type Ib or by a tendency for type Ib to occur in dusty, obscuring regions of galaxies is not yet known, but in either case it is clear that supernova searches have discriminated against type Ib relative to type Ia, owing to their faintness. Type Ib supernovae are rare among the supernovae that have been discovered, but they may really occur almost as often as type Ia. Various models have been suggested for type Ib, but none has yet won general acceptance among astronomers.

IV. The Origin of Type II Supernovae

A. Supergiant Progenitors

Type II supernovae occur in spiral arms and therefore must be produced by stars more massive than $8M_\odot$. Their optical spectra correspond to a hydrogen-rich composition, and so type II explosions occur in stars that still have their outer layers in place. Two-thirds of the observed type II supernovae have light curves of the plateau variety, with shapes unlike type I light curves and not consistent with the nickel–cobalt radioactivity model. Type II plateau light curves can be explained in a different way.

A massive star in its late evolutionary phases becomes a supergiant, having a small, dense, heavy-element core surrounded by an extended hydrogen-rich envelope, which may attain a radius of 1000 times the solar radius, approaching 10^{14} cm. If some instability in the core releases a large amount of energy in a short period of time, a shock wave may form and transport the energy outward, heating and ejecting the stellar envelope. Because the envelope initially is so extended, it can attain a radius of 10^{15} cm while still hot (10^4 K) and become optically bright without the need for any delayed input of energy. Hydrodynamical computer simulations of the response of a supergiant star to the sudden deposition of energy at its center predict light curves and other properties that are in good agreement with observations of type II-plateau (II-P) supernovae. The plateau is explained as a phase of diffusive release of the thermal energy deposited in the envelope by the shock. To account for the observed absolute brightness, the expansion velocity, and the duration of the plateau phase, the initial supergiant envelope needs to have a mass on the order of $10M_\odot$ and a radius of 10^{14} cm, consistent with observations of supergiant stars, and the energy deposited on a short time scale at the center needs to be on the order of 10^{51} ergs.

B. Core Collapse

The sudden energy deposition at the center of a massive star follows the inevitable gravitational collapse of its highly evolved core. In a star more massive than $8M_\odot$, carbon is nondegenerate when it ignites and fuses nonexplosively to a mixture of oxygen, neon, and magnesium. Nuclear burning of these and still-heavier elements will not be able to disrupt the increasingly tightly bound stellar core, and so the core is destined eventually to collapse. When the initial mass of the star was less than $10M_\odot$, the oxygen–neon–magnesium core contracts to become electron degenerate, and the subsequent capture of electrons by neon and magnesium nuclei decreases the degenerate pressure and causes gravitational collapse. In a star of more than $10M_\odot$, the contracting neon–oxygen–magnesium core ignites to fuse to still-heavier elements, ultimately to iron. No nuclear energy can be extracted from iron; instead, high-energy photons characteristic of the high core temperature photodisintegrate the iron endothermically, decreasing the energy and pressure in the core and leading to core collapse.

The collapse of the core occurs on a time scale of a tenth of a second. As the density of the core approaches nuclear densities, 10^{14} g/cm^3, neutrons become degenerate and form a neutron star, and the pressure rises enormously to resist further compression. At this point a gravitational energy of 10^{53} ergs has been released; most is carried off in the form of neutrinos, which hardly interact with matter and therefore stream freely out of the star. To the extent that the collapse is asymmetric, some energy should be radiated in the form of gravitational waves. To produce a type II supernova, on the order of 1% of the gravitational energy somehow must go into an outgoing shock wave. This process is not yet completely understood. A promising possibility, at least for stars not much more massive than $10M_\odot$, is a hydrodynamical explosion. Owing to its infall velocity, the collapsing core overshoots its equilibrium nuclear density, rebounds, or "bounces," and transfers energy mechanically to material falling in from above, leading to an outgoing shock wave (Fig. 8). In more massive stars the energy put into the shock wave by the bounce tends to be consumed in the photodisintegration of nuclei just outside the core before the shock can propagate into the stellar envelope. In these stars, a second mechanism, involving the neutrinos, may be critical. If the star is sufficiently massive, the neutrinos may be trapped at the outer edge of the core, where they deposit their energy and momentum and revive the stalled shock wave. The hydrodynamical explosion, if successful, immediately follows the bounce of the core and is referred to as a prompt explosion, while the neutrino depo-

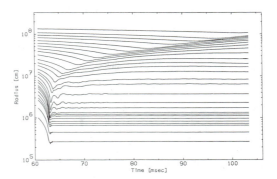

FIG. 8. Radius versus time for selected zones in a computer simulation of the gravitational collapse and bounce of the core of a star of $10M_\odot$. Only the innermost $1.4M_\odot$ is shown. Time (in milliseconds) is measured from the beginning of core collapse. The bounce occurs at 66 msec. [Reproduced with permission from Hillebrandt, W. (1982). *Astron. Astrophys.* (*Lett.*) **110**, L3.]

sition occurs a second after the bounce and is a delayed explosion mechanism.

C. PAIR-INSTABILITY EXPLOSIONS

A star with initial mass greater than $100M_\odot$ is predicted to encounter a catastrophic instability when its core is composed of oxygen. At the temperature expected, 2×10^9 K, γ rays may spontaneously produce electron–positron pairs. The energy to provide the rest mass of the particle pair comes at the expense of the thermal energy of the gas, and the core collapses. Theoretical simulations predict that in a star in the range $100M_\odot$ to $300M_\odot$ oxygen fusion during the collapse is able to reverse the implosion and explode the entire star. For a mass in excess of $300M_\odot$, the oxygen burning cannot overcome the star's gravity and the whole star is predicted to continue to collapse to form a black hole.

The existence of such very massive stars is open to question. There are observational and theoretical reasons to think that stars more massive than $100M_\odot$ do not form. Nevertheless, one observed supernova, SN 1961v in NGC 1058, showed properties that suggest that it may have been the explosion of a very massive star. Supernova 1961v is the *only* supernova to have been observed as a stable star before it exploded. The estimated absolute brightness of the presupernova star corresponded to a minimum of several hundred solar masses. The light curve of the supernova was very slow compared to

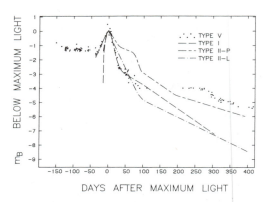

FIG. 9. The light curve of SN 1961v in NGC 1058 compared to the mean light curves of type I, II-L, and II-P supernovae. Supernova 1961v took at least a year to reach maximum light. [Reproduced with permission from Doggett, J. B., and Branch, D. (1985). *Astron. J.* **90**, 2303.]

ordinary supernovae (Fig. 9); it took a year to reach maximum light and was followed optically for at least 6 yr after maximum light, while no other supernova has been followed optically for more than 2 yr. The long time scale for the light curve has been interpreted as a consequence of a long diffusion time characteristic of thousands of solar masses of ejected material. Theoretical simulations suggest that a star so massive could explode rather than collapse to a black hole only if the presupernova was rapidly rotating. The case for SN 1961v having been a very massive star is not compelling, but astronomers are now watching for further extraordinary supernovae, which will be observed and modeled in greater detail.

D. TYPE II-LINEAR SUPERNOVAE

Spectroscopically, type II-linear (II-L) supernovae resemble type II-P supernovae; the outermost layers of the ejected matter are hydrogen rich. The physics of a type II-L explosion may be like that of a type II-P, the principal difference being that, owing to extensive presupernova mass loss, the envelope of the star is depleted and much less material is ejected in type II-L than in type II-P, accounting for a reduced diffusion time scale and therefore for a short, inconspicuous plateau phase. However, numerical simulations have not yet shown that the observed shape of the type II-L light curve can be accounted for in this way.

Figure 4 shows that the shape of the type II-L

light curve is more like the shape of type I than like type II-P. Another possibility, therefore, is that a type II-L explosion is physically like that of a type I explosion, powered by the decay of radioactive nickel, but with a small amount of hydrogen on the outside.

V. Supernova Nucleosynthesis

Nuclear reactions during the first few minutes of the Big Bang are thought to have produced an initial universal composition of 75% hydrogen and 25% helium by mass. Only small quantities of lithium and beryllium, and negligible amounts of heavier elements, were produced in the Big Bang. Lithium, beryllium, and boron also are produced in the interstellar medium by energetic collisions between cosmic-ray nuclei and heavy elements in the interstellar gas. In a fundamental paper published in 1957, E. Margaret Burbidge, Geoffrey R. Burbidge, and William A. Fowler of the California Institute of Technology and Fred Hoyle of Cambridge University proposed that carbon and all of the heavier elements are formed by nuclear reactions inside stars. Some of the heavy elements may be injected from stars into the interstellar medium by stellar winds, planetary nebulae, and nova explosions, but the main way that stars transfer heavy elements to the interstellar medium is by supernova explosions. A supernova ejects heavy elements formed during the presupernova evolution of the star, as well as elements synthesized during the explosion itself.

Each type I supernova apparently ejects $1.4M_\odot$ of heavy elements, including carbon, oxygen, and other elements up to iron. The elements heavier than oxygen were formed during the explosion. The amount of nickel needed to power the light curve by radioactive decay is enough to ensure that type I supernovae are significant if not dominant producers of the nickel decay product—iron. Type I supernovae, therefore, certainly are important for nucleosynthesis, but they cannot be the principal source of all of the intermediate-mass elements.

Type II supernovae, if they do eject heavy elements from their deep interiors, hide this material from our view by shrouding it in a hydrogen-rich envelope. Thus the evidence for nucleosynthesis by type II supernovae is primarily theoretical rather than observational. Supernovae from stars having masses near $10M_\odot$ are predicted to eject negligible amounts of elements heavier than oxygen, but those from more mas-

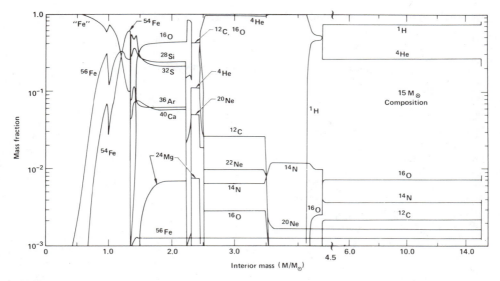

FIG. 10.   The predicted composition of a star of $15M_\odot$ at the onset of core collapse, based on a computer simulation of the star's presupernova evolution. The amount of mass interior to the point in question is plotted along the horizontal axis; note the changes in scale at $4.5M_\odot$. [Reproduced with permission from Woosley, S. E., and Weaver, T. A. (1986). *Annu. Rev. Astron. Astrophys.* **24,** © 1986 by Annual Reviews, Inc.]

sive stars are expected to eject several solar masses of heavy elements (Fig. 10). If so, type II and I supernovae may be of comparable importance for nucleosynthesis.

VI. Observations across the Electromagnetic Spectrum

Observations of extragalactic supernovae outside the optical region of the spectrum began to be made regularly only in the early 1980s. Radio and infrared observations are made from the ground, while ultraviolet observations can be made from orbiting observatories. The study of X rays from supernovae has hardly begun. No γ rays from supernovae have been detected, nor have neutrinos or gravitational waves.

A. RADIO OBSERVATIONS

The study of supernova remnants in our galaxy has been based mainly on observations at radio wavelengths. Supernova remnants are among the brightest radio sources in the sky. The radio emission is a continuum of synchrotron radiation produced by relativistic electrons accelerated either by a central pulsar (in the Crab Nebula, for example) or in the region of

interaction between the ejected matter and the swept-up interstellar medium (in Cassiopeia A, Tycho, and Kepler).

The first supernova to be detected at radio wavelengths during its explosive phases was the type II supernova 1970g in M101. Since then, with the advent of more sensitive radio telescopes, a few more radio supernovae have been detected, including the type II supernovae 1979c in M100 and 1980k in NGC 6946 and the type Ib supernova 1983n in M83. The form of the radio "light curve" varies from one event to the next; generally it is slower to rise to its peak than the optical light curve and also slower to fall. The fraction of a supernova's total emission of electromagnetic energy that is in the form of radio waves is only on the order of 10^{-5}.

The radio emission from supernovae is synchrotron radiation from relativistic electrons thought to be accelerated in the hot (10^9-K) region of interaction between the ejected matter and a shell of circumstellar gas. The circumstellar shell is generated by nonexplosive mass loss from the presupernova star, that is, by a stellar wind. Future observations and theoretical modeling of the radio emission from supernovae are expected to provide new, detailed information on the mass-loss processes that stars undergo shortly before they explode.

A supernova remnant becomes a strong radio source only after decades or centuries, after it has swept up a mass of interstellar gas comparable to the ejected mass. Radio supernovae, on the other hand, fade on a time scale of years or decades, as their circumstellar shells become swept up entirely. Recently, radio emission has been detected from several supernovae that were discovered optically decades ago. The first such recovery of a historical extragalactic supernova was the detection in 1981 of radio emission from SN 1957d in the bright southern galaxy M83. Future, more detailed observations of radio emission from such supernovae of intermediate-age will provide information on how a radio supernova, interacting with circumstellar gas, makes the transition to a supernova remnant, interacting with interstellar gas.

B. Infrared Observations

Each supernova emits infrared radiation simply as a long-wavelength extension of the thermal continuous spectrum from its photosphere. The shapes of the infrared light curves of type Ia and Ib supernovae have been found to be distinct, despite the similarity of their optical light curves. The differences in the infrared are thought to be caused by a broad absorption feature in the infrared spectra of type Ia, perhaps produced by lines of neutral silicon; recall that silicon lines are identified in the optical spectra of type Ia supernovae, but not in type Ib supernovae.

Some supernovae have been observed to emit large excesses of infrared radiation, beyond the moderate amounts expected to be radiated by their photospheres. The infrared excesses are attributed to thermal radiation from cool (1000-K), small (10^{-5}-cm), solid particles, or dust grains, in a circumstellar shell. The grains absorb optical and ultraviolet radiation from the photosphere, are heated, and then emit a thermal continuous spectrum in the infrared. Infrared observations provide information on the nature and distribution of the dust grains. The grains surrounding most supernovae apparently condensed from gaseous matter lost nonexplosively from the presupernova stars, but the infrared observations of a few supernovae also are consistent with the possibility that the grains formed from the ejected supernova matter as it cooled. [See ASTRONOMY, INFRARED.]

C. Ultraviolet Observations

Ultraviolet observations have established that a type I supernova emits much less energy at short wavelengths than would be expected on the basis of an extrapolation of its optical–infrared continuous spectrum; that is, type I supernovae are ultraviolet deficient. The few spectral features that have been observed in type I ultraviolet spectra may be produced by lines of singly ionized iron. Type II supernovae have, instead, mild ultraviolet excesses. The ultraviolet spectra of a type II supernova contains numerous narrow absorption lines formed along the line of sight in the interstellar gas in our galaxy and in the supernova host galaxy. The presence of the interstellar lines provides a valuable probe of the nature of the interstellar medium, but it complicates the study of the ultraviolet spectra emitted by the supernovae. Spectral lines of carbon, nitrogen, and oxygen, detected in the supernova ultraviolet, provide an opportunity to determine the abundances of these important elements, which have only weak spectral features in the optical spectrum. [See ASTRONOMY, ULTRAVIOLET SPACE.]

VII. Supernovae as Distance Indicators for Cosmology

Two goals of observational cosmology are to determine the present rate of expansion of the universe and the rate at which the expansion is decelerating. The first requires the determination of *absolute* distances to galaxies at distances of tens of megaparsecs. The second requires *relative* distances to more remote objects, at distances comparable to the radius of the observable universe. Supernovae, because they are so luminous and, therefore, detectable from afar, are promising indicators of both absolute and relative extragalactic distances.

A. The Extragalactic Distance Scale

In 1929 Edwin Hubble of the Mount Wilson Observatory discovered that the universe is expanding. The radial velocities of galaxies with respect to us are proportional to the distances of the galaxies from us; thus, the universal expansion can be represented by the Hubble law: $V = Hd$, where V is radial velocity, ordinarily expressed in kilometers per second, d is distance in megaparsecs, and H is the Hubble constant in kilometers per second per megaparsecs. The reciprocal of the Hubble constant gives the Hubble time, the age of the present expansion of the universe if deceleration has been negligible. Radial velocities of galaxies are directly measured by means of the Doppler effect, and so the determination of the Hubble constant reduces to the problem of determining absolute distances to galaxies. In practice, because galaxies also have random motions, it is necessary to determine distances to galaxies whose radial velocities exceed 1000 km/sec, corresponding to a distance between 10 and 20 Mpc. The classical approach to the determination of the Hubble constant is to estimate distances to the nearest galaxies by comparing the apparent brightness of stars contained in them to the absolute brightness of similar stars of known distance in our galaxy and then to estimate distances to more remote galaxies by comparing their observed properties to those of the nearer ones. Attempts to apply this bootstrap method lead to distance scales that differ by a factor of 2, corresponding to values of the Hubble constant in the range 50 to 100 km/sec/Mpc. These values give a Hubble time of 20 to 10 billion yr.

A straightforward way to use supernovae to determine the distances to galaxies is to assume that all supernovae of a given type or subtype have the same absolute brightness and to calibrate that brightness on the basis of a supernova in our galaxy or in one of the nearest galaxies whose distance is known. Unfortunately, type II supernovae do not all have the same peak absolute brightness, but the more conformist type I supernovae do tend to cluster about a common value (Fig. 11). In our galaxy, both Tycho's and Kepler's supernovae are suspected to have been type I supernovae, primarily on the basis of their optical light curves as reconstructed from contemporary reports of the apparent supernova brightness with respect to the planets and bright stars. However, the absolute brightness of Tycho's and Kepler's supernovae is not well known, owing to uncertainties in *their* distances and in the amount of extinction produced by dust in the line of sight. Furthermore, one or both may have been type II-L rather than type I, considering the similarities of the two kinds of light curves (Fig. 4). Among the nearer galaxies of known distance, only the Andromeda Galaxy has produced a supernova observed at peak brightness. The distance to Andromeda is known to within 10%, but the supernova was a

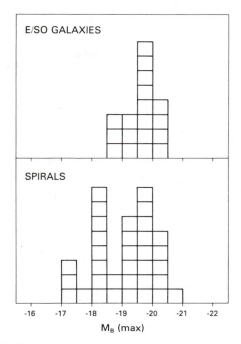

FIG. 11. The number of type I supernovae versus absolute blue magnitude at maximum light. Magnitudes have not been corrected for interstellar absorption, which is negligible in E/S0 (elliptical and lenticular) galaxies, but not in spirals. [Adapted with permission from Panagia, N. (1985). *Lect. Notes Phys.* **224,** 14.]

perculiar type I with an unusually rapid light-curve decline, and so there is no guarantee that its absolute brightness was the same as that of ordinary type I supernovae.

A second approach to the distance scale is to calculate the absolute brightness of an extragalactic supernova from its observed properties on the basis of an understanding of its physical behavior. The absolute brightness of a supernova depends on the temperature at, and radius of, its photosphere. The temperature can be estimated from the shape or color of the continuous spectrum, and the radius is given by the product of the expansion velocity at the photosphere and the time elapsed since the explosion. The expansion velocity can be derived from the widths and blueshifts of the spectral features, as discussed in Section II,B. This method, first applied to supernovae in 1973, has the very attractive feature that it gives the distance to a supernova directly, independent of other astronomical objects at intermediate distances. The method tends to give large distances, corresponding to a Hubble constant nearer to 50 than to 100 km/sec/Mpc. The main uncertainty is in the assumption that the surface of a supernova radiates like an ideal blackbody of the same temperature. Detailed numerical models of supernova atmospheres now under construction may accurately predict the emissivity and strengthen confidence in the method.

The recognition that the peak luminosity of a type I supernova is provided by the decay of radioactive ^{56}Ni and ^{56}Co leads to another independent way to infer the absolute luminosity of a type I supernova. Recall that the kinetic energy of the explosion comes from the nuclear fusion of carbon and oxygen to heavier elements, including nickel, and that the peak luminosity is determined by the amount of nickel initially synthesized. Expansion velocities inferred from the spectrum fix the kinetic energy, which in turn fixes the amount of nuclear burning that must take place and the initial mass of ^{56}Ni. If type I supernovae do completely disrupt and eject $1.4M_\odot$, the observed velocities imply kinetic energies of 10^{51} ergs. The nuclear fusion needed to produce this kinetic energy would also produce about $0.5M_\odot$ of ^{56}Ni. This leads to a luminosity greater than 10^{43} ergs/sec, corresponding to a Hubble constant of about 60 km/sec/Mpc. Only if, contrary to most indications, type I supernovae fail to disrupt completely, and thus eject less than $1.4M_\odot$, can the kinetic energy, ^{56}Ni mass, and peak luminosity be low enough to be consistent with a value of the Hubble constant as high as 100 km/sec/Mpc.

A fourth approach to supernova distances is provided by radioastronomy. Very long baseline radio interferometers can be used to resolve the angular sizes of extragalactic radio supernovae several years after the explosions take place. Comparison of the measured angular size with a linear size, calculated as the product of the expansion velocity and the time elapsed since the explosion, yields the distance. The main uncertainty associated with this method is that the velocity of expansion inferred from supernova spectra refers to the region in which the optical spectrum is formed, but the velocity of the material that emits the radio waves is needed for the method. To improve the accuracy of the distance determination, a more detailed understanding of how the interaction of the ejected matter and the circumstellar shell produces radio emission is needed.

All four methods for determining distances to supernovae lead to values of the Hubble constant in the range 50 to 100 km/sec/Mpc, and the present accuracy of the supernova approaches appears to be comparable to the accuracy of methods based on galaxies. The supernova method based on ^{56}Ni radioactivity points to a value nearer to 50 than to 100 km/sec/Mpc, provided that type I supernovae are produced by white dwarfs that explode completely.

B. The Deceleration Parameter

The expansion of the universe is expected to be decelerating owing to self-gravitation. If the deceleration exceeds a critical amount, the universe is closed and its expansion will come to a halt and subsequently contract; otherwise it is open and will continue to expand forever. None of the many attempts to determine the deceleration have been convincing. One direct approach is to compare the expansion rate in the past to the present, local expansion rate. The expansion rate in the past is obtained by measuring velocities and relative distances of a sample of extremely remote objects. This has proved to be difficult to accomplish for galaxies because the distances to the remote galaxies are based on the assumption that all galaxies of a given morphological type have the same absolute luminosity; the assumption is invalid because the properties of galaxies evolve with time.

With the Hubble Space Telescope, supernovae can be measured at distances sufficiently large for this purpose. If the absolute brightness of at least a subtype of type I supernovae can be proved to be within a narrow range, if the absolute brightness does not change with time, and if type I supernovae at remote distances can be discovered in sufficient numbers to obtain a good statistical sample, the question of whether the universe is open or closed may be answered in this way.

VIII. Prospects

The study of extragalactic supernovae began with the discovery of the explosion in Andromeda in 1885, but it was not until 50 yr later that the grand scale of such events was recognized and the term supernova began to be used. The following 35 yr were primarily an observational period, when the basic characteristics of the various supernova types were recorded and statistical information was gathered. Since 1970 our understanding of supernovae has advanced rapidly. Observations with modern detectors, primarily in the optical region but increasingly in other regions as well, have provided high-quality data for comparison with the results of detailed numerical models. A basic understanding of the nature of supernova spectra and promising models of the explosion mechanisms have emerged from the interplay between observation and theory.

Further observational advances are expected. Ground-based optical astronomy will contribute automated searches with dedicated telescopes, to increase the supernova discovery rate. New powerful instruments, including the Hubble Space Telescope for optical and ultraviolet astronomy, the Very Large Baseline Array (the VLBA) of radiotelescopes, the Advanced X-Ray Astronomical Facility (AXAF), the Space Infrared Telescope Facility (SIRTF), and the Gamma-Ray Observatory (GRO), will provide observations of more and fainter supernovae, all across the electromagnetic spectrum. New sensitive detectors may also detect neutrinos and gravitational waves.

The flow of observational data will stimulate further theoretical work. Improved physical data from atomic and nuclear physics will be fed into new generations of computers to produce increasingly realistic models of the explosions. Astronomers will intensify their efforts to understand how supernova explosions are produced, to evaluate the role of supernovae in nucleosynthesis and in the evolution of galaxies, and to use supernovae to probe the present and future expansion of the universe. These are grand goals, and elusive ones, but they are not hopelessly out of reach.

BIBLIOGRAPHY

Bethe, H. A., and Brown, G. (1985). How a supernova explodes. *Sci. Am.* **252,** 60.

Clarke, D. H. (1984). "Superstars: How Stellar Explosions Shape the Destiny of the Universe." McGraw-Hill, New York.

Clark, D. H., and Stephenson, F. R. (1977). "The Historical Supernovae." Pergamon, Oxford.

Shklovskii, I. S. (1968). "Supernovae." John Wiley and Sons Ltd, London.

Trimble, V. (1982). Supernovae: Part I. The events. *Rev. Mod. Phys.* **54,** 1183.

Trimble, V. (1983). Supernovae: Part II. The aftermath. *Rev. Mod. Phys.* **55,** 511.

Wheeler, J. C., and Nomoto, K. (1985). How stars explode. *Am. Sci.* **73,** 240.

Woosley, S. E., and Weaver, T. A. (1986). The physics of supernova explosions. *Annu. Rev. Astron. Astrophys.* **24,** 205.

VARIABLE STARS

Steven N. Shore   *New Mexico Institute of Mining and Technology and DEMIRM, Observatoire de Meudon*
    and Radioastronomie, Ecole Normale Superieure

GLOSSARY

Hertzsprung–Russell diagram (HR diagram):
Plot of surface, or effective, temperature versus luminosity for a star. Observationally, the locus of the observed stellar population in a color-magnitude plane. Theoretically, the description of the path covered by the surface of a star with time.

Instability strip: Portion of the HR diagram, spanning the temperature range from about 3000 to 8000 K, and from the main sequence to the supergiants, in which most periodic stellar pulsational instabilities are detected. The strip is characterized by a period–luminosity (P–L) relation, with increasing period corresponding to higher intrinsic brightness.

Overstability: Growing, but periodic, instability characterized by a complex frequency.

Population I and II: Characterized mainly by age and metallicity differences. Population I is the young, stellar galactic component, with metallicities about the same as the sun and dynamical confinement to the plane of the Galaxy. Population II, found in the galactic halo and globular clusters, consists of older, low-mass stars with below solar metal abundances.

Secular instability: A nonperiodic, or steady, growth of the pulsation with time.

Units: Solar luminosity, $L_\odot$, 4×10^{33} ergs^{-1}; solar mass, $M_\odot$, 2×10^{33} g; solar radius, $R_\odot$, 7×10^{10} cm; mean solar density, $\langle \rho_\odot \rangle$, 1.4 g cm^{-3}.

Variable stars are stars that vary in photometric output over any period of time. The manifestation of periodic behavior is rare among stars, resulting from resonance phenomena in the interior. The primary mechanism for intrinsic stellar variability is pulsation, which may be either radial or, in rapidly rotating stars, nonradial. Magnetic fields also play a role in structuring the variations. Mass loss and shock phenomena are also associated with large amplitude pulsators, which may affect stellar evolution. This article deals exclusively with pulsating variables.

I. Introduction

Like most astronomical discoveries, the first observations of stellar variability were accidental. Fabricius noted, in 1596, that the star o Ceti showed a range of variability from invisibility to second magnitude. In honor of this, he named the star Mira, the wonderful. This discovery served as confirmation of the view, first forcefully promulgated in the Latin west by Tycho Brahe following the "new star" in Cassiopea of 1572, that the heavens are not static.

A major step was achieved in the late eighteenth century with Goodricke's discovery, announced in 1782, of the variations of β Persei, also called Algol. The variations were observed to be periodic and stable, and Goodricke explained this by the model of a close, unresolved binary star system undergoing eclipses. This is the first explanation for stellar variability, and one which served as an effective stimulus to search for other binary stars. The statistical reality of binarity and the application of Newtonian mechanics made a potent combination. In 1784, Goodricke discovered the variability of δ Cephei (δ Cep); in the same year, Piggot discovered the variability of η Aquilae (η Aql). The eclipse model was applied at the time to both of these stars, whose periods are

TABLE I   Early Discoveries of Intrinsic Stellar Variables

Star name	Discoverer	Year
o Ceti	Fabricius	1596
ψ Leonis	Montanari	1667
κ Sagittarii	Halley	1676
χ Cygni	Kirch	1687
δ Cephei	Goodricke	1784
η Aquilae	Piggot	1784
R CrB	Piggot	1795
α Herculis	Wm. Herschel	1796

similar to Algol. Both δ Cephei and η Aql are now known to be pulsating variables, unrelated to the Algol class of binaries, but the general explanation of the variations did not become clear for more than 150 years after this discovery. Table I, adapted from John Herschel's 1847 edition of "Outline of Astronomy," gives a list of some of the earliest discoveries of what are now known to be intrinsically variable stars.

During the next century, the search for stellar variability grew in importance to astronomy. The work of Argelander at Bamberg, the establishment of the Bonner Durchmusterung star atlas, and the international efforts to construct the Carte du Ciel, a photographic all-sky atlas, stimulated the discovery and cataloging of variable stars.

Perhaps the most important group working at the end of the nineteenth century was that at the Harvard College Observatory. S. Bailey, in observing globular clusters, discovered a class of periodic variables that showed the same characteristics as the field variable RR Lyrae, and the search for such stars was extended to the southern hemisphere with the establishment of the Bruce telescope at the Harvard Southern Station in Peru.

The most important discovery in this period was connected with observations of variable stars in the Large and Small Magellanic Clouds, most of the work being performed at Harvard by H. Levitt. She realized, around 1905, that these stars were generally of the same type as the variable δ Cephei. The extraordinary discovery was the relation between the period and the apparent brightness of the stars, the so-called period–luminosity relation. While the zero point for this relation was not then known, the relative brightness of these stars, with the same period as δ Cep and other variables in our galaxy, led to a singularly important cosmological breakthrough—the realization of the great distance that must characterize the Magellanic Clouds.

In the meantime, theoretical work on the cause for variability continued. Russell pointed out several paradoxes resulting from the assumption of binarity as the cause for the variations of many stars. Eddington suggested the pulsation model for stellar variability and derived the linearized, adiabatic pulsation equation. Subsequent work by Rosseland and Cowling further clarified the nature of the instabilities. The culmination of this epoch of study was the encyclopedic review article by Ledoux and Walraven, in 1958, which still serves as an important starting point for all discussions of stellar variability. The role played by stellar convection zones in driving pulsation and producing the radial velocity variations in Cepheids was understood by Cox around 1963. The first machine calculations of nonlinear, nonadiabatic pulsation models were accomplished in the mid-1960s. Around the same time, the first large-scale grids of stellar interior models were produced. Schwarzschild and Harm discovered the thermal instability of double shell sources in 1965.

This article deals exclusively with pulsating variable stars, those which are mechanically and thermally unstable because of intrinsic internal structural processes. Several other classes of variable stars are dealt with in the articles "Supernovae" and "Binary Stars" (which includes a discussion of the RS CVn stars, novae, and eclipsing binary stars).

II. Observations of Stellar Variability

Photometric and radial velocity variations are the basic data for the study of stellar variability. Light curves are the most obvious manifestation of variations, but because of the problems associated with interpretation, they cannot be deemed sufficient to pinpoint a mechanism responsible for the observed changes. Spectroscopic observations, specifically radial velocity variations, when coupled with color and light variability, serve to identify the causative agent behind most stellar variability.

The detailed behavior of velocity and light changes of a Cepheid variable or RR Lyrae star are usually very different from those observed in an eclipsing binary, although both will show both light and radial velocity changes. Most variable stars show asymmetric light curves, usually steepest on the ascending branch. Many are not strictly periodic, some are genuinely aperiodic. There is often a lag between maximum brightness and maximum radial velocity. Most telling is that there are often both color and spectral changes, which are gradual through the photometric cycle, indicating alterations in the physical structure of the stellar atmosphere.

The presence of radial velocity variations means, for a pulsating star, that the radius of the star is physically changing with time. The scale of this change is the time integral of the radial velocity curve. Changes in the sign of the radial velocities take place at extrema of the radius changes, so that they can be thought of as a bounce or recontraction. If the pulsations are strictly periodic, the radial velocity curve will be as well. The relation between the phase of maximum light and maximum radial velocity is most important in specifying the driving mechanism for the pulsation. One application of the radial velocity variations, in conjunction with the light curve, is the determination of temperature by the Baade–Wesselink method. This technique assumes that at two phases for which the colors are the same, the temperatures are also identical. From the comparison of luminosities at these points with the radial velocities, the radii can be obtained and the amplitude of pulsation can be specified. Additional information is available using the Barnes–Evans calibration of surface brightness versus intrinsic red color [$(V - R)_0$ color, in magnitudes]:

$$F_V = 3.956 - 0.363(V - R)_0 \qquad (1)$$

This provides an angular diameter, ϕ (in arc seconds), for the star using the apparent (reddening corrected) visual magnitude V_0:

$$F_V = 4.221 - 0.1V_0 - 0.5 \log \phi \qquad (2)$$

The radius thus obtained can be compared with that derived from the radial velocities.

For Cepheid variables, there is an empirical, as well as theoretical, period–luminosity–color relation that relates the period of pulsation of the star to its absolute magnitude and surface temperature. In combination with the Barnes–Evans relation, this can help place the star in its evolutionary state, and it also serves as a fundamental calibrator for the extragalactic distance scale (see discussion in Section IV).

III. Basic Theory of Stellar Pulsation

Stellar pulsation is a phenomenon resulting from departures from strict hydrostatic and thermal equilibrium. Such a state can be reached either by internal, that is to say evolutionary, avenues or by the presence of external perturbers. For instance, the tides observed on Earth are the result of the periodic forcing on the oceans and crust, on a rotating planet, by the gravitational pull of the sun and moon. On the other hand, the relaxation of the planet following an earthquake is internally driven.

For stars, the problem is that the precise nature of the mechanical and thermal couplings of different parts of the interior is difficult to specify. Stellar interiors are far from uniform or isothermal. There are steep temperature gradients near the surface, which is defined by the photosphere, and chemical composition changes because of the processing of the interior mass by nuclear reactions near the core. In addition, the ionization of the gas is a function of depth, as is the opacity, and therefore both the heat capacity and radiative losses are strongly depth dependent.

Pulsation serves as one of the few direct probes of the state of the interior of a star. The way in which an instability, a temperature and pressure variation at some zone in the stellar envelope, manifests itself at the stellar surface from which point it can be studied by an external observer is a complex resonance phenomenon that is not completely understood.

There are several time scales associated with a stellar interior. One is the characteristic time it takes for a zone of some physical size to achieve pressure equilibrium, the sound crossing time. This is given by $t_s = l/a_s$, where a_s is the sound speed that is a function of both the equation of state, through the ratio of specific heats, and the temperature and is therefore strongly depth dependent. Another is the time scale for radiative losses, which depends strongly on the opacity and thus on the equation of state, density, and temperature. Finally, and most critically, there is the time scale for a parcel of matter to oscillate in a gravitational field, given its local density. This last time scale, the so-called free-fall time scale, is given by

$$t_{ff} = (G \langle \rho \rangle)^{-1/2} \qquad (3)$$

where $\langle \rho \rangle$ is the mean density of the star. For a star with a mean density of 1 g cm^{-3}, this time scale is about 10 min, essentially the collapse time for matter from any distance to the center of mass. Stars have large density gradients and this mean density is heavily weighted toward the core, so in general, to account for the degree of central concentration, one can take the pulsation time scale to be related to the free-fall time through:

$$Q = P(\langle \rho \rangle / \langle \rho_\odot \rangle)^{1/2} \qquad (4)$$

where $\langle \rho_\odot \rangle$ is the mean density of the sun, about 1.4 g cm^{-3}, and P is the pulsation period. The value of Q can be calculated from any interior model once the period has been determined from the pulsation equation.

In the first approximation, a star can be assumed to be spherical. While rotation may play a role in the internal structure, resulting in polar flattening and equatorial expansion to an oblate spheroid, in general this is not important. The equation of hydrostatic equilibrium can be written as a balance between the outward pressure gradient and gravitation:

$$\ddot{r} = -\frac{1}{\rho}\frac{\partial p}{\partial r} - \frac{Gm}{r^2} \tag{5}$$

Here r is the radius, ρ is the density, p is the pressure, and m is the mass interior to a point r.

Energy loss is assumed to be due to radiation, which takes place diffusively because of the large optical depth of the stellar envelope:

$$L = \frac{16\pi acr^2T^3}{3\kappa\rho}\frac{\partial T}{\partial r} \tag{6}$$

Here, L is the luminosity, a and c are the Stefan constant and speed of light, respectively, κ is the opacity coefficient (a function of the density and temperature), and T is the temperature.

The energy balance of the stellar interior is given by the change in the entropy of the matter as a result of internal energy generation and radiative losses. The second law of thermodynamics

$$T\frac{dS}{dt} = \frac{dE}{dt} - \frac{P}{\rho^2}\frac{d\rho}{dt} \tag{7}$$

when supplemented by the equation of state $p = p(\rho, T)$, becomes

$$T\frac{dS}{dt} = \frac{p}{\rho(\Gamma_3 - 1)}\left[\frac{d}{dt}\ln p - \Gamma_1\frac{d}{dt}\ln\rho\right]$$

$$= \varepsilon - \frac{\partial L}{\partial m} \tag{8}$$

where S is the entropy; $\Gamma_1 = (d\ln p/d\ln\rho)_{Ad}$ and $\Gamma_3 - 1 = (d\ln T/d\ln\rho)_{Ad}$ are the adiabatic exponents for the pressure and temperature, respectively; and ε is the rate of energy generation per unit mass. This latter quantity, like the opacity, is a function of the density and temperature.

Finally, the explicit assumption of sphericity of the star enters in the equation for the mass interior to any radius:

$$\partial m/\partial r = 4\pi r^2\rho \tag{9}$$

It should be noted that all of these equations will be more complicated in the event that the star is distorted from sphericity, because of the coupling in the momentum equation between the various components of the gravitational acceleration on a distorted star and possible effects of rotation in both the momentum and energy equations.

In this enumeration of the equations of stellar structure, there are two time dependent equations, for the acceleration and the energy generation, and two equations of constraint, the diffusion approximation for the energy loss and the mass conservation equation. Thus, we would expect that the final equation will be third-order in time, and that there will be both stable and unstable solutions for many possible combinations of the parameters of energy generation and opacity. In fact, one point is already manifest in these equations—there are three critically important parameters in the pulsation characteristics of a star: the ratio of specific heats, the opacity, and the rate of energy generation. These govern the stability of the star.

Since the radius is changing with time in a pulsating star, the radial coordinate is not the best one to employ for the analysis of the instability. Instead, the Lagrangian approach is used, in which a mass zone is chosen and followed in its oscillations, rather than using the radial coordinate. In this way, the interior model can be calculated in equilibrium and perturbed, and thus the perturbation is only a function of time. For example,

$$r = r_0\left(1 + \frac{\delta r}{r_0}\right) = r_0(1 + \xi) \tag{10}$$

where r_0 is now the equilibrium position of a mass zone $q = M(r)/M$, M being the stellar mass.

The calculation of pulsational properties proceeds in a deductive way. One assumes an interior structure and asks what the response will be for the mass if this configuration is infinitesimally perturbed. The usual argument is that if the mass is intrinsically unstable any perturbation, no matter how small and irrespective of its origin, will be sufficient to begin pulsation. Consequently, it is assumed that all physical parameters, T, p, and ρ, are perturbed from equilibrium values by a small component.

The combined effect of the perturbations of the equations of stellar evolution is that a single equation can be obtained for the temporal variation of the pulsation radial amplitude:

$$\frac{\partial^3\xi}{\partial t^3} - 4\pi\dot{\xi}r_0\frac{\partial}{\partial m}[(3\Gamma_1 - 4)P_0]$$

$$- \frac{1}{r_0^2}\frac{\partial}{\partial m}\left[16\pi^2\Gamma_1P_0\rho_0r_0^6\frac{\partial\dot{\xi}}{\partial m}\right]$$

$$= -4\pi r_0\frac{\partial}{\partial m}\left[\rho_0(\Gamma_3 - 1)\delta\left(\varepsilon - \frac{\partial L}{\partial m}\right)\right] \tag{11}$$

which is also known as the linearized wave equation (although not really a wave equation). This equation, first derived by Eddington, is one of the best studied tools of stellar interiors. There is an exact solution for the adiabatic version of this equation, which gives secular instability if $\Gamma_1 \leq \frac{4}{3}$ (for a perfect gas, this ratio is $\frac{5}{3}$). The effects of partial ionization, which occurs in convection zones, decrease the star's stability. In consequence, the convection zones, already mechanically unstable, tend to drive the envelope pulsation.

The reason for dwelling for so long on the derivation of the pulsation equation is that there are many classes of instability. The pulsation is driven from different regions of the star depending upon the state of the stellar interior. For instance, as first emphasized by Eddington, a star can be thought of as a self-gravitating analog to a steam engine. If the star is to be pulsationally unstable, or what he called overstable, then the driving region must heat up on compression and overcool on expansion. Thus, the restoring force for the pressure rises on compression, and the rate of energy dissipation is most efficient at the lower temperature, so the matter overcools. The combined effects of the heat capacity and the opacity throttle the rates of heating and cooling, and thus there are specific zones within the star that are identified with the initiation and maintenance of the oscillatory behavior of the envelope.

There are few stars that are driven into instability by the nuclear burning, unless that burning is not central. For example, in the case of stars in the transition stage between main sequence and red giant, hydrogen burning occurs in a thin shell zone surrounding the hot but inert helium core. The temperature dependence of the CNO reaction is not, though, by itself, sufficient to induce the instability observed in stars in this stage of their lives. Instead, once the helium has undergone consumption in the stellar core, the double shell source that surrounds the red giant core is unstable because of the extreme temperature sensitivity of the He and CNO processes combining to produce the observed luminosity. This double shell source instability, first recognized by Schwarzschild and Harm in 1965, is one of the few, and still best studied, examples of nuclear-driven instabilities.

The primary cause of stellar pulsation is that the convection zones of the envelope, which correspond to ionization zones, are thermally unstable. When the zones expand and cool, the ionization drops and so does the opacity. When compressed, they heat up and the ionization goes up thereby increasing the heat capacity (the value of γ drops). Detailed nonlinear modeling of both Cepheids and RR Lyrae stars confirms this expectation. Because of the large energy associated with the ionization of neutral (He I) and ionized helium (He II) and because these zones lie fairly deep in the stellar envelope, they serve as the

sources for pulsational driving in most classes of variable stars. The explanation lies in another aspect of the driving of the instability. If the mass overlying the driving zone is too large, the mechanical impulse of the instability in the convection zone will be damped before it can raise the surface layers substantially, and the star will be stable, or of very low amplitude. Should the mass above the zone be too small, however, the inertia of the zone will be small, and it will not be able to drive the compression sufficiently on fall-back to continue the pulsation. In addition, if the convection zone is too near the surface, the layers will be optically thin and the energy of the pulsation will be too easily dissipated by radiative processes. Thus, a delicate balance exists between the depth of the pulsational driving zone and the conditions in the zone, and this serves as the simple explanation as to why most stars do not vary.

Intuition would lead one to expect that, when the star is smallest, the luminosity would be greatest since the decrease in surface area should be more than compensated by the increase in the surface temperature. The observations show, however, that maximum luminosity usually occurs at maximum expansion velocity, the phase lag being about 0.25 of the period. This effect is produced by the presence of the convection zones, mainly the surface hydrogen ionization region, which stores the wave of luminosity coming from the driving He II zone deep in the interior and slowly releases it on expansion. This nonadiabatic behavior is largely due to the variation in both the opacity and specific heat ratio in this zone as a function of phase in the pulsational cycle.

Nonradial pulsation will result if the star is not spherical, for instance because of rapid rotation or the presence of a close perturbing companion in a binary system. Such pulsations are often nearly adiabatic, involving only the stellar atmosphere and, consequently, of short period and low amplitude. In the case of the sun, for example, there is no organized resonant radial pulsation, but the envelope shows a rich spectrum of modes because of the mechanical instability of the convection zone. Nonradial modes are driven by gravity rather than thermal instabilities and are often more sensitive to the equation of state than to the opacity (the β Cep stars may be an exception to this).

IV. Classification of Intrinsic Variable Stars

Classes of variable stars are generally named after the first discovered, or prototypical, member of the class. Subsequent to the recognition of a given phenomenology, it sometimes appears that the namesake is not the best representative of the class, but the name persists. As in all taxonomy, there is a dichotomy between the "splitters" (those who relish the proliferation of subclasses for every departure from the precisely defined properties of a class) and "lumpers," and in this article, the latter presentation is followed. In this article, a broad separation is made between the radial and non-radial pulsators, driven in large measure by the differences in the theoretical models required to understand their behavior.

Variable stars are named according to a standard international rule. The stars are named in order of discovery. The first star discovered in a given constellation is named R (the first letter that had not previously been used in star catalogs during the nineteenth century), if not already named with some other designation (such as Greek, or Bayer names). The single letters are exhausted with Z, and the sequence restarts with RR through ZZ, then goes from AA through QZ before starting as V335 and continuing. The central clearing house for all variable star designations and properties is maintained by the International Astronomical Union in Moscow, which publishes the "General Catalogue of Variable Stars." The current edition (the fourth, complete through 1982) contains about 28,450 stars, listed by constellation.

A. RADIAL PULSATORS

Most radial pulsators are confined to a nearly vertical strip on the Hertzsprung–Russell (HR) diagram, called the instability strip. In the range of temperature represented by the strip, the stars have thermally and mechanically unstable envelopes.

1. Delta Scuti Variables

The Delta Scuti (δ Sct) stars are main sequence and slightly evolved, intermediate mass stars, A-type stars that show small amplitude (0.003 to 0.9 magnitudes) pulsations in a period range from about 0.01 to 0.2 days. Their luminosities range from 10 to 100 $L_\odot$. They are primarily driven by the surface convection zones, and only a small amount of mass is involved with the pulsational activity of the star. Several classes of stars coexist in the HR diagram with these stars, the δ Del stars, Am stars, and Ap stars (see Section IV, B, 3). The maximum radial expansion velocity closely matches the phase of maximum light to within about a tenth of the period. These variables are numerous but difficult to detect because of their short periods and small amplitudes; however,

because many are found in clusters, their evolutionary status is well understood. They form the lower luminosity (main sequence) end of the Cepheid instability strip.

2. RR Lyrae Stars

The RR Lyrae (RR Lyr) stars are identified with a unique stage in the life of a low-mass star, the horizontal branch, helium core burning period of evolution. As such, they show a small range in luminosity, the most variable property being their effective temperature as a function of mass. The locus of stellar masses on the horizontal branch is determined by helium core mass. The RR Lyr stars range in spectral type from A3 to A6, and have absolute magnitudes of about $0.^m5$, about $10^2 L_\odot$. They are low-mass stars, 0.5 to 0.7 $M_\odot$. The pulsation is driven by instabilities in the He II convection zone. The RR Lyrae stars display several different light curves, which are correlated with their periods. The RR(b) stars show both fundamental and first overtone pulsations, the typical period ratio being about $P_1/P_0 = 0.75$; the prototype of this class is AQ Leo. For those RR Lyr stars with only a one observed period, Bailey originally divided the "cluster variables" into two main subclasses. The RRab stars have periods ranging from 0.3 to 1.2 days and amplitudes typically from 0.5 to 2 magnitudes. They are characterized by "sawtooth" (steep ascending branches) light curves. The prototype of the class, RR Lyr, is a member of this subclass. The RRc stars show small amplitudes, generally less than 0.8 magnitudes, short periods from 0.2 to 0.5 days, and essentially sinusoidal light curves. As in the δ Sct and δ Cep stars, the maximum expansion velocity corresponds to maximum light.

The RR Lyr stars are best studied as members of globular clusters, where they appear in large numbers because of the sizable population of the horizontal branch in these evolved stellar systems. They can be classified according to a metallicity index, first proposed by Preston, which compares the spectral type assigned from the metal lines, especially Ca II, to that obtained from the hydrogen Balmer lines. This ΔS index is a useful indication of metal abundance and shows that the RR Lyr stars are low-mass Population II stars, related in their pulsational properties to the W Vir and RV Tau stars, but of very low mass.

3. Delta Cepheid Variables: Classical Cepheids

These are the prototype variable stars and the first class for which the period luminosity relation was discovered. They are generally moderately massive stars, being above 2 or $3M_\odot$, evolved, and in the stage of helium core burning. They reside in the instability strip, in fact defining the position of the strip on the HR diagram, and have single modes of pulsation with periods between about 1 day and 2 months.

Most δ Cep stars have periods ranging from 1 to about 140 days, with amplitudes ranging from 0.01 to 2 magnitudes. In general, their variations are greater at shorter wavelength and longer period. The stars are usually spectral type F at maximum, and G or K at minimum. Their luminosities range from $5 \times 10^2 L_\odot$ to about $2 \times 10^4 L_\odot$, depending on their mass and evolutionary status. The classical Cepheids are Population I stars, occasionally residing in open galactic clusters from which their approximate masses of 3 to $10 M_\odot$ are obtained. Several have recently been shown to be members of binary systems, notably SU Cas. Their radial velocity variations are characteristic: with only a small phase shift (about 0.1 in phase), the maximum expansion velocity corresponds to maximum light. δ Cep light curves are usually distinctive, being steep on the ascending branch with little structure to the curve.

A subset of the δ Cep stars, called Cep(B) or bump Cepheids, show multiple periodicity, usually with $P_1/P_0 = 0.7$ and P_0 between 2 and 7 days. TU Cas and V367 Sct are typical of this subclass.

Finally, the DCepS subclass shows amplitudes that are always less than 0.5 magnitudes and almost symmetric light curves. Their periods rarely exceed 7 days. The well-studied star SU Cas is typical of this subclass. These stars seem to be pulsating in the first harmonic (hence, the smaller amplitudes and shorter periods), while classical Cepheids appear to be pulsating in the fundamental mode.

The Cepheids arise from a high-mass population, and as such they do not represent a unique stage in the life of a star. Following hydrogen core exhaustion, stars greater than about $3M_\odot$ will cross the instability strip at least once. The higher mass stars, greater than about $7M_\odot$, will traverse the strip several times, including the post-helium-core burning stage, each time with increasing luminosity; as many as 5 crossings have been calculated for stars above this mass.

Theoretical models provide the following calibration for the Cepheid period–luminosity relation:

$$P_F \sim \left(\frac{L}{L_\odot}\right)^{0.83} \left(\frac{M}{M_\odot}\right)^{-0.66} T_{\text{eff}}^{-3.45} \quad (12)$$

where P_F is the fundamental period and T_{eff} is the effective temperature. Turning this relation around, one can derive a mass for the Cepheid from theoretical models, called the pulsational mass, if the lumi-

nosity, surface temperature, and period are known. Few masses are well known, but the pulsational mass determinations are quite sensitive to the temperature determinations for these stars, while the evolutionary masses seem to agree with the few stars for which masses have been obtained. For example, for SU Cyg, the Cepheid mass is about $6M_\odot$, which agrees with the evolutionary mass.

The δ Cep stars are also important as distance indicators, in large measure because of their place in the instability strip; as a result, an enormous effort has, in the past 40 years, been put into the determination of intrinsic properties of these stars, especially the study of their masses and luminosities. The discovery, in the past 5 years, of large numbers of binary stars among the Cepheid variables has been an important clue to these properties. Many of the companions are still on the main sequence and are of sufficiently early hot spectral type that they must be moderately massive. They are best studied using ultraviolet spectra, in which the spectrum of the Cepheid variable does not appear. Masses can be determined from a comparison of the Cepheid radial velocity curve, obtained from the optical spectrum, with that for the companion, obtained from the UV. In consequence, the lower mass of the Cepheid can be understood, and the time scale for the evolution of the variable specified within limits. The binaries will also serve as useful calibration of the evolutionary and pulsational mass determinations, once a large enough number of them have been identified.

An additional determination of intrinsic properties is provided by the use of the infrared. For stars of surface temperatures less than about 8000 K, the peak of the spectral distribution falls in the red. Therefore, the use of the IR, wavelengths of 1 to 3 μm, can give a clear indication of the bolometric variability of the star and is not sensitive to changes in the shape of the spectrum. Further, since the infrared is far less sensitive to the effects of interstellar reddening than the optical, the zero point for the magnitudes can be more reliably determined. The slope of the relationship depends on the wavelength at which the variations are observed, but the mean stellar magnitude as a function of log P appears to be invariant between galaxies, independent of stellar metallicity, for classical Cepheids.

4. W Virginis Stars: Population II Cepheids

These are the Population II analogs of the classical Cepheids. They arise from a lower mass population, from about 0.5 to $0.8M_\odot$, but being similarly internally structured, they display the same mechanical and photometric properties as the more massive

Population I stars. They range in period from 0.5 to 35 days and have amplitudes in the range 0.3 to 1 magnitude. These stars also display emission in hydrogen Balmer lines. There are two subclasses identified among the W Vir stars. The CWA stars, like W Vir, have periods greater than 8 days; CWB stars, like BL Her, have shorter periods. In BL Her, which has also served as the prototype of a subclass of W Vir stars, bumps are observed after maximum on the descending portion of the light curves.

The W Vir stars are fainter than classical Cepheids at the same period by between 0.7 and 2 magnitudes, a fact that has important cosmological consequences. Since the extragalactic distance scale rests in large measure on the Cepheid variables as distance calibrators for relatively nearby galaxies, the period–luminosity relation is used to determine the distance modulus, $m_j - M_j$. Here, m_j and M_j are the apparent and absolute magnitude in the jth wavelength band. With only the period and apparent magnitude in hand, one can determine the distance to a galaxy knowing, from the P–L (or period–luminosity–color or P–L–C) relation, the intrinsic luminosity of the star, hence its distance. In the late 1940s, the inclusion of the W Vir variables in the relation produced an overestimate of the magnitude of the Hubble constant because of an underestimate of the distance to the external galaxies. The discrepancy between the age of the globular clusters and the apparent age for the universe was the leading spur for the creation of the steady-state theory of cosmology. With the removal of this problem and the recalibration of the P–L–C relation, the empirical foundation for this cosmological model collapsed.

5. Long-Period Variables

a. Mira Variables. Mira variables, named after o Cet, are emission-line cool giants, generally arising from a population of low to intermediate mass (1 to $2M_\odot$) stars. Miras typically have luminosities of about $10^3 L_\odot$ and temperatures less than 4000 K. They are found among the Me, Ce, and Se stars, often displaying time variable emission lines of both atomic and molecular species. These stars have some of the largest amplitudes observed among pulsating variables, ranging from 2.5 to 11 magnitudes in the optical, but generally considerably smaller in the infrared (at a wavelength of a few micrometers, the amplitude is less than one magnitude). The periods range from 80 to over 1000 days, although many of the longest period systems are quite irregular in their light curve behavior. A characteristic of the light variations is the change in the light curve structure with time. Miras in general do not appear to be rig-

orously periodic, with amplitudes that do not repeat from one cycle to another at any wavelength and considerable evidence for mass loss and shock processes accompanying the pulsation cycle.

The Miras are perhaps the best studied examples of nonlinear pulsators, having amplitudes large enough to produce the ejection of matter and strongly nonadiabatic behavior of the envelopes. One interesting aspect of these stars is that their pulsation periods are sufficiently long, and their amplitudes large enough, that time-dependent convection may be important in modeling their internal structure. Considerable theoretical attention is currently being paid to these stars.

b. RV Tauri Stars. The RV Tau stars are intermediate temperature (F and G) supergiants with luminosities of approximately $10^4 L_\odot$, highly evolved, and consequently of long period. They are often associated with old Population I or Population II. Their periods range from 30 to 1000 days, and their amplitudes are about 3 to 4 magnitudes.

The RV Tau stars are distinctive because of their light curves, which on the short-period end resemble classical Cepheids, but which for the long-period stars are quite distinctive. The longer period subclass, called RVb stars, shows variable mean magnitudes, with a variation in the amplitude of maximum between successive cycles in a periodic way. The RV Tau star is the prototype of this class. The RVa stars, typified by AC Her, show no variation in the mean magnitude.

c. R Corona Borealis Stars. The R CrB stars are hydrogen deficient, helium rich, and often carbon rich, high-luminosity stars, characterized by a semiregular behavior, which is unique among variable stars. They undergo quasi-periodic fadings during which time they can decrease in brightness by more than 1 to about 10 magnitudes on a time scale of 30 to several hundred days. The intervals between these events are irregular; some low amplitude periodicity of less than a year may be present as well. These episodes are marked by the ejection of considerable amounts of dust, as determined from ultraviolet satellite observations of their spectra. Optically, they display high polarization. The mechanism responsible for this behavior is not understood presently. The R CrB stars, which are poorly understood, also span much of the temperature range of the stellar population, occupying spectral types Be to the carbon stars. These stars show very high luminosities, about $5 \times 10^4 L_\odot$.

d. Semiregular Variables. Finally, there are high-luminosity stars that show photometric and radial velocity changes on time scales of months to years that do not appear to be periodic, although there is a characteristic interval of time during which the brightness of the star is observed to change. The supergiants are among the best examples of this class of variable, perhaps the best known being the M supergiant Betelgeuse (α Orionis). In many ways, these stars resemble the RV Tau and Mira variables, but without the regularity of variability associated with these other stars.

The SRc, or supergiant semiregular variables, have amplitudes of up to one magnitude and periods from tens to thousands of days. Their luminosities range from 10^4 to $10^5 L_\odot$, and temperatures are generally in the range of 2000 to 4000 K. The SRc's overlap with the R CrB stars. The OH/IR stars, red supergiants that display OH maser emission and large middle and far infrared luminosities, may be related to these stars. Other subclasses of the semiregular variables are the SRa and SRb stars, giants with similar periods but lower luminosities, and PV Tel stars, which are helium-rich, long-period supergiants of low amplitude (several tenths of a magnitude). The SRa and SRb stars overlap with the Mira variables.

6. Luminous Blue Variables

The luminous blue variables (LBVs) are a recently recognized class of massive supergiants. Also called S Dor variables, after their Large Magellanic Cloud prototype, these are the most luminous stars in a galaxy and easily identified in extragalactic systems. Their luminosities are often about $10^6 L_\odot$, and inferred masses are in excess of $30 M_\odot$. Their amplitudes range from less than 1 magnitude to over 10. Also called Hubble–Sandage variables, they display long (years) time scales for temperature and luminosity changes, often ejecting shells that form pseudo-photospheres, causing their spectral types to change from O to as late as F. The galactic supergiant P Cyg, well known for its high rate of mass loss and irregular light variations, is a galactic member of this class. Others recently recognized as LBVs are HD 269858f in the LMC and AE And and AF And in M 31.

B. NONRADIAL PULSATORS

1. β Cephei Stars

The name of this class is dependent upon which side of the Atlantic one is on. The Europeans typically refer to these as β Canes Majoris stars, while the North Americans typically use β Cep as the prototype.

These stars are the prototypes of nonradial pulsa-

tion. They do not show simple mode structure, but often pulsate in several different modes at once. Their periods are extremely short, about 3 to 6 h. Several Be stars (emission line B stars that are generally rapid rotators) are also in this class, the most notable being λ Eri. The best studied subclass, the short-period group, lies around spectral type B2-3 IV-V, displays periods from 0.02 to 0.04 days, and has amplitudes between 0.015 and 0.025 magnitudes. The short period and the high-mode number are indicative of atmospheric or shallow envelope pulsation, but the details of the driving mechanism for these stars are not currently understood.

The β Cep stars show an instability strip unrelated to that for the classical Cepheids. It is roughly parallel to the main sequence, and at luminosity class IV, characteristic of post-hydrogen-core exhaustion stars. A recently suggested mechanism, by A. N. Cox, is that the envelopes of these stars may have time-dependent convective instabilities as the driving mechanism. An alternative, suggested by R. F. Stellingwerf, is that the pulsation is opacity-driven and that models will require an enhanced opacity at temperatures of about 10^5 K.

Apparently related to these stars is a class of massive pulsators, which generally lies between O8 and B6, covering the range in luminosity from main sequence to supergiant. These nonradial pulsators show periods between 0.1 and about 1 day and small amplitudes, usually less than 0.3 magnitudes. Maximum brightness corresponds to minimum radius, indicative of adiabatic pulsation but distinctly different from the normal behavior observed for the radial pulsators. Spectral lines in these stars show the effects of distortion of the photosphere, varying in shape during the pulsation period in a fashion that indicates distortion of the stellar surface.

2. White Dwarf Stars

Several classes of variables are found among degenerate stars. The hottest are called GW Vir stars, or PG 1159-035 objects, after their prototype. These stars are extremely hot helium white dwarfs, typically with temperatures of about 10^5 K. They show small amplitude pulsation, which is usually observed in many nonradial modes. Because of the dependence of the period on the temperature of the atmosphere, the star's period can change during the course of centuries as the dwarf cools. The pulsation periods are about 500 sec, indicative of nonradial modes; radial pulsation periods for degenerate stars are several orders of magnitude smaller than this. Several of the central stars of planetary nebulae have been found to be members of this class, with long periods up to about 2000 sec, indicating extensive atmospheres and temperatures that may be as high as 150,000 K.

The cooler class, which lies in the instability strip and was one of the few predicted classes of variable stars, is the ZZ Ceti stars. These are DA-type white dwarf stars, with temperatures between 10^4 K and 2×10^4 K. Here, the driving is due mainly to the hydrogen convection zone. Periods are about 500 to 1000 sec, and masses are about $0.6 M_\odot$. The DBVs, helium-rich analogs of the ZZ Cet stars, have temperatures between 2×10^4 K and 3×10^4 K but otherwise similar properties. In both cases, the amplitudes are between a few tens and hundreds of millimagnitudes. Among the cataclysmic systems, there are many white dwarf systems that show multiple periodicities, but it is not clear whether these should be ascribed to pulsation or not.

3. The Rapid Magnetic Pulsators

These recently discovered stars have been shown to be main-sequence stars with strong magnetic fields that inhabit the instability strip at the same place as the δ Sct stars. The difference between these and the normal variables of the main sequence is that they display time variable signatures of pulsation, which are due to the site of surface displacement. They provide a unique chance to study the interaction between pulsation and stellar magnetism.

The periods range from about 5 to 25 min, indicative of atmospheric pulsation, and they show very small amplitudes (typically less than $0.^m01$. The pulsation appears to be active only in the vicinity of the magnetic poles. Thus, as the star rotates, the polar region crosses the line of sight and the pulsation is observable. When the magnetic equator crosses the observer's sight line, the pulsation disappears. The mode of the pulsation is very high, typically $l = 20–40$. The presumed driving is essentially the same as for the δ Sct stars, that is, the hydrogen convection zone is unstable. According to Shibahashi and Cox, the modification introduced by the magnetic field is to suppress pulsation at the magnetic equator by the addition of a strong restoring force, while producing an overstability at the poles. To date, about a dozen such systems are known, all confined to stars with magnetic fields of several hundred to several thousand gauss.

V. Pre-Main-Sequence Variables

The variability of stars in the stage before the onset of hydrogen core burning is not presently well understood. Intrinsic variations are observed in both the T Tauri stars, which are the young, late-type stellar population associated with, but not imbedded in, their parent molecular clouds, and the FU Orionis stars. The latter are notable for large amplitude outbursts, sometimes of more than 5 magnitudes, taking place over a period of months to years. These have been recently explained as arising from an accretion disk instability in the environment of the forming star. Neither of these classes of stars appears to be pulsationally unstable, although detailed modeling is difficult. All stars in evolutionary stages are characterized by emission line spectra and strong stellar winds. One subclass, the YY Ori stars, shows reverse stellar wind profiles, with absorption on the red side of the emission line suggestive of mass accretion.

VI. Concluding Remarks

As a tool for the study of stellar interiors, pulsation theory and observations of stellar variability are only now maturing. Increased instrumental sensitivity is allowing the study of very small scale, nonperiodic oscillations in solar-type stars, the field of helio- and asteroseismology; and it is now possible to compute physically interesting nonlinear, nonadiabatic pulsation models for a wide variety of stellar interior conditions. The effects of magnetic fields and rotation are being included in theoretical computations. Improvements in detectors, especially the widespread use of CCDs, are quickening the pace of discovery and study of faint variables. Improved model atmospheres allow for the study of stellar abundances with increased precision. The advent of true flux calibration spectra, again using CCD detectors, allows for direct comparison with models and much improved radial velocity determinations at many different wavelengths.

The field of variable star research is also a fruitful one for both amateurs and professionals with access to small telescopes; it is a field in which every newly determined light curve makes a substantive contribution to our knowledge of the cause and behavior of pulsating stars.

BIBLIOGRAPHY

Cox, J. P. (1974). *Reports Prog. Phys.* **37,** 563.
Cox, J. P. (1980). "Theory of Stellar Pulsation." Princeton Univ. Press, Princeton, New Jersey.
Cox, A. N., Sparks, W., and Starrfield, S. G., eds. (1987). "Stellar Pulsation." Springer-Verlag, Berlin and New York.
Hansen, C. J. (1978). Secular stability, *Annu. Rev. Astron. Astrophys.* **16,** 15.
Hoffmeister, C., Richter, G., and Wenzel, W. (1985).

"Variable Stars." Springer-Verlag, Berlin and New York.

Johnson, H. R., and Querci, F. R., eds. (1986). "The M-Type Stars." NSA SP-492, NASA, Washington, D.C.

Kholopov, P. N., ed. (1985). "General Catalogue of Variable Stars: 4th Edition" (3 vols.). NAUKA Publishing House, Moscow.

Ledoux, P., and Walraven, Th. (1958). *In* "Handbuch der Physik" (E. Flügge, ed.), Vol. 51, p. 353. Springer-Verlag, Berlin and New York.

Madore, B. F., ed. (1985). "Cepheids: Theory and Observations." Cambridge Univ. Press, London and New York.

Percy, J., ed. (1986). "The Study of Variable Stars Using Small Telescopes." Cambridge Univ. Press, London and New York.

Petit, M. (1987). "Variable Stars." Wiley, New York.

Rosseland, S. (1949). "The Pulsation Theory of Variable Stars." Oxford Univ. Press, London and New York.

Unno, W., Osaki, Y., and Shibahashi, H. (1979). "Nonradial Oscillations of Stars." Tokyo University Press, Tokyo.

X-RAY ASTRONOMY

Jonathan E. Grindlay *Harvard University*

GLOSSARY

Accretion: Process of matter being drawn near and down onto an object by the gravitational attraction of the object. It is the process by which matter falls onto compact objects such as neutron stars to produce luminous X-ray emission.

Accretion disk: Thin disk structure through which the gas spirals inward to reach the compact object at its center.

Alfvén radius: Inner edge of an accretion disk, or accretion flow, where the magnetic field from the central compact object exerts a pressure on the inward flow of gas that is comparable to the dynamical pressure (from gravity of the compact object) in the flow itself.

Black hole: Compact object, usually resulting from the end point of the evolution of either a very massive star or a very dense star cluster, which has collapsed under its own gravity to such a small size that not even light can escape its intense gravitational field.

Compact Object: Very dense, and therefore very small, object with mass (typically) of at least the mass of the sun (1 $M_\odot$) that has collapsed to this extreme state by processes at the end points of stellar evolution. In order of decreasing size for an object with mass 1 $M_\odot$, the possible compact object states are white dwarf ($\sim 10^4$ km radius), neutron star (~ 10 km radius), and black hole (~ 3 km radius).

Coronas Extended clouds of hot X-ray-emitting gas that can surround ordinary stars (e.g., the sun), accretion disks, and entire galaxies, where they are more commonly referred to as galactic *halos*.

Dark matter: Apparently major constituent (90%) of mass in the universe, detected by the gravitational motions of luminous matter not directly detected itself as a source of visible light or X rays.

Energy band: Soft, intermediate, and hard X-ray energy bands refer to, respectively, X rays at low, medium, and high energies. The detection techniques and the astrophysical questions each band can address are overlapping but different.

Neutron star: Collapsed remnant core of a relatively massive star that is sometimes produced in supernova explosions. The neutron star has a density comparable to that of the nuclei of atoms ($\sim 10^{14}$ g/cm^3) and therefore a radius of only ~ 10 km for a typical mass of 1.4 $M_\odot$.

Supernova remnant: Debris of an exploded star, a supernova, which emits thermal X rays for a period of perhaps 5×10^4 years after the explosion.

White dwarf: Collapsed remnant core of a less massive star (such as the sun) that is the normal end product of stellar evolution. The white dwarf has a density some 10^5 times that of lead (or $\sim 10^6$ g/cm^3) and is thus only the size of the earth ($\sim 10^9$ cm in radius), although its mass is typically ~ 1 $M_\odot$.

X-ray astronomy is the study of cosmic X-ray sources using detectors and telescopes above the atmosphere of the earth. As one of the newest fields of astronomy, it has already become as rich in scope and promise as the more traditional fields of optical or radio astronomy. Virtually all classes of astronomical object (from planets to stars to galaxies) have been found to be emitters of X rays. It is the study of these emissions and the physical nature of the objects that their X-ray emission reveals that is the main province of X-ray astronomy. X-ray astronomy also allows the study of compact objects such as neutron stars and black holes, as well as the study of the most energetic processes in nature. Because of the great penetrating power of X rays, cosmic X-ray sources can be studied in regions as obscure as the nuclei of galaxies and as remote as the most distant quasars.

I. Overview

A new window on the study of celestial objects and cosmic phenomena has been opened by measuring and interpreting the X-ray emissions (or absorptions) they produce. Both the observational study of cosmic X-ray sources and the astrophysics of their interpretation are included in the field of X-ray astronomy. (Throughout this article we use the term ''X-ray astronomy'' to describe both the observational and the theoretical studies carried out on cosmic X-ray sources in the more general field of high-energy astrophysics.) A popular misconception is that somehow X-ray astronomy proceeds by ''X-raying'' the objects to be studied, implying that the X-ray astronomer somehow sends out an X-ray beam to study the objects or phenomena of interest. Not only would such a procedure be doomed by the immense distances and sizes of the astronomical objects for study, but it would also pale in comparison with the rich variety of totally natural X-ray radiation emitted by cosmic objects such as stars and galaxies. Whereas these objects are traditionally studied by measuring the optical or infrared light they produce (or, again, absorb), at still shorter wavelengths—beyond the violet and ultraviolet regions of the electromagnetic spectrum—lies the vast region of the X-ray (and, finally, the gamma-ray) portion of the spectrum.

The X-ray band may be defined for photons with energies in the approximate range 0.1–100 keV. This may be contrasted with the relatively narrow band of optical photons visible to the human eye: from approximately 4000 to 7000 Å, or equivalently 2–3 eV = 0.002–0.003 keV. The unit of energy, the kilo-electron-volt (keV), is more frequently used in X-ray astronomy than the inverse units of wavelength (the angstrom,

or Å, where 1 keV corresponds to 12.4 Å). Nevertheless, the short wavelengths of X rays remind us that their production and propagation processes are closely related to the inner-shell electrons of atoms and ions and to the interactions of high-energy electrons with matter.

X rays are the natural form of electromagnetic radiation emitted by matter raised to temperatures in excess of about a million degrees. While such high temperatures may seem extreme by terrestrial standards, they are in fact relatively common on an astronomical scale in the so-called astrophysical plasmas. The plasma state of matter, where most (if not all) the electrons have been stripped from the constituent atoms of matter by virtue of its very high temperature, is in fact the most common state of matter in the universe. Thus X-ray emission is a particularly natural process for astronomical objects over a broad range of object types and masses.

X rays are also produced by the interaction of high-energy particles, such as electrons, and fields. When high-energy electrons with energies so high as to be relativistic—that is, for total energies (greatly) in excess of the rest mass energy of the electron—traverse a magnetic field, the electrons lose energy to radiation by the synchrotron process. If the electron energies are high enough, for a given strength of magnetic field, the synchrotron radiation is emitted in the X-ray region of the electromagnetic spectrum. Relativistic electrons can also produce X rays by scattering lower-energy photons of radiation, from the radio or optical regimes, into the X-ray portion of the spectrum. For both this ''inverse Compton'' process and the synchrotron mechanism, electron energies typically in excess of a few hundred times the electron rest mass (i.e., Lorentz factors $\gamma \gtrsim 10^2$) are needed to produce X rays under typical cosmic conditions. In many cases, however, the necessary electron Lorentz factors are in excess of $\sim 10^6$ for such nonthermal production of X rays. Thus X-ray astronomy also allows the study of the vast realm of nonthermal or high-energy phenomena in the universe.

II. Historical Background

A. THE FIRST X-RAY SOURCE DISCOVERIES

The first discovery of a cosmic source of X rays was made as recently as in 1962, when a sounding rocket equipped with several small

Geiger counter type detectors found an unexpected intense source of X rays in the constellation Scorpius. In keeping with the tradition already established by the radio astronomers, the source was accordingly named Sco X-1. The discovery was totally unexpected, since the strongest sources of cosmic X-ray sources anticipated were X rays from individual stars like our sun, which would be far too faint at their much greater distances to be detectable. Instead, the rocket had been programmed to scan across the moon to search for the fluorescence X rays expected due to the bombardment of the solar surface with the (then) recently discovered solar wind of high-energy particles. Both the lunar fluorescence and ''normal'' stellar X rays were in fact measured years later with much more sensitive investigations (in the lunar case, made possible by the Apollo moon landing program), but the discovery of Sco X-1 had already opened up a rich new field of high-energy astrophysics. The intense X rays from Sco X-1, which were progressively better localized on the sky over the next 5 years and finally identified with a moderately faint blue and variable star-like object, were eventually recognized to arise from the intense accretion of matter down on to the surface of an incredibly collapsed state of matter: a neutron star. Neutron stars, which had been predicted theoretically by Oppenheimer in the late 1930s and which were identified with the radio pulsars upon their discovery in 1967, were thus first ''observed'' with the discovery of the intense X-ray emission from Sco X-1 in 1962. [See PULSARS.]

Over the next decade, the number of bright X-ray sources discovered grew to more than 30 as a series of rocket and high-altitude balloon flight investigations were carried out. Most of the sources were found to be time-variable, even within the short 5-min observing times available for observation above the earth's atmosphere on a sounding rocket. Thus, although these early measurements were only able to locate source positions on the sky to a fraction of a degree, it was evident that the sources were point-like and probably involved compact objects. It was not until the first all-sky survey for X-ray sources was carried out with the first satellite devoted to X-ray astronomy, the Small Astronomy Satellite-1, or SAS-1, which was launched by the National Aeronautics and Space Administration (NASA) in December 1970, that the nature of the compact X-ray sources began to become clear. SAS-1, called UHURU by its creators

(from the American Science and Engineering Company in Cambridge, Mass., the same group that had previously discovered Sco X-1) and the scientific community, found that several of the bright X-ray sources were not only variable but were periodic in their X-ray output on several different time scales. Pulsations were detected with X-ray periods of a few seconds, while the overall intensity of the source was in turn modulated on a longer time scale of a few days. The overall interpretation was obvious: the X rays were produced (somehow) by matter falling down on to the magnetic poles of a strongly magnetized and spinning neutron star orbiting a companion star from which the matter was supplied. This general picture and its many variations and interesting physical implications are described in much more detail in Section IV.

B. THE ALL-SKY SURVEYS

UHURU enabled the discovery of some 400 X-ray sources over the whole sky during its several years of operation. In addition to the luminous X-ray binaries, X-ray sources were identified with supernova remnants (the Crab nebula remnant was in fact the first source to be optically identified, by the group at the Naval Observatory, from rocket flight observations prior to UHURU), white dwarf binaries, entire galaxies (our closest neighboring large galaxy, M31), clusters of galaxies, and active galaxies and quasars. The discovery of X-ray emission from galaxy clusters and from active galaxies were particularly significant and comparable in importance to the discovery of X-ray binaries. A map of the apparent positions of the sources in the UHURU catalogue of 2- to 10-keV X-ray sources is shown in Fig. 1. Source positions (dots) are shown plotted in the astronomer's system of galactic coordinates (for which the galactic center is at zero degrees galactic latitude and longitude as viewed from the solar system), and the relative X-ray intensities are coded by the size of the dot. Several of the most prominent X-ray sources and representatives of a number of classes of both galactic and extragalactic sources are individually labeled.

The many discoveries made with UHURU motivated the scientific community and NASA to plan for a series of much larger and more sensitive X-ray satellite missions in the late 1970s: the High Energy Astronomical Observatory (HEAO) program. The latter half of the 1970s also saw important X-ray astronomy satel-

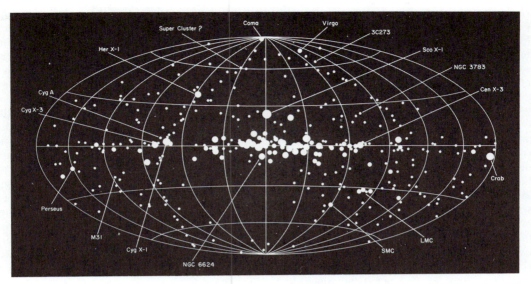

FIG. 1.   Map of cosmic X-ray sources plotted in galactic coordinates and showing the relative intensity and source type for several major classes of source.

lites launched and operated by the Dutch (and United States, in a joint effort), the British, and the Japanese. These programs enhanced the activity and scientific discovery of a number of groups of X-ray astronomers around the world.

The HEAO-1 mission carried out a much more sensitive all-sky survey, with energy coverage extending (with a series of different instruments) from approximately 0.1 to 200 keV (one instrument even had response extending into the low-energy gamma-ray range, to about 3 MeV). The total number of sources was increased to approximately a thousand, and source locations were possible to determine to accuracies of ~30 arcsec in many cases (versus the typical UHURU source locations of ~30 arcmin). Comparably precise X-ray positions had already been obtained for a number of sources by similar detector systems on the SAS-3 satellite, which studied selected X-ray sources during 1975–1977. The improved source positions enabled the optical identifications of many more sources, and new classes of objects were found: from certain classes of "normal" binary star systems (i.e., not containing a collapsed star such as a neutron star or white dwarf) to the first examples of relatively low-luminosity active galactic nuclei. At the sensitivity limits reached with the HEAO-1 survey, approximately half of the sources found were distributed along or near the plane of our galaxy, indicating their galactic nature, while half were found at higher galactic

latitudes and toward the poles of our galaxy, indicating their extragalactic nature.

In November 1978, the HEAO-2 satellite was launched by NASA and the era of imaging X-ray astronomy was ushered in. Renamed the Einstein Observatory by the scientific community, HEAO-2 carried out a 2½-year program of pointed observations of a wide variety of objects and found that virtually all classes of celestial objects are X-ray emitters. Its sensitivity was nearly a thousand times better for the detection of low-energy (0.1–4 keV) X rays than any previous X-ray detector system, and its grazing incidence optics telescope and position-sensitive X-ray detectors at the telescope focal plane allowed the first images of cosmic X-ray sources to be obtained (all previous X-ray satellites and rocket or balloon experiments had incorporated only nonimaging detector systems). The angular resolution of the telescope and detector system was ~2 arcsec, so that the X-ray images obtained could be nearly as sharp as optical images obtained with ground-based optical telescopes.

C. THE EINSTEIN OBSERVATORY: X-RAY ASTRONOMY IMAGE ESTABLISHED

The Einstein Observatory had a profound effect on observational and theoretical studies of many of the most pressing problems in astrophysics. Among its many discoveries summarized in this article, the Einstein Observatory opened up a new field of investigation of the hot coronas of stars and allowed the pervasive dark matter in the universe to be traced indirectly by the hot gas it confines in the halos of galaxies. Despite remarkable discoveries such as these, X-ray astronomy in the United States has had no flight opportunities since, due to budget restrictions at NASA. A modest X-ray telescope and a nonimaging higher-energy X-ray detector was launched by the European Space Agency (ESA) and operated for several years until it failed in April 1986. This European X-ray Observatory Satellite (EXOSAT) was much more limited than Einstein in its imaging capability and provided its most significant results from its nonimaging study of the time variability of X-ray sources.

III. Techniques and Telescopes

A. Energy Band

Cosmic X-ray astronomy has achieved most of its results in the energy band of approximately 0.5–10 keV. The upper end of this band, the "intermediate" band from 2 to 10 keV, is the band most successfully explored in all the X-ray satellites and experiments prior to the Einstein Observatory, whereas the "soft" X-ray band from approximately 0.5 to 4 keV was the range covered by Einstein. These somewhat overlapping bands have differing limitations in both the types of objects they typically detect and the detection techniques that can be used. Soft X rays are preferentially absorbed by matter (due to the photoelectric effect), so that higher sensitivity is needed to detect soft X-ray sources at great distances in our galaxy through the absorbing gas and dust. This same photoelectric absorption is what makes X-ray astronomy at soft and intermediate energies impossible from even the highest-altitude balloons (at altitudes of some 40 km), which can be used for "hard" X-ray astronomy (at energies above about 20 keV), and instead requires that these experiments be flown on satellites in earth orbit or beyond. The soft band enables the X rays to be focused by specially constructed mirrors so that telescopic images are possible, while the intermediate band is more difficult in this respect and has thus far only been explored with nonimaging detectors.

B. X-Ray Detectors

The most commonly used detector for cosmic X-ray astronomy, in both the soft and intermediate bands, is the proportional counter. These de-tectors may be configured to record simply the detection and energy of individual X rays or the X-ray position (on the detector) as well. Position sensitivity is necessary if the detector is used with an imaging telescope. The basic proportional counter detector is a sealed chamber filled with an inert gas (usually argon or xenon) in which an X ray interacts with the gas by the photoelectric effect and is detected by the amplification of secondary electrons produced in a strong electric field (typically 2000 V). The field is established between a cathode and anode wire, or grid of wires, and the charge pulse due to each X ray detected is sensed on either or both of these electrodes. The detector is usually constructed as a flattened box with the X rays incident through a thin window (typically beryllium, which, because of its low atomic number, is relatively transparent even to soft X rays) on "top." The depth of the detector is kept minimal to reduce the non-X-ray background levels detected, as well as to keep the detection plane as sharply in focus as possible in the case of imaging proportional counter detectors used at the focus of X-ray telescopes. However, because the X-ray interaction depth increases with increasing energy, this means the energy range of maximum sensitivity for proportional counters is limited to the soft and intermediate bands (although some response is possible up to energies of 20–30 keV).

A proportional counter is not only a detector but also a spectrometer. That is, it can enable measurement of the spectrum of X rays it detects, since the total electronic charge produced by the X ray interacting in the detector gas is directly proportional to the incident X-ray energy—hence the name "proportional" counter. The spectrum of X rays is nearly as important as the mere detection of X rays within a particular band, since it allows the physical nature of the cosmic X-ray source to be probed. In practice, the pulse heights of each X-ray event detected are usually telemetered down with the data from an X-ray satellite, and spectral analysis is carried out after the fact in computer analysis of the distribution of pulse heights recorded on a given source.

The UHURU and HEAO-1 detectors were proportional counters with, respectively, very coarse and moderately fine spectral resolution, or numbers of spectral energy channels included in the data. Thus, although UHURU allowed most of the initial discoveries of new types of sources by virtue of conducting the first all-sky survey, the HEAO-1 detectors were able to carry out more detailed follow-up studies of individual sources. The detectors on both missions were nonimaging, and X-ray arrival directions could only be reconstructed by the relative intensity of a source as it was scanned across the restricted angular field of view of a collimator (with angular widths in the range 1–5 degrees in both cases). HEAO-1 also included a specialized "modulation collimator" design on several of its detectors, which, by recording the time-varying shadow modulation of a wire grid above the detector, allowed the ~30-arcsec X-ray source location uncertainties mentioned above to be derived. Nonimaging proportional counters employing collimators have been the detectors used on all other X-ray astronomy satellites launched by the United States (NASA), Europe [European Space Agency (ESA)], and the Japanese Institute for Space and Astronautical Science (ISAS), except for the Einstein Observatory and EXOSAT.

Only Einstein and EXOSAT have thus far made use of direct imaging techniques for X-ray astronomy. In this case, the proportional counter (or other detector) also registers the detected X-ray position so that an image of the X-ray source or sources can be built up, photon by photon, for later analysis on the ground. Since the pulse height of each X ray is also recorded in an imaging proportional counter, the images are "in color;" that is, the spectral distribution of X rays across the image can also be measured. Both Einstein and EXOSAT also employed another type of imaging X-ray detector with much higher spatial resolution on the detector face, which in turn allowed much higher angular resolution on the sky. These detectors [called the high-resolution imager (HRI) on Einstein] incorporated a microchannel plate array of electron multipliers (consisting of a close-packed array of microscopic glass tubes, each only some 25 μm in diameter) in which the primary X-ray interaction produced a greatly amplified cloud of secondary electrons, which were detected by a position-sensitive charge detector. Since the spatial resolution on the detector could be about 30 times better than with an imaging proportional counter, correspondingly better angular resolutions (about 2 arcsec) in the image could be derived. However, no spectral resolution was possible with the versions of these detectors used thus far.

Finally, the imaging capability of an X-ray telescope (cf. Section III,C) also means that the physical size of a detector to detect a single point source need only be very small. This, in turn, means that the internal sources of detector background can be greatly reduced and that higher sensitivities can be achieved for spectrometers with very high spectral resolution than in a larger detector not at the focus of a telescope. The Einstein Observatory included two such high-resolution spectrometers, which allowed the spectra of individual sources to be studied in much greater detail than with the comparatively coarse spectral resolution achieved in proportional counters. In one case, the spectrometer was a solid-state device in which the total number of electron–hole pairs produced in a silicon detector was recorded with energy resolution of ~200 eV. In the other case on Einstein, the spectrometer employed a Bragg crystal and proportional counter detector (with position resolution along the dispersion direction) for much higher spectral resolution (about 10 eV) but much lower throughput or sensitivity. Unfortunately, neither of these spectrometers was itself position-sensitive, so that spatially resolved high-resolution spectroscopy must await future X-ray observatory missions now planned.

C. X-RAY TELESCOPES

Both the Einstein Observatory and EXOSAT were based on focusing X-ray telescopes, with finely polished mirrors that reflect X rays at very shallow grazing incidence angles onto a focal plane with (typically) a position-sensitive proportional counter detector. The grazing incidence angles are required to reflect X-rays, since an X ray with energy of 2 keV can only be reflected from a polished surface if it strikes the surface at an angle less than about 4 degrees. In addition, the surface must be coated with a high-atomic-number metal (such as gold) to ensure good reflectivity up to the highest energies possible. The high-energy limit of a grazing incidence X-ray telescope is governed by the inverse relationship between the incidence angle at which X rays can be reflected (and not transmitted or absorbed by the mirror surface or substrate material) and the X-ray energy; at energies of 8 keV, for example, this angle is only about 1 degree. Since the X rays must be detected at the common focus of the telescope mirror(s), as will be described, the shallower the grazing incidence angle, the farther back the focal point must be behind the mirror. On the Einstein Observatory, the focal length was about 3 m behind the front of the mirror assembly; this

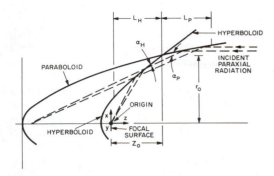

FIG. 2. Focusing geometry of a Wolter type I grazing incidence X-ray telescope constructed by a paraboloid followed by a hyperboloid cylindrical mirror.

limited the energy response of the telescope to energies below about 4 keV.

Several different configurations of curved grazing incidence mirror segments are possible to form a telescope, which in turn brings X rays collected over a relatively large area to a common focus where they can be detected with a relatively much smaller detector. The most common configuration, and the one used for both Einstein and EXOSAT, is the so-called Wolter type I geometry, whereby the telescope is con-

structed with two cylindrical and coaxial mirrors: a paraboloid followed by a hyperboloid, as shown in Fig. 2. The paraboloid or hyperboloid curvature is polished into the long (axial) direction of the otherwise round cylindrical mirror. The actual mirror segment (either the paraboloid or hyperboloid) is constructed from a cast and then finally polished glass substrate material onto which is finally deposited the high-atomic-number metallic coating for the actual reflective surface. The collecting area of such a two-segment mirror is governed by the radius r_0 at the rear of the paraboloid and the length of the paraboloid L_P (which then determines the radius of the paraboloid at its front edge), as shown in Fig. 2.

In order to gain as much total collecting area of the telescope as possible, which is required to maximize the number of X rays that can be imaged and detected per unit time, the cylindrical mirror segments are nested one inside the other so that the entrance aperture (which has a radius of approximately the r_0 value for the outermost mirror) is as nearly filled as possible. In the case of the Einstein telescope, four such nested mirrors were used. The entrance aperture diameter was 0.6 m in diameter, and the overall collecting area of the telescope was about 150 cm^2. This

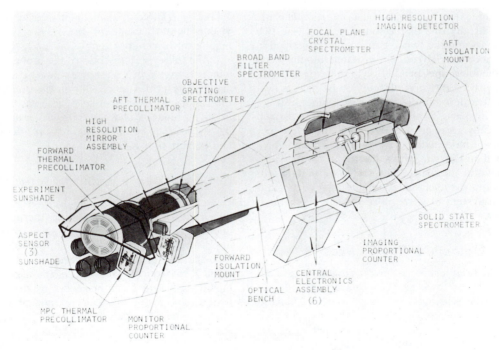

FIG. 3. Overall layout of the Einstein X-Ray Observatory.

total effective area for X-ray imaging with a 0.6 m-diameter telescope is seen to be much smaller than the equivalent number (~2800 cm²) for a normal incidence optical telescope of the same outer diameter. Thus, X-ray telescopes not only have to be polished much more smoothly than optical telescopes because of the much shorter wavelengths being imaged, but they must also be polished over much larger total areas. For example, the largest and most precise X-ray telescope now planned (the AXAF, which is described in the final section of this article) will have a total effective area for X-ray imaging (with 0.5 arcsec resolution) of about 1600 cm², an entrance aperture of 1.2 m, and will have six nested grazing incidence mirrors with a focal length of 10 m. The total amount of glass substrate material for the mirrors, which must be polished to an overall surface roughness of only about 10 Å, is more than eight times the surface area of the 2.4-m-diameter single mirror of the Hubble space telescope for optical/ultraviolet astronomy.

The overall construction of an X-ray telescope consisting of both the nested mirror assembly and the focal plane detector assembly is shown in Fig. 3, which depicts the layout used for the Einstein Observatory. Although the mirror assembly was rigidly fixed and aligned relative to the focal plane, the detectors were mounted on a rotating carousel at the focal plane so that any one of the four detectors (the two imagers and the two spectrometers described in the Section III,B) could be rotated into the beam of the telescope. In addition to these major components, the X-ray telescope must have a system for determining its precise pointing direction in space, so that images, which are formed typically in exposures lasting a thousand seconds or more, will not be blurred. This was provided on Einstein with a star tracker system also shown in the figure. The telescope pointing direction was itself maintained relatively constant by a system of gyros, also indicated schematically.

IV. Galactic X-Ray Sources

The brightest X-ray sources in the sky are objects in our own galaxy. This can be inferred from the distribution of these sources (the largest dots in Fig. 1), which is sharply clustered around the directions toward the center of our galaxy as well as the plane of the galaxy (the horizontal "equator" line of the coordinate grid in Fig. 1). These sources, which include the pro-

totype galactic X-ray source Sco X-1, are predominantly binary systems containing a neutron star accreting matter from a relatively normal binary star companion. Intense X-ray emission is produced in these systems by virtue of the fact that the energy released in the (near) free-fall of matter in the intense gravitational field of a neutron star is liberated with great (~10% of the rest mass energy) efficiency at the hard stellar surface and is thermalized into radiation in the X-ray band (cf. discussion of blackbody emission in Section IV,A). Over a wide range of accretion rates, measured in solar masses per year that reach the neutron star surface, the resulting X-ray luminosity is directly proportional to the accretion rate. The total X-ray luminosities of these systems may be estimated from their X-ray fluxes and spectra and their approximate distances (relatively precise distances are known for only a fraction of the galactic sources, as will become clear in the following subsections). The luminosities of the brightest sources appear to not exceed values of about 2×10^{38} erg/sec, which is nearly 10^5 times the total luminosity of the sun or most "normal" stars, and which corresponds to an accretion rate of about 2×10^{-8} $M_\odot$ per year. Thus, cosmic X-ray sources in the galaxy are individually spectacular objects, radiating vast quantities of energy in X rays. However, their collective total energy output in the galaxy is relatively small compared to that in visible light, because of the relatively small number of bright sources.

In the following subsections, the processes by which X rays are produced in galactic sources are first described more fully in the context of the major classes of sources. The mass transfer in X-ray binaries and accretion disks are then discussed, followed by a summary of the origin and evolution of X-ray binaries.

A. THERMAL X-RAY EMISSION AND SOURCES

As mentioned in Section I, X rays are produced either by very hot matter (thermal production processes) or by high-energy particles interacting with either magnetic or photon fields (non-thermal production processes). Most galactic X-ray sources and most of the luminosity in X-ray emission from the galaxy are thermal in nature. The major thermal emission processes may also be described in order of their typical luminosity.

Thermal bremsstrahlung from stellar coronas is the process by which stars with surrounding high-temperature coronas radiate X rays. The solar corona, the outermost and thinnest layers of the solar atmosphere with a temperature in excess of 2×10^6 K, is a typical example and was discovered to be an X-ray emitter in rocket experiments in 1948. The Einstein Observatory was sensitive enough to detect this same type of emission, which is typically only a small fraction ($\sim 10^{-7}$–10^{-2}) of the total energy output of the star, from a wide range of stellar types. The X-ray emission arises from the braking (hence the radiation process name, which translates as braking radiation) felt by an electron with high temperature (and thus velocity) when it passes near a proton (more massive and hence more slowly moving) in a hot gas. The process is relatively inefficient, but the rate of bremsstrahlung cooling of a hot gas depends on the product of the gas density and temperature as $n^2 T^{1/2}$. Thus, study of the stellar coronal emission discovered with the Einstein Observatory for stars with a broad range of mass, age, and temperature (much lower than the temperatures of their surrounding coronae) provides a unique probe of the density and temperatures in the outermost regions of stellar atmospheres. The X-ray luminosities of stellar coronas are, once again, strongly dependent on the type of star but are in the range $\sim 10^{26}$–10^{31} erg/sec.

Thermal bremsstrahlung from supernova remnants is a more luminous source of cosmic X rays, mainly because the mass of radiating gas is much larger than in the stellar coronal case, although the temperatures and densities of the radiating gas can be similar. A supernova is the final end point in the evolution of a relatively massive star when its nuclear fuel has been exhausted and the stellar core collapses. The bulk of the stellar mass explodes back into space with very large velocity when the collapsing core "bounces." The supernova remnant (SNR) is then an expanding gas cloud, or shell, and is heated to X-ray emission by shock waves set up as the shell traverses interstellar space. A beautiful example of the X-ray image of such an object is shown in Fig. 4. The SNR in the figure is referred to as the Tycho SNR, since it is the remnant of the supernova seen visually by the noted Danish astronomer Tycho Brahe in 1572. The total mass of radiating gas is uncertain but is at least 2–4 $M_\odot$, indicating the presupernova star was much more massive and short-lived than the sun. The composition of the gas in the SNR provides vital information on the physics of the explosion and the stellar evolution leading up to the explosion. The X-ray spectrum of the SNR provides not only the bremsstrahlung temperature and densities but also, from the emission lines detected with the spectrometers on Einstein, the composition and ionization states of the gas in the SNR. With future X-ray observatories such as AXAF, it will be possible to spatially resolve the X-ray spectrum across the image and to then carry out much more detailed studies of the remnant. The X-ray luminosities of thermal emission from SNR are typically in the range $\sim 10^{34}$–10^{36} erg/sec. [*See* SUPERNOVAE.]

Thermal emission from accretion sources is the mechanism primarily responsible for the bright X-ray sources discovered in the UHURU and HEAO-1 surveys. The X-ray binaries, including the brightest sources in the galaxy mentioned above, are all powered by accretion of gas onto a compact object (a white dwarf, neutron star, or black hole) from a binary star companion. This process will be described later in this section on galactic X-ray sources. The actual production of thermal X-ray emission occurs when the gas spirals down through an "accretion disk" onto the compact object due to its intense gravitational field and is heated by the "friction" (or viscosity) of the gas as it tries to rush onto the relatively tiny compact object. The emission mechanisms are complex and depend on the details of the still poorly understood gas flows, but they involve thermal bremsstrahlung from relatively diffuse gas in a corona surrounding the central regions of the accretion disk, as well as thermal emission by the "black body emission" process (cf. Section IV,B) from the accretion disk surface as well as the surface of the compact object itself if it is a white dwarf

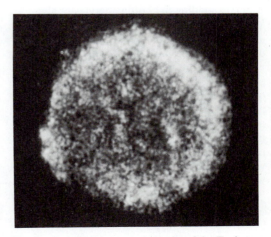

FIG. 4. Einstein Observatory image of the Tycho supernova remnant. (Courtesy S. Murray, Smithsonian Astrophysical Observatory.)

or neutron star. The typical X-ray luminosities from accreting neutron stars or black holes in binary sources are in the range $\sim 10^{35}$–10^{38} erg/sec.

Blackbody emission from the surface of a neutron star yields the highest-luminosity X-ray emission from sources in the galaxy. The blackbody emission process is the maximum emission possible from an object at a given temperature T. It is achieved (instead of the corresponding bremsstrahlung luminosity) when the radiation emitted by individual particles is scattered many times before it can escape the source; thus it is the "natural" thermal radiation process for solid objects or very dense gases but not very diffuse gases heated to a given temperature T. For an object (e.g., a neutron star) with an emitting surface with radius R and temperature T, the total blackbody luminosity is given by the expression

$$L_{\text{tot}} = (4\pi R^2)ST^4$$

where S is the Stefan–Boltzmann constant. The actual shape of the spectrum is a thermal spectrum with a broad peak at a frequency (or energy) directly proportional to the emitting temperature T. The emitting temperature T, in turn, is fixed by the total luminosity and object size R in accordance with the equation above. Thus, a neutron star radiating at its maximum luminosity possible from the accretion process, the so-called Eddington limit luminosity

$$L_{\text{Edd}} = 1.4 \times 10^{38}(M/M_\odot) \quad \text{erg/sec}$$

where the outward force of the radiation pressure exactly counterbalances the inward force of gravity on the accreting matter, will have an equivalent blackbody temperature of about 10^8 K corresponding to photon energies of about 2.7 keV. Even for more modest rates of accretion and thus X-ray luminosities well below the Eddington limit, the blackbody radiation spectrum produced at the neutron star surface is still expected to be primarily in the X-ray band because of the weak ($\frac{1}{4}$-power) dependence of the temperature on the luminosity. This is the essential reason why accretion onto neutron stars in compact X-ray binaries produces copious X rays, rather than ultraviolet light or gamma rays, which are (fortunately) relatively penetrating as well as easy to detect.

The most dramatic evidence for a neutron star radiating X rays at near its Eddington limit luminosity occurs during an X-ray burst. This phenomena, discovered by the author in 1975 and now understood as being due to the gargantuan thermonuclear explosion of some 10^{21} g of material that has accreted onto the surface of a neutron star since a previous burst, is shown in Fig. 5 in the time profile of X-ray emission from a particularly interesting X-ray binary in a globular star cluster (NGC 6624). The count rate of individually detected (with the Dutch–U.S. X-ray satellite ANS, which carried two X-ray detectors: the "HXX" and "SXX" detectors referred to in the figure) X-rays is seen to rise abruptly from about 10 counts/sec to more than 100 counts/sec in only about 1 sec. The count rate before (and well after) this sudden rise is due to the persistent X-ray emission from the thermal X-ray emission process from accretion, as described in the preceeding subsection, whereas the extra emission in the burst is due to blackbody emission from the sudden release of energy at the surface of the neutron star. The blackbody nature of the spectrum of X rays during the burst can be measured from comparisons of the burst counts recorded in different energy bands such as the intermediate versus soft bands shown in panels (a) versus (b) of the figure. Careful analysis of data such as these reveals that in X-ray bursts the blackbody temperature cools throughout the decay of the burst profile; this is due to the cooling of the neutron star surface after the sudden heating in the thermonuclear explosion, which is the energy input for the burst.

B. NONTHERMAL X-RAY EMISSION AND SOURCES

The most conspicuous nonthermal X-ray source in the galaxy is the Crab nebula, which is an SNR in which high-energy electrons are accelerated by a central rapidly rotating neutron star, that is, a pulsar. The Crab pulsar spins with a period given by the 33-msec period of the pulsations detected in the radio through the gamma-ray energy bands. The neutron star in the Crab nebula is the best evidence that neutron stars can be produced in the supernova event itself, which, in the case of the Crab, was actually recorded as a "guest star" by the Chinese court astronomers in 1054 A.D. During the collapse of the stellar core to form the neutron star, the original magnetic field present in the star (e.g., the sun has a surface magnetic field strength of about 1 G) is amplified enormously as the stellar core collapses to only about a millionth the diameter of the original star. The magnetic field emerging at the neutron star surface is accordingly some $(10^6)^2$ times bigger than the original

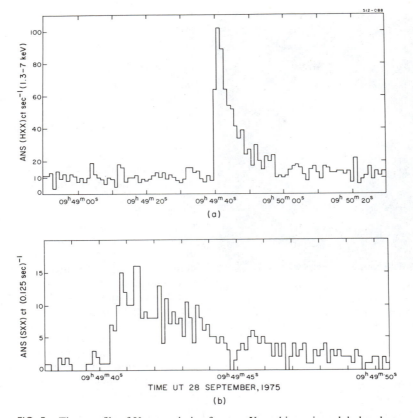

FIG. 5. Time profile of X-ray emission from an X-ray binary in a globular cluster, showing the first X-ray burst discovered from a known source. (a) ANS (HXX) observation of x-ray burst from NGC 6624 (4U1820-30). (b) ANS (SXX) observation of X-ray burst from NGC 6624 (4U1820-30).

surface field and so increases to $\sim 10^{12}$ G. When such a powerful magnet is spun around rapidly (some 30 times a second in the case of the Crab pulsar!), extremely large electric fields are created as in a generator. It is these strong electric fields that accelerate the electrons into the Crab nebula with energies in excess of 10^{15} eV. The spectrum of cosmic ray electrons so produced radiate nonthermal X-ray emission (actually a spectrum from the radio through the hard X-ray bands) by the synchrotron process as they traverse the relatively weak magnetic fields (about 10^{-4} G) that thread through the whole SNR. The X-ray image of this diffuse synchrotron emission from the Crab nebula is shown in Fig. 6.

The contrast between the Crab and Tycho SNR X-ray images is striking. In the Crab, the emission is relatively smooth and increases toward the source of the nonthermal energy, the bright central pulsar. In the Tycho image (see Fig. 4), the thermal emission is much more confined to the spherical shell of hot debris from the exploded star. No corresponding thermal X-ray component has yet been detected for the Crab. Synchrotron X-ray nebulas have been detected by the Einstein Observatory around other isolated pulsars (not obviously still in SNR, which themselves are expected to be visible only for some 10^5 years), indicating that the high-energy particle production by magnetized neutron stars (pulsars) is widespread. Recent radio studies of compact X-ray binaries such as the brightest galactic sources discussed above indicate that the neutron stars or accretion disks in these systems are also capable of accelerating high-energy electrons. However, the accompanying nonthermal X-ray emission that might be expected from these systems has not yet been detected.

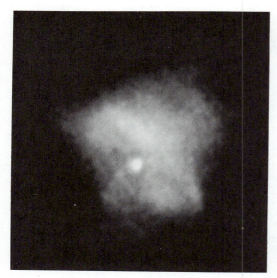

FIG. 6. X-Ray image of the Crab nebula, recorded with the Einstein Observatory, and also showing the Crab pulsar (neutron star) as the bright spot of emission. (Courtesy F. Harnden, Jr., Smithsonian Astrophysical Observatory)

C. MASS TRANSFER AND ACCRETION PROCESSES AND CONSEQUENCES

For accretion to occur onto a neutron star in an X-ray binary, mass transfer must occur from the companion star in the binary system. This can occur by either of two principal mechanisms: overflow of the atmosphere of the companion star directly onto the neutron star, if the stars are close enough together so that the companion star overflows its "Roche lobe," or loss of mass from the companion star via a stellar wind, some of which is intercepted by the neutron star. In either case, it is critical that the companion star be close enough to the compact object that mass transfer can occur. Applying Kepler's laws of motion to the binary system, with consideration of the two processes in more detail, shows that the corresponding binary periods must typically be shorter than a few days or a few weeks, respectively, for most types of companion stars. In the case of X-ray binaries with the most massive companion stars, which can lose matter via intense stellar wind outflows for a relatively short period of their evolution, the binary periods are typically 2 weeks (as in the case of the peculiar X-ray binary and source of cosmic jets SS433, which may contain a black hole for the compact object). At the other extreme, most (if not all) of the X-ray burst sources, such as that in the globular cluster NGC 6624 already discussed, have binary companions that are probably very-low-mass stars (with masses much less than the neutron star mass they orbit around) with correspondingly small radii. The binary periods in these systems are typically less than a few hours, and in the case of the source in NGC 6624 as short as only 11.4 min (!), the shortest binary period known.

For both wind-fed and Roche lobe overflow mass transfer, the gas from the companion star has appreciable angular momentum, which must be lost if it is to be able to accrete onto the compact object and produce X rays. The angular momentum is lost by the gas shearing against itself in a thin, but relatively dense, accretion disk into which the gas settles as it orbits the compact object. The angular momentum loss rate, and thus the rate at which gas is able to spiral inward through the disk, is dependent on the viscosity of the disk. This, in turn, is difficult to model exactly, since the effects of both complex turbulence and magnetic fields generated in the disk are probably important. Therefore, accretion disks are thus far described with a free parameter α, which includes the unknown physics factors in the model (the so-called α-disk models).

Once the gas reaches the inner portions of the disk, it may begin to feel the effects of the magnetic field of the neutron star. The radius from the neutron star where the magnetic field pressure is equal to the dynamical pressure of the incoming gas is called the Alfvén radius, and its value increases with increasing magnetic field and/or decreasing rate of accretion. For a neutron star such as that in the Crab pulsar (which does NOT accrete gas, since it has no binary companion) with a magnetic field of perhaps 5×10^{12} G, if the mass transfer rate were sufficient to produce an X-ray luminosity of 10^{37} erg/sec (i.e., a typical compact X-ray binary source), then the Alfvén radius would be at about 100 times the radius of the neutron star, or at $\sim 10^8$ cm. In this case, where the Alfvén radius is substantially greater than the radius of the compact object, the accretion disk is disrupted and the matter can instead plunge directly onto the compact object following its lines of magnetic field down to the stellar surface. Since most of the magnetic field lines reach the stellar surface near the magnetic poles, the accretion of matter is primarily onto the magnetic poles, and two hot spots of X-ray production occur. If the neutron star is rotating, as would almost certainly be the case (since it is receiving angular momentum from the gas it accretes), then the resulting X-ray emission will be seen as pulsations and an accretion-powered pulsar is produced. Exactly this sort of phenomena is observed in about 20 of the galactic compact X-ray binaries, which are in fact accretion-powered pulsars. They differ, once again, from the rotation-powered pulsars, such as that in the Crab nebula, in that the latter do not accrete matter but instead expel high-energy particles by virtue of their rotating magnetic and electric fields.

In the case of the X-ray burst sources (about 30 known in the galaxy), as well as many (perhaps most) of the most luminous X-ray binaries in the galaxy (e.g., Sco X-1, and some 10–20 other sources), the magnetic field is probably so weak that the Alfvén surface is not appreciably above the neutron star surface. In this case, the accretion disk may extend down to near the stellar surface, where it is disrupted in a "boundary layer," and intense X-ray emission is produced. The detailed structure and emission from this boundary layer as well as from the accretion disk itself are only now beginning to be explored in detail and compared with observations from both the Einstein and EXOSAT detectors.

D. ORIGIN OF X-RAY BINARIES

The creation of an X-ray binary with a neutron star or black hole orbiting a normal stellar companion and fuel supply is not well understood in detail, although the basic ideas are becoming clear. In the case of the massive X-ray binaries (i.e., systems with companions stars more massive than the compact object), it is likely that the neutron star forms from the supernova of a preexisting binary member without disrupting the binary in the supernova explosion. The fact that these systems are found preferentially near the plane of our galaxy indicates that they are the result of the evolution of normal massive stars, which were themselves formed in binary systems. The case of the low-mass X-ray binaries such as the X-ray bursters and the most luminous X-ray sources in the so-called galactic bulge is not nearly as "obvious." Here the problem is at its most extreme when

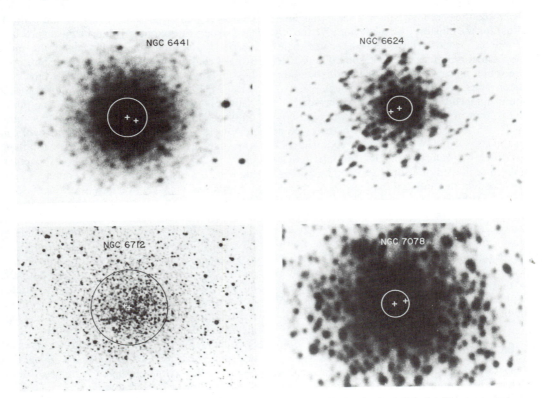

FIG. 7.   Positions of X-ray source versus cluster center and core radius for four globular clusters containing luminous X-ray sources imaged with the Einstein Observatory.

the origin of the X-ray bursters with neutron stars and apparently very low mass but *unevolved* companion stars is considered. The problem is to form a neutron star in such a close binary system without either disrupting the system or otherwise finding that the companion star has itself evolved. [*See* BLACK HOLES (ASTRONOMY).]

Only for those members of this class of sources that are found in globular clusters, such as (once again) the burster in the cluster NGC 6624, is the origin reasonably well understood. In this case, the neutron star and (initially) normal low-mass companion star probably capture each other into a bound orbit by virtue of a relatively close encounter in the very high stellar density core of the cluster. Direct evidence that the luminous X-ray sources in globular clusters are low-mass binary systems involving a neutron star and not a massive black hole, which was once thought likely to have formed in the collapse of the cluster core, is contained in the relative positional offsets of the sources from the centers of their parent clusters. In Fig. 7, the relative positions for four of the globular cluster sources are shown relative to the centers of each cluster and the characteristic size scale (the core radius) for the cluster. The more massive the X-ray source, the closer it would be expected to lie to the cluster center, where it should sink if it were much more massive than the surrounding cluster stars. Analysis of eight such cluster sources imaged by the Einstein Observatory indicates that the source masses are approximately 1.5 $M_\odot$, or consistent with their being neutron stars with low-mass binary companions. This further supported the interpretation that these were indeed neutron star systems by virtue of the fact that most of them are X-ray burst sources. Although the capture process almost certainly works in the dense cores of globular clusters, it does not in the much lower stellar density of the galactic bulge, and so these sources remain a mystery. One possibility is that they too were formed in globular cluster cores but that the parent globulars have since been disrupted by encounters with giant molecular clouds or other massive structures in the galaxy.

V. Extragalactic X-Ray Sources

The UHURU survey firmly established the existence of extragalactic X-ray sources (several of which were hinted at before) with the discovery of X-ray emission from individual sources in the Magellanic Cloud galaxies, the integrated X-ray emission of our closest neighbor large galaxy (M31), X-ray emission from clusters of galaxies, and emission from active galaxies and quasars. However, once again it was with the more sensitive HEAO-1 and, especially, the Einstein Observatory that detailed study of the spectra and images of these sources were carried out. In this section and the next, we briefly summarize some of the more recent results (from Einstein observations), as well as the major implications of ongoing studies.

A. GALAXIES AND CLUSTERS

Although UHURU observations established the existence of individual compact binary sources in the Magellanic Cloud galaxies (our closest neighboring galaxies), it was the Einstein observations of M31 that really initiated the study of X-ray properties of galaxies. In Fig. 8 we show the positions of some of the individual sources detected in this galaxy overlayed on an optical image of the galaxy. Many of these sources are identified with globular clusters in M31, and higher-resolution observations of the central region of the galaxy resolve the centralized group of sources visible in the figure into nearly 100 sources in total. One of the most striking features of this source distribution is how much more strongly peaked about the center of M31 it is than the distribution of compact X-ray binaries is in our own galaxy. One possibility is that M31, which contains a larger number of globular clusters than does our own galaxy, may also have had many more of them disrupted in its nuclear bulge, thereby populating it with more luminous X-ray sources.

Studies of the X-ray emission from individual galaxies have now been carried out for a wide range of types of galaxies. It has been found that the X-ray emission is often correlated with both the visible-light and radio emission from spiral galaxies, suggesting that the distribution of compact X-ray binaries is closely tied to both the distribution of mass and high-energy particles (cosmic rays) in these galaxies. The total luminosity of their resolved sources (i.e., not including their nuclei, which may be "active"—see Section VI) are typically in the range $\sim 10^{39}$–10^{41} erg/sec. The actual emission mechanisms and types of sources are probably similar to those discussed above for our own galaxy.

In the case of the elliptical galaxies, a remarkable discovery was made: the halos of these gal-

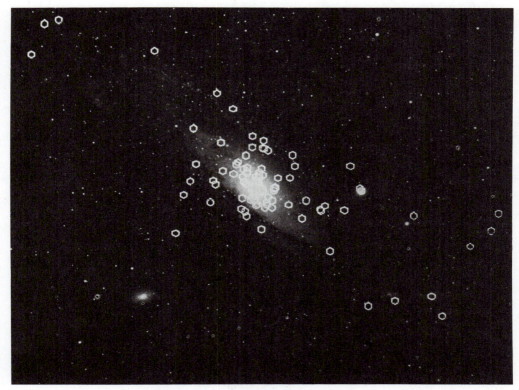

FIG. 8.   Positions of X-ray sources in the galaxy M31 (the Andromeda nebula) superimposed on an optical photo of the galaxy. The angular scale of the picture is about 3 degrees across. (Courtesy L. Van Speybroeck, Smithsonian Astrophysical Observatory)

axies, which are thought to be deficient in gas (and regions where stars are being formed), are glowing in soft X rays. This diffuse emission very likely is produced by hot gas expelled by evolved giant stars in the galaxy and trapped by the gravitational attraction of the mass of the galaxy itself. When the mass necessary to confine this hot gas is calculated, it is found that it greatly exceeds the mass in visible stars in the galaxy. Thus the hot X-ray gas provides independent support for the claim that the halos of galaxies contain a substantial amount of "dark" matter, that is, matter that is felt gravitationally but is not accounted for in the visible stars. Studies of this dark matter are among the most actively pursued in astrophysics today and can be expected to be greatly enhanced by future X-ray observations.

Similarly, the study of entire clusters of galaxies is now greatly enhanced by X-ray astronomy. Galaxy clusters were discovered with UHURU to contain diffuse hot gas filling the space between the galaxies and readily detect-able in X rays. The X-ray spectrum of this emission shows that it is clearly thermal emission, with temperatures typically of about 3–8 keV and total luminosities as great as $\sim 10^{45}$ erg/sec. The Einstein observations have imaged this emission and enabled the corresponding gravity of the cluster to be mapped out in a way analogous to that for individual galaxy halos. The mass necessary to bind the hot cluster gas is once again much greater than that visible in the light from the component galaxies, so that dark matter is again required but now on a much larger scale.

Therefore, diffuse X-ray emission appears to be the ideal tracer of mass, and therefore missing mass, in both galaxies and clusters. It also allows the origin of the gas in clusters to be constrained: if the spectrum of the gas shows emission from heavy elements such as iron, then it must have been expelled from the cluster galaxies themselves. Absence of iron line emission from the hot gas far from the constituent galaxies in the cluster, however, would indicate that the hot gas might be primordial and possibly related to the gas clouds from which the galaxies themselves may have formed. Although the HEAO-1 (and other) X-ray detectors had already firmly established that the sum total X-ray emission from galaxy clusters contains substantial iron emission, spatially resolved spectra (i.e., imaging with spectral resolution) are needed. Unfortunately, the Einstein Observatory was not able to do this, since its images had only coarse (or no) spectral resolution and also its energy response did not extend to the 6.7-keV iron lines.

B. ACTIVE GALACTIC NUCLEI

The study of the nuclei of galaxies is of great interest and importance in astronomy and astrophysics today. It is now generally believed that the most luminous objects known in the universe, the quasars, are extreme examples of active galactic nuclei (AGN). These objects are very luminous X-ray sources and in fact produce a large fraction of their total energy output in the soft through hard X-ray bands. The brightest AGN, which can be either the closest or the most intrinsically luminous, were already de-

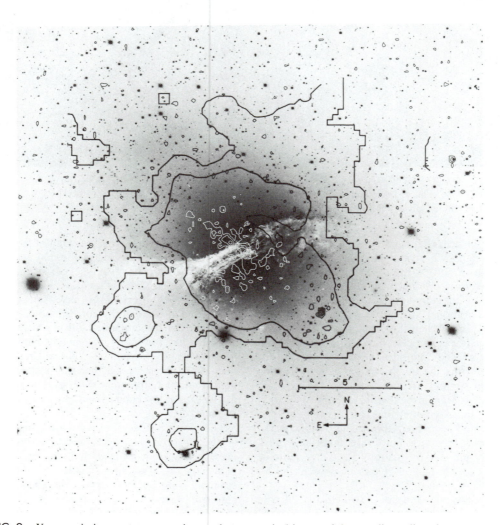

FIG. 9.   X-ray emission contours superimposed on an optical image of the peculiar radio galaxy Cen A (NGC 5128). This is the closest AGN and may be a system triggered into nuclear activity by the collision and merger of two galaxies in the relatively recent past.

tected as X-ray sources by UHURU. However, it was not until the Einstein Observatory that the sensitivity was achieved (largely because of the virtually nonexistent background in a grazing incidence soft X-ray telescope) to study much fainter AGN in detail. This has allowed studies of both the point source (at the actual galactic nucleus) radiating most of the X-ray energy as well as much fainter surrounding sources, either in the associated galaxy or in cosmic jets produced by the central source. An example of such a complex configuration of X-ray emission in an active galaxy is shown in Fig. 9, where the X-ray emission intensity contours are superimposed on an optical image of the nearby peculiar galaxy Cen A. [See GALACTIC STRUCTURE AND EVOLUTION; QUASARS.]

The detailed X-ray studies of AGN allow the regions closest to the central source to be probed. This is because the X-ray emission from these objects is typically observed to be variable in time, indicating that the X-ray emission is somehow produced near a compact object with source sizes only light-minutes to light-hours across (corresponding to the fastest variability timescales). Although variability is also seen at radio through optical wavelengths for many AGN, it is generally on much longer time scales, indicating that the X-ray emission indeed arises from closest to the central source, which is obscured by gas and dust at other wavelengths. The most widely accepted physical model of the nature of the central sources in AGN is that they are massive black holes: for so much energy (up to $\sim 10^{47}$ erg/sec, or equivalent to $\sim 10^{14}$ solar luminosities!) to be produced in such a small volume (roughly comparable to the size of the solar system) would inevitably lead to the collapse of the source or sources into a single central massive black hole. The mass of the black hole inferred is in the range $\sim 10^7$–10^9 $M_\odot$, as compared with the value of only ~ 10 $M_\odot$ for the largest mass black hole suspected in an X-ray binary system in our own galaxy.

The X-ray spectra of AGN are almost certainly nonthermal (although some contribution of thermal sources is also still possible) and may be closely connected to the production of high-energy particles in the region near the massive black hole. The sources can be described by models in which a spectrum of high-energy electrons is produced near the black hole and that in turn radiate synchrotron photons in the radio through soft X-ray bands. The relativistic electrons responsible for this emission are somehow confined to a relatively small source region in which the density of synchrotron photons they produce is very high. Therefore, the electrons have an appreciable probability of scattering (by the inverse Compton process) the synchrotron photons or also photons produced from thermal sources (such as hot clouds) farther out in the system. In this case, hard X-ray photons are produced and an extension of the X-ray spectra of AGN up to energies in excess of 100 keV is expected. Unfortunately, hard X-ray observations have thus far been of limited sensitivity and the predictions of this so-called synchrotron–self-Compton model have not yet been possible to check in detail. Hard X-ray observations of AGN, therefore, offer a particularly powerful probe of nonthermal phenomena around massive black holes.

It is likely that the integrated emission from "all" AGN is the major contributor to the pervasive X-ray background radiation detected from all directions of space. This X-ray background was in fact discovered on the pioneering rocket flight experiment that discovered Sco X-1 and effectively launched the field of X-ray astronomy. It remains nearly as much of a mystery today, despite more than two decades of work. Understanding the origin of the X-ray background as well as the role of AGN will require determining the luminosity and spectra of quasars out to very great distances. Because quasars are so bright in X rays, their study at very great distances can also provide clues to the physical conditions in the early universe. This work has just begun with the Einstein observations of quasars, which appear to show that the X-ray properties of AGN evolve strongly but at a different rate than the optical properties of the same objects. Understanding this will require much more sensitive work in the future with the next generation of X-ray telescopes.

VI. Future Directions

X-ray astronomy is now in the mainstream of modern astrophysics. It allows many of the most pressing questions about the universe to be investigated from a new perspective. The advent of imaging and high-resolution spectroscopy with the exploratory Einstein Observatory revealed the rich promise of the field. The next major cosmic X-ray telescope mission planned for space is a soft X-ray telescope built by the Germans but including a high-resolution detector and launch (in $\sim$1989) into orbit from the

United States. It is expected to reveal hundreds of thousands of new cosmic X-ray sources and to allow the study of many of the classes of problems discussed above. However, the major step will occur with the launch of the AXAF telescope in the mid 1990s. This will allow much higher-resolution studies to be undertaken, with sensitivities greatly improved. It should allow the frontiers in many of the most sought-after astronomical problems to be pushed back significantly and may well, along with the other Great Observatories to be launched (for studies in the infrared, optical, and gamma-ray bands), change our view of the cosmos in profound ways.

BIBLIOGRAPHY

Fabbiano, G. (1986). *Publ. Astronom. Soc. Pacific* **98,** 525.

Giacconi, R. (1980). *Sci. Am.* **242,** 80.

Gorenstein, P. (1983). *In* "Astronomy from Space" (J. Cornell and P. Gorenstein, eds.), pp. 171–192. MIT Press, Cambridge, Mass.

Grindlay, J. E. (1983). *In* "Astronomy from Space" (J. Cornell and P. Gorenstein, eds.), pp. 141–170. MIT Press, Cambridge, Mass.

Overbye, D. (1979). *Sky and Telescope* **57,** 527.

Seward, F., Gorenstein, P., and Tucker, W. (1985). *Sci. Am.* **253,** 88.

INDEX